PROCEEDINGS OF THE

FOURTEENTH INTERNATIONAL MACHINE TOOL DESIGN AND RESEARCH CONFERENCE

held in Manchester
12–14 September 1973

Edited by

F. KOENIGSBERGER
Professor of Machine Tool Engineering
University of Manchester Institute of Science and Technology

and

S. A. TOBIAS
Chance Professor and Head of Department
Department of Mechanical Engineering
University of Birmingham

MACMILLAN

First published 1974 by
THE MACMILLAN PRESS LTD
London and Basingstoke
Associated companies in New York
Dublin Melbourne Johannesburg and Madras

SBN 333 14913 0

© The Macmillan Press Limited 1974

Published in the U.S.A.
by Halsted Press, a Division
of John Wiley & Sons, Inc.
New York

Library of Congress Catalog Card No: 73–16545

Printed in Great Britain by Thomson Litho, East Kilbride

PROCEEDINGS OF THE

FOURTEENTH INTERNATIONAL
MACHINE TOOL DESIGN AND RESEARCH
CONFERENCE

CONTENTS

OPENING SESSION

OPENING ADDRESS

by

A. G. M. GALLIERS-PRATT*

When I was approached, in the unavoidable absence of Lord Stokes who, as you know, was to have addressed you this morning and to have opened this important conference, I must confess that I felt some hesitation and trepidation in accepting the most flattering invitation to perform this ceremony today. There were two reasons for my hesitancy. The first is that I am conscious that Lord Stokes would, in his inimitable style, have made a most informative and interesting address to you. The second is that as I stand here this morning I must confess, before you discover it for yourselves, that although I am here as President of the Machine Tool Trades Association of Great Britain, I am not myself an engineer in the sense that you would consider such a precise definition to be attached to a person.

It is true that the whole of my working life has been spent within the machine tool industry, but I can claim no formal recognition from any engineering institute so that, and without further apology, I shall be speaking to you as a layman and one who has seen machine tools in use, has experienced their operation both in my company's own factories and in those of other installations that I have visited, but one who, perhaps from the sidelines, sometimes sees more of the game—or struggle—in which we all have to contend in meeting the needs of humanity without destroying irrevocably its ability to obtain truly full satisfaction from the results.

I am particularly pleased that this morning gives me an opportunity to say from our industry, the British machine tool industry, how delighted we are that you, Lord Bowden, Professor Koenigsberger and others associated in bringing about and organizing this distinguished gathering, have so framed the proceedings as to further cement the communication links between industry and the academic world.

In this day and age of progressively faster moving technological development, it is essential that these communications between, on the one hand the universities and other technological establishments, and industry should be fostered as closely as possible.

When I refer to 'industry' I am embracing both those that manufacture for the user and the user himself, who ultimately produces on the machine the component which is required in the process of building whatever the end product may be, whether it is a motor car or a nuclear power station!

Naturally, close though these tripartite links may be, in my view we still have ground to cover in order to establish more fully, and hence satisfactorily, the necessary understanding which is essential if the right technology is to be applied at the correct moment in time in order that manufacturing industry can have the benefits which it requires to satisfy the markets of the world commercially and to improve the standard of living, which must be our constant endeavour.

To this end such conferences as this mounted here this year in Manchester, provide the correct forum and the necessary background for education—in the broadest sense—and exchanges of views on the problems which we all must surmount if, as I have indicated, better understanding is to be built up so that metalworking machine tools can continue to make a maximum contribution to the well-being of peoples throughout the world.

May I, for a moment, speak purely from the benches of the machine tool industry—if I may adopt such a phrase—and use this opportunity to say something of which you are all aware but which, nevertheless, cannot be said too frequently. That is that the cyclical pattern of investment which has particularly bedevilled the industry which I represent, over at least two and a half decades, is one of its most difficult problems which up to the present time has proved to be insoluble. As I speak today this industry has emerged from the longest and most severe downturn in its fortunes in the post-war era. Unless some method can be found for ironing out these very low troughs, it is going to continue to be an inhibiting factor in the technological advancement which must, in the next ten to twenty years, contribute so much to bringing to the user the new and better, more efficient, more reliable and thereby most cost-saving machine tools and production techniques which must herald in the 1980s. Even the advancement of group technology—which I am glad to see receiving attention during the seminar—requires attention to the

---

* President, The Machine Tool Trades Association

ironmongery concerned if it is to make a maximum contribution.

The factors which new generations of machine tools will embody will include the all-important question of noise, of safety and of pollution. But, at the same time as we consider these and many other technological aspects, there is a further factor which I should like to comment upon as we cast our minds into the future.

This is the human one.

So far I have been briefly touching upon a few points that strike me at this moment in time and which I believe are appropriate to the opening address of this conference. But whilst, in the next few days, you will be hearing, reading and discussing many important and necessary technological aspects concerned with machine tools, may I ask you to remember the men and women who operate the machine, the men and women in the design offices and all those others, whether they be in sales. marketing, administration or other sections which all go to build up a successful company or a forceful team within a university or technical establishment.

We may be able, and indeed have proved that we can reduce machine cycle times, cut down on noise and increase the safety in operation of machine tools. But, at the end of a day, when we go home, what have we to look forward to and what can we expect? The increasing number of cars on the roads not only pollute the atmosphere but cause frustration to those drivers who, tired after a day's work, have to suffer the irritation of being delayed in getting back to their home and to their families.

Greater leisure time is good in that it can refresh us mentally for the exacting tasks which we all have to perform. Greater affluence in our society allows us to have the benefits of a modern technical age. But one thing which we cannot recreate as quickly as we tend to destroy it is our environment.

Great Britain, this tiny island of some ninety thousand square miles, cannot be enlarged. It is not possible to stretch out into the seas which surround us and take in more land. Our population has expanded by 4·6 per cent over the past decade. Today, despite increasing modern facilities which allow us to move from country to country more easily, we have not the will nor have found the way to provide any substitute for the country into which we were born or in which we have lived and which is our home and

our heritage. All and everyone of us must remember this and fight to preserve it.

New motorways may ease some of the frustrations to which I have referred. Better communication links are vital to industry. They make it easier for us to reach parts of this country which previously were almost inaccessible. We can, therefore, in our leisure time, find new places to visit and new unspoiled countryside to spend time in. But we must beware that what has been unspoiled is not now despoiled.

Sadly, in some instances in the guise of providing more recreational and other facilities, we see the countryside being ruined by over-commercial development. Caravan sites for holiday accommodation should be strictly controlled or else they will ruin the very reasons for people visiting them. New holiday homes in some of the least-spoiled countryside in this country, are being built, or applications to build are flooding in to our planning authorities. Just as we live in an era of two cars per household—or more—so we are gradually coming to see a large proportion of our countrymen acquiring a second leisure home away from the place in which they work. Do not imagine that I would wish to deny these facilities, but I would say to you 'Beware' for, in creating them, we may be destroying the very heritage which we wish to give us the necessary artistic and cultural and environmental refreshing which our metabolism requires.

Coupled with this, I would too make a plea for our architectural heritage which gives so much enjoyment and so much pleasure to so many. We are rightly proud of the buildings which our forebears created. We are the envy of the world in the magnificent richness of our architectural past. We must not let this be swept away or allowed to be desecrated in the name of progress.

If we are not careful and if we do not nurture all these pleasures of real living, we will find, at the end of the day, that whilst we have achieved much advancement in technology, when we return in the evening we have little in life to have made this worth while.

May I, in conclusion, say again how delighted I am to have had the privilege of addressing you this morning. I would thank you Lord Bowden, Professor Koenigsberger and your colleagues, for the excellent arrangements which you have made for this important conference which it now gives me much pleasure to declare open.

MACHINE TOOL ENGINEERING AT U.M.I.S.T.

by

F. KOENIGSBERGER*

Teaching and Research in technology are unashamedly tendentious and their tendentiousness has not been mellowed (as it has for medicine and law) by centuries of tradition. Technology is of the earth, earthy: it is susceptible to pressure from industry and government departments: it is under an obligation to deliver the goods. (Sir Eric Ashby, now Lord Ashby, *Technology and the Academics,* Macmillan, London, 1959).

This statement, which was made on the subject of technology in general, becomes even more appropriate when applied to teaching and research in the field of machine tool and production technology. As far as the community is concerned the design, manufacture and utilization of machine tools is not an end in itself but a means to an end, namely the production of consumer goods. In order to assist those who make and those who use machine tools—that is, in order to deliver the 'goods' mentioned by Lord Ashby—the universities who work in these fields must carry out three tasks.

(1) The education and training of men;
(2) the study and development of new ideas concerning manufacturing processes and procedures as well as of machine tool design, manufacture, investment, installation, maintenance and operation;
(3) investigation of specific problems arising in industrial applications, which may also lead to contributions under (1) or (2) or both.

In order to satisfy the requirements of these tasks not only the manpower but also the necessary equipment must be at the disposal of the university. While it is appreciated that the best equipment in the world has little value if men qualified to use it are not available, the study and teaching of industrial production problems will be, to say the least, hollow and incomplete if it cannot be backed by research and by operational demonstrations and experience.

In addition to the teaching of students there is an obligation to convey to the engineer in industry the latest developments in his field, to provide him with knowledge of and develop in him the ability needed to apply modern techniques of designing the necessary equipment, as well as planning and controlling its application. Here our tasks must be broken down into detailed topics so that each can be treated within the short period during which industry can spare those engineers who are later to apply the new techniques.

The tasks which we have set ourselves at UMIST thus cover the education of undergraduate and postgraduate students and of mature engineers. The breadth and depth to which the various topics should be treated vary, of course, because not only the prevailing knowledge and experience of the 'students' but also their available time must be considered.

While 'short courses' are concentrated on two to four days of intensive specialized work, the introduction of undergraduates to the subject may be spread over the academic year in doses of some one or two hours per week. The postgraduate studies and research which ultimately form the basis of the teaching activities require detailed work in great depth and may thus have to be extended over longer periods.

Last year we were given the opportunity of equipping our laboratories with the new plant and machines which were needed for carrying out our work comprehensively and effectively. We are dividing our research laboratories into five sections.

(1) Metal machining research laboratory (Fig. 1);
(2) grinding research laboratory (Fig. 2);
(3) electrical machining research laboratory (Fig. 3);
(4) the laboratory for general purpose conventional machine tools (Fig. 4a);
(5) the laboratory for NC machine tools (Fig. 4b).

In addition we are expanding the laboratories in which researches are carried out into machine tool structural elements, with particular reference to

---

* Department of Mechanical Engineering, Division of Machine Tool Engineering, U.M.I.S.T.

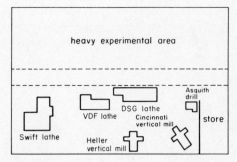

Figure 1. Machine tool layout in the metal machining research laboratory.

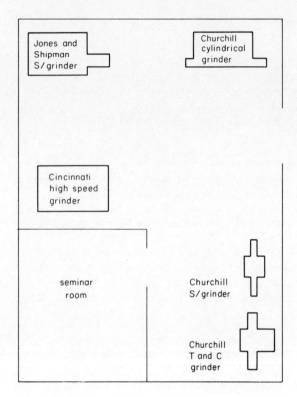

Figure 2. Machine tool layout in the grinding research laboratory (occupied late 1973, some equipment being earlier accommodated in the general purpose machine tool laboratory).

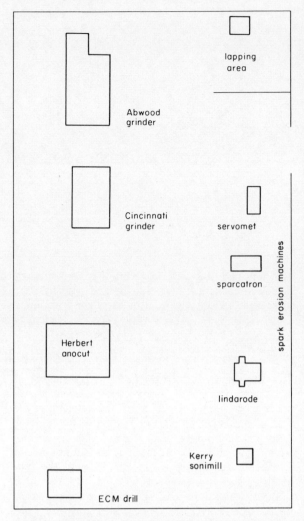

Figure 3. Equipment layout in the electrical machining research laboratory (equipment accommodated in the general purpose machine tool laboratory and installed in its final location in late 1973).

computer-aided design, and into machine tool control.

Although our fields of activity are thus clearly defined both technologically and geographically, the guiding principle of our work is the correlation of all our efforts so that none of the above-mentioned activities is allowed to take place in isolation. In this manner, for example, some work on drive and control elements is carried out directly on the machines in the main laboratory, while machining data for the work on NC machines is prepared in the metal machining section and tested in the main laboratory. Thus the 'goods' in Lord Ashby's definition can be given their right place within the overall framework of machine tool design and operation and delivered within the right context to industry.

One of our less spectacular projects concerns the performance testing of machine tools, in particular NC machines. Here we go beyond the boundaries of the laboratory and gather experience in factories in Manchester and in other areas both in this country and abroad[1].

In this respect as well as in many others it is most important to maintain contacts with other researchers and we are indeed fortunate that through the International Institute of Production Engineering Research (CIRP) we are able to take part in a number of cooperative efforts. Two projects among these may be mentioned.

(1) *Research into metal machining processes*, with particular reference to the establishment of cutting data banks. These are of great importance in the efficient programming of NC machine tools, and especially when adaptive control is to be applied effectively[2].

(2) *Computer-aided design of machine tool structures*. In this connection the application of the finite element method for determining static and dynamic behaviour under load is being studied and developed. The finite element method has also been used successfully for predicting the conditions in and behaviour of sliding and bolted joints in machine tools. Both the normal and the shear characteristics as well as the body deformations around the joint have been studied[3].

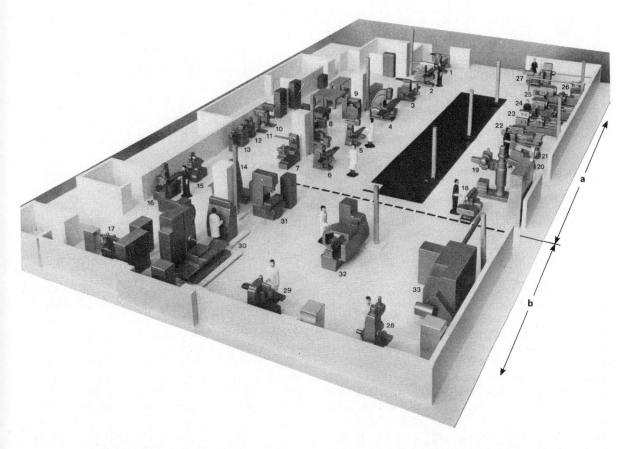

Figure 4. Model of the general purpose and NC machine tool laboratory (in its final form after the establishment of the laboratories in Figs. 2 and 3 in late 1973).

1. Adcock and Shipley horizontal milling machine
2. Parkson horizontal mill
3. Kearney and Trecker horizontal mill
4. Kearney and Trecker vertical mill
5. Beaver copy milling machine
6. Elliott vertical mill
7. Kearns horizontal boring machine
8. Southbend bench lathe
9. Sykes gear hobbing machine
10. Startrite drill
11. Elliott pillar drill
12. Dormer drill grinder
13. Abwood pillar grinding machine
14. Elb surface grinding machine
15. Jones and Shipman surface grinding machine
16. Jones and Shipman universal grinding machine
17. Brierley tool and cutter grinding machine
18. Butler shaping machine

19. Asquith radial arm drill
20. Startrite band saw
21. Rapidor 12 in saw
22. Dean Smith and Grace 1307 lathe
23. Elliott MIS lathe
24. Dean Smith and Grace 1910T tool room lathe
25. Dean Smith and Grace lathe with digital readout
26. B.S.A. No. 48 single spindle automatic lathe
27. Ward No. 3 autoward plug programmed capstan lathe
28. Newall 1520 jig boring machine with airmec PTL 100 control
29. Hayes Tapemaster 1380 with Plessey NC 4300 control
30. Herbert De Vlieg 75 JMC with Plessey 7300 control
31. Middlesex NC 200T turret drill with Plessey NC 11 control
32. Mastiff Commander lathe with Plessey 4600 control
33. Herbert B.S.A. Batchmatic 75–250

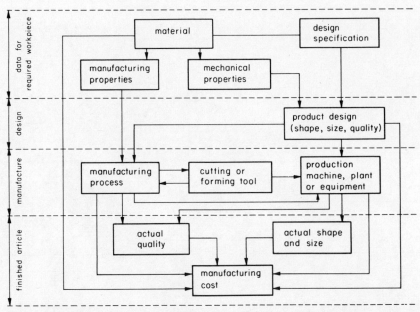

Figure 5. Interrelation between the components of the industrial production process.

Apart from the more fundamental research activities, in which international cooperation has been not only possible but also most fruitful, cooperation with industry has led to investigations into specific problems. These resulted in the study, evaluation and development of topics which concerned processes[4], procedures[5] and design applications[6].

However, we are not limiting our efforts to internationally planned or industrially inspired or supported research projects. Some considerable efforts are concerned with investigations into the fundamental processes for which basic information is not yet fully available. This applies in particular to electrical machining methods. These have been investigated for some time[7], and we are establishing a special laboratory (see Fig. 3) with up-to-date equipment for electrochemical and spark erosion machining, electrochemical grinding and ultrasonic machining.

The machine tool is an important part of a manufacturing system. Its purpose is the technically and economically efficient manufacture of workpieces which are, within specified limits, of a given shape, dimensional accuracy and surface quality. While in the case of other machines such as IC engines, turbines, motors, pumps or compressors the performance can be measured by comparing input and output energies, machine tool performance depends on a combination of technological, economic and human aspects. Any research and teaching in this field must, therefore, be geared to cover and interrelate all these factors. In particular, the combination of the influences created by the design of the product, the specification of the production process, the synthetic approach to the design of the production equipment, including the machine tool, and the analytical approach towards studying its performance must not be overlooked.

In other words the problems and the interests of the product designer, the machine tool designer, the machine tool builder and the machine tool user are closely interrelated. Figure 5[8] shows this interrelation

between the various components of the industrial production process, starting from the establishment of design data specification and leading via product design and manufacturing process to the completion of the finished product.

Engineering in general and machine tool engineering in particular is never static[9] and we can contribute towards its progress only if in addition to our own research and development work we enable all concerned to keep up with the work of others who work in the field. This is the purpose of the conferences at Birmingham and Manchester.

The three tasks, mentioned at the beginning, which we at UMIST have set ourselves are thus geared both to the needs of the machine tool manufacturer and to those of the machine tool user, both terms being employed in their widest sense.

For this reason we maintain close links with industry, because we will be able to make our contribution to the benefit of all concerned only if we keep our feet firmly on the industrial ground, even if our heads may sometimes be in scientific clouds.

The importance has already been mentioned of considering those engineers and managers in industry who wish to become acquainted with the latest developments, and the principles of their application. These men are not interested in obtaining degrees or other formal qualifications, in fact, many may not be qualified for admission to university courses. Moreover, as they cannot afford to spend lengthy periods of time away from their posts we provide intensive courses of short duration, say of two to four days, which deal with specific subjects such as

— metal removal processes
— computer-aided design of machine tool structures
— spindles, bearings and guideways for machine tools
— feed drives for numerical control
— numerical control
— accuracy and assessment of NC machine tools
— group technology.

In order to illustrate and practise in the laboratory those subjects which have been dealt with in the course lectures, full use is made of our facilities, the results of our investigations and studies playing a prominent part in the lectures and demonstrations.

In industry it is usual practice to distinguish between *research* and *development*, and in doing so to consider the former a more or less academic luxury with the latter as a practical yet necessary evil. This attitude is not only wrong but also harmful to progress because research—at least technological research—may be useless unless it is combined with or followed by development, and development which is not based on sound research is either trouble shooting or the perpetuation in a more efficient manner of antiquated ideas.

The fact is not often recognized that efforts in manpower and money can be reduced if purposeful research is used for preparing sound development. The operative word is, of course, 'purposeful'. This must always be borne in mind, not only by those who plan and carry out research but also by those who ask for it to be carried out in order to obtain answers to their problems.

Frequently, however, there may be a difference in the manner in which engineers in a university and those in industry approach a problem. In the latter an economical solution is required within a minimum of time, whereas academics have the reputation of deliberately ignoring expediency in order not to interfere with the serenity of their mind and the purity of their thinking. While some years ago this reputation may have been justified, it is certainly not applicable universally any more. On the other hand, a carefully planned, programmed and progressed research operation can be carried out to the benefit of all concerned if it is done independently of pressures which are unavoidable in industry, but with initiative and cooperation of all concerned. When the research worker has investigated a problem, the results of his investigation must be evaluated by designers and checked by production engineers in the workshop, whose experience will disclose new problems for the research worker, and so on. This cycle represents a mental servo-loop which must be a closed one in order to provide satisfactory results. Moreover, considerable reluctance—to say the least—is often encountered among those who would have to apply these results in industrial practice. Amongst such people there are still many—in fairly senior positions—who, before taking radical decisions tend to look in two directions, in one at men below their level whom they do not wish to disturb by upsetting their customary routine, and in the other at men above their level who may criticize their initiative if a newly introduced idea is not immediately successful.

In order to support his initiative and back his determination, the engineer in industry must be supplied with all the relevant information which he needs, and for this purpose suitable means and methods of communication must be created.

The art of communication between research and industry consists, therefore, of keeping a good balance between the minimum information which the men in each group require in order to carry out their work and the maximum necessary for stressing the justification or necessity of actions to be taken. We have tried to make this aspect one of the guiding principles in planning this conference, because we are convinced that the ability to 'deliver the goods', in the words of Lord Ashby, is the hallmark of the research worker needed by industry: the 'academic' who makes himself understood and appreciated by the industrialist, in fact the man who helps to advance industrial production towards its tasks within technological progress on all fronts.

The challenge lies in the research worker delivering the right goods and the engineer in industry being able to accept and make good use of the goods so delivered. If both accept this challenge their co-operation is easy and success is bound to follow.

REFERENCES

1. (*a*) R. WADSWORTH, A. COWLEY and J. TLUSTY (1970). Theoretische und experimentelle dynamische Analyse einer Horizontalbohr-und-fräsmaschine, *Fertigung*, 4/70.
(*b*) J. TLUSTY and F. KOENIGSBERGER (1971). New concepts of machine tool accuracy, *Annals of the C.I.R.P.*, **19**.
(*c*) F. KOENIGSBERGER, A. COWLEY and M. BURDEKIN (1971). Assessing the capabilities of metal cutting machine tools, *The Production Engineer*, June.
(*d*) C. P. HEMINGRAY, A. COWLEY and M. BURDEKIN (1972). Positioning accuracy of numerically controlled machine tools, *Proceedings of the 12th Int. M.T.D.R. Conference*, Macmillan.
(*e*) M. BURDEKIN, A. COWLEY and J. TLUSTY (1972). Establishing standard cutting conditions for performance testing of universal metal cutting machine tools, *Proceedings of the 12th Int. M.T.D.R. Conference*, Macmillan.
(*f*) E. J. GODDARD, A. COWLEY and M. BURDEKIN (1973). A measuring system for the evaluation of spindle rotation accuracy, *Proceedings of the 13th Int. M.T.D.R. Conference*, Macmillan.
2. (*a*) G. BARROW (1972). Tool life equations and machining economics, *Proceedings of the 12th Int. M.T.D.R. Conference*, Macmillan.
(*b*) I. YELLOWLEY and G. BARROW (1974). The stress-temperature method of assessing tool life, p. 000 in the present work.
3. (*a*) A. COWLEY and M. A. FAWCETT (1968). The analysis of a machine tool structure by computing techniques, *Proceedings of the 8th Int. M.T.D.R. Conference*, Pergamon.
(*b*) A. COWLEY and S. HINDUJA (1971–2). The finite element method for machine tool structural analysis. *Annals of C.I.R.P.*, **19**.
(*c*) S. HINDUJA and A. COWLEY (1972). The finite element method applied to the deformation analysis of thin walled columns, *Proceedings of the 12th Int. M.T.D.R. Conference*, Macmillan.
(*d*) A. COWLEY and S. HINDUJA (1972). Computer-aided design of machine tool structures, Conference on the application of finite elements in mechanical engineering design, The Institution of Mechanical Engineers, London.

(*e*) N. BACK, M. BURDEKIN and A. COWLEY (1973). Review of research on fixed and sliding joints, *Proceedings of the 13th Int. M.T.D.R. Conference,* Macmillan.

4.   (*a*) C. E. DAVIS and C. RUBENSTEIN (1972). Reciprocating plunge surface grinding with diamond abrasive wheels, *Int. J. Mach. Tool Des. Res.,* **12.**
(*b*) V. PACITTI and C. RUBENSTEIN (1972). The influence of the dressing depth of cut on the performance of a single point diamond dressed alumina grinding wheel, *Int. J. Mach. Tool Des. Res.,* **12.**

5.   (*a*) R. CONNOLLY and A. J. P. SABBERWAL (1971). Management structure for group manufacture, *C.I.R.P. Annals,* **19.**
(*b*) R. CONNOLLY, G. H. MIDDLE and R. H. THORNLEY (1971). Organising the manufacturing facilities in order to obtain a short and reliable manufacturing time, *Proceedings of the 11th Int. M.T.D.R. Conference,* Pergamon.
(*c*) F. KOENIGSBERGER, F. W. CAUDWELL, E. A. HAWORTH and H. H. LEVY (1972). Verbesserter Produktionswirkungsgrad durch eine computerunterstützte Produktionssteuerung im Zusammenhang mit Teilefamilienfertigung, *Fertigung,* 5/72.

6.   (*a*) M. E. MOHSIN and B. HODGSON (1964). A hydrostatically lubricated ram for a horizontal milling and boring machine, *Proceedings of the 4th Int. M.T.D.R. Conference,* Pergamon.
(*b*) J. F. JACKSON and R. BELL (1971). An electrohydraulic stepping motor for numerically controlled machine tools, *Proceedings of the 11th Int. M.T.D.R. Conference,* Pergamon.
(*c*) B. ENNIS and A. COWLEY (1971). Application of sealed hydrostatic bearings to the vertical slide of an horizontal drilling machine, *Proceedings of the 11th Int. M.T.D.R. Conference,* Pergamon.
(*d*) J. G. M. HALLOWES and R. BELL (1973). The dynamic stiffness of antifriction roller guideways, *Proceedings of the 13th Int. M.T.D.R. Conference,* Macmillan.

7.   (*a*) C. S. KAHLON, H. J. BAKER, C. F. NOBLE and F. KOENIGSBERGER (1970). Electric spark toughening of cutting tools and steel components. *Int. J. Mach. Tool Des. Res.,* **10,** 95–121.
(*b*) M. M. SFANTSIKOPOULOS and C. F. NOBLE (1973). Dynamic and geometric aspects of vertical spindle E.C.G., *Proc. 13th Int. M.T.D.R. Conference,* Macmillan.
(*c*) C. F. NOBLE and S. J. SHINE (1973). Electrochemical drilling using alternating current, *Int. Conf. on E.C.M.,* Leicester University, March.

8.   (*a*) F. KOENIGSBERGER (1970-1). The teaching of machine tool technology at university level, *Proc. I. Mech. E.,* **185,** 8/71.

9.   (*a*) G. A. B. EDWARDS and F. KOENIGSBERGER (1973). Group technology, the cell system and machine tools, *The Production Engineer,* July/August.

A RESEARCH AND DEVELOPMENT UNIT IN THE MACHINE TOOL INDUSTRY

by

L. K. LORD*

SUMMARY

This paper describes the circumstances which led to the setting up of a research and development unit to meet the needs of a machine tool manufacturing group, and some of the factors considered when doing so. The size of the unit is touched upon, and examples given of the wide range of topics dealt with during its twelve years' operation. Other sources of research endeavour used are briefly mentioned, and in conclusion the cost of the unit and economic considerations are discussed.

INTRODUCTION

In May 1959 the Research Council of the then Department of Scientific and Industrial Research considered a report on the research and development requirements of the British Machine Tool Industry presented by the DSIR's Economics Committee. Among its recommendations were that efforts be made to persuade the industry to take more responsibility for research and development. The Machine Tool Trades Association was the point of contact between DSIR and the industry, and also the vehicle by which many of the recommendations were given effect. Very early in 1960 the culmination of the MTTA's examination of the affairs of the industry and its future (which had in fact antedated the DSIR efforts by some two years, the MTTA's Research sub-Committee having been formed in 1956/57), was the acceptance by the industry of the notion that an extension of research activity within it would be beneficial to it.

One very important outcome was the formation that year of our Machine Tool Industry Research Association (MTIRA) but in addition stemming from that point in time there was an increasing awareness on the part of individual companies in the industry that independent research and development facilities properly used could greatly improve a company's capacity to develop new designs and so improve its competitive edge.

Whilst there are of course a number of R & D units in our industry the reasons for the existence of one particular department and its underlying ethos may be worth examination.

THE PARENT ORGANIZATION

It is probably desirable first to define the structure and size of the group that it serves, in order to enable one to determine the extent to which it may or may not be typical. Although the Group I represent has interests in the field of small tool manufacture as well as machine tools, the small tool companies have their own facilities, and so for the purposes of this paper we are concerned with a group of some five machine tool manufacturers with a total of rather less than 2000 employees (neglecting of course personnel in selling or marketing companies not involved in production) and a total turnover in the region of £7.5 m. In other words we represent maybe 4 per cent of the British machine tool industry. The greater part of the R & D endeavour (although certainly not the whole) is in support of the nearby principal machine tool plant in the Group, which accounts for about one half the number of employees and machine tool turnover quoted.

ETHOS

Although our own Research and Development Department was not set up until June 1961, about a year after the industry's collaborative venture MTIRA was established, this should not be interpreted as indicating that we did not agree that it could be uneconomic and wasteful for an individual approach to problems which are of universal application. We are of course in membership of MTIRA (and indeed I am currently the chairman of its Research Committee) and two other research associations as well,

---

* Group Director of Engineering, Wickman Ltd., Coventry

but there were two main factors which I think prompted us to set up our own facility.

Which work is done where?

Firstly we recognized the existence of different types or kinds of problem to the solution of which different kinds of research organization could well be better suited than others. From the earliest days and still today our policy has always been that fundamental and long-term applied research should go outside, and product development should stay within, the organization. The distinction between fundamental and applied research will be clear to all at this Conference. For the purpose of convincing boards of directors perhaps less familiar with the distinction we defined fundamental research as an attempt to advance the frontiers of one's knowledge although it is not even known what is going to be found. When something is found, it must then be turned via engineering to commercial advantage. In fact the company we are discussing, at least in the machine tool division, has never undertaken or sponsored fundamental research. However, the case of applied research, with which we are of course concerned, differs in that here one does know what one is looking for, and the search is for the 'best' way of achieving it.

Now it seemed to us that long-term research probably required academic qualifications of a kind more plentiful in Universities and Research Associations than in industry, and moreover the very nature of much long-term work rendered of little importance (or certainly of very much less importance) the question of commercial confidentiality or secrecy, so that to put such work outside is logical. On the other hand, product development, which many of our projects comprise, requires more practical skills and experience in the appropriate industry than are sometimes found in academic environments (although this would not to the same extent be true of many RA's); some of the problem-solving necessary is what may be called a 'fire-brigade' operation which often demands a speedy solution, whereas Universities for very good reasons are commonly not organized to deal with many problems urgently; and finally commercial secrecy is almost always vital to us in this area. For these reasons those aspects of our total work-spectrum are, in our view quite properly, retained in our own R & D department.

Thus virtually all our in-house work is in reality product development, although we may as a company be guilty, occasionally and unintentionally, of referring to it as research. In this paper, whether the word 'research' or the word 'development' is used, in relation to our own endeavours on our own site, applied research or product development is meant.

Human considerations

The second factor, which was an important one in our decision to have our own development capability, is concerned with technical manpower. Even the most ardent protagonists of the co-operatively organized centrally available research facility for the industry will agree that in order to benefit from research, wherever it is done, the individual employer must have personnel who are able to digest, interpret and apply such research results to their own organization.

Of course the MTTA had itself (in fact in 1959 before its efforts which culminated in the establishment of MITRA) recognized that the achievement of a position of leadership for the industry *vis-à-vis* overseas competition in applying the newer technologies to machine tool design would beforehand require soundly trained design engineers, and had therefore established Design Scholarships at this Institute to enable competent and ambitious engineers to attend a two-year full-time course specially aligned to the needs of machine tool design. Moreover, somewhat later, the MTTA established first a Higher Degree Scholarship scheme to enable Science and Engineering graduates employed by member companies to undertake full-time post-graduate studies in a field related to machine tool technology and leading to an M.Sc. degree. These have been tenable at a variety of institutions. Then, in 1967 the MTTA set up the unique Integrated Graduate Programme at the University of Birmingham which provides for honours graduates a specially selected two year course comprising not only advanced specialized technological study but also a period of meaningful industrial participation leading to M.Sc. qualification.

We have as a Company greatly benefited from our participation in all these schemes, but able technicians and technologists would not for long, we feel, be content to continue merely to apply the results of the efforts of others. If our industry is to attract and then retain the most able minds and workers in these vitally important technological fields, then it must create an environment in which such men will flourish, and the work task must be scientifically satisfying. They must have the opportunity of pursuing worthwhile investigations themselves, and indeed failing from time to time, in order that they are stretched intellectually. Moreover, I personally believe they should be encouraged to discuss their work with others, and that they should not be prevented from publishing some of their work. To associate a man's name with some specific piece of work, and expose it to the critical scrutiny of his national and international fellow-investigators, whether the work is of a research kind or of a more directly creative kind, should, I think, stimulate accuracy and strict objectivity in the case of research and success in the case of design. That is my personal view. Developing a capacity then, for quickly applying research results wherever the research is done, was the second factor influencing our decision to establish a research and development department of our own.

THE DEPARTMENT

For convenience I will refer to the unit by the initials of the title by which it has always been known—CRD or Central Research Department—'Central' to distinguish it from our hard metal company's research department which is close at hand.

Size

The size of CRD is very modest, 15 000 sq ft (1400 sq m) and with a total staff of twenty-four, a figure which has been as high as thirty-two. The facilities comprise a large Projects Area organized on an open-plan basis to encourage interaction and the cross-fertilization of ideas between the various project teams involved, surrounded by supporting services such as an Engineering Office, Model Shop, Instrumentation Laboratory, Mechanical Laboratory, Electrical/Electronics Laboratory and a Technical and Patents Library.

ASPECTS OF ITS WORK

Our success, like that of any research unit, can only be judged in retrospect, and twelve years is a convenient milestone from which to look back and see what has been achieved.

Hydrostatic bearings

Possibly the first in importance and certainly the first chronologically (for the subject was that covered by our very first Research Report in January 1962) is our work in the field of hydrostatic bearings. Initial investigations were devoted to oil-hydrostatics applied to slides in February 1962 culminating in a successful application of air-hydrostatics including auto-compensation for asymmetrical loading and table droop and to a large planer table used in the grinding mode in May 1964. The application to spindles (both journal and thrust bearings) commenced in 1963 with air bearings successfully run on a small surface grinding machine in November of that year, and on a larger machine in June 1964. While these early attempts at air-hydrostatics were relatively primitive, by March 1966 the spindle of the larger machine had been optimized in the CRD using computer methods with the very satisfactory result that we achieved a greater than four-fold resultant stiffness at the point of grind, and well over five times the axial stiffness, all with rather less than one quarter of the original airflow.

Following the application of oil-hydrostatics to a then new internal grinding machine in July 1964 and initially of air-hydrostatics to centreless grinding machines in November that year, it was realized that for versatility and applicability to a wide range of duties, including the wide-wheel centreless grinder where substantial forces are called into play, oil-hydrostatics provided the most rational approach, and so in October 1965 oil was adopted in lieu of air throughout. A considerable step forward was made when, in the following month, a simple but efficient double diaphragm valve form of variable capillary resistance for oil-hydrostatic journal and plane bearings was developed in CRD so that when the first hydrostatic bearing wheelhead version of our internal grinding machine was exhibited at Olympia in June 1968 it was easily capable of consistently grinding within 30 millionths (0·8 μm) for roundness.

EDM work

Another quite different area in which CRD has consistently contributed to the company's knowledge from its earliest days is that of EDM, formerly known as electro-erosion or spark-machining. It was in May 1962 that CRD started to prepare designs for an 80 A split output EDM generator, and in August 1964 the prototype, fully tested on a variety of spark-erosion machine tools, was passed over to our Electro-Mechanical Department for series production. Today, new versions of EDM generator are entering production, the development of which has used the wide experience of many of the same team which designed and built our very first electro-erosion machine more than ten years before CRD was set up (in fact in 1951/52) and several since. Considerable development work was also done in the related areas of electrolytic and electro-chemical machining, and the breadth of our experience in these fields can be judged on turning the pages of CRD reports, for we see work on electrically conductive grinding wheels, tests on graphite electrodes and much effort on moulded graphite electrodes, parallel work on reformable spark-erosion electrodes, the development of an electrolytic form grinding machine (for carbide form tools), and many more exciting projects up to the present day.

Centreless grinding

To revert for the moment to research in support of grinding machines an early and important project was the Mechanics of Centreless Grinding, already briefly mentioned, which we commenced in November 1961 here in UMIST. The first phase of that work, which for the first time laid bare precisely how round work is generated in the centreless process, and established a number of important criteria to be satisfied if the machine is to be capable of rapid rounding-up—design criteria that we have of course been quick to exploit throughout our range—was completed in November 1964. In a related field are CRD studies into the stiffness and dynamic performance of machine structures, which originated with a vibration analysis of a small Centreless Grinding Machine in April 1963. This particular machine study was not completed until January 1966. In the intervening thirty-three months we had by analytical methods reduced the two principal modes of vibration by no less than 80 per cent, and the lessons learned thereby have made every subsequent centreless grinder of our manufacture superior in performance, but as a matter of interest the cost of finding this out was £7000 in labour and computer charges alone. Although the cost is not insignificant, our customers of course benefit.

Machine control developments

A development which has proved highly popular in relation to our single spindle bar automatics has been the photo-electrically actuated punched-disc programme control equipment also designed and engineered in CRD. The extra-cost option, which provides pre-programmable control over spindle speed range, spindle direction, selection of feeds and fast motion, and such auxiliary functions as coolant, is particularly valuable in conjunction with pre-set tooling, also optionally available with all our automatics. Requested in June 1963 the design was completed in the early part of the following year and built into a prototype machine then of $2\frac{5}{8}$ in capacity in October 1964. A full test programme was undertaken with the result that on passing this equipment to production in March 1965 the CRD's work was, at least for the moment, complete.

Projects not proceeded with

Of course, not every development project comes to fruition, either for technical or economic reasons, perhaps for political ones, or maybe simply because the development is several years premature. To be seen in our project reports are such fascinating topics as electrostatic precipitation for dust collection on grinding machines; automatic line following—as a development at one time considered for our optical profile grinding machine; ultrasonic cleaning of grinding wheels in use; alternative machine tool base and bed constructional materials, including for example concrete and granite; a coolant-driven turbine for powering internal grinding machine spindles; and infiltrated silver dispersions in abrasive wheels for electrolytic grinding. None of these was proceeded with, but quite considerable development effort was expended on every one of them.

Side benefits

Some development paths are long and tortuous, and not infrequently problems met with en route have to be solved first. Such situations result in quite useful development items, usually pieces of hardware, which are really peripheral to the central problem, but which nevertheless are necessary as a means to an end. Examples of this kind include the design and development of a range of interchangeable static switching function modules, including a single-phase 100 VA static switch, a 3 ph static contactor for a 1 hp ac motor, a static latching circuit, a static time-delayed switch and a heavy-duty (400 VA) static switch, all engineered in 1965 and 1966 for use with our range of Automatic Assembly machines then in design, a tacho-generator speed indicator capable of measuring spindle speeds from 300 to 5000 rev/min to within ±2 per cent for use with a planetary high-speed spindle drive developed in 1965 for a Template Milling Machine not in the event proceeded with; transistorized peak-load measuring equipment for use on coining presses; and a range of capacitive displacement transducers developed in 1966 for a heavy single spindle chucking automatic then in its early stages of design. These transducers were developed to meet our own need of equipment to compare the performance of hydraulically- and mechanically-actuated slides then under consideration for that machine. There are many other such examples which have resulted in a useful piece of instrumentation or equipment which fills, or at one time filled, a gap in what is commercially available.

Persistent projects

Mention has been made of our first Research Report early in 1962 on the subject of hydrostatic bearings, and remarkably enough some of the topics we were investigating twelve years ago are to be found among our current projects today, some 240-odd Research Reports and Technical Notes later. They may not have been the subject of continuous endeavour, but set aside on the completion of one phase, and picked up again more recently as new opportunities for exploitation emerge, or perhaps as developments in, for example, materials technology and production techniques make feasible in the 1970s something impossible or uneconomic in the 1960s. Thus, among the topics looked at in the past we are today still investigating hydrostatic bearings, now in relation particularly to the demands of high-speed grinding with its peripheral speeds up to 12 000 ft/min (60 m/sec); EDM generators, the first of a new generation of these, a single 30 A and a single 60 A unit, with the development and engineering of which CRD have been very closely involved, finding a place at the International Machine Tool Exhibition at London, last year; the Mechanics of Centreless Grinding, the subject of research sponsored originally jointly by us and the Science Research Council, originally, as mentioned, at this University but now at the Lanchester Polytechnic, Coventry; and one or two investigations facets of which are being pursued specifically in relation to new design projects in the Group.

Investigations 'extra to programme'

Informal or ad hoc investigations also occupy many R & D man-hours. These are investigations which are 'extra to programme' in that they are not primary projects, but they are nevertheless studies for the completion of which CRD, with its access to expertise in a variety of disciplines, is admirably equipped. Instancing a few at random we see the automation of a tool and cutter grinder for automatic continuous diamond-wheel testing; an investigation into isolation mountings for a precision grinding machine; the examination of service failures in proprietary equipment; planer table thrust measurements; investigations into electric motor balance; and noise measurements of many kinds, on ball mills in our Hard Metal Division, on natural-gas-fired boilers in the Machine Tool company's main works, cyclones at another of our Coventry factories and, of course, tests on the Group's machines, such as multi-spindle hydraulic chuckers, internal grinders with high-frequency spindles, and high-speed coining presses to name a few potentially noisy classes of machine.

Some current work

Other projects on which CRD are currently engaged, by no means as long-established as some mentioned earlier but nevertheless of extreme importance, and on some of which many thousands of research man-hours have already been spent, include rig and life testing in support of designers throughout the Group. It would not be appropriate at this stage to reveal the nature of most of this current work, but an area of importance to all of us, manufacturers and users alike, is noise, briefly referred to above. Excellent progress has been made with investigations into means of reducing multi-spindle auto noise, mainly from stock tubes, main-drive transmissions, and hydraulic elements. Perhaps the aspect of this with which the multi-spindle auto user will be most concerned is stock tube noise, and here we have pursued a programme of stock tube development to see whether the use of various forms of double-walled construction could bring about improvement. A variety of fillers has been tried, such as mineral fibre, granulated or foamed plastics, proprietary sound barrier mattings, sound deadened steel, resin-bonded laminated steel sheet and so on. Recent developments in this and other directions are showing very great promise, so that our contribution to noise reduction in auto shops will, we believe, prove to be no mean one.

Relationships with other research bodies

It has already been mentioned that we are in membership of collaborative research associations, and it is through these that we endeavour to keep abreast of parallel developments in the machine tool, mechanical and production engineering research fields, and indeed in the case of two well-known RA's we are actively co-operating in certain of their research programmes to which some of our own domestic endeavours are complementary. Use is also made as appropriate of Government establishments such as NEL, East Kilbride and RAE, Farnborough. Close relations are also maintained with the leading British centres of University research in machine tools, namely Birmingham and Manchester (to name them in alphabetical order) and we have worked with other establishments such as the Cranfield Institute, The Universities of Cambridge, Aston, Sheffield, Southampton and Warwick, as well as of course Coventry's Lanchester Polytechnic.

ECONOMIC CONSIDERATIONS

Cost

Of course, R & D is expensive, but we have proved that it need not be all that expensive. We invested a little short of £100 000 in establishing CRD in 1961, covering structural alterations to premises we already owned and which fortunately were available, new plant, furniture and office equipment, and the transfer of existing equipment from the several establishments in Coventry in which separate development activities had formerly been undertaken. To salaries and personnel charges and so on, must be added the cost of research rigs, materials, instrumentation, hire of computer time etc.—charges directly concerned with the R & D activity, which amount to about 20 per cent of salaries. General expenses such as depreciation and maintenance, standing charges and sundries account for a further sum amounting to about 15 per cent of the salaries. All three major headings comprise the gross annual cost of the facility, and in the case of the unit described, excluding the cost of prototypes, admittedly a very considerable sum but one which varies from year to year, our in-house endeavour excluding design-development, the cost of which is very heavy, and

currently totals around 1 per cent of the aggregate turn-over of the companies it serves. By comparison the aggregate annual cost of subscriptions to the three RA's to which we belong is about one twelfth of this (although to the RA's themselves it is worth substantially more as it attracts a Government grant).

Justification

Can expenditure on in-house research and development be justified? The cost effectiveness of it is certainly not easy to measure. At one extreme it may be argued that expenditure at a level of only one or two per cent of turnover is, like advertising, something no company in a technological field can afford not to do, and so, again like advertising or for that matter exhibitions, its cost effectiveness is perhaps irrelevant. The other extreme is to expect an R & D operation to be capable of standing on its feet and be a profitable area like any other cost centre. For this policy to succeed it would be necessary for a research department to 'sell' its expertise and solicit a programme by encouraging divisions of the organization to sponsor projects and pay for them. The CRD operates a policy between these two extremes, and its aim is not only to see that a wide range of technological developments are exploited for the benefit of the Group, but also to provide a capability for dealing with product development problems, product (or technique) evaluation, and for consultation.

CONCLUSIONS

We believe that subsequent events and achievements have vindicated the decision to establish our own capability, and justified the investment which has amounted to no mean sum over the twelve years existence of the department. Quite apart from the tangible products of our endeavours there are intangibles which could be added to the benefit side of the equation, the training value of in-house R & D, the contribution made by the facility to a Company's 'image', and so on. These are benefits to the company. And on a wider scale there are of course corresponding benefits to the industry and hence the nation. For example when wastage occurs, and some labour turnover is inevitable, the leavening effect of this is of benefit to the industry.

However, for the reasons discussed the possession of one's own facilities has not been exclusive, and we continue to use, as appropriate and from time to time, the special capabilities of other sources of expertise such as RA's, Government Establishments, Universities and academic centres.

MACHINE TOOL UTILIZATION
AND PERFORMANCE

AN ANALYSIS OF MARKET REQUIREMENTS FOR NC TURNING

by

E. F. MOSS*

SUMMARY

This paper outlines one machine tool manufacturer's rationalization of requirements for a new range of N.C. turning machines for the majority user. Specific applications are described and conclusions are given based on the results obtained.

INTRODUCTION

This paper sets out the basic marketing, design and production policy of Hydro Machine Tools Limited with regard to N.C. turning. The whole question of user requirements in the next decade was analysed against a group marketing philosophy which, historically, has been the volume production of machine tools for world markets.

The volume market for turning machines is already clearly established; namely, the smaller and medium sized firms which form the bulk of engineeering companies throughout the world. The problems associated with these companies are:

(1) limited investment capital.

(2) rising wage rates.

(3) scarcity of skill and labour.

(4) relatively simple production set-up and know-how.

The conclusions to be drawn from these factors are that there is a definite market for N.C. turning machines of relatively low cost, but with an effective range of performance and flexibility to meet a wide variety of requirements. At the same time, such machines must be simple to understand, operate, and maintain. This marketing specification has proved highly successful with the centre lathe which is, according to world statistics, still the most widely used machine in engineering. This overall requirement, and the economic and production factors detailed above, provide a very reliable guide to the market available for the volume production of N.C. turning machines.

USER REQUIREMENTS

In the United Kingdom at least one hundred manually operated centre lathes are sold every week—on a world-market basis, the weekly sales total well over one thousand. Most of these machines are bought for use in small or medium batch production.

The problem now facing these users is that rapidly rising labour costs and standards of living are making it less and less feasible to continue producing by such methods with such a high dependence on skilled manual labour. Of necessity, the trend is definitely towards greater automation of the production process. Having said that, most automatic machines lack the essential flexibility for small and medium batch production and are often too expensive for the majority of smaller firms. As a consequence, there remains a heavy reliance on the centre lathe.

In view of the above it is not surprising to find a rapidly growing interest in the lower-cost N.C. turning machine-system combinations now emerging in the price range of £10 000–£25 000.

With these lower unit costs has come a growing awareness on the part of the lathe user that he can no longer afford to put off an in-depth appraisal of N.C. turning, with regard to his own specific requirements, in order to remain industrially competitive.

However, very little information is available on the significance and criteria relating to the potential for N.C. turning in the majority of companies where both capital and engineering resources are limited.

PRODUCT SPECIFICATIONS

The first task was to establish essential parameters of N.C. turning for the majority user and optimize performance requirements against cost. Apart from

---

* Managing Director, Hydro Machine Tools Limited, and Engineering Director, 600 Group Machine Tool Division. Previously Technical Director, Colchester Lathe Company Limited.

the United Kingdom, where capstan lathes are still popular, most production turning is still done on centre lathes of varying degrees of sophistication, but, in general, retaining an inherent flexibility for use as a chucker or bar machine, a between-centres lathe, or screw-cutting machine. It was clear that this general purpose facility should at least be maintained if not extended further towards the full machining centre concept. It was also accepted that the essential re-education of users towards N.C. would be more readily achieved by maintaining a comparatively simple and familiar configuration of the basic elements.

Once basic specifications were established, selective use was then made of appropriate and already well-proven components available from The Colchester Lathe Company Limited with its high weekly output of some two hundred lathes. Although the degree of rationalization with the high volume standard centre lathe was deliberately limited, it nevertheless provided a significant contribution towards a favourable optimization of overall performance and cost.

The designs were finally rationalized to two basic models with the following specifications:

The Hydro NC 390

Swing 390 mm, 1250 mm between centres, $7\frac{1}{2}$ hp 16 speed spindle drive. The machine has a point-to-point two axes numerical control system with all conventional manual centre lathe controls and facilities retained to provide high versatility. It has been designed for the production of small/medium batch turned components with the emphasis on ease of programming and setting. The basic specification is simple and offers special appeal for advanced educational purposes. Additional features are provided to extend the capability of this machine to production purposes.

Both machine axes are driven by high performance electro-hydraulic stepping motors under the control of a Pratt 21 L100 system.

The input medium is punched paper tape in E.I.A. or I.S.O. coding and a simple tab sequential programme format allows full straight cut and taper facilities with tape control of feed and rapid traverse rates.

All features of a manual centre lathe have been retained and changing from numerical control to manual control for setting purposes is achieved by simply switching off the numerical control system.

Pre-set quick-change tooling is standard.

Optional features include a continuously variable spindle drive, four-position eight station indexing turret, four pairs of tool offsets, hydraulic chucking, threading and profile copying—all under tape control, where appropriate. Variable speed control to the spindle is effected by means of an electronically actuated Kopp Variator giving 50 speeds in each preselected gear over a range of 4:1 at any one setting at virtually constant horse power.

The hydraulically actuated indexing turret is provided with two Dickson interfaces on each of four faces, allowing the mounting of eight tools, if required, and precision of engagement is ensured by use of a hardened and ground face-tooth coupling. Each face is provided with a coolant supply distributed from the central turret stem.

The profiling system enables contours outside the scope of the normal tape programme to be performed by means of a conventional high-level template. The price for production versions (fully tooled) is about £13 000.

The Hydro NC 540

Larger and more sophisticated than the Triumph Commander NC, with a swing of 540 mm, and 1525 mm between centres, with continuous path control of two axes. An infinitely variable 15 hp spindle drive is provided, together with tape controlled 8-station indexing turret. The control system itself is very comprehensive in its basic form, incorporating buffer storage, manual data input and tool offsets as standard features.

The main machine axes are driven by D.C. servos controlled by an Allen-Bradley/Plessey 4600 Series system, giving a full slope and arc facility (linear and circular interpolation), and including buffer storage, and absolute or incremental programming as standard.

The input medium is punched paper tape in E.I.A. or I.S.O. coding and the word address programme format.

The D.C. servos are of the permanent magnet type and are driven by an S.C.R. system. Motion to each axis is via 1:1 gearboxes to preloaded ball-screw assemblies. The spindle drive consists of an electronically actuated Kopp variator input driving the spindle via four manually selected gear ranges. By this means, four overlapping ranges, each of 6:1 ratio, constitute the speed range ($7\frac{1}{2}$–1400 rev/min) of the machine. Hydraulic clutches allow for stop/start and forward/reverse operation under full tape control and a hydraulic disc-brake acts directly on the spindle when the drive clutches are disengaged.

The automatic indexing turret mounted on the cross slide consists of a four-position indexing member carrying on each face two quick-change interfaces for attachment of eight toolholders. Indexing and clamping are effected hydraulically, and a precision hardened and ground face-tooth coupling interposed between turret and base ensures precision and rigidity of clamping.

All machine slideways are faced with a PTFE based anti-friction material with a one-shot system which automatically injects a supply of lubricant to the slides on operation of the machine 'ON' control and also when a lubrication push button is depressed.

Manual controls on the control console include 'turret index', feed override, spindle override, brake 'off', jog, return to datum, and a remote control unit on the machine sliding guard gives the facility for emergency stop, auto lubricate, cycle start, feed hold, spindle hold.

There is an option of single or twin-turret configurations. Drilling, threading, and light milling facilities are also provided. Quick change, pre-set tooling is standard. Price range: £20 000–£24 000, fully tooled.

CASE STUDIES

The following applications of the machines described above provide a basis for evaluating the viability of the general purpose N.C. turning concept.

1. Lathe spindle

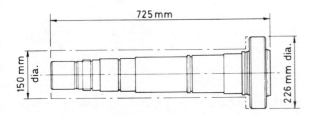

Figure 1. Lathe spindle (En9 forging).

Prior to introducing N.C., this lathe spindle was produced by means of one automatic copying lathe and two conventional centre lathes.

The spindle flange face was first rough turned by centre lathe and then turned to finish size, apart from grinding allowances and a limited number of grooves plunge-formed. The spindle was then handed to the second centre lathe for threadcutting and additional grooving.

The times for the above operations, including loading, unloading, and checking with set-up times shown separately were as follows:

	Operation times min	Set-up times min
Auto copy lathe	26·50	240
First centre lathe	14·50	30
Second centre lathe	6·75	15
	47·75	285

Therefore, total machine operation time, per spindle, for a batch of one hundred components, including set-up time = $47·75 + \frac{285}{100} = 50·60$ minutes.

For a batch of one hundred components, the total operator time, including work handling, was 3125 minutes, i.e. 31·25 minutes per component.

By introducing the N.C. lathe, the three previously separate operations of finished turning, threading, and grooving were completed in one operation on one machine only with one part-time operator in thirty minutes processing time, including loading, unloading and checking.

The related set-up time, including changing chuck jaws, adjusting tailstock, changing support centre, resetting z-axis datum, changing the control tape, and corrections to first-off component required a total time of 10 minutes. Therefore, total time, per spindle, for a batch of one hundred components, including set-up time = 30·1 minutes. The operator time per component was 10 minutes.

Summarizing, the introduction of one N.C. lathe effectively eliminated the need for three separate machines and related operators. Throughput time for the one hundred batch was reduced by 40·5 per cent and direct operator time by 68·0 per cent. Similar gains were established for other sizes of lathe spindles for which the same Hydro NC 540 was used.

The following case studies cover the application of the Hydro NC 390 Straight-Line N.C. lathe to work previously done by centre lathes, capstans, and chucking autos.

2. Speed selector dial

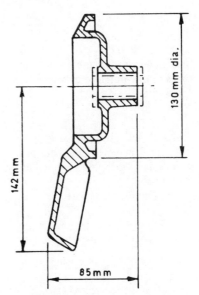

Figure 2. Speed selector dial (L.M.4 aluminium gravity die casting).

Previously machined on a capstan lathe in batches of five hundred against a requirement of thirty per week.

Operations
(1) Load to special jaws on main diameter and angled face.
(2) Rough and finish bore.
(3) Turn face.
(4) Reverse in chuck and turn second face.

Results

	Capstan lathe min	N.C. lathe min
Set-up time	120	10
Cycle time	7·05	2
Total run time for batch, including set-up time	3645	1010
i.e. 73 per cent saving on previous throughput time		

3. Spindle bearing covers

Figure 3. Front cover (L.M.4 aluminium).

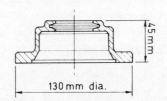

Figure 4. Rear cover (L.M.4 aluminium).

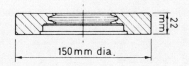

Figure 5. Front cover (grade 14 cast iron).

Front bearing cover–Fig. 3
Previously machined on a manually operated centre lathe in batches of one hundred against a requirement of eleven each week.

Operations
(1) Load to hard jaws gripping an inside diameter. Finish turn face, diameter and chamfer.
(2) Load to soft jaws gripping on outside diameter. Finish turn bores. Plunge turn oil collector groove.

Results

	Manual centre lathe min	N.C. lathe min
Set-up time	60	10
Cycle time	7·75	3·0
Total run time for batch including set-up time	835	310

i.e. 63 per cent saving on previous throughput time

Rear bearing cover–Fig. 4
Previously machined on a chucking auto in batches of two hundred against a requirement of eight each week.

Operations
As previous component.

Results

	Chucking auto min	N.C. lathe min
Set-up time	90	10
Cycle time	1·25	1·55
Total run time for batch including set-up time	340	320

i.e. 6 per cent saving on previous throughput time

Front bearing cover–Fig. 5
A similar component to Fig. 3, but in this instance previously machined on a chucking auto in batches of three hundred, against a requirement of eight per week.

Operations
As for previous component.

Results

	Chucking auto min	N.C. lathe min
Set-up time	90	10
Cycle time	4·9	3
Total run time for batch including set-up	1560	910

i.e. 42 per cent saving on previous throughput time

The above four cases have a low weekly usage, but relatively large batch sizes were necessary to justify using machines with long set-up times. With N.C., the set-up time is very low and economic runs on much smaller batches can readily be achieved.

4. Leadscrews

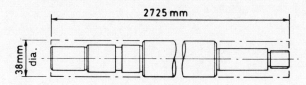

Figure 6. Leadscrews.

Previously produced on capstan lathes, with a separate set-up for each end, in batches of one hundred against a requirement of twenty per week.

Operations–N.C.
(1) Load to collet.
(2) Face off.
(3) Centre drill.
(4) Turn diameters.
(5) Cut recesses and chamfer.
(6) Screwcut.
(7) Programme stop and reverse bar in collet.
(8) Face off.
(9) Drill and tap $\frac{3}{8}$ UNC x $\frac{3}{4}$ deep.
(10) Turn diameters.
(11) Cut recesses and chamfer.

Results

	Capstan with copy unit min	N.C. lathe min
Set-up time	180	15
Cycle time	14·50	14
Total run time for batch including set-up time	1630	1415

i.e. 13·2 per cent saving on previous throughput time

Advantages over the previous method were that a flow could be created in the production line because there was no longer any need for batching; one operator could attend the N.C. lathe as well as a grinding machine, a thread whirler and a spline shaper and set-up times between each leadscrew were cut drastically, with operator time reduced from 14·50 to approximately 2·0 minutes.

This application highlights the potential of N.C. turning for reducing production times, work handling, and also stocks, by combining the first and second capstan operations into one N.C. lathe operation, stopping only to reverse the component. The same combined operation could not be done on a capstan because of the limited number of turret stations which can be made available.

CONCLUSIONS

Prior to N.C., companies with small to medium batch production have generally chosen to use either:

(1) low cost centre and/or capstan lathes with the inevitable commitment to high labour costs.
(2) relatively high cost single or multi-spindle automatics, the economics of which dictate large

batch sizes resulting in low stock turnover—often no better than two or three times a year.

The case studies cited in the paper illustrate the considerable advantages to be gained for small/ medium batch production by the use of N.C. turning machines against conventional methods.

(1) Faster set-up and throughput times, facilitating smaller batch sizes and consequent higher stock turnover figures. With N.C., there is every reason to expect stock turnovers as high as ten or even fifteen times a year.
(2) Reduction of work handling and stocking between operations.
(3) Higher labour utilization and a reduction of direct labour costs.
(4) Savings achieved by elimination of special tooling.
(5) Increased machining efficiency as a result of continuous improvement exerted by programme control—eliminating setter and operator bias.

For the future, there is a very large potential market for N.C. turning machines within the price and specification parameters outlined in this paper. As the demand increases to allow really high volume production of such machines, then the whole pattern of industrial turning is likely to be transformed.

DISCUSSION

Q. L. Wood. There seems to be two schools of thought in the design of this class of turning machine. One design uses a tape for control, the other a form of plug board whereby the operator sets up the control functions of the machine. Would the author comment on these two features and explain why the tape controlled machine was preferred.

For shaft work much quicker positioning of the tailstock was desirable with in built scales for position.

A. Both schools of thought are accepted as being valid customer requirements. As a manufacturer we have chosen to first market the design using a tape for control because the market requirement for this is much more clear than any form of plug-board programming.

A hydraulic tailstock is available for faster quill positioning. Agreed built-in scales could improve correct positioning.

Q. P. C. Hagan, Eaton Ltd. Axle Division. Your market analysis appears to have been very extensive, however, the machine design appears very orthodox in construction i.e. it has manual controls for feed and speed, etc., as well as a long bed type of construction with a tailstock. Bearing in mind the results of Tinker at M.T.I.R.A. on machine tool utilization of machine features and Opitz's work on component statistics and the success of the short bed lathe, 'why did your company'

(1) Not fit automatic speed and feed change to have a more complete control of the processes involved.
(2) Change the basic machine design to fall in line with the results of component statistics published to date (it is understood that these statistics will vary according to the individual companies products).

A. 1. Automatic speed and fitted change controls have in fact been provided as standard on the basic full continuous path machine and on an optional basis on the smaller straight line machine.

2. The volume market for N.C. turning is to be found in the small to medium general purpose companies, who in our experience demand a highly versatile machine capability generally in accordance with the specifications as briefly defined in the Conference paper. Note: In many cases the small general purpose sub-contract man often requires extra machine capability or features simply as insurance against possible minority order requirements.

NC—YOUR CHOICE

by

C. G. SCARBOROUGH*

HISTORICAL

The year 1958 saw the introduction of the first NC machine into Rolls-Royce. This was a three axis machine to which a control system had been retrofitted. Its main use was in the manufacture of former plates and three-dimensional model blocks from which die blocks for engine blades were copied. This combination demonstrated to the company the potential value of NC.

Although a computer had been installed in 1954, it was used exclusively for calculations relating to blades stressing and detailing. It was, in fact, too rudimentary for NC purposes, and as a consequence, tapes for this early NC system were obtained from a bureau.

The installation of a more modern computer in 1959 meant that work could be started to enable NC control tapes to be produced 'in-house'. Two basic systems were developed (PROCONSEL and COCO-MAT) and these two systems, which have been subject to considerable development, are still in regular and profitable use today. These systems showed clearly the benefits of linking NC with computer support.

From 1958 to 1962, a number of point to point machines were introduced into various parts of the company, while in 1962 a new 3-axis milling machine was introduced. This was a Newall Contimatic fitted with a Ferranti control system. The result was a higher quality of work than had been achieved on the retrofitted machine previously used.

Progress until the end of 1966 was steady but unspectacular. A rapid increasing involvement in NC began in 1967, a year which also saw the first NC lathes and machining centres in the company.

RECENT TRENDS

After the crisis of 1971, Rolls-Royce decided to take a more critical look at NC and its application to that date, and one of the first major decisions was to create an NC steering committee which was given the responsibility of formulating policy recommendations on a company basis.

Committees had existed in the past, but these were divisional in nature and so had not been successful in producing a company coordinated policy.

The first task of the committee was to survey the company's NC investment and activities. The survey revealed that, although the NC machines represented a profitable activity, there was little doubt that they could be made more efficient and more profitable.

By the end of 1972, considerable information had been gathered and a number of policy recommendations made. The future function of the committee will be to review performance and policies in an endeavour to extract the absolute maximum of benefit.

THE NC MACHINE TOOL

The key factor in any NC machining operation is obviously the NC machine tool. The technological factors are largely the same, no matter whether one uses a conventional or NC machine. It is the introduction of the control system which produces the additional problems which are experienced in connection with the need for different requirements, such as supporting services or training requirements.

Thus, while acknowledging the prime importance of the machine tool, this paper will largely be concerned with such differences.

MACHINE/CONTROL SYSTEM OPTIONS

One aspect of the machine tool which will be considered is the range of options offered to the potential purchaser of a machine tool, because these are often confusing to the new-comer. Inevitably, these options, if taken up, cost money. As a result, they may be rejected, or alternatively purchased on the grounds that they may prove useful in the future.

Before any decision is made, a very careful appraisal is required to ascertain whether they really are desirable and do, in fact, offer some real benefit. In assessing benefits resulting from the purchase of a particular option, the answers will vary according to

* Chief Applications Engineer, NC Computer Group, Rolls-Royce (1971) Ltd, Derby.

B

whether manual or computer assisted part programming is to be used. A brief examination of some
typical examples follows.

Interpolation

This is a feature which allows detailed points on a
cutter path to be calculated from a basic minimum of
information. For instance, two points may be used to
define a straight line and an interpolation system is
used to calculate all the necessary intermediate points
on that line. This type of interpolation is referred to
as 'linear', two other commonly offered systems
being 'circular' and 'parabolic'.

By their very nature, interpolation facilities are of
use only on continuous path systems, but when they
are correctly used, they reduce part programming
time considerably.

Interpolation may be carried out in the control
system or by means of a computer. Obviously, with
manual part programming it can only be carried out
in the control system, and thus this facility is really a
'must'.

The position is not so clear cut when computer
assisted methods are used, because there is a choice of
where it shall be done. If it is carried out in the
control system, a capital spend is incurred. If in the
computer, then there is a considerable increase in the
amount of computer time used. The deciding factor
will be the amount of contouring which is necessary
to make the company products. If the amount of
contouring required is considerable, then interpolation in the control system will be better choice,
while, if there is a minimum of contouring, then the
computer will be better.

Canned cycles

These are routines which permit a series of machining
operations to be carried out as a result of one
instruction from the tape.

The potential purchaser needs to study the operations which will be carried out by the canned cycles
offered and to decide whether these match the
machining techniques used in his factory, particularly
if he is working in sophisticated materials.

Assuming that the canned cycles offered in the
control system meet the company's requirement, and
that manual part programming is to be used, then
they will be worth while. The position is not so
obvious when using computer assisted programming.
Each case calls for separate assessment, but the
following considerations concerning the use of
canned cycles in the control system should be borne
in mind.

Advantages of canned cycles in the control system
   (1) Shorter tape lengths.
   (2) Easier manual programming.
   (3) Dependent on the speed of the tape reader, or
the amount of buffer area in the control, the actual
machining cycle may be reduced.

Disadvantages of cycles in the control system
   (1) Canned cycles in the control cost money for
additional hardware, whilst post processor routines
can be developed to produce machining cycles at
minimum cost.

   (2) Inflexibility.
   (3) Canned cycles in the control system may
restrict block by block movement.

Incremental–absolute

This feature determines the form in which positional
information is fed to the control system. With an
incremental system the position of the next point to
be approached is specified relative to the current
position. While in an absolute system, the positions of
all points are specified relative to a common datum.

It therefore follows that when manual part programming is used, the amount of work involved in
writing a part programme will be considerably increased if the piece-part drawing is dimensioned in
absolute units with the control system requiring
incremental units and vice versa. The necessary
conversions are not only tedious, but error prone.

Imperial–metric

With the advent of metrication and Britain's entry
into the EEC, it is fairly obvious that metric units are
going to be the future standard. Newcomers to NC
will therefore opt for metric control systems.

Companies who are already involved in NC may
need to consider this a little more carefully if they are
contemplating the purchase of a machine, which with
its control system, is identical in all other respects to
an existing installation. In such a case, it may be
deemed essential to have complete interchangeability
of tapes between the new and the old, in which case
the units will need to be the same.

Some controls are fitted, as an option, with a
switch which can be used to change the units of the
control, and, despite the potential risk of operating
with a wrongly set switch, such a method would give
some, but not necessarily full, interchangeability of
tapes.

Such problems do not arise with computer produced tapes since post processors can be devised
which will produce tapes to either or both units at
very little extra cost. These tapes can be positively
and clearly identified by the computer as being
'Imperial' or 'Metric' and thus there is absolutely no
risk of confusion on the shop floor.

Feed rate codes
Feed rates can be supplied to the control system in
one of two forms.

   (1) A feed rate code such as the 'magic three'.
   (2) The true feed rate value.

Neither method causes any problem if computers
are used, because they take true feed rates from the
part program and convert them into the form required
by the control system with no difficulty.

Using manual methods of part programming there
is a distinct preference for the use of true feed rate
values because this eliminates the need for tedious
conversions which again can be a source of error.

Feed rate computer
An option which is useful when manual part programming is used, is the feed rate computer. This part
of a control system accepts feed rate information in

inches a minute or millimetres a minute and converts to the form required by the control system. Such a feature is completely unnecessary when computer assistance is used.

PART PROGRAMMING

An essential part of the supporting services required for successful NC operation is the part programmer. Many views have been expressed on this subject. Some have gone as far as to say that girls straight from school could carry out this task, while others have claimed the need for a top class production engineer. This diversity of opinion would seem to stem from a failure to produce a job description with the result that everyone is referring to a different job.

In Rolls-Royce (1971) Limited, the part programmer is, in general, a planning engineer who may have the responsibility for

(1) deciding how the piece part will be machined
(2) deciding the cutting tools required
(3) designing any holding fixtures
(4) producing the control tape
(5) co-operating in tape proving.

These responsibilities determine the prerequisites of a candidate for part programming work.

(1) experience of production engineering
(2) knowledge of material and cutter properties
(3) knowledge of machining capabilities
(4) ability to design holding fixtures, etc.
(5) mathematical ability.

Part programmer training

As in all areas of NC, training for part programmers is essential. Having satisfied the pre-requisites, the part programmer should learn the characteristics of the NC machine tools for which he will have part programming responsibility. He should become fully conversant with the function of the control system.

A formal course in manual part programming should be undertaken, even though he is going to be 100 per cent involved in the use of a computer. It is often forgotten, when dealing with a computer produced tape which requires minor modification, that rapid modification can be achieved by the use of a Flexowriter type machine. In such cases there is a need to fully understand post processor output. Of course, disciplines need to be established to ensure that the computer file is updated, so that, in the event of a future design change, the computer file can be accessed to produce the correct data for the basic piece-part.

For the part programmer who is to use a computer, a further course is necessary to provide instruction in the part programming techniques required for the processors and post processors to be used.

An interesting point has emerged as the result of experience in training. With some exceptions, it would appear that staff over 35 have some difficulty in assimilating the contents of these training courses which generally tend to be intensive. A probable reason for this is that such people have not been in such a formal and intensive learning environment for some considerable years. Two possibilities are open to help in this situation—either an extended course or additional tutorials after the course.

It is important to note that, no matter how good the standard of tuition on the course, the new part programmer can only learn the basic rules of the craft. Ability can only result from involvement and experience.

Ideally, when a man returns from a course to his work situation, he should be given the opportunity to put into practice what he has learned. The longer it is between the actual course and starting to write part programs, the less satisfactory will be the results.

COMPUTER ASSISTANCE

Newcomers to NC frequently ask the question 'Does a computer really help the NC process?' Apparently the thought of the additional computing costs predominates in the mind of the questioner.

This concern about costs must be encouraged and, unless the benefits produced more than outweigh the costs involved, then the latter should not be incurred.

Fortunately, experience has shown that benefits do, in general, outweigh costs and include

(1) reduced part programming time
(2) correct tapes offered to the shop floor
(3) reduced proofing costs.

Further benefits accrue from access to additional computer facilities.

(1) Inspection information
(2) mathematical facilities
(3) technological files.

REDUCED PART PROGRAMMING TIME

When computer assisted part programming is used in preference to manual part programming, it has been found that the time required to write the part program may be reduced by as much as 75 per cent, and, in a large organization, this represents a considerable saving of skilled and scarce manpower.

To achieve such savings the assumption is made that the component being manufactured is itself suitable for NC computer programming. The difficulty is to assess the validity of this assumption.

No hard and fast guide lines can be laid down. It is quite easy to select extreme cases and to say, for example, that a simple cylindrical shape is best tackled by manual part programming whilst a complicated three-dimensional surface is best tackled by computer part programming, but where is the switch made from one method to the other?

The answer to this can only lie in the experience and know-how of one's staff in respect of

(1) critical lead times
(2) the capabilities and problems of manual part programming
(3) the capabilities and problems of computer part programming
(4) the capabilities of the machine tool
(5) the nature of the component.

There is no easy way of acquiring such experience, but a firm basis on which to build can be obtained by proper training.

TAPES FOR THE SHOP FLOOR

When a tape arrives on the shop floor, it would be ideal if it could be placed in the tape reader and machining commenced immediately. In real life, this is not always the case, particularly when using manual part programming which is subject to the normal error rate for human beings. Computer produced tapes, however, have a much lower error rate because of the reliability of computers, together with their ability to carry out checks on data much more easily.

Often after an operator has been used to manually produce tapes and is then introduced to tapes which have been produced by computer, he has commented on the increased reliability of these tapes compared with manual tapes. The result is not only greater confidence in the tape information but increased confidence in his machine tool.

PROOFING COSTS

In the manufacture of aero engine components, piece parts requiring machining have often undergone so much preliminary processing that they are already expensive when they reach the machine tool. Production engineers, fully aware of this, have ordered blank forgings at a much lower cost, in order that the tapes supplied can be proved. Even these blanks have proved comparatively expensive.

The computer has been used in many instances to eliminate the need for blanks. This has been achieved by using the computer in conjunction with a draughting machine to produce drawings of the information contained on the tape.

The simplest application is to compare the drawing of the cutter movements with a drawing of the finished component which has been drawn at a similar scale. A simple examination will show up large sources of error and indicate potential collision paths. Naturally, minute discrepancies will not be revealed, but disasters, such as a cutter trying to shear through the workpiece will be revealed.

In Rolls-Royce (1971) Limited, considerable use is made of this facility with a highly significant reduction in the use of test pieces. An excellent example of this is to be seen in the application of EXAPT 2 where drawings are produced via an appropriate post processor to show

(1) the blank from which the piece part is to be machined
(2) the finished part
(3) the holding device
(4) the cutter path.

The latest software developments permit each tool movement to be plotted separately showing the condition of the blank at the commencement of the operation. In fact, this system can be used to produce Operation Sheets without manual intervention, thus saving tedious drawing work.

These drawings also indicate whether the feed is at rapid or programmed feed rate, but do not, and cannot indicate whether the programmed feed rate is the optimum for the piece part configuration. The latter can only be ascertained when the first piece is cut. However, the operators are now becoming confident that a computer produced tape, whilst not necessarily being optimum for speeds and feeds, will not produce scrap.

The most obvious advantage of subscribing to systems such as APT and EXAPT is that development and maintenance costs, which are quite substantial, are spread over a number of companies and organizations. In this way, costs are not onerous or prohibitive. On the other hand, in-house systems are costly to develop and maintain.

The processor is not the only source of expense. There is still the cost of post processors to be considered. Rolls-Royce (1971) Limited hold the view that efficient post processors are essential for most effective use of M/C Tools and to this end, now write all their own.

Previous experience has shown that, generally speaking, bought-in post processors present many problems.

(1) They are costly to implement on a specific computer.
(2) Maintenance is a problem and to train staff to become competent in maintenance is nearly as costly as writing the post processor.
(3) The output does not conform to a definite standard and the resulting profileration of output presentation is liable to cause confusion and possible scrap on the shop floor.

Post processor writing is a specialized job requiring training and experience. To minimize the amount of training and experience required, and to reduce the lead time in post processor writing, modular post processors are the order of the day. This means that each routine in a post processor is designed as a separate building brick, such that individual building bricks can be assembled to produce an efficient post processor in the shortest time and at the lowest cost.

THE RIGHT SUPPORT

In Rolls-Royce (1971) Limited, there has been a continuous appraisal of the processors which are available. As a result of this, the current preferred processors are

–EXAPT 1 for drilling and boring
–EXAPT 2 for turning
–APT for complicated machining and draughting.

These systems have been selected as they are considered best suited to the general requirements of the company. Apart from these well-known processors, additional specialized processors have been developed in-house to satisfy special requirements.

The power of APT is used not only for complicated milling, but for the writing of part programmes to drive a draughting machine. To produce even greater benefits, system macros or special purpose routines within APT have been developed with considerable success.

A large number of such successes could be quoted, but a couple may suffice as examples.

The manufacture of impellors have been a problem for many years, involving a lead time of the order of nine months and, whilst the geometry tended to become more complicated, the lead time could not be shortened. These impellors are manufactured on a multi-axis copymill with the aid of templates, and, although an NC machining centre might seem to be the answer, the simple fact was that the small number of impellors required made the purchase of such a machine uneconomic.

The answer lay in producing the definition of the basic shapes from aerodynamic data by means of a special computer program. The resulting geometric data was fed into APT system macros to produce tapes to machine templates and inspection layouts.

This approach helped to reduce the lead time to weeks—a worthwhile exercise by any standards.

Another problem area lay in the production of templates and inspection layouts used in the manufacture of broaches for the dovetail roots of engine blades. In the conventional approach there was always a large amount of desk calculation required, with consequently the possibility of error. Moreover, it was found in practice that considerable dressing of the finished gauges was required.

Once again, the system macro technique was used. The benefits expected in the reduction of man involvement was achieved, together with a cost saving of £200 per operation per broach. Two additional benefits were achieved. Firstly, the need for dressing disappeared. Secondly, because the computation carried out by the computer was dependable, it was possible to demonstrate that some problem in the past had been caused by a malfunction on the machine. With this rectified, extremely gratifying results are being achieved.

The EXAPT systems were adopted largely because of their potential. Technology appeared to be a valuable tool because of the standardization it could bring to the process. Standardization of tools, etc., could obviously lead to high cost savings, whilst feed and speeds could be used to give optimum cutting conditions. One benefit, which only became apparent as the system was developed, was the further reduction in the time required to write a part program, because, for instance, in EXAPT 2, automatic stratification of the piece eliminates the need to detail the cutter path.

IN-HOUSE VS. BUREAU SERVICES

The foregoing discussion has made the claim that in many cases a computer is an essential tool for a part programmer, but some companies may contend that these arguments are invalid because they do not have a computer or do not have a sufficiently powerful computer.

It must be realized that it is not necessary to have an in-house computer, because various computer bureaux offer facilities for NC processing. In fact, any company, even those with in-house facilities would be wise to investigate the services available from such bureaux before making a decision to use in-house services. Once again, the cost and benefits of each approach must be considered.

In-house services
The use of a company's own computer has many points in its favour.

(1) It can normally offer faster service.
(2) It is easier to bring pressure to bear to ensure completion of urgent jobs.
(3) There is less problem in transportation of input and output between user and computer. This is particularly important where lead time is critical.
(4) If there is adequate computer capacity available, processing does not produce a cash flow.

Bureaux services
When investigating the services offered by computer bureaux, it is necessary, first of all, to find out where these are sited. A good source of reference is the 'Numerical Control Directory' published by the British Numerical Control Society.

Application to the bureaux will rapidly produce details of the services offered and the charges which will be made for these services.

Naturally, the bureaux must be capable of offering a processor to meet the technical requirements so far as product geometry is concerned, as well as post processors to suit the machine tool/control system combinations in one's factory.

If these criteria are met, then one company is not confronted by the cost of maintaining processors and post processors, etc., even though so many of the advantages of in-house services are lost.

Having conducted adequate research on the two methods of support available, it should be a simple matter, knowing one's requirements for the future, to carry out an economic appraisal and so determine the best line of action.

There are, however, a number of disadvantages, some of which have already been mentioned.

(1) The use of processors such as APT and EXAPT requires membership of the appropriate association or 'club'. This costs money.
(2) The purchase or writing of post processors costs money.
(3) The maintenance of processors and post processors costs money.

The computer must be of sufficient core capacity to handle the processor(s) and post processors which are necessary to the company's requirement. It must also have sufficient time available to carry out the NC processing, bearing in mind that if the computer is dedicated to the payroll, etc., such items will have priority.

A further requirement will be to ensure that sufficient trained staff are available to carry out, at least, first line software maintenance.

BATCH PROCESSING

Whether one chooses a bureau or an in-house service, one mode of operating which will be available is that known as batch processing. In this mode, the part programmer codes his program on sheets specially designed to suit the particular processor, and computer. When complete, and preferably checked, these sheets are sent to the computer department by mail, special messenger, etc. On arrival at the computer department, the information is transferred onto punched cards or punched tape and then fed into the computer for processing. Subsequently, the output in the form of punched paper tape and printed output is fed back to the part-programmer, together with the original part-program.

Experience has shown that a part programmer may require three or more such passes before he achieves a successful computer run which actually produces valid tapes.

This may sound to be a quick cycle of operations, but in practice it is time consuming and a source of frustration and delay on the shop floor.

STANDARDIZATION OF LANGUAGES

When writing post processors, every effort is made to conform to current BSI and ISO standards, and to ensure that vocabulary words used in part programming languages always have the same meaning.

The format used for input and output have also been standardized, to avoid the confusion which would otherwise arise.

The question of standardization has much wider implications than in the software aspects and one of the benefits of EXAPT processors is that the use of tool library files tends to induce standardization of cutting tools and holders. Here, there is an obvious cash benefit as a result of a reduced variety of types.

One of the main areas where standardization has been applied is in the definition of preferred machine tools to cater for each class of work piece, as well as recommendations of preferred types of control systems for point to point, turning and milling.

As early as 1966, it was realized that some form of standardization or rationalization was desirable because the proliferation which then existed in connection with point to point machines was causing problems. As a result, certain control systems were recommended for specific applications in Rolls-Royce.

Early in 1971, the state of the art was again reviewed, and a report was published detailing the observed performance of every machine tool control system combination. When this report was studied in depth, it was very apparent that some combinations were sound whilst others had failed to live up to their expected performance. This failure was due to different causes. In some instances, there was a failure to achieve the predicted metal removal rate, in some, a high incidence of breakdown, and so on.

These recommendations will be reviewed from time to time with a view to continually improving the return on invested capital.

Apart from trying to ensure that all machines installed are capable of producing a high performance, standardization of machine tools and control systems produces benefits in respect of

(1) reducing the number of post processors required
(2) ensuring interchangeability of work between machines with a minimum of effort.

All aspects of NC benefit from standardization and some of these are mentioned elsewhere in this paper.

TERMINALS

To overcome the delays encountered in batch processing, the use of terminals provides an answer. Of course, like all solutions to problems, this solution also costs money, the amount of which can vary considerably.

These terminals can be divided into two classes.

(1) Intelligent
(2) non-intelligent.

Intelligent terminals are really small computers which reside in a particular area and are connected by land line to a major installation some distance away. These tend to be very expensive, but they can be used in the local area to carry out other computing tasks which do not require a large core. Examples of this type of terminal have been installed at Rolls-Royce (1971) Limited factories in Coventry, Bristol and Glasgow for transmission of NC work to the large installation in Derby.

Unintelligent terminals, on the other hand, can consist of a simple inexpensive teletype and tape-punch connected via modems and a GPO line to a central computer. Whilst they do not have the flexibility or capability of the intelligent terminal, they are relatively inexpensive, and it is to this kind of terminal that Rolls-Royce (1971) Limited are moving.

With either type of terminal, it is possible to introduce sophistication by adding high speed printers, plotters etc. Of course, sophistication means increased cost.

Terminal facilities can be provided by many commercial bureaux.

TAPES

Tapes have been mentioned many times but have not so far been subject to any detailed discussion. They are, however, most important. They are an essential part of the tooling kit for a given piece part, yet how often are they treated as just another piece of paper? Personnel handling tapes should have impressed upon them the fact that tapes are a valuable collection of data, not just a piece of paper costing a few coppers.

Careful consideration should be given to methods of storage and transport. It is strongly recommended that, wherever possible, tapes should be stored on spools and contained in a reasonably rigid box for transportation purposes, and that adequate storage cabinets should be available.

Magnetic tapes are not in wide use and will not be dealt with.

As is the case in most aspects of NC a code of standards for tapes is highly desirable, and the first aspect of standardization to be considered concerns the material from which they are made. The choice facing a user is very wide, ranging from plain paper, through what are known as 'durable' tapes, mylar to metal coated mylar. Not only is there a wide choice of materials, but there is a wide range of prices ranging from a few pounds to £20 per roll according to the type of material. Where computers are used for tape production, as in Rolls-Royce (1971) Limited, ordinary paper tape is a useful medium. This can be used as a 'master' tape and copies in 'durable' tape produced on a copying machine which also acts as a comparator. Both the paper tape and the 'durable' tapes are despatched to the workshop where the master paper tape is retained in store while one 'durable' tape is used as a working tape and a second 'durable' tape is kept in reserve in case the first one becomes damaged.

This arrangement has been found to work extremely well on all optical tape readers as well as on many mechanical readers, although it has sometimes been found necessary to use a tape which will stand extremely rough treatment in the reader. In such cases mylar has been used. The use of mylar is avoided wherever possible because of its high cost and the heavy wear which is induced in the anvils of the tape punch.

When ordering tapes, manufacturers should be asked to guarantee that their product conforms to BS3880, which lays down standards relating to tape width and thickness. It further defines standards relating to hole positioning and hole sizes, and these standards should be accepted as 'in-house' standards.

Before any manufacturer's tape is acceptable to Rolls-Royce (1971) Ltd., it must satisfactorily undergo rigid tests in respect of

(1) resistance to oil and water
(2) resistance to abrasion, hole tear and delamination
(3) stability with changing temperature
(4) weight per unit length
(5) tensile strength
(6) flammability
(7) thickness
(8) opacity
(9) conformance to BS3880.

When testing tensile strength, virgin tape is not used, instead, tape which has actually been punched is used. This is because the results from virgin tape are misleading.

In the workshop environment, tape identification is extremely important. Each tape should be clearly identified to show at least

(1) the operation it will perform
(2) the drawing standard to which it was prepared
(3) whether it is a proven and accepted tape.

Of course, with computer produced tapes, the beginning or leader of the tape can be punched in man-readable characters so that the relevant information is an integral part of the tape.

In addition to this information on the tape itself, it is essential to introduce a documentation system which clearly shows the identification of the current production tape. It follows then any obsolete tapes should be destroyed.

Factors which tend to make tapes obsolete are design changes which demand a new version of a tape. Specific procedures must be laid down relating to the production of such new tapes and also to their adequate documentation.

INSPECTION

Rolls-Royce (1971) Limited experienced difficulty in producing casings for the Spey engine in the required quantities. The bottleneck was identified as being in the machining department, and after due consideration Omnimil machining centres were installed. This action eliminated the bottleneck of piece parts awaiting machining but did not solve the overall problem. A bottleneck developed in the inspection department. A solution was found by using more sophisticated inspection equipment, but this was just another example of the need to consider carefully the supporting services required for NC. It also demonstrates the need for even closer liaison between production engineering and inspection.

Investigations in 1972 showed that

(1) despite the claims made for increased accuracy from machine tools, there had been no significant reduction in the inspection process;
(2) practice in inspection tended to be to detect scrap and not so much to prevent it occurring.

The possibility of reducing the inspection process was discussed by the interested parties and it was agreed that reduction was a long term objective on any aspect other than positional measurement.

So far as positional measurement was concerned, a reduction of inspection procedures was feasible once the performance capability of the particular machine tool had been established. This proviso led to the regular calibration of NC machines by means of laser equipment. The result is that on hole positions, once a tape has been proven, sampling techniques can be used.

Attention to prevention of scrap has led to a very serious approach to in-process gauging, and this is expected to lead to very substantial cost savings.

It is interesting to note that the inspection departments are making more use of the computer to produce inspection graticules and also dimensional checking information.

MAINTENANCE

Another supporting service which requires careful consideration is that of maintenance. This question should arise when purchase of an NC machine tool and control system is under consideration. The manufacturers reputation for after sales service should be investigated together with his policy on the stocking of spares. These items, together with the geographical situation of the manufacturers nearest service depot and service engineers can be most important.

The level of spares, both for the machine tool and control system which need to be carried 'in-house' for first line maintenance, should also be determined and implemented prior to the installation of the machine.

Some companies take the view that their own maintenance department need not be involved with a new machine during the period of guarantee. This is a completely short sighted outlook because

(1) no matter how efficient a manufacturer is in after sales service, there can always come a time when his engineers are committed and cannot tackle your urgent requirement;

(2) whenever 'in-house' maintenance engineers becoming involved in supporting a particular new machine, there is always a learner curve.

One factor often overlooked when planning a new machine is preventive maintenance. From the outset, this should be scheduled into production plans to prevent the situation arising when maintenance fails to get done because of the need to meet an excessive production plan. Failure to carry out such maintenance can result in large production losses at a later date.

The nature of the NC machine tool means that the maintenance problem is more complex than for a conventional machine, due to the additional electronic equipment.

Rolls-Royce have always had 'in-house' maintenance staff, but the actual responsibilities have been divided: for example, millwrights for the machine tool, and electronic engineers for the control system. Each of the branches of responsibility were the responsibility of centralized groups. There is now a move to introduce into some work centres a new job of NC maintenance engineer. This new type of engineer will be responsible for the maintenance of both machine tool and control system. It is confidently expected that this new approach will produce more effective maintenance than was experienced previously.

The personnel for the new role have been selected from men with experience in depth of one aspect of maintenance. Training in the other required disciplines has been initiated. It is this training which will delay the full implementation of this system of maintenance.

UTILIZATION

While it is important to consider support services in some detail, sight must not be lost of what happens to the machine tool itself.

In most companies, it is quite a common practice to prepare a justification for the purchase of a machine tool. A valid question to ask is whether these same companies look back at the justification once the machine tool has been installed. Such a retrospective look after the machine has been in production can be very revealing. It can demonstrate

(1) whether the machine is achieving the objectives set out in the justification
(2) whether the application has changed
(3) whether there is spare capacity
(4) whether the machine and control system are as effective as was claimed

(5) whether the reliability is good enough.

Obviously such information can be of use to management in a number of ways, for example

(1) it can provide guide lines to future buying policy
(2) it can lead to better machine utilization
(3) it can demonstrate whether maintenance is adequate.

Rolls-Royce are carrying out such reviews more than once during the life span of a machine.

In order to produce a more realistic review, the performance of each machine is monitored in detail. Elements of performance are identified, e.g. awaiting repair, no tapes, no material, productive time, etc., etc. Each element is so constructed that responsibility for that element can be identified easily.

Simple data sheets are completed, shift by shift and returned to the computer depot each week. At the end of each four-week period, the whole of this data is processed and analyses produced for use by each level of management up to board level. These analyses are continuously reviewed with the firm objective of obtaining optimum utilization.

In considering ways of improving utilization in an organization with a large NC population, many ideas occur. For instance, it might be thought possible to improve utilization by grouping all machines of a similar type together, since this would mean that variations in load could be catered for more easily with less machines. Such grouping would also give better back-up facilities in the event of a machine failure. Many other potential advantages could be listed, but, on the other side of the coin is the important question as to whether such an arrangement would suit the work flow for the organization. Therefore, whilst the idea is worthy of consideration, no general answer can be given. Like so many other things, each situation must be considered on its merits—in the words of the punter, 'horses for courses'.

An examination of many NC situations seems to suggest that setting up the workpiece often takes a disproportionate amount of time, suggesting that if one workpiece can be loaded while the next is being machined, considerable benefits might be gained. Obviously for some machines, this is impracticable. However, the possibility should always be considered.

CONCLUSION

NC can be a very effective tool if properly used. The newcomer needs to make a proper study of the capabilities of NC and to identify the benefits he aims to obtain.

Steps must be taken to ensure that any staff who are to be involved are adequately trained and that the organization is arranged so as to cater for the new requirements of these machines.

Once the machines are installed, backed up by the proper services their performance must be monitored if the best results are to be obtained.

DISCUSSION

Q. J. A. Stokes. If a company has started using NC with manual programming, what criteria should that Company use in deciding whether to employ computer aided programming?

A. The costs for the manual programming of tapes would be known to the company which had to make the decision whether or not to use computer assisted programming. It has been shown by many companies that the latter method produces savings, although opinions vary as to the amount of saving. Therefore, assume the savings to be 50 per cent, which is a conservative figure. If the cost of using a bureau service is less than this saving, then, computer assisted part programming becomes commercially viable. The company, because of the conservative estimate used, can employ a bureau with benefit. The sums which need to be done in respect of 'in-house' computer services is much more complicated.

SOME EXPERIENCES ON NC SHAFT TURNING

by

G. D. WARD and L. WOOD*

SUMMARY

This paper describes the introduction of NC shaft turning into a heavy engineering organization manufacturing shafts for electrical machines often in batch quantities of five or less. It describes the initial considerations leading to the capital investment, and the effects of this type of plant on the organization from design to manufacture of shafts.

PLANT

There are two lathes involved. Figure 1 shows the larger machine. It is a VDF M1120B N/C which has a capacity of 7000 mm between centres, 1110 mm diameter and will carry $12\frac{1}{2}$ tons without steadies. It has a DC thyristor drive and a GE 100S NC system (metric).

Figure 2 shows the small machine which is a DS & G with a capacity of 2438 mm between centres, $431\cdot8$ mm diameter and will carry 3 tons between centres. It also has a GE 100S NC system (metric). The swing is reduced to 355 mm with the milling unit. The first machine requirement was evolved in 1967 and was destined for the Stafford Works of English Electric Company where the whole organization had been carefully nurtured through "Digital Read-out' to the point where NC was the next logical step. Before delivery of the machine, GEC and English Electric merged in May 1969 and the machine was diverted to the Rugby Works of GEC which was to become the headquarters of electrical machine manufacture. The capacity of the machine remained satisfactory in spite of the changed production requirement, but the organization had not been conditioned to accept such a modern piece of new plant and intensive effort had to be made in design, DO and planning areas in the short time left before delivery in September 1969. Shafts were mainly in batch quantities of one or two, but could repeat in the year.

The opportunity arose in September 1970 to acquire the DS & G from our manufacturing research and development centre where it had been extensively tested in conjunction with the manufacturer and was not considered suitable for release to a production unit. This was a suitable machine for our small shaft range where batch quantities were quite high and a greater degree of shaft standardization existed.

PLANT INVESTMENT

A family group had been set up at Stafford Works in 1966 for the manufacture of frames and shafts for electrical machines; this group included a shaft lathe, vertical borer and a travelling column horizontal milling and boring machine with bedplates and a rotating table. The lathe was an existing lathe on which we had done our experimental work on 'Digital Read-out'. This group had made substantial strides towards the philosophy of 'mass producing a batch quantity of one' and as a result attention was turned towards further improvements.

Although shaft standardization was desirable it was not practical in the business of making heavy machines, the designers wishing to retain a very high degree of design flexibility. It was therefore argued that a shaft was a list of features connected by metal and if these features could be standardized then the whole process from design to manufacture could be speeded up by using the computer, linking to a computer program for preparation of tapes for NC

---

* GEC Machines Limited.

Figure 1. VDF M1120B N/C lathe.

Figure 2. DSG NC lathe.

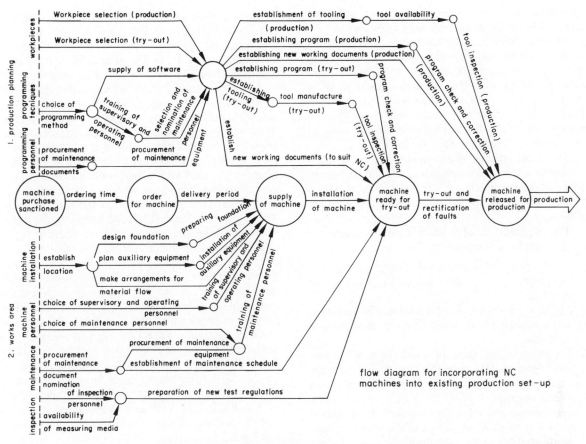

Figure 3. Flow diagram for introduction of NC turning.

machining. Thus we began to investigate the possibilities of NC shaft turning which led to the purchase of the VDF.

It is important to note that the basic argument for NC arose out of a consideration of family grouping as a means of producing a critical path component for electrical machines and that it was the only possible technique for getting near to the philosophy 'mass producing a batch quantity of one'.

PREPARATION FOR MACHINE

For all new major plant which represents a radical departure from existing practices it is advisable to prepare a network of things to do so that all the facets of the necessary work can be seen. Figure 3 shows a network prepared by VDF which is quite comprehensive.

TAPE PREPARATION

The most important aspect is of course programming which embraces all the manufacturing problems in the machining of the components. In the case of the VDF we are faced with low batch quantities of high cost shafts where scrap cannot be tolerated, principally because of difficulty in replacement. The DS & G produces larger quantities of lower cost shafts which repeated frequently, so that a scrap shaft can be more easily replaced.

The reliability of the tapes was therefore of paramount importance, both in the information on the

tape and the reading performance on the machine. We have had our share of tape punching errors, and tape stretching problems; suffice to say that the back-up service for tape preparation is vital and that any attempts to do this on the cheap will become more expensive in the end. In the case of the VDF a new tape is prepared from the master each time it is required by the machine even though it is a repeat order. The tape material used is Waterlo A.

For the DS & G a more expensive tape, Syntosil, is used. This can be stored and reused up to ten runs, but a careful note has to be kept of condition, to safeguard against tape stretching and damage in a shop environment.

Initial investigation showed that for large shafts approximately 1000 blocks of information would be required, with a manual part programming time estimated at two man-weeks. The projected order book indicated that an average of three shafts would require to be finish machined per week and this would fully occupy a trained part programming staff of six; this was clearly unacceptable. The alternative was to use an outside programming bureau service or computer assistance; the former was accepted and initial components part programmed by the supplier of the machine. The proliferation of NC languages available seems to increase day by day so it was thought that no problems would arise in having a suitable system working in due course as, in the long term, a bureau service was thought not to be satisfactory. Experience with outside programming showed that the turn round time of three weeks was not acceptable for production requirements, and that tapes could not be guaranteed to be error-free.

The frustration and confusion caused to personnel trying to familiarize themselves with the new and to them, completely novel machine tool; to satisfy a manager by ensuring that the lathe produced three shafts per week (one if lucky); and to explain to Production Control why one shaft could not be substituted by another as on a conventional lathe at short notice, was trying to say the least.

All of this highlighted the importance of an efficient backing service for a machine of this type, which, when working at maximum efficiency, was capable of producing three times the output of a conventional lathe.

The requirements of a suitable part programming service now became apparent.

(1) It must be capable of being run on the company's in-house computer (Honeywell H2015).
(2) In the event of errors these must be corrected and a new tape made available to the shop floor in half a day.
(3) Every effort must be made to supply tapes free from major geometric errors so that the machine could be run on fully automatic with unproven tapes.
(4) Part programming time should not take longer than one week from receipt of drawing to a tape being available to the shop floor.
(5) Capable of being used by a low level of programming skill.
(6) Imperial–Metric Conversion.

Consideration was also given at this stage to the philosophy of designing and manufacturing shafts from a collection of 'bits'. If a typical electrical machine shaft is examined (Fig. 4), it can be seen to have two bearings, coupling, exciter, commutator fit and armature fit; all or some of these would occur, depending on an AC or DC design. All of these shaft details are basically the same and differ only in length and diameter, and in detail are very often dependent on the whim of a particular draughtsman. Taking plain journal bearings as a test case, it was found that designers could be persuaded by logical and economic arguments to condense twenty or thirty designs down to a standard three journal designs. These were then issued to the drawing office, with instructions that any deviations from the preferred standards would have to have the authorization of the chief engineer. The first requirement (1) placed a limit on the size of processing language which could be used due to the size of the computer available, and ruled out high level languages such as EXAPT or 2CL which would have been first choice. Some existing programs could have been adapted but they were too comprehensive and would have been very costly to run in production as well as requiring a post processor. Because our shaft requirements were fairly specialized and relatively simple, a special purpose program would be much shorter and more economical to run and use. It was decided therefore to write our own processing program. This was justified as the post processor cost of EXAPT for the VDF lathe was quoted at £2000, and the writing of our own eventually came to half this figure.

As all our shafts were 'the same but different' and only varied basically in lengths and diameters, even the work material being nominally the same, a fixed format input was decided on, which would require a low level of skill. Feeds and speeds could be standardized and also, to some extent, depth of cut. In practice this has proved successful on the VDF; two

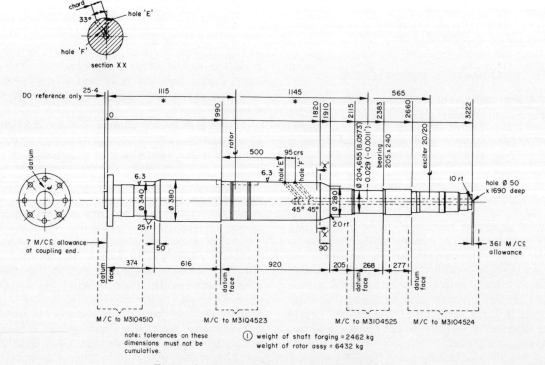

Figure 4. Typical shaft for an electrical machine.

surface speeds are used for machining standard shafts, and feeds are restricted to 1 mm per rev. roughing, 0·5 mm per rev. for semi finishing and 0·25 mm per rev. for finishing. Exceptions do arise when the part programmer has to rely on experience but most eventualities have now been covered. In designing the program, consideration was given at this stage to using a 'pack of cards' technique whereby a library of standard shaft features could be held by the design office and from these the majority of shafts built up, thus achieving savings in design and draughting. These standard design features could be matched by turning programs: Fig. 5 shows a standard design detail for a

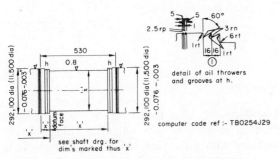

Figure 5. Method for turning journal for grinding.

plain journal, Fig. 6 shows a shaft drawn by this method. The running costs for a machine tool control tape are approximately £20 per computer run, producing a control tape of from 800 to 1000 blocks, from 300 lines of input information. The average cost per block is £0·02.

The second requirement (2) has been achieved as access to the computer is available at half an hour's notice so that a new control tape can be produced and made available to the shop within two to three hours of discovering an error. The majority of this time is consumed in tape copying and transport. Requirement (3) is everybody's dream but 99 per

cent success has been achieved. Initially the first tapes to be produced all had errors which were not detected before actual cutting took place. Simple checks had been built into the main processing program, and the machine tool operator would discover the majority, but there was always one that eluded discovery until actual cutting was taking place when it was too late. The operator would only run new tapes block by block. When machining large shafts, using tapes of up to 1000 blocks in length, running block by block increases the cycle time by a factor of four. It was more by luck than by good judgment that no shaft was scrapped during this period. The answer seemed to be to use graph plotting techniques to simulate the tool path. This was tried and proved a major step forward, as it was found that by laying the full size tool plot on to a full size outline drawing of the component, geometric errors greater than 0·5 mm could be easily detected. The part programmer now uses a full size component drawing and a tool path plot produced by a graph plotter (Fig. 7). Since the plot is full size, it is easier to visualize a large shaft, which could be 23 feet long, than would be possible with a small scale drawing. An absolute system of dimensioning in the required English or metric units is used, with the tolerances calculated by the computer in the mid point system.

The tool path plot shows cutting modes in black and fast traverse in red, this has proved invaluable as complicated details can appear as a forest of black lines which are extremely difficult to interpret. The system will not detect errors due to incorrect tooling being specified or incorrect feed and speed, but as a large degree of standardization has been achieved in this respect, errors of this nature are extremely rare and can usually be rectified by the machine tool operator. Only 3 basic tool geometries are used on the large lathe and these are capable of producing all designs of shaft to date; changes are continually being made in the grade of carbide used, and/or the

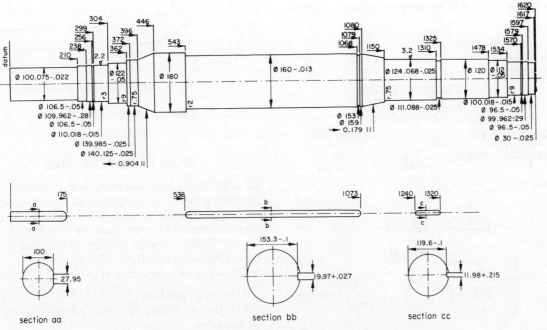

Figure 6. Shaft drawing produced on a graph plotter.

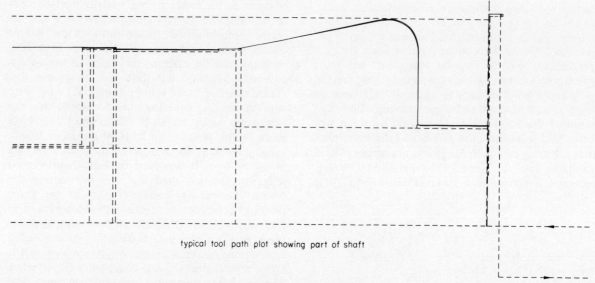

typical tool path plot showing part of shaft

Figure 7. Typical tool path plot.

supplier, as with the closely controlled cutting conditions tool material evaluation can be carried out with ease. It may be of interest that the average tooling cost for large shafts is under £2. With regard to requirement (4), the system enables one part programmer easily to handle two lathes on a part time basis, and a complete tape has been produced in three days for a large shaft. Programs have been produced by a girl with little knowledge of turning. The main processing program has limitations in that it must be modified for a change in machine as the machine parameters are included in the main program. It will only cope with external shaft turning and internal work would have to be programmed manually if this was ever required; this has not proved a disadvantage so far. the large VDF machine produces 80 per cent of the large machine shafts required, the other 20 per cent being mainly outside its capacity. The DS & G lathe can produce 60 per cent of the medium machine requirements and spare capacity is still available on both machines to allow for increased orders.

Possibilities of using EXAPT are still being kept in mind, as in the long term with changes in computer hardware within the company it will be possible to run a program of this size on site in due course. Whether it could compete for a simple and standardized type of work on an economic basis with the present system would seem doubtful. The same programming technique is used for the DS & G lathe which has a similar NC control system

The complete programming system is shown in Fig. 8.

Figure 9 shows the graph of tape production costs demonstrating the learning curve of the part programmer.

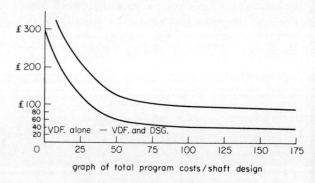

graph of total program costs / shaft design

Figure 9. Graph of program costs.

SOURCES OF LOST TIME

In order to reduce and eliminate if possible the penalties incurred in operating with a batch quantity of one, all possible sources of lost time in changing from one shaft to another must be examined. In this we also include possible sources of lost time during the machining of a shaft.

These can be enumerated as:

(1) Use of steadies.
(2) Use of offsets for close tolerance work and measuring.
(3) Obtaining fine surface finishes.

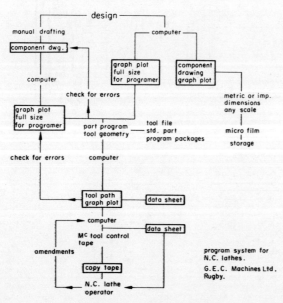

Figure 8. System flow line.

(4) Tool setting and tool changing during machining cycle.

(5) New set up.

(6) Shaft preparation—centring and control of length.

Although the two machines have similar control systems with EIA codes they represent quite different problems in production technology. Cycle times can be up to one week for the VDF and one hour for the DS & G. The smaller machine was concerned with a family of shafts where it was consdered that they could be turned without the use of steadies and where tooling variety could be reduced to the point of tip changing so that no tool setting was involved. This was not possible in the larger machine so that tool setting equipment had to be provided for the operator, with sufficient tool holders to allow tool setting to be done in parallel with other activities.

Both machines have a common requirement for shafts to be centred and controlled to length. It is obvious that standardization of angle of centre is required but depth also has to be controlled, particularly on the small machine where tool clearances at the tailstock centre are of the order of 0·25 mm. If the raw material is a forging or a casting proof turned by the supplier, then this could be part of the supply specification. Where the material is bar stock, then it is usual to have a preliminary operation on a manual centre lathe, if this fits into the family grouping. Other methods include ending and centring machines, which however becomes progressively more expensive with increasing diameter and length of bar.

It seemed rather pointless to saw bar to approximate lengths and then face up on the lathe, and it was discovered that in many cases this was done out of custom and practice rather than a design requirement, so that the best answer seemed to be a saw which could cut accurately to length combined with a centring machine. Unfortunately, not a great deal of attention has been paid to this aspect of cutting off and this would seem to be quite an area for improvement by designers of cutting-off machines.

Setting up for the DS & G has been reduced to adjusting chuck jaws, moving the tailstock and inserting the tape, and possibly changing two tool tips. In order to achieve this, strict control of the tool library is essential and sometimes it is necessary to compromise on metal removal in order to achieve tool standardization.

Driving a shaft by means of a three-jaw chuck means that the shaft must be turned round for finishing the other end. Experiments with face drivers were partially successful but on some of the shafts in the family, severe vibration occurred. Changing a chuck for a face driver would lose as much as would be gained so chuck driving has been retained as the standard method of work driving.

Since the use of steadies can double the operation time on a shaft, every effort was made to avoid a situation where they become necessary. Shaft work can be divided into three classes. When the shaft is uniform and stiff without excessive reduction in cross section it is obvious that there is no need for steadies. There is also the opposite case where turning could not be done without using a steady because of a high reduction in cross sectional area. In between there is a group which could fall into either class. This can only be determined by development work on the particular machine tool involved. In cases where vibration occurred, every effort was made to find the cause and the cure by trying tool geometry overhang, tailstock position, and headstock rpm.

Answers were found in all cases; these often meant compromise in metal removal rates in that these could be less than the theoretical optimum. This aspect can apply particularly in the use of ceramic tools. If the sole criterion was metal removal then these would be used, but there is a maximum rpm at which a shaft will start to throw without applying a steady; if attempts were made to increase the cutting speed the net result could well be a longer floor to floor time even with a shorter metal removal time.

Close tolerance work is also a source of lost time, particularly if measuring and use of offsets is involved. The lathes can hold nominal tolerances of ±0·4 mm

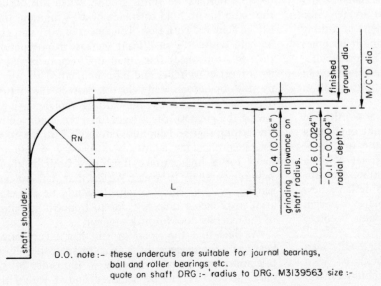

size	Rn	L	D.O. ref. nom. corner rad. of mating part.
I	I	4	1,5 and 2
2	1,5	6	2,5
3	2	6	3 and 3,5
4	2,5	6	4
5	3	8	5
6	4	8	6
7	6	8	8
8	8	8	10
9	9	10	12
10	12	10	16
II	15	15	20
12	20	15	25

D.O. note :– these undercuts are suitable for journal bearings, ball and roller bearings etc.
quote on shaft DRG :– 'radius to DRG. M3139563 size :–

Figure 10.

without the use of offset switches; thus shafts can be roughed out without stopping the machine. Tolerances for grinding can be held after setting the first offset switches for repeat shafts in the batch. In general tolerances up to ±0·1 mm can be repeated without any major sources of lost time, but tolerances finer than this require repeated measuring and use of offset switches and result in a steep increase in lost time. This particularly happens in the case of the large lathe where tolerances of 0·025 mm are required on diameters of up to 200 mm.

A typical breakdown of the floor to floor time on large shafts with close tolerance requirements is one-third of the time roughing out, two-thirds of the time finishing. From this it is apparent that a successful answer to the 'in-process' gauging problem would lead to a significant increase in productivity, and it is to be hoped that development will eventually produce the answer. Surface finishes of 100 microinches are obtained but anything better than this requires the usual manual lathe methods of broad tooling, roller burnishing, and the like. It is usual to finish journals by grinding; turning of journals is done to a special detail, as shown in Fig. 10, so that there is no need to grind into corner radii.

FURTHER OPERATIONS

It is quite usual for shafts to require keyways. The DS & G is fitted with a keyway unit, and separate tapes are produced for this operation.

The foundation for the VDF has been prepared so that a lathe bed can be lined up at the rear of the machine. This lathe bed can carry a milling unit for keyways and other light work. The spindle of the lathe is capable of being locked.

CONCLUSION

This paper demonstrates what is believed to be the correct and sensible approach to the use of NC. This is in our case, to see what technology can be used to design and manufacture electrical machines as quickly as possible with the minimum work in progress. Shaft turning represents a considerable part of the critical path of the manufacture of an electrical machine and any improvement in this area is of far greater value than a simple floor to floor improvement in turning time. With a concept of family grouping it becomes easier to see that NC can provide the best technological answer to arrest or slow down deterioration in turning performance resulting from labour inflation and reducing skills.

Whilst variety reduction is always a worthwhile exercise, it was shown that this need not apply to a whole shaft but could be applied to the functional pieces of a shaft. Ultimately there must come a situation where it is physically impossible even for the most individualistic of designers to design a new shaft, so that in due course of time whole shafts begin to repeat, and the need for new programs becomes progressively less.

The advantage of family grouping is readily seen in the amount of standardization achieved with simple tool library and programming techniques with built-in technology, to the point where preproduction aspects of NC turning are a fairly simple activity dealt with as normal aspects of planning; not, as is very often the case, something special set apart from the normal activities of the business.

DISCUSSION

Q. R. T. Webster, Midcast. What scrap level has been experienced on the large expensive shaft billets during NC turning; particularly in relation to machine malfunction and operator setting errors?

A. It is true to say that no shafts have been scrapped on the VDF lathe. We have had several near misses particularly before the graph plotting technique was developed.

We took the view that error-free tapes must be produced before they arrive at the lathe and the necessary techniques developed to eliminate both the human errors introduced by the programmer and sources of errors introduced by tape preparation equipment. The economics of tape proving on the lathe could not be tolerated, even to the extent of running block by block. In this respect it is obvious that the geometry was the most important aspect. Errors in feeds and speeds could be corrected on the machine.

The graph plotting technique will detect errors greater than 0·5 mm. Tool setting errors can occur in spite of this but the operator has the tool setting dimensions programmed in the tape. Roughing or semi-finishing stages usually indicated these errors if present. At this point there is usually sufficient material left on to overcome such errors.

The majority of errors occurred during the first six months of starting up, these comprising programming or tool setting faults. This is part of the learning curve of all concerned and will always be present when a company installs its first NC machine tool.

Q. C. F. Turner. What degree of accuracy are you able to maintain on the shaft lathe? How is it possible to produce a one-off workpiece without proofing tape?

A. Normally tolerance diameters on shafts such as bearing journals are finish ground. When this occurs the lathe has to hold tolerance of ±0·4 mm and this can be achieved with ease. Tolerances of ±0·1 mm can be held without a significant increase in production time but tolerances finer than this require repeated use of offset switches and measurement of the shaft by the operator. Many factors affect the actual diameter produced by the cutting tool and those having the greatest affect seem to be the condition of the cutting edge and depth of cut used. With the close tolerance–diameter is usually associated a requirement for a high surface finish and both of these requirements have to be met. We feel that an in-process gauging system which is reliable would be a useful asset in reducing time lost due to frequent stopping of the machine tool for measurement.

We feel that the operator still has a very large bearing on the success of the machine tool when close tolerances and surface finishing in the order of 32 microinches have to be achieved and it would be many years before the operator's skill can be completely discounted.

ECONOMICAL INTRODUCTION OF NC MACHINES IN A MACHINE TOOL FACTORY

by

D. GUENTHER*

SUMMARY

The following facts resulted in knowledge which originated from the introduction of NC machine tools in a machine tool factory. Due to the difficulties experienced with the very first NC machines questions arose concerning the economy of these machines, of the programming and tooling used, and about the increase in economy of production.

INTRODUCTION

The numerically controlled machine tool is today regarded generally as the most economical means of production for the manufacture of workpieces in small and medium batch sizes. It has a very profitable effect when individual batches are repeated at certain intervals. With few exceptions the production of machine tools in middle-class industry today meets, to a large extent, the above conditions and represents an almost ideal introduction for NC machine tools.

FIRST EXPERIENCES WITH NC AT GILDEMEISTER

In many manufacturing firms the first NC machine to be obtained was a boring machine with point-to-point control. However, economic points of view played no role at all in this respect. It was more important to become better acquainted with, and to obtain first-hand experience with, the new technique and to compare this relatively simple machine with other processing machines.

At Gildemeister however, we adopted a different policy. Our main product, the multi-spindle automatic turning machine, consists of a number of differently formed levers, slide bars, pull rods, etc. with milled flats, which were originally finished on two conventional knee-type milling machines. In 1962, the necessity arose to replace one of those two machines. At that time the decision was made not to purchase a new conventional machine, but instead to purchase a $2\frac{1}{2}$ axes numerically controlled knee-type milling machine with manual toolchanger and a device suitable for planetary milling. The purchase followed in 1963; two years ago the machine was abandoned.

Looking back one can say that this very necessary step of purchasing this NC machine was an incorrect one for the time and the conditions that existed. We should have started with a simple NC machine because we had to ascertain that there could be no possibility of an economical utilization during the first year. Quite the contrary, former production on the conventional machine was still partly more economical. However, this was not only the fault of the new NC machines, but deficiencies also existed in the control and the mechanics. The precision of the machine was unsatisfactory, there were problems with programming which had to be executed by three highly qualified engineers. Furthermore, the required organizational standards were only gradually reached; for example with regard to the preparation of the tools.

ECONOMIC CALCULATIONS FOR THE INTRODUCTION OF NC MACHINES AND PRACTICAL RESULTS

However, the setbacks already mentioned did not discourage us. The next step was the forming of a project group, the task of which was to study the questions which arose about the economic comparisons of the conventional machine tools and the NC machine tools. The computer program, namely Invest, must be mentioned as an essential result of the study carried out by this group. This computer program serves as a help in deciding whether to buy new machine tools, especially those with NC. Today this program enables the comparison to be made of up to four different units. Based on the calculation of the hourly rate of the machine, the setting-up costs of the respective machines are determined, and this also allows the calculation of the production costs per workpiece to be ascertained.

Figure 1 shows the result of an economic calculation according to this program, in which an NC turning machine is compared with a drum-type turret

---

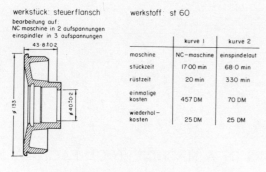

werkstück: steuerflansch   werkstoff: st 60

bearbeitung auf:
NC maschine in 2 aufspannungen
einspindler in 3 aufspannungen

	kurve 1	kurve 2
maschine	NC–maschine	einspindelaut
stückzeit	17·00 min	68·0 min
rüstzeit	20 min	330 min
einmalige kosten	457 DM	70 DM
wiederhol–kosten	25 DM	25 DM

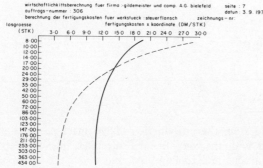

Figure 1. Economic comparison NC turning–turret turning.

turning machine for the machining of a control flange. As is seen, the NC turning machine takes batch sizes of up to 20 pieces, and is thus more economicial than the turret turning machine.

From the results obtained it can be established which type of machine is more economical when applied in different cases. For example, for the production of a large number of threads, etc. and forms which are not cylindrical, the NC turning machine would normally be superior to the conventional machine and to the turret turning machine. On the basis of these findings we have replaced, in the course of the last four years, thirteen turret turning machines, for the production of flange-formed pieces and gear blanks, by four 2-axis and two 4-axis turning machines with numerical continuous path control (Fig. 2). The reason for this action was the high proportion of setting-up, namely the time-consuming tool-settings on the machine as well as the very expensive special tools required. Each of the two axes turning machines which replaced the turret turning machines, is equipped with a hexagon turret for internal machining, and a disc turret for external machining (Fig. 3). At the production of the workpieces with complicated external and internal con-

Figure 2. NC turning section.

Figure 3. Tool post RN.

tours, the application of the two 4-axis NC turning machines enabled us to achieve an increase in output of about 40 per cent. This relatively high percentage was obtained due to the fact that external and internal machining of the workpieces could be executed simultaneously. Figure 4 illustrates the tool position of such a machine.

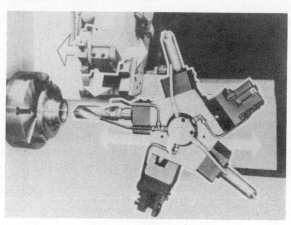

Figure 4. Tool post RSA.

The six NC turning machines, previously mentioned, run in a 2-shift action and achieve an average production rate of 85 per cent against about 55 per cent obtained with the conventional drum-type turret turning machine, which was replaced. This, it is believed, that for the economical running of a manufacturing plant, today as well as in the future, a more considerable amount of capital, and this includes NC machines, is absolutely vital. The reason for this statement, is shown in the results obtained by a study executed by the Union of German Machine Manufacturers, VDMA, on cost increases in the Federal Republic of Germany up to 1980 (Fig. 5). Wages were considerably increased during this period than for instance, the cost for automated plant.

At present there is a total of twenty-two NC turning machines in action at our works. The following gives a general idea of the types of machine.

two 2-axis centre lathes with continuous path control
four 2-axis turret turning machines with continuous path control
three 4-axis turret turning machines with continuous path control

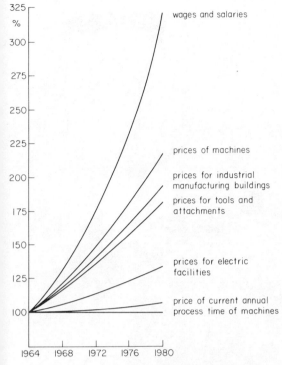

Figure 5. Cost increase up to 1980 (according to VDMA).

four vertical milling machines with linear path control
one turret boring machine with linear path control
one 8-spindle boring automatic with linear path control
two jig boring machines with point to point control
three horizontal boring machines with linear path control
one cam boring machine with point to point control
one cam milling machine with continuous path control

The latter two machines are special machines. They are used for the manufacture of cam drums and cam discs for our multi-spindle turning automatics.

For the twenty-two NC machines, five types of control are used. At first very little value was placed on the actual control systems. This is now considered as a disadvantage as in the early years the manufacturers of control devices developed their products at a relatively high pace. However, there were hardly any difficulties with service. We ourselves produce NC machines and thus disposed of the need for skilled personnel. For users only of machine tools, these facts were treated differently, especially with regard to maintenance and service and consequently a limit on the number of control products appeared.

Loss at present is about 3 per cent for the total of twenty-two machines. However, this amount could only be kept up after running periods of between three and six months. At the introduction of these machines, it was naturally much higher. 80 per cent of the total losses are due to the condition of the controls or the electrics, e.g. limit switches. The amount of purely mechanical failures is very low indeed, about 20 per cent.

QUESTIONS OF PROGRAMMING

The first NC machine tools we obtained were manually programmed, where, as already stated, at first highly qualified engineers from the production planning departments had to be employed. Later it was our intention to change over to mechanical programming via computer by means of the programming language EXAPT. Due to the drawn-out period of development, however, particularly of EXAPT 2, we had to choose another way. In the meantime, we ourselves produced NC turning machines, and our customers expected our assistance in the supply of the control tape for the machines delivered by us. This led to the development of our programming department NPZ (Fig. 6), consisting of a typewriter

Figure 6. Programming place NPZ.

with tape reader and perforator and a small computer. Contrary to the manual programming the programmer only has to establish the geometrical data of the workpiece, as well as the required cutting conditions. These data are printed into the tape and, together with a post processor, which contains the specifications of the machine, are processed in the small computer, and thus the actual control tape for the NC machine is made. By means of a drawing table, which may be connected to the programming place, the geometrical content of the control tape can be checked.

With regard to the expenditure and economy, this lies between the manual and computer type of programming. For a workpiece and a control tape of 100 blocks we have established 5·5 min/block for the manual programming at 2·32 DM/block and for the programming via small computer, 3·17 min/block at 1·21 DM/block. This is equal to a saving in time of 42·4 per cent and a cost saving of 47·8 per cent when computer programming is used.

At present fourteen of the twenty-two NC machines are being programmed by means of a small computer. Consequently, the number of required programmers per machine could be lowered to two, as opposed to three programmers. After the change-over of all machines to this procedure, we hope to lower this number to one and a half. Today, the programming is basically carried out by technicians.

PREPARATION OF TOOLS

Naturally for the economical use of NC machines the setting up of a tooling organization is required. All tools of the NC machines are registered in a card index. Along with a picture of the tool, each index sheet contains its main dimensions. For the tools which are necessary for the machining of a work-piece, the programmer fills in a card (Fig. 7), which contains the order No., designation of the tool and

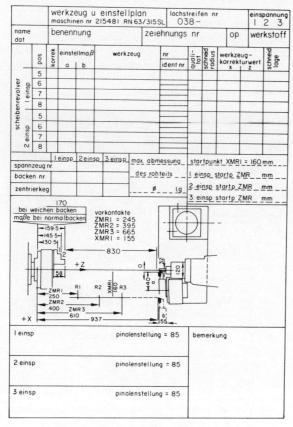

Figure 7. Tooling and adjustment diagram.

the machine, as well as the dimensions of the cutting tools regarding the fixed point of the tool support and the station of the tool support. This card goes to the central tooling pre-setting, where the individual tools are set up on optical pre-setting instruments (Fig. 8). This shows the pre-setting of a complete disc turret. We have at our disposal three of such optical pre-set instruments. The central pre-set employs four persons. In general the tool set is in triplicate: one is always on the machine, a second is being pre-set for the next order, and the third serves as a reserve. After pre-setting the tools, together with the setting-up cards and the control tape, the programming sheet and the blank to be operated are placed in readiness at the machine.

SETTING AND OPERATION OF THE MACHINE

In the NC turning section trained workers are assigned to operate the machines. The setting of the machine as well as the control of the first finished workpiece is carried out by a toolsetter. This tool-

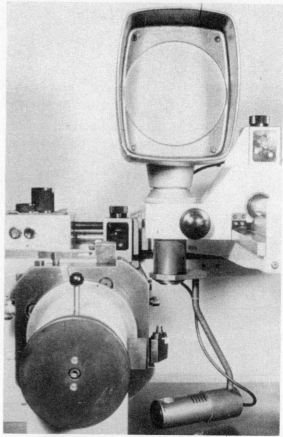

Figure 8. Presetting-up of a disc type turret.

setter is also responsible for the elimination of difficulties which may occur at the machine. Up to now, we have not yet introduced the multi-machine operation, because the machine running times of many pieces is far too short. However, there are plans for a multi-machine operation trial on two NC turning machines, where long pieces are being machined. The operation of the NC milling machine and the horizontal boring machine is carried out by skilled personnel, and they also set-up the machines by themselves.

In this connection it is important that our operation personnel of the NC machines neither works on a time wage nor premium wage basis, as is usual. Studies in our factory revealed that the share of time allocated to chucking and changing of the tools on machines with manual tool changing system, may amount to between 20 and 40 per cent of the machining time. For this reason we have introduced piecework rates, and this has proved a great success.

GENERAL INCREASE OF ECONOMICS BY USE OF NC MACHINES

As is well known, it is very difficult, when calculating the economy to predict precisely the amount of savings which result from the use of NC machines. There are effects which are revealed only after a certain length of time. Some effects are more noticeable than others, e.g. travelling of the parts through the works, and the savings of storage. In this, we could hardly see any improvements. However, this is not surprising, if one takes into account that against the twenty-two NC machines there are about 350 conventional tooling machines. Certainly a further advantage is that one has to study the organization of production and to adapt it to the needs of the NC production. Here, however, it must be pointed out that this adaptation is also good for the production on conventional machines.

Real savings of personnel could be accomplished in the elimination of marking, especially for large parts. Furthermore, the personnel cost for quality control was reduced. Another remarkable advantage of NC production is the reduced scrap rate. Today, we have a scrap rate of about 1·5 per cent. In addition, a more proportionate production rate is achieved, which, for example, led to a reduction of overtime for assembly and thus less difficulty in the assembly procedure.

CONCLUSION

I conclude that we are convinced of the advantages resulting from NC production and that in future we will improve the economy of our works by the use of further NC machines.

CRITERIA FOR PLANNING AND EVALUATING INTEGRATED MANUFACTURING SYSTEMS

by

H. J. WARNECKE and P. SCHARF*

SUMMARY

This paper concerns manufacturing systems, which consist of one or several machining stations, integrated by workpiece storage areas, workpiece handling systems, and control systems. Corresponding to the hierarchy, shown in Fig. 1, it deals primarily with manufacturing systems of grade 2, which are multi-station systems. In relation to different manufacturing tasks different structures of manufacturing systems are possible. For determining the optimal structure in the totality of all possible structures it will be necessary to evaluate the structures of manufacturing systems in respect to their performance, their costs, and their behaviour in comparison with changing manufacturing tasks.

This paper gives a survey of evaluation criteria and of the problems in forming only one utility (or performance) index, which represents an equalized relation between all advantageous and disadvantageous attributes of a manufacturing system.

CRITERIA FOR INTEGRATED AND FLEXIBLE MANUFACTURING SYSTEMS

So far as one-off and mass production are concerned, the automation of the production process has attained a high level. For one-off production, numerically controlled single machine tools with their great versatility, known as NC machining centres, may be a successful method of automation. On the other hand, in mass production, transfer lines characterized by their very specialized equipment are an optimal way of automation with regard to technical and economic conditions. Between these two fields of production there is a big gap in the state of automation in the field of small- and medium-sized batch production. The solution of the problem lies in the principles of the so-called 'flexible manufacturing systems'.

The main criteria of 'flexible manufacturing systems' are:

(1) Ability to machine different workpieces with different ranges of shape, and/or production, similarity. The versatility of the machine equipment depends on the width of the parts spectrum and on the frequency of batch changing.
(2) Adaptability of the machining stations to the changing geometrical and technological conditions of the different manufacturing tasks.
(3) System integration by automated workpiece handling and tool changing.

This necessarily means:

(1) Automation of the technological and the organizational information flow.
(2) Ability to enlarge a step-by-step installation.
(3) Compatibility with other systems of the same grade and with manufacturing systems of higher grades (see Fig. 1).

BASIC NOTES FOR THE EVALUATION OF ALTERNATIVE STRUCTURES FOR A MANUFACTURING SYSTEM

For the evaluation of different types of manufacturing system it is necessary to determine:

(1) the benefit for the system developer and producer, and
(2) the benefit for the user of manufacturing systems.

Here, we shall only deal with the system performance and the benefit to the user.

Up to now investment decisions have been made on the basis of economy, considering only those factors which are quantifiable in terms of money. This procedure is practicable if the investment objects considered are equal in technical merit.

In the planning and evaluation of complex integrated manufacturing systems, there is a very wide

---

* Institut für Industrielle Fertigung und Fabrikbetrieb, der Universität Stuttgart

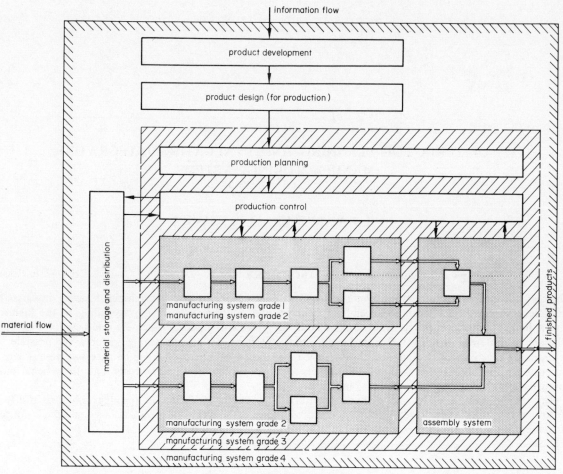

Figure 1.

range of possible system types and system structures for a given spectrum of parts: but they are not equal in all their technical and economic attributes. Therefore it is not possible and not reasonable to compare only factors which are quantifiable in terms of money. A comparison of alternative types of manufacturing system must take account of non-quantifiable as well as quantifiable attributes if it is to be complete.

An evaluation of alternative projects or systems can be carried out only when there are previously defined criteria of desired objectives. The different attributes of a system have to be measured against the corresponding objective.

OBJECTIVES FOR THE EVALUATION OF MANUFACTURING SYSTEMS

The totality of all objectives forms a hierarchical structure. There is one main objective, the rest being secondary objectives which are generally interdependent. In the dependence of two different objectives we can distinguish between:

(a) a complementary dependence, and
(b) a conflicting dependence.

For a direct comparison only, independent objectives or conflicting dependent objectives apply. With regard to different decision fields we need to distin-

guish between three groups of objectives characterized by the following main objectives:

1. Maximize the benefits of the whole plant

Under this main objective, all objectives which are influenced by a manufacturing system and which affect the performance of the whole plant are subordinate (Table 1).

2. Maximize the benefits of the manufacturing system

Under this main objective, all objectives which affect the performance of the manufacturing system are subordinate (Table 2).

TABLE 1

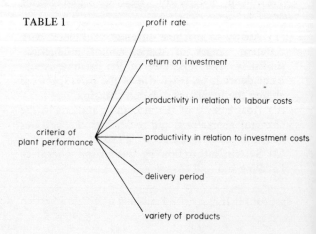

TABLE 2

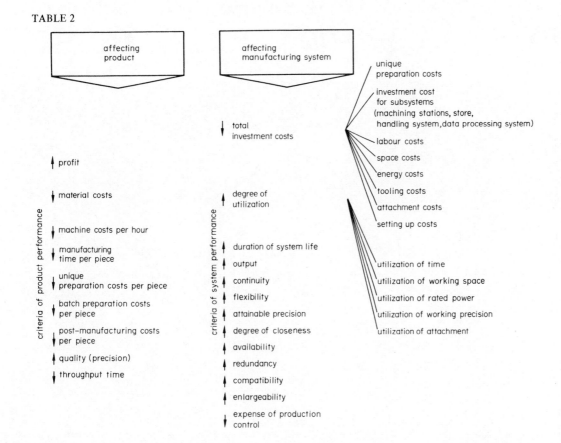

3. Maximize the benefits of the product

Under this main objective, all objectives which affect the benefits of the product are subordinate. They are mostly dependent on characteristics, especially cost, of the manufacturing system (Table 2).

The directions of objectives listed in the tables are indicated by arrows:

$\uparrow$ = Maximize the criteria of the objective.
$\downarrow$ = Minimize the criteria of the objective.

EVALUATION OF PERFORMANCE INDICES

For a clear comparison of alternative manufacturing systems we need, ideally, only one performance index, which includes all the positive and negative qualities of the main objective. Therefore a method of performance index analysis is suggested. The performance index is shown as a dimensionless number, which determines the weighting given to the various factors which affect the main objective.

The procedure for evaluating the performance index is shown in Fig. 2. The important steps are:

(1) Definition of the objectives.
(2) Weighting of the objectives.
(3) Evaluation of the different objective factors.
(4) Valuation of these objective factors by attaching a value V_i from a value scale (which is for example divided in units from 0 to 1).
(5) Calculation of the performance index.

The performance index may be illustrated in a disc diagram, called the Profile of Performance Index (shown in Fig. 2).

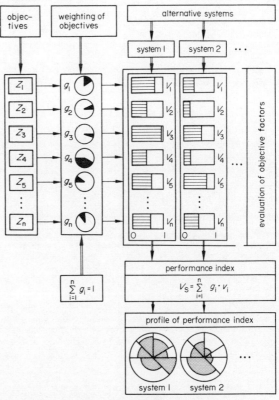

Figure 2.

THE USER'S VIEW OF THE MACHINE TOOL INDUSTRY

by

A. J. TAYLOR*

SUMMARY

This paper outlines the requirements of the user with particular reference to machine tools used in the production of large quantity, high precision components for diesel fuel injection equipment.

The customer's experience must be fully integrated with the machine tool supplier's specialized knowledge. Package deals, including automatic handling, in-process or post-process gauging, together with guarantees covering component output rate and quality, must be considered. As well as price and delivery, after-sales service and availability of spare parts are of major importance.

Examples of specific areas where the British machine tool industry has failed to meet requirements in comparison with foreign machine tool manufacturers and vice versa are given. Examples are concerned with technical competence, risk taking, attitude of mind to productivity, and after-sales service and delivery.

The needs for the future are summarized with particular emphasis on improving communications between supplier and user, and more joint co-operation between the user, the supplier, research organizations, and universities.

THE USER'S REQUIREMENTS

The following comments are generally restricted to machines for high volume production of precision components as used for fuel injection equipment, although in many cases they can be applied more widely.

1. Machine utilization

Machine tools must be designed to work five days a week at least two shifts per day at an effective machine utilization of between 70 and 80 per cent depending upon their complexity. To achieve this, there are many factors which must be observed.

(a) Reliability must be built into the machine in all areas to keep time lost owing to breakdown to a minimum. An overall target of 5 per cent loss caused by machine breakdown is set at the CAV Injector factory. This percentage will vary from machine to machine obviously depending upon the complexity, but even on the most sophisticated special purpose machines the maximum breakdown time is set at a target of 8 per cent.

(b) Automatic fault finding should be built into the control system to reduce the investigation time required by the maintenance engineer.

(c) The machine should be of such construction as to enable complete spare units to be fitted easily in case of breakdown to keep the down time to a minimum.

(d) Plug-in modules should be used wherever possible in the electrical, hydraulic, and pneumatic circuits.

(e) Tooling should be constructed in such a way as to facilitate removal for maintenance or component changes.

(f) One of the major factors affecting machine utilization is the unreliability of work handling equipment 'hung on' to the machine tool. The handling must be an integral part of the machine design and not just another item that is subcontracted to a 'specialist'. The machine tool supplier must be the 'specialist'.

2. Quality

To obtain the required component quality, the fundamental machine tool design must be correct, i.e. slides moving in a straight line, repeatable positional accuracy of slides and indexing tables, elimination of vibration, high rotational accuracy and rigidity of spindle, etc. In addition, more emphasis must be put onto in-process gauging.

There have been vast advances in gauging technology in the field of grinding but only very limited work appears to have been done in soft stage machining on single- and multi-spindle autos, profile turning machines, boring machines, etc.

---

* Company Production Engineer, C.A.V. Limited

3. Productivity

There have been rapid strides in both turning and grinding to obtain a higher metal removal rate. This has been achieved in turning by the introduction of machine tools capable of using optimum feeds and speeds, and by the wider use and development of carbide and other tipped tools.

In grinding, more companies producing external cylindrical and centreless grinding machines are offering the facility of high speed grinding (60 or 90 metres per second). However, grinding wheel development does not appear to have advanced at a rate sufficient to use the full potential of these machines.

Adaptive machine control available in the precision internal grinding field enables cycle times to be reduced whilst at the same time producing components to improve accuracy, geometry and surface finish. There is a need to widen the field to cover external grinding, turning, and drilling.

The reductions in actual cutting times in turning and grinding have emphasized the lost time in getting the components in and out of the machine. In a number of grinding operations on diesel nozzle manufacture, more than 50 per cent of the floor-to-floor time is used in handling the component. More concentration is required by the machine tool designer on reducing this wasted time.

4. Environment and safety

Much publicity is being given to environmental control and pressure is being applied to machine tool suppliers to reduce noise levels, reduce or contain atmospheric pollution, and improve machine guarding and safety features.

Lucas Standards on noise are issued to all potential suppliers of machine tools but many are reluctant to give guarantees. There is no doubt that in many cases considerable developments and therefore on-costs will be involved. A shortsighted attitude must not be taken, particularly bearing in mind the probability that legislation will eventually be introduced in the U.K. requiring provisions for the protection of employed persons against the harmful effects of noise. In the U.S. and Soviet Union very tight standards already exist. More initiative by the supplier is required to meet these needs.

5. Guarantees

When placing an order, the user should not simply expect to be supplied with a piece of machinery but instead with a complete package. Guarantees should be given on output rates, quality standards, tool usage, reliability, and after-sales service. An area leaving much to be desired is the supply of spare parts. Many companies give a good service but there are others who fail to meet their commitments. It is accepted that the user has a responsibility to hold a stock of fast-moving spares but the machine tool suppliers and their agents must play their part.

6. Co-operation

There is a need for closer co-operation so that the user's product experience and know-how are blended with the supplier's specialized knowledge of the machine tools. Data obtained on machine breakdown with the emphasis on areas of weakness shown by repeated faults should be fed back from the user to the machine tool designer.

Over the past two years there has been much more direct contact between the British machine tool companies and the Lucas Organization through formal presentations, where there has been a two-way discussion on new developments and user requirements.

HOW HAVE THESE REQUIREMENTS BEEN MET BY U.K. AND FOREIGN SUPPLIERS?

The policy adopted by the Lucas Organization is to buy British plant wherever possible, providing it is economically justified and meets the foregoing requirements.

In the field of general purpose machines, the British machine tool industry generally meets these requirements. However, when it comes to special and 'off-beat' standard machines, Britain still lags behind her overseas competitors.

The following examples are representative of the experience of CAV Ltd when a decision has been to buy foreign machines.

1. Special purpose machines for grinding bore and seat of diesel injection nozzles (see Fig. 1)

When investigations started, no British machine existed capable of working to the required tolerances.

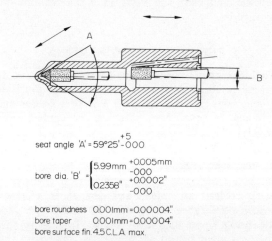

seat angle 'A' = 59°25'$^{+5}_{-000}$

bore dia. 'B' = $\begin{cases} 5.99\text{mm} & ^{+0.005\text{mm}}_{-000} \\ 0.2358" & ^{+0.0002"}_{-000} \end{cases}$

bore roundness   0.001mm = 0.00004"
bore taper   0.001mm = 0.00004"
bore surface fin. 4.5 C.L.A. max.

Figure 1. Section of diesel injector nozzle.

A manually loaded American machine was used for about ten years. In 1959, British suppliers were approached, without success, to start developing an automatic machine in conjunction with CAV. Development started with a continental machine tool company in 1959 and by 1962, as a result of joint effort, a fully automatic machine was produced, with in-process gauging and automatic wheel dressing. As a result of further developments, the machine cycle time has been reduced by 30 per cent and further improvements are planned as a result of modifications to the control system.

There are now fifty of these machines in use at CAV on various fuel injection components. The value of this investment is approximately £1¼m. Numerous similar machines have been supplied to fuel injection equipment manufacturers in Western and Eastern Europe, the United States, and Japan. Knowledge gained from this development is now being used in machines for the bearing industry in all parts of the world.

2. Fully automatic precision centreless grinding machine for grinding diesel injector needles (see Fig. 2)

There was a need to replace old machines used in grinding diesel injector needles. The specification required was:

Needle outside dia.:	6·0 mm ± 0·005
Length:	56 mm
Surface finish:	20 μ inches CLA
Roundness:	0·002 mm
Eccentricity of outside dia. to the 60° angle:	0·002 mm TIR
Production:	2 components per cycle
Dressing:	Automatic diamond roller

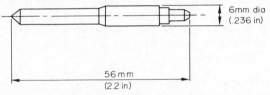

Figure 2. Diesel injector needle.

Six U.K. and continental machine tool suppliers were invited to quote (see Table 1).

The continental supplier who received the order gave guarantees on quality, output rate, after-sales service, and delivery. There are now 8 of these machines in use at CAV and further machines are on order.

To the credit of one U.K. supplier, although there was considerable delay, the failure to get into a very important market has not now been taken lightly. During the past eighteen months considerable work has been carried out to prove that a British built machine will meet the strict specification set for the

manufacture of fuel injector needles. Of course, it has been understood by them that it will be difficult to justify moving away from the continental supplier unless distinct advantages can be shown.

3. Special purpose transfer machine for aluminium rotary pump bodies

A machine was required which was small by motor industry standards, but the accuracy necessary was very critical.

Six U.K. and continental machine tool suppliers were invited to quote. The order was placed with a continental supplier not because of price but because of the very professional approach regarding details prior to quotation, excellent logical technical documentation, and a package deal quotation including swarf removal and mechanical handling. In addition, Lucas Group experience was available on similar machines. All the guarantees which were given prior to delivery were met.

THE CASES WHERE U.K. MACHINES WERE SELECTED

There are a number of examples where U.K. machine tool suppliers have shown that they can compete with foreign suppliers.

The following are examples where British special purpose or 'off-beat' standard machines have been used at CAV with success.

1. Size control match grinder for grinding of nitraloy pump rotors

There was a need to purchase machines to meet a plannned increase in programme. Previously, for fifteen years, German or Swiss machines had been used to grind these components.

Four continental and U.K. suppliers were approached and it was decided that, although their experience was very limited in the area of match grinding, the British company would be given the opportunity to prove the capability of their machine. This was carried out as a joint project in conjunction with the Ministry of Technology.

TABLE 1 Comparison of quotations on centreless grinders

Supplier	Two components per cycle Rough	Finish	Cycle time (seconds) (incl. dress) Rough	Finish	Fully auto. load and unload	Diamond roll dressing	Delivery	Price of machine (without spares)	After sales service
Continental	Yes	Yes	5·5	5·5	Yes	Yes	12 months (with penalty clause)	£18 000	24-hour guarantee clause
Continental	Yes	Yes	6·5	6·5	Yes	Yes	12 months (no penalty clause)	£20 000	Good
U.K.	No	No	12·5	15	Yes	Yes (rough only)	7–8 months (no penalty clause)	£10 000 (finish) £11 000 (rough)	Good
Continental	Yes	No	5·7	11·4	Yes	Yes	6 months (no penalty clause)	£9000	U.K. service poor
U.K.	Yes	No	6	8·5	Yes	Yes	7 months (no penalty clause)	£13 000	Fair
Continental	No	No	20 (one per cycle)	20	No	No	Enquiry stopped	£9000	Enquiry stopped

The specification required was:

Nitraloy rotor to be
ground:	19 mm dia.
Length:	38 mm
Surface finish required:	3 μ inches CLA
Roundness required:	0·002 mm
Clearance between ground dia. and lapped bore:	0·001 mm to 0·002 mm
Stock removal:	0·020 mm
Cycle time:	30 seconds

After a period of development during which there was very close co-operation between the machine tool suppliers and the CAV engineers, the machine met all the requirements and further orders have been placed.

2. Vibratory deburring, metal finishing and cleaning plants

There are a number of applications for which this type of plant has been successfully supplied by U.K. manufacturers. They have been very competitive on price and have met the quality and cycle time specifications, although in some instances the engineering has caused problems. In all cases these problems have been overcome by concentrated joint efforts.

3. Numerical controlled plant

British NC milling and drilling machines have been proved as good as, if not better than those supplied by continental rivals, and ther are a number of very successful applications now running within CAV. The continental suppliers generally lead the field in the area of NC turning, although as a result of recent developments a number of British machines have been ordered by CAV.

LOOK TO THE FUTURE

What does the user consider should be done to improve the image of the British tool industry? This can be summarized as follows:

(a) The representative of U.K. machine tool suppliers must be more knowledgeable about his products. The initiative must be taken by the supplier to suggest improvements on the customer's initial proposal.

(b) Liaison between the user's and supplier's engineers and designers must be at a level at least as good as, or even better than that currently experienced with foreign suppliers.

(c) Full use must be made of knowledge available from research organizations and universities. Co-operation between the user, the supplier, and research organizations or universities has proved very successful in several major projects at CAV. A number of very sophisticated precision machine tools are now in production as a result of this approach.

(d) More effort should be made to give a package deal with integrated handling, gauging, and control systems. Get away from the 'hang on' concept.

(e) More emphasis should be laid on modular design techniques to improve reliability and reduce down time. Particular attention must be paid to machine control systems.

(f) A better after sales service must be provided.

It is most important to let the receptive user see more initiative coming from the machine tool industry to determine the future needs. Let the industry show that this is being done by regular top level communication meetings aimed at giving the user details of future developments and finding out their views.

DISCUSSION

Q. F. W. Craven, Herbert Machine Tools Limited. I disagree with the first questioner's objection concerning criticism by users. In the common interests of the machine tool supplier and the machine tool user it is important that a continual dialogue of criticism should continue. If both parties have a better understanding of the other's problems there will be a better chance of achieving mutual solutions.

Mr Taylor's lecture demonstrated some first class and courageous production engineering. His argument for the supply of (for example) extremely precise internal grinders from the UK is not likely to be taken up because of the firm foothold that certain Continental suppliers have already established in this relatively limited market.

A. I agree with Mr Craven's comments relative to 'precise internal grinders'. However, the object of my comments was to illustrate how the British Machine Tool Industry has failed in the past to recognise the potential of projects where a certain degree of risk has to be taken. I believe that the Machine Tool Industry must show by example that they can compete on the more sophisticated projects. They must be prepared also to provide a 'package' as outlined in my paper.

Q. P. Brooks, Giddings and Lewis Frazer Limited. A previous contribution (B. T. Evan, Wickman) has taken exception to Mr Taylor's comments concerning the British Machine Tool Industry and given the impression that such a paper is not welcomed.

On the contrary, and also speaking as a machine tool manufacturing representative, I feel that constructive criticism of this nature is beneficial in enabling the user and the manufacturer to see and understand each other's problems.

I do not wish to dwell on this general theme but would be grateful if Mr. Taylor could expand a little on the aspect of poor aftersales service. In what specific areas, from his point of view, is the Industry failing to meet user requirements?

A. Generally, the reaction to the request to Machine Tool Suppliers for after sales service is good. However, two major failings are:

(i) The delivery time for spares is in many cases bad. There is a reluctance on the part of a supplier or his agent to hold sufficient stocks.

(ii) The information provided relative to spares is often inaccurate and in fact, changes in design are frequently made without prior consultation with the user or without notifying the user of changes made.

A USER'S CRITICAL ANALYSIS OF HEAVY MACHINE TOOLS

by

L. WOOD*

SUMMARY

The paper examines present machine tool designs and concepts for the machining of heavy workpieces in the light of their contribution to the problem of work in progress and their tool/workpiece contact time; and suggests that the time has come for fresh thinking for the machine shops of the future.

INTRODUCTION

In the heavy engineering industry with a high jobbing content, the machine tool plant changes very slowly and is expected to have a life of some twenty years or even more. In this length of time the product can change more rapidly and it is important that parameters of any new plant are chosen with this in mind. Although it is true that engineers and designers of the product can very often design within the limits of existing plant it would be much more desirable if these limits could be changed from the plant to the building.

In general, transport limitations are a safer guide to size and weight but these can be ignored if the factory is located at a port with direct access to shipping. But even this has to be carefully considered as the practice is sometimes to assemble parts together for machining operations, which are later dismantled for transport to site.

Machine tools with fixed envelopes therefore impose a physical limitation on workpiece size which could be a handicap within the life of the plant. In the paper entitled 'Time critical path analysis charts' it is argued that heavy engineering in particular was prone to high working capital investment and that this could be significantly reduced, by eliminating or combining operations and secondly by improving the time that the cutting tool was in contact with the workpiece. It is thus necessary to get away from a layout consisting of single purpose machine tools with its excess handling, queuing and lengthy setting time to a situation where the workpiece is placed in position and the cutting tools brought to the workpiece without disturbing the setting.

TRADITIONAL PLANT

The traditional plant used by heavy engineering has the basic characteristics of being single purpose with fixed envelope and weight capacity, and these are the very characteristics which should be avoided. Such plant includes, vertical borers, planers, plano mills, table type horizontal borers, slotting machines, radial arm drills and even centre lathes.

Another point to bear in mind is that, with the exception of the plano mill, these machines apply single point cutting tools to the workpiece, and in most cases crane handling is an integral part of the floor to floor time.

Planers

This type of machine tool has a fixed table size, a fixed width between the columns and a fixed maximum height under the tool post. There is also a maximum weight which the table can carry. It does not seem to make mechanical engineering sense to traverse heavy loads at high speed backwards and forwards underneath a piece of high speed steel or carbide in order to remove metal. Many efforts have been made to improve planer performance by cutting in both directions and by using multiple tool boxes but in the end it must stand comparison with milling cutter performance if it is to remain a viable tool.

Plano-mills

These suffer the disadvantages of the fixed envelope of the planer, even though they use a multi-point cutting tool. Good metal cutting performance can be obtained provided that the milling heads are a direct drive type. Vibration is the enemy of high speed milling, and geared heads should be avoided if maximum performance is to be obtained.

The plano-mill can have the additional advantage of being able to bore and drill holes and, if provided with readout, marking out can be eliminated. The major disadvantage in this type of machine is that the floor-to floor time consists of the sequence: cleaning

---

* GEC Machines Ltd.

C

the table, crane lift of workpiece on to the table, manhandling the workpiece into position, fitting clamps before any attention is paid to the spindle in terms of fitting cutting tools, and so on. In a heavy engineering workshop cranes are not always available at any precise moment. The percentage of cutting time to total time is very often 20 per cent or even less.

Some machine tool designers, recognizing the problem, have designed plano mills with tandem tables which can be used together for long work or independently for short work so that work can be set up on one table while the other is being used.

Swarf is also a major problem on these machines as it collects either on the table or on either side of the table making it difficult to use conveyors or collecting bins. They usually need the man with the wheelbarrow, and this type of labour is fast becoming difficult to find.

Horizontal borers (table type)

It is quite usual to find a situation where this is the most expensive machine tool in the shop to operate. Labour is of the highest of skills and the machine is the least productive in terms of metal removal time. Apart from the disadvantages of a fixed envelope and table weight it is single point cutting. In addition there is a lot of wasted time in stopping the machine for gauging and measuring. The more accurate the work, the more lost time in this aspect.

Swarf is not an acute problem because the machine tool does not make enough. The floor-to-floor sequence also includes lost time for handling and setting. Although indexing tables are a normal fitment, sometimes there is extra setting involved if machining is required on the horizontal planes.

Modern designs incorporate milling and it is important to recognize the difference between a machine tool basically designed as a boring machine which, it is claimed, can do light milling as against a milling machine which can bore. The price will usually indicate the difference and there is no doubt that many people have been influenced by the price differential and by doubtful claims on milling performance by a boring machine.

Vertical borers

It would be difficult to imagine a more complicated mechanical engineering problem than the one caused by suspending a turning tool at the end of a long ram hanging from a beam supported by two vertical structures. Perhaps it is because of the mechanical engineering problems inherent in such designs that a great deal of talent is devoted to improving the mechanical performance of this type of machine. If machine tool designers really studied a breakdown of the floor-to-floor times it should become apparent that such design energy was perhaps misplaced and would be better employed in looking at ways and means of recovering operator lost time or better still questioning the whole concept of a vertical borer as a viable method of removing metal.

The first consideration of the workpiece is one of envelope size: will the workpiece swing between the uprights and underneath the rams? If not, then a horizontal boring technique must be used. For some workpieces there is a choice between vertical and horizontal boring, and this will depend upon accuracy specified, size of bore and available table rpm to obtain the necessary cutting speeds. There is also the reverse envelope problem of a small workpiece on a larger vertical borer table, where even the maximum table rpm will not give the necessary cutting speed.

These types of envelope considerations have given rise to machine shops with a range of vertical borers in steps to cover what is considered to be any eventuality in workpiece size. Each of these machines is manned, each is single-purpose and the organization spends all its energy in keeping these machines working and the operators happy—all at the expense of work in progress.

In heavy engineering the workpiece is either a unique forging or casting or a unique fabrication; in other words the workpiece can have characteristics which differ from those that have been experienced before, even though it may be a repeat order.

Individual forgings and castings present individual problems in dimensions and machinability; fabrications present problems of setting up in order to get the best out of them. This will always involve a learning curve which will depend on familiarity with the workpiece, how well the job has been planned and the degree of planning detail given to the operator.

An overhead crane will then pick up the workpiece and drop it on to the table, the workpiece being guided to an approximate position on the table by suitably prefixed stops or clamps. If the workpiece is circular and there is a lot of metal to be removed, then it could be that no further positioning is required and the operator can proceed to clamp up. If the workpiece is not circular then more precise positioning is required; first we have to check, either by clocking or by a sweep gauge from the table centre, or visually by concentric rings turned on the table, and then some form of manhandling is required to push the workpiece in the desired direction. For a deep workpiece it is then necessary to check the vertical plane and correct this by using packing and shims. This handling may require the crane or, as is more usual, the use of jacks or crowbars. It is now possible to start clamping; for a workpiece near enough in balance this is not too difficult, but for other workpieces additional clamping is necessary to resist the centrifugal forces. For fabrications the cutting speeds on vertical borers are usually below the optimum because of centrifugal forces generated by out of balance and the difficulty of clamping; there is an additional problem of possible distortion due to a combination of clamping and out of balance.

It is now possible for the operator to turn his attention to the rest of the machine tool; the bridge requires positioning to give the maximum ram rigidity, the tool requires inserting in the tool post, and feeds, speeds and depth of cut have to be chosen and a trial cut taken. The machine is stopped and the operator has a mountaineering exercise in climbing on the workpiece together with his measuring equipment in order to measure size. For castings and forgings it is necessary to get under the skin; this is not always possible due to a variation in the depth of cut, and very high tool wear and breakage is common. The usual practice is to use heavy cuts at slow speeds. On fabrications light cuts have to be taken as the workpieces very often lack rigidity, and since the rpm limitations referred to previously can apply, the metal removal rates on the whole are below the optimum which can be obtained with modern tool materials.

Recent innovations in vertical borer design, such as constant cutting speed capability (infinitely variable rpm) preloaded table bearing designs eliminating table lift, have improved accuracies which can be obtained in finishing. Nevertheless, even with digital readout whereby a tool tip can be positioned to within 0·001 in of the centre of the table and the same accuracy above the table, the resulting face or diameter on the workpiece is very different, because of tool push-off.

There is a continual stopping of the machine for measuring and gauging, the amount of which largely depends on the confidence of the individual operator.

Attempts have been made to reduce this time lost by fitting in process gauging, consisting of roller devices; these are very useful for work with uninterrupted bores or outside diameters, but they are complicated to fit and do not really suit a general engineering environment with a constantly changing workpiece. After machining we have the inevitable secondary operations where the workpiece has to be removed by the crane, taken to a convenient location, turned over and reloaded on the machine. Meanwhile there is a large quantity of swarf to be swept up and handled to a collecting point.

In short, vertical borers, say from 6 ft table diameter and over, would fail on the following counts in relation to a policy which regarded work in progress and investment in working capital as its main objective.

(1) Fixed envelope requiring a range of sizes.
(2) Single purpose. Other operations such as milling, drilling, etc., would require special design considerations.
(3) Crane handling and setting of work pieces in sequence with the cutting.
(4) Single point cutting.
(5) Accuracy of tool setting does not equate to accuracy of workpiece, necessitating constant gauging and stopping of the machine.
(6) Inability to operate at optimum cutting speeds because of unbalanced workpiece, table speed range, interrupted cutting conditions.
(7) Swarf not easily collected.
(8) Operator working in blind conditions when boring.

Centre lathes

The basic fundamental technique of turning shafts between centres has not changed since the days of Whitworth and Watt. Size and weight have brought problems, but these have been solved quite admirably by succeeding generations of machine tool designers all working on the basis of the same thing but bigger and better.

A lathe has the same disadvantage of a fixed envelope as previously described, and an error of a fraction of an inch in specifying the height of centres and/or length between centres can have serious repercussions. For this reason lathes are usually purchased with plenty of safety margin over the best predicted maximum workpiece size, resulting in a lathe which can do all of the workpieces but the majority inefficiently because it has become too big. Thus there is a tendency, as for vertical borers, to have a range of lathes in a general engineering environment. It is not uncommon within the life of a large lathe for it to have had an increase in bed length and/or to have had the headstock and tailstock height packed up.

Heavy shaft lathes are usually quoted with two maximum workpiece weight capacities, one for shafts between centres and one for shafts supported by one or two steadies. The raw material is either hammered bar, a casting or a forging, which requires machining before one can apply a steady. The purchaser is faced with a problem of spending a lot of extra money for capacity between centres when he only requires this operation for turning steady bands, the remainder of the turning being done on steadies. Castings and forgings are usually proof machined by suppliers, and one way out is to pass the problem to him. The forgemaster usually has machine tools designed for removing metal and not necessarily for producing steady bands with roundness accuracies of 0·0003/0·0004 in. There is need therefore for simple techniques of machining steady bands on shafts based on rotating tool rather than rotating workpieces. Mounting of shafts between centres, and then trying to adjust and tighten four jaws without pulling the shaft out of line, is almost an impossibility. This is not too important if the shaft is rigid enough to be machined at one setting, but can cause trouble in the machining of flexible shafts where the practice is to machine at the tailstock end and turn the workpiece end-for-end for machining the other shaft end.

Lathe manufacturers seem to have spent a lot of design and research energy on headstock and bed design and not enough on tailstock and steady design. If dead centres are used, the rpm must be kept low in order to minimize wear in the component centre; it is usual practice to recut such centres before commencing finishing turning. If live centres are used, finish is only as good as the vibration from the tailstock resulting from the bearing design. Improved quality of bearings and preloading techniques have improved tailstock performance in recent years, but they still lack the necessary rigidity for turning at speeds where the operator can use modern cutting tools. Since it is desirable to keep the width of the steady bands as narrow as possible, the load on the steady pads can exceed permissible loading on anti friction metals. The use of a roller bearing design creates problems in finishing through vibration and shaft marking. It would seem that obvious developments in the fields of hydrodynamics and hydrostatics could well be applied, but the problem is complicated by a desire for open steadies and variable steady band diameters.

The problems of positioning and moving steadies and tailstock along the bed of a lathe seem to have been totally neglected. In setting up these units have to be positioned on the bed to suit the shaft being turned, and the addition of a rule fastened to the bed would work wonders. The use of comparatively simple air lift units would ease considerably the problems of pushing these units into and out of position.

It is usual to find in heavy engineering that the lathe is expected not only to finish-turn the shaft but somehow or other to produce close tolerance fits and finishes. There are many devices from roller burnishing, toolpost grinding, broad tool finishing, each of which are lengthy operations depending on how well the finish turning has been carried out. Finish turning and trying to obtain size usually destroys the finish. If size can be obtained while retaining a reasonable depth of cut, then good finishes can be achieved which would not require much effort to polish up. D.R.O. shows up to great advantage in this respect and is an essential feature on any large lathe. Finishing can be as high as 75 per cent of the total time to turn a shaft and in process gauging would be a great asset.

It is rare for large shafts, say 10 tons and over, to be machined at anything like the performance at which cutting tools are capable, and the majority of the reasons for this can be traced to trying to machine a shaft between centres.

The ultimate engineering purpose of most shafts is the transmission of rotary motion, and they are, to say the least, rarely if ever mounted between centres. They are usually supported in bearings or by couplings, or in one bearing and one coupling. The particular engineering design will take into account the rpm and the natural deflection. Why therefore must it be turned between centres? Would it not be better if the shaft were captured in its natural environment for the purposes of turning?

For a shaft which will ultimately be mounted in two bearings it would seem logical to machine the bearing diameters first, with the shaft stationary using a rotating tool, and then to use these diameters to hold the shaft for further machining. In this sense the headstock is only required to rotate and stop the shaft, which should be done through a flexible coupling. It is then only necessary to provide means of taking up the feed force of the cutting tool. The resulting layout would still look like a lathe, but with the emphasis on holding the shaft and not steadying it. In general large shafts rotate at much faster speeds when in service than when being turned, so that it would be possible to use the job bearings for this purpose. For shafts with separate couplings the end of the shaft can be used; for shafts with integral couplings a slave diameter would have to be created near to the coupling back face.

An interesting example of combining turning and grinding into one machine is now appearing; this shows that machine tool designers are beginning to appreciate the advantages of combining operations. With this example, however, it is suggested that endless belt techniques would give a better answer. Most shafts need keyways, and very often this is the only extra operation required. It is most annoying to have to remove shafts from a lathe, mark them out, and set them up again for a milling operation. Lathes should be designed so that the spindle can be locked and to be suitable to take keyway milling operations either by a separate machine tool located back to back, or by a separate accessory.

THE FUTURE

The machine shop of the future must surely be designed around multi-point cutting tools which can be applied with the minimum envelope restriction in any linear or rotary plane. This inevitably leads to the milling cutters applied to the workpiece by a travelling column horizontal spindle. With the layout the column height can be close to the clearance under the overhead crane, and the bed the length of the shop. One or two travelling rotating tables can be chosen as necessary for rotary work, and the floor covered with bedplates (Fig. 1). The only fixed part of this envelope which cannot be increased is the height of the spindle above the floor plate. The bed length of both column and rotary table can be increased at any time, and if the table is properly designed both to take the weight which can be lifted by the overhead crane capacity and to take sub tables for extra large work, it can be said that with the layout the envelope restriction has been shifted away from the machine tool to the building.

The layout has also the inherent advantage that workpieces can be set up while the column is working on another workpiece. With proper planning this layout can achieve as high as 80 per cent tool workpiece contact time, and in general there is little work at present done on traditional planers, plano mills, borers, drills, etc., which could not be machined by such a travelling column set-up.

There are, of course, many examples of travelling column type milling and boring machines available, but it is suggested that the enormous potential of this layout is not being realized because both user and machine tool manufacturer are looking at the hardware as just another individually operated machine tool instead of it being the basis for an integrated machine shop. This basis would consider the following main features.

(1) The machine shop would be one integrated foundation block independent and isolated from the building structure. The block would be designed to take the bed for a number of travelling columns and any necessary tables.
(2) The bedplate area would be designed for a load of 2 tons per square foot and fitted with accurately machined bedplates so that work could be set up without using the column as a checking device, and for back up machine tools for supporting ancillary work. The bedplates would be mounted on jacks set in epoxy resin with quick levelling facilities.
(3) The tables would be capable of separate operation for setting purposes and for use with other machine tools, but would be operated as an integral unit with the column when required.
(4) Datum stops, independent of the bed of the column, would be fitted such that the cutter centre line can automatically be located to the centre line of the table.
(5) The fastest possible first traverse speeds on the

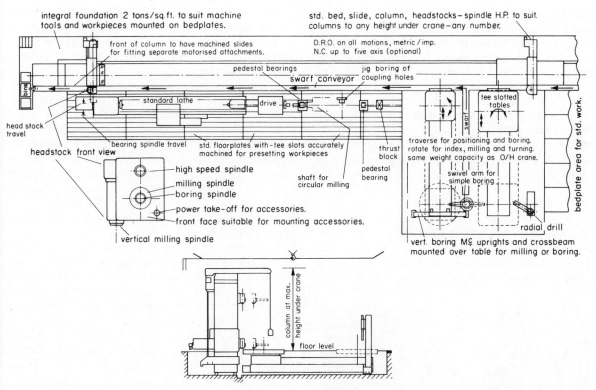

Figure 1. Open side layout.

long travel of the column with the delicacy of control to position the tool to a random job datum. From job datum to any other position, automatic positioning to a preselected figure to 0·01 mm.

(6) A machining capability which will develop and use the available hp at all vertical positions on the column.

There are two recent variations of this layout theme with the same objective of minimizing working capital. The first one is shown in Fig. 2. This has

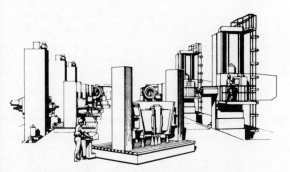

Figure 2. Air pallets cut down N/C down time.

fixed machining positions with a facility for setting up elsewhere in the shop and rapid moving of the set-up workpiece into and out of the machining location by the use of air pallets. It is by Ingersoll USA. The pallets are 12 ft by 18 ft with T slot surfaces, and can be moved by one man to a precise dowel location within 0·001 in. The workpieces measure approximately 14 ft diameter x 12 ft long and weigh 40 tons. It is claimed that this method can reduce down time from 60 per cent to only 10 per cent. The only arguable disadvantage is the fixed

envelope size of the pallet and the fixed machining stations, each of which is manned.

Figure 3 shows another variation known as a gantry type machining arrangement. If overhead crane handling is used it is difficult to see the advantages of

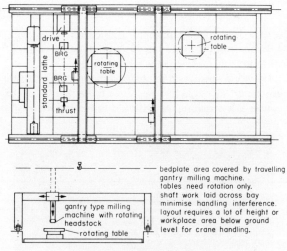

Figure 3. Gantry type.

this layout, as the bridge will put an effective materials handling barrier across the shop, and the envelope will be considerably less than the building capability. It could work in conjunction with the air pallet handling system providing enough room was left for manoeuvring. Although this layout may well be appropriate in special cases, it is not so flexible as the open sided layout.

Rotating tables

Modern design of rotating tables has given us many interesting possibilities in machining practice. The tables have longitudinal motion for boring and milling and can be indexed accurately so that work can be bored from both sides. They can be indexed to 0·001° and have rotary milling feeds. Rotary turning feeds can be fitted if desired. These tables allow the user to take a close look at his vertical borer type operations to see if such work could be done firstly by a milling cutter and, if not, by using the system as a vertical borer.

Circular milling

This combination of a table and column enables a great deal of circular milling to be done on faces and O/dias of all kinds of components and shallow bores of ring type components. Figure 4 shows face milling at 11 mm depth of cut, 290 mm machined band width,

Figure 4. Face circular milling.

with a feed of 680 mm/min. The cutter is removing 1·46 cu. in/hp (80 to 85 hp) completely without vibration and leaving the workpiece cold.

Figure 5 shows milling the periphery with a cutting efficiency of 1·73 in^3 per hp (63–8 hp). This of course is a roughing cut with a finish in the form of a

Figure 5. Periphery circular milling.

scallop where the peaks and valleys are a function of the speed of rotation of the table and the rpm and number of blades in the cutter. This can be predetermined to examine whether the finish is acceptable on a particular workpiece.

This peripheral circular milling technique can be further extended when very fine finishing is required. The set up is the same as with roughing, except that the cutter is given an additional vertical feed so that the machining path is a helix. As an indication the periphery of the ring was finish machined with a cutting speed of 200 m/min (656 ft/min), depths of cut 0·15 mm (0·006 in), table rotation 0·07 rpm, head vertical feed 8 mm/min (0·315 in/min) in 1½ revolutions producing a surface finish of 40–110 RMS[1].

As a comparison, Fig. 6 shows the same set-up, but using the table as a turning device. Although it was possible to almost achieve the same metal removal rates with a turning tool, considerably more vibration was present and the interrupted cutting broke the tool. Equivalent surface finishes were not obtainable even with this set up which would be much more rigid than that which could be obtained on a vertical borer. A summary of comparisons between this method and vertical boring can be detailed, as in the appendix.

Figure 6. Turning on circular table.

There is no fundamental reason why this circular milling technique could not be applied to shafts, and the lathe of the future could well take the form of independent units such as bearing supports, drive motor, tailstock mounted on the bedplate in front of a travelling column. It obviously makes good sense from an engineering and safety point of view to move workpieces at low speed and to have the hp in the cutter rather than the other way round.

No assessment of circular milling is complete without considering circular interpolation of the spindle. For workpieces which have small bores so that it is not possible to reach from the headstock when the workpiece is flat on the rotating table, the usual practice is to use some form of boring and facing slide attachment mounted on the headstock for boring horizontally. Not only are these attachments expensive to purchase but they require putting on and taking off with consequent loss of spindle time, and the question must be asked whether a standard milling cutter plus circular interpolation will not give a more economic answer.

Most of the large milling/boring machines at present on the market are driven by Ward Leonard sets feeding one feed motor for each axis. An essential condition is that any feed motor has a minimum rpm which is set at some safe margin above the rpm at which the motor becomes completely unstable. In theory therefore such a machine tool will produce a circle with four flats. In practice if all the implications are studied for bores of, say, 36 in and over, with high feed rates the flats are negligible[2]. If the bore is for a mechanical fit rather than a running fit flats can be allowed. The same arguments can be used for linear interpolation of slopes. Obviously a form of NC control is required to monitor and continually correct the position.

For such work it is imperative that each axis should have its own feed motor and although the best examples of these machine tools can be controlled through a Ward Leonard set by a computer, it is suggested that the Ward Leonard set is now due to be discarded in favour of the integrated feed drive unit. This comprises a permanent magnet DC motor with tacho generator and resolver all mounted on one shaft, with high torque, low inertia characteristics which allow the elimination of gear boxes. The result could be a much simpler and we hope cheaper machine.

CONCLUSIONS

This paper has attempted to apply the lessons of group technology to machine tools used in the heavy engineering industry and suggests that some of the traditional designs are suspect when attention is focused upon the use or misuse of working capital. Milling is suggested as the process most worthy of machine tool design effort in meeting the needs of the future, and if the paper does nothing more than make a designer stop and ask himself why he is spending so much effort in spinning heavy workpieces at high speeds it will have served its purpose.

ACKNOWLEDGMENT

Figures 4, 5 and 6 are reproduced by permission of Innocenti and C. W. Berthiez.

REFERENCES

1. C. W. Berthiez Patent.
2. Specification and tests of metal cutting machine tools, UMIST.

APPENDIX: CIRCULAR MILLING AND VERTICAL BORING

	Circular milling	Vertical borer
Group technology	Readily groups with plain milling, drilling, boring, tapping, etc.	Does not easily group with other operations. Usually separate operation.
Accuracy	Feed force is vertical on to the table, therefore no tool push off. Component accuracy can reach the guaranteed accuracies of the table.	Feed force against ram. Tool push off considerable. Impossible to bore and turn accurately as ram extends.
	Face milling (feed force parallel to top of table). Flatness within rotary travel of table 0·0004 in in 80 in.	Table lifts at high turning speeds result in taper.
Finish	32 μin can be achieved.	100/250 μin, depending on position of ram.
Interrupted cutting	Does not cause any serious problems.	Increases tool push-off and vibration with reduced metal removal performance and accuracy.
Alloy steels	On materials with difficult machinability castings, hard spots in forgings, etc., milling can cope.	Can become difficult if not impossible.
Tooling	More expensive. Throwaway tips can be used, but for accurate finishing brazed tips with tool and cutter grinding would be required.	Comparatively cheap. Throwaway indexable tips.
Chip removal	Milling produces short curly chips, and this, combined with the layout of the machine tool, lends itself to mechanized handling of swarf.	Chips can be continuous type which is awkward to handle. A vertical borer does not lend itself to mechanized chip handling, and more manual labour is required.
Unbalanced workpiece	Because speed of rotation is low, these do not set up unequal centrifugal forces, and clamping of the workpiece can be simple.	Requires complicated clamping to overcome unbalanced centrifugal forces. Speeds have to be reduced below the cutting ability of the tool because of safety and wear on the bearings.
Workpiece temperature	Remains cold. All heat is in chips. Measurements can be taken with confidence.	Rises in temperature *pro rata* with rate of metal removal.

APPENDIX: CIRCULAR MILLING AND VERTICAL BORING—*continued*

	Circular milling	Vertical borer
Special forms	Cannot perform special forms other than straight tapers.	Can perform any form, taper or radii.
Maintenance	If table inoperable, column can be moved to continue work on bedplate. If column inoperable, table can be used with portable machines for drilling, boring, marking out, etc.	The whole unit is inoperable if any part of the machine fails.
Obsolescence	Less danger of obsolescence as cutting speeds can be increased fairly easily.	Any break through in machinability requiring increased cutting speeds cannot always be used on a vertical borer as the workpiece rpm would require increasing.
In-process gauging	The variation in dimension from the cutter setting would be minimal; within the normal tolerances allowed the machined dimension is likely to be within the cutter setting dimension thus reducing the need to measure.	Because of tool push-off, actual component dimensions vary from tool setting dimensions larger than normal tolerances. Measuring is therefore essential.

DISCUSSION

Q. F. W. Craven, Herbert Machine Tools. Mr Wood's clear exposition of the many problems concerning design and utilization of heavy machine tools rightly recognizes that in this field multi-tooth cutting is of paramount importance. The relatively small number of heavy machines, their high capital cost and the opportunity of designing machines for a limited range of heavy components make such considerations into a special case. In this field more than any other the purchaser cannot expect the supplier to simply quote a machine tool to cover nominated components. An iterative process of knowledge exchange is required with both parties gaining as much information as possible on each other's problems and possible solutions.

A. Mr Craven is quite right in that this kind of plant involves close cooperation between the machine tool maker and the customer. This cooperation usually only begins at the stage when the basis for the machine tool process has been settled and very often the customer is requesting tenders for say vertical borers when he should be discussing the broader problem with the machine tool makers of 'how to make this range of components'. At this stage it must be borne in mind that the components can very often be redesigned to suit a proposed new piece of capital plant and new machining processes.

The criticism implied in the paper is of a situation to which both customer and machine tool maker have led us, but in which the customer must take the greater responsibility. The emphasis has been on mechanical engineering aspects in feeds, speeds and accuracies with the problems of working capital a poor second. Once the purchaser realises that the use of working capital is the most important aspect of any business then he will devise layouts very different from our present shops and will begin to demand machine tools which can be arranged to suit as described in the second part of the paper. I have no doubt that the machine tool industry, once given the challenge to provide this type of machine, will more than adequately meet the demand.

Q. B. J. Davies. (1) In Fig. 1, does Mr Wood advocate a separate simple machine on the same bed for drilling holes in large workpieces, or a drilling spindle or attachment on an expensive machine such as a ram borer?

(2) Is achieved workpiece accuracy with the large machines shown adequate?

(3) In the next decade, will components become so large that some different machining approach other than building larger and larger machine tools be essential?

A. (1) The layout in Fig. 1 shows a layout developed from machine tools designed in modules. The details of the layout would vary according to customer requirements. Since drilling is always a secondary operation the question of whether one purchases a drilling machine would be a secondary decision. The first decision would be the milling facility, and it is obvious that drilling, etc., would be incorporated in the first machine.

If the capacity required to be increased then the next step could be a lighter machine on the same bed or installed separately to serve the rotating table. If the lighter machine were installed in the same bed this may restrict the flexibility of laying out workpieces on the bedplate. In general I would be against the purchase of a single purpose machine tool such as a drilling machine except as a supporting tool for the main machine, i.e., either for use with the table or for temporary installation on the bedplates.

(2) Accuracy on large workpieces is a complex subject depending on whether it is boring, milling, point-to-point accuracy, linear or circular interpolation. It depends a great deal on the tooling and on the way the workpiece is set up and clamped, particularly when machining fabrications. The best examples of modern large milling and boring machines and tables are more than adequate for workpiece accuracies. In general they are more

accurate than the means of inspection and are within the limits where one would begin to consider the temperature of the workpiece in conjunction with the tolerance.

(3) Designers of large capital plant must still design within the transport limitations of weight and cube size, so that in general these can be accepted as the limits for machine tool design.

The exception to this is when there is an assembly to be machined. Such machining is generally done because of the inadequacy of machining the parts of the assembly, and good design with modern machine tools can make such operations unnecessary. These aspects should be considered very carefully before investing in what would be a very special, expensive and often single-purpose machine tool.

Q. J. A. Stokes. It is not clear whether Mr Wood is complaining that he cannot get the machines he wants, or whether he is not prepared to pay the costs of meeting his particular requirements. In general, all that he wants is available from various sources. Indeed, an installation having some of the elements was installed in AEI Rugby over 25 years ago.

If Mr Wood's machines show detectable flats in milled bores, there is certainly something wrong. Modern equipment of good quality keeps such flats to less than the equivalent of the surface roughness of the milling cut.

A. It is not the intention of the author to complain about the machine tool industry in the sense of the question, but rather to test the validity of the evolution of traditional machine tool concepts against the economic needs of the heavy engineering industry.

The evolution of machine tool design is as much the fault of the purchaser as it is of the designer; perhaps the purchaser is really the one at fault since it is he who creates the market.

Once the purchaser obtains a better understanding of business economics, and from this deduces that reduction in work in progress is vital, then the machine tool requirements will surely change from traditional designs to the concepts illustrated in the paper.

The general elements of the layouts shown in the paper are available from various sources, but only of recent years are there any signs that these are being used in a broader sense for machine shop layouts. For this, machine tool designers must design standard modules which can be assembled to suit differing layouts for different customers.

The flats referred to are those generated by circular interpolation which requires a NC system for control. If the feed motors are part of a Ward Leonard, then even with the restriction of a minimum speed flats are negligible in most conditions. The introduction of permanent magnet type motors with brake and tachogenerator will result in improved performances in this respect.

STRATEGIES FOR INCREASING
THE UTILIZATION AND OUTPUT OF MACHINE TOOLS

by

P. C. HAGAN* and R. LEONARD†

SUMMARY

As the cost of machine tools and manufacturing systems escalates, the attainment of an acceptable level of utilization and hence output becomes crucial. The present work describes certain procedures for measuring and subsequently improving the utilization of machine tools. A case study of the improvement in utilization of an NC lathe is detailed and it is emphasized that all machine tool purchase decisions require a prediction of the output of the new plant. Finally, an analysis concerning batch sizes indicates that, in general, utilization, as defined in this work can be markedly improved by increasing the batch size.

INTRODUCTION

Machine tool utilization

'The output from a machine tool directly depends upon its utilization'

While initially the above statement appears correct, its validity depends upon how utilization is defined. If a machine operates with a utilization of 70 per cent, the vital points of consideration are

(1) is it efficient?
(2) what is its output?

A definition of utilization is required which is both meaningful to management and aids efforts to increase productivity. Initially, a number of alternative definitions are considered and these are subsequently reviewed and improved.

(a) Definition I
'Machine tool utilization is the percentage of the shift that the machine tool is actually cutting'

With this definition, a machine could operate at very low values of feed or speed and, in consequence, achieve 100 per cent utilization and yet produce little output. In addition, how would the following be considered?

(1) In-process gauging
(2) Tool positioning during drilling
(3) The return stroke on a shaping machine
(4) Loading and unloading the workpiece
(5) Initial set up and re-setting.

(b) Definition II
'Machine tool utilization will be taken to be identical to operator utilization'

This approach appears to have definite merits since it includes many of the items categorized within Definition I which could not be accounted for or at least explained in the figure for utilization obtained. The machine and the operator are taken as being a single unit; thus it is assumed that if the operator achieves a 90 per cent output in terms of measured daywork, then the utilization of the machine must also be 90 per cent. Regrettably this is not the case. If one considers a large boring machine, the operator may be asked to perform an initial setting up operation taking two hours. However, although the operator has correctly performed his duties and achieved two hours of measured daywork, what has happened to the machine? It has remained completely passive and no output or utilization has been achieved. Hence, operator utilization is not a measure of effective machine tool utilization, although operator utilization is important in correct circumstances, such as low cost plant environments where depreciation on equipment is less than the operator's wage. For batch and jobbing production, a significant part of each shift will be consumed by obtaining special tooling, inspection procedures and numerous other delays associated with this type of production.

In spite of the above difficulties, in the past machine tool utilization has been loosely defined in terms of operator utilization or more generally in terms of operator output when compared with imprecise piecework times. The prime reason that this

---

* Eaton Ltd, Axle Division
† U.M.I.S.T.

definition had hitherto been acceptable was the relatively low depreciation of the machine when compared with an operator's wage. This, however, is not the case with many machine tools today. The price of certain machines may result in a rate of depreciation

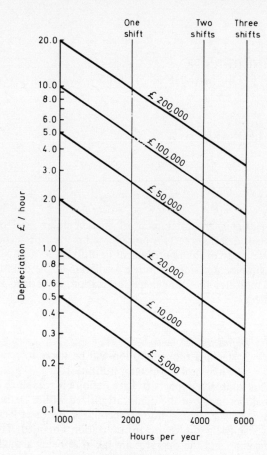

Figure 1. The depreciation/hour for machine tools.

which exceeds the wages of the operator by a factor of 10 (Fig. 1). Such machines cannot remain inactive for one minute more than necessary. Coupled with this is the need for the output and utilization to be defined in terms of the capacity to produce and not the level of utilization of the operator.

Proposed definition
Assumptions:

Before an acceptable definition of machine tool utilization can be given, it is necessary to state a number of assumptions which are implied in that definition.

(1) The component is correctly allocated to the machine most suitable for its production. Due account must be taken of component size and machine features available.

(2) The feeds, speeds and tooling for machining the component are both correctly chosen and acceptable within today's climate of industrial relations. It is essential at this stage to discuss and evaluate any implications or advantages which may rise from the use of jigs, fixtures or multiple tooling arrangements.

(3) The operation times are obtained by work measurement techniques within the production environment, preferably by the use of standard synthetic data.

(4) Multi-shift working is in operation where this is economically justifiable.

Case I—numerically controlled machine tools
The utilization of an NC machine tool is solely the tape running time per job, multiplied by the number of jobs successfully machined per shift, divided by the time available per shift. Analytically this becomes

T_R = tape running time, hours
S_t = shift time, hours
N_s = number of successful jobs produced/shift
U_m = utilization of the machine

Then

$$U_m = \frac{T_R \times N_s}{S_t}$$

Case II—conventional machine
The utilization of a conventional machine tool is taken as the time between the job being clamped in position before machining and the job being ready for unclamping after the operation, multiplied by the number of parts successfully produced per shift, divided by the time available per shift. However, this definition can be influenced by clamping conditions and equipment such as two machine work tables or automatic loading and unloading features. Analytically this becomes:

N_o = operation time between job clamping and subsequent unclamping
N_s = number of successful jobs produced/shift
S_t = time available per shift
U_t = machine utilization

Then

$$U_t = \frac{N_o \times N_s}{S_t}$$

Discussion
The above definitions appear severe, since they do not include any allowances for initial set up, component set up, tape proving, operator relaxation or scrapped work. However, in a strict financial sense the company does not produce saleable goods during setting up times, operator relaxation periods or through scrapped components; indeed during these periods of lost production time, storage costs on components are being incurred. A machine tool engaged in batch production cannot reach 100 per cent utilization when judged by the above definition since initial setting up times and part loading will occupy a significant percentage of the time available. However, direct use of this definition will suggest where improvements may be obtained if investigated correctly. Systems involving pallet loading, pre-set tooling, in-process inspection and efficient production control will improve machine tool utilization and hence output. The utilization of a machine can be measured

and stated in terms of the above definition. However, it must be carefully considered what value of utilization is acceptable or attained within various departments.

If department A has a utilization figure of 50 per cent and department B one of 40 per cent, it is not necessarily true that A was more efficient than B. Due consideration must be given to batch size, new tape proving, the size of components machined and the type of machines used. Can management therefore decide if utilization in their workshops is acceptable? This factor can be accomplished by analysing a normal week's operation for any machine tool engaged in batch production, the result in general will be as follows:

$$T_T = (U_t \times T_T) + (S_u \times N_u) + (T_p \times N_t) + (N_p \times T_1) + (M_n + T_d)$$

Now

$$N_p = N_u \times \text{average batch size } (Q_b)$$

where T_d = time lost by such factors as absenteeism, lack of work, tools etc.
T_T = total time available/shift/week, hours
S_u = average set up time, hours
N_u = number of set ups/week
T_p = average tape proving time
N_t = number of new tapes proved/week
N_p = number of parts/week
T_1 = time to load and unload average part
M_n = maintenance time, hours

hence

$$U_t = 1 - \frac{(S_u \times N_u) + (T_p \times N_t) + (N_u \times Q_b \times T_1) + M_n + T_d}{T_T} \tag{1}$$

Thus, if the average set up time is small or zero, no new tapes are produced and machine loading is accomplished quickly, then utilization can approach 100 per cent if maintenance is performed outside the regular working period. In order to validate this analysis the authors will discuss what values for such items used in the above expressions are likely to occur in practice.

Tape proving

When a new product is initiated, the maximum use should be made of existing component designs with the necessary modifications made to fulfil the new products function. However, some new tapes will still be required. This involves correction, modification and production of the tape to eliminate errors common in such methods of machine control; this may involve the production of three or four tapes before an acceptable tape is available.

The time required for tape proving can be markedly reduced if a computer language, such as EXAPT, is used but such systems are not universally acceptable for various reasons. Once the tapes are produced and the product design is established, few new tapes are required and so the time spent 'tape proving' is reduced considerably. One factor which should not be

overlooked with numerical control is that as experience is gained, this knowledge should be used to reduce the tape running and proving time by the development of suitable aids. At each stage in the life of a product, management should develop an analysis technique which results in a figure per machine corresponding to the amount of tape proving. This should be extended to evaluate tape preparation costs for similar types of components, say all types of crownwheels or differential cases. A simple study of a company's component statistics or parts groups will illustrate the implied conditions here. Marked advantages occur if the purchase of a new machine coincides with a new product, because, for this case, no existing jigs or tapes will be duplicated.

Machine maintenance

Equation (1) shows that machine maintenance, during the normal shift, must reduce machine tool utilization. Hence, wherever possible, routine maintenance should not take place during shift time.

Machine loading and unloading

Management must attempt to reduce the time necessary for loading and unloading components. Direct aids to such strategies should include consideration for pallet loading, duplicate machine tables, special lifting devices and faster operating jigs and fixtures. This approach to material handling at the machine will result in significant improvements; however, a finite time will still exist for loading and unloading components. The duration of the time spent loading and unloading is not only dependent upon the fixtures used, but is also greatly influenced by the size and complexity of the product and the operation itself. Bright steel bar can be speedily located in lathes whether small or large; conversely a large casting may require considerable time before it is correctly positioned. The nature of the operation influences the load time since 'roughing' operations are often carried out on separate machines from the final operation which may be working to close tolerances. Thus, accurate, positive location methods are required with cleanliness of the working area being of considerable importance. In all cases the lifting, positioning and locating times should be obtained by synthetic data or correct work measurement techniques. The use of such times results in an average value for loading and unloading each machine or machine type.

Initial setting up times and batch sizes

The majority of metalworking production in the world today is produced by batch production techniques (Merchant[1]), with the result that initial setting up times provide a very significant percentage of each shift. Setting up times amounting to 30 per cent of the time available are not uncommon. This is both costly and non-productive and as such the frequency of setting up should be critically examined: especially when high cost plant is involved, where depreciation on machinery and the associated investment is of major importance. Set up costs can be reduced by two methods.

(1) The first is to improve the efficiency of the setting up process by the use of standard tooling or various pre-set tooling techniques. Included in this category are improvements in management and production control techniques to ensure that all tooling, drawings, components and other documents are available at the machine when required.

(2) The second method of reducing setting up costs is to increase the batch size, with the results that for a given quantity of production per year, fewer batches are required, hence fewer set ups are needed. Strategies for increasing the batch size should be of a twofold nature, namely the economic batch size can be increased after a component classification and hence a variety reduction exercise has reduced the number of individual components. Secondly the question should be considered if the batch size produced is the most economical in terms of total cost per component and the effective utilization of the plant.

With reference to the latter statement, it is shown in Appendix 1 that the batch size necessary to minimize the total production cost per part is given by:

$$Q_{opt} = \sqrt{\frac{2R \cdot S_u \cdot O_h}{(C_p \cdot I_r + C_s)(1 + 2R \cdot t_p)}}$$

It is also shown in Appendix 1 that this quantity is not critical and for an optimum batch size of 63, batches between 13 and 313 components are within 5 per cent of the minimum cost batch size. However, this ratio does become critical for small batches with relatively lengthy initial setting up times when compared to machining times. The economic batch criterion is not met when after a two-hour setting up time, four components are machined at 1/4 hour per part. The formulae given in Appendix 1 enable economic batch considerations to be tested; however, a general rule of thumb approach gives the economic batch size as

$$Q \times \frac{t_p}{S_u} \geqslant 10 \quad \text{or} \quad Q \geqslant 10 \cdot \frac{S_u}{t_p}$$

Thus the batch size should be such that the total production time per batch is not less than 10 times the setting up time.

EXAMPLES OF MACHINE TOOL UTILIZATION

A comprehensive study of the utilization of conventional machine tools was performed by Tinker[2] and a summary of the results is given by permission of the M.T.I.R.A. in Table 1. This shows that in general a figure of 46 per cent utilization is achieved according to the proposed definition when the cutting time is added to the in-process gauging time.

It should be noted that in this context 'cutting' is synonymous with 'spindle rotating'. Initial setting up, loading and unloading consume, on average, 37 per cent of each shift; this is an area where significant improvements are possible. The third major category, operator being absent for no known cause accounts for 11 per cent of the shift; this latter element must be seen as a problem for management, specifically closer supervision and control of the work force.

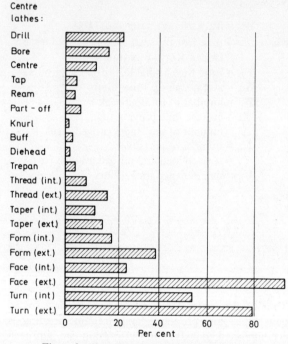

Figure 2. The utilization of a centre lathe.

Tinker further sub-divided machines into their features and usage by a study of workpieces produced on each type of machine. A sample of the results are shown in Fig. 2: these cover the machining requirements of work produced on centre lathes.

TABLE 1.   The utilization of conventional machine tools

Activity Percentage	Mill	Centre lathe	Turret lathe	Planing	Drill	Capstan	Horiz. bore	Vert. bore	Cylind. grind.	Surf grind.	Total
Loading and unload	9.1	7.4	7.3	9.3	16.9	7.7	12.7	12.8	18.1	14.4	10.7
Idle loaded operator absent	16.2	17.9	16.0	12.0	12.1	16.7	17.9	14.8	11.2	11.9	15.1
Idle operator receiving instructions	1.3	2.1	1.2	1.2	1.1	1.3	1.4	1.0	2.8	1.0	1.4
Miscellaneous	0.7	0.3	1.1	0.7	1.1	1.2	0.7	0.4	0.4	0.6	0.8
Gauging	2.8	6.0	4.2	2.7	1.2	3.6	2.7	5.6	7.5	6.4	3.9
Setting and handling	24.2	25.4	27.8	26.8	32.5	32.0	28.6	24.3	15.0	21.0	26.7
Cutting	45.7	40.9	42.4	47.3	35.1	37.5	36.0	41.1	45.0	44.7	41.4
	100.0	100.0	100.0	100.0	100.0	100.0	100.0	100.0	100.0	100.0	100.0
Proportion of m/c time usefully emp.	84.0	76.8	82.9	89.2	69.2	79.8	80.8	88.8	64.3	68.7	78.7
Total observations	1,867	1,638	1,353	1,241	1,551	1,317	998	580	787	709	12,041

A study of workpiece statistics by Opitz[3], and machine capability by Moll[4], led to the proposal that a short bed lathe be produced; this can be seen as a forerunner of the Herbert Batchmatic, a CNC machine specifically designed to machine the majority of turned components occurring in medium sized batches, with a minimum set up time between batches.

NC machine tool utilization

The results of a survey[5] concerned with the utilization of NC machines produced results which were similar in magnitude to those given for conventional machine tools. The highest average utilization figures occurred in company A and these were 50 per cent, generally an average level of utilization of 40 per cent existed. Set ups, loading and unloading accounted for about 20 per cent of the shift time, tape and tool proving (6 per cent), no tape or tools available (5 per cent), machine breakdown and maintenance (5 per cent), no work or awaiting inspection (5 per cent), and no operator available (19 per cent).

The output of machine tools

The output of a machine tool is its capacity to produce multiplied by its utilization level obtained by inplant operation. The capacity of an NC lathe exceeds that of a centre lathe; hence if the utilization of both are similar, the NC machine will produce more components. This does not of necessity mean a cheaper product since no account has been taken of relative machine operating costs.

Reduction of machining time by using NC

Machine/Part	Conventional time (hours)	NC time (hours)	Reduction ratio
OMNIMIL			
Casing	22·67	7·2	3·15
Casing	23·85	5·12	4·66
MOLINS TWIN SPINDLE MILL			
Workpiece 'A'	0·5	0·15	3·3
Workpiece 'B'	2·26	0·28	8·1
CINTIMATIC DRILL			
Workpiece 'C'	1·05	0·40	2·6
Workpiece 'D'	1·33	0·23	5·8
WEBSTER & BENNETT VT LATHE			
Workpiece 'E'	2·75	0·92	3·02
Workpiece 'F'	6·34	1·27	5·6

Reduction of tooling costs using NC

Application	Conventional tooling costs	NC tooling costs
Casing	£22 200	£12 853
Casing	£20 705	£12 777
Drilling ops. (some)	Up to £150 per plate	Not required
Proving turning ops. (some)	£300 (plus) (Two blanks)	Not required

Reduction of setting up times using NC

Application	Conventional ops.	NC ops.
Casing	56 hours	2 hours
Casing	14 ops.	4 ops.

Reduction of scrap rate and rectification using NC

Year	Scrap rate (£ per STH) Conventional	NC	Rectification (£ per STH) Conventional	NC
1968	0·410	0·162	1·986	0·647
1969	0·376	0·090	2·304	0·702*
1970	0·378	0·275	1·804	0·875

*The 1969 figures are regarded as nearer to the norm, as 1968 and 1970 figures contain 'learner' elements for new parts.

Reduction of implementing design changes via NC

Application	Design change	Conventional	NC
Casing	Change made to one fine limit bore and 3 holes added	£285	£85
Casing	2 additional bored bosses plus 4 setscrew holes	£90	£45
Gear casing	Scallop added	£70	£45
	Spanner clearance added	£150	£10
Casing	New profile added	£200	£50
Disc	New roughing operation	£84	£8

METHODS OF IMPROVING MACHINE TOOL UTILIZATION AND OUTPUT

These strategies should be of a twofold nature, namely improving utilization by reducing nonproductive parts of the day and in addition increasing output directly by improving the efficiency of the production process.

Improving utilization

(1) Improve the efficiency of production control techniques to ensure that components, data and tooling are available at the machine when required.

(2) Standardize tooling, machinery and, where possible, control systems to give the best advantages while at the same time maintaining the flexibility of the batch production environment.

(3) Improve the methods of work handling, not only loading and unloading the machine, but also critically discuss with machine tool manufacturers and materials handling companies, techniques to give minimum handling time or handling at minimum cost to the component produced, i.e. every time a component is moved it should have value added to it in the way of extra produced features.

(4) Test the possibility of using duplicate machine tables or pallet loading.

(5) Consider increasing the batch size in order to reduce the number of set ups, hence reducing component cost.

(6) Whenever possible, perform maintenance duties outside the normal working shift.

(7) Minimize the time lost by first-off inspection; this may be achieved by the use of special templates or possibly by use of the machine features.

(8) Consider the economic advantages of multi-shift working, but, at the same time take due regard of the increased management control needed.

(9) Apply comprehensive work study practices, not only work measurement techniques, but also the total aspects of the work study function. These range from an efficient layout of plant and equipment at the industrial engineering level to the layout of the working area by suitable method study of the job.

(10) To reduce the percentage that the machine is inoperative through no known cause.

(11) It is extremely important with any analysis to measure and record data, thus the percentage time spent on each activity should be obtained. It is of importance that the data measured is valid and accurate since if it is not, areas for improvement will not become evident.

(12) Utilization cannot commence until the machine is fully commissioned. For the case of an NC machine, a considerable time period can elapse between ordering a machine and it operating in a productive capacity. A specific installation procedure is suggested (Fig. 3).

(13) Design for production and, where desirable, exploit the advantages available when NC machines are used, for example, multiple operations on one machine, repeated accuracy and controlled contouring facilities.

Improving the production process

(1) Verify that the correct feeds and speeds are being used and that these are acceptable to all parties.

(2) Ensure that the optimum tool path is obtained.

(3) Consider changing the type or quality of the cutting tool or consider a multiple tooling arrangement.

(4) Ensure that adequate service facilities are available to avoid delays that would be experienced by waiting for clamps, vee blocks, fixtures and measuring instruments.

(5) Consider the use of digital readout facilities, especially on older machinery or where an initial survey of the machine's operation indicates excessive time spend moving and positioning slides or workheads.

Utilization improvements possible with an NC lathe

Any reference to utilization should include adequate practical results which justify the time spent improving the system. In this respect the results achieved with a Warner Swasey S.C. 28 numerically controlled turret lathe are summarized. In order to illustrate these results a brief history of the development, installation and operation is given because this had a direct bearing on the final output.

The machine was installed in April 1970, but with inadequate back-up and service facilities. As is the case with most new technology many requirements were not evident or emphasized until a crisis arose. The machine was introduced into the plant as the pride of the machine shop. Unfortunately, it was treated, for a while at least, as a very much protected item and was not released to production until a considerable time afterwards.

During this period questions were asked, especially 'what level of output and utilization is being obtained from the NC lathe?' This question could not be adequately answered. No system was in operation which recorded the following:

(1) machine operating time
(2) operator available or absent
(3) time spent tape proving
(4) machine maintenance

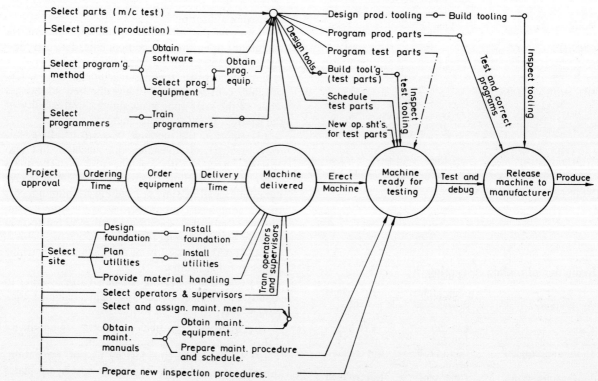

Figure 3. Installation procedure for an NC machine.

or other machine downtime features which would affect the overall machine utilization figure. The answer was clear: some system or method of recording data had to be established.

In order to introduce any recording techniques into the company, management had to prove to the shop floor personnel that the machine was not achieving or approaching the output needed to justify the capital expenditure on such a machine. This was done by a graphical technique, shown in Fig. 4, which illustrates very clearly the time that appeared as excess when the machine was not productive. The excess time over that of production time utilized was obtained by establishing the tape running time, and adding to it extremely generous loading and unloading times, gauging and tooling allowances obtained from standard data in conjunction with the operator's daily production figures to determine the lost time per day.

T_R = tape running time, min
T_l = load and unload time, min
T = tooling allowance, min/component
G = gauging allowance
P = daily production, number of components per day
T_T = total time available per day, min
E = excess or lost time

Thus, the excess time per day (E) is given by

$$E = T_T - (T_R + T_1 + G + T)P$$

These results plotted on the graph were the simplest method of illustrating to the operators concerned that something had to be done to justify the return on the investment. However, it should be mentioned that the sound personal relationships between the NC

machine operator and the NC engineer were very significant in the production and evaluation of the data obtained—indeed similar systems are not as yet introduced elsewhere within the factory.

In order to develop the graph into a system which can give an indication of the level of machine utilization, more items than those given in equation (1) have to be included. The list of factors involved was prepared initially by the NC engineers, who asked the NC operators to suggest any additional factors; these were added to the parameters to be measured and evaluated. A complete list of each of the items is shown by Fig. 5. The operators themselves completed the record sheet and these were analysed and evaluated by the NC engineers to determine areas for improved effort. A comparison of Figs 4 and 6 show the improvements made in production capacity by such a scheme. In brief, the result is that the production quantities obtained approach those required to justify the investment in the machine. Further work on the analysis sheets has enabled other improvements to be made, namely:

(1) New programming aids and method of tape preparation were developed to eliminate the time lost due to tape modifications.
(2) Development of realistic tool life trials and adequate feedback of data.
(3) From the frequency of the maintenance visits, a simple introduction to a planned or preventative maintenance scheme was developed.
(4) The need for improved inspection techniques became very evident, especially when checking off the first component or sample.
(5) The introduction of quality control techniques to the machine eliminated unnecessary tape running time and other negative elements were avoided by a careful evaluation of results.

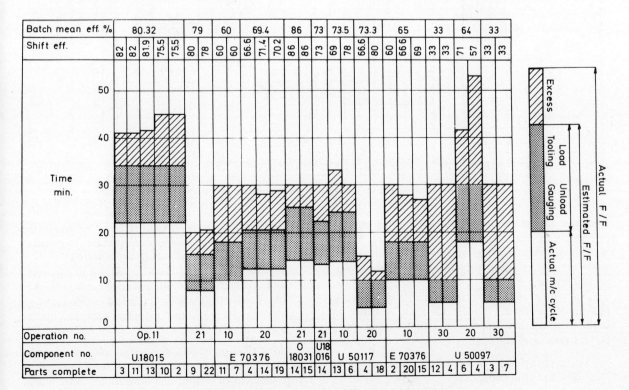

Figure 4. The utilization of an NC lathe prior to improvement.

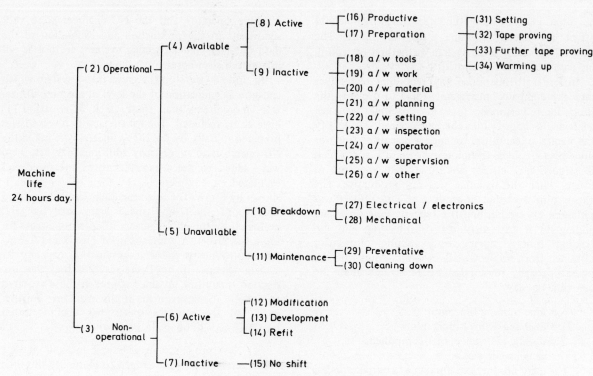

Figure 5. Parameters to be recorded.

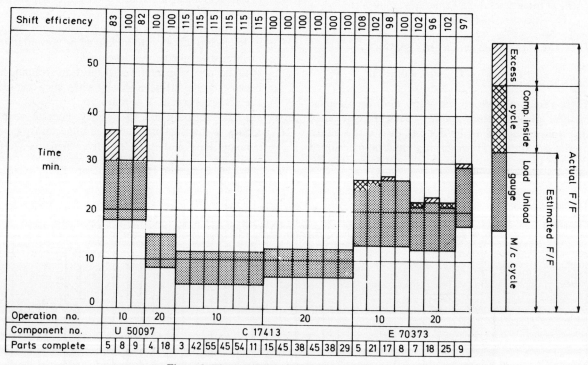

Figure 6. The utilization of an NC lathe after analysis.

Excessive downtime was reduced and further analysis in these particular areas listed previously (1–5) enabled improved utilization figures to be obtained. The work content required to achieve these results was proved justified.

This section is only a brief summary of the work performed; however, it indicates that, in order to improve utilization and hence output, some basic factors must be established.

(1) To obtain some details of the present level of output and utilization.

(2) To involve the operators with the work that has to be done to enable improvements to be made and explain why such results are required.

(3) To develop and install a method of recording and evaluating the results obtained.

(4) To use the results obtained to enable the overall objective of the planning function to be attained.

As a final comment, the results obtained on the NC lathe have indicated that, because of the ease of measurement of output and utilization, future investment in equipment will include a place for advanced technology with program controlled machine tools as the basis.

CONCLUSIONS

As the cost of plant continues to rise, the attainment of an adequate level of utilization becomes of increasing importance. A number of suggestions have been made in this work on how to increase the output of machine tools. The authors are of the opinion that the most profitable areas of investigation include a consideration of increasing the batch size, improving part loading and better production control systems. A pre-requisite of any improvement is the measurement of the present utilization and suggested parameters for investigation have been detailed.

ACKNOWLEDGMENTS

The authors wish to express their thanks to Mr Cyril Scarborough, Mr Wilf Smith and Mr Frank Davenport of Rolls Royce (1971) Ltd., for their most valuable advice during this work.

Specific thanks are also due to Eaton Axle Division for their co-operation with this program and for allowing the results of the NC lathe investigation to be published.

The Machine Tool Industry Research Association is particularly thanked for their permission to publish some of the results of their investigation.

REFERENCES

1. M. E. MERCHANT (1969). Trends in manufacturing systems concepts, *10th M.T.D.R. Conference*, Manchester.
2. G. C. TINKER (1968). The utilisation of machine tools, M.T.I.R.A. Research Report No. 23, April.
3. H. OPITZ (1964). Workpiece statistics and manufacture of 'parts families', *VDI Zeitschrift*, **106**, No. 26, September, M.T.I.R.A. Translation T146.
4. M. MOLL (1963). A user's viewpoint of the machine tool of the future, *Proceedings of the 4th M.T.D.R. Conference*, Manchester, Pergamon Press.
5. F. DAVENPORT (1972). Numerical control at Rolls Royce (1971) Ltd., The Operation of Numerically Controlled Machine Tools: Short Course: U.M.I.S.T., April.

APPENDIX 1
OPTIMUM BATCH SIZE

Notation

C_p = Capital price of component (£)
C_s = Storage costs/component/year
D = Delay time/production in works
I_r = Company interest rate
K_c = $C_p . I_r + C_s$ (£/year)
L_t = Lead time = $D + T_p$
O_h = Over head rate (£/hour)
Q = Batch size
Q_m = Minimum stock level

Q_o = Optimum batch size (minimum cost/part)
R = Annual rate of usage of components (parts/year)
S_u = Setting up time (initial) hours (total for all operations)
T_c = Total cost/component while in factory
T_p = Machining time (total/component for all operations) hours.

Stock level in stores

(see Fig. 7)

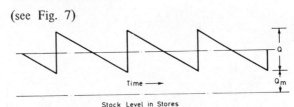

Figure 7. Typical stock level diagram.

Average number of components in stores

$$= \frac{Q}{2} + Q_m$$

Average time spent in stores

$$= \left(\frac{Q}{2} + Q_m\right) . \frac{1}{R}$$

Cost of keeping component in stores

$$= K_c . \left(\frac{Q}{2} + Q_m\right)\left(\frac{Q}{2} + Q_m\right)\frac{1}{R}$$

Cost/component

$$= K_c . \left(\frac{Q}{2} + Q_m\right)\left(\frac{Q}{2} + Q_m\right) . \frac{1}{\left(\frac{Q}{2} + Q_m\right)} . \frac{1}{R}$$

$$= \frac{K_c}{R}\left\{\frac{Q}{2} + Q_m\right\} \tag{1}$$

Stock level in works

Lead time

$$= L_t = D + Q . T_p$$

Number of batches in works

$$= \frac{R}{Q} . L_t$$

Interest charge for batches in works

$$= \frac{R}{Q} . L_t . L_t . K_c . Q$$

Interest charge/part in works

$$= \frac{R . L_t^2}{R . L_t} . K_c$$

$$= K_c . (D + Q . T_p) \tag{2}$$

Production cost/part

Production cost/part

$$= T_p . O_h + \frac{S_u . O_h}{Q} \qquad (3)$$

Total cost/part

$$= (1) + (2) + (3) = T_c$$

Therefore

$$T_c = \frac{K_c . Q}{2R} + \frac{K_c . Q_m}{R} + K_c . D + K_c . Q . T_p + T_p . O_h$$

$$+ \frac{S_u . O_h}{Q}$$

$$= \left(\frac{K_c . Q_m}{R} + K_c . D + T_p . O_h \right) + Q . K_c . \left(\frac{1}{2R} + T_p \right)$$

$$+ \frac{S_u . O_h}{Q}$$

To obtain a batch size which makes the total cost

(T_c) a minimum

$$\frac{dT_c}{dQ} = 0,$$

$$\frac{dT_c}{dQ} = K_c . \left(\frac{1}{2R} + T_p \right) - \frac{S_u . O_h}{Q} = 0$$

therefore

$$Q_{opt} = \sqrt{\frac{2R . S_u . O_h}{K_c(1 + 2R . T_p)}}$$

Hence optimum batch size is given by

$$Q_{opt} = \sqrt{\frac{2 . R . S_u . O_h}{(C_p . I_r + C_s)(1 + 2R . T_p)}}$$

Now T_p = total machining time/part
therefore

$$T_p = \sum_{1}^{m} T_{pm} \text{ for 'm' operations}$$

S_u = Total initial set up times

therefore

$$S_u = \sum_{1}^{m} S_{um}$$

Example:
  R = 20 parts/year; I_r = 10 per cent = 0·1
  C_p = £100; O_h = Overheads = £10/h
  C_s = Storage costs £4/year
  S_u = Sum of set ups = $\frac{1}{2}$ + 2 + 2 + 3 + 1 + $1\frac{1}{2}$
      = 10 h
  T_p = Sum of machining time/part
      = 1 + $\frac{1}{2}$ + $\frac{1}{2}$ + 1 + 1 + 2 = 6 h
      = $\frac{1}{4}$ day
      = $\frac{1}{1500}$ year

therefore

$$Q_{opt} = \sqrt{\frac{2 \times 20 \times 10 \times 10}{(100 \times 0·1 + 4)\left(1 + \frac{2 \times 20}{1500}\right)}}$$

$$= \sqrt{\frac{4000}{14 \times 1·03}}$$

$$= \sqrt{280}$$

$$\simeq 17$$

To obtain the limits on the batch size for the actual component costs to be within an amount Δ of the minimum cost

$$T_c = T_p . O_h + \frac{S_u . O_h}{Q} + Kc(D + Q . T_p)$$

$$+ \frac{K_c}{R} . \left(\frac{Q}{2} + Q_m \right)$$

$$= \left(T_p . O_h + K_c . D + \frac{K_c . Q_m}{2R} \right) + \frac{S_u . O_h}{Q}$$

$$+ K_c . T_p . Q + \frac{K_c . Q}{2R}$$

Writing

$$K = \frac{T_p . O_h}{S_u . O_h} + \frac{K_c . D}{S_u . O_h} + \frac{K_c . Q_m}{2R . S_u . O_h}$$

$$\frac{T_c}{S_u . O_h} = K + \frac{1}{Q} + \frac{Q . K_c}{S_u . O_h}\left(T_p + \frac{1}{2R} \right)$$

and

$$\frac{K_c}{S_u . O_h} . \left(T_p + \frac{1}{2R} \right) = \frac{1}{Q_o{}^2}$$

Hence

$$\frac{T_c}{S_u . O_h} = K + \frac{1}{Q} + \frac{Q}{Q_o{}^2}$$

T_c optimum is obtained by putting $Q = Q_o$. Hence

$$\frac{T_{co}}{S_u . O_h} = K + \frac{1}{Q_o} + \frac{1}{Q_o}$$

$$= K + \frac{2}{Q_o}$$

Giving

$$\frac{T_c}{T_{co}} = \frac{K + \dfrac{1}{Q} + \dfrac{Q}{Q_o{}^2}}{K + \dfrac{2}{Q_o}}$$

Defining the ratio

$$\left(\frac{T_c}{T_{co}} \right) = R_c = \text{Relative cost}$$

$$= 1 + \Delta$$

Then

$$Q^2 + Q(K . Q_o{}^2 - R_c . K . Q_o{}^2 - 2R_c . Q_o) + Q_o{}^2 = 0$$

Solving the quadratic in Q yields

$$Q = \left(\frac{2R_c \cdot Q_o + Q_o^2 K(R_c - 1)}{2}\right) \pm \frac{Q_o}{2}$$

$$\times \sqrt{[Q_o^2 K^2(1 - 2R_c + R_c^2) - 4R_c \cdot Q_o \cdot K(1 - R_c)}$$
$$+ 4(R_c^2 - 1)]$$

Now

$$R_c = 1 + \Delta \quad \text{and} \quad R_c^2 = 1 + 2\Delta + \Delta^2$$

this gives

$$Q = \left(Q_o + \Delta\left(\frac{2Q_o + Q_o^2 K}{2}\right)\right)$$

$$\pm \frac{Q_o}{2}\sqrt{(Q_o \cdot K \cdot \Delta^2 + 4\Delta)(Q_o \cdot K + 4) - 8\Delta}$$

Put $Q_o = 63$; $K = 1$; $\Delta\ 0.05$

$$Q = 163 \pm 150$$
$$= 13 \rightarrow 313$$

Hence batches of between 13 and 313 are within 5 per cent of the optimum cost.

Example:

$S_u = 24$ h;	$T_p = 12$ h
$O_h = £10/$h;	$C_p = £1000$
$I_r = 0.1$;	$C_s = £20/$year
$R = 100/$year;	$D = \frac{1}{4}$ year
$Q_m = 10$;	$\Delta = 0.05$

$$Q_{\text{opt}} = \sqrt{\frac{2R \cdot S_u \cdot O_h}{(C_p \cdot I_r + C_s)(1 + 2R \cdot T_p)}}$$

$$= \sqrt{\frac{2 \times 100 \times 24 \times 10}{120 \times \left(1 + \frac{2 \times 100 \times 12}{365 \times 24}\right)}}$$

$$= \sqrt{\frac{48\ 000}{120 \times 1.3}}$$

$$\simeq 17$$

Now determine K

$$K = \left(\frac{1}{2}\right) + \left(\frac{120 \times 1}{24 \times 4 \times 10}\right) + \frac{120 \times 10}{2 \times 100 \times 24 \times 10}$$

$$K = 0.5 + 0.125 + 0.02$$

$$K = 0.645$$

Now determine Q

$$Q = (34 + 0.05(34 + 300 \times 0.645))$$
$$\pm \frac{17}{2}\sqrt{[(17 \times 0.645 \times 0.0025 \times 0.2)}$$
$$\times (17 \times 0.645 + 4) - 0.2]$$

$$= \frac{34 + 11.5}{2} \pm 8.5\sqrt{0.23 \times 15 - 0.2}$$

$$= 23 \pm 8.5\sqrt{3.25}$$

$$= 23 \pm 8.5 \times 1.75$$

$$= 23 \pm 15$$

$$= 8 \text{ or } 23$$

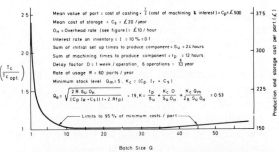

Figure 8. The effect of batch size on component costs.

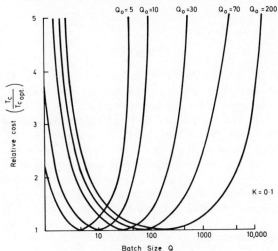

Figure 9. Batch size results for $K = 0.1$.

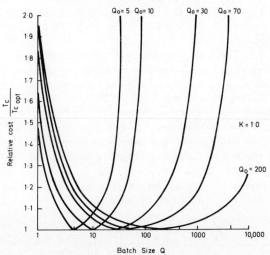

Figure 10. Batch size results for $K = 1.0$.

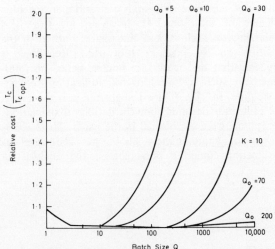

Figure 11. Batch size results for $K = 10.0$.

Figure 8 depicts a typical batch size/cost per part relationship. It is seen that for batches in excess of 8 components, the cost per part is steady until 45 parts per batch is reached. Figures 9, 10 and 11 show the general relationships plotted logarithmically for various values of optimum batch size, Q_o, and constant costs, K. Clearly, when the constant costs, K, are small ($0 \cdot 1 \rightarrow 1 \cdot 0$) large batches are indicated. However, if the constant costs are large, small batches may be produced. It must be emphasized that in the case of small batches, output will be reduced because of set up time.

DISCUSSION

Q. F. W. Craven, Herbert Machine Tools Ltd. Mr Hagan's paper demonstrates some first class analytical techniques. There are three questions.

(1) His comparison of percentage utilization of NC lathes and conventional lathes is suspect. The NC lathe figure of 40 per cent is substantially lower than actual achievements in the field and it is suggested that the data backing his figure of 40 per cent is taken from too limited a sample. In addition, ref. 5 relates to a special application of such machines.

(2) The comment on the high cost of contouring facilities used only occasionally is open to doubt. Firstly, such facilities as a percentage of total machine/system cost are relatively low, and secondly the decision must be made on sound judgement based on adequate component analysis at the procurement stage.

(3) The statement that GT automatically means smaller batches thereby leading to increased premiums of set-up costs is not automatically true. GT, properly applied using the right machine tools, particularly NC lathes where relevant, plus good production engineering, can give the opportunity to reduce batch sizes. Such reductions in batch sizes are a matter of judgement based primarily on cost structures. Batch size reduction is certainly not essential to GT application, but can be a further added benefit in certain circumstances.

A. (1) Regarding your question on the utilization of conventional lathes and NC lathes, I did not make any comparison on the degree of utilization of these machines. The figures which were quoted in the paper did not compare NC and conventional machine figures of utilization. If, however, Mr Craven is referring to Table 1 and the figures quoted in the section on NC machine tool utilization, then it appears that the figures are similar, hence his reference to such a low level of utilization for NC machines. In light of this I might add that the degree of utilization depends upon the definition and various proposals were established in the paper. Mr Craven's figure of approximately 80–85 per cent level of utilization could be obtainable if the analysis included such items as tape proving, machine tool setting and loading and unloading—and in fact, a further analysis by the author's company indicated levels of approaching 100 per cent on an NC lathe, but in this case a different definition was used to establish the figure.

In reply to Mr Craven's remark concerning ref. 5 as being a special application, I totally disagree: surely the company concerned are the country's prime users of NC and have been for a considerable time. What exactly Mr Craven means by 'something special' is open to discussion, but if he is referring to the type of machine purchased and the work it is used upon, then a visit to the company will illustrate that most of the machines are manufactured standard machines, although some are expensive and look 'something special'; they are freely available on the open market, and are used on products which are similar in complexity in the industry concerned throughout the world.

(2) Mr Craven is exactly right with this question, and he mentioned the very point which was meant to be illustrated in the paper: that of adequate component analysis that must be made prior to machine tool purchase. Indeed, I believe Mr Craven's own company must have had such ideas in mind when they built a short bed lathe and later developed this to the Batchmatic series of machines.

Regarding the comments on the high cost of the contouring equipment, used only occasionally, is from the authors' point of view open to discussion, but the point that was meant to be made was a correct selection of products for the machine to make adequate use of the machine's capability. It is conceded, however, that if a company has purchased a machine with such features then, no matter what products have to be machined, effective machine utilization must be achieved, and in order to gain a contribution from those NC machines they must be kept working for as long as possible, even if they are used only for straight line work or even point to point, in order to avoid the machine being left idle, which may in itself psychologically be a bad thing for shop floor morale, not to mention management disillusionment in the capability of numerical control.

(3) Mr Craven makes a valid point when stating that group technology does not of necessity mean smaller batches. However, one of the main aims of group technology is to reduce 'work in progress', both in the works and the rough and finished stores. If the batch sizes are not reduced, little gain will ensue from the works stores. If the batch sizes are not reduced, little gain will ensue from the works stores and considerable material will still exist at the machine. The point the paper was trying to make was specifically, 'If batch sizes are significantly reduced, set-up costs may become disproportionately high, and in addition, machine utilization low.'

TIME CRITICAL PATH ANALYSIS CHARTS

by

L. WOOD*

SUMMARY

This paper states that the successful business is one that makes the best use of working capital and demonstrates a technique for examining this, principally in a general engineering environment with low batches and a high design engineering content. The technique is used to develop a manufacturing efficiency ratio, and methods of improving this ratio are discussed.

INTRODUCTION

In order to appreciate some of the basic rules used in both construction and appreciation of a Time Critical Path Analysis (TCPA) it is necessary to discuss briefly 'cash flow' and communications.

CASH FLOW

An accountant's appreciation of the subject is not the aim of this paper which is a deliberate simplification in order to demonstrate the relationship between this and the production control function. Cash flow can be regarded as a 'closed loop' system with 'drain holes' fitted with valves which can be opened and closed by management (Fig. 1).

There are two main parts to the diagram in Fig. 1, 'fixed capital' and 'working capital', the important

Figure 1. Cash flow.

point being that 'in' and 'out' should be in balance from the point of view of cash flow.

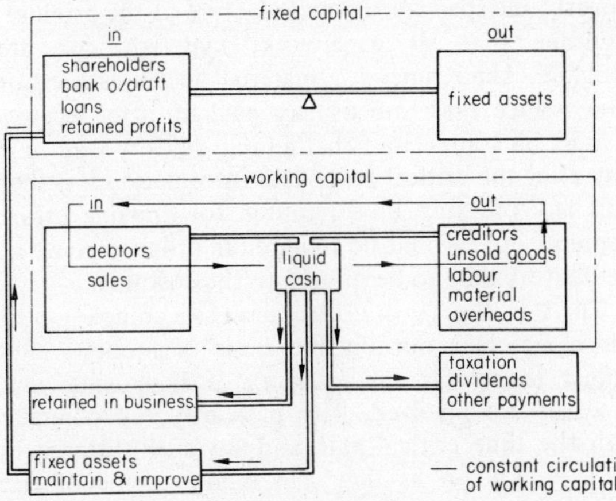

Fixed capital is not liquid cash—it is a sum of money which has been spent to provide buildings, land and plant, and is not generally available for transfer to liquid cash should the company run short of working capital. The circulation of working capital is the important part of the diagram, with the volume of 'liquid cash' being the key to the state of the business. In practice this is expanding and contracting daily but there should be an overall growth to allow management to draw off for various payments; for maintaining the plant; and for putting aside for the purchase of new plant and equipment.

There are many pressures on the volume of liquid cash, such as increased costs for labour, materials and services which cannot be recovered in the price of the goods, reduction in profit margins due to lack of competitive position, credit squeezes etc., but the matter that concerns the manufacturing organization is the one generated by a sluggish or spasmodic speed of circulation which can be said to be typical of a 'jobbing' (small batch) type of business.

The overall growth of the volume of liquid cash is controlled by the profit on the sale of goods. If the goods are sold infrequently the rate of incoming cash will be infrequent and the overall growth depends on infrequent opportunities of making profit. The business is therefore on a higher risk than in mass production where turnover is faster with smaller profit margins and a more regular and predictable growth.

An examination of Board of Trade statistics for various companies reveals that a league table based on return in capital shows that companies engaged in

* GEC Machines Ltd., Rugby

retailing or with a high degree of automation are at the top of the table with ratios of 80 per cent fixed to total assets, i.e. low work in progress values, and those engaged in heavy plant manufacture are at the bottom with 30/40 per cent ratio of fixed to total assets, i.e. high work in progress values.

There is therefore a sound economic argument for the adoption of mass production principles in 'jobbing' and 'batch' production.

COMMUNICATIONS

Working capital consists of raw material stocks, finished part stocks, work in progress, unsold stocks of saleable units, and it is the speed at which these are turned over which is relevant. Since stocks of all kinds tend to increase or decrease in sympathy with the organization's performance in controlling work in progress, it is this performance which is important. Controlling work in progress is a problem of communication between work stations, which is fixed by component routing and flow lines or as is generally termed 'Factory Layout'.

A functional type of layout, i.e. grouping operations according to type, results in a complex routing problem, with complex progressing problems. In a 'batch' and 'jobbing' factory the usual practice is to adhere to functional type layouts, and the resulting complex progressing problems are usually resolved by employing extra overheads in the shape of paperwork and people making efforts to reduce the work in progress. 'Progress' people must have progress meetings and even further efforts have been employed in the shape of computer systems and even operations research specialists to cure the problem. A simpler answer must lie in the techniques used in mass production so the phrase is coined that the fundamental objective of a manufacturing organization must be to 'Mass Produce a Batch Quantity of one Unit of Sale'. The following rules are given which may help to summarize the communication problems.

(1) Work in progress increases directly as the distance between operations and with the number of operations.
(2) Work expands to fill the available floor space.
(3) Work expands to fill the available labour force.
(4) Overheads increase with extended lines of communication.
(5) Progress must be seen to be apparent.
(6) Storage is an unnecessary evil.

In its simplest form this problem may be measured by the ratio of direct labour hours to total throughput time of the critical path.

TIME CRITICAL PATH ANALYSIS CHARTS

A sample chart (Fig. 2) shows the critical path analysis to be the sequence of events laid out on a time scale (not calendar scale). The chart can be for one particular model or it can represent a range of products which follow similar manufacturing sequences within a specified group of shape, size and weight.

In constructing the path, care should be taken to represent what is actually happening rather than what should happen. It may be that in planning production, thirty weeks is allocated for manufacture; if there is a permanent overdue situation it must be established whether this is due to a particular resource bottleneck or due to random causes. Any operations which are regularly double shifted should be shown in parallel rather than sequential; this should also be the case when there are a number of duplicate items which can be made in parallel.

Although it is fairly easy to establish the time of an operation it is not so easy to establish the time between operations when the workpiece is not being worked on. For this, rules have to be established and tested for validity by checks on the shop floor. The suggested rules are based upon the premise that work in progress increases directly as the distance between operations and also with the number of operations, as well as with the change of administration. The rules are

(1) No penalty for operations linked by conveyors or other automatic handling devices.
(2) No penalty for operations performed in a family group where the group is responsible for completed units.
(3) A penalty of two days for operations performed in the same bay under the same foreman where these are not inside a group as rule 2.
(4) A penalty of two days for operations performed in the same bay under different foremen.
(5) A penalty of three days for operations performed in different buildings or under different superintendents.
(6) A penalty of a week for sub-contracting.

In the later stages of the chart, it is natural that more progress effort is made as the delivery date gets nearer so that the penalties for changes in administration and location can be halved.

Having constructed the chart and tested its validity then the critical path becomes apparent as being the longest time span for the manufacture of the product, assuming that all paperwork, materials etc. are available. Lead times for material deliveries are not shown since these are usually paid for after receipt. It must be appreciated that adding delivery lead time may alter the critical path. This information together with any pre-shop time required for drawing office planning, etc., would be required in order to arrive at the delivery date to be quoted to the customer.

There is also a cost critical path as a companion to this chart, laid out on the basis of money value against time; this can be done if high value raw materials are involved. This path may not coincide with the time critical path and any such differences must be known as they can modify some of the philosophies and action taken in improving the path. The time critical path analysis is therefore a visual aid which can demonstrate the efficiency of the organization in creating saleable goods from raw materials on a time basis from which can be arrived at the ratio.

Sum of operations on manufacturing critical path to throughput time expressed as a percentage

For a completely automatic factory this ratio is 100 per cent. Hence the phrase previously used of 'mass producing a batch quantity of one unit of sale' is the goal of the manufacturing organization. In working towards this goal it should be obvious that the resources of plant investment, production engineering manpower etc., should be concentrated on the critical path as any reduction achieved here is worth more to the company than in an area not on the path. The battle never ends as there is always a critical path.

If the critical path changes then it is possible that the new ratio of Sum of operations on critical path to throughput time can change adversely since the new path may have many more delays. It is essential therefore to always identify this ratio with the particular path turnover time.

Another more positive way of looking at this ratio is to state that we are only working on this particular range of contracts 17 per cent of the time or for a sixteen week cycle we are only working on this job 6·4 hours each working week. There should only be two other basic reasons for spending resources of either money or manpower in areas other than the critical path. These are:

(a) The value of the material is high as shown by the cost critical path.

(b) There is a problem of a bottleneck in plant resources in that total output is being limited by a piece of plant or tooling or quality problems.

ANALYSIS OF TIME CRITICAL PATH ANALYSIS CHART

The sample chart shown in Fig. 2 would be a representative chart for a general engineering jobbing business. For a ratio of approx. 17 per cent it can be seen that the real problem is one of lost time between operations, i.e. a problem of layout. Problems of feeds speeds etc. at individual operations do not greatly affect the situation. For a business with more mass production and flow lines the ratio improves and what happens at an individual operation becomes more relevant. The first examination of the chart looks for opportunities of eliminating operations, combining operations, and moving operations closer together physically and administratively, and attempts to formulate a shop layout based on the flow line of the critical path. This examination must not accept any of the conventional restrictions imposed by traditions of a functional organization, but follow a line that on the critical path we are trying to maximize the speed of travel of a component or assembly and that utilization of labour and plant take second place.

A sequence colour system will assist the examination. Basic colours for the main groups such as fabrication, machining, sub-assembly, assembly, test, shipping with variations of the basic colours to show movement between locations and of foremen in the same group, i.e. if red is for fabrication we may have flame cutting, shearing, bending, rolling, assembly, welding, stress relieving, painting all in functional locations each with a different foreman, and it is necessary to ensure that work is not going backwards and forwards between the groups. Similarly for the machine shop for a functional layout grouping lathes, drills, mills, grinders etc. together.

Operations not in the colour sequence then stand out visually and can be examined. Obvious situations which break the colour sequence would be a machining operation in the middle of an assembly or sub-assembly sequence, intermediate machining operations inside the fabrication sequence or vice versa. In such cases attempts are then made to design

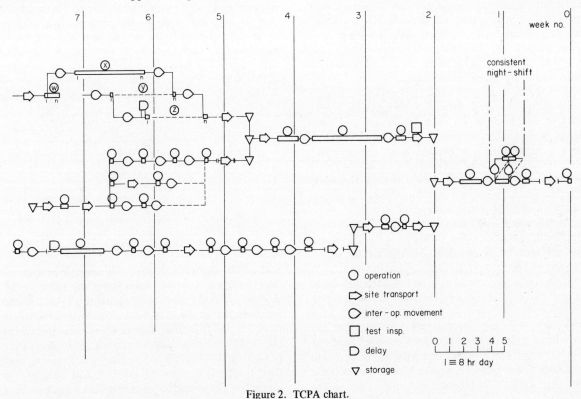

Figure 2. TCPA chart.

out the offending operation or put the necessary machine tools in the line or conversely put the fitting operations in the machine shop, at the same time using production engineering expertise to simplify the operation so that it can be done by an operator with training rather than an operator with skill. Choice of control points along the critical path is important; these should coincide with an appropriate level of supervision, major control points coinciding with senior management levels such as the manager, superintendent etc. and sub-control points with next level of management such as the foreman. Control points are defined as 'points of no return', and it follows that all paperwork should be designed to suit a control point. Each control point can be regarded as a business unit selling a product to the next control point, each area of control adding value and having work in progress. Thus we have the basis of an added value accounting system, an incentive system for supervision, a production control monitoring system, and a progressive costing system all of which can be computerized.

Assuming that we now have a chart which has been corrected for obvious faults in routing and which can be said to be true for a product range, we examine the full product range to see if they all have the basic sequencing similarity, and put on one side those that are different. These are then examined to see why they are different and whether they can be made the same.

This part of the exercise questions some generally accepted views of what is meant by standardization. Standardization is practised in an effort to reduce the total costs of manufacture, this being the pre-shop (design and drawing office, purchasing, planning jig and tool etc.) plus hardware.

Standardization concerned with raw material standards, fasteners etc. is probably necessary and worth while but engineering standards solely concerned with reduction in engineering and drawing office activity, can very often be in conflict with the shop floor requirements. The shop floor view of standardization is to standardize on the time critical path analysis chart within the specified envelope.

It is worth bearing in mind that within a complex engineering business a designer who bases his economics of manufacture on material, labour and overheads will send components all over the factory chasing areas with low overheads. A design which accepts the discipline of the critical path is probably a better design for production even though the traditional accounting system based on material, labour and overheads may prove otherwise.

At this point let us assume that the chart (Fig. 2) represents a product range within an agreed envelope; the envelope being physical, determined by shape, size and weight.

Each operation should now be considered in detail to test its contribution or otherwise to the product.

Any operation consists of the combination of a workpiece, tool and operator (Fig. 3). The tool can vary from a simple hand tool to a complex machine tool. The influence of the operator on output will vary with the degree of control which he can exercise over the cycle. Management can only control the

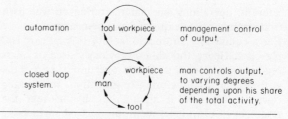

Figure 3. The operation.

output from the operation by excluding the operator, as in true automation. If an operator is necessary it is he who controls output to a greater or lesser degree depending on whether he is simply loading and unloading, to the situation where he is in complete control as in a fitting operation or a manual turning operation. The output of the operator will vary according to his estimate of the value of what he does in return for his conditions of employment which include other things beside wages.

The operation can be recorded on a cost basis and on a time basis. Normal costing procedures do not recognize the increasing value of the workpiece as is demonstrated in Fig. 4, and it is necessary to guard

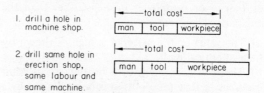

Figure 4. Cost of operation.

against decisions by engineers, draughtsmen and planning engineers based on attitudes created by such systems and their derivatives such as machine hour rates. 'Drill to suit on assembly' is a common drafting technique. If this operation is moved to the detail part at an early stage in the critical path or even better off the path altogether and this means an investment in tooling, or even a numerical controlled drill thereby saving four days' work in progress, is the truly economic decision to take. The operation is shown on a time basis in Fig. 5 and defines that

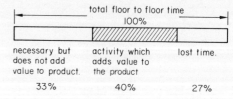

Figure 5. Time.

part of the cycle which adds value to the product as being only the activity which alters the shape of the workpiece. For a metal cutting operation this would be when the tool was in contact with the workpiece as in Fig. 6. It further defines those activities which are considered necessary but do not add value to the product and the rest which is lost time. It can be seen that a lot of engineering effort could be spent in improving the metal removal rates on a machine tool

necessary, but does not add value to product
transport workpiece to work area.
crane lift to machine or work area.
remove any protections on workpiece.
study drawings and bills.
set up machine.
prepare equipment, process or plant.
make and fit temporary equipment to workpiece.
regrind and reset tools, or replace.
move cutting tools to workpiece, index etc.
rotate workpiece.
test.
inspect.
check workpiece.
communicate with nightshift.
move finished workpiece aside.

should be removed from critical path

approx. 33%

lost time due to other causes
confer or wait for supervision.
confer or wait for inspection.
confer or wait for ratefixer.
confer or wait for progress.
confer or wait for drawing office.
obtain drawings.
obtain P.W. bills and make out, etc
obtain material.
obtain necessary tools, jigs etc.
obtain setting up equipment.
obtain own kit.
go to stores.
maintenance.
rectifying work and waiting for decisions on what should be done.
wait for transport.
wait for crane.
cleaning machine or work area.
necessity to work out of own shop.
await availability of machine.
should be eliminated or minimised.

approx. 27%

Figure 6. Breakdown of direct activity.

activity which adds value to product	
fit and assemble:	the joining of parts together.
paint:	from when the paint is applied to the workpiece, until it is completely painted as required.
process work (heat treatment, plating etc.)	
fabrication:	the fusing together of parts.
flame cut:	from when the flame is introduced to the sheet until the workpiece is completely cut out.
machine:	the removing of metal from the workpiece.
bends:) press:) shear:)	from when the operator motivates the machine until the cycle has been completed.

critical path activity

approx. 40%

by a small percentage, say 5 per cent, which in terms of overall gain is 5 per cent of 40 per cent. It could be that the same effort expended in the other areas would yield greater savings, and that more attention should be paid to sources of lost time and ancillary work such as setting up material handling, inspection, tool service etc. It follows that all the operations have to be tailored to suit the philosophy of 'mass producing a batch quantity of one' but the unit of one can be part of a family of components of a specified shape, size and weight.

PRACTICAL APPLICATIONS

The foregoing philosophy is best demonstrated by working through the critical path and posing the kind of problems which arise when one is trying to improve the speed of work in progress.

Starting at the beginning, the problem may arise as to the use of a fabrication, a casting, or forging. Although the use of a casting or forging would apparently shorten the path by the whole of the fabrication cycle, the same aim can be achieved by stocking fabrications. In making a decision therefore the true cost of a casting must include additional factors such as; the overhead cost purchasing, and progressing a unique component as against the same effort for a general purpose material such as plate, section etc. more readily available; the unique cost of goods receiving inspection; the variability in machinability, particularly of forgings; the dislike of machinists working in cast iron, the havoc to a machine tool caused by cast iron dust, mill scale etc. and above all the ordering and stocking of excess quantities to make the work attractive to a supplier. In short, in comparing fabrication and casting prices, the latter should be weighted with an agreed penalty to cover the aspects above. Probably the first operation in fabrication would be flame cutting. The greatest source of lost time in this operation is a change of plate thickness, and it is usual to organize the work into groups according to the plate thickness. This practice can also result in lower plate utilization as there is less tendency to use offcuts for making other parts. In our 'batch quantity of one unit of sale', if all the plates we require are the same thickness, then we have achieved an ideal situation. If not, then there will be compromise depending on the time span of one day, one week etc. within which the operator is working. The longer the time span the greater is the work in progress. Before compromising the design of the fabrication should be examined to see if material thicknesses could be standardized, as it is quite common for draughtsmen to work on the basis of minimum material cost only, without considering these other aspects. Fabrication is the putting of parts together, tack welding and final welding, with possibly stress relieving, some intermediate machining, shot blasting and painting.

In a fabrication shop there is probably some grouping according to weight and physical size enforced by the buildings but grouping should also take into account shape and method of construction so that it is possible to arrange work stations for workpieces with the same basic erection and welding techniques. A stress relieving oven instead of being a source of pride can also be a source of embarrassment. An oven has a fixed cubic envelope and a fixed charge for the treatment cycle so that the tendency is to try and fill the oven to capacity to minimize the cost per piece. This can form a queue so that the process is shown on the chart as a time interval of one week. This process does not fit the philosophy of 'mass producing a batch quantity of one' and the necessity for the operation must be closely examined. If essential, there are other techniques now available which more readily fit our philosophy.

These remarks also apply to shot blasting and painting in that much more flexible methods can be found than rooms with fixed cubic capacities and specialist plant in embarrassing layout positions. These operations are however usually much quicker, forming less of a queue than stress relieving. Where there are separate machining and fabrication shops the chart heavily penalizes a machining operation in the middle of a fabrication sequence. Drillers and plate edge planers are acceptable machine tools to a fabrication shop, but the introduction of other machine tools particularly for circular work could cause problems. There would seem to be quite an area for development of machine tools, as used in the pipe-line industry for field work, for more sophisticated work such as facing and boring. These would be more flexible and more acceptable to a fabrication environment than V/borers, H/borers etc. particularly since it is more of a problem of squareness of faces than one of surface finish or close tolerances.

A machining sequence on the critical path is best examined out of context of the general machine shop since in this path we are maximizing the speed of work in progress. In the general machine shop we can relax this a little and compromise because of a specialist labour or plant situation.

Work in progress is best reduced by firstly eliminating, secondly combining and thirdly grouping the operations which may consist of turning, milling, drilling, boring and grinding. This is the first attack and the second attack is eliminating setting, operator learning curves, materials handling, gauging etc. The technology used to arrive at the right answers depends very much on the materials handling problem.

Basically any component should stay in one position to have as many operations carried out as possible. This is true whatever the size of the component. It is now fairly common to see milling, drilling, boring combined into one machine tool but care must be taken to recognize the difference between a milling machine which can drill and a drilling machine which is supposedly able to do milling operations. It is perhaps less common to see turning combined with milling and/or drilling, and even less common to see turning combined with grinding. In this respect there are, of course, other finishing techniques available which would eliminate the grinding machine.

If it is not possible to combine all the operations into one piece of plant, the family of components must be transferred in a disciplined way from one operation to the next. Acceptable methods would be conveyor, pallets etc. assuming discipline of a line production where each station is manned and a line balance can be obtained, a grouping of the necessary plant where operators become skilled in manufacturing the family of components and are not necessarily restricted to, or employed as, skilled turners, millers etc. In any such group there is probably a key operation which would control the output; this operation should be given the closest attention in order to maximize the output in terms of tool/workpiece contact time. This means investigating all the sources of non-productive time as shown in Fig. 6. The supporting operations may well total up to a fraction of an operator; this fraction requires removing by speeding up one or other of the supporting operations or increasing the work content by doing something inefficiently in order to balance the labour force.

As the materials handling problem becomes more difficult and we become involved with cranes, layouts based on moving the machine tools to the component become more logical and it is most important that in heavy engineering a good look is taken at the traditional methods and machine tool designs of the past.

Erection areas are very often in a state of confusion particularly if a great deal of fitting has become the accepted practice. It is here that particular skills develop around particular people, and work very often waits the availability of these special skills.

Any erection is only as good as the individual parts and the amount of fitting is pro rata with the dimensional quality of the individual parts. Many of the problems in this area can be traced to the lack of jigs, fixtures and the operation of old plant in the machine shops and other feeder shops. The grouping arrangement in the erection areas should be around the family of products so that each erector is working on an erection with the same kind of sequence and problems so that a work station can be developed complete with the necessary tools and spaces for component parts. Although this is fairly standard practice for bench work it is not so for large work, where very often more than one man is involved. In such cases it is obvious that if a close knit team at a work station could be developed where each man is concerned with the finished product and can carry out any operation necessary irrespective of functional type casting, then the work would move faster than with a labour force organized into traditional functional skills.

CONCLUSIONS

The paper has deliberately avoided the use of a lot of terminology introduced in recent years such as 'group technology', cell production, value analysis, value engineering etc. Each new phrase seems to be covered in the publicity and proliferation of technical papers in an effort to make industry interested in them. In consequence some of these techniques become practised for their own sake with the very opposite result of what is intended. This paper states the view that the use or misuse of working capital is the fundamental problem in manufacturing and suggests a method of controlling this. In working to improve the techniques we can use any of the tools which become available, whatever name they are given. This is not to decry the excellent work done by many people in the field of group technology and other derivatives or forms of grouping. In fact probably the most important factors of all in unlocking some of the money tied up in working capital is to make it available for purchasing new plant.

THE SIGNIFICANCE OF ACCEPTANCE TEST RESULTS

by

H. TIPTON*

SUMMARY

The purpose of the paper is to point out the need for proper statistical assessment of the results of acceptance tests. A machine tool is a variable system and without knowledge of the variability, characterized by the standard deviation of properly repeated tests, no satisfactory comparison can be made of machines on the basis of these tests.

INTRODUCTION

The purpose of this paper is to point out the need for proper statistical assessment of the results of acceptance tests—in particular those kinds which lead to a single figure of merit as an indication of the quality of the machine.

With two exceptions this requirement is largely ignored in such acceptance testing as is now practised. These exceptions are tests at the two extremes of accuracy of measurement. With cutting tests of the dynamic performance of machine tools the variability of the test results is well known and is often quoted as an objection to the use of such tests. Work at M.T.I.R.A. on dynamic performance tests[1] and reported by Stone[2] has placed these tests on a firm statistical foundation and has shown how, with proper care, repeatabilities as good as ±10 per cent can be obtained.

The other example is in the evaluation of the accuracy of NC machine tools. Accuracies of 1 part in 10^5 are claimed and to check this, measuring instruments with accuracies of the order of 1 part in 10^6 are used. This high accuracy immediately shows up the variability of the results and has led to the requirement, in both the N.M.T.B.A. and V.D.I. specifications for evaluating the accuracy of NC machine tools, for a statistical criterion to be applied. This is that a range is to be specified within which 99·7 per cent of all readings of a repeated positioning will fall in practice. Statistically this is equivalent to a range of width ± three standard deviations about the mean value obtained and this is usually known as the ±3σ range. This means that when the measuring system indicates a certain value the actual position will be within a range of ±3 standard deviations quoted from it in 99·7 per cent of cases, or alternatively that it will only be outside that range in three measurements out of a thousand, on average.

However, statistical considerations of this kind apply to all tests, and the accuracy of any single result depends on the standard deviation or scatter of the results which can only be determined by repeated tests. Repeated tests will yield a mean value which has an improved precision and is a more representative measure of the quantity under investigation.

In this kind of situation, before any conclusion can be drawn on which is the better of two machines a proper statistical analysis of the evidence must be made, and this will not always substantiate what appears to be an obvious conclusion.

THE SIGNIFICANCE OF A SINGLE RESULT

All measurements are subject to a combination of systematic and random errors, of which the systematic errors can be reduced by an amount depending on the time and money available. The errors due to the measuring instruments are usually known and if we are measuring a system which can be regarded as being both linear and time-invariant then we can say that the value obtained represents that particular quality of the system to the accuracy of the measurement. This is the case when, for example, we measure the capacitance of a capacitor.

More complex systems, however, such as machine tools, which are made of many components, and of which the quality to which we wish to put a figure or figures depends on the cooperation and interaction of many parts, cannot be regarded as linear time-invariant systems. Any single measurement will be representative only of the machine in

* The Machine Tool Industry Research Association

the state that it was in at the time of measurement. A repeat measurement on another occasion will, in general, yield a different result from the first one, the difference being due to the inherent variability of the system as well as that of the measuring instruments.

The variability of the system itself will also have random and systematic components, of which the latter can be reduced depending on our knowledge of the various causes, by holding constant the factors which produce these systematic variations in the result. There will, however, always be some irreducible random variation in the value of the measured quantity which must be regarded as a property of the system being investigated.

All this is evident from the two examples previously quoted. We can obtain more consistent results of cutting tests by, for example, carefully controlling the workpiece material, and a major feature of the second type of test—of the accuracy of a numerically-controlled machine tool—is the determination of the magnitude of the random variations in the setting of the measuring system which is a prime measure of its quality.

Without a knowledge of the magnitude of the random variations of the results of repeated tests on a complex system (even though all those factors known to affect the result have been thought to be closely controlled), the results of any single measurement are virtually meaningless. We may know the accuracy of the measuring instruments, i.e. we may know the accuracy of the actual measurement of the state of the machine tool at that time. Our ignorance is of how representative the current state of the machine is of its general or average condition. All we know is that the desired value lies somewhere within a range of results which would be obtained if the tests were repeated indefinitely. Such a series of tests would produce a distribution curve approximating to the well known 'normal' or 'Gaussian' bell-shaped curve. The chances are that a value obtained is likely to be near the mean value but it could be anywhere within the range covered by the distribution. Without knowing the width of this curve, characterized by the standard deviation, we do not know how reliable or representative any single value is. It is thus futile to compare such single measures of two machine tools to decide which is the better. The fact that one value is bigger than the other has no significance unless the respective standard deviations are known, and even then it does not imply that the machine tool from which the 'better' value is obtained is in fact better than the other one.

The best measure of systems of this type is the mean value determined from a number of repeated observations. The mean value of a set of values has a scatter—a standard deviation—which is reduced by the square root of the number of observations compared with the standard deviation of the single observations.

These considerations apply to all tests on variable systems. In particular it is not logical to reject dynamic acceptance tests of the cutting variety because of their inherent variability and then to neglect proper statistical treatment of instrument-type tests. Many of the sources of variability within the machine tool are common to both tests, and each test method introduces further systematic and variable factors peculiar to the test method.

So far we have discussed variability within a single system (or machine tool). Machine tools are made in batches and when testing any single machine for the first time we may already know something of measurements on other nominally-identical machines. These results can provide useful information with which to assess measurements on machines of the same type. These measurements however are affected by a further source of variability: that between machines. This points to an alternative type of test: that to determine whether an individual machine is accepted as a satisfactory member of the class of machine tools to which it belongs.

THE VARIABILITY OF MACHINE TOOLS

The work on dynamic acceptance tests on lathes at M.T.I.R.A.[1,2] has shown that, at the 95 per cent confidence level, (equivalent to ±1·96 standard deviations) the machine variability can in certain cases be as small as ±10 per cent for a single machine and about ±30 per cent for a batch of nominally-identical machines. Although much work has been published on instrumentation tests, and the effects of various factors which affect the results, such as preload, spindle rotation, and feed rates, have been pointed out, no results appear to have been published on the variability of repeated measurements of this kind. Because of this it is worthwhile giving some details of some forced vibration measurements made by M.T.I.R.A. on radial-arm drills in 1967. These consisted of a series of repeat measurements on a single radial-arm drill and also of measurements on a small batch of nominally-identical drills.

It is worth mentioning here that the real interest in repeated measurements is to know the validity of such tests if single tests of this type were to be used as acceptance tests, so that we are really interested in the variability that results when each test is conducted as a separate job; that is, when each test includes setting-up the machine and apparatus for the test from scratch. In this sense the mere repetition of a set of readings immediately after a previous one, such as might be done to check the results, does not count as a repeated test.

The tests reported here are not the results of experiments specially done for the present purpose and so they fall short of the ideal in many respects, in particular the first set of measurements did not include setting-up the machine and apparatus from scratch before each frequency response measurement, but they still provide useful information. In the first tests the apparatus and drill were undisturbed except that between the second and third measurements the coupling between the apparatus and the spindle was temporarily disconnected to allow the spindle to be run continuously for 1 hour at 1550 rev/min. Repeated measurements were made of direct receptance with an r.m.s. force of 47 N (10·6 lbf) at intervals over a period of 94 hours which included a weekend. Eight sets of readings were taken at the various time intervals listed in Table 1 in which the values of the maximum negative in-phase receptance are given. This

TABLE 1   Repeated receptance measurements on a
           radial-arm drill

Time h	Maximum negative in-phase receptance 10^{-6} m N^{-1}
0	0.226 ± 0.007
3	0.200 ± 0.006
Spindle run at 1550 rev/min	
4	0.183 ± 0.006
69	0.166 ± 0.005
72	0.160 ± 0.005
74	0.186 ± 0.006
76	0.184 ± 0.006
94	0.166 ± 0.005
Mean	0.184
Standard deviation	0.022
95% confidence limits	0.184 ± 0.041

TABLE 2   Receptance measurements on a batch of
           radial-arm drills

Machine	Maximum negative in-phase receptance
1	0.193 ± 0.006
2	0.200 ± 0.006
3	0.080 ± 0.002
4	0.194 ± 0.006
5	0.110 ± 0.003
6	0.059 ± 0.002
Mean	0.139
Standard deviation	0.064
95% confidence limits	0.139 ± 0.125

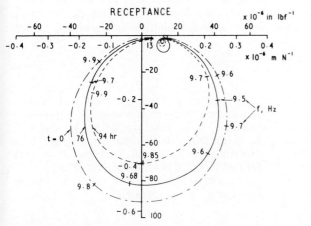

Figure 1. Receptance loci for a radial-arm drill measured at times t = 0, 76 and 94 h.

occurred near the major resonance at about 10 Hz and three typical response curves are shown in Fig. 1. The measurements were made in the normal laboratory environment where the temperature control is probably better than that in a typical workshop. Even though the apparatus was virtually undisturbed throughout, the variability of this quantity, which is that normally taken as a figure of merit for radial-arm drills, is 22 per cent at the 95 per cent confidence level. The range quoted with each figure represents the estimated accuracy of a single measurement of ±3 per cent.

Measurements were also made, on the manufacturer's premises, of a batch of five nominally-identical radial-arm drills of a different make. The resonant frequencies were about 12·5 Hz and the corresponding results are given in Table 2.

These results clearly indicate variabilities of the same order as those found in the cutting tests, and that they are inherent in many machine tools as at present manufactured. We need to know their magnitudes in all kinds of performance tests on machine tools if they are to give a useful and reliable index of performance of whatever kind. This is particularly true of short-duration tests; there are some long-term tests, such as, for example, the determination of

metal-removal rates on E.D.M. machines in which there will be some averaging effect, which will reduce the scatter of the results in the same way as repeated short-term tests.

DEDUCTIONS FROM THE RESULTS OF ACCEPTANCE TESTS

In the absence of any prior information the result of a single test on a variable system has little significance since we have no indication of the reliability of the result. For this we need some statistical measure, such as a standard deviation σ, of the repeatability of tests on the system, obtained from the results of repeated tests.

If n values x_i are obtained from repeated tests the most probable value which best represents the average behaviour of the system is the mean value $\bar{x}$ given by

$$\bar{x} = \left\{ \sum_{i=1}^{n} x_i \right\} \bigg/ n$$

The mean value is that value from which the mean square deviation of the n values is a minimum. The mean square deviation

$$S = \left\{ \sum_{i=1}^{n} (x_i - \bar{x})^2 \right\} \bigg/ (n - 1)$$

is known as the variance. Strictly the true mean should be used for this calculation, but since this is not usually known the mean value of the set of results has to be used instead. Because this is not independent of the measured values the number of independent variables reduced by one; hence the use of $n - 1$ in the denominator rather than n.

The standard deviation is the square root of the variance

$$\sigma = \sqrt{\left\{ \left(\sum_{i=1}^{n} (x_i - \bar{x})^2 \right) \bigg/ (n - 1) \right\}}$$

If we assume that the distribution of the results about the mean value is 'normal' or 'Gaussian', which usually gives a good approximation to the truth, we can say from a set of readings that any future readings are likely to be within the range $\bar{x} \pm 2\sigma$ in 95 per cent of cases, i.e. these are the 95 per cent confidence limits, or that they will fall into the range $\bar{x} \pm 3\sigma$ in 99·7 per cent of cases.

Statistically values near the mean value $\bar{x}$ are most likely to occur and the mean value of a set of results represents a better measure of the system than any individual reading in the set. The precision of the mean is increased by the square root of the number of values from which it is calculated, and the standard deviation of the mean σ_m is given by

$$\sigma_m = \sigma/\sqrt{n}$$

To determine whether one machine tool is better than another it is necessary to compare both the mean values for each machine and also the individual sets of results from which the mean values were obtained, since differences between sets of results with narrow distributions (small standard deviations) are clearly more significant than those between sets with larger standard deviations. Tests to determine whether differences of this kind are statistically significant consist of tests to determine the probability that the two sets of results are not different, i.e. tests of the probability that all the results might have come from the same set. The fact that the results were obtained on different machines does not invalidate the test, however.

These standard procedures, involving the use of tables which allow the significance of the results to be determined, are given in standard textbooks[3] and are not repeated here. There are two methods depending on the total number of measurements involved. If, as is likely, this is less than thirty the small sample method is used and this assumes that the standard deviation of both sets of measurements is the same. If the test shows that the probability that there is no difference between the results is say 5 per cent, then there is a 95 per cent chance that there is a difference between them, or the statement that there is a difference 'is true at the 95 per cent confidence level'.

It cannot then be safely predicted that the machine that yields the higher value is the better machine. Further simple statistical analysis yields a range within which the true difference will lie in, say 95 per cent of cases, when it may be found, for example, that in 30 per cent of cases the difference could equally well have been the other way round. In cases like this the only safe thing to infer is that there is insufficient data on which to make a judgement.

ACKNOWLEDGMENTS

This work was undertaken as a part of the research programme of the Machine Tool Industry Research Association and is published by courtesy of the Director.

REFERENCES

1.   A dynamic performance test for lathes, *M.T.I.R.A. Report,* Ministry of Technology Contract No. KJ/4M/150/CB 78A, March 1971.
2.   B. J. STONE (1971). The development of a dynamic performance test for lathes, *Proc. 12th Int'l. M.T.D.R. Conf.,* Manchester, 229 pp.
3.   K. A. BROWNLEE (1947). *Industrial Experimentation,* 2nd ed., H.M.S.O.

THE NOISE EMITTED FROM POWER HACKSAW MACHINES

by

G. J. McNULTY*

SUMMARY

Investigations into the parameters which affect the output noise level of power hacksaws have been carried out. The discussion includes suggestions whereby the noise level can be reduced by careful selection of cutting force and blade pitch without recourse to elaborate acoustic enclosures. The philosophy of noise reduction recommended in this paper whereby cutting parameters may be controlled may with success be adopted in other machine tools.

NOTATION

dB	Decibel (see Appendix 1)
S.P.L.	Sound pressure in decibels
ϕ	$\left(\dfrac{\text{Longitudinal tension in hacksaw blade}}{\text{Thrust force of blade on workpiece}}\right)$
N	Speed of stroke (times per minute)
P_b	Blade pitch (teeth per inch)
K	Kasto machine
W	Wicksteed machine

INTRODUCTION

In view of the *Code of Practice*[1] problems involving the noise emitted by machine tools are of contemporary interest. For example the problem of noise in manufacturing areas and its concomitant nuisance to factory personnel has received attention by several authors[2-5]. In these cases the authors base their criterion for the nuisance on the *Code of Practice* and cite case studies where reduction in sound pressure levels has been achieved by enclosures and vibration isolation. Also the background noise in a production unit initiated from ventilating systems has been adequately covered[6,11].

In this paper attention has been given to the control of noise from two different makes of hacksaw machine. Emphasis has been directed to certain cutting parameters which affect the noise output. The machines are modified as research tools and the cutting forces are controlled through a hydraulic system, and measured by means of a dynamometer.

The main reason for undertaking this work is to establish a philosophy of noise control in machine tools through control of cutting action, rather than restricting the exercise to screening-off the source by bulky enclosures, although the latter method is also necessary in certain cases.

A subsidiary aim of the work is to reduce the noise of each machine tool in the workshop, so that the limits specified in the *Code of Practice* will be satisfied. When each machine is treated for noise reduction there will be a resultant diminution of the overall S.P.L. See Appendix 1 for the definition of the decibel, and Appendix 2 for a brief explanation of the *Code of Practice*.

CONDITIONS OF THE EXPERIMENT

The sound pressure levels were made in accordance with British Standards[7] which formed a basis for the test procedures. The measurements were taken at several points round the machine and averaged according to ref. 7a. However the aim was to establish noise level trends when cutting parameters were varied rather than consider the machine itself as a high level nuisance.

Narrow band (6 per cent) frequency analysis were taken from tape recordings made during the investigation. A precision sound level meter was used which complied with ref. 8.

The machines under test were situated in a quiet room free from reflections ensuring that the environmental background noise was negligible as shown in Figs. 2 and 3.

To fulfil requirements of the investigation, near field sound pressure levels were adequate and detailed determination of sound power would not be justified.

---

* Principal Lecturer, Department of Mechanical and Production Engineering, Sheffield Polytechnic

THE HACKSAW MACHINES

Power Hacksaw machines are classified according to the method used to develop the load between blade and workpiece during stroke; this is fully discussed in ref. 9. The makes of the two machines used were a Wicksteed and a Kasto. Both machines are classified as hydraulic machines, because the resulting thrust force between the blade and workpiece is developed by a hydraulic device.

NATURE OF NOISE SOURCE

The predominant noise is of a fluctuating nature during the cut giving a peak value at approximately maximum thrust force. In Fig. 1 the thrust force is plotted over the stroke length and superimposed in the

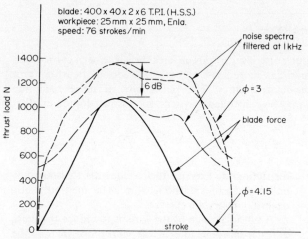

Figure 1. Noise spectra over the cutting stroke of the blade—filtered at 1 kHz showing the relationship of blade force and noise spectra.

curve is the noise spectra for the blade cutting action.

The spectra for two loads are given for $\phi = 3$ and $\phi = 4\cdot15$; it will suffice at this juncture to consider these as relative values. To obtain the noise curve the spectra were analysed from a tape loop where the start and finish of the stroke was monitored on the tape. The peaks of the spectra are shown for filtered frequencies of 1 kHz—the main peak in the noise spectra, which by coincidence, is also the preferred frequency for measurements recommended in British Standard[10].

Another source of noise in the machine is the rotating parts of the motor which is a continuous broad band frequency hum; this contrasts with the blade cutting action. To mask this noise a substantial screen was built round the motor of the Kasto machine, thus enabling the blade noise to be studied in detail.

CUTTING FORCE MEASUREMENT

The cutting force was measured by a dynamometer clamped in the workpiece vice of the machine. The measuring element was a Kistler piezo electric dynamometer with a charge amplifier. The displacement of the blade was measured by a linear transducer and meter.

At slow rotational speeds the output from the instruments were displayed on an X-Y plotter, giving load against displacement. For high speeds the outputs were displayed on a multi-channel oscilloscope and recorded with a polaroid camera.

Parameters controlled

Several parameters in the test were measured against the sound pressure level (see Notation), i.e. ϕ, P_b and N.

RESULTS

Microphone placed at distances recommended by ref. 7 from the machine and the results averaged according to ref. 7a.

TABLE 1

Octave band centre frequency (Hz)	63	125	250	500	1000	2000	4000	8000
1. P_b=6, ϕ=3 S.P.L. N=72, W S.P.L.	73	73	72	72	78	81·5	78	72
2. P_b=6, ϕ=4·15, N=72, W S.P.L.	71·5	71	71	71·5	79	78·2	72	74
3. P_b=8, ϕ=4·15, N=72. Blade and workpiece damped S.P.L.	71	71	70	70	71	71	70	67
4. P_b=10, ϕ=3, N=72, W	72	71	72	74	80	81	·73	68
5. S.P.L. P_b=10, ϕ=3 N = 144, W	71	71	72	74	79	81	84	71
6. S.P.L. P_b=6, ϕ=3 N=72, K	81	81	81	83	85	85	83	79
7. S.P.L. P_b=6, ϕ=3, N=72, K Motor screened	67	71	72	77	80	77	77	75
8. S.P.L. P_b=6, ϕ=3 N=72, K. Motor enclosed workpiece and blade damped	67	71	71	74	74	73	72	70

DISCUSSION

Blade force and blade pitch

Table 1 shows the variation in sound pressure levels (S.P.L.) for various parameters, of which blade pitch, and blade forces are two. The table indicates that there is no significant change in S.P.L. when the blade pitch is altered, this is consistent throughout the tests, i.e. rows 1 and 4 show only 0·5 dB difference at the peaks. An increase in S.P.L. is shown when the blade thrust force is increased. Figure 1 shows for example that the effect of increasing the thrust force can cause a rise in S.P.L. of 6 dB at peak frequency bands.

Time of cut

When considering the attenuation achieved by the reduction in blade force, the factor of cutting time must be considered. Table 2 shows the time of cut for a 3·0 in diameter specimen. Two blade forces and two blade pitches are compared for the time of cutting operation. It is shown that for the higher blade force

TABLE 2   Times of cut for different blade pitches and cutting forces. 3·00 in diameter EN44 1 bar

Blade force (ϕ)	Blade pitch (P_b)	Time (min)
3	6	3
3	10	5
4·15	6	10
4·15	10	15

the cutting time is reduced by a factor of 3. Thus, although the S.P.L. is increased for the higher blade force, the time for the lower force is trebled, which results in an increase in the noise energy emitted in the latter case. It would appear from the above figures, therefore, that it is advisable to use a coarse blade and as high a thrust force as possible. This will serve the dual purpose of reduced noise exposure coupled with a faster throughput.

Workpiece and blade damping

From Figs 2 and 3 the effect of damping the material of the workpiece and blade is shown to attenuate the noise level by about 5 dB. It is hot always practicable to apply this technique, due mainly to the time involved and also to physical limitations of the operations in many machine tools. However, if the time of cutting operation is long, compared to the time of applying the added damping, it would appear to be a useful expediency.

Screening

Screening an offending noise source is generally costly and cumbersome. Added to this, the important parts of the machine tool may become inaccessible because of a screen; making the whole operation intractable. The Kasto machine in the tests lent itself to the screening of the motor, resulting in a drop of 8 dB as given in Fig. 2. Figure 2 also shows that the

noise spectra is then sensitive to changes in blade force.

Blade cutting speed

Table 1, columns 5 and 6, show the comparison of S.P.L. when the cutting speed is doubled. There is no significant increase in the S.P.L. The table also shows that there is a frequency shift of the peaks.

The speed alteration therefore, increases the occurrence of the peak shown in Fig. 1. From Fig. 1 it can be calculated that the high peak in the spectrum during the stroke lasts approximately 0·6 s; doubling the speed does not alter this time interval but increases its frequency. If for any reason it was desirable to spread the main noise peak, alteration in cutting speed may prove efficacious.

The blade cutting action due to wear and its effect on the noise

When a hacksaw blade wears, a flat is produced at the tip of each tooth and the outer corners of the teeth become rounded. This wear phenomenon has been discussed in ref. 9. It tends to increase the thrust load for a given depth of cut.

When operating under the action of a light load a high pitched single tone noise is observed when the blade wears. It is suggested that this is caused by a rubbing action on the side of the flat produced by the worn blade against the workpiece. Measurements were taken of the noise caused by this rubbing effect; it was found to increase the peak S.P.L. by as much as 18 dB at 1 kHz making conversation in the proximity of the machine quite inaudible. For light loads this may last about 4 min when cutting a 3·0 in diameter mild steel specimen.

Comparison of noise levels in the Kasto and Wicksteed

There is only a small difference in the noise levels for the two power hacksaws considered. The Kasto is marginally higher for the same conditions of cut as given in Table 1, rows 1 and 6; at peak levels this amounts to 3·5 dB. The spectra suggests similar characteristics as exhibited by Figs 2 and 3.

Thus these tests should be typical of most power hacksaws. The one advantage from the noise aspect is

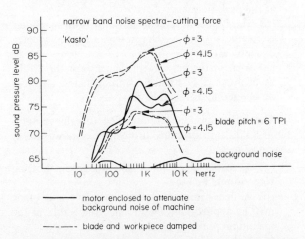

Figure 2. Narrow band noise spectra (Kasto) showing effects of damping the blade and workpiece and screening of motor.

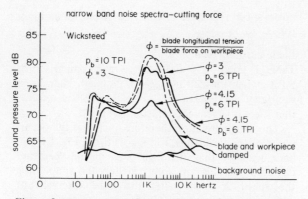

Figure 3. Narrow band noise spectra–cutting force (Wicksteed) showing effects of blade and workpiece damping.

that the Kasto can be partially enclosed by screening. Whereas in the case of Wicksteed, to screen the motor would have prohibited the stroke action.

CONCLUSIONS

The work of noise reduction in the two hacksaw machines described in this paper demonstrate principles which may be applied in the general situation of machine noise. The following points summarize the findings of the investigation.

(1) A narrow band frequency analysis (6 per cent) of the noise spectra should be made to establish the position and source of the highest peaks.

(2) The reduction of the thrust force from tool to workpiece may reduce the peak sound pressure level (S.P.L.). In this investigation the S.P.L. at the peak frequency was reduced by 6 dB. However, this was offset by a trebling of the time to complete the cut, resulting in a net increase in noise energy emitted for the lower blade force.

(3) Where possible damping material should be added to the workpiece and blade. In the tests this reduced the peak level by 5 dB.

(4) The coarser blade pitch had no effect on the noise level but had the advantage of speeding the cutting time.

(5) Screening sections of the machine to attenuate the S.P.L. will in general achieve the desired effect. However, this process is cumbersome and expensive. The screening of the motor in the Kasto machine causes attenuation of sound about 5 dB at the peak frequency.

(6) The type and level of noise of both machines were similar, and may represent a typical spectra of such devices.

(7) Wear of the hacksaw blade can raise the S.P.L. by as much as 18 dB at peak frequencies. This was observed only for light loads.

The following are recommended to achieve minimum S.P.L. for optimum throughput of power hacksaw machines.

(a) A coarse blade pitch and maximum cutting force should be used.

(b) Added damping of the workpiece and blade may be applied, and may reduce the S.P.L. at the maximum peaks by as much as 5 dB.

(c) Light loads accompanied by worn blades must be avoided in cutting operations as this combina-

tion produces dangerous high screeches for relatively long periods.

(d) When the noise level of the manufacturing area remains above the critical level and all other machine tools have been treated for noise reduction, then screening of the offending source should be considered.

REFERENCES

1.   HMSO (1972). *Code of Practice for Reducing the Exposure of Employed Persons to Noise.*

2.   K. HEALLIS (1972). Designing to reduce product noise, *P.E.R.A. Conference on Industrial Noise,* September.

3.   K. HEALLIS and E. BOYES (1972). Avoiding noise nuisance to the public, *P.E.R.A. Conference on Industrial Noise,* September.

4.   J. D. WEBB (1972). Noise assessment in and around the factory, *P.E.R.A. Conference on Industrial Noise,* September.

5.   R. S. ALLSOP (1972). Fundamentals of noise control, *P.E.R.A. Conference on Industrial Noise,* September.

6.   I. J. SHARLAND (1972). The control of noise in ventilating systems. *Noise Control and Vibration Reduction,* November.

7.   BS 4196: 1967. *Guide to the Selection of Methods of Measuring Noise Emitted by Machinery.*

7a.   Appendix A of ref. 7.

8.   BS 4197: 1967. *Specification for a Precision Sound Level Meter.*

9.   P. J. THOMPSON and M. SARWAR, Power hacksaws. Paper to be published.

10.   BS 3593: 1963. *Preferred Frequencies for Acoustical Measurements.*

11.   I. J. SHARLAND (1972). *Wood's Practical Guide to Noise Control.* Chapter 5.

APPENDIX 1

Definition of the decibel (dB)

$$dB = 20 \log_{10} \frac{P_1}{P_{ref}}$$

P_1 = sound pressure level in N/m^2
P_{ref} = reference level = $2 \times 10^{-5} \ N/m^2$
For example if the sound pressure is doubled 6 dB is added to the sound pressure level.

Some typical sound pressures for comparison are given below:

Riveting at operator position	130 dBA (intolerable)
Boiler shop	120 dBA (intolerable)
Automatic lathe shop	100 dBA (very noisy)
One power hacksaw	83 dBA (noisy)
Two power hacksaws	86 dBA (noisy—just tolerable)

APPENDIX 2

The Department of Employment's recommendations *Code of Practice*[1] limits the noise exposure to employed personnel to 90 dB (A)* for an 8 h day. The other limits are based on exposures to equal

_____
* This 'A' stands for a weighting network which is incorporated on all noise level meters, given in ref. 8.

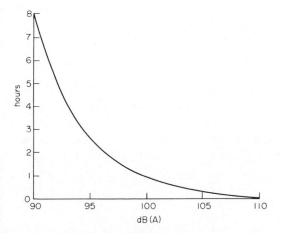

Figure A.1. Noise levels–daily exposure allowance, from Department of Employment's *Code of Practice.*

amounts of energy shown in Fig. A.1. The characteristic of the curve shows a steep drop between 90 and 93 dB (A) which represents a small part of the day. It is thus desirable to maintain the level within the 90 dB (A) limit. Consideration, therefore, must be given to reduce the level of every noise source in the manufacturing area. This would then help to lower the overall noise and furthermore is a necessary, incipient step in all noise reduction planning.

ACKNOWLEDGMENT

Much gratitude is extended to Mr P. J. Thompson for his help in the measurement of thrust forces in the hacksaw. Mr Thompson (Sheffield Polytechnic) is engaged in power hacksaw research and it was his research effort which provided the instrumentation for the blade forces.

DISCUSSION

I thank Dr Wasiukiewics (DIXI, Switzerland) for his encouragement and helpful comments. He suggests two main improvements in power hacksaw machines for noise reduction:

(i) provide automatic feeding for the workpiece
(ii) redesign the machine.

Several makes of power hacksaws provide automatic feeding. However, the varied geometry of the workpieces in this type of operation obviates automation and for this reason the loading with be predominantly manual.

An important improvement would be the redesign of the machine to minimize noise. This could be achieved either by remote positioning of the motor or by locating the motor integrally with the body of the machine. In the latter case the motor should be substantially baffled by the machine structure, resulting in extra capital cost.

I agree with Dr Wasiukiewics that a well designed screen built round a remote motor may be cheaper than a redesigned machine. Nevertheless, future legislation may demand that all machine tools do not exceed certain sound power levels, and the design aspect would be given serious consideration.

DISPERSION IN MACHINE TOOL CAPABILITY TESTING

by

B. STEEN*

SUMMARY

This paper discusses results from machine capability testing of turning and grinding machine tools. Long time testing and repeated tests in different machines give an indication of the dispersion involved in the test method. That dispersion is compared with statistical laws of significance.

This report also contains results from dispersion in measurement with the normal equipment which is used in shop practice and it also presents an acting programme for machine capability tests.

INTRODUCTION

A metal cutting machine tool very often gives the possibilities for machining in a very wide volume, but in the workshop a great percentage of the work is details of the same shape and only a small part of the machining volume is used. Moreover there is in some workshops an effort for using group technology. Under those conditions machine capability tests could be used for describing accuracy. This paper describes a test method preferably related to machining details with limited variations in shape and size. The application is for more or less automated machine tools.

The scope of this paper is to present statistical methods in describing the uncontrollable dispersion for a machine tool. We think that it is better to write about dispersion and possibilities of meeting tolerances than accuracy. The uncontrollable dispersion is the standard deviation when the production process is in control; otherwise it has to be calculated with deviations. The standard deviation is estimated from one measurement taken from each of a number of workpieces produced in one batch under controlled conditions. According to confidence limits, number of workpieces and the possibilities of dividing the amount of measurements into groups of the same size a factor can be calculated which is $>1 \cdot 0$. This factor assures the calculated value to be as small as possible but bigger than the real value of the uncontrollable machine tool dispersion. In order to achieve a given tolerance we should know the systematic deviations according to material, cutting data, and thermal effects, operators' skill and the dispersion of the measuring method besides the uncontrollable machine tool dispersion. The systematic deviations will sometimes be described as difference between the programmed target and the average value. The standard deviation can be calculated in many ways. In the workshop it is very easy to use the range method. In German V.D.I. recommendations (VDI 3254) that method is used for the determination of 'Positioning tolerance' and the German association of quality control has made a proposal for an acceptance test with a similar theory.

STATISTICAL THEORY FOR THE TEST METHOD

The application of the method of estimating standard deviation from the range value of small samples will be described.

The test is carried out with one single measurement from each workpiece and the values are divided into subgroups. Only the change of workpieces has to be done with the machine when machining those workpieces whose measurement values form a subgroup. Between two subgroups it is possible to make adjustments of the machine or even change of the tool. An advantage with this method is that only the pieces within a subgroup need to be consecutive. The number of pieces in each subgroup and the number of subgroups will be discussed later.

From each subgroup there is calculated a range value (R). The standard deviation S_R is obtained as the mean of all the range values ($\bar{R}$) divided by a factor d_n given in statistical literature and Table 1/Fig. 1 shows the calculation schematically.

* The Swedish Institute of Production Engineering Research, Göteborg

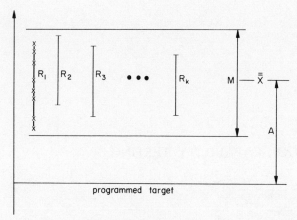

$$\frac{\Sigma R}{k} = \bar{R} \Rrightarrow \frac{\bar{R}}{d_n} = S_R \Rrightarrow S_R \cdot C \cdot 6 = M$$

Figure 1. Schematic description of the calculation for a process in control.

Confidence limits for σ

The real standard deviation is a value between the confidence limits given by equation (1).

$$\frac{\bar{R}}{d_n} \cdot \frac{1}{1 + F.\beta/(d_n . \sqrt{k})} < \sigma < \frac{\bar{R}}{d_n} \cdot \frac{1}{1 - F.\beta/(d_n . \sqrt{k})} \tag{1}$$

where

σ is the true value of the standard deviation

F is a probability unit

d_n and β are statistical factors depending on number of pieces in each subgroup

k is number of subgroups

$\dfrac{\bar{R}}{d_n} = S_R$ is a calculated value of the standard deviation.

If we transform (1) in order to get the factors for the confidence limits we will get

$$\frac{1}{1 + F.\,\beta/(d_n . \sqrt{k})} < \frac{\sigma}{S_R} < \frac{1}{1 - F.\,\beta/(d_n . \sqrt{k})}$$

TABLE 1   Factors for calculating the confidence limits

Number of values in each sub-group	3	4	5	6	7	8	9
d_n	1·693	2·059	2·326	2·534	2·704	2·847	2·970
β	0·888	0·880	0·864	0·848	0·833	0·820	0·808
Statistical probability		90%	95%	99%			
F		1·65	1·96	2·58			

If we accept 95 per cent statistical probability we will get the limiting lines in Fig. 2. The diagram shows the ratio between σ and S_R as a function of the number of testpieces. Between these limiting lines of the factors it is possible for a calculated value to fall.

It is now possible to calculate a value for the uncontrollable dispersion (M) of a machine tool

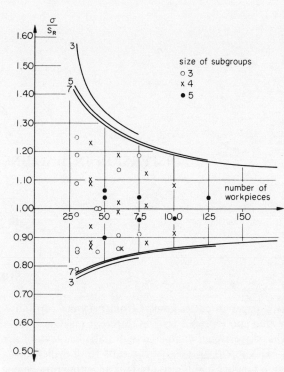

Figure 2. The scatter of test results from shorter and longer dispersion tests with different size of subgroups. The test is carried out on an automatic grinding machine.

which is larger than the true value with the following formula

$$M = C \cdot 6 \cdot S_R$$

if C is a factor equal with the upper limit of the confidence interval. The factor 6 is given for the statistical probability of 99·7 per cent that the measurement of the workpieces will fall in that interval.

Grubbs and Weaver[1] have shown that it is advisable to divide observations into groups of 5 to 10, the best size being 8. David[4] has calculated the ratio of standardized range to its normal-theory expected value for different types of frequency distributions. He has found that the differences are only a few per cent.

TEST PERFORMANCE

Planning the test

The theory described shows that the best group size is 8 but there is little difference between subgroups of 5 to 9 values. The bigger the subgroup the faster is the calculation. The number of pieces for the whole test ought to be more than fifty according to the slope of the confidence limits.

In order to obtain enough details for a test it could be necessary to gather details of different size and form. For this choice classification systems of group technology could be used and the dimensions might not differ more than 30 per cent between batches. Clamping device has to be of the same type in each case with the same number of points of support etc. Tool geometry, tool clamping and cutting data might not differ during the test. When machining one subgroup there has to be no other changes but the change of workpiece.

It has been found necessary to machine 25 per cent more workpieces than needed for the acceptable security of the calculated value. There are measurements which cannot be used according to the group size and the machine adjustments and tool change are not in the same intervals at the time.

Before starting the test it is very important to inform all personnel involved in running the machine to undertake it exactly as normal. In the test-record machining and workpiece data should be noted and how the measurement has been carried out.

Form defects, e.g. of roundness and cone-shaped, must be checked before the test starts. Normally those defects are systematic errors. When machining cylindrical workpieces and checking the dispersion in diameter dimension it is necessary to measure the maximum or minimum size in order to get the machine tool dispersion if there are form defects. Another check on the machined workpiece is the surface roughness. Bad surface could be an indication of failure in machining data.

During the test it has been found that it is important that the supervisory engineer has a good control over the machine tool in order to notice all adjustments. If measurements taken before and after an adjustment of the machine are put together in the same subgroup the R value might be bigger than that which depends on the machine itself.

Control of dispersion in measurement

The measuring equipment has to be checked before the test starts. It has been found suitable to use the range method for describing the standard deviation of the measurement equipment. This check is made with twenty-four pieces measured twice at the same point. From the difference between the two measurements a

standard deviation devoted to the measuring method is calculated with the range method (S_{Rm}). That deviation has to be less than 20 per cent of the preliminary deviation calculated form one of the two rows of measurements from the pieces; otherwise the measuring technique could not be accepted.

The significance of dispersion in measuring equipment is then given by

$$S_R{}^2 + S_{Rm}{}^2 = S_{R\,tot}{}^2$$
$$S_{Rm} < 0 \cdot 20\, S_R$$
$$S_{R\,tot}{}^2 < (1 + 0 \cdot 04)\, S_R{}^2$$

This means that the measurement has an influence of the total value of less than 8 per cent.

RESULTS

Figure 2 shows test results from a grinding machine during a long period of time. The dispersion of the machine has been calculated for short and long time tests with 30 to 125 workpieces in each test. This has also been made for turning and other grinding machine tools and the results have a similar shape.

Table 2 shows test results from an automatic lathe for turning rings in ball-bearings, as an example of test results from a machine tool over a long period. The test results are taken during a time of one and a half years. Test results are mean values of tests made with 200–400 workpieces. The dispersion result of the machine tool follows the state of machine tool wear (see Fig. 3).

Another test is made on a cylindrical grinding machine for two operations on the same workpiece. In the first operation the feed of the grinding wheel is

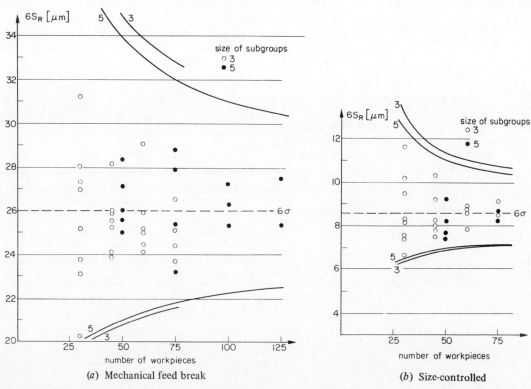

(a) Mechanical feed break   (b) Size-controlled

Figure 3. Machine capability test results from a cylindrical grinding machine operated with a mechanical feed break (a), and size-controlled (b).

broken with a mechanical break and the other operation is size-controlled. The test results are shown in Fig. 3. In each operation there are a few tests in order to check the possibilities of meeting confidence limits.

The measuring equipment used for the tests has been checked according to the dispersion method described above. The indicator with a scale reading of 0·5–1·0 μm has to be used in most cases. Dispersion of this ordinary measuring equipment with a skilled operator is between one to three μm. When testing a grinding machine special care has to be taken of the measurements because of danger of too big a dispersion with the measuring process.

An indication of the value of these tests according to other workpieces and dimensions is given by the small differences in dispersion in Table 2 when machining workpieces of different diameters under the same machine tool condition. On the other hand if the setting of the machine is changed like the grinding machine in Fig. 3, there will be differences in dispersion. It might be possible to use this performance test when checking the capabilities of machine tools for different groups in group technology.

It has been shown that it is possible to use this method for calculating the uncontrollable dispersion of a machine tool. The test method is used for delivery control of machine tools, and in planning and maintenance departments of some workshops with good results.

TABLE 2   Test results from an automatic turning machine tool

Machine tool condition	Test value (M) (mm)	Actual dimension (mm)
Normal	0·180	ϕ110
	0·174	ϕ 98
Worn out before repair	0·218	ϕ 89
After repair*	0·292	ϕ 82
Newly repaired and	0·167	ϕ 94
controlled	0·155	ϕ 89
	0·169	ϕ119

* Failure in assembling the spindle.

REFERENCES

1. F. E. GRUBBS and C. L. WEAVER (1947). The best unbiased estimate of population standard deviation based on groups ranges, *J. Amer. Stat. Ass.*, **42**, 224–41.

2. A. HALD (1952). *Statistical Theory with Engineering Applications*, New York.

3. F. KOENIGSBERGER *et al.* (1971). Assessing the capabilities of metal cutting machine tools, *The Production Engineer* (June).

4. SARIN and GREENBERG (ed.). Contributions to order statistics, chapter 7 in H. David, *Order Statistics in Short-Cut tests*.

DISCUSSION

Q. J. Peters, Leuven. I question your method of computing the standard deviation of your measurements S_R from the range R. Doing this by the mean of the mean of the coefficient d_n, issued from the χ^2 distribution you introduce an additional uncertainty factor, by which your expected results are widened. Using available mini or micro computer programs there is no need any more to compute the standard deviation indirectly from the range; it is easily done directly from the measurement results.

A. Yes, d_n is the mean of the range of the observations from a standardized normally distributed population. The ratios between root-mean-square (s) and range (S_R) estimators for some types of populations are shown in ref. 4. For the actual size of subgroups the differences are only a few per cent. The confidence limits are calculated by Hald[2] for s and S_R. The factor for the estimators differ with one unit in the second decimal.

One important reason for using the range method is that we can avoid 'outside' factors like adjustments in the calculated value. Between subgroups it is possible to make adjustments without any influence on the estimated deviation. That is why we are speaking about uncontrollable dispersion as a 'fingerprint' of the machine tool itself, the machine elements, its assembly, and the steering system.

Using mini and micro computers we get some s-values which we have to add with the 'square' method in order to get enough measurements and avoid systematic deviations like adjustments.

Q. C. P. Hemingray. (i) The fundamental point of using small order statistics is to reduce computation labour; but in the situation quoted this is at the expense of additional measurements. It can roughly be said that twice as much data is necessary to obtain similar reliability of the results with small order statistics as with more conventional statistics.

(ii) With the imminent widespread distribution of small, cheap, advanced electronic calculators, any objctions on the ground of complexity of use to conventional statistics must rapidly disappear.

A. We have tried the test method on some production machines with different shop-engineers carrying out the test. When we have examined the test procedure afterwards, we have found that the total loss of workpiece measurements is about 25 per cent. That figure includes mostly practical reasons and not theoretical reasons according the calculation method. The first reply contains the discussion of the theory.

MACHINE TOOL NOISE

by

J. B. DAVIS* and R. J. GRANGER†

INTRODUCTION

The increasing attention being paid to environmental conditions in industry means that the machine tool builder will have to give even greater heed to the control of noise. Research into noise control has now developed to a point where levels of acceptable exposure can be defined accurately and so engineering capability in noise control must be acquired to meet the requirements of these exposure limits.

Noise control engineers must be trained to

(1) develop measuring techniques;
(2) develop company standards using known standards as reference;
(3) compile a 'catalogue of noise levels' of company products;
(4) provide a consultative service at the design stage; and
(5) undertake research into methods of noise reduction.

These five areas of activity have been practised for some time by our own company to develop proper noise control techniques. They have arisen as a result of examining the present stage of research into noise levels (particularly that of Burns and Robinson[1]), and present and proposed restraints on noise emission. The authors define the three types of noise produced by machine tools and recommend general acceptance of the concept of Equivalent Continuous Sound Levels. Methods of measuring noise related to existing standards are noted, and control of noise through reduction at source is illustrated by an example.

STATE OF THE ART

The five-year research programme undertaken in the UK by Burns and Robinson has shown that the maximum noise level to prevent noise induced hearing loss in industrial workshops is 90 dB(A). The level arrived at by independent research in the USA and now incorporated in the 1970 OSHA Act is also 90 dB(A). Both UK and USA levels are based on 8 hours exposure.

The concept of Total Noise Emission, as proposed by Burns and Robinson, provides a relationship between the permissible level and the duration of exposure (Fig. 1).

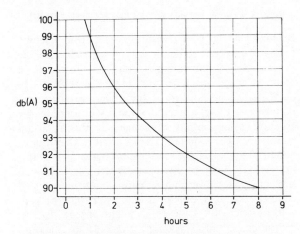

Figure 1. Permissible noise level and duration of exposure.

To ensure overall levels within 90 dB(A) individual levels must be below this figure. It is expected that the maximum level of 90 dB(A) will be reduced over the next few years to 85 dB(A) or even lower.

Data given above reflects the present position in the UK and the USA. Maximum noise levels vary from country to country. Currently there is no legal requirement in UK relative to noise control in machine tools, but the Code of Practice[2] issued by the British Department of Employment in 1972 advises on maximum permissible levels of noise exposure for employed persons. The machine tool builder must be continuously aware of pending legislation and standards in existence at company, national and international levels.

TYPES OF NOISE

Machine tool noise comprises one or more of three basic types of noise classified as steady state, cyclic and impulse.

* Cincinnati Division, Cincinnati Milacron Limited
† HME Division, Cincinnati Milacron Limited

Steady state

Noise that is continuous or fluctuates over a range of 5 dB or less. Standard sound level meters are suitable for measuring this.

Cyclic

Noise levels that change in excess of 5 dB during the machine cycle, the noise pattern being similar for each machine cycle. Standard sound level meters are used for measurements but are usually combined with some form of recorder.

Impulse (impact)

Noise in discrete pulses of less than 1 second duration, having peak pressures at least 10 dB greater than the r.m.s. sound pressure level. For practical purposes, if the repetition rate is greater than 10 per second, the noise may be considered as steady state.

Impulse sound level meter or oscilloscope techniques are required for measurement of the noise levels.

EQUIVALENT CONTINUOUS SOUND LEVEL

Existing standards for measuring noise from machine tools requires tests to be carried out with the machine in its noisiest operating condition. However, in practice, many machines run in this condition for only a small proportion of their operating time and therefore a more logical method of evaluating the effective noise output of a machine during a full working shift is required. The Equivalent Continuous Sound Level (ECSL) has been derived for this purpose. It gives the energy equivalent of a fluctuating or cyclic noise and is defined as the level of continuous sound which, in any given time interval, will produce the same total sound energy. Instrumentation is now becoming available to measure ECSL.

If the operating cycle of a machine is known or can be reliably estimated, then noise measurements should be taken to determine the level for each portion of the cycle. The ECSL should then be calculated as shown in the Appendix. An example is also given.

By developing the ECSL concept, national and international standards for machine tool duty cycles could be derived to form a comparator for machines of like function. Additionally, standardized loading conditions could be developed.

METHODS OF MEASUREMENT

Present standards for measuring noise emission are many and varied, and occur at industrial, as well as national and international levels.

International and national standards

General requirements for the 'Preparation of Test Codes for Measuring the Noise emitted by Machines' are included in ISO R495 which sets out guidelines for methods of measurement of noise from machine tools.

More detailed national standards are available, for example, BS4813 'Methods of measuring noise from machine tools'. Other examples are listed below.

National standards and codes of practice

Australia:	SAA 1217-1972–Noise measurement of machine tools.
Austria:	OAL-Richtlinie Nr. 1–Measurement of machine noise.
Czechoslovakia:	CSN 01 1603–Methods of noise measurement, 1967.
France:	S30 006–General rules for drawing up test procedures for the measurement of noise emitted from machines.
Germany (DBR):	DIN 45635–Measuring the airborne noise emitted by machines.
Hungary:	MSZ 11131-69–Methods of noise measurement.
India:	IS:4758-1968–Methods of measurement of noise emitted by machines.
Japan:	JIS B 6004: 1962–Method of sound level measurement for machine tools.
South Africa:	SABS 19/60/3–Code of practice for measurement in the field of sound emitted by machines.
USA:	NMTBA–Noise measurement techniques, 1970.
USSR:	GOST 11870-66–Machine noise characteristics and their determination.
	ENIMS N89-40–Limiting sound spectra levels for metal-cutting machine tools.

Industrial standards

Many standards established by individual companies now exist which detail methods of measurement and list permissible levels.

The diverse nature of these standards has prompted us to formulate a standard procedure to be used throughout the company's European Division. It was developed after examination of many national and industrial standards combined with our own experience in practical noise control.

This standard includes

(1) scope
(2) units of measurement
(3) equipment
(4) test conditions
    (a) machine operating requirements
    (b) test area suitability
    (c) ambient noise levels
(5) measurement positions
(6) measurements
(7) presentation of results
(8) types of noise
(9) equivalent continuous sound level.

The test sheet that combines the results obtained is shown in Fig. 2.

NOISE CATALOGUE

In order to develop effective noise control techniques, it is necessary to catalogue the noise levels of

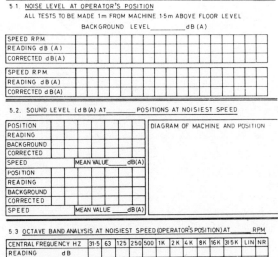

5.1 NOISE LEVEL AT OPERATOR'S POSITION

ALL TESTS TO BE MADE 1m FROM MACHINE 1·5m ABOVE FLOOR LEVEL

BACKGROUND LEVEL_____dB (A)

SPEED RPM													
READING dB (A)													
CORRECTED dB(A)													
SPEED RPM													
READING dB (A)													
CORRECTED dB(A)													

5.2. SOUND LEVEL (dB(A) AT_____POSITIONS AT NOISIEST SPEED

POSITION				DIAGRAM OF MACHINE AND POSITION
READING				
BACKGROUND				
CORRECTED				
SPEED	MEAN VALUE____dB(A)			
POSITION				
READING				
BACKGROUND				
CORRECTED				
SPEED	MEAN VALUE____dB(A)			

5.3 OCTAVE BAND ANALYSIS AT NOISIEST SPEED (OPERATOR'S POSITION) AT_____RPM

CENTRAL FREQUENCY HZ	31·5	63	125	250	500	1K	2K	4K	8K	16K	31·5K	LIN	NR
READING dB													
CORRECTED dB													
BACKGROUND dB													

5.4 OCTAVE BAND ANALYSIS AT NOISIEST SPEED (NOISIEST POSITION)

CENTRAL FREQUENCY HZ	31·5	63	125	250	500	1K	2K	4K	8K	16K	31·5K	LIN	NR
READING dB													
CORRECTED dB													
BACKGROUND dB													

Figure 2. Test sheet.

samples of each existing product. From this, an appreciation of troublesome areas is developed and a programme of noise control work set out.

NOISE CONTROL

In a machine tool noise effects can be reduced either by reduction at source or by shielding.

Reduction at source

Machine tool noise can be generated from
– machine
– machining process
– work handling equipment.

In order to reduce noise, the source making the largest contribution to the overall noise level must be identified and attention paid to its reduction.

For example, reduction of noise level at source in the design of a machine tool spindle drive, such as illustrated in Fig. 3, will be influenced by

design parameters
manufacturing tolerances

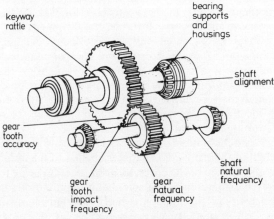

Figure 3. Gear train.

Design considerations include:
(1) pitch line velocity of gears
(2) number of teeth in contact
(3) shaft deflections
(4) bearing deflections—rigidity of housing
(5) choice of bearings
(6) choice of materials.

Manufacturing considerations include such factors as:
(1) Involute profile error
(2) tooth to tooth pitch error
(3) pitch circle diameter concentricity
(4) misalignment of bores.

The example, shown in Fig. 4, is for a 16 speed gearbox having standard outputs at spindle from 25 to 1500 rev/min. All gears are hardened to give required load and impact capacity, and ground to provide accuracy.

Conventionally produced cluster gears cannot be ground. Such gears have to be joined together mechanically or electron beam welded after completion of the grinding operations.

All bores containing bearings are in a rigid casting designed such that they can be machined with sturdy boring bars having a low length-to-diameter ratio.

Gears are also mounted on involute splines which assist in quiet running.

Additionally, shafts are run on axially spring loaded ball bearings. These shafts are designed to have a maximum deflection of 0·05 mm under full load conditions, generally resulting in shafts with length-to-diameter ratios of about 5 to 1.

The gearbox should be designed with the minimum number of gear contacts.

Finally, the gear casing is designed to give high static and dynamic rigidity and also to minimize thermal effects which could give centre-to-centre errors and increase backlash.

The noise levels of this gearbox measured are as shown in Table 1.

TABLE 1

Speed rev/min	Position	1	2	3	4	5	6
1500		77	78	77	78	78	78
290		68	68	69	70	70	68
25		65	65	66	68	67	65

Noise levels dB(A)–see Fig. 5.

Another example of reduction at source on which we have done work has been concerned with tooth belt drives. Investigations into the dynamic properties of the belt combined with pulley accuracy has yielded considerable noise reduction for some applications.

Shielding

This second method of noise control can take the form of either enclosures around the noise source or protection for the hearer. Shielding the source can be effected by simple enclosures around noisy parts of a machine, such as drive systems or hydraulic tank units or, in the extreme case, by total enclosure of

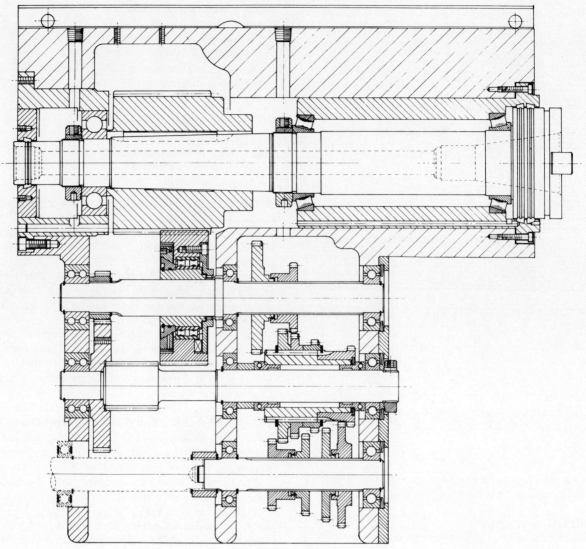

Figure 4. Gearbox.

the machine and ancillary equipment. To be effective however, the principles of noise absorption and insulation must be fully understood by the designer and great attention must be paid to detail during design and construction.

The typical thin walled enclosure often encountered as a sheet steel guard on machines has few

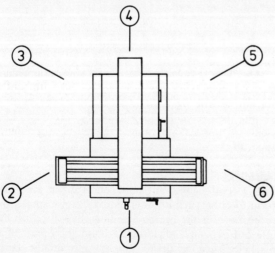

Figure 5. Positions of measurement.

absorptive properties and may even increase the sound level through the effects of resonance and coincidence. Resonance occurs when the natural frequency of the sheet material is matched by a principal frequency in the sound being produced. Coincidence occurs when the projected plan of the wavelength of the sound coincides with the wavelength of surface bending waves in the material.

Resonance and coincidence effects can be dramatically reduced by using a material with high damping, high density and low stiffness; lead being an excellent material in these respects. However, lead is expensive and possesses little structural strength.

Sheet steel coated with a proprietary damping compound is much stronger and cheaper than lead and, since it is as good acoustically as lead, offers the best compromise. The proprietary damping compounds may be applied as a liquid paste or glued on in sheet form. Applying it as a liquid takes longer but permits greater control over thickness.

Aesthetic considerations aside, it matters little whether the damping material is applied to the inside or the outside of the cover, its main function being to prevent vibration of the steel. Any sound absorbing properties of the compound itself are relatively small by comparison. To reduce vibration of the cover further, it is necessary to isolate it from the noise source. This is best achieved by making the cover a free standing unit having no contact with the equipment inside, or by insulating the cover from the equipment using neoprene or rubber.

Noise absorption is achieved by making the sound energy expend itself by doing work against friction— that is, the sound energy is degraded into heat energy, though the temperature increase is practically negligible. Sound waves trying to pass through tiny holes or fibrous materials is an effective way of converting sound into heat. Polyurethane is a good and relatively inexpensive example of a noise absorbing material. The best results are obtained if the sound waves repeatedly encounter the absorbing material by means of multiple reflections.

When the above techniques are incorporated into sound insulating enclosures, the most important design consideration is that there must be no leakage of air through or around the extremes of the enclosure. Any holes in the enclosure or lack of sealing between the enclosure and the machine structure have a drastic effect on the noise reduction achieved. For example, if there is leakage through an area equal to 1/100th of the total surface area, then the best sound reduction possible is 20 dB.

Great attention is also necessary in the detail design of outlets in the enclosure for pipes and cables, and where there is total enclosure of a machine element the problem of adequate heat dissipation often arises. This necessitates ventilation systems being incorporated into the enclosure, the cooling air being drawn through acoustic ducting. Where visual inspection of the machine or element is required, safety-glass panels have to be provided which should be kept as small as possible.

Shielding economics

The additional cost of acoustic treatment of sheet metal panels by the application of damping materials is negligible.

Where further acoustic treatment is required (for example around noisy machine elements or around the machine process), the cost obviously varies with the size and type of application, Fig. 6. For example, enclosures of hydraulic units could cost around £300 to £400 whilst partial enclosures of machines such as high-speed Blanking presses would cost £1500 to £2000. This last consists of hinged acoustic shields, 100 mm minimum thickness, fitted to the area of the press at the front and rear. It would be complete with inspection panels, together with acoustic treatment provided at the sides of the machine. Such an application to a blanking press reduces the sound

pressure level measured in front of the machine by 10 to 15 dB.

If further noise reduction is required, total enclosure of the machine and, probably, ancillary equipment is necessary. Several manufacturers now market acoustic enclosures in a range of standard sizes, a typical cost being from £5000 upwards.

CONCLUSION

Noise reduction on existing plant may require expensive re-design, or expensive modifications to incorporate shielding techniques. Expertise must be developed to ensure that cost of reduction does not become disproportionally high compared with the results obtained. Some source reduction techniques will not add significantly to product cost particularly when included in the original design.

For true economic noise control, expertise must be developed by the machine tool builder to minimize noise at the initial design stage. A planned method of training engineers to develop measuring techniques, develop company standards, catalogue existing product noise levels, provide consultative skills at design stage and undertake research into all methods of noise reduction are recommended as a positive step to reduce noise emission in machine tools.

REFERENCES

1.   W. BURNS and D. W. ROBINSON. *Hearing and Noise in Industry.*

2.   *Code of Practice for Reducing the Exposure of Employed Persons to Noise,* H.M. Stationery Office.

APPENDIX: EQUIVALENT CONTINUOUS SOUND LEVEL

Mathematical basis

Equivalent continuous sound level (L_{eq}) is that sound level which, in the course of an 8-h period, would cause the same A-weighted sound energy to be received as that due to the actual sound over the actual working day. L_{eq} is expressed in terms of A-weighted sound levels, and is defined mathematically by:

$$L_{eq} = 10 \log \frac{1}{8} \int_0^T \frac{p_a(t)^2 \, dt}{p_0{}^2}$$

where L_{eq} = equivalent continuous sound level, normalized to 8 h shift

T = total working period in hours

p_a = instantaneous A-weighted sound pressure in N/m^2

p_0 = reference (r.m.s.) sound pressure (2.10^{-5} N/m^2)

t = time

A Nomogram is available for easy calculation in the Code of Practice.

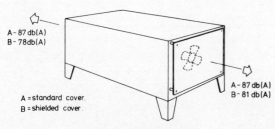

A – 87 db(A)
B – 78 db(A)

A – 87 db(A)
B – 81 db(A)

A = standard cover.
B = shielded cover.

Figure 6. Acoustic shield.

Example
During a machine cycle of 2 min duration, the sound level varies as follows: 94 dB(A) for 25 s, 83 dB(A) for 40 s, and 87 dB(A) for 55 s. Each cycle is followed by 45 s with the machine running idle, at a sound level of 80 dB(A). Total duration time of machine running is $6\frac{1}{2}$ h out of an 8-h working shift.

Calculation of the ECSL gave a value of 87 dB(A) (to nearest decibel).

ADAPTIVE CONTROL OF AUTOMATIC ROLL GRINDERS

by

P. F. AINSCOW*

SUMMARY

This paper sets out to show the unique qualities of the control system which adapts itself to the condition of the workpiece on a traversing wheelhead carriage type grinding machine used for regrinding steel mill rolls.

The adaptive electrical control system, which is completely automatic, detects and eliminates roll taper and uneven roll wear while grinding a given shape of roll.

The control is used to achieve matching of rolls, amount of stock removal and final sizing. Control of automatic balancing of the grinding wheel and automatic loading of the roll is also discussed.

INTRODUCTION

The operation necessary to reestablish the shape and surface finish of worn steel mill rolls has, in the past, been carried out by manual operation of the roll grinding machines by highly trained and skilled operators. In the steel industry in particular, there has arisen a demand to increase the tonnage of finished product in relation to the labour force. The availability of this skilled force can also be a major problem.

The mill rolls are of high value and thus roll life is a significant factor in the overall production costs.

The ability of the roll grinding machine to remove the minimum amount of stock necessary to produce the required shape and finish to the roll is also of major importance.

It was to achieve these objectives that the automated roll grinder was introduced, and this paper describes the control system within each area of automation in the order in which it is used for the automatic regrinding process as applied to a Churchill grinding machine.

The complete process of regrinding is fully automatic. A punched paper tape or punched card reader is used to give information to the system. This programmed information will include

(a) stock to be removed
(b) final size of roll
(c) number of semi finish passes
(d) number of finish passes
(e) final size of matched roll
(f) auto wheel balancing, auto wheel truing, etc.

Once the system has been set up and the roll loaded into the machine the whole automatic cycle is then initiated by pressing a button and then follows the areas of automation which are described:

(i) roll alignment in the machine
(ia) constant peripheral wheel speed, etc.
(ii) the magnetic amplifier
(iii) optimum method of re-shaping the roll form
(iv) roll matching, types of gauging, etc.
(v) grinding cycle.

In conclusion, the loading and unloading of the roll into the machine are described which are a further automation processes of the regrinding cycle.

ROLL ALIGNMENT IN THE MACHINE

Basically two types of roll alignment have been used.

The first was by measurement of wheelhead motor current on the roll body ends by establishing an identical plunge cut at identical points at each end of the roll. This was achieved by having one adjustable power operated steady (at the work-head end) and correct alignment relied upon the ends of the roll face being undamaged and parallel. If there was taper in the roll this method would not eliminate it.

The second and more sophisticated method of roll alignment is by probing the roll necks with transducers (see Fig. 1). This involves two identical adjustable power operated steadies and two centring devices and this method ultimately detects and eliminates taper on the roll face.

This method is designed to ensure accurate location of the axis of the roll, not only parallel to the

---

* Herbert Machine Tools Limited

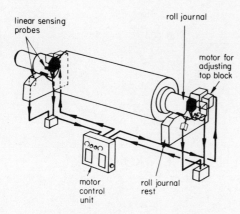

Figure 1. Automatic roll alignment by probing the roll necks.

wheel carriage traverse but located in a specific position relative to the machine bed. The accuracy is also unaffected by variations in diameter of the roll necks.

We have two pairs of capacitive transducers, one pair at each end of the roll and mounted on a simple unit.

The pair of transducers situated at one end of the roll are arranged to measure the out of alignment of that end of the roll. The two transducers are driven from oscillator 1. The output from these transducers (see Fig. 2) is connected together and arranged so that for equal displacement of each transducer the resultant signal is zero (i.e. the signals cancel each other). If the roll is out of alignment the transducers will not be deflected equally and a signal will be produced which is proportional to the amount of misalignment. This signal is amplified by Amplifer 1 and fed to Comparator 1.

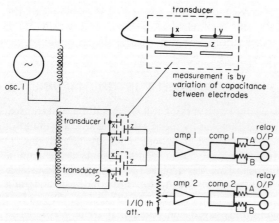

Figure 2. Diagram showing roll alignment transducers.

The purpose of Comparator 1 is to measure the input signal from the amplifier and to operate a relay when this signal reaches a certain level preset on Control A. A second identical circuit in Comparator 1 detects the input signal when it is of the opposite polarity and at a level set by Control B. The two relays controlled by Comparator 1 are used to drive the centring motor to aligh the roll.

The accuracy of the centring will depend on the speed of the motor being sufficiently slow to prevent

overrun when signalled to stop. The speed of the motor is controlled, fast or slow by the signals obtained from Amplifier 2 and Comparator 2.

The signal into Amplifier 1 is also supplied to Amplifier 2 but reduced in amplitude by a factor of 10. If this signal is sufficiently large in spite of being divided by 10, to operate one of the detecting circuits in Comparator 2 then the roll is considerably off centre and the motor is switched to fast. As the roll becomes more central the signal becomes insufficient to operate the comparator circuits and the motor is switched to slow.

In order to assist in setting up, there is a push-button on Oscillator 1 which introduces a signal into the circuit corresponding to a transducer displacement of 0·0025 inches (0·0625 mm).

The 'pad wear' transducer is driven from Oscillator 2 (see Fig. 3). The signal from the transducer is

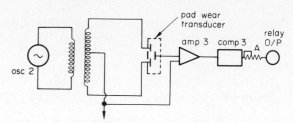

Figure 3. Diagram showing pad wear transducers.

amplified by Amplifier 3 and then used to operate the detector circuit in Comparator 3. This detector then operates a relay to indicate that the pad is worn.

All the above modules are mounted in one rack and there is an identical rack for all the modules associated with the transducers at the other end of the roll.

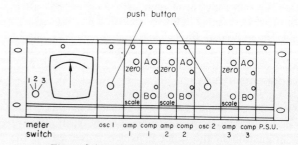

Figure 4. Typical layout of module rack.

The setting up procedure is as follows (see Fig. 4).

(1) The meter switch is used to connect the meter across the output of Amplifiers 1, 2 or 3. With switch set to 1 and zero control on Amplifier 1 set to mid-position, adjust the roll centring transducers *mechanically* to give as near as possible zero reading when the roll is centred. When the best mechanical zero position has been obtained the meter should then be set exactly to zero by means of the zero control.

(2) Operate pushbutton on Oscillator 1 and adjust the 'scale' control on Amplifier 1 so that full scale deflection on the meter is obtained. When the push-button is released the zero setting may need slight correction by use of the zero control again.

Repeat this procedure until the meter indicates exactly full scale with the button pressed and zero with the button released. (The meter scale will then represent 0·0001 in (0·0025 mm) per division and 0·0025 in (0·0625 mm) full scale.)

(3) Using the zero control to deflect the meter, in a clockwise direction, to the reading corresponding to the trip point of the control relay, adjust control A on Comparator 1 until the red light next to the control is illuminated.

It is important to first set the control A fully clockwise and then adjust until the lamp lights. Check the setting by varying the zero control either side of the setting and note the reading of the meter when the lamp lights.

Repeat the above using control B (for the setting on the opposite side of zero (anticlockwise). (These two points should be the same amount either side of zero.) We have now set the point at which the roll alignment motor will be de-energized and allowing for over run (however slight) we hope the meter will in fact finish on zero (i.e. once we have reset the zero).

(4) Set the 'meter switch' to 2.

Repeat sections (2) and (3) but adjust control on Amplifier 2 and Comparator 3. (Because of the 1/10 attenuation of the signal into Amplifier 2 the 'scale' setting should be adjusted to give 1/10 of the full scale deflection of the meter when the pushbutton is pressed on Oscillator 1.

When adjusting A and B on Comparator 2 which are the fast/slow change points, note that the meter at full scale represents 0·025 in (0·625 mm).)

The setting up procedure for the 'pad wear' transducer follows a similar pattern.

CONSTANT PERIPHERAL WHEEL SPEED, ETC.

A constant surface wheel speed is desirable throughout the life of a grinding wheel in order to achieve the optimum finish on the roll.

The theory of operation is as follows.

As the radius R of the grinding wheel reduces and thus its axis moves towards the roll, the axis displacement X can be utilized to displace a rectilinear potentiometer by the same amount and this can be utilized to maintain the peripheral speed of the grinding wheel as follows:

$$\text{Peripheral speed}(s) = 2\pi R N$$

where R is the instantaneous radius, N is the number of revolutions per minute

$$\therefore N = S/2\pi R \propto 1/R$$

As the radius reduces the rpm must increase to keep the peripheral speed constant.

We can thus use a linear potentiometer to give a measurement of radius R and divide a reference voltage by the output of the potentiometer.

This linear potentiometer is mounted on a slide which carries

(a) a unit consisting of a group of proximity probes

(b) the main coolant nozzle.

The particular probe that is used for the peripheral speed control of the wheel actually switches the

motor driving the slide backwards when the gap between the probe and the roll reduces below a certain distance. As the wheel wears the whole wheelhead moves forward, the gap reduces, the proximity probe signals the motor to drive the slide back which moves the linear potentiometer and the wheel speed increases accordingly. Depending on the differential of the proximity probe, which is low, the slide moves back in small steps as the wheel wears, the wheel speed increases and the peripheral speed is kept constant.

When a wheel change is necessary a simple mechanical interlock prevents a new wheel being fitted without first bringing the slide to its fully forward position.

Because the slide also carried the main coolant nozzle, once this is set directed at the point of contact of the roll and wheel it will remain directed at this point of contact throughout the life of the wheel.

The single proximity probe already mentioned acts as a reference for the unit as a whole in relation to the face of the wheel.

It maintains a fixed relationship in terms of distance of the proximity probe unit behind the face of the wheel regardless of the diameter of the wheel.

We use this feature to give the 'automatic wheel approach system'.

When the roll is correctly aligned in the machine, it is first necessary to obtain contact between the grinding wheel and the roll face in as quick a time as possible and this is achieved by the action of two of the other proximity probes on this unit. (See Fig. 5.)

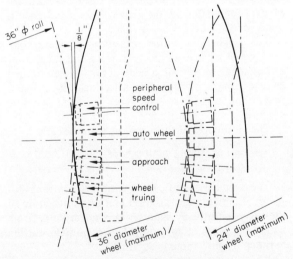

Figure 5. Layout of proximity probes.

One probe controls the changeover point from wheelhead rapid approach to slow traverse and the other probe controls the changeover point from slow traverse to a final programmed auto feed rate, and these points occur at given preset distances of the wheel from the roll face regardless of change of diameter of roll or of change of diameter of wheel.

Similarly for automatic truing of the wheel from a diamond mounted on the tailstock these two proximity probes are again used along with a third to bring the wheel in very close proximity to the diamond.

The wheelhead carriage is then traversed, a predetermined timed end feed is given to the wheelhead slide and the diamond dresses the face of the wheel.

Automatic wheel balancing is achieved by using an extremely sensitive electronic device to determine the amplitude of the grinding wheel spindle vibration caused by an out-of-balance grinding wheel.

The signals from this device are monitored to control a balancing mechanism mounted on the front of the grinding wheel collet. The device disconnects when the amplitude is at a minimum.

The advantage of being completely automatic is that the mechanism can be actuated at any point in the machine cycle. At the appropriate time, as programmed, a check is made and if an increase of 'out of balance' is detected the mechanism reengages and corrects this out of balance. On completion of correction the cycle continues.

THE MAGNETIC AMPLIFIER

On completion of roll alignment (i.e. the more up-to-date method of probing the necks) the carriage moves into position and the wheel is brought into close proximity with the roll (automatic wheel approach system). The automatic grinding cycle then commences.

The sensing of the shape of the roll and in fact the whole grinding cycle is controlled by the magnetic amplifier.

The magnetic amplifier is the heart of the adaptive control system.

There are two magnetic amplifiers in the system and two are necessary because for certain conditions, as explained later in the paper, two simultaneous levels of load current are required.

The magnetic amplifier measures very accurately the level of wheelhead motor current and the output can be made to operate a relay at any predetermined level of motor current by selecting, via external potentiometers, a certain level of bias winding current.

The advantages of the magnetic amplifier are

(i) ease of mixing input signals
(ii) rugged construction.

The original specification was: 'DC current differential between energization and de-energization of the relay with the 72 amp series connection is to be within 0·5 amps at any control potentiometer setting between 6 and 72 amp load current'.

We will now consider the operation of just one magnetic amplifier (see Fig. 6).

There are four windings.

(1) The signal winding—this carries the main wheelhead motor load current, the load we want to measure. This is the heavy winding consisting of a single turn or two.

(2) The bias winding—this winding preselects the operating point of the amplifier by introducing throughout the grinding cycle different potentiometers which give different values of bias winding current and hence different levels of operating points of the amplifier.

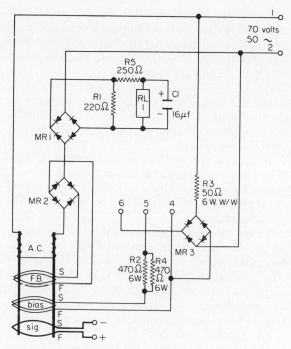

Figure 6. Diagram of one magnetic amplifier.

(3) The AC winding.
(4) The feed back winding.

The AC winding is in series with the feed back winding (DC) and the relay RLI (DC) which performs the switching action.

Operation of feed back winding

(i) Connection of AC supply to terminals 1 and 2 causes an alternating current to flow through the AC winding of the amplifier. With no DC signals applied to the amplifier windings the impedance of the AC winding is fairly high (point O) (see Fig. 7) and initially a low current flows through rectifier MR1, which is insufficient to energize relay RL1.

(ii) The AC current, however, also flows through rectifier MR2 and produces a DC current in the feedback winding, which causes a slight reduction in the impedance of the AC winding (point A).

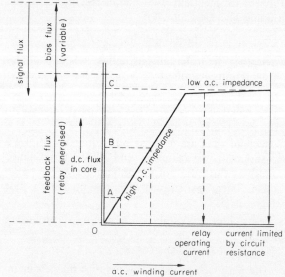

Figure 7. Characteristics of magnetic amplifier.

(iii) This again causes an increase in the current through the AC winding and hence, via MR2, an increased DC current in the feedback winding (to point B).

(iv) This effect causes a continuous increase in the AC and feedback currents up to point C, when the current becomes limited by the resistance of the circuit and no further increase occurs. This value of AC current is sufficient to cause relay RL1 to energize.

(v) The circuit is now in a stable state, with relay RL1 energized and the flux in the core due to the feedback winding above point C, i.e. the amplifier is 'saturated'.

Effect of sign 1 current

(vi) If a DC current is now passed through the main signal winding it will create a flux opposing that produced by the feedback winding. When this signal current reaches a sufficiently high value, the net flux in the core falls below point C and the impedance of the AC winding increases, causing the AC current to decrease.

(vii) The decrease in AC current is magnified by the action of the feedback winding, which further reduced the net flux so that the AC current falls rapidly and relay RL1 de-energizes. The feedback winding gives 'fail-safe' operation of relay RL1 and assists any change in AC winding impedance giving a 'snap' action and reducing the differential between energization and de-energization of RL1.

Effect of bias current

(viii) With relay RL1 energized and the core flux above point C, a fixed amount of DC signal current will cause relay RL1 to de-energize.

(ix) DC current can be passed through the bias winding from rectifier MR3 and The magnitude of the current is adjustable by means of an external potentiometer connected across terminals 4, 5 and 6.

(x) The bias current creates a flux which opposes that due to the signal winding. Thus a greater signal current is required to bring the net flux in the core below point C. In this way, the operating point of relay RL1, in relation to the magnitude of the signal current, can be varied by adjusting the remote potentiometer.

(xi) A tapped resistor is provided and this is connected in series with the remote potentiometer. The range of the amplifier can be pre-set by means of this resistor. For different values of tapped resistor, typical amplifier working ranges are

0–36 amps
0–72 amps
0–125 amps
0–175 amps (resistor shorted out).

For the above conditions the relay RL1 coil

(a) energizes at 2·1 V DC
(b) de-energizes at 2·0 V DC.

We, therefore, have a device which can switch its output very accurately, i.e. with a very low differential, at any given point within the range of the amplifier by means of selecting one of several poten-tiometers available which are themselves preset at the optimum load levels for the different stages of the grinding cycle.

OPTIMUM METHOD OF RESHAPING THE ROLL FORM

The wheel is auto fed into the end of the roll, an initial cut is established and this is determined by a preset potentiometer. When this level of load is reached the auto feed is stopped and then the wheelhead carriage is traversed towards the centre of the roll. The system is now looking to see what shape the roll is.

Although it is usual for a strip mill roll to be worn most heavily in the centre, there are instances when the camber requires reducing. The control system caters for this requirement and detects any variation from true form that exists in the roll shape of the worn roll and this determines the type of reshaping cycle.

As the grinding wheel traverses towards the centre of the roll face the magnetic amplifier detects any significant increase or decrease in cut.

A decrease in cut indicates a roll worn in the centre and end diameter grinding is automatically selected.

An increase in cut indicates a roll that is high in the centre and centre grinding is automatically selected.

End diameter grind

If the system senses the shape to be 'worn in the middle' (see Fig. 8), the wheel traverses to the centre of the roll after having established an initial cut at the tailstock end.

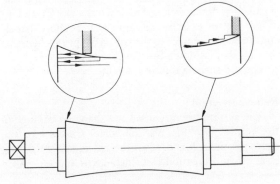

Figure 8. End-diameter grind.

A 'load loss' potentiometer is now brought into circuit and this determines the current level to which the cut has to decrease before the carriage changes direction.

This is a preset dwell on reversal (end diameter reversing timer) when auto feed is applied (normal roughing feed rate or fast roughing feed rate depending upon material of roll). The amount of cut which is applied is determined by another potentiometer (reversing). When this level is reached the feed stops and the carriage traverses towards the tailstock. During this pass if the load on the wheel should fall below a preset value then the feed is automatically selected again to maintain this value (wheel compensation potentiometer). This then compensates for wheel wear when necessary.

The carriage stops at the tailstock end of the roll for a preset dwell (auto-dwell carriage reverse dwell timer).

A preset timed infeed is applied (fast wheel break up and diameter infeed timer or normal wheel break up and diameter infeed timer).

The carriage traverses towards the centre of the roll, there is no wheel wear compensation because now the system is looking for the worn part of the roll again. The load loss potentiometer determines the point of reversal and the sequence continues as above until the wheel arrives at the centre of the roll.

The number of end diameter passes required depends only on the amount by which the roll is worn. The number of passes, therefore, varies from roll to roll but is always the minimum necessary to achieve the correct shape.

When the wheel arrives at the centre of the roll a switch is operated and the wheel then continues towards the workhead end of the roll. The load loss potentiometer is taken out of circuit and the wheelhead is then automatically set to retract in steps, controlled by the magnetic amplifier until it reaches the workhead end.

When the load increases at the high unground workhead end to a level preset by 'retract protentiometer', the carriage stops and the wheel retracts for a pre-set timed period (workhead end retract timer) or *until the load decreases to a preset level*, whichever is the longer.

The carriage continues in steps until the workhead end is reached. The carriage stops at the workhead end of the roll.

The wheel moves forward until a preset cut is established identical to that at the tailstock end. The wheel then proceeds to grind down the workhead end of the roll in exactly the same way as the tailstock end until the centre of the roll is again reached this time by shaping down the workhead end.

The roll shaping is now complete.

Roll centre grind

The system having sensed the shape to be 'high in the centre' (see Fig. 9), the wheel traverses to the centre of the roll and when the cut reaches a pre-set level set by potentiometer 'high centre retract'—Amp

2, the carriage stops and the wheel retracts until the load decreases to a preset level.

The carriage continues towards the centre of the roll and this sequence of stopping the carriage and retracting the wheelhead is automatically repeated whenever necessary until the wheel arrives at the centre of the roll.

The carriage traverses past the centre of the roll until the 'load loss' potentiometer signals the carriage to change direction.

There is a preset dwell on reversal (End diameter reversing timer) when auto feed is energized. The amount of cut applied is determined by potentiometer 'reversing'. When this level is reached the feed stops and the carriage traverses back towards the centre of the roll. During this pass up to the centre of the roll, wheel wear compensation takes place. Beyond the centre of the system is looking for load loss and, therefore, no wheel wear compensation is in circuit.

This sequence is repeated as often as necessary until the full length of the roll is covered.

Shaping is now complete.

ROLL MATCHING, TYPES OF GAUGING, ETC.

Duplex roll measuring gauge—this is an in-process gauging system.

Because with the latest roll alignment techniques we are physically placing the roll in a definite horizontal plane, i.e., it is in a specific position relative to the machine bed we are able by means of a single point gauge to monitor the barrel diameter of the roll.

The device is used for controlling the final diameter of the roll and to produce a digital read-out and record the roll barrel diameters. It also plays a vital function when automatically matching a pair of rolls.

The gauge system is brought into action on completion of roll shaping and just prior to the rough grind, and consists of a measuring head positioned at the longitudinal centre of the roll.

Electrically there are two transducers on the gauge head assembly (see Fig. 10). A long range transducer on the base slide gives a digital display of roll diameter which can be recorded on a ADDO printer machine. A short range transducer on the top slide

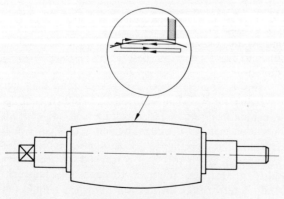

Figure 9. Roll-centre grind.

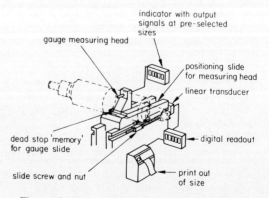

Figure 10. Layout of in-process gauging system.

gives a digital indication of stock to be removed, a series of oversize positions and a zero size datum.

The method of operation is as follows

The gauge head traverses forward on its top slide to a mechanical dead stop position.

A signal is given to index the tape.

The bottom slide (carrying top slide) moves forward at a fast rate after the tape has been indexed to the correct position.

During this movement the long transducer is measuring and indicating diameter (we assume this has already been calibrated) which is reducing as the slide comes in.

The gauge probe makes contact with the roll face—still moving forward fast.

As the gauge probe is depressed the remote set indicator indicates movement of the short range transducer (see Fig. 11). When this figure (on digital readout) coincides with the numbers on the tape a signal is given for the fast rate to change to slow (movement of head forward).

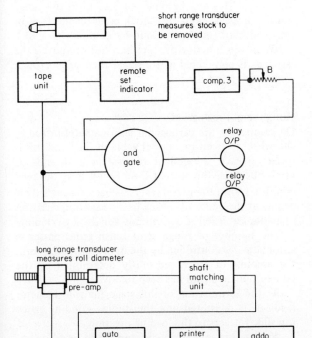

Figure 11. Block diagrams of short and long range transducers.

When slow selected, the tape also indexes bringing a new set of numbers into circuit.

The gauge probe is depressed further as the bottom slide moves forward at a slow rate until a point is reached when the preset stock removal indicated by new numbers on the tape coincide with the reading from the transducer.

The interface relay contacts close—the gauge positioning motor is dynamically braked bringing the gauge head to rest quickly. Print out of size is carried out because the long transducer is now in its final position. This will be the size of the roll when the short transducer is released to its zero position (i.e. when grinding stops).

The gauge head is now in a position to control the actual stock removal (it has been displaced from its zero position by an amount equal to the amount of stock to be removed). This stock removal is carried out during the rough and rough semi finish grind stages.

The machine prepares to rough grind and at the same time the tape indexes to present a number equivalent to the stock to be removed during roughing.

The rough grind passes are carried out. Metal is removed. The transducer probe is gradually released.

A position is reached where the preset rough grind stock removal of the transducer coincides with the preset rough grind stock removal indicated on the tape. (This position is probably 0·005 in above zero size.)

The interface contacts close, rough grind completes and commence rough semi finish grind. Signals are given. The tape indexes to give zero size information.

The gauge probe is further released until a position is reached where the rough semi finish grind stock removal output of the transducer coincides with the preset rough semi finish grind stock removal indicated on the tape (this position is 'zero' on gauge transducer and tape).

The interface contacts close. The rough semi finish grind is complete. At the end of the roll face the semi finish grind commences. The tape is indexed and the next number presented is the number of semi finish grind passes.

At the same time the gauge head is withdrawn on the top slide only. (The bottom slide remains stationary.)

The first roll is completed by counting the number of passes on the semi finish grind (number of passes on the tape coincide with number of passes given by the counter) and also counting the number of passes on the finish grind in the same way. The tape is indexed to the start position.

In ROLL MATCHING a second roll is loaded into the machine and the sequence restarted.

On completion of roll aligning and shaping (end diameter) at the tailstock end the carriage traverses towards workhead end of the roll.

As the roll centre switch is operated, the gauge head traverses forward on its top slide to its mechanical dead stop position. It is now in a position to match the second roll to the first roll.

The tape indexes giving 'end of rough grind 0·005 in above zero'. On completion of the end diameter grind at the workhead end, the rough grind commences and the gauge circuits associated with the stock removal are brought into operation.

The gauge probe is slowly released until the stock removal on the gauge coincides with that on the tape. Rough semi finish grind commences. The tape indexes to give zero size information.

The gauge probe is slowly released until a position is reached where the rough semi finish grind stock removal output of the transducer coincides with the preset rough semi finish grind stock removal indicated on the tape (this position is 'zero' on the gauge transducer and tape).

Figure 12. Roll grinding machine and automatic loader.

Semi finish grind.

The tape indexes to give the number of semi finish grind passes.

The gauge head is withdrawn on the top slide.

The gauge positioning motor moves the bottom slide to the back position at a fast rate and is then dynamically braked bringing the bottom slide to rest quickly.

The in-process gauging system has not completed its cycle of operation.

GRINDING CYCLE

Once the correct roll shape has been achieved it is necessary to remove marks and blemishes from the surface. There are usually five stages of grinding to achieve the required finish after reshaping the roll:

> rough grind
> rough semi finish grind
> semi finish grind
> finish grind
> feedling elimination.

At the completion of the end diameter grind (or roll centre grind) the carriage stops at the roll end for a preset dwell period. The special duplex roll measuring gauge is now automatically positioned at the centre of the roll.

Rough grind

The carriage now traverses the full length of the roll repeatedly and during this process the load on the wheel is kept up to a certain level determined by the 'rough grind' potentiometer which applies an in-feed should the value of current fall below this preset value. Wheel wear compensation is automatic throughout the rough grinding cycle during which the primary function is to remove stock.

Should there be a rapid fall off in load (faulty condition) below a preset value determined by 'rough grind' AMP 2 potentiometer then wheel wear compensation does not take place. This low value of current indicates low roll surface and at this stage of the cycle we do not want the wheel to follow.

The number of rough grind passes depends on the amount of stock to be removed, i.e., until the gauge signal corresponds to that stock code on the tape or punched card (originally done by counters on desk).

When this amount of stock has been removed the last pass is completed and a signal then selects rough semi-finish grind.

Rough semi finish grind

On reversal of the carriage an end diameter infeed to the wheel is given to a preset value of time and this is given on each subsequent reversal at both ends. If after this time the current level is still low, the feed continues until the preset level has been reached. This occurs at the ends only. No wheel wear compensation or minimum level of current is given during grinding.

The number of rough semi finish grind passes is automatically controlled by the in process gauge until the amount of stock previously determined by the tape or punched card has been removed.

At the termination of this stage the roll diameter read-out is registered and the duplex grinding gauge top slide retracts to the loading position.

At the start of the rough semi finish grind operation the carriage speed, feed rates and workspeed are adjusted automatically.

Semi finish grind

Stock removal during the preset number of semi finish and finish grind passes although minimal has been previously established.

The wheelhead is only fed forward at the ends of the roll by a timed amount of infeed preset on a timer potentiometer. In fact the conditions of feed are exactly as those for the rough semi finish grind operation but the preset level of current is at a lower value of amps.

Carriage speed, work speed, cut and feed rates are adjusted automatically.

The carriage traverses continuously between the carriage reverse limits until the number of passes indicated on the tape coincide with the number on the counter.

Finish grind

On completion of the semi finish grind the tape is indexed to give the number of finish grind passes. The counter is reset.

The wheelhead is retracted at a fast rate until the current has been reduced to a value set by FG Amp 2 potentiometer.

On reversal if the wheelhead motor load is below its preset finish value (determined by FG Amp 1 potentiometer) an infeed is applied to the wheel. No wheel wear compensation takes place.

The carriage traverses continuously between the limits until the number of passes indicated on the tape coincide with the number on the counter.

Feedline elimination

The feedline elimination process is designed to remove any trace of feedline which may remain on a roll face on completion of the finish grind stage.

The wheelhead motor current level is maintained between two close limits. The bottom limit is set up on magnetic amplifier 1 and the top limit on magnetic amplifier 2.

The sequence of operation depends upon the shape of roll that is being ground, but in each case the physical disposition of the wheel relative to the work is such that the trailing edge of the wheel is taken away from the roll.

For a parallel roll the workhead steady top block is retracted away from the wheel and then a single pass is made starting from the workhead end and finishing at the tailstock end.

For a *convex roll* the feedline pass is made in two halves. Retracting the workhead steady, a feedline pass is made from the workhead end to the centre of the roll. The wheel then continues to the tailstock end but clear of the roll. The workhead steady is then brought right forward and a second feedline pass is made from the tailstock end to the centre of the roll where the wheel is completely retracted.

For a *concave roll* the feedline pass is again made in two halves. Retracting the workhead steady, the wheel traverses to the centre of the roll but clear of the roll. From the centre the wheel is brought into contact with the roll and a feedline pass is made to the tailstock end. The workhead steady is then brought right forward, the wheel traverses to the centre clear of the roll and a second feedline pass is made from the centre of the workhead end where the wheel is completely retracted.

AUTOMATIC LOADING AND UNLOADING OF THE ROLL IN THE MACHINE

Normally the roll shop electric overhead crane is used for loading rolls into the machine before grinding and unloading rolls out of the machine after grinding.

It is almost impossible to utilize fully the high production potential of a modern, heavy duty roll grinder due to non-availability of the cranes. It was for this reason that the steel industry showed interest in a form of automatic loading mechanism, the controls of which I will now briefly outline.

The device consisting of two main units, the roll transfer carriage and the roll stillage.

Transfer carriage

This is the dynamic section of the loader. It is a structure running on rails and transports the rolls to and from the roll grinder. It is propelled by a two speed motor drive system to the rear wheels.

There are two pick-up units at the ends of the twin lifting jibs operated by a common lifting beam. The jibs are raised and lowered by a system of motor driven (three separate speeds) screw jacks incorporating twin hydraulic balance units to provide optimum load/speed characteristics.

Roll stillage

The stillage consists of a series of fabricated steel mounts on which the rolls are positioned. The type of mount depends on whether chocked or dechocked rolls are to be accommodated.

The automatic loading system is fully integrated into the roll grinder cycle. Normally the stillage would be part or fully loaded with paired rolls from the mill.

The system relies completely on proximity switches for positional control. There are, of course, conventional safety override limit switches where necessary, particularly on the carriage traverse where a direct series connected three-phase limit switch prohibits the fast traverse speed of the carriage beyond a certain point in both directions.

Once loaded into the stillage the number of rolls to be ground is selected on a multiposition selector switch.

On selection of full auto and with all functions in their start position, the 'grinder auto start pushbutton' is depressed.

The loader, starting from its parked position with arm raised and jaws open, accelerates and moves forward at a speed of 17 in/s.

As the loader moves forward along the track the location proximity switches which are positioned at each stillage are operated by a target on the carriage.

Depending on the number of rolls to be ground, that is, as selected, the appropirate stillage proximity switch brings into circuit the traverse proximity switches which are located on the travelling carriage and rely on targets at each stillage to signal them.

As the first traverse proximity passes the target at the selected station the carriage slows down to 1 in/s.

When the second traverse proximity reaches the target the carriage stops and a brake is applied to lock the carriage in position over the selected station.

The arm lowers at top speed, slows down just prior to its lowest position, the lifting jaws close and contact is made with the roll when the arm is raising at slow speed.

When the roll is lifted from its stillage the arm accelerates to a medium (third) speed and then slows down again at the top position.

There are brakes on all motions when de-energized. The carriage moves forward again to a position over the machine where the roll is lowered into position (again using all three speeds). Proximity switches determine the different heights and also the open/close position of the jaws.

The roll is loaded into the machine, the loader moves to its top position with jaws open and then the machine is given the signal to start its own cycle. The loader moves to its parking position.

The loader waits in this position until the machine has finished grinding the roll. The loader then moves forward, picks up the roll from the machine and returns it to its original stillage.

The other roll of the pair is then selected by the loader, loaded into the machine, ground and then returned.

This action is repeated until all the rolls have been ground.

You will appreciate that this is essentially a sequence scheme and multi-bank uniselectros are used to control the sequence.

A series of interlocks (proximity switches) ensure correct orientation of chocked roll in stillage and of roll driving plate on the end of the roll neck.

The roll has to be correctly radially positioned in the machine before unloading can take place. This is achieved by positioning the roll radially via the workhead (again using proximity devices) so that any risk of collision is avoided.

Both the machine and the loader can be controlled manually for setting-up purposes. The loader and stillage areas are completely enclosed by a safety fence with electrically interlocked gates.

There is little doubt that development will continue in this field of automation. In the future we can foresee a bank of roll grinders entirely controlled by direct line to the mill computer. The rolls will be changed in the mill by auto roll changers and the supply of rolls to and from the mill will be by automatic transfer systems. the present automated roll grinders will be part of this overall system.

DISCUSSION

Q. H. E. Hefford, A. P. Jones and Shipman Ltd. Concerning the auto wheel approach mechanism:
(1) What is the fast wheel approach speed?
(2) How close may the wheel be allowed to approach the unground workpiece at the fast approach speed?

A. (1) The fast approach speed of the wheel to the roll is 100 in/min.
(2) The changeover point from fast to slow approach speed takes place at a distance of approximately $\frac{1}{8}$ in between wheel and roll.

DAM
DATA COMMUNICATION MANUFACTURING SYSTEM
OF ADAPTIVE CONTROLLED MACHINING CENTRE

by

H. TAKEYAMA*, T. HONDA*, K. INOUE*, H. SEKIGUCHI*, K. TAKADA*, H. SUZUKI†, M. SATO†,
K. SUZUKI†, K. KOBAYASHI‡, N. YOSHITAKE‡ and K. ICHIJO‡

SUMMARY

A data communication manufacturing system has been developed aiming at both automating to the extreme, routine works of comparatively low value, and centralizing engineering functions, including data, know-how, production planning, optimizing data processing, etc. in an integrated manufacturing system, as a cooperative research work between the Mechanical Engineering Laboratory, Makino Milling Machine Co., Ltd. and Fujitsu Fanuc Ltd.

Communications between the computer and the factory engineer by a typewriter, prior to each lot and during machining is a fundamental feature of this system. The computer calculates the optimum machining conditions based on the information obtained through the dialogue between the engineer, the computer and the filed data, indicates the job specification and the *a priori* optimum machining conditions to the adaptive controlled machining centre, and immediately sends out the actual job result to the factory as a table.

The system model was constructed and experimentally run in February and May 1972, and it was revealed that programmed manufacturing costs and times were within an accuracy of several per cent.

INTRODUCTION

Considerable progress has been made towards the automation of factories in recent years. In order to implement this, however, there are a number of problems to be solved not only at the level of the factory but also the enterprise as a whole.

An image of an on-line automatic production system of the future is illustrated in Fig. 1. A large central computer at the head office of an enterprise is connected in an on-line fashion with the technical department, production control department, purchasing department, sales department, etc. for each factory which is located far from the head office. Moreover, each factory is equipped with a computer at factory level, and the production engineering department of the factory is directly connected with the on-site machine tools. Process control, tool control, machine tool control, and other similar functions inherent to the individual factory are handled by the computer at factory level, and the overall production planning, production control and data bank, etc., which are regarded as common and highly intelligent functions, are handled by the main computer.

The machine tools employed in this type of system are expected to optimize the total system, quickly and accurately by feeding back the work data hierarchically to the computer and at the same time to perform safely and efficiently at all times. For this purpose, adaptive control machine tools are most promising.

One of the objectives of this system which has to be especially stressed in to automatize to the extreme routine works and to centralize highly intelligent functions such as technical know-how, optimizing data processing, production planning, production control, data bank, etc. by means of a data communication system.

System DAM, which is the abbreviation of Data communication Adaptive control machining centre Manufacturing system was jointly developed and experimented by Mechanical Engineering Laboratory, Makino Milling Machine Co., Ltd. and Fujitsu Fanuc Ltd. In this paper the characteristic features of this system is outlined and the result of the model test is described.

SYSTEM INTEGRATION

An experimental model of this system is shown in Fig. 2. A process computer (TOSBAC-3000, 16KW) assumed to be the main computer, is located at the Mechanical Engineering Laboratory and an adaptive

* Mechanical Engineering Laboratory
† Makino Milling Machine Co., Ltd
‡ Fujitsu Fanuc Ltd.

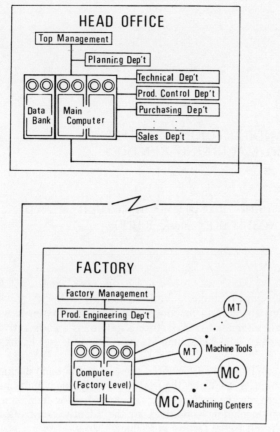

Figure 1. One future vision of automated manufacturing system.

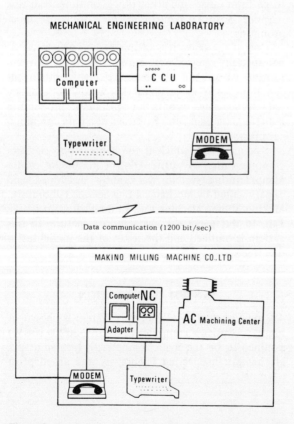

Figure 2. Integration of data communication manufacturing system.

control machining centre model MCC100-A100-AC equipped with FANUC-250A computer NC is located at the Makino Milling Machine Co., Ltd. are connected over a distance of 45 km through the Japan Telegraph and Telephone Public Corporation's limited telephone circuit (toll, 1200 bit/sec).

A monitoring typewriter is connected to the main computer and the data from the computer is sent to the modulator-demodulator (MODEM) through a communication control unit (CCU). The signals from the MODEM at the factory are sent to the NC control unit through an adapter. If a signal is related to the technical data of machining, it is processed to operate the machine, and if it is a conversational mode, it is printed out by the typewriter connected to the NC control unit. The data from the factory, on the other hand, is sent to the main computer through the adapter, telephone circuit, MODEM and CCU to be processed by the main computer.

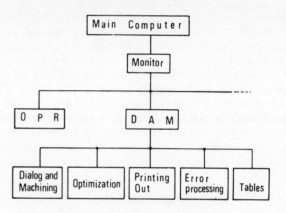

Figure 3. Software of main computer.

The hardware of the main computer (approx. 16KW) is illustrated in Fig. 3. The main program of DAM consists of four subprograms, that is, dialogue and machining, optimizing, printing out and error processing programs and tables under the control of the monitor.

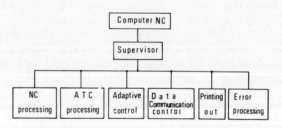

Figure 4. Functions of computer NC.

The adaptive control machining centre in the factory, which is equipped with a tool magazine for 100 tools and an automatic tool changer, is capable of self-determining the feed in reference to the torque limitation and the cutting speed for each tool so that the specified tool life can be satisfied. The computer at the factory level, which is represented by the computer NC, FANUC-250A, has software as shown in Fig. 4. The geometrical processing of parts is performed by NC tapes on the factory side.

FUNCTION OF DAM

Outline

The functional features of DAM are summarized as follows.

(1) The system lays stress on the technological aspects at the shop floor level, which is very complicated and critical, rather than on the managerial or I.E. aspect, which has been handled frequently in various DNC systems.

(2) Highly intelligent functions are centralized at the head office through a data communication system.

(3) The best suited solution in view of the factory situation can be selected through communication between the computer and the factory engineer. This will be very helpful to satisfy capable engineers or operators.

(4) A computerized numerical control (CNC) system is utilized for high flexibility and adaptive control.

(5) Optimization in the case where a variety of tools are utilized is satisfactorily realized.

(6) A sophisticated tool life control is performed.

System flow

The system flow is outlined in Fig. 5. The factory is connected to the main computer by keying in at the factory, and the conversation between the factory and the main computer commences with printing out the data and a message 'Computer is OK'. Keying in DAM, which the factory is ready to use now, the inquiry and answer step QA1 starts with the conver-

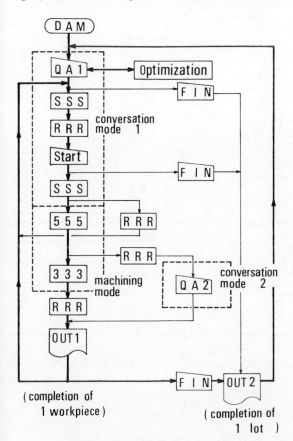

Figure 5. Flow diagram of DAM.

sation on the job specification such as machine number, part number, lot size, etc.

When the answer to these inquiries is correctly keyed in at the factory, the main computer performs an optimizing computation by picking up the necessary data from the machine file, part file, tool file and other files. At this point, six feasible criteria are printed out at the factory. At the same time the number of tools required for the job is printed out, and the factory engineer who knows the actual plant situation or requirements selects the optimum criterion out of the six. When the number of the selected case is keyed in, the manufacturing cost, time, set-up of the tooling required, tool life, and predicted machining time are printed out.

At this time, the tool life can be revised in accordance with the factory situation such as tool inventory, grinding schedule, etc. When the tool life is revised, the tool number and revised tool life are keyed in and the necessary data of cost, time and the like are calculated again until the factory engineer is satisfied. Thus, keying in the final decision, the step QA1 is over.

Now the machining mode starts with 555. Then the numerical controller sends the signals of machine start and the tool number to be used to the main computer. The main computer starts to count the machining time and sends the tool data and the mathematical model in reference to cutting speed and feed corresponding to the optimal tool life to the numerical controller.

The adaptive control machining centre performs machining while self-adapting the cutting speed and feed to the real situation, satisfying the aforementioned mathematical model within the predetermined threshold of cutting torque.

Within the period of tool change, after finishing the process with the tool, the actual cutting time of the tool just used is notified to the main computer. By utilizing this data the remaining life of the used tool is calculated and filed in the tool file. If a tool has reached the end of its life, a signal is sent to the numerical controller to change the tool. Since all these processes are performed during ATC, they do not affect the machining time at all.

When the machining of one workpiece has been finished, the numerical controller sends a signal to notify this to the main computer, and the machining mode finishes. The main computer stops counting the machining time, and it summarizes the machining time and the actual cutting time of the tools used. The result is then printed out on the main computer and factory sides, at the stage of OUT 1 after 333 and RRR.

When a certain tool reaches the end of its life during machining, a signal to change the tool is sent out at the stage of ATC, and the tool table stored in the computer NC is revised. If the same type of tool has to be used again, the same type of spare tool is picked up.

If any trouble occurs with the workpiece or tool during machining an alarm signal is sent out and the machining mode is interrupted by RRR which is actuated by the adaptive control function or by the operator who checks the tool or workpiece.

At QA2, the main computer inquires the state of the tool and workpiece. When the operator answers this inquiry, the main computer calculates the time of machine interruption and prints out the result.

If tool breakage or a machining defect has occurred, the cost of the tool or material waste is added to the machining cost, the result being also printed out at OUT 1. In addition, if a message that machining is incomplete is printed out, the workpiece is distinguished from the completed ones, and it is sent to be reprocessed, if possible.

When one program is finished, FIN is keyed in from the factory typewriter. The main computer then summarizes the whole results for the lot and prints them out in comparison with the predicted cost and time at OUT 2.

This is one cycle of the system, and one example of the system flow is shown below.

Notes	Main computer	Factory
(k): Requested keying in Conversational mode 1 Start of aux. time counting	1972. 2. 13 COMPUTER IS OK.	→ → ← (k) DAM
	MANUFACTURING NO. 12345678 1. MACHINE NO.?	→ → ← (k) MCC100-A100 -AC-1
	2. PART NO.?	→ ← (k) BMC5-1059-A
	3. PREPARATORY TIME?	← (k) 20
	4. LOT SIZE?	→ ← (k) 35
	5. COST LIMITATION? (YEN/PIECE)	→ ← (k) 5200
	6. TIME LIMITATION (HOUR/LOT)	→ ← (k) 30
	* CHANGE LINE NO.?	→ ← (k) 4
	4. LOT SIZE?	→ ← (k) 31
	* CHANGE LINE NO.?	→ ← (k) NOTHING

$$ DAM 1972. 2. 15. MCC100-A100-AC-1 BMC5-1059-A→

CASE	COST YEN/PIECE	TIME HOUR/LOT
1	3900.0	1.02
2	3950.0	0.97
3	---- -	- --
-	---- -	- --
6	---- -	- --

TOOL CODE NO.	CASE 1 2 3 4 5 6
T0404	1 1 1 1 1 1
T1124	2 2 2 2 2 1
-----	- - - - - -

	Main computer	Factory
C: Change	* WHICH CASE?	→ ← (k) 1 or 2 or ... 6 or C

CUT: Net machining time	TOOL NO.	LIFE MIN	CUT MIN/LOT	
	T0404	120	58	→
	T1125	60	60	
	T1125	60	26	
	T5146	35	17	
	T----	--	--	
	- ----	--	--	

	Main computer	Factory
	* ANY TOOL LIFE CHANGE?	→ ← (k) YES or NO
	* LINE NO.?	→ ← (k) 4
	TOOL LIFE? T5146	→ ← (k) 50
	LINE NO.?	→ ← (k) NO

$$ DAM 1972. 2. 13. MCC100-A100-AC-1 BMC5-1059-A →
LOT SIZE 31
MANUFACTURING TIME 28 HOUR/LOT (56 MIN/PIECE)
MANUFACTURING COST 5010 YEN/PIECE (155310 YEN/LOT)

```
TOOL NO.   LIFE   CUT
T0404      120    58
T1125      60     60
T1125      60     26
T5146      50     23
T----      --     --
-----      --     --
```
                                                    ← SSS (START)
                                                    ← RRR (RESET)
                                                    ← SSS (START)
Machining mode   ← 555 (M50)
Start of machining
and counting net
machining time   ← #04040000
                    T-Data, J-Data, 0   →
1125: TOOL NO.   ← #11250056
0056: Actual cutting
      time of T0404
                    T-Data, J-Data, 1   →
Completion of 1st   ← 333 (M30)
workpiece   ← RRR (RESET)
OUT 1 $$ DAM BMC5-1059-A   →
                       PIECE NO.   1
                       MACHINING TIME 63 MIN
                       MACHINING COST 4358 YEN
2: The 2nd spare   CHANGED TOOL T1125-2
   tool
                                                     ← SSS
Machining mode   ← 555
                                                     ← #04040000
                    T-Data, J-Data, 0   →
                                                     ← RRR
                                                     ← #0039
Conversational   TOOL OK?   →
mode   ← (k) YES or <u>NO</u>
                    WORK OK?   →
                                                     ← (k) <u>YES</u> or NO
                    $$ DAM BMC5-1059-A   →
                       PIECE NO.   2
                       MACHINING TIME   50 MIN
                       MACHINING COST   2251 YEN
                       UNFINISHED REPROCESS
                       CHANGED TOOL T3792-2
                                                     ← SSS
                                                     ← 555
Start of machining
and counting net
machining time
                                                     ← #37920000
                    T-Data, J-Data, 0   →
Completion of   ← 333
2nd workpiece
Conversational   ← RRR
mode
                    $$ DAM BMC-1059-A   →
                       PIECE NO.   3
                       MACHINING TIME   33 MIN
                       MACHINING COST   3532 YEN
                                                     ← SSS
Machining mode   ← 555
                                                     ← #04040000
                    T-Data, J-Data, 0   →
                                                     ← RRR
                                                     ← #0072
Conversational   TOOL OK?   →
mode   ← (k) YES or <u>NO</u>
                    WORK OK?   →
                                                     ← (k) YES or <u>NO</u>
                    $$ DAM BMC5-1059-A   →
                       PIECE NO.   27
                       MACHINING TIME   17 MIN
                       MACHINING COST   2528 YEN
                       UNFINISHED
                                                     ← SSS
Machining mode   ← 555
                                                         .
                                                         .
                                                         .
```

E
```

Conversational mode   ← (k) FIN

Completion of 1st lot

$$ DAM 1972. 2. 13. MCC100-A100-AC-1 BMC5-1059-A →
MANUFACTURING NO. 12345678
PRESCRIBED
LOT SIZE   31
MANUFACTURING TIME 28 HOUR/LOT
MANUFACTURING COST 5010 YEN/PIECE
MACHINING

PIECE NO.	MACHI. TIME	MACHI. COST	NOTE
1	63 MIN	4358 YEN	
2	50	2251	UNFINISHED
3	33	3532	REPROCESS
4	..	....	
.	..	....	
27	17	2528	UNFINISHED
..	..	....	
32	53	4628	

TOTAL	1728 MIN	156241 YEN	UNFINISHED .. 2

AUX. OPERATION

AUX. TIME	27 MIN
AUX. COST	1057 YEN

MANUFACTURING

MANUFACTURING TIME	29.3	HOUR/LOT
MANUFACTURING COST	157298	YEN/LOT
FINISHED LOT SIZE	30	
AVE. TIME	58.5	MIN/PIECE
AVE. COST	5244	YEN/PIECE
1972. 2. 13.	→	

Conversational mode 1   COMPUTER IS OK.   →

PROCEDURE OF OPTIMIZATION

The performance indexes for the six alternative criteria mentioned in the previous section are taken as the manufacturing cost per workpiece C (¥/piece) and time per lot t_w (h/lot), which are calculated by the following equations, respectively.

$$C = M \left[\frac{t_{prep}}{N_L} + t_e + N_T t_{ATC} + \sum^{N_T} \left(\frac{L_P}{P} + \frac{L_r}{F_r} \right) \right] + M_c \frac{t_c}{N_L}$$

$$+ \sum^{N_T} \left[\frac{D_c L_c}{318 f_{ave} Z} \left\{ \frac{M}{V_{ave}} + \frac{f_{ave}^{m/n} V_{ave}^{1/n-1}}{K_c^{1/n} \zeta} \right. \right.$$

$$\left. \left. \times \left(\frac{C_c}{k_1 + 1} + G t_g \right) \right\} \right] \tag{1}$$

$$60 t_w = t_{prep} + N_L \left[t_e + N_T t_{ATC} + \sum^{N_T} \left\{ \frac{L_p}{p} + \frac{L_r}{F_r} \right\} \right.$$

$$\left. + \sum^{N_T} \frac{D_c L_c}{318 f_{ave} Z V_{ave}} \right] \tag{2}$$

where

M	labour cost + overhead (¥/min)
M_c	computer cost (¥/min)
C_c	tool cost (¥)
G	tool grinding cost (¥/min)
t_{prep}	preparation time (min)
t_e	work loading time (min)
t_{ATC}	automatic tool changing time (min)
t_g	tool grinding time (min)
t_c	computer time (min)
N_L	lot size
N_T	number of tool
k_1	number of tool grinding times
L_p	positioning length (mm)

L_r	adaptive-approaching length (mm)
L_c	machining travel (mm)
p	positioning rate (mm/min)
F_r	adaptive-approaching rate (mm/min)
D_c	cutter diameter (mm)
Z	number of teeth
ζ	peripheral length of tooth engagement/πD_c
f_{ave}	average feed (mm/tooth)
V_{ave}	average cutting speed (m/min)
T	tool life (min)
m, n and K_c	constants of tool life equation $f_{ave}^m \cdot V_{ave} \cdot T^n = K_c$

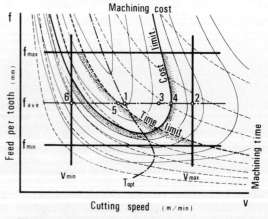

Figure 6. Optimization with regard to cost and time.

Reducing the cost and time to those of each tool, they can be represented as illustrated in Fig. 6 in reference to feed and cutting speed, in which the solid and dotted lines represent equi-cost and equi-time curves, respectively. The permissible area in view of cutting speed and feed is within the constraints denoted by straight, thick solid lines f_{max}, f_{min}, V_{max} and V_{min}. The cost and time limitations which are specified by the factory are shown by the curved, thick solid line and curved, thick dotted line, respectively.

The average feed f_{ave} is picked up from the data file on the main computer. The f_{ave} is the feed permissible in view of the torque constraint for the tool and the average stock allowance given by the part programming data, that is,

$$f_{ave} = \text{Const} . (\text{Torque})_{max}/$$

$$\{(\text{Depth of cut})_{ave} . (\text{Width of cut})_{ave}\}$$

On the line of f_{ave} six sets of optimal solutions such as cutting speed and tool life are calculated by the main computer from the standpoints of six criteria, that is, (1) minimum machining cost, (2) minimum machining time, (3) compromised point between (1) and (2), (4) minimum machining time within the cost limitation, (5) machining time limitation, and (6) maximum machining time.

The optimum cutting speed and tool life are determined for each tool, and then the total machining cost or time, which is the sum with the whole number of tools used for one workpiece and under the above respective cutting conditions, is assessed.

Taking up the case 'minimum machining cost', for example, the optimum cutting speed for each tool $V_{opt,it}$ can be obtained by putting $dC/dV_{ave} = 0$. Therefore,

$$V_{opt,it} = \left\{ M \frac{n}{1-n} . \frac{K_l^{1/n} . \zeta}{f_{ave}^{m/n} \left(\frac{C_c}{k_1 + 1} + G . t_g \right)} \right\}^n \quad (3)$$

If $V_{opt,it}$ is outside the constraints $V_{max,it}$ and $V_{min,it}$ the finally selected speed will be set to be equal to one of the constraints.

In the case (2), $V_{max,it}$ which is stored in the filing table is utilized for each tool to give the minimum machining time. The case (6) is the reverse of the case (2).

In the case (4), starting with the data obtained in the case (1) the cutting conditions are modified so that the cost allowance when comparing between the cost calculated in the case (1) and the cost limitation can be balanced and distributed to each tool in proportion to the machining cost with the respective tools, and the cost limitation can be satisfied as a whole.

Making the final decision on the choice out of the six criteria, the mathematical model in reference to the cutting speed and feed, is transferred to the numerical controller. The adaptive control machining centre performs machining, while self-adapting the cutting speed and feed to the real situation and satisfying the equation.

TEST RESULT

In order to evaluate the feasibility of the system a test run was performed while machining test parts. The work material is Meehanite cast iron, and the dimension is 144 x 150 x 140 mm. The tools used are carbide end mills (40 mm ∅, 4 mm ∅), face milling cutters (5 in ∅) for rough cut and finish cut, respectively, boring bar (60 mm ∅), drills (50 mm ∅, 6·8 mm ∅), centre drill (19 mm ∅), tap (M8-1·25) and bevel cutter (20 mm ∅). The machining time of one workpiece is roughly 15 minutes although it depends upon the performance criteria and the conditions of tool and work material.

The summarized result for the case (1) 'minimum machining cost' is shown in Fig. 7 when the lot size is two.

As seen in the figure, the actual performance aiming at 'minimum cost' is very close to the predicted result.

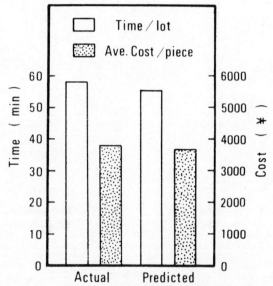

Figure 7. Comparison of predicted and actual performances.

CONCLUSIONS

A technology-oriented data communication system, in which highly intelligent functions are centralized, has been implemented and experimented in the light of industrial needs. A considerable amount of information and know-how related to the data communication control has been obtained through this experiment. Moreover, an approach for optimization when a variety of tools are used has been demonstrated.

ACKNOWLEDGMENTS

The authors thank Mr Y. Imai in Makino Milling Machine Co., Ltd. and Dr S. Inaba in Fujitsu Fanuc Ltd. for their guidance and cooperation.

AUTOMATIC, COMPUTERIZED OPTIMIZATION OF MULTI-SPINDLE DRILLING WITH PROBABILISTIC TOOL LIFE

by

J. L. BATRA* and M. M. BARASH**

SUMMARY

An algorithm is presented for the planning and optimization of multiple drilling operations performed simultaneously on a multi-spindle drilling machine. The mathematical relationships are formulated and solved using the techniques of geometric programming and Monte Carlo simulation. The optimal cutting parameters and a suitable tool change scheme are determined simultaneously and automatically with the aid of a digital computer. A package of computer programs has been prepared which permits the optimization of machining operations according to one of the following criteria: minimum production cost, maximum production rate, or minimum production cost for a stipulated production. Results are obtained for both deterministic and probabilistic tool life. In the latter case each tool may be considered as obeying one of the following life distributions: normal, lognormal, Weibull, or gamma.

Analysis of the information generated about the optimal selection of machining parameters and tool change scheme indicates that the optimum cutting speed for the 'probabilistic case' is either lower than that for the 'deterministic case' or equal to it. Further, it is shown that the current arbitrary practices of reducing speed or terminating tool life before the computed value is attained are unsatisfactory and often lead to economic waste.

INTRODUCTION

Optimization in machining is achieved by the selection of the correct machining parameters for either minimum cost, maximum production, or maximum profit. Whatever the criterion, the fact remains that the variables influencing tool life represent a major subgroup in the overall system to be optimized. Tool life variables affect the determination of both costs and production rates of machining operations for which about 50 billion dollars are spent yearly in the United States alone.

In order to establish a suitable basis for calculating the cutting parameters required by the planning engineers, well known empirical formulae[1-8] are extensively used. The analyses are strictly deterministic. Such formulae, defined specifically for particular cutting conditions, assume different terms in the various expressions derived, and the parameters calculated from them differ essentially from practical experience. Such discrepancies occur because a rigid functional relationship between the cutting parameters is provided in the formulae, whereas in reality the relation between them has a probabilistic character. It is well known to anyone connected with machining that the actual tool life very rarely coincides with the predicted value. The life of individual (even similar) tools, due to instability of their prop-

erties, heterogeneity of the workpiece material and several other factors, is of random magnitude, often with wide scatter. At present, two general practices are being followed in the industry for the selection of cutting parameters for a given workpiece-tool combination. One is to use the recommended values given in the standard metal-cutting handbooks in the form of tables or nomograms, and the other is to determine optimum cutting parameters using either cost, production time, or profit models, generally ignoring the effect of tool life variability. To ensure that very few of the tools which fail prematurely have tool lives shorter than the optimum value calculated, the general practice is to increase the average value of the tool life by an arbitrary factor and then determine the corresponding cutting speed. A similar suggestion was made by Radford and Richardson[9] in their recent article on the optimization of parameters in turning with a single tool. A thorough search of the literature concerned with the optimization of machining operations shows only one instance where the investigators considered tool life to be probabilistic in nature and accounted for this fact in determining optimum machining parameters. Berra and Barash[10,11] have tackled the problem by imposing a probabilistic constraint on the predicted tool life. The predicted tool life is treated as a probabilistic value and the lower confidence limit of the

* Assistant Professor, Indian Institute of Technology, Kanpur, India.
** Professor, Purdue University, W. Lafayette, Indiana 47907, U.S.A.

tool life at a desired probability level is used as the tool life instead of its mean value. It should be pointed out that this analysis relates to simple, single tool machining operations only.

In a multi-spindle drilling machine when more than one drill is cutting simultaneously and the drills have a probabilistic tool life, it is an involved problem to find at what speeds and feeds to operate the multi-tool assembly.

Some authors[12,13] have suggested that the cutting conditions for the multi-tool assembly should be selected on the basis of the critical tool, i.e. the tool which has the shortest tool life. But in actual practice a tool with long mean life and larger scatter may often cause failure of an assembly, rather than a tool with short mean life and small scatter. It is a well-known fact that the unpredicted failure of a single tool in a multi-tool assembly, or in a large transfer line, can be very costly. Thus it is necessary to establish a tool replacement scheme for limiting the percentage of such failures to an acceptable value. The availability of large computers and developments in optimization theory have encouraged the authors to develop a technique and a computer program for the planning and optimization of multiple drilling operations performed simultaneously on a multi-spindle drilling machine.

Nature of variation of drill life

The type of the distribution which fits the drill failure data closely depends on the drill size, drill material, cutting conditions, and the selected criterion of drill life. The literature indicates that the following failure distributions have been successfully fitted to the drill life data by various investigators.

(1) Normal distribution[14-17]
(2) Lognormal distribution[14,18]
(3) Weibull distribution[18,19]
(4) Gamma distribution[18]
(5) Alpha distribution[20,21]

Of all the distributions listed, the normal distribution is the easiest to work with from the computational standpoint. When the tool life distribution indicates high positive skewness—lognormal, Weibull, gamma and alpha distributions fit the drill life data more closely. Again from computational point of view, lognormal and alpha distributions are easier to handle and so should be used whenever possible. Weibull and gamma offer the advantage that they are two-parameter distributions and can be adjusted to cover a wide variety of shapes. The numerical evaluation of distribution parameters and random deviations is easier for the Weibull distribution than for the gamma distribution. Consequently, when the empirical distribution function is accepted equally well by either a gamma or a Weibull, the Weibull fit is preferable.

MATHEMATICAL MODELLING FOR THE OPTIMIZATION OF MULTI-SPINDLE DRILLING MACHINES WITH PROBABILISTIC DRILL LIFE

As pointed out earlier the optimization in machining involves the selection of correct machining parameters for either minimum cost, or maximum pro-

duction, or maximum profit rate. Armarego and Brown[12] and Hitomi[22] have compared the three criteria. They found that for high feeds, the maximum profit rate conditions lie somewhere in-between the conditions found by using minimum cost and maximum production rate criteria. Although the maximum profit rate criterion is most desirable to use, it suffers from the following disadvantages:

(1) The analysis is complex.
(2) Some knowledge of the selling price of the product is needed, and in case of multiple operations, the allocation of selling price to each operation is required.

Because of the above-mentioned difficulties, the general practice is to select the cutting parameters based on either minimum cost of production or maximum production rate or minimum production cost with production rate restriction (stipulation). Mathematical formulation of the problem involving multiple machining operations performed simultaneously on a production type multi-spindle drilling machine is presented with the following assumptions:

(1) The spindles of the multi-spindle drilling machine may be rotated simultaneously at different speeds when the drill feed for all the drills is restricted by the rate at which the drill head is fed. Further, each spindle has no more than one tool.
(2) A tool life relationship of the form $VF^{n}f^{m} = C$ [Taylor Tool Life Relationship[7]] only applies for the mean tool life of the drill. In this relationship 'V' is the cutting speed, 'T' the mean tool life, 'f' the feed rate, 'C' a constant, and 'n' and 'm' are the exponents. This assumption is based on the experimental results of Sinpurwala and Kuebler[14].

Planning and optimization of multiple spindle drilling machines with probabilistic drill life

Consider the machining of a component with 'J' tools working simultaneously on a multi-spindle drilling machine.

H = the total distance travelled by the drill head to machine the component (inches)

L_i = length of the ith machining operation (inches) $(i = 1, 2, \ldots, J)$

$L_k = (L_i)_{max}$

Then

$$H = L_k + \Delta_k$$

where Δ_k is the distance between the lower tip of the kth tool and the starting point of the machining operation performed by the kth tool.

The total time needed to perform the machining operations is given by the following equation.

$$T_p = T_a + T_e + T_t + \sum_{i=1}^{J} T_{di} \left(\frac{T_{ci}}{T_i} \right) \qquad (1)$$

where

T_p = production time per part, minutes
T_a = sum of the tool approach time and actual machining time of a tool in the setup, minutes

T_e = sum of load, unload and setup time per part, minutes

T_t = retract time for the drill head, minutes

T_{di} = tool changing time for the ith tool, minutes

T_{ci} = cutting time for the ith tool, minutes

T_i = tool life of the ith tool, minutes.

It should be noted that the tool change time is the sum of the following items

machine start and stop time, t_1;
average tool load and unload time, t_2;
average time to search for the defective tool, t_3;
average tool procurement time, t_4.

The production time may be expressed in terms of the cutting parameters through the following relationships

$$T_t = H/R_{tr} \tag{2}$$

$$T_{ci} = \frac{L_i}{H_f} \tag{3}$$

$$\bar{T}_i = (C_i)^{1/n_i}\left(\frac{\pi D_i N_i}{12}\right)^{-1/n_i}(f_i)^{-m_i/n_i} \tag{4}$$

$$T_a = \frac{H}{H_f} \tag{5}$$

L_i = length of the ith machining operation, inch

N_i = spindle speed of the ith spindle, rev/min

f_i = feed rate of the ith tool, inch/rev.

$\bar{T}_i$ = mean tool life of the ith tool, minutes

C_i = Taylor constant for ith tool

D_i = drill diameter for the ith tool, inch

n_i = tool life exponent (Taylor tool life equation) for ith tool

m_i = feed exponent (Taylor tool life equation) for ith tool

H_f = feed rate for the spindle head, inch/min

R_{tr} = rapid traverse rate for the drill head, inch/min

For a multi-spindle drilling machine in which the drills are fed at a rate corresponding to the feed rate of the drill head, it is necessary that the following relationship be satisfied:

$$f_i = \frac{H_f}{N_i}; \quad \text{for all values of } i \tag{6}$$

Substituting equation (6) into equation (4), we have

$$\bar{T}_i = (C_i)^{1/n_i}\left(\frac{\pi D_i}{12}\right)^{1/n_i}N_i^{(m_i-1)/n_i}H_f^{-m_i/n_i} \tag{7}$$

The variable cost of manufacturing one component (C_p) is given by the following equation:

$$C_p = X(T_p) + \sum_{i=1}^{J}\left[Y_i\left(\frac{T_{ci}}{T_i}\right)\right] \tag{8}$$

X = operating cost per minute of machine and operator

Y_i = cost of ith tool per cutting edge.

Depending upon the desired optimization criterion, either equation (1) or equation (8) is to be optimized subject to the various technological constraints imposed by the machine-tool and the workpiece. For the case when the desired optimization

criterion is the minimum cost for a stipulated production rate, there will be an additional production rate constraint.

Metal cutting constraints

1. *Production rate constraint.*
If T_{pd} is the permissible production time on the basis of production rate stipulation, then

$$T_p \leqq T_{pd} \tag{9}$$

2. *Horsepower constraint*
Kronenberg[26] suggests the following relationship for cutting force, P_c, for drilling

$$P_c = C_p(f)^a(\text{BHN})^b(D)^c$$

where C_p, a, b and c are constants and BHN is the Brinell Hardness Number.

The total horsepower required by the drills (HP$_t$) can be calculated from

$$\text{HP}_t = \sum_{i=1}^{J}\frac{P_{ci}V_i}{33000}$$

where V_i is the cutting speed in ft/min.

If HP$_{max}$ and η represent the maximum horsepower and efficiency of the machine tool, then

$$\text{HP}_t \leqq \eta(\text{HP}_{max}) \tag{10}$$

3. *Cutting speed constraint.*
The Taylor tool life relationship of the form given in equation (3) is generally valid for a range of cutting speeds called the 'Taylorized speed range'. Let V_{itmax} and V_{itmin} represent the Taylorized speed range for the ith tool of diameter D_i, then

$$N_{itmax} = \frac{12V_{itmax}}{\pi D_i}$$

$$N_{itmin} = \frac{12V_{itmin}}{\pi D_i}$$

where N_{itmax} and N_{itmin} are the maximum and minimum values of the Taylorized spindle speed for the ith tool.

The optimal cutting speed selected for each tool should be within the Taylorized speed range for the tool and the speed range available for the spindle which rotates the tool. Therefore

$$N_i \leqq (N_{imax}, N_{itmax})_{min} \quad \text{for all values of } i \tag{11}$$

$$N_i \geqq (N_{imin}, N_{itmin})_{max} \quad \text{for all values of } i \tag{12}$$

4. *Feed constraint.*
The maximum feed with which a drill can be fed is restricted by its maximum permissible shear stress. Bhatacharya and Ham[29] suggest that for standard geometry drills, the maximum permissible feed rate (f_{max}) is given by the relationship

$$f_{max} = C_s K_l D^{0.6}$$

where

C_s = a constant depending upon tool–workpiece material combination

K_l = a ratio of drill length to drill diameter.

Therefore

$$f_i \leqq f_{imax} \quad \text{for all values of } i.$$

Since the selected feed rate of the drill head (H_f) should be within the drill head feed range available on the machine, H_f should satisfy the following constraints.

$$H_f \leqq [H_{fmax}, (N_{imax}f_{imax})_{max}]_{min} \tag{13}$$

$$H_f > H_{fmin} \tag{14}$$

where H_{fmax} and H_{fmin} are the maximum and minimum drill head feeds for the machine tool.

The problem of determining optimum cutting parameters involves the minimization of an objective function represented by either equation (1) or equation (8) subject to various constraints represented by equations (10) to (14). The constraint represented by equation (9) is included only when the desired optimization criterion is minimum production cost with production rate stipulation. For a setup consisting of 'J' tools working simultaneously, the optimization involves the determination of ($J + 1$) variables considering that one of the quantities in the objective function, i.e. tool life, has probabilistic characteristics and is a function of the variables to be evaluated.

Literature on optimization techniques studied has indicated that, at present, there is no efficient mathematical technique for solving problems of this type. Moreover, even if a solution to the problem as formulated above could be obtained, it would be based on the fact that the tools are allowed to fail at random. Since the unpredicted failure of the tools can be very expensive, it is desirable that the failure rate is kept below a prescribed value. The random tool failures can be kept below a preselected value by adopting a suitable tool change program. Therefore, the problem of optimization of multi-spindle drilling machines involves the determination of optimal cutting parameters and the generation of an optimal tool change scheme, both solutions being obtained simultaneously. It needs to be pointed out that for any given set of cutting conditions, one can evaluate an optimal tool change scheme. This would mean that the optimization of multi-tool assembly would require:

(1) the determination of optimal tool change schemes for all the speeds and feeds available on the machine tool.
(2) the evaluation of an overall optimum condition for the optimal solutions found in (1).

Such a solution would involve an enormous amount of computing time. Therefore, it is necessary to have an approximate solution for machining parameters before analyzing the various tool change schemes.

If the tool life is assumed to be deterministic despite its probabilistic nature, and mean values of tool lives for various tools in the set up are substituted in equations (1) and (7), the problem reduces to optimizing a nonlinear objective function with respect to both linear and nonlinear constraints. The solution of the deterministic nonlinear economic model can serve as an approximate solution for the determination of optimal tool change scheme when the tool life is considered probabilistic.

TOOL CHANGE SCHEMES

For the selection of an optimal tool change scheme, the following policies are compared:

(1) Series Tool Change Scheme. In this scheme only the tool which fails is replaced.
(2) Parallel Tool Change Scheme. If one of the tools in set up fails, all the tools are replaced.
(3) Scheduled Tool Change Scheme I. All tools are replaced at fixed intervals. The tools which failed during the interval and were then replaced, are also finally replaced.
(4) Scheduled Tool Change Scheme II. All tools are replaced at fixed intervals except those tools which failed during the interval and were replaced.

It should be borne in mind that a policy of replacing the tool as it fails cannot be implemented literally for obvious reasons. The tools have to be removed prior to imminent failure. Unless suitable sensors for sensing the imminent failures of the tools are used, tool change policies 3 and 4 should be preferred. In tool change policies 3 and 4, an optimum value for the tool change interval is determined by keeping the ratio of unscheduled tool changes to the total number of tool changes below an acceptable value.

The literature survey reveals that at present there is no mathematical relationship for evaluating the mean life of a multi-tool assembly when the tools are replaced at a predetermined tool change interval. Moreover, the removal of tools prior to imminent failure poses additional problems for the mathematical formulation for mean assembly life in all the four tool change policies. Keeping these in view, a simulation by Monte Carlo method was used for determining the

(a) most economical tool change scheme
(b) optimal tool change interval.

The cost and production time models of the various tool change schemes were developed for use in the simulation runs. These models are presented later in this paper.

OPTIMIZATION TECHNIQUE FOR DETERMINISTIC ECONOMIC MODEL

There exist several techniques which can be employed to solve an optimization problem involving a nonlinear objective function, a set of linear constraints, a set of convex separable inequality constraints, and nonnegative variables. These include Lagrange's Method, Fiacco-McCormick's Sequential Unconstrained Minimization Technique (SUMT), Rosen's Gradient-Projection Method, Zontendijk's Method of Feasible Directions, and Zener-Duffin's Geometric Programming. However, either nonlinearities in the constraints or an objective function of more than second degree introduce serious problems when determining an optimal solution. In such cases, the most efficient method appears to be the recently developed technique called geometric programming. Zener and Duffin called their method Geometric Programming, since it is based on a generalization of the arithmetic–geometric mean inequality. The basic theory and formal proofs can be found in refs. 23, 24.

Reklaitis[25] developed a Differential Algorithm for Posynomial Programs of non-zero degrees of difficulty via a transformed auxiliary problem using a modified version of the Differential Algorithm of Wilde and Beightler[24]. For determining an optimal solution of the multi-tool economic model, a computer program based on Reklaitis' algorithm is used. Since the details of Reklaitis' algorithm are beyond the scope of this paper, the readers are referred to ref. 25 for the complete information.

As stated earlier, the solution of the deterministic nonlinear economic model serves as an input for the determination of optimal tool change scheme when the tool life is considered probabilistic. On the basis of the values of Taylor tool life exponents reported by Kronenberg[26], one concludes that an increase in cutting speed (V) is far more detrimental (three times or more) to the tool life than an equivalent increase in the feed rate (f). Therefore, the optimum feed rate values obtained from the solution of deterministic nonlinear economic model can be assumed to be very close to the optimum feed rates for the determination of the optimum tool change scheme. For all practical purposes, the feed can be maintained at the same value as that obtained for the deterministic case, and the simulation of the probabilistic behaviour performed on speed values only.

COST AND PRODUCTION TIME MODELS FOR VARIOUS TOOL CHANGE SCHEMES

The optimum cutting parameters obtained for the deterministic case serve as an input for the selection of an optimum tool change scheme. The optimum values of the various spindle speeds are determined by using the Golden Section Search[24] technique. The cost and production time models for the various tool change schemes are developed in terms of the 'shop parameters', i.e. *spindle speeds* and the *number of parts* produced by the tool before it is 'worn' out. For simulation purposes, the use of 'shop parameters' instead of cutting speed and tool life offers the following advantages:

(1) If the 'shop parameters' are used, the tool life will be expressed not in duration of time but in the number of parts which will be produced during the life of the tool. If that number is a non-integer it can be rounded off to an integer, thus producing a meaningful solution.
(2) The simulation of tool life expressed in time does not provide an immediate answer as to the total number of parts to be produced. However, if the number of parts to be produced is stipulated in advance, the use of shop parameters will terminate the program when the required number has been simulated and this will result in somewhat better optimum conditions.

For a fixed value of feed rate f_i, the Taylor's modified tool life relationship can be expressed as

$$N_i P_i^{A_i} = B_i \quad \text{for all values of } i \quad (15)$$

where

N_i = spindle speed of the ith spindle in rev/min

P_i = number of parts produced by the tool on the ith spindle

A_i and B_i = constants for the tool on the ith spindle

A derivation for the above relationship is given in the Appendix. Referring to the Appendix, we obtain

$$A_i = \frac{n_i}{1 - n_i}$$

$$B_i = \left(\frac{12C_i}{\pi D_i}\right)^{1/1 - n_i} (f_i)^{m_i - n_i/n_i - 1} (L_i)^{n_i/n_i - 1}$$

Therefore, from the various parameters of the Taylor's modified tool life relationship, one can develop an expression in the form shown in equation (15).

The cost and production time models for the various tool change schemes are given in the following paragraphs:

1. Series tool change scheme

The production time and production cost per part are given by the following equations

$$T_p = T_a + T_e + T_t + \frac{1}{P_s} \sum_{i=1}^{J} T_{di} N_{tci} \quad (16)$$

$$C_p = X(T_p) + \frac{1}{P_s} \sum_{i=1}^{J} Y_i N_{ti} \quad (17)$$

where

P_s = simulation run size, number of parts. In case of limited production, it would represent the stipulated production size.

N_{tci} = unscheduled tool changes performed on the ith spindle

N_{ti} = number of tools required on the ith spindle for manufacturing P_s parts.

2. Parallel tool change scheme

The equations for production time and production cost per part for the parallel tool change scheme are similar to those for the series tool change scheme.

3. Scheduled tool change scheme 1

The production time and production cost per part for the scheduled tool change scheme (all the tools are replaced at fixed intervals) are given by the following relationships:

$$T_p = T_a + T_e + T_t + \frac{1}{P_s}\left[\sum_{i=1}^{J} T_{di} N_{tci} + N_{stc} K_s \sum_{i=1}^{J} T_{di}\right] \quad (18)$$

$$C_p = X(T_p) + \frac{1}{P_s}\left[\sum_{i=1}^{J} \{N_{stc} + N_{tci}\} Y_i\right] \quad (19)$$

where

N_{tci} = unscheduled tool changes of the ith spindle
N_{stc} = number of scheduled tool changes
K_s = a factor by which the sum of tool change times for all the tools is reduced due to the

fact that all the tools are being replaced at the same time

N_{stc} = number of scheduled tool changes required to manufacture P_s parts

If z represents the value of the predetermined tool change interval, then

$$N_{stc} = \frac{P_s}{z}. \qquad (20)$$

The value of factor K_s is determined by examining the various time elements which comprise the tool change. A listing of these time elements is given earlier.

4. Scheduled tool change scheme 2

In this scheme, those tools which were not replaced during the preceding interval, are replaced at the fixed interval. Therefore, it is necessary to:

(a) determine a new value of the factor K_s at each tool change interval and

(b) keep an account of the various tools which were replaced before the predetermined tool change interval.

The production time per part and the production cost per part are estimated by

$$T_p = T_a + T_e + T_t + \frac{1}{P_s}\left[\sum_{i=1}^{J} T_{di}(N_{tci})\right.$$

$$\left. + \sum_{p=1}^{N_{stc}} K_{sp} \sum_{i=1}^{J} \delta_i T_{di}\right] \qquad (21)$$

where

$$\delta_i = \begin{cases} 0 & \text{when tool } i \text{ is not replaced for scheduled tool change number '}p\text{'} \\ 1 & \text{when tool } i \text{ is replaced for scheduled tool change number '}p\text{'} \end{cases}$$

K_{sp} = a factor by which the sum of tool change times for all the tools replaced at the pth scheduled tool change interval is reduced to account for the fact that the tools are being replaced simultaneously.

$$C_p = X(T_p) + \frac{1}{P_s}\left[\sum_{i=1}^{J} (N_{stc} + N_{tci})Y_i\right] \qquad (22)$$

GENERAL METHODOLOGY FOR THE SOLUTION OF AN OPTIMAL TOOL CHANGE SCHEME

The selection of an optimal tool change scheme in conjunction with the determination of optimum cutting parameters is carried out with the following procedure:

(1) Using the optimal cutting parameters for the deterministic case, the values of A_i and B_i for the relationship used in equation (15) are calculated.

(2) The tool with the lowest deterministic cutting speed is considered as the controlling tool which is held in the controlling spindle.

(3) Preliminary studies involving several tens of computer simulation runs for minimum production cost and maximum production rate restriction, indicate that the optimal deterministic speed can be treated as the upper bound of the speeds examined for the probabilistic case. This is well in line with what could be expected from mathematical considerations.

(4) Lower limits on the speeds to be examined when the controlling spindle is arbitrarily set at 0·75 of deterministic optimum speed. If the lower limit of the Taylorized speed for the controlling spindle is exceeded, the lower speed is set equal to the lower limit of the Taylorized speed range.

(5) In case of series and parallel tool change schemes, the optimum values of the speeds for each spindle are calculated by the Golden Section Search and the Monte Carlo simulation techniques. Golden Section Search is used to determine the experimental speed for the controlling spindle. The speeds for the non-controlling spindles are modified to ensure that the product N_{ifi} is the same for all the spindles. Simulation is performed either for the given batch size or a duration which results in at least 25 failures of each tool. Optimum values of speed, production cost and production time are recorded. In order to ensure repeatability of the results, the simulation is carried out for 5 runs. A preliminary study was conducted to determine the number of runs required for good reproducibility of results. It was found that, in general, the number of runs should be a minimum of 5.

(6) For scheduled tool change schemes, the optimum values of speeds are selected on the basis of the desired optimization criterion and the acceptable ratio between unscheduled tool changes to the total number of tool failures. As in the case of series and parallel tool change schemes, the experimental speed for the controlling tool is selected by the Golden Section Search technique. For the calculated set of spindle speeds (corresponding to an experimental value of speed for the controlling spindle), a search for an optimum tool change interval is also carried out to ensure that with the selected tool change interval the ratio between the unscheduled to scheduled tool changes is below an acceptable value. Spindle speeds for the controlling spindle are searched for an optimum value. For the Monte Carlo simulation, the duration of run and the number of runs are determined as in the case of series and parallel tool change schemes. Figure 1 is a general logic flow chart for the optimization of multi-spindle drilling assemblies. A description of the general flow chart is given in the following section.

GENERAL DESCRIPTION OF COMPUTER PROGRAMS

Based on the techniques discussed in the previous section, a computer package has been prepared and fed to a CDC 6500 computer. The computer package can handle up to 10 tools working simultaneously. In its present level of implementation, it needs 90K memory of 60 bit words. Extension of the capability of the program will require much greater storage. A detailed description of the package is given in refs. 27 and 28. The general logic incorporated in this package is given in Fig. 1.

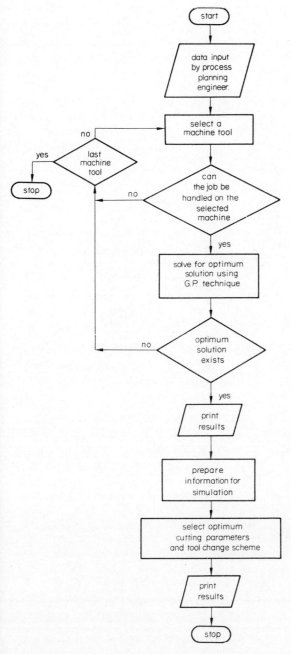

Figure 1 General logic flow chart for the optimization of multi-tool set-ups.

For the current computer package the process engineer provides all the necessary information concerning the workpiece, tool material and the machine tool or a group of machine tools from which the machine will be selected.

Table 1 gives typical input data provided by the process engineer. This information is programmed by the process engineer for a casting (Fig. 2) in which

four holes are to be drilled on a multi-spindle drilling machine.

Once all the necessary information is submitted, the computer program will carry out the following functions:

(1) Selecting one machine out of the given group of machines.

(2) Checking the compatibility of tool motions for the particular machine. In other words, it will confirm whether the machine tool can drive the tools the way the process planning engineer has specified. If the selected machine tool cannot drive the tools, this machine is eliminated from the group of machines and another machine is selected.

(3) Computing the spindle head feed and the spindle speeds for the group of simultaneously working tools based on the specified criterion of optimization considering deterministic tool life relationships. If no feasible solution exists for manufacturing the job on the selected machine, this machine is eliminated from the group of machines. Steps 1 through 3 are repeated until an optimum solution is obtained. It should be noted that the current package does not select the most economical machine from the group of machines. This limitation was imposed by the authors to conserve computer time. However, with slight modifications this feature can be incorporated in the program.

(4) Printing out the cost and production time for the selected machine (tool life considered deterministic).

(5) Computing feeds and speeds for the selected machine using the specified criterion of optimization and considering the tool life to be probabilistic in nature. Simultaneously it examines the various tool schemes and selects the best tool change scheme.

(6) Printing out information about the selected machine tool, optimum cutting parameters, the tool change scheme and the cost and production time.

One of the various examples solved using the current program is given below.

Example

Figure 2 shows a casting in which four holes are to be drilled on a multi-spindle drilling machine. Table 1 gives all the necessary information found by the process engineer. It is emphasized that in this case, the machine tool group comprises of only one machine tool.

The results of the analysis are given in Table 2. The results indicate that the optimum cutting parameters (the speeds for the various spindles) are lower in the probabilistic tool life case as compared to when the tool life is deterministic. Further it should be noted that out of the various tool change schemes, the scheduled tool change scheme 2 results in the minimum cost and time of production. However, both the cost and the production time are greater as compared to the case when the tools are replaced as

TABLE 1.   Input data for example on multi-spindle drilling

Data on Workpiece

Work material = cast iron
Brinell hardness no. = 165·0
Total no. of operations = 4

Operation no.	Type of operation	Length (in)	Diameter (in)
1	Drilling	1·50000	0·50000
2	Drilling	1·00000	0·37500
3	Drilling	1·00000	0·37500
4	Drilling	0·50000	0·25000

Data on Cutting Tool

Tool no.	Material	Taylor tool life		
		Life exponent	Feed exponent	Constant
1	H.S.S.	0·102	0·500	8·52
2	H.S.S.	0·102	0·500	8·52
3	H.S.S.	0·102	0·500	8·52
4	H.S.S.	0·102	0·500	8·52

Tool no.	Drill Length (in)	Helix angle	Point angle	Taylorized speed range (ft/min)	
				Upper limit	Lower limit
1	3·2500	26·0	118·0	88·00	16·50
2	3·2500	26·0	118·0	88·00	16·50
3	3·2500	26·0	118·0	88·00	16·50
4	2·2500	26·0	118·0	88·00	16·50

Tool no.	Type of tool life distribution	Ratio of mean and standard deviation	Total tool change time (min)	Tool load unload/time (min)
1	Lognormal	3·000	1·00	0·75
2	Lognormal	3·000	1·00	0·75
3	Lognormal	3·000	1·00	0·75
4	Lognormal	3·000	1·00	0·75

Average time to search a tool (min) = 0·000
Average time to procure a tool (min) = 0·250
Average tool procurement time factor = 0·800

Data on Machine Tools

Total number of machine tools in the group = 1

Spindle no.	Speed range Upper	(rev/min) Lower
1	1000·00	27·00
2	1000·00	27·00
3	1000·00	27·00
4	1000·00	27·00

Spindle head feed range (in/min)

Upper limit	Lower limit
8·00000	1·00000

Horsepower = 50·000
Efficiency = 0·800
Operating cost (cent/min) = 20·000
No. of spindles = 4
Total necessary travel of multi-spindle
head to machine the workpiece = 2·50000
Start stop time (min) = 0·000
Set up load and unload time (min) = 0·250

TABLE 1—*continued*

Cutting force relationship

Cutting tool no.	Force relationship constant	Feed exponent	Drill dia. exponent	BHN exponent
1	0·174	0·8000	0·8000	1·0000
2	0·174	0·8000	0·8000	1·0000
3	0·174	0·8000	0·8000	1·0000
4	0·174	0·8000	0·8000	1·0000

Batch size = 1000
Shear strength const. for standard drills = 0·0175
Production rate = 30·0 per hour
Non-simultaneity restrictions = operations none

Optimization criterion is minimum production cost

Tool Change Scheme to be Examined

Parallel Tool Change Scheme
Scheduled Tool Change Scheme 1
Scheduled Tool Change Scheme 2

TABLE 2.   Summary of results for example given in fig. 2

Tool Life Deterministic

Operating no.	Rev/min	Feed rate (in/rev)	Cost (cents per 100 parts)	Prod. time (min per 100 parts)
1	300·15	0·00530	2981·10	156·32
2	405·13	0·00393		
3	405·13	0·00393		
4	600·29	0·00265		

Tool Life Probabilistic

Series tool change scheme—results as in tool life deterministic

Parallel tool change scheme

1	275·34	0·00530	3324·08	168·82
2	370·51	0·00393		
3	370·51	0·00393		
4	550·70	0·00265		

Scheduled Tool Change Scheme 1

1	289·00	0·00530	3120·00	160·01
2	391·00	0·00393	Optimum tool change interval = 65 parts	
3	391·00	0·00393		
4	578·00	0·00265		

Scheduled Tool Change Scheme 2

1	295·50	0·00530	3080·00	159·00
2	397·00	0·00393	Optimum tool change interval = 55 parts	
3	397·00	0·00393		
4	591·00	0·00265		

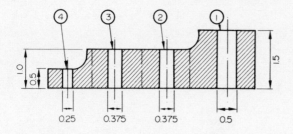

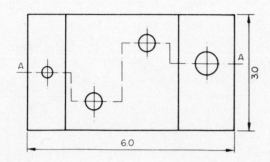

cast iron.

dimensions in inches.

Figure 2. Example workpiece for multiple spindle drilling operations.

they fail and the tool replacement in the middle of a cut is permissible (deterministic tool life).

On the basis of the results of this example and many other examples solved by the authors, the authors find that the current practice of reducing the cutting speed by an arbitrary factor disregarding the tool life distribution for the various tools working simultaneously is wrong because this factor cannot be prefixed. (There are specific instances when the optimum cutting speed for both the probabilistic and deterministic cases is the same. Complete details of this conclusion are given in refs. 27 and 28.)

CONCLUSIONS

The analysis of the information generated on the optimal selection of machining parameters and tool change scheme, indicates that the existing practice of reducing the cutting speed or terminating the tool life before the computed value is unsatisfactory, and often leads to economic waste. Such economic waste may not be significant in the case of single tool–single operation machining. However, in the case of multi-spindle drilling machines which involve number of tools working simultaneously, the running of the whole machine on the basis of the truncated tool life of a tool with the shortest mean life, will lead to enormous loss of productivity and high cost of production. Therefore, for operations in which relatively large number of components are produced, the tool changing scheme should be considered as an integral part of the process planning phase and should be generated simultaneously with the selection of optimum cutting parameters.

As regards various tool changing scheme, it is found that the most economical one is to allow each tool to run until it fails. Since this is not a practical

proposition, an equivalent scheme is to run the tool until it is approaching imminent failure. In many cases there already exist relatively simple tool sensors which can be used to indicate imminent failure. The availability of such sensors will permit the running of the machining system in a much more efficient manner. The utilization of sensor signals would require the availability of a control facility, i.e. a process computer. Since the introduction of on-line computers is already a reality, the findings of the present study should be taken into account when planning such systems.

Conceptually the problem could be solved fully automatically in the sense that only the final description of the component has to be given. The computing machine could produce the overall optimum solution. However, it does not appear to be feasible for the following two reasons:

(1) The capacity of the existing computers does not appear to be sufficient for carrying out such planning except for very simple parts.
(2) Such automation will completely eliminate the process planning engineer who may have certain creative ideas. Even if the computing capabilities were available for fully automatic planning, it is doubtful that one should subscribe to the philosophy of completely eliminating the human element. The appropriate course is to incorporate the human creative ability by maintaining interactive mode of programming as the computer technology continues to develop in capability.

The computer program developed for the optimization of multi-tool assemblies with probabilistic tool life, is of considerable magnitude and requires use of a large computer. It is, therefore, not advisable to use the package for small batches of production in which the cost of running the program will not be offset by increase in production. However, multi-tool operations (e.g., multi-spindle drilling), in the sense treated by this program are specific for large volume production where even significant amounts of computing time (several minutes compared to several seconds for small jobs) are offset many times by the optimization of the process.

ACKNOWLEDGMENTS

The authors wish to express their gratitude to Dr Theodore J. Williams, Director of the Purdue Laboratory for Applied Industrial Control for the technical assistance obtained in the Laboratory.

REFERENCES

1. A. NOVAK (1971). A method for optimization of cutting data, *C.I.R.P.* Paper No. CPA-16, presented at the Third International Seminar on Optimization of Manufacturing Systems, Pisa, Italy, June.
2. R. C. BREWER and R. RUEDA (1963). A simplified approach to the optimum Selection of machining parameters, *Engineering Digest*, 24 (9), 133–51, September.

3. B. COLDING and W. KONIG (1971). Validity of the Taylor equation in metal cutting, *C.I.R.P. Annals*, **19** (4), 793-812.

4. B. N. COLDING (1959). A three-dimensional tool-life equation—machining economics, *Trans. of A.S.M.E., Journal of Engineering for Industry*, **81**, 239.

5. M. KRONENBERG (1970). Replacing the Taylor formula by a new tool life equation, *International Journal of Machine Tool Design and Research*, **10** (2), 193-202.

6. M. J. C. MATTHIJESEN (1965). The economic aspects of high cutting speeds and short tool lives, *C.I.R.P. Annals*, **13** (1), 31-8, August.

7. F. W. TAYLOR (1907). On the art of cutting metals, *Trans. of A.S.M.E.*, **28**.

8. S. M. WU, D. S. ERMER, and W. J. HILL, (1966). An exploratory study of Taylor's tool-life equation by power transformations, *Trans. of A.S.M.E., Journal of Engineering for Industry*, **88** (1), 81-92, February.

9. J. D. RADFORD and D. B. RICHARDSON (1970). Optimization of parameters in Turning, *The Production Engineer*, **49** (4), 197-201, May.

10. P. B. BERRA (1968). Investigation of automated planning and optimization of metal working processes, Unpublished Ph.D. Thesis, Purdue University, June.

11. P. B. BERRA and M. M. BARASH (1968). An algorithm for the planning and optimization of a rough turning operation, Paper FP 6.3, presented at 34th National Meeting, *ORSA*, Philadelphia, November.

12. E. J. A. ARMAREGO and R. H. BROWN (1969). *The Machining of Metals*, Prentice-Hall, Inc.

13. R. H. BROWN (1962). On the selection of economical machining rates, *International Journal of Production Research*, **1**, 1-22.

14. N. D. SINGPURWALA and A. A. KUEBLER. A quantitative evaluation of drill life, *A.S.M.E.* Paper 66-WA/PROD-11.

15. M. N. LARIN (1961). Quality control for cutting tools, *Russian Engineering Journal*, **41** (7), 58-62.

16. J. G. WAGER (1967). The nature and significance of the distribution of H.S.S. tool life, Unpublished Ph.D. Thesis, Purdue University, January.

17. C. J. OXFORD (1959). The evaluation of cutting tools, Research Report, No. 5906-1, National Twist Drill and Tool Company.

18. J. DUNCAN (1968). Tool replacement scheduling, Technical Report, General Parts Division, Ford Motor Company, February.

19. C. J. OXFORD, (1970). Personal communication, letter dated December 8.

20. P. G. KATSEV (1968). *Statistical Methods in the Study of Cutting Tools* (in Russian), Mashgiz, Moscow.

21. E. S. VYSOKOVSKII (1970). Reliability of cutting tools in automated production, *Russian Engineering Journal*, **50** (3), 63-7.

22. K. HITOMI (1971). Studies of economical machining, *Bulletin of the J.S.M.E.*, **14** (69) 294-301.

23. R. J. DUFFIN, E. I. PETERSON and C. M. ZENER (1967). *Geometric Programming*, John Wiley.

24. D. J. WILDE and C. S. BEIGHTLER (1967). *Foundations of Optimization*, Prentice-Hall, Inc., Englewood Cliffs, N.J.

25. G. V. REKLAITIS (1969). Singularity in differential optimization theory, differential algorithm for posynomial program, Unpublished Doctoral Dissertation, Stanford University.

26. M. KRONENBERG (1966). *Machining Science and Application*, Pergamon Press.

27. J. L. BATRA (1972). Computer-aided planning of optimal machining operations for multiple-tool setups with probabilistic tool life, Unpublished Doctoral Dissertation, Purdue University, January.

28. J. L. BATRA and M. M. BARASH Computer-aided planning of optimal machining operations for multiple-tool setups with probabilistic tool life, Report No. 49, Purdue Laboratory for Applied Industrial Control, School of Engineering, Purdue University, Lafayette, Indiana 47907.

29. A. BHATTACHARYA and I. HAM (1969). *Design of Cutting Tools—Use of Metal Cutting Theory*, American Society of Manufacturing Engineers, Dearborn, Michigan.

APPENDIX: TAYLOR'S TOOL LIFE RELATIONSHIP IN TERMS OF SHOP PARAMETERS

For a tool 'i' let

$$N_i P_i^{A_i} = B_i \tag{1}$$

where

N_i = cutting speed, rev/min

P_i = number of satisfactory parts produced by the tool

A_i = an exponent

B_i = a constant

The relationship (1) can be derived from the Taylor's modified tool life relationship which, for tool 'i', can be represented as

$$V_i \bar{T}_i^{n_i} f_i^{m_i} = C_i \tag{2}$$

where

V_i = cutting speed, ft/min

$\bar{T}_i$ = mean tool life, min

f_i = feed rate, in/rev

n_i = tool life exponent

m_i = feed rate exponent

C_i = Taylor constant

For a drill of diameter D_i,

$$V_i = \frac{\pi D_i N_i}{12} \tag{3}$$

If L_i is the length of the hole drilled with tool (drill) 'i', then the machining time,

$$T_{ci} = \frac{L_i}{N_i f_i} \tag{4}$$

and

$$P_i = \frac{\bar{T}_i}{T_{ci}}$$

therefore

$$\bar{T}_i = \frac{P_i L_i}{N_i f_i} \tag{5}$$

Substituting equations (3) and (5) into equation (2) and rearranging the terms we obtain

$$N_i(P_i)^{n_i/1-n_i} (f_i)^{m_i-n_i/1-n_i}$$

$$= \left(\frac{12C_i}{\pi D_i}\right)^{1/1-n_i} (L_i)^{n_i/n_i-1} \tag{6}$$

For a known value of 'f_i',

$$(f_i)^{m_i-n_i/1-n_i}$$

will be constant.

Therefore equation (6) can be rewritten as

$$N_i(P_i)^{n_i/1-n_i}$$

$$= \left(\frac{12C_i}{\pi D_i}\right)^{1/1-n_i} (f_i)^{n_i-m_i/1-n_i} (L_i)^{n_i/n_i-1}$$

$$= \text{constant} \tag{7}$$

Comparing equations (1) and (7) we obtain

$$A_i = \frac{n_i}{1-n_i}$$

and

$$B_i = \left(\frac{12C_i}{\pi D_i}\right)^{1/1-n_i} (f_i)^{n_i-m_i/1-n_i} (L_i)^{n_i/n_i-1}$$

THE AUTOMATION OF SELECTIVE ASSEMBLY PROCESSES

by

W. S. BLASCHKE and B. REED*

SUMMARY

In many assemblies the clearance tolerance between mating parts has to be smaller than the sum of the tolerances which can be achieved in the manufacture of the components. In these cases matching of components, viz. selective assembly, is common practice.

Advances in metrology and transducers permit the rapid gauging of components to the required accuracy. This has made the automation of selective assembly feasible.

In the construction and control of the necessary flow lines, interesting problems of synchronization and interlocking arise. Solutions based on decentralized control and computing units, which have been implemented, will be described.

Selective assembly has statistical repercussions on the control of the machining processes which produce the components.

Automation of the assembly allows easy data logging. This opens the way to on-line control of the combined manufacturing and assembly process.

INTRODUCTION

Selective assembly of matching parts is used where the clearance tolerance is smaller than the sum of the part tolerances.

The paper concerns itself with the problems of automating the process. These problems fall into three categories.

(1) Gauging of components to the required accuracy and at the required rate.
(2) Control and synchronization of flow lines for the mating parts and administration of buffer store.
(3) Statistics of matching and rules to be applied at the production stage so as to minimize the accumulation of unmatchable components.

A survey of industrial applications might allow some deductions as to the area of greatest difficulty. It might also indicate common features which would allow the development of generalized techniques. Our own conclusions are that only a few generalizations are of value. Beyond these, details of components and processes dictate the solutions.

Selective assembly is concerned with precision assemblies and therefore the gauging problem will almost always be difficult.

The organization and control of flow lines and storage of parts is a laborious exercise but one that can always be solved. The techniques described in Section 2 of this paper should help to simplify the problems.

In Section 3 some simple statistical results of a general nature are given. In most cases an empirical approach based on data logging from an actual process will be necessary.

AUTOMATIC GAUGING

Methods of gauging industrial components fall into three classes.

(1) Contacting gauges using mechanical probes.
(2) Noncontacting gauges based on optical, inductive or capacitive phenomena.
(3) Indirectly contacting gauges such as air gauges.

Type 1 techniques must generally be ruled out for dimensional work, since they cannot be performed at high speeds. Stylus tips wear and it is difficult to safeguard against damage. Components with omitted machining operations can occur and locating devices can fail.

Type 2 techniques are better but are often of limited resolution. The proximity required for induc-

---

* National Engineering Laboratory, East Kilbride, Glasgow

tive or capacitative gauges is often inconvenient. Optical methods are preferable. Particularly important here is the use of lasers which can provide a highly collimated beam of high intensity and small cross-section. An example is a technique described by Parkinson[1] and illustrated in Fig. 1.

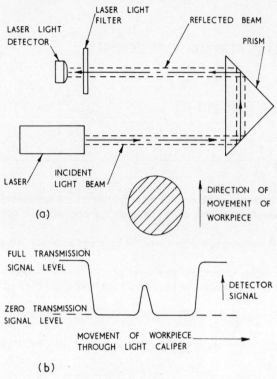

(a)

(b)

Figure 1. A laser caliper gauge due to Parkinson[1]: (a) schematic arrangement; (b) detector output.

A narrow cross-section laser is turned through 180° by a right angled prism. The forward and return beams are separated by a distance which is determined by the position of the apex of the prism. A cylindrical workpiece moves through the beam at right angles to its axis, first obstructing the incoming beam then the return beam. In the central position light passes at either side of the component and the relative height of the output pulse from the detector is a gauge of the diameter. An accuracy of 4 μm in diameter determination can be achieved by this method.

The measurement is in analogue form and there are particular difficulties with specular reflection from metallic components. The gauge can only be used in an accept/reject mode and is unsuitable for grading.

An alternative approach is to use direct shadow determination with a photoarray as the detector. Photoarrays are designed using integrated circuits with element pitch as low as 25 μm. A magnification of one order of magnitude will thus suffice for practical diameter determination. However the problems of decollimation, diffraction and specular reflection are aggravated. The range over which a threshold value of the shadow boundary moves linearly with diameter change, is still limited.

Further work is required in internal bores which present many additional problems.

In all these methods a vital factor is the accurate location of the component with respect to the optical system.

Type 3 techniques, in particular air gauging, will overcome the problem of component location since the gauge aligns itself to the component. This explains why the technique is widely used in the manually operated selective assembly.

The automatic guidance of an air gauge into a bore or air ring over a cylinder is made difficult by small stand-off clearances. However, with some elaboration of the gauge using a lead-in taper and some method of drawing the gauge onto the workpiece, these difficulties could be overcome.

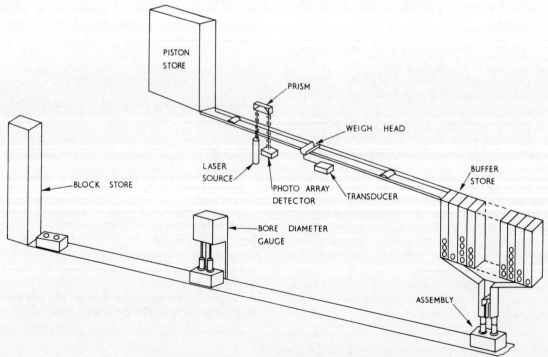

Figure 2. Test bed rig for selective assembly built at NEL.

Provided small stand-off clearances can be tolerated, adequate resolution can be obtained. The gauging rate, though obviously slower than for Type 2 gauges, is sufficient in many applications.

CONTROL OF FLOW LINES AND BUFFER STORES

Most of our work has been done in this section. Figure 2 illustrates schematically a test bed rig for the gauging, signal communication and control algorithm which would be needed in a typical selective assembly process.

Piston pairs are to be inserted into the two bores of a cylinder block so that diameters are matched to the bores and weights of the pistons are identical.

Pistons are released from the piston store, diameters and weights determined, and accordingly channelled into a buffer store. From the block store cylinder blocks are released, bore apertures determined and pistons selected, so that assemblies satisfying the above criteria are produced.

To simplify the problem, measurements are limited to a 1:4 resolution. The two parameters of the piston which are gauged (diameter and weight) determine one of 16 buffer store locations to which the piston is to be routed.

The control block diagram is shown in Fig. 3. This also illustrates the buffer store in matrix form in which columns are determined by diameters, rows by weights. The two columns to be accessed are known as soon as the bore diameters have been gauged. However there still remains the choice of row, provided it is the same for both pistons (equal weights).

A simple algorithm is used for the selection. As the contents of each store are known, the row which has the maximum of minimum content for the pair is chosen. This assures that empty stores will occur as infrequently as possible. If two selected columns have for every row a zero content in either column, the requirements of that cylinder block cannot be satisfied.

Much of the detail of the block diagram of Fig. 3 is unimportant in the context of this paper, but a few points of principle are worth explaining.

Contrary to much current practice, the control is not centralized in a computing system. It is considered artificial to centralize a process which is by its nature decentralized. For example the piston line and block line are isolated by the very existence of a buffer store. Their only interaction is taken care of by the interlocks 6 and 7 shown in Fig. 3.

The decentralization is carried further. Since two pistons are present on the piston line at any time, two cycles are distinguished. In one, pistons are released, diameter and weights determined. In the other, measurements are recorded, appropriate channels opened and the arrival of piston in the buffer store registered.

One main problem exists: since the cycles are asynchronous and time independent, there must be some guarantee that the two cycles are completed one for one. Interlock 2 of Fig. 3 assures this by requiring two stages, one from each cycle, to commence simultaneously.

A similar subdivision of the block line into 3 cycles—Block Measure, Piston Selection and Piston Insertion—with appropriate interlocks, is also shown in Fig. 3.

Since most stages in the flow lines have durations of the same order and yet are subject to variation, an on-line computer would be presented with a bewildering permutation of interrupts. The complexity of the interrupt handling program would be considerable. When it is considered that upon receipt of an interrupt very little computation is needed before the next action is engendered, it is obvious that a central computer control would become unnecessary.

The most suitable function for a computer is the data logging and statistical analysis of the results which are of great importance for a quantity production process.

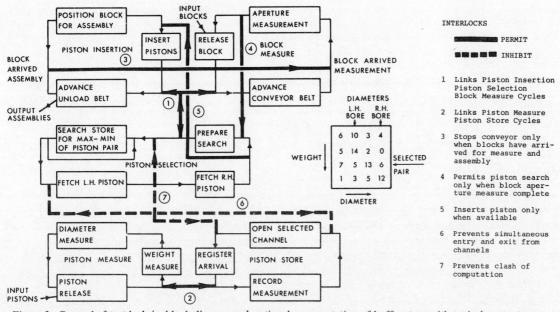

Figure 3. Control of test bed rig; block diagram and notional representation of buffer store with typical contents.

STATISTICAL RULES

The problem considered here is the accumulation of unmatchable components. Only one dimension from each component is considered, the external diameter of a male component and internal diameter of a female component. Nothing is lost by assuming that matching requires identical diameters, since a non-zero clearance can easily be allowed for by substituting a nominal diameter.

The analysis proceeds from the following premises.

(1) The male component is produced by a process A, such as centreless grinding, which can be controlled to produce components whose mean dimension μ_m remains constant.

(2) The female component is produced by a process B, such as broaching, where the internal diameters produced have a mean value μ_f which decreases as the tool wears.

(3) Both processes produce dimensions assumed normally distributed.

(4) The standard deviations are characteristic of a process on a particular machine and do not vary throughout a run.

(5) The standard deviations σ_m, σ_f, characteristic of the male and female processes respectively, do not differ widely and are in fact adjustable within small limits. By improving mechanical stabilities, reducing feeds, changing the geometry of tools and other means the standard deviations can be reduced to some extent.

It is required to consider

(1) The ratio σ_f/σ_m.

(2) Over what range should μ_f be allowed to drift before the female process is stopped and the tool replaced? If μ_{f1} and $\mu_{f\lambda}$ are the initial and final values of μ_f, the drift $2a = \mu_{f\lambda} - \mu_{f1}$ is required.

Figure 4(a) shows the distributions and excesses of unmatchable components for the case of equal standard deviations ($\sigma_m = \sigma_f = \sigma$) and a drift of $4\sigma(a = 2\sigma)$.

The fact that excess female components of large and small dimensions are inevitable can be seen by inspection. The drift of 4σ is obviously too large.

At the same time near the central dimension μ_m excess male components are produced. Thus for example the dimension μ_m will be produced from the complete integral under the female distribution ($\pm 2\sigma$ excludes only 4 per cent of the components). The excess male components follow from the relation

$$\frac{1}{\sqrt{(2\pi)}\sigma} > \frac{1}{2a} \quad \text{when } a = 2\sigma.$$

This result can be interpreted by subdividing the range $\mu_{f1} - \mu_{f\lambda}$ into $2a$ subranges of unit width into each of which a normally distributed proportion of components fall. The central subdivision (centred on μ_m) sees each of these subranges in turn as the mean μ_f drifts from μ_{f1} to $\mu_{f\lambda}$. The contributions thus integrate effectively to 1. If production is one for one $2a/\sqrt{(2\pi)}\sigma_m$ male components of dimension μ_m will

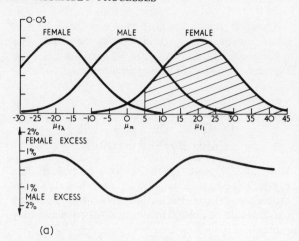

(a)

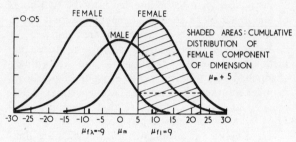

(b)

Figure 4. Statistical distributions of component dimensions and unmatchable excesses: (a) case of $\sigma_f = \sigma_m = 10, a = 20$; ($b$) case of $\sigma_f = 8, \sigma_m = 10, a = 9$.

be produced to accompany them. The male excess is then

$$\frac{2a}{\sqrt{(2\pi)}\sigma_m} - 1.$$

Central male excesses can be avoided only by reducing the drift range and making $\sigma_m > \sigma_f$.

A semiquantitative argument leads to the conclusion that a satisfactory solution is always near the conditions

$$\sigma_f/\sigma_m \approx 0.85 \quad \text{and} \quad |a| \approx \sigma.$$

Figure 4(b) illustrates the case of $\sigma_m = 10, \sigma_f = 8$ and $|a| = 9$. The resulting excesses are all within 0·1 per cent and cannot be shown on the scale used in Fig. 4(a).

Let

$$G(x) = \frac{1}{\sqrt{(2\pi)}} \int_{-\infty}^{x} e^{-t^2/2} \, dt$$

denote the cumulative standard normal distribution, and

$$g(x) = G'(x) = \frac{1}{\sqrt{(2\pi)}} e^{-x^2/2}$$

denote the standard normal density function.

Subscripts f and m will be used for the non-standard distributions with standard deviations σ_f and σ_m respectively. To simplify the expressions $\mu_m = 0$ is assumed. The excess female–male distribution is

$$E(x, a) = \frac{1}{2|a|} \{G_f(x - a) - G_f(x + a)\} - g_m(x).$$

$$(1)$$

Its derivative

$$e(x, a) = \frac{1}{2|a|} \{g_f(x - a) - g_f(x + a)\} - g_m'(x). \quad (2)$$

The degenerate case of $\sigma_f = \sigma_m$ can be seen from

$$\lim_{a \to 0} \frac{1}{2|a|} \{G_f(x - a) - G_f(x + a)\} = g_f(x),$$

to give $E(x) = 0$ for zero drift. Equal distributions require equal means and will produce excesses for any drift of μ_f from μ_m.

To arrive at a more useful result for $\sigma_f < \sigma_m$ which allows a drift of μ_f, conditions are required which, in the first place, make $E(0,a) = 0$. Before determining these it is more fruitful to consider $e(x,a)$.

From equation (2) it follows, since g_f is symmetrical, that $e(0,a) = 0$. But $g_m''(0)$ is a maximum. Therefore $e(x,a)$ will remain small near $x = 0$ when

$$\frac{d}{dx} \left[\frac{1}{2a} \{g_f(x - a) - g_f(x + a)\} \right]_{x=0}$$

is also a maximum. This divided difference will change most rapidly if the extremes of the interval lie on the point of inflexion of the curve, i.e. at $x = \pm\sigma$. It is thus reasonable to assume that $|a|$ should not differ greatly from σ.

Returning to equation (1) for $a = -\sigma_f$ at $x = 0$

$$\frac{1}{2\sigma_f} \{G_f(\sigma_f) - G_f(-\sigma_f)\} = \frac{0.34}{\sigma_f}$$

while

$$g_m(0) = \frac{1}{\sqrt{(2\pi)\sigma_m}}.$$

Therefore

$$E(0, \sigma_f) = 0 \text{ implies } \frac{\sigma_f}{\sigma_m} = 0.85.$$

One further general result shows agreement with this ratio for $|a| = \sigma$.

$$\frac{1}{2\sigma} \{g(0) - g(2\sigma)\}/g'(\sigma) = \frac{\sqrt{e}}{2} (1 - e^{-2}) = 0.71$$

since

$$\frac{g_m'(\sigma_m)}{g_f'(\sigma_f)} = \frac{\sigma_f^2}{\sigma_m^2},$$

$$e(\sigma_f, \sigma_f) = 0 \text{ implies } \frac{\sigma_f^2}{\sigma_m^2} \approx 0.71, \text{ or } \frac{\sigma_f}{\sigma_m} \approx 0.84.$$

The fact that $x = \sigma_m$ and $x = \sigma_f$ are not coincident does not invalidate the result since $g'(\sigma)$ is at a stationary value.

In a practical application the units of Fig. 4 might be microns. Figure 4(b) shows the case $\sigma_m = 10$, $\sigma_f = 8$, $|a| = (\sigma_m + \sigma_f)/2 = 9$. If the total range of acceptable dimensions is $\mu_m \pm 30$ only about 0.2 per cent of components will fall outside it. Subdivision of this range into 60 subranges of unit width would require a 1:60 resolution to separate the components. In the extreme case of measurement error, two mating components would fall into the same subdivision while their real dimensions are at the extremes of adjacent subdivisions. The tolerance on the clearance is thus ±1.5.

The tool producing the female component is set initially so that the distribution of the output has a mean $\mu_{f1} = \mu_m + 9$. A drift of μ_f to $\mu_{f\lambda} = \mu_m - 9$ is allowed before it is reset.

Selective assembly has achieved a clearance tolerance of ±1.5 μm using processes whose outputs are subject to standard deviation of the order of 9 μm and a cutting tool has been used until its diameter has reduced by 18 μm before resetting.

CONCLUSIONS

Selective assembly of matching components is frequently the only possible solution to the manufacture of precision assemblies The continuing demand for closer tolerances on clearances is expanding its use. The technique is labour consuming and costly. The automation is an important task for modern technology.

ACKNOWLEDGEMENT

This paper is published by permission of the Director, National Engineering Laboratory, Department of Trade and Industry. It is Crown copyright.

REFERENCES

1.   G. J. PARKINSON (1972). New applications of multi frequency lasers to alignment and gauging, *Proceedings of the NELEX 1972 Conference on Metrology*, Paper No. 24.

GROUP TECHNOLOGY

THE FORMATION AND OPERATION OF A CELL SYSTEM

by

C. ALLEN*

SUMMARY

This paper describes the steps leading up to the re-organization of a machine shop, engaged in the batch manufacture of avionic equipment, into a mainly cellular system. This involved the setting-up of machining cells varying in complexity from a simple turning cell to an advanced mixed cell based on numerically controlled machines. The paper describes the latter cell in detail and also covers the design of parts to suit numerically controlled machines; the automatic derivation of control tapes from the design geometry; and inspection by means of a coordinate inspection machine with coupled computer. Mention is also made of production control and machine loading by means of a computer system, quality considerations, difficulties encountered in setting up the system, and probable future trends.

INTRODUCTION

The main aim of the technique of grouping machines into cells, sometimes known as 'Group Technology', is to try to get the advantages of flow line production in batch production. Many claims are made on the advantages of this technique but what is not always made clear is that each situation tends to be different; that other considerations may influence the result; that benefits will vary widely from cell to cell and company to company. In this paper it is assumed that the basic principles of cellular production are familiar to the reader[1,2].

In this particular case history several other major changes were being made in parallel with the change to a cellular structure. These were as follows:

(1) Several machine shops including a previously separate NC one were being combined.
(2) Computer techniques for machine loading were being applied.
(3) Work was being transferred from conventional machines to numerically controlled machines.
(4) Design departments were being encouraged to specify parts which could be machined economically from the solid.
(5) Techniques were being developed for automatic preparation of control tapes for machining and inspection.
(6) Efforts were being made to reduce the cost of inspection by getting the machine operators, foremen, etc., to accept that 'quality' was their responsibility.

(7) There was an increasing need to reduce the dependence on skilled men and also to reduce the amount of indirect labour.

It must be borne in mind that each of these factors exerted some influence on the evolution of the organization. Because so many changes were made at the same time it is not possible, except in a few obvious instances, to apportion credit for the improvements to the particular changes. Hence, only a few examples of savings are quoted in this paper.

TYPE OF WORK

Clearly, the type of equipment being produced, the production rate, and the total quantity required are of the utmost importance when considering a production system. In the case of Ferranti, Edinburgh, much of the equipment is developed from scratch and because of this, there tends to be a relatively high number of new parts in the flow through production. The equipment produced is, in the main, avionic and tends to be electronic combined with high precision optical and electro-mechanical. Usually, after a lengthy development stage in which up to fifty development models, probably all different, may be required, production means the making of several hundred pieces of equipment at most, spread over several years. This leads to a pattern of manufacture of small batches of a wide variety of parts, delivered for assembly at frequent intervals, in order to minimize the effect of modifications and also to keep down the cost of the 'work in progress'.

* Ferranti Ltd., Edinburgh

The range of components being produced varies from small shafts, screws and gears, to complex castings and sheet metal fabrications. Materials, although standardized, are varied. The range in use includes stainless steels, carbon steels, bronzes, brasses, aluminium alloys, glass fibre reinforcements, ceramics, and many others.

Many of the castings are of such a complex shape that they require many machining operations and the provision of many jigs, fixtures, and tools. Some are aircraft, class 1 type and require special care, including full traceability against a serial number through manufacture, heat treatment, and inspection.

Components turned from round bar comprise the largest proportion of the work, about 46 per cent. Of this, 70 per cent is made from bar stock of 1 inch diameter or less. Many of the round components are gears.

The remainder of the work comprises components produced from slabs by machining, parts made from sheet materials, and sheet metal fabrications.

FACTORS GIVING RISE TO THE NEED FOR CHANGE

1. Skilled effort is in short supply and is becoming increasingly expensive.

2. Availability of numerically controlled machines which are capable of producing parts from solid at a high rate.

3. The potential of the computer in design, planning, and production control.

4. Long and uncertain delivery of castings from suppliers.

5. In the more complex parts many operations are required and this usually means short operation times and comparatively long setting times. This leads to a need for large batch size in order to keep the cost down.

6. The small quantities required of most components means that many movements have to be made and recorded. At any given time there are some 5000 different batches in work, and about 600 operations are completed daily.

7. Long make times. In the functional organization each operation is likely to take a week even on very small operation times. This means that a component made in thirty operations could easily take thirty weeks. The ratio of cutting to cycle time is often as low as 1 per cent[8].

8. Difficulty in controlling the movements between operations of so many parts and the complex flow pattern between machines.

9. Increasing need for shorter and more certain delivery times in order to beat the competition.

ORGANIZATION OF THE FIRST MACHINE CELL (Ref. – Fig. 1 – Cell 'T')

Obviously, the answer to all of these factors does not lie in making a cellular structure. Our main hope was that by making such a change, the calendar time taken to make the parts would be reduced and that more reliable dates for delivery to assembly would be obtained. It was also hoped that, due to having a

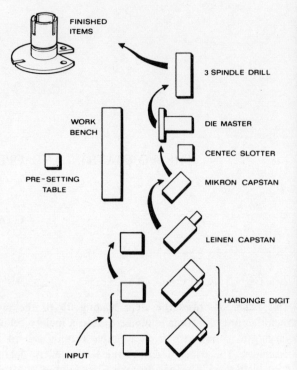

Figure 1. Turning/milling/drilling cell T (early).

clearer line of responsibility and greater job interest, an improvement in the quality of work and a more contented work force would result.

Our first attempt, which was made around 1967, was to form a simple group to deal with turned components which, in some cases, required subsequent milling, drilling and tapping operations[1]. It was clear that many parts were being machined from bar stock of 1 inch in diameter or less. Hence a machine group was arrived at empirically by analysing the operations on all such parts. It was found that the machines, shown in the figure, allowed us to manufacture about 300 different components. Two turret lathes, each fitted with digital display and equipped with pre-set tools, were used as the prime machines in the cell. Work was fed to these in a sequence based on a simple tooling code, in order to minimize tool changing. After the first operation, the operator moved the part on to its next operation on the milling and drilling machines as indicated by the arrows on the diagram. The results of this initial work may be summarized as follows:

(1) Work movements in the cell were simple, queues of work being held at the machines. It proved satisfactory for the cell leader to supervise work movement. No progress effort was required, other than to note batches 'in' and 'out'.

(2) Throughput time was reduced by a factor of 5. However, it should be noted that this can only be achieved by ensuring that the cell is not overloaded. If it is, delays will occur until a balance of capacity and load is achieved.

(3) It proved possible to group the work into sub-families, each of which was allocated a simple code which referred to a given arrangement of pre-set tools on the turret of the lathe. It has been found that four basic turret set-ups cover the work. By arranging the work flow so that all parts

which require the same basic turret are loaded in sequence, the set-up time is minimized. In the case of Cell 'T', setting time was reduced to about a third of what it was previously.

(4) The secondary machines were only partially loaded. This meant that the men had to be flexible, moving from machine to machine. The underload on these machines was of little importance because the machines were of low capital cost.

Since it was first set up this cell has now evolved into the form shown in Fig. 2.

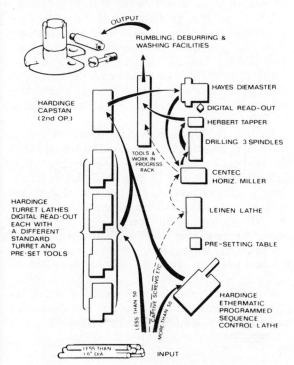

Figure 2. Turning/milling/drilling cell T (current).

Initially, work flow was matched to the capacity of the cell, any surplus being manufactured in the old way. Later, all suitable work was fed to the cell and its capacity augmented to suit. Hence there are now four prime machines in place of the previous two. This permits minimum turret changing as each machine is fitted with a different one of the four standard turret arrangements.

The only other change is that another lathe has been added. This is a sequence controlled machine which is used to deal with larger batches (about 50 per cent and over).

The secondary machines are now more fully loaded.

This cell is proving very effective and tends to be used as the pattern on which all the others are based.

Since the results were so encouraging it was decided that the proposition of extending the system to the remainder of the shop should be studied.

EXTENSION OF THE CELL SYSTEM

Classification

It should be noted that, although the detail drawings carried a code related to the geometry of the part, it was only necessary to use a simple turret number to control the flow to the first cell. On examination of the remaining components it was clear that a very sophisticated code would be necessary if machine groupings were to be obtained automatically. For example, a part with holes could require punching, drilling, boring, or reaming dependent on the tolerance and the nature of the part. This obviously applies to many other features on the parts and, in addition, in the case of machined parts, the method could vary with quantity, availability of jigs and fixtures, and also the availability of machine tools. Hence it was decided that the most promising method of forming the groups was to use the existing machine routing for parts and to work backwards in order to see how many parts could be made with certain machine groupings. This technique is known as 'Production Flow Analysis' (PFA)[2,6,7] and the continued assistance of G. Edwards and his team at UMIST was enlisted to carry out the work.

Production flow analysis

Since the use of this technique involves the analysis of the route and the machines used in making each part, it is essential that this information is available as an input. In our case, complete operation sheets showing machine numbers and values of setting and make time were available. This formed the basic input, the only doubtful area being the level of the workload which, of course, is required in order to fix the number of machines in a group. This could only be given roughly, as 25 per cent variation can occur over relatively short periods. Hence there could be no expectation of a precise result. Components were sorted, initially, in machine numbers in ascending order, irrespective of the sequence on the operation cards. This resulted in family groupings of components requiring a certain combination of machines. It proved possible after some empirical adjustments to arrange about 80 per cent of the machines into cells. Some of these cells, like the first one described, are simple, flow being sequential and the same for all parts. In other cells the flow is complex and there is more than one machine which can be described as prime. In the former, virtually no progress system is necessary, whereas in the latter a good progress system is necessary[4] (see 'Casting Cell').

At the start of the PFA exercise it was clear that there already existed an obvious group of parts, viz. castings. A separate analysis was carried out for this type of component and a group of machines was set up to deal with them in parallel with the work on the PFA (see 'Casting Cell').

Overall, the result of the PFA and the previous work has been that thirteen groups have now been set up, ranging from simple to complex. Machine groupings are shown in the Appendix.

Compromise was necessary in several instances: for example when one operator looked after, say, three machines, then these machines could not readily be separated into different groups. Similarly, where only one machine of a type was available this could only be put into one group. Of course if the machine was inexpensive it was duplicated, e.g. tapping.

Overall work flow, from raw material to finished part, was studied with a view to further reduction of movement. For example, movement to the plating area was reduced by arranging for degreasing and de-burring facilities to be available locally and accessible to the cells on demand. Similarly, raw material, stampings, etc. are now carried in or near the cells where appropriate.

In all cells the supervisor, assisted where necessary, is responsible for both progress and quality.

The re-organization has necessitated a considerable re-layout of the plant.

Much of the existing planning required to be changed to suit the new groupings and the additional processing facilities. A new discipline of planning to suit the groups has been introduced, and in many cases standard planning is being adopted.

Where practicable, a planning engineer has been allocated to a group and made responsible for all planning for that group. This provides better communication between the cell leader and the planner.

The casting cell (see Fig. 3)

As mentioned earlier it was decided to set up a cell to machine the 500 different castings[5]. From the list of machines in Fig. 3 it can be seen that many of the

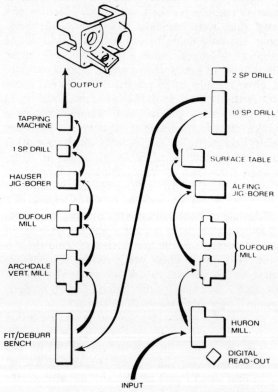

Figure 3. Casting machining cell C (early).

machines are expensive and that no one machine can be termed 'prime'. The work flow pattern is complex, the parts starting at various machines for their first operation and subsequently following their own special route. Because of this, and the variation in operation times, some queuing is inevitable.

Because of the complex flow pattern it is not possible for the group leader to monitor the flow through the cell. Initially a progress engineer was attached to the section and he controlled progress by

means of a manual load board. This system has been superseded by a computer system, the two being run in parallel for a six-month changeover period. The computer system is up-dated twice a week, at present, the operation cards now being used as input for both wages and progress. The information is fed to the computer via punched cards. Every two days printouts are available of 'queue lists' for each machine, complete 'state of the batch' schedule, 'batches late' schedule and some other useful data.[4] This machine loading system is used on all complex groups.

This cell has functioned effectively since it was set up. The principal benefit derived is that the throughput time has fallen to about one-third of what it was previously. No saving has yet been made on setting times but this aspect is being studied. One of the main factors acting against change is that proven tools, fixtures and methods are available for the parts and it does not make sense to bring in changes which could cause unnecessary trouble (refer to 'Development of NC Cell').

Initially the machines in this cell did not include an NC mill although one was specified for several of the castings. Instead, a number of hours per day on an NC machine in another area were allocated to the casting cell. The reason for this was that it was believed to be safer to allow specialists to supervise the NC equipment. In practice this arrangement proved to cause long delays and gave rise to divided responsibility. Subsequently therefore, an NC milling machine was allocated to the cell and this cleared the difficulty, although it tends to carry a lighter load than it should. Later still, an NC jig boring machine was attached to the group. This was due to the discovery of a novel method of tooling which allowed accurate

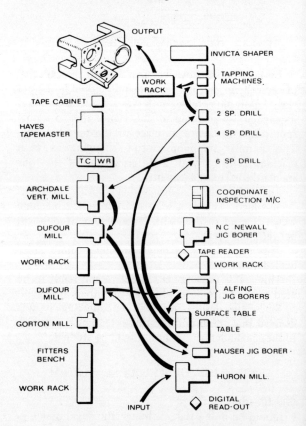

Figure 4. Casting machining cell C (current).

hole diameters to be produced by means of special boring bits. The present machine complement of this cell is shown in Fig. 4.

Due to variation in work load it is sometimes necessary for the cell supervisor to take action. If the work input drops and he gets too far ahead of schedule he will arrange to transfer some of his work force. When the work load rises and delays occur, as shown on his computer output and work queues forming at the machines, he must arrange to get more effort, work longer hours, or put the work elsewhere. However, he carries the responsibility for all work booked into his cell and he must ensure that it is kept moving. Of course he is expected to seek advice from a superior in the event of some difficulty which is outside his control.

The development of an NC cell[5,8]

In parallel with the work described in the paper so far, work was going on in the use of NC machines. Our aim in transferring work to such machines was to get higher productivity from the machine and to reduce our dependence on skilled effort. The first NC machine was brought into use in this factory around 1956 and later NC acquisitions tended to be grouped together. In earlier years these machines were used mainly for the manufacture of specialized components such as waveguide blocks. In recent years, with the advent of machining centres it became clear that provided design department could change their ideas to suit, many more components could be produced economically by machining from solid. This applied particularly to our quantities; tooling could be minimal and many of the awkward castings could be replaced on new designs.

It was a small step from thinking in terms of an NC machining centre to arriving at a group based on NC machines which would allow parts to be made from bar to finished component. This resulted in a

cell being set up in which parts could be machined from light alloy billets of sizes up to 12 x 12 x 6 in (305 x 305 x 153 mm). The formation of this cell[5] is as shown in Fig. 5. This cell is indicative of the way batch manufacture will be organized in the future as a greater range of numerically controlled machines comes on the market capable of making economically the full range of components used in industry. The other cells may be regarded as transient although the changeover time will be many years.

The amount of special tooling required in this cell is minimized by the use of standard pallets, ref. Fig. 6.

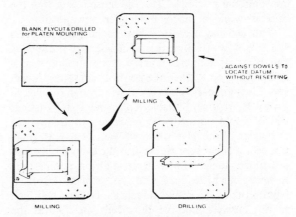

Figure 6. Palletization.

Work is mounted on these pallets in known standard positions and the pallets can be transferred from machine to machine in the cell, tape and machine datums being lined up automatically to suit a pre-determined component position.

All parts made in this cell are machined from billets of a standard pre-stretched aluminium alloy which is held in about ten standard thicknesses, all 12 in (305 mm) wide. A supply is kept within the cell. Initially a special alloy was specified but due to supply difficulties a gradual change is being made to BS 1470-HS 30TF.

When a batch is issued the blanks are cut from the billet on a high speed saw which can cut the thickest section in about 30 seconds. The blanks are then faced on a milling machine, a Vertical Maserati, fitted with a large toothed cutter. In one or more passes a smooth flat surface is obtained and the blank is reduced to the required thickness. Fixing holes are then drilled in the blanks by either a standard drill jig mounted on a manual drilling machine or an NC turret drill. The blanks are then mounted on pallets and are passed to the other machines as specified on the operation list. In general, efforts are made to carry out as much of the machining as possible on a Molins twin spindle milling machine. This machine has automatic pallet loading and automatic tool change, and has high speed spindles of adequate power. Hence setting time is low and, due to the high rate of metal removal, the operation time can be very low. Since the machine is nearly automatic the operator need not be very skilled, and times are nearly independent of the operator. Another NC milling machine, a Hayes Tapemaster, is included because the Molins is restricted to a thickness of 6 in (153 mm) and sometimes milling is required on the side faces of

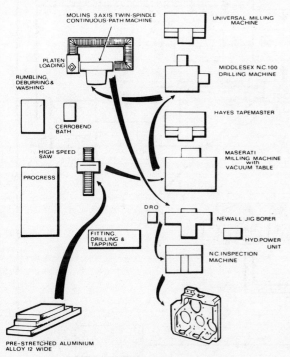

Figure 5. NC machining cell (current).

the 12 x 12 in (305 x 305 mm) block. Drilled holes are machined on an NC turret drill, a Middlesex NC 100, and accurate holes are machined on an NC jig borer, Newall 2 axis. Fitting, cleaning, and de-burring facilities are available in the cell and inspection is carried out on a coordinate inspection machine.

Progress is by means of the computer system mentioned earlier, and one man is able to deal with this cell and also two other smaller NC based cells. It should be borne in mind that the number of parts produced in this cell is many times that which could be produced in a cell with the same number of conventional machines. This of course means that the progress work per machine is higher in proportion.

This cell has been in action for about two years and it is now effective. It is of course very productive and parts of suitable design can be produced at a competitive cost. Once the control tapes are proved, and this is not easy, parts can flow through this cell in a very short time. Suitable replacements for castings which would normally be machined in the casting cell, taking probably twelve weeks, can pass through this cell in about four weeks, but the average throughput time is longer at present. In addition, setting time is very much reduced and less skilled effort is required than in a conventional cell. Hence a cell such as this answers most of the needs specified at the beginning of this paper. The savings are of course not only due to the machine grouping and, in fact, are due mainly to the change to numerical control and the design of parts to suit the technique.

'Costing' in such an NC cell

It is said by some that company costing can be simpler and more accurate in an organization based on the cell system. In our case, and for this particular cell, this has not been found to be the case. Briefly, the reasons for this are as follows:

(1) Operation lengths on each machine cannot be matched and this leads to some queuing and not a simple through flow. Setting times and operation times are required for each part and some comparison must be made between 'times allowed' and 'time taken'. It is not easy to see how an average bonus could be paid in the cell taking into account variation in performance and the demands of individual members of the team. We have in fact maintained an individual bonus system which necessitates booking 'on' and 'off' for each operation.

(2) The amount of planning required per machine is up roughly in proportion to the increased output of the NC machine. In addition, part-programming is required in order to transfer the drawing geometry to control tape. This causes an increased overhead and varies from machine to machine.

(3) Capital cost of an NC machine is high and depreciation must balance this. Multi-shift work helps to some extent but other charges go up due to this.

(4) Repairs and maintenance charges are high.

(5) Progress charges per machine are high.

(6) More consumable tools are used.

Because of these factors, the overhead on the NC machines in the cell is much higher than that on conventional machines (up to four times in some cases) and varies significantly between one machine and another. In our case, we believe that it is necessary to charge the work against the particular machine used rather than against an average for the cell. By doing so, the cost of making the part is

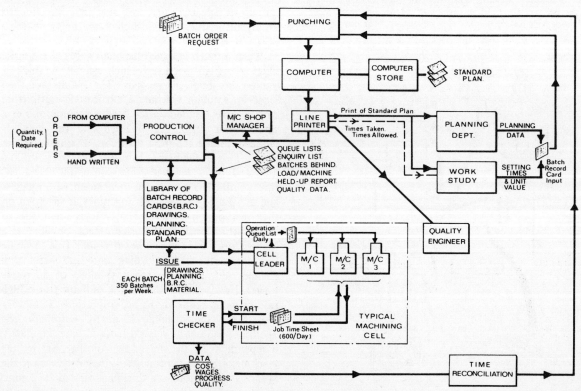

Figure 7. Information flow; production control.

known to some accuracy and this information can be used by design and planning to reduce costs. An overall machine rate could hide the high cost operations. In practice three levels of rate per machine hour are used, these including planning, programming, and progress in addition to the usual overhead content. The highest rate is four times the lowest. It should be noted that, provided the NC machine produces an adequate output, it can compete favourably with its conventional equivalent.

Preparation of data for this cell

Although this cell has been set up to take over, in due course, the work of conventional cells such as the casting one, it does not mean that all work can simply be transferred. In fact, some 90 per cent of it cannot be moved in some cases due to its unsuitability for machining by NC, and in other cases due to the fact that tooling exists and the NC method could not compete. Hence, on existing designs no appreciable transfer is possible and re-design is not possible because of cost and other reasons. However, on new designs, the picture is different. On these our designers are using parts which are suitable for manufacture by NC. This will mean eventually that all milled, bored, and drilled parts will be made in such NC cells. Cost of 'work in progress' will be greatly reduced, raw material stock being standard and throughput time being under, say, a month for all parts.

However, once design has specified a shape, there is still the difficult stage of detail drawing and control tape making to be passed through. In this connection, we have been using automatic draughting equipment for many years and its use is described fully elsewhere[3,10]. See Fig. 8. The principle is that all geometry and cutting data are digitized at the design stage, transferred into alpha-numeric form, fed to a computer, and all information required by production produced automatically by means of the computer and its peripheral equipment. Outputs include detail drawings with list of coordinates, scale drawings for checking, etc., machine control tapes, and inspection tapes. This system has meant that drawing office work is being combined with planning and part-programming. In families of parts, such as flat panels, once the production planning has been laid down it can be copied for all similar parts automatically, the only requirement being to feed in the new geometry. Advantages of the system include speedy availability of control tapes and piece parts and a possible rapid build up of parts for production. On the debit side of course, there is the cost of writing computer programs and the additional capital investment on the computer system.

Automatic inspection in the cell

Although not situated in the cell, an inspection machine with a coupled computer[9] is available for the use of the cell, ref. Fig. 8. Initially when a part is digitized, two types of tape are produced, one for manufacture, the other for inspection. Firstly, the automatic inspection machine is used to check that the item produced by the control tape is correct. Where necessary the tolerances on the inspection tape can be reduced in order to allow for drift in sizes during the manufacture of a batch. In our case we think that we should aim for 50 per cent of the

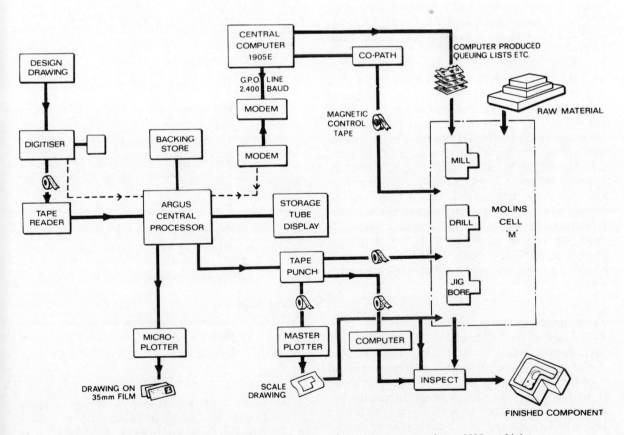

Figure 8. System diagram: draw, make, and inspect using computer processing and NC machining.

tolerance in certain cases. Once the tapes are proved, subsequent batches can be checked to the inspection tape either fully or to any extent decided by the inspection/quality organization. Programs are being developed at present to enable us to process all flat plates and $2\frac{1}{2}$D parts. At present for example, a part with up to 2000 holes can be checked in the following way:

After digitizing the information is passed to the Argus computer (ref. Fig. 8), by means of punched tape. The first output is from a micro-plotter and is in the form of a 35 mm film showing all holes to scale. This is obtained a few minutes after input and can be checked by the draughtsman and any obvious error eliminated. A second tape is obtained, fed to a master plotter and a master drawing to an accuracy of about ±0·002 in (0·05 mm) produced. This can be on film and can be used as a checking overlay on the design drawing and later on the part. At this stage, the draughtsman should be fairly sure that the part is specified correctly. Next, the machining tape and an inspection tape are produced. The part is then drilled or jig-bored in the cell and passed to the automatic inspection machine. Here it is simply mounted on the table, there being no need to line up the datum lines on the part to the axes of the machine. The tape is read into the computer store (in less than a minute) and the inspector places a taper probe in each hole in turn. Once probed, a correction is made by the computer for the inclination of the work to the table, and the resultant coordinates compared with the information which has been provided by the inspection tape (takes about 4 seconds a hole). The program has been written so that the holes can be probed in random order and only errors are printed out. At the completion of the job, the inspector can request a check to see if he has probed all the holes. If he has missed any, the coordinates of the holes missed are printed out on the teleprinter. The accuracy of the machine is ±0·0003 in (0·0075 mm) and, if a zero tolerance is fed in, all the coordinates are printed out. This method may be used to arrive at the amount of drift on the machine. It is not necessary to check all holes every time; any random selection may be made. The diameters of the holes are checked by means of plug gauges. When the inspector wishes to know the size and tolerance of any hole he is inspecting he can simply ask the computer to provide the information via the teletypewriter. By means of this system a saving of inspection time of up to six times over a manual inspection machine is made. In addition, it is now feasible to make and inspect without the need for a drawing.

Future developments
In the case of the NC based cell, the stage has now been reached where work is taken through from design to final inspection with a very much reduced need for intervention by skilled effort. In batch work, it is doubtful if we can move any nearer to the automatic factory than this in the foreseeable future. Unfortunately this type of cell can only be applied to a limited family of parts, wider application to other

families being dependent on the development of new machines.

The other basic machine groups have now settled down and are unlikely to vary over the next few years. Some detail changes may be necessary, including changes to match capacity against the changing load but, since the groups are tailored to suit fairly general types of components, requirements will only change slowly to suit any variation in our products in the future. Much work is still required on tooling and methods of sequencing in order to reduce set-up times.

CONCLUSIONS

The application of 'Group Technology' (or whatever you like to call it) is proving beneficial in reducing throughput times in all cases and setting times in some, and is creating a clearer definition of responsibility for both operator and supervisor. In the change to numerical control, increasing savings are being made as new designs come into production. NC based cells are proving effective and an increasing use will be made of computer aided drawing for tape preparation and computer aided inspection.

However, it must be emphasized that each factory has a different set of problems, and anyone who is considering re-organization should analyse his problems carefully and be prepared to tailor a system to suit his own needs. Basically GT is a simple common sense idea, somewhat confused with coding, etc., occasionally, but it has much to offer to those engaged in batch production in medium and large size factories.

APPENDIX: MACHINE GROUPS

Conv.

C Casting group

D
E } Turning, milling and gear-cutting

F Milling, turning and drilling; closely related to L

G Turning, grinding and milling

L Grinding and turning; feed to F

T Turning followed by simple milling and drilling

NC

M NC—parts under 12 x 12 x 6 in (305 x 305 x 153 mm) in light alloy

A NC remainder

H Punched, drilled and bored

Sheet metal

P Guillotines, presses and benches

Q Benches, milling

R Benches only

REFERENCES

1.   F. R. E. DURIE (1969). A survey of group technology and its potential for user application in the U.K., I. Prod. E., George Bray Memorial Lecture.

2. G. A. B. EDWARDS (1971). *Readings in Group Technology*, The Machinery Publishing Co. Ltd.

3. J. HOLLINGUM (1971). Design–production gap is bridged by auto-draughting, *The Engineer*, 2/9/71.

4. J. HOLLINGUM (1971). Computer helps plan plant layout to speed batch flow, Machine shop work scheduling plan puts first things first, *The Engineer*, 16/9/71, 30/9/71.

5. J. J. MARKLEW (1971). Advanced ideas for production in the 70's at Ferranti, Edinburgh, The cell system—a variety of applications in the same factory, *Machinery*, 4/3/71, 24/3/71.

6. J. L. BURBIDGE (1969). Production flow analysis, I.L.O. Group Technology Seminar.

7. M. CROOK (1969). The investigation and development of production flow analysis as a method of introducing group production to engineering production, M.Sc. Thesis, U.M.I.S.T.

8. D. T. N. WILLIAMSON (1968). A new pattern of batch manufacture, *The Chartered Mechanical Engineer*, July.

9. J. J. MARKLEW (1972). Computerised inspection—some possibilities and benefits, *Machinery*, 15/11/72.

10. C. ALLEN (1966). Computer aided drawing and design, *The Production Engineer*, August.

DISCUSSION

Q. J. L. Sheldon, ICL. How has the formulation of groups in the machine shop affected the quality of the product?

How has it affected the organization of inspection and the amount of inspection necessary?

What problems have arisen over changes of responsibility of individuals as a result of the re-organization, and how has group versus departmental organization affected general morale?

A. At the same time as the organization of the machine shops was being changed to a cell structure, the concept of quality/inspection was also being changed. The altered quality concept was that more attention should be paid to quality at all stages from design through procurement, manufacture and test. This meant that in the machine shops the foremen and operators were asked to make the parts correctly instead of making parts and having them checked in detail by inspection. Line inspection has now been withdrawn although, in certain cases, specialized effort is available to a group leader but under his control.

This change has had more effect on inspection cost and quality than the change to cell structure. Throughout the machine shops—i.e., either in the cells or in what is left, which is still organized on functional lines—the percentage of inspection has gone down from over 30 per cent direct inspection per direct operator to less than 15 per cent per direct operator. The quality is now better than it was previously and the present level of scrap, is about 7 per cent.

The cellular structure will help in the longer term when arrangements are being made to collect scrap figures for each cell. Although a different level of scrap may be the norm for each cell, trends should become obvious and will be a useful guide to management.

Several problems have arisen over changes of responsibility, some supervisors being asked to do more and others less. However, the job of each supervisor is now clear. It has proved to be most important to make everyone aware of the objectives and reasons for the changes. Supervisors are being asked to do more, i.e., to control both progress and quality in addition to providing direct supervision. Some may have lost some status and others may have fewer buffers between themselves and upper management. However, the cells and the men are settling down and show much promise for future development.

F

CONSIDERATIONS FOR THE FORMATION OF CELLS IN GROUP MANUFACTURE

by

M. Y. MALIK*, R. CONNOLLY* and A. J. P. SABBERWAL*

SUMMARY

This paper reviews the main existing methods of bringing workpieces together for group production and discusses briefly the limitations and disadvantages of these approaches.

A method is proposed whereby the manufacturing families and the system design are interlinked: this is illustrated by means of an actual example of the creation of component/machine families for erratic work loads of essentially non-repeating small batch production. The approach adopted shows the means by which are met the constraints of existing machines, minimum extra machinery investment, space restrictions, and the number of operatives in a group, to achieve the maximum labour and machine utilization. The method used avoids the limitations of existing schemes and is essentially a quicker and cheaper way of forming family group cells.

INTRODUCTION

It has been recognized for a long time now that large scale benefits can be gained if conditions are created whereby batch-production can be made to follow mass-production methods of manufacture. Although the benefits of group manufacture have been written about extensively in the past decade, the system is still not widely adopted anywhere. One basic reason for this can be attributed to the fact that various protagonists of this new manufacturing concept often emphasize only a particular aspect, e.g. a certain system of parts classification or a certain area of analysis, and the overall system approach has in general been narrowed to the machine shop.

If, however, a total production system is considered from the outset study, then the questions whether or not to apply a part classification system and the extent of the analysis and collection of meaningful data can be answered in the light of overall management requirements.

A review of the existing methods of creating component families has revealed that the approaches adopted are narrow in concept (i.e. machine technology orientated), costly in application, and lengthy in execution. This paper illustrates how, by considering the system, management, and environmental constraints, the need for extensive component classification exercises is made unnecessary. The approach also simplifies the analysis necessary for cell formation since it is related to the overall requirements.

VARIOUS APPROACHES FOR THE CELL FORMATION

Figures 1(a) and 1(b) show the two kinds of production situations which can be met in practice, i.e. implosive or explosive. The former describes a situation starting with a number of components (made in or bought out) which ends up with a product or multi-products (e.g. electric motors), whilst the latter is concerned with the case where an essentially one-off mix or type of material leads to a number of end components or products (e.g. a range of twist drills or brake linings). In both types of production situations, a machine shop is encountered at some stage in the manufacturing cycle as indicated by the shaded areas in Fig. 1. The extent of the machine shop activity in relation to other facilities varies from firm to firm and from product to product, but it remains just one of many facilities in a complete manufacturing system. Hence it is of special interest to note that, in general, the classification systems are mainly concerned with this one activity. Since it is held in various circles that a classification system is a pre-requisite for a group manufacture system, it is necessary to discuss its shortcomings in view of the point made above.

Classification has been claimed to be necessary for a multiplicity of purposes, e.g. retrieval, grouping of parts, workpiece statistics, and tooling requirements. Each major system claims to fulfil most of these purposes but some are essentially workpiece statistics

* Ferodo Ltd., Chapel-en-le-Frith, Derbyshire

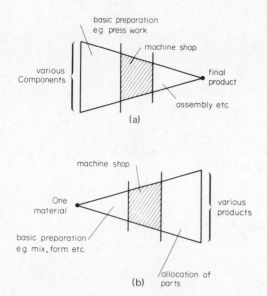

Figure 1. (*a*) Implosive production system; (*b*) explosive production system.

based[1], whilst others are design retrieval orientated[2].

Systems proposed by Opitz[1], IAMA[3], PGM[4] are essentially similar and suffer from the same shortcomings, viz. each assumes that the total number of components in a shop can be divided into groups for machining purposes. It does not follow that because the components can be grouped together in the machine shop (thus optimizing the sub-system), (i) their end use is similar in nature, and (ii) the preceding and following operations are such that components are going to be grouped there as well in the same way. In fact, in using these systems, little thought is given to the overall flow in a company producing multi-products either to the customers' designs or the company's own designs. As Edwards[5] very rightly observes, 'Preoccupation (by Mitrofanov, Opitz, Czech, Yugoslav, French, etc.) is with engineering and technology to the exclusion of management, systems and economics of the total company'. Zimmerman has produced a classification system[6] which has 30 digits. The length of this code alone makes it a mammoth task to compile data. The Brisch[2] system of classification is essentially a design-based system although a production-orientated code can be supplemented to the design code. The system comes as a package and is more complete than the systems mentioned above. It has not, however, proved very successful in producing component families once one extends beyond the established type of product (e.g. rotational)[7]. This is given as one reason why Ferranti went over to the Production Flow Analysis method. Before moving to the next aspect of component grouping methods, it is worth repeating that classification and coding in themselves produce nothing and a great deal more information is required than given by workpiece statistics and classification schemes[8].

Other well-known methods of component groupings are Production Flow Analysis[9] and Component Flow Analysis[10]. PERA have developed a method[11] but there is a lack of published information about the actual methodology involved and of the totality of the approach.

PFA is based on three major steps, (i) Factory Flow Analysis, (ii) Group Analysis, and (iii) Line Analysis. The end result can be shown in Fig. 2(*a*) where it is noted that the factory can be divided into certain areas, usually those that already exist, e.g. machine shop, assembly; and that grouping takes place in each area. Burbidge assumes various items, e.g. the flow, can be modified sometimes drastically and that nothing leaving one area will have to visit a similar facility again, i.e. sub-optimization can be obtained in each area. Since the system essentially does not make any allowance for different kinds of end products, one can imagine problems where sub-optimization of setting-up times occurs. For example, two parts belonging to two completely different products with a very different lead time are processed together simply because these two products happen to group at a certain operation. Although such a system can attain group control it does not achieve overall system control (inter-group). While the parts of the idea proposed are pioneering and are widely used, the total concept is limited, and this may be one reason why, to the authors' knowledge, no successful example of a PFA based total manufacturing system exists.

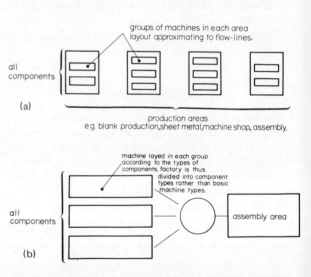

Figure 2. (*a*) Production flow analysis method for factory division and grouping; (*b*) component flow analysis method for factory division and grouping.

Component Flow Analysis is a method developed by El-Essawy in conjunction with multi-product firms, notably Ferranti, when their classification system could not cope with the casting type components[12]. The method developed is more thorough and the essential features are shown in Fig. 2(*b*). In spite of claims by various people[13] that it is no different from PFA, it is believed that the difference is very fundamental and a move in a more practicable and right direction as Fig. 2 shows. From the written evidence available it seems that the assembly is not considered as the major criterion in the cell formation. However, it can be expected that this system will suffer from some kind of imbalance at the assembly stage and, as Allen of Ferranti wrote, 'with overload, the progress system tends to break down'[12]. What is not comprehended, however, is the manner in which the 'overload' occurs in the first place, because a planning system should ensure it does not happen. But with the system as it appears, the planning system will have to be fairly sophisticated to avoid imbalance.

The survey would be incomplete if the simulation methods being developed were not mentioned. They are due to McAuley[14] and Crookall[15] where the former considers a computer method for grouping by PFA. The example given is on a very limited scale (12 machines, 10 parts) and assumptions such as 'no backward flow', etc., cast doubt on the practicability of the method. Crookall uses the known data to simulate group manufacture conditions, and the method is being claimed as a means for finding the benefits 'on paper' before committing the overall efforts. This is feasible only if all the necessary data are freely available in the correct form (which is rarely the case); otherwise collecting the data for computer simulation may take longer than collecting the data for a full analysis. Crookall quotes the results of a survey wherein it is stated that a detailed classification system is not necessarily required in order to implement group manufacturing methods.

OVERALL SYSTEM APPROACH

The approach described here is based on the following major considerations:

(1) the reduction of the work input (i.e. supply) to each production section to a minimum, ideally to one supplier;

(2) the splitting of production into either material groups (for processing industry) or product groups (assembly industry);

(3) the creation of as much product (material) group autonomy as possible—in the reporting structure, span of control, and physical layout considerations, e.g. transport routes consistent with the economical and technological constraints;

(4) the completion of the highest amount of finished work for the least number of sectional (or departmental) visits, i.e. the reduction of the number of exit points to a minimum;

(5) the siting of each section to be geographically compact and compatible with other physical constraints, e.g. roof heights, availability of primary services, and

(6) the splitting of work in each section into simple/complex types to overcome the problem of extra machinery investment and to allow for manufacturing flexibility.

With regard to the above, the total production can be split into sections which are then given the cellular or flow line layouts based on the production routes taken and the machine utilization obtained from analysis of orders. For the purpose of analysis, the sample taken can be based on either (i) input (actual orders placed) or (ii) output (actual production achieved), both requiring to be related to a sales forecast. In effect, the input or output sample is used for establishing details of product (component) mix and the sales forecast to determine the level of volume. An output sample has the advantages that it records the actual situation as it happened (some-times not known) off the shop floor, but it has the disadvantages that it misses out on what never emerged from the system (in the given time) or what was never recorded to have been completed. The output sample is also 'smoothed' by the facilities available. It does, however, represent a solution. The analysis of the input sample has the advantage that the total input requirement can be ascertained. In a recent study, the order input was taken as a basis for the sample, although in an earlier analysis of another industry[16] the output was the criterion for data collection. Clearly, both analyses would enhance the end result—but experience has shown that either analysis modified to counteract its limitations achieves much the same end result.

It is important to appreciate at this point that, whatever analysis technique is used, the whole activity is with historical data, both suspect in its recording of what happened and even more suspect in relation to what might happen after a major system change. Thus, excessive sophistication in analysis will not improve its reliability in relation to predicting the outcome. The main objective is for the technique to place the solution in the 'right street', subsequent onsite modification using live data to produce the optimum solution. Thus the major requirement is to find the best overall solution ('the right street'), the details of the 'families' being more easily definable once the overall constraints are fixed.

The Overall System Approach (OSA) embodies the total search for optimizing the production system and not just one facet, e.g. machine tool utilization. (This later idea—another example of sub-optimization—is given by Frost-Smith[17] where he advocates the rescheduling of machines every time a new order is put on the machine shop. It is not difficult to imagine that if the advice were carried through it would create chaos in conventional layout situations and it would be impossible in group layout.)

The OSA can be applied in a batch production situation—either implosive or explosive (Fig. 1). An example of an explosive manufacturing situation serves to illustrate the technique. The batch sizes for finished components (in this case the product) are given in Table 1 which illustrates that over 65 per cent of the orders were below the quantity of 100: not an uncommon situation in manufacturing industry in general. Table 2 shows that a variety of orders are fulfilled at various exit points in relation to three basic material groups, i.e. some orders are fulfilled at the end of operations in a press shop, the

TABLE 1   Batch sizes of finished components

Batch Size	Percentage of orders	Cumulative percentage of orders
1–10	30·75	30·75
11–50	26·38	57·13
51–100	8·76	65·89
101–200	9·75	75·64
201–300	5·86	81·50
301–500	3·4	84·90
501–1000	8·3	93·20
1001–5000	4·4	97·60
5000 +	2·4	100

TABLE 2   Exit points for all orders in three product groups

Exit Point	Product Group A		Product Group B		Product Group C	
	Per cent of orders	Per cent of components	Per cent of orders	Per cent of components	Per cent of orders	Per cent of components
Semi-finished stock	25·2	8·95	10·6	0·24	50·1	18·0
After trimming, etc.	28·2	20·9	55·0	71·5	19·4	1·30
After general machine shop	35·7	60·6	28·5	24·5	30·5	80·7
After special treatment 1	–	–	5·3	3·28	–	–
After special treatment 2	8·5	1·4	0·6	0·28	–	–
After special treatment 3	1·2	7·3	–	–	–	–
After special treatment 4	1·2	0·85	–	–	–	–

others requiring special treatment sometimes after the machine shop. This again is typical of manufacturing industries.

It is tempting, when analysing an existing production system, to sectionalize the existing situation on the group manufacture principles, i.e. Fig. 2(a). This may have limited benefits, but remains unsatisfactory because of the interaction of the various product groups being manufactured and the resultant cross-flow between groups. Hence greater efficiency in the production planning and control would be sacrificed for the sake of grouping, control at inter-section level also being of major consideration. In relation to this, Edwards[5] has reached a similar conclusion when he states that 'G.T. is not just a parts and tooling exercise, but also a system exercise which concerns itself with that most complex of systems—the production planning and control system'. Hence, when investigating an existing production system, it is essential that some conceptual decision concerning the formation of product (or material) groups should be made at a very early stage. This can save a lot of time in reaching an optimal solution because data collection and analysis becomes subjective to a known end result. Data collection is so often the first major concern, it being assumed that the collected data will make the fundamental decisions—which it does not. In any industry, the product groups would emerge automatically, e.g. in a processing industry it could be the material types, whereas in an assembly industry it could be the product types. The underlying principle is the same. In an example derived from the processing industry, three basic material types existed (although the actual materials were over 150). One basic material produced only flat sheets at the press shop stage and had essentially different constituents from the other two; the primary services systems required were also different. This was endorsed since it also served a particular section of the market. The other two materials were essentially similar for primary service systems, but one material not only produced flat sheets, but also rotational bars and rods. A consideration of the total activity showed that it was equally divided between sheet

type and rotational type products in these two materials. Hence three essential product groups were created, viz:

Product Group A—all activity on material 1.
Product Group B—all work on material 2 and all sheet work on material 3.
Product Group C—all rotational work on material 3.

Again, the sheet type and rotational work in materials 2 and 3 served particular sections of the market.

The split into product groups clears the way for fuller subjective analysis of each group and produces the conditions conducive to the first requirements in the OSA methodology—i.e. reducing the inputs to each section to a minimum, ideally one.

Figure 3 shows the traditional situation where two materials are shown to share the same facilities, and it can be seen how difficult it is to have the minimum number of inputs and good intersectional control. Figure 4 shows the solution based on the product group concept. It will be noted that the material flow is simplified and the semi-finished material stores become the responsibility of planning, hence unifying the management function. The management structure proposed to operate such a system is shown in Fig. 5. The structure is based on providing the maximum amount of product group autonomy (e.g. maintenance, examination functions) at an economical cost[18]. This autonomy creates group efforts at a larger micro-scale. It also gets control by commitment and accountability and brings about the most significant contribution to production control. Each product manager has his clear span of control and deals with all concerned about matters arising from his product group. The planning supervisor and foreman are linked to provide a service which would (i) maximize the production, and (ii) minimize the disruptions. The three product groups mentioned earlier showed that for each product group, the number of production sections (or departments) varied in accordance with the exit points. In one product group a section was too small to warrant

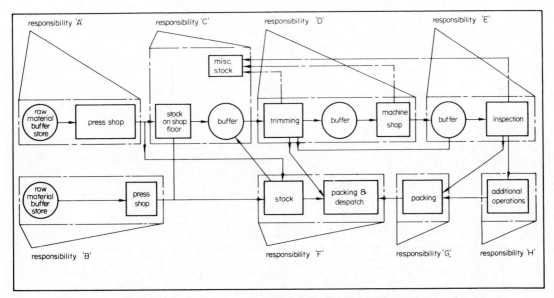

Figure 3. Traditional manufacturing system for Product Group A, and part of Product Group B.

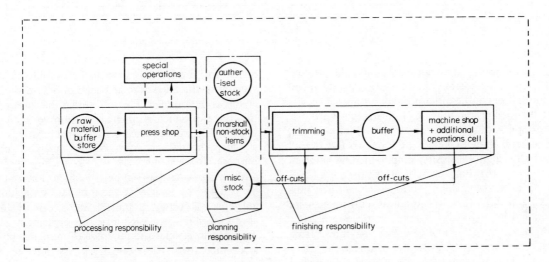

Figure 4. Product Group A manufacturing system based on product group concept.

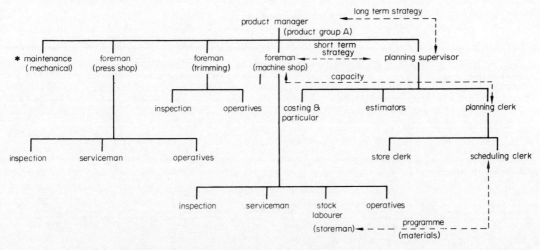

Figure 5. Organization structure for Product Group A.

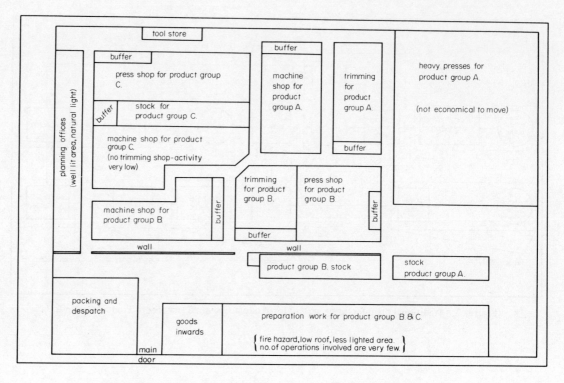

Figure 6. Proposed geographical location for all three product groups.

separate existence, although as a combined production facility that section did exist as a separate entity (as it still does) in the other product groups.

If the total production area is divided into product groups and each group is under a product manager, then careful thought has to be given to the overall layout so that it ensures (i) simplified flow between the sections, (ii) the geographical compactness of all the sections in a product group for supervision purposes, and (iii) that geographical constraints are considered beforehand and taken into account. For example, it may be uneconomical to move large presses or ovens with expensive foundations, or a confined area may not be used for environmental reasons, or the low roof height may preclude the installation of certain machinery. Other considerations such as the roof structure or insurance require-

ments for fire hazards have to be considered when specifying the areas for each section before the final layout is determined. Figure 6 shows the solution for the three product groups in relation to such constraints, the layout indicating sectional locations.

Once the data is collected and analysed for the purpose of machine layout, it usually provides answers showing that the best possible flow can only be achieved by increasing the number of machines in existence. This may be possible in the case of simple, cheap machines, e.g. a circular saw or a single spindle drill, but in general the purchase of more machines is a dis-incentive. Therefore it is essential to keep investment in new machines to a minimum level. This is possible if the principle of simple/complex split of the work is adopted. This principle is applicable to all types of section since it retains the same flexibility as

TABLE 3(a) showing all the sequences of operations in a product group

Machine Types	A	B	C	D	E	F	G	H	J	K	L	M	N	P
No. of machines available	1	2	5	1	1	1	2	1	5	1	2	1	1	1
Sequence of operations { 1 2 . . . n			(−) hours of work in each sequence	−		−	−		−		−			
Total time (hours)	x	x	x	x	x	x	x	x	x	x	x	x	x	x
Grand total					X hours									

TABLE 3(b) showing the division of work into 'simple' type based on machine's availability

Machine types	B	C	G	J	L
No. of machines required for 'simple' work	1	3	1	3	1
Sequence* of operations	1	–			
	2		–		
	.	–		–	
	.		–		–
	.		–		–
	n_1				
Total time (hours)	x_1	x_1	x_1	x_1	x_1
Grand total			X_1' hours		

* These sequences are derived from the total sequence in (a).

already existed. For example, consider a machine shop in a particular product group:

The first exercise has been to allocate machines to each product group, since this is the overall system requirement. Each type of operation (normally a machine) is analysed against each sequence for its work content as shown in Table 3(a): essentially this is flow analysis.

The idea is to split this facility into small manufacturing groups (up to 10 operatives each) if it already exceeds this figure. In general, it has been found that splitting by product group produces fairly small units which only require sub-splitting into (say) 3. A further requirement is to retain flexibility of manufacture after splitting. This is achieved by withdrawing the 'simple' components from the total (Table 3(a)). 'Simple' is defined as components manufactured solely on machines of which there are more than one. Reference to Table 3(b) would suggest 'simple' as components using machines B, C, G, J, L. The next step is to determine the work content associated with the 'simple' group. This may produce the required split in relation to size of groups. A further sub-division of the 'simple' group can be undertaken on the same lines, if required. It is not important to have an even split between 'complex'

and 'simple' components ('complex' being the items remaining after completing the above exercise) since by definition 'simple' components can be manufactured on the 'complex' component facilities. This mechanism allows for meeting the market fluctuations[19]. Figure 7 shows the division of work into simple/complex types for the three product groups mentioned previously. The previous machine shop, however, can now be seen to comprise 6 groups, the important feature being that the groups are now flexible, controllable, and aligned to the actual market.

CONCLUSIONS

(1) The Overall System Approach is necessary to avoid the pitfalls of narrow machine shop approaches and meaningful results can thereby be achieved, thus optimizing the total manufacturing system.

(2) The method developed for cellularization of the facilities is simple but effective in operation, whilst the number of sub-divisions into simple/ complex work types is not limited. As well as being simple, this method ensures that the investment on new machines can be minimized without loss in production capacity and flexibility.

(3) Divisions into product groups based either on material type (processing industry) or product groups (assembly) at the conceptual stage simplify the subsequent analysis by applying certain overall constraints.

(4) Production control is simplified and the system described is superior in that the management responsibility and accountability is firmly established, the management structure being an integral part of the system.

(5) It is implicit that no formal classification system is necessary for the formulation of component/machine families. Although classification is necessary, it is subjective in relation to conceptually predetermined factors.

(6) Having established the basic framework of a manufacturing system, particular areas such as tooling (e.g. a multiple jig) and composite workpiece creation within a particular cell can be investigated, since gain in this area now contributes to the overall efficiency. Similarly, it can now be decided on much more rational grounds whether different machine tools (e.g. NC) are necessary and, if so, the workpiece statistics available from the program of work load would indicate the parameters to be fulfilled.

ACKNOWLEDGMENTS

The authors wish to thank the Board of Directors of Ferodo Ltd. for permission to publish this paper.

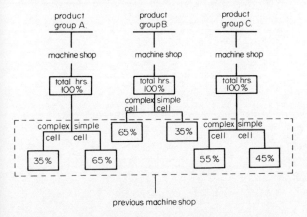

Figure 7. Division of work between simple and complex cells for all three product groups machine shops.

REFERENCES

1. H. OPITZ et al. (1969). Workpiece classification and its industrial application, *Int. J. Machine Tool Design and Research*, **9**, 39–50.
2. J. GOMBINSKI (1969). The Brisch classification and group technology, *Proceedings of International Seminar on Group Technology*, Turin, September.

3.   V. B. SOLAJA and S. M. UROSEVIC (1969). Optimisation of group technology lines by methods developed in the Institute of Machine Tools and Tooling (IAMA), *Proceedings of Int. Seminar on G.T.,* Turin, September.

4.   F. HELLSTROM and C. SELJEE (1966). Shape classification as an aid to production control, *Teknisk Ukellad,* **113** (23), June.

5.   G. A. B. EDWARDS (1971). *Readings in Group Technology,* Machinery Publishing Co. Ltd., Brighton.

6.   D. ZIMMERMANN (1968). Design—the focal point of rationalisation, *Engineers' Digest,* **29**, (5 and 6).

7.   J. HOLLINGUM (1971). Computer helps plan plant layout to speed batch flow, *The Engineer,* 16/9/71.

8.   G. H. MIDDLE, R. CONNOLLY and R. H. THORNLEY (1970-71). Considerations for an industrial classification system, *I. Mech. E. Proceedings,* **185**, Paper 45/71.

9.   J. L. BURBIDGE (1969). Production flow analysis, *Int. Seminar on Group Technology,* Turin, September.

10. I. F. K. EL-ESSAWY and J. TORRENCE (1972). Component flow analysis, *The Production Engineer,* May.

11. P.E.R.A., Report Nos. 207, 243.

12. C. ALLEN (1972). Group Technology short course at U.M.I.S.T.

13. (1972). Letters in the *Production Engineer,* July.

14. J. McAULEY (1972). Machine grouping for efficient production, *Production Engineer,* February.

15. K. I. BALDWIN and J. R. CROOKALL (1972). An investigation into application of grouping principles and cellular manufacture using Monte Carlo simulation, *Manufacturing Systems (C.I.R.P.),* **1** (3).

16. R. CONNOLLY, G. H. MIDDLE and R. H. THORNLEY (1970). Organising the Manufacturing facilities in order to obtain a short and reliable manufacturing time, *Proceedings of the 11th MTDR Conference,* Birmingham.

17. E. H. FROST-SMITH (1971). Optimisation of the machining process and overall system concepts, *C.I.R.P.,* **19**.

18. R. CONNOLLY and A. J. P. SABBERWAL (1971). Management structure for the implementation of group technology, *C.I.R.P.,* **19** (1).

19. R. CONNOLLY and A. J. P. SABBERWAL (1973). Group Manufacture. Presented at the C.I.R.P. General Assembly in Bled, Jugoslavia, August 1973.

DISCUSSION

Q. F. W. Craven, Herbert Machine Tools Ltd. What is the definition of simple and complex, and what is the precision of this definition?

A. For simple and complex work piece definition please refer to the text of the paper. It will be seen from the earlier notes that the precision of the definition depends upon accuracy of route cards, which in general are assumed to contain correct information.

THE ECONOMICS OF GROUP TECHNOLOGY

by

G. M. RANSON*

SUMMARY

There is little value in talking about the economics of Group Technology without reference to the economics of the conventional way of operating a batch manufacturing business. It is well known that, under the conventional practice of batch manufacturing, excessive stocks and work in progress are inevitable. Despite these stocks, deliveries are invariably bad and throughput times lengthy and unmeasured. In simple terms, if a company has stockholding representing anything up to 55 per cent of annual sales value, there is the added annual burden of supporting that investment with high operating rates. In addition to this there is the further burden of excessive space utilization to house these stocks, heating and lighting, constant movement, and the ever present danger of obsolescence.

There are a number of indisciplines which also attend the conventional practice such as the ever-open door stores policy which enables materials to be withdrawn for the purpose of keeping machines and work-people busy irrespective of whether or not there is a need for the parts created, at least within a reasonably immediate future. The inability to measure properly and, therefore, predict manufacturing throughput times always endangers a correct delivery promise with the consequence of continued overdue deliveries often repetitive for the same product. A strong emphasis is placed upon the use of high volume capability machines in the belief that costs are reduced in an isolated place to the benefit of the company as a whole, whereas much simpler machines, properly arranged, would provide much lower investment with a far better reward to the company when all the facets are taken into account. Even with the items mentioned above, which are far from the total which can affect the conventional method of manufacture, the sum total of operating is far lower than the rewards to be gained by the application of Group Technology.

The accompanying short description of Group Technology and its application at Serck Audco Valves amply demonstrates the economic benefits to be derived.

GROUP TECHNOLOGY

Production engineers have long realized that the most economic method of manufacturing is mass or flow-line production. This rationalization has resulted in manufacturing facilities tailored to produce a clearly defined range of components against an established demand.

Unfortunately a large number of organizations never really attain this ideal and, faced by the nature of their products and the fluctuation in market demand, consider the manufacture of component parts in less than desirable batch sizes.

When this situation occurs, invariably an 'economic batch quantity' approach to component manufacture is adopted in order to reduce machine costs per unit. This in turn results in large stocks of finished components.

Group Technology, as it is called, is a technique which allows the production of components normally produced in small batches to achieve similar economic advantages to those associated with continuous flow-line production.

Getting started
The approach is to analyse a company's total product line and identify those families of components that are related by a similarity in shape and/or type of production facilities required to manufacture them.

To perform this analysis properly, it is essential that each component part be examined according to predetermined standards; the shape and manufacturing requirements should be clearly identified and related to other components within the total product group. The use of functional descriptions of component parts, however, is not satisfactory.

---

* Serck Audco Valves

The key is to establish a classification and coding system that identifies the shape and manufacturing requirements by assigning a specific digit to each significant feature.

Before using this approach, a basic division can be made to separate rotational (i.e. components normally rotated during their machining) from non-rotational components and also to identify their shape features.

The general information content that a classification system suitable for GT applications should carry is as follows:

(1) Geometric definition of external and internal shape.
(2) Additional features, such as holes, slots, splines, etc.
(3) Material type and, where practicable, the initial material form (bar forgings, castings, etc.).
(4) Size envelope.

This information can be supplemented by other features such as accuracy and weight where the product type requires them.

The preferred approach is to use a code number classification system which contains the necessary levels of information from drawings and production planning documents.

The next step—having identified all of the families of components—is to determine production quantities required for each component for a given period, including setup and machining times. An assessment of machine-load content per component family can now be established. However, if the work content is not sufficient to load a group of machines economically, the parameters of the family should be extended in either size, shape, or material until a proper load is developed.

Having reached this point, an examination should be made of the processes used to manufacture the components. These may also warrant changes, considering that now, in effect, production consists of large batches of components with a high degree of similarity.

The final stage in this process of GT planning involves the examination of each component within a family to define its tooling requirements. Once this is done, a group of dissimilar machines can be brought together capable of manufacturing a complete component family.

What to expect

It is expected that initiation of this manufacturing system will result in a radical reduction in total component manufacturing time. Secondly, a drastic reduction in setup time should occur as machines will only require adjustments rather than being completely reset for each new component. Work in progress will be reduced as production reflects more accurate production figures for production control. Finally, the queuing time associated with inter-operational machine loading will be eliminated, thus further reducing back-up.

By establishing a classification-based data retrieval facility, a new management tool becomes available for estimating production times and costs for each new product. Each component of the new part is coded. Then a search of the data files is conducted to determine into which 'family' each new component will be placed.

Finally, the documented relationships between component shape and manufacturing requirements, used to establish machine groups, define the type and size of various machine tools required to produce them. Thus future machine tool purchases can be based upon these clearly identified needs.

An investigation of machine capacities in a workshop usually reveals an imbalance between the size of components produced and the capacity of the machines used to produce them. In other words, the machine tools being used are larger and more expensive than necessary.

In establishing groups of machine tools capable of producing a defined family of components, it is necessary to accept a disciplined approach as to their control. Each new order must be exploded into its component parts, coded, and then examined against the existing machine group's parameters. Decisions must be made as to the acceptability of a new component within an existing group, and if necessary, the group's parameters and tooling extended to accept it.

Applying Group Technology

Group Technology has been adopted by a number of engineering companies in Britain, but as J. A. Harris, project leader of the Ministry of Technology's GT Centre points out:

'Companies are generally reluctant to let others, especially competitors, know of their internal activities, particularly while they are passing through the development stages of a changed system'.

However, there is one company that is not hesitant to discuss their application of GT—Serck Audco Valves, valve manufacturers of Newport, Shropshire, England. This company is now being used by MinTech's GT Centre as a showplace to illustrate what can be attained with proper implementation of the group technology concept.

In June 1966, the Government-inspired National Economic Development Office called a conference on production planning and control. At this conference it singled out Serck Audco Valves as an outstanding example of a company that had found a successful way to increase production, cut stocks, and achieve better and more reliable deliveries.

When Serck Audco Valves began to evaluate their company, it became evident that there were problems involving excessive stock and indifferent delivery performance, and many interdepartmental frustrations. In total, these problems could only be successfully overcome by a co-ordinated company effort.

Results indicated that a total integrated approach was necessary: first, to 'plan the work', from incoming orders or sales forecasts, through the various departments of the company to final despatch; second, to 'work the plan', allowing management to deal with those problems that fail to meet the scheme.

Sales, purchasing, accounting, etc., as well as the

production departments, were all involved in initial planning activities such as work study, production engineering, cellular grouping of machines, job evaluation, and production control. They also made additional contributions to the overall plan.

Serck Audco Valves concluded that it was more advisable to have £150 000 of machine tools under-utilized than £½ million of stock. A report summarizing what had been achieved up to 1965 shows:

(1) Sales up by 32 per cent.
(2) Stocks down £500 000, or 44 per cent.
(3) Stocks/annual sales ratio down from 52 to 25 per cent.
(4) Average manufacturing time reduced from twelve to four weeks.
(5) Past-due orders down from six weeks output to one.
(6) Wages per employee per annum up from £700 to £900.
(7) Increased despatches per employee, £2200 to £3100.

The task had proved a big one, but the company did not rest on the basis of this report. Impressive as the above figures are, the position at 1971 shows even greater progress. Sales are up 67 per cent; stock annual sales ratio is down to 22·5 per cent; average manufacturing time down to 2½ weeks; wages per employee up to £1520; and average despatches increased to £5700.

Capital investment of the plan, which took three years to achieve, was recovered four-fold by stock reduction alone and enabled the company to move ahead with its planned expansion.

Failure to satisfy a delivery promise is becoming more and more the charge levelled against British industry, particularly in the export field, and becomes even more serious when our records are compared with overseas competitors. These charges are equally valid for home sales.

Therefore, the real and ultimate objective is to deliver products on time and arrange the operation so that the time required is truly the minimum necessary. In order to attain this objective it is necessary to take a cold hard look at the total operation. Often it is believed that merely by introducing a new production system the problem is solved. However, the total operation must be studied and an integrated plan—from receipt of an order to the despatch of a product—formulated and then acted upon with determination and proper implementation.

Studies show that invariably far too much of the control of business is vested in people who are departmentalized and generally act in isolation, without semblance of coordination, each keeping his or her own records—often duplicated elsewhere—which are either not complete or contain irrelevant information.

Following a thorough examination of their production activities and manufacturing operations, Audco's management found they were faced with a nine-fold task:

(1) Create facilities to provide accurate information for delivery dates and maintain them.

(2) Provide a coded system for identification of all parts, assemblies, and products.
(3) Create facilities yielding the shortest processing time from receipt of order to despatch of product.
(4) Reduce stocks and work in progress to a minimum commensurate with efficient operation.
(5) Examine the product line and develop a rationalized approach to provide a more acceptable product line in total.
(6) Base the whole activity on measurement so that facts can be used for computations.
(7) Institute tight controls on all material movements and record keeping.
(8) Create materials supplier relationships to ensure reliability.
(9) One of the major benefits from classification and coding was the enforced study of the products and component designs. Not only did the exercise reveal very many instances of closeness which enabled a substantial rationalization to be made, which amounted to more than 20 per cent of component parts, but the opportunity was afforded to study closely the design features of both products and piece parts to enable as many common parts as possible to be used within the product range.

The point concerning materials supplier relationships was felt to be particularly important, since suppliers were now aware that Audco demands are very real and broken promises from them could seriously upset the operation.

To ensure that the new scheme was planned as efficiently as possible, the company engaged a team of six work study engineers who were unfamiliar with any of the former work traditions of the plant. The team found that the greatest difficulty encountered in the implementation of the system was the resistance to it.

Cellular group production

The term 'cellular group' is described as a number of flow-line cells of plant equipment arranged alongside conveyors so that piece parts are begun at one end and completed at the other. Each group is designed to process a specific type of piece part, material, and size range; each cell is also limited to a certain size. This system involves a considerable number of methods changes in order to balance, as nearly as possible, the elemental times of each piece part.

Each cell contains the required number of machine tools to deal with the variety of piece parts within the size range. The rate of output is generally based on the number of machine tools used; an accumulation of work in progress is virtually eliminated.

All concerned, including machine operators, were invited to study the model layouts and were given an explanation of the advantages of the new system versus the existing one.

A completely new wage structure was developed in accordance with the new plan of operation. The result was nine hourly rates of pay throughout the plant. This replaced some ninety-three different rates which previously existed.

The storage system was also revised to ensure faster material movement and inventory control. Assembly lines, which were formerly located in three areas, were consolidated into one. Consequently, this rationalization of personnel and equipment provided much greater labour flexibility.

The new operation was based on logic, measurement, and planning. Knowing what was required and when, enabled management to concentrate on whatever failures occurred. This could be termed management by exception. Moreover, the function of management at all levels became clear-cut and objective. Able to concentrate on true functions, management turned its energies to forward planning and better practices.

All in all, once the new principles and philosophies were established, they become self-generating by creating a progressive-thinking, progressive-working environment illustrating that when group technology is put into action—it works!

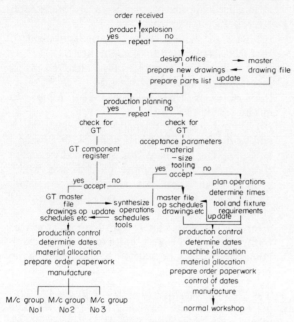

Figure 1. Diagram illustrating Serck Audco's processing under Group Technology planning—from receipt of order to manufacturing.

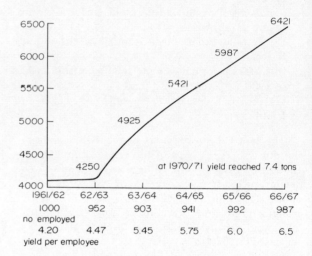

Figure 2. Serck Audco total annual output in tons and yield per employee.

TABLE 1   Serck Audco Valves performance record

(1)	(2) Net despatches	(3) Value added	(4) Average no. employed	(5) Total wages and salaries	(6) Despatches per employee	(7) Value added per employee	(8) Average income per employee
	£	£		£	£	£	£
1961/62	2·220	1·615	1001	0·714	2218	1613	714
1962/63	2·184	1·580	952	0·706	2294	1660	742
1963/64	2·585	1·872	903	0·752	2863	2073	833
1964/65	2·922	2·007	941	0·844	3105	2133	897
1965/66	3·363	2·303	992	0·951	3390	2320	953
1966/67	3·768	2·646	987	0·979	3818	2681	992
1967/68	3·576	2·519	906	0·951	3947	2780	1050
1968/69*	4·984	3·117	1130	1·338	4411	2759	1184
1969/70	6·008	3·786	1181	1·578	5087	3206	1336
1970/71	6·727	4·374	1179	1·787	5706	3710	1516

* During this period the progress in column (7) was retarded by the problems associated with the transfer of £1·3 m of Ball Valve production to Newport.

REFERENCES

Information about Serck Audco Valves application of GT is to be found in the HMSO 1966 publication *Production Planning and Control* and the UKAE *Atom* bulletin number 158 published in 1969.

DISCUSSION

Q. J. R. Green, Alfred Herbert. In assessing the annual cost of holding work-in-progress the multiplying factor should be $(v + u)/(1 - t)$ times the capital tied up.

In this expression v is the rate of return on shareholders' equity (capital plus reserves), u is the annual rate of inflation and t is the corporation tax rate.

The corresponding expression for the annual cost of *plant* depends on writing down factors and depreciation but will usually be less than the cost of holding work in progress for the same amount of capital tied up.

This is in conflict with the relative values of £150,000 for machine tools and £½ million for stock and WIP mentioned at the top of the left-hand column of page 3 of the paper.

A. The £150,000 was the cost of the total reorganization, of which £75,000 was used for the purchase of balancing plant.

The consequence of the reorganization was a reducting of stock of £550,000 and the direct support cost then running at 8 per cent borrowing rate. We computed a cost of 15 per cent which specialists advised us was too low and should have been 18 per cent. Thereafter stocks were turned over four times a year. 50 per cent of the work space was released; delivery performance was accurate and 97 per cent on target; throughput times 5 to 12 times faster; no progress chasing; proper accent on improving performance with immediate results; return on capital employed at best 37 per cent and worst 27 per cent. 'Normal' batch manufacturing performance cannot match these results.

Q. J. A. Stokes. Mr Ranson rightly comments that we are seeking to maximize the return on our capital employed. We have heard much of techniques which reduce work in progress and something of improving machine tool utilization. Since both work in progress and machine tools represent capital employed, does Mr Ranson know of any work directed towards establishing the criteria for minimum capital employed, i.e. for hitting the right balance between fully loaded machines with lots of WIP and lots of spare machining capacity with rapid throughput and minimum WIP?

A. The best answer I can give is the revelation of results: See answer to previous question.

Q. P. C. Hagan, Eaton Ltd, Axle Division. Mr Ranson mentions in the paper a new wages structure; could he give a brief outline of his scheme in answer to the following questions?

(1) Is his pay scheme one of payment by results, measured daywork or otherwise, and is this scheme based on the individual performance of the operator, or the group as a unit, or the total plant output?

(2) He mentions the reduction of wages rates from 93 down to 9 throughout the plant: in the light of this:

(a) Are people of different rates employed in each group, or are the operators of one grade of labour rate only employed in that unit?

(b) Obviously this reduction of the number of wage rates indicates some analytical work was performed, i.e., some form of job evaluation. If this is so, how did you classify between different traditional skills which in most companies, apart from, say, toolmakers are all paid equivalent wage rates?

(3) Is transfer of operators possible between machine Groups? or does the classification of jobs prevent this? As an extension to this, can an operator move from one grade of labour payment to another (whether higher or lower), and whose responsibility is it to recommend this?

(4) GT will mean new training programmes for operators and management alike; could Mr Ranson give a brief summary of his company's operator training schemes, and give the opinion of the unions in his plant on the mobility of labour?

A. (1) (*a*) Group payment in cells.

   (*b*) Individual bonus earning on the larger machines.

(2) (*a*) In the main the simpler cells have uniform payment as the skills are common. Certain more skilled requirements within some cells command a higher rate. All operators are classed as skilled and belong to one of three categories of skill, and are so paid.

   (*b*) Job evaluation was implemented based on the BIM points rating system and formulated jointly with management and unions.

(3) Operators are mobile and move between cells. If an operator is called upon to carry out the next category of skill and remains for a period of three weeks, he is automatically upgraded and stays in the new grade.

(4) A training programme of versatility with the use of machine tools was carried out. A training school was established and is still in use where new personnel can undergo induction in the use of a variety of machine tools.

The unions describe the old method of conventional machine shop practices as 'the bad old days'.

Q. W. H. P. Leslie. In this paper Mr Ranson appears to be saying that production control is at least equally important to group technology in obtaining the improved financial performance of his company. There would be no point in producing different components quickly in their particular cells if they did not arrive for assembly at the correct times.

A. The deduction is absolutely correct. The philosophy of the production of pieceparts and storing them hopefully to assemble into products gives way to the planning of product requirements and exploding into the requisite components which form the manufacturing programmes of two-week frequency.

SOME CONSTRAINTS, FALLACIES AND SOLUTIONS IN GT APPLICATIONS

by

F. W. CRAVEN*

SUMMARY

This paper, in identifying the importance of Group Technology and Cellular Manufacture, specifies some of the problems to be recognized and dealt with in the preparatory stages. Attention is drawn to the need to avoid the development of general purpose solutions from successful narrow based achievements. Reference is made to the contribution that modern NC lathes can make in this field.

INTRODUCTION

During the last few years, a great deal has been written on the subjects of Group Technology and Cellular Manufacture, both in the technical press and in academic publications; if the degree of conviction were related to the volume of argument, the manufacturing world should, by now, be fully dedicated to GT as the answer to most of its problems of batch production. Much of the literature, however, is unapplied theory or, and possibly more misleading to the wide product range, general engineering industry, based on actual applications in untypical narrow product range environments. This is not to imply that many such 'single product' applications have not been relevant to a particular company's problems, nor is the competence with which they have been carried out in question. Rather is notice drawn to the danger of developing general solutions based on simple applications.

It is also suggested that many articles may have dealt with the benefits of cellular manufacture without adequate attention having been paid to the constraints and the disciplines necessary in applying GT or cellular manufacturing techniques to obtain the most beneficial results.

In some cases, assumptions have been made empirically, or upon narrow based data, leading to fallacious general conclusions. As an example, much of what has been written on set-up costs falls into this category. On the other hand, a great deal of energy has been dissipated by protagonists of particular aspects of GT in defining their particular solutions as the best way (with strong implications of being the only way), whether their arguments have concerned coding, production flow analysis or whatever.

It is neither the purpose of this paper to support a particular approach, nor to define a ready-made solution technique, but rather to draw attention to some of the principal elements that should be considered when deciding to launch a GT project. It should be clearly understood that the items dealt with in this paper by no means make up a comprehensive list of matters requiring consideration.

The matter is of vital importance to British industry; there are great benefits to be gained which can be maximized if the subject is thoroughly understood, the pitfalls recognized and adequately dealt with at the planning stage. In the writer's company, in one factory alone the whole of the machine shop is part way through a complete reorganization, moving from a functional layout to a cellular layout of twenty-six cells. The benefits so far realized with the cells already established are substantial. The level of achievement is directly related to the thoroughness of the preparatory work.

We now go on to define and comment on some of the most common reasons used to justify GT applications.

1. Popular reasons for implementation of GT and some guiding comments

Statements of the main advantages to be gained when moving from functional layouts to cellular manufacture often include the following:

(a) Reduction in setting time and costs
(b) Increased machining capacity
(c) Reduction in tooling investment
(d) Reduction in handling costs
(e) Reduction in throughput time
(f) Reduction in work in progress
(g) Unification of responsibility.

It is suggested that several of these statements require further investigation as to their validity and

* Planning and Development Director, Herbert Machine Tools Ltd

significance against a particular company's operations, prior to setting course towards the conversion of manufacturing systems to GT or cellular manufacture.

Thorough consideration needs to be given to the prime purpose in making the change so that the best solution can be determined and the problems in achieving this end can be identified and solved before implementation.

As indicated in the introduction, because the possible effects have been so widely and persistently proselytized, there can be an assumption that wholesale improvements in many factors will be automatically achieved. There is little doubt that this can be done, provided objectives are clearly defined and the preparatory stages of the operation are carried out in a thorough manner. Let us examine the items listed above in more detail.

(a) Reduction in setting time and costs
In a paper of this brevity, it is difficult to draw continual attention to the differences between GT and cellular manufacture at the various stages of discussion.

It is stressed that the following comments will not apply in all circumstances. In many cases, the grouping of components into families will not, in itself, reduce actual setting time but may increase it. In the section in this paper on the construction of cells, reference is made to the opportunities for grouping similar set-ups in functional layouts, which could, in certain circumstances, lead to actual setting times being less in functional layouts due to easier matching of component requirements to machine facility. It is probably true to say that in the case of products which fall into natural families there may often be a greater setting cost benefit in cellular layout than in functional layout. Much of the literature deals with examples of this type. In the general engineering industry where there is a wide variety of component type, unless adequate production engineering work is carried out, actual set up times may increase, particularly in turning operations.

There is a considerable difference in the nature of work holding and setting on turning machines, particularly combination lathes, compared with drilling machines and milling machines where set-up varieties are more easily condensed. Inevitably, in general engineering situations, product geometry and size within families can still display considerable variety unless cells are made so small as to be virtually unworkable, principally because of unacceptably low utilization of second operation machines. In some cases, the effect of component size is of importance; with components of similar geometry, it is generally found that the ratio of set-up to operation time is considerably more adverse for small components.

Many statements on GT applications in the turning field refer to the composite component or some similarly termed device, with the automatic implication that the use of such techniques substantially reduces or eliminates alterations to setting. Whereas this may be true to some degree with certain cross sliding lathe applications (and it is comprehensively true with NC lathes) it is certainly questionable on turret lathes and capstan lathes. It seems hardly necessary to state that different sizes of collets are required for different sizes of bar, different sizes and settings of roller boxes are needed in turning different finished diameters and that a whole range of end working tools are required to produce a variety of different sized holes.

Combination lathes have to be set for turning to specific diameters and lengths and unless the features and dimensions of sequential components are identical and not just approximately similar, many alterations in tooling or stop setting have to take place. There is a fundamental difference between turning operations and milling and drilling operations, principally because of the effect of the three main elements of set-up, namely:

(a) Holding the workpiece
(b) Tool capability/size
(c) Relative movements, in several planes, of work piece/tool.

Considerations of these features lead to the conclusion that set-ups can be better and more simply standardized on milling and drilling operations, rather than on turning, particularly with combination lathes.

The development of consolidated set-ups should be slanted towards producing set-ups which are low in cost to operate and which are either comprehensive, that is, capable of producing a variety of components with no tool change (a limited solution), or flexible, that is, able to produce a variety of components with minimum tool change. It is possible in a cell environment to engineer such set-ups since the work is presented to the Production Engineering Department in more manageable parcels; the opportunity is afforded to define preferred set-ups and preferred planning methods which if rigorously applied, can substantially reduce on-going planning costs. Using standardized set-ups, one can afford to have a higher provisioning cost to obtain reduced set-up application time, because overall, there can be a substantial reduction in the number of such facilities. It must be borne in mind, however, that this is an expenditure which has to be undertaken in the construction of effective cells.

In the writer's company, work has been done in identifying the ratios of set-up to operation time through various production flow sequences with a view to establishing the most adverse ratios in such sequences and to determine their permanence, in order to schedule operations in such a manner as to optimize the sequential use of set-ups with the most adverse setting/operating ratio. In this work it is necessary to grade elements of the alterations to set-ups in order of significance and frequency. This is a most complex subject and it is prudent in such work not to attempt to develop a fully optimized solution. Consideration needs to be given to the method in which the scheduling information is transmitted to the cell supervisor and to the level of scheduling which is prescribed to him, together with the degree of options which should be left to his own immediate choice.

It is of great importance in a production limited operation to achieve minimum set-up costs because excess costs of this type will, in the extreme, have to be converted into additional capital investment. In conclusion, bearing in mind that batch sizes are often influenced by existing machine facilities, the efficiencies obtained from reduced set-up costs can give the opportunity for further reductions in work in progress by batch size reduction.

(b) Increase in machine capacity

Generally, this is not an automatic benefit and in some cases the implementation of GT may lead to an overall reduction in effective machine capacity. It is accepted that certain machines will be under-utilized, and a prime factor in the construction of cells is the necessity of ensuring that the higher capital cost machines are fully utilized. The internal balance of machine loading is somewhat dependent on the scale of family range and variation of batch size. In designing cells, there is a conflict between large cells with a well-balanced work load with their attendant drawbacks and smaller cells based on narrower families which, in the ultimate, can lead to gross under-utilization, particularly on second operation machines. It is unfortunate that, when the question of under-utilization is discussed, the cell proposals are based on a comprehensive analysis using data which has been calculated with some accuracy. Conclusions from this are frequently compared with what may be a false assumption of previous individual machine loading. Where a section is already seriously production limited, particular care needs to be paid to the validity of the decision for moving to GT. Steps can be taken to counteract specific reductions in capacity but such problems need to be recognized and solutions determined at an early stage.

(c) Reduction in tooling investment

There can be considerable deception in this statement. If a plant is being established from scratch with properly designed cells, the total tooling investment could certainly be less than that required in a functional layout. This stems from the consolidation and standardizing of set-ups. This 'green field' situation is rarely met in practice. Most concerns will be moving from a functional layout with its existing tooling investment to a GT layout which initially will probably make some existing equipment obsolete and which will certainly require additional expenditure on tooling as indicated in the section on set-ups. Furthermore, even in the best run batch production shops, equipment and tooling are frequently developed and used on the shop floor unknown to the Production Engineering Department; the disciplines required by GT cannot accept this situation and provision has to be made for the withdrawal, identification and formalization of the use of such equipment or, in some cases, agreed replacements. The situation is somewhat similar to the purchase of, for example, an NC drill where existing drill jigs are made obsolete and have to be replaced by improved provision of tooling, equipment and better disciplines, and initially this costs money.

(d) Reduction in handling costs

In most cases, the proximity of sequential machines in cells reduces the length of the material flow line. It is generally necessary in establishing cells to provide improved handling facilities or at least, a restructuring of existing facilities, particularly for larger components.

(e) Reduction in throughput time

In the writer's own company, in the many cells already established, a substantial reduction in throughput time has been achieved and this clearly is the general case. The prime benefit from this is, of course, the reduction in work progress which is dealt with separately, but throughput time reduction, per se, is of great importance in handling urgent work, spares orders and in rectifying omissions.

(f) Reduction in work in progress

The correct application of GT procedures should generate substantial reductions in work in progress, and this is the prime reason for most companies' activity in this field. The opportunities for lowering work in progress are several. There is firstly the initial reduction that stems from quicker throughput time which hardly needs explanation. Secondly, and particularly in complex businesses with a wide product range, if the whole of the machine shops is of a cellular nature, there can be a far closer coupling of the production programme to the constantly changing order input. The controls stemming from this facility enable changes in overall levels of business to be reacted upon much more quickly and, of even greater importance, random changes in demand for particular products can be recognized and dealt with more speedily. The third opportunity for reduction in work in progress can be obtained from a combination of lower stock holding in finished part stores and the manufacture of smaller batches. The quicker throughput time enables changing market needs, errors, random demands etc., to be quickly dealt with and for stock-outs to be corrected more quickly. With regard to reductions in batch size, it is most important that proper judgements are made on the savings to be obtained from reduced stock holding compared with the increases in cost associated with smaller batches and more frequent set-ups. This point endorses the need to develop lower cost set-ups, together with the facility for sequential scheduling, to obtain minimum set-up change.

In passing, it is interesting to note how much work has been done in the machining area, compared with the little attention that appears to have been paid to the importance of the time for building sub-assemblies and complete units.

(g) Unification of responsibility

There are many benefits to be gained from the corporate nature of cellular manufacture. The close involvement of work people with one another's activities, particularly in the interaction of interdependent operations such as turning for grinding etc., enable common problems to be solved quickly. The need for flexibility of labour gives the opportunity for work people to gain additional skills. The self-containment of the cell produces a better morale which, among other things, leads to quicker correction and reduction of faults. A direct operator contribution is the improvement of quality and other positive attitudes stemming from the feeling of belonging to a reasonably small unit.

There can be problems in obtaining the right supervisor; in the functional layout, the foreman is generally very experienced in a small number of operations such as turning or milling, whereas in a cellular layout, the supervisor requires a much broader experience and such people may not easily be found. The supervisor's tasks are somewhat different from those of the conventional shop foreman; progressing functions, for example, either disappear or reduce, or are modified, and one of the prime qualities a supervisor needs is good production engineering experience.

2. Construction of cells—avoidance of loss of functional layout benefits

It is not the purpose of this section to deal with details of the analytical work necessary to synthesize particular cells; rather to draw attention to some of the problems that may arise in moving from a functional layout to a cellular layout, so that they can be recognized at the planning stage and effectively dealt with. The following argument is developed on a typical turning group, as the lathe, in its various versions, is the most common machine tool in general engineering plants. In the Coventry works of the writer's company, for example, 69 per cent of all components have turning operations.

A conventional layout of a typical lathe section can contain:

(a) A number of types of machine, for example, turret lathes, PSC lathes, NC lathes etc.
(b) A number of sizes of some or all of these machine types.
(c) A number of machines of some or all of each type and size.

Such a section will handle a variety of turned components, all of which are within the capabilities and capacities of the turning machines in the group. We could say that the process planning work has automatically classified the components, insofar as the turning operations are concerned, to fit the machine group. On the shop floor, the balancing of machine size/type with component needs of size, geometry and volume leads to easily recognized and relatively simply executed opportunities for matching component requirements to the best facilities available. There is a wide spread of type of demand and type of solution which generally, because of this width of requirements and capability, enables more or less ideal matching to be made. There is a total overall matching of capabilities and capacities but more important, there is a significant degree of optimum matching of detailed requirements with capability.

The functional layout foreman, by his experience, can as a matter of course improve the actual matching of immediate requirements with immediate capability and his span of control in such judgements can extend forward over short periods of time to further optimize this matching. Examples of such opportunities include having certain machines with pre-

ferred collet size or chuck size or the normal use of one machine with a small roller box and another machine with a larger roller box etc. The foreman of such a section is generally a specialist with considerable detailed knowledge of the premiums he has to pay for changeovers.

In addition to physical matching of requirements with capabilities, the foreman also has the opportunity of dealing effectively with quantity variations where smaller or larger batches of similar components can be put on the appropriate type of machine. The functional layout foreman, therefore, has a continuing opportunity of reducing the number of set-up changes because of the wide span of facilities which are capable of handling the variable work pattern in size, geometry and volume. He can optimize the operation of his section so as to keep set-ups to a minimum and by putting different batch sizes on the right types of machines. Furthermore, he can meet abnormal or random production requirements, even though in a sub-optimal manner, by carrying out work on machines which are not necessarily the ideal type. Provided the variety of machine dimensional capabilities is not too broad, there is already a family within these capabilities. The features of sub-families within this range are easily visually appreciated and dealt with. If there is a reasonable degree of spare machine capacity, the foreman can leave set-ups for short periods and move his labour so as to balance the load, and at the same time, reduce machine change-over time.

In such a section, the effect of variations in load and absenteeism can be handled relatively easily. Let us assume that this turning section contains say thirty machines of the types defined earlier. In reorganizing these turning machines into cells, influenced by the family definitions, which are dependant to some degree upon subsequent operations, the cellular layout consists of say, ten cells with an average of three lathes per cell (this example is probably extreme and over-simplified). Certain of these cells would still contain relatively broad turning families within the capacity of the chosen machine types but the machine type variation has inevitably been narrowed. Unless a great deal of production engineering expertise is applied beforehand, the supervisor may have considerably less ideal solutions because his requirements are still relatively broad when compared with the reduction in machine capabilities. This argument can lead to having even smaller families with narrower definitions being handled by a larger number of smaller cells. With a narrow product range, this can even lead to the single machine approach but whereas this might be suitable in some cases, it can only be regarded as an exception and certainly not as a general rule. Extremely small cells can lead to machine breakdowns having disastrous effects, gross under-utilization of second operation machines or transfer of work between cells, which is generally unacceptable. With the reduced number of machine types in the cell, problems can arise from batch size variation with less options available to handle such demands and finally there are greater load balancing problems in smaller cells because of reduced flexibility. There are certainly GT applications which contain limited conventional machine variety but their acknowledged success is dependent upon the narrow range of components which they are designed to handle. It is suggested that the majority of batch production shops would not find this a satisfactory solution.

The purpose of this section of the paper is to draw attention to the importance of making the right decisions with regard to the size of cell, its flexibility and capability of handling work loads which vary in quantity, variety and batch size. These problems can be solved, but there is no ready-made general purpose solution. In the writer's company, there is considerable variation in the size of cells and the number of operators in them. Each opportunity has been considered on its merits and upon the considerable amount of case history already established.

3. Principles of ordering

The type of ordering system often associated with Group Technology is one of Period Batch Control where components are ordered in balanced product sets from a series of short term programmes. This type of system is afforded by the generally lower lead times and because dispatching rules of this type are easier to administer where manufacture is divided into a number of smaller production units as in the case of cells. Period Batch Control applies greater control to W.I.P. reduction, lower risk of obsolescence, reduced ordering cost and simplification of production ordering documents. For example, it is sometimes possible to release orders to a fixed standard listing which can be used over and over again. Being a fixed cycle system, the order quantity from each cycle is variable.

This type of system has proved successful and advantageous to a number of organizations employing Group Technology layouts. However, Group Technology itself may not automatically dictate the decision to employ Period Batch—it depends as much as anything on the type, size and variability of inventory demand. Where the number of piece parts and the standardization between products is relatively low, Period Batch control could well offer the ideal ordering system and at the same time provide good 'flow control'.

Where the products are complex and the number of piece parts is very high, the number of batches derived under a pure Period Batch system can be prohibitive. Certainly under Group Technology, the effect of increased ordering and setting-up costs by smaller batches can be counteracted but unless assembly and handling services are as efficient as the cells themselves, the sheer volume of the number of batches can present a real problem in physical and administrative control. Fluctuating demands such as spares call-off and scrap also are a problem when parts are ordered selfishly to short time programmes. Under these conditions, a fixed quantity variable cycle system derived from inventory analysis is likely to be more successful. A suitable compromise between systems is a hybrid of period batch and stock control. Break-points between methods can be set by the total number of orders required per annum imposed on a sliding scale according to usage value.

This means that period batch will be used for the high usage value items and stock control for the smaller consumable type of items, having pre-selected the desired number of batches to be processed through cells.

4. Influence of GT on machine tools

(a) Turning

As outlined earlier in this paper, the main problems with set-up variation lie in the turning field. With conventional turret lathes, the use of an identical set-up for two components necessitates either exactly identical features and identical dimensions or composite set-ups which are capable of machining a family of components with certain tools being left on the machine, some of which are omitted from the machining cycle. The opportunity to leave machines completely set is not often met in practice. It is certainly possible to develop sub-families with preferred and clearly identified categories of set-up, recognizing that even if this is done, in the majority of cases, tools will still have to be changed which can lead to a strong case for greater use of pre-set tooling and quick change tooling. This approach has been successfully developed and in some applications has been further improved by the use of read-out which is particularly advantageous for cross-sliding machines.

The natural development from the cross-slider, fitted with read-out, and the direct and total solution to the turning set-up problem is the NC lathe. This is because the geometry and dimensions of the components are generated by tape command and a limited range of standard tools (excluding minor exceptions such as narrow grooves). The application of NC lathes can be further refined by the development of sub-families which can be machined with minimum or no tool change (apart from end-working tools) and this technique can lead to a reduction of programming costs. A possible restriction, however, with many NC lathes, stems from the inherent caution built into the programme by the part-programmer. The part-programmer, who is generally removed from the NC lathe operation, both in time and distance, has to allow for possible hazards such as surplus material on forgings, variations in casting size etc. With the majority of conventional lathes, even PSC machines, the operator can take short cuts and improve the machine performance, particularly where problems allowed for by the process engineer do not arise. With most NC lathes, as all instructions are generally on the tape and not quickly capable of being altered, the operator does not have the facility to make adjustments of any magnitude with the result that the machine may be cutting at lower conditions than are ideally possible. This point has been recognized in the development of the HMT mini computer system so that the part-programme can be modified on the machine to improve the cutting conditions and the tape permanently modified at leisure.

A common advantage claimed for NC machines is that of improved management control. Because NC lathes have approximately double the percentage of cutting time of conventional machines, it is important to increase the effectiveness of this cutting time. If this can be done on the spot, so much the better, even if some control is lost. One has only to see the exasperation of the operator of a conventional NC lathe who has not got this facility, who is completely constrained by the tape and is prevented from raising the cutting conditions to those which he knows are possible.

(b) Machining centres

The use of machining centres goes one step further towards solving the question of sequential set-up/scheduling problems by direct elimination of most of the set-ups. Whereas on certain operations, machining centres may frequently be no quicker than conventional machines in metal removal, the elimination of set-ups and the telescoping of operations substantially reduce throughput time. It automatically follows that cells based on machining centres generally contain smaller numbers of machine tools, and as a corollary, machining centres will generally be better utilized in a cell environment.

(c) Effect on work in progress

For some years, one of the advantages claimed for NC machine tools has been their ability to reduce work in progress. This statement is often true but its validity may depend upon the position of the NC machine in the operation sequence. In the past, the majority of NC machines have been drilling machines and as drilling operations are frequently towards the end of the component cycle, such machines have been effective in reducing work in progress. As the NC lathe generally is at the start of the operation sequence, the reverse can be true. NC lathes, properly used, are most effective generators of work in progress if subsequent operations are not dealt with adequately. The NC lathe with its many other advantages makes an ideal marriage with cellular manufacture and in such an environment, the high output from this type of machine can be more quickly turned into saleable goods.

(d) Conventional machines

A popular belief appears to be developing that lower cost machine tools will be required in cellular manufracture because fewer features and functions are needed, as the machine specifications are closely defined against particular cell parameters. It is doubtful if this argument is wholly valid.

Many standard machines, when purchased with less facilities than normally offered as standard, are generally only marginally cheaper because of the cost to the supplier, not in fitting the unrequired options, but in taking them off. This comment probably applies more to lower cost machines which are manufactured and sold in higher volumes. The situation could change, if a worthwhile demand arose for such

simpler machines. More important is the probability that when people purchase machine tools for a particular cell, there is the possibility that in the future, there will be a re-layout of the shop, or that their design department will call for components with different features or dimensions, with the possible result that a machine may ultimately have a different application. Natural conservatism will probably cause people not to forgo low cost features which may not be required immediately, but the absence of which could be a serious draw-back in the future.

CONCLUSIONS

As indicated in the introduction, it is not the purpose of this paper to present a case study. The paper is not intended to illustrate the many sequential and partially interdependent steps necessary in planning and implementing a particular GT or cellular application. Rather is comment made on but a few aspects of the subject.

The writer's company which is fully committed to GT is multi-product based with 40 000 individual made-in piece-parts in one site and 30 000 in another. The whole of these have been coded, flow routes analysed and, supported by comprehensive production engineering and scheduling work, twenty-six cells have been engineered on the first site and eighteen cells on the second. Substantial gains have been made in the many cells already implemented. This is a major task of engineering and re-organization. As an indication of its magnitude, when the work is complete, some 1050 machine tools will have been repositioned. Clearly, considerable conviction of planned benefits is necessary before launching such a project. We have been gratified that the forecast improvements have been, and continue to be, achieved in practice.

Such a major physical and social upheaval in a company's machining operations has to be comprehensively and professionally engineered. Of equal importance is the need for a 'company approach' so that a complete corporate strategy can be developed. In the conceptual and planning stages, consideration should be given not only to the manufacturing operation but to many other matters, including improved methods of value engineering, design retrieval/variety reduction, simplified payment systems, reduction in paperwork, simplified costing procedures etc. The interaction of such activities across normal management structures is considerable and the opportunities of changes for the better can be many. It is, therefore, most important that the chief executive of the company is in full support from the outset. GT is not a nine-day wonder—let us make sure that we generate the environment in which it will flourish and give us the greatest yield.

THE IMPROVEMENT OF MANUFACTURING CONTROLS IN A MEDIUM-HEAVY ENGINEERING COMPANY

by

F. W. CAUDWELL*

SUMMARY

This paper gives an account of how a newly formed division within a large engineering group improved the use of its manufacturing facilities by adoption of improved controls over a five-year period. The improvements involved the introduction of group technology concepts in both machining and assembly areas, improved quality control disciplines, introduction of computer data shop scheduling and inventory control systems, reassessment of time standards and incentive payments for operators, development of better method cutting standards and introduction of new methods of process planning.

INTRODUCTION

The Mather & Platt Group consists of 24 companies with operations in 25 countries and with a total of 34 factories, which are located in the U.K., various parts of Europe, Africa, Asia, South America and Australia. These companies manufacture and market a wide range of products which include all types of fire protection equipment, security systems and bank safes, specialist electronic equipment, textile finishing machinery, food processing and packaging machinery, centrifugal pumps, electric motors and anti-pollution equipment.

The main factory is located in Manchester and its products are fire protection systems, textile machinery, centrifugal pumps and electric motors.

Five years ago, a decision was taken to divisionalize into product groups, as a result of which the power division was formed with responsibility for development, marketing and manufacture of centrifugal pumps and electric motors, foundries and more recently, anti-pollution equipment. This paper is concerned with the subsequent development of manufacturing facilities in Manchester for this power division.

Traditionally, the company has manufactured pumps and motors over the widest range to meet general industrial requirements and, consequently, it has been concerned primarily with the supply of 'tailor made' units produced on a jobbing shop basis.

Pump and motor production has been maintained over a period of ninety years and requirements for spares parts, particularly for pumps for both old and new designs, calls for substantial manufacturing facilities.

The policy of the new division was aimed towards the improvement of sales; ratio of sales to capital employed and profit; and these objectives were seen to be achievable by some rationalization of existing product groups, introduction of some new designs for standard and special pumps and motors, reorganization of the manufacturing facilities available and the use of improved production methods in both manufacture and control.

PRODUCTION RESEARCH

When the new division was formed, stock, work in progress and machine tools employed in the production of pumps and motors were located and controlled separately in two machine shops and the value of these items amounted to a substantial part of the capital employed by the division.

Each separate machine shop had a traditional disposition of machine tools by machine type; these dispositions were, therefore, duplicated to serve the requirements of two separate types of products.

A production research section was formed to examine the methods of control and manufacture within the electrical and pump departments and to study the problems associated with their organization into one efficient divisional manufacturing unit.

This section was led by a senior engineer with experience of special project work and he was supported by staff specializing in machine tools, production engineering and production control. This section also received support from post graduate students and staff of the Department of Mechanical Engineering at UMIST.

* Mather & Platt Ltd.

The introduction of group technology was considered to be the first essential stage of reorganization, as this would reduce material movement, simplify material control and reduce the time lost in transit during the period required to machine the component parts.

It was also considered necessary to support the introduction of group technology by investment in some more sophisticated plant, improve methods of production and stock control and develop assembly flow lines, to deal with the new designs of standard products.

The investigation and planning of these various aspects of reorganization ran in parallel and allowed implementation to take place on a broad front.

Part classification

The application of group technology and the formation of 'families' of parts for manufacture is greatly facilitated by the use of a component identification code. A pilot project covering the classification of all stock drawings for modern designs of pumps and motors was carried out with support and guidance of a staff member in the Machine Tool Engineering Division of UMIST.

The classification system used in the project was that developed in Germany at Auchen University. This is a production biased five digit classification code identifying a component firstly by its general shape and then by its various production features.

Other relevant component data recorded while coding were drawing numbers, stock numbers, material used, component dimensions and a product code.

The sample of components used in this pilot project classified 1500 machined items. Analysis showed that 82 per cent were rotational parts; approx. 50 per cent were of the short disc type with a length to diameter of less than 0·5; approx. 65 per cent were made of ferrous alloys, the remainder being of bronze or copper alloys, and over 60 per cent of all items were castings.

The work carried out on this pilot project helped to create a general interest in and understanding of the general principles of group technology and provided a source of information which was useful in the ultimate determination of machine tool groups.

Machine utilization

Concurrently with the pilot project, studies were made of all machine tools available to the division to determine the age, value, quality and degree of utilization of individual machines employed in both the electrical and pump machine shops. Machining methods were also examined with a view to making more effective use of the overall availability of machine tools.

These investigations revealed that many machine tools were surplus to the requirements of the Division and were occupying valuable floor space.

After evaluation of these various studies, plans were developed for the disposal of approximately one third of all machine tools and for their replacement by a small number of modern machines more suited to the division's requirements. The plans covered the relocation of all of the remaining machine tools on group technology lines in such a manner as to reduce the area occupied to 75 per cent of the original separate pump and electrical machine shops.

The relocation of machine tools was planned to take place over a period of eighteen months and the plan required, in some cases, that machines be moved two or three times in order to avoid dislocation of production.

GROUP TECHNOLOGY—MACHINE SHOP

The reorganization of the machine tool facilities into one single machine shop was completed early in 1970 and consists of ten groups of machine tools. The arrangement of these groups is such that components falling into particular categories are planned for process in one machine group only. Material and instructions are issued, via Production Control, to the appropriate machine group and the material is processed to a finished state ready for movement to a parts store or an assembly area. Each section has its own inspection area and its supervision is, therefore, responsible for the quality of the finished component.

Four of the ten groups provide production facilities for small and medium batch quantities (1 to 20) of large family groups of components, i.e. shafts and body bolts etc.; sleeves and bushes; impeller and fans; and horizontally split pump casings. Five of the groups provide for miscellaneous components dependent on quantity and size (100 mm to 6 m overall diametral size). The last group covers the large quantity (over 20) production of components ranging up to approximately 1250 mm diameter by automatics and NC machine tools.

These ten groups are arranged as follows:

Group 1—shaft section

Bar preparation equipment and material stock followed by profile and centre lathes, grinders and milling and keyseating machines. This section produces all spindle-like components required for both pumps and motors including shafts and body bolts, etc. Some of these machines have been fitted with digital read out equipment.

Group 2—sleeve section

Centre and combination lathes, grinders, millers, slotting and honing machines. All types of sleeves, or sleeve-like components other than large quantity stock, are produced in this section.

Group 3—impeller section

Centre and combination lathes, broaching and slotting machines and dynamic balancing equipment. The section provides facility for complete machining of pump impellers of all types up to approximately 450 mm dia. and certain types of motor fans.

Group 4—split casing section

Plano-miller, various horizontal boring machines and radial arm drills. This section is arranged to meet primarily the machining requirements of one basic product, i.e. horizontally split pump casings.

Figure 1. Machine group.

Figure 2. Machine group.

Machines utilizing digital read out equipment and pre-set tooling are included in this section.

Group 5–apprentice section

Various types of lathes, milling and drilling machines. All the components processed in this section are under 100 mm dia. with batches of less than twenty. New apprentices joining the power division commence work in the section following their one-year period of module training in the practical training workshops which are located away from the main workshops. This section is, therefore, the first contact that a new apprentice machinist has with production work. After a period of six months to a year they move on to large machine tools in the other machine groups.

Group 6–small component, small batch section

Combination lathes, milling and drilling machines. These provide facilities for the production of small miscellaneous components ranging from approximately 100 mm to 450 mm diameter for both pumps and motors, e.g. bearing housing, caps, brackets, shrink rings, fans, etc.

Group 7–medium size, small batch section

Various sizes of small vertical borers, milling and drilling machines. This section caters for miscellaneous pump and motor components, motor and shields, pump suction covers, middle bodies, larger bearing housings, etc. The components range in size from approximately 450 mm to 1200 mm diameter.

Group 8–large component, small batch section

Vertical boring machines ranging from approximately 1200 mm to 3000 mm, horizontal borer, open-sided milling and drilling machines. This section provides a facility for the complete machining of pump bodies, larger motor frames, large impellers, etc.

Group 9–heavy machine tool section

This section includes vertical boring facilities up to 6 m diameter and large plano-millers, planers and drilling machines. It also includes a 150 mm spindle horizontal boring/milling and drilling machine fitted with digital read out on three axis and capable of machining completely, components of sizes within a 5 m cube. Components can be machined on an indexing table or mounted on floor plates.

Group 10–automatic section

(a) Small chucking automatics supplemented with milling machines, drills and an automatic indexing drill.

(b) Medium chucking automatics equipped with pre-set tooling facilities.

(c) Pre-set Tool Room with mechanical and optical pre-set equipment to meet the needs of the medium and larger turning and boring automatics and the NC machines. This tool room also provides a pre-set service to the horizontal boring machines.

(d) Large horizontal chucking automatic which works in tandem with an NC Turret drill and light milling capability.

(e) Large vertical automatic boring machine with plug board control which works in conjunction with a conventional vertical boring machine and an NC machining centre. This section provides for the production of parts in batch quantities from approximately twenty for components ranging in size to approximately 900 mm.

QUALITY CONTROL

In all manufacturing sequences, the control of quality is vitally important if disruption to production programme is to be avoided. Delays in reproducing or repairing defective components can have financial repercussions which far exceed the cost of repair or replacement, particularly in the case of castings produced in special alloys.

Work is processed within the ten machine groups and is subject to inspection procedures applied within each group by inspectors who are responsible to a quality control manager.

Defects found at this stage fall into one of the following categories:

(a) to be scrapped and replaced

(b) to be acceptable by concession

(c) to be acceptable following rework or repair.

Defective work which falls into category (b) clearly minimizes financial loss and delays to production programmes, and implementation of the necessary procedures invariably involve appraisal by engineering departments.

Those falling into category (c) are almost always subject to concession either by technically competent people within the organization or by the customer and, although incurring delay and expense, can be more advantageous in all respects than rejection as in category (a).

To improve the quality of work carried out within the ten machine groups, two disciplines were introduced within the quality control department:

(1) Routine collection of data about every item found to be defective within the machine groups irrespective of whether scrapped, found acceptable after concession, or subject to rework. A programme was developed which enables the data to be analysed with regard to reason for rejections, material type, machine tool, operator, work centre, etc., and for application of control measures to minimize the total loss due to defects to reasonable limits.

(2) Development of a specialist machine group to rework components which would otherwise be scrapped. This 'hospital' group is staffed by highly competent operatives under the overall guidance and supervision of a quality engineer. The section has the capability of undertaking almost all forms of repair normally considered appropriate to the division's class of product. It has production facilities for both replacement and removal of metal and this includes air and gas welding, powder welding, dot welding, flame spraying and fusing, shot blasting, electro plating and spark erosion. The machining facilities within the 'hospital' section are supplemented by a small number

of allocated machines within the ten machine groups. These machines are available for priority reworking of components under quality control supervision.

By these means, indiscriminate, unapproved and irresponsible repair operations are avoided, product costs are reduced and delays to the production programme kept to a minimum.

PRODUCTION CONTROL

Work centres

Each machine tool or number of similar machines located in any one of the ten machine groups is designated by a work centre number, and non-machining operations complementary to tooling processes have also been allocated work centre references, e.g. heat treatment, marking out, inspection, dynamic balance, etc.

By this means, any component sent for machining to a particular machine group can be routed as directed on a planning document for process through a defined sequence of work centres located within that group.

Manual control limitations

Before divisionalization, the electrical and pump machine shops had independent production control sections which used 'Kardex' files but with procedures and documentation peculiar to their own mode of operation. Both these systems failed to give a clear appreciation of the basic problem of relating production commitments for a given period of time to the production capacity available; consequently, neither system was capable of allocating the capacity available in such a way as to ensure the maximum production of completed pumps or motors in order of priority or to cater for the frequent feedback of accurate information to the commercial departments.

During the course of the initial feasibility studies into the application of group technology, investigations into the production control activities resulted in the recommendation of one single improved system of control of pumps, motors and the associated spares and repairs orders. These recommendations were adopted before the integration of the machine shops took place and so enabled the two production control sections to be formed into a single control unit located in a central area.

Computer aided shop scheduling system

Although the reorganization of the machine shops was intended to reduce the movement of materials during the course of manufacture and also to simplify the problems dealt with by the production control staff, the introduction of a more sophisticated system of planning and production control was seen to be a necessary part of the process of improving the manufacturing facilities of the division.

The use of a computer was considered necessary to give an improved production control system by which means a frequent statement might be obtained on

(a) the current position of each total order
(b) the forward load on each work centre
(c) a work-to list in order of priority for each work centre within each machine group.

The required information was obtained by using a package type program and processing the information available from the process planning sheets on a service bureau computer located in Birmingham. This program was ultimately modified for use on a new computer installed at the Manchester works.

This 'computer aided' scheduling and loading system was first introduced, as a pilot operation, on the first machine tool section formed on group technology lines, i.e. the shaft section. To do this, the machine group was sub-divided into fourteen work centres and work scheduled for the shaft section was issued weekly to the supervisory staff on a computer 'print out'.

This 'print out' provided a summary of all work scheduled for process in such a way as to maximize the use of plant and labour available on that section, provided that the sequence of tasks set down was observed.

This method of control was tried for a three month period and many problems were encountered, but the comparatively small size of the pilot operation enabled the systems problems to be solved without serious effect to the overall production programme.

This method of control was then extended, section by section, as the machine groups were formed on group technology lines.

GROUP TECHNOLOGY—ASSEMBLY

Flow-lines

Concurrent with the machine shop and production control reorganization, plans were prepared for the development of assembly flow lines for batch produced motors, which would connect with and extend from the 'automatic' machine group of the reorganized machine shop.

Approximately 90 per cent of the motor components for motors in the range of 20 to 350 hp are produced in the Automatic/NC machine group where the disposition of machine tools is such as to provide for a flow line of batches of parts between automatics and NC machines and into an adjacent finished parts store. This store is flanked on one side by a Goods Received section for all incoming materials other than machined parts, and on the other side by a conveyor system allowing a continuous sequence of sub-assembly process to be carried out within twenty-one work centres located along the conveyor track terminating in test, paint and despatch or warehouse sections.

Work centres

The work centres on the assembly flow line can be joined with the 208 machine shop centres for purposes of scheduling and loading by the same production control computer program.

Stock control program

Plans are now being implemented to introduce a complementary program for the control of stock to meet both production requirements and market forecasts. These two programs will be used to provide ready access to production capacity, stock levels and

Figure 3. Flow-line.

gross or nett requirements of parts to meet actual order commitments or market forecasts and for the evaluation of cost and technical data.

This will allow manufacturing management to modify production plans quickly in response to changes of design, marketing policy or out of pattern demands of sales.

PROCESS PLANNING

Limitations

Although the method of production control described was subsequently extended, section by section, as machine groups were finished on group technology lines, it became quite clear that the quality of data available to the planning department and the methods used for processing planning information was not adequate for use with a sophisticated production control procedure.

This was due to the use within the factory, by all divisions, of an outdated piecework system which, having been applied with varying degrees of flexibility prior to divisionalization, had resulted over many years of practice in widely differing levels of payment for efficiency assessed from the ability to work within planned time allowances.

Effect on incentive payments

These time allowances for work were not in effect a reasonable criterion for the assessment of the time required by a competent operator to complete a given task because

(a) existing forms of process planning were not sufficiently explicit;

(b) machining allowances assessed from drawings were frequently found to be quite inaccurate;

(c) time allowances were invariably based on low definition statements of work content;

(d) machining data available was out of date and too generalized, and in some cases, quite inconsistent in application to the various product groups within the division, and also between divisions operating within the same factory complex.

Effect on scheduling

The quality of the process planning information diminished the effectiveness of the sophisticated control system due to the inaccurate time allowed, for process of manufacture, through successive operations.

The inconsistencies of these time allowances were reflected in bonus earnings and were thus a source of concern and dissatisfaction to both management and work people alike.

PRODUCTIVITY AGREEMENT

Objective

The negotiation of a productivity deal, two years ago, provided the opportunity to improve shop floor earnings pending the introduction of a new incentive scheme. It was agreed that this would be based upon use of new data suitable for the calculation of accurate time allowances and suitable also for use with a sophisticated computerized shop scheduling system.

Requirements

The new data resources required by the process planning department were therefore:

(a) work measured manually controlled work data
(b) verified cutting data
(c) negotiated allowances
(d) product know-how

To provide these resources, three teams were set up with the guidance of consultants.

(1) Work study team recruited from within the factory and including shop stewards, operators and ratefixers. This team was trained in Work Study methods by consultants.

(2) Agreements committee including representation of senior management and shop representatives to negotiate suitable allowances and procedural agreements.

(3) Machine process team recruited from within the power division and including operators, supervisors and planning engineers selected for their skills and experience.

WORK STUDY

Manually controlled work elements

Approximately 1500 elements were observed within the various machine groups and statistical analysis was carried out on the studies made.

These elements were subsequently arranged into larger data blocks to suit various processes considered by planners and all elements and data blocks were identified uniquely.

Machine processing parameters

The initial investigations of cutting data available from tool manufacturers and other agencies suggested a need for carrying out a programme of supervised cutting tests on a variety of machining operations carried out at our own works and under normal workshop conditions.

Over 2500 tests were carried out at the factory in Manchester and these tests enabled data to be compiled for each material group at varying feeds and speeds ranging up to the limitations of the machine tools used, the cutting tool life and the machining tolerances on accuracy and surface finish imposed by the specification detailed on the engineering drawings. Data compiled from each test included, component description, material code (84 items), machinability code (84 items in 5 groups), machine tool references, machine tool work centre, type of cutting tool (e.g. titanium coated throwaway tip etc.), cutting conditions (e.g. intermittent, continuous etc.), rigidity, surface cutting speed code (alpha A to P, e.g. $A = 1000/1200$ ft/min), cutting feed code (Numeric 1 to 9, e.g. $5 = 0.015$ in/rev.).

The results obtained from these tests were placed onto a computer file to provide access to the cutting conditions appropriate to components by their type and material and for comparative assessments with similar components.

Cutting data is specified by alpha-numeric code form which can be transposed, by operators, into spindle speed and feed rates by use of charts unique to individual machine tools. By this means, planning can be carried out for machine tools within the same group which, whilst having similar capacities, may have dissimilar settings of available spindle speeds and feeds.

Where similar operations could be analysed from these tests, mean feeds and speeds were calculated and considered in relation to their range of variation. This sort of appraisal indicates some of the problems involved when using a single value of feed or speed to cover all applications within the range of samples taken.

PROCESS PLANNING

Planned method

A planning 'method sheet' is developed for each component or group of similar components.

This method sheet provides a detailed specification of the work to be carried out and includes a sketch to illustrate tooling sequences, tool details, operation description, data block references, feed and speed codes and the frequency of cuts required.

Computation of time

By reference to the method sheet, a planner can then prepare a 'planning computation' sheet which provides a detailed statement of how a particular 'time standard' has been developed from the method sheet for a particular component.

The method sheet is readily accessible to the operator and the computation sheet can be made available in the event of a dispute arising on the shop floor.

Planning conversion problems

The task of converting planning sheets from 'old time' standards to the new time values has proved to be formidable and possibly the most difficult problem encountered during all of the changes made since the formation of the division. This has been due to the very large number of component drawings in common useage (about 100 000), the number of machining operations required per component, the high definitions of work content required to be considered for the production method used for each operation and the small number of components which have a high degree of repeatability over a short time span.

Some simplification of the planning conversion task has been achieved by analysis of family groups, but this has not proved to be as helpful as had been anticipated originally and in consequence, a large team of temporary planning staff has been used for over one year. This temporary staff has been obtained by secondment of supervisors and experienced and suitable operational personnel from manufacturing areas of the division at Manchester and from the two subsidiary factories.

Routing file

A routing file has been established to provide a storage bank of component information relative to tooling operations on the computer.

This information includes drawing, stock material and pattern references. Also available are product codes, Opitz number and other component descriptions along with operation sequences, work centre references, machine groups, machine types and both old and new time standards, where these are appropriate.

The file provides a means of holding on file a large volume of information on planned tooling operations in such a way as to provide for rapid updating of work values, comparison between 'old' and 'new' time standards, changing of work centre routings etc.

Eventually the file will provide an opportunity to issue shop floor paperwork from the computer from input data relevant to the parts list issued for a particular order.

MANUFACTURING CONTROL

A department has recently been established to maintain overall control of progress of all orders from date of receipt to date of despatch. This department comprises three sections with the following general responsibilities.

Coordination section

(a) Schedule incoming orders and produce a plan for their completion on or before the contractual commitment date.

(b) Establish target completion dates for each order for every department involved in completion of the order, e.g. contracts, design, drawing office, planning department, foundry, machine shops, assembly and test areas and for receipt of 'bought out rough' or 'finished' components.

(c) Monitor subsequent progress of all orders by product group in accordance with the established target dates.

Planning section

(a) Carry out our routine process planning.

(b) Carry out advance planning connected with the application of work study or method study.

(c) Maintain computer files which provide access to technical and manufacturing data on all manufactured items and to evaluate such data when required.

Production control

(a) Operate the computer aided shop scheduling program for preparation of weekly reports.

(b) Operate the inventory control programme for preparation of stock control reports (currently operational on electrical products and in course of implementation for pump products).

(c) Load control within all work centres.

(d) Evaluation of forward load data to determine limitations of capacity (machines and operators) and initiation of corrective measure, e.g. redeployment of labour, recruitment or sub-contracting activity to send out or bring in work for overloaded or unloaded areas.

The manufacturing control manager is in a position, each week, to review the production position of any particular order in the light of information made available to him from staff located in the three sections of his department. He is in a position to authorize, if necessary, the use of the external priority control facility which is built into the scheduling programme. These facilities allow the finite manufacturing resources to be matched, at weekly intervals, with the total known demand. This matching is carried out automatically by evaluating the natural priorities of scheduled dates but with provision for orderly manual intervention at a senior level where routine manual monitoring of product groups demonstrates a need for such intervention.

SAVINGS AND BENEFITS

The improvements carried out to-date have produced the following identifiable savings or benefits:

(1) A 25 per cent reduction of area occupied by the machining facility of the division at the main works in Manchester. The area saved was used to the benefit of one of the other divisions which operate at the main works.

(2) A very substantial reduction in the movement of material which has contributed to a reduction of indirect labour employed and to a simplification of control disciplines.

(3) The formation of 'Group' work units together with the issue of weekly 'work-to' lists for each work centre has simplified the tasks of supervision and increased its responsibility for the quality of finished components. This has contributed to a reduction in the number of Supervisory staff on the shop floor.

(4) Development of multi-shift operation of selected machine tools.

(5) An improvement in communications has been made due to provision of better documentation, the availability of detailed method sheets, where necessary, in support of the new 'time' standards and the use of microfilm to permit issue of disposable drawings for each job operation.

(6) Availability of forward load assessments for each work centre in each machine group which provide better opportunities for avoiding crisis situations due to under or over loading of the manufacturing facilities.

(7) Improvement in the quality of work processed through the machine shops.

(8) Extensions of the same procedures and disciplines at two subsidiary factories which allows their work centres to be loaded by the production control programme processed at the Manchester works—this has produced reductions in administrative, supervisory and indirect labour.

(9) Determination of better 'time' standards with consequent benefits of a better system of incentive payment and a better measure for purpose of production control.

(10) A steady improvement in the use of direct labour.

(11) A steady improvement in the ratio of turnover to stock plus work-in-progress.

(12) A steady improvement in the delivery of goods on schedule or ahead of schedule.

CONCLUSIONS

The manufacture of medium to heavy electrical rotating machinery (400 to 10 000 hp) was terminated during 1969. The facilities used on those particular products and much of the labour was transferred to other divisions operating within the main factory complex for the development of textile machinery and fire protection equipment. By the end of 1969, the power division was operating with approximately 30 per cent of the machine tools, also factory area and labour force which was made available to it in 1967.

During 1969 the effect of the reorganization carried out in the machine shops and assembly areas for motor production and the various aspects of improved manufacturing control, became effective on a rationalized group of pump and motor products. The effect of these diminished resources applied to the U.K. manufacture of the rationalized products relative to the 1969 quarterly production average is illustrated in Fig. 4.

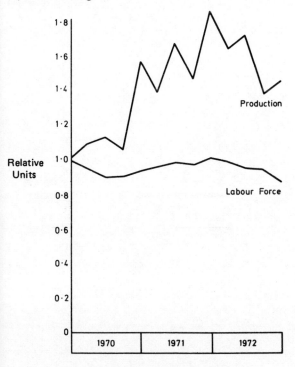

Figure 4. Variation of production/labour.

The budget level of production for the last quarter of 1972 was twice that for the 1969 performance and failure to achieve this level was due to the reduced availability of commitment.

The pattern of orders obtained for a production commitment in 1973 now suggests that the trend established between the beginning of 1970 and the second half of 1972 will be resumed.

The reducing labour force employed to achieve the production trends shown in Fig. 4 is illustrated in Fig. 5, along with the content of direct effort used in the machine shops which level has remained substantially level over the four-year period.

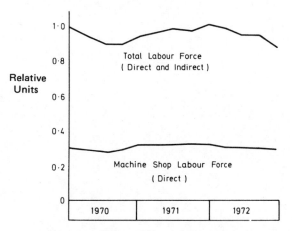

Figure 5. Variation—direct and indirect labour.

The value of production achieved over the four-year period in relation to the value of stock plus work-in-progress is shown in Table 1.

TABLE 1   Ratio of value of production to value of stock and work-in-progress

1969		1·0
1970	(Jan./June avge)	1·2
	(July/Dec. avge)	1·4
1971	(Jan./June avge)	1·5
	(July/Dec. avge)	1·5
1972	(Jan./June avge)	1·6
	(July/Dec. avge)	1·3

Although a significant improvement is indicated this improvement has been limited by the retained value of stocks associated with slow moving products of old designs and the need to generate additional stock in support of fast moving products of new designs.

More significant reductions in the investment in stocks and work-in-progress will occur in future.

Reorganization has been carried out on a broad front over the past five years and with a modest but steady improvement in performance.

During the next five years emphasis will be placed on the development of improved methods of production in the foundries, machine shop, assembly and test areas, and on the development of products designed for ease of manufacture within the resources available to the Division.

ACKNOWLEDGMENT

The author wishes to thank the directors of Mather & Platt for permission to publish this paper, and Professor Koenigsberger and those members of his staff who have provided valuable advice and much practical help during the development period described.

SOME FACTORS AFFECTING THE DESIGN OF PRODUCTION SYSTEMS IN BATCH MANUFACTURE

by

F. A. LEWIS*

THE PRODUCTION SYSTEM FOR COMPONENTS

The planning and design of the production system for components is probably the most important task of management in an established engineering enterprise. Investment in machine tools is the largest item of capital expenditure in most engineering companies, and the selection and layout of plant effectively prescribes the manufacturing capability of the company. The decisions involved therefore are primarily the concern of top management.

In a viable company the production system is subject to continual modification in order to keep pace with market growth and technological advance. Machine tools may be added here and there to increase capacity, ageing equipment may be replaced by new and more productive machine tools, improved tooling may be introduced, and attempts may be made to raise efficiency by the application of scientifically derived processing data. Very rarely however is the basic concept of the production system changed from that laid down when the manufacturing unit was first established. Thus, although machines and tooling have changed substantially during the past twenty-five years, the production systems used today have remained essentially unchanged since the beginning of the century.

The production system of the company, together with its operating procedures, can be regarded as management's solution to a problem of variety. Variety commences with the market requirement for product diversity, and is reflected in the size of the component drawing file, the type and layout of production plant, manufacturing time, and the levels of stocks and work-in-progress inventories. It will be evident that variety cannot be simply classified as low or high: there is a scale from low to high variety and individual companies can find themselves at any point along the scale.

Two basic solutions to the problem of variety, characterized by the layout of plant and machinery, were recognized until recently: (1) layout by processes and (2) layout by products. In layout by processes, machines are grouped together by similarity of type and function to create what is now commonly referred to as a functional layout. In layout by products, machines are arranged in groups or 'flowlines' for the high volume production of specific components.

The technical and economic factors involved in the design of product-specific flowlines are well understood by production engineers. The social factors involved in the specialization and de-skilling of human activity on product-specific flowlines are not so well understood, but the classical example of the assembly line is becoming increasingly suspect as an acceptable form of production system in terms of human motivation and satisfaction. In the area of engineering component manufacture however, management decisions about the form of production system are still governed largely by consideration of the technical and economic factors. Management evaluation of these factors has resulted in an overwhelming preponderance of production systems based on functional layout, although in many cases the original reasons for choosing this form of layout are no longer valid.

Recently a new form of production system based on the concept of manufacturing component families in small groups or cells of machines has received increasing attention as an alternative to the functional system for batch manufacture. Known as group technology or cell manufacture, the technique has been implemented by a few companies in this country with impressive results. Early applications have been in relatively simple areas of manufacture or were carried out on a very limited scale. As experience has been gained, there have been one or two notable examples of larger scale applications to the small batch manufacture of a widely varying range of highly complex components. Publication of the results of this more recent work has undoubtedly led to a revival of interest in group technology, and a greater willingness on the part of managements to investigate its relevance to their own manufacture.

In the published literature of the past twenty years, numerous developments have been heralded as

* P.E.R.A.

the long awaited solution to the problem of efficient batch manufacture, starting with numerical control in the early 1950s. Although some spectacular advances have been made in limited areas, the overall improvement in terms of the specific criteria of production efficiency, namely, machine and labour productivity, rate of stock turnover and delivery performance, has been disappointing.

It is now realized that industry cannot afford to wait for the widespread introduction of fully integrated batch manufacturing systems based on numerical control and computer technology. Such systems may provide the ultimate solution, but economic factors will inevitably limit their application for many years to come. In the meantime there is a compelling need for a re-organization of batch manufacturing methods in which the machines and equipment available today are used more effectively to contain rising costs and improve the return on industrial investment.

THE NATURE OF THE PROBLEM

It has been stated previously that the production system represents a solution to a problem of variety. Also it has been suggested that different companies can be visualized as operating in variety situations represented by points on a continuous scale from low to high variety. We might infer from this that there should be a great variety of solutions to match the many different variety situations encountered by companies in the engineering industry. Yet we have seen that except for clear cut situations of high volume/low variety manufacture, most companies have continued to operate with production systems based on a functional layout of plant. The decision to divide a company's set of machines by type or function effectively permits only one solution to the production systems design problem: it follows that the variety of solutions available cannot possibly match the variety of situations encountered. Thus from elementary considerations alone we are led to suspect that the functional layout will be ill-adapted to the requirements of many of the companies which use it.

Nevertheless, the functional layout has been with us for a long time and we should examine the factors which have resulted in its widespread adoption, because it is only when these are shown to be no longer relevant that company managements are likely to feel justified in embarking on a fundamental and broadly based restructuring of the production system. The factors to be considered can be discussed under three headings:

(1) Company History
(2) The Traditional Division of Skills
(3) The Availability of Planning Information.

Many companies started life as a result of the enterprise of one or two highly motivated people with little more than a good idea for a product and a possible market in view. In the early days of the company, higher management activity will have been concentrated primarily on product design, development, and selling. The manufacturing unit will have consisted of a small number of universal type machine tools manned by a handful of skilled tradesmen. Survival will have been dependent on flexibility and the ability to improvize in a situation of rapid product development and great market uncertainty. Because the overall company operation could be perceived and comprehended by one man and communication was direct, a quick response to unforeseen opportunities and emergencies was possible.

Successful growth will have been accompanied by a stabilization of the relationship between the company and the market. The product range will have become established, and management will be in a position to undertake long term planning of the company activities. The manufacturing unit will have expanded considerably in size and output and a formal system of production planning and control will of necessity have grown up within the company. Machines will have been added piecemeal type by type to satisfy the demand for increased output, and there will be a labour force of skilled and semi-skilled operators working within functional divisions of machines and supervision.

It is characteristic of companies which have grown up in this way that higher management attaches great importance to extreme flexibility of manufacture, although there may no longer by any cause for doubt about the nature of the long term manufacturing programme. Over-emphasis on a requirement for flexibility may also be coupled with unawareness of the cost penalties associated with it.

The traditional division of skills by principal types of process, i.e. turning, milling, drilling, etc. is often given as a strong reason for preserving a functional layout. It is argued that the various machining skills are more easily developed in functional groups and that the supervisory task is more easily defined, i.e. the job goes to the best operator. This argument may once have been true, but it is based on a rather narrow 'craft' view of industrial tasks. Developments in machine tools and changing attitudes on the part of the entrants to industry have rendered this view largely invalid.

The strongest case for adopting a functional layout can in fact be based on the availability of planning information. If a company cannot make useful predictions about its future manufacturing programme, or is unwilling to assemble product information in a form suitable for analysis, the planner has no rational alternative but to take a decision that leaves the largest number of options open. In other words he will design a production system which has the smallest number of constraints, i.e. one based on a functional layout. The situation is one of a pay-off between the investment cost of planning and a continuing production on-cost due to a mismatch between requirement and manufacturing capability.

VARIETY, INFORMATION, AND SYSTEM

Running through the previous discussion have been three concepts of great generality and usefulness, namely, variety, information, and system. We are concerned firstly with the variety of components deriving from the range of products that a company makes; secondly with the comprehension of this variety in some form of information system; and finally with the design and operation of a physical system of machines which will generate the requisite variety in the most economically advantageous way.

The variety of the component set is formally measured by the number of distinguishable items it contains, or more conveniently for some purposes, by the logarithm to base 2 of the number of items, i.e. in 'bits'. The variety of components in different companies may vary from say 500 to 500 000, or from 9 to 19 bits, depending on the type of manufacturing activity. Except in those cases where the variety is small, it is physically impossible to comprehend the component variety by listing the items as, for example, in a serial drawing file. Some trick has to be found which will enable the variety to be conveniently handled and analysed.

A method of universal utility is to dimension the variety in terms of the relevant variables and attributes of components. Thus some of the variety will be taken up by material, some by shape, some by overall dimension, and so on. The diagram in Fig. 1 shows how the information about components can be laid out conceptually in the rows and columns of a data field. The component axis extends horizontally to the right and the information axis extends vertically downwards. On the information axis are listed all those information items or attributes considered relevant to manufacture. A mark in the intersection of a row and column denotes the presence of a bit of information, e.g. '8 in long' or 'holes on a PCD' against a component item.

The data field is a very powerful device for handling variety. For example, with just 20 attributes listed on the information axis, 2^{20} or more than 1 million components items can be distinguished. Thus the variety encountered in any company can be comprehended and analysed with a mere handful of properly chosen information attributes.

The relationship between the data field as presented here and component classification and coding systems generally is worthy of comment. Underlying every classification system is the data field of Fig. 1.

The various classification systems differ in the attribute sets chosen, in the way attributes are grouped to create classes, and in the number coding used. In an attempt to compress the code number, many classifications reduce the variety which can be distinguished to the point where the classification can no longer cope or it becomes difficult to use. At PERA we believe that a data field embodied in a punched card file suitable for rapid manual, mechanical, or electronic sorting provides the most satisfactory solution to the immediate requirements of the company for analysis. Classifications can then be created as required for the specific needs of the company.

Few companies have component data of the type shown in Fig. 1 assembled in the form of a data file. Most companies do however have process data on planning sheets or a master job card file. In particular, the process route data constitutes an encoding of the component in terms of the machine tools used to produce it. This existing information on process routes can be used as a starting point for the analysis of component variety, or more strictly, the variety of component processes in the company.

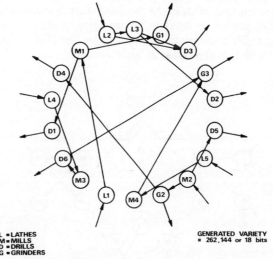

L = LATHES
M = MILLS
D = DRILLS
G = GRINDERS

GENERATED VARIETY
= 262,144 or 18 bits

Figure 2. Work flow—random layout of machines.

Using the process route data for components, we can examine the relations amongst the machine tools comprising the production system. The diagram in Fig. 2 shows 18 machines scattered at random around the periphery of a circle. This corresponds to a machine shop in which the machines are placed in any convenient space that may be available. No assumptions are made about work flow: it is assumed that a job leaving a machine has an equal chance of moving to any other machine. The lines joining machines represent a flow of work between them; an arrow at each end of a line would signify that flow can take place in both directions.

Figure 3 shows the same set of machines rearranged around the circle to bring machines of the same type together. This corresponds to the conventional functional layout of machine shop. The lines of work flow criss-crossing the diagram show the same general dispersion of movement as the previous diagram; grouping by machine type has failed to bring any detectable order to the work flow.

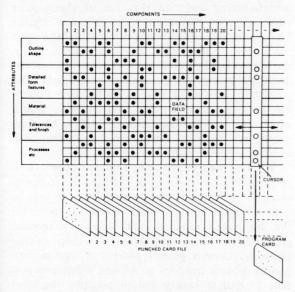

Figure 1. Structure of a component data file.

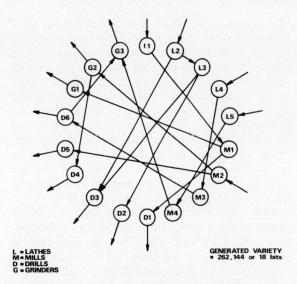

L = LATHES
M = MILLS
D = DRILLS
G = GRINDERS

GENERATED VARIETY
= 262,144 or 18 bits

Figure 3. Work flow—layout by machine types.

Figure 4 shows the same set of machines again, but this time the machines are grouped according to their relatedness in terms of the jobs which flow between them. The grouping of machines now reflects the pattern that exists in the component set, and exploits this pattern to shorten the workflow paths and to segregate components into sub-sets or 'families' which remain in limited areas or 'cells'.

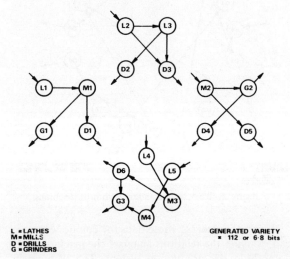

L = LATHES
M = MILLS
D = DRILLS
G = GRINDERS

GENERATED VARIETY
= 112 or 6·8 bits

Figure 4. Work flow—layout by relatedness of machines.

The essential difference between the production systems shown in the diagrams can now be quantified. The layout of the 18 machines in Figs 2 and 3 is effectively based on an expected process route variety of 2^{18}. Thus, although the actual variety of process routes to be handled in the component mix might be no more than 50, the functional layout in Fig. 3 would provide a huge variety of rather more than a quarter million. The group layout of machines in Fig. 4 on the other hand has been constrained to cope with an expected variety of $(2^4 + 2^4 + 2^4 + 2^6)$ or 112. Unwanted variety has been destroyed to enable a much more efficient production system to be formulated.

Earlier it was suggested that the nature of batch manufacture should lead to a variety of production systems solutions capable of matching the variety of manufacturing situations encountered. Figure 4 indicates the form that these solutions may take. It turns out therefore that the group technology concept provides the basis for generating the variety of solutions needed in practice.

The problem of production systems design for batch manufacture is illustrated graphically in Fig. 5. As the variety of the component mix to be produced decreases, components can be brought together into increasingly specific families; we move from 'once-off' jobbing manufacture at one end of the scale to the continuous high volume production of single components at the other end. At the top of the diagram is shown a continuous spectrum of solutions ranging from functional layout to fully mechanized flowlines. Of primary interest to the present discussion is the wide band of group technology solutions in the middle.

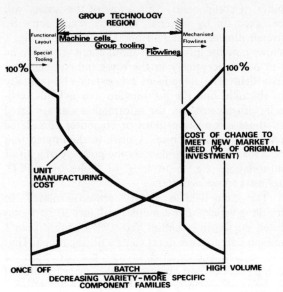

Figure 5. Relationships between form of production system, manufacturing cost, and investment risk.

The two curves on Fig. 5 show respectively unit manufacturing cost and the cost of changing or adapting production plant to meet a new market need. In the case of functional layout with little or no special tooling the cost of change is minimal, but in the case of fully mechanized flowlines it may be necessary to lay down completely new plant. The problem of production systems design lies in how to specify component families and machine groups which enable unit manufacturing cost to be minimized, but which do not burden the company with unacceptably high plant renewal costs.

The first step therefore in a rational approach to production systems design is to locate the company along the component variety axis. This can only be done if information is available on market trends for the company's products and higher management plans for the long term business activity of the company. Once a general pattern of future component demand has been established it becomes possible to analyse variety and to prescribe the most suitable form of production system.

THE DESIGN OF CELLULAR PRODUCTION SYSTEMS

It has been shown that production systems based on a layout of machines in group technology cells are the logical outcome of an attempt to improve manufacturing efficiency by achieving a closer match between the generated variety of the system and the expected variety of the component mix. There is ample evidence that such systems substantially reduce throughput times and work-in-progress, raise machine and labour productivity, and improve delivery performance. Contrary to popular belief, the principles of group technology have also been demonstrated to apply equally well to both simple and complex work.

Any company wishing to implement group technology however may find itself faced with two obstacles to rapid progress, one of which we can call the information barrier and the other the combinatorial trap. The information barrier arises from the need to assemble information on upwards of perhaps 10 000 manufactured components in a form suitable for analysis. The combinatorial trap lies in wait for the unwary when attempts are made to analyse the component data in order to specify machine groups and associated families.

Both obstacles can be overcome by the application of common sense and modern data processing techniques. There is no one method of tackling the problem which is equally suitable for every company. The approach to be adopted will depend on the type of product, the number of components, the data already available in a convenient form, whether the quantities of any one component can always be assumed small by comparison with the total throughput, and so on. A principle which should always be applied however is that the analysis should be comprehensive: it should start with the whole component range across all the products of the company. The results of the overall analysis establish at an early stage the scope for group technology and the form it should take.

Reference has already been made to the relationship between process route information on components and the relatedness of machines. Burbidge recognized the usefulness of this relationship in 1963 and proposed a method of analysing the information on route cards to find the division into group technology cells and associated component families. Known as production flow analysis, the method was not widely applied because of computational difficulties. Development at PERA has overcome these difficulties and the method can now be rigorously applied to any number of job operations on any number of machines.

The method of analysis by process routes is particularly attractive because many companies possess well maintained process route information for all their components on planning sheets or a master job card file. In some companies the information may already be on a computer file, but if not it can easily be converted into a form suitable for analysis in the computer. However, the importance of the method chiefly reposes in its emphasis on work flow and the relatedness of machines as the first criterion for establishing group technology families. Within a pro-

duction or machine group family it will usually be possible to reduce set-up times by creating sub-families of jobs which can be produced with common tooling, but this follows as a secondary stage of group technology analysis and planning.

As developed at PERA, flow analysis sets out to achieve a division of components into families, and a corresponding division of machines into groups, such that the members of a component family can be machined complete in one machine group. The primary constraints on this division are:

(a) that a component family should constitute a sufficient volume of work to justify the establishment of a machine group;

(b) that the composition of a component family should permit a satisfactory load situation on the machines in a group;

(c) that the size of the machine groups should be chosen to enable all aspects of processing within the group to be directly controlled by one man;

(d) that technologically incompatible processes are kept apart, e.g. a precision grinder would not be grouped with a power press.

As a first step it is assumed that the company's existing process routes will allow such a division to be made, although the method of analysis recognizes that the divisions may not be at all clear cut. Usually the underlying pattern of work flow is obscured by numerous connections between machines due to the presence of a proportion of jobs which do not conform to the pattern.

The problem to be solved is illustrated in Fig. 6. The components are assumed to be divisible into sets which can be processed on the machines divided into groups A, B, C, etc. When the attempt is made to divide components and machines, applying the constraints (a) to (d), it is found that some of the components must go to machines in more than one machine group. These components are shown in the shaded intersections of the sets A, B, C, etc. In a favourable case a high proportion of components will be found in the non-intersecting regions of the diagram. These components will constitute the group technology families.

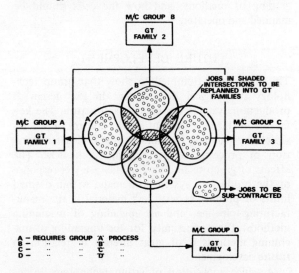

Figure 6. Division into component families and associated machine groups.

An important part of group technology planning is to move components from the intersections and into families by replanning jobs and modifying the composition of the machine groups. Components which cannot then be assigned to a specific family may have to be completely or partially machined in a miscellaneous machining area. Sometimes it may be desirable to sub-contract manufacture of the non-conforming components.

Where the component variety is large and the batch sizes are uniformly small, a re-allocation of machines into group technology cells is possible solely on the basis of the incidence of components. The existing numbers of machines of various types can be used to calculate the balance of machines in the cells. More usually, however, it will be necessary to introduce information on quantities and machining times so that the analysis can be performed in terms of the machining hours represented by component families.

An aspect of group technology planning which sometimes causes concern is the utilization of machine tools. Critics of the cell system on this count usually ignore the fact that many of the machines in a conventional machine shop are under-utilized. There is no evidence to suggest that the effective loading of machines in a properly designed group technology cell system is any more difficult than in a conventional system. On the contrary, what evidence there is suggests that the overall utilization of machines is improved. The extent to which the load on machines can be smoothed is a function of the number of jobs freely available for loading. Although there may be three or four times as much work out on the shop floor in the conventional system, the number of jobs actually available for selection in a given period of time may well be less than in the cell system.

At the completion of the overall analysis, the composition of viable component families and the general layout of machine cells will have been established. The detailed design of the production system and its implementation can then be carried out in convenient stages. Further work will be concerned with determining the best arrangement of machines in the cells, the creation of tooling families and upgrading of methods, and how the cells should be manned and operated.

FUTURE DEVELOPMENT

This paper has attempted to show that group technology is a logical development in the design of production systems for batch manufacture. The key factors in this development are the control of process variety by planning for similarity, and closer integration of processes by simplifying work flow. The effects of group technology include a faster response of the production system to demand within defined limits of product variety, less material in the manufacturing pipeline, and an upgrading of machining methods. The opportunity for the upgrading of machining methods is of great interest in relation to future development.

A wider application of group technology in industry will almost certainly stimulate developments in the field of machine tools and production equipment which have remained dormant because they could not be made to work effectively in the conventional production system. For example, the grouping of components into technological families is a prerequisite for the application of advanced machine tools such as numerically controlled lathes and machining centres; and the economic use of preset tooling is bound up with the identification of tooling families amongst the components to be machined. Any company therefore which undertakes group technology development will find itself (a) defining areas of application for new equipment and methods; (b) generating the information required to evaluate them; and (c) creating the system conditions in which they can be used.

The availability of comprehensive survey data on component statistics and the concept of technological grouping of components have already led to the development of a few machine tools well adapted to the new production systems, but machine tool development for group technology is still in its infancy. In a comprehensive survey of industrial machining requirements carried out some years ago, a proposal was put forward for the development of 'machining complexes' consisting of a number of numerically controlled machining units, each specialized towards a particular set of machining operations and linked together by a simple work transfer system. It was intended that the complex should be made up of standard modules arranged to suit the component families and particular needs of individual companies.

The results of the survey suggested that between 30 and 40 per cent of non-rotational work in the general engineering industry could be produced advantageously on the 16 in cube capacity, four-station machining complex shown in Fig. 7. The four machining stations of the complex consist of one station designed primarily for plane milling and precision boring operations, and three identical stations designed primarily for drilling, reaming, tapping, and end milling operations. This ratio of machining stations corresponds closely to the average balance of machining effort on non-rotational work, and since all but the more specialized operations can be done at any station, there is in-built flexibility to allow operation times to be balanced over the four stations.

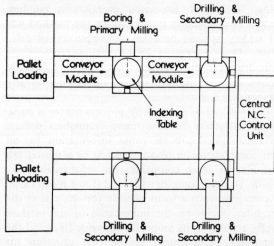

Figure 7. Four station NC machining complex.

Outline specifications of the machining stations and sketches of the units are shown in Figs 8, 9, 10 and 11.

It may be expected that developments such as the machining complex will receive greater attention from machine tool builders in the future as more companies convert to production in group technology cells.

A further development intimately connected with the cell system is the possible use of mini-computers for the real time scheduling of work within the cells. One of the important objectives of group technology is to transform the complex overall batch scheduling situation of the conventional machine shop into a

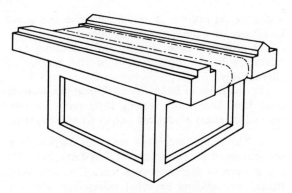

Figure 11. Conveyor module.

series of simpler group schedules. This opens up new possibilities for the close scheduling of work loads within the group technology cells. With a mini-computer it becomes feasible to optimize the performance of a cell using a flexible system of job priority rules which can be worked out to suit best the requirements of the company.

Station	Machining Range	Accuracy Capability	Details of Tool Head	Work Positioning	Main Features of NC Control
Drilling and secondary milling	1¾in. max. drill size. 1in. max. tapping size. 1¼in. max. milled slot. Machining in steel and cast iron.	Positional accuracy to IT10	6-position tool turret. Quick change over from horizontal to vertical mode of operation. Speed range: 100–2000 rev/min	Indexing table about vertical axis	*Tool Head* Variable rate position control of three axes; simultaneous control of any two axes. Automatic speed change. Automatic turret index. Automatic table index.
Primary milling and boring	8in max. dia. face mill. Max. dia. bore— 3in. Machining in steel and cast iron.	Positional accuracy to IT8 Diametral accuracy to IT6.	Single horizontal spindle. Speed range: 90–2160 rev/min.	Indexing table about vertical axis.	*Tool Head* Variable rate position control of three axes, one axis at a time. Automatic speed change. Automatic table index.

Figure 8. Outline specification of machining stations for 16 in machining complex.

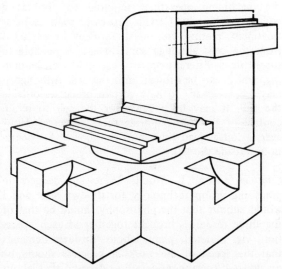

Figure 9. Machining station for machining complex.

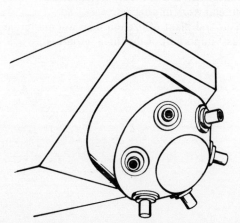

Figure 10. Universal spindle drive unit capable of horizontal or vertical operations.

CONCLUSION

Group technology can be seen as the logical outcome of a need to design production systems which are more closely matched to the product variety of the individual company. Such systems are more efficient and more responsive to demand within defined limits of product variety than conventional systems based on a functional layout of machines.

The design of group technology cells entails an investment in data analysis and planning as compared with the minimal information required for functional layout. The analysis of component data can be carried out rapidly and effectively by computer, using existing process route information or specially developed component data files.

The wider adoption of production systems based on group technology cells should stimulate further developments in machine tools and operating methods. Machining complexes for the production of component families are likely to receive increasing attention from machine tool builders, and mini-computers may be used for the real time scheduling of work in production cells.

ACKNOWLEDGMENTS

The author wishes to thank the Director and Council of PERA for permission to publish this paper.

DISCUSSION

Q. J. R. Crookall. My observations are made in the light of some experience with simulation studies of GT systems of manufacture carried out at Imperial College. It would have been of interest to have available in the paper details of the simulation which were given on slides during presentation of the paper.

If my memory serves me correctly, the average utilization for an average work load on the system was 79 per cent and in one case I observed a machine utilization of 93 per cent. Surely such a high level of utilization in practice is unrepresentative and could be achieved in a simulation study only if:

(1) utilization was deemed an overriding consideration in the optimization of the system; or alternatively

(2) the range of flexibility of throughput assumed must have been small, leaving little room for variations (increases) in required output from the system.

A. The results presented were not derived from simulations of the operation of GT systems, but from the analysis of aggregate work loads. The utilization figures are therefore expected values. The next step will be to submit production period work loads to the systems to determine their ability to absorb fluctuations in the work mix. The cell systems described permit a substantial amount of flexibility in the assignment of jobs. Also there is reserve capacity available through overtime or two-shift working.

Although we would not give undue weight to machine utilization in attempts to optimize a cell system, in practice we find that companies are unwilling to accept low utilization figures, especially in the case of expensive machines. The problem of course is to secure the right balance between the cost of work-in-progress and the cost of under-utilized machines.

Q. G. F. Purcheck, Cranfield Inst. of Tech. The amount of variety of selective information in M objects or attributes is 2^m inclusive of the zero vector. With regard to Figs 2, 3, 4, the links should be removed and the numerical values altered to read 264143, 264143, and 108 respectively, i.e. $2^m - 1$. If the links and direction arrows remain then the numerical values are obtained from

$$m \sum_{n=0}^{K} (m-1)^n$$

where m = no. of machines

K = summation index

n = route length $- 1$

This excludes follow-on machines.

Also Fig. 5—decreasing variety: more generality in component families

increasing variety: more specificity.

A. The diagrams in Figs 2, 3 and 4 represent three different arrangements of machines and the effect of these arrangements on the flow of a typical set of jobs: the lines and arrows have no significance other than to illustrate the effect of machine arrangement on work movement. It is evident from the text that the measure of variety in the diagrams relates to the total number of machine combinations that can be generated from the different arragements. The zero combination corresponds to those processes which do not have the machines of the set, and the author can see no reason why it should be excluded from the measure of variety. However, exclusion or otherwise of the zero combination has no effect on the argument presented in the paper, which is that the capability of conventional production systems to generate variety is vastly in excess of that needed for efficient production.

The author cannot agree with Mr Purcheck's comment regarding the variety axis of Fig. 5. For a given volume of production, a reduction in variety must mean the production of larger quantities of identical or nearly identical parts. Thus a component family will contain fewer component types, i.e., it will have less variety or be more specific.

Q. Prof. Warnecke, Stuttgart. It is said that in the different cells certain group families can be manufactured 100 per cent. How do you solve the problem of, for instance, heat treatment, sand blasting or plating?

A. Usually it is possible to combine in the cell all the machining operations required to produce a component family. Certain processes such as heat-treatment, plating, etc., must normally be excluded from the cell, although sometimes it is possible to substitute a different process, e.g. induction hardening, which can be brought into the cell. Although it may be necessary to have certain processes outside the cell, it may be possible for the cell to retain control of the jobs requiring these processes. However, a longer lead time must be allowed for operations outside the cell.

Comment from L. Wood. There are dangers in practising group technology for its own sake and I would submit that the philosophy should be that of the three resources brought together at each operation, viz. material, plant, labour. Business demands that the speed of the work in progress should be maximized; social considerations demand that labour be consulted, and the only compromise is between labour and work in progress.

Plant and plant utilization does not matter, as it is the least important of the three resources.

MANAGEMENT MOTIVATION TO APPLY GROUP TECHNOLOGY

by

D. BENNETT* and W. MACCONNELL†

SUMMARY

This paper deals with the areas within a manufacturing organization that could be effected by the introduction of group technology techniques and a survey of 150 companies to determine the reasons for their contemplating introduction.

INTRODUCTION

The Group Technology Centre was founded in late 1968 with the prime objective of assisting industry to apply group technology techniques to solve their manufacturing and management problems. Its task was to promote a broad based 'educational' campaign coupled with a back-up advisory and consultancy service for industrialists wishing to apply GT. In essence this has involved general and detailed discussions with the managements of many engineering companies to identify their potential for GT development. These discussions have led to the following results:

(1) The company and the Centre have decided there is little to be gained from the implementation of GT.
(2) The company has decided that the application of GT would be of advantage and has used the Centre's services to assist them in implementation.
(3) The company has decided that the application of GT would be of advantage, and implemented the technique without outside assistance.

This paper examines the reasons for 150 of these companies spending considerable time and effort in the detail examination of the effects of applying GT to their manufacturing organization.

In analysing data of this type the results can be no more than subjective, but as relatively few people were involved in collating the data the level of variability of individual interpretation is low.

The spread of industrial groupings of the companies is shown in Fig. 1, and it should be noted that of the 150 companies contributing to the analysis, 90 have to date committed themselves to the implementation of GT.

type of company	no. in analysis	no. using 'GT'
general engineering	34	16
valve and pump	20	13
machine tool	14	9
electrical products	12	5
instruments	11	10
tool making	9	4
aircraft	9	5
motor car accessories	8	6
paper converting and printing machinery	8	5
civil engineering equipment	6	3
textile machinery	4	3
sheet metal	4	2
computers	3	2
gears	2	2
ship building and equipment	2	2
office machinery	2	2
agricultural equipment	2	1
	150	90

Figure 1. Industrial groupings of companies used in the analysis.

THE DEFINITION OF GT

Why do companies make the decision to implement group technology? When this question is posed it is often difficult to obtain a rational and clear answer, especially when the answer has to fit into an analytical framework. Even company personnel who were responsible for the decision to implement GT may not be able to give a clear and unambiguous reply.

* Project leader, The Group Technology Centre Reading
† Project engineer, The Group Technology Centre, Reading

The reason for this is not simply 'People do not know what they are doing or what they have done', but initially they see the potential solution to a particular problem by applying GT. When the design and production data etc. has been identified, manipulated and grouped in a logical manner, further uses of the data can be envisaged and the solutions obtained to other equally pressing problems. It is at this stage that the original reasons to implement GT often becomes clouded over or forgotten.

Another reason is possibly the lack of understanding or interpretation in application of the technique. This may be attributed to the current definitions of group technology since these are often criticized as being too narrow or too broad and neglecting or excluding some particular field of management.

One definition put forward in the early days was that GT was a conglomerate of established management techniques and if a new one was developed it could be automatically incorporated. This definition, however, does not consider the application of the particular management technique to the jobbing and small batch production environment coupled to the philosophy of grouping information, data, machines and men. When one considers that some of the techniques were developed and used for large-scale production, the latter point is especially important.

For this paper, therefore, it is preferable to use the following definition:

'Group Technology is a technique which by using grouped information and data with grouped human and physical resources, together with established management techniques, achieves improved control and manufacturing performance in jobbing and small batch production.'

Within the context of this definition it is possible to envisage the application of an established management technique using 'grouped' data, components machines and operators in order to achieve improved control and performance.

FIRST STAGE OF THE ANALYSIS

The analysis of the various company's reasons for investigating the implementation of GT was carried out in two stages. The first was to define the main areas of interest under the broad headings—production engineering, design, management and manufacture. The result of this is shown in Fig. 2 and it can be seen the greatest interest is in production engineering.

It is significant that the improvement in managerial control is not most companies' initial major interest. Since this is only 34 per cent of that for production engineering, it could be considered there is a reversal in the true order of importance. However, the reason for the lower interest is often due to the company specifically limiting the areas in which to implement GT and not necessarily a lack of realization of the technique's full potential to ease manufacturing management problems.

Because of this it is important that the initial approach to implementation is not carried out in a

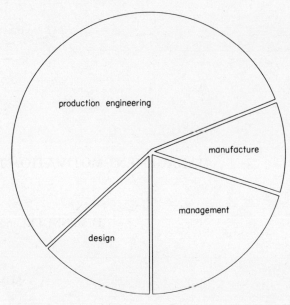

Figure 2. First stage analysis—main areas of company interest in GT.

way that it restricts the development of GT as a complete manufacturing system within the company at a later date.

SECOND STAGE OF THE ANALYSIS

The second stage of the analysis was to sub-divide each of the main areas into specific fields which were related to the company's interest or reason for implementing GT. An example of the analysis form is shown in Fig. 3.

The original or prime interest in GT is shown ⊘ and secondary interests shown ✓. The assessment of a company's reason for investigating the implementation of GT was often difficult since there was no black or white but many shades of grey. In many cases the company's interest or reason changed after discussion or when a feasibility study or the initial stages of implementation had been completed. An example of this was a company's interest in more efficient use of the machine tools in their workshop

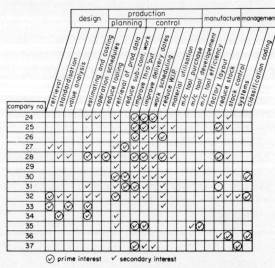

⊘ prime interest ✓ secondary interest

Figure 3. Example of analysis form.

which resulted in the re-appraisal of their sub-contract work. Analysis of their components quickly showed that the existing low efficiency was mainly caused by attempting to produce components on unsuitable machines with respect to capacity, machine facilities and accuracy. The end result of this exercise was an almost complete reversal of workload, components previously sub-contracted are now made in-house and the in-house components sub-contracted.

GT IN PRODUCTION ENGINEERING

The result of the second stage analysis of the use of GT to provide solutions to production engineering problems is shown in Fig. 4. The retrieval and re-use

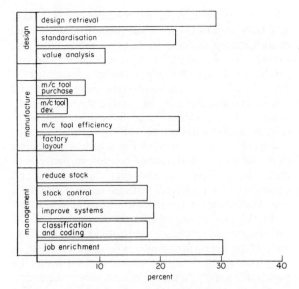

Figure 5. Second stage analysis—design, manufacture and management.

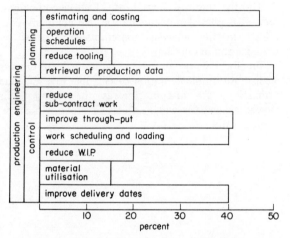

Figure 4. Second stage analysis—production engineering.

of existing production data shows as the prominent interest (50 per cent of companies), closely followed by estimating and costing, improved through-put, work scheduling and improved delivery dates (40–45 per cent of companies).

GT IN DESIGN

The second stage analysis of the interest in design is shown in Fig. 5. In the past, the relevance of design retrieval, standardization and value analysis to GT has often been discussed. The results of the analysis show that industrialists certainly see these techniques within the context of GT, especially when the interfaces and cross relationship of 'grouped' components and data from design with other areas are considered as a basic part of the company's total manufacturing system.

GT IN MANUFACTURE

In manufacturing, the main interest has been the more efficient use of machine tools (25 per cent of companies). Two of the other interests or reasons, i.e. purchase of machine tools and factory layout, are closely associated with the main interest of efficient use of machines (see Fig. 5).

In recent years, production engineers, accountants and management have become very aware that increasing cutting speeds and feeds is not necessarily the current answer to improved production efficiency. Research and studies into machining efficiency are proving that the actual metal cutting time of most machines working in a batch production environment is small in relation to the total manufacturing time. The greater time is spent on setting, waiting material, tooling, information and other delays. It is in this area of work organization and system that GT can and does make a major contribution.

A number of companies look to GT to help an improved factory layout usually when planning a new factory. The aims of most being smoother work flow from the raw material to assembly stage, greater output from the same floor space and improving working conditions. Where this is a primary interest it is usually coupled with little understanding of the wider concepts of GT.

GT IN MANAGEMENT

The use of GT in management is mainly concerned with systems reappraisal or establishment of control systems except for the area termed 'Job Enrichment' (see Fig. 5). This area will be discussed in more detail later. The use of GT in the areas of stock reduction and control has been fully described in a number of papers in the past. Substantial reductions in stock levels are claimed by companies and have been verified by independent bodies. During the last few years of economic difficulty it is no surprise that 20 per cent of all companies making the decision to implement GT have included these two areas as of prime interest.

One other area of interest is that of classification and coding. This, as a management technique within the context of GT, has been much maligned and caused major controversies. However, this does not deter managements from implementing the technique as a basic step to GT or as the foundation for a total manufacturing system. In both, a meaningful and logical identification of components, data, and end products is an essential step for either manual or EDP processing systems. Alternatives to classification and coding are available for the manipulation of data to form machine groups or to perform other basic steps in the implementation of GT, but these do not necessarily produce a situation which significantly improves total manufacturing control. Experience to date, though, has indicated that classification and coding is the preferred method of establishing a sound data base for across-the-company activities.

In the field of job enrichment 30 per cent of the various company management teams expressed a keen interest in improving worker job satisfaction. It is, therefore, clearly being recognized that GT offers the potential to apply to the manufacture of components in a job shop situation, the type of work structuring being instituted with success in the mass production assembly plants. However, it is felt that the actual level of response in this particular area may be suspect due to heavy publicity of job structuring programmes at Volvo and Saab in Sweden which in some references have wrongly been associated with GT applications.

CONCLUSIONS

A number of conclusions may be drawn as a result of the analyses of the types of industrial company and their reasons for considering the implementation of GT.

Firstly, the interest in implementing the technique is spread over companies producing a wide range of products and includes manufacturing units employing from 50 to greater than 2000 people.

Secondly, group technology is perceived by many as part of a basic step towards a new and total manufacturing system. The technique being used to establish a sound data base. Because of this, many companies are interested in approaches to GT involving the creation of product information in a form that is compatible with systems use and manipulation.

Thirdly, the introduction of GT into a company is seen as a phased implementation. No company contained in the analysis contemplated a total reorganization in one step.

Finally, many of the 'traditional' arguments against the implementation of GT, like low machine utilization, are disappearing as the technique is viewed against the wider aspects of superior management control.

THE CELL SYSTEM OF PRODUCTION EMBRACES GROUP TECHNOLOGY AND ALSO CONCERNS MANAGEMENT, TECHNICAL AND SOCIAL CHANGE

by

G. A. B. EDWARDS and G. M. FAZAKERLEY*

SUMMARY

To many, new to the field of Group Technology, there seems to be some kind of special mystique which surrounds components, machines and tooling. There is no human factor of any significance when 'technical' factors such as these are discussed. This paper suggests that in order to understand the overall systems concept of GT, it is necessary to look at management, unions, the community of operators on the shop floor and the individual.

In order to do justice to these important and too often ignored factors it has been necessary to give the authors' interpretation of the nature of GT, the cell system and its development so far, before dealing with the human factors.

The current over-concentration of much of the talk surrounding GT with techniques of analysis justifies a paper which attempts to deal with some of the 'people' factors.

INTRODUCTION

Much of our work in the Department of Management Sciences at UMIST has been concerned with designing cells and critically examining existing and emerging tools to assist with such design. We have been very critical of a variety of different systems of component classification, particularly those of the universal or panacea type, not only because they fail to do the job for which they were intended but more importantly because the job itself has been defined imprecisely or entirely misunderstood. Because we are still in the early development phase of the cell system the delegates at this conference should expect that some speakers will present papers dealing with the technical aspects of group technology to the exclusion of human, organizational and system aspects.

Although this particular contribution to this conference is meant to remedy this, it cannot do so if the human aspects are considered as 'additional factors'. It is most important, at this stage of cell system development, to recognize the integration of human and technological development. Too often the minor importance of the techniques have been over emphasized at the expense of the people they serve. The novelty of the technology of the computer, the interesting equations of linear programming and, in the field of GT, the intricacies of component classification have consumed the interest of those new to the field. Of course, these techniques have an important place in the general scheme of things but machines are to serve men, not men machines. Unless we are clear about our objectives and define our ends before our means we will reach a stage at which technological development will outstrip man's capacity to handle it.

At a time when 'in place of strife' has given way to an Industrial Relations Act which brings forth conflict and strife at almost every turn and where men are demanding that they are led by informed management who have their interests at heart, it is not only opportune but mandatory that human aspects of production are given priority when those from industry meet to talk about the present and the future.

The delegates at this conference will, like those on most others, go back to their firm and report that the non-formal sessions around the bar, the lunch table and at the hotel were at least as important to them as that being preached from 'the platform'. Although this contribution may well be referred to as a kind of gospel, it is towards the informal organization that the human factors should be directed. Organizational change has not, so far, been brought about by politicians nor those charged with vast industrial responsibility, neither on the management side, at group board level or by the CBI, nor on the union side, at general secretary level or by the TUC. It is significant that, in common usage, we refer to the term 'both sides of industry'. Such demarcation lines

* Department of Management Sciences, University of Manchester Institute of Science and Technology

must surely disappear if the people of industry are to produce a new kind of organization which places teamwork and co-operation at the head of the priority list. British industry has developed more of an interdependent relationship than some countries. We rarely 'bow to superiors' or 'click our heels' as a mark of respect. This does not mean that we are 'better' or 'worse' than others, rather that we are different and at a different stage of development socially and industrially than, for example, Zanzibar, Cyprus or Germany. Such factors influence the kind of production system which can exist in the glove and shoe factories of Zanzibar or the gear-box maker in Huddersfield.

In Britain, there is a different hierarchical structure than in Germany or France and the kind of production system needed in Britain is one where interdependence, teamwork and people are more important, and seen to be so than machines.

The flow-line system is no longer acceptable to many workers. Volvo have announced that they are spending £20 million on their movement away from lines and into cells. The reasons for the change at Volvo are many and various but the tedium of the assembly line seems to be responsible for the kind of unrest and militancy that is evident in Britain. For similar reasons Philips began a similar movement in TV assembly in 1966 and Friedland-Doggart in Stockport have had assembly cells operational for more than two years.

STAGES IN GT AND CELL SYSTEM DEVELOPMENT

The idea that Group Technology is a technique is to underrate its significance. There is little in Group Technology perceived as a technique for improving machine and labour utilization by the process of introducing a rough coding system which supposedly forms component families. Unfortunately Group Technology is fast taking over the centre of the fashion stage from value engineering, OR, MBO, and this detracts from the possibility of it being fully utilized in a broader context.

This paper is not about GT as a technique; it is about the philosophy of GT, the result of which is a new production system called the 'cell system', which can make production planning and control in batch manufacture both possible and relatively simple. The significance of the research in GT conducted in the Department of Management Sciences at Manchester University is not so much because it has produced the practical use of GT at Ferranti[1], Platt International, Stibbe-Monk, G.E.C.-E.E., and Whittaker Hall[2], but rather that it has produced a new analytical framework around which batch manufacturing firms can re-examine themselves and begin to design new and improved production systems. This not only leads to the design of cells of machines dealing with families of components, but also changes the social structure of the workshop[3], which in turn, changes planning and control, wage payment and methods of performance measurement.

Although production can be seen as a system involving inputs, a transformation process and outputs, J. L. Burbidge[4] has suggested that it is the material flow analysis which should be of prime concern. His view was ignored for about ten years, partly because of the system of analysis which he advocated. His papers on production flow analysis (PFA) were based upon samples. Other researchers[5] have since pointed out, the statistical pathfinding method gives no practical solution; nor does the approach of factory, group and line analysis, which are three essential steps suggested within PFA. It is now common to hear a further criticism, namely that because the basic data collection document is the planning card, this is another reason for criticizing this approach. As with all factors concerned with the manufacture of components, there are individual differences and preferences. The plant available often dictates the play. Although the planning document itself is perhaps unreliable, the use of a flow analysis using drawings and a good plant code by a competent production engineer, who is assisted by a specialist in GT, can prove to be most useful. El-Essawy[6] has explained this so ably in a recent Ph.D. thesis.

The importance of flow analysis leading to production system design has now been shown to be of great significance in the theory and practice of production management. It is in production analysis itself therefore that the analysis of flow, developed during GT research, has been helpful, since it enables us to clarify major segments in production system design. The use of this research can best be illustrated in the context of the following:

(1) It provides a clearer definition of the factors operating in the transformation process of material into products.
(2) It enables us to create a framework for the analysis of these factors.
(3) It suggests an order for the analysis and the places for synthesis.
(4) It has given rise to the development of new or improved methods of analysis.
(5) It has highlighted the method and importance of collective syntheses, leading to production system design.

The general value of this work to those interested in management and production control is, however, to be found in the way it illuminates and differentiates between INFORMATION: Analysis of Information: Synthesis AND Production System Design as work stages in the process *leading* to production planning. It is only after planning that production control becomes a reality. The realism of this control is that which brings it close to the whole concept and practice of management control. The stage that seems almost entirely under-developed is production system design, whereas it is in the analysis area that the techniques have been developed for their own sake rather than within some general framework[7].

FUNCTIONAL AND LINE LAYOUT

For many years engineering manufacturers have looked with jealousy upon automobile, television, and refrigerator assembly plants. They have said 'If only we could take advantage of the flow-line system'. Because of the variety of different products made to the precise specification of many customers, and because of irregular and unpredictable market forecasts, these firms have been forced to use systems of production which, like Topsy, 'just growed'.

The most common system of plant layout dates from the early part of this century and is called functional layout. It is characterized by all similar machines being brought together. It exists because works managers perceived the importance of linking machines and associated operator skill when they set about creating their first systems of organization in workshop and factory. In engineering manufacture today, functional layout is the rule rather than the exception. Systems of labour organization, payment, measurement, cost allocation, budgeting and production control have all been developed around a method of plant layout which has been predominantly machine and skill orientated, but which has neglected the developments in thought and analysis that have resulted from the rapid growth in education.

When we look at such plants we observe that the machines are frequently 'on parade' like guardsmen at Buckingham Palace, but when we look at the work, it is piled up in makeshift stores, or queues before and after each machine. In short, the system of work flow is disorganized. There are progress chasers running about, trying first to find the overdue component, and secondly, to hasten its flow through the production cycle. The machines are impervious to this activity since they stand 'in order', ready for anything to be 'thrown' their way.

Some people observe, when seeing workers running from the works at 'finishin' time', that this is extraordinary behaviour. 'Why should they run?' they ask. To be a normal operator under the chaos of functional layout *is* to run away at 'finishin' time' or, hopefully, somewhat earlier!

Factory work is becoming generally less acceptable, but the particular frustration and low efficiency of batch manufacture offers the skilled man nothing other than 'a dull, boring job which demands skill, training, alertness and accuracy...'[8] It appears that the early promise of computers has turned sour. In Europe, the computer is still seen as a brain, but in the United States the reality of working alongside them for a longer period has modified the analogy of brain and computer. It is 'old hat' to say that the computerized chaos of functional layout costs more than the chaos that now exists, but it is true.

Functional layout is wasteful, inefficient, and chaotic in most places where it is the basis for the production system. Usually it is adopted because engineering companies have been unaware of the need to examine the organic nature of production. Any production system only becomes 'a system' in the true sense of the word, when an analytical process is begun in the four areas of components, demand, machines and people. The very nature of production emphasizes that it is a dynamic environment. Even those authors who have become household names seem to have failed to appreciate the full complexity of that which they have supposedly taught us to plan and control. Group Technology frequently makes our existing machine shops look inadequate. More importantly, once it has been implemented some of the pillars of production wisdom can be seen to be narrow precise men with little real insight into the nature of production, for it asks us to recognize the obvious. Like the discovery that the world is round, it is really quite simple when we get used to it.

The approach of group technology should first be seen as an analytical process which begins in the following four areas.

(1) Component analysis
(2) Machines available (in the factory or on the market)
(3) Demand for the components—a level further than products, and two levels from product range
(4) People—operators with the skill and willingness to work machines.

At present, Group Technology is perceived as the production of families of components on groups of machines. It has not yet been understood as being a production system which results from an analysis of the above followed by their collective synthesis.

The production system which arises from this total analytical approach can be a single-machine concept, a functional-layout concept, or a cellular concept, or any combination of these.

The early development of Group Technology resulted from component orientation in the workshop. For example, at English Electric in Bradford, England, where electric motors are made, the early work in GT was aimed at improving machine and operator performance by reducing the setting-up and change-over times between different components. Components were analysed, and a family was found which, with revised and specific family tooling, reduced setting time by about 80 per cent. The first machine was operated by the lowest bonus earner, who, after GT was implemented, became the highest earner.

The diagram (Fig. 1) shows that a component family was found. Each of its components left the functional layout for the turning operation, but returned for subsequent operations. As a result the turning operations showed a tremendous productivity improvement but there was no such improvement for other operations. Automation had come through component family selection and a clever tooling arrangement. The year was 1966. Previously published work had pointed to setting time improvements from round parts produced on those machines (lathes), where labour is most scarce and expensive because of its high skill content.

SIMPLE SAVINGS FROM THE SINGLE–MACHINE APPLICATION

The advantages of the simplest forms of GT (Fig. 1) are reduced setting-up time, improved machine utilization, increased operator efficiency and earning potential on one machine. The companies which have conducted work along this pattern, e.g., GEC–English Electric, Bradford, and GEC–Elliott Control Valves Ltd., Rochester, England. These companies have produced figures of their financial savings which

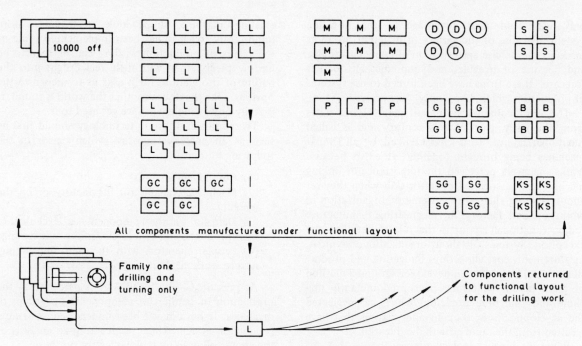

Figure 1. The simple form of GT adopted at English Electric, Bradford.

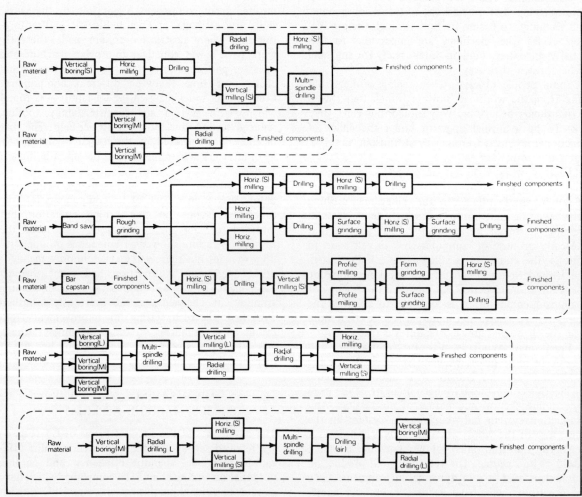

Figure 2. The layout at one of the manufacturing sites of the Stibbe Knitting Machine Company.

exhibit 80 per cent plus reductions in setting-up time as well as improvements in the output of each machine of $1\frac{1}{2}$ to 2 times.

This approach to the improvement of production efficiency has been attractive because of its simplicity and the lack of need for any substantial changes in long-established methods of wage and bonus payments, systems of production ordering and cost centre specifications.

EXPERIMENTAL GT CELLS

The early GT experiments at Ferranti, Edinburgh, Jeumont (France) and at the Elland works of Hopkinsons Valves, serve to emphasize that similar approaches to GT, involving Brisch systems of component classification, lead to similar but wider benefits than those derived from the single-machine method. The benefits are comparable when setting time and machine utilization are considered but the reduction in work-in-progress and increase in rate of component production output are additional and staggering benefits from the adoption of cells.

Each of these three firms introduced GT as a pilot scheme in order to test the idea under practical circumstances. Ferranti, in association with the University of Manchester, has continuously developed its approach, so that it has now conducted a complete GT analysis and more than 70 per cent of component manufacture is in cells. The benefits from the first Ferranti cell were different from those of the single-machine approach, because they allowed for the importance of the speed of component throughput. Consequently, a vast reduction in work-in-progress, as well as in reduced setting times and enhanced machine utilization, was obtained.

Figure 2 shows the results of similar co-operation between a university and industry, and is the actual layout adopted at one of the manufacturing sites of the Stibbe Knitting Machine Company. The Hopkinson cell proved successful for its limited purpose. The first published French case (Jeumont)[9] was impressive enough for restrictions to be placed upon the publication of results.

These firms were examined by the research group at UMIST. It became clear that although financial advantages had been achieved, it was because of the improvements in job satisfaction, performance measurement, shop-floor attitudes and production control systems, that GT had become part of a fundamentally new approach to management.

FINANCIAL SAVINGS FROM A 'TOTAL' APPROACH

The Serck Audco Company in Newport[10] were dissatisfied with their performance in the early sixties, but found only traditional ideas of improvement in published work. The idea of an economic batch quantity was unviable. Their prime concerns were work-in-progress and delivery reliability.

Serck Audco attacked their problems on a broad front. They worked alone at producing a solution to their problems. The effects of their changes, which centred upon GT, were:

(1) sales up 32 per cent;
(2) stocks down 44 per cent;
(3) average manufacturing time down from twelve to four weeks;
(4) overdue orders down from six weeks to under one;
(5) despatches per employee up from £2200 to £3105 p.a.;
(6) average earnings up by £182 per person p.a.

The management research group at Manchester University began to study what had been achieved at Serck Audco and found a parallel in the work of the companies mentioned earlier. It was a parallel, however, which had to be perceived from a management rather than from an engineering viewpoint. The integrated philosophy was close to the practicality found at Serck Audco. Thus it became clearer that in production management few authors had understood either the integrated nature of production or the integrated nature of management. An understanding of GT seemed to bridge the gap between theory and practice. An acceptance of this had made it possible for the practising managers at Serck Audco to understand better what they did in those pioneer years, and through their continuing association with others, to produce a useful body of knowledge which enables others to achieve improvements by quicker, well-established routes.

THE CELL SYSTEM AS A PEOPLE-CENTRED SYSTEM

The cell system results from the analysis and synthesis which are implicit in the total managerial or integrated approach to group technology[11]. Although it is the production rather than the assembly areas towards which this paper is directed, the cell concept is appropriate in both. As yet it remains to be seen whether or not there is as much comparability in technical analysis as in social analysis. The assembly cells at Philips in Holland[12] and Friedland in Stockport, seem to have a close affinity with the manufacturing cells at Ferranti, Audco and Whittaker-Hall in the 'people sense' even though they differ in the technical sense.

Most of the components in the majority of workshops travel between two or more different machines. Those that require, or might require, production on only one machine will be ascertained in the component and machine analysis stages of (GT) production system design. These 'single machine cells' will be quite different from those of the GT single machine concept which was described earlier. This is because they will be created as a result of conducting an overall analysis of all components based upon the entire manufacturing sequence and not just that part dealing with one machine type. Firms with this type of family will adopt an approach to machine selection and tooling design which will be very similar (in nature) to that suggested by the early Russian researchers and put into practice at English Electric, Bradford (Fig. 1).

The analysis leading to production system design is conducted when certain quantities and types of machines are available in a firm. Usually it is found that new machines are on order and that these must always be embraced within any new production system suggested. At this stage a major problem invariably occurs. The work leading to the design of a new system provides the researcher with a much clearer picture of the relationships between machines and components because of the cell concept covering the entire component manufacturing cycle. This places the machine-tool buying policy on a much firmer footing for those firms who examine their existing machines and seek to invest in more appropriate machines in the future.

In many firms, fashions dictate play frequently. Machines of higher technical performance than is required are purchased. Although they may be far superior to those which they replace, they make a relatively tiny impact upon the business. This is because the system in which they are required to operate is not a system at all, but a chaotic, functional-layout based, disorderly, non-system. Analysis leading toward the cell system is important primarily because machines can then be selected to suit components in a more specific way. Because of component family selection, machines are bought to fit the entire component manufacturing system, namely, the cell. This makes the measurement of returns from an investment policy more easily and more accurately quantifiable. The selection of machines from their competitors can and should be calculated in the framework of each cell.

Engineering managers need to recognize that Group Technology has to be introduced into an existing factory which has its own organizational structure. Inevitably problems will have to be overcome. One of the most basic is getting people to consider change as the rule rather than the exception.

Some of the main problems likely to be faced by a manager acting as a change agent are as follows:

(1) Existing functions and their current role
Those functions of management which are different from, but dependent on the production process such as marketing and works accountancy are affected by the introduction of GT. Their particular functions will be subject to modification when group technology is implemented. Within the sales and marketing functions, for example, it will be necessary for the senior executive to be convinced of the need to generate historical and current information so that trends and forecasts can be prepared for use in a useful (to production) production forecast.

Equally, in production management itself, changes will be required. The production director has to originate and develop his planning and control functions with the new and (usually) improved information. Such information may well change his whole approach to production forecasting and production planning and control—a fact demanding a high degree of co-operation between the individuals and their respective staff.

To originate changes without the involvement of the works accountant is industrial suicide not only because of the importance of his function but also because of his role in budgeting and financial control. This is particularly true with the introduction of GT because it is usual both for the basis and the centres of works costing to change. Those concerned with industrial relations, wages, incentives, productivity agreements, etc., will also have to be orientated towards the acceptance of change.

(2) The position of works management
In Britain the works management function is sometimes not represented at board level. Such an absence demonstrates that the production function is sometimes omitted at the decision-making stage. In such companies the works manager may be given impossible tasks, such as impossible delivery dates, simply because he was 'not in' at the decision-making stage.

Not only are there differences in training and specialization at board level, frequently there are differences in socio-economic background and resultant career-patterns. The works director is likely to have reached the top by a different route from his colleagues in finance and marketing. Thus not only the established packing order but subtle nuances in expression lead to difficulties in understanding each other's problems.

(3) Parochialism
In general, firms are organized into various functions with a senior executive at their head. Although some of these functions will undoubtedly be well managed within themselves, there may well be a significant amount of friction between different functions, leading to reduced actual effectiveness of each function. Traditional jealousies between functions such as 'Accountants' and 'Engineers', or 'Design' and 'Production' are much in evidence. This tends to reduce the speed of change.

If the cell system is to succeed, an overall approach must be taken to the total operation of managing the company. It is essential that the parochial thinking within the organizational structure is broken down. 'Tramline thinking' is a phrase which has been coined to describe the propensity of some managers to place the interest of their function above that of the company as a whole. Most 'tramline' thinking is not deliberate or planned, but an attitude of mind which has evolved over many years of separatist thinking.

(4) Fashions in management techniques
Companies sometimes introduce, or pretend to introduce new techniques in order to 'keep up with the Joneses'. For these companies GT may be perceived as little more than the latest in a long line of new fashionable 'gimmicks'.

(5) Difficulties of change in the production system
Some things are easier to change than others. A change in the production system is very difficult, first because it demands a relatively large amount of concentrated thought and effort, and secondly because it involves so many different people at different levels in the firm. For a manager, time available to devote to concentrated thought is at a

premium. Managers find themselves dealing with immediate and specific problems. Usually they are unable to find enough time to devote to the long periods of concentrated thought necessary to achieve success in the medium and long term.

(6) Resistors

Many resistors to change have no understanding of what GT is but are absolutely certain that it cannot be utilized by their factory since it threatens their existing position and security! It is common to hear managers state, 'we already have GT and the cell system', as an excuse for retaining the *status quo*. It happens also to be true that one 'prime mover' can be the sole reason why GT has been introduced.

(7) Traditional attitudes

The people that carry out workshop and factory planning, particularly machine loading, are often those who have worked at the factory and grown with it 'man and boy'. Their local knowledge is useful in certain respects but in others it is a positive hindrance. Certain behaviour patterns exist due to habit more than anything else. In many factories the idea of keeping machines running is a form of divine law and almost the sole criterion upon which one determines the efficiency of the works. Return on investment, meeting delivery dates, the levels of absenteeism and speed of throughput are, however, all relevant factors which should be taken into account when assessing performance. Acceptance of this wider view is implicit in the cell concept of production.

(8) Product-orientation

Most companies who manufacture a wide range of products in the engineering field are organized on a product basis. Consequently most of the information available on paper and in people's heads is related first and foremost to products. The first requirement of the change agent handling GT is to ignore 'product-thinking' at some stages, whilst using it intelligently at others.

(9) Information level prior to change

In some companies, preparation for GT is hampered by poor basic data. The change agent has to know the best route using the best data or, in some cases, the best people, since the best data is in their heads!

(10) Education

Because the cell system implies an overall approach it is necessary for the change agent to plan an 'educational programme'. It is necessary for all functions and working groups to have some inkling of the changes afoot and to be involved in the change.

For example, the marketing manager needs to appreciate why he should produce information in a more detailed manner than hitherto. It is necessary for him to understand that sales forecast information is required for each component being sold irrespective of whether it is a spare part, a standard, or a purpose designed product. The educational problem is important because not only must the sales manager, production manager, production controller and the production planning engineer be brought together in the interests of providing more useful information for group technology, but changes in paperwork, foremanship and lines of responsibility have to be made if one is to make progress in the right direction.

Detailed discussions and co-operative decision-making is essential between those already mentioned and those concerned with the management information system such as the computer section. Such meetings are the foundation stone of a good GT plan which is the first phase of the implementation process. Obviously such meetings, organized as a programme of courses, some of which could be away from the factory, are vital if one is to be sure that the attitude of the company is ready for the changes which must result from the implementation of the cell system.

(11) Trade Unions

Within a cell or unit, men will be required to work as a team integrating the traditional skill patterns of lathe-operator, milling-operator and drill-operator, who have different degrees and types of skill, and this presents problems which frighten some managers. However, in more than one British Group Technology application such problems have been overcome and changes have been welcomed at a time when industrial relations have tended to be one of the major national problems.

(12) Time

The length of time to implement Group Technology properly is bound to deter some companies but time is required at all stages. Time for thought, time for planning, time for training, time for working together in new ways with new people with different skill patterns, and time for entering into and hearing about other people's problems. With sensible planning and an appreciation of the long term aim it is possible to make major savings in the short run of around six months.

(13) Financial considerations

One thing that characterizes the too simple manager is that he asks, 'what is it going to cost and what will be my return?' The question itself seems at first glance to be reasonable and capable of receiving a quick and sensible answer. There is some difficulty in quantifying savings because answers are often required in too narrow a framework. For a total system change, total system benefits need to be calculated and since the total system is the firm it is towards the many management ratios that the good change-agent usually refers. The significance of people, their job interest, involvement in decisions, cannot be ignored by those who would succeed with the cell system.

In the past decade the distance between management and worker seems to have grown much wider. Instead of the face-to-face relationship that once existed, there is now a 'formal communication channel' through which all disputes are supposedly dispatched. However well oiled or systematic these channels are, there will be no industrial peace without trust and mutual confidence.

As we enter the seventies, it is not only management which will wish to increase the rate of change. Within the trade union movement there are those whose aims are the same as those of forward looking management, namely that a co-operative and participative environment shall be created in industry.

On the shop floor traditions and skills are well established but there has been a progressive erosion of the fixed lines between skilled, semi-skilled and unskilled men. Developments in machine tool technology, such as numerical control, have had the effect of changing the status of the skilled man in the workshop hierarchy. The introduction of the flow-line, for example, has shown that the semi-skilled can earn more than the highly skilled. Emphasis has changed from skill brought to the job to skill in designing the job.

The pattern of change, which brought about developments in technology is unlikely to be reversed. In the future it is likely to gather momentum. To date, the worker on the assembly-line is usually working in a different company with different management and different union representation from the worker on component production. In component production not only the high volume assembly lines are supplied but also the intricate one-off special products.

Group Technology will create as great a revolution as the flow-line. The problems are different, the people affected are of a different skill grouping and the factories deal with a smaller volume and higher variety of products. The ability to produce almost any product to satisfy any customer is a constant requirement of the men in these firms. Organizational structures and systems have evolved as a result of the constraints such an approach imposes.

SOME COMMON PITFALLS WHEN INTRODUCING GT

A characteristic of the work conducted in The Department of Management Sciences at UMIST has been to try to marry academic quality to shop floor reality. The shop floor has become the laboratory of the GT research group. Leading engineering firms have enthusiastically supported the testing of hypotheses within their factories.

When the government-sponsored GT centre was established, it was not surprising that they approached those who had been working some years on GT to look at areas of some significance to the nation. One of the projects at Manchester was to examine the various methods of proceeding to GT and the other was concerned with the impact of GT upon the men on the shop floor.

The research group found it necessary to explain in the first project that the GT centre and others had perceived GT in a limited way. Consequently the use of a universal code (Opitz) was counter-productive when the cell system is defined as a new production system.

In the second project the work done by Mrs G. M. Fazakerley's participant observers and interviewers was most helpful in explaining the importance of the more general cellular definition used by UMIST which emphasized the role of people in production systems. The work indicated that a new manufacturing system was evolving which needed technical, social and systems definition.

Work is now progressing to this end but attention has yet to be paid to preparing the earth rather than forcing sickly plants. Although as yet there are uncertainties, the chances of ultimate rejection can be reduced if attention is paid to the methods and timing of implementation.

Of the factories visited which had implemented group technology only one seemed to have reached the stage where the system had become an integral part of the informal as well as the formal organization of the factory. In all other cases group technology was a foreign element which had been grafted on to the existing culture. In such circumstances it is to be expected that there would be reserve about its value.

Innovation is probably a way of life and therefore tends to characterize some companies much more than others. It is unlikely, for example, that companies which are sensitive and responsive to developments in one of their fields of activity are total laggards in others. As a result the problems of change are constantly with them and they develop strategies which they find suitable for dealing with such problems.[13] Even amongst these progressive organizations there had been problems connected with the introduction of Group Technology and therefore it seems worth while to spell out some of the lessons learned from these companies.

There is a skill in presenting change so that it is seen as an opportunity rather than as a threat, and the following points cover some of the sub-skills required in implementing Group Technology.

(a) It is necessary to determine which definition of Group Technology has been accepted as relevant for a particular company's needs. The definition will help determine not only objectives, but the necessary strategies for achieving these objectives and the criteria against which achievement can be measured. The accepted definition will also help determine what kind and degree of benefits can be expected from the implementation.

(b) The objectives of the company must be clearly and accurately stated. Far too frequently objectives are stated in terms which cannot be concretely assessed. Phrases such as 'better utilization of labour', 'increased share of the market', from constant repetition become unconsciously mistaken for objective realities. This conventional phraseology is passed from hand to hand like a smooth worn coin. Because it is so much 'of course' it scarcely demands proof and so offers no criteria for efficiency. Far too many companies leave their objectives unstated. They believe them to be so self-evident that they can be left implicit but objectives which are not stated tend to get overlooked and consequently stand little chance of being achieved.

(c) It must be accepted that every individual will want to know 'what does this mean to me?' Having accepted this the company must explain exactly what the introduction of Group Technology is likely to mean in personalized terms. However much men may identify with the company they are not going to be so altruistic that changes which imply threats to security, income and livelihood are going to be accepted willingly. If such threats exist, personnel have a right to know of them; if they do not the company can scotch the rumours which always arise when any change is suggested.

(d) The company must establish the right climate for change. What this will be depends very much upon individual situations, but certainly it will mean that top management understands the principles behind Group Technology. Without this the company can neither plan the application nor control it once implemented. Equally it will require a readiness to respond to local conditions and what may be thought to be quite irrational reservations. In two cases, for example, there was some resistance by the men on the shop floor to the term 'cell'. Rather than being seen as a biological analogy it conjured up comparison with prison cells. Obviously in these companies some other term ought to be utilized, even though in other companies the term 'cell' is perfectly acceptable.

(e) Just as the company must establish the right climate for change it must select appropriate strategies. Examples of such strategies have been given in detail above; however it must be appreciated that these are subject to change and a good company will allow new strategies to evolve as appropriate as well as taking part in their initial selection.

(f) The timing of the introduction of Group Technology is very important. The best time for innovation is at the beginning of a likely upswing brought about by other factors, e.g., the general market situation. This will allow for any difficulties to be absorbed more readily and some of the benefits obtained from other sources to be psychologically credited to the innovation. In contrast if change is introduced when other factors are depressed or beginning to be so then at best the innovation will have an uphill struggle for acceptance and at worst could be totally rejected as it may be seen to have brought calamity to the company.

(g) The men selected for the first groups should be the best the company has, not those most easily spared. If this is done it helps to build up the prestige of the innovation and helps to establish the group as something with which other operatives can identify the most positive aspects of their own motivation. Group Technology is asking the men to change and thus threatens their identity. Those things which can mitigate the threat should be attended to.

(h) Opportunities for social contact should be provided. Conscious effort to increase interaction appears to lead to the better achievement of the company's objectives.

(i) Attention must be paid to communications. The content of communications must be meaningful to the recipient as well as to the sender and the flow of communications must not be impeded. Even in companies where the communications between management and shop floor and that within the cells appears to be good, communication between cells tends to be poor. There tends to be a feeling that if work is not actually in the cell it is somebody else's responsi-

bility. Cell leaders need training in how to handle this kind of situation.

(j) The selection of the right change agent is very important even though there are no specific criteria against which this requirement can be measured. Essentially it is a matter of matching a man to a situation. The actual implementation of Group Technology should be carried out by a company's own management team. Although external people may be brought in to plan, or to gather data, or to work closely with management, they should be treated as advisers. The shop floor tends to resist people from outside coming to tell them what to do.

(k) The co-operation of management with the shop floor can reveal great potential savings. Often managers are not prepared to listen to or to be advised by the man in the shop but practical day-to-day experience can be of inestimable value. Although senior managers are concerned with policy-making, middle and junior management are more concerned with problem-solving. Such a function makes co-operation with the shop floor essential. Management training really requires much more stress to be placed upon this aspect of a manager's role.

(l) Although orientations to the introduction of Group Technology can vary within an organization and this be healthy for the company concerned there must be a basic consistency in attitudes towards Group Technology otherwise problems are inevitable. It is management's responsibility to bring about such cohesion.

(m) When introducing Group Technology a company must encourage all personnel to look at its functions. If this is done it will be more readily accepted that the introduction of Group Technology will mean change in areas which are tangential to production. In practice the implementation of GT cannot be confined merely to the shop floor. This applies no matter which definition of Group Technology has been determined upon.

(n) The planning department should be closely involved with the implementation of Group Technology. Although it may be thought that only the most shortsighted of companies would separate production from planning, this is not the case. Very often planners appear to have no idea of what actually happens on the shop floor nor do they design with ease of production in mind. Production orientated design tries to make components as simple as possible to manufacture whereas planning which is not production orientated often appears to serve as a vehicle for the self-expression of the planner.

(o) Component rationalization is not synonymous with product rationalization. The latter is not crucial to the implementation of Group Technology and is a decision for a company's policy makers.

(p) Product change may mean cell obsolescence; a well-planned layout should allow for such changes, particularly if the company is involved in an industry with a rapid rate of obsolescence.

(q) The introduction of Group Technology has not been completed once the new methods appear to be operating satisfactorily. Only when appropriate norms and behaviour patterns have evolved can a change be said to have been completed. In the interim period constant feed back is required to ensure that the old values and behaviour patterns which are irrelevant to the new system do not re-emerge, and that inappropriate norms and behaviour patterns related specifically to the new system do not emerge either.

(r) Social awareness is in an organization's self-interest. It is, of course, sometimes difficult to determine what is important in a particular situation. For example some men may say that they are concerned that the introduction of Group Technology will mean deskilling and yet their behaviour patterns demonstrate no concern over this matter at all. Other operatives may not even mention this as an area of concern unless asked about it specifically and yet their behaviour indicates great concern over this issue. The breaking down of jobs into their sub-elements serves to minimize still further the gap between the skilled, time-served man and the semi-skilled, and is a factor which may require great skill in handling.

(s) Important innovations require suitable recruits who can tolerate ambiguities and confusion of identity. If the changes are very important as with the introduction of Group Technology they might also require an ideology. Only the total concept of Group Technology can provide this.

(t) An approach to innovation based solely on rationality will fail. Many of the problems which emerge have an irrational basis because people's emotions are involved. Unless the basic approach to innovation allows for this it will prove too rigid to cope with the problems of change.

(u) The most useful perspective on the shop floor is to see it as an open socio-technical system. It is not on technical variables alone that differences in behaviour and attitudes depend, rather it is the interaction of those variables with the history and traditions of the factory and the external environment which bring about changes in men.

Perhaps the only caveat that can be entered against the possible benefits arising from the introduction of Group Technology is that not necessarily all the benefits come from Group Technology itself. The very process of looking at the organization critically could reveal where improvements and savings could be made. In any company eventually effective but inefficient methods evolve, paperwork multiplies and people lose their enthusiasm. A fresh approach reveals those things which are being done in a way determined by tradition rather than by logic. A good example of this can be seen in machine utilization—although some managers are horrified to find that under the new system it is planned to keep some machinery idle for as much as 20 per cent of the time, they are even more horrified when they find that previously it was idle for 30 per cent of the time when it had been planned for 100 per cent utilization. This is a problem of which companies are becoming increasingly aware and it is interesting to note that some research is currently being implemented to measure the degree to which actual and predicted machine utilization coincide.

REFERENCES

1. C. ALLEN (1973). *The application of Group Technology in the Ferranti Machine Shops—A Case Study,* Ferranti Ltd., Edinburgh, 25th January.

2. J. J. MARKLEW (1973). The cell system of manufacture. Report of work conducted at Whittaker-Hall by W. J. Hancock in association with UMIST, *Machinery and Production Engineering,* 9th and 23rd August.

3. G. M. FAZAKERLEY (1972). *The Impact of Change on the Shop Floor with Particular Reference to the Introduction and Implementation of Group Technology,* Department of Management Sciences, UMIST, March.

4. J. L. BURBIDGE (1962). *The Principles of Production Control,* Macdonald and Evans.

5. W. K. HOLSTEIN and W. L. BERRY (1970). Work flow structure—an analysis for planning and control, *Management Science,* **16,** (6), February.

6. I. F. K. EL-ESSAWY (1971). Component flow analysis. PhD thesis, Department of Management Sciences, UMIST.

7. J. P. SCHMITT and G. A. B. EDWARDS (1972). *La Production L'Enterprise Moderne* (H. Dougier Ed.), Les Sciences De L'Action, Paris (French).

8. D. T. N. WILLIAMSON (1972). The anachronistic factory. A paper presented to the Royal Society, March.

9. P. A. SIDDERS (1962). The flow production of parts in small batches. *Machinery and Production Engineering,* 31st January and 25th April.

10. G. M. RANSON and B. H. TOMS (1964). *Production Planning and Control.* NEDO, HMSO.

11. G. A. B. EDWARDS (1971). Readings in group technology—cellular systems, *Machinery.*

12. H. G. VAN BEEK (1966). The influence of assembly line organization on output, quality and morale. *Occupational Psychology,* **38.**

13. F. KIRKMAN (1970). *Coping with Change,* Department of Management Sciences, UMIST.

VARIETY CONTROL OF A SEMI-FINISHED PRODUCT WITHIN A GROUP TECHNOLOGY MANUFACTURING SYSTEM

by

T. M. GIBSON* and R. H. THORNLEY†

GENERAL INTRODUCTION

Stock levels and stocking policy to meet manufacturing and customer service requirements often lead to large sums of money having to be invested in slow moving items and large quantities of work in progress, much of which is designed for one specific need. Rationalization and standardization reduce the manufacturing and stocking costs, provided they are carried out with due consideration of re-order cost and effect on customer service. Within a Group Technology manufacturing system at Ferodo Ltd. consideration has been given to rationalization of semi-finished products for the manufacture of brake linings. The rationalization has not only led to a reduction of basic manufacturing varieties but also provided the basis for a new stocking policy.

In the particular investigation reported here, the function of the semi-finished product, hereafter termed semi-finished (SF) pad, is to serve as a curved sheet of friction material from which the linings can be machined. The SF pad serves a purely internal function and as such the design can be controlled by Ferodo Ltd. and not, as in the case of the final product (brake lining), by the customer. Ferodo Ltd. can be said to be a process industry and is peculiar in that it manufactures its own material from which the final product is machined. Thus, as regards the brake lining group, the factory can be split into two natural divisions, Stages 1 and 2 manufacture. Stage 1 deals with the manufacture of the SF pad and Stage 2 with the manufacture of the lining from the SF pad. In Stage 2 a GT manufacturing system is employed[1-3].

Between Stage 1 and Stage 2 manufacture there is a buffer stock of approximately one week's work, this being in the SF pad form. Similar orders are then grouped together from this buffer stock and fed into the basic shape GT machining cell.

PROBLEM DEFINITION

The rules regarding the creation of a new SF pad size were ill-defined and generally speaking the situation appeared to be degenerating to such an extent that, regardless of order quantity, a new SF pad size was created every time a new lining size was demanded. In consequence a large variety of SF pad sizes resulted, many of which were almost identical in size. This had the following effects:

(a) Many small orders for Stage 1 manufacture causing inefficient use of machines and material.
(b) Many 'overmakes', causing redundant stock to appear in the buffer between Stage 1 and Stage 2 manufacture, despite the policy to manufacture SF stock only for specific customers' orders.

Customer service for small orders of linings outside the authorized stock range was poor simply because SF stock had to be manufactured, i.e. standard lead time was usually quoted. It was felt that a rationalization exercise coupled with a stocking policy for SF pads would solve the attendant problems.

An initial survey showed that eight different materials generated 3500 different pad sizes; this quantity was increasing at the rate of approximately 300 per annum. A Pareto analysis was carried out over a four-month period and the results shown in Fig. 1. The analysis indicated that approximately 4·3 per cent of the total number of live specifications represented 90 per cent of the total cash value of pads demanded in the four-month period. It appeared

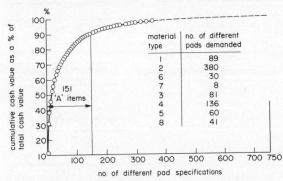

material type	no. of different pads demanded
1	89
2	380
6	30
7	8
3	81
4	136
5	60
8	41

analysis of S.F. pad demand over a four month period for the materials shown above

Figure 1.

* Production Planning Manager, Ferodo Ltd.
† Professor and Head of Production Engineering, University of Aston in Birmingham

that equal attention was being paid to all the specifications whether high or low usage. Examples were found where, from the material point of view, high usage pads were being used in an uneconomical manner.

ORDER OF QUANTITIES INVOLVED

Production at the factory is in one-week cycles; thus it is convenient to think of the batch quantity for SF pads as the quantity manufactured in Stage 1 in any specific week. This quantity can be as high as 5000 or as low as 2. In a typical week's production 239 different pad sizes were demanded. Appendix 1 shows a breakdown of the batch sizes.

METHODOLOGY

Figure 2 shows the method of grouping for rationalization. It was decided to concentrate on the major groups of material since these constituted the bulk of the turnover. Material type 2 was chosen to establish the rules for rationalization because this was the largest in terms of both turnover and variety. Referring to Fig. 2, all linings manufactured in the same material were grouped together, and then grouped according to forming radius. Linings are moulded to their correct radius on formers rather than machining, the radius of the former being a function of the finished lining radius and thickness. Additionally several slightly dissimilar lining radii can be considered to be nominally identical, within a small tolerance range. Hence the parameter to group on is

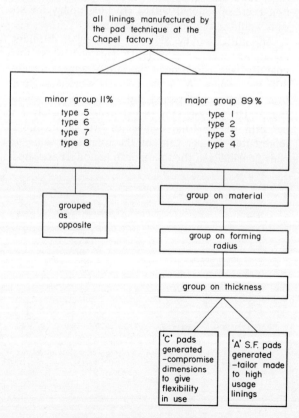

method of grouping for rationalisation

Figure 2.

forming radius and not finished lining radius. The linings manufactured on the same forming radius were then arranged into groups of 0·005 in thickness increments (see Fig. 3). Thus it was decided to 'tailor make' SF pads to cater for high usage linings—these to be designated 'A' type pads and would cater for 90 per cent of the turnover. The remaining 10 per cent

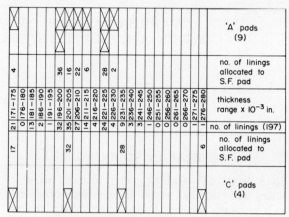

rationalised system no. 3 for linings grouped into material type 2 and forming radius 3.813 in.

Figure 3.

would be manufactured from a limited range of SF pads, designated 'C' type pads. C type pads would be low usage pads and would cater for small orders, material economy being far less critical in this range than in the 'A' range. Thus, by careful design of the dimensions of the 'C' range of pads it would be possible to vastly reduce the number of SF pad specifications and at the same time increase the average batch size for SF pads, since many linings would be allocated to one 'C' type pad. It was decided to carry this philosophy one step further and introduce a stocking policy for these 'C' type pads. The implications of this will be discussed later.

RATIONALIZATION OF SF PADS WITHIN THE MATERIAL TYPE 2 GROUP, FORMING RADIUS 3.813

No. of SF pads prior to rationalization 58
No. of SF pads after rationalization   13

Figure 3 shows the final number of SF pads that were designed after rationalization, the four 'C' type pads being held as stock items. Thus, in arriving at this finalized system, the following questions had to be answered.

(a) What decides the generation of an 'A' pad?
(b) How many 'C' pads should be generated?
(c) What should be the dimensions of the 'C' pad?
(d) How many of each 'C' pad should be stocked?

A Pareto analysis was carried out on the previous twelve months sales and any item appearing in the top 80 per cent was classed as an 'A' lining for which an 'A' pad was automatically generated. Low usage linings having similar dimensions were also allocated to the 'A' pad. Thus any 'A' pad has one specific lining allocated to it, for which it is designed and additionally other linings similar in size but of low usage are also allocated to it.

Clearly, a fairly close control system is necessary when operating in a dynamic market. This is described later.

The number of 'C' type pads to be generated in each material forming radius group had to be decided. The cost of material wastage increases as the number of 'C' pads generated decreases, as shown in Fig. 4, but due to variety reduction and rationalization, the number of pads stocked also reduces. Thus, by balancing the cost of excess material to be machined (unrecoverable) against the cost of storage, it was possible to arrive at the optimum number of 'C' pads that should be generated (see Fig. 4).

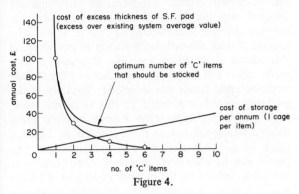

Figure 4.

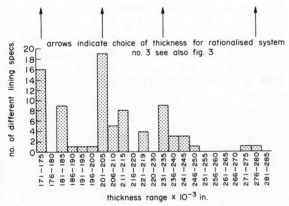

'C' item distribution for linings grouped into material type 2, forming radius 3.813 in.

Figure 5.

In deciding the dimensions of the 'C' pads, the thickness is the critical dimension since the material in this plane is unrecoverable. Figure 5 shows a distribution of the lining thickness in the group under consideration; thus in the absence of a quantity forecast for each part number, the choice of the four thickness dimensions becomes self evident. Regarding the width dimension, this is governed by a maximum tooling dimension of 26·75 in. The length of each 'C' pad is governed by the length of the longest lining allocated to the pad.

The quantity of each 'C' pad to be stocked caused concern because of the high risk of carrying obsolescent items. Figure 6 shows the batch size to be ordered (economic batch quantity—EBQ) when replenishing the stocks, i.e. orders for SF pads to Stage 1 manufacture; this formula is derived from the Camp formula, details of which are given in Appendix 2. For the group considered the number of pads/cage was approx. 200 and the demand rate, found by simulation over a ten-week period, was 17/week.

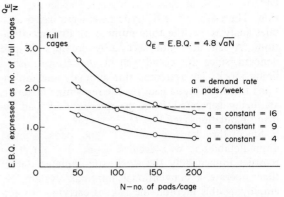

graph showing variation of E.B.Q. with no. of pads/cage and demand rate

Figure 6.

From Fig. 6 it is seen that $Q_E = 1·5$ cages. Bearing in mind the acute shortage of both space and cages at the factory it was decided to round off the EBQ to the nearest whole number of cages, in this case 1, i.e. $Q_E = 200$ pads.

The whole range of material type 2 SF pads has now been rationalized along these lines and rationalization is proceeding for the other materials.

CONTROL OF THE ESTABLISHED SYSTEM

Having rationalized the range of sizes it was essential that a system be developed to adjust according to the needs of a dynamic market. This was achieved using a computer to produce on a quarterly basis an 'ABC' analysis (see Fig. 7). In addition a tabulation was produced listing all lining orders allocated to each SF pad. These two tabulations enable decisions to be taken by management on a quarterly basis, e.g. item number 151, Fig. 7, is losing popularity; by reference to the previous quarter tabulation it is possible to

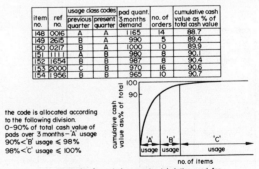

example of quarterly computer tabulation used for monitoring S.F. pad demand

Figure 7.

take a decision whether to leave it as an 'A' type pad or to relegate the once high usage lining to a 'C' type pad and eliminate the 'A' type pad reference.

DISCUSSION ON THE OVERALL ADVANTAGES AND DISADVANTAGES OF THE SYSTEM

(a) Disadvantages

Six months after establishing the full range of 'C' type pads, approximately 20 per cent of the sizes remained unused. This was not totally unexpected and the following reason is suggested as an explanation. The sizes of the 'C' type pads were based on a sales analysis, i.e. the total number of 'live' specifications. A sale does not necessarily create a SF pad demand, since this could well be met from finished lining stock. To overcome this anomaly the dimensions for the 'C' type pads for the remaining materials are being generated from the quarterly SF pad demand tabulations (see Fig. 7).

Since the demand for 'C' type pads is most unpredictable, it was decided from the outset to re-order immediately the stocks are depleted, rather than operate a conventional min/max system. By employing this method the risk of carrying obsolescent stock is reduced and cage utilization is maximized. In order to minimize the delay when a demand is created for a 'C' pad for which there are no stocks available, the system shown in Fig. 8 was devised. When a demand is created for a 'C' type pad this immediately takes preference over any work queuing to be loaded into the Stage 1 manufacturing system. Since the quantities involved are small this does not seriously disrupt the production schedule.

This system tends to stabilize the overall lead time for linings manufactured from 'C' type pads, since the variable element in the overall lead time is eliminated.

The cost of the clerical effort involved in the rationalization exercise was estimated to be £1000. The cost of computer effort for the control system was estimated to be £750.

(b) Advantages
(1) Improved customer service and more efficient use of tooling on GT lines. It is suggested that these two factors are interconnected as shown in Fig. 9. The SF stock acts as a buffer for the start of the GT manufacturing system. Similar sizes of linings are grouped together to reduce machine set ups (or increase average manufacturing batch size). The average period allowed for grouping is one week. If this period is increased, then the likelihood is that more orders can be grouped together and the average manufacturing batch size increased. Similarly, if the grouping period is reduced to less than one week, then one would expect the average manufacturing batch size/machine set up to be reduced. By having a stock of SF pads available ('C' pads) the grouping period can be effectively increased. On occasions it has been possible to deliver small quantities of linings within a week of receiving the order simply because SF pads were available. Admittedly in these cases material economy was not good, but the effect of good customer service was immediately apparent and considered of high importance. The economics of this action are extremely difficult to quantify, but this kind of flexibility is vital in a highly competitive industry.

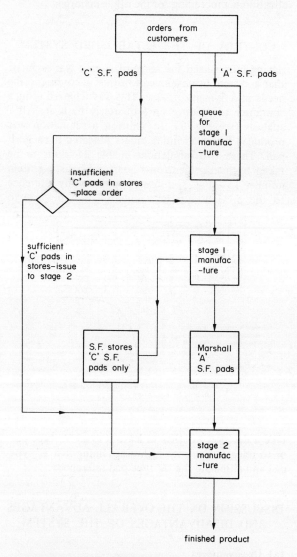

proposed method of achieving a quick consistent delivery of items allocated to 'C' S.F. pads

Figure 8.

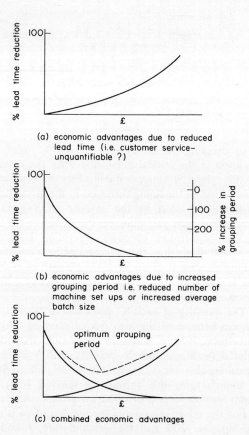

(a) economic advantages due to reduced lead time (i.e. customer service-unquantifiable ?)

(b) economic advantages due to increased grouping period i.e. reduced number of machine set ups or increased average batch size

(c) combined economic advantages

Figure 9.

It is almost impossible to quantify the effect of reducing the standard lead time by one week. Thus Fig. 9(a) indicates a probable trend. Work is being carried out at the moment to attempt to quantify the reduction in machine set ups associated with effectively increasing the grouping period; the probable relationship is indicated in Fig. 9.

(2) The reduction in SF stock levels due to this exercise was approximately 32 per cent and 700 cages costing £5600 were released for other work.

(3) The effect on Stage 1 manufacture was to increase the average batch size, see Appendix 1. Due to several developments taking place simultaneously in the Stage 1 area, quantification of benefits directly attributable to rationalization are extremely difficult.

(4) Overmakes (in SF pad form) for small orders have been eliminated. For example, for an order demanding one SF pad only, it was necessary under the old system, to manufacture a minimum quantity of two pads. Under the rationalized scheme one 'C' pad can be issued from stock.

CONCLUSIONS

The authors consider that the advantages of rationalization clearly outweigh any disadvantages.

Technology has itself produced a demand for greater variety. Engineers are constantly making very precise specifications to meet specific circumstances. Manufacturers have met this demand and so expanded their product range. Companies are constantly satisfying the demand for all orders of non-standard products and are being coerced by customers into expanding their product range. Conversely, scientific management is taking over within manufacturing companies and these are becoming more cost conscious. Customers usually require quick, reliable deliveries at low cost. To effect a quick, reliable delivery of a large product range it is necessary to hold large stocks of material in one form or another. The quantity of these stocks is generally proportional to the variety. Thus, a quick delivery of a large variety of products is usually synonymous with large stocks and this is incompatible with low product cost.

A rationalization exercise coupled with a sensible SF stock policy, while not being a panacea, could well be a step in the right direction to solving this syndrome.

ACKNOWLEDGMENTS

Some of the work described here formed part of a research contract between Ferodo Ltd. and UMIST* and the authors would like to thank Prof. F. Koenigsberger for allowing the work to be carried out in the Machine Tool Division at UMIST. Also the Ferodo board of Directors are thanked for their permission to publish this paper, and include information obtained in their company.

* University of Manchester, Institute of Science and Technology.

REFERENCES

1. R. C. PARKER (1970). University and industry collaborate, *Advance* No. 8. UMIST.
2. R. CONNOLLY, G. H. MIDDLE and R. H. THORNLEY (1970). Organising the manufacturing facilities in order to obtain a short and reliable manufacturing lead time, *11th M.T.D.R. Conference*, Birmingham, September.
3. R. CONNOLLY and A. J. P. SABBERWAL (1971). Management structure for the implementation of Group Technology, *C.I.R.P.*, **19** (1).
4. S. EILON (1962). *Elements of Production Planning and Control*, Macmillan.

APPENDIX 1

ANALYSIS OF THE BATCH SIZE OF SF PADS (STAGE 1 ORDERS) BEFORE AND AFTER RATIONALIZATION

Batch size (S)	Percentage of orders	
	Before rationalization	After rationalization*
$S < 49$	62·5	36·5
$99 \geqslant S \geqslant 50$	9·5	28·5
$200 \geqslant S \geqslant 100$	11·5	12·5
$S > 200$	16·5	22·5

* This analysis covers all materials and illustrates the effect after only one material has been rationalized (2602 material).

APPENDIX 2

DERIVATION OF THE FORMULA

$$Q_E = 4 \cdot 8 \sqrt{(a N)} \quad \text{(see also Figure 6)}$$

The EBQ for 'C' pads is determined from the formula given below (see ref. 4), this being a modification of the CAMP formula.

$$Q_E = \sqrt{\left(\frac{2 a S}{I + 2 B}\right)}$$

Q_E = economic batch quantity (EBQ)
  a = weekly consumption rate
  S = preparation cost/batch, estimated to be £2/order
  I = i.c. = interest payable/pad/week
  i = interest rate (10 per cent)
  c = unit pad cost
  B = unit storage cost/week
  N = number of pads/cage

$$I + 2 B = \frac{£0 \cdot 173}{N}$$

$$Q_E = \sqrt{\left(\frac{4 a N}{0 \cdot 173}\right)} = 4.8 \sqrt{(a N)}$$

DISCUSSION

Q. B. J. Davies, Staveley Machine Tools Ltd. From Fig. 5: would not the amount of grinding to produce finished pads be reduced if the choice of thickness had been 181–185 instead of 171–175 and 211–215 instead of 201–205?

A. Figure 5 shows a distribution of existing thicknesses for small demand items. The demand for such items is very unpredictable and it would therefore seem rational to generate thickness sizes corresponding to the most popular thickness ranges.

DATA PREPARATION FOR NUMERICALLY CONTROLLED
MACHINE TOOLS

THE WORK ON NC AT NEL

by

W. H. P. LESLIE*

SUMMARY

The paper describes the wide ranging scope of the work at NEL which is directly related to numerical control, tracing its development since 1960. Currently a staff of forty are involved. The major aim has been to encourage industry, by the use of standards, to use NC machines efficiently.

INTRODUCTION

Since 1951 NEL has had an interest in both digital (numerical) measurement and control, although in the 1950-60 period this interest was confined to speed, time, fluid flow, pressure, torque, angular position, linear position, surface finish, and temperature. During this period NEL was also developing the radial grating and applying it to gear cutting and measuring machines with accuracies of a few seconds of arc. Work was also going on at its sister establishment, the National Physical Laboratory, on the development and application of linear gratings to position measurement and control in machine tools. This work has subsequently come to NEL where linear and radial gratings and their applications continue to be developed.

Since 1960 the staff at NEL have built up an expertise on most aspects of numerical control of machines. Currently there is a staff of forty working on this topic.

MANUFACTURING SYSTEMS GROUP

The Manufacturing Systems Group is one of the four main NEL Groups; it employs a staff of eighty on separate subjects of which those directly related to NC and Production Control are shown in Fig. 1. Any attempt to break a research interest into smaller units inevitably results in the smaller unit having links to others in the wider group, and in Fig. 1 the full lines indicate active links whilst the dotted lines are links currently being established. The seventeen subjects shown in Fig. 1 have staffs of between two and eight working on each, and for geographic and administrative purposes in NEL they are dealt with in two separate NEL Divisions (approximately thirty staff per Division) known as the Computer Aided Manufacturing and the Control Systems Divisions.

The subjects on Fig. 1 which are directly related to NC are shown separately in Fig. 2 where the incomplete links show how these NC subjects benefit from and contribute to associated Production Control work.

This paper concentrates on these NC subjects (shown separately in Fig. 2) which have recently been highlighted by being considered within the scope of the work on NC at NEL within MS Group.

NC MEASUREMENT AND CONTROL

This is box 206 in Fig. 2 and is dealt with first because it was one of the first NEL interests in NC of machines twelve years ago. (The use of 206 and similar numbers is related to a numbering system for research projects at NEL.)

Metrological tests were carried out on the more accurate measuring systems then available and the various control systems which went with them. It was decided that there was no need to develop new measuring methods but that it was worth while continuing to develop the use of linear and radial gratings.

In 1965 it was felt that there was a need for an NC measuring system which could retain the high accuracy available from optical gratings, but provide the absolute indication required for many machine tools, for which it was desirable to indicate the machine position after switching on without following a setting up procedure. As the result of a development program, a system was obtained which had three decimally related grating patterns on a grating scale for the three least significant figures, combined with a three decimal place digitizer for the three most significant figures. This provided a 100 in scale with a resolution of 0·0001 in whose basis is shown in Fig. 3. By subdividing the grating cycle into ten parts, each pitch of a fine grating of 'm' lines/in formed 10

* NEL

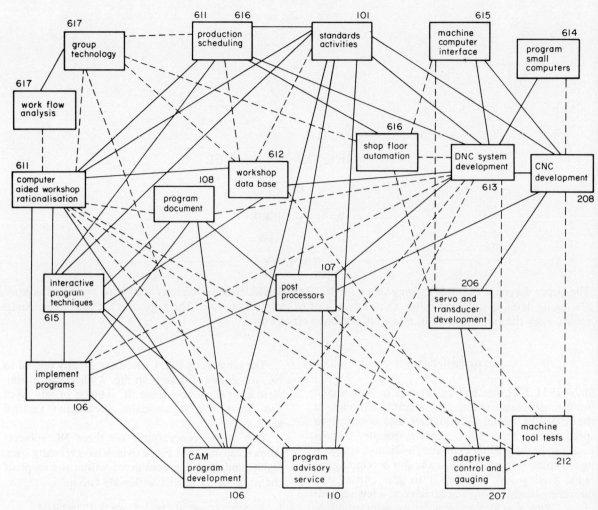

Figure 1. NC and production engineering projects.

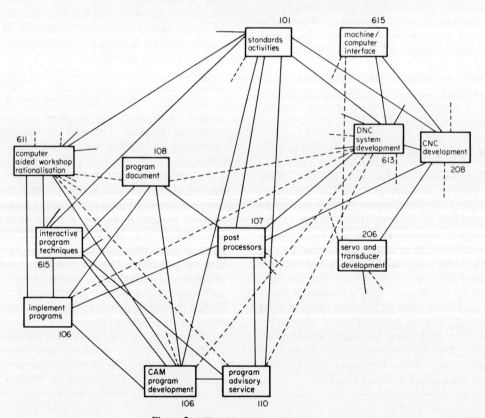

Figure 2. NEL NC projects, 1973.

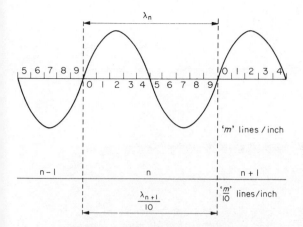

Figure 3. Absolute grating system.

units (each of $m/10$ in) which were the least significant digits of an absolute decimal number. The next coarser grating had $m/10$ lines/in and the third grating $m/100$ lines/in and each was similarly sub-divided (each subdivision being equal to ten of those on the finer track). The three digit digitizer completed this process for the three coarser decades. By this method each track was 'covered' by a coarser track until the whole absolute quantity was built up.

Two licences were granted for this system which featured eight patent applications. The first licensee was Whitwell Developments Ltd who engineered linear and angular measurement and control systems and made substantial sales. This Teletrak system is shown in Fig. 4. The second licensee, Moore Reed

(Industrial) Ltd, has engineered a rotary version incorporating their coded disc with grating tracks, which has made considerable inroads in the military field (the Auto-scale).

A further development has been the NEL Optosyn Transducer whose sponsor (Newall Engineering Company) required an accurate grating system which would be compatible with servo elements such as the resolver, Inductosyn and other a.c. devices. The outcome of the development project was a grating transducer (Fig. 5) with a multi-cell self-scanned photodevice array as the reader of the Moiré pattern of light.

The a.c. reference signal 'f' of the resolver system is fed to a phase-locked loop digital multiplier which generates clock pulses at the rate 'fn' where n is the number of cells in the array (up to 256 cells but more generally 64 or so). These pulses cause the array to be scanned at the frequency f. The output change-of-phase of the Moiré signal is proportional to the grating displacement and is compatible with other a.c. phase transducers. The errors encountered in normal d.c. grating systems are removed as the new system is not amplitude-conscious and, by combining the Optosyn transducer with resolvers for coarse measurements, absolute systems can be arranged.

Several new techniques have been developed to exploit fully the features of the Optosyn and one of these (Fig. 6) is a constant frequency technique which ensures that the transducer output frequency remains constant irrespective of the direction and magnitude of travel of the machine member. Filters can then be introduced without following errors.

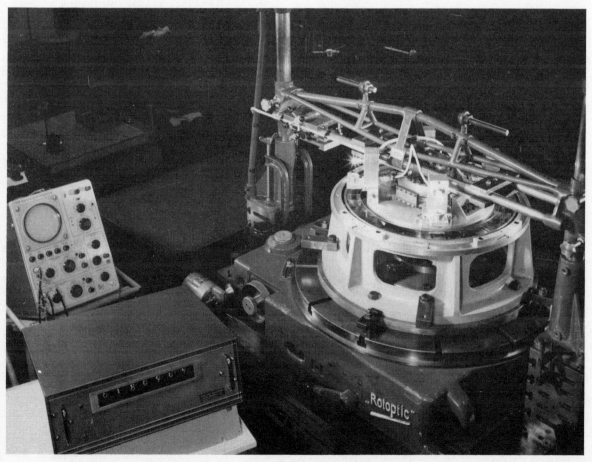

Figure 4. Teletrak system.

H

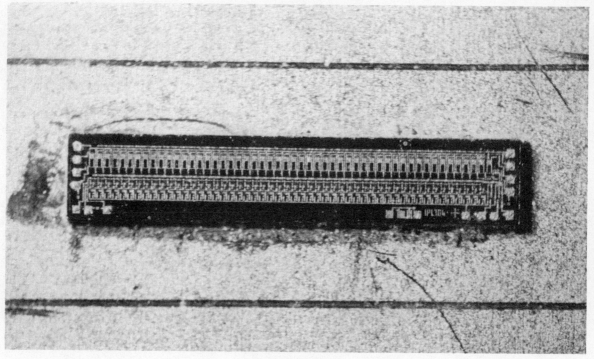

Figure 5. Optosyn.

An example of the application of these techniques is the servo-controlled rotary table shown in Fig. 7, which combines a constant frequency grating phase transducer with a frameless torque motor on a single

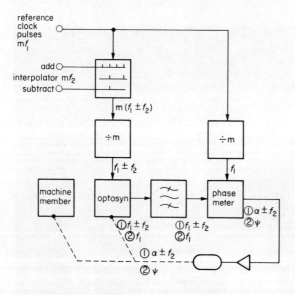

Figure 6. Constant frequency servo technique.

shaft and bearings. The prototype has been constructed to evaluate the design philosophy and can be scaled for any particular requirement. Positional information is picked off at four points on a 2160 (10 arc min) line grating and provides an extremely stable 1 arc s system resolution. The omission of gears in the drive has removed hysteresis and the 11 in table has a holding torque and following torque of 5 lbf ft. The maximum traverse rate with a table load of 50 lb is 0·3 rad/s. As in all servo systems 'trade offs' can be

made to increase the load or traverse rate when the following error is less important. Light workloads can be withstood and clamping can be arranged for heavy workloads.

TESTING MACHINE TOOLS

Although not directly related to NC, NEL has had a long standing interest in certifying the accuracy of machine tools. The main types of machine where this has been important have been jig boring machines and large gear hobbing and grinding machines.

Since purchasing its first NC machine in 1961 NEL has displayed an interest in evaluating the cutting accuracy of NC machines, and has co-operated with PERA, BEAMA, and MTTA in the design of suitable cutting tests for NC machines. This is a problem not yet solved.

ADAPTIVE CONTROL

Work is also done at NEL on the evaluation and development of adaptive control and in-process gauging. This applies to all types of machine tool, although adaptive control can be applied most economically to machines having existing control of feedrate and spindle speed, as is often the case with Numerical Control. Again, if the NC is obtained by using a small computer (see CNC below) it is possible to carry out the necessary calculations for optimizing the adaptive control function at little extra cost.

COMPUTER NUMERICAL CONTROL (CNC)

Box 208 in Fig. 1. With the advent of cheap minicomputers there is a techno-economic race between providing NC functions by programming a minicomputer which is cheap because it is mass produced

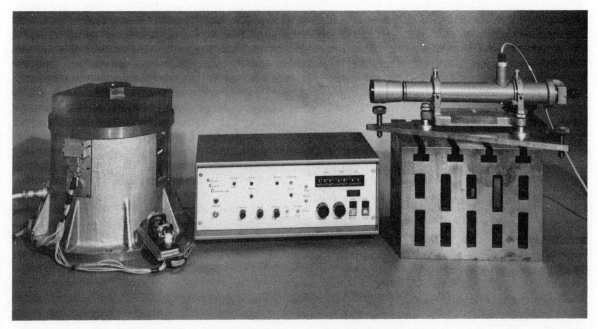

Figure 7. Servo controlled rotary table.

for many other purposes as well, or by taking advantage of the complex circuit functions which can be obtained on a small silicon chip (LSI circuits) and designing modern NC systems with a minimum number of cheap components.

In order to keep in touch with the CNC possibilities, NEL has developed a system on the lines of Fig. 8 which is modular in both software and hardware and is, basically, a contouring system for three or more axes. All conventional ISO standard NC functions are available with the addition of axis calibration, control tape storage and editing, and adaptive control, in the near future. Interpolation may be implemented in hardware or software. A generalized software interpolator of NEL design is available which can cope with straight lines, slopes, and conic sections such as circles, ellipses, and parabolas. An extended CNC system can be shared between a number of similar machines with, in favourable applications with short control tapes, the control tape program stored in computer memory. This

eliminates all but one tape reader and reduces tape reading to one initial input operation. A significant saving in control system cost can therefore be realized in multi-machine systems. The modularity of the NEL CNC system ensures that widely differing specifications can be met. A generalized control program can be readily assembled and defined in terms of a specific machine by completing a parameter table containing data on table limits, maximum feedrate and spindle speeds and acceleration coefficients etc.

The software modules from which CNC systems can be assembled have been designed as generalized routines. The structure and organization of a typical system is shown in Fig. 9, although specific applications may differ in detail.

The NEL CNC software is available to industrial users either on the basis of user employee training at NEL or in terms of NEL undertaking its implementation at user plant. It is therefore possible to utilize it in other computer-based industrial systems and a versatile library of modules is being developed.

DIRECT NUMERICAL CONTROL (DNC)

Boxes 613 and 615 in Figs. 1 and 2. Due to the reducing cost of small computers it has become practical in the last few years to consider distributing control tape information to a number of NC machines in a workshop from a magnetic disc store attached to a small computer.

Such a system has become known as a Direct Numerical Control system and the immediate advantage is the elimination of tape handling and of the repeated use of rolls of punched tape in the relatively hostile workshop environment. It is also easier to ensure that an up-to-date version of a control tape program is used if only one copy exists. The program can also be edited temporarily (oversize casting or a batch of harder material to be machined), or permanently if the tape program is being tried out before starting a production run.

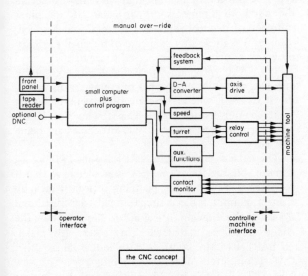

Figure 8. The CNC concept.

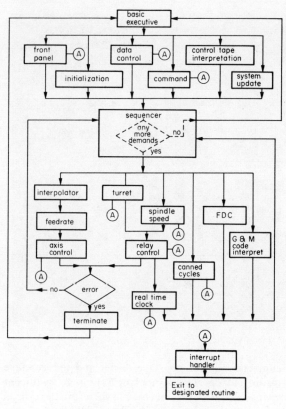

Figure 9. CNC system structure.

It is only likely to prove economical to pay for this extra computing system if it is purchased along with perhaps 10–20 new machine tools so that economies can be made in individual NC systems (no tape readers, circular interpolators, cutter compensation, feedrate override on each, for example). More commonly the DNC system will be introduced to control information flowing to existing NC machines, with perhaps one or two new machines added. The DNC system then connects to the NC systems behind each tape reader (BTR), and project 615 is concerned with this aspect.

The cost of the DNC system must then be recovered from reduced operating costs and by the improved work flow resulting from the collection of valuable workshop management information. The DNC system in this case can be used to collect information as well as to distribute. It then has an up-to-the-minute picture of the progress of each batch of parts, recording for example how far through its cutting cycle a particular part in a batch is, whether feeds or speeds have been modified manually or by adaptive control, whether a machine is idle (and why).

When adaptive control is successfully applied to an NC machine such a DNC system will almost be a necessity because the time to machine a batch of parts will no longer be known from the programmed cutting conditions, but will depend on variations in blank size and in machinability of the workpiece. To cope with this and to ensure that the machine does not lie idle after an easily machined batch, large reserves of workpieces will have to be scheduled at each machine or else, with DNC, short term rescheduling will have to be carried out to determine a new

pattern of work flow. This will then also help if a machine is not available due to breakdown or an additional part is required due to a flaw being discovered in a workpiece after machining.

Such a DNC system also offers an ideal test rig for the acquisition of metal cutting data or the development of programs handling workshop information, because so much of the necessary information can be acquired automatically and quickly. Work at NEL on DNC is aimed at making a system available to industry to help users determine the value of DNC in their particular area.

CAM PROGRAMS

Box 106 in Fig. 1. The work at NEL on Computer Aided Manufacturing programs was started in 1965, after an industry based committee investigation, to deal with computer programming aids for NC tape production.

At that time, it was believed that it would be necessary to encourage industry to adopt a standard part programming language for the description to computers of the cutting which had to be done by NC. The committee recommended that an NC computer program be developed by NEL and made freely available to industry, so that users could employ it without being tied to a particular machine tool control system, or computer manufacturer's products.

It was then correctly forecast that this part programming language should conform to that developed with Government Funds in the USA and known as the APT system. It was recommended that NEL should develop computer programs which responded to the parts of the APT language required by UK industry.

With the assistance of control, computer programming, and computer manufacturing firms, NEL has developed a set of program modules which can be combined to form separated NC programs for turning (2C), drilling (2P,L), milling (2C,L), flame cutting, design, etc, or they can be combined to form the complete NELAPT processor. Users find that one approach to part programming is then possible for all types of NC machining and that they may forget all the detail requirements of particular NC machines (number of decimal digits for X, Y, Z, feed, speed, tool, etc.) as these are taken care of by small auxiliary programs known as post processors (see below).

The 2C,L program is available at computing bureaux, and is on most types of computer, including the ICL 1900, where there are fifty users, including fourteen universities and colleges who use it for teaching purposes, and a further 60 copies are being studied.

The current work on NELAPT is aimed at making it more suitable for use in the design office, so that it becomes possible in the planning office to pick up from a computer file a shape already described and drawn in the drawing office and to describe how the shape is to be machined. This work also makes NELAPT easy to interface with industrial or other CAD (Computer Aided Design) programs. The new NELAPT version with this added capability is now available from NEL and is ready to be adopted by industry.

Program implementation is an associated activity which makes tested versions of NELAPT available to registered users. This is done by having a small team who accept proposed NELAPT modifications, assemble these into a complete system with each card image identified, and test the system before making it available to users.

PROGRAM DOCUMENTATION

Box 108 in Fig. 1. No programming system is complete without documentation. This is available in five levels.

Some 700 copies of the NEL NC News are sent three times a year to interested recipients. The NC News contains information about: new features; notes on avoiding problems not yet solved in NELAPT; meetings and courses available; news of developments in NC standards, etc.

Users can also register for a fee to obtain a loose-leaf version of the Part Programming Reference Manual, regular up-dates, and occasional users' notes which deal in detail with a particular facility which beginners find difficulty in exploiting fully. To date these notes have dealt with using Tabcyls (Interpolated curves), Motion Statements (to cause desired tool movements in some complicated situations) and fully Three-Dimensional Machining using MATRIX and TRACUT.

From time to time the Part Programming Reference Manual is printed as a laboratory report which is freely available but is not updated for a few years.

A small number of looseleaf Computer programming Manuals are distributed for a fee to give details of the working of NELAPT from the point of view of a computer programmer who may implement it on a different computer.

Producing the two reference manuals takes time and until printed revisions are available there are usually hand-written or typed pages available.

In addition to these activities papers are written giving details of particular aspects of the NELAPT system, technical memos for internal and limited external circulation are written and occasional NEL reports are produced.

PROGRAM ADVISORY SERVICE

Box 110 in Fig. 1. A major task since 1965 has been to make available to industry advice on using NELAPT and APT. This is usually provided free for a one-day visit to NEL. Longer visits will involve a fee where the benefit is mainly to the visitor. NEL has 10 NC machines ranging from a 2-axis drill through machining centres, jig borers, and milling machines to a 6-axis milling machine and two lathes. On these it is possible to machine many of industry's requirements (although for 6-axis work it may be necessary to demonstrate by cutting a small-scale part due to the limited size of the machine, 300 × 300 × 150 mm).

It is also possible for a part programmer who has attended an initial part programming course to arrange to spend time at NEL part programming beside experienced men to become familiar with some of the more complicated aspects of the subject. NEL retains membership of the CAMI organization in the USA which develops APT and the EXAPT Association in Germany so that it has an up-to-date experience of these two NC programming systems. In a joint exercise with the Machine Tool Institute in Aachen a joint processor, 2C,L–EXAPT 1, was developed to test the practicability of combining parts of each programming system. 2C,L is the contour milling part of NELAPT, and EXAPT 1 is the drilling part of EXAPT.

The program advisory service attends one or two NC exhibitions each year demonstrating NELAPT at work, and also arranges in-house demonstrations of NELAPT to potential users.

COMPUTER AIDED WORKSHOP RATIONALIZATION (CAWR)

Box 611 of Figs. 1 and 2. A study of two ICL NC programs which had been developed for specific purposes, and a test carried out jointly with a local firm on the use of EXAPT 1, convinced NEL in 1968 that it would be necessary to do more research before developing the section of NELAPT which would automatically supply feeds and speeds and sequences of tools to obtain desired features on a workpiece (e.g. a tapped hole). A feasibility study was commissioned by NEL with ICL, and PERA were brought in as advisers.

This resulted in recommendations in a report, given at a Conference in NEL on 'NC Programs with Technology' in October 1970, that a scientific approach which attempted to reduce workshop practice to a series of curves or formulae would not be acceptable to the average workshop. Instead it was proposed that the NC program should consult tables of data filled in by individual users.

This approach has been cautiously developed at NEL and has led to a new concept. The CAWR system can be used in a stand-alone form to aid the planning and estimating departments in any workshop. It provides a program to enable the chief planner to enter or modify workshop data relevant to particular materials when cut on particular tools.

A separate interactive or conversational program allows the planner to enter details of the feature to be produced, for example a hole of given diameter, accuracy, and depth. The program immediately returns the sequence of tools, tool holders, feeds, and speeds for the material specified on the chosen machine. The planner can accept or modify this information and then enter the number of holes of this type on the part. The process is repeated until the work on the part is all described. If the number of parts in the batch is then entered the time and cost for making the batch is obtained, together with the list of operations to be done.

Used in conjunction with NC the planner can also supply the PARTNO (the reference number of the part) and leave the cutting information for later use.

When the part program is later presented to NELAPT the details of cutters and cutting conditions are fetched automatically from disc and used in the part program. In this way the planner can have ensured that the machine tool can cope with the range of tools and operations required whilst only using CAWR, ensuring no hitches from this cause when using NELAPT.

On the other hand, if the part programmer wishes he can cause NELAPT to automatically consult the CAWR system without any preliminary planning run, in much the same way as other NC programs. The risk then is that the whole NELAPT and CAWR run can be wasted because too many tools are chosen for the machine tool magazine or turret, or because a suitable tool or tool holder is not available for a given operation.

CAWR is currently emerging from its development stage and is now ready for industry testing.

INTERACTIVE PROGRAMMING

Box 615 in Figs. 1 and 2. There is often a nett saving in cost in having a programming error pointed out and corrected when it is generated, rather than waiting up to 24 hours (local computer) or up to a week (posting to and from a remote computer) for the result of a batch processing run.

In the last three years conversational, or interactive, versions of NELAPT and CAWR have been developed and tested. These offer immediate response to each line of programming information entered by teletype or alpha numeric display. A dialogue then occurs between user and computer in the course of which the user need know nothing of the special computer requirements such as how to assign files. Much of his response can be 'Yes' or 'No' to questions such as 'Have you an existing part program filed?' and he can respond with a proposed name for a file when this is requested.

Such interactive techniques can save both time and money and can be used over standard telephone lines. They are also ideal for teaching or learning purposes.

The systems developed at NEL have benefited from a systems engineering approach. Interactive and batch processing can be carried out at will on the same part program, and indeed a suitable part program (which calls only for facilities available in APT) can be developed using conversational NELAPT and, still using the terminal, transferred to batch processing by APT. As explained above, CAWR and NELAPT are also completely compatible.

POST PROCESSORS

Box 107 in Fig. 1. As mentioned earlier, the post processor is a small computer program (usually supplied by the machine tool maker) which converts the standard output of the NC Program (CLDATA) to the special form required on a particular machine tool.

Until NEL took an interest, and since, the art of post processor writing had been regarded as valuable industrial information which is only hinted at in any technical article.

NEL has published reports, chapters in books, and technical articles, making the means of writing post processors more widely known. Complete coding of sample post processors have been purchased or written by NEL and are available for reasonable charges. Most of these post processors are assembled from a library of sub-routines available at NEL.

Incidentally, NEL predicted in 1965 the likely form of the ISO standard for CLDATA (contrary to some well informed US and German opinions) and have seen it progress in 1973 to the status of a standard closely conforming to NEL predictions. Because of this the output of NELAPT, APT, and IFAPT (a French program) all conform to the ISO standard, and the machine tool manufacturer is in the happy position of being able to write one post processor which will work with all three NC programs. In Germany however, an unfortunate guess at future standards in 1967 has meant that most German post processors must have different versions for use at home with EXAPT and abroad with the other APT-type systems, although the EXAPT Association have agreed to conform to the ISO standard in the future.

STANDARDS ACTIVITIES

Box 101 in Fig. 1. This is the most important of NEL's activities in the NC field and has been so regarded since 1965. NEL has attempted to alert industry to the trends, first of all with the punched tape code (ASCII and subsequently recognized as the ISO codeset) when industry generally adopted the now obsolescent EIA codeset, then axes nomenclature (particularly +Z away from workpiece), and EIA format which has become the ISO format.

Current discussions are on revising the tape programming format and addresses to accommodate the needs of DNC without interfering with existing control systems. NEL has maintained a consistent support of BSI and ISO on these topics and also on the developing NC computer programming languages and CLDATA where the NEL predictions are being vindicated eight years later. Some users who, in the meantime, have ignored these trends are to some extent embarrassed by large stocks of part programs in other languages which make it more and more difficult to make the more and more necessary change to the ISO programming language. However when the change is made we can expect some years of peaceful development of the use of a language which has a complete capability to develop with user needs.

In Europe, NC program developers and users have cooperated for some years on the development of NC languages and CLDATA on the Unified Numerical Control Language (UNCL) Committee with UK, French, German, Belgian, Dutch, and Swedish representatives, together with the European representatives of the CAMI organization of USA. This committee has met about three times a year and helped to iron out differences which would otherwise have taken time at ISO. One valuable task which UNCL has undertaken is a register of CLDATA record numbers so that post processor writers can quickly check, through their national representatives, whether a record number already exists for their purpose and if so what it is. If there is not a suitable record type a number can be allocated for the proposed function. Currently NEL maintain this register on behalf of UNCL, having recently taken this over from CAMI who had operated it for some years.

Not directly a standards activity, but closely related, is the organization of International Conferences. NEL has played a prominent part in organizing the first two PROLAMAT (Programming Languages for Machine Tools) Conferences in Rome in September 1969 and Budapest in April 1973. The third PROLAMAT Conference will be held at Stirling in June 1976 with NEL backing. These are IFIP/IFAC conferences. On a smaller scale and on a wider range of topics, NEL and Strathclyde University are holding a Computer Aided Manufacture Conference (CAM 74) in Glasgow in June 1974, at which it is hoped a number of European contributions will be made.

CONCLUSIONS

It has not been possible within the scope of this paper to deal technically with so wide ranging a topic as numerical control and its applications. The author can only apologize and suggest that if any particular aspect is to be followed, contact be made with NEL where an enquiry will be directed to the appropriate member of staff.

DISCUSSION

Q. R. T. Webster, Midcast. 1. Has the development of NELAPT over 8 years been worthwhile?

2. How much has it cost?

3. Will the number of users continue to expand?

A. I am convinced that Industry has benefited from the work on NELAPT and APT. Many users have been spared the expense of pursuing non-standard programs up blind alleys only to have later to retrace their steps. Again the machine tool industry has been able to produce post processors for NELAPT which can be sold abroad for use with APT and IFAPT, and with modification for use with EXAPT. The user of NELAPT is working with a program which is being developed into an integrated CAD/CAM design and manufacturing system.

The cost would be difficult to estimate because the staff have worked on many projects in parallel but perhaps the direct effort on NELAPT would be about 120 man years.

As the trend in NC machine tools is away from the simple 2 axis drilling machine with the manual setting of feeds, speeds and depths, to the 2C,L type machine contouring in two axes and line milling on the third and to contouring lathes and milling machines the need for computer programs increases and we see a continually widening field for the use of NELAPT.

Q. M. A. Sabin. In some circles the clubbing of profile data is regarded as a very mixed blessing. Particularly for simple work the simpler language had great advantages.

A. Everyone is agreed that programs written for a particular purpose can be more attractive for that purpose than general purpose programs. They can, however, be a liability later when a user finds that in addition to the special purpose need (2C,L type milling in the case of PROFILE DATA) he now has other needs which his special purpose program cannot cope with (2P,L type drilling, 2C type turning and 3C-5C type multiaxis milling). Since 1963 NEL has been pointing out that such special purpose programs would become obsolescent and that the earlier users recognised the advantage of using processors which accepted part programs in APT-based language the less eventual trouble they would be in. In the last few years most industrial users have got this message, despite another wave of salesmen trying to cause industry to learn the hard way again by extolling the immediate apparent advantage of the special purpose non-standard program.

Q. W. Budde. Likely form of the ISO standard for CL DATA: In 1965 there was no ISO standard for CL DATA because of the different versions of APT, of the Computer on which the system is used, post-processors, etc. Using technology in NC language makes it necessary to standardize CL DATA records, because they are generated automatically (not 'coding' of part programmer input). So EXAPT did a step forward to standardize CL DATA in UNCL ISO etc. Using the new proposal of CL DATA standard (ISO) all systems have to be changed slightly. (With special input routine we use today postprocessor of other APT like languages together with EXAPT.)

A. The point I was making in the paper was that in 1965 we were advised in the U.S. that a fairly new APT4 CLDATA would become an eventual ISO standard. After careful consideration we came to the conclusion that this proposed APT4 CLDATA would not be economically viable. In 1965, and consistently since then we advised industry that APT3 CLDATA would eventually become an ISO standard and we designed 2C,L and NELAPT to this APT3 CLDATA. We also invested effort in helping to tidy up an untidy situation in the U.S. where the APT language aimed at the post processor and the APT3 CLDATA record numbers were not properly looked after. Eight years later ISO is at the final stages of confirming a CLDATA standard which conforms completely with our predictions. I can state categorically that APT3 systems and NELAPT conform to the new ISO standard and I have been assured that IFAPT does, but I have no user experience of IFAPT.

CONVERSATIONAL NELAPT

by

D. G. WILKINSON*

SUMMARY

A computer aid to N.C. programming is described which allows N.C. control tapes to be produced on a Teletype connected through the normal telephone network to a time-sharing computer. The computer is used in a conversational manner in which control is affected by YES or NO answers to simple questions. Errors made by the programmer are detected by the computer which replies with an explanatory diagnostic and allows the programmer to correct his mistake. Simplified file handling is included together with versatile editing facilities. Examples of typical costs are given.

INTRODUCTION

The programming barrier which for a long time has impeded the effective use of the more sophisticated N.C. machines is being eroded from two directions. N.C. control systems are making manual programming easier and computer-aided programming systems are also becoming very much easier to use. Conversational NELAPT is in the vanguard of N.C. computer programs which, using the internationally accepted N.C. language, allow the user to interact with the processing of a part program, correct mistakes, look at the output and produce a control tape within a few minutes of sitting down at a Teletype and being connected to a computer by means of a telephone call.

All the features of the NELAPT programming system are available when using Conversational NELAPT: terse spellings to keep the amount of input small; calculation facilities, including mathematical functions; a wide range of geometric definitions, including a powerful curve-fitting facility; and repetitive programming using MACROs or loops. To these facilities, with which many people are now familiar, Conversational NELAPT brings flexibility; immediate detection of errors with a diagnostic indicating the type of mistake; the ability to print the results of a geometric or mathematical calculation as the part program is being developed; storage of a new part program or re-use with editing of an existing part program. All these operations can be achieved without using any of the usual computer jargon.

BACKGROUND TO CONVERSATIONAL NELAPT

The NELAPT system is one of the products of NEL's long involvement in N.C. programming techniques which started in 1965 with a study of APT the prototype of most modern N.C. programming systems. The program 2C,L for N.C. milling[1], was introduced in 1968 and after several years of active development has formed the basis of the modular NELAPT system which is suitable for all types of N.C. programming. In the batch version of NELAPT a complete part program, prepared either on punched cards or punched tape, is processed through each of the NELAPT modules in turn. These are Input, Decode, Geometry, Tool-selection and Motion. A diagram of the flow through this system is shown in Fig. 1. If mistakes are detected during the processing through any one of the modules the processing is terminated at that point and a listing of the processing so far, together with diagnostic messages is output from the computer. The part program is then corrected by the part programmer and re-submitted for processing. Depending on the computer installation this turn-round time may be hours or even days and if several error are made—however trivial—the time between first submission of a part program and a

* National Engineering Laboratory, East Kilbride, Glasgow

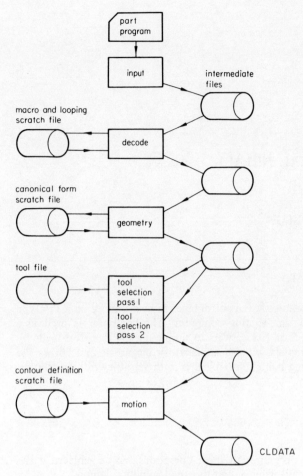

Figure 1. NELAPT batch processor.

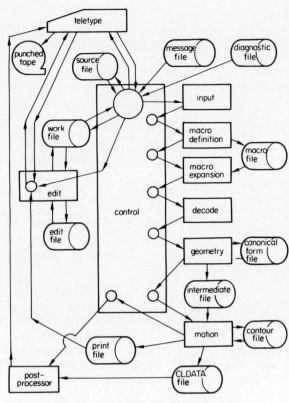

Figure 2. Conversational NELAPT system.

control tape being produced can be quite lengthy. This system is typical of all batch programming and is perhaps a major deterrent to would-be users of computer-assisted programming for N.C.

THE CONVERSATIONAL NELAPT SYSTEM

The program modules developed for the batch version of NELAPT are used in the Conversational System[2] with slight modifications. The principle difference between the two systems is in the way in which they are organized and the addition of a control program.

Figure 2 shows the communication paths in the conversational system and reveals the large organizational differences between it and the batch version of NELAPT. The only sequential intermediate files remaining are between GEOMETRY and MOTION and the CLDATA file. The CONTROL module contains buffer areas of storage which are used to transfer information between the various modules and external files. In addition to the new CONTROL module there is an EDIT module and two other modules, MACRO DEFINITION and MACRO EXPANSION, which have been extracted from the DECODE module. Of the new files, the diagnostic and message files are random access, the addressable quantum being 28 computer words stored on drum on the Univac 1108. The other new files are all sequential files written using a normal FORTRAN write statement with a 12A6 format. This ensures compatibility with the DATA file structure standard on the Univac

1108 under the EXEC 8 operating system and allows these files to be conveniently handled independently of the NELAPT processor.

System operation

The Conversational System has three main functions: file handling, part program processing, and part program editing. The CONTROL module dynamically assigns files. It reads source program statements either from the Teletype, or a previously established file and stores completed part programs on a new file, or replaces the source part program with an up-dated version.

Interactive section

Part program statements from whatever source are processed individually through the INPUT, MACRO DECODE and GEOMETRY modules. A mistake discovered by any of these modules is signalled to the CONTROL module which prints out an error message and ignores the incorrect statement. If the source statement was obtained from the Teletype a corrected statement is requested from the Teletype, but if the source was a part program file, then control is transferred to the EDIT module. The signal to terminate the interactive phase of part program analysis is the FINI part program statement.

Motion

The sequential file output from GEOMETRY is now available to be processed by MOTION and assuming that the user decides to initiate this phase, a CLDATA file and print file are generated. The print file may be scanned by the EDIT module for selective listing by the user and, if required, it may be routed to the line printer for a complete listing. Post-processing is initiated by CONTROL after the user has indicated whether an output listing is required from the post-processor. (This feature is necessary because the output from the post-processor may not be formatted to suit Teletype operation.)

After completion of post-processing, which is a separate operation, control passes from the post-processor back to the computer operating system. To produce a control tape the user must then type the control command:

@TAPE.PUNCHT   PUNTAP.

Editing module

Because of the standardized file structure adopted for both the source part program data and the print file, the editing module is utilized at all stages of processing in which the user is allowed access to files. The commands handled by this module are:

(a) LIST/SOME LIST/ALL PUNCH
(b) INSERT REPLACE DELETE
(c) RUN RESUME QUIT

The commands (a) provide facilities for inspection of a part program or having it punched on tape at the terminal in use. Commands (b) allow the stored part program to be modified, and commands (c) provide exits from the editing mode.

Error handling

No program as complex as Conversational NELAPT can be entirely error-free or able to handle every conceivable user response. Considerable effort, however, has been put into making the system foolproof and providing a tidy exit if a situation should get beyond rectification. Error messages guide the user on how to correct mistakes and if an error is made in reply to a question affecting the logical flow of the program, the user is normally given three chances to type an acceptable answer before the system apologetically terminates.

USE OF CONVERSATIONAL NELAPT

In contrast with batch processing, the user of a conversational system is in constant communication with a computer which is sharing its time with many other users. Interaction with a computer in this manner usually requires a knowledge of the computer operating system and the commands associated with it. Conversational NELAPT takes care of the computer jargon and presents the user with questions, the answers to which are generally YES or NO. Each statement is examined as it is typed (this is the essential difference between interactive and batch processing) and if an error is detected the response to the user is appropriate to the mistake which has been made. For example if the expected response from the user is YES or NO and any other reply is typed, Conversational NELAPT replies with

**PLEASE ANSWER YES OR NO TO NEXT
QUESTION**

followed by a repetition of the question asked previously.

Example of a 'Conversation'

As with all interactive computer systems the computer must first of all be instructed which computer program is to be used. With the NEL system this is done by typing the statement

@XQT   NELAPT.NELAPT

as in Table 1. From this statement onwards Conversational NELAPT controls the computer operations and translates the replies of the part programmer into commands understood by the computer's executive system.

Table 1 shows a very simple example of the use of Conversational NELAPT in which a new part program is being built up, hence the reply of NO to the question

*DO YOU WANT TO USE AN EXISTING FILE?

(In the examples, replies typed by the programmer have been underlined.) The statement FULIST/OUT is an addition to the normal 2C,L language and its use here is to instruct NELAPT not to type back each part program statement as it is processed, thus saving time and reducing the amount of printing. The word MACHIN has been mis-spelt; this is detected and the incorrect statement is ignored. Notice that the alternative spelling of the word POINT, 'P' is used in the definition of P2 and that P1 and P2 are defining points with identical co-ordinates. The attempt to define a line L1 passing through these two points is immediately rejected as a geometric impossibility. The command EDIT switches Conversational NELAPT to the editing mode and allows the point P2 to be replaced, whilst the command RESUME switches back to inter-active processing of part program statements. Following this, the line L1 is re-defined and the mathematical parameters of the line as stored by NELAPT are printed as a result of the PRINT/3,L1 statement being typed. The FINI statement terminates this part program, which is stored for future reference under the name MTDR.

Another example of the use of Conversational NELAPT is shown in Table 2 which this time illustrates how a part program stored as a result of a previous use of the system may be added to and edited to produce a new part program. After the contents of the file 'MTDR' have been listed some editing is performed and interactive processing resumed. The statements RAPID and GOTO/5,5,2 are added to the part program, which this time contains simple motion statements directing the cutter to explicit co-ordinates, and after the part program is terminated by means of the FINI statement it is stored in its updated form under the new file name 'MTDR1'. The previous file 'MTDR' will still be saved, although had it been required, the updated part program could have been stored under the same name, thus over-writing the original part program. The option of having the cutter positions calculated has been accepted this time. The opportunity to have a section or all of these co-ordinated listed on the Teletype is declined, but the complete listing of part

TABLE 1

```
@XQT NELAPT.NELAPT
  **CONVERSATIONAL NELAPT - VERSION 006 - N E L
  * ENGLISH? FRANCAIS? DEUTSCH? SHORT?
ENGLISH

  *DO YOU WANT HELP?
NO

  *DO YOU WANT TO USE AN EXISTING FILE?
NO

  *NOW*
PARTNO MATADOR
             1     PARTNO MATADOR
  *NOW*
FULIST/OUT

  *NOW*
MACHINE/NELCNC,1
        WORD MORE THAN 6 CHARS.
  **ERROR IN STATEMENT          3
  *NOW*
MACHIN/NELCNC,1

  *NOW*
P 1=POINT/0,0

  *NOW*
P 2=P/0,0

  *NOW*
L 1=LINE/P1,P2
        SOLUTION IMPOSSIBLE
  **ERROR IN STATEMENT          6
  *NOW*
EDIT

  **EDITING**
  *CHOOSE*
REPLACE

  *FROM WHICH STATEMENT?
5

  *HOW MANY STATEMENTS?
1

  *NOW*
P 2=P/1,1

  *CHOOSE*
RESUME

  *FROM WHICH STATEMENT?
5

  *NOW*
L 1=L/P1,P2

  *NOW*
PRINT/3,L1

  STATEMENT NO    7     PRINT OF CANONICAL FORMS
  NAME  TYPE     CANONICAL FORM
  L1 LINE  .707107 -.707107  .000000  .000000
  *NOW*
FINI

  *DO YOU WANT TO SAVE YOUR PART PROGRAM?
YES

  *PLEASE GIVE FILE NAME*
MTDR

  *CUTTER MOTION CALCULATIONS?
NO

**CONVERSATIONAL NELAPT NOW TERMINATED - CHEERIO**
```

TABLE 2

```
@XQT NELAPT.NELAPT
  **CONVERSATIONAL NELAPT - VERSION 006 - N E L
  * ENGLISH? FRANCAIS? DEUTSCH? SHORT?
ENGLISH

  *DO YOU WANT HELP?
NO

  *DO YOU WANT TO USE AN EXISTING FILE?
YES

  *PLEASE GIVE FILE NAME*
MIDR

  **EDITING**
  *CHOOSE*
LIST/ALL

     1 PARTNO MATADOR
     2 FULIST/OUT
     3 MACHIN/NELCNC,1
     4 P1=POINT/0,0
     5 P2=P/1,1
     6 FEDRAT/8.5
     7 GOTO/P1
     8 GOTO/P2
  *CHOOSE*
INSERT

  *AFTER WHICH STATEMENT?
5

  *HOW MANY STATEMENTS?
1

  *NOW*
SPINDL/1500

  *CHOOSE*
RESUME

  *FROM WHICH STATEMENT?
8

  *NOW*
RAPID

  *NOW*

GOTO/5,5,2

  *NOW*
FINI

  *DO YOU WANT TO SAVE YOUR PART PROGRAM?
YES

  *PLEASE GIVE FILE NAME*
MIDR1

  *CUTTER MOTION CALCULATIONS?
YES

  **INITIATING CUTTER MOTION CALCULATIONS**
  **NO. OF CUTTER LOCATION RECORDS =     1R
  **NO. OF ERRORS DETECTED =      0
  *DO YOU WANT TO EXAMINE CLPRNT?
NO

  *DO YOU WANT CLPRNT VIA LINE PRINTER?
YES

  *DO YOU WANT TO USE A STANDARD POST PROCES:
YES

  *DO YOU WANT TO EXAMINE PPRINT?
YES

**CONVERSATIONAL NELAPT NOW TERMINATED - CH
```

TABLE 3

POSTPROCESSOR LISTING

MATADOR

N001G80   S715   M03

N002G01X00000 Y00000   F485

N003   X01000 Y01000   F485

N004   X05000 Y05000   Z02000   F999

PAPER TAPE LENGTH 6 FEET

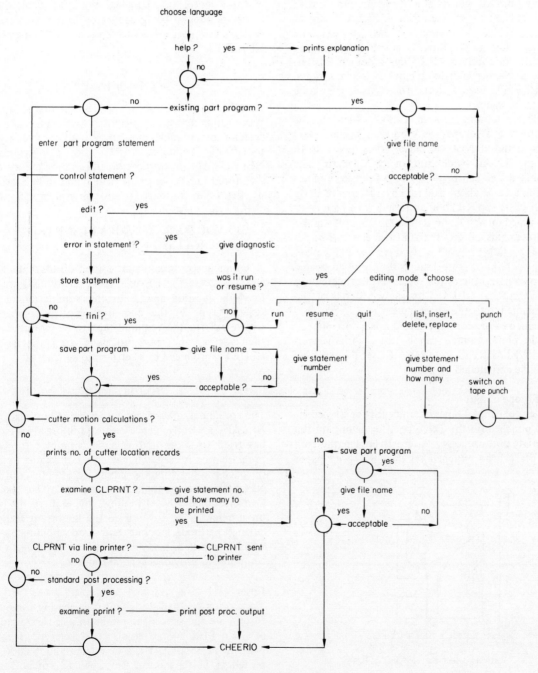

Figure 3. User's flowchart.

program input and tool co-ordinates has been asked for on the line printer, which is situated adjacent to the computer itself. If this part program had been submitted from a Teletype connected to the computer through the GPO telephone network, then this listing would have had to be sent by post. The use of a post-processor, which is called for by the MACHIN statement, is required and in this case the output will be tailored to suit the NEL CNC drilling and milling machine. The option of having the post-processor printout typed on the Teletype is also accepted and control is then passed from Conversational NELAPT to the post-processor, whose output is shown in Table 3.

A detailed flowchart of the system from the user's point of view is shown in Fig. 3.

Graphical output

Two methods of graphically displaying the cutter path calculated by Conversational NELAPT are at present available. Post-processors which tailor the output to suit graph-plotters can be called up by using the appropriate MACHIN statement. Alternatively a selective plotting facility in Conversational NELAPT can be used with the Tektronix storage tube computer terminal.

a plotter, costing around £4000 which can be attached to a Teletype and used over normal GPO lines, is in use at NEL. It can produce a plot up to 300 mm (12 in) wide and of any length. The Tektronix device can be used like a Teletype and has a viewing area of about 200 mm by 150 mm (8 in by 6 in). It costs about £2000 and can have a hard copy unit attached for about another £2000. Figures 4 and 5 are photographs of the Tektronix screen showing a plot of a part program output for a N.C. plate burning operation. Figure 4 shows the complete nested plate and Fig. 5 utilizes the selective plotting feature of Conversational NELAPT to show a detail of the plate enlarged to fit the screen area.

Attached to the computer at NEL is a 740 mm (29 in) wide plotter which can be used with Conversational NELAPT and for a remote user the resulting plot can be despatched by post.

Punched tape output

Machine-tool control tapes in either ISO or EIA codes can be punched on the Teletype after calling up the appropriate post-processor. Alternatively lengthy control tapes can be punched on the high-speed paper

standard test part program std 04

Figure 4. N.C. plate burning example.

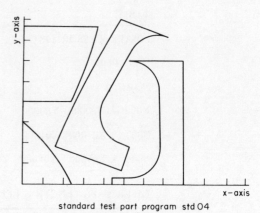

standard test part program std 04

Figure 5. N.C. plate burning detail.

tape punch attached to the computer and the tape despatched by post to remote users. This facility of having a control tape punched on the Teletype within a few minutes of being connected to the computer is perhaps the most attractive feature of Conversational NELAPT.

POST PROCESSING

At NEL a wide range of post-processors for many types and makes of machine tools and control systems are available and can be used by a variety of processors including APT and Conversational NELAPT. The post-processors in common use at NEL are stored on the magnetic drum and can be accessed by the Conversational System when a user answers YES to the question.

'DO YOU WANT TO USE A STANDARD POST-PROCESSOR'.

Other post-processors are stored on magnetic tape and to access these a user must answer NO to the previous question and, following termination of the system, type in the computer control commands which call for the required tape to be mounted and then cause the post-processor to process the CLDATA produced by Conversational NELAPT.

CONCLUSIONS

The principal advantages of using Conversational NELAPT are convenience and ease of use. This paper has tried to show that using a computer for N.C. programming is easy, what it does not do due to lack of space is to describe the simplicity and power of the NELAPT part programming language. The use of Conversational NELAPT allows the skill of the part programmer to be focused on the important areas of cutting methods, tooling and work holding, areas which, in spite of many attempts, are still defying the computer to provide a general solution.

Costs

The capital costs of using any time-sharing computer system are the same. The minimum is a GPO modem at about £100 a year rental, and a Teletype, second-hand from about £450, a new from about £600, or rented at £30 a month. Any new user of N.C. will have to provide himself with tape punching facilities and would be well advised to ensure that this equipment can be used with a GPO modem at the outset.

Operating costs are made up of telephone charges and computer charges. Telephone call charges are the normal ones published by the GPO in the telephone directory and are dependent on time of day, distance and whether the call is STD or manually connected. At present the range of costs is from 50p to £5.10 an hour, but for distances over 50 miles connected by STD in the afternoon, the charges would be £3.60 per hour.

Computer charges depend on a number of factors; the time the central processor of the computer is being used (this is only a small fraction of the time connected); the size of the computer program being used; the number of times files are accessed; the amount of file space taken up; whether magnetic tapes have to be mounted and also the connected time. To give an example a lathe component was cut recently on NEL's Dean, Smith & Grace N.C. lathe for a customer. It was programmed using Conversational NELAPT at an overall commercial cost of £5.44. This was for 106 blocks of control tape taking 13 minutes maching time which included roughing and a final contouring cut. The program was prepared on punched paper tape using the Teletype off-line and then read into the computer, processed by Conversational NELAPT and a control tape produced on the Teletype. The connected time was $18\frac{1}{2}$ minutes and the computer time used was 54 seconds. Printout was obtained from the line printer and the control tape was listed on the Teletype as it was being punched. A photograph of the tool path taken on the Tektronix terminal is shown in Fig. 6.

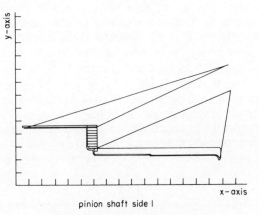

Fig. 6. N.C. lathe cutter path.

One important factor which must be considered before Conversational NELAPT can be used, is the provision of a post-processor for the machine tool to be programmed. NEL have a large number of post-processors, but if a firm is buying an N.C. machine tool and is contemplating using computer-aided programming, it should find out the availability of a post-processor before the order is placed. NEL will be happy to give advice on this point whatever programming system is being considered.

Further developments

NEL have a team of highly skilled programmers and engineers who are employed full time on developing computer aids to manufacturing. Conversational NELAPT is just one facet of this work and is complemented by many other projects, some of which, like links to computer-aided design and the selection of tools and machining methods by computer, are already being incorporated into the NELAPT system. This activity will ensure that Conversational NELAPT remains in the forefront of technological progress and that users will never find that suddenly the computer programming system they are using has become obsolete.

ACKNOWLEDGEMENT

This paper is published by permission of the Director, National Engineering Laboratory, Department of Trade and Industry. It is British Crown copyright.

REFERENCES

1. National Engineering Laboratory. 2C,L Part-programming reference manual. Second Revision. *NEL Report* No. 543. East Kilbride, Glasgow: National Engineering Laboratory, 1973.
2. ELDER, I. C. S. Conversational NELAPT, a user's guide. NEL N.C. *Program Notes* No. 4, East Kilbride, Glasgow: National Engineering Laboratory, 1973.

DISCUSSION

Q. I. R. Wasiukiewicz, Dixi II Switzerland. Firstly, I would like to congratulate the research team at NEL on the idea of allowing the programmer to do his job in one of three languages. Being personally involved in the introduction of APT in my company in 1969, I was faced with a great reluctance on the part of French and German production staff to use English-like terms.

It is a common practice nowadays in Switzerland (as is shown by a survey of employment offers in technical periodicals), for companies seeking programmers to employ experienced machine tool operators and train them. At the same time it is very unlikely that these people have a good knowledge of English. As we must increasingly bring the job to the man and not the man to the job, we should simplify the task of the programmer by allowing him to use his mother tongue:

1. Because of the better efficiency. The computer assisted preparation of the tapes is only justified when time savings are obtained. The search for the right word in English or reflection time can considerably influence the overall efficiency of the tape preparation process.
2. Because the files must be clear to the machine operator.
3. Finally, because the knowledge of some specific terms as applied to programming, cannot contribute to a real knowledge of the English Language.

Therefore it would be very desirable, if one of the ISO Committees were to establish a sort of International vocabulary for the programming of machine tools. This vocabulary must have national supplements in different languages. Because of the modular structure of most programming systems, there would be no major technical problem involved in inserting a translation-module in them.

A. The choice of languages available to the user of Conversational NELAPT are at present English, German, French or a shorthand version of English; 'Short'. These options ensure that control of the processor and all diagnostics are in the chosen language. The language of the part program, however, is still that of APT and no provision has been made for alternatives to the vocabulary words other than the 'terse' spellings, e.g. P for POINT, C for CIRCLE. There is no logical reason why the principle of language selection should not be extended to the part program vocabulary, but to do this for NELAPT would mean alterations to standard modules and is not contemplated at present.

Q. M. A. Sabin. When we first had access to a multiaccess terminal, some 5 years ago, one of the first things I implemented was a conversational version of the geometry calculation part of Profile-data. After one day's playing our part programmers, who are all 'accustomed users', started using the new facility by proxy, sending up their punch girl to talk to the machine. Since then the whole of my experience has been that it is the two-minute response to a complete part program which saves money, not the two-second response to each line. What is your experience?

A. We have tried to provide the best of both worlds by giving the user the choice of entering his part-program line by line, or by first establishing a part-program file from cards or punched paper tape and then processing the complete part-program by means of the editing command RUN. We believe that the first method is ideal for the learner or student, whilst the accustomed user would invariably use the second method.

Q. P. Aughton, B.A.C. Have NEL tackled the problem of post processor output (i.e. control tapes), in particular the problem of checking these tapes for noise generated on the telephone line or the possibility of more sophisticated terminal hardware.

A. The NELAPT system allows the user to have his control tape punched out on his own terminal if he wishes, or alternatively to have it punched on the high-speed punch attached to the computer at NEL. We are at present writing a control tape checking program which will accept as input the control tape and tooling details. The control tape is checked for errors and the tool path is plotted. This program is machine tool dependent of course, but capable of being easily modified to suit other machine tools.

Q. A. R. Kimber, M.O.D. The language appears rather 'verbose'. Does the option 'short' at the beginning of conversation enable a shorthand version to be used.

A. The SHORT language option does give shortened questions and allows replies to be given as Y and N instead of YES and NO. The terse spellings such as P for POINT and C for CIRCLE are available at all times. The latest version of the processor gives the accustomed user the option of bypassing the 'language' and 'help' questions to further reduce the verbosity.

Q. W. A. Carter, Plessey N.C. Ltd. The editor commands REPLACE, RESUME, etc., do not carry operational parameters. The parameter must be entered in a separate command to indicate statement number. For experienced users, it would be more efficient to have the operational parameters in the same statement as the command. Does conversational NELAPT intend to provide this facility?

A. This is a valid comment and we intend to alter the processor so that it is possible to enter the parameters immediately following the editing commands. When the parameters are not given following the commands the system will prompt the user as at present.

THE PROGRAMMING AND USE OF NUMERICAL CONTROL TO MACHINE SCULPTURED SURFACES

by

C. BELL, B. LANDI and M. SABIN*

SUMMARY

Shapes traditionally thought of as being difficult to program for numerical control can be and are being cut economically in hard materials to close tolerances. This has been made possible by the joint development of the machining technology and the computer programs through close collaboration between workshops, programmers and mathematicians.

INTRODUCTION

The manufacture of wind tunnel models is not an everyday task, but the demand for more detailed and more accurate models in stronger materials to a shorter time-scale has given it many aspects which are shared by other, more familiar, engineering activities. The making of moulds and dies, in particular, necessitates the handling of awkward shapes and the working of tough materials; close tolerances have to be maintained and a good surface finish has to be achieved and the production run is short, single items being the rule rather than the exception.

Many other tasks too, are characterized by some of these problems and a description of how numerical control has been applied successfully to overcome these difficulties may therefore be relevant.

The authors propose to examine first each of the problem aspects and then the various contributions to the total solution.

THE PROBLEM ASPECTS

Shape complexity

Shape-related difficulties arise from two sources. One is that with aircraft-like shapes it is far from obvious how the shape can best be specified and represented numerically. Several methods have been tried, but all have well-defined disadvantages. The method used by the British Aircraft Corporation (B.A.C.) in their Numerical Master Geometry (N.M.G.) system has proved to be very practical from the machining point of view. It has a fully defined surface which can be offset precisely to allow for cutters of arbitrarily complex shape and it is suitable for the calculation of cutter paths defined in almost any way conceivable. The value of this will appear below, in the examina-tion of the types of cutter path required to overcome the other problems. For the moment, it is sufficient to say that the N.M.G. system can handle surfaces made up from many bicubic 'patches' slightly better than APT can handle cylinders.

The second shape-related difficulty arises where a large number of independent shape elements have to be considered together. This complication is far more difficult to handle automatically than that of having faired surfaces. All current commercial programming languages leave it to the part programmer to specify which surfaces need to be taken into account at each part of the cutter path.

Fortunately, these two complications do not often arise simultaneously. Although pathological situations can be contrived, it is seldom necessary to consider even the 'multiple check surface' case when sculpture milling.

The typical finish cut operations are:

(1) clearance of a surface by scanning with closely spaced parallel strokes, possibly within some boundary (Fig. 1).

(2) clearing cusps left along the boundary by cutting along the intersection of two surfaces (Fig. 2).

(3) cutting detail superimposed on a surface. Some wind tunnel models have grooves in the surface in which pipes for sensing pressures are laid (Fig. 3).

Of these three operations, the first usually domi-nates in the total cost and lead time of a model, although on some models the third will be more significant.

The roughing operations are:

(1) pre-roughing, in which some 60–70 per cent

---

* British Aircraft Corporation Limited, Commercial Aircraft Division, Weybridge, Surrey

of the swarf is removed, and which involves only a crude representation of the shape finally to be cut.

and

(2) roughing, which brings the surface down to within 2–5 mm of the finished shape, and which needs to take the final shape into account quite closely.

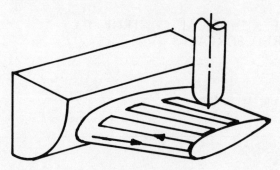

Figure 1. Area clearance.

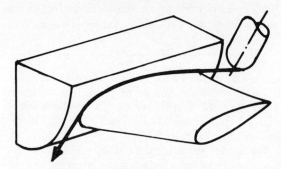

Figure 2. The intersection of two surfaces.

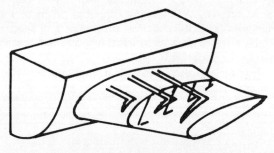

Figure 3. Detail on a surface.

Tough materials

The metals currently used for wind tunnel models are not usually exotic. Steels in the 50–100 tons U.T.S. range can be and are being machined in many workshops throughout the country. The high strength of the material does, however, compound the other problem aspects and prevents the simplest solutions to the shape problem from being used economically. Hardened steel ball-end cutters, driven only along the definition curves of a surface, would wear too fast and cut too slowly to be really useful.

B.A.C.'s answer has been to develop a cutter with cylindrical carbide cutting inserts, which gives very good performance in terms of metal removal rate and cutter life.

The effective cutter shape is fairly complex because the disks are mounted with an appreciable rake angle (negative). Furthermore, like all cutters

without teeth underneath, the maximum sink angle is severely limited when cutting into block material. When roughing, it is essential either to cut uphill all the time, which can result in wasted machine time or else to drive the cutter along horizontal paths only. Both these factors complicate the calculation of the tool centre paths.

Also, when having hard material to work with, it is important to leave less on for hand finishing. The optimum cusp height between the strokes is lower on steel than on aluminium alloy or even wood, which were used for all models until a few years ago. The factors defining the optimum cusp height and stroke density are examined in Appendix 1.

Tolerances and finish

In this context, the economically achievable tolerances are tight by normal engineering standards, though not uniquely so. A demonstrably consistent accuracy of approximately ± 0·05 mm is achieved only by continual care and by paying attention to all potential sources of error. The cutter design mentioned above is substantially better than a ball-end cutter, because the disks can be ground accurately circular off the cutter. They are mounted on precise locations not subject to wear.

Workpiece deflection has to be expected and catered for. Two problems arise here. The first is distortion of the material due to stress relief when a large fraction of the original billet has been cut away. This is dealt with by careful measurement after the roughing stage, followed by re-machining of the datum faces and setting holes so that the final shape is well within the roughed out billet. If the distortion has been excessive, the job is scrapped and the roughing started again with a new billet. As a further safeguard, the finishing cuts are taken in two stages, with inspection between them. The tapes for the first finishing cuts are calculated with an extra 1 mm of offset all over, because merely running 1 mm high does not give adequate clearance round a leading edge. They may also have a larger spacing between strokes to reduce the machining time at this stage.

The second problem is deflection of the job under cutting loads. This is most extreme on slats which may be over 1 m long, with a chord of perhaps 20 mm and a thickness of only 3–4 mm, but it is always considered.

The final surface finish obtained is not determined by the machining process, but by the hand finishing which has to follow any scanning. The only special contribution made here is in the avoidance of dwell marks. The computer program notes where momentary low feed rates will occur and arranges to withdraw the cutter slightly there. This may leave a 'pip' on the surface, which is relatively easily polished off. Blending out a dwell mark is much harder work.

Small runs

'One-offs' are not generally considered suitable for numerical control. Programming costs are quoted as making numerical control uneconomic for very short runs and this may also apply to parts which are either simple to make or which are so detailed as to need an inordinate amount of part programming. Surface sculpturing, however, is not easy by any other method and part programming is considerably less detailed than typical '2½D' work. The part programming costs are not really a dominant factor.

The principal problem which does arise from the individuality of every model is that it is not possible to start machining from a forging. The only way of producing models quickly enough to be useful is to work from the solid billet. This involves machining away quite large volumes of material, especially as the billets available may not be very close to size. Efficient cutting patterns for pre-roughing and roughing are therefore absolutely essential for economic use of the numerical control machine. Running the finishing tapes high is just not adequate, because the number of passes required to get down to depth would double or triple the cost. For aircraft models, good roughing patterns can get down to within 2–5 mm in about a quarter of the machining time taken for the finish cut. Furthermore, the roughing pattern will allow clearances in plan form as well as in depth.

The pre-roughing, because of its very crude surface representation, can actually overlap the setting up of the surfaces in elapsed timescale, or even the detailed design of the model.

SOLUTION TECHNIQUES

The successful solution of these problems has involved the development of a combination of workshop technology, part programming techniques and numerical representation of shape.

Workshop technology

The principal development here relates to special carbide insert cutters. The ball-end cutter traditionally used for surface scanning has two great disadvantages: it is difficult to achieve really accurate grinding and it cuts very inefficiently on surfaces of low inclination because the cutting point is so close to the axis of rotation (Fig. 4). The sharp-cornered

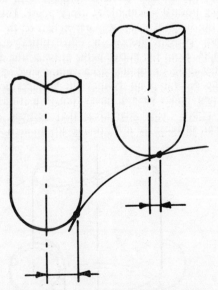

Figure 4. Variation of effective cutting radius.

end-mill or side and face cutter is much better from both points of view, but it has the disadvantage that,

when travelling across the slope of the surface, it leaves extremely large cusps between the cutting strokes (Fig. 5).

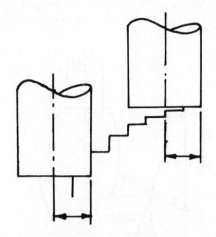

Figure 5. Large cusps between strokes.

The disk insert cutter combines the good features of both types of cutter (Fig. 6). Its disadvantages are that it cannot sink at all steeply into block material without fouling underneath and that, because of the negative rake on both side and base of each cutting edge, the effective profile has a rather difficult equation. These disadvantages are readily overcome, however, by the appropriate combination of part programming technique and computer algorithms.

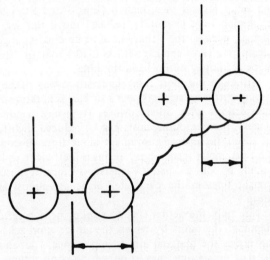

Figure 6. Inserted disk cutter.

The carbide inserts give good cutting performance and high metal removal rates. Development of cooling, cutter mounting and accurate setting has complemented that of the cutter itself. The difficulty arises in setting, because the cutter axis is often inclined by a few degrees to give predictable cutter movements on almost flat regions of the surface.

Part programming technique

This is the art of knowing what cutter movements can be generated and which are appropriate to various situations.

For example, when hogging out of a solid billet using a cutter with no teeth underneath, it is necessary to drive the cutter horizontally. If, however, a narrow flange is being finish machined which has appreciable twist, it is better to drive the cutter so that the cutting point travels horizontally (Fig. 7). If the cutter moves horizontally, the cuts quickly leave the surface.

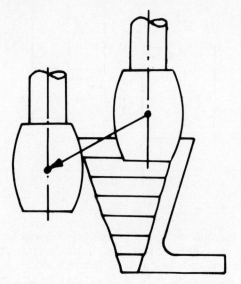

Figure 7. Skew cutter movement to keep the cuts on the surface.

Again, when finish machining a surface of light slope, the hand finishing time can be reduced substantially by driving the cutter directly up and down the slope instead of around it (Figs 8 and 9). The machining time cannot be reduced much this way, though, because the strokes have to be close enough together to leave enough witness marks to define the shape accurately during hand finishing.

Although the most easily understood way of describing cutter movements over a surface is in terms of intersections with other surfaces (typically parallel planes), the computing costs can be reduced slightly by specifying instead movement along lines of constant parameter (generators). If this is inefficient, as it would be on a sharply tapered surface, general straight lines in the parameter plane are almost as cheap.

Not only the actual cutting strokes need careful planning: the paths by which the cutter approaches and leaves the material are of importance, too, and the part programmer has to be able to control them. For example, a cutter which cannot sink when

cutting has to approach each cut by sinking clear of the billet and then moving in horizontally. Unless each cut is to be programmed individually, the approach movements have to be specified, both in length and orientation, in generic terms. At the end of each cut, a simple vertical lift is usually adequate, but situations can easily occur (for example when undercutting) when more complex paths are essential. Again, the quickest safe path from the end of one stroke to the beginning of the next depends largely on the shapes being machined.

A substantial part of this technique has been incorporated into the computer programs, because the program suite has been developed in-house, jointly by the part programmers concerned and the mathematicians.

Numerical representation of shape

The N.M.G. system was originally developed by B.A.C. as a system for the design and lofting of fair surfaces. It is therefore fortunate that the representations adopted have proved to be so well suited to the machining context.

The N.M.G. system has two principal types of data items, surfaces and curves. Surfaces are represented by rectangular arrays of bicubic 'patches' or 'tiles', which have continuity of slope and curvature over all boundaries and are controlled in shape by the co-ordinates of the points at the corners. Design of such a surface is analogous to pinning down a stiff, but stretchy membrane at points in space by three-dimensional drawing pins. Slope constraints can be imposed at some or all of these points if required, but the continuity of curvature is then lost.

This type of surface description is not particularly biased towards aircraft shapes. It has been used 'in anger' for ships; ships' propellers; cars and aircraft. It has also been applied on a demonstration basis to shoe lasts, jugs and easy chairs.

The first important property of this representation is that the surface is fully defined, with a unique surface normal calculable at every point. This means that the shape is in no way dependent on the cutting pattern chosen; hence, no difficulties arise, for example, from the cutter paths crossing the data at a shallow angle, or running off a leading edge.

The second point is that the surfaces are shape invariant under general linear transformations of the data points. This guarantees that vertical, or near vertical, surfaces do not provide any numerical prob-

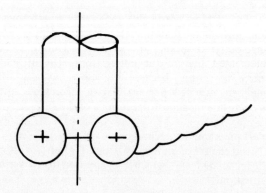

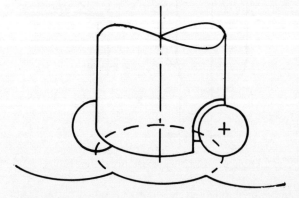

Figures 8 and 9. Variation of effective stroke radius with cut direction.

lems, although the part programmer obviously still has to concern himself with cutter access to all parts of the surface.

The third point is that precise offsets can be calculated for arbitrarily complex cutter shapes. Variation of cutter orientation has only been implemented for stepwise variation, because continuous variation provides very little advantage on surfaces which have to be scanned anyway. There is no unsurmountable technical obstacle to driving four or five axes continuously and simultaneously.

Curves are represented by lists of points, the points being close enough together for linear approximation to be within a defined tolerance. Other information, such as the tangent vector to the curve and the normals of the surfaces by whose intersection the curve was generated, can optionally be stored as well as the coordinates of the points.

Having the curves as well as the surfaces stored in a well defined intermediate interface allows any curve, however defined and calculated, to be used in any of the several output procedures. It also allows—and this is important for the flexibility of the system—a wide range of editing processes to be applied to a curve before it is used.

It may be truncated at a plane, or at some other surface, or at the point where a surface goes beyond the vertical. The cutter offset can be applied at this point, or, alternatively, during the actual calculation of the curve. It is even possible to flatten a curve out, or to apply it to some other surface.

Consequently, there is great freedom in specifying where the cutter will go, which is put to good effect in cutting models accurately and rapidly.

CONCLUSIONS

The techniques described provide a cost-effective way of meeting close tolerances and tight time-scales in the manufacture of wind tunnel models. They might also be applicable to other areas in which faired surfaces have to be machined.

ACKNOWLEDGMENTS

We wish to thank the Management of B.A.C., Ltd. for permission to publish this paper, and all those colleagues who have contributed to the work described.

APPENDIX 1

Optimum stroke density

The cost of making a model can be divided into three parts: (i) the constant part, independent of stroke density, which includes the billet cost, the part programming and all the rough machining and inspection; (ii) the part which varies linearly with the number of strokes, which includes the computing and finish maching costs; and (iii) the part, due to hand finishing, which varies with the amount of metal left on to be removed by hand.

Now the amount of metal left on varies inversely with the square of the number of cuts (N) and the cost equation may be written thus:

$$C = C_1 + C_2 N + \frac{C_3}{N^2}$$

If we choose N to minimize C, we have

$$\frac{dC}{dN} = 0 = C_2 - \frac{2C_3}{N^3}$$

and thus

$$C_2 N = \frac{2C_3}{N^2}$$

Therefore, for a minimum total cost, computing and finish machining should together cost roughly twice as much as hand finishing.

The equation for total time taken is of exactly the same form, with different coefficients. For minimum lead time, therefore, the finish machining time should be roughly twice that of the hand finishing.

These theoretical optima have to be constrained in some circumstances by having enough cuts to provide an adequate set of witness marks to retain accuracy and in any case the costs are not exactly linear or quadratic. They may often vary with the surface form, concave surfaces being more difficult to hand finish than convex ones. The choice of cutting pattern and stroke density may even be made after the +1 mm cuts have been made. This analysis is, however, a reasonable first approximation and can provide a basis for a more detailed optimization.

APPENDIX 2

N.M.G. facilities

Surface definition

Surfaces can be:
(1) defined by data points.
(2) modified in shape.
(3) extended.
(4) combined.
(5) shifted to new coordinate axes.
(6) developed into the flat.
(7) distorted to allow for flight or tunnel deflection.

Parts of surfaces can be abstracted for separate manipulation. Surfaces can be offset to allow for cutter dimensions or skin thicknesses.

Curve calculation

Curves can be:
(1) defined by data points.
(2) calculated as the intersection of two surfaces.
(3) calculated as the intersection of a surface with a plane.
(4) calculated as the intersection of a surface with a circular cylinder.
(5) calculated as the silhouette line of a surface as seen from any point.
(6) calculated as any straight line in the parameter plane of a surface.
(7) calculated as any finite straight line in space.

Curve editing

Curves can be:
(1) truncated by planes.

(2) truncated by surfaces.

(3) truncated by silhouette lines.

(4) offset to allow for cutter dimensions or skin thicknesses.

(5) re-evaluated on a developed surface.

(6) interpolated at intermediate points.

(7) shifted to new coordinate axes.

(8) reflected in a surface.

(9) dropped perpendicularly on to a surface.

Output

Curves can be:

(1) listed in full computer sheet form, with many angle facilities.

(2) drawn out in any view, in true view or perspective.

(3) put out as geometric definitions or as cutting sequence in the Profiledata numerical control language.

(4) used directly as tool centre paths, all pickfeeds and feedrate changes being set up automatically. This is the particular facility used for surface machining.

REFERENCES

1. Meeting on curved surfaces, *Proc. Roy. Soc. Lond., A.*321, 1971.

2. *Proceedings of Curved Surfaces in Engineering Conference, April, 1972,* IPC Science & Technology Press, London.

3. M. GARDNER. The machining of wind tunnel models, *Machinery,* 16 March, 1972.

DISCUSSION

Q. A. R. Kimber, MOD. The B.A.C. system is in-house-unique-non-standard. Do you not consider that APT or any other STANDARD GENERALISED SYSTEM could be used for this work to the greater future benefit of users in the future.

In the opinion of the questioner the continual development of in-house unique systems is detrimental—in the long term, to future programming development.

A. The BAC system is in-house only in the sense that it was written by one company rather than by a consortium. It is being used in various versions by all the major centres in BAC, by at least two other companies who run it on their own computers (it is available commercially), and by at least two more who use us as subcontractors. Not very long ago 2CL had only seven users.

Could an existing system have done the job? At the time certainly not; even now it is doubtful whether any 'standard' system has anything like the same competence in terms of ease of part programming and effectiveness of use of machine tool time. Logically Mr. Kimber should oppose the development of further systems in this field.

In one particular respect we have been very concerned with standardization. From the start we used our range of standard Profiledata postprocessors,

and now that APT postprocessors are slowly becoming available we expect to convert to use the standard CLFILE format.

Q. W. Carter, Plessey N.C. Ltd. Some of the APT IV facilities use similar ideas to the NMG. There could have been some leapfrogging.

A. One would expect APT IV to have picked up some of the things which have been learnt about sculptured surfaces in the last seven years, but I would like to raise one warning here. My impression from reading the APT IV sculptured surfaces specification is that either it uses theory no hint of which has ever been published, or else the author did not understand the nature of the mathematical problems. I suspect the latter and therefore expect the package to be unreliable and unpredictable.

Q. R. T. Webster, Midcast. Does the author see a future for the programme in the general die sinking industry.

A. There is a *prima facie* case that using N.C. with this type of program in diesinking has a future, and what I'm saying here today is that somebody in that industry should be looking at it. There's a lot of money to be saved by the first to make it work.

Q. W. Carter, Plessey N.C. Ltd. 1. The intersection of the surface of, for example, the wing and the fuselage may be some form of toroidal or similar surface and not just a simple surface which can be generated by a single cutter pass. Can your system have the facility to deal with surfaces of intersection?

2. All the examples shown in your paper use 3 axis machining methods only. What is B.A.C.'s opinion of the use of 5 axis machines for this work.

A. The first question has two answers, both of which are 'yes'. I have never come across a situation in which the problem was anything but that of getting a small enough cutter to clear out the intersection adequately. Often one has to drive along the intersection with a sequence of cutters of smaller and smaller tip diameter to approximate adequately the required sharp concave corner. However, if a flaired out fillet is required it can be defined as a surface in its own right and cleared in the same way as any other surface. Obviously, if you don't want what is generated by a single pass of a cutter you will have to define what you do want.

The second question I cannot answer. My own opinion, a slightly different thing, is that five-axis machines have little advantage in this situation, where you have to scan anyway, and they have disadvantages of cost and lower rigidity.

We also considered five-axis machining for bevelled frames, where it would have reduced the machining time considerably, but as in the case we were examining the total time and costs would have been increased we elected to use scanning instead. In both cases, though, we were looking at specific situations, and it is quite possible that in the next case we look at, five-axis machining will be the best solution.

APT PART PROGRAMMING VIA REMOTE TYPEWRITER TERMINALS

Part Programmers from several areas of the country are using the computer facilities at B.A.C. Filton for APT part programming. This paper describes how users are able to update and schedule their part programs remotely. A facility for returning full diagnostics to the user is described, also a proposal for accurate transmission of machine tool control tapes.

The part programming work for large B.A.C. contracts such as Concorde cannot all be done at a single location. In the case of Concorde the work is distributed firstly to other locations of B.A.C. and secondly to numerous subcontractors throughout the country. In all the latter cases, and some of the former, the processing is done by the computer facilities at B.A.C.'s Filton works.

In this situation, where a large number of part programmers are working at a location remote from the computer, it makes good sense to try and build a system which gives these outside locations remote access to the Filton computer. Such a system would have to satisfy a number of requirements:

(1) It must be based on APT processing. APT and ADAPT have been used extensively for Concorde components, and APT has now been accepted as the standard NC processing language for B.A.C. To introduce a radically different processor would therefore be highly undesirable.

(2) The system must be genuinely remote, i.e. driven through the GPO standard telephone network and not through any specialized communications media.

(3) It must be possible to multiprogram. Any number of users should be able to use the system at any one time, sharing the same area of core. As extra users are accommodated it should not be necessary to supply extra hardware at the mainframe computer end.

(4) Modularity and Flexibility. It should be possible to add or replace new versions of processors and post processors, to take full advantage of any new software (or hardware) which may become available.

(5) Low cost. Cost justification is the final limiting factor for any system, many of B.A.C.'s subcontractors have limited resources and any expensive hardware is out of the question.

These five requirements have been met by the NC terminal system in use. It has proved that processing APT through typewriter terminals is a practical proposition. It has led the NC computer applications group to look at more difficult problems such as the transmission of post processor output. This problem involves not just the transmission of paper tapes but the checking of them to ensure that they do not contain errors produced by noise on the telephone line.

HARDWARE

The sections on hardware and software refer to computer equipment and computer systems, they may be omitted by readers interested only in the part programming content of the paper. Subsequent reference to foreground and background processing will be made however, and the subheading 'Foreground and Background Processing' under the heading 'SOFTWARE' should be read and must be clearly understood to appreciate the functioning of the system properly.

Mainframe
The mainframe computer at Filton is an IBM System 370 Model 155. It has one and a half megabytes of core, sixteen 3330 disc drives, seven tape decks (six 9

* Engineering Computing Services, British Aircraft Corporation

track, one 7 track) plus card reader, card punch, two line printers, and two IBM 1018 paper tape punches.

Offline are two Calcomp plotters, one Versatec electrostatic plotter and a Copath digital to analogue NC tape converter.

Terminals
Up to twenty-three remote terminals can be in use at any one time. At the time of writing these consist of fifteen IBM 2741's, five Itels and three teletypes. Five of these terminals are used for NC applications.

SOFTWARE

NC processing
The standard NC processing language is a B.A.C. version of APT known as APT140. It was developed by the B.A.C. Military Aircraft division with the prime object of reducing the core requirement to 140K, a number of features extra to the standard IBM APT processor have been added. All new part programs are written in APT140 but several other processors are supported by the system to enable modifications to be made to old part programs. These processors are IBM APT, ADAPT and PROFILE DATA.

NC post processing
Post processing to produce tool centre path plots is done via a Calcomp post processor.

A second post processor produces control tapes for paper tape control systems. BR3100, Plessey NC33 and Marwin 302 output is available.

A third post processor provides digital tapes for input to Copath, control tapes for Ferranti IV B, IV B* and IV D control systems can be produced.

These three post processors are for APT and ADAPT systems, a similar set of post processors is available for PROFILE DATA systems.

Interactive software, Time Sharing Option (TSO)
The original interactive software was written to run on CPS (Conversational Programming System). Many desirable features are included in CPS but it is not ideally suited for an interactive NC system. A second terminal system called TSO (Time Sharing Option) contains additional features which make it more attractive for the NC applications but on the other hand TSO does not contain some of the more useful features of CPS. This dilemma was eventually overcome by B.A.C. systems programmers who were able to make CPS run *under* TSO and we were therefore able to get the best of both worlds.

Part program filing scheme
The part program filing scheme is an essential building block in the overall system. Over five thousand part programs are stored on the filing scheme as card images on magnetic tapes. About six hundred current part programs are stored as card images on disc and can be accessed directly by TSO.

Housekeeping programs are run daily to retrieve data from tape and copy it to disc and all part programs are indexed and backed up by at least two other copies. The filing scheme is also used to log accounting information and to supply statistics for management information purposes. It is a suite of programs with two years of production usage behind it, and it was used prior to the development of terminal usage for NC (although terminal input was envisaged when the filing scheme was originally designed).

Foreground and background processing
TSO contains an area of core, called a partition, for on-line processing, all the TSO users being allocated in turn time slices in this partition, called the foreground partition. In addition to foreground processing, users are allowed to schedule jobs to be run other partitions, i.e. other areas of core in the main computer. After scheduling one of these jobs, it enters the batch queue and is thereafter under the control of the computer operator as with an ordinary batch job—this is called background processing. Thus the foreground can be used for highly interactive processing and the background reserved for the non-interactive work such as post processing.

OPERATING THE NC
TERMINAL SYSTEM
There are two modes of operation. The first mode is for part programmers who want rapid turnround and who will want to run jobs in the foreground. They would normally operate the terminal themselves to do on-line debugging. The second mode is to submit batches of part programs, with updates and post processing instructions, to the background. In this mode of operation a clerk or typist can be used to input the data because no knowledge of part programming is necessary.

Foreground processing under TSO (Fig. 1)
Figure 1 shows schematically the steps involved in this mode of operation. The user dials into the computer and logs onto TSO and he issues a command to the computer to say that he intends to process NC.

The user is then asked to type in an identifier for the part program he wishes to process. The specified part program is copied to a work area where the user can make alterations by inserting, deleting or updating statements without changing the original part program. These alterations can be made either by searching for character strings and editing on a character basis or by specifying card numbers and editing on a card basis.

The user can schedule his edited part program to be processed through his choice of NC processor. There follows a short delay whilst the part program is actually processing. The length of this delay depends on the complexity of the job being processed—if the job is known to be very long or very complicated then, in the case of the APT140 processor, it can be broken down into groups of statements. The number of statements in a group can be specified by the user.

After processing, the whole of the line printer output is held by the system and can be scanned by the user. Normally the user is only interested in any errors thrown up by the processor. An assembler

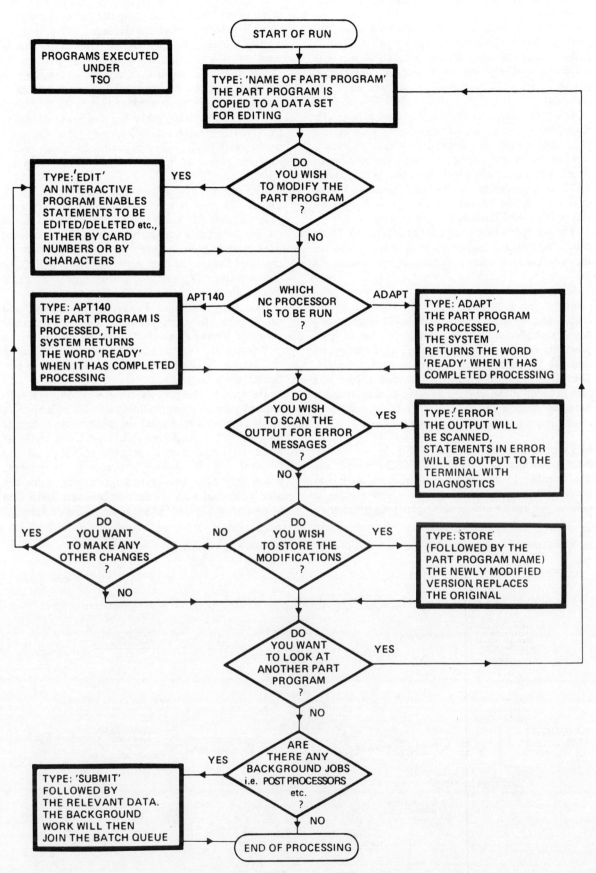

Figure 1.

program is therefore provided for the purpose of scanning the output and passing error messages and the relevant statements to the terminal for inspection by the user.

If the error message is insufficient to diagnose the problem then it is possible to scan the output for any string of characters that the user thinks will help with his diagnosis. Every line in the output containing a specified string of characters can be written at the terminal.

When the user is satisfied that he has correctly diagnosed all his errors, he can return to the editing phase and correct the statements which are wrong. The processing cycle can be repeated any number of times until a successful processing run is obtained and a CL file is produced. At this stage the user can type in an instruction to run the post processor of his choice from the CL file he has just created.

The post processor is run in the background. There is little advantage in making it interactive because of the very low frequency of failures at this stage in the job.

Sometimes the user is prepared to type in a complete part program at the terminal as this would enable him to get from coding sheet to machining tape in the same day. In practice, however, part programs are too long to be typed in directly and they are punched off-line by punch card operators. The part program is then stored on disc as part of a part program filing scheme. As soon as it is stored it becomes possible to use the TSO foreground processing techniques described in this section.

Background processing using CPS (Fig. 2)

Prior to the implementation of the TSO system it was not possible to run processors such as APT and ADAPT in a foreground region. Consequently a system was built which enabled part programmers to prepare a file of data to be used as input to a sequence of batch programs as indicated in Fig. 2. The batch programs were then scheduled from the terminal and run by the computer operators when a suitable partition became available.

This method of scheduling NC work is still widely used at B.A.C. because, although it is generally slower than the foreground system, it can be used to stack up a file of NC work to be run overnight. This batch of work is compiled late in the afternoon and scheduled at the end of the day shift just before the terminal system closes down. The batch of work, which tends to be a large amount of processing, can then be run at the convenience of the computer operating staff.

The various stages of the system are shown in Fig. 2. The user types in the name of the first part program and if it is not one of the 600 or so stored on disc the system will tell him it must first be copied from tape. It gives the user an opportunity to check that he has typed the right identifier before he requests that the part program be transferred from tape to disc.

The user continues by typing in updates and deletions to the part program together with card numbers. Having made all the changes, he then types in a 'MACHIN' card which tells the system he wishes to execute the part program and to call up a particular post processor.

He repeats the process for the second and subsequent part programs, building up a file of input to a batch program also a small file of identifiers for part program to be transferred from tape. When his file of input is complete it is possible to list it at the terminal and, if necessary, to edit any statements which may have been typed incorrectly. Once the user is satisfied with the data he issues an instruction to execute a sequence of programs in the background

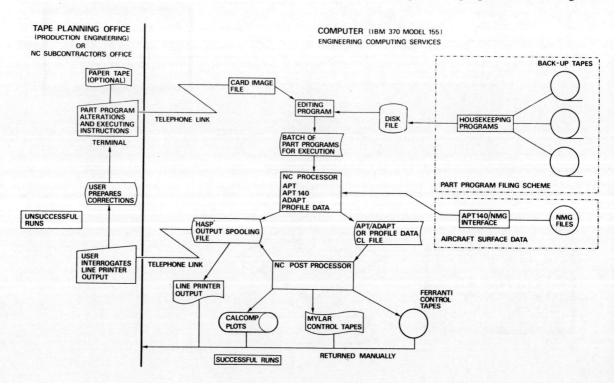

Figure 2.

partition. The system will ask him if he wants his output to be retained for his inspection (a facility described below) or printed off on the line printer as with normal batch work.

The first program in the sequence looks at the identifiers of the part programs to be retrieved from tape. Frequently this program has no work to do because all the current work will be already available on-line, and only the older jobs coming up for modification will have to be copied across to the disc.

The second program does all the editing, i.e. the updates, deletions and insertions specified by the user. It also puts the part programs to a file for execution with the correct 'MACHIN' statement as supplied by the user.

Thirdly the NC processor is run. This program controls the various post processors, and the processors and post processors between them produce line printer output, plots and the various machine tool control tapes. These control tapes and other output are returned manually to the user.

If the user has told the system that he wishes to retain the line printer output for inspection then it will not be printed off immediately. The user can log-in when he thinks his run is completed (a facility exists for interrogating the system to determine whether or not the job has finished). He is able to scan the output for error messages and also for specified strings of characters to help him with his debugging. This scanning facility, however, is now largely superseded by the foreground processing under TSO which gives the same facilities with a faster turnround.

FUTURE DEVELOPMENTS

The work done on terminal developments to date has shown that it is a promising and worthwhile development for NC, but using a basic typewriter terminal it soon becomes obvious that there is a limit to the level of development. It is possible to purchase terminals with paper tape facilities and these are indeed used by some B.A.C. customers, so it is possible to punch out a machine tool control tape using such a terminal. This raises a number of problems. Firstly, it is obvious that the control tapes must be perfect in every character transmitted and the length of the average control tape is such that there is a definite possibility of incorrect characters due to noise on the telephone line. Secondly, the time taken to transmit the average tape is about thirty minutes so that transmitting half a dozen tapes would occupy the terminal for half a day and would almost certainly interrupt other work.

The next stage of development will therefore involve more sophisticated hardware, consisting of a mini-computer driving paper tape equipment and a storage tube. The mini-computer will then be used to check the paper tape data before punching. A block of data will be stored in core at the terminal end and will not be committed to paper tape unless the data satisfies certain checking procedures. If these checking procedures show that there is an error then the mini-computer will instruct the main frame to transmit the block again. The existence of processing equipment at the terminal end will also enable tapes to be transmitted in a compressed format, making use of every piece of information, and it should be possible to reduce the average transmission times to the order of five to ten minutes per tape.

Some experiments with storage tubes have shown that the tool centre path plots can be transmitted down a telephone line and a plot produced with reasonably good definition. It is anticipated that on-line plotting of this nature would be a useful debugging aid, especially as it is possible for the part programmer to watch the plot building up on the screen.

There is evidence that the cost of on-line graphics processing will fall drastically in the next year or two. Combined with the development of more sophisticated software this will make the use of on-line NC systems a very feasible proposition. The next few years will show more and more companies following the same lines as B.A.C. in trying to bring their computer facilities right into the part programming office.

DISCUSSION

Q. D. G. Wilkinson, NEL. APT 109 was mentioned as an interactive version of APT in which control and editing of the part program is done within the APT system. I would like to have some more details of the language used for control and editing.

A. Using the TSO foreground processing system described in the paper it is necessary to leave the APT processors and to enter the 'EDIT' program in order to make alterations to part programs.

With version 1 of APT, however, editing of part programs can be done within the processor. Two methods of editing are available, one is based on the IBM EDIT facility, the other on the IBM utility IEBUPDTY. The language used by APT140 Version 109 is similar to that used in these IBM programs.

Q. W. H. P. Leslie, NEL. It is always interesting to follow how names of programs develop. Could Mr. Aughton explain how the systems he discussed jump from the more usual APT3 and APT4 first to APT140 and then to APT109?

A. The name APT140 is derived from the core requirement of this processor which is 140 kilobytes as against 256K for IBM APT.

There is no APT109 as such but Mr. Leslie is referring to APT140 *version* 109, the 109 refers simply to the numbering system employed by the authors.

Q. C. F. Turner, Is it possible with computer programming to replace the skilled operator for turning applications?

A. It is possible for computer programs to pick up elementary mistakes and to make basic checks on materials, feed and speeds, etc. and the use of computers can therefore make savings in this area. There is no evidence to show that the skilled operator will be replaced in the forseeable future.

THE EXAPT SYSTEM:
PRESENT STATE AND CURRENT DEVELOPMENT WORK

by

W. BUDDE* and H. WEISSWEILER*

INTRODUCTION

With numerical control a tool has been found for performing the most complicated manufacturing tasks and for automating the manufacturing process from the smallest to medium batch size. Today, the numerical control unit and NC machine can be considered to have matured. The number of NC machines in use shows that the advantages of its use are recognized. If the number of machines installed today is lagging behind the expectations of 1965, the main reason for this situation must be looked for in the difficulties of the organizational incorporation of the NC machine into the general operational process of a plant, and in a certain reserve towards the more rigid type of organization required when using NC machines. However, the developments in the machine tool industry, in control and manufacturing techniques continue to progress: today we speak of ACC, ACO, CNC, DNC and of flexible NC manufacturing systems. To enable industry to use this new manufacturing equipment economically in the future, we must today start developing new methods and systems to bring company organization to a higher level as regards adaptation to the automation level in manufacturing.

Considering the background of today's situation and the long-term development this presentation gives a summary of the present-day state of development of the EXAPT system.

PRESENT STAGE OF THE EXAPT SYSTEM

The use of numerically controlled machine tools shifts the work, expenditure and responsibility for the workpiece and the manufacturing equipment from the workshop to the operations planning department, as the manufacturing process must be predetermined in great detail by the part programmer and be transferred to punched tapes. This activity entails a great amount of tiring routine work and must be performed with maximum concentration.

The use of automatic programming languages makes the computer an effective aid in the technical range of an industrial enterprise. Automatic programming systems make it possible to transfer the manufacturing specialist to the operations planning departments of the companies and to entrust him with the detailed preliminary planning of the manufacturing and cutting process. Automatic determination of the machining technology by the computer leads to a reduction in the demands on the programmers to be employed. This is a great contribution towards solving the shortage of personnel, as a few specialists with particular knowledge of the manufacturing process will be in a position of support and effectively employ a greater number of part programmers.

EXAPT philosophy

In the past, a great number of different programming languages were developed. Characteristic of all these developments is the fact that all these languages were conceived for solving geometrical problems of programming and/or machining. However, if one considers the path of a workpiece from product design to the manufacturing process, as is shown in Fig. 1, it becomes obvious that the decisions to be taken in this process are mainly of a technological nature. When using NC machines, the user is compelled to plan the machining cycle for the automatically operating machine tool very accurately in great detail. For this, the part programmer will be principally required to deal with problems of technology planning and is,

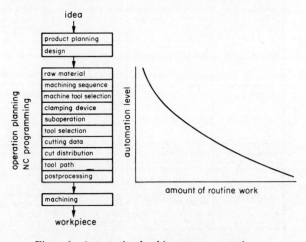

Figure 1. Automation level in part programming.

* EXAPT, Aachen

therefore, more a manufacturing specialist than a programmer. The geometry-oriented programming systems simplify the tool path determination and the post-processing. All extensions towards a higher level of automation lead to an extension of the technological part of programming systems.

Going out from these considerations, the development of the EXAPT system was started in 1964. The name EXAPT stands for 'extended subset of APT'. This designation reflects the characteristic features of EXAPT. The EXAPT system belongs to the family of 'APT-like languages' and represents a subset of the APT system as far as solvable geometrical tasks are concerned. On the other hand, the facilities for programming technology were considerably extended. Assuming that the facilities of APT are sufficiently known, the following explanations are limited to the technological aspects of EXAPT.

The EXAPT system

The differences between EXAPT and some other languages of the APT family are shown in Fig. 2. The figure makes it clear that the different parts of EXAPT among one another and in relation to APT are not competing systems, but that they complement each other and offer specific advantages for different machining problems.

BASIC-EXAPT is the basic language part of the EXAPT system. BASIC contains the geometrical processing facilities of the EXAPT system, but is not tailor-made for a specific manufacturing method. Thus, BASIC-EXAPT is at the same time the most universal language part of the system, permitting the programming of manufacturing problems that are new to NC control, such as nibbling, electro-discharge machining, flame-cutting, etc.

EXAPT 1 was conceived for machining tasks on point-to-point and straight line controlled boring, drilling, and milling machines. It allows automatic

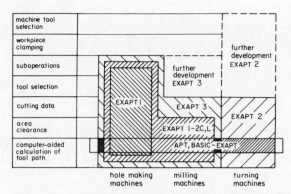

Figure 2. Computer-assisted programming systems for NC machine tools.

determination of work cycles, that is, the different machining operations such as centre drilling, pre-drilling, boring open, tapping, etc. are determined automatically until the final state is reached. For this, EXAPT 1 selects the required tools, determines the necessary tool motions and cutting values to be used.

EXAPT 1.1 contains all facilities of EXAPT 1 for automatic determination of technological data. The following essential extensions can be noted: variable work cycles make it possible to solve company-specific manufacturing problems. Motion statements allow $2\frac{1}{2}$-dimensional milling operations. Boring bars with several cutting edges can be programmed easily. Tool selection is based on optimization criteria.

EXAPT 2 was specially conceived for the programming of turning operations. The system performs the cut distribution and cutting value determination automatically. For this, the clamping conditions, the geometrical and technological properties of tool, machine tool, and workpiece are taken into consideration, in order to ensure a trouble-free, optimum-cost machining process on the NC machine tool.

Figure 3 gives once again a summary of the facilities of the EXAPT system. For further informa-

language part possibilities	BASIC	EXAPT I	EXAPT II	EXAPT 2	EXAPT 3	2 C.L – EXAPT I
geometrical part	point, circle contour description (blank and finished part) point pattern transformation matrix	point definition point pattern auxiliary definitions line, circle transformation matrix	point definition point pattern auxiliary definitions line, circle transformation matrix	point, line, circle contour description (blank and finished part)	point, line, circle point pattern tabulated functions contours transformation	point, line, circle point pattern tabulated functions contours transformation
technological part		centre drilling drilling counter boring end milling spot facing reaming tapping milling	drilling counter boring end milling spot facing reaming tapping milling boring bar operations	drilling counter boring reaming tapping straight line turning operations contour–parallel turning operations	drilling counter boring end milling spot facing reaming tapping contour milling area clearance milling	drilling counter boring end milling spot facing reaming tapping contour milling area clearance milling
executive statements	single statements	single statements	single statements	single statements	single statements	single statements
technological possibilities	points, where spindle speed is changed for facing	cutting speed feed tool selection work sequence tool path determination	cutting speed feed tool selection work sequence tool path determination	cutting speed feed collision check cut distribution	cutting feed, speed collision check tool selection (drilling) cut distribution for milling operations work sequence (drilling)	cutting feed, speed collision check tool selection (drilling) cut distribution for milling operations work sequence (drilling)
field of application	all operations ($2\frac{1}{2}$ dimensional)	drilling	drilling milling ($2\frac{1}{2}$ dim.) operations with boring bars	turning	drilling milling ($2\frac{1}{2}$ dimensional)	drilling milling ($2\frac{1}{2}$ dimensional)

Figure 3. Facilities of the EXAPT system.

tion on the different EXAPT language parts as well as on the overall system extensive information material is available that can be obtained from EXAPT.

Figure 4 shows the general flow of computer processing. The programmer describes the manufacturing problems (for instance as a global description of the work sequence) in a part program by using the EXAPT language. The EXAPT processor that processes the part program input by punched cards, teletype terminal, etc. consists of two separate program parts, in which the geometrical and technological computations are made in two steps. The results are stored on an interface, the CLDATA 2, and are adapted to the specific features of the machine tools and control units with the aid of post-processors. Besides the control information for NC machine tools lists are also prepared for the operator at the NC machine tool as well as further information for the operation planning and the workshop organization.

The EXAPT processors can be used uniformly for all NC machine tools and are not related to a specific company. For adapting the manufacturing data to the equipment of an individual company and to the production-technical know-how, the EXAPT processors have access to files in which the company-specific information is stored. The following files are used:

— a tool file that contains a geometric description of the tool and technological information;
— a material file containing all characteristics of the materials to be machined that are required for automatic determination of cutting values;
— a machining and machine tool file containing both the characteristic data of machine tools and data for influencing the machining operations.

For simplifying the introduction of the system, the users have as far as possible EXAPT standard files at their disposal to use freely.

Examples and cost situation

Figure 5 shows the input and output of the EXAPT 1 system. Beside general technological data concerning the workpiece (material, initial condition), the input comprises the description of the different machining positions and their position in correlation with one another (point patterns) as well as relative to the workpiece and the description of the different machining operations to be carried out at these machining positions. The lower part of the figure shows the results of an automatic planning process.

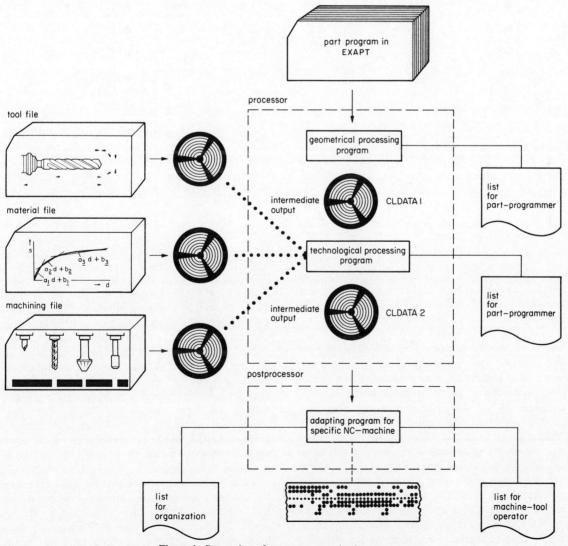

Figure 4. Processing of a part program in the computer.

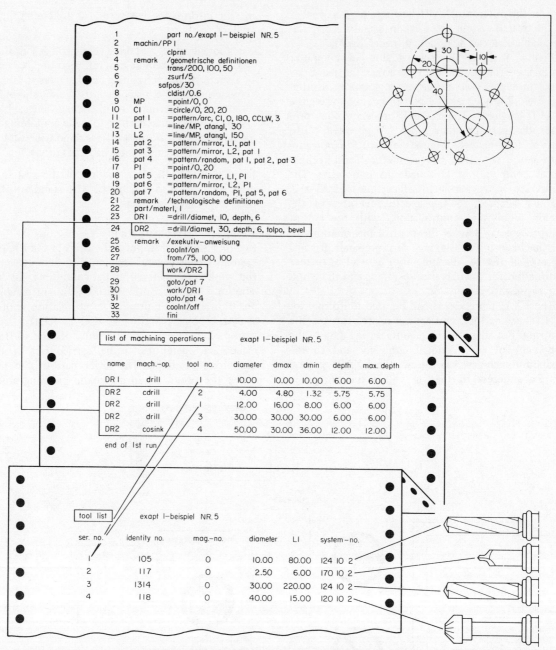

Figure 5. Computing an EXAPT 1 part program.

The descriptions of the machining operations have been split into a sequence of single operations to which appropriate tools from the tool file were selected or designed automatically. After determination of the single operations and tools, the detailed production planning process is continued in another program section, where tool paths and cutting values are determined and where the post-processor prepares the control paper tape for the NC machine tool.

Contrary to EXAPT 1, EXAPT 2 part programs contain a complete description of the blank and finished part and, at the present stage of development, an aimful pre-specification of the different machining operations and tools. Figure 6 shows some plotter drawings which were prepared automatically from the output of the EXAPT 2 system for checking the machining cycle for the shown workpiece. All tool motions drawn were determined automatically by area clearance routines. As the planning process is based on the description of the blank and

finished part, the tool motions can be drawn for every workpiece clamping in relation to the finished part and current blank. This has resulted in excellent graphic checking facilities which, as practice has proved, make test runs with new control paper tapes on the expensive NC machines superfluous. These checking facilities have also for active display units already been developed with appropriate correcting facilities for the paper tape.

Figure 7 shows the time and cost advantages when using BASIC-EXAPT in comparison with manual programming of the workpiece shown on the left-hand side of the figure.

Figure 8 shows with the aid of a practical example for turning operations the economic advantages when the technological values are determined by the computer. By the use of a geometry-oriented programming language the programming time was reduced from 70 to 35 hours, i.e. by 50 per cent when

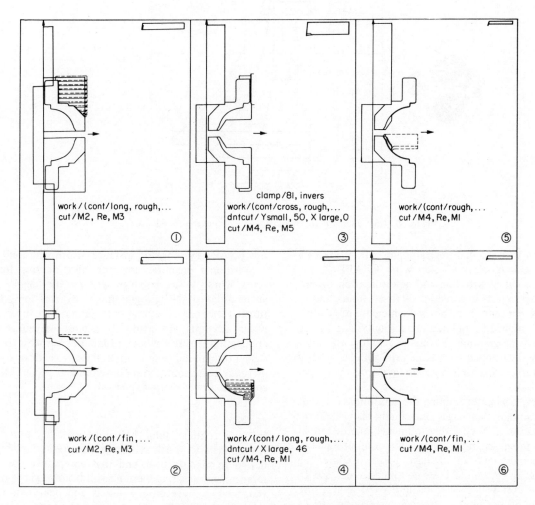

Figure 6. Plotted sequence of turning operations.

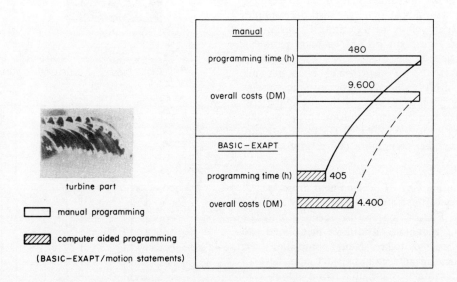

Figure 7. Cost and time comparison: manual programming–BASIC-EXAPT.

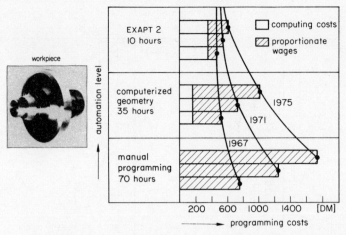

Figure 8. Cost increase for manual and computer-assisted work.

compared with manual programming. The utilization of the technological efficiency of EXAPT 2, i.e. automatic cut distribution and determination of cutting values results in a reduction to 10 hours, i.e. to about 14 per cent. This shows the increase of productivity of individual programmers when using highly developed programming systems. Owing to automatically determined cutting data of a higher quality the manufacturing time was also reduced by 14 per cent.

However, when examining the development trends of programming languages one should also take into consideration the cost developments mentioned at the beginning. In the example, the computing costs remain almost constant or have a slightly increasing tendency, the personnel costs continue to increase. In 1967 the cost saving when using EXAPT 2 at a rate of DM 10·80 per hour, personnel costs amounted to DM 288 as against manual programming. In 1971 the cost saving amounted already to DM 720 at a rate of DM 18 per hour, and in 1975 the saving is expected to amount to DM 1140 at an average increase rate of personnel costs of about 8 per cent per year.

Figure 9 shows the same situation for an example programmed with EXAPT 1. The list of examples of application as well as the time and cost advantages when using the different EXAPT language parts can easily be continued.

This proves the efficiency and economic advantages of highly automated programming systems containing technology. Owing to the prevailing cost situation, their significance in enhancing the productivity is going to increase further. The more tasks are transferred to the computer, the greater are the savings.

Present and future EXAPT development projects
So far a survey of the features of the already available parts of the EXAPT system has been given. The technical and economic advantages mentioned can already be realized today. Many companies have recognized these advantages and make use of them.

Practical examples show that computer-assisted programming becomes all the more economic, the more functions are transferred to the computer and the more the programming method is automated. Therefore, besides the propagation of the facilities of

the EXAPT system, for instance also in the field of conventional manufacture, the aim of our long-term work is the enhancement of the degree of automation of NC programming. By shifting additional functions of operations planning to the computer attempts are made to extend the range of application to the whole production process from design to manufacture. In the following sections some of today's and some future development projects that go in this direction are explained.

Modular system
It is easy to understand that an integrated system that is extended to the whole range of production, from design to manufacture, and that enters into the different manufacturing methods will become very large. But, owing to specific products and manufacturing methods of different companies, only parts of large,

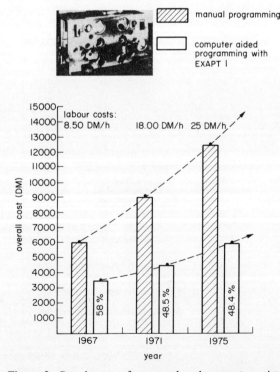

Figure 9. Cost increase for manual and computer-assisted work.

integrated systems are of interest to the individual user. He is faced with the alternative of using large-sized computers or of carrying out company-specific and product-dependent developments of his own. This decision entails of course a great number of organizational, personnel, and financial problems.

A way out of this situation is provided by the modular structure of the processing programs and by the development of a modular NC system. The user is then in the position to compile the system that fulfils his requirements and technical facilities best. Furthermore, such a program concept allows him to make extensions for further ranges of operation planning such as design, etc. by way of modules.

What does modular structure mean? In the scope of a project that we started working on in the year 1970, 'modular structure' means an exact assignment of individual functions of a computer-assisted system to corresponding computer programs, the 'modules'. Figure 10 shows the structure of a complex NC-Processor divided into such modules. The left-hand side in the figure shows the sequence in processing an EXAPT part program. This process can be divided into four phases:

— input translation phase;
— standardization of (workpiece) geometry;
— planning of machining operations; and
— specification of the machining steps.

These four phases are followed by the post-processing for adapting the CLDATA to the required control information on the shop floor.

During input translation the information of the input language is transformed into a computer-internal storing method. In the subsequent standardization phase the geometric definitions are processed and geometric elements, such as lines and circles are linked to contours.

In the third phase the operations sequence is determined. For hole-making with EXAPT 1, for

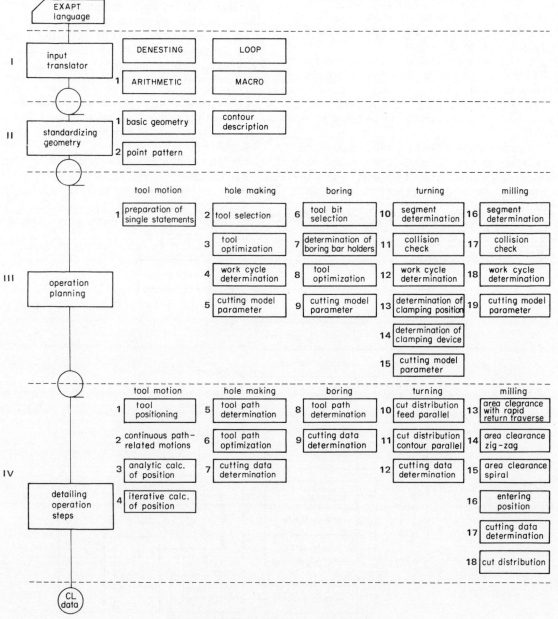

Figure 10. Modular EXAPT system structure.

instance, the work cycles, the tools as well as the cutting value parameters, such as object function and constraints on the part of the machine tool, the tools, and the material are determined. The further development of computer-assisted programming methods is carried out principally in this phase.

In the fourth phase the specification or detailing of the machining steps which were determined automatically at an earlier stage or pre-specified by the part program is carried out. On the right-hand side of the figure the EXAPT modules are shown with which the individual tasks are solved by the computer. The linkage of the modules directs the processing from input language to control tape. Essential characteristics of the modular order and of the modular structure are

(1) clear structure of the overall system
(2) unified data structure of all modules
(3) reapplicability of modules
(4) multiple combination facilities
(5) small maintenance expenditure.

This results in a great number of technical, organizational, and personnel-related advantages both for the user and developer of the system.

Figure 11 shows some general types of application. For the input different media are available (terminal, punched card, screen, CRT, etc.)

Different input languages (APT-like, fixed format, etc.) can be used if appropriate input-translators are available.

The input program can call for standard programs, the so-called system-macros, that are stored in a pre-translated form in external libraries. When the translation is terminated, the technological data, such as material, tool, machining, work cycle and clamping device files, are made available from external data stores.

The processing of part programs and the information flow for NC and AC machines are shown by arrows. When using AC machine tools, the determination of cutting values in the NC processor can be dropped. This applies also to the determination of cutter paths provided that the AC system calculates automatically the tool paths. Graphical display on a plotter or screen can be omitted, but may be useful for verification of the part programming as a replacement of test runs on the NC machine tool or for the preparation of working papers for conventional manufacturing.

The processing of the part program is to take place pass by pass. In Fig. 11 each pass is represented by a modular block. The advantages of this structure are that each pass can be easily overlooked, as all input and output data are stored on external storage media, easy exchangeability of individual modules, and the possibility to generate processors for the various tasks involving little expenditure. For instance with a specific master program it is possible to use the modules for building up a dialogue-processor that will permit a data flow with modification and interruption facilities on the different interfaces. For the practical use of NC processors it is important that this system is conceived in such a way that it can be handled easily by the user. This requirement is fulfilled by a system structure covering the following topics.

(1) A general concept for passing the data on between the different modules;
(2) a special data-handling-routine which is available in each module;
(3) improved facilities for syntactical checking of statements;
(4) a flexible solution for the use of postprocessor words by the part programmer;
(5) avoiding array limitations within wide ranges of the processor;

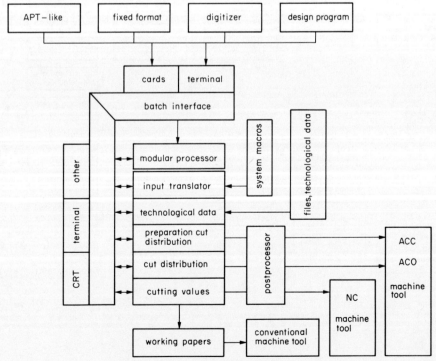

Figure 11. Modular processor system.

(6) the MASTER organization which is set over the modules for controlling the flow (process) within the module processor;

(7) Programming of the geometry-orientated modules taking account of the fact that they can be used for all language parts without having to be modified and that in their size they can be adapted to the size of the technological modules.

User-dependent work cycles in hole-making

Since the beginning of 1972 the test version of the EXAPT 1.1 processor is available on different computers (CDC 6400/6600, UNIVAC 1108 and IBM 360/370). The concept of EXAPT 1.1 took place in close co-operation with industry in EXAPT working groups. Experience values that had been gathered in the practical use of EXAPT 1 were included in the concept. Contrary to EXAPT 1, the language part of the EXAPT system that is meant for programming point-to-point and straight-line-controlled machine tools, EXAPT 1.1 was developed for programming straight line and continuous path controlled drilling and milling machines as well as machining centres.

The experience that was gathered in the practical use of the EXAPT 1 technology for the drilling and milling process made it obvious that in many cases standard work cycles are not flexible enough for practical use. This led some of the users to waive automatic determination of work sequences and tool selection and to prespecify the work sequence and tool in the part programs themselves.

By describing the work and tool cycles in decision tables and by their handling as a file, it was possible to make available a flexible method that allows company-specific requirements to be taken into consideration without having to change the processing programs. An analysis of possible work and tool cycles revealed the possibility to give the user standard cycles as a data base. Building up on these standard cycles, additional cycles required by the individual companies can be entered in this machining file.

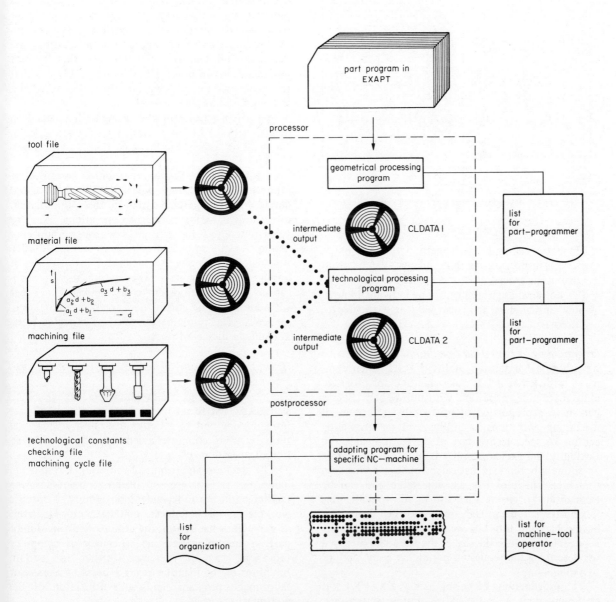

Figure 12. Processing of a part program in the computer.

The dislocation of the program logic from the technological processor to the work cycle file gives greatest possible flexibility and adaptation to company or branch-oriented requirements and 'know-how' without processor modifications.

The determination of work cycles as shown in Fig. 12 is based on

— a standard computer program,
— the work sequence file,
— a file for technological constants which is referred to by the work sequence file, and
— a check file for the formal accuracy of the machining definitions in the part program.

Figure 13 shows in comparison the determination of a standard cycle (left) and a user-dependent work cycle for the same machining task. In the sequence shown on the right, one cycle and one additional tool are saved by using a special tool for predrilling and chamfering in one operation.

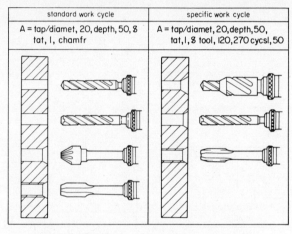

standard work cycle	specific work cycle
A = tap/diamet, 20, depth, 50, 3 tat, 1, chamfr	A = tap/diamet, 20, depth, 50, tat, 1, 3 tool, 120, 270 cycsl, 50

Figure 13. User-dependent work cycle generation.

Essential advantages of this method are

(1) better economy, particularly with bigger batch sizes, as manufacturing time and costs can be saved when using special tools and adapted cycles;
(2) universal applicability, as the individual know-how of the user plant can be stored and reproduced.

Programming of boring bar operations
An additional extension of EXAPT 1.1 when compared with EXAPT 1 is a new method for programming machining operations with boring bars. Boring bars have a proportion of up to 50 per cent in all machining operations in drilling and on machining centres. A special advantage of the boring bar is the flexibility and exchangeability of the tool bit. Figure 14 shows a boring bar that can be equipped with two tool bits. The tool bits are adjustable and can be exchanged. To ensure that also in the tool file this flexibility is guaranteed, separate file sheets for boring bars and tool bits were developed. This made it possible to reduce the size of the file to be prepared for describing the requested boring bar combinations to a minimum.

The programming of boring bars in EXAPT 1.1 is at present done with the aid of motion statements,

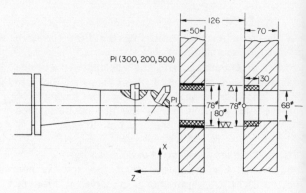

Figure 14. Boring bar operation.

but this method and the use of the tool file simplify part programming considerably, as the tools are selected by the program, the tool dimensions are stored in files, and the presetting dimensions are determined automatically and are taken into consideration during the cutter path calculation.

But the programming becomes much easier if the technological data are determined automatically. In the next development step of EXAPT 1.1 which is in preparation, boring bars, cutting values and cutter paths will be determined automatically. The computer programs are ready except for the cutting data determination. During testing it proved that the part programming can be simplified considerably and that the programs are working quickly and reliably and that good results are obtained.

A particular problem when using boring bars is the determination of cutting data. A great difficulty can above all be seen in the fact that boring bars tend to chatter.

The EXAPT Association met with a great deal of interest from industry when in 1970 it started developing a computer program for the determination of cutting data for boring bar operations. For two years now there has existed a working group consisting of eight important companies and the EXAPT Association dealing with a systematic investigation of boring bar cutting data. The aim of these investigations is the development of a practical model that will determine the cutting data for boring bars automatically. The result of this work is expected to be available in the near future.

Utilization of group technology criteria in NC programming languages
The development of NC programming languages such as EXAPT led to an important automation of part programming and to a better utilization of machine tools by using computer-aided automatic determination of operations planning.

The standard algorithms of the EXAPT system are based on an analysis of different manufacturing methods and are independent on groups of specific workpieces, machine-tools, materials, etc. However, as experience has shown, the degree of automation of part programming and the efficiency of computer processing can be increased further for certain functions of the manufacturing process by the use of tailor-made programs and with utilization of the group technology.

The EXAPT processors perform tasks of operations planning and detailing of the work sequence automatically. Operations planning in general means assignment of technology, such as work cycles, tools, etc., to the workpiece geometry.

This user-dependent assignment rule can be established easily for function-related shape elements. For instance, the hole-making operations in EXAPT 1.1 are adapted with the aid of decision tables to the user requirements. Using this method, a high degree of automation can be reached accompanied with a great flexibility. The advantages of this method which is built up according to group-technological aspects are also applicable to other machining operations if a

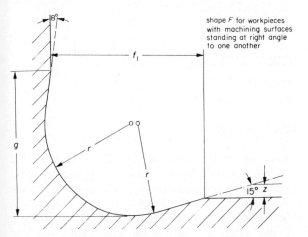

Figure 15. Standard shape element according to DIN (Deutsche Industrie-Norm = German Standardization Committee)

standardization of the shape or form elements at workpieces can be reached. This is also of great importance for reasons of exchangeability of workpieces. Figure 15 shows an undercut called type *F* according to DIN 509 with two rectangular machining surfaces. The type of the undercut as well as the workpiece diameter determine precisely the different dimensions of this shape element. For this, standardized and/or user-dependent machining cycles can be stored in decision tables similarly as in EXAPT 1.1.

The utilization of this method is presently being prepared for the EXAPT 2 system as well. The advantages are

— reduction of part programming time and cost
— user-adapted machining cycles via decision tables
— reduction of the number of required tools
— feedback from production to design to use standardized form elements.

For utilizing the advantages on a wider basis, the EXAPT Association is cooperating in different standardization committees.

Besides this, it is of course possible in all EXAPT parts to increase further the productivity of part programming for specific part families by the use of MACROs. The adaptation of the machining operations to the modified workpiece dimensions is done automatically by the technological processor.

System for file maintenance and evaluation

Technological planning models for automatic NC programming, such as are contained in the EXAPT system, must be adaptable to the specific manufacturing conditions and requirements of industrial companies. The technological algorithms in the computer programs must not be programmed rigidly, but there must be the possibility for every user to influence it by his own manufacturing data.

For this reason, the EXAPT processors that have a technological part use files in which the data of the manufacturing equipment are stored which do not necessarily change with every new manufacturing task. Such data are

— conditions of use and organizational data of the machine tools
— conditions of use and dimensions of existing tools and toolholders
— conditions of use and dimensions of clamping devices and fixtures
— technological characteristics for the determination of machining sequences and tools to be used
— material data for the calculation of economic cutting data.

To put this data into the computer and to maintain it throws up some organizational problems. Great advantages are offered by service programs with which the file data are maintained on external files of the computer and are made available for EXAPT processor access. For maintaining the EXAPT files the universal file handling and maintenance program (MAPEX 2) was developed. The structure of the service program is shown in Fig. 16. This program must be seen in addition to the 'information centre for cutting values—INFOS' project of the Machine Tool Laboratory, Aachen. One of the many-sided tasks of INFOS is the preparation of the EXAPT material file. The administration of the data stock of tools, materials, etc., that is tailor-made for the EXAPT system is done by MAPEX 2.

The manufacturing data contained in the files are of course not only valid in the field of NC manufacture, but they can also be made use of by the work planning department for conventional manufacture, provided that suitable aids are available to evaluate the stored data. Thus, a test program has been available for some years already that was developed at the Technical University in Aachen with which cutting values for the different types of turning operations can be calculated. For this, the program uses the

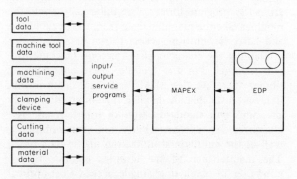

Figure 16. EXAPT file handling system MAPEX.

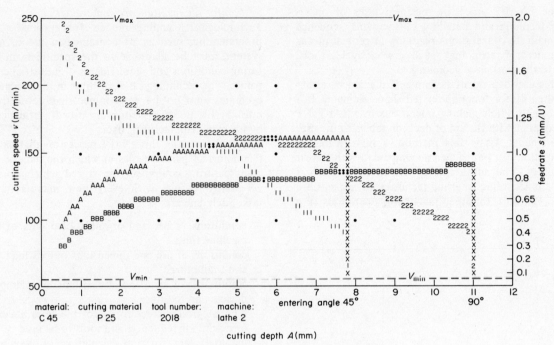

Figure 17. Cutting data calculation EXAPT 2.

same files, i.e. the cutting values are adapted automatically to the technological conditions of the tools and of the machine tool. The program can be output graphically as is shown in Fig. 17 or in the form of cutting value tables. This program is a component part of MAPEX 2.

Similar evaluation programs were developed by users of the EXAPT 1 system with a view to utilizing the EXAPT 1 files for the preparation of cutting data catalogues for conventional manufacturing.

As mentioned earlier, one of the aims of the future development of the EXAPT system is to provide further facilities for the evaluation of the stored machining data files according to different aspects. The above-mentioned range of planning for conventional production, economy comparisons of alternative manufacturing equipment, etc., represent the starting-off points for such developments.

System utilization

Figure 18 shows the structure and the facilities for utilizing the modular EXAPT system. Normally, the user has to produce a limited part spectrum. In the course of time a certain know-how has accumulated for this. The shop floor is tailored to the manufacturing profile and requirements. The user selects the program blocks for his different manufacturing tasks from the program library. The universally applicable modules take the planning logic from the cycle library and have, in addition, access to the data file library containing the characteristics for tool, machine tool, and material. For the module, cycle and data file library there are the EXAPT standard software and dataware available. A detailed documentation of the programs and standardization of interface and files guarantee the exchangeability and extendibility as well as the multilateral utilization of the data block. The maintenance of the libraries is done by the computer in batch or dialogue access. Cost-optimal, user-oriented systems are obtained by combining

company-specific software and dataware with the standard system. Thus it becomes possible to utilize the advantages of data processing to a greater extent also by the largest possible group of companies. Naturally, the facilities of the system can be used also in the range of conventional work preparation and manufacture.

What practical benefits do different companies have from the use of the EXAPT system? The essential advantages are shown in Fig. 19.

The first five points were the expressed aim of the development of EXAPT and were realized by the different language parts of EXAPT, as is shown in the many practical examples, which are:

reduction of the programming time and costs by shifting the repetitive work to the computer
storing the know-how in the technology-planning-models of the processors and the file data to which the computer has access

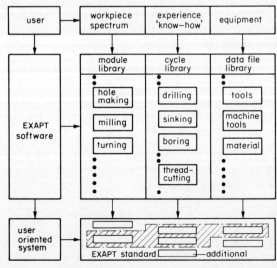

Figure 18. Structure and utilization of a modular NC system.

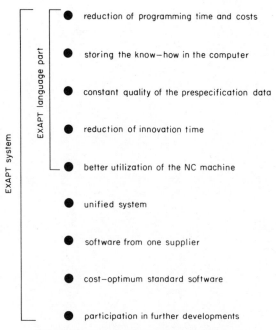

Figure 19. Advantages of EXAPT utilization.

for this reason each part program has a constant quality of results (same object function and constraints)

direct utilization of new technologies after the updating of the stored algorithms and files

better utilization of the NC machines (to meet the aims of company policy).

Besides these advantages which are already offered by every language part, the overall system EXAPT has features that must be of the same interest to the user.

EXAPT is not fixed on one manufacturing method, but provides a high automation level for different, frequently recurring manufacturing methods.

For this reason, the user can solve his multilateral NC problems without changing the system. The unified programming system makes it possible to keep the number of system and programming experts in the companies small and to avoid costs for the introduction of new systems, e.g., for other post-processors, the preparation of new files, etc. The fact that the whole software is made available by one central body (the EXAPT Association) should in addition solve many problems of competence and questions of updating and maintenance.

The development of such efficient NC processors causes high costs which are not justified for individual firms. With the EXAPT standard software the benefits can be taken advantage of at very low investment costs by the individual companies. The competition situation between the different users is not affected by this, as the processors are adapted by files to the production-technical know-how of the individual company (and not *vice versa*).

The EXAPT user participates in projects for the development of further technology-models, that are used in new or extended processors.

The work of EXAPT-Verein is financially borne by its members. Membership with EXAPT-Verein is at the same time the prior condition for industry leading to the utilization of the services of EXAPT-Verein and the EXAPT system. The EXAPT Information brochures give more information on the legal, organizational, and financial aspects of membership as well as on the specific facilities of application of the EXAPT system.

Summary
From the point of view of work preparation, control paper tapes for NC-machines represent detail work plans. Therefore, it is desirable that in automatic programming the greatest possible number of reproducible steps of work planning are carried out by the computer. Beside the required geometrical specifications, whose calculation and machine coding are natural component parts of all NC programming systems, the use of technology-oriented algorithms for the determination and optimization of work sequence, as well as for the selection of tools and calculation of cutting values provides the possibility of a successive automation of the entire manufacturing process.

The advantages of the determination of work sequences and tool selection for drilling, turning, and milling operations, as well as the use of algorithms for collision checks of the tool paths and the calculation of cutting values in the EXAPT-System were explained. A central factor in such work is the making available of company-specific empirical values and technological data. The structure of company-oriented files for storing machine tool, machining, tool, and cutting data and their main parameters, respectively, were also explained. Another subject was the use of files and also their effect on conventional manufacture. Finally, the organizational effects and facilities for the enhancement of productivity when using the developed procedures were explained. It was shown that through the construction of a modular NC-System both the interests and concerns of the small-sized manufacturer can be met and that through universal extensibility construction of an integrated system is feasible.

NC COMPUTER PROCESSING AT A GOVERNMENT RESEARCH AND DEVELOPMENT ESTABLISHMENT

by

A. R. KIMBER* and H. R. JOHNSON*

SUMMARY

This paper outlines the development of computer processing for production of numerical control tapes over the past decade at AWRE, Aldermaston. It describes how certain philosophies about computer processing have resulted, and how these have effected the organization of the AWRE NC computer processing system. Examples of aids to machine tool part-programming that have been developed are described, and ideas for the future of NC computer processing are put forward.

INTRODUCTION

Numerically controlled machine tools and computers to produce control tapes have been used at Aldermaston for over ten years. It is not intended that this paper should be an historical report, but an indication will be given of the way in which our approach to, and philosophies about, numerical control have developed over the years. While it is appreciated that AWRE work differs in many respects from general machine shop production, there are principles and policies which apply in general once computer production of control tapes is necessary.

We believe that the principles we have adopted through involvement with the complexities of computer processing may be of general use.

EARLY DAYS

Numerically controlled machine tools were first purchased for AWRE to produce work which no conventional machine tool could manufacture within an acceptable time scale. This was the economic justification. With the complex geometry of the components it was essential that control tapes for the machines were produced using a computer. At that time we had no experience of numerical control, and no available NC processing system. We did however have available a scientific computer, so special purpose computer programs were devised for the production of control tapes, one for each type of numerically controlled machine tool. This early work was the subject of an important policy decision which has been continued throughout our work with numerical control; all numerical control computer programs would be developed by applications staff.

In other words our early special purpose programs, and subsequently all our postprocessors, have been written by Engineers rather than computer specialists. A central team of computer staff has been available to provide advice when required.

With our early processing systems the Engineers involved had as their priorities the need to cut metal on the new machines—accurately, efficiently and economically. The niceties of programming were unimportant providing the work was done. An early example of the success of these efforts with NC processing was the reduction in production time of one exceptional family of components from fifty weeks with a conventional jig-borer to six hours with a special purpose NC machine.

With several single purpose processors being used for NC tape production it became desirable to simplify the processing procedure and aim at standardizing those parts of each system which were common to all the machine tools. While we had been introverted in our approach to NC processing development, other organizations had cooperated and had made great progress towards the development of general purpose geometric processors and towards the setting of standards for the writing of the machine tool dependent parts of a system. Most of this work had been done by the Americans under the APT umbrella—this was 1964—and had been published by the IIT Research Institute.

THE APT PROCESSOR

APT is a generalized processing system which has been developed into the most geometrically comprehensive system available. It lays down the principle of a Processor to handle the geometric

* Machine Tool Development, AWRE, Aldermaston

problems of NC tool positioning and all general functions of part-programming, and a standard interface with Postprocessors which handle the specific machine dependent functions of one or more machine tools. When we first obtained a version in 1965 there was only limited experience of NC processing in this country and we followed very closely the guidance of American experts. We believe that the basic grounding we received in American software at that time has been invaluable in enabling us to standardize our own NC software, and has greatly simplified the computer changes we have made since then. It has always been our policy to write our own postprocessors, and while initially this produced problems and a need for following rules laid down— sometimes blindly—we have now developed an expertise in the subject which can easily satisfy our own postprocessor requirements, and which has been used by industry as a postprocessor writing service for several years[1].

SYSTEM ORGANIZATION

The APT system, the Processor and Postprocessors, are stored on a computer and are used to process completely part-programs, each written to enable one particular type of machine tool to manufacture one particular component. It would be ideal if, once the system had been set up, it could be forgotten for a few years, but this is not the case. Numerical control computer systems are complex logical creations; there are millions of different ways in which they can be used and complete testing is impossible. So there are errors, and there are improvements and developments and changes in method and other reasons why the system must be regularly updated. The software is in fact living. It lives to keep up with technological progress, with the increasing demands made by man's creative ingenuity and with the slow progress towards perfection.

To cater for this living requirement a computer system needs to be organized in a way that makes it practical for regular updating and modification. It may also need to be interrogated in the case of errors occurring. There is in fact a demand for a considerable amount of work and organization to run a numerical control computer system.

In our experience to use a computer, to keep a systems team to implement and run such a system, and to provide a control tape production service, is very expensive. Certainly the advantages gained cannot alone justify the cost of a computer. Economic justification of any NC computer system is very difficult, and we have found it essential to organize processing to minimize the major costs in running our system. It can be argued that to keep a system of the complexity and size of APT is wasteful and unnecessarily expensive, especially if one's work is predominantly simple and could be served by a less capable processor. This may be true, but it can be appreciated that the savings in part-programming effort are greater the more complex the component. This principle is illustrated in Fig. 1 where it is suggested that, whereas the usage of the more complex areas of a geometric processor is limited, the cost savings on

these few occasions in man effort and time are proportionately far higher.

We therefore rely on an APT processor to satisfy our geometric requirements for the foreseeable future. We know from past experience that the construction of this processor and the use of a standard source language simplifies and reduces the cost of moving the processor from one computer to another. In the past ten years our system has been on three different computer types and seven different computer models.

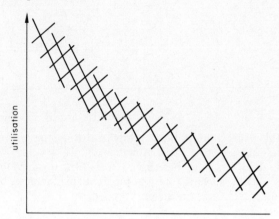

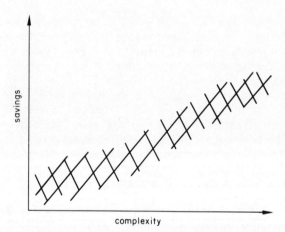

Figure 1. Utilization and savings vs complexity.

We keep the organization and use of the APT system as simple as possible, while allowing for the necessary regular updates which are made to both processor and postprocessor areas. The cost of development of software is so high that the system must include back-up copies to protect against both accidental erasure in the computer and the possibility of fire. When updates are made all copies of the system must be included, so to simplify this we use a remote terminal and let the computer control the updating procedure. In this way we obtain quick modifications with minimal risk of error. The use of a terminal in this way is our only refinement in an otherwise simple system. We have considered the use of interactive terminals for part-programming, we have researched into the development of a so-called 'technological' system for lathe programming[2], and we have flirted with the use of graphical input devices for part-programming[3]. We have found that the expense of implementing, organizing, testing and running all these types of 'advance', especially graphics, is far above any economic advantages they may bring to the production of components on numerically controlled machine tools in our organization.

In order to avoid multiplying the implementation and system organization costs we use the flexibility and versatility of APT to satisfy all the varied machining requirements across our whole range of numerically controlled machine tools. We obtain the maximum versatility by fully developing and tailoring the postprocessor parts of the system.

POSTPROCESSOR AIDS TO MACHINING REQUIREMENTS

As an indication of the ways in which we have utilized our full control of the postprocessor capability to advantage, the following sections describe some of the use we make of postprocessor facilities.

1. Lathe programming

The problems of programming for NC lathes are unique, and hence the development of special NC processors to overcome these problems[4]. We have also been faced with these problems, receiving our first NC turning machine in 1962 before any special processors were available.

With turning, to enable a reasonable maintenance of constant cutting conditions, strict control of the spindle speed is essential, but, although facilities for constant cutting are features of more recent special lathe NC control systems, on some early machines the spindle control was a stepped gear change which could not be operated during cutting operations. The responsibility for changing spindle speed could not therefore be built as a fully automatic procedure into the postprocessor. Within this limitation the selection of cutting feeds and speeds has been made as simple as possible. The surface speed of the workpiece, and the inches per revolution feedrate are more useful parameters than conventional feedrate and spindle speed. A part-programmer is required to specify his surface cutting speed (surface feet per minute) and the feedrate per spindle revolution (inches per rev). He then has only to program the word 'SPINDL' in suitable places in the part-program and the post-processor will automatically select and output the correct spindle speed for the cutting conditions and radius of cutting at that time—if necessary. Feedrates for all subsequent cutting movements are then provided automatically from the specified inches per revolution. More modern NC lathes are capable of changing speed under cutting loads and, with this control of the spindle speed, changes can be completely handed over to the postprocessor if required.

Geometric aids to part-programming have been developed which give immediate benefits in reduction of part-programming effort. One example is the process of roughing a blank down to a form by describing the finished shape. Working from a conical or cylindrical blank the postprocessor will generate all the necessary clearance cuts and spindle speed changes to rough the described form. The procedure will leave a smooth contour rather than a stepped finish so that finish machining will involve a constant metal removal rate (Fig. 2).

Another turning process is that of threading. Those who have seen an NC lathe cut a thread will know that it is a rapid procedure, far shorter than the time taken to program it by successive cuts. This applies particularly if the thread must be side cut—the tool advancing down the thread angle—or if the thread is multi-start. The calculations involved for either of these cases are repetitive and inclusion of such a feature in the postprocessor is straightforward. Our part-programmers can cut threads in our NC lathes by programming one statement, giving any

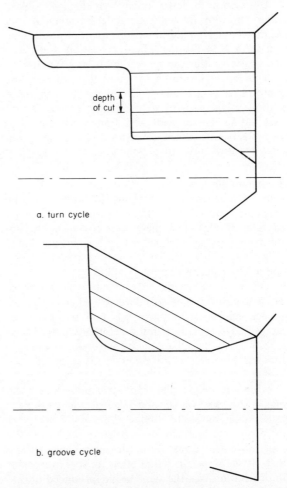

Figure 2. Lathe roughing cycles.

thread, any size, any number of roughing and finishing cuts, any number of starts, plunge or side cut, left or right handed, parallel or up a taper if required, at any cutting speed.

Tooling for NC turning is another problem. To machine one component there may be call for several different tools, all of which differ in size and shape, each mounted in a different position relative to the workpiece. The relationships between the different tools and the machine movements necessary to allow for changeover from one tool to another can be complex, and a logical system of tool dimensioning, setting and programming is necessary to avoid positioning and collision problems. The use of a computer means that such a system can be laid down, and the information about each tool fed into the part-program for the postprocessor to organize and keep account of the relative positions of tool cutting edges. A part-programmer need then only select a new tool and its different relative position will be compensated for. This illustrates one of the basic principles of computerization of any process. The major point of the tooling procedure is the organization of tools and tool geometry made by the part-programmer before his program is written. If that organization is bad or in error so will be the postprocessing of the information. The computer demands that a system to be implemented must be well organized (and occasionally after it has been organized the computer is no longer necessary).

Our system for lathe tooling uses standard tools, a library of tool drawings, a standard system of tool geometry, and a rigid procedure for tool positioning and setting. In this way the organization is efficient and postprocessing can proceed without difficulty.

The facilities offered by our lathe programming postprocessors are not as comprehensive as those within special purpose lathe processors, but we find that within the standard APT framework they provide practical solutions to many of the major problems and difficulties of programming for NC turning.

2. Step milling

One of our earliest problems to solve was the machining of accurate free form cams for copy turning. At this time there were no accurate contouring jig-boring machines available so we purchased a jig-borer with a card fed point-to-point NC control and machined cams using step milling.

Other organizations have done work to make point-to-point machines contour[5]. Their usual approach has been to move the tool point-to-point along a zig-zag path within a defined tolerance zone (Fig. 3). This may be done by moving one axis at a time or by using simultaneous positioning of both axes.

With the requirement for machining cams the major aim was to cut the edge of the workpiece rather than move the tool along a path. We were therefore able to make use of relative freedom of the tool movement away from the component. Rather than define a tolerance limitation for tool movement the limitation was the cusp height produced on the machined surface. Steps were calculated to give a certain cusp height with a certain diameter tool (Fig. 4).

This approach for cam machining can show a reduction in number of steps of 10:1 over the alternative method, and makes practical the machining of free form cams using a point-to-point paper tape control, tape length being a major problem.

The original jig-borer has recently been replaced with another. There is still no cost comparable contouring system available which could give us the accuracy we require on cam forms and also leave the machine available for easy manual operation when required, so we still use a point-to-point system for accurate cam production. With new equipment we machine to give approximately 80 steps per inch, at a rate of about 20 a minute, and we consistently mill cam forms to within a 0·0002 in tolerance band.

The computation for step calculation is done in the postprocessor. A part-programmer defines the

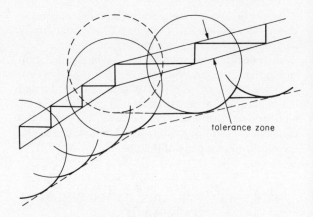

Figure 3. Contour point-to-point milling.

required height of machined cusp, the diameter of tool to be used, and the side of the workpiece to be machined. He then takes the cutter round the cam form by conventional APT motion statements, and all calculations of X and Y steps are made automatically.

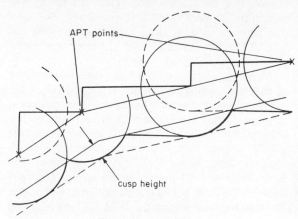

Figure 4. Step-milling.

3. Arc machining

As has been stated, a common requirement in our work is to machine curves which are free form rather than of geometric origin. The usual way to machine these is to break the curve into many small straight line cut vectors, provided by APT, and machine using linear interpolation. This results in many very small steps, and very long lengths of control tape. The associated heavy usage of paper tape readers on control systems has caused a noticeable decrease in reliability. A reduction in both tape length and in strain on tape readers has been obtained by using the postprocessor to fit circles to the free form curves within the available tolerance band, and to machine using circular interpolation (Fig. 5). This has been very successfully applied to both contour milling and turning work.

4. Inspection

The APT system is used with appropriate postprocessors to provide data necessary for use with electronic inspection equipment. In some cases separate part-programs are written, while in others a link between the manufacturing machine and inspection postprocessors has been made so that the provision of inspection data is made through the machining part-

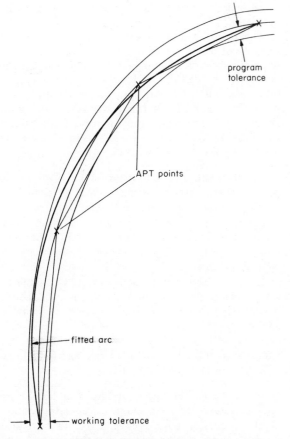

Figure 5. Arc machining.

program. In this way part-program effort is mini-mized.

The accuracy of Inspection equipment is higher than most manufacturing equipment. The tolerances with which one is working for measuring a component must of necessity be at least an order better than those of the component to be measured. When trying to use the APT system to provide the 'ideal' figures for a component from which error is to be detected, one is therefore using the system at an accuracy level for which it has not been designed, and to which it has not been thoroughly tested. This has presented problems. By using the system with very small tolerance limits in the more complex geometric areas we have found inaccuracies which could effect the efficiency of our inspection processes.

THE FUTURE

This paper has briefly discussed the development and use made of computer processing for numerical control tape production at AWRE, Aldermaston. It has indicated that due to the expense of computer techniques, and to provide a standard approach to tape production across our whole range of numerically controlled machine tools, we have settled on a production facility centred around one very capable NC processor. We use simple and straightforward system organization with full development of postprocessors to aid the part-programming of diverse machine tool types.

The basis of a processor–postprocessor system is the reliance on a strictly maintained interface between the general and machine dependent areas.

However the considerable variations in machine tool types and methods of operation and production, and the infinite variations in component geometry, mean that attempts to confine all tailoring of user requirements into the postprocessor stage are limited by the relative position of the interface within the whole system. The adopted standard interface is known as the APT III interface, and its position within an overall system is as important as its detailed format if the viability of hundreds of existing postprocessors is to be maintained. Although certain processors may claim to maintain the APT III interface it is sometimes in format only and the shifting of the interface within these systems has created new breeds of postprocessors and a breakdown of standardization.

We have for some years recognized a need, within our policy of one general processor for all NC processing, for some way to bridge this barrier—to extend the influence of the user not only into the postprocessor, but also into the processor while still maintaining a strict recognition of the interface between the two. With current processors this can only be done by creating non-standard user-orientated processors which are contrary to the overall philosophy of a generalized system.

There will soon be available a new generation of generalized NC processing systems which would seem to satisfy this need, and also fit ideally into our ideas for NC processing[6]. This type of system will give the ability for a user to work in a area behind the interface—on the processor side—from the part-programming level. The opportunities for user development within this area, creating flexibility of use unknown in NC processing up to the present time, seem to us to be unlimited. Use could vary from the field of lathe programming to the inclusion of full technological data banks and automatic machining features.

In our opinion the method of advance in numerical control processor development in the past has been limiting.

Complex programs have been created at great expense which admirably satisfy the problems of a limited number of users and go part way to helping a large number of users, but which do very little to aid most users either because their problems are different or because the cost of supporting development is too high.

The new type of processor will enable each user to develop his own techniques, dictated by his requirements and by available finance. Users will not be committed financially and organizationally to one expensive philosophy over which they have limited control. Because the user developments will take place on the processor side of the interface they will of necessity be standard, and by encouraging collaboration and exchange of user techniques as they develop we should see gradual progress on a wide front instead of the present 'big strides' on a rather limited front. In addition all development should be in the hands of the users—those who are face to face with day to day metal cutting problems. It is not unknown for 'big stride' progress to lose sight of and be out of line with the developing needs.

The above is a simplified view of systems to come,

they are not here today. They require full development, and for operation will require modern computers with advanced storage techniques.

However the facility for tailoring the system to meet user requirements through the part-program lends itself to use by many varied users through one computer, and no doubt bureau use will be encouraged.

We look forward with interest and expectation to a time when systems offering this type of part-programming facility become current.

REFERENCES

1. Blacknest Centre, Brimpton, Nr Reading, Berkshire.
2. A. R. KIMBER (1969). M.Sc. Thesis, Cranfield Institute of Technology.
3. (1968). *Harwell Engineering Review*, 7, No. 7.
4. Autopol—IBM.
   Exapt II—Exapt Verein, Aachen.
   2C—NELAPT.
5. T. C. HARRIS. *Analysis of Contouring on a Point-to-point Machine Tool*, General Electric, New York.
6. IBM New APT.

DISCUSSION

Q. Sartorio, D. E. A. Italy. In the paper, mention is made of problems met in inspection. I would like to know more about these problems and how you intend to overcome them in the future.

A. The type of work associated with these problems is the inspection of very accurate free form curves on special purpose inspection equipment. Where other people may manufacture to tolerances of ·001 in we are often an order tighter with tolerances in tenths. (·0001 in). For inspection purposes we require the ideal theoretical data to an order better than this, i.e. ·00001 in. Using APT to provide the interpolation and stylus offset calculations at these very fine tolerances has highlighted the fact that the system is not designed and tuned to work for this class of processing.

The solution to the problem, which is basically a systems problem, should be to have the processing retuned to cater for the higher orders of accuracy. We would hope that it is a problem which will affect a very few people.

Q. W. H. P. Leslie, NEL. Mr. Kimber claims that it is better to keep the APT, or other, N.C. processor development frozen and that development to suit each particular user's requirements should be confined to the post processor. This claim is based on the value of standardizing the APT 3 CLDATA which

is, he claims, upset by adding new features into the processor.

The opposite appears to be the case. Each user feature specially handed by the post processor requires that corresponding information be passed via new types of CLDATA to the post processor. On the other hand, if the feature can be dealt with by the processor it is likely that only standard information about the consequent tool moves needs to be transmitted to the post processor and this can be handled by the standard CLDATA.

A. On the contrary one of the main arguments in the paper is that software is living and that developments of the processor are very important. The paper is differentiating between general facilities and use-oriented facilities. The first should, in the long term, be processor facilities developed through a control organization.

In the short term, and for use orientated requirements, some user development is essential. We would agree that new CLDATA is very undesirable and we have never found the need to create new types of APT III CLDATA. Our post processors are compatible with, and are used on, many standard APT systems including NELAPT.

Q. W. A. Carter, Plessey N.C. Ltd. You stated that your preference to keep the main geometric section of the processor unchanged and not to introduce special features to that section and that users should introduce their own special features mainly in the post processors.

While this may be true for large and competent organizations such as A.W.R.E. there are many more smaller users who do not have the staff or funds to carry out this work economically. Therefore, such features must be introduced in general terms on the processors, leaving only the minimum of special software to be written specially for the customers requirements.

Could you comment please.

A. I would agree in principle, but there is always a point where general features finish, and user oriented features begin—it is these that we have put in our postprocessor. I do not necessarily advocate postprocessor development as a long term method of solution of this problem but at AWRE we had problems to solve and resisted development of a non-standard processor. As indicated in the paper we accept there are difficulties and undesirable factors in the development of sophisticated postprocessors, and this is why we look forward to systems where user oriented facilities may be developed with ease in an area which interacts directly with the standard generalized processor.

MACHINE TOOL ACCURACY AND
WORKPIECE INSPECTION

THERMAL BEHAVIOUR OF NC MACHINE TOOLS

by

G. SPUR and P. DE HAAS*

SUMMARY

Thermal influences can deteriorate the working accuracy of NC machines. The generated heat influences the measuring system and the machine. The loss of accuracy of a screw spindle with stepping motor used as feed drive in a turret lathe has been tested. The thermal influences on the accuracy of a jigmill equipped with a linear measuring system are discussed. The different reasons for the deformations and improvement methods are described. The heat transfer between hot chips and the machine bed and the insulation by a layer of cool chips have been investigated.

INTRODUCTION

The working accuracy of a machine tool is determined by the accuracy of the relative motion between workpiece and tool. If one or several parts of the machine being responsible for the relative motion are deformed the working accuracy deteriorates. This chain of deformations consists of the machine tool itself, of the workpiece and the tool. For NC machines the deviations caused by the measuring system must be added. These deformations can be referred to forces and temperatures. The head line 'Thermal behaviour' means all heat-effects biasing the accuracy, no matter whether they result from internal or external heat sources.

THE LOSS OF ACCURACY OF NC MACHINES

The relative motion between tool and workpiece of NC machines is determined by the control unit, where the stored information is transduced into positions. The error of the working accuracy is composed of different components (Fig. 1). One component of the deviation of the set position is the error at positioning time. During the cutting process the thermal deformations of the machine will increase the error. The accuracy of positioning is deteriorated by systematic and by random errors, composed of the control system errors and the measurement system errors. In the following the thermal influence on the errors after positioning and on the measurement errors will be discussed.

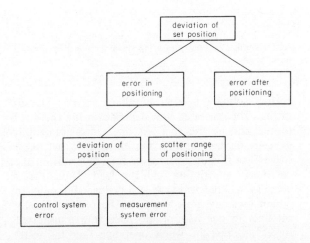

Figure 1. Thermal influences on the errors of NC machines.

THERMAL BEHAVIOUR OF A SCREW SPINDLE AS FEED DRIVE

For reasons of economy the feed drive often will be used at the same time as measuring system. The investigated feed drive of a turret lathe consists of a screw spindle with screwed roller bodies between spindle and nut. The electro-hydraulic stepping motor determines by the number of angular steps the position of the nut. This measuring system works indirectly in the form of an open loop system. The positioning is only accurate if the condition of a constant screw pitch is fulfilled. A deformation of the spindle by the increase of temperature will cause a

* Lehrstuhl und Institut für Werkzeugmaschinen und Fertigungstechnik, Technische Universität Berlin

positioning error. Because of the two functions of the spindle, this error is an error of the measuring system[1,2].

The heat sources consist of the frictional heat of the bearings, of the rolling friction of the roller bodies against the screw and the nut, and the friction of the slideways (Fig. 2). The heat of chips and

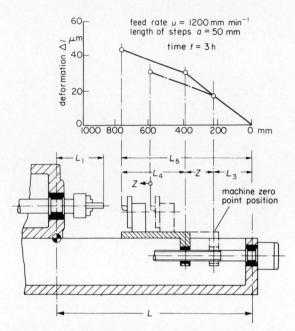

Figure 2. Deformation state of a screw spindle used as both feed drive and indirect measuring system, caused by thermal influences.

cutting solution is not considered in this investigation. The generated heat depends on the frictional torque and on the rpm. At linear displacement it depends on the force of friction and on the linear speed. The frictional torque or force depends on the coefficient of friction and the load. Nuts, thrust bearings, and slideways are used preloaded in order to obtain great dynamic stiffness. The preload usually amounts to one half of the maximum transmitted forces[3]. During the whole operating time the roller bodies are submitted to high pressure due to their preload, although the proportion of cutting time involving high forces is small.

The turret lathe with the feed drive is shown in Fig. 2. The relative position between tool and workpiece is given by the lengths L_1 to L_5 and their errors. The thermal error of the feed drive (length L_5) has been tested by measuring the repeating accuracy of positioning. The length of the steps between two positions was constant. The length from the thrust bearing to the tool was measured with a laser interferometer, fixed upon the machine frame. The mirror to reflect the laser beam was clamped instead of the tool.

Figure 2 shows the deformation state measured 3 hours after the start. The length of steps between each position was $a = 50$ mm, the feed rate $u = 1200$ mm min^{-1}. The total error is composed of the deformation of the screw spindle and of the slide L_4. The length responsible for the spindle deformation is

the sum of the distance L_3 between thrust bearing and machine zero point position and the variable length Z. The increase in temperature of the spindle was about 5·5°C, greater than the temperature of the slide. Therefore the deformation line is buckled. If the feed rate increases, the deformation increases too. Because of the simultaneous influence of the feed rate on the coefficient of friction a linear dependance has not been found.

For metrological reasons the feed drive of the longitudinal slide has been investigated although the error in this direction is not so important. The effect on the working accuracy can be neglected if the raw material is machined at one chucking and if machining time is short in relation to the deformation rate. The deformation mechanism is the same for the deformation of the cross feed drive which is very important because the diameter error is doubled.

THERMAL BEHAVIOUR OF A LINEAR MEASURING SYSTEM

The thermal error, caused by deformations of a direct measuring system, has been investigated in a horizontal boring and milling machine. The vertical motion of the headstock is controlled by a linear measuring system. The optical measuring head, clamped at the headstock, moves along the measure rule which is composed of several parts mounted at the machine frame (Fig. 3). If these parts are deformed or displaced, the set position will not be reached because of a thermal measuring system error. The influence of the mode of mounting the parts of the measure rule on the machine frame has been investigated.

error at positioning time t : $e_1(t) + e_2(a,t) + e_3(t) + e_5(t)$
additional error at time Δt

after positioning time t : $e_4(a,t,\Delta t) + e_5(t,\Delta t)$

a : nominal position

e : error

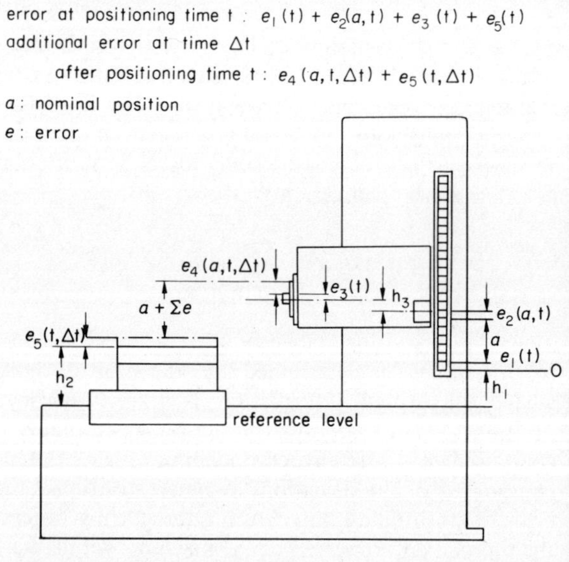

Figure 3. The chain of deformations for the vertical axis of a jigmill.

In order to determine any given length the measure rule is subdivided into standardized parts, which are mounted on brackets on the machine frame (Fig. 4). For adjustment the parts are shifted along a guide pin clamped in the brackets and fixed by pressing screws. When the machine frame warms up during operation it will be deformed, and the distance between the brackets will increase. Because of the small heat flow through the brackets into the parts of the rule, and their large surfaces for heat removal, the increase in temperature and the elongation of the parts are smaller than those of the frame. This difference of elongation causes a tensile force in the chain of the rule parts. The force is greater than the clamping force of the pressing screws, so that the parts slide and the joints between neighbouring parts move apart.

The thermal error of the measuring system is shown in Fig. 4. The error is not a continuous function of the length, there being steps at the joints. The upper curve shows the thermal error for the described mode of mounting A. In order to obtain a uniform distribution of gaps, the rule parts are clamped only at the left hand end. If both sides are clamped one cannot forecast which side will slide, and the positions of the gaps are accidental. The increase in temperature of the machine frame was 10°C, and the error greater than one hundred micrometres in one metre length.

In order to improve the thermal stability of the measuring system, the mode of mounting was changed as shown in Fig. 4. In the first place the

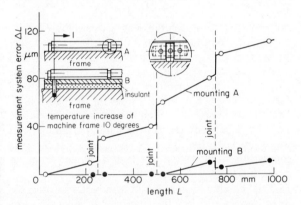

Figure 4. Thermal influence on the measure rule of a linear measuring system.

brackets were fixed to a separate plate and not to the machine. In the second place this plate was fixed to the machine through an intermediate layer of insulation in order to diminish the heat flow from the machine to the measure rule. Furthermore the relative position of the plate and the machine was fixed by one locking pin so that the starting point and the direction of the elongation could be determined. The lower curve in Fig. 4 shows the thermal error of the improved mounting B which amounts to only 10 micrometres, less than 10 per cent of the initial error.

THE CHAIN OF DEFORMATIONS

The error which diminishes the working accuracy is composed of a number of particular deformations. The deformation components in the vertical direction of a horizontal boring and milling machine are discussed below (Fig. 3). Deformations of the tool and workpiece are not considered. It is assumed that the machine bed remains undeformed, and the upper

edge represents the reference level. The real relative position between worktable and milling spindle is the difference in their distances to the reference level changed by thermal influences. The distance to the spindle is composed of the height h_1 from the reference level to the starting point of the measure rule, of the nominal position a and the height h_3 between the optical measuring head and the spindle. The addition of the corresponding errors gives the error at the moment when the working position is reached and the headstock is clamped at the machine frame. From this moment the position is independent of the measuring system and the increase in the error is determined by the deformation of the assembly of headstock and machine frame.

THE TIME CURVE OF THERMAL DEFORMATIONS

The electrical energy fed into the machine tool is completely transformed into heat energy. The electrical and mechanical losses can amount to 50 per cent of the initial power, so that only the rest can be used for cutting. If the places of heat transformation are known, it is possible to evaluate the heat flow into the different sections. The principal places are the motor, the transmission train and clutches, the bearing of the main spindle, and the hydraulic system. In the cutting zone the rest of the mechanical power is transformed into heat, flowing into the workpiece, and the tool, chips, and cutting solution. The influence of hot chips falling onto the worktable is discussed later on.

If a heat flow runs into a part the increase of temperature with time depends on the heat storage capacity and the heat emission ability. The energy balance for an ideal volume element is the basis of the equations obtained for the curves of temperature, shown in Fig. 5. It is an exponential function, characterized by the values of the time constant and the steady-state temperature or deformation. This function is approximately in accordance with the measured curves for the machine tools[4]. The time constant is determined by the ability of the part to

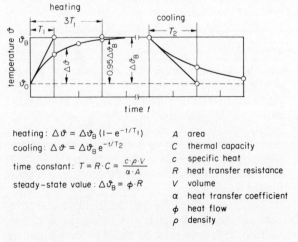

heating: $\Delta \vartheta = \Delta \vartheta_B (1 - e^{-t/T_1})$

cooling: $\Delta \vartheta = \Delta \vartheta_B e^{-t/T_2}$

time constant: $T = R \cdot C = \dfrac{c \cdot \rho \cdot V}{\alpha \cdot A}$

steady-state value: $\Delta \vartheta_B = \phi \cdot R$

A	area
C	thermal capacity
c	specific heat
R	heat transfer resistance
V	volume
α	heat transfer coefficient
ϕ	heat flow
ρ	density

Figure 5. Time curve of increase and decrease of temperature for an ideal thermal volume element.

store and remove heat. If the conditions of removal of heat change, the time constant changes too. This happens when machine tools cool down and there is no longer a ventilation effect and oil circuit, so that the time constant of cooling is greater than the time constant of heating. Whereas the time constant is independent of the quantity of heat flow, the steady-state temperature depends on it, and on the heat emission ability.

In order to improve thermal stability there are several possibilities. The difference between steady-state and ambient temperature can be reduced by cutting down heat generation. Some heat sources (the hydraulic system and the gear set) can be placed outside the machine so that the heat generated does not influence the parts responsible for the working accuracy. But there are some heat sources remaining (for instance the heat generation of the spindle bearing) which are parts of the deformation chain. The effect of these unavoidable heat sources can be diminished by increasing the efficiency, for instance by improving the frictional conditions.

The second way to diminish the inertia temperature is the cooling. That means all operations with or without auxiliary energy to increase the removal of heat.

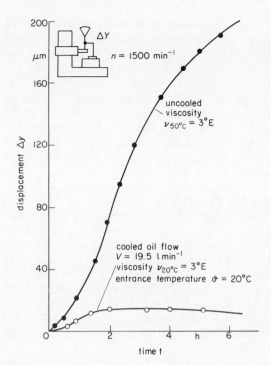

Figure 6. Cooling influence on the vertical displacement of a jigmill.

IMPROVEMENT OF THERMAL BEHAVIOUR BY A COOLING SYSTEM

The jigmill described above (Fig. 3) has been equipped with a cooling system[5]. The main heat source is the headstock including the heat generation of the spindle bearing, the gear set, and the direct-current motor. The heat will be carried away by conduction to the machine frame and by convection and radiation to the ambient air. The removal of heat has been improved by splashing the inside of the headstock walls with oil. In this way the inside of the walls can also be used as a heat emission surface. By splashing the oil the flow velocity was great and the film thickness small, giving a good heat transfer coefficient. The lubrication oil was used also for cooling, but the delivery was one-tenth of the delivery of the coolant pump. The hot oil flowing away was cooled to the ambient air temperature (20°C) and pumped back into the headstock. By measuring the flow and the difference in temperature it was possible to calculate the heat flow removed from the machine.

The influence of the cooling system on the vertical deformations of the jigmill is shown in Fig. 6. The error in the distance between worktable and spindle in the clamped headstock has been measured as a function of the idle running time. The deformation curve without cooling has a similar shape to the exponential function shown in Fig. 5. This proves that the deformation is caused mainly by linear elongation of the machine frame. The steady-state of the deformation has not been reached after a running time of 6 hours.

This curve enables the user to calculate the movement of the spindle after the headstock has been clamped. This error must be added to the positioning error (Fig. 3). The cooling system allows the removal of heat before it reaches the column. Thus thermal deformation of the column is almost negligible.

The time constant is also influenced by the cooling process. After 2 hours a practical steady-state can already be observed. Furthermore the cooling process can reduce the generation of heat. The frictional losses depend among other things on the viscosity of the oil. If there is a cooling system, the temperature and the dependent viscosity remain constant. In this way a low viscosity oil can be chosen, so that the heat generation during the first few hours can be diminished.

This cooling system has the advantage of great effectiveness and low installation cost in relation to the machine installation cost. However there are some disadvantages. The removal of heat is only possible if the temperature of the headstock is greater than that of the coolant. Thus there will always remain a thermal deformation if the temperature of the inlet cooling oil is equal to the temperature of the ambient air. In order to avoid the increase in temperature of the headstock despite heat generation the coolant temperature must be less than the ambient temperature. For this task a controlling system is needed which ensures that the cooling power never exceeds the heat generation. If the cooling power is too great, the temperature decreases below the ambient temperature, so that the machine shrinks. Furthermore the air humidity will condense on the outside surfaces which subsequently may rust. Investigations into improved cooling systems are being made at present at the Institute of Machine Tools of the Technical University of Berlin.

COOLING OF DISC CLUTCHES

Disc clutches and disc brakes are also important heat sources in the headstocks of machine tools. If it is not possible to place the gear train outside the machine, the temperature level can be reduced by cooling. In electromagnetic disc clutches without slip rings the heat is generated by friction losses in the bearing, between the discs, and by current losses in the magnet coil (Fig. 7). If the outside of the disc clutch is splashed with cooling oil, the depth of penetration into the discs caused by the centrifugal forces is very small. By using a hollow shaft it is possible to feed the cooling oil into the interior of the disc clutch. But

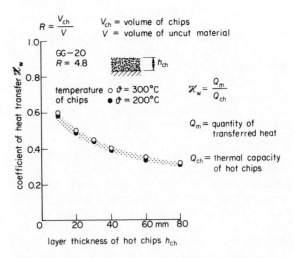

Figure 8. The heat transfer from hot chips to the worktable of a machine tool.

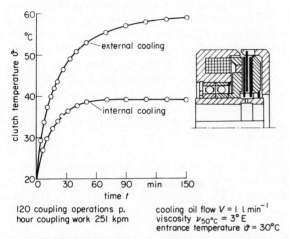

Figure 7. Comparison of external and internal cooling of a disc clutch (coupling torque 25 kp m).

even this method of cooling does not allow complete wetting of the discs. The oil film bursts caused by the increase of flow area and centrifugal forces as a function of increasing radius. In Fig. 7 a comparison between external and internal cooling of a disc clutch without slip rings is shown[5]. The difference between the inertia temperature of the clutch and the ambient temperature could be reduced by up to 50 per cent.

HEAT TRANSFER OF HOT CHIPS

The thermal deformations discussed up to now were caused by internal heat sources. The cutting process causes a heat flow into the workpiece, the tool, the chips and into the cutting solution. The influence of dry hot chips falling down on the worktable has been investigated[4].

The ratio of the stored heat quantity of the hot chips to the quantity of heat transferred to the worktable is the coefficient of heat transfer $\mathscr{H}_w$ (Fig. 8). An artificially heated mass of chips was poured onto a steel test plate. The temperature of the chips can be calculated by calorimetric investigations into the stored heat in the chips after cutting. The temperature depends on the cutting parameters.The chip temperatures for milled steel (St-50) range from 350 to 450°C, and for milled grey casting (GG-20) from 200 to 250°C. In Fig. 8 the coefficient of heat transfer is shown for grey casting chips as a function of the layer thickness. Only 30 to 60 per cent of the quantity of heat stored in the hot chips is transferred to the worktable. It was found out that the heat transfer is not much affected by the temperature of the chips, by their form, or by the manner of pouring (all at once or continuously). The form of the chips

can be characterized by the relation between the volumes before and after cutting.

There are two possibilities for reducing the influence of the chips on the thermal behaviour of the machine. Firstly the chips can be removed, e.g. by sucking away dry chips. Without doubt this is the best way, but auxiliary energy is needed. Secondly parts of the machine, for instance the worktable, can be insulated in order to reduce the heat transfer coefficient. This point of view led to the idea of investigating the insulating effect of layers of cool chips. Figure 9 shows the influence of the layer

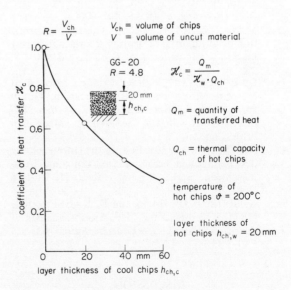

Figure 9. The influence of a layer of cool chips on the thermal insulation.

thickness of cool chips on the coefficient of heat transfer $\mathscr{H}_c$. This coefficient represents the ratio of the heat transferred from the hot chips through the layer of cool chips to the machine, to the heat transferred without cool chips. Because of the good insulation effect a layer of chips should be removed only when necessary.

CONCLUSIONS

The working accuracy of the cutting process can be adversely affected by thermal influences causing deformations of the machine tool, of the tool and of the workpiece. The error resulting from the thermal deformation of the NC machine is composed of an error at the time of positioning, and an additional error developing after the tool has been clamped in position.

The positioning error can be attributed partly to a thermal error in the measuring system. The thermal influence on an indirect and on a direct measuring system are shown in the examples of a screw spindle with a stepping motor used as feed drive in a turret lathe, and of a linear measuring system for the vertical axis of a jigmill.

The different reasons for the deviations and the influence of the conditions of heat emission on the temporal curve have been analysed. The effects of improving heat emission by cooling on temperatures and deformations are discussed, for the cooling system of a headstock of a jigmill and for cooling disc clutches.

The heat generated by the cutting process influences the deformations if the hot chips fall onto the machine. The rate of heat transfer has been investigated for different layer thicknesses. It was found that a layer of cool chips gives a good insulating effect.

REFERENCES

1. G. SPUR and P. DE HAAS (1972). Thermisches Verhalten eines Transrollspindel-Vorschubantriebes, *Zeitschrift für wirtschaftliche Fertigung,* **67** (8), S. 403-406.
2. Y. Sh. ZBARSKII (1965). Thermal deformation of differential measuring screws in jigboring machines, *Machines and Tooling,* **36** (10), S. 28.
3. G. ENGEL (1972). Special design features of NC contouring machines, Discussion, Query from Wasiuktiewicz, *Proceedings of the 12th International MTDR-Conference,* Macmillan Press, London.
4. H. FISCHER (1970). Beitrag zur Untersuchung des thermischen Verhaltens von Bohr- und Fräsmaschinen, Berlin, Techn. Universtät, Dr.-Ing. Diss.
5. G. SPUR, P. DE HAAS and H. KAEBERNICK (1971). Verbesserung der Arbeitsgenauigkeit durch Kühlung, *Werkstattstechnik,* **61**, S. 530–535.

DISCUSSION

Q. Dr Schultschik, E.T.H., Zurich. Considering Fig. 2 it seems that there is a connection of two problems. One of them is the thermal behaviour, the other is the spindle stiffness. The influence of thermal errors could be minimized in moving the machine zero point into the spindle reference point on the left in your figure. On the other hand the spindle stiffness would be reduced in that new machine zero point.

Consequently the step motor should drive the spindle from the left end.

I can imagine that the realization would bring a lot of troubles in designing, but I think this aspect could be a considerable possibility.

A. The chain of deformations is valid for deformations caused by temperatures and also by forces; however, there is an important difference. The deformation between tool and workpiece caused by forces is totalled from the sum of the individual deformations of the parts, the thermal deformations of the parts may compensate each other.

Considering the turret lathe shown in Fig. 2, there is no question that the rigidity would be improved if the stepping motor were placed at the left hand side below the main spindle. With regard to the improvement of the thermal behaviour the time curve of the single deformations must be known at different operating modes as a function of the running time. The superposition of these deformations gives the deformation between tool and workpiece, which has to be minimized.

Q. Ir. Wasiukiewicz, DIXI II, Switzerland. Always with the same great interest we follow the work in progress of TU Berlin on the thermal behaviour of the machine tools. Being involved personally for many years in similar work (applied to the jig boring machines), I would like to add some comments on the paper from the industrial point of view.

From the very beginning DIXI have equipped machines with the optical measuring system based on the graduated steel rule. There were three reasons for doing this:

(1) It allows the absolute measurement of the moving parts both during their displacement and at rest with very high precision.

(2) The rule is not exposed both to the wear and the self-generating heat due to friction as it is in the case of the ballscrew (Fig. 2).

(3) The thermal expansion coefficient of steel is nearly the same as for cast iron. Therefore it is possible to reduce the influence of the ambient temperature on the working accuracy. Our philosophy is to have a good thermal conductivity between the precision rule and the rest of the machine structure. The temperature of the rule has to follow the machine temperature, which in turn has to follow the ambient temperature. At the same time it is necessary to stop all heat penetration from the heat sources into the machine structure, but not exclusively to insulate the rule as is proposed in Fig. 4. It can also be dangerous because of the heating of the rule by the lamp.

Now some words about cooling systems which can be classified as follows:

(1) Oil lubrication. Cooling of oil by ambient air.

(2) Lubrication with the oil of constant temperature (the constant cooling power).

(3) Lubrication with the variable cooling power to ensure the thermal stability of the machine. Only this system can catch up with variable heat generation due to the variable machine work programme. Taking the appropriate measures there is no danger because of 'shrinking' of the machine (System DIXI).

Finally, it has to be said that the power equilibrium of both heating sources and cooler is only valid in the steady state conditions (which occurs seldom for these types of machines, with very short programme sentences).

I hope it would be possible for me to develop this subject further during the next MTDR Conference in Birmingham.

A. Measure rules. Let us consider two examples:
(1) You are using a big machine tool with very small internal heat sources and without ambient air temperature control. Assuming equal thermal expansion coefficients of the machine tool, of the measure rule and of the workpiece, the accuracy depends on their average temperatures. The temperature in the different parts does not immediately follow the changing air temperature. The propagation of temperature changes from the outside of a part into the interior depends on the thermal diffusivity and the depth of penetration. Consequently, the machine tool, the workpiece and the rules may be deformed differently because of their different average temperatures although they have the same thermal expansion coefficient.
(2) You are using a machine tool of which the internal heat sources are not negligible—the regular case. The warming up of the machine tool is quite different from the warming up of the workpiece. Consequently, the error of a rule in good thermal contact with the machine body is much greater than the error of an insulated rule as shown in Fig. 4.

The best way will be to have temperature controlled measure rules and to compensate the dilation of the workpiece.

Cooling systems. The term 'cooling systems' means cooling where one or several parameters of the process are controlled. A classification has to be adjusted to the controlled variable.

The cooling power of an oil flow through a machine tool is

$$\dot{Q} = \dot{V} \cdot \rho \cdot c(\vartheta_{out} - \vartheta_{in})$$

where $\dot{V}$ = oil flow, ρ = density, c = specific heat, ϑ = temperature.
Assuming a constant oil flow, the increase in temperature depends on the heat generation of the machine tool and the heat transfer conditions. Consequently, the cooling power is not constant if the lubrication oil temperature is constant (controlled variable = ϑ_{in}) as assumed by Mr Wasiukiewicz.

The described cooling system of the jigmill had a controlled inlet oil temperature. The set-point was adjusted to 20°C, equal to the temperature controlled ambient air. In this way there was no danger of shrinking, even if the machine tool had no heat generation. At the present time investigations are being made about cooling systems of which the controlled variable is the outlet oil temperature which improves the cooling effect by reducing the temperature level. Furthermore cooling systems having a certain thermal deformation as controlled variable are being investigated.

ANALYSIS OF THERMAL DEFORMATION OF MACHINE TOOL STRUCTURE AND ITS APPLICATION

by

T. SATA, Y. TAKEUCHI, N. SATO and N. OKUBO*

SUMMARY

A general computer program for the design and analysis of machine tool structures by the use of the finite element method has been completed. The system consists of three parts; the input program for the automatic mesh generation in idealizing the structure, the program for the analysis of the static, dynamic and thermal deformations of the machine tool and an output program to draw the configuration of these deformations and to print out the numerical data requested. A sample computation is carried out on the thermal deformation of a jig grinding machine in the steady and non-steady state as well as in the on–off operation of the machine.

The program is also used to determine the optimal operation mode of the machine or the optimal control mode of the artificial source if installed in order to minimize the transient period at starting or changing the machine operation. The method to identify the intensity of heat sources from temperature measurement has been developed by modifying the equation of heat conduction based on the finite element method.

INTRODUCTION

The thermal deformation of a machine tool is an important factor influencing the shape and the accuracy of the workpieces and hence must be properly understood in order to pursue the measures to preserve the dimensional accuracy in production. Possible measures to prevent thermal deformation of machine tools would not only be advantageous to remove the heat sources from the structure, or to take off by cooling any heat which is generated, but also to design the structure with high thermal rigidity or to compensate thermal deformation easily. It is thus desirable to examine the thermal behaviour of the structure in the design stage, before any prototype of the structure is built.

Analysis of thermal deformation was first made by considering the simplified thermal expansion and bending of the structure based on temperature measurements[1]. The finite element method has been recently introduced to give a better idealization of the structure, not only for static[3,4] and dynamic[5] analysis but also for the analysis of thermal behaviour[6].

It has been the aim of our research to develop a general computer program for design and analysis of the machine tool structure by using the finite element method and to apply the program to explain and to control the behaviour of the thermal deformation.

The general computer program developed consists of three parts. The input program for the automatic mesh generation in idealizing the structure, the program for analysis of the static, dynamic and thermal deformation of machine tool and the output program to draw the configuration of these deformation and to print out the numerical data requested. A sample computation is carried out on the thermal deformation of a jig grinding machine in the steady and non-steady state as well as in the on–off operation of the machine. The program is also used to determine the optimal operation mode of the machine or the optimal control mode of the artificial source if installed, in order to minimize the transient period at starting or changing the machine operation. Application of the finite element method leads to a method to identify the intensity of heat sources from a limited number of temperature measurement.

THE BASIC CONCEPT OF COMPUTER AIDED DESIGN SYSTEM

The finite element method enables one to analyse the static, dynamic and thermal behaviours of the machine tool structure by idealizing the structure with an assembly of finite elements. Thus the structure of a machine tool can be determined from the drawings in the design stage, to meet various requirements for

---

* Department of Precision Engineering, Faculty of Engineering, Tokyo University, 7-3-1 Hongo, Bunkyoku, Tokyo, Japan

the working characteristics of the structure. According to this idea a computer aided design system for machine tool structures is under development at our laboratory.

The flow chart of the system is shown in Fig. 1.

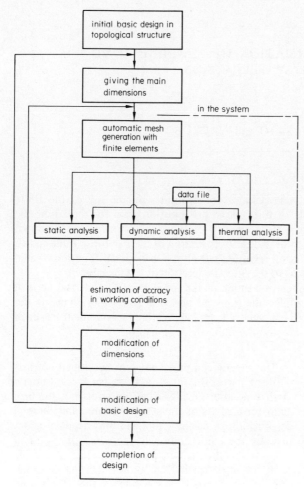

Figure 1. The computer aided design system of machine tool structure.

The designer first considers the basic topological structure and then gives the main dimensions of the structure. The geometrical configuration of the structure and the height, width, wall thickness and other dimensions of each component in the structure thus given should be described in the input to the computer program, in which the finite element mesh is automatically generated. It is necessary to describe the material of the structure, the properties of the foundations, interfaces between components, the external static and dynamic forces, and heat sources. The data necessary for the computation are prepared in the system by referring to the data file. With these preparations, the static, dynamic and thermal deformations of the structure are computed and evaluated from the standpoint of various requirements of design, and these results are printed or plotted for the designer. When the results do not meet the requirement of design, the designer can modify the dimensions of the structure. The procedure is repeated until all requirements are satisfied.

THE ANALYSIS OF THERMAL DEFORMATION

For the analysis of thermal behaviour, the temperature distribution and thermal deformation of the machine tool structure in both steady and non-steady state are calculated in the system. In this method the elements are idealized automatically, the equilibrium equations of heat conduction for the whole system are formulated and then the temperature distribution is obtained by solving these equilibrium equations under boundary conditions given. The thermal deformation of the structure can be obtained by calculating the thermal strain under this temperature distribution and replacing the strain by equivalent nodal forces. Triangular, tetrahedral or rectangular elements are used in the system to idealized the machine tool structure.

The temperature distribution in the steady state is obtained by minimizing the equation (1) derived from heat conduction and Euler's theorem.

$$X = \int\int \left[\frac{1}{2}\left\{kx\left(\frac{\partial\phi}{\partial x}\right)^2 + ky\left(\frac{\partial\phi}{\partial y}\right)^2\right\} - Q\phi\right] dx\, dy$$

$$+ \int\int q\phi\, dx\, dy + \int\int (\tfrac{1}{2}\alpha\phi^2 - \alpha\phi_\infty\phi)\, dx\, dy, \quad (1)$$

in which kx and ky are anisotropic conductivity coefficients in x and y directions respectively, the function Q is the rate of heat generation, ϕ is the temperature, q is the heat flux per unit of surface, α is heat transfer rate and ϕ_∞ is ambient temperature.

The temperature in an element, $\{\phi\}$, is expressed as a function of the nodal temperature, $\{\phi\}^e$ by

$$\{\phi\} = [N]\{\phi\}^e. \quad (2)$$

By differentiating equation (1) and substituting equation (2), the equilibrium equation in matrix form assembled for the over-all system is expressed as follows:

$$\{Q\} = [H]\{\phi\}. \quad (3)$$

where $[H]$ is called the over-all thermal stiffness matrix.

The temperature distribution in a non-steady state is obtained in the same way as in steady state

$$[P]\left\{\frac{\partial\phi}{\partial t}\right\} = -\{Q\} - [H]\{\phi\} \quad (4)$$

in which $[P]$ is the matrix assembled from the components of $[p]$ for an element

$$[p] = \int\int c[N]^T[N]\, dx\, dy \quad (5)$$

where c is the thermal capacity per unit volume.

We can solve equation (3) by unit partitioning or the conjugate gradient method and equation (4) by conjugate gradient or Runge Kutta's method.

Once the temperature distribution is determined, the thermal deformation due to the equivalent force is calculated by solving the following equation

$$\{F\} = [K]\{X\} \quad (6)$$

where $[K]$ is stiffness matrix, $\{X\}$ is the nodal displacement and $\{F\}$ is the equivalent force assembled by a force, $\{f\}^e$, for an element

$$\{f\}^e = v[B]^T[D]\{\epsilon_0\} \quad (7)$$

in which v is the volume of the element, $\{\epsilon_0\}$ is thermal strain and $[B]$ and $[D]$ represent the relation between strain and nodal displacement and between stress and strain respectively.

The thermal behaviour of a machining centre with a single, thin-walled column is first analysed with the computer program. Figure 2(a) shows the geometrical configuration of the structure whose height is 2 metres and its idealization with finite elements. The main heat sources in the structure exist at the front and rear bearings of the spindle and oil sump in the head stock. The computed temperature rise with time at the location a, b, c and d in Fig. 2(a) are shown with the solid lines in Fig. 2(a) which can be compared with the experimental data in the figure. One can see good agreement between computed and experimental results. The thermal deformation of the structure in the steady state is shown in Fig. 2(b),

where the displacements at the front bearing of the spindle in the head stock are -2, -58 and $9\ \mu$ in x, y and z direction respectively and the difference of these values from measured ones are found to be less than 20 per cent. Comparison of the deformation of the machining centre having a single column with that having double column with the heat source of the same intensity shows that the deformation of the latter structure is much smaller and simpler than that of the former due to the less torsional deformation.

On the shop floor the machine tool is operated by repeated switching on and off. The temperature rise in this case can be easily calculated by applying the analysis to the non-steady state. Figure 3 shows the

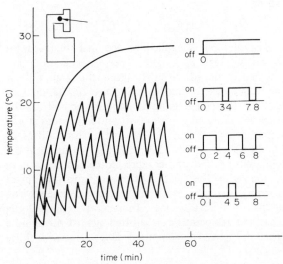

Figure 3. The temperature change of jig grinding machine in on–off operation.

temperature change of a jig grinding machine during on–off operation of different intervals. It is seen that the longer the off time, the lower the ultimate temperature level. Analysis of the thermal deformation of the machine shows that the head stock displaces mainly in a forward direction and that the displacement is proportional to the temperature rise of the head stock.

ESTIMATION OF THE INTENSITY OF HEAT SOURCES

By applying the matrix partitioning method, a method has been developed to estimate easily and accurately the intensity of heat sources from the temperature measurements at a few points concerned with such heat sources in machine tools. This method can be applied to both the steady and non-steady states.

The relation between the intensity of heat sources, $\{Q\}$, and temperature, $\{\phi\}$, equation (3), can be reassembled and subdivided into two parts, for nodes concerned with heat sources and the others as follows,

$$\left\{\frac{Q_1}{Q_2}\right\} = \left[\begin{array}{c|c} H_1 & H_2 \\ \hline H_3 & H_4 \end{array}\right]\left\{\frac{\phi_1}{\phi_2}\right\} \qquad (8)$$

in which Q_1 and ϕ_1 are concerned with heat sources, and Q_2 and ϕ_2 are not concerned with them.

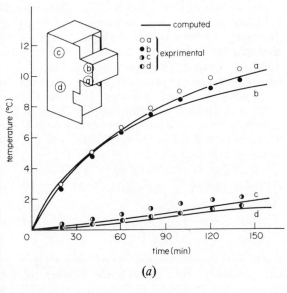

(a)

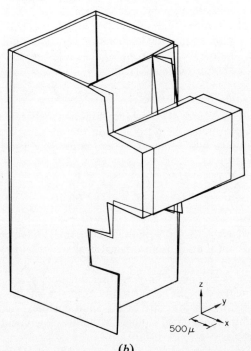

(b)

Figure 2. The temperature change and thermal deformation of machining centre.

Rewriting equation (8), we have

$$\{Q_1\} = [H_1]\{\phi_1\} + [H_2]\{\phi_2\} \qquad (9)$$

$$\{Q_2\} = [H_3]\{\phi_1\} + [H_4]\{\phi_2\} \qquad (10)$$

Equating $\{Q_2\}$ to 0 from the assumption, we obtain

$$\{\phi_2\} = -[H_4]^{-1}[H_3]\{\phi_1\} \qquad (11)$$

From this equation it is seen that if the temperature, $\{\phi_1\}$, of the nodes concerned with heat sources are measured, the temperature, $\{\phi_2\}$, of the other nodes can be calculated and/or the temperature distribution of the over-all structure can be easily evaluated. Substituting $\{\phi_2\}$ into equation (9), the intensity of heat sources, $\{Q_1\}$, can be obtained as follows,

$$\{Q_1\} = [H_1]\{\phi_1\} - [H_2][H_4]^{-1}[H_3]\{\phi_1\} \qquad (12)$$

The method is applied to the machining centre shown in Fig. 2(a). The three heat sources are assumed in the idealized structure, the front wall of the head stock for the bearing of the spindle, the rear wall for the rear bearing and the lower wall for the lubricant oil resting in the bottom of the head stock. Experiments are carried out with three different spindle speeds of 280, 560 and 1670 rev/min and the temperature of 8 points representing the nodes concerned with the heat sources are measured after reaching a steady state for each experiment. The intensity of heat sources, thus computed from the measured temperature, is plotted against the spindle speed in Fig. 4. The computed values agree with those

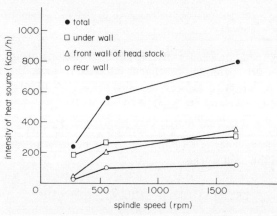

Figure 4. Intensity of heat source in machining centre.

estimated from the traditional method. The temperature distribution, which is computed by using equation (11) from temperature measurement at the nodes of heat sources, is seen to agree very well with that measured.

A similar analysis is made in a non-steady state to estimate the intensity of heat sources. The latter method is found to be also effective and more desirable than the former because the temperature measurements are carried out in a rather shorter period without waiting to attain the steady state temperature distribution.

The method proposed here is most valuable for predicting the quantitative value of various heat sources involved in machine tools such as the gear box, bearings, motors, and lubricant oil whose heat

generation has been considered difficult to be estimated.

THE CONTROL OF THERMAL DEFORMATION

To obtain high accuracy from the machine the change in the thermal deformation between tool and workpiece must be as small as possible during operation. Large changes of thermal deformation take place in the warming-up period after starting the operation of the machine, while the deformation becomes less with increasing operating time and approaches the steady state value. Thus a higher rate of change of the deformation is desirable at the initial stage to reduce the warming-up time of the machine. To attain this goal two methods have been developed which utilize either additional or existing heat sources.

The former method consists of installing an artificial heat source at a suitable location and supplying the energy to the structure during a certain period after starting an operation. In the latter method the spindle is operated at higher rotation speeds than required for a certain period to give the excess energy to the structure in the initial stage of operation.

In controlling the additional heat source, the temperatures at several representative points are aimed to be brought to the same value as that of the steady state at each point.

The objective function J to be minimized is expressed as follows

$$J = \int_0^\infty \sum_{i=1}^n (\theta_i(t) - \theta_i)^2 \, dt \qquad (13)$$

where $\theta_i(t)$ is the temperature of the ith representative point at time, t, n is the number of the representative points and θ_i is the temperature of the ith representative point in steady state.

The temperature of the ith point, $\theta_i(t)$, is expressed by adding the temperature contribution from the original heat sources, $\theta_i^s(t)$, and from the additional heat source, $\theta_i^c(t)$, as

$$\theta_i(t) = \theta_i^s(t) + \theta_i^c(t). \qquad (14)$$

By the use of weighting function between the temperature at the ith point and the intensity of the additional heat source, Q^c, the temperature, $\theta_i^c(t)$, is described as

$$\theta_i^c(t) = \int_0^t g_i^c(t - t)Q^c(t) \, dt \qquad (15)$$

in which g_i^c is the weighting function. The temperature contribution from original heat sources, $\theta_i^s(t)$, is obtained by the method for the non-steady state above mentioned. The weighting function, g_i^c, can be estimated by the temperature response of the ith point for unit intensity of the additional heat source in unit time interval obtained either by experiment or analysis.

Substituting equations (14) and (15) into (13) we get the optimal intensity of the additional heat source, Q_0, to minimize the objective function, J. In the case of supplying a certain heat energy, Q, during a certain period to the additional heat source, the optimal intensity, Q_0, is obtained by differentiating equation (13) and equating it to zero thus:

$$\mathrm{d}J/\mathrm{d}Q = 0. \qquad (16)$$

The solution of the equation is given by

$$Q = \frac{\sum_{i=1}^{n} \int_0^\infty (\theta_i - \theta_i^s(t)) \left(\int_0^t g_i^c(t - \tau)\,\mathrm{d}\tau \right) \mathrm{d}t}{\sum_{i=1}^{n} \int_0^\infty \left(\int_0^t g_i^c(t - \tau)\,\mathrm{d}\tau \right)^2 \mathrm{d}t} \qquad (17)$$

In controlling the spindle speed the spindle is operated at higher rotation speed for a certain initial period and then switched to the required rotation speed. Denoting the temperature change of the structure at the initial operation by $\theta(t)$ and that after stopping the spindle rotation speed at time $t = t'$ by $\theta'(t)$, one can get the expression for $\theta'(t)$

$$\theta'(t) = \theta(t) - \theta(t - t'). \qquad (18)$$

When the spindle rotational speed is switched to the required value from a higher speed for an initial period, T, the temperature change of the ith point, $\theta^i(t)$, is written by using equation (18) as follows

$$\theta^i(t) = \theta_1^i(t) \qquad \text{for } 0 \leqq t < T$$
$$\theta^i(t) = \theta_1^i(t) - \theta_1^i(t - T) + \theta_2^i(t - T) \quad \text{for } T \leqq t, (19)$$

in which $\theta_1^i(t)$ is the temperature change in the initial stage by higher spindle rotation and $\theta_2^i(t)$ is that for the required spindle rotation.

Thus the objective function to be minimized is expressed as

$$J = \sum_{i=1}^{n} \left[\int_0^T (\theta_1^i(t) - \theta_i)^2 \, \mathrm{d}t + \int_T^\infty (\theta_1^i(t) - \theta_1^i(t - T) \right.$$
$$\left. + \theta_2^i(t - T) - \theta_i)^2 \, \mathrm{d}t \right]. \qquad (20)$$

As equation (20) is the function of variable T, we can determine the optimal time of switching from higher speed to required speed by differentiating equation (20),

$$\mathrm{d}J/\mathrm{d}T = 0. \qquad (21)$$

Both controlling methods are applied to a jig grinding machine. An additional heat source is installed close to the spindle head of the structure, where five reference points are adopted as shown in Fig. 5.

In the former method the optimal intensity of the additional heat source is computed on the jig grinding machine against the heating time by using equation (17). The result is shown in Fig. 5. When the capacity of the additional heat source is limited to 62·5 W, the optimal heating time is seen to be about 65 min from Fig. 5.

Figure 6 shows how fast the temperature measured at the reference points approaches each final value by

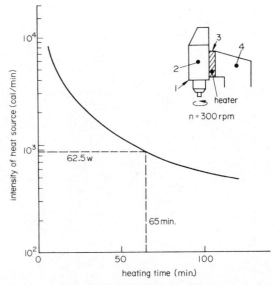

Figure 5. The optimal intensity of additional heat source against heating time.

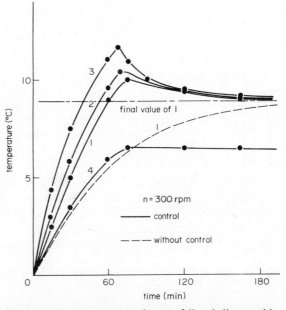

Figure 6. The temperature change of jig grinding machine with and without control of the additional heat source.

the controlled additional heat source, compared with the broken line in this figure for the reference (1) without control.

The thermal deformation of the spindle in the y direction with control of the additional heat source and without control is shown in Fig. 7 with a solid and a broken line respectively. It is seen from this figure that displacement with control approaches the final value earlier than that without control.

In controlling the spindle speed, the higher speed is taken as 300 rev/min and the required one 150 rev/min. From equation (21) the optimal switching time from higher speed to required one is calculated to be 52 min. Experimental result under these conditions is shown in Fig. 8. It is seen that the spindle displacement with control approaches the final value much earlier than without control.

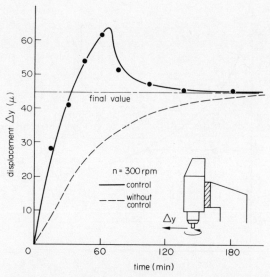

Figure 7. The thermal displacement of jig grinding machine with and without control of the additional heat source.

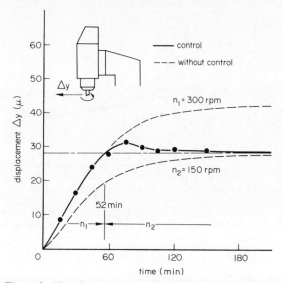

Figure 8. The thermal displacement of jig grinding machine with and without control of the spindle speed.

The objective function in controlling either using an additional heat source or changing spindle speed is taken to minimize the temperature difference from that in the steady state. The procedure is effective in as much as the pattern of the displacement with control is similar to that without control. The displacement of the jig grinding machine with control of

additional heat source shows overshooting from the final value. This would be due to the slight difference of the deformation pattern in both controlled and uncontrolled cases.

CONCLUSION

(1) A computer aided design system for machine tools by using the finite element method has been developed. The system consists of routines for automatic mesh generation, analysis of static, dynamic and thermal behaviour of machine tools and evaluation of these behaviours for design requirements.

(2) The routines for thermal analysis are used to compute the change of temperature and deformation of the structure in the steady, non-steady and in the on–off operation of the machine.

(3) A new method is proposed to estimate the strength of heat sources involved in machine tool from the temperature measurement at the limited number of points.

(4) The change of temperature and deformation of machine tools can be reduced by controlling either the additional heat source or the spindle speed. The optimal condition of these controlling methods is determined and proved experimentally.

ACKNOWLEDGMENT

The authors wish to express their appreciation to Nippon Univac Co., Makino Milling Co. and Seiko Seiki Co. for their assistance.

REFERENCES

1. Y. YOSHIDA (1964). Thermal deformation of a knee-type vertical milling machine, *Proc. of the 5th International M.T.D.R. Conference.*

2. A. COWLEY and S. HINDUJA (1970). The finite element method for machine tool structural analysis, *Annals of the C.I.R.P.*, **18.**

3. A. COWLEY (1971). Co-operative work in CAD in the CIRP, C.I.R.P. Ma Group, January.

4. T. SATA and N. TAKASHIMA (1971). Dynamic analysis of machine tool structures by the finite element method, Preprint C.I.R.P.

5. T. SATA, N. SATO, Y. TAKEUCHI and N. TAKASHIMA (1972). Analysis of thermal deformation of machine tool by finite element method, C.I.R.P.

6. G. SPUR, H. FISCHER and P. DEHAAS (1972). Thermisches Verhalten von Werkzeugmaschinen, C.I.R.P.

SOME ASPECTS OF THE ACCURACY EVALUATION OF MACHINE TOOLS

by

C. P. HEMINGRAY*

SUMMARY

The major elements of errors in numerically controlled machine tools are discussed, and the likely order of magnitude of errors is considered, some typical results being used in illustration. The problems associated with a realistic assessment of overall machine errors are also briefly mentioned, but the lack of proven techniques means that the major effort is directed towards the prediction of likely workpiece errors from the known individual error values. It is hoped that this will introduce some much needed realism into the way machine accuracy is described.

INTRODUCTION

The whole purpose of accuracy evaluation of machine tools is to obtain a 'quality assurance' of finished component accuracy. For NC machines this is particularly important, for the very advantage of such a machine—its consistency—means that, if one component is out of tolerance, they all will be, and the possible means of rectification are few. However, the very nature of the subject is so complex that few, if any, producers of such machines are prepared to give any realistic figure for the overall behaviour of general purpose machines. Of course many will quote (and generally achieve) specific levels of accuracy for given production components or test workpieces, but the problems of a guarantee of some form of total accuracy has not really been attempted. The existing specifications and evaluation techniques are, even in the most advanced state, almost entirely restricted to two dimensions. Consequently, especially for machines designed for producing essentially three-dimensional components, considerable confusion exists as to the way in which 'accuracy' is stated. This paper will attempt to clarify this area, not by theoretical considerations, but by descriptions of some of the individual error elements, with typical results being presented. It is hoped this will introduce some much needed realism into the description of the accuracy of numerically controlled machine tools, and to assist in this aim the likely overall accuracy is derived from the known values for the various elements of error.

This paper will concern itself with the largest class of machine tools, for which the accuracy evaluation presents the greatest problem, i.e. numerically controlled machines designed for three-dimensional work (e.g. machining centres, milling machines, boring machines, etc.). Two-dimensional work includes turning and many drilling machines, and is subject to much fewer errors and fewer problems in their evaluation. Three classes of accuracy of machine are common, despite some individual variations high precision machines, typically with alignment tolerances of 3 μm/300 mm, medium precision (10 μm/300 mm), and low precision (20 μm/300 mm). The results cited and estimates of accuracy derived are for medium precision machines, but available evidence leads to the conclusion that the accuracy of the classes can readily be scaled in the given ratio of 3:10:20.

POSITIONING ERRORS

Most manufacturers of numerically controlled machines quote values for the positioning errors of their machines, usually either ±0·01 mm or ±0·02 mm. Only rarely is the fact mentioned that this is related to distance travelled, but most machines have traverses of one metre or less. However, the same values of positioning errors are often quoted for machines with traverses of several metres and a value of ±0·01 mm over traverses of 10 m is not unknown. In this connection, environmental factors put a practical limit to attainable accuracy, especially over long distances. It is generally recognized that temperature stability of closer than ±1°C is impossible to guarantee under shop conditions, even over distances

* Staveley Engineering and Research Centre, Worcester

K

of 1 m. As ±1°C represents about ±0·01 mm/m, any claim for a higher precision than this must be unlikely without extreme precautions to ensure temperature stability. This is reinforced by the available test data. No conventional machine tool so far examined has bettered ±0·01 mm/m, although a significant number were little worse than this.

Repeatability is also frequently given, but normally only for uni-directional positioning. Two values of repeatability are common, i.e. ±5 μm and ±12·5 μm, and these are realistic values for uni-directional repeatability (i.e. for positions approached from one side only). However, the presence of dead zone, which is typically of a magnitude between 5 μm and 30 μm (±3 to ±15) does increase the realistic positioning repeatability to between ±6 μm and ±25 μm, with most systems being round about ±10 μm. There is some indication, both from test data and specifications, that results towards the lower end of the ranges quoted are likely for machines equipped with a linear measuring system, and towards the

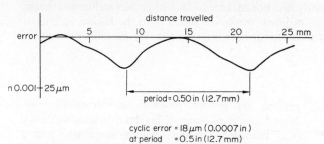

cyclic error = 18 μm (0.0007 in)
at period = 0.5 in (12.7 mm)
no other cyclic errors of any period

Figure 1. Example of cyclic error.

higher range with resolvers, although the systems do overlap.

Another very important parameter is that of cyclic error. This normally repeats at the pitch of interpolation of the system, i.e. once per revolution of the leadscrew and/or resolver for a leadscrew measuring system, once per revolution of the pinion for a rack system, and at the pitch of the grating or inductosyn for a linear system. These can vary up to ±25 μm, and are not normally much less than ±3 μm, with linear systems being, if well set up, of around this value and others being typically between this value and ±9 μm. A typical cyclic error of this magnitude is shown above (Fig. 1); the relatively smooth variation is quite normal.

GEOMETRIC ERRORS AND POSITIONING ERRORS

The previous section assumed implicitly that the machine geometric errors did not affect the positioning accuracy of the machine tool. This is, in general, only true for cases where the measurements are taken very close to the axis of the linear transducer. The further away the actual line of interest is from the transducer the more the machine geometrical errors can and do influence the observed errors. Indeed, for machines of over 300 mm capacity the geometrical errors of typical machines contribute more than the inherent errors of most positioning systems. This is illustrated below (Fig. 2), where the positioning

errors shown are of a magnitude approaching 100 μm at around 1000 mm away from the spindle nose, but only half this near the nose. This was caused by geometrical errors within the normal tolerance for this class of machine. By such simple tests as these,

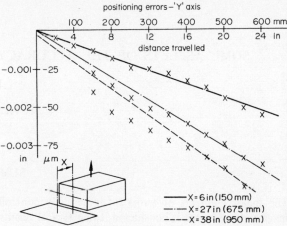

all cases:–dead zone =0.00025 in (6 μm) uni. –dir'n rep Y=0.00012 in (3 μm) bi.–dir'n rep Y=0.00025 in (6 μm)

Figure 2. Effect of displacement of measuring axis on observed errors.

even without calculations on the effects of geometrical errors, it is apparent that no single measurement of errors along an axis is of much use. This is especially true for the three-dimensional case, as is demonstrated below.

THERMAL ERRORS

Undoubtedly the largest errors observable on machine tools are those caused by the effects of temperature on the structure. 'Normal' errors for small machines are 0·05 mm (measuring total movement from cold to stabilization at top speed), but for large machines under the worst conditions (i.e. sustained running at top speed) 1 mm movement is not uncommon. Fortunately, this is normally in the 'insensitive' direction (along the spindle axis), although some high-speed machines have been known to move by amounts approaching this in transverse directions. These movements are an inherent feature of the combination of heat sources with large physical dimensions. Even though, on well-designed machines, the maximum temperature rise under 'normal' conditions is usually held to 5–10°C, which is barely perceptible to the touch, this is, assuming a dimension of 1 m, 0·05–0·1 mm movement. By their very nature, these movements are long term, which is the reason for the relative unimportance of these effects on manual machines, as re-datuming under such conditions is routine and frequent. However, particularly on large machines, the production system and programming often does not lend itself to frequent drift correction, and hence over a working day substantial errors have occasionally been observed. Warming up does reduce these effects, and many complex machines are normally run continuously or have heaters and thermostats to ensure thermal stability. Otherwise, three hours running or more is often necessary, as can be seen in Fig. 3. This shows movements typical of those observed under 'normal' operating conditions; in this case, the machine was run at 315 rev/min for 5 min on/5 min off, and displacements in the two 'sensitive' directions were only 12 μm or so at most, but the machine quite obviously took three hours to warm up. At higher speeds it took even longer, at top speed six hours or more being necessary.

In general, for most machines, at speeds typical of those normally used (as demonstrated by surveys conducted by MTIRA), thermal displacements perpendicular to the spindle are roughly the same as those produced by most other geometrical errors. Parallel to the spindle they are somewhat higher, and the spindle attitude only rarely stays within the nominal alignment limits at half top speed. The larger

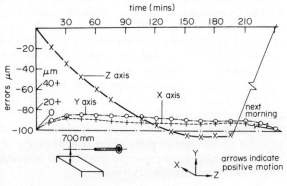

Figure 3. Typical thermal errors.

the machine, in general, the more severe the problem, and indeed the general workshop temperature instability of, ±2°C or so frequently causes significant movements.

MEASUREMENT OF COMBINED ERRORS

It has become increasingly apparent that for complex numerically controlled machines, control of errors as is normally practised does not guarantee the accuracy of workpiece production, at least to limits as close as customers frequently specify. For the common case of a specified workpiece, the machine can be set up to produce the required tolerances, although frequently with some difficulty. But control of alignments and 'positioning accuracy' are not by themselves sufficient in many cases. Indeed, although numerical control offers many advantages in terms of tooling and productivity, very few machines so equipped can better the best levels of accuracy attained using jigs and fixtures.

Three approaches to the problem are being developed. The first is to measure the various error components of the machine and correct the tape of the component by use of a suitable sub-routine in the post-processor. The second is to measure the actual errors of the machine throughout the workpiece zone. And the third is to use some form of external reference framework to ensure machine accuracy independent of the machine structure. However the first approach does not correct for machine instability, and the third is likely to be viable only for very large machines.

The second method, that of determining the required component accuracy assurance, is a metrological problem that has not yet been adequately solved for workpiece dimensions above 400 mm or so. As has already been stressed, it is insufficient to determine positioning accuracy along one line on each axis, and the only feasible technique requires measurements along many of the extreme lines in the workpiece zone within which quality assurance is required. For the two-dimensional case, the techniques are not difficult. Even manufacturing and measuring a workpiece is quite feasible, and is the only commercially practical way of determining the behaviour of a contouring machine.

A typical set of data for a simple 9-hole test is shown in Fig. 4, with an accuracy of within ±13

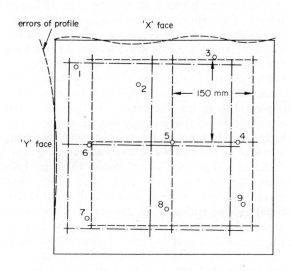

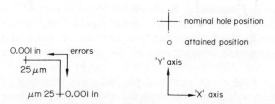

Figure 4. 'Normal' workpiece errors in a single plane.

μm/300 mm demonstrated. The results are shown with both the unmodified datum axes, using hole '5' as zero, and also datum axes modified to represent the functional effects of hole centre distance errors. This is not the place to discuss the problems of selecting datum location and axes, but for a three-dimensional investigation the origin and attitude of these axes is critical to obtain realistic assessment of machine errors, especially for fairly high precision machines.

Investigating the errors of a basically cuboid workpiece zone does present substantial problems. Including thermal effects but excluding dynamic control errors, there are twenty-six components of error for even the simplest machine capable of machining a three-dimensional workpiece. Prediction of typical likely errors will be given in the next section, but the measurement techniques are not yet available to give enough data for confirming the assessments. Errors have been calculated in practice from the individual error components, and compared with tests measuring various combinations of these errors. The agreement was within ±7 μm over a 600 mm cube with total errors of ±50 μm, excluding thermal effects, but the process is slow, laborious, and requires computer processing. Some form of optical technique shows the most promise, but further work is clearly needed on the methods.

LIKELY MACHINE ACCURACIES

Despite superficial differences, the large majority of machines are built to levels of alignment accuracy which are broadly similar, and indeed have changed little over the last eighty years. Hence it is possible to generalize on probable machine accuracy on the basis of available knowledge of these limits, even though measurements are not yet available to confirm these in detail. The problem is amenable to statistical analysis, for a study of a large number of alignment test results have revealed an equal probability for the measured value to be anywhere within the tolerance band, as shown below for 115 such tests (Fig. 5). Combining the errors applicable to a cuboid workpiece, using geometrical analysis and statistical techniques, leads to the discovery that alignment errors alone lead to location accuracy four times worse than each individual error. Hence, assuming alignment tolerance of 15 μm for a 500 mm side working zone, errors of ±60 μm can be anticipated, to 95 per cent limits (i.e. at most, one machine in twenty will exceed this limit in going across the diagonal of the zone). Adding arithmetically the previously disclosed values of ±15 μm for positioning accuracy, ±10 μm for repeatability, and ±15 μm for thermal instability, the worst case became around ±120 μm across the diagonal of the zone. All the evidence would be that for any machine a figure of ten to twenty times the quoted values of uni-directional repeatability would give a realistic confident level to those in search of a component assurance value, for a working zone of around 300 mm–600 mm size. Larger machines do not have errors increased *pro rata*, but doubling of these values can be expected over a 2 m working zone. In addition, for those few contouring machines where the actual contour is critical, allowance will

have to be made for dynamic control errors, which do not, however, normally exceed 50 μm under most conditions.

It cannot be too strongly emphasized that these error values are for fairly large three-dimensional movements, and relate to operations where re-datuming is not performed, and also the errors are measured diagonally. For a 300 mm sided workpiece machined in one plane and with adequate programming, both statistical analysis and the evidence of workpieces machined and subsequently measured, indicates that errors of ±20 μm can be hoped for, using more normal single-axis error measurements. These errors can be expected to be divided by 3 for high-precision machines (i.e. jig-boring) and multiplied by 2 for low precision (drilling) machines.

It is considered that the values given above are realistic 'worst case' estimates, the few machines for which comprehensive data is available exhibiting errors only about one-half of that estimated. Only one machine in twenty can be expected to be close to the limits, and it could be argued that routine measurement of actual errors would render feasible a 'quality assurance value' of substantially less than the worst estimate given above. This is because any machine approaching the limit would do so because of a whole series of machine errors all adding, instead of some errors counteracting others, and discovery of this would make it very easy to modify the machine alignments to reduce the observed errors.

CONCLUSION

Some of the more important elements of the accuracy of numerically controlled machines have been considered, and realistic values of both the individual and overall errors presented. It is hoped that some realism has been introduced into the area of attainable component accuracy, although the values cited can at the moment be only estimates, albeit well founded on a considerable amount of test data.

That there are other facets of machine tool behaviour which can cause component errors should not be forgotten. Workpiece weight deformations have given trouble in the past, although modern machines are much less prone to trouble in this respect than their predecessors. The control system itself can introduce dynamic errors which have only very cursorily been mentioned, and pre-set tooling can introduce many pitfalls for the unwary. It is considered, however, that the combination of geometric, positioning, and thermal errors presents by far the worst problems of measurement and control, and it is therefore on those aspects that this paper has concentrated.

ACKNOWLEDGEMENT

The author thanks the General Manager of the Staveley Engineering & Research Centre for permission to publish this paper. He also would wish to acknowledge the assistance of Dr A. Cowley of UMIST, under whose supervision were derived many of the results on which this article is based.

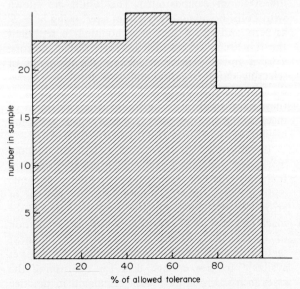

Figure 5. Distribution of alignment errors.

TESTING AND EVALUATING THERMAL DEFORMATIONS OF MACHINE TOOLS

by

J. TLUSTY* and G. F. MUTCH*

SUMMARY

On the basis of previous work in the development of metrology of NC machine tools, error functions to be measured for thermal effects are pointed out with the use of several examples of machine tools. Typical features of thermal deformations in machine tools are illustrated by means of computations of transient thermal fields and transient thermal deformations of simple typical structures. This discussion is concluded by the formulation of rules for the specification of conditions for thermal test cycles. Finally, the way of evaluating the error functions is explained and illustrated by examples.

FORMULATION OF THE TASK

In a number of previous publications we have concentrated on the metrological part of the problem of testing and evaluating the accuracy of NC machine tools [1-3]. The main conclusions of these investigations consisted in formulating:

(a) The form of the tolerance rule to be applied for evaluating the errors in a two- or three-dimensional working zone of a machine tool. This tolerance rule is of the form

$$|\delta_{m_2} - \delta_{m_1}| \leqslant A_{mn} + K_{mn}|n_2 - n_1| \qquad (1)$$

where $m = x, y, z$, $n = x, y, z$, and x, y, z are the individual coordinate motions as related to the individual coordinate axes of the working zone. It was shown that, in order to cover the whole working zone, the tolerance rule (1) has to be applied to errors obtained at extreme off-sets in the working zone as they affect the basic 'positioning' and 'straightness' errors due to the angular deviations connected with the motions of the individual moving bodies (pitch, roll, yaw). The rule (1) refers to 'positioning' for $m = n$ and to 'straightness' for $m \neq n$. It was shown that each individual error function $\delta_m(n)$ as evaluated along any line in the working zone may be taken separately for checking against the corresponding tolerance rule.

The tolerance rule (1) when expressed graphically has the form of a simplified NMTBA 'tolerance template'. For the latter, see ref. 4.

The error functions $\delta_m(n)$ may either be directly measured as 'positioning' and 'straightness' errors along the lines at the corresponding extreme offsets, or they may be derived from a combination of such translative-type error measurements with angular error measurements.

(b) Included in the statement (a) is the important notion that it is sufficient if the error functions $\delta_m(n)$ taken at extreme offsets satisfy the tolerance requirement in order to be assured that this requirement is satisfied throughout the whole working zone. This results from the recognition of the fact that differences between error functions $\delta_m(n)$ taken at different offsets are caused by angular deviations in the motions and are, therefore, proportional to the size of the offset. Any error taken along a line between extreme offsets is a linear interpolation of errors taken at extreme offsets.

In an analogous way, as far as weight effects are concerned, such as those of the load on the table of a horizontal boring machine, it was suggested to carry out those error measurements which are affected, both at no load and at extreme load. If both the results satisfy the tolerance it will be satisfied for any load between the two extremes. This is based on the assumption that differences in the error functions caused by the load increase all the way through increasing the load.

For the thermal effects it was suggested to carry out those measurements which are likely to be affected repeatedly through a range of thermal states of the machine such as result from a typical thermal working cycle of the machine. This recommendation is, obviously, far less exact than those related to space and weight effects. It is the purpose of this paper to elaborate this recommendation into a more precise specification. This concerns which measurements to

* McMaster University, Hamilton, Ontario, Canada

repeat, and also through which range of thermal states. For the first part of this question illustrations will be given for several examples while, simultaneously, the problem of the minimum number of measurements in the working zone will also be briefly discussed. For the second part concerning the typical thermal cycle the recommendation will be based on the analysis of the character of thermal deformations of machine tools.

MEASUREMENTS TO BE TAKEN

Here, taken means 'to be considered for evaluation' and it can also mean 'to be carried out' depending on whether we carry out translative error measurements only or combine them with measurements of angular deviations. In ref. 3 an exact way was given of determining all error functions necessary to express the accuracy of a complete working zone. In many instances less than the whole working zone is used and, consequently, the number of error functions required decreases.

As an example, a type of machining centre is considered in Fig. 1. The tools used are all of

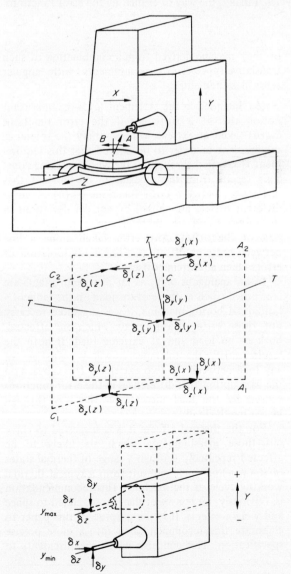

Figure 1. Errors to be measured on a type of machining centre.

practically the same length and, therefore, as regards X and Y motions machining is practically carried out in a single (X, Y) plane instead of in a three-dimensional working zone. This does not exclude the workpiece being three-dimensional with machined surfaces through the whole range of the maximum possible cube. This, first, is reduced to one-half by machining in that half-cube only which faces the headstock and subsequently bringing the opposite half into action by indexing the table. For the rest, the motions Z, A, B are used to bring the machined surface into the above mentioned single X, Y plane. In this respect motions Z, A, B are used for setting only and not for working. An exception is the boring operation which is carried out by the Z working motion while X and Y are the setting ones.

It is obvious that a sensible approach to this case is to consider the errors associated with the individual motions in two separate groups: X, Y as regards work in one (X, Y) plane only which passes through the tip of the tool of average length, and Z, A, B as regards straightness and depth of boring holes.

Let us first specify the necessary error functions for the latter group: there are (1) straightness errors $\delta_x(z)$, $\delta_y(z)$ along the line C_1 and additionally, for the effect of the roll $\epsilon_z(z)$, also $\delta_x(z)$ along the line C_2; (2) positioning error $\delta_z(z)$ along C_1 and, for the effect of pitch $\epsilon_x(z)$, also along C_2; (3) a tolerance is set for the setting motions A and B and it can practically be limited to $\delta_a(a)$ and $\delta_b(b)$ errors of angular positioning.

For the X, Y group the necessary measurements are limited for the X motion to straightness errors $\delta_y(x)$ and $\delta_z(x)$ along A_1 and $\delta_z(x)$ along A_2 (for the effect of the roll $\epsilon_x(x)$) and to positioning error $\delta_x(x)$ along A_1 and along A_2 (for the effect of the pitch $\epsilon_z(x)$) and for the Y motion to straightness errors $\delta_x(y)$, $\delta_z(y)$ and positioning error $\delta_y(y)$ along the line C.

As another example a bed-type milling machine is diagrammatically shown in Fig. 2. The rather small number of measurements indicated is based on the assumption of using this machine for flat workpieces only, i.e. for machining low above the table surface and of a single spindle. It can be imagined that for a truly three-dimensional working zone and, especially, for a three-spindle machine the number of the required error functions increases considerably.

As the last example the configuration of a knee-type vertical spindle machining centre is given in Fig. 3 and the necessary error functions are indicated for a case assuming the use of tools of fairly uniform length.

So far we have not considered thermal effects. We implement them now for all the three preceding cases. First, it is necessary to briefly discuss and comment on possible heat sources. This problem has been thoroughly considered in ref. 5. There are three classes of heat sources which will not be considered in our further discussions.

The first one is the surrounding air. Even if the environmental effect is often not negligible and the usual shop air conditioning systems with corresponding high local air flow rates do not solve the problem[6], it is an effect which may rather easily be

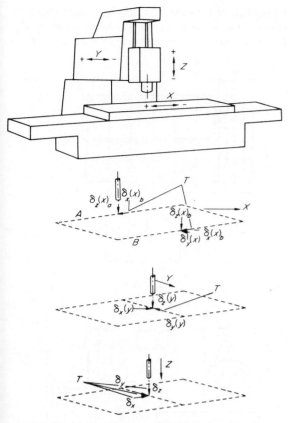

Figure 2. Errors to be measured on a bed type milling machine with limitation to flat workpieces.

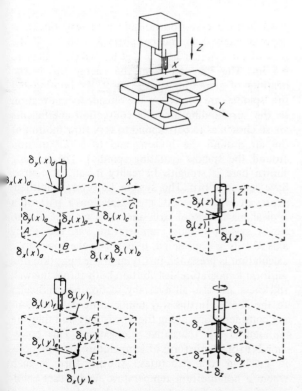

Figure 3. Errors to be measured on a small machining centre.

considered separately from and superimposed on the effects of internal heat sources. The latter constitute a more complex variety and we shall limit the discussion to them.

The second class neglected here is the heat generated in the cutting process. It is assumed that for precise workpieces rough machining is carried out first and finish machining is done separately later. The heat generated in finishing is small, and if chips are carefully separated from the machine this effect becomes negligible.

The third class omitted here is the effect of the coolant. The omission is justified by assuming that in most operations no coolant is used and where it is used and high precision considered it should be temperature controlled, and this is a special case.

All three omissions mentioned are made not for denying the omitted effects but for limiting our discussion to those many cases in which these omissions are well justifiable.

The remaining sources are those which are really internal to the no-load work of the machine. They may roughly be divided into two groups: constant heat sources like a hydraulic power supply with constant delivery pump and relief valve, or heat generated in the electric motor and blown on to the structure; and variable heat sources like, typically, the spindle bearings and the gearbox (possibly with hydraulic clutches) in the spindle drive with heat rate depending on spindle speed or friction in feed drives with heat rate depending on feed rate.

It is usually possible to estimate well which are the heat sources of this type in a given machine. Furthermore, it is usually possible to estimate well which of these sources affect which parts of the structure, and it is even possible to estimate well which of the error functions will be affected.

We now consider the above mentioned examples and try to localize the individual heat sources and their effects.

For the machining centre in Fig. 1 the case may be such that there are no significant heat sources located in the transversal (Z) bed. Heat generated by sliding in the guideways during Z motion is negligible and we shall consider one possible effect only: that of the heat generated between nut and leadscrew, especially during the rapid traverse withdrawals in boring, provided that positional feedback is derived from the leadscrew. Thus, in the group associated with the motion Z, the $\delta_z(z)$ tests only have to be repeated.

In the given situation the heat sources and effects of the Z group and of the X, Y group may be considered separately. For the X, Y group, first of all the heat generated in the spindle bearings and in the gear transmissions and clutches located in the headstock must be considered. This heat, obviously affects all error functions associated with the Y motion, $\delta_x(y)$, $\delta_y(y)$, $\delta_z(y)$, and these have to be repeated during a corresponding thermal cycle. This heat does not influence errors associated with the X motion. The $\delta_y(y)$ error is affected depending on how the positional transducer is attached to the column and how it follows the thermal deformations of the column. If X displacement is measured by an Inductosyn or another type of linear transducer practically no measurements associated with the X motion are thermally affected. If displacement is derived from leadscrew $\delta_x(x)$ has to be repeated during a cycle of motions heating the leadscrew.

Summarizing: In the case of linear transducers $\delta_x(y)$, $\delta_y(y)$, $\delta_z(y)$ measurements only have to be repeatedly measured through a heat cycle of the headstock. These are denoted T in Fig. 1. In the case of positions derived from leadscrews, $\delta_z(z)$ and $\delta_x(x)$ must also be repeated through a thermal cycle heating the leadscrew.

It is often satisfactory to replace the repeated measurements of the error functions by a single measurement of such functions complemented by repeated measurement of the corresponding errors at the ends of the corresponding travel. In this way the repeated $\delta_x(y)$, $\delta_z(y)$, $\delta_y(y)$ measurements may be replaced by repeated measurements of deviations δ_x, δ_z, δ_y in the positions y_{min} and y_{max} as shown in Fig. 1b.

For the machine Fig. 2 there are again no significant heat sources in the bed. Heat generated in the motor, the gearbox and the spindle mounting affects the ram (motion Y) and the headstock (motion Z). Consequently, measurements $\delta_x(y)$, $\delta_y(y)$, $\delta_z(y)$ are affected. Of the whole motion Z, with respect to the assumed case of flat workpieces, in practice the bottom end of travel only is of interest. Therefore, error functions associated with Z motion are reduced to deviations δ_x, δ_y, δ_z to be measured at the end of the spindle as they vary during the thermal cycle. These variations are sometimes called 'drifts'.

If on that same machine the whole Z travel is used for machining (i.e., the limitation to flat workpieces does not apply), the functions $\delta_x(z)$ and $\delta_y(z)$ and also $\delta_z(z)$ should be measured instead of the drifts alone. In that case, in a simplified way and in analogy to Fig. 1b, instead of $\delta_x(z)$ and $\delta_y(z)$, drifts at top end as well as at bottom end of the travel are measured.

In the case in which positional signal is derived from leadscrews, error functions $\delta_x(x)$, $\delta_y(y)$, $\delta_z(z)$ are again affected by heat generated by friction in the nut.

The case of the machine according to Fig. 3 is similar to the two preceding ones. Heat generated in the headstock affects errors associated with the Z motion and this effect is measured as shown in Fig. 3b by measuring drifts δ_x, δ_y, δ_z at the bottom end and also drifts δ_x, δ_y at the top end of a mandrel clamped in the spindle. In this machine feed drives are by means of hydraulic cylinders. The positional feedback is derived from special gauging system inside of these cylinders. In this way the hydraulic oil heated in the power supply affects all error functions associated with both X and Y motions.

Having illustrated which are the measurements to be repeated through the thermal cycle it is now necessary to discuss and specify the form of the thermal cycle to be used in the tests and, further on, how the results are evaluated. These two questions are dealt with in the following two sections, respectively.

THE CHARACTER OF THERMAL DEFORMATIONS IN MACHINE TOOLS: SPECIFYING THE THERMAL TEST CYCLE

The background for the discussion in this section is presented in the appendix, where calculations are given of transient thermal fields and of transient

thermal displacements of a simple typical structural system. The diagram of the system is shown in Fig. 4a. It consists of a 6 in long steel cylinder attached to

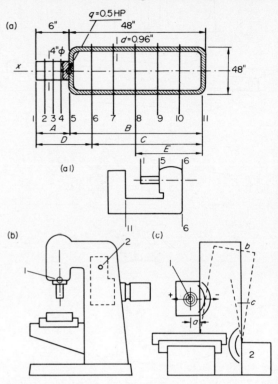

Figure 4. Diagrams for the discussion of main features of thermal deformations.

a 48 in long cast iron housing with wall thickness approximately 1 in and periphery $P = 48$ in. At the root of the cylinder heat q is generated equivalent to 0·5 hp. This may represent the energy loss in the bearings of a rotating spindle. Both the housing and the 'spindle' are cooled on the outside by convection to the surrounding air. The convection coefficients are so chosen as to correspond to very slow motion of the air around the housing and to a fast motion around the spindle (rotating spindle). The housing shown here is straight. In reality it can, of course, have various forms. This system, however, simplified, expresses with a good approximation a part of a typical machine tool structure. It is a single dimensional system because temperature variation along axis X only is considered. In part 1 of the appendix a calculation is presented justifying the assumption of a uniform temperature distribution both across the wall thickness and also all over any section perpendicular to the X axis. In this way temperature deformations are considered also along the X axis only and they are represented by translative elongation. This corresponds to one basic form of temperature deformations in machine tool structures. Another form is such where a non-uniform temperature field causes bending of a part of the structure such as the column in the case of Fig. 4c which will be discussed later on. It will also be shown later that both these fundamental modes of deformations have common features.

The case of Fig. 4a is calculated in the appendix. Here we present some of the results of the calculation. In Fig. 5 the variation with time is shown of the transient temperature field in the system during the

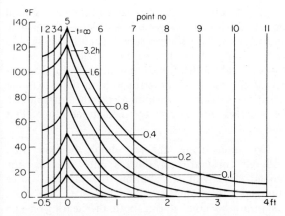

Figure 5. Transient temperature field for heating-on period for system 4a.

heating period at the beginning of which the heat source q was switched on and kept constant further on. The temperature field consists of data computed for sections 1 to 11 distributed at regular distances of 0·125 in for the spindle and 8 in for the housing. In Fig. 6 the transient temperature field in the system is

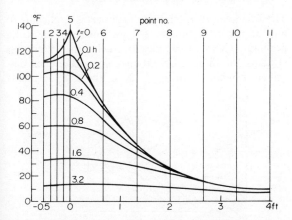

Figure 6. Transient temperature field for cooling-off period for system 4a.

shown for the cooling-off phase following a long (infinite) period of heating at the end of which a steady state was reached and then at the start of cooling-off the heat source was shut off. It can be seen how the temperature first rises and then decreases at point 5 where the heat source is. Temperature at other points follows these changes with a certain attenuation and a certain time delay which is the greater the more the point is distant from the heat source. All these variations are, at every point, composed of variations in individual 'modes' with various time constants, as they are depicted in the appendix (Fig. 17).

These temperature variations produce thermal deformations between the various points of the system. In Fig. 4a various parts of the system are selected and denoted A, B, C, D, E. The variations of these distances for both the above-mentioned heating-on and cooling-off periods are shown in Fig. 7. First, curves A and B show how the end points of the spindle and of the housing move with respect to the heating point 5. The directions of the graph are such that motion to the right from point 5 in Fig. 4a corresponds to the positive displacement in Fig. 7.

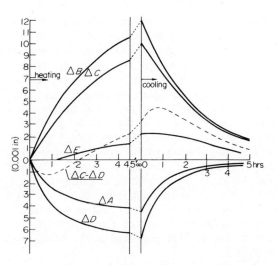

Figure 7. Transient thermal deformations in system 4a.

The curve E shows that the deformation of a section distant from the source is delayed in both the heating-on and cooling-off periods.

The sum of deformations A and B representing the variation of the distance between the end points 1 and 11 shows that this distance increases throughout the heating-on period (with a rate decreasing with time) and decreases throughout the cooling-off period.

An arrangement could be imagined in such a way that the relative displacement of two points would result from the difference of changes of two parts of the system. Such a case is diagrammatically represented in Fig. 4a1. There the distance between points 11 and 1 increases with the expansion of the distance C between points 6 and 11 and decreases with the expansion of the distance D between points 1 and 6. Because the prevailing time constants in part D are shorter than those in part C, the difference $C - D$ which expresses variation of the distance between points 1 and 11 first decreases and then again increases. This is shown by the broken line curve in Fig. 7. In the cooling-off period this distance first increases and then decreases.

The case just described illustrates some very basic characteristics of thermal deformations in machine tools. Our approach is similar to that of McClure[6], who has stated in a very general way that basic feature which in this paper is expressed in the appendix by equation 35 for a very simple part and by equation 45 for the system of Fig. 4a. It can be formulated as follows.

Assuming a particular structure with a particular configuration of cooling by heat convection, the variation of a thermal deformation along a particular path (between two particular points) is obtained as a sum of particular exponential time functions each of which is characterized by a time constant. Given a particular location of particular constant heat sources in the structure, each of the time functions is also characterized by an amplitude. The effect of several heat sources is obtained by superposition of the effects of the individual heat sources.

Our contribution to this statement is its detailed illustration for a simple typical case. McClure has also shown how the deformations of a structure can be modelled using model constants obtained from simple experiments. A very interesting contribution to the general picture of thermal deformations is presented in refs. 7 and 8 where, especially, the effect of periodic switching on and off of a heat source is treated. In the same references one practical example of a milling machine is described with measured deformation curves of all the varieties presented in Fig. 7. In such a milling machine (see Fig. 4*b*), the vertical displacement of the spindle end above the table may have the form of our curve $C - D$ (Fig. 7), both for the reasons explained in relation to Fig. 4*a*1 and as a result of opposing influences of the expansion of the headstock due to heat generated in the spindle mounting 1 and of the expansion of the column due to heat generated in the spindle drive gearbox 2.

A number of case histories showing curves similar to those in Fig. 7 have been described in ref. 5. Among them is the case represented in Fig. 4*c*. The heat generated in the headstock 1 of a horizontal boring machine will first produce the positive displacement of the spindle because the distance *a* from spindle to column grows. Later, also, the column side is heated which is close to the headstock. This produces bending of the column and reversing of the displacement of the spindle into the negative direction. The effect of the hydraulic power source 2 located at the root of the column causes a different bending of the column which is opposite to the preceding one.

We now summarize those features of the described phenomena which are relevant for our problem of specifying the thermal test cycle:

(*a*) As distinct from the effects of the offset in the working zone or the weight effects, as have been discussed above, the thermal effects depend on two parameters: the intensity of the heat source and the time. The former of these parameters may be a complex one because more than one source are acting simultaneously. However, we first consider such cases only where, apart from constant heat sources, there is only one variable source, or where there are two variable sources, the effects of which are separated (this was well illustrated in relation to Figs 1, 2, 3).

(*b*) The effect of time on any error is either such that the error increases (decreases) continuously during heating on (cooling-off) or the variation is reversed after a certain initial period; see Fig. 7.

(*c*) The effect of a heat source increases with the intensity of this source. For an intermittent source the average intensity increases with the increase of the ratio of on time to off time (provided the periods of changes are shorter than the decisive time constants of the system).

From these statements the following conclusions may be drawn:

(1) Every error-constituent of the whole accuracy formulation (see above) must be measured over a time period long enough and the whole range of its variation during that period must be established. The 'period long enough' is that during which a sufficient steady-state is reached, or the usual working period, whichever is shorter.

(2) The effect of every variable heat source should be tested with two extreme intensities of this source and with constant sources simultaneously acting. The two extreme intensities of the variable source are source on at specified maximum intensity, and source off. If the range of the error in both these extreme situations satisfies the required tolerance, it will also be satisfied during any intermediate action of that source.

(3) Summarizing: the test should be done first with all constant heat sources only, and for the second time with, in addition, the variable source running at maximum specified intensity. The latter test must be done during both the heating-on and cooling-off periods as regards the variable source. Only those measurements are made which may be affected. If one and the same error function may be affected by two sources tests must be made under extreme combinations of these sources.

EVALUATING THE THERMALLY AFFECTED ERRORS: EXAMPLES

The tolerance rule, relation (1) above, is applicable to any of the error functions—constituents of the total accuracy check. Examples of such error functions systems for three different machines have been described in relation to Figs 1, 2, 3. Figure 8 represents

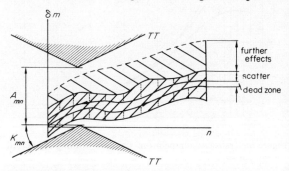

Figure 8. The general form of error functions and tolerance template.

the general form of the result of measurement of any error function $\delta_m(n)$. The error function has a certain range (shaded field) due to dead zone and scatter. This range may be further extended by weight effects or by thermal effects. The whole range of the error function must fit into the tolerance template TT.

The range of the error function as obtained during the thermal test is determined by the field between the extreme curves of the error function concerned as they have been obtained through both the time and intensity of heat source extremes of the thermal test.

There is, however, one provision which may reduce the range of the error. In the evaluation, the possibility of 'zero-shift' of the coordinate system must be admitted.

We now describe some examples. The examples we give here represent a small selection of a great number of tests and measurements we did during the last year. They were not carried out exactly according to the recommendations specified above because these have only now been formulated on the basis of our experience gained during the measurements. We shall use our recommendations for out next measurements. Nevertheless, these examples, though imperfect, help us to understand the recommendations.

The first example is given in Fig. 9 and it concerns a particular measurement carried out on the machine

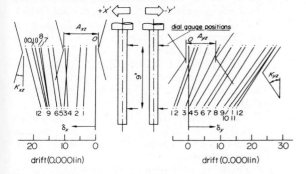

Figure 9. The $\delta_x(z)$ and $\delta_y(z)$ drifts on machine according to Fig. 2.

represented by Fig. 2. In the test we did not apply the limitation of low work-pieces only and, therefore, error functions $\delta_x(z)$, $\delta_y(z)$, $\delta_z(z)$ have been measured as well. For the thermal test the simplified approach was chosen analogous to that explained in relation to Fig. 1b. Displacements δ_x and δ_y were read periodically during the thermal cycle both at the bottom and the top ends of a mandrel in the spindle. The thermal cycle was an extreme one: all constant heat sources plus a continuous run of the spindle at 1800 rev/min. Measurements were taken every 30 min for a total period of 6 h of heating on; none, however, were taken during the cooling-off period (spindle off, all constant heat sources on), which does not comply with our present recommendation. Making an allowance for zero-shift after every 2 h the corresponding tolerance templates which include the worst 2 h (4 measuring intervals) error ranges are drawn into both the $\delta_x(z)$ and $\delta_y(z)$ error fields. Both the δ_{xz} and δ_{yz} tolerance templates are shown twice because there are two different 2 h ranges for each of the A_{xz} and K_{xz} or A_{yz} and K_{yz} constants respectively:

$$|\delta_{x2} - \delta_{x1}| \leqslant 0.001 \text{ in} + 0.0008 \text{ in}/12 \text{ in } |z_2 - z_1|$$

$$|\delta_{y2} - \delta_{y1}| \leqslant 0.001 \text{ in} + 0.0015 \text{ in}/12 \text{ in } |z_2 - z_1|.$$

The next example (Fig. 10) concerns a similarly simplified error concept for the machine depicted in

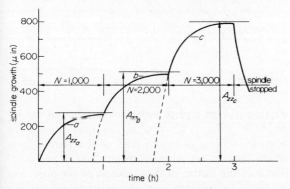

Figure 10. Spindle growth in machine according to Fig. 3.

Fig. 3, or, in other words, it is the 'spindle growth'— with respect to table—graph. In this case the whole

error function is replaced by a single value δ_z corresponding to the top position of the Z coordinate. Correspondingly, instead of a tolerance template δ_{zz} the constant A_{zz} only is applied for evaluation:

$$A_{zza} = 0.0003 \text{ in}, A_{zzb} = 0.0005 \text{ in}, A_{zzc} = 0.0008 \text{ in}.$$

The test was done by continuous spindle run at three different speeds subsequently in one hour periods and a continuous record of δ_z was taken. A small part of the cooling-off period is shown as well and it can be seen that there is no 'overshoot' like in curve $\Delta C - \Delta D$ (Fig. 7). Thus, there is no real necessity for showing the whole cooling-off period.

The next example concerns positioning error function $\delta_x(x)$ for the machine according to Fig. 3. As mentioned before, this machine uses a system of hydraulic gauges for positional signal and these, together with the feed driving hydraulic cylinder, are rather affected by the temperature of the oil. The diagram shown in Fig. 11 is the record of the

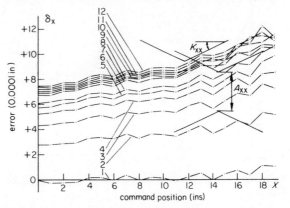

Figure 11. The positioning error in machine according to Fig. 3.

positioning error as recorded from a laser interferometer repeatedly in intervals of 15 min up to position 9 and then in 30 min intervals. It is seen that the variation of the error with position x preserves its character, but the actual distance positioned increases almost uniformly over the whole travel length. In this case it is difficult to decide whether the first two readings should not be disregarded and half-hour warm-up period permitted. In order to answer this problem correctly it would have been necessary to interrupt the cycle and see how the beginning of the variable heat source cool-off period looks. The heat cycle consisted here of continuously moving the slide with slow feed. It may be assumed that stopping the feed should not start a sudden reverse of the drift with a rate similar to the beginning of the heat-on period.

The usual fast initial reversals of thermal deformations following the shut-off of the heat source make it impractical to allow for warm-up periods concerning variable heat sources. This is different for the effect of a constant source which is probably the present case. Therefore, in evaluating, we disregard the range between curves 1 and 3 and, assuming a 1 h period for zero-shifting, we obtain the value A_{xx} for the applicable tolerance template from the distance of curves 3 to 6 and the slope K_{xx} from the curve 12:

$$|\delta_{x2} - \delta_{x1}| \leqslant 0.003 \text{ in} + 0.005 \text{ in}/12 \text{ in } |x_2 - x_1|.$$

Figure 12 is the diagram of the positioning error $\delta_x(x)$ as obtained during a very severe cycle of

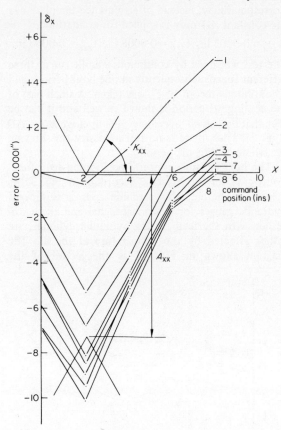

Figure 12. The positioning error in machine according to Fig. 2.

continuous rapid traverse (50 in/min) table motion of the machine in Fig. 2 for $3\frac{1}{2}$ h. The individual graphs 1 to 8 are records from the laser interferometer taken every 30 min. The heat source effective here is the heat generated between the ball nut and screw, and the effect is the elongation of the leadscrew to which the feedback synchro is attached. Again, the records should correctly have been taken after having stopped this motion. Probably the same great change in reverse would have been obtained during the first 30 min as is observed between 1 and 2. It is seen that in this case the effect consists mainly of a zero drift and could be eliminated by a zero shift. If such a zero shift is admitted every hour and no warm-up period is found justified the tolerance template as shown must be accepted:

$$|\delta_{x2} - \delta_{x1}| \leqslant 0{\cdot}0007 \text{ in} + 0{\cdot}0025 \text{ in}/12 \text{ in } |x_2 - x_1|.$$

ACKNOWLEDGMENTS

This work is a part of a project supported by the National Research Council of Canada. The help of Dr R. Judd in classical solutions of heat transfer and of Mr W. H. El Maraghy in programming of eigenvalue problems is gratefully acknowledged.

REFERENCES

1.   J. TLUSTY and F. KOENIGSBERGER (1971). New concepts of machine tool accuracy, *Annals of the C.I.R.P.*, 24.

2.   J. TLUSTY (1972). Accuracy testing of numerically controlled machine tools. *12th Int. M T.D.R. Conference*, Macmillan.

3.   J. TLUSTY (1972). Testing and evaluating accuracy of numerically controlled machine tools, S.M.E. Conference on Manufacturing Technology, Chicago, S.M.E. paper MS72-164.

4.   N.M.T.B.A. (1972). *Definitions and Evaluations of Accuracy at Repeatability for N.C. Machine Tools*, 2nd Ed. (August).

5.   J. TLUSTY and F. KOENIGSBERGER (1970). Specifications and tests of metal cutting machine tools, conference proceedings, U.M.I.S.T. (February).

6.   E. R. McCLURE (1969). Manufacturing accuracy through the control of thermal effects, D.Eng. Thesis, U.C.R.L. 50636, Livermore, California.

7.   S. SPUR, H. FISCHER and P. de HAAS (1972). Thermisches Varhalten von Werkzeugmaschinen, presented to C.I.R.P. (August), Stockholm.

8.   G. SPUR and H. FISCHER (1969). Thermal behaviour of machine tools, *10th Int. M.T.D.R. Conference*, Pergamon.

9.   V. S. ARPACI (1966). *Conduction Heat Transfer*, Addison-Wesley.

APPENDIX: CALCULATING TRANSIENT TEMPERATURE FIELDS AND THERMAL DISPLACEMENTS FOR SIMPLIFIED TYPICAL CASES OF MACHINE TOOL STRUCTURE PARTS

1. Steady state temperature field in a simple structural body with constant heat flux input and convection cooling over entire surface

The cast iron body in the situation as in Fig. 4a may rather well be considered in an approximation as a plate of thickness d (see Fig. 13a), and of width β, with a steady heat input q at the left-hand end, with no cooling on one side (the inner wall of the body) and convective air cooling on the other side. Then the problem can be easier to formulate, as in Fig. 13b, if

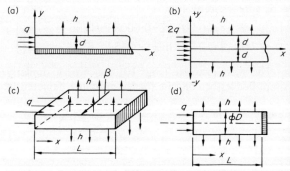

Figure 13. Diagrams for the thermal calculations of simple bodies.

we take instead a plate with thickness $2d$ and cooling on both sides, with double the heat input. The original problem is exactly half of this latter one in which there is no temperature gradient in the direction y at the centre of the thickness of the plate. For guidance in solution, see for example ref. 9. For steady state we have

$$\frac{\partial^2 T}{\partial x^2} + \frac{\partial^2 T}{\partial y^2} = 0 \qquad (1)$$

where T is the temperature difference above environmental temperature which is considered constant.

The boundary conditions are

$$-k\frac{\partial T}{\partial x}(0, y) = q'' \qquad (2)$$

$$\frac{\partial T(x,0)}{\partial y} = 0 \qquad (3)$$

$$T(\infty, y) = 0 \qquad (4)$$

$$-k\frac{\partial T(x, d)}{\partial y} = hT(x, d) \qquad (5)$$

The heat flux $q'' = \dfrac{q}{2bd}$.

Condition (3) is valid for the centre of the thickness of the wall and condition (4) is an assumption of no heat loss at the sides of width B which corresponds to the fact that this plate is an enclosed wall of a body. This condition also leads to the consequence of no variation of temperature along the width of the plate.

Separating the variables:

$$T(x, y) = X(x) Y(y) \qquad (6)$$

$$\frac{d^2 X}{dx^2} - \lambda^2 X = 0, \qquad \frac{d^2 Y}{dy^2} + \lambda^2 Y = 0 \qquad (7)$$

A general solution for Y is:

$$Y = C_1 \cos \lambda y + C_2 \sin \lambda y$$

where from condition (3) it is

$$C_2 = 0$$

and from condition (5):

$$k\lambda_n \sin \lambda_n d = h \cos \lambda_n d \qquad (8)$$

Equation (8) allows us to calculate the characteristic values λ_n for the complete function:

$$T(x_1 y) = \sum_{n=1}^{\infty} a_n e^{-\lambda_n x} \cos \lambda_n y \qquad (9)$$

where the coefficients a_n may be obtained using the condition (2) as:

$$\sum_{n=1}^{\infty} \lambda_n a_n \cos \lambda_n y = \frac{q''}{k}, \qquad (10)$$

where

$$a_n = \frac{2\dfrac{q''}{k\lambda_n} \sin \lambda_n d}{\lambda_n d + \sin \lambda_n d \cos \lambda_n d} \qquad (11)$$

Let us now calculate a typical case by choosing: wall thickness $d = 0.96$ in $= 0.08$ ft, thermal conductivity $k = 30$ Btu/h ft $^\circ$F, and a rather high value of the heat convection coefficient $h = 2$ Btu/h ft^2 $^\circ$F.

Thus, Equation (8) becomes:

$$\lambda_n \tan 0.08\lambda_n = 0.0666 \qquad (12)$$

The first three characteristic values of λ_n and of the corresponding $d\lambda_n$ and a_n are obtained as in Table 1.

TABLE 1

n	λ_n	$d\lambda_n$	$ka_n/2q''$
1	0.908	0.07264	0.551145
2	39.290852	3.14326	−0.00001356
3	78.539816	6.283185	0.0000017

It is obvious that the distribution of temperature in the direction Y for any particular value of x, as it is expressed by Equation (9) is practically given by the term for $n = 1$, while all a_n coefficients for $n > 1$ are negligibly small. Neglecting them, Equation (9) becomes

$$T(x, y) = 1.102 \frac{q''}{k} e^{-0.908x} \cos 0.908y$$

$$T(x, y) = X(x) \cos 0.908y \qquad (13)$$

where $X(x)$ is the temperature in the centre of the thickness of the wall (i.e. at $y = 0$). Towards the surface it drops to

$$T(x, d) = X(x) \cos 0.908d = 0.997X(x).$$

This calculation of a typical structure shows that temperature across the thickness of the wall may be considered constant.

An analogous calculation for the cross-section of the 'spindle' of Fig. 4a revealed a temperature variation of less than 2 per cent.

2. Transient temperature fields in simple bodies with constant heat flux input and convection cooling over entire surface

The two typical bodies considered correspond to the 'housing' and 'spindle' respectively of Fig. 4a.

In this paragraph we consider them separately. For this purpose they may be simplified as shown in Fig. 13c and d. In (c) the housing is considered as a plate of length L, width B and thickness $2d$ (for the same reason as in Fig. 13b) which has a steady heat flow input at $x = 0$ and is insulated at the sides, which corresponds to the housing being enclosed. Heat is conducted from the surfaces of the plate. Similarly, the spindle represented by a solid cylinder is considered insulated at the end $x = L$. This may be understood in such a way that the surface at the face is added to the circumference which is correspondingly lengthened.

With respect to the result of the calculations in the preceding paragraph temperature is considered constant through each section (for any particular value of x). Thus, as regards geometry, the problem is single dimensional in the coordinate X. Considering transients, the other coordinate of the problem is time t. The situation is described by:

$$\frac{\partial^2 T}{\partial x^2} - m^2 T = \frac{1}{\alpha}\frac{\partial T}{\partial t} \qquad (14)$$

where

$$\alpha = \frac{k}{\rho c_p} \quad \text{is thermal diffusivity,}$$

and

$$m^2 = \frac{hP}{kA},$$

where P is the periphery of the section and A is its area.

The solution of Equation (14) is considered as a superposition of a steady state temperature Φ and the transient one ψ:

$$T(x, t) = \psi(x, t) + \Phi(x) \tag{15}$$

For the transient ψ no heat input is considered and it is assumed that its initial state is the negative of the steady state Φ so that $T(x)$ is zero at $t = 0$ and as ψ vanishes with time $T(x, \infty) = \Phi(x)$. Thus, we have

$$\frac{\partial^2 \psi}{\partial x^2} - m^2 \psi = \frac{1}{\alpha} \frac{\partial \psi}{\partial t} \tag{16}$$

and

$$\frac{d^2 \Phi}{dx^2} - m^2 \Phi = 0 \tag{17}$$

with the initial condition

$$\psi(x, 0) = -\Phi(x) \tag{18}$$

and the boundary conditions

$$\frac{\partial \psi(0, t)}{\partial x} = 0 \tag{19}$$

$$\frac{\partial \psi(L, t)}{\partial x} = 0 \tag{20}$$

$$k \frac{d\Phi(0)}{dx} = -q'' \tag{21}$$

$$k \frac{d\Phi(L)}{dx} = 0 \tag{22}$$

The transient function $\psi(y, t)$ is solved by separation of variables

$$\psi = X(x)F(t) \tag{23}$$

$$\frac{1}{X} \frac{d^2 X}{dx^2} - m^2 = \frac{1}{\alpha} \frac{1}{F} \frac{dF}{dt} = -\lambda^2 \tag{24}$$

The solutions for the three functions have the following forms:

$$X(x) = \sum_{n=0}^{\infty} A_n \cos \mu_n x, \tag{25}$$

where

$$\mu_n = \frac{n\pi}{L}, \quad n = 0, 1, 2 \ldots \tag{26}$$

$$F_n(t) = C_n e^{-\alpha \lambda_n^2 t} \tag{27}$$

and

$$\lambda_n^2 = m^2 + \mu_n^2 \tag{28}$$

$$\Phi(x) = B_1 e^{-mx} + B_2 e^{mx} \tag{29}$$

Using the above boundary conditions the following is obtained:

$$\Phi = \frac{q''}{km} \frac{\cosh m(L - x)}{\sinh mL} \tag{30}$$

and

$$\psi = \sum_{n=0}^{\infty} a_n e^{-\alpha \lambda_n^2 t} \cos \mu_n x \tag{31}$$

where using the initial condition (18) it is found:

$$a_0 = -\frac{q''}{kLm^2}, \quad a_n = -\frac{2q''}{kL\lambda_n^2}, \quad n = 1, 2, 3 \ldots \tag{32}$$

Equation (31) shows that the temperature field along x is obtained as a sum of 'modes'. These are functions of x. Then $n = 0$ ('zero' mode) is constant over the whole length of the body and all the other modes are cosine functions of x: the first ($n = 1$) one extending over the interval $(0, \pi)$, the next one over the interval $(0, 2\pi)$, etc. The 'amplitudes' of these modes are equal to a_n for $t = 0$ and they decay exponentially with time with corresponding time constants

$$\tau_n = \frac{1}{\alpha \lambda_n^2} \tag{33}$$

for each mode.

The thermal deformation between two points x_1 and x_2 is obtained as

$$\Delta x_{1,2} = \alpha_s \int_{x_1}^{x_2} T \, dx = \alpha_s \int_{x_1}^{x_2} \Phi \, dx + \alpha_s \int_{x_1}^{x_2} \psi \, dx$$

$$= \Delta_{s1,2} + \Delta_{t1,2} \tag{34}$$

where α_s is the coefficient of thermal expansion and Δ_s denotes the steady state ($t = \infty$) deformation and $\Delta_t(t)$ is the transient one. With the use of (31) it is

$$\Delta_{1,2} = \alpha_s \int_{x_1} \left(-\frac{q''}{kLm^2} - \frac{2q''}{kL} \sum_{n=1}^{\infty} \frac{\cos \frac{n\pi}{L} x}{\lambda_n^2} \right) e^{-\alpha \lambda_n^2 t}$$

$$\Delta_{1,2} = \alpha_s \left[\frac{q''}{kLm^2} (x_1 - x_2) e^{-\alpha m^2 t} \right.$$

$$\left. + \frac{2q''}{kL} \sum_{n=1}^{\infty} \frac{L \left(\sin \frac{n\pi}{L} x_1 - \sin \frac{n\pi}{L} x_2 \right)}{n\pi \lambda_n^2} e^{-\alpha \lambda_n^2 t} \right]$$

$$\Delta_{1,2} = \alpha_s \left[D_0 e^{-\alpha m^2 t} + \sum_{n=1}^{\infty} D_n e^{-\alpha \lambda_n^2 t} \right] \tag{35}$$

For any particular path $(x_2 - x_1)$ over which the integration (34) is carried out the individual D_0, D_n terms are constants. In this way such a thermal deformation is a sum of modal terms, each one of which is an exponential function of time with a corresponding time constant. This is in full agreement with what McClure has obtained[6].

For our particular cases we obtain:

(a) For the housing type structural part: the parameters are $d = 0.08$ ft, $B = 4$ ft, $L = 4$ ft, $h =$

1·666 Btu/h ft² °F, k = 30 Btu/h ft °F, α = 0·666 ft²/hr, the actual heat flow input q = 1260 Btu/h which corresponds to 0·5 hp. According to our double wall thickness formulation we have to calculate with q'' = 2 q/A = 3940 Btu/ft² h. Using these data we have m^2 = 0·69444 and the amplitudes and time constants for the individual first six modes are obtained.

TABLE 2

n	$-\psi_n$ °F	$\alpha\lambda_n^2$	τ hours
0	47·28	0·463	2·16
1	50·078	0·874	1·14
2	20·77	2·1	0·474
3	10·51	4·164	0·24
4	6·22	7·043	0·14
5	4·08	10·743	0·093

(b) For the 'spindle' the following parameters are chosen: diameter D = 4 in = 0·3333 ft, L = 0·5 ft, h = 4 Btu/h ft² °F, k = 30 Btu/h ft °F, α = 0·570 ft²/h, and m^2 = 1·6. In order to obtain the same temperature for x = 0, t = 0 as for the housing the values of q'' = 3383 Btu/ft² h and q = 295 Btu/h are used. The resulting modal parameters are as in Table 3.

TABLE 3

n	$-\psi_n$ °F	$\alpha\lambda_n^2$	τ hours
0	140·7	0·912	1·1
1	10·96	23·4	0·04
2	2·82	90·9	0·01
3	1·26	203·4	0·005
4	0·71	361	0·0028
5	0·46	563	0·0018

It is obvious that only a small number of the lowest modes is sufficient to express the temperature field accurately enough. For the spindle, the first two or three modes are sufficient.

The variation of the temperature fields during the heating on for the above given parameters is shown in the diagram (Fig. 14). Diagram (a) applies to the housing and diagram (b) to the spindle. They are

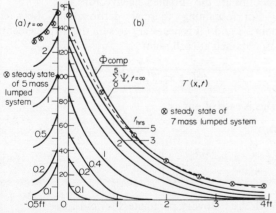

Figure 14. Temperature fields during heating-on in the 'housing' and in the 'spindle'.

separated here because only the steady state temperatures are matching at points x = 0 and not the transients. This was expected with the given specifications of the cases. The variation of temperature of a number of points of the housing is given in Fig. 15.

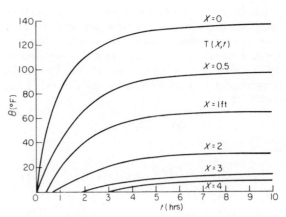

Figure 15. Temperature variation of various sections of the 'housing'.

3. The transients of temperatures and displacements in lumped-mass systems

In the preceding paragraph it was shown that in these typical cases the internal resistance to heat flow is rather small compared to that for convection of heat from the surfaces of the bodies. Therefore, the temperature gradients in these bodies are low and a small number of 'modes' is sufficient to describe the temperature fields with good accuracy. It is therefore tempting to try and treat them as consisting of a small number of lumped masses. While the above given exact solutions become very difficult for even slightly more complicated systems or boundary conditions, it is not so for the lumped mass system treated by means of the computer. Here we do not apply the complete finite differences technique or finite elements technique and treat the time variable as continuous. For cases with constant boundary conditions this seems to be of advantage.

We shall compute our spindle-housing system as a whole and represent it by 11 discrete masses as shown in Fig. 16. The spindle length of 0·5 ft is divided in

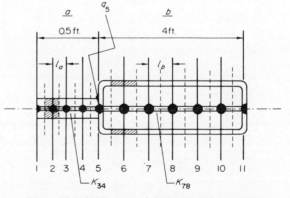

Figure 16. The diagram of the lumped-mass system.

four parts and the housing length of 4 ft in six parts. Each part is centred in points 1 to 11 and it extends half-way towards the neighbouring one. The masses at the ends are half those in between and point 5 consists of a half-unit mass of the spindle and half-unit mass of the housing. Within each mass temperature is assumed constant, each mass has a surface S_i and volume V_i and the masses M_i, M_{i+1} are interconnected by heat conducting channels with section areas $A_{i,i+1}$ and lengths $l_{i,i+1}$.

Similarly, as above, we solve both temperature fields and displacements by superposition of steady state and of transient state. There is a single geometrical coordinate. It is discrete and is replaced by the number i of a point. We denote by T_i the temperatures at the individual points. For the steady state it is:

$$\Sigma q_i - hS_i T_i = 0$$

or

$$K_{i-1,i}(T_{i-1} - T_i) + K_{i,i+1}(T_{i+1} - T_i) - hS_i T_i$$
$$= -q_{ex,i}$$

or

$$-K_{i-1,i}T_{i-1} + (hS_i + K_{i-1,i} + K_{i,i+1})T_i$$
$$- K_{i,i+1}T_{i+1} = -q_{ex,i} \qquad (36)$$

where

$$i = 1, 2, 3, \ldots 11, \qquad K_{i,i+1} = \frac{kA_{i,i+1}}{l_{i,i+1}}$$

and $q_{ex,i}$ is the external heat input.

In our case it is

$$A_{i,i+1} = A_a \text{ and } l_{i,i+1} = l_a, \quad i = 1 \text{ to } 4,$$
$$= A_b \text{ and } \qquad = l_b, \quad i = 5 \text{ to } 10,$$

$q_{ex,i} = 0$ for all values of i except $i = 5$.

Solving the steady state case means solving the system (36) for the eleven values T_i. The system has the form:

$$\begin{bmatrix} hS_1 + K_{12}, -K_{12}, 0, 0, \ldots \\ -K_{12}, hS_2 + K_{12} + K_{23}, -K_{23}, 0, 0, \ldots \\ 0, -K_{23}, \\ \vdots \quad \vdots \end{bmatrix} \begin{Bmatrix} T_1 \\ T_2 \\ T_3 \\ \vdots \\ T_{11} \end{Bmatrix}_{st}$$

$$= \begin{Bmatrix} 0 \\ 0 \\ 0 \\ 0 \\ -q_{ex5} \\ 0 \\ 0 \end{Bmatrix} \qquad (37)$$

where the subscript 'st' stands for steady state.

For the transient case it is

$$\Sigma q_i - hS_i T_i = \rho c_p V_i \frac{dT_i}{dt},$$

or,

$$C_i \frac{dT_i}{dt} - K_{i-1,i}T_{i-1} + (hS_i + K_{i-1,i} + K_{i,i+1})T_i$$
$$- K_{i,i+1}T_{i+1} = 0 \qquad (38)$$

where

$$C_i = \rho c_p V_i \qquad (39)$$

For the cases of constant boundary conditions it is possible to assume the solution

$$T_i = X_i \, e^{-\beta_i t} \qquad (40)$$

$$\frac{dT_i}{dt} = -\beta T_i$$

Inserting (40) into (38) and rearranging, we obtain

$$\beta \begin{Bmatrix} X_1 \\ X_2 \\ X_3 \\ \vdots \\ X_{11} \end{Bmatrix} = \begin{bmatrix} A_{11}, A_{12}, 0, 0, 0, \ldots \\ A_{21}, A_{22}, A_{23}, 0, 0, \ldots \\ 0, A_{32}, A_{33}, A_{34}, 0, 0, \ldots \\ \vdots \\ \end{bmatrix} \begin{Bmatrix} X_1 \\ X_2 \\ X_3 \\ \vdots \\ X_{11} \end{Bmatrix}$$

$$(41)$$

The system (41) yields a characteristic equation solving for eleven eigenvalues β_n and eleven eigenvectors X_{in}. These eigenvectors represent the 'mode shapes' of distribution of temperature on the individual masses.

Finally, it is

$$\begin{Bmatrix} T_1 \\ T_2 \\ T_3 \\ \vdots \\ T_{11} \end{Bmatrix}_{trans} = \sum_{n=1}^{11} a_n \, e^{-\beta_n t} \begin{Bmatrix} X_1 \\ X_2 \\ X_3 \\ \vdots \\ X_{11} \end{Bmatrix}_n \qquad (42)$$

The coefficients a_n are obtained by means of the initial condition equating the solution of (42) for $t = 0$ to the solution of (37):

$$[X_{in}]\{a_n\} = \{T_i\}_{st} \qquad (43)$$

where $[x_{in}]$ is now a square matrix (11 by 11) whose columns are the obtained eigenvectors X_{in}.

For obtaining thermal deformations between points i and j it is necessary to sum the displacements Δl_i obtained at each point:

$$\Delta l_i = \alpha_s \cdot T_i \frac{l_{i-1,i} + l_{i,i+1}}{2} \qquad (44)$$

This is carried out as:

$$\Delta_{ij} =$$
$$\alpha_s \sum_{n=1}^{11} a_n \, e^{-\beta_n t} \begin{Bmatrix} X_i \\ \vdots \\ X_j \end{Bmatrix}_n \begin{Bmatrix} \frac{l_{i,i+1}}{2}, \frac{l_{i,i+1} + l_{i+1,i+2}}{2} \cdots \frac{l_{j-1,j}}{2} \end{Bmatrix} \qquad (45)$$

With the following parameters for our case:

i	C_i	hS_i	$K_{i,i+1}$
1	0·3	0·25	21
2 to 4	0·6	0·5	21
5	9·6875	4·25	26
6 to 10	18·775	8	26
11	9·3815	4	26

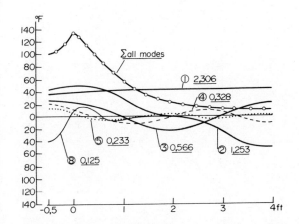

Figure 17. The 'mode' shapes and time constants of main components of the temperature field of the system of Fig. 16.

the following solutions have been obtained. For steady state cases for separated systems 1 to 5 and 5 to 11 as entered for comparison with exact solutions in Fig. 14. The time constants are compared in Table 4. For the whole system as regards the eigenvalues β and the eigenvectors X_{in} as shown in Fig. 17. The transient temperature fields are given in Figs 5 and 6 and thermal deformations between various pairs of points are plotted in Fig. 7.

TABLE 4   Time constants τ hours

(a) Housing Mode	0	1	2	3	4	5
System continuous	2·16	1·14	0·474	0·24	0·14	0·093
System lumped	2·35	1·26	0·55	0·31	0·22	0·18

(b) Spindle Mode	0	1	2	3	4	5
System continuous	1·1	0·04	0·01	0·005	0·003	0·002
System lumped	1·2	0·06	0·04	0·01	0·008	

AN AXIS OF ROTATION ANALYSER

by

P. VANHERCK and J. PETERS*

INTRODUCTION

The principal motions in most cutting operations are rotations: mainly the rotation of the workpiece in turning, cylindrical grinding, etc., and the rotation of the tool in boring, milling, etc.

Under ideal chip-formation conditions, a complete suppression of the unwanted relative motions between the axis of rotation and the tool-point for a rotating workpiece or between the axis of rotation and the workpiece for a rotating tool, would result in a perfect geometrical form of the machined workpiece.

Due to imperfect bearings and structural deformations, caused by dynamic forces, it is impossible to avoid completely these relative motions and, as a consequence, the machined workpieces or surfaces will present deviations with respect to the ideal expected geometrical form. Consequently the measurement of the relative motion of the axis of rotation, or generally referred as the error motion, is present as one of the most important items of the geometrical acceptance tests of machine tools.

In several previous papers, it was proven that the classical methods of testing, presented by Schlesinger, Salmon, etc. are not able to give a good picture of this error motion phenomena.

One of the most important limitations is due to the limited response of the mechanical gauges, so that these methods cannot be applied at the real cutting speeds, though it is evident that the behaviour of the bearings, the vibration level and the picture of the structure, are strongly speed dependent. These methods only give the run-out of components.

On the other hand measuring the errors on a machined workpiece does not provide a complete picture, because of the influence of the chip formation on both the machine structure and the machined surface. Interference of cuttertraces, build up edge, and damping due to flank wear dissimulate most of the data. Therefore several authors and research institutions[1-5] have proposed or discussed new measuring techniques, aiming to produce on a cathode ray tube screen the shape of the generated surface excluding the perturbances due to the cutting operation itself.

Recently the ASME, submitted a provisional draft standard on the 'Axis of Rotation' including definitions and characteristic accuracy parameters. Unfortunately, up to now no commercial instrument is available to measure to this standard.

At the University of Leuven, a prototype of a universal instrument was developed which is easy to operate. A comprehensive description of this instrument will be given in this paper.

THE AXIS OF ROTATION ANALYSER

The aim of the instrument is to express quantitatively the different error motions of the axes of rotation, as defined in previously mentioned standards. In order to obtain good correlation between the error motions and the geometrical shape of the machined surfaces, the instrument must detect the error motion due to the structural deformations of the machine as well as the error motions due to the imperfect bearings and spindle.

Two cases must be considered: Firstly the case of rotating workpiece, and stationary tool, as in a lathe for plunge, cylindrical turning or facing. Secondly the case of rotating tool and stationary workpiece, as in boring. Both cases are handled by the instrument, by means of two different circuits, respectively described in the following sections. Before going into the detailed description of each of these circuits, let us summarize here the advantages of the proposed methods.

The first circuit for detection of the error motions in rotating workpieces is an extension of the methods of Lawrence Radiation Laboratories and that of VUOSO. It has the following advantages.

(1) Reference circle is always perfect and centred without re-adjustments (no zero shift, and a constant scale over the circumference are guaranteed).
(2) No cams are needed.
(3) Only a single contactless pick-up (no contact noise) is used.

---

* Instituut voor Werktuigkunde, Katholieke Universiteit, Leuven

(4) It practically responds without frequency limitations of the error signal.

(5) Using a very light synchro (total weight of only 55 g) the mechanical influences of the synchro are negligible.

The second circuit, for detection of the error motion in rotating tools, is an extension of the VUOSO method, with respect to which it has the advantage of providing a possible phase adjustment. When the measuring method is properly used a direct correlation exists between the image on the screen and the shape of the machined workpiece, not only qualitatively but also quantitatively.

It should be emphasized that each type of cutting operation needs its appropriate measuring procedure; otherwise no correlation can be achieved.

ROTATING WORKPIECE

Radial error motion

Principle
Supposing a groove-plunging operation on a lathe (Fig. 1a): a relative motion between the tool and the axis of rotation, parallel to this axis of rotation, will have no influence upon the geometrical form of the cylindrical part of the groove, while the influence of the relative motion in the cutting speed direction will be of second order.

It is clear that, under ideal chip-formation conditions, the geometrical error of this cylindrical part will only be due to the relative motions in the radial direction along a line, passing through the tool point, perpendicular to the axis of rotation. This direction is called the 'sensitive direction'. In general, one calls 'sensitive direction', the direction along a line through the tool point and perpendicular to the considered generated workpiece surface.

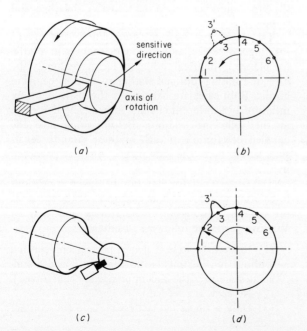

Figure 1. Radial plunge turning: (*a*) plunging operation; (*b*) section of the workpiece; (*c*) measuring set-up; (*d*) figure on the screen.

Figure 1(*b*) represents a steady state image of the cross section of the workpiece. At the moment $t_1, t_2, \ldots t_n$, the points $1, 2, \ldots n$, will be cut by the tool.

If the error motion between the tool and the axis of rotation is reduced to zero, the geometrical form of the workpiece will be a perfect cylinder. However, if at the moment t_3 the distance between the tool and the axis of rotation increases suddenly, the machined workpiece will present a bump at this particular place, that is point 3.

The principle of the method is to generate a rotating vector by means of a synchro fixed at the spindle and to modulate its length by the error signal detected by means of a capacitive pick-up fixed on the tool post and measuring the gap between its end and a spherical master mounted on the spindle (Fig. 1c).

If no error motion is detected the image on the screen will be a perfect circle. In order to get the true

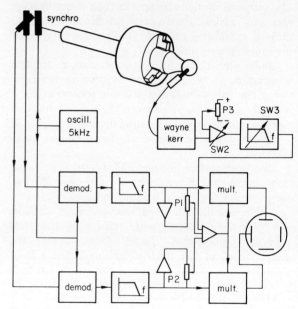

Figure 2. Block diagram: axis of rotation analyses; radial error motion.

image of the machined workpiece the rotation of the vector on the screen must be inversed with respect to the rotation of the spindle. It means that on the screen the workpiece is represented stationary whereas the tool is rotating. If the gap between the spherical master and the pick-up varies, a corresponding deviation from the circle is marked on the screen. This point is easily shown by means of the block diagram (Fig. 2) showing the rotor of the synchro fed by a 5 kHz oscillator. If ω is the rotating speed of the spindle, one yields after phase linked demodulation and filtering two pure harmonic signals ($A \sin \omega t$) and ($A \cos \omega t$).

By linking these two signals to the x and y deflection plates of the oscilloscope, a vector of magnitude A will be created on the screen, rotating in synchronism and opposite direction with the spindle rotation.

The length of this vector must now be modulated by the output signal of the pick-up. This can be done by multiplying the sine and cosine signals of the synchro with the pick-up output by means of a multiplier integrated circuit, before linking them to the deflection plates.

Error compensation

In constructing the instrument care must be taken to choose a master ball with a shape error an order of magnitude less than the error to be measured. It is possible to select balls with spherical errors less than 0.1 μm. If higher measuring precision is required, e.g. for testing measuring spindles of roundness measuring instruments, special techniques are used, described by Donaldson[4]. The main error arises from the lack of centring of the ball on the spindle.

It should be pointed out first of all that even when the rotation axis of a spindle does not coincide with the geometrical axis, still a cylindrical piece is machined, although possibly not centred with a chosen reference axis. As a general rule we can see that constant eccentricity does not produce deviation from the cylindrical shape, it can give a dimensional error depending from the reference taken.

Although the spherical master is mounted on a special rig on the spindle that makes a mechanical adjustment possible, it is practically impossible to centre the sphere perfectly on the rotational axis just by mechanical adjustments.

This eccentricity will produce on the screen a Lissajous figure with a period equal to the rotational period, instead of a circle (although the machined piece will be cylindrical).

As this deformation could be misinterpreted as an error motion, it must be reduced as far as possible. For that reason, an electronic 'centring' circuit was built into the instrument, to compensate all harmonic signals with the same periods as the rotation itself (fundamental).

The first approximation showed that the component of the output signal of the capacitive pick-up, due to this eccentricity, can be expressed by

$$B \cos (\omega t + \theta)$$

where B represents a value proportional to the product of the mechanical offset and the gain of the pre-amplifier, and θ represents the angle between the reference direction of the synchro and the direction of the off-set.

The above expression furthermore can be expressed as follows:

$$B \cos (\omega t+\theta) = B \cos \theta \cos \omega t - B \sin \theta \sin \omega t$$
$$= C \cos \omega t - D \sin \omega t$$

where $C = B \cos \theta, D = B \sin \theta$.

If one can add signals of $-C \cos \omega t$ and $+D \sin \omega t$ to the signal of the pick-up, the harmful effect of the offset of the sphere can be suppressed completely.

In order to achieve this, two potentiometers P_1 and P_2 allow to derive from the synchro signal components consisting of two compensating signals adjustable respectively between $-P \cos \omega t$ and $+P \cos \omega t$ and on the other hand between $-P \sin \omega t$ and $+P \sin \omega t$.

P represents the saturation output from the pick-up amplifiers so that the compensation can be made over the full range of the instrument. Both compensation signals are added to the pick-up signal before entering the multipliers.

Range adjustment

Generally the error motion is much smaller than mean gap between the pick-up and the spherical master, e.g. with a safety gap of 20 μm and a motion error of 0.5 μm the error of radius ratio would only be 2.5 per cent. In order to increase this ratio the signal due to the average gap (20 μm) is partially compensated by a d.c. signal, adjustable by a potentiometer P_3. On the other hand an adjustable gain of the instrument SW_2 provides the mean to match the sensitivity of the instrument with the magnitude of error in different machines.

Further filtering units SW_3 are provided in order to isolate low frequency bearing motions from high frequency structural motions, whenever needed.

As a conclusion the scale of the instrument results from the previous adjustments. Let (S) be the overall sensitivity in millimetre spot deflection per micrometer error motion:

$$S = S_1 . S_2 . S_3$$

where S_1 represents the pick-up sensitivity (V/μm),
  S_2 the gain of the electron unit (V/V),
  S_3 the sensitivity of the oscilloscope (mm/V).

Angular error motion

In cylindrical turning (Fig. 3d) the angular error motion (α_t) must be considered besides the radial error motion (x_t). This can be measured using two spherical masters separated by a distance L, of which the radial error motion is measured by two pick-ups, with exactly the same sensitivity, fixed in the same direction and at the same tool post.

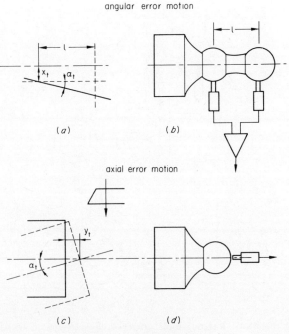

Figure 3. Error motion.

By adjusting the function-selector in the appropriate position, the difference of the outputs of the two pick-ups is taken, and further processed as explained in the previous section.

The overall sensitivity of the instrument S_α expressed in mm spot deflection per milliradian can be calculated as follows:

$$S_\alpha = S_1 . S_2 . S_3 . L$$

where S_1, S_2, S_3 have the same meaning as above,
L is the distance between the centre of the two spherical masters in millimetres (mm).

Axial error motion

In Fig. 3(c) the form error can be seen in facing results from both the angular error motion α_t and the axial error motion y_t. For measuring this axial motion error the capacitive pick-up is placed along the axis of rotation.

In the measurement of the axial error motion the possible angular or radial error gives only a second order result whereas the axial error is of the first order. The output signal of the pick-up is treated as explained above except as to the compensation for the eccentricity that must be switched off by putting the selector switch in the appropriate position.

ROTATING TOOLS

In cutting operations with rotating tools as in boring operations the sensitive direction is no longer steady in space, but it rotates with the tool as shown in Fig. 4(a). The same basic instrument as described above can be used provided that two pick-ups are detecting the spherical master motion in two perpendicular directions. Two different measuring procedures can be applied according to the manner in which the reference circle is generated.

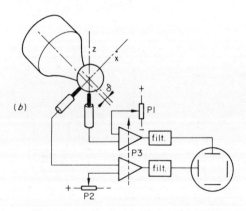

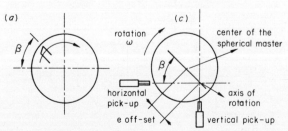

Figure 4. Rotating tool: reference circle generated by an offset of the spherical master.

In the first method the reference circle is generated by giving to the spherical master an offset in the direction of the cutting edge of the tool. No synchro is used. In the second method the reference circle is generated by a synchro to which the error motion signal is superimposed.

First method: reference circle generated by offset

Let the spherical master be mounted on the spindle with an offset e in the direction of the cutting edge. The two capacitive pick-ups are placed stationary in horizontal and vertical direction and linked respectively with the H and V axis of the oscilloscope.

Let the spherical master have an offset e in the direction of the tool point and assume that this makes an angle β with the x axis at the time t_0 (Fig. 4c) and further suppose that no error motion is present. Two signals V_H and V_V are generated when the spindle is rotating with rotational speed ω:

$$V_H = S'_1 \, |E_H - e \cos(\beta + \omega t)|$$
$$V_V = S'_1 \, |E_V - e \cos(\beta + \omega t)|$$

where S'_1 is the sensitivity of the pick-ups (mV/mm); both sensitivities must be the same,
E_V and E_H represent the mean gap width between pick-ups and master.

It may be noted here that the second harmonic is negligible and can only produce a second order error.

It is easily seen that when no error motion is present a rotating vector is generated on the oscilloscope screen, of which the end produces a circle; however, the centre of this circle is a point of which the co-ordinates are proportional to $S'_1 \, E_H$ and $S_1 \, E_V$. This circle can be centred on the screen by adding compensating d.c. signals by means of the potentiometers P_1 and P_2 to the output of the pick-up signals. This centring is not critical and does not produce any shape error in the reference circle.

On the other hand every error motion of the spindle axis from its theoretical position causes a variation of the gap value, and consequently produces a deformation of the generated circle. It images the shape of the generated hole in the workpiece, provided that the direction of the eccentricity e coincides with the cutting edge direction, as said above.

Let us emphasize the fact that the adjustment of the offset of the spherical master precisely in phase with the tool point is critical in order to obtain an exact correlation between the shape of the figure on the screen and the geometrical form of the machined surface. One should realize that if this adjustment is not kept the image is not only rotated but completely distorted. Suppose for example that a phase shift of 90° would exist between the spot and the tool point when this is along the horizontal axis; it is known that a possible vertical error motion of the tool produces only a second order error on the piece whereas it would produce a first order deviation on the screen.

On the other hand the error to radius ratio cannot be arbitrary, if the offset is too small with respect to the error, the image of the error on the screen will cross the centre of the screen, if the offset is too large, the amplification factor, which is common to the eccentricity and the error motion, must be kept small in order to keep the image on the screen. Practically the radius can only be 2 at 3 times the amplitude of the error motion.

Consequently, the main disadvantage of the method is the very critical setting of the spherical master in phase and magnitude. As in the preceding method filters can be used for separating structural errors, and the gain can be adjusted with the selector P_3.

Second method: reference circle generated by synchro

Principle

The block diagram of this method is shown in Fig. 5. As in the preceding case the spherical master is mounted on the spindle and two capacitive pick-ups are stationary perpendicular to each other. But contrary to the preceding case, the spherical master must

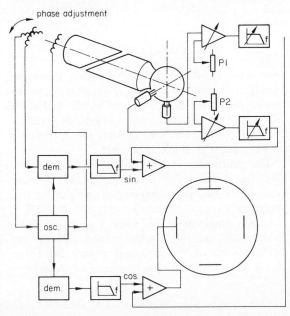

Figure 5. Rotating tool: reference circle generated by a synchro.

be centred on the rotational axis. It will be shown afterwards what happens when the centring of the spherical master is not perfect. On the other hand a synchro is rotating with the spindle and as the signals are fed to the deflection plates of a cathode ray tube, a perfect circle is produced, when no error motion is present. The signals produced by the pick-ups are now *added* respectively to the signals produced by the synchro before being fed to the cathode ray tube.

When the error motion is zero the reference circle is again perfect although not centred on the screen, due to the signals produced by the average gap widths. This can be compensated by the potentiometers P_1 and P_2.

For the same reason as explained in the previous section, the spot on the screen must be perfectly in phase with the tool motion, otherwise a distorted image is produced. This adjustment can be made by

rotating the stator of the synchro over the wanted angle. Any error motion will produce a vertical and horizontal component that causes the spot to deviate from the reference circular path so that the image of the machined workpiece will appear on the screen provided the spot and the tool point rotation are adequately adjusted. The gain factor for the error signal can be set with selector P_3.

Effect of a residual eccentricity of the master and its compensation

Until now the spherical master was assumed to be perfectly centred. However, it is very likely that the mechanical adjustment will leave some residual eccentricity of the same order as the error motion to be measured.

It must be emphasized that this does not produce a limaçon as in the case on page 2 (where the pick-up signal modulated the synchro signal); in the present case, when no error motion is present, a vertical and a horizontal component are *added* to the circle tracing vector on the screen of which the values are:

$$d_v = S\,e\,\sin(\omega t + \beta)$$
$$d_h = S\,e\,\cos(\omega t + \beta)$$

where S is the sensitivity factor of the overall instrument in mm spot deflection per micrometre error motion,

β is the angle between the vector generated by the synchro signals and the offset direction,

and e is the offset of the spherical master.

The spot on the screen is now the endpoint of a vector resulting of two components: AB due to the synchro and BC due to the offset, as shown in Fig. 6.

Still a perfect circle is generated by point C, that can be used as reference circle. The trouble however is that the length of the component BC and consequently the angular location of the tracing point C with respect to the synchro reference B are dependent on the phase angle β as well as of the amplifica-

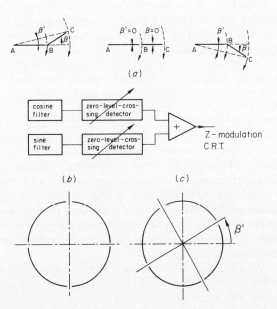

Figure 6. Compensation of the angular shift of the spot rotating tool: error motion superimposed on reference circle.

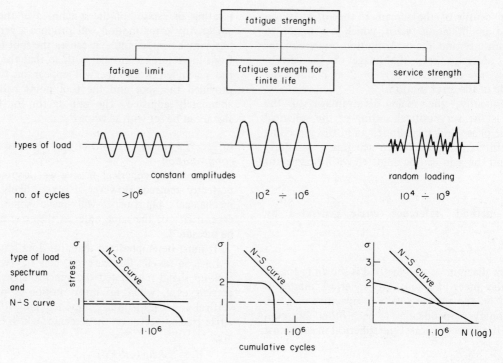

Figure 7.

tion factor S. Consequently the spot on the screen will be not perfectly in phase with the tool point, when the synchro is adjusted with respect to the tool point, and as explained above, a distorted image of the machined surface will be seen.

It is theoretically possible to compensate this phase shift between tracing spot and tool point, by adjusting the angular position of the stator of the synchro. The practical problem however is to know exactly the amount of the phase compensation to be introduced, particularly as this depends upon the selected sensitivity.

This problem has been solved by introducing a z axis modulation derived from two zero level crossing detectors, respectively linked in the sine and cosine branch of the synchro signal. These devices work as follows (Fig. 7a)—whenever the sine or cosine wave is passing the zero level a negative signal is applied to the first grid of the cathode ray tube and consequently suppresses the spot.

If the stator of the synchro is adjusted in phase with the horizontal position of the tool point, and if low error amplification factors are used so that vector BC is small or zero, the spot will be suppressed on the screen at the intersections with the vertical and horizontal axes (Fig. 7b). If however the amplification factor is increased the spot follows point C and deviates from the AB direction, it will be suppressed at a certain angle β' from the intersections with the vertical and horizontal axes. A slight rotation of the stator of the synchro makes it possible to readjust the phase of spot C so as to make it cross the horizontal axis at the same moment as the tool crosses the horizontal axis.

As a conclusion it can be stated that this second method, although somewhat more elaborate, allows to set the error motion to radius signal ratio almost independently from the offset of the spherical master and from the gain of the amplifiers.

CONCLUSION

The 'Axis of Rotation Analyser' was developed, making it possible to express quantitatively, on the base of the partial draft proposed by the ASME, the different error motions of the axis of rotation. Special attention was paid to keep the operation of the instrument as simple as possible.

The instrument is universal. It is provided with measuring circuits for the detection of the error motions for rotating workpieces as well as for rotating tools.

It was emphasized that only a good correlation between the measured error motion and the geometrical form of the machined surface can be obtained when the appropriate measuring assembly is used.

ACKNOWLEDGMENT

This study was supported by the C.R.I.F. (Research Center of the Metalworking Industry, Brussels).

REFERENCES

1. BRYAN, CLOUSER and HOLLAND (1967). Spindle accuracy, *American Machinist*, Spec. Rep. no. 612.

2. VANEK (1969). Measurement of accuracy of rotation of machine tool spindles, V.U.O.S.O.

3. GODDARD, COWLEY and BURDEKIN (1973). A measuring system for the evaluation of spindle rotation accuracy. *Proc. 13th MTDR-Conference*, Macmillan.

4. DONALDSON (1972). A simple method for separating spindle error from test ball roundness error, *Annals of the C.I.R.P.*, 21(1), 125–6.

5. KOENIGSBERGER and TLUSTY. Specifications and tests for metal cutting machine tools, U.M.I.S.T.

6. U.S.A. Standard Axes of Rotation. Draft standard, proposed by the American Society of Mechanical Engineers.

7. RAEKELBOOM and THIJS (1973). Fouten op rotatieassen. Thesis 73E4, Instituut voor Werktuigkunde, Leuven.

8. P. VANHERCK and J. PETERS (1973). Digital axis of rotation measurements, *Ann. C.I.R.P.*

DISCUSSION

Q. H. R. Taylor. Have the authors considered using hall probes? If a polar plot is required, two stationary hall probes and a rotating magnet provide a simpler and less expensive method of generating the base circle than the syncro method described in the paper. A description of this device is to be published.

A. We did some experiments with stationary hall-elements and a rotating permanent magnet. However, we did not obtain the same precision of the sine and cosine signals as with a synchro. The precision of these signals has to be very high, expecially when large eccentricities have to be compensated with a limited deformation of the polar plot. We agree with Dr Taylor that high precision rotary transducers, based on hall-elements, would simplify the circuit.

THE DETERMINATION OF THE VOLUMETRIC ACCURACY OF MULTI AXIS MACHINES

by

W. J. LOVE* and A. J. SCARR†

SUMMARY

An analysis is made of the displacement, planar and volumetric errors of multi-axis machines by determining the combined effects of errors in the measuring system and errors in geometric features such as linearity of motions and orthogonality of axes of movement.

INTRODUCTION

The final accuracy of a machined workpiece is compounded of many contributing factors. Among the more important of these are

(1) The machine tool, e.g. positional accuracy, thermal distortion, load distortion.
(2) Cutting conditions, e.g. tool wear, cutting force deflections.
(3) Environment, e.g. temperature differences from 20°C.
(4) Workpiece, e.g. clamping distortion, stiffness.

This paper is concerned with the first of these factors, namely the positional accuracy of the machine tool. The analysis of the problem is considered in three stages commencing with the one-dimensional case, i.e. displacement errors; this is then extended to the two-dimensional case, i.e. planar errors and finally considers the three-dimensional case of volumetric errors in a multi-axis machine.

Two alternative approaches are possible to establishing displacement, planar and volumetric errors. In the first instance, a direct measurement is made of the true position of the cutting tool (or probe, in the case of a measuring machine) in relation to the nominal position for a suitable number of points within the working capacity. In the second approach the true position of the cutting tool or probe is established as a synthesis of the contributing accuracy factors. These factors include the measuring system errors, linearity errors of movement in pitch, yaw and crosswind and errors in the orthogonality of axes. This latter approach of determining the combined effect of the various factors affecting the positional accuracy is the one considered in this paper.

ANALYSIS

Position error

Position error can be defined as the error between the desired position of the tool or probe and the actual position when the tool or probe is moved from a reference point to any other point in a specified axis.

There are a number of possible causes for this error. It may be caused through geometric errors such as pitch, yaw, crosswind or lack of straightness in the specified axis or in perpendicular axes. It may be caused through non-parallelism between the specified axis and the axis of the built-in measuring system, or it may be caused through the Abbé offset error due to the distance between the specified axis and the axis of the built-in measuring system. The position error is most likely caused by a combination of all these errors.

Displacement error

Displacement error can be defined as the error between any two points along a specified axis. It can be found by taking a reading of displacement from a reference point at various positions along the specified axis, and comparing each reading with the corresponding reading from the built-in measuring system. The differences (position errors) can be plotted on a graph of error against measuring system displacement readings. A better estimate of the position errors can be found, however, by taking a number of readings at each position point by approaching that point first from one direction, then from the opposite direction (Fig. 1). At each point the readings obtained in one direction will be dispersed about a mean μ_1, and those obtained in the opposite direction will be dispersed about another mean μ_2. The difference between μ_1 and μ_2 is backlash or hysteresis error. The

---

* Research Assistant, Cranfield Institute of Technology
† Senior Lecturer, School of Production Studies, Cranfield Institute of Technology

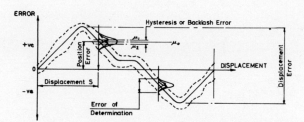

Figure 1. Displacement accuracy—single specified axis.

mean values of μ_1 and μ_2 can be arithmetically averaged to give a mean positioning error PE. Assuming the two distributions about each mean are normal, their variances may be pooled, and the standard deviation for all the errors at the particular displacement point may be found. This can be denoted as σ_0. The error of determination for the mean position error can then be taken as $\pm 3\sigma_0$ which is the spread of the pooled distributions. Hence the value of the positioning error for any given displacement S is

$$PE = \pm PE(S) \pm 3\sigma_0 \qquad (1)$$

The displacement error between any two points a and b along the specified axis is given by:

$$(PE_a - PE_b) \qquad (2)$$

and the maximum displacement error is given by:

$$(PE_{max} - PE_{min}) \qquad (3)$$

Planar error

If a second specified axis is introduced which is perpendicular to the first axis, then a planar error can be found. Planar error can be defined as the error between any two points in a specified plane or area set up by the two perpendicular axes.

One method of determining this error is by taking measurements of individual parameters and synthesizing all the errors.

There is no difference in the resulting planar error values obtained by arriving at any one point from any other point in the specified area, along preferential routes. This is because for a given point in that area the errors of angular pitch, angular yaw, angular crosswind and true straightness of carriage motion in each direction are constant for that point, irrespective of the size of the area. At any given point a in the specified area the normal orthogonality error (N.O.E) between the two axes intersecting at point a, one of which is parallel to the reference axis, is always constant (Fig. 2). At any other point b in the area apparent changes in orthogonality are due to the

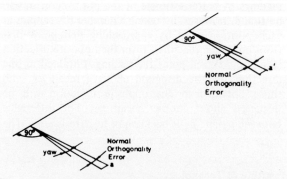

Figure 2. Orthogonality between axes in a plane.

effects of angular pitch and/or angular yaw along the intersecting axes. Hence if the angular pitch and/or angular yaw along the intersecting axes are known at the point in question, the normal orthogonality error for any point can be calculated.

Some part of the errors in straightness of carriage motions in the given plane can be attributed to the effects of angular pitch and/or angular yaw in the directions of the reference axes. It is necessary to separate these errors out in order to find the true value for carriage motion straightness in each direction.

Errors in displacement along each of the two reference axes will occur through a combination of geometric errors, non parallelism and the Abbé offset error, as previously mentioned.

Consider some point O in a specified area bounded by the x and z direction axes, and take this point as a reference for subsequent work. Take one axis through the reference point as the reference axis, and let angular errors in the clockwise direction be negative. Let all errors at point O be zero.

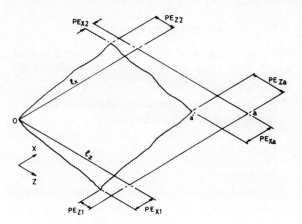

Figure 3. Positional errors in a plane at point a.

Now consider some point a at distances of l_x and l_z from the reference point (Fig. 3). Let angle γ = angular yaw. Then

Measured orthogonality error θ_{xz}
= Normal orthogonality error NOE_{xz}
+ error due to yaw in the x direction
+ error due to yaw in the z direction

i.e.

$$\theta_{xz} = NOE_{xz} + \gamma_x + \gamma_z \qquad (4)$$

Hence

$$NOE_{xz} = \theta_{xz} - (\gamma_x + \gamma_z) \qquad (5)$$

When moving the probe or tool point a distance of l_z in the z direction positioning errors occur, due to the angular errors and the magnitude of l_z, in the x direction:

$$PE_{x1} = l_z \sin(NOE_{xz} + \gamma_x + \gamma_z) \qquad (6)$$

and in the z direction:

$$PE_{z1} = l_z - l_z \cos(NOE_{xz} + \gamma_x + \gamma_z) \qquad (7)$$

Also when moving the probe or tool point a distance of l_x in the x direction, positioning errors also occur in the x direction:

$$PE_{x2} = l_x - l_x \cos(NOE_{xz} + \gamma_x + \gamma_z) \qquad (8)$$

and in the z direction:

$$PE_{z2} = l_x \sin(NOE_{xz} + \gamma_x + \gamma_z) \qquad (9)$$

Equations (6), (7), (8) and (9) are not strictly true since they rely on the assumption that the true position and the actual position of the probe or tool point are on a circular arc whose centre is at the axis intersection point. Since it is assumed that θ will be in terms of seconds of arc and l_x and l_z in terms of metres, the difference is so small that it has been neglected.

The total errors due to angular errors in moving the probe or tool point from point O to point a in the area are

(in x direction) $PE_{x1} + PE_{x2} = PE_{xa}$

(in z direction) $PE_{z1} + PE_{z2} = PE_{za}$

These errors can also be found by considering the Abbé offset principle. From Fig. 3,

$$PE_{x1} = l_z \sin \theta_{xz} \qquad (10)$$

$$PE_{z1} = l_z - l_z \cos \theta_{xz} \qquad (11)$$

and

$$PE_{x2} = l_x - l_x \cos \theta_{xz} \qquad (12)$$

$$PE_{z2} = l_x \sin \theta_{xz} \qquad (13)$$

Again

$$\theta_{xz} = NOE_{xz} + \gamma_x + \gamma_z$$

and again the total errors due to angular errors in moving the probe from point O to point a' in the area are:

$$PE_{xa} = PE_{x_1} + PE_{x2}$$

and

$$PE_{za} = PE_{z_1} + PE_{z2}$$

as before.

If the values of the N.O.E. and the yaw errors are known, then the planar error for any given point a from the reference point O can be expressed as

$$\text{Planar error} = \sqrt{(PE_{xa})^2 + (PE_{za})^2} \qquad (14)$$

Now consider the probe or tool point to be moved to some other point b in the specified area. Further errors have been introduced into the system due to this movement. The magnitude of these errors depend on the magnitude of the probe or tool movement from the reference point. Again by consideration of the Abbé offset principle these errors are (from Fig. 4):

$$PE_{x3} = (l_z + \delta l_z) \sin \theta_{xz} \qquad (15)$$

or

$$PE_{x3} = PE_{x1} + \delta l_z \sin \theta_{xz} \qquad (16)$$

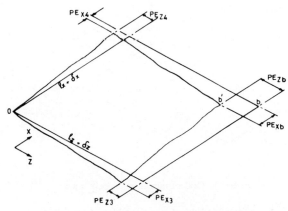

Figure 4. Positional errors in a plane at point b.

$$PE_{z3} = (l_z + \delta l_z) - (l_z + \delta l_z) \cos \theta_{xz} \qquad (17)$$

or

$$PE_{z3} = PE_{z1} + \delta l_z (1 - \cos \theta_{xz}) \qquad (18)$$

and

$$PE_{x4} = (l_x + \delta l_x) - (l_x + \delta l_x) \cos \theta_{xz} \qquad (19)$$

or

$$PE_{x4} = PE_{x2} + \delta l_x (1 - \cos \theta_{xz}) \qquad (20)$$

$$PE_{z4} = (l_x + \delta l_x) \sin \theta_{xz} \qquad (21)$$

or

$$PE_{z4} = PE_{z2} + \delta l_x \sin \theta_{xz} \qquad (22)$$

and the total errors in moving the probe or tool from point O to point b' are:

$$PE_{xb} = PE_{x3} + PE_{x4}$$

and

$$PE_{zb} = PE_{z3} + PE_{z4}$$

N.B. Equations (16), (18), (20) and (22) are only true if γ_x and γ_z are equal at both points a and b.

The planar error between any two points a and b in the specified plane can be expressed as

$$\sqrt{(PE_{xa} - PE_{xb})^2 + (PE_{za} - PE_{zb})^2} \qquad (23)$$

If the geometric errors are measured along a number of axes in each direction in the form of a grid (Fig. 5) and the positional errors plotted for each, a possible maximum planar error occurs when either

$$\sqrt{(PE_{xamax} - PE_{xbmin})^2 + (PE_{za} - PE_{zb})^2} \qquad (24)$$

or

$$\sqrt{(PE_{zamax} - PE_{zbmin})^2 + (PE_{xa} - PE_{xb})^2} \qquad (25)$$

is a maximum, whichever is greater.

However, it is possible that the second squared term in either equation (24) or (25) could give a zero value, and the actual maximum planar error for the plane or area may occur where either of the squared terms in equation (23) gives a value slightly less than maximum while the other squared term gives an appreciable value.

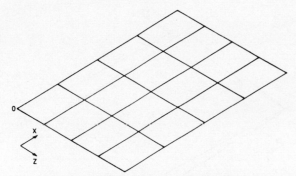

Figure 5. Grid of measuring axes in a plane.

In order to find the maximum planar error it is necessary to find the values of the positional errors at any points a and b which will maximize equation (23) and to find the coordinates of the points a and b. This can be readily done on a computer by presenting to the computer all the positional error data for each point in the grid, in the form of x and z direction arrays. A program has been written which will, in turn, take the x and z direction errors at each point in the grid, together with the errors, in turn, at each other point in the grid to solve equation (23) for each pair of points. The computer program then demands print-out of the maximum value of equation (23) together with the matrix coordinates of the errors which produced that maximum.

Straightness errors

The actual yaw straightness of the carriage motions at any point can be found by subtracting the errors due to all the angular changes from the appropriate

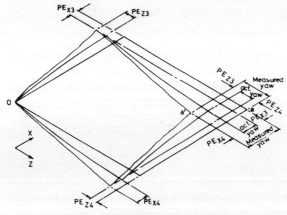

Figure 6. Straightness errors in a plane.

straightness measurement taken along any axis parallel with the carriage motion (Fig. 6). E.g. the measured yaw straightness of x carriage motion,

$E_{z(MXY)}$ = actual yaw straightness of x carriage motion
+ positional errors due to angular errors in the z direction
+ positional errors due to Abbé offset in the z direction.

i.e.

$E_{z(MXY)}$ = actual yaw straightness + PE_{z3} + PE_{z4}

(26)

∴   Actual yaw straightness of x carriage motion

$$= E_{z(MXY)} - (PE_{z3} + PE_{z4}) \qquad (27)$$

By a similar method,
actual yaw straightness of z carriage motion

$$= E_{x(MZY)} - (PE_{x3} + PE_{x4}) \qquad (28)$$

Volumetric error

If a third specified axis is introduced into the system which is mutually perpendicular to the previous two specified axes, then a volumetric error can be found. Volumetric error can be defined as the error between any two points in a specified working volume within which the machine probe or tool point can normally operate.

As for planar error, the method used here for determining the volumetric error is by taking measurements of individual parameters and synthesizing the results.

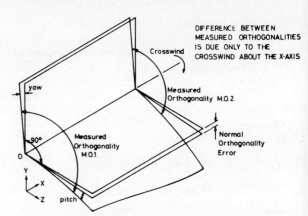

Figure 7. Orthogonality between planes in a volume.

Again there is no difference in the resulting volumetric error values obtained by arriving at any one point from any other point in the specified volume, along preferential routes. As before this is because for a given point in that volume the errors of angular pitch, angular yaw, angular crosswind and true straightness of carriage motion in each direction are constant for that point, irrespective of the size of the volume. At any given point in the specified volume the normal orthogonality error (NOE) between any two planes intersecting at that point, one of which is parallel to the reference axis, is always constant (Fig. 7). At any other point between the same two planes apparent changes in orthogonality occur due to the effect of crosswind about the mutually perpendicular axis at the intersection of the two planes, and due to the effects of angular pitch and/or angular yaw in both planes in question. Hence if the relevant angular pitch and/or yaw of the two planes that intersect at that axis are known, then for a given position along the axis the orthogonality between the two planes can be calculated, i.e. changes or apparent changes in orthogonality are not separate errors, but are pitch, yaw and crosswind errors.

Straightness measurements are taken at the probe position whereas the pitch, yaw and crosswind angles are measured directly on the carriage whose motion is being checked. Some part of the straightness errors can be directly attributed to the effects of crosswind about an axis in the direction of carriage motion, and to the linear effects of angular pitch and angular yaw in the other two mutually perpendicular directions. It is necessary to separate all these effects out to give a true reading for the straightness of carriage motion.

Consider some point O in a specified volume bounded by the x, y and z direction axes, and take this point as a reference for subsequent work. Take one axis through the reference point as the reference axis, and let angular errors in the clockwise direction be negative. Let all errors at point O be zero.

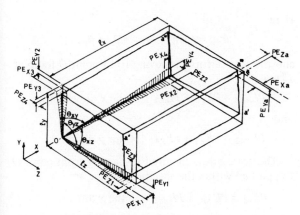

Figure 8. Positional errors in a volume at point a.

Consider some point a at distances of l_x, l_y and l_z along the axes from the reference point, and consider each plane separately (Fig. 8).

Let angle α = angular crosswind

$\qquad \beta$ = angular pitch

$\qquad \gamma$ = angular yaw

For the XZ plane:

Measured orthogonality error θ_{xz}

$\qquad$ = Normal orthogonality error NOE_{xz}

$\qquad$ + error due to crosswind about $y = \alpha_y$

$\qquad$ + error due to yaw in the x direction = γ_x

$\qquad$ + error due to yaw in the z direction = γ_z

i.e.

$$\theta_{xz} = NOE_{xz} + \alpha_y + \gamma_x + \gamma_z \qquad (29)$$

or

$$NOE_{xz} = \theta_{yz} - (\alpha_y + \gamma_x + \gamma_z) \qquad (30)$$

When moving the probe or tool point a distance of l_z in the z direction positioning errors occur, due to the angular errors and the magnitude of l_z in the x direction

$$PE_{x1} = l_z \sin (NOE_{xz} + \alpha_y + \gamma_x + \gamma_z) \qquad (31)$$

and in the z direction

$$PE_{z1} = l_z - l_z \cos (NOE_{xz} + \alpha_y + \gamma_x + \gamma_z) \qquad (32)$$

Also when moving the probe or tool point a distance of l_x in the x direction positioning errors occur in the X direction

$$PE_{x2} = l_x - l_x \cos (NOE_{xz} + \alpha_y + \gamma_x + \gamma_z) \qquad (33)$$

and in the z direction

$$PE_{z2} = l_x \sin (NOE_{xz} + \alpha_y + \gamma_x + \gamma_z) \qquad (34)$$

Equations (31), (32), (33) and (34) are not strictly true, but the differences have been neglected on a previous assumption.

The total errors due to angular errors in moving the probe from point O to point a' are

$\qquad$ In direction x, $\quad PE_{x1} + PE_{x2} = PE_{xa1}$

$\qquad$ In direction z, $\quad PE_{z1} + PE_{z2} = PE_{za1}$

Now consider the YZ plane:

Measured orthogonality error θ_{yz}

$\qquad$ = Normal orthogonality error NOE_{yz}

$\qquad$ + error due to crosswind about $x = \alpha_x$

$\qquad$ + error due to yaw in z direction = γ_z

$\qquad$ + error due to pitch in the y direction = β_y

i.e.

$$\theta_{yz} = NOE_{yz} + \alpha_x + \gamma_z + \beta_y \qquad (35)$$

or

$$NOE_{yz} = \theta_{yz} - (\alpha_x + \gamma_z + \beta_y) \qquad (36)$$

When moving the probe or tool point a distance of l_z in the z direction positioning errors occur, due to the angular errors and the magnitude of l_z, in the y direction

$$PE_{y1} = l_z \sin (NOE_{yz} + \alpha_x + \gamma_z + \beta_y) \qquad (37)$$

and in the z direction:

$$PE_{z3} = l_z - l_z \cos (NOE_{yz} + \alpha_x + \gamma_z \beta_y) \qquad (38)$$

When moving the probe or tool point a distance of l_y in the Y direction, errors in the Y direction are

$$PE_{y2} = l_y - l_y \cos (NOE_{yz} + \alpha_x + \gamma_z + \beta_y) \qquad (39)$$

and in the z direction

$$PE_{z4} = l_y \sin (NOE_{yz} + \alpha_x + \gamma_z + \beta_y) \qquad (40)$$

The total errors due to angular errors in moving the probe from point O to point a'' are

$\qquad$ In direction y, $\quad PE_{y1} + PE_{y2} = PE_{ya1}$

$\qquad$ In direction z, $\quad PE_{z3} + PE_{z4} = PE_{za2}$

Now consider the XY plane:

Measured orthogonality error θ_{xy}

$\qquad$ = Normal orthogonality error NOE_{xy}

$\qquad$ + error due to crosswind about $z = \alpha_z$

$\qquad$ + error due to pitch in the x direction = β_x

$\qquad$ + error due to pitch in the y direction = β_y

i.e.

$$\theta_{xy} = NOE_{xy} + \alpha_z + \beta_x + \beta_y \tag{41}$$

or

$$NOE_{xy} = \theta_{xy} - (\alpha_z + \beta_x + \beta_y) \tag{42}$$

When moving the probe or tool point a distance of l_y in the y direction, errors in the X direction are

$$PE_{x3} = l_y \sin(NOE_{xy} + \alpha_z + \beta_x + \beta_y) \tag{43}$$

and in the y direction

$$PE_{y3} = l_y - l_y \cos(NOE_{xy} + \alpha_z + \beta_x + \beta_y) \tag{44}$$

When moving the probe or tool point a distance of x in the x direction, errors in the x direction are

$$PE_{x4} = l_x - l_x \cos(NOE_{xy} + \alpha_z + \beta_x + \beta_x) \tag{45}$$

and in the y direction

$$PE_{y4} = x \sin(NOE_{xy} + \alpha_z + \beta_x + \beta_y) \tag{46}$$

The total errors due to angular errors in moving the probe from point O to point a''' are:

$$\text{In direction } x, \quad PE_{x3} + PE_{x4} = PE_{xa2}$$
$$\text{In direction } y, \quad PE_{y3} + PE_{y4} = PE_{ya2}$$

$\therefore$ The total errors incurred in moving within the specified volume from point O to point a^* are

In the x direction, $\quad PE_{xa} = PE_{xa1} + PE_{xa2} \tag{47}$

In the y direction, $\quad PE_{ya} = PE_{ya1} + PE_{ya2} \tag{48}$

and in the z direction, $\quad PE_{za} = PE_{za1} + PE_{za2} \tag{49}$

These errors can also be found by considering the Abbé offset principle. From Fig. 8

$$PE_{x1} = l_z \sin \theta_{xz} \tag{50}$$

$$PE_{z1} = l_z - l_z \cos \theta_{xz} \tag{51}$$

$$PE_{x2} = l_x - l_x \cos \theta_{xz} \tag{52}$$

$$PE_{z2} = l_x \sin \theta_{xz} \tag{53}$$

Here

$$\theta_{xz} = NOE_{xz} + \alpha_y + \gamma_x + \gamma_z$$

and the total errors in moving from point O to point a' are

$$PE_{xa1} = PE_{x1} + PE_{x2} \quad \text{in the } x \text{ direction}$$

and

$$PE_{za1} = PE_{z1} + PE_{z2} \quad \text{in the } z \text{ direction}$$

Also from Fig. 8

$$PE_{y1} = l_z \sin \theta_{yz} \tag{54}$$

$$PE_{z3} = l_z - l_z \cos \theta_{yz} \tag{55}$$

and

$$PE_{y2} = l_y - l_y \cos \theta_{yz} \tag{56}$$

$$PE_{z4} = l_y \sin \theta_{yz} \tag{57}$$

Here

$$\theta_{yz} = NOE_{yz} + \alpha_x + \gamma_z + \beta_y$$

and the total errors in moving from point O to point a'' are

$$PE_{ya1} = PE_{y1} + PE_{y2} \quad \text{in the } y \text{ direction}$$

and

$$PE_{za2} = PE_{z3} + PE_{z4} \quad \text{in the } z \text{ direction}$$

Also from Fig. 8

$$PE_{x3} = l_y \sin \theta_{xy} \tag{58}$$

$$PE_{y3} = l_y - l_y \cos \theta_{xy} \tag{59}$$

$$PE_{x4} = l_x - l_x \cos \theta_{xy} \tag{60}$$

$$PE_{y4} = l_x \sin \theta_{xy} \tag{61}$$

Here

$$\theta_{xy} = NOE_{xy} + \alpha_z + \beta_x + \beta_y$$

and the total errors in moving from point O to point a''' are

$$PE_{xa2} = PE_{x3} + PE_{x4} \quad \text{in the } x \text{ direction}$$

and

$$PE_{ya2} = PE_{y3} + PE_{y4} \quad \text{in the } y \text{ direction}$$

The total errors incurred in moving from point O to point a^* within the specified volume are

$$PE_{xa} = PE_{xa1} + PE_{xa2} \quad \text{in the } x \text{ direction}$$

$$PE_{ya} = PE_{ya1} + PE_{ya2} \quad \text{in the } y \text{ direction}$$

and

$$PE_{za} = PE_{za1} + PE_{za2} \quad \text{in the } z \text{ direction}$$

as before.

If the values of the NOE for each plane pair and all the pitch, yaw and crosswind errors are known, then the volumetric error for any given point a from the reference point O can be expressed as:

$$\text{Volumetric error} = \sqrt{PE_{xa}^2 + PE_{ya}^2 + PE_{za}^2} \tag{62}$$

Now consider the probe or tool point to be moved to some other point b in the specified volume. Further errors have been introduced into the system due to this movement. The magnitude of these errors depend on the magnitude of the probe or tool point movement from the reference point. Again by consideration of the Abbé offset principle these errors are (from Fig. 9):

$$PE_{x5} = (l_z + \delta l_z) \sin \theta_{xz} \tag{63}$$

or

$$PE_{x5} = PE_{x1} + \delta l_z \sin \theta_{xz} \tag{64}$$

$$PE_{z5} = (l_z + \delta l_z) - (l_z + \delta l_z) \cos \theta_{xz} \tag{65}$$

or

$$PE_{z5} = PE_{z1} + \delta l_z (1 - \cos \theta_{xz}) \tag{66}$$

and

$$PE_{x6} = (l_x + \delta l_x) - (l_x + \delta l_x) \cos \theta_{xz} \tag{67}$$

or

$$PE_{x6} = PE_{x2} + \delta l_x (1 - \cos \theta_{xz}) \tag{68}$$

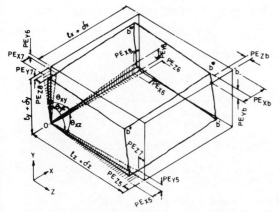

Figure 9. Positional errors in a volume at point b.

$$PE_{z6} = (l_x + \delta l_x) \sin \theta_{xz} \qquad (69)$$

or

$$PE_{z6} = PE_{z2} + \delta l_x \sin \theta_{xz} \qquad (70)$$

and the total errors in moving the probe or tool from point O to point b' are

$$PE_{xb1} = PE_{x5} + PE_{x6} \quad \text{in the } x \text{ direction}$$

$$PE_{zb1} = PE_{z5} + PE_{z6} \quad \text{in the } z \text{ direction}$$

N.B. Equations (64), (66), (68) and (70) are only true if α_y, γ_x and γ_z are equal at both points a and b. Since these alternative ways of writing the equations can also apply in the other two planes no further reference will be made to them.
Also from Fig. 9

$$PE_{y5} = (l_z + \delta l_z) \sin \theta_{yz} \qquad (71)$$

$$PE_{z7} = (l_z + \delta l_z) - (l_z + \delta l_z) \cos \theta_{yz} \qquad (72)$$

and

$$PE_{y6} = (l_y + \delta l_y) - (l_y + \delta l_y) \cos \theta_{yz} \qquad (73)$$

$$PE_{z8} = (l_y + \delta l_y) \sin \theta_{yz} \qquad (74)$$

and the total errors in moving the probe or tool from point O to point b'' are

$$PE_{yb1} = PE_{y5} + PE_{y6} \quad \text{in the } y \text{ direction}$$

and

$$PE_{zb2} = PE_{z7} + PE_{z8} \quad \text{in the } z \text{ direction}$$

Finally, also from Fig. 9

$$PE_{x7} = (l_y + \delta l_y) \sin \theta_{xy} \qquad (75)$$

$$PE_{y7} = (l_y + \delta l_y) - (l_y + \delta l_y) \cos \theta_{xy} \qquad (76)$$

and

$$PE_{x8} = (l_x + \delta l_x) - (l_x + \delta l_x) \cos \theta_{xy} \qquad (77)$$

$$PE_{y8} = (l_x + \delta l_x) \sin \theta_{xy} \qquad (78)$$

and the total errors in moving from point O to point b''' are

$$PE_{xb2} = PE_{x7} + PE_{x8} \quad \text{in the } x \text{ direction}$$

and

$$PE_{yb2} = PE_{y7} + PE_{y8} \quad \text{in the } y \text{ direction}$$

L

The total errors incurred in moving the probe or tool point from point O to point b^* within the specified volume are

$$PE_{xb} = PE_{xb1} + PE_{xb2} \quad \text{in the } x \text{ direction}$$

$$PE_{yb} = PE_{yb1} + PE_{yb2} \quad \text{in the } y \text{ direction}$$

and

$$PE_{zb} = PE_{zb1} + PE_{zb2} \quad \text{in the } z \text{ direction}$$

The volumetric error between any two points a and b in the specified volume can be expressed as

$$\sqrt{(PE_{xa} - PE_{xb})^2 + (PE_{ya} - PE_{yb})^2 + (PE_{za} - PE_{zb})^2}$$

$$(79)$$

If the geometric errors are measured along a number of axes in each direction in the form of a lattice (Fig. 10) and the positional errors plotted for each, a

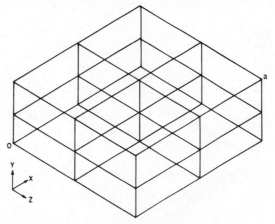

Figure 10. Lattice of measuring axes in a volume.

possible maximum volumetric error occurs when either

$$\sqrt{[(PE_{xa\max} - PE_{xb\min})^2 + (PE_{ya} - PE_{yb})^2 + (PE_{za} - PE_{zb})^2]} \qquad (80)$$

or

$$\sqrt{[(PE_{ya\max} - PE_{yb\min})^2 + (PE_{xa} - PE_{xb})^2 + (PE_{za} - PE_{zb})^2]} \qquad (81)$$

or

$$\sqrt{[(PE_{za\max} - PE_{zb\min})^2 + (PE_{xa} - PE_{xb})^2 + (PE_{ya} - PE_{yb})^2]} \qquad (82)$$

is a maximum, whichever is greatest.

However, it is possible that either or both of the last two squared terms in expressions (80), (81) and (82) could give a zero value, and the actual maximum volumetric error for the specified volume may occur where any of the squared terms in expression (79) gives a value slightly less than maximum while the other two squared terms give appreciable values.

In order to find the maximum volumetric error it is necessary to find the values of the positional errors at any points a and b which make equation (79) a maximum and to find the coordinates of the points a

and b. This can be readily done on a computer by presenting all the positional error data for each point in the lattice in the form of x, y and z direction arrays, similar to the case for the planar error. An extended program has been written which will, in turn, take the x, y and z direction errors at each point in the lattice, together with the errors, in turn, at each other point in the lattice to solve equation (79) for each pair of points. The program then demands a print-out of the maximum value of equation (79) together with the matrix coordinates of the errors which produced that maximum.

Straightness errors

The actual pitch and yaw straightnesses of the carriage motions at any point can be found by subtracting the appropriate errors, due to the angular

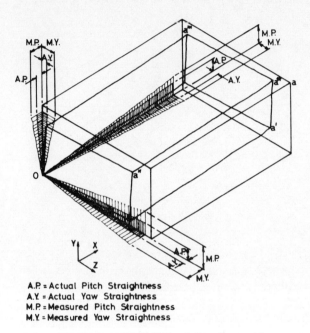

A.P. = Actual Pitch Straightness
A.Y. = Actual Yaw Straightness
M.P. = Measured Pitch Straightness
M.Y. = Measured Yaw Straightness

Figure 11. Straightness errors in a volume.

changes, for the appropriate straightness measurement taken along any axis parallel with the carriage motions (Fig. 11). The measured pitch straightness of x carriage motion, e.g.

$E_{Y(MXP)}$ = actual pitch straightness of x carriage motion

+ positional errors due to angular errors in the other two directions

+ positional errors due to the Abbé offset in the other two directions

i.e.

$E_{Y(MXP)}$ = actual pitch straightness + $PE_{y7} + PE_{y8}$

∴ Actual pitch straightness of the x carriage motion

$$= E_{Y(MXP)} - (PE_{y7} + PE_{y8}) \qquad (83)$$

By similar methods:

Actual yaw straightness of the x carriage motion

$$+ E_{Z(MXY)} - (PE_{z5} + PE_{z6}) \qquad (84)$$

Actual pitch straightness of the y carriage motion

$$= E_{x(MYP)} - (PE_{x7} + PE_{x8}) \qquad (85)$$

Actual yaw straightness of the y carriage motion

$$= E_{Z(MYY)} - (PE_{z7} + PE_{z8}) \qquad (86)$$

Actual pitch straightness of the z carriage motion

$$= E_{y(MZP)} - (PE_{y5} + PE_{y6}) \qquad (87)$$

Actual yaw straightness of the z carriage motion

$$= E_{x(MZY)} - (PE_{x5} + PE_{x6}) \qquad (88)$$

Control and measuring system errors

These are the repeatability of positioning of any carriage by the control system, and the accuracy with which the lines are engraved on the scales. The repeatability of the positioning can be found by experiment, and the accuracy of engraving can be found from the manufacturer. These errors are direct additions to the positional errors in the respective directions.

Accuracy statements

When stating a displacement, planar or volumetric accuracy it is recommended that the following statements should also be given.

(1) The axis, plane or volume concerned.
(2) The straightnesses of motions of the machine carriages.
(3) The coordinates of the two points between which the error occurs.
(4) The N.O.E. of all the plane pairs.

CONCLUSIONS

It is possible to measure the individual parameters of geometry of a machine tool, such as pitch, yaw, crosswind and orthogonality errors in each of the three dimensions, and to synthesize these errors in order to find an overall volumetric error between any two points in the specified working volume of the machine tool. Furthermore the position, sense and magnitude of the maximum possible volumetric error can be found.

From the parameters measured it is possible to calculate the true straightness of motion of all the machine carriages.

This method of analysis will enable a machine tool manufacturer to pin-point the parameter which contributes most to the volumetric error. The necessary measures can then be taken if the volumetric accuracy is outside acceptable limits. The single error value can also be used to compare the accuracies of machines of a similar type.

ACKNOWLEDGMENTS

The authors gratefully acknowledge the advice and help given by Mr P. A. McKeown, Director of the Cranfield Unit of Precision Engineering in the preparation of this paper and to the Science Research Council for their sponsorship of the research project from which this paper was derived.

DISCUSSION

Q. J. J. Ashton. Has an analysis of the uncertainty of determination of volumetric accuracy been made (further to Dr Hemingray's earlier remarks)?

The exchange of ideas for the direct measurement of volumetric accuracy should be encouraged.

What problems arise from our lack of knowledge of volumetric accuracy?

A. The uncertainty of determination of the volumetric accuracy has been examined and is, in fact, calculated within the computer program. This information can easily be given with any volumetric accuracy specification.

There do not seem to be many technical problems due to the lack of knowledge of volumetric accuracy. The main problems will lie in the education of machine tool manufacturers and users in the use of this parameter.

Q. W. J. Wills-Moren, Cranfield Unit for Precision Engineering. I have not studied the mathematics of the paper but it is my understanding that the input data relating to pitch, yaw, crosswind and orthogonality errors consists of *actual measured* information from a specific machine. If this is so the computed errors between specific points in the volume are actual errors and not theoretical maximums as Dr Sartorio suggests.

Am I correct?

A. The computed errors are true errors, statistically derived from actual measured information.

Q. B. J. Davies, G.M.T. (1) Individual errors in roll, pitch, yaw, do not contribute equally to volumetric error of a machine tool so it is not certain that volumetric error will be satisfactory if ever individual slideway errors do not exceed limits.

(2) The sum of individual maximum errors gives a very pessimistic view of the working volumetric error of a machine tool.

A. On the first. point, it is true that individual errors do not contribute equally to volumetric accuracy. It is not intended that volumetric accuracy of a machine tool should replace all other specifications, but should complement them. The overall accuracy or volumetric accuracy will give a comparison between the accuracy of similar machines, so that, for example, the production engineer can decide more easily what work should be allotted to each machine.

On the second point, the value of volumetric error for a given machine need not be so pessimistic; again, it is a matter of education. A volumetric error of 0·002 inch would indicate a good machine in fairly new condition, whereas a workshop machine about ten years old may have a volumetric accuracy of 0·030 inch.

Q. P. Brooks, Giddings and Lewis Frazer Limited. The paper is a valuable contribution to the concepts and problems of specifying the geometric performance of a machine in terms of volumetric accuracy parameters. Unfortunately the ensuing discussion has placed us on the point of concluding that because machine accuracy is not specified in terms of this relatively new concept components are unwittingly being produced without tolerance requirements! This is not the case. The present methods of specifying accuracy are directly related to both the geometric configuration of the machine and the functional requirements of the components being processed.

If any particular manufacturing alignment checkout procedure using conventional methods allows error accumulations between the various parameters such that functional requirements are not met, then there is merely an error in procedure which can be corrected.

If volumetric accuracy can be shown to be a functional requirement of components, then it should be specified, but I suspect that its main value will lie in ensuring that the compounded errors, which are known to result from the existing orthogonal approach, have been fully considered.

A. The questioner's conclusions are wrong. The volumetric accuracy does not, in any way, indicate that components are produced without tolerance requirements. The volumetric accuracy is derived from the tolerances set on individual parameters and compounded together in such a way that a comparison may be drawn between similar machines. Machines of higher volumetric accuracy will produce components of an inherently higher accuracy than those produced on a machine which has a lower volumetric accuracy. It may be possible for a designer to specify which grade of machine a specific component should be made on.

NUMERICAL INSPECTION MACHINES

by

H. OGDEN*

SUMMARY

The advent of NC Machine Tools produced a considerable pressure on conventional inspection facilities and led to the development of inspection machines working with co-ordinate displays and eminently suitable to inspect first-off parts from milling, drilling, and boring machines.

Inspection machines have now been associated with mini-computers and a range of software programs has been developed to simplify decision making in first-off inspection of complex parts. Numerical inspection machines working in point-to-point or continuous scanning mode have been developed.

The paper will cover the progression of development of inspection machines to data and focus attention on the computerized inspection machine and the developed N.C. capability.

INTRODUCTION

Through the 1950's my Company was deeply involved in the development of Numerical Control Systems and their application to Machine Tools. A key factor in achieving accurate control of Numerical Control Machines was the need to develop a linear measuring system, and it was the development of Moiré Fringe Measuring Systems, in the case of Ferranti, which provided this essential feed-back element.

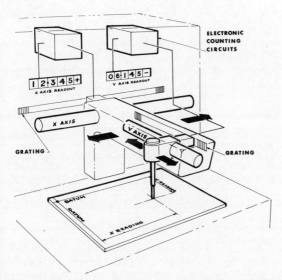

Figure 1. Isometric view—inspection machine working arrangement.

The successful development of a 3D Continuous Path NC System and its application to Vertical Milling Machines produced complex components with remarkable economy in machining time. Proving the accuracy of such parts by conventional inspection methods was shown to be completely inadequate for this new emerging technology of machining, and led to consideration on how a corresponding improvement could possibly be achieved in the inspection of the products thus created.

A solution to inspection was sought and led to the concept of a Coordinate Inspection Machine, using mechanical kinematic principles, in association with Moiré Fringe Measurement and Digital Display. This concept is as illustrated in Fig. 1.

MACHINE DEVELOPMENTS

The original Size 1 Co-ordinate Inspection Machine had a capacity of 24 x 15 in (610 x 380 mm), with 10 in (254 mm) of daylight (Fig. 2). The mechanical design of the machine is based on kinematic principles, the endeavour from the outset to be able to move the machine freely with low friction in its working area, and use a range of mechanical probes to pick off dimensional information from the component.

The Size 1 Machine was developed into the Size 2 and Size 3 Co-ordinate Inspection Machine, having 24 x 15 in (610 x 380 mm) of travel but with a daylight of 14 and 24 in (356 and 610 mm) respectively. There was a need also for an increased size

* Ferranti Limited, Dalkeith, Scotland

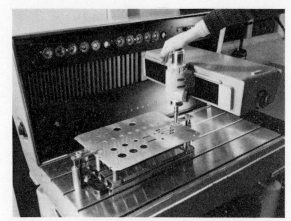

Figure 2. Original Size One Inspection Machine.

Inspection Machine to cater for larger manufactured components in the capital goods industry, which led to the design and development of the Size 4 and 5 Co-ordinate Inspection Machines (Fig. 3), which were based on mechanical principles of the Size 2/3 Machines and, indeed, used a number of common

Figure 3. Size 4 inspection machine. C.O.I.D. award.

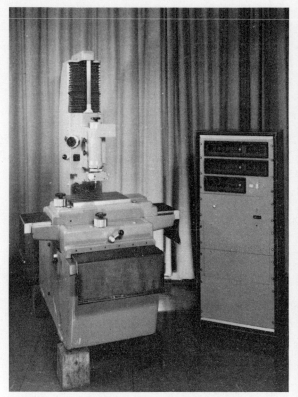

Figure 4. Size one measuring machine.

mechanical parts. It was a Size 4 Co-ordinate Inspection Machine which won a C.O.I.D. Design Award for capital goods in 1967.

Coincident with the development of the Size 4/5 Inspection Machine it was considered that there would be the need to make a high precision 10^3 in Machine to suit the needs of high precision parts manufactured by Jig Boring and Jig Grinding techniques. This led to the development, in conjunction with Coventry Gauge & Tool, of a compact vertical column machine with a compound table movement as illustrated in Fig. 4.

Subsequent development of the Size 1 Measuring Machine established 0·00005 in (0·0013 mm) resolution and a planar accuracy of ±0·0001 in (0·0025 mm), in conjunction with a novel tram-a-matic probe (Fig. 5).

The tram-a-matic probe greatly enhanced the facilities of the Size 1 Measuring Machine. When measuring along the X and Y axes, it serves as a sensitive indicator of contact, but its main use is as a device for trammelling holes. The tram-a-matic probe provides an extra axis of measurement (probe radius

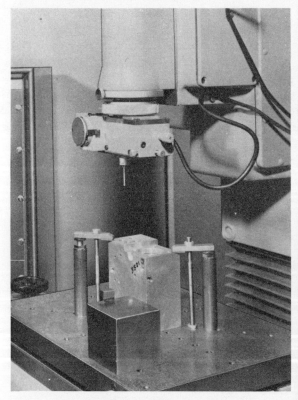

Figure 5.   Tram-a-matic probe.

P) and, in conjunction with an optical switching system in the quill of the Measuring Machine, provides signals for the logic system giving the read-out of the positional error of holes from nominal.

If the probe datum is positioned at the actual centre of the hole (by moving the table until the readings on the X and Y error indicators is zero), and the hole is re-trammelled at this new position, digital

Figure 6. Hydrocord machine.

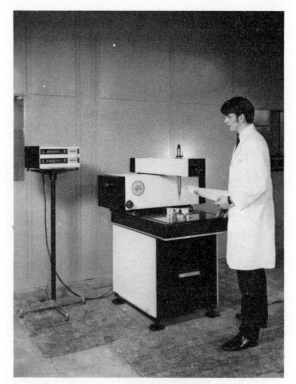

Figure 7. Mercury machine.

indicators show the radius of the hole and any out-of-roundness.

In the search for large capacity, high accuracy performance based on the principles of constraint inspection already well established in the Size 2, 3, 4, and 5 Co-ordinate Inspection Machines, the Hydrocord Machine was developed, having a capacity of 30 x 20 in (762 x 508 mm), with a daylight of 20 in. The machine geometry was that of a fixed bridge design to provide maximum machine stiffness and maximum accuracy on the X and Y axes of the Machine (Fig. 6). In order to minimize friction, hydrostatic pads replaced rolling bearing elements, and use was made of feed-back in order to maintain a constant oil film gap under changing load conditions.

An interesting design feature was incorporated in the Hydrocord Machine for use of an inclined weight to counter-balance the Z Axis of the Machine. In order to compensate for changes in weight of ancillary units fitted to the probe column of the machine, provision has been made to change the angle of the inclined plane and thus change the effective weight balance on the vertical column.

More recently my Company has introduced other Inspection Machines into the product line, including the Mercury Inspection Machine, which is a new concept in modern design which offers the end user a basic Machine which will fit into an existing surface table, together with additional items which form the elements to fit together into convenient and compact Inspection Centres (Fig. 7).

ACCESSORIES

The initial concept of Ferranti Inspection Machines was to use the principle of constraint inspection. A

range of interchangeable taper probes which fit into the probe column are used to automatically locate hole centres and other physical features of the component being inspected. It was found by experimentation that remarkably rapid and accurate checks were available by this means with repeatability of the order of the digital resolution.

It was recognized that certain measurements needed to be undertaken which required modified taper probe techniques. Bore locations, for instance, of substantial diameter are measured by means of range extenders as illustrated in Fig. 8. Ball Bushes are used in conjunction with taper probes in order to locate centres of small bores of high precision from within the bore itself.

Inspection Machines initially were concerned with measurement in the X-Y co-ordinate planes, and as the need arose to measure in the Z axis, accurate axial relocation of probes in the Z axis was required, which

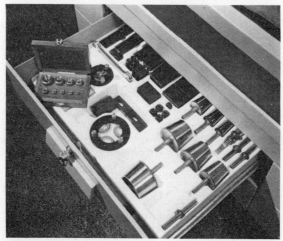

Figure 8. Inspection machine ancillaries: probes; range extenders; ball bushes.

led to the development of the probes which were subsequently used on the Hydrocord Machine and the 5*. The need to measure in the Z axis led to the development of Z axis units for the range of Inspection Machines, and typical of these designs is that of the 5*, illustrated in Fig. 9. The unit has 20 in (508 mm) of travel and is counter-balanced by means of a Tensator spring. This particular unit makes use of glass Moiré Fringe Scales.

Figure 9. 5 star Z axis bearing arrangement.

The need to trammel bores led to the development of a rotating probe, whilst the emerging need to measure three dimensionally led to the development of four-way probe devices so that measurement could be made on the exposed five faces of a cube located on its sixth face. The need to align work pieces precisely to the X and Y axes of the inspection machine led to the use of a swivelling sub-table in early inspection machines, and later to the swivelling of the inspection machine work table. Accommodation of large components led to the development of the variable daylight machine by means of a power elevating table which has also swivelling capability. Thus we see the evolution of mechanical devices that facilitate the mechanical process of making three dimensional measurements on manufactured components.

DATA PROCESSING

Alongside the developing mechanical features of Inspection Machines there was an increasing activity in the electronic capability of these Machines. The electronic measurement display was extended from two axes to the three axes display of machine position.

Initially, a fixed format strip printer was offered to provide two axes printout of the measured dimensions, together with a sequence number to identify the specific measurements recorded. Interfacing with the electronic display counter was simple but the move to three axes display could not readily be accommodated in the standard printout units available at that time. Further, with this parallel type printer small machine movements during the printout sequence could result in printout errors of large magnitude and movement detection circuits were incorporated to indicate possible printout errors.

The interfacing of a teletype unit to the electronic display gave greater flexibility of operation and removed limitations imposed by the simple strip printer. Further, operator instructions and line by line printing of the nominal dimensions are provided by the use of the teletype tape input facility.

A further unique development of this printout called 'AUTOCHECK' indicates out of tolerance dimensions measured by the Co-ordinate Inspection Machine when the printout is provided with the upper and lower dimensions of each feature being measured (Fig. 10).

Figure 10. Printout unit with autocheck and Teletype 33ASR.

In the U.S.A. attention was paid to more universal ways of data processing and comparison; use was made of the mini-computers then coming onto the market to carry out a data-processing function which compares in detail the nominal information with actual information and makes automatic GO—NO GO decisions. The real capability of the mini-computer became apparent when programmed to carry out calculations derived from actual measurements which gave additional inspection information not readily available by other means. For instance, cartesian co-ordinate measurements can be converted readily into polar co-ordinate measurements; hole centre location and diameter measurements of components can be carried out by three or four point routines within the bore, which simplifies the measurement routine.

The inspection of colour television masks in the

U.S.A. led to the development of area scanning programs, using the mini-computer; these programs were further developed for scanning of profiles such as cams used in high speed packaging machinery and the automobile. A listing is given in Fig. 11 which illustrates the emerging range of software programs which are currently available for computerized inspection machines, which gives some idea of the rapidly developing power of this technique.

<u>MEASURING MACHINE COMPUTER PROGRAMS</u>

1. Measurement Monitor

    a) Automatic Alignment Computation

    b) Difference from Nominal Computation

    c) Out-of-Tolerance Computation

    d) Co-ordinate Conversion (Cartesian to Polar)

    e) Polar Deviation (True Position)

    f) Three Point Radius/Diameter

2. Linear Scanning

3. Angular Scanning

4. Random Check

5. Electronic Probe

6. Computer Controlled Measuring
    Machine Programs

    a) Positioning
    b) Measuring (includes 1 to 3)
    c) Statistical

7. NC Tape Generation

Figure 11. Software list: standard, linear, angular, CNC and NC tape output.

N.C. INSPECTION

The mini-computer effectively complements the mechanical configuration of the Inspection Machine and presents the possibility of being able to measure, inspect, and verify complex components which are produced by N.C. machinery. The Conquest Mark II Inspection Machine illustrates the class of machine which is now available for these parts, being interfaced with the DEC PDP8E Computer as a standard package (Fig. 12).

To complement the facility of the mini-computer electronic probes have been developed which enable precision touch measurements to be taken by the machine without the necessity for interchanging of probe tips, etc.

Complex components produced by modern N.C. Machine Tools require a rapid inspection procedure. This now can be achieved with Numerical Inspection Machines.

Programs have been made available for computerized inspection machines which enable components to be physically measured and provide a paper tape output which is acceptable as the input to a number of established N.C. Systems.

CNC Inspection Machines are now available which work in either point-to-point mode or in a continuous path mode. In the case of automatic point-to-point

Figure 12. Conquest Mark II with PDP8.

inspection an early development was to use digital stepping motors on the axes to drive the machine to the computer demanded position in the X-Y axes, and to use the Z axis to locate on the feature being inspected with a conventional taper probe with the drive mechanisms declutched. Thus the machine would automatically centre in the actual position giving automatic comparison with the nominal derived from the computer input. Automatic routines on complex point-to-point type components have been carried out in this way. Automatic contour following has been achieved in conjunction with a servo following Z axis unit. A machine of this type is used for the automatic inspection of three dimensional profile on television masks as illustrated in Fig. 13.

The Saturn Machine is an example of a large capacity (2 metres x 1.25 metres x 1 metre) Inspection Machine designed for automatic N.C. Inspection. The machine has servo drives which facilitate rapid traverse, creep and incremental digital feed. It is designed to work with electronic probes

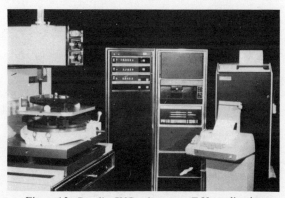

Figure 13. Bendix CNC colour type T.V. application.

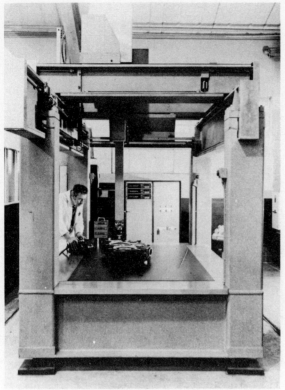

Figure 14. Saturn.

which sense the work piece and provide control signals to drive the axes of the machine and the electronic measuring system. An essential feature of such a machine is the use of a mini-computer suitably programmed to carry out inspection routine calculations which maximize the effectiveness of the measuring machine remarkably.

CONCLUSIONS

The challenge to the inspection process as presented by the components produced by N.C. Machine Tools led to the development of the Co-ordinate Inspection Machine. Starting from the relatively simple 2 dimensional measurements it has developed in capacity and in mechanical capability to accommodate components produced in industry by these means. Additional electronic developments, probing techniques, and data-processing have complemented its mechanical measuring capability and has resulted in machines being developed from a manual machine to semi-automatic and automatic in function. These developments complement the progress which has been made in the development of N.C. machine tools and provide a powerful and economic inspection facility to backup the manufacture of N.C. components in industry.

A SPECIAL CNC MACHINE FOR SHAPE INSPECTION OF WANKEL ENGINE CASES

by

J. H. M. LEDOCQ*

SUMMARY

The control of the shape of Wankel engine bores makes a requirement for a special machine which must be able to generate the theoretical trochoidal profile and to measure with a fairly high resolution the departure between this reference and the actual profile. The author shows the reasons why a valuable solution to this problem can be reached by use of a computer controlled digital machine provided it is of an original and appropriate design. The specialized grinding machine used in the production field must use a purely mechanical process to generate directly the profile to be machined; on the contrary, due to the high precision level required, the metrology machine needs very sophisticated guiding mechanisms in spite of the low level of the loads involved; it seems difficult to associate those mechanisms giving the proper degrees of freedom, with mechanical boundary insuring the profile generation as it can be done in a production machine. Moreover, the use of digital control and of computer numerical control (CNC), leads to an easy way of adjusting the parameters, thus insuring a full versatility which was completely unnecessary in the grinding machine. Industrial requirement for such a control machine exists now, this requirement should be a generalization of the well-known roundness control machines. The author wants to present here the project in its early stage, of such a machine, the design of which has been based upon the theoretical analysis of the trochoidal movement generation and its possible discrepancies with practical machining.

INTRODUCTION

The control of the shape quality of the cross internal section of a Wankel Stator rests upon the generation of the theoretical profile by means of a suitable mechanism. The measuring transducer must be given such a movement that in the ideal situation where its tip would follow a perfect bore, its deviation would remain zero all around the profile. Moreover, the axis of the transducer must be the normal to the theoretical profile at the contact point in such a way, that the deviation between the reference and the actual profiles is measured along this normal. The reason for this assumption is that a family of trochoidal profiles which are to differ by one single dimensional parameter, are to be uniformly distant from each other. All the profiles of such a family are said to have the same 'Shape'. This is true for every non-circular continuous profile of the functional surface of a machine part surface, through which this part is in relation with an alternative[1].

Those conditions, relevant to the profile generation, are assigned to the mechanism of the grinding machine too, but, over and above, there is one more compulsory condition: the normal to the point of contact between the trochoidal profile and the wheel circumference must remain parallel to a fixed direction. This condition results from the fact that it must be possible to generate any profile of the family defined here over (that is uniformly distant) by a single adjustment on a cross slide of the machine, and this must be possible when the grinding process is underway. The fact that the stator itself is a relatively flat and light-weight workpiece with regard to the complete wheel head of a production grinder, make it preferable to give all the generating movement to the workpiece table and keep the wheel head stationary[2]. The control machine can be designed more freely because its 'tool' is a light-weight small displacement transducer which can be given any movement except a continuous revolution.

The automobile industry production rate, makes it possible to use one complete generating mechanism for each stator geometrically machined, without any adjustment capabilities that ought to be fitted. A good condition of precision can thus be reached with a comparatively small and low cost machine, including a simple, purely mechanical, generating device, involving the exact number of necessary elements which appear in the cinematic definition of

---

* Lecturer, Department of Mechanical Engineering, Faculté Polytechnique de Mons, Belgium

the modified trochoid[3]. To define this curve, three geometrical parameters must be used

the mean radius R_m
the excentricity E
the equidistance A.

The two former parameters lead to a reference true trochoid of which the theoretical machined profile is equally distant of the length A. The theoretical analysis of the shape irregularities happening in such multiparameter profiles shows[4] that it is possible to control a machine profile by generating the reference trochoid (R_m, E) and calculate the shape error according to any particular assumption from the measured variations of A, whichever are the primary errors on R_m or E in the production machine.

Two ways of designing the control machine can then be anticipated: the first one proceeds from the design of the grinding machine and consists of building several pure mechanical systems, each of which is able to generate one theoretical profile only, using the fact that the level of loads involved by the measuring process is one order of magnitude less than that of the machining process; the building cost of such a machine would be comparatively low, but it must be taken into account that the precision level required is very high (similar to that of a roundness control machine) and such a machine will be able to generate one theoretical reference only.

It is our opinion that a good solution from the point of view of economical feasibility versus precision performance compromise, is given by a versatile machine of original design where the reference generation is performed through a computer numerical control system.

This machine is in fact able to control any bore which profile is a non-circular but a continuous closed curve, at the only condition that the polar radius never exceeds a range which is an indication of the machine overall capacity.

The workpiece is a heavy but round-shaped hollow piece and it can easily be located on a table which only has to be given a rotating movement. So far only the workpiece is concerned, the machine will be quite similar to a roundness control one.

The other two movements which are necessary to generate the non-circular profile having regard to the normal orientation of the transducer, will be given to the transducer support as they have comparatively small amplitude because the profile which can be ground technically, always lies close to a circle.

The mechanical part of the measuring machine only provides a guiding mechanism giving either the workpiece or the transducer the proper degrees of freedom[5] and the drive is assured by high precision servomechanisms having a numerical input from a tape reader or better still a computer.

CNC SYSTEM

Those non-circular but continuous profiles to which this measuring system is relevant have a geometrical or cinematic definition. The computation of the coordinates of the generating point are easy if one uses a parametric form of the equations.

A small computer is able to calculate, at a fairly high speed, the three displacements associated to any point of the profile, provided it is given the geometrical parameters (R_m, E, A). Those displacements which will be called hereafter the 'coordinates' of the point, are a function of one single independent variable κ which can be indexed step by step.

It must be noted that it would be impossible to compute two coordinates fast enough as functions of the third one—for instance the angle of rotation of the workpiece table—such a calculation would have to be iterative owing to the transcendental form of the implicit equation system.

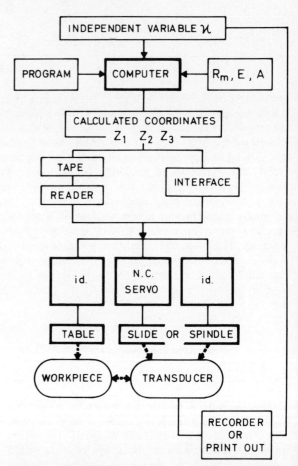

Figure 1. General layout of the complete CNC system for inspection machine drive.

Figure 1 gives the block diagram of the on-line CNC system required to drive the control machine: three servos are used to produce the movements of the table and of the transducer, the signal provided by the transducer can either be directly plotted versus the independent parameter κ or digitized and printed out. The main reasons why a specialized machine has to be designed for this CNC purpose is that there is no obvious opportunity of interposing an interpolation system, though a continuous profile has to be inspected, the high resolution required allows a point to point process which is of the required accuracy. In fact, the particular shape of the profile and the proper choice of the necessary motions, will be shown to need very low band-pass for such an interpolator, should it be foreseen.

On the other hand, very few metrology machines can be automatically controlled at high speed, by digital devices, and they are very expensive due to their universality. Moreover, existing machines are not able to measure properly the shape error following the direction of the normal to the theoretical curve as it will be seen here that the coordinate system which has to be carried out is no conventional one.

In the system of Fig. 1 the independent variable is automatically indexed by a clock or by the computer itself as it receives a strobe pulse from the recording or print-out device. The computer programs can solve the equations of the profile if the computer is supplied the parameter R_m, E, A of the reference profile. The calculated coordinates Z_i can be stored on a punched tape or a magnetic cartridge but for laboratory purposes the on-line process is far more interesting as it allows research about the effect of a slight variation of one parameter and makes full use of the machine versatility.

Input of the coordinates blocks (Z_i) activates the servos to position the table and the transducer at the point which is related with the particular value of κ parameter. The error at this point in the normal direction is then sensed and printed out versus κ value.

CONTROL AXIS

The table sustaining the workpiece, is rotating about a fixed axis just as in a roundness control machine. Figure 2 shows a scale reduction of the R080 profile

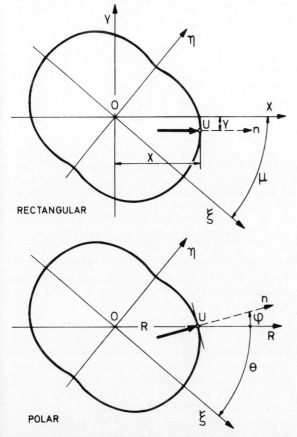

Figure 2. Definition of the two possible coordinates systems: (a) one table and two slides (μXY), (b) one table one slide one spindle ($\theta R\phi$).

tilted at an angle. Two ways of positioning the transducer according to our assumptions can be considered. The first system is derived from that of the grinding machine, and one can see how the normal keeps the constant direction of the absolute axis OX. This situation is a compulsory one in the grinding machine but can be disregarded in the control rig as no adjustment in that direction is necessary during the profile generation. Hence, the second system is such that the tip of the transducer rests on the point of the profile lying on the fixed radial axis OR, but the transducer axis itself has of course to be tilted by an angle ϕ, to remain parallel to the normal $\vec{n}$.

In the first case, the two motions given to the transducer support are linear and crossed, parallel to absolute axis OX and OY, while in the second system, one is a linear motion along R, the other being a rotation about an axis perpendicular to the profile plane and going through the theoretical point U. These two systems might look quite similar, though it seems easier from a mechanical point of view to design the second one: it involves a spindle (ϕ) located in a carriage (R) and this can be made more accurate and compact than two cross slides.

With the object of selecting the best of both systems, we shall compare the variations of the 'coordinates' in either case versus the independent parameter κ.

This geometrical variable is an angle reaching the value 2π for one complete revolution of the workpiece. In the particular case of the Wankel trochoid according to its order 2 symmetry, one elementary cycle of the measurement process is completed for half a revolution of the stator ($\kappa = \pi$). For comparison purposes, we have used the R080 profile whose definition parameters are:

$$R_m = 100 \text{ mm} \qquad E = 14 \text{ mm} \qquad A = 2 \text{ mm}$$

Furthermore, for velocity evaluation, the duration of the measurement cycle has been made equal to 60 s.

The calculation has been completed, according to the first steps of Fig. 1 and, deleting print-out delay, the time lapse required to calculate the three coordinates for each step of κ variation is very short. This shows that, with the type of computer used, the control can be performed in one minute, while sensing the profile at a great number of different points. Figures 3(a) and 3(b) give the computed variations of μ, X, Y and θ, R, ϕ, for the mean measuring frequency of 1 rev/min. One can at a glance notice that the six functions are rather closed to sinus sums, involving low order harmonics, as it has been earlier pointed out.

A keener examination of the curves shows that the X motion, which can be compared to R oscillation is somewhat more distorted: a plateau is observed about the minimum value; such a quasi-stop of the X slide makes this variation more difficult to be closely followed that the even curve of R versus time.

The oscillation along the Y slide can be compared with the spindle swivelling ϕ: as a matter of fact, those additional degrees of freedom are required owing to having to make the transducer axis parallel to the normal. Though it can seem somewhat steep to compare a linear amplitude and a swivelling angle,

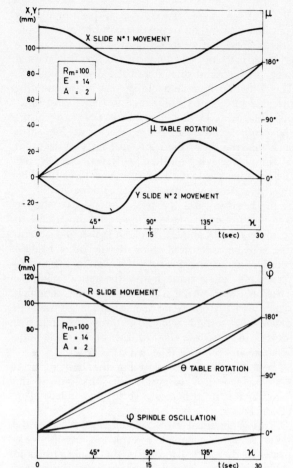

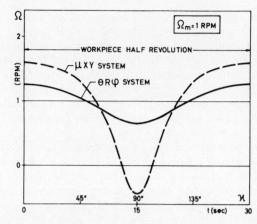

Figure 4. Variation of the angular velocity of the table for the two machine types.

Figure 3. Variation of the coordinates for generation of R080 profile at one RPM mean frequency: (a) (μXY) machine, (b) ($\theta R\phi$) machine.

one can easily realize that the ($\theta R\phi$) system is again of greater interest as the Y motion has more overtones than the ϕ rotation. At last the table movement is in either system close to a constant velocity rotation but the departure from this mean motion is of quite a different order of magnitude: in the ($\theta R\phi$) system, the θ motion is slightly speeded up during the first half cycle and slowed down by the same amount for the second half cycle. At the opposite end, the angle μ overshoots 90° before κ reaches $\pi/2$ and moves backward for a short time after κ is over $\pi/2$.

The tremendous difference between the two systems, making the selection almost unquestionable, is strongly emphasized by Fig. 4 which is a plot of the angular velocity variation during one cycle for the two cases; it can clearly be seen that the two systems are quite different with regard to the movement of the workpiece. The overall relative variation of the angular velocity is scarcely 60 per cent in the better case—($\theta R\phi$) system—and reaches some 200 per cent in the other one. Moreover, one can notice that the reverse motion about $\kappa = \pi/2$ leads to a negative velocity reaching more than 40 per cent of the mean value.

Such an irregular alternating movement must of course be strictly prohibited in a precision machine, and the ($\theta R\phi$) system has thus been adopted; it involves one slide and spindle system in addition to the rotating table.

GUIDING LAYOUT

The particular three axis control system is now selected. The workpiece table must be allowed to rotate about a fixed axis while the transducer support must be given two oscillating movements. Those degrees of freedom are to be given by means of a sophisticated bearing, owing to the precision level to be reached but this can be achieved by a fairly simple general layout. Figure 5 shows this system composed

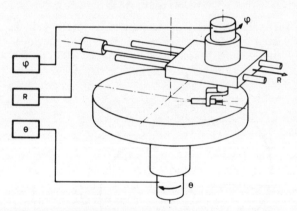

Figure 5. Schematic layout of a machine for point to point control of non-circular profile.

of one spindle for the table, one slide and another spindle for the transducer. Such a design could be used for both internal and external profile inspection but in the latter case the transducer would be rotated by 180° as regards the drawing of Fig. 5. It can be seen that the transducer must be swivelling on a spindle the axis of which goes past its tip; this requires some kind of an L-shaped device as a support.

With the object of achieving an accurate and stiff layout and of simplifying stator locating operations, it has been preferred to restrict the versatility of the designed machine to bore checking only. Hence, the layout of Fig. 6 has been retained: it can be noticed at a glance that the table, as well as its bearing spindle and ancillary equipment, are ring-shaped so that the transducer guiding system could be located wholly inside the main frame, underneath the table itself. Such an arrangement has the main advantage of avoiding the need for a bridge, or C-shaped frame.

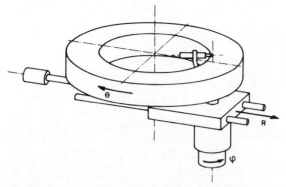

Figure 6. Functional design of machine for control of W.E. bores.

Higher stiffness can thus easily be reached and several joints can be omitted, which is the best way of obtaining the highest possible accuracy. The whole machine is thus concealed inside of a square frame of small height, topped by the only hollow table where the engine stator is laid. The peculiar features involved, especially in the table bearing system will be discussed in the last section of the paper, let us but point now that all the three degrees of freedom needed for the non-circular profile generation are achieved by means of hydrostatic air bearings and slides.

SERVO DRIVE

These three motions, allowed by proper freedom are caused by looped servo drive of mixed digital-analogue design. We shall hereafter only describe the driving and positioning loop of the θ axis (workpiece table) as the two other ones are of similar design and their accuracy are not as high to ensure a correct measuring working.

The external signal supplied by the tape reader (the computer interface in full CNC configuration) is of course of digital form, some eventual code conversion device is needed, according to the output code of the computer (i.e. binary, BCD, ASCII). The feed-back signal is also digital, because it is impossible to consider an analogue position transducer for the required resolution (5×10^{-6} rad).

Moreover, the table has a continuous rotating movement requiring a direct drive from the feed-back signal generator, as well as the shaftless table spindle. The digital encoder supplies a digital signal which is compared with the input order in a remotely programmable preselection counter, allowing the use of a comparatively cheap incremental encoder instead of an absolute one.

The digital error signal can be easily translated into analogue voltage by means of a low cost converter: the point to point check out is progressive and the error signal $\Delta\theta$ is the difference between the polar angle of two successive check points and is thus of minute magnitude. This is one more reason for using a digital feed-back allowing to locate the D.A.C. downstream of the comparator (this is true also for the two other channels).

The power output of the driving loop is provided by a conventional servo amplifier driving a differential torque motor, directly coupled again to the

moving parts in such a way to avoid inaccuracies due to back-lash.

Those three power elements are of special design owing to the desire of integrating them closely within the machine structure, this point will be briefly discussed in the section related to mechanical design.

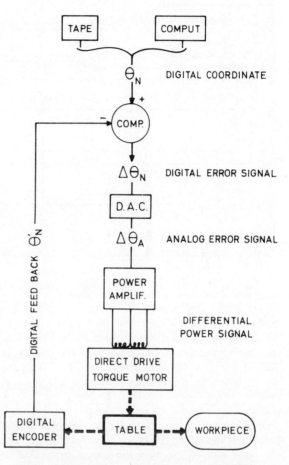

Figure 7. D.A. servo drive system for table orientation.

This whole CNC servodrive is described in Fig. 7 for θ axis. No interpolator has been interposed as the computer is able to calculate the three coordinates of check points very close to one another even in on-line configuration. The numerical control system can be completed with logical devices ensuring the printing out of the measured error only when the proper position has been reached on either axis. Furthermore, proper damping of the servo mechanism, in its analogue part, can be provided to avoid overshoot and realizing an approximate interpolation insuring the motions are progressive enough to keep the transducer displacement within its restricted mechanical range.

PROSPECTIVE MECHANICAL DESIGN

Figure 8 shows a part of a prospective drawing of the control machine in its very early design. Essential mechanical features which have not been discussed as yet will be outlined in this section with sketch support.

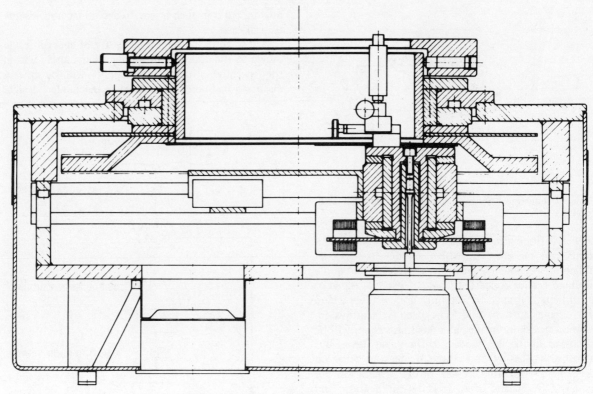

Figure 8. Prospective design of machine prototype.

Bearings

All the bearings used in the two spindles and the slide for linear motion are of the hydrostatic gas type. Special attention has been given to build an isostatic layout where exact necessary thrust pads and radial bushings have been provided to avoid redundant boundaries.

The axial–radial composite bearing of the table placed on the upper flat face of the main frame and the fixed rails of the cross slide, are secured also to this frame by means of adjusting supports allowing proper centring of the transducer spindle.

Incremental encoders

Two optical linear ·episcopic encoders are used for table and slide indexing while the second one is conventionally mounted, the first one uses a steel engraved tape wrapped around a disc secured to the table hub.

The third encoder, a standard rotary one is used for spindle swivelling motion control, it is driven by the spindle shaft end.

Power elements

The direct drive torque motor of the table is composed of a flat ring-shaped rotor secured to the encoder disc and rotating within the gap of the magnetic system of the stator. A similar disposition of smaller diameter is used for the spindle. The linear motor of the slide uses a similar stator in the gap of which moves a U-shaped aluminium beam. An a.c. version of the driving elements has been considered with the purpose of avoiding the torque pulsations promoted in the d.c. system by brush commutation. All three motors are of the differential, push-pull design, which is standard in electro-hydraulic servo

systems. Mechanical damping is caused by permanent magnets creating eddy currents within aluminium moving parts[6].

Frame

The main frame is a single stiff steel welded part for the prototype but it could easily be made of cast iron for batch production. Only two parallel flat faces must be machined to receive the table bearing and the slide rails.

Measuring transducer

The eccentrically mounted transducer is a standard lever transducer as used on roundness checking machine; it is mounted on an auxiliary mechanical slide for coarse mechanical zero adjustment.

CONCLUSION

The problem of shape check out of non-circular profile is not solved as yet by conventional measuring machines. A good solution is provided using a CNC machine of appropriate design, the relatively high cost of which is compensated by its versatility and ease of adjustment. The feasibility of such a machine has been demonstrated and the main topics of its original design have been pointed out. Quantitative analysis related to the well-known example of the R080 stator can be translated to any non-circular grindable profile: conclusions are general, as the layout which has been considered is imposed by the continuous variation from a constant radius profile.

The solution which has been carried out ensures greater operating ease as no adjustment is necessary on the machine itself. The profile equations are contained in the software of the computer and the geometric parameters can be stored by keyboard instructions.

The measured error is immediately printed, or better punched on a tape in digital form thus readily available for subsequent analysis however sophisticated it can be. Immediate storing into the computer memories opens the way to a detailed analysis, which is one reason more of building a CNC system for on-line operation.

REFERENCES

1. B. VISEUR (1970). Signification et mesure de l'erreur de forme sur les éléments d'un assemblage 'Polygon', *Revue M.*

2. A. LEROY and J. LEDOCQ (1971). Rectifieuse de génération pour surfaces trochoïdales, 1st World meeting on Machine tools, Milan.

3. A. LEROY and J.-M. FLAMME (1972). Génération des surfaces trochoïdales, CIRP General assembly, Stockholm.

4. A. LEROY and J.-M. FLAMME (1973). Contrôle morphologique des surfaces trochoïdales, CIRP General assembly, Ljubjana.

5. A. LEROY and J. LEDOCQ (1971). L'intervention des composants dans la conception des machines outils, 1st World meeting on Machine tools, Milan.

6. Magnetic memory disc drive, *Hewlett Packard Journal.*

NEW METHODS AND APPARATUS FOR CHECKING THE KINEMATIC ACCURACY OF GEARS AND GEAR DRIVES

by

KAREL ŠTĚPÁNEK*

SUMMARY

A definition of kinematic accuracy is given, and kinematic errors are divided into 'slow', 'fast' and cyclic ones. With intricate gear drives the necessity of harmonic analysis is emphasized. It is stated that the graph and amount of kinematic errors and their harmonic spectrum are the main values for the classification of the quality of gears and gear drives. Therefore the activities of the Research Institute of Machine Tools and Machining Technology in Prague consist in this field mainly in the research and development of apparatus and devices for checking kinematic accuracy. The TOSIMO-JS and -DS devices for checking the kinematic index drives of gear cutting machines, and TOSIMO-JK and -DK apparatus for checking the kinematic accuracy of gears, are described. Finally the application of checking units in conjunction with a mechanical correction device for improving the accuracy of gear cutting machines, and with correction servo-mechanisms for master gear cutting machines with high accuracy, is given.

INTRODUCTION

The kinematic accuracy of a gear drive is given by the graph of errors of the gear ratio of a certain gear as compared to that of an ideal one. The errors are checked in angular seconds (sec), microradians (microrad) or micrometres (μm) on a given diameter and are related to the initial or final member of the gear drive.

A characteristic graph of the kinematic error Δ_k of a plain gear drive (consisting of two members) is seen in Fig. 1. The graph contains the so-called 'slow' errors Δ_p (low harmonic components, resulting from the errors of the position of tooth flanks, and the so-called 'fast' errors of the shape Δ_r of tooth flanks. The error which is repeated for each tooth is usually the most important and is called the cyclic error Δ_c.

The graph of the kinematic error of a complex gear drive (consisting of several members) is shown in Fig. 2. Generally, it includes various harmonic components. The best way of investigating this sort of gear drive is the harmonic analysis of the graph of the kinematic error and the determination of its harmonic spectral components.

The magnitude of the kinematic errors of a gear drive is the basic quantity characterizing its properties from the point of view of the accuracy, vibrations, noise, service life, and so on. Consequently, the checking of kinematic accuracy (also called the single-flank test) is the basic checking operation, and the equipment for checking the kinematic accuracy can

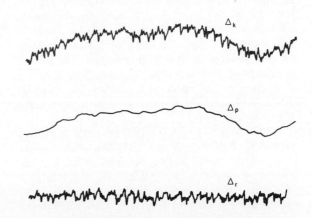

Figure 1. Kinematic errors of a plain gear drive.

be considered as the basic aid for checking the quality of gears. In spite of that, the existing units of this sort find only a limited field of application, mainly because of their complexity, lack of versatility and poor accuracy.

In the following text the TOSIMO apparatus and devices for checking the accuracy of gears, developed in Czechoslovakia, will be described.

SPECIFICATION OF TOSIMO APPARATUS AND DEVICES

The basic principle of TOSIMO apparatus and devices is the TOSIMO-J gear checking method for measuring high transmission ratio gears ($i > 20$) and the

---

* VUOSO, Prague

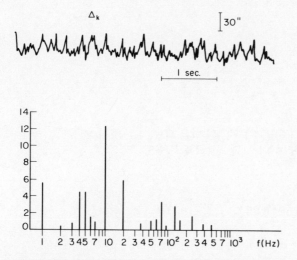

Figure 2. Kinematic errors of a complex gear drive and their harmonic spectrum.

TOSIMO-D checking method for measuring low transmission ratio gears (i = 20–1:20). A detailed description of these methods is given in the magazine *Czechoslovak Heavy Industry*, 7 (1969) and 3 (1971).

TOSIMO apparatus and devices include the TOSIMO-JS and TOSIMO-DS devices for checking the kinematic accuracy of high and low transmission ratios (indexing gears of gear-cutting machines, indexing units, gear boxes, etc.) and the TOSIMO-JK and TOSIMO-DK apparatus for checking the kinematic accuracy of gears (by means of a master worm or master gear). The main advantage of TOSIMO apparatus and devices is their high precision, reliability, possibility of statistical checking, versatility and ease of operation. The checking is carried out quickly, continuously and the results are recorded by means of a recording device.

The TOSIMO-JS and TOSIMO-DS units consist of individual TOSIMO-JSP and TOSIMO-JSR sensors fitted on the initial and on the final element of the gear to be checked. The TOSIMO-JK and TOSIMO-DK units are equipped with built-in sensors and enable the clamping of gears and master worms or master gears. A specification of TOSIMO apparatus and devices is seen in Fig. 3. In this specification the terms of the individual types of units have the following signification:

TOSIMO-J: Static phase checking method for measuring the accuracy of high-ratio gears (approx. 20–2000)

TOSIMO-D: Differential static phase checking method for measuring the accuracy of low-ratio gears (approx. 1:20–20:1)

TOSIMO-JE: Electronic part of the apparatus for checking the accuracy of high-ratio gears

TOSIMO-DE: Electronic part of the apparatus for checking the accuracy of low-ratio gears

TOSIMO-JS: Devices for checking the accuracy of high-ratio gears (indexing gears of gear hobbing machines, worm gear boxes, indexing tables and heads, etc.)

TOSIMO-DS: Devices for checking the accuracy of low-ratio gears (indexing gears of gear shaping machines, gear boxes, etc.)

TOSIMO-JK: Single-flank checking apparatus for gears by means of master worms

TOSIMO-DK: Single-flank checking apparatus for gears by means of master gears

TOSIMO-JSP: 'Slow' sensor

TOSIMO-JSR: 'Fast' sensor

TOSIMO-DSR: Dual fast sensor

TOSIMO-JSP: 10, 16, 25, 40—Slow sensors dia. 100, 160, 250 and 400 mm

TOSIMO JK 200: Single-flank checking apparatus for medium size and large gears by means of master worms (axial distance 200–1100 mm)

TOSIMO DK 40: Single-flank checking apparatus for medium size gears by means of master gears (axial distance 62–400 mm)

TOSIMO-JS AND TOSIMO-DS CHECKING DEVICES

The TOSIMO-JS checking device is suitable for measuring the kinematic accuracy of high-ratio gears (higher than 20). It consists of a TOSIMO-JSP slow sensor, a TOSIMO-JSR fast sensor and the electronic part TOSIMO-JE.

The TOSIMO-DS unit is intended for checking the kinematic accuracy of low-ratio gears (20:1–1:20). It consists of two TOSIMO-JSR slow sensors, two TOSIMO-JSR fast sensors (or a TOSIMO-DSR dual fast sensor) and the electronic part TOSIMO-DE. The TOSIMO-JSP slow sensors are made for various applications in four sizes: 10, 16, 25, 40. According to the slow sensors employed the TOSIMO-JS 10, 16, 25, 40 and TOSIMO-DS 10/10, 16/16, 25/25, 40/40 checking devices are distinguished. The TOSIMO-JSP slow sensors and the TOSIMO-JSR fast sensors are shown in Fig. 4.

The TOSIMO-JS checking device enables the measuring of the kinematic accuracy of index gears of gear hobbing machines, gear grinding machines, worm gear boxes, indexing tables and heads, etc. A TOSIMO-JS 16 checking device is seen in Fig. 5.

An example of checking the kinematic accuracy of a gear hobbing machine by means of the TOSIMO-JS 25 device is shown in Fig. 6. In this case the TOSIMO-JSP 25 slow sensor is fitted to the machine

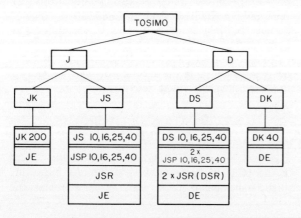

Figure 3. Specifications of TOSIMO checking apparatus and devices.

Figure 4. TOSIMO-JSP 16, 25, 40 slow sensors and TOSIMO-JSR fast sensors.

Figure 5. TOSIMO-JS 16 checking device.

Figure 6. Checking the kinematic accuracy of a gear hobbing machine by means of the TOSIMO-JS 25 checking device.

table and the TOSIMO-JSR fast sensor on the tool spindle. The results of checking are seen in Fig. 7.

The TOSIMO-DS checking device is well suited for measuring the kinematic accuracy of the index gears of gear shaping machines, bevel gear hobbing machines, gear boxes, etc. The TOSIMO-JS and TOSIMO-DS checking devices are intended for checking in laboratories as well as under service conditions. The checking is done very quickly while the gear to be checked is rotating and the kinematic errors are recorded by a recording device. On these units any gear ratios and irrational ratios can be checked. The TOSIMO-JSP 10 and 16 slow sensors can also work in horizontal position.

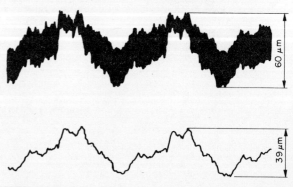

Figure 7. Results of checking the kinematic accuracy of gear hobbing machine.

Specifications of TOSIMO-JS checking device

TOSIMO-JS		10	16	25	40
Diameter of slow sensor	mm	100	160	250	400
Height of slow sensor	mm	150	205	205	205
Maximum measured gear ratio		1000	2000	4000	8000
Speeds of slow sensor	rev/min	0–4	0–2	0–1	0–0·5
Frequency of measured errors	c/s	0–25	0–25	0–25	0–12
Maximum measured error	sec	1600	800	400	200
Measuring accuracy of slow errors	sec	± 2·5	± 1·5	± 1	± 0·5
Measuring accuracy of fast errors	sec	± 1	± 0·7	± 0·5	± 0·3
Weight of slow sensor	kg	3	10	15	20

The specifications of the TOSIMO-DS checking devices are equal to the specifications of the respective TOSIMO-JS devices.

CORRECTION EQUIPMENT

An active application of the results obtained when checking the accuracy of index gearing may be acquired by means of correction equipment. This includes correction equipment with fixed memory and correction servo-mechanisms.

KZ mechanical correction equipment

A mechanical correction device for the correction of kinematic errors of index worm drives of gear hobbing machines, gear shaping machines and gear grinding machines has been developed. This device is based on the principle of acceleration and slowing down of the worm.

The equipment consists of two correction systems—a slow and a fast one—working simultaneously. The period of the slow system is given by the worm gear revolution, the period of the fast

Figure 8. KZ mechanical correction device.

system by the worm revolution. The memory consists of four cams. Two cams of the slow system are made according to the graph of the slow kinematic error of the worm gear drive in one and the other direction of rotation. Two cams of the fast system are made according to the measured graph of the mean fast kinematic error of the worm gear drive in one and the other direction of rotation. The checking of the worm gear error is performed on the TOSIMO-JS device. The KZ mechanical correction device is seen in Fig. 8. The influence of correction is shown in Fig. 9.

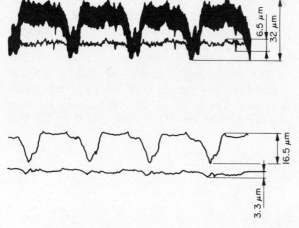

Figure 9. Influence of the KZ correction device.

The mechanical correction device is used on machines of higher accuracy and is marked by ease of operation as well as reliability. It helps to reduce especially the slow kinematic error of worm gears.

Machines fitted with the mechanical correction device work in the fourth class of accuracy according to DIN 3962.

Correction servo-mechanism AEC

The AEC automatic correction servo-mechanism developed for gear hobbing machines works according to the scheme in Fig. 10. The index gear drive between the tool N and the table O consists of the bevel and spur gears P_n, the differential D_s, the index change gears a, b, c, d and the worm drive P_o. The device is driven by the electric motor M through the gear box P_m.

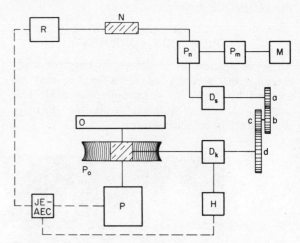

Figure 10. Scheme of the AEC correction servo-mechanism.

The accuracy of the index gear drive between the tool N and the table O are checked by means of the TOSIMO-JS device consisting of the fast sensor R on

Figure 11. OF 45 gear hobbing machine with AEC correction servo-mechanism.

the tool spindle N, of the slow sensor P under the table O and of the electronic part JE-AEC. The error signal produced in the electronic part from the inaccuracies of the index gear drive controls the valve of the hydraulic servomotor H. The rotation of the servomotor H, which is transmitted into the indexing gear drive by means of the correction differential D_k, acts against the arising of the error signal and thus corrects the inaccuracies of the index gear drive.

The OF 45 gear hobbing machine with the AEC correction servo-mechanism is seen in Fig. 11. The

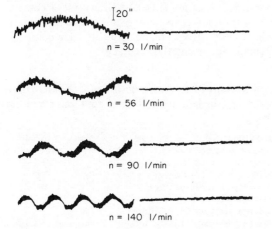

Figure 12. Influence of the AEC correction servo-mechanism.

influence of correction for various speeds of the tool N is shown in Fig. 12. Tests have proved that the automatic correction device substantially increases the accuracy of the machine and enables the building of high precision or master gear hobbing machines for the production of gears with the highest claims on accuracy. The correction is done with an accuracy of 2 sec so that the machines with automatic correction servo-mechanism work in the first class of accuracy according to DIN 3962.

TOSIMO-JK 200 CHECKING APPARATUS

The TOSIMO-JK 200 checking apparatus is intended for checking the kinematic accuracy of gears in diameters ranging from 200 to 2000 mm. It is employed for single-flank rolling tests on medium size and large spur and helical gears by means of master worms. These worms are of the envolute single-pitch type, with one or several threads. It is also used for checking the kinematic accuracy of work gears before assembly into gear cutting machines and gear boxes. The TOSIMO-JK 200 apparatus is illustrated in Fig. 13.

The TOSIMO-JK 200 apparatus is suitable especially for precision laboratory measurements and enables a perfect classification of gears and gear drives according to their quality. Very important is the finding out of the optimum relative position of the

Figure 13. TOSIMO-JK 200 checking apparatus.

worm and worm gear in worm gear drives. The checking proceeds very quickly and the results obtained are recorded by a recording device.

Specifications of TOSIMO-JK 200

Axial distance of measured gear and worm	mm	200–1100
Range of measured modules		any module
Vertical travel of headstock	mm	600
Headstock swivels		± 90°
Maximum length of worm	mm	600
Maximum weight of measured gears	kg	3000
Range of measured gear ratio		20–4000
Frequency of measured errors	c/s	0–7
Maximum measured error	sec	400
Measuring accuracy of slow errors	sec	± 1
Measuring accuracy of fast errors	sec	± 0·3
Weight of equipment	kg	3500

TOSIMO-DK 40 CHECKING APPARATUS

In its basic design the TOSIMO-DK 40 apparatus is intended for checking the kinematic accuracy of spur gears the axial distance of which ranges from 62–400 mm. It is employed for single-flank rolling tests of small and medium size spur and helical gears by means of master gears. The unit is also used for checking the kinematic accuracy of spur gears before assembly into machines and gear boxes. The TOSIMO-DK 40 unit is seen in Fig. 14. In conjunction with additional devices the apparatus can be employed for single-flank testing of pinions, bevel, spiral, worm and internal gears or bevel, spiral and worm gear drives (Figs 15, 16, 17, 18).

Additional devices enable the dual-flank rolling tests as well as the checking of main quantities of the geometrical accuracy of gears such as the accuracy of the envolute and pitch of helix, accuracy of indexing and running untrue (run-out), accuracy of tooth thickness and amount of play. The TOSIMO-DK 40 apparatus is intended for precision particularly laboratory measurements and allows for a complex classification of gears and gear drives according to their quality from practically all viewpoints of kinematic and geometric accuracy.

Specifications of the basic design of TOSIMO-DK 40 apparatus (single-flank testing of spur gears and gear drives)

Axial distance of measured gears	mm	62–400
Range of measured modules		any module
Range of measured gear ratios		20:1–1:20
Maximum weight of measured gear	kg	150
Frequency of measured errors	c/s	0–12
Maximum measured error	sec	800
Measuring accuracy of slow errors	sec	± 3
Measuring accuracy of fast errors	sec	± 1·5
Weight of unit	kg	500

Figure 14. TOSIMO-DK 40 checking apparatus.

Figure 15. Pinion checking unit.

Figure 16. Bevel gear checking unit.

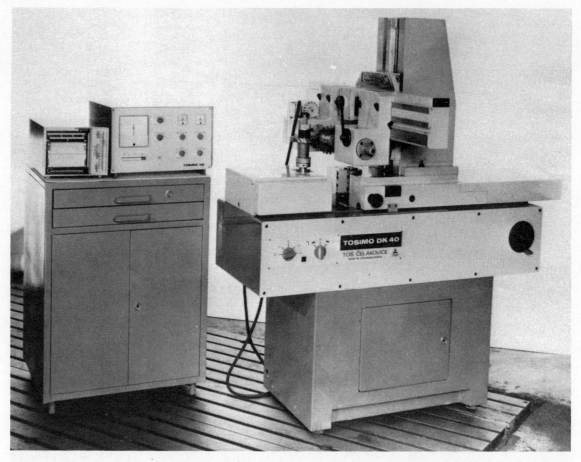

Figure 17. Spiral gear checking unit.

Figure 18. Worm gear checking unit.

CONCLUSION

This paper details work done in Czechoslovakia in the field of checking the kinematic accuracy of gears and gear drives. The TOSIMO-JS and TOSIMO-DS devices for checking the kinematic accuracy of gears are also described and some examples of application are given. The units enable a complex checking of the accuracy of gear-cutting machines, gear boxes, indexing tables, etc.

The above-mentioned units make possible a special application in conjunction with correction mechanisms. The KZ mechanical correction equipment and the OF 45 gear hobbing machine with the AEC correction mechanism is described. The correction servo-mechanisms enable the production of high precision gear-cutting machines.

The TOSIMO-JK 200 and TOSIMO-DK 40 for checking the kinematic accuracy of gears are described. These units make possible a complex checking of the kinematic as well as geometrical accuracy of spur, bevel, spiral and worm gears and gear drives.

All the above-mentioned checking units have been developed by cooperation of the Research Institute for Machine Tools and Machining Technology in Prague and the TOS Čelákovice works, the latter being the maker of them. They enable a wide range of problems to be solved in the field of the production and checking of precision gears and gear drives.

DISCUSSION

Q. J. J. Ashton. (1) What uncertainty of measurement is achieved?

(2) Has the weaving-trace type of instrument (with data retrieval improvements) been utilized for this work?

A. (1) It is very difficult to answer the first question; the first point is that the automobile manufacturer says that they want accurate checking equipment, but when they are asked to give a figure—they can't. Secondly, what I have presented is the first design of a prototype which is not yet built, and it is difficult to assess the resultant accuracy that can be achieved. What we have tried to manage is a level of accuracy equal to that of a roundness control machine for the main spindle which is of primary importance, error on the two other axes being much less important.

(2) I do not know of this instrument, but if it is a two axis (*XY*) type of measuring machine, it cannot be used for this problem as the measurement of errors *must* be made along the normal to the theoretical curve, especially because evenly distant curves ought to be measured starting from a single reference.

NUMERICALLY CONTROLLED MACHINE TOOLS

MACHINE DRIVEN BY A MINICOMPUTER FOR THE AUTOMOTIVE STYLING

by

F. SARTORIO*

The machine illustrated in the film I had the pleasure to present is a system for high speed scanning, continuous digitizing, and milling of soft material surfaces. This machine is a 5 axis machine with continuous movement control, directly driven by a minicomputer.

The need for the development of such a strange machine arose from a typical problem in automotive styling, particular to those companies that during styling use clay for the creation of a full size car model. This model is very heavy (15/18 tons) and so fragile that its transport is not recommended. Therefore when the General Management of a company decide that a certain model should pass from the styling to the body creation phase there is the problem of scanning the model completely to obtain all the geometric data without altering it. This involves two requirements:

1. a self-propelled machine capable of moving to the model;
2. a machine to continuously scan the surface with such a light pressure so as not to leave any trace, even on fresh clay.

This is precisely the task of the machine presented here that we called 'Leonardo 02'. The Leonardo combines a new machine configuration with a totally new computer. The Leonardo is a totally self-contained, self-propelled, fully automatic scanner, capable of moving to the model, aligning itself, and scanning the model, quickly and accurately. The operator can drive the machine from either end for maximum ease of positioning. Power steering, power brakes and smooth speed control simplify this operation. This machine is driven by a computer designed by DEA, called DEAC 1001, it is stored in a self-propelled cabinet. This is an 'interpretative' computer with 12K memory plus 2, 5 K ROM with 16 bit. In order to provide proper planes of reference for the dimensions taken from the model, the machine may be aligned either to a gravity reference or to reference points on the model. Both gravity reference or to reference points on the model. Both gravity orientation and orientation to the model are fully automatic

sequences. The servo-driven pads are computer-adjusted to provide the correct orientation.

The three axes of movement generate a measuring envelope 2650 x 1400 mm in the horizontal plane and 1460 mm in the vertical plane. Machine displacement along the three cartesian axes are continuously displayed by means of rear projection read-outs mounted on a portable support.

The system must operate accurately on ordinary studio floors even though they deflect under the shifting loads that could result as the column is traversed along the base (approximately 1, 2 tons that move from one side of the machine to the other) or as the ram is extended. As the ram extends, a counter balance moves out on the opposite side to match the load exactly and thus keep the centre of gravity in a constant position. A more sophisticated system is used to counterbalance the mass of the travelling column. Here mercury is pumped between ballast tanks under computer control.

Various tools can be mounted on this machine and the probes are all interchangeable. Some of them are:

> a scribing device
> a fixed probe
> a probe for continuous scanning (H1)
> a sensor
> a milling head

These tools are mounted on a rotary support which is mounted at the end of the machine's horizontal arm. This horizontal arm, called 'multi-axis arm', has two degrees of freedom: a rotation on a vertical axis up to 270° and a rotation on a horizontal axis up to 225°. With the multi-axis arm, freedom from probe attitude restrictions is obtained since the tool tip is held to a fixed point in space as the tool altitude is altered. Tremendous time savings result from the ability to rotate in the scanning position without having to withdraw the probe from a part and reposition it each time an index is required.

The rotation of the tool can be decided by the operator or by the program itself depending on the space position of the contact zone. This is a first adaptable control capacity.

---

* DEA, Turin

M

Figure 1. Leonardo 02: portable machine for continuous scanning and milling. On the left side one of the handle bars.

Figure 2. DEAC 1001 computer: stored in a self-propelled cabinet for driving the Leonardo 02.

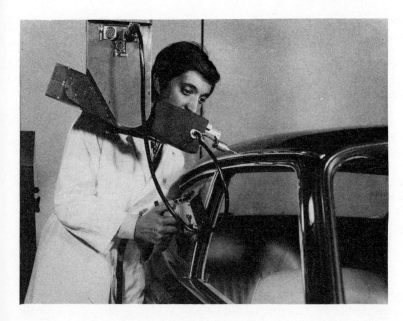

Figure 3. Multi-axis arm with the H1 probe.

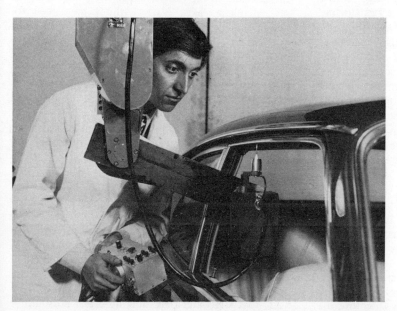

Figure 4. The multi-axis arm allows the rotation in the vertical plane of the tool around the contact zone of the piecepart.

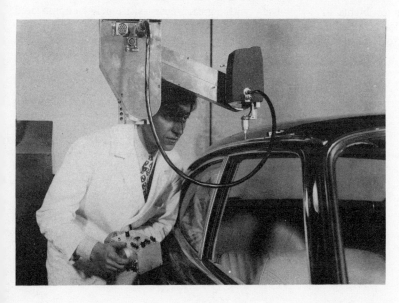

Figure 5. The H1 probe, with only a few grams of contact pressure, works on very soft surfaces without leaving a mark.

The H1 probe, with a low inertia servoed tip, can scan a surface precisely at a speed of more than 2500 mm/min with a very low pressure (3/4 gr). These two characteristics allow it to scan fresh clay or other very soft materials without leaving a mark. To obtain this capability the H1 probe has a supplementary movement (fourth axis), servo-commanded, with a 50 mm stroke. Due to the very small masses involved, this fourth axis can reach very high speeds and accelerations, driven by a high-sensitivity inductive contact probe. The movement of the fourth axis is also measured electronically and the values of its movement are reported to the machine's three main axes on the base of the fourth axis direction in the space, so as to have at any time the coordinate values of the contact point with the partpiece. Dimensional information is continuously output by means of a high speed tape punch fed from a buffer memory in the computer.

Data may be taken at equally spaced, or at points selected automatically by the computer, on the basis of minimum data needed to define the profile within a pre-defined chordal error. Chordal error is the distance between the true curve and the straight line segment that approximates the curve. Operator intervention is unnecessary even for highly irregular surfaces because of a special feature that we call 'learning capability'. If, as the machine scans a model, it encounters a sharp change in contour, it will automatically slow down to ensure gathering sufficient accurate data. If the machine cannot slow quickly enough, it will retrace its path at a reduced speed to obtain the correct record. Obviously much time would be lost if it had to retrace its steps each time it crossed a sharp contour. The learning capability feature means that the control memorizes the location of the problem area and anticipates it on the next scan by slowing down just prior to reaching it. The system automatically runs at the maximum speed that will allow gathering sufficient data to define the profile. Top speed is set by an operator override, after which acceleration and deceleration are fully automatic. This is an example of the learning capability of the machine by experience. This is therefore the second case in which the machine shows adaptive capabilities.

Years of experience in solving the problems of getting complete data for such difficult items as character lines led to the development of this special probe that we call the sensor. This unit combines the features of an operator control and a sensitive measuring probe. The probe tip is free to move in all directions, but if released will return to its null position. Moving it up, causes the machine to move up; moving it down, causes the machine to move down; movement to the left or right is similarly followed by the machine. Signals from the sensor are correctly interpreted by the machine regardless of the direction in which the probe happens to be pointing. This allows the operator to lead the machine by the probe to any point he chooses to locate. Once there, he releases the probe and the machine comes to rest with the probe at null, and with its tip touching the desired point. He then signals the control by means of a pushbutton on the control box, and the coordinates in three axes are recorded.

The Leonardo is also equipped with a light milling head, which runs at a speed of 6/18 000 r.p.m. on 0·5 hp. The large degree of freedom of the multi-axis arm, permit milling on parts normally approached with difficulty, such as undercuts, cavities and so on. In particular the rotation in the vertical plane takes place without any need for programming retraction from the workpiece, this is due to the fact that the rotation stabilizes the position of the tool's tip and a considerable increase in the speed of milling is obtained. The milling program permits movement modes in rapid positioning, linear interpolation, parabolic interpolation, and circular interpolation at the operator's choice.

Surface data is transmitted to the computer through a punched tape or directly to another remote computer via telephone. Therefore Leonardo is also an advanced example of CNC and DNC.

Leonardo has been operational for one and a half years at the Design Center of the Ford Motor Company in Detroit (U.S.A.) and is currently used during two shifts proving that:

1. modern technologies allow the realization of systems as complicated as the one described with a degree of reliability sufficient for an intensive use in industrial environments;

2. that CNC and DNC now have real possibilities of industrial use;

3. that it is possible to train average operators to use with great ease extremely complicated systems, even those equipped with computers, hence achieving huge increases in production.

NEW MANUFACTURING SYSTEMS FOR SMALL AND MEDIUM BATCH PRODUCTION

by

K. TSCHINK*

MANUFACTURING PROBLEM, PRODUCTION PROCESS, AND AUTOMATION APPROACH

Small and medium lot sizes and their variation are characteristic wide areas of mechanical engineering. They derive from the rapid development of science and engineering for satisfying the increasing social needs. This process is reflected by the new and continued development of technology, aimed at the permanent new and continued development of technology, aimed at the permanent new and continued development of technical goods with new and improved capabilities.

Efficient manufacturing production equipment with a high flexibility, is therefore necessary to enable automated one-off, small, and medium batch production to be achieved economically.

In determining and designing flexible manufacturing equipment we must always set out from the specified production job. The amount of flexibility i.e. the work capacity and actuating ranges required, will essentially be determined by:

the form, dimensions, material, weight, number, etc. of the workpieces to be produced, and; the capacity of the manufacturing means itself.

The greater number of identical workpieces required, the longer and time used per workpiece on a specific machine; and the smaller the capacity of the manufacturing equipment, the smaller will be the number of the various workpieces required for the rate of utilization and hence the necessary flexibility.

The time used per workpiece and batch of identical workpieces respectively, as well as the capacity of the manufacturing means are thus essential criteria.

Modern automation technology offers a possibility to meet effectively the requirements for a high flexibility, at the same time ensuring high productivity and accuracy with minimum cost. In this connection the attractive solutions are those which take account of all operations including the technological production planning, the scheduling, management, and control of production. Thus identical information and memories, identical programming systems and methods and a uniform device system are required.

The basis of such integrated automation is supplied by numerical control technology, the digital computer, and electronic data processing.

All endeavours in modern production engineering are aimed at machining an assortment of defined parts with high labour productivity as far as possible completely in one run on one manufacturing installation. And here it is not only a matter of applying simple technological processes, but rather such processes that can be easily realized and automated under given conditions and which can be inserted into the total automation approach of the factory. The use of conventional machine tools as well as N.C. machines and production equipment with adaptive control alone does not as yet enable a continuous, technological process of high efficiency to be realized.

The shaping process as a whole is carried out both on N.C. machines and conventional machine tools in such a way as to use the individual machines successively. The down time of the workpieces for a particular lot as well as the handling time between the machining units are not reduced.

Though adaptive control reduces the time for production planning and the basic time considerably, there is no connection between the machines, not to speak of the machines required for performing the total process on a workpiece. Direct computer control is the first step towards accomplishing this connection.

DIRECT NUMERICAL CONTROL

In direct numerical control (DNC) already several N.C. machines are interfaced to a process computer (Fig. 1) so that the program provision too, can be automated through the direct access to a program libarary (bulk store). The preparation, proofing, storage, as well as the exchange of the punched tapes at the machine are eliminated. The on-line connection of the computer with the correspondence devices, the

* Fritz Heckert A.G., Karl-Marx-Stadt, D.D.R.

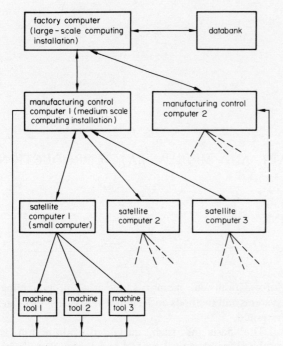

Figure 1. Computer hierarchy for control of machine tools.

memory, and the large computer on the one hand and the various N.C. machines on the other hand allow

(a) the direct control of the machine by the computer after calling;
(b) the automation of control data managing and supplying;
(c) the feedback of certain operating conditions;
(d) the evaluation of the operating data and perhaps their processing, using the connection to the large computer.

Thus also the problems of management and control of production, important questions of operating statistics, etc., are automated. In this way can be

logged: the utilization of the machine capacity; the susceptibilities to trouble of particular machines and groups of machines; the tool life; the workpiece run times.

The flexibility and the functional possibilities of numerical control increase decisively by using small digital computers as integrated components of the controls.

For this type of control the English designation CNC has become usual. A suitable computer is, for example, the KSR 4100 small control computer of VEB Carl Zeiss, Jena, with the following data: 12-bit word length, 8k words, 2·5 μs cycle time. The function of the control system is no longer realized through different control devices and a special wiring, but by a stored program.

In the D.D.R. the CONCEMA system (computer N.C. for machine tools) was developed both for DNC and CNC applications. CONCEMA 16 N.C. peripherals (NCP) can be connected to the standard terminal of the computer. The NCP can be used as an N.C. ramp control or output buffer.

If the NCP is used as a simplified N.C. control for a machine tool, the computer can control for example 12 axes of a machining centre. 80 per cent of the control functions are managed by the computer, the comparison between the reference and actual values and the information input and output are handled by the N.C. peripherals. The CONCEMA system comprises a computer with standard terminal system, transmission facilities for up to 1000 m cable length, several NCP modules in different modifications, monitors for the communication of the operators with the computer, and computer programs (software).

The state of DNC and CNC engineering in the D.D.R. was represented by the common exhibition of 13 DNC and 3 CNC machine tools from six member countries of the Council for Mutual Economic Aid.

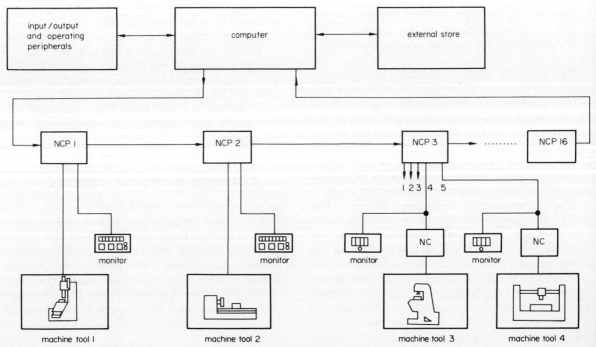

Figure 2. CONCEMA System.

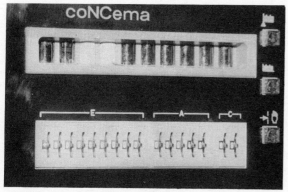

Figure 3. Monitor.

This show demonstrated the possibility of a flexible production installation for different components, which is for instance required for the manufacture of sub-assemblies. Programming, controlling, and monitoring are done by a hierarchy of computers. It comprises an external data processing unit, the central ODRA 1204 control computer, and small control computers of the KSR 4100 type.

The CONCEMA system has matching facilities for control systems of a great variety of types and manufacture. This permits machines of different operational modes to be compatible and to be incorporated into flexible systems.

Computer controlled production systems can be controlled on-line by a process control computer or off-line via an external computer to and from which data transmission can be performed.

In the second case an open path exists between computer and production systems, which must be bridged by man (Fig. 5).

Via keyboards or punched tape readers the computer receives input information about the situation in the machine system and the manufacturing tasks planning for the next day.

It calculates the optimum machine and workpiece store utilization and optimizes the operations sequence. The results are printed out in lists and a punchtape for workpiece transport is punched out.

On-line control is also referred to as real-time control, as there is no time delay (Fig. 5a). In this mode the process control computer is fully utilized for the production system and serves for storing and managing the N.C. programs, controlling the workpiece, flow, monitoring machines, and quick trouble shooting in case of failures. The direct numerical control system is the system at a higher level in which a further external computer is used for programming, run scheduling, accounting, etc.

Direct numerical control as realized by the types known today does not cover the process of manufacturing a workpiece or assorted workpieces. This requirement is met only by the machining centre and production system.

MACHINING CENTRES

The machining centre is designed for finish machining the workpieces of a component family. It must allow machining from all required sides and with all

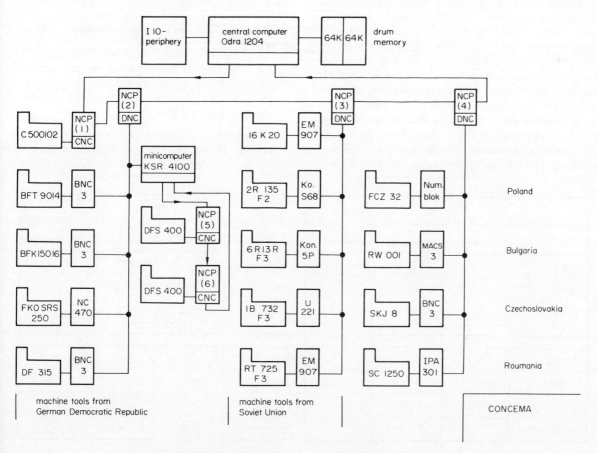

Figure 4. CONCEMA at International Leipzig Spring Fair 1973.

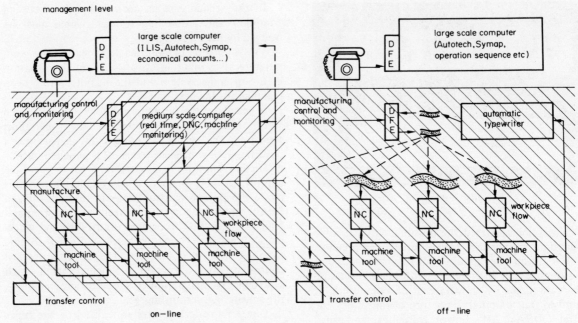

Figure 5. Types of computer controlled manufacturing systems.

required techniques. These conditions can scarcely be carried out when, during the manufacture of a workpiece, the basic movements of the tool or the workpiece change from one operation to the other. The prismatic workpiece (Fig. 6), which has to be machined by milling, drilling, boring, etc., is much more suited for a machining centre than, for example, the rotationally symmetrical workpiece, which must be machined by such basic operations as cutting-off to length, centring, turning, grinding, etc., because the basic movements of the tools and the workpiece are different. That is why for the machining of rotationally symmetrical workpieces there are frequently machining centres available only for one machining technique, for example turning centres, or such machining centres on which a combination of a few techniques with identical basic movements can be made. An essential criterion for the decision to what extent a machining centre or manufacturing system will be used is the expenditure for workpiece and tool changing. On the machining centre the workpiece is set up once and the whole process will then be carried out on this set-up, using tool changing.

Workpieces which are difficult to be set up—extremely small or unwieldy, large, and heavy workpieces—are, therefore, particularly suited to being machined on machining centres, especially when the expenditure for tool changing does not exceed double the expenditure for workpiece changing. Therefore, easily exchangeable workpieces are especially suitable for machine systems.

Important and often conclusive in deciding the use of a machining centre or machine system is the ratio between:

the total time volume required for producing the workpieces of a component family per annum and the capacity available per annum of one or several machining centres or

the average expenditure per workpiece of a component family at the bottle-neck machine or bottle-neck unity of a machine system and the

	machining centre			machine system			
operation	primary tool movement	secondary movement	workpiece movement	operation	primary tool movement	secondary movement	workpiece movement
milling		↓↑⊢		sawing	⟳	⇄	—
drilling		←		centering	⟲	⇄	—
boring		⊤⊢←		turning	⇄	⊤	⟳
counterboring	⟲	⊢		gear cutting	⟲‖	↓⊢	⟮
reaming		←		milling	⟲	⊢	—
tapping		←		hardening	←	—	—
facing		↑ ⊢		grinding	⟳	⊢↓↑	⟳
plunge-cutting		⊤ ⊢		finish honing	↓⊤	⊢	⟳

Figure 6. Basic movements of tools during machining.

number of pieces per annum of a component family, which necessitates the utilization of the machine system in more than two shifts.

If we try to contrast the breakdown of the machining times for the machining centre

$$MZ_{MC}$$

$$= mt_A + n\left(t_{WSt} + \sum_1^i t_P + \sum_1^i t_G + \sum_1^{i-1} t_{WZ} + \sum_1^k t_K\right)$$

$$(1)$$

to those for the machine system

$$MZ_{MS} = mt_Z + n\left(t_{WSt} + t_P + t_G\right)_{max} \qquad (2)$$

we get a difference of

$$t = n\left(\sum_1^{i-1} t_P + \sum_1^{i-1} t_G + \sum_1^{i-1} t_{WZ} + \sum_1^i t_K\right) \qquad (3)$$

Legend:

MZ	actual running time
m	number of lots per annum
n	number of pieces per annum of a component family
t_A	set-up and shut-down time of the machine
t_G	basic time
t_{WSt}	workpiece changing time
t_{WZ}	tool changing time
t_P	time share for positioning
t_K	time share for testing and checking
i	number of machining operations
k	number of checking operations

The capacity gain of the machine system is quite remarkable because of the parallel machining and becomes the more important, the greater the number of pieces 'n', the more machining operations 'i' and checking operations 'k' in machining the individual workpieces will be required.

As can be seen in equation (3) for the machining centre, the tool changing time is contained within the total machining time on equal terms with the basic time and positioning time. As the two last-mentioned times cannot readily be changed, special attention must be paid to a maximum reduction of the tool changing time by a proper automation of the tool flow. High flexibility is achieved by an arrangement of the tools in the magazine independent on the operation sequence (tool coding). A store location coding on the other hand requires an exact storing of the tools by the operator. Dead travels in the magazine are eliminated in this way.

To further reduce the machining time on machining centres a reduction of the workpiece changing time is necessary. In Fig. 7a for example two rotary tables are located on a slide, one of which will be loaded during the basic time. Figure 7b shows a similar approach where two separate rotary tables are used, and finally a particular advantage is attained when the worktable of the machining centre is loaded by means of workpiece pallets. This method also forms the basis for incorporating machining centres in production systems.

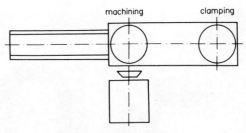

a) one slide has two rotary tables

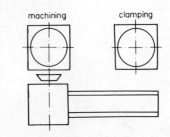

b) two rotary tables

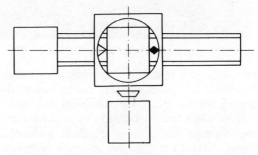

c) one rotary table – pallet clamping

Figure 7. Workpiece flow of NCMC.

To illustrate my viewpoint I would like to describe to you a machining centre produced in the D.D.R.

At present about 20 various types of machining centres for machining prismatic parts are built in the D.D.R. the manufacturer is VEB Werkzeugmaschinen-kombinet 'Fritz Heckert' Karl-Marx-Stadt.

The C 101 N.C. numerically controlled machining centre consists of a column mounted on a cross-bed and a rotary table receiving the pallets, a separate tool magazine and two workpiece loading stations.

The vertically arranged, flexibly usable chain-type magazine is space-saving and can accommodate 60 or 120 tools with 4 to 200 mm diameter. While 60 tools are in the work cycle, the machine can be loaded with 60 new tools.

A double-arm gripper carries the tools from the chain-type magazine into the tool carrier unit, which is an integrated part of the main drive assembly.

The work area of the machine allows the machining of workpieces with maximum dimensions of 630 x 630 x 800 mm. The rotary table with its 4 x 90° positions together with the horizontal work spindle permits a 4-side machining of the workpieces.

Automatic workpiece changing can be realized by a workpiece handling device. The transport of the pallets loaded with workpieces is made on special trucks. Resetting-up on a second pallet enables the workpieces to be machined from all sides. The

Figure 8. Machining centre C 101 N.C.

extremely high rigidity of the machine, which results from the chosen design and the optimum dimensioning of the main subassemblies, the preloaded anti-friction bearing guideways in all three axes, and the mounting of the work spindle in preloaded precision anti-friction bearings ensure a high working accuracy.

The performance of the machining centre is two to four times as high as that of conventional machine tools. A reduction of machining cost up to 50% is possible.

MANUFACTURING SYSTEMS

If few machining centres do not meet the requirements in manufacturing a component family on account of their capacities, or the machining conditions cannot be realized on a machining centre because of too many diverging motions in the various successive operations, a manufacturing system will be suitable. The manufacturing is an installation, integrated with regard to control technique systems, or various automatic individual machines provided with conventional N.C., which participate in the manufacture of a component family and between which the workpiece flow, too, has been automated. This workpiece flow, which is optimized according to a predetermined program or the utilization of the machine system, is thus an essential part of the machine system.

The problems of automating the workpiece flow are far more difficult to handle than those of the tool flow on the machining centre, because there must be a continuous deposit of the workpieces after the individual operations, an identification and transport to the next machine according to the program, another storing, etc.

To enable the clamping of the workpiece with still greater safety, accuracy, and speed, devices such as chucks, clamping pallets or similar devices are used so that the workpiece can be clamped outside the actual running time. Heavy pallets are moved by means of roller tracks provided witith switches or guides with drive elements, intermediate and light pallets by fork-lifts or overhead conveyors.

In the D.D.R. an air-cushion type transport system with linear motor drive was designed for pallets having a base surface of 1000 x 1600 mm. It includes clamping and reclamping stations, junctions, storage places, etc.

Automatic tool exchange—blunt against sharp—favours the reduction of non-productive times in the manufacturing system. This tool change can be done for each machine with a tool magazine of its own or for many machines provided with identical tools through central magazines. The ways to the machine and to the tool changing unit are then of considerable length.

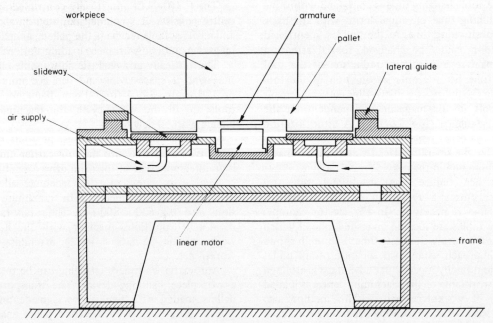

Figure 9. Pallet transport system.

A difficult question, is the actual utilization of such systems. Either the component group should be structured so that the various operations on the individual workpieces do not take very different times, which will permit a maximum utilization of the capacities of the individual machines, or for certain operations parallel machines must be used when an optimization of the capacity becomes possible. Eventually universal machines may partly be used, which for certain operations serve to relieve the bottleneck machines.

Two basic types are feasible:

manufacturing systems with machining stations *complementing* one another;
manufacturing systems with machining stations *substituting* one another.

The benefits of both types are listed as follows:

Manufacturing system with machining units complementing one another	Manufacturing system with machining unit substituting one another
simple, cheap machines	high redundancy of machining
great stability of machining stations	good utilization with respect to time
good setup	far-reaching independence on a particular parts assortment
good performance	
high accuracy	little effects in case of failure of individual machining stations
economic manufacture, provided that the parts assortment is strictly defined	simple possibility of extension

Actually constructed manufacturing systems represent a combination of both design possibilities.

In the D.D.R. four basic types of systems, both for machining rotationally symmetric workpieces and for machining prismatic parts, have been developed and built. The first example is the ROTA F 125 N.C. manufacturing system of WMW-Kombinat '7. Oktober' Berlin. In this system 4 N.C. lathes, 2 N.C. milling machines, and 1 N.C. chucking grinding machine are arranged under an annular workpiece storage unit so that the workpieces can be removed from the storage rings by means of elevators and are immediately received by a chuck of the machine for machining. The capacity of this storage unit, comprising nine rings, is 540 workpieces. There is an additional clamping and unclamping station where the workpieces are accommodated in the chuck and after having been machined will be cleaned and released. The control of the machines is achieved off-line, i.e. an external computer produces the control punched tape for the system machines and optimizes the capacity of the system and hence the sequence of motions of the workpieces.

In practical application the following characteristic values have currently been reached:

increase of labour productivity up to 300%;
reduction of production area down to 50%;
reduction of manpower down to 30%;
reduction of programming time by 10–20%;
availability of the system 90%;
output: 135 000 workpieces per annum.

Figure 10. Manufacturing system ROTA FZ 125 N.C.

ROTA FZ 200 manufacturing system

The ROTA FZ 200 manufacturing system of VEB Werkzeugmaschinenkombiant '7. Oktober' Berlin, which was shown at the Leipzig Spring Fair 1972, provides for the soft machining of spur gears up to module 4, comprising turning, hobbing, deburring, and shaving, with cleaning as an intermediate operation. Besides three DF 200 N.C. chucking lathes of VEB Werkzeugmaschinenfabrik Magdeburg, which are computer controlled by the CONCEMA control system, two gear hobblers, one tooth edge chamfering and deburring machine, one gear shaving machine, and a metal cleaning installation are used, which operate by program control.

Figure 11. Manufacturing system ROTA FZ 200.

At a pallet loading station the workpieces are loaded on pallets in three storeys (Fig. 12), temporarily stored at the entrance and exit of storage

Figure 12. Workpiece-changer.

locations, and transported to the machines. A stacking crane provided with a target control device and receiving its commands from a KSR 4100 transport computer carries out the transport (Fig. 13). The computer is also connected with all machines, the washing station, and the pallet loading station.

Figure 13. Workpiece transfer and magazine system.

Figure 14 shows a DF 200 N.C. technological station with workpiece change unit. During setting-up operations or in case of computer failures the system can be controlled from a manual control panel.

Characteristic values of the system are:

increase of labour productivity up to 310%;
reduction of production area down to 45%;
reduction of number of production workers down to 32%;
average reduction of prime cost 13%;
output: about 200 000 spur gears per annum;
assortment: about 2000 gears comprising 64 different groups;
average lot size: 40 gears.

Figure 14. Computer controlled DF 200 CNC.

Figure 15. Manufacturing system M 250/02 CNC.

Figure 16. Horizontal machining centre C250/01 N.C.

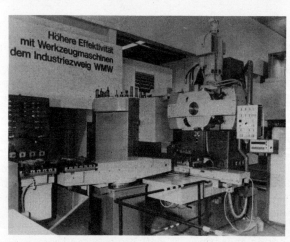

Figure 17. Vertical machining centre C 250/02 N.C.

M 250/02–CNC manufacturing system

The M 250/02 manufacturing system (Fig. 15) is designed for 5-side machining of prismatic workpieces with 400 mm edge length. It was designed by VEB Werkzeugmaschinenfabrik Auerbach, a branch plant of VEB Werkzeugmaschinenkombinat 'Fritz Heckert'. Two machining centres C 250/01 N.C. (Fig. 16) and C 250/02 N.C. (Fig. 17) as well as the tool flow and workpiece flow are controlled through two drive units of the CONCEMA system by the Polish ODRA 1204 process control computer.

The transport system consists of two pallet change units and an inserted loop conveyor. A third pallet change unit is connected with a workpiece store. An ingenious combination of the elements of the work-piece transport system permits the connection of further MC machines and machining centres. Through the use of this DNC machine system an increase of productivity up to 300%, compared with machining on individual machines, is attained.

SUMMARY

The rapid development of new products with con-tinuously improving user-qualities also requires an adaptation of subassemblies and components in machine building. Small- to medium-batch quantities can be economically machined only on very flexible automated manufacturing equipment. A prerequisite to the development of flexible machine systems is the use of numerical controls and the far-reaching adap-tation of the process sequence to the existing machining conditions. For automating entire techno-logical processes in the field of shaping, partly in-cluding handling processes, the combination of all operations required in shaping and handling on one machine, the machining centre, or in a system with integrated control technique, the machine or manu-facturing system, is therefore a necessity. The question whether or not a full utilization is possible will be decisive in the use of machining centres or machine systems; in all cases a high economic profit has to be ensured.

REFERENCES

1. TSCHINK, K. (1971): Problemlosungen des Werkzeugmaschinenbaues zur komplexen Bear-beitung von Werkstucken in Klein- und Mittel-serien (Problem approaches in machine tool building for the complex machining of work-pieces in small and medium batches) *Fertigung-stechnik und Betrieb* **9**, 515-526.
2. TSCHNIK, K. and KUHN, R. (1972): Numerisch gesteuerte Bearbeitungszentren und Werkzeugmaschinensysteme für die Klein- und Mittelserienfertigung (NC machining centres and machine tool systems for small and medium batch production) *Die Technik* **7**, 445-452.
3. HERMANN, J. (1971): Auswahlkriterien für Fertigungseinrichtungen im Bereich der Klein-serien (Criteria for the selection of manu-facturing equipment in the field of small batches). *Fertigungstechnik Kolloquium* **70**, VDI Verlag, 11-21.
4. ANON (1973). CONCEMA-System zur direkt rechnergefuhrten Werkzeugmaschinensteuerung (CONCEMA system for direct numerical con-trol of machine tools) *Forschungszentrum des Werkzeugmaschinenbaues, Karl-Marx-Stadt,* 2nd enlarged edition.

COMPUTER NUMERICAL CONTROL SOFTWARE

by

W. A. CARTER*

SUMMARY

With computer numerical control the design emphasis changes from being hardware orientated to software orientated. While the user will not see any radical external changes, merely because the machine tool is being driven by a computer numerical control (CNC) system, the design changes are considerable. User demand for a cheap, but reliable system places stringent constraints on the software design.

A typical software system is described in the paper to provide an understanding of the software requirements. Features and problems associated with CNC design, field applications and maintenance are also described in this paper.

INTRODUCTION

Latest in the series of major developments that have taken place in numerical control systems is the use of a mini-computer. Software programs for the mini-computer take over the essential NC functions which were previously executed by hardware. The basic NC functions are tape decoding, interpolation, servo control and miscellaneous function control. With the CNC (computer numerical control) system it is possible to introduce additional control by taking over some of the functions that were previously performed by the machine tool builder's interface cabinet. These are the machine tool functions such as spindle, overtravel limits, tool changes etc.

Which functions will be controlled by software and which will remain in hardware will depend on several factors. The most important being the amount of computer memory required and speed at which the computer can service the NC functions. A NC machine tool has several functions that must be serviced in relatively small time periods and these requirements tend to stretch the small computer to its limits. The hardware system does not suffer such severe timing problems since it is possible to perform most of the functions by parallel processing. Unfortunately, computers are basically serial processors, such that the set of instructions which constitutes the software NC program must be executed in a serial manner. When the computer has to perform tape decoding, circular interpolation and control the machine tool magnetics there is very little idle time on the computer if none of these functions are allowed to suffer a timing deterioration compared with a conventional hardware system.

The timing problem concerning the execution of the machine tool magnetics and control panel functions can be reduced by using word manipulation of what are essentially bit operations. Unfortunately, a penalty of greater core usage has to be paid which can increase to the point where the CNC system becomes uneconomic. Good facilities for bit manipulation are important in CNC design, but many mini-computers do not provide such facilities in an economic manner.

Moving from hardware based technology to a software technology in control system design and maintenance has also introduced new problems of organization and control. The implications of these are discussed at the end of the paper.

TYPICAL SYSTEM

Although the methods used in CNC software design will differ from system to system, some of the problems are fundamental to the NC operation and to the computer architecture. By examining in detail a typical working system it is possible to gain an appreciation of what is expected in CNC software design.

The system described here has been designed to control, by software, up to six axes in simultaneous contouring, all tape decoding, servo position loop control, all the machine tool dependent logic and the control panel operation. This machine dependent logic includes the control of spindle, tool changing,

* Software manager, Plessey Numerical Control Ltd.

canned cycles, and machine slide overtravel limits. Control panel operations including jog controls, tape reader controls and manual data inputs are also software controlled. Unusual for most CNC systems, the AB 7300 system described uses a Visual Display Unit in conjunction with a manual keyboard for all information display and manual data input. Direct interface to the machine tool is via solenoid drivers and contact monitors.

The computer used is the Hewlett Packard 2100 having a 1 μs memory cycle time. The word size is 16 bits and the NC system requires an 8K memory in its basic form.

Software organization

The main system software flow chart is shown in Fig. 1. The system operates under a main controller routine which establishes the system mode and calls the appropriate auxiliary routines as required.

At 10 ms intervals the hardware clock interrupts and control is temporarily diverted to the timed interrupt controller which initiates those functions which need to be serviced on a real time basis. These are functions concerned with the interpolation, the servo and the machine dependent software. In essence the system can be divided into two sections; those routines operating under the main controller routine and those that operate under the interrupt control.

The timed interrupt controller (Fig. 2) is called upon to interrupt every 10 ms by the system clock and then proceeds to carry out certain functions on a real time basis. The servos are checked to see that they were correctly serviced in the previous 10 ms. The front control panel contacts are monitored for any change of state. All status flags are set for use by the machine dependent software, and all the contact

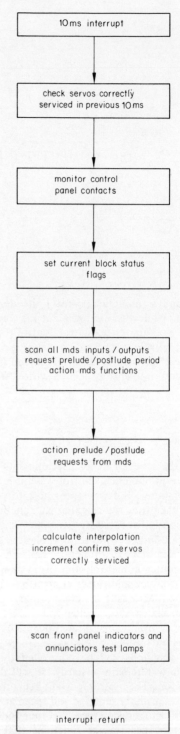

Figure 2. Time interrupt controller.

inputs and solenoid drive outputs are scanned, the MDS (Machine Dependent Software) then requests prelude or postlude period, e.g. clamping, and actions all the MDS functions. The interpolation increments are then calculated and commands passed to the servo. The servo routines are again checked that they have been satisfactorily serviced and finally the front panel indicators and annunciators are checked for satisfactory operation and the interrupt then returns control to the main controller routine.

Interruption priorities

Certain functions such as tape reading need to be carried out on a priority basis and the interrupt

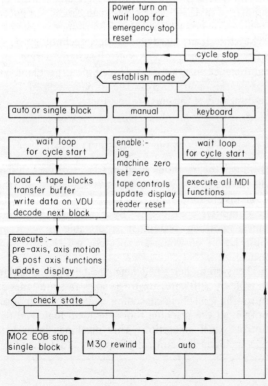

Figure 1. System software chart.

system provides a means of temporarily diverting control from the main controller to the device which needs to be serviced. Since several devices may wish to transfer data at random times it is necessary to lock out all transfers except the one being serviced. The input/output priorities are established by the enabling line which runs in series through all the device interface cards. A device in the process of transferring data essentially breaks this line thus disabling all devices of lower priority. However a higher priority device can break in at any time and temporarily disable the device that was being serviced. The interrupted transfer will continue when the higher priority device has completed its task.

The order of priority starting with the highest is as follows.

(a) Power fail interrupt with automatic restart. A power failure causes an interrupt to a trap cell which defines the address of the power failure routine. 500 μs are available for this routine to perform an orderly system shutdown. A resumption of power is detected and causes an interrupt to this trap cell. The difference between power coming on or going off is defined by the state of the flag at the time of the interrupt. When power supply resumption is detected the main control is entered at system start up entry point, this then causes the system to take up the emergency stop state.

(b) Memory parity error. Memory parity is checked and if an error is present an interrupt occurs in the trap cell which defines the address of the shut down routine for memory parity error.

(c) Tape reader. A tape reader request again causes an interrupt and entry into the tape reader driver routines.

(d) 10 ms clock. This is third in the order of priority which causes the interrupt to the trap cell and entry into the timed interrupt controller.

(e) Keyboard. An interrupt from the keyboard does nothing more than set a flag which is later monitored by the main controller.

Main controller software

Probably the best way of obtaining an understanding of the main controller software is to examine the logical sequence of events for system start-up, program start-up and then operation in the various modes such as, auto, block by block, keyboard and manual.

System start up—This area of the main controller performs the necessary initialization operation, details of which are shown in Fig. 3. It is executed on two occasions.

(1) When the system tape is loaded and the start up address is set on the computer;

(2) on having previously executed the system and undergone a power shut down or power start up.

In order that the initialization procedure shall be properly executed the interrupt system is disabled. The initialization procedure is necessary to set up the NC system in a defined status where the state of the following error on all axes and condition of the major system parameters is set. The appropriate action to be taken during the power fail, power up, parity error and clock interrupt is dictated on execution of an

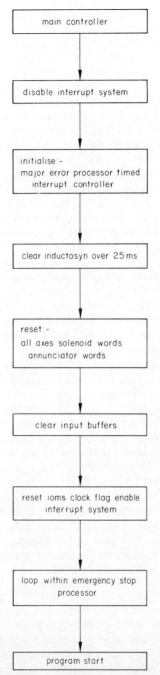

Figure 3. System start-up procedure.

initiator section within these areas which places an executable statement in the appropriate trap cell locations for the select codes. The executable statement causes a jump in the continuator part of the interrupt handler. The first part of the interrupt handler causes storage of the data in the main registers and the remaining part determines which action is to be taken.

The first time the system start up is entered after the system is loaded, entry is made to a configuring routine that sets up such machine dependent values as maximum traverse speeds, following error suppression points and home limit positions. This routine is by-passed on subsequent system start-up calls.

Initialization of the select codes allocated to the axis inductosyn takes place over a 25 ms period to ensure correct phasing coordinated through 2 ms pulses from the clock. All words set aside for carrying the bit status of all the solenoids and annunciators in the system are zeroed.

Flags are used by the basic system which on interrogation describe the state of hardware and software functions. Some are initialized to zero but others, mainly those system state parameters, are required to have specific values to cope with logic conditions throughout the system. Such parameters are those relating to g modal words, absolute or incremental programming etc. The part program information when entered via the tape reader for a given component will dictate future values. To ensure the correct operation of the hardware associated with the servo area, it is essential that the flags reflect true dynamic status. The servos are updated with the command increment every 2 ms to remove possible system ripple. If in that time the servos have not been addressed a flag is set to cause a non-retrievable emergency stop condition. A similar protection for the machine dependent functions is used to ensure that there is no continuous looping within one area. A check is carried out by monitoring the state of the flag, which should be clear by the time the 10 ms clock interrupt occurs. If the flag is not cleared then a non-retrievable emergency stop condition is initiated.

Machine control information read from the tape reader or the keyboard is stored in respective buffers. A tape reader buffer consists of 24 words, each 16 bit word is capable of holding 2 characters. The system has 8 of these buffer areas or bins plus an extension for the tape reader, the keyboard has a buffer for 200 characters. Once the buffers are full, loading can only take place when spare buffer bins become available.

Finally in system start-up the clock flag is initialized and the interrupt system enabled. A closed loop is maintained until the clock interrupt occurs causing entry into the timed interrupt controller where it is realized that the system should be forced into an emergency stop state. The state is maintained until the emergency stop pushbutton is depressed.

Program start up. This section of the program is entered after pressing the emergency stop pushbutton. The VDU screen is cleared and the system flags such as jog retract, active and fixed cycles are initialized. The machine control unit resets switches monitored and actioned which forces the system into the defined state.

After turning on the absolute contouring lights the mode of operation switch is monitored to decide which of the three defined areas, auto, keyboard or manual needs to be actioned.

Auto. The first block is decoded and set up as active while tape data is loaded into the buffer bins. If an invalid character has been loaded into the buffer a pointer to that particular buffer is set and causes an error message to be displayed when the buffer becomes active. The operator has the option to cycle stop during execution of the current block, otherwise the block execute subroutine organizes the necessary logic for completion of execution.

The operator is made aware of which block is active by the appearance of an asterisk on the left of the block, which in fact will usually be the second line of the buffer display area, the first line being the previously active block. The display area will also give updated axes coordinates with spindle speed feedrate required and tool designation.

Throughout interpolation the operator has the option of cycle stopping. From a cycle stop position the operator has the choice of continuing the block via cycle start, aborting the current block by active reset or initiating the jog retract sequence. On cycle start after a retract the retracted axis will return to its last position at a slow feedrate and then continue at the previous feedrate. Any canned cycle active within the block will be executed according to its definition. Within the block routine there is a wait loop for completion of postlude functions. An end of block stop can be actioned to request the system to hold in a cycle stop state whereby a change of mode could be achieved. All optional stop, program end, end of tape, auxiliary codes are continuously monitored. The end of tape condition monitored at buffer stage causes a rewind status before the block becomes active. Tape data is loaded as sufficient bins become available for reloading.

Block by block. This is essentially identical to the auto sequence of operation, with the exception that an end of block condition is forced for each block necessitating a cycle start before execution of the current block.

Keyboard. This mode of operation is available to the operator for entering commands and accompanying data into the VDU associated buffer in readiness for execution. The centre of the keyboard has a transmit pushbutton which sets a flag to be monitored by the basic software when it is ready. The commands available to the operator are, MDI (Manual Data Insert), Block Edit, Tool length compensation, and Tape search.

Manual. This mode of operation allows the operator to jog any axis in a required direction with feedrate override allowances with the option of setting zero at that point or to a machine zero. He can also select tape scan whereby, depending on the type of scan, that is end of block, any reference, slew and direction can be selected.

If on searching a block a parity error is discovered an attempt is made to re-read the character three times. If after the third attempt the error remains the reader halts on the offending character and the reader light on the front control panel is lit.

MACHINE DEPENDENT SOFTWARE

The CNC system has a capability of dealing with additional functions to those that are normally covered by hardware control systems. This is the area of the magnetics interface between the control system and the machine tool. It is possible through MDS software to construct a series of programs which will simulate the logic previously covered by the machine tool magnetics. The software receives inputs from contact monitors and output signals to solenoid drivers, and performs the interlocking and sequencing logic by routines in the MDS.

Because the MDS is a machine dependent area, a software interface exists between the basic control software. Information is passed in either direction

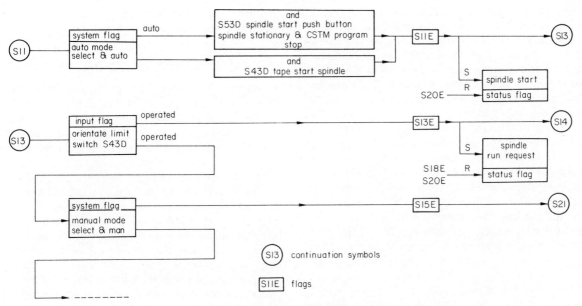

Figure 4. Section of logic for spindle start.

from the basic system to the MDS via a set of flags which can be made standard to all CNC systems of one type.

To ensure that the magnetics are updated at a regular interval the MDS is entered once during every pass through the timed interrupt controller, that is they are updated once every 10 ms. Before the MDS is entered a routine sets up flags to indicate the state of the basic control software at this point of time. These flags are then monitored by the MDS which in turn sets up its own flags dependent upon the state of the machine tool. The flags form two types. There are those which are conveyors of information and others which are requests from the basic software for some function to be carried out by the machine. These latter flags are monitored by a routine when the MDS is complete. When the flag has been acknowledged it is reset. The MDS itself cannot reset these flags since other areas of the basic software may use them.

The output from the MDS software and indeed its input are in the form of hardware signals, this interface can consist of 115 V a.c. signals or 24 V signals. Each signal is transformed through hardware to become 1 bit of a 16 bit word. This 16 bit word is allocated a select code for use in the MDS software. The interface between the basic system hardware and the computer is such that it is not allowed to pass data in both directions at the same time, and therefore common select codes are used between input and output words.

One of the major problems in interfacing a CNC system with a machine tool, if it is intended to control the machine magnetics by software, is to obtain a concise statement of the logic so that the software can be written. In order to get these concise logic definitions they must be specified by the machine tool builder in conjunction with the control system builder. It is in this area that often a problem exists since the logic of the machine tool interface is something that has usually grown over several applications of similar machines and a completely integrated logic statement may not exist for any given machine tool build, even though it may be a new machine

tool. Machine tools having simple relay logic are more easy to define since the timing characteristics are probably better known. Those machine tools employing solid state logic for the interface or a mixture of solid state and relays, present a more difficult problem. In general the solid state logic operates at such a fast speed that the machine tool builder may in fact use the speed of the logic to overcome sequencing problems. The computer, of course, cannot have this advantage where every logic sequence must be examined in a serial mode.

Two methods are available for specifying the logic, these are by logic diagrams or flow charts. Logic diagrams are probably the easiest for the machine tool builder to understand, but these of course do not define the system completely unless some element of timing is introduced. Timing considerations cannot be shown as logic symbols but have to exist as a separate description of the system.

The flow chart is probably less familiar to the customer but from the software system design, it is much more convenient to use since the sequencing of systems is more readily observed. Similarly jump instructions are not readily derived from logic diagrams but are obvious from flow charts. Such instructions are important in MDS software in order to make most efficient use of time and core. It is possible to use a compromise approach using a combination of logic diagram and flow chart techniques. A simple example of spindle start logic is shown in Fig. 4.

FACTORS INFLUENCING THE SOFTWARE DESIGN

Having examined one proven system in detail it is now convenient to examine other factors that could influence the design of the CNC system. One of the major problems that besets the design of the CNC system is of course the choice of computer. The major factors are speed of operation and cost.

Technically the computer must be capable of performing the NC functions including the machine dependent software in a 10 to 25 ms time cycle.

Therefore the power of the instructions, their speed and the access time, are of vital importance.

Probably the worst feature of all general purpose mini-computers is that they are not ideally suited for machine tool control since they are basically word orientated, or at best byte orientated. Their capability for addressing and manipulating bit logic leaves a lot to be desired. Unfortunately in operating the machine dependent software it is logical manipulation of bits that is the prime function. In most computers, in order to process the machine dependent software in the right time cycle a bit has to be manipulated as one part of a complete computer word. A bit manipulation subroutine is of course feasible, but it takes ten times as long to manipulate a bit. Unfortunately operating the machine dependent software by word manipulation instead of bit manipulation is extravagant on computer memory which pushes up the cost.

Software interpolation, does not take up an undue amount of core but does aggravate the timing problem. Timing conditions become quite critical when one attempts to control more than one machine from the same computer. If the machine tools are circular interpolating on different programs at the same time, there is not much time left for decoding, machine dependent software, and so on. Three machines circular interpolating on different components is probably at the top limit of any present-day mini-computer. Apart from the critical time considerations in the machine dependent software and in the interpolator areas, it is of course, by software, quite possible to add several interesting features. These can include such items as format compatibility with other control systems, tape editing, program storage, axis calibration and a host of other things. The premium the user has to pay is increased core size, more complex software, probably additional peripherals and certainly a more complex control panel, and probably a faster computer than the one that could be used for straight NC functions.

There is no great problem in getting the information to the computer from the solenoid output or from the contact monitors or indeed the control panel, but the method chosen will influence the software design. It is possible to design the system such that all the information, from whatever source, is entered through direct memory access channel to a set location in memory such that the information is immediately available at the start of the MDS logic program. The problem is the manipulation of the data within the computer.

One area that requires extremely careful thought in the software design is the operation of the control panel and the operation of the machine tool in manual mode. In automatic mode or block by block mode then the conditions of input and output are much more predictable. When the machine tool is under manual control and in particular with the modern control systems, the operator has a large selection of options available to him, such as jog, slewing modes, entry of manual data, cutter compensation, values, feedrate override. All these events have to be taken as possible inputs at any time, and in any order. Only valid sequence must be allowed. Invalid sequences or fault conditions must be rigorously examined and appropriate action taken.

APPLICATION AND MAINTENANCE

Of course it is possible to use a CNC as a conventional NC control, that is, the hardwired interface to the machine tool is duplicated by the software control and no attempt is made to perform the machine magnetics software. However, when the CNC system applied to a new machine tool and a machine magnetics are controlled by software then the CNC becomes much more an individually tailored device to the machine tool. It can be argued that the CNC system is more flexible and the machine magnetics is after all only a software problem, but the fact that is often forgotten is that the design of the software logic is no simpler than a new design of hardware logic and initially will probably cost more. When the machine dependent software is incorporated into the basic control system then this may be considered as a special in the field. It is highly important then that the basic control software has a clean interface between machine magnetics, which is the dependent part of the software and the basic control. This allows the special part of the system to be defined more readily for any new integration.

A heavy and sometimes unseen burden that the builder has to carry is one of maintenance. When these systems are in the field then the builder must have a record of what was in the basic system, what level it was and what was defined in the machine magnetics software. Since software is somewhat flexible and can be updated in the field it is important that such changes are recorded and known to the builder. Otherwise, as is quite often the case, a machine tool modification is made, maybe one or two years later requiring some additions to the software. It could be possible that a copy of that particular software for that machine does not exist with the machine tool builder and the original basic software has been upgraded or modified from the days when the system was first integrated. Another departure from normal maintenance practice is that the service engineer now has to carry a teletype with him if he wishes to examine faults and run diagnostics on the system, or to make system modifications. Luckily the suitcase type teletypes are now available. The system designer faces another problem in the fast moving technology of mini-computers. He may at any time wish to change the computer he is currently using in the light of cheaper or more powerful systems coming on to the market. Since almost all the software will be written in basic assembler for the particular computer, then software change for a new computer requires a complete re-write from flow diagrams or at least a complete change of all instructions. It is believed that the computer manufacturers will recognize this problem and the design of computers will be very much more flexible with their instruction sets on micro-processors, allowing software changes to be made more easily. We will not achieve a complete standardization or capability, since not only do the instructions change with a new model, but the system architecture changes, which often faces the user with a problem of having to change the whole structuring of the programming logic or, even worse, a system design change.

CONCLUSIONS

We expect to see a continued increase in the use of mini-computer or computer technology driven control system for NC, with increasing capability of controlling more than one machine from one controller.

Current costs are such however, that the mini-computer system cannot compete economically with the lower end of the market but becomes more effective with complex lathes and machining centres. Even as computer prices drop so do the prices of hardware systems, and it must be remembered that with a computer based system there is still a great deal of hardware connected with it, such as tape reader, drives, input/output logic, control panels etc. No amount of cost cutting in the mini-computer will ever affect this.

An added problem is that the mini-computer carries its own power supply in addition to the power supplies already needed for the control system, which in turn means that more heat is being dissipated by the system with an added factor that the memory of the computer is most sensitive to high temperatures and becomes the most vulnerable part of the system.

Apart from these considerations computer based technology will be increasingly used in control system design. Development in micro-processor technology will speed this process since they can be more directly applied to specific control functions. Computer controlled systems will increase the capability of NC systems, by offering such features as tape edit, tape storage and axis calibration. It will also increase the scope of control by providing digitizing capability, monitoring and data logging. When CNC becomes generally used it will undoubtedly change the concept of NC computer programming by providing at least some of the post processor functions and may change the now established processor interface and control tape format.

REPORT ON A SPECIAL CNC SYSTEM

by

T. PFEIFER and T. DERENBACH*

SUMMARY

This report deals with system components which were developed in the setting up of CNC control. The process to be controlled is a lathe with an automatic tool-changer and stepping-motor drive. The system software and hardware components described combine to form a flexible and extensible control unit. Future scope of the extended system-application will be mentioned.

INTRODUCTION

Numerical control systems using a digital mini-computer (CNC-systems) are increasingly being considered when setting up automatic methods of control of conventional or new manufacturing processes. Until a few years ago, all methods of numerical control which could in general be understood to be information systems[1], were hard-wired. The primary task of the information system is to combine logically the control data and the process data and determine from these the actual guiding values for the production system. Since logical combinations in Boolean algebra can best be solved with digital computers, it was logical to apply suitable computers in the area of the information system. Although initially the high price and size of computers at that time impeded the application of such equipment, computers are now on the market which are universally applicable, small and inexpensive. Thus many manufacturers of control systems find it necessary to reconsider their previous concept of hard-wired systems and investigate the possibilities and justification for application of control systems based on mini-computers.

On the occasion of the 14th Machine Tool Conference at Aachen, a Computerized Numerical Control-system (CNC) of a two-axis, contour-controlled lathe was presented in the Laboratorium für Werkzeugmaschinen und Betriebslehre at Aachen University (Fig. 1). This paper will deal with the setting up, structure and future area of application of this system.

CNC EQUIPMENT

Figure 1 shows the complete production equipment. In the left part of the picture two units are shown; the one at the edge of the picture contains a display and input panel for operational data acquisition. The tasks of the CNC computer in on-line data acquisition will only briefly be dealt with in the final section. The adjacent unit contains the actual CNC equipment. The tape reader at the top is followed by the tape puncher, the mini-computer MINCAL 523 and the process interface with power supply. The external operating elements of the control system include a function board which can be seen in the lower-right part of the foreground, and a teleprinter which does

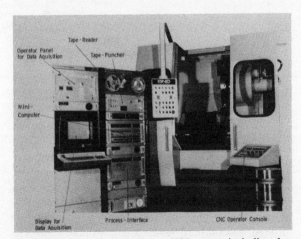

Figure 1. View of the built up CNC-system including data acquisition.

* Laboratorium für Werkzeugmaschinen und Betriebslehre, University of Aachen, West Germany

not, however, appear in the picture because after completion of all setting up and correcting operations, it is not placed in the operator's working area. The system to be controlled, an NC automatic lathe with drum tool-storage and stepping-motor drive, can be seen in the background.

DEVELOPMENT OF THE CONTROL SYSTEM

With computerized control systems it is necessary to examine all the tasks and establish whether the hard-wired or programmed solution is more suitable, so as to produce an efficient task-distribution. For this reason, the types of solution chosen will be presented separately for the software and hardware areas.

CNC software

Process controls of the type treated here are so-called programmed control systems, that is, all work sequences up to the finished product are stored as control data in the workpiece program. The control data contain all necessary geometric and technological information. They are listed in program sequences according to agreed syntax rules. If there is no access to mass storage, the workpiece program is usually in the form of punched tape.

Controlled operation of the workpiece program is carried out sequence by sequence. Since for economic reasons it is not advisable to make the whole workpiece program core-memory resident, the information must be fed in sequentially as required. As the input of a character is much slower than the computer handling of a character, alternate buffers are built up in the system for program sequences—so as to avoid computer waiting-times. Two sequence-buffers have proved to be sufficient for refilling the alternate buffers in multiprogramming operations.

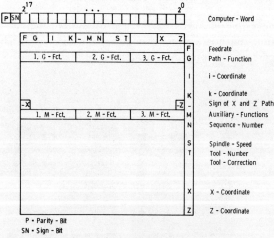

P · Parity - Bit
SN · Sign - Bit

Figure 2. Matrix used for one data-sequence.

For the processing of controlled data sequences by the control system, the form in which the data are offered is important. As invariable storage configurations are the simplest for program interpretation, data presentation in matrix form[2] was chosen. Figure 2 illustrates the layout of the quadratic matrix. Each matrix can hold the contents of one sequence only. In the matrix heading, a bit position, reserved for

function called up in the corresponding control data sequence, is entered, whereby the appropriate value can be found in the matrix line allocated to the bit position of the function. The order of the function positions in the matrix heading is arranged so that they occur in the logical operational sequence required for the task.

The interpretation of the control data matrix and the carrying out of necessary measures is a task of the control program. As shown in Fig. 3, it is the central

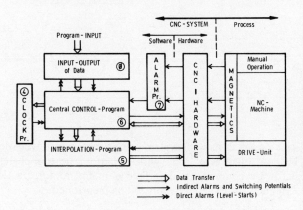

Figure 3. Basic components of the built up CNC-system.

program component of the CNC system and can be described as a flexible, logical form which, with dynamic consideration for all accompanying parameters and alarm signals, must make decisions with the process in real-time operation. The interpretation schema is very simple. For each stage, a given question sequence for function call-up is run through once. The order of the function interrogation corresponds to the function layout in the heading of the control-data matrix. If the function call-up has been established, the program leaves the decision sequence and branches off according to the type and extent of the measures to be taken, to the temporary or final conclusion of the operational stage. If the sequence function in operation does not prohibit the simultaneous occurrence of further functions, the program automatically returns to the next stage of the decision sequence. The advantage of this simple interrogation scheme is that with eventual changes of individual functions or addition of newly defined functions, program changes or expansion can be made in blocks. Thus it is possible to arrange the central program package modularly and in a clear way.

Contrary to positioning control and line-motion control, contouring control requires an interpolator. This has the task of dividing into sections of fixed length a function analytically described by the interpolation parameters. In Fig. 3, the program module responsible for handling the contour interpolation and which generates the data flow to the production-system over the process interface, is identified as an interpolation component.

The task of an external interpolator lies in the transformation of the contour function into a digital, geometric representation. The computed results are stored and recalled on demand. Apart from this, the internal interpolator has a second task: the representation of the transformation result as a function of time. The result is a coordinated movement of the controlled axes. This is insured either by the computing frequency of the interpolator itself, or by suitable data-output equipment. The task of the internal interpolator is, therefore, to obtain as accurately as possible digital, geometric information from the analytically given function and to present the information in real time mode or synchronously to the process events.

Apart from this, interpolators differ with relation to the mathematical process in effecting the transformation. If the contour to be interpolated is given in the form of difference equations, the interpolators are referred to as digital differential analysers[3]—DDA-interpolators for short. This type of interpolator should now be mentioned.

With two-axis contouring control, two functions have priority because of their frequent occurrence: the straight line and the circle.

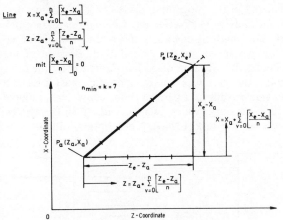

Figure 4. The linear DDA-interpolation.

The straight-line difference equation whose mathematical formulation and graphic interpretation are reproduced in Fig. 4, states that with each interpolation computation the value to be calculated $\sum_{v=0}^{n}$ increases by one summation term $(X_e - X_a)/n \leqslant 1$ of constant value, in which the summation term can only be considered if it is greater than or equal to 1. Reaching a unit value is synonymous with increased movement in axis direction of increment size. The so-called divisor is identified as n. Its value must be greater than or equal to k, which is produced by the cutting of the straight-line contour to be followed, into rounded up increments. With two-axis movement this corresponds to the cutting of the hypotenuse into path-increments (Fig. 4).

The circular equations (Fig. 5) are mathematically set up in the same way as the straight-line equations. Typical differences are that the terms of the sum are no longer constant and their value is calculated from the coordinate value already reached for the other axis. The divisor n is subject to relevant laws, as with straight-line interpolation, although its value is fully independent from the length of the contour to be followed. The reason for this is that in setting up the differential equation system for linear motion, that time is chosen as reference point which is needed for following the given course. With circular motion, how-

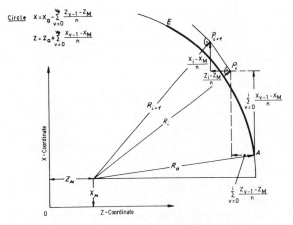

Figure 5. The circular DDA-interpolation.

ever, a period of time is chosen as reference-time, which will be needed for running through the angle $\alpha = 1$ to $57°$. From that the divisor n takes on a value in its circular movement which is greater than or equal to k, which results from the cutting of the relevant radius into rounded-up path increments. Figure 5 illustrates the geometrical relationship to the circular equations. In addition, Fig. 5 shows a circular expansion. It is caused by the computer process which determines tangents to the circle, whose 'points of contact' do not, however, lie on the circle, with the exception of the initial tangent. It follows from elementary geometry that the relative error R_v/R_0 is proportional to the computing steps v and conversely proportional to the square of the divisor n. For large values of n the radius error can therefore be made as small as required.

The form of the difference equation for the contour representation is particularly accessible by means of reproduction through digital subtraction and addition operations. The interpolation calculation is therefore chiefly practicable using fast running, simple programs in the CNC-computer. It was estimated that the software internal-interpolation alone for the computer would require considerably more computing time especially with high-speed movements. However, the low load on the computer in handling solely the control operations within the turning process described, would permit much additional computing time. The choice of effecting a purely software internal interpolation was not only justifiable, but could even be considered as an interesting experiment. This experiment could answer the question: which interpolation-data rates in terms of software only, can be reached as maximum for a given computer? In filing the interpolation program in a core memory with a cycle time of $1·5$ μs, interpolation frequencies of approximately $9·6$ kHz could be reached for both linear and circular interpolation. If semiconductor memories are used, interpolation frequencies can be reached which lie above this by a factor of two to eight. A software interpolator is useful for the relatively simple task of controlling a turning process, without the computer having to handle more complicated and intensive operations. Not only can it be built up simply and quickly, but special requirements of the user can later be met. This is not, however, the case when the control system must also handle process adaptation, process surveillance or operational data-acquisition. In such cases it is necessary to make curtailments on the interpolation side. There is a solution which allows the system to preserve, for the greater part, the advantage of having a flexible effect on the interpolation process and in addition, decidedly assists the computer, especially in high-speed movements: that is the reduction of the software interpolation-routine to the plotted points of the contour, and carrying out the fine interpolation by means of additional, external hardware linear-interpolations. Since such interpolators are small, inexpensive and easy to assemble, this type of job-distribution will prevail in future CNC-systems.

The alarm program (Fig. 3) which handles the reading and releasing of most program-interruption signals is normally on standby and has absolute priority if activated, so that a decision can be made immediately whether measures should be taken and what these should be. As the necessary operational routine must first be determined on the basis of a programmed decision process, an indirectly working interruption-system is used. These systems are by nature slower than direct interruption-systems which avoid the software decision process, whereby the necessary programs are directly activated by particular alarms or alarm groups. The indirect interruption system is, however, necessary in order to be able to make the alarm operation parameter-dependent. Flexibly carrying out these tasks using the alarm program requires detailed and methodical organization of the identification routine, using very simple and speedy command sequences.

The clock program should be briefly explained, to complete the description of the program components described in Fig. 3. Its function is to register elapsed periods of time and store them for simple read-out. It also acts as a programmable alarm release which, after a predetermined time has elapsed, restarts that program level which was originally set. The clock program is triggered at a rate of 100 ms by an external generator or short-interval alarm clock.

In controlling a machine tool, problems of sequence generally occur. This means that during the operation, events take place which require the functioning of certain parts of the equipment or the starting of particular test routines. These occurrences can take place simultaneously as well as consecutively, and in the latter case in such a way that one or more events take place in the operations phase of a previous event. Thus the computer must allocate certain priorities to all types of events or related groups of events and carry out the alarm operation according to the priorities allocated. A corresponding number of program levels is allotted to the number of priorities distinguishable by the computer in hardware. The program components mentioned in Fig. 3 were allocated priorities which increase with process proximity. The various computing levels are marked with circled figures, whereby zero means the lowest and seven the highest priority.

The total storage requirement for the CNC basic software could be maintained with 2·52 K 18 bit

words in a suitable dimensional arrangement. Because the standard mini-computers available have a storage of 4 K words, the standard model provides enough latitude to enable the computer to handle additional tasks in process surveillance, operational data-acquisition or adaptive process-control, without expensive storage extension.

CNC hardware

The interpolation carried out has been characterized as internal interpolation. This is not quite correct. In order to allow the control computer the possibility of carrying out important alarm operations even at high interpolation rates, without having to interrupt the interpolation data-flow, a small amount of data was stored intermediately. Contrary to the external interpolator however, the whole of the contour information was not stored, but only small contour sections measuring only a few increment lengths. In coordination with this, not every interpolation datum calculated in the program is passed on to the drive unit immediately. Only after determining one data block of the length of one computer word, is the information passed on to the data output. Because of the serialized processing of path information by the drive unit, the hardware for data output must have the characteristics of a parallel series converter. A simple data memory which works with a shift register, assumes this task (Fig. 6). A description follows.

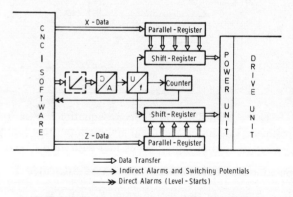

Figure 6. Output circuit for interpolation data.

The parallel register and shift register are generally filled with data in the following way. Initially the interpolation program produces one computer word with interpolation data for each moving axis and hands it over to the parallel register allocated to the axis. After takeover of the data into the shift register, the interpolation program is restarted directly by the hardware so as to have the second data word determined, whilst in the meantime the shift register passes its information serially to the drive unit. If the second data word has been established and passed on to the parallel register, the interpolation program pauses until the second word has been taken over by the shift register for output, and a command is given for provision of the third word, etc.

The data output to the moving axis is automatically coordinated because the shift register is given its shift rate by a central pulse generator, whose frequency is proportional to the contouring speed. The correct setting of the voltage controlled generator was obtained by the control computer calculation of a digital value proportional to the programmed contouring speed, and by passing it on to a digital analog converter, which in turn provides the generator with the appropriate voltage level.

As the frequency of the stepping-motor control may only fluctuate up to the so-called 'start stop-frequency', all nominal frequencies which exceed this typical value must be guaranteed a slow increase or lowering of frequency in the acceleration and braking phases respectively. Therefore an integrating component could be placed in front of the D/A converter if required.

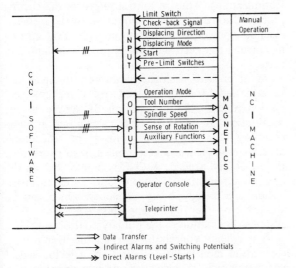

Figure 7. Process interface and control panel.

Apart from the hardware control for the interpolation data output, connected to the driving side of the process by means of data channels with only one information direction, there is also an exchange of data and signals between control and process. On the process side, the magnetics unit is a direct signal receiver and transmitter. On the control side, the input–output interface performs this function (Fig. 7). The whole input–output logic was built up word by word with hardware modules which nowadays form part of the standard equipment of every computer manufacturer.

To guarantee a desired production sequence, the operator must have control over the process events. On the other hand, there must be a situation reporting facility on the production equipment side. These process communications are as much an orientation aid for the operator to carry out any necessary measures, as a confirmation of the correctness of his decisions. The dialogue between operator and production equipment is carried out over two terminals which can be regarded as one unit because of their function (Fig. 7).

The unit marked 'Operator console' provides an optical indication of binary process-situations and computer-register contents. Additionally, the contents of this register which can be interrogated are newly defined and the machine set to one operation-mode. Finally, the operator is able to reduce all feed rates continuously down to 20 per cent of their programmed value, alter the spindle speed and set all computer programs to a defined starting point.

Although it may not be absolutely necessary, it is very useful with CNC systems if the staff have equipment available with alpha-numeric input and output. In this way, for example, it is easy to carry out input and output as well as changes of co-ordinate values for tool-length correction. Such equipment is valuable if technological errors in the workpiece program are to be eliminated. Since such program errors do not occur until a trial workpiece is run, it is advantageous to make the correction on the production side. Workpiece programs can quickly be expanded, shortened, corrected and listed with the aid of a small correction program and using an alpha-numeric input or output.

Future scope for application

Running a workpiece program by means of the control computer is achieved in a kind of fine analysis of elementary settings and control signals as well as internal sequence regulations. The volume of data to handle increases noticeably. An increase of data characterized as data explosion[4] inevitably occurs with interpolation. The time periods between information demands to the control system, which normally last fractions of seconds, become about one thousandth shorter through software interpolation. A useful task distribution in the form of plotted-point calculations carried out by software with consecutive, hard-wired linear interpolation allows the data volume, to be generated in the computer, to increase not thousandfold but by about ten times. It follows that the increased free time created by distribution of the computer load will not be fully available in the near future, through increase in further control and test functions. Evidence that this opinion is generally becoming accepted is that consideration is being given to a mini-computer taking over the control of several production processes—that is, building up a kind of mini-DNC system. This should not now, however, be covered in detail for the immediate interest is in most promising, future operations which a CNC system can additionally carry out.

In the so-called magnetics of a machine tool, decoding, blocking and logical combinations etc. are realized: e.g. functions intended to be carried out by a digital computer. An extraction of the logic area from the magnetics through its reproduction in the CNC computer implies a shifting of the signal boundary nearer to the process (Fig. 8). In this way

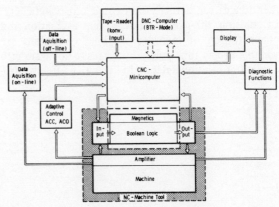

Figure 8. Extended CNC-concept.

the control gains a greater, immediate influence on the internal functions of the process. This magnetics software should be easily interchangeable and stored in read-only memories. After formulation of the special machine logic in Boolean equations, it is conceivable that program generating as well as program test and recording in the read-only memory could be run automatically. Such a set-up is of primary benefit to the system manufacturer, as the magnetics unit itself is considerably smaller and cheaper, its realization is speeded up and the manufacturer can deal more flexibly with equipping different types of machines or special designs.

In taking over the logic area from the machine magnetics, the computer has direct access to internal machine-conditions. This information can be of advantage to the user in machine supervision and fault diagnosis. Such diagnosis routines can be put to good use in system changes, so as to shorten as far as possible the test phase by using the interactive method of testing or improving testing. Additionally, such routines can run either as a cycle during the process sequence in the form of preventive machine-supervision, or can be activated on demand in the event of machine failures so as to locate and eliminate the cause of breakdown as fast as possible. In the latter case, the diagnosis program does not need to be core-memory resident. It can be set out correspondingly amply and comfortably and can provide the operator with the results found—in alpha numeric form for example.

Success in the unceasing and strenuous efforts to increase productivity largely depends on internal operations being scrutinized in detail. With respect to the individual production process, it becomes necessary to determine, collect and evaluate operational data such as productive and idle times, quantity of good and rejected parts, technical and organizational causes of machine down-times etc. Some of these data can only be automatically registered on-line, that is during the production sequence. Other data which can usually be fed in off-line, and which can therefore be subjectively manipulated by the operator, can be made available for automatic registration. Besides the general operation-dependent machine condition, situation queries on individual and important functions of the machine are made possible by the control system. With CNC systems this kind of machine similarity control-configuration close to the machine is not only possible but (as has already been mentioned) can actually be carried out beneficially. In addition, the parameter-dependent evaluation and concentration of recorded machine conditions are variable at any time purely by means of software into explicit operational-data. In using a freely pro-grammable control computer, not only can data-acquisition be carried out more easily, distinguishably and objectively, but additional tasks in part-preparation of data—in the sense of background calculations—can be dealt with by the computer through collecting, sorting and concentration.

There is practically no doubt that the coming generation of computer control will decidedly be influenced by efforts to create adaptive systems. This includes the simple, marginal control of technological values (Adaptive Control Constraint) as well as the difficult determining of technological nominal values from adaptive optimization models (Active Control Optimization). Whilst a control improves the cutting process through greater exploitation of the installed machine's power, an optimizing control of the cutting parameters can keep the cutting process at minimum costs.

As at present, analogue hardware-techniques will be preferred for continuous control of technological process values. If the addition of automatic cut distribution is required, the calculation strategy necessary for automatic control of the tool-movement sequence is best applied in software with the flexibility needed. A CNC computer has all the basic requirements. Methods of automatic cut distribution are already being developed which can be applied to the mini-computers used for CNC systems—in addition to the primary control tasks[5].

It is equally conceivable that in the near future, calculation models for optimization of technological parameters will be developed and implemented on a CNC computer.

To summarize, Fig. 8 shows the area of tasks mentioned which can very effectively be applied in a CNC system. Only a system layout in this or in a similar form provides the user with the extensive flexibility and capability of the independent freely programmable computer.

REFERENCES

1. G. SPUR (1972). Analyse des Fertigungssystems Werkzeugmaschine, *Zeitschrift für wirtschaftliche Fertigung*, **67** (1).
2. W. REHR (1972). Echtzeitsteuerung von Fertigungseinrichtungen, Dissertation TH Aachen.
3. E. GÖTZ (1961). Digital arbeitende Interpolatoren für numerische Bahnsteuerungen, *AEG Mitteilungen*, **51**.
4. BJORKE (1971). On-line numerical control systems, *CIRP*.
5. E. GIESEKE (1972). Automatische Schnittaufteilung beim Drehen, *Industrie-Anzeiger*, **94** (14).

THE MTIRA COMPUTER NUMERICAL CONTROL (CNC) SYSTEM

by

H. TIPTON* and J. I. ROBERTS*

SUMMARY

The main features of the MTIRA CNC system are described and illustrated by an application to a contouring lathe with control facilities similar to those found in current hard-wired control systems.

INTRODUCTION

The MTIRA CNC system is a computer numerical control system which uses a dedicated minicomputer to perform the decoding, arithmetic, logic and control functions of an NC system. It has been designed to form the basis of a general-purpose CNC system applicable to control of up to about six axes on any of the usual types of machine tool, or to a multi-machine system controlling independently several machines (not necessarily of the same kind) with the same total number of axes. The system accepts conventional part-program tapes, whether prepared manually or by computer, and can thus replace the hard-wired controllers on one or more conventional numerically-controlled machine tools.

As well as offering an attractive alternative to hard-wired systems as a general-purpose system which uses common hardware but gains flexibility from ease of changing the control configuration by programming (software), this same flexibility can offer other advantages not readily available with conventional NC control systems. For example, either whole or commonly-used parts of part-programs can be held in store for use as required, or a generalized program for a family of parts can be stored so that only a minimum of input data to define the part is required. The system is inherently compatible with further forms of control, including DNC on the one hand and adaptive control on the other, and the computer can also handle and process information about the operation of controlled machines as required for management purposes.

HARDWARE

Figure 1 shows the various items of equipment making up the system, together with an indication of the flow of information between them.

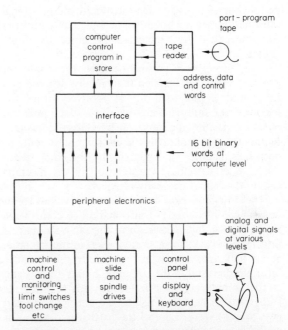

Figure 1. Block diagram of the MTIRA CNC system.

The heart of the system is the minicomputer—typically a 16-bit word computer with 4000 words of core store. The computer communicates with most other parts of the system via a general-purpose interfacing system specially developed by MTIRA with machine-tool requirements in mind. The tape reader, however, is shown as being separately connected. This is because the addresses allocated by the computer manufacturer for customers' peripherals do not encompass the tape reader address used in the manufacturer's software. There is no other reason why the tape reader (or further tape readers) should not be connected via the general interface.

The interface is, in effect, an extension of the

* The Machine Tool Industry Research Association

computer and handles information in computer words and at the voltage and current levels used in the computer. It allows the computer to pass information to a number of 'write-only' registers each holding one computer word and to receive information from a number of 'read-only' registers. Here the terms 'read-only' and 'write-only' refer to the way in which the computer communicates with the particular register. The MTIRA interface contains sixteen 16-bit read-only registers and sixteen 16-bit write-only registers. Each register is identified by an address and the interface routes information between the computer and the appropriate register.

The different parts of the system which control or are controlled by the computer—push buttons, limit switches, transducers etc.—are referred to as 'peripherals' and are connected to the computer via these 'read-only' and 'write-only' registers. In some cases only one-way communication is involved and a peripheral is connected via one register, but in other cases connection to both types of register is necessary. For the operation of a switch only a single bit is required and thus one 16-bit register can control sixteen peripherals of this type. On the other hand positional information needs about 20 bits and thus a position transducer will be connected to more than one register.

The information passing between the computer and the registers is handled sequentially (a word at a time) and is presented to, or received from, the machine tool through a number of parallel inputs and outputs via the registers. Each register acts as a temporary store and holds the information it contains until it is modified or 'updated'—by the computer in the case of write-only registers, or by the machine tool in the case of read-only registers.

The interface can output up to sixteen 16-bit words (256 bits in all) of information to control the machine tool and receive up to the same amount of information back; for a single 2-axis contouring lathe about half this capacity is required. In the interface information is still in the computer binary format and at the computer voltage level, typically 0 to 5 volts and it has to be converted to the correct form and level to operate or control the particular devices for which it is intended, and in the same way the output responses and various control signals from the peripherals to the computer need to be converted from their various levels and formats to binary signals varying between 0 and approximately 5 volts. Both these functions are carried out in the block labelled 'peripheral electronics'. In practice much of the hardware for this is on a set of printed circuit boards mounted in a rack but some of it may be elsewhere, on or near the controlled devices.

It will be clear that a computer and an interface are essential parts of all CNC systems whatever type of machine tool is controlled. Moreover, the computer/interface combination is general-purpose in nature in that it can be used to control any type of machine tool. Some peripheral electronics are also essential with every system but their nature will depend on the type of machine being controlled. In particular, they will depend upon the kind of servo drives and measuring systems used, the type of tool-changing facilities, etc. Nevertheless most of the necessary peripheral electronics for any actual system can still be selected from a limited number of standard circuit boards containing, for example, D–A (digital-to-analog) converters, triac switches for controlling solenoids, etc.

SOFTWARE

Method of operation

In a conventional NC system the various functions and all the individual details of the system (such as, for example, the type(s) of part-program control tape format that can be accepted) are determined by the wiring of the system which is then called a hard-wired system. A CNC system on the other hand is controlled by a computer program or by software, and the control configuration of the system can be changed by reading in a new CNC control tape. In the MTIRA system the program for the CNC control tape is written in modular form, that is, as a set of sub-routines each of which deals with a particular aspect of system operation such as decoding the information on the part-program tape or working out the tool path (interpolation). Both these sub-routines are applicable to any type of contouring machine tool and this is true of many of the other individual sub-routines so that much of the software is general-purpose in nature.

This same flexibility makes it easy to provide particular features required by individual users. For example, alternative decoding sub-routines for either EIA or ISO part-program tape coding, or both, can easily be provided.

The basic operations performed by the CNC control program are:

(a) decoding the input information from the part-program control tape or the manual data input;
(b) controlling the geometric shape of the part produced by interpolation of the data provided and ensuring that the feed drives move the tool along defined linear or circular paths within an acceptable tolerance;
(c) controlling the auxiliary motions and logic such as sequence control, spindle speed, tool changing, interlocks and limit switches, etc.;
(d) detecting and indicating any error conditions and taking any necessary protective action.

The first three of these functions are shown on the block diagram in Fig. 2.

A hard-wired controller has a number of special electronic circuits to perform various functions which can, if necessary, operate at the same time, i.e. in parallel, and each of which, when in operation, works continuously, i.e. when it is fully occupied. A computer on the other hand, deals with information one 'word' at a time but at a very high rate, each basic instruction or step typically being completed in a few microseconds. The computer is thus dealing with separate processes sequentially, i.e. one after the other. To work in this way a computer uses a store which not only holds the CNC control program which decides how and in what order it services the various

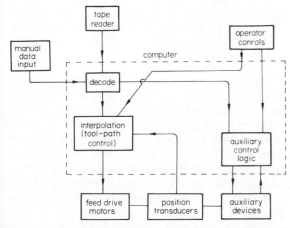

Figure 2. The basic operations performed by the CNC control program.

processes but which also holds the data on which it operates. These data consist of input information, i.e. the part-program describing the part to be manufactured, which the program manipulates, and also the results of these manipulations which determine the actions of the controlled machine. The results are placed in the store by the control program and are held there until they are required for use.

In general, there will be at any one time several different things for the computer to do, e.g. up-dating the information to the servo drives, acting on control signals from the operator, checking that the machine is in a correct state to perform the operations demanded, etc. Although the computer can deal with only one task at a time it operates so quickly that any delay that may arise in dealing with a particular requirement is unimportant. It is, however, necessary to determine in advance the order in which the computer deals with or 'services' the various control operations. This could be done in a fixed sequence but since the actions of a machine tool vary widely depending on the part to be made only a limited number of operations are required at any one time and it would be wasteful to use a fixed sequence. Instead the various sub-routines are called according to the requirements of the part-program. The time required by the computer to deal with different parts of the program varies depending on the part being made and the rates or times at which the various peripheral devices such as the tape reader, the feed drives, etc. need controlling.

In many conventional NC systems the feed drives are fed with continuous (but varying) analog signals. This can still be done with a CNC system by using 'hard-wired' interpolators (as are used in conventional NC systems) external to the computer but this tends to defeat the object of making the fullest possible use of the fast, sophisticated, but relatively cheap, data-processing facilities of the computer. Optimum use is made of the computer when the position-control signals are calculated at the minimum rate necessary to produce adequately smooth operation of the feed drives. The theoretical minimum rate is twice the system bandwidth but in practice a factor of five–ten times may be required. A sampling rate of 200 Hz has thus been selected although tests indicate that any-

thing above 70 Hz is acceptable on this particular servo system. The servo mechanisms cannot respond to stepwise changes in position at this rate so that effectively the position-demand signals appear to be changing smoothly. This is the principle used in the MTIRA CNC system and the sampling rate is determined by a 'real-time clock'—a pulse generator operating at the desired sampling rate. As its name implies, the real-time clock also provides a timing reference for the system which is used for controlling the actual feedrates and dwells etc. With closed-loop feed drives the actual slide position is compared with the demanded slide position to provide a position-error signal, so that the position loops of the servo drives are effectively closed within the computer.

The interpolation routines, linear and circular, calculate the demanded positions to an accuracy a fraction of the resolution of the measuring system and this is then rounded to the nearest whole number so that demanded positions are always accurate to better than the system resolution. The circular interpolation algorithm uses second-order terms to determine a series of chords which will follow the circular arc to the same accuracy. If the chord is shorter than the distance moved in a clock pulse at the demanded feed rate, then the feed rate is reduced to maintain the desired accuracy.

It is clear that the operation of servicing the servo drives must be performed after every real-time clock pulse and must receive priority over most other functions except those safety functions—such as emergency stops—which require the servo drives to be stopped. This is achieved by using an interrupt system whereby each of the routines is allotted a priority or order in which they are to be serviced, and which allows a high-priority routine such as feeding information to the servo drives to halt or interrupt a lower-priority routine and thus take precedence when necessary.

The tape reader also operates under interrupt control at a lower priority. The information on the tape is read into a buffer store, a block at a time and in advance of being required. This means that the machine tool does not have to wait while a block of part-program instructions is being read in from the tape and a high-speed tape reader is not necessary. In fact, as many blocks as required can be held in the computer store (in the current implementation it is six) which the computer can use whenever it is ready, and these are kept filled up from the tape reader during the time when the computer is not servicing routines of higher priority.

Control program philosophy
The CNC control program is written in modular form with the majority of modules or sub-routines being of a general nature and applicable to any kind of machine tool. It is intended that a control program for any particular machine tool should be prepared by assembling the appropriate modules with the minimum modification of the standard modules. All the modules are designed to be relocatable and re-entrant. Re-locatable means that the sub-routines can be assembled in any order, completely independently of the position in the computer store of the

other sub-routines with which they are used, and re-entrant means that the same sub-routine can be used repeatedly by any part of the program as necessary.

In an interrupt environment if a sub-routine, for example that for decoding a tape, is being used at low priority to fill up a buffer store on one machine tool it must be possible to interrupt this in the middle of its operation in such a way that the same sub-routine can be used again immediately in full by the interrupting program. In this case the interrupt may, for example, have been caused by a second machine which is in much more immediate need of decoded data. Having completed operation at the higher-priority level it must then be capable of resuming the first process where it left off as though the interrupt had not occurred. With re-entrant sub-routines this process can occur several times—the sub-routine could be interrupted and re-entered several times in succession. If re-entrant sub-routines are not used then separate copies of each sub-routine must be held in store for use separately. In a multi-machine CNC system each machine would need its own copy of every sub-routine so that an n-machine system would need core storage space for n times that required for a single machine. With re-entrant sub-routines one copy of a sub-routine suffices for all machines.

The current implementation of the MTIRA CNC system to control a two-axis contouring lathe uses about 4000 16-bit words of store. It has time available to control up to three similar machines. With re-entrant sub-routines the one control program would suffice for all three machines with the addition of a small executive program to determine in what order the three machines were serviced. Without re-entrant programs the same sort of additional executive sub-routine would be required together with three copies of the main program taking a total of about 12 000 words of store. Even where three different types of machine are controlled simultaneously, considerable saving of computer storage results by using re-entrant sub-routines since many of them, e.g. linear and circulation interpolation, decoding, are still common to all the machines.

Program organization

The program uses a set of executive routines which determine which one of a number of modes is currently active.

The various modes incorporated are:

Datum-set For use with incremental measuring systems after the power is switched on. This sends the slides to a datum position to set the position counters.

Single A single block of a part-program will be executed.

Continuous A complete program will be executed block by block.

Hold The servo drives are effectively clamped (i.e. the demanded position remains at its current value) and the system remains in a waiting condition.

Manual This allows the machine slides to be moved under manual control.

Keyboard (Manual data input). This allows data to be entered manually—insert, delete or modify blocks, tool offsets etc. via a keyboard.

Each mode of operation has its own control routine which calls on those sub-routines which it requires to perform the particular functions. Certain modes, e.g. 'single' and 'continuous' use the same sub-routines and in some cases a sub-routine can call other sub-routines. The actual mode in use is determined in the first place by the operator but may be modified later by the program in response, e.g., to error signals, which under certain conditions switch the control system into the 'hold' mode.

Other facilities such as the alpha-numeric display for the display of information on the control panel are controlled in the appropriate way for the particular mode as will become evident in the description of the implementation of the system on a lathe.

A PRACTICAL APPLICATION OF THE SYSTEM

The MTIRA CNC system is largely independent of the particular types of machine tool to be controlled and of the type of control equipment—servo drives, transducers, etc.—used on them. Naturally the detailed design of the interface and peripheral electronics will depend to some extent upon the type of equipment used on the machine tools but it will depend to only a limited extent upon the types of machine tool. It is, therefore, possible to make a standard CNC system that can be used, with very little adaption, to control almost any machine tool or group of machine tools within the capacity of the system.

In principle, many different types of minicomputer could be used, but both the efficiency of operation of the system and the design of the interface are influenced by the choice of minicomputer. As developed at MTIRA, the system uses a PDP 11 computer, and the following description of an application of the system to control an HPE 2-axis contouring lathe is based upon the system developed for this purpose around the PDP 11 computer. Particulars of the main features of the equipment, which is now fully operational, are given below but it is again emphasized that these details are not fundamental to the system; other types of drive, transducer, etc. could equally well have been used.

The relevant details of the lathe are:

X-axis traverse — 400 mm (inclined at 10 degrees to vertical)
Z-axis traverse — 300 mm
Feed drives — 1·75 kW, 800 rev/min TENV Lucas Hyperloop motors and drive amplifiers

Each axis incorporates a light-emitting diode/phototransistor transducer for datum-setting purposes and also dual limit switches, the inner ones operating via the computer, the outer ones tripping the feed power contactor directly.

Measuring transducers—twin-track optical digitizers generating 2500 pulses per revolution of the leadscrew giving a resolution of 0·0025 mm (0·0001 in) (improved digitizers now available with a separate track giving one pulse per revolution will enable the datum-set transducer to be replaced by a simple limit switch).

Spindle drive—11 kW variable-speed d.c. motor with s.c.r. drive. A forced-lubrication system is incorporated for the spindle bearings.

Two toolholders are fitted to the cross-slide.

GENERAL SYSTEM DETAILS

The format used for the part-program tape is the British Standard word-address variable-block format as defined in BS 3635 Part 1: 1972 Specification for the Numerical Control of Machines. TAB characters can be used for format control of the tape print-out but are ignored by the control system. Sub-routines for either ISO or EIA coding can be used. The input can be in imperial or metric units and in incremental or absolute coordinates. The detailed format specification is

$$\text{N3.G2.X}\overset{+}{-}24.Z\overset{+}{-}24.I\overset{+}{-}24.K\overset{+}{-}24.F3.S3.T3.M2*$$

in imperial units, and

$$\text{N3.G2.X}\overset{+}{-}33.Z\overset{+}{-}33.I\overset{+}{-}33.K\overset{+}{-}33.F3.T3.M2*$$

in metric units.

The plus sign is optional and trailing zeros are not required.

All interpolation, linear or circular, is done by the computer which feeds out the axis-control information, as discussed previously, at a rate determined by the real-time clock, and, with the closed-loop servo drives used, the position feed-back loop is closed within the computer at the clock rate. The drives and displays etc. are serviced on a time-sharing basis. It is estimated that the system uses about 10 per cent of the available computer time, leaving the remainder of the time free for controlling more axes or more machine tools.

Operation of the system

The operation of the system is best described with reference to the control panel shown in Fig. 3. All control operation are via push buttons, most of which are illuminated when they are active, and which thus display the current control state. No decade switches or variable potentiometers are used; information on tool offsets, manually entered blocks, etc. is entered via the keyboard, and the word currently being entered, including the appropriate address (N, G, M, F, X, etc.), is shown on the display.

The various active modes of control are initiated by the row of push buttons on the top right-hand side of the control pannel. The active modes are DATUM, SINGLE, and CONTINUOUS and are mutually exclusive. Pressing any of the buttons lights the button but causes no action, and can be cancelled by pressing one of the others. Action takes place in the mode last entered (and shown by the button being lit) only when the EXECUTE button is pressed. The two buttons remain illuminated until the action is complete when the system enters (and indicates) the HOLD mode.

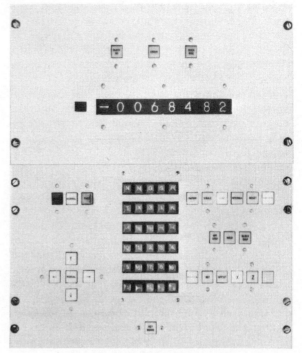

Figure 3. The machine control panels.

The DATUM mode is necessary because an incremental measuring system is used, and it is intended to be used immediately after switching on the power in order to set the position counters on each axis at reference points determined by the datum-set transducers. Entering this mode sends the machine slides to the right and then downwards, setting up the Z and X axis position counters in turn.

The CONTINUOUS mode causes normal operation of a part-program up to the end-of-tape character. If the tape includes 'optional stop' or 'block skip' characters these can be activated by pressing the corresponding illuminated buttons at each side of the hold button.

The INTERNAL button is used to switch the input from the tape reader to an internally-stored program (held in the computer store).

In both SINGLE and CONTINUOUS modes the part-program is entered a block at a time into a buffer store, the extent of which can be varied but which is, at present, six blocks. The RESET button is pressed when it is desired to restart the tape from the beginning; pressing this button clears out the input buffers and initializes the tape reader but leaves the 'operator store' containing tool offsets, coordinates of the part origin, etc. intact.

The display consists of a 16-segment display to show address letters, followed by 8 'Nixie' tubes which display a sign followed by seven digits. In the normal mode it displays in the first three digits the number of the block currently being obeyed and the percentage feed-rate override (normally 100) in the last three digits; the 'block number and override' lamps above it are lit.

Feed-rate overrides are entered by the three buttons—SLOW, NORMAL and FAST—at the top left of the control panel. Pressing the FAST button causes the feed rate to be increased above the programmed value at a given rate which might, for example, be 1 per cent per second, the actual percentage increase at any time being displayed in the last three digits of the display whilst the button is pressed, even if the display is set to display other information. Similarly, pressing the SLOW button causes the feed rate to be reduced at the same rate; pressing the NORMAL button immediately resets it to 100 per cent—the programmed rate. These buttons are active during either of the 'cutting' modes—'single', 'continuous'—when the drives are working and also in the 'hold' mode when they are not.

Alternative information can be displayed by using the bottom right-hand row of buttons. Simultaneously pressing the POSITION and X buttons causes both buttons to be illuminated and the display shows the current X position in the part coordinates in 2·4 or 3·3 format respectively, depending upon whether the system is set to work, in inch or metric dimensions, at the same time extinguishing the 'block number' and 'override' indicators above the display. Pressing both buttons simultaneously, a second time, extinguishes the button lights and the display reverts to the base mode showing block number and override. Similarly pressing REF and X or Z shows the position of the part origin in the machine coordinates and pressing OFFSET and X or Z shows the offset of the current tool in use.

The 'manual' and 'keyboard' modes can be entered only from the 'hold' mode. Pressing the MANUAL button illuminates it and allows the slides to be moved by the arrowed buttons, either singly or both simultaneously. Invalid commands, such as simultaneously pressing the left and right arrowed buttons, are ignored. The normal feed rate in the manual mode is set at 10 per cent of the maximum cutting rate of 1500 mm per minute (i.e. 150 mm per minute) and is displayed as 10 in the override display. This can be ramped up or down by the FAST and SLOW buttons, as in the cutting modes, whether the drives are moving or not.

Pressing the KEYBOARD button allows data such as part origins, tool offsets etc. to be entered into the 'operator store' and also allows the part-program to be modified by inserting, deleting or editing blocks; on running a part program the system will obey the modified instructions entered via the keyboard instead of the original ones. The particular function required is indicated by first pressing the key labelled f and the appropriate digit—see below—, followed by the data to be entered which are automatically displayed, a word at a time, on the display as they are entered. The address letter is displayed by a 16-segment display character at the left of the numerical display. At any time during which data are being entered the last digit or character entered can be deleted by means of the delete key. Successive deletes can be used to erase as many characters as required (in reverse order) and when a whole word has been deleted the previous word is then displayed. The various functions provided for at present are:

f0   insert a tool offset
f1   insert part origin (in X or Z or both axes)
f2   delete a block

f3   insert a block after the block indicated
f4   edit the block indicated
f5   re-enter cut
f6   jog X and/or Z by the distance(s) entered
f7   clear operator store to begin a new part–program

Exciting and re-entering the keyboard mode allows the last entries to be modified but exciting and entering this mode a third time deletes the whole of the last f entry.

The computer detects various error conditions and indicates them in the last three digits of the display and lights up the 'error' lamp above the display as soon as they are detected. This operation takes precedence over whatever is currently being displayed. Further action depends on the kind of error and whether it is considered 'fatal' or not. If the error is considered serious then the system immediately enters the 'hold' mode and no further action can be initiated until the fault is cleared. Non-fatal errors include, for example, a tape syntax error. This will be detected immediately the block containing it is read into store. The error light is illuminated and the appropriate error number shown in the display until the system comes to use that block; at this point the error, if not previously corrected, becomes a fatal error. After initial detection and before the block is to be used, i.e. while the error is still non-fatal, the operator can enter the hold mode and rectify the error, by, for example, editing the block containing the syntax error by entering an f4 instruction via the keyboard.

The errors currently incorporated are:

Feed power off
Spindle lubrication off
Spindle power off
Limit switch activated
Servo error out of range
Keyboard syntax error

The first five errors are all immediately fatal in the cutting modes but may not be in other modes such as the 'manual' mode. It is, of course, necessary to enter the manual mode to correct a limit-switch error.

Tapes are entered via a 300 c/s bi-directional photoelectric tape reader and, as previously mentioned, the input passes into the buffer store which is filled up from the tape reader as necessary under program control.

Although the basic philosophy is to make the maximum use of the sophisticated, high speed central processor and to use the computer to do all the data processing, this cannot include the major safety features, which must be independent of the computer. The emergency stop button acts directly to remove the power from the whole system. Duplicate limit switches are used on all axes, the inner ones acting via the computer. If, for any reason actuation of one of these is ineffective then the outer limit switch of the pair removes the power from the feed drive.

ACKNOWLEDGMENTS

This work was undertaken as part of the research programme of the Machine Tool Industry Research Association and is published by courtesy of the Director.

The software was developed by R. Kissach and B. G. Evans and the hardware by D. J. Stanton and T. Brigg.

SOFTWARE DEVELOPMENT FOR BATCHMATIC COMPUTER NUMERICAL CONTROL SYSTEM

by

I. W. SMITH, D. A. HEARN and P. WILLIAMSON*

SUMMARY

The paper describes the development of computer numerical control (CNC) software for Herbert Batchmatic numerical control lathes—the first production lathes to be offered with CNC. The paper deals with the topics of system cost, system flexibility, machine logic, error diagnostics, tape preparation and prove-out, feedrate and interpolation. In each case the advantages and problems of implementing the features in software are described.

INTRODUCTION

In mid-1971, the Lathe Divison at Herberts decided that the new range of NC lathes would be produced with CNC controls instead of NC controls. These controls made their debut at Olympia 1972, and are now in production with orders worth over £2 000 000. The principal reason for switching to CNC was the realization that in the quantity production of the lathe market, that is about 150 machines a year, CNC gave lower cost control systems and that CNC also gave the following advantages.

(1) A general purpose control system can be developed which may be interconnected to any one of a number of different machines, changeover being obtained by priming the system with an executive program tape. This reduces work in progress and spares stocking costs, for Herbert Machine Tool Co. For the customer, as for HMT, it reduces the cost of part programmer, operator, and service engineer training.

(2) The machine logic can be built into the system, so that unlike conventional systems, commonality of control system also includes the machine logic.

(3) In the event of a program or operator error, or machine failure, the computer can give a more detailed indication of the error than conventional systems.

(4) During tape prove-out, correction to the tapes may be stored in the control system and used for the rest of a batch of components, without recourse to cutting a new tape. This significantly reduces tape prove-out time.

This paper discusses the effect these objectives had on the software design and on the CNC controller.

COSTS

It must be emphasized that we are aiming at a quantity production market, so that on the one hand development costs can be amortized over a substantial number of machines and, on the other hand, production costs must be kept to a minimum. Software is obviously well suited to this market with comparatively high development but negligible production costs. In fact, software development costs have not been excessive, only three people full time for eighteen months have been used to produce the software for the Batchmatic range. Secondly, it has been worth optimizing the software to obtain the maximum number of features for a given core size. Batchmatic machines for production with 8K of eight bit core, have more features than many systems with 8K or more of sixteen bit core. Later on in the paper the effect of this on the modularity of the software will be discussed.

Again, to keep the number of hardware boards to a minimum as much as possible was put in the software. Though this obviously increases the core storage taken by the executive program, once the basic cost of computer and interface has been accounted for, core itself costs comparatively little. So the decision was taken to put not only feedrate generation and axis interpolation in the software, but also to handle all the machine logic by software.

---

* Herbert Machine Tool Co. Ltd.

FLEXIBILITY

Consideration of flexibility leads to the same conclusions as that of cost-implement as much as possible in the software, rather than the hardware. This means that differences such as

—different machine logic
—different slide strokes
—two axis or three axis with independent feedrate
    and simultaneous motion
—bar machine, or chucking machine
—different start up procedure
—automatic datum shift and size change on switching
    turrets on the two turret machine

can all be accommodated by just repriming the core store with a different executive tape.

It must be said that the software is not the only reason that a very flexible design has been obtained. In a CNC system the computer communicates with the outside world by what is known as the 'computer highway'. To transfer information the software calls for an input to output transfer instruction. This first sends down the path a code unique to a particular interface board, say the X-axis servo board. In the clock pulses immediately after this, information is then transferred down the path to and from the computer with only this particular board responding. The implication is that it is relatively easy to add extra boards into the system, for instance a fourth axis, by assigning an address code to the board and just plugging in the board. Contrast this with many conventional systems where boards are linked together by extensive back board wiring.

While to the software designer, input and output are just 'bits' transferred by input and output instructions, each 'bit' must eventually drive or be driven by the machine or the control panel. The power capability of the interface hardware can itself limit flexibility and both the interface and the magnetics of the control system were carefully designed to try to preserve its flexibility.

An indication of the success of this approach was that though the system was designed for lathes, it is now being quite successfully applied to the DeVlieg 75J five axis jigmil.

MACHINE LOGIC

As stated above the machine logic functions are implemented in software. On the lathes these include
—spindle speed control with intermediate speeds
—coolant
—turret indexing
—bar feed
—chuck open and close
—workguard
—slide hold.

Whereas, in a conventional system each of these would have their own circuitry dedicated to that particular function and operating simultaneously with one another, a CNC system behaves rather differently. The central processor can only carry out one operation at a time and only gives the appearance of handling many things at once by moving rapidly from one to the next. To do this the software at regular intervals scans the machine inputs and the machine states, and if any change from the last time is noticed,

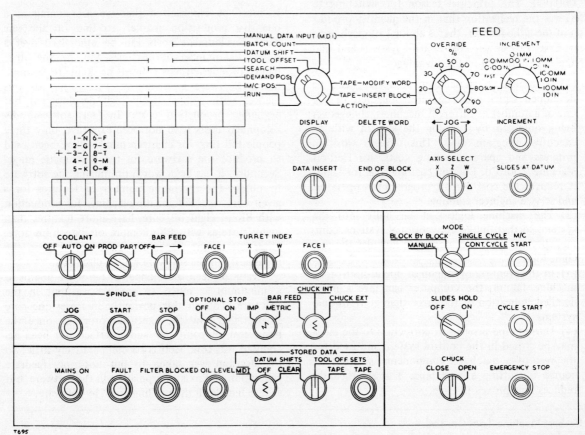

Figure 1. Control panel

it adds the change into a software 'queue'. The queue is served by the 'sequencer' which picks up each queue entry in turn, and dispatches to the software routine for the entry change. To give as an illustration the start of a turret index, the turret index routine sets the bit in the output corresponding to the unclamp solenoid. The routine returns to the sequencer which then continues handling other routines entirely unrelated to turret indexing until the movement of the turret itself activates the turret unclamp limit switch. This changes the input slightly and causes an entry into the queue.

When the turn comes, it is dispatched to another routine for setting the bit for activating the index motor. The sequencer thus forms a natural software interface to the main machine logic.

The question of modularity of software is now discussed. Many of the companies developing CNC software are aiming at the machining centre end of the NC market, where either sales of a particular machine are limited in number or there are so many options that each machine becomes a 'special'. For this market the cost of software design and commissioning of machine logic may well be significant, and as a result, a great deal of effort has been put into making this area as modular as possible. Indeed some companies claim that the logic can be specified by the mechanical designer without knowledge of software and that computer programs will then automatically generate the appropriate software routines.

In the Batchmatic software design, modularity was not a primary objective. This was for two reasons. Firstly, we were aiming at a quantity selling market where development costs are less significant than production costs. Secondly, we took the view that until we had written the software for at least two systems, we could not assess what facilities general purpose software would have to provide. However the pressure of core space, trying to squeeze as much as possible into 8K of store, forced us, wherever possible, to share software routines for a variety of functions until by now a number of standard table driven routines have emerged, allowing us to do the simple machine logic in a highly standardized modular fashion, and indeed these same routines are used in decoding many of the M and G codes.

However, our experience is that the more complex machine logic still requires some special coding. An illustration with turret index routines is given: the simple example of starting the index motor when the turret unclamp switch is made was discussed earlier. Consider the case when the operator carried out an emergency stop procedure during a turret index, so that the turret has not clamped properly. The unclamp switch is already made and when he switches on and operates the turret index control, no signal saying that the unclamp switch has changed state will occur and the turret will not index. So, a timer is set at the same time as the turret is commanded to unclamp. If this timer times out before a turret unclamp signal appears in manual mode the turret indexes regardless, but in auto it is regarded as a systems fault and shut down is commenced. Interlocks with the other turret, problems of switching in and out of auto in mid-turret index, rapid adjustment

of the turret control all cause further complications. In fact, about six different approaches to this logic were tried before settling on the present form, a combination of special coding linking together standard basic routines.

Similar problems with M codes such as MO2, or spindle speed changes or tool change could be illustrated. And so to conclude, while we think we have made some progress towards modularity of software without loss of flexibility, we believe that machine logic writing, like for instance post processor writing, will remain a skilled job for some time yet.

ERROR DIAGNOSTICS

A significant portion of software is devoted to error diagnostics. Some of these are associated with tape information areas (Table 1), the objective is to indicate the nature of error so that it can be quickly identified and corrected. This is a feature proving very valuable during tape prove-out.

TABLE 1   Part program errors

Error no.	
*10	turret out of range
*11	tool offset out of range
*12	dwell duration not specified
*13	invalid G code
*14	invalid tape character
*15	too many digits after address
16	threadchasing error—spindle not rotating
17	dwell in use
*18	invalid M code
*19	invalid S code
*20	no M code in first block
*21	x axis out of limit
*22	z axis out of limit
*23	w axis out of limit

* Active during dry run.

Another range of errors are associated with operator errors (Table 2) and another range with system errors (Table 3). The control system has a single row of 8 'NIXIE' display tubes, on which a wide range of information can be displayed, and in all cases if an error condition occurs, the displayed information is replaced by the error number shown on the table. At the same time the red fault light comes on. The operator can then look up the error number on his error list and take the appropriate action. The error indication can be removed by pressing the 'display' button, when the display resets to information it was displaying before the error occurs.

An additional feature of the software is that any word in the core store of the computer can be displayed on the 'NIXIE' tubes. the display continually being updated. This feature can be used in testing some machine faults. For instance, with a turret sequence fault there is one word in core which represents the states of the turret limit switches and another which represents the demanded state of the solenoid outputs to the turret, so that monitoring the two words enables the turret sequence to be followed

TABLE 2   Operator errors

Error no.	
50	256 identical characters: tape may be missing or broken
51	bar feed not selected: M80 on tape
52	slide not on datum
53	tape and panel dimensions not the same
54	chuck open: switched to auto
55	searched for N and not found
56	input tool offset out of range
57	modification buffer full
58	turret still indexing: switched to auto
59	machine start when not in manual
60	action buffer meets mods buffer
61	attempt to delete N word
62	axis select switch off position
63	turret not clamped or not on datum face when machine switched to auto
64	workguard not shut, machine to auto
65	x not on correct start
66	z not on correct start
67	w not on correct start
68	clear select switch between positions
70	insert action when not at end of block
71	invalid feedhold release

TABLE 3   System error

Error no.	
100	datum limit switch made when in auto
101	emergency stop input
102	tape reader fail
103	corrupt system tape error Q empty
104	corrupt system tape error Q full
105	invalid queue entry
106	end of cycle error
107	end of cycle/end of bar
108	other turret not clamped (BM 350 only)
109	turret on wrong face
110	corrupt system tape: list increment zero
114	end of block error
115	end of block, end of bar
122	end of cycle, end of block
123	end of cycle, block, bar
129	x servo excess following error
130	z servo excess following error
132	w servo excess following error

in detail. Similarly such functions as servo following error or last character read to the tape reader can be easily monitored.

Let us consider tape reader parity failure. A typical tape for a lathe may contain 5000 characters, that is 100 000 characters over a batch of twenty components. It is our experience that if a tape reader develops a fault with say 1 in 10^5 failure, it is sometimes difficult to isolate the fault and correct it, so we have added a small feature of software, whereby if a tape parity failure is detected, the system will back step the tape reader, and re-read the character. Only if it also fails the second time will it display for example error No. 14—invalid tape character. This feature enables the tape reader to continue to work with an intermittent fault, until the fault becomes hard enough to be easy to isolate and correct.

As an option, a 16K store can be supplied so that the whole program is read in once, stored in the computer memory, and accessed direct from there throughout a batch of components, thus reducing the load on the tape reader.

So far I have discussed error diagnostics associated with the program in store during normal operation. However for servicing, this program can be replaced by special diagnostic programs, which can be used to give a complete check out of the machine function, servo area performance, electrical interface, control panel and the tape reader. Indeed the computer plus interface provides an excellent monitoring system for a machine tool.

If the computer itself goes down, special hardware has been developed to check out the computer at the individual instruction and individual micro instruction level.

TAPE PREPARATION AND PROVE-OUT

A major restriction on the growth of the NC market has been the problem of tape preparation and prove-out. The Batchmatic has all the normal facilities for manual part-programming—absolute or incremental, inch/metric, G92 datum shift, mm/min feedrate etc., and post processor support for computer assisted part-programming, but it is in the tape prove-out area that the present system has made a significant impact. The problems of tape prove-out are now considered further.

When a new part-program is first tried out on the machine, a number of faults may occur. Some of the spindle speeds and feedrates will probably need changing, part programmer slips or typing errors may occur and finally some of the movements may result in collision. The conventional procedure is to run the tape until an error occurs, which prevents going any further and then re-typing the tape. Retyping the tape sounds easy, but in many firms, it means returning to the part-program department and waiting for a teletype to be free before repunching the tape and then returning to the machine—typically at least an hour's delay, and more than one retyping may be necessary.

Now for most firms, the organization of preparing an alternative batch and changing over the tooling to enable the machine to be cutting usefully while the part-programmer is away, is not possible. For the whole of the prove-out period the machine will be idle. The problem is most acute in a machine like the Batchmatic 50/2, where many components have small cycle times, and rapid set up enables small batches to be economic. Indeed in the first year, a customer purchasing a Batchmatic 50/2 may typically have to prove-out 500–800 new part-programs.

Let us now consider the tape prove-out with the Batchmatic consoles, bearing in mind the twin objectives—rapid identification of part-program faults and rapid correction of the faults.

First of all there is a dry run facility enabling the system to run the tape at full tape reader speed, but without the tape commands being transmitted to the machine. During this the system checks out the tape for format errors, or movements outside axis limits (those marked with an asterisk on Table 1) and if it finds an error the system stops and the tape displays the appropriate error number. The operator dials in the correction, which is stored in the computer memory, and then continues the dry run.

The machine can then be switched to the normal auto mode and cycled through the part-program again. The control panel can be set to display the axis position for any axis relative to component datum, or alternatively the distance to go to the end of that block. If the operator feels a collision is imminent he can apply a feedhold and from the display determine whether he has only a few thousandths of an inch to go and hence no collision, or a considerable distance so that a correction must be dialled in to modify the part-program. Of course in some cases the collision can only be avoided by a major change in the machine process, or by a change in tooling.

Finally, if on cutting the metal, the operator decides a speed or feed needs changing, the correction can be dialled in for this also, and like the others stored in core, overriding the information on tape for the rest of the batch. Thus in most cases tape prove-out can be completed and the batch machined without the need to re-cut the tape. A correct tape may later be produced for future batches by the part-programming department without keeping the machine standing idle or as an option punched out at the time by the system on a 40 char./sec paper tape punch.

While an edit package is a recent feature for an NC control system, in the computer world it is very much a standard. However, we opted for writing an edit package especially for an NC tape format, rather than using a standard package so that the operator can identify the point at which the correction is to be made, by defining the block number and the word address within the block and is working in terms familiar to him. In the edit package one or more words in a block can be added, or modified, or alternatively a complete block can be inserted.

FEEDRATE AND INTERPOLATION

In the Batchmatic system both feedrate generation and axis interpolation are software functions. In practice it is the feedrate generation that takes the most software, both in number of instructions and stops used, and in percentage of computer time. At the start of each movement, the feedrate is 'ramped up' to its required value, that is accelerated linearly. If the feedrate override is changed or a feed hold applied, the feedrate must be ramped to its new value. During all these changes, up or down, the system must monitor the feedrate and estimate the position at which to start the ramp down to the final end point. As this ramp down distance varies as the square of the feedrate, the start point, especially in rapid must be accurate if long creep times at the end of ramp down are to be avoided.

Similar difficulties occur in Threadchasing. To simplify slightly, in hard-wired controls threadchasing is accomplished by replacing the pulses from a clock by pulses from the spindle transducer, and these pulses are then modified by a factor proportioned to the required pitch on the thread. This procedure

meets problems if the slide is to be ramped up to its correct speed for approx. the same acceleration is needed whether the spindle is operating at 30 or 3000 rev/min. A scaling factor has to be introduced into the acceleration area. This turning out to be inversely proportioned to the square of spindle speed so has a range of 10 000 to 1 for a box of range 100–1.

Now the basic difference in using hardware feedrate generation and interpolation, to software is that hardware can make simple decisions very quickly, while software is several orders of magnitude slower, but can handle more difficult decisions.

For instance multiplication can be handled much more easily with software and also a great deal of setting up and scaling can be done especially at the start of the block. For the Batchmatic lathes, the all software solution turns out to be cheaper than an all hardware solution, but much more information is required before the optimum point is known, at which to move from software into hardware.

Feedrate and interpolation is carried out entirely by software. To be precise the calculation of demanded increments for each axis is done by software, but the servo loop itself is closed by hardware. The demanded increment is picked up from store and added into the servo hardware error store for each axis by micro-program as is the check for excess following error, this being a process which has to be carried out at such frequent intervals that it would require an unjustified proportion of total computer time if done by software. It has been a useful feature of the Minic computer to have been able to turn to micro-programming to give extra speed when necessary.

CONCLUSIONS

The development of the Batchmatic Control System has shown that CNC gives cost advantages over conventional NC systems for anything over $2\frac{1}{2}$ axis point to point in complexity. The effect of this cost reduction will be to increase the batch size at which NC is economic at the expense of the plugboard machines.

Secondly, already important features such as error diagnostic and part-program editing have emerged, which are not possible with conventional systems except at considerable cost. There is clearly great potential for further features to be incorporated in software to increase the overall productivity of the machine.

Thirdly, the concept of implementation by software, and the concept of linking different boards of the hardware by the data highways, mean that it will be possible to update systems in the field with the new developments, by only increasing core storage, or replacing subunits of the system. Thus the system should be much less prone to obsolescence than previous systems.

The impact of CNC will alter not only our view of the NC control system but also of NC processors and post processors, group technology, DNC and machine design. It has been a very interesting two years for the software team and the next two look even more interesting.

APPENDIX: STANDARD FEATURES OF BATCHMATIC GENERAL PURPOSE CNC SYSTEM

Listed below are the standard features of the Batchmatic CNC System for the 50, 75–250 and 350. The system is capable of extensions to cover larger machine departures, and a more accurate output resolution.

1. Self-contained general purpose computer with 8K words of storage, with the option of 16K words to facilitate complete storage of the part-program.
2. Linear Industosyn feedback.
3. Two axis contouring with simultaneous line motion control of the third axis.
4. Tape input to EIA. RS-244 or EIA RS-358 (ISO) character code, including automatic recognition of the particular code on the tape.
5. Tape format:
    (a) Metric:
        N31. G2. X-43. Z-43. W-43.
        I43. K43. F41. S2. T3. M2*
    (b) Imperial:
        N31. G2. X-34. Z-34. W-34. I34.
        K34. F32. S2. T3. M2*
6. Absolute or Incremental input, selectable within the part-program.
7. Resolution
    (a) Input   0·001 mm
                0·0001 in.
    (b) Output 0·0025 mm.
8. Maximum departure:
        1999·999 mm
        199·9999 in.
9. Direct mm/min or in/min feedrate programming.
10. Maximum programmed feedrate:
        4500 mm/min
        180 in/min.
11. Manual feedrate override: 0–100 per cent in 10 per cent steps.
12. 100 c/s bi-directional tape reader with $5\frac{1}{4}$ in spools.
13. Linear and circular interpolation.
14. Programmed dwell.
15. Buffer storage—6 blocks.
16. Electro-hydraulic servo outputs—9 V dc.
17. Axis jog/incremental jog.
18. Power supply failure protection.
19. Feed hold.
20. Automatic tape search to selected sequence number.
21. Programmable datum shift—G92.
22. Up to 99 M codes B.C.D. output.
23. Voltage input range 220–575 V.
24. Ambient temperature range 0°C up to +45°C.
25. Totally enclosed cabinet.
26. Manual data input.
27. Display of tape information.
28. Display of absolute slide position.
29. Sequence number display.
30. Display of batch count.
31. Error indication display.
32. Part-program editing with storage of modified or inserted tape blocks.
33. Full length datum shift.
34. Twenty tool offsets in pairs, three digits.
35. Optional stop.
36. Manual machine controls.
37. Machine tool magnetics.
38. Machine logic (software).
39. Threadcutting from spindle transducer.
40. Imperial/metric input, switchable.
41. Automatic acceleration and deceleration.
42. Parity check.
43. Display of demanded incremental movements.
44. Automatic axis jog to position slides at the end of the current block.

GENERAL DISCUSSION
(Answering points raised in the preceding four papers).

Q. M. A. Sabin. Most of the papers in this session have mentioned tape editing. Is not this a very dangerous option in contents where the product itself is evolving? A tape gets changed at the machine—nobody tells the part programmer, so when he makes a new tape the fact that the editing needs doing again probably gets discovered the hard way by scrapping the first part.

A. Whether tape editing is required at the machine tool or not depends on the user, particularly the environment in which the machine tool is used, and the type of machining.

With aircraft components where a high degree of tool path computation has been done, it is difficult and undesirable to edit at the machine tool. Moreover, with machining centre work, where hole position, depths, feed rates, machining sequences, etc., can be changed, then editing at the machine tool can be very useful.

The main danger with editing tapes at the machine tool is when a source program has been prepared by computer and does not get up-dated—as you stated in your question.

Q. Dr A. E. Middleditch. I would like to endorse Mr Sabin's remark that machine control tape editing is a dangerous procedure since the source check remains in error. However, tape editing will be a necessary evil until part programming/NC systems become more efficient. Let me draw the analogy that no one would consider editing the assembly code of a FORTRAN or APL program.

My questions relate to the timing aspects of CNC systems and are addressed to all speakers. Firstly, I would like to address the selection of sampling rate. Mr Roberts suggests that a sampling rate of twice the servo bandwidth is theoretically necessary but that the extended slope of the Bode plot necessitates a rate 5 to 10 times the bandwidth. I would like to know where the theoretical factor of 2 originates. The suggested factor of 5 to 10 seems reasonable but this means that Mr Smith's sampling rate of 1 sample/second implies a servo bandwidth of about 150 Hz. Perhaps Mr Smith could comment on this high bandwidth.

I would like all the authors to comment on the restrictions which must be placed on the transport delay in the position control loops of a CNC system and how these restrictions effect the software. Also, since the position control loops must be serviced sequentially, reference path errors occur. How do the allowable path tolerances limit the lack of synchronism which can exist between the servicing of the different axes.

Could Mr Carter outline the algorithms used in his control to implement circular interpolation and comment on the reference path accuracy.

A. I. W. Smith. Beat frequencies associated with feedrate generation and axis interpolation mean that the servos should be updated as often as possible, so that the demand increment update and the beat frequency amplitude are as low as possible.

In the H.M.T. system the servo loop is closed outside the computer so that problems of transport delay are avoided. After calculation of demand increments, the demands are fed to the axis within 10 microseconds of each other, so that reference path errors are negotiable.

A. W. A. Carter, Plessey NC. It is not necessary to have a sequential delay in the position control loops for each axis, since the interpolation increment can be calculated and delayed until all axes can be serviced in parallel. The small delay that occurs in holding the increment is insignificant and can be catered for in system design.

The algorithms used in the control system for circular interpolation are on DDA principle. The reference path accuracy is such that over one quadrant the accuracy is better than one programming unit. The minimum radius is one programming increment and the maximum ten million programming units.

A. J. I. Roberts. There seems to be some confusion in Dr Middleditch's endorsement of Mr Sabin's comment on tape editing rather than editing the part program. Machine control tape editing is equivalent to changing the wiring on a hard-wired NC controller and we agree that this should only be done by changing the source tape.

Editing a part program is a different matter and this can be done at the machine tool when the part program is held in some form of read-write store. This is preferable to several trips to a planning office to produce a modified input tape. There is currently a trend in NC systems to enter the part program manually and also to check and edit it manually at the machine.

The theoretical factor of two in determining the sampling rate arises in a simple analysis which considers the servo as an ideal low-pass filter and then applies Shannon's criteria that a continuous signal of frequency ω can be specified by samples at a rate of 2ω.

It is obviously desirable to minimize transport delays in a feed back system and these can be minimized in a sampled-data system by calculating the demanded position beforehand so that only a subtraction is necessary between reading the current axis position and outputting the position error signal.

In development we have run our system with delays of up to 1 millisecond with no apparent ill-effects. Many current NC systems use d.c. servo motors with thyristor drives and the (variable) delay which these introduce can be appreciably longer than this.

In the M.T.I.R.A. system the delay between servicing the two control loops is approximately 2·5 microseconds. The resulting maximum 'lack of synchronism' corresponds to 2 per cent of the minimum position increment of the system when the servos are running at the maximum feed rate.

Q. Sartorio, D. E. A., Torino, Italy. Since nothing is mentioned in their papers, I would like to ask all the authors if they have found the same type of numerous problems we found in introducing CNC into the shop. The 'blue collar' cannot be substituted by a professor when introducing a computer on a machine tool. We found the problems: one related to the efficiency of the system, the second to its reliability.

The first: we have seen the operator under stress when he has to decide to press the start push button ('have I forgotten anything?')–therefore we have been compelled to develop a rather sophisticated man–machine communication system and to develop programs to aid the operator in the initialization phase where the computer lists all the parameters that it must receive to perform its task.

The second: the commercial computers are not designed to work in a shop environment and under the fingers of a 'blue collar'. we have been compelled to kick it and to give to the operator the bip knobs, levers, etc., to which he is used.

What is the opinion of the authors?

A. I. W. Smith. The control system has been designed so that the operator need not be aware that a computer is present at all. He is using a machine tool control panel designed in a similar form to a conventional NC control panel. The feature of diagnostic error numbers makes the control in many ways easier to use than a conventional control and is proving very popular with operators.

In our system a process control computer designed to withstand a normal work environment is used. The computer front panel is not supplied with the equipment, the operator communicating with the computer by the control system panel, as explained above.

A. J. I. Roberts. We think that there should be no essential difference in the use of CNC and conventional NC systems in the shop. The CNC system should not appear to be different to a conventional system and there should be no need for the 'blue collar' worker to be aware that the system contains a computer. There is equally no reason why a CNC system should not be just as reliable as a conventional system which these days are made using the same components.

A. W. A. Carter, Plessey NC. It is not necessary for the operator or the user to be aware that he has a computer in his system. The computers used are designed to work in hostile environments. Our own systems are designed to work in total enclosed non-ventilated atmosphere at temperatures and humidities at the extremes associated with machine shop environments.

To the operator the control panel is similar and perhaps more simple than the control panels on normal NC systems using heavy duty switches and push buttons. If a keyboard is used then it is again a heavy duty keyboard designed for process operations. Our system uses the CRT which is again designed to operate in hostile environments and the information presented has received approval from operators and it is a comparatively simple matter to add more or less information to the display.

For operators who can neither read nor write we supply symbols on the control panel rather than letters.

Throughout all our design considerations we would like to underplay the fact that the computer is in the system since it is put there to ease the manufacturing problems and to provide more facilities on the system and not for any reasons of gimmickry.

Q. B. J. Davies, Staveley Machine Tools. Please can the authors of these three papers comment on field experience of reliability and service problems, particularly those involved with split responsibility for the computer.

A. The computer within its fundamental calculation functions is today of remarkable reliability. It may be considered as a mass-product. A certain part of the control hardware on the other hand is designed individually which often deteriorates the system-reliability and makes service tasks sometimes difficult. Well designed modular hardware makes us forget those problems so that CNC-systems one day will have greater reliability and much easier maintenance than pure hardware configurations.

Q. Roger Gettys Hill. Do I understand correctly that Herbert went to CNC to *save* money when all the other speakers emphasized that CNC would cost *substantially* more? If CNC is cheaper and better why have any hardwired NC?

A. I. W. Smith. Our experience at H.M.T. is that for systems more complex than 2-axis point-to-point, CNC does reduce costs. Reasons for this include absence of back board wiring, use of a mass produced article (computer core), elimination of special hardware for machine logic and the ability to use a common controller across all machines with differences only in Software. This commonality enables quantity buying, reduces spares stocks, work in progress costs, training and servicing costs.

Q. E. F. Moss. Could Mr Smith provide any information on feedback of actual and user reaction to the Matchmatic CNC concept?

A. I. W. Smith. As machines have only recently been delivered to customers, it is too early to assess customer reaction. However, the reaction of our demonstrators who have had to learn to use the new control has been very favourable.

Q. Carbonato, D.E.A., Torino. As Mr Pfeifer and Mr Smith said, there are two different approaches to the CNC problem:

(1) The first one makes use of the computer to control completely the machine, including the interpolation task. It requires probably more memory in the computer and certainly a higher operating speed of the arithmetic unit.

(2) The second one leaves to an inexpensive external hardware the task of the fine position interpolation allowing the computer either to be less complex or to spend its time in other tasks like adaptive control or other machine's control.

Already knowing the opinions of Mr Pfeifer and Mr Smith on this argument, I would like to know those of Mr Carter and Mr Roberts.

A. W. A. Carter, Plessey NC. The major design intent in using a minicomputer in a CNC control is to replace by software those special hardware modules which the conventional NC required. Options can also be written in software; thus the remaining hardware is simple and common for many types of machine tool. This is an advantage to the control system builder since he only has to build one basic system. Because the NC functions and options require a computer, then it is desirable to utilize the computer as much as possible for these tasks. For example, software interpolation which does not use large amounts of memory even up to 5 axes simultaneous should be a computer task. It is true that a high speed arithmetic is required for this task but it is also true that other options such as cutter compensation, axis calibration, etc., which are very desirable in the CNC, also require a high speed arithmetic unit.

It is possible to construct a combination of hardware and software control where a cheaper computer would be used, but it is doubtful whether the total system would be less expensive.

A. J. I. Roberts. We do not think there is a well-defined dividing line between what should be done by software and what should be done by hardware in a CNC process. In the early days our intention was to have the division biased towards hardware, but we subsequently decided to do as much as possible with software. This obviously increases the amount of computer core and computer time. The only exception we have made is the keyboard where the thirty keys would have consumed almost two complete words of read-only interface buffer. We therefore decided to encode the keyboard into an eight bit code.

Q. G. V. Bloomfield, A.H. Ltd. Mr Smith mentioned his ideas of the future development in CNC; would the other authors give theirs?

A. W. A. Carter, Plessey NC. Developments in CNC will probably move in two directions. The first is where additional capability is required over the hardware system as we know them today. For example, the introduction of features such as axis calibration, part program storage, connection to a DNC system, and greater operator control through CRT display etc. In these applications even greater computer power in the CNC system will be required.

Secondly, at the other end of the scale we can see major developments in silicon chip technology which will provide control and processing ability in small special areas such that the NC functions could be designed on special chips and provide basic system building blocks. Unfortunately at this point in time arithmetic calculations on chips are quite slow and thus the minicomputer has the advantage in this computational area. Generally developments will move more and more towards computer technology in control systems design.

A. J. I. Roberts. In the immediate future it looks as though a 'buyer's market' will exist as a result of the flexibility of CNC systems, and it will be interesting to see which features NC customers are prepared to request and pay for.

It looks as though the cost of minicomputers is going to decrease still further and that they will incorporate more LSI (large scale integration), which could mean 'processors on a chip'. What happens to CNC depends mainly on developments in the computer field—we may get central processors on a chip or a number of smaller, special-purpose processors performing different functions in parallel.

Most of the newer NC systems, if not incorporating a minicomputer, show the influence of computer technology in their design and the future development of NC is going to depend on developments in computer technology. Such developments may not always provide the user with additional features apart from reduced cost or maintenance.

Q. B. J. Davies, Staveley Machine Tools. Please can the authors of these papers comment on field experience of reliability and service problems, particularly those involved with split responsibility for the computer.

A. W. A. Carter, Plessey, NC. There are approximately forty 7300 systems integrated on machine tools in the U.S.A. Reliability of these systems is at least equal to modern hardware systems and probably better. There is less special hardware in the NC system since the basic control and most options are software controlled.

The computer is able to provide comprehensive diagnostic checks, reducing the service engineer's task when locating faults. Naturally, if the computer memory or the CPU fails, then the diagnostic routines are of little use.

The service engineer has an additional requirement over that of the hardware engineer in that he must understand at least the basic elements of the computer and to be able to make simple software modifications. On the question of staff responsibility the Plessey Company takes on complete responsibility for servicing the system including the computer; thus the end user does not suffer from the problems of divided service responsibility.

A. J. I. Roberts. M.T.I.R.A. has no field experience with CNC, but in $2\frac{1}{2}$ years of development work on CNC we have only had one fault which could be attributed to the computer and that was in the early stages. Our main problem has been in the lack of reliability of paper tape readers.

DESIGN OF HUNGARIAN NC MACHINE TOOLS WITH PARTICULAR REFERENCE TO THE MACHINING CENTRE FV-6-41

by

L. M. REVESZ*

NC technical development started in Hungary a few years later than in other developed countries. In 1965 the Hungarian Foreign Trade Company, Technoimpex, together with the Csepel Machine Tool Factory, entered into cooperation with the West German firm Krupp in Essen. This firm together with Aachen Technical University carried out a survey in numerous factories of the Krupp Group. This survey was carried out statistically on parts classification and grouping, on the dimensions of the components, their surface roughness, the character of their surfaces and other data regarding the technology of the machine parts. From this Krupp's found that the machine required, e.g., for machining disc-like components, is a lathe to machine up to 200 mm diameter and 200 mm length with an accuracy of IT7 surface roughness, R_a = 1·6 microns, and with 8–10 tools. The statistical survey also showed that the surfaces are: 4 per cent conical, 5 per cent other curves and about 20 per cent threaded. However, the company wanted at that time straight-cut control. Accordingly, the first Hungarian machine designed as an NC machine, was the ERS200, a machine suitable for straight-cut controls. There were a number of NC controls adapted to this machine, such as Grundig, Masing, AEG, Siemens, Plessey and the Hungarian Vilati.

After the first successes of Hungarian NC machines development started on a wider basis. In 1968 the Ministry of Metallurgical and Machinery Industry initiated an agreement for research and development of NC machine tools among the SZIM (machine-tool works), the Csepel Machine Tool Factory, the Vilati (Electrical Automation Institute) and the GTI (Industrial Technology Institute). Since then a number of new types have appeared on the market. SZIM bought a licence from the French firm Forest-Ratier for the manufacture of large size, bed-type continuous-path controlled milling machines in different sizes (V800, 630, 500) and a machinery centre. (CU9). As their own development they have started the manufacture of EV630 type lathe, both turret and tool-changer versions. This machine is also suitable for continuous path controls.

Based on the design of the ERS200 mentioned

earlier, the Csepel Machine Tool Factory has built a new type, ERI250, using d.c. motors (Hyperloop type) for the auxiliary movements and applying continuous path control. From these types Hungary has exported more than 80 sets, most of them to West Germany, with some pieces working in Austria, the Soviet Union, Switzerland, Australia and Sweden. In 1973 Csepel presented the ERI400 which is suitable for adaptive control.

Csepel has also developed a new type of machining centre, the FV6, a new bed-type milling machine the M6, the MFI400 knee-type milling machine also working with continuous path control, and an NC controlled radial drill, the KFS50. Vilati has already developed the Unimeric family, as its own design for NC controls, and has bought the licence and know-how to build up-to-date third generation NC controls.

For the time being there are two factories in Hungary building machining centres: the Csepel Machine Tool Factory and the Milling Machine Factory of the SZIM. Csepel builds a machine of its own design, the machining centre manufactured by the SZIM Factory is type CU9 designed by the French Ratier-Forest company. A detailed description of Csepel's FV6 machine will be given.

Hungarian design engineers make the most of the results of the statistical surveys and take due consideration of the constructional effects of the adaptation of modern technology. By considering the statistical surveys the designer treats the consumer's requirements as primordial. Statistics have shown that demands of the customers cannot be met by one single machine of optimal parameters. It follows that there is a need for various models with different dimensions, capacity and constructional features.

With the machining centre FV6 the first question arising was if it was worthwhile at all to manufacture such a machine. The answer was given by marketing experience, i.e. the number of units sold up to the present time and continual increase of sales. The size of workpiece to be machined is indicated at the same time by the main dimensions of the machine. Again the answer lay in the statistical surveys.

The findings showed that a machine with a table

---

* Technoimpex, Hungarian Machine Industries, Budapest V

of 630 by 800 mm (max. length of workpiece 800 mm) and with a cutter heat moving by 550 mm vertically can handle 55 per cent of the workpieces most frequently encountered. So these parameters became the main dimensions of the machines. To be sure another survey showed that the manufacture of machines having these sizes was a very economical proposition. This meant that pricewise the machine could very well compete with machines of similar size on the international market. As to technical aspects the machine will prove to be a worthy competitor of similar machining centres.

The next question for the design engineer was: what primary and secondary motions are necessary on the machine in order to handle the workpieces that, according to statistics, are the ones most frequently encountered? This question is rather complex, the following factors having to be taken into consideration:

(1) The proportion of the drilling and milling operations to be carried out on the workpiece— this is from 75 to 25 per cent.

(2) With milling operations, the division of work by surfaces to be machined (given in frequency and the time involved) presenting the following picture:

	per cent
profiled surfaces	4
one plane surface	19
several plane surfaces	19
perpendicular surfaces	38
surfaces joining at an angle other than 90°	20
Total	100

(3) Division by number of bores

	per cent
uni-directional	29
two, forming a right angle	17
two, in angular arrangement	1
three, forming right angles	32
three, in angular arrangement	2
four forming right angles	7
other	12
Total	100

(4) Division by bore dia. to be machined

	per cent
ϕ 5 to 10	1
10 to 20	2
20 to 50	8
50 to 100	17
100 to 200	26
200 to 400	31
over 400	15
Total	100

The decision of the designer, based on statistical data on the above four items are given: The machine will have four axes, three constituted by the longitudinal (x), and transverse (z) movement of the table, and the vertical travel of the milling head (y). It might be necessary to apply a face plate to the machine (see analysis of the bores), to form the fourth axis (w).

It is important to note that the statistics referred to earlier can be used for further designing information, specially for the determination of the tooling. In turn, the latter is closely related to decisions on the power of the drives of the primary and secondary motions. This raises another constructional problem concerning the power that the drives should have. Here the designer is assisted by the technologist who analyses the technological aspects of the possible machining operations (cutting speed and feed) for different materials. In the course of the investigations relating to the machine and with regard to tools of various dimensions and quality calculations have been made for a medium quality steel, a cast iron and an Al-silicon material, for the following operations:

— drilling from the solid
— drilling with twist drill
— reaming
— tapping
— face milling (roughing f = 8 mm)
— finishing
— milling with slotting mill
— boring
— fine boring

Because of the accuracy and precision required from NC machining centres, these machines rarely carry out roughing operations. From this fact, in the evaluation of the above technological information, a value of 6 kW proved to be the optimum milling spindle performance (the output of the main motor at 3000 rev/min is 18 kW). The required torque for the feed actuating d.c. (Hyperloop) motors is from 1 to 2 m kg. At the same time the technological analysis determined the main spindle speed range. In the construction of the individual machine elements, the correct decision on machining accuracy plays an important part. Statistical data with regard to the required machining accuracy gives:

	per cent
Class IT4	3
IT5–6	18
IT7–8	47
IT9–11	18
IT12	14
Total	100

For the machining centre the accuracy, class IT6-7 for milling and class IT5-6 for boring, may be regarded as the maximum requirement.

On the FV6 class IT6, when drilling or boring (but not with a twist drill) and class IT7 when milling can be guaranteed.

In the design of the machine the slide system and the columns are arranged on a T-shaped baseplate. The slide system is composed of a cross slide, longitudinal slide and, as a basic type, of a four-position table. The upper parts of the two columns are connected by a beam. The four-position turret head is mounted on the bridge hydraulically relieved of load, moving vertically together with the head on hardened and ground slideways. The turret head takes four spindles of which only one moves at a time that one being in the working position. For two of the four spindles automatic tool change from the tool magazine is possible.

The tool change system serves the economical utilization of time. With one spindle in the working position and machining in the opposite one, tool change is carried out simultaneously.

In one of the other two spindles a cutter is mounted, in the second a facing head for bigger boring with higher tolerances for siding, or chamfering. The versatility of the machine is enhanced by the digital actuation of the tool in the facing head.

The spindle in the working position is driven by a thyristor type d.c. motor through a transmission gear with electromagnetic clutches. From the punched tape values between 45 and 2200 rev/min may be selected.

The machine has a magazine holding twenty-four tools which can be changed automatically by their code rings. No separate manipulator is used for tool change, the magazine itself functioning as a manipulator. With the twenty-four tools the most intricate workpieces can be machined, especially when using the faceplate, and given the further possibility to employ large-sized cutter heads. To sum up, the machine is equally suited for drilling, boring, reaming, tapping and milling operations. The slide system, the bridge and the faceplate, are driven each by separate d.c. motors. Other types of drive may be applied if required e.g. hydraulic or stepping motors.

In conclusion, the main technical characteristics of the machine, type FV6, are:

table dimensions	630 x 800 mm
longitudinal travel	800 mm
cross travel	700 mm
vertical travel	550 mm
feed rate	1 to 1400 mm/min
rapid traverse	4500 mm/min
main spindle speeds	45 to 2200 rev/min
tool taper	ISO 40
facing head travel	± 40 mm

facing head speed	45 to 600 rev/min
main motor rating	
(at 3000 rev/min)	18 kW
number of tools	24 + 3
overall dimensions	2·3 x 2·3 x 2·3 m
weight approx.	10 000 kg

Main characteristics of the control system, and optionals. The control system is a four-axis, so-called mixed control in which on two axes simultaneously continuous-path control is possible. The equipment contains third generation TTL integrated circuits.

Resolution accuracy for all the axes is 0·005 mm. Data input in the basic design; may be in the form of: $n3, g2, x\pm7, y\pm7, z\pm7, i7, j7, k7, h2, f2, s2, t2, m2$.

In addition to the standard services, the following optionals are suggested:

automatic zero-points: for four axes
absolute position read-out, ± 7 positions
manual data input
automatic punched tape reeling
linear-circuit interpolar
symmetry switch for x, y axis
boring cycles $g80 \pm$ to $g89$
8 decade switches ± positions
cutter radius compensation

DISCUSSION

Q. Sartorio, D.E.A. Italy. 1st Question: The best machining centre shown has a 4 position turret for the tool. You described the purpose of 3 positions. What about the 4th?

2nd Question: About the facing and boring tool with vertical NC controlled movement. Is the accuracy given by NC positioning sufficient for boring?

A. 1st Question: The 4th position is for applying any kind of special tool which cannot be applied to the tool magazine, for example, a big diameter face-milling cutter.

2nd Question: Yes. Some accuracy can be achieved as with any other tool on the machining centre.

LOW-COST OR HIGH-PERFORMANCE NC-TURNING MACHINES?

by

H. G. ROHS*

THE PROBLEM

Looking at the tendencies of recent years in the development of NC-Turning Machines, there are two decidedly opposite trends which have been introduced into industrial application. One is the so-called low-cost NC-turning machine and the other is the high-performance NC-turning machine.

Both tendencies are based on the idea of achieving an economic production. The low-cost machine is to answer that requirement by keeping the capital investment as low as possible at the expense of productivity, whereas the high-performance machine is primarily expected to render a high productivity which in turn entails a relatively high investment of capital.

These different means towards the same end give rise to the question whether each way is justified, and which are the criteria to be observed. These are the problems to be investigated in the following study.

First of all, both types of turning machines, and their underlying philosophy of development shall be described in detail. The next step will be some basic views on the similarities and differences in application. An analysis of the cost components is the basis for an economic study and comparison, and the conclusions drawn from our arguments are the criteria for the economical ranges of application of both types of machines.

Figure 1 shows the most important features and characteristics of the two types of machine, as well as

construction	toolroom lathe engine lathe	low−cost machine	high−performance machine
bed arrangement	horizontal	horizontal	in the rear
working area	open	open	closed
chip fall	obstructed by bed	obstructed by bed	free
chip conveyor	no	no	no
driving capacity	5−10 kW	5−10 kW	20−30 kW
progression ratio φ	1.41 (1.25)	1.41 (1.25)	1.12
speed selection	manual	manual	automatic
speed range	50	50	250
maximum speed	2000−2500 RPM	2000−2500 RPM	3500−4500 RPM
tool change	manual	manual	automatic
tailstock	manual	manual	automatic
rapid traverse	120−200 IPM	120−200 IPM	400−500 IPM
chip removal rate	2.2 lb/min	2.2 lb/min	8.8 lb/min
NC−control system	none	2−axis,straight−line or cont.−path	2−axis continuous−path
total investment	£8500	£17000	£45000
productivity relative	I	~2	~4

Figure 1. Comparison of some characteristics between low-cost/high-performance NC-turning machines, and hand-operated universal lathes (swing 16−20 in).

a comparison with a conventional universal lathe. The following descriptions are based on this chart.

THE LOW-COST NC-TURNING MACHINE

As mentioned already, this machine aspires to keep the capital outlay as low as possible in order to ensure an economical application. We do not deny that one of the aims is also to remain under a certain financial barrier for gaining access to a new market range. However, this aspect can be disregarded for the purpose of our discussion.

A typical example in the category of low-cost NC-turning machines is shown in Fig. 2.

The low purchase costs are achieved by using to a large extent the major assembly groups of conventional lathes produced in batches, such as bed, headstock, and tailstock. This, however, means that such a machine looks basically like a hand-operated lathe, i.e. the machine has a bed, with the cross slide arranged horizontally; the working area is open or can only be slightly closed; and the fall of chips is obstructed by the bed. As a rule, such machines are not equipped with a chip conveyor.

The driving power, too, is that of a hand-operated machine; the spindle speeds are manually selected; progression ratio, speed range, and maximum speed are similar to those of a hand-operated lathe.

Tool change is also controlled by hand in most cases; these machines are usually equipped with either a manually indexing fourway turret or quick-change toolholders.

The tailstock again is manually operated, the quill being traversed by means of handwheel and screw. Tailstock movements and clamping are hand-operated.

The rapid traverse rate is relatively low— approximately 3-5 m (120-200 in) per min. Higher rapid traverse rates would call for more expensive feed drives which are dispensed with because of the costs involved.

THE HIGH-PERFORMANCE NC-TURNING MACHINE

With such a machine endeavour is made towards a maximum utilization of the inherent high basic costs of the NC control system. The design concept has been specifically adapted to the use of NC-control. All measures taken towards an increase in performance are consistent with the ultimate aim of lowering

* Gebr Boehringer GmbH.

Figure 2. A low-cost NC-turning machine.

Figure 3. A high-performance NC-turning machine.

the costs per piece produced, in spite of the initially higher capital investment. An example of such a machine is shown in Fig. 3.

The overall concept of the high-performance machines answers the requirements for high metal removal rates. In most cases, the machines are equipped with an inclined bed in the rear for a free fall of chips, with a totally closed working area in order to permit the application of the latest high-speed cutting methods. The large volume of chips produced is taken care of by a conveyor transporting the chips out of the working area to avoid any standstill of the machine due to the necessity of removing the chips.

The driving power is twice or even three times that of a hand-operated machine of similar size. Speed change is automatic and controlled from the tape; a finely stepped geometrical progression is available for optimum surface speeds and, consequently, short machining times. If desired, equipment for constant speeds enables infinite variation of the spindle speeds.

In order to permit the machining of rather small diameters with high economic cutting rates, the maximum spindle speed is usually higher than on hand-operated machines, and so is the speed range in most cases.

The application of extremely high cutting rates results in a reduction of the main machining times to such a level that special attention has to be paid to considerably reducing the idle time. Therefore, the tools are changed automatically during the operating sequence by quickly indexing turrets which, in addition, are more accurate, more robust, and more reliable than tool change magazines.

All functions of the tailstock have been integrated in the automatic sequence, and the rapid traverse rate is 10 m/min (400 in/min) or even higher.

The endeavour to meet the requirements of a high performance and production rate did, of course, not preclude considerations of realizing all these technical properties at costs as low as possible.

It goes without saying that the two machines described here represent the two extremes under consideration. But they are typical for the principles which governed their design.

PRE-REQUISITES FOR A COMPARISON OF THE TWO TYPES OF MACHINES

There are, of course, numerically controlled turning machines on the market which do not always answer all aspects of the tendencies shown. Nevertheless, a distinct trend to either the low-cost or the high-performance philosophy can be discerned. The following comments are, therefore, applicable to more or less all types of numerically controlled turning machines offered today.

It is taken for granted that both types of machine have a numerical control system with punched-tape input, and that pre-set tools are used. There are, of course, other control systems with different information input such as decade switches or cassettes with magnetic tapes. If these possibilities were to be included in the comparison, this would only confuse the issue and might even lead to concealing the underlying tendencies. Moreover, these types of controls have been used very sparingly on the two machine concepts under consideration.

Pre-set tooling is a must in conjunction with numerical information input, and neither the one nor the other machine concept can do without it.

When investigating the economy of both machine concepts, the production costs per piece should be the criterion, and for this it is necessary to include all cost components properly assigned in accordance with their nature of origin. The generally accepted procedure is to split all cost components into three groups: the production costs, the appropriate share in order repeat costs, and the appropriate share in preparation costs. Figure 4 is a summary of the different cost components.

The preparation costs can be taken as being the same for both machine concepts because programming is, in fact, identical. Although the speed and tool commands are not contained in the tape for the low-cost machine, these values have to be included in

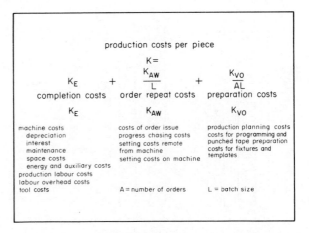

Figure 4. The cost components for the comparison of economy.

the program sheet for the operator to know when to select them manually for correct application. In addition, the programmer must have these values and consider them when establishing the coordinates and feed commands.

Similarly, to simplify matters, it can be assumed that the order repeat costs are the same. These are essentially the setting costs on the machine, and the costs for pre-setting the tools. Setting-up on the machine, and the costs for pre-setting the tools. Setting-up on the machine comprises changing or re-setting of the chuck, changing the tools and, if necessary, the finding of a new reference point.

All these procedures, except for the tool change, are almost identical on both machine concepts. The same can be said of pre-setting the tools.

It is possible, therefore, to restrict the investigations to the comparison of the production costs (completion costs in Fig. 4) K_E. These are primarily determined by the time per piece, and the machine burden costs per hour, including the wages for the operator. By way of simplification it is possible to concentrate the investigations to the interrelation of investment costs and productivity, without being excessively inaccurate.

THE STRUCTURE OF COSTS ON BOTH TYPES OF MACHINES

Compare first the machine burden costs of both machines. These are based on the values for the Federal Republic of Germany because of the author's better knowledge of the situation here but conditions are likely to be very similar in the U.K.

Figure 5 shows the burden costs for both machines, composed of the major cost components. The left-hand column in each case is applicable for one-shift operation, the right-hand column for two-shift operation. It can clearly be seen that in spite of the great difference in the capital costs, the machine burden costs as such are not widely different. The graph shows further that a two-shift operation of the low-cost machine results in a smaller cost reduction than on the high-performance machine.

From that it can be derived that any standstill of the high-performance machine is much more detrimental to cost considerations than a standstill of the

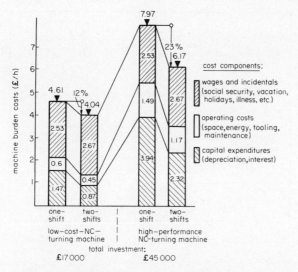

Figure 5. Comparison of machine burden costs among various NC-machines.

low-cost machine. There are a number of reasons why a machine might not be in operation. Firstly, there might not be enough workpieces to be machined for a full utilization of the production capacity of the machine. Secondly, there might be shortcomings in the organization, primarily in production control, which may lead to standstills. For instance, the correct material is not available, or the tool is not there, or the tape is not yet prepared, or the operator is absent, and so on.

As to the comparison of economy, the following is an established fact: If workpieces are produced on both machines at the same costs, the high-performance machine must yield a production rate which is 72 per cent higher when working in single shift or 50 per cent higher when working in double shift, than the low-cost machine. Inversely, the low-cost machine is still a more economical proposition if its production rate exceeds 58 per cent of that of the high-performance machine, when working in single shift or 67 per cent when working in double shift.

THE PRODUCTIVITY ON BOTH TYPES OF MACHINES

From there, one must go on to the comparison of the production rate yielded by both machines. This is only possible by a comparison of the actual production time of a particular workpiece on both machines. Considering the great variety of workpieces which could be selected for this purpose, any choice of examples will always be an individual one and will, therefore, be only valid with certain restrictions. For this reason, a few general aspects may first be outlined.

A primary criterion for the productivity is the metal removal rate, which can be calculated from the driving power of the machine. Given that approximately one kilowatt driving power is capable of removing a quantity of 130 g of medium tensile strength steel then the low-cost machine, having a driving power of 7·5 kW, can remove a maximum of 1 kg (2·2 lb) of steel per minute in contrast to the high-performance machine, having a driving power of

30 kW, which can remove a quantity of 4 kg (8·8 lb) per min.

This high metal removal rate, however, can only be achieved if certain conditions are fulfilled. These again can be obtained better on a low-cost machine than on a high-performance machine. The first, a very trivial one, is of course that the machine is cutting at all. From this it follows that the idle time should be as short as possible. Furthermore, it must be possible to select the cutting conditions such as depth of cut, feed, cutting speed in such a way that the metal removal rate is used to its utmost. This is surely not attainable for all workpieces and operations.

As a rule, a driving capacity of 30 kW can be fully utilized during roughing cuts even on small workpieces. When using form tools, or during finishing cuts, this is possible only occasionally although a further development of the tooling technique can be expected in this field.

At the Olympia Show last year for instance, the finishing of cast iron with ceramic tools was demonstrated at a surface speed of approximately 1600 m/min (5000 ft/min), at a feed rate of 0·20 mm/rev (0·008 in/rev), and a cutting depth of 0·5 mm (0·02 in). This calls for a driving power of nearly 30 kW. Incidentally, similar cutting speeds are being applied in actual production at a German automobile plant.

The extent to which the high driving power of a high-performance machine and, consequently, the high metal removal rate can be utilized depends primarily on four conditions, namely:

(1) low handling and setting times (idle times)
(2) the spectrum of workpieces to be produced
(3) the quality of programming
(4) the efficient organization, particularly the production control.

These four factors shall be looked into more closely: Low idle and setting times are ensured by a suitable design of the machine. High rapid traverse rates, quick switch-over times in the main drive, short tool change times, and an automatic operation of all possible functions are the most important measures to be taken. A decisive factor for short setting times is, however, also an efficient organization, mentioned under point 4.

The spectrum of workpieces to be produced can hardly be influenced by the respective user. A careful analysis must be made, however, of its suitability for high-performance turning. A few examples show the great differences which might exist among the various workpieces.

Figure 6 depicts a workpiece quite suitable for machining on a high-performance turning machine. It

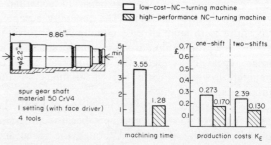

Figure 6. Economy study for NC-turning machines. Workpiece: spur-gear shaft.

is produced in one setting in 1·28 min. Its production on the low-cost machine takes 3·55 min, also in one setting, i.e. almost three times as much.

The production costs per piece on the high-performance machine, with single shift operation, are thus £0.18 or 37 per cent lower than on the low-cost machine. With double shift operation, the production costs on the high-performance machine are lower by even 45 per cent.

Quite a different picture is obtained on the workpiece shown in Fig. 7. This is a casting of grey cast

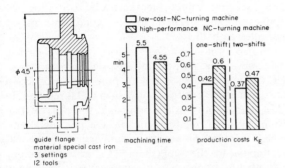

guide flange
material special cast iron
3 settings
12 tools

Figure 7. Economy study for NC-turning machines. Workpiece: guide flange.

iron having rather poor machinability, and very little machining allowance. This prevents the high driving power of the high-performance machine being utilized. The many recesses can be machined only at cutting speeds which are just as well possible on the low-cost machine. For this reason, the machining time on the high-performance machine of 4·55 min is only 17 per cent lower than that on the low-cost machine of 5·5 min, and is definitely not sufficient to compensate for the high hourly burden cost of the former machine.

As illustrated by the columns in the graph, the low-cost machine is more economical in that case, the workpiece is produced in single shift operation at £0.43 on the low-cost machine, compared with £0.61 on the high-performance machine. Even when working in two shifts, the low-cost machine remains more economical although the advantage, costs per piece £0.38 compared with £0.47 for the high-performance machine, is no longer quite as impressive.

Similar examples could easily be given in great numbers but this would not assist anybody in making a specific decision because the workpieces to be machined differ from one plant to the other.

The quality of programming is a decisive factor for ensuring that the workpieces are really machined in the shortest possible time. An important requirement is the fact that the programmer must be conversant with the technology of modern high-speed turning, and able to apply it consistently. Some sort of reluctance may be experienced here from the workshop personnel but this must be gradually reduced by the production management.

The influence of a good organization on the down times of an NC-machine and, consequently, on its economy has been mentioned before.

CONCLUSIONS

Summing up it can be said that both philosophies are obviously justified in entering into the design of an NC-turning machine. For both, the low-cost machine and the high performance machine, there exist ranges of application where one type of machine is more economical than the other. A more difficult task is to establish which of these machines is the more suitable for a given application. In this paper it was only possible to give a few criteria for making a decision, derived from the aspects as explained above. These criteria have been consolidated in Fig. 8. For a decision in a particular case, they require a careful weighting of their relative merits.

criteria:	no	yes
is a two-shift operation possible ?	no	yes
is a sufficient production volume available to ensure full utilization of the machine ?	no	yes
is the majority of the workpieces in the upper working range of the machine ?	no	yes
are many parts being machined from solid ?	no	yes
are there large machining allowances on the pre-shaped blanks ? (castings or forgings)	no	yes
is the material easy to cut ?	no	yes
will technologically difficult tools be used only occasionally ? (profile, tools, slender boring bars,etc.)	no	yes
are the workpieces stable and easy to hold ?	no	yes
are qualified programmers available or could they be trained ?	no	yes
is an efficient production control system available ?	no	yes
is a quick and efficient service available for the machine ?	no	yes
is adherence to close tolerances required ?	no	yes

low–cost–NC–turning machine

high–performance–NC–turning machine

Figure 8. Criteria for the choice between low-cost and high-performance NC-turning machines.

In general, it can be said: if the majority of the questions have been answered by 'yes', the high-performance machine should be preferred. If the majority of the answers is 'no', then the low-cost machine is the better choice. In some cases, however, just one single answer can be decisive. For instance, if the machine load for a high-performance machine is by far insufficient, its purchase is not justified no matter how many times the answer 'yes' can be given to the other questions.

DISCUSSION

Q. E. F. Moss, Hydro Machine 100 LS. I would like to ask Professor Rohs a question on market potential for NC lathes but first it is essential to be clear on what we mean by high performance and low cost. My own company is currently selling NC lathes with specification in which the basic high performance NC features mentioned by Professor Rohs are included— plus others.

For example,

full continuous path control
high performance servo drives
variable spindle speed
constant cutting speed
multistation indexing turret
hydraulic tailstock—all under full tape control

In addition: a tool presetting capability which is believed to be more precise than any other system, solid state interface integrated into basic console for maximum reliability.

The selling price for this machine—with 540 mm swing capacity is about £22 500 fully tooled, i.e. effectively half the price stated by Professor Rohs for a high performance machine. In this respect very much a low cost machine. However, the term 'low cost NC' appears firmly associated with lower-spec semi-automatics. Therefore, a better term would be lower-cost high performance M/C. Turning to the low-cost concept, a more realistic price for the specification stated would be in the region of £12 000 to £13 000 and not £17 000.

A point to remember is that an overgenerous provision of spindle power is a way of making a high cost machine appear more favourable i.e. metal removal rates are increased to show a lower cost per price on specific components but here metal removal rates could well be too high for the general requirement.

I would welcome any comments Professor Rohs may have on these points and on the potential market for each of the three machine categories as follows

1. High cost/ultra high power/performance m/c—the Rolls-Royce-type
2. lower cost, high performance m/c—the Jaguar-type
3. Low cost-semi-automatic—the Ford Escort-type.

A. 1. *To the Machine Concept mentioned in the query.* In the course of the lecture it was pointed out that the two machine concepts on which the comparison was based represent two extremes which have been chosen for the very purpose of distinctly showing the economical ranges of application of both conceptions. Although most of the machines at present on the market do not conform to these extreme cases with regard to their features and equipment, they still can be related, to a greater or lesser degree, to these categories.

For instance, an NC-turning machine of the size mentioned in the lecture would not be considered a high performance machine by me if having a horizontal base, minus a fully enclosed working area. Such a machine can be applied only for cutting techniques inherent with a manually operated universal turning machine but never for high capacity metal cutting.

2. *To the Prices mentioned in the lecture.* It is extremely difficult to make price comparisons over a longer period on the international market if there is such a pronounced fluctuation of the rates of exchange in currencies as has been the case in recent months. Any values mentioned in £-Sterling are applicable and related to the conditions having prevailed early in 1973.

3. *To the installed Power on NC-turning machines.* The purpose of installed power on an NC-turning machine is not to prove low production costs per piece but to enable a high metal removal. In this connection it is impermissible to refer to traditional general requirements in the considerations, when introducing a new technology—viz: the NC technique—in the workshop. This new technology permits the application of latest cutting techniques and for this reason, a high performance machine should be equipped with the necessary power for these techniques. From our experience we can say that there are only a few cases where the high metal removing capacity cannot be fully utilized, and with the progress being made in the field of tool application, these few exceptions will become even less in the future. Of course, if any company should be unwilling to go all the way in accepting the improvements in the techniques, then such a company would be better off with the low-cost NC turning machine for an economical production.

4. In my opinion, there is a considerable market potential for both the low-cost and the high-performance machines, and I would like to restrict myself to these types, for the reasons outlined above. In Germany, the share of installed NC-turning machines is approximately 40 per cent of all NC-machines in operation. Their share in the machines delivered in 1972 is even higher so that the number of NC-turning machines within the total of all NC-machines installed will still increase. This big share of NC-turning machines can be contributed to the fact that NC control for turning operations results in especially high rationalization effects since it permits not only a remarkable reduction of setting and handling times but also of the main machining times due to the high metal removal rate. Conseqently, the majority of the NC machines installed in Germany can be said to answer the concept of the high-performance machine.

NUMERICAL MACHINE CONTROL FOR PROCESSING LARGE WORKPIECES

by

P. NEUBRAND*

MACHINE CONCEPT FOR PROCESSING LARGE PIECES

The degree of rationalization and automation in the metal-cutting production of large-series products is more and more approaching a limit whereas the technical possibilities of medium-series, small-series and single-piece processing, and for processing bulky and heavy pieces, are not yet fully used. The most different machine concepts are already being offered for medium-series, small-series and single-piece manufacture of prismatic pieces with up to 1 m cube length. The numerical control of such systems has become an integral part of the machines. It ensures quick and high-accuracy access to the individual machining points and guarantees at the same time a uniform quality.

In most of the cases, the machine concepts available on the market are so designed that the workpiece is movably arranged on a travelling machine or a circular table. For processing, the workpiece is positioned in one or more axes or it is moved during the manufacturing process.

Workpieces which, however, due to their dimensions and weight exceed the measures of standard machines require a completely different machine concept for their treatment. A change of the position of large pieces having a height-to-length ratio of 1:1·5 up to 1:10 is difficult or even impossible.

If it is suitable to move a workpiece during the machining process, all movements are to be carried out by the machining tool. This basic requirement leads to a machine concept according to which the arrangement of the workpiece during the machining process is stationary.

The mounting base consists of one or several face plates with T-slots fastened in the foundation. The size of the plates depends on the workpieces to be machined (Figs 1 and 2).

If several plates are used it is possible to machine the workpieces in a pendulum rhythm. The preparation time for a new workpiece or a second setting may then coincide with the main period.

For multi-face treatment, a stationary circular table with fixed or continuous partition or a reversible clamping device also with fixed or continuous partition can be provided instead of the face plates.

The linear travelling axes X, Y and Z are on the machine side. The machine itself is arranged on a long bed of the X axis. The cross slide movable on the long bed with guide ways for the tool feeding axle (Z axis) carries the machine stand with the vertical axle (Y axis), main spindle and all units required for single-tool change including a chain store for single tools.

Due to the use of units out of the standard series of machining centres, problem solutions can be offered which allow the economic production of large-size pieces in small-series and single-piece manufacture.

MACHINE DESCRIPTION

Long bed

The long bed as X axis represents the basic element of the machine. In order to obtain a standard production series, the whole machine bed is subdivided into parts of 4000 mm and 2000 mm length.

According to the combination of these partial lengths, the following bed lengths and working strokes of the X axis are obtained:

Machine type	Bed length approx. (mm)	Working stroke X axis (mm)
TC 2–SO 4	6000	4000
TC 2–SO 6	8000	6000
TC 2–SO 8	10 000	8000
TC 2–SO 10	12 000	10 000

Three guide ways are provided on the machine bed (Figs 3 and 1).

The guide ways themselves consist of hardened, solid partial pieces which are screwed to the machine bed. The centre guide way serves at the same time as the base for the cross head guide of the machine axle. Rack sections for driving the machine axle are mounted in the machine bed.

At the longside of the machine bed, flat spaces are provided for attaching the measuring system. They are provided on the side turned towards the workpiece. Inductosyn scales (for direct measurement) are used for this purpose.

* Ludwigsburger Maschinenbau GmbH

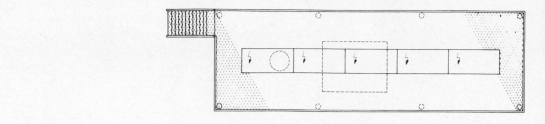

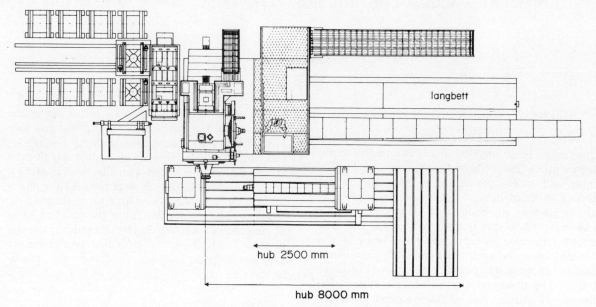

langbett

hub 2500 mm

hub 8000 mm

Figure 1. Machine TC 22 SO (view from top).

Cross slide

The cross slide is the unit representing on one hand
the slide of the X axis and on the other hand the base
for the mobile stand unit (Fig. 3). It is supported by
roller elements on the three guide ways of the long
bed.

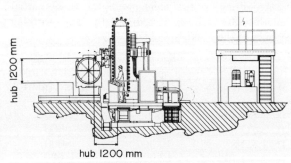

hub 1200 mm

hub 1200 mm

Figure 2. Machine TC 22 SO (side view).

The rigidity between the cross slide and machine
bed is increased by pre-stressing of the roller
elements.

Similar to the flat guide the cross head guide is
also provided with roller elements at the central guide
way.

On the upper side of the cross slide, guide ways
and cross head guides are arranged at 90° to the
travelling axle of the machine bed for providing the
tool feeding axle (Z axis). The guide ways and the
cross head guides are solid, hardened steel bars
screwed to the cross slide.

Since in the case of this machine concept the
feeding forces in the Z axis may reach several MP due

to the use of multispindle units, the ball rolling
spindle is arranged as a transmission element in the
middle between the guide ways. The forces are thus
centrally transmitted to the stand unit.

Two hardened, solid bars, arranged symmetrically
to the central axis, are fixed by screws as cross head
guide for this axis. This arrangement guarantees perfect guiding conditions and a high degree of cross
rigidity. The axial rigidity of the ball rolling spindle is
additionally increased by pre-tensioning.

The cross slide is moved in the X direction on the
long bed by a hydrostatically supported worm
attached to its bottom which slides in the worm rack
of the long bed. The hydrostatic worm can be driven
according to the equipment of the machine by a
hydraulic servo-motor or an electric d.c. motor. A
low-friction transmission element pre-tensioned in
travelling direction is obtained by the both-side pressure application on the worm flanks by pressure oil.

Column

In order to obtain an appropriate distribution of
forces, good guiding properties and favourable deformation behaviour at temperature variations, the
machine stand is of symmetrical design.

The double-walled frame structure ensures
vibration-free machining even in case of the heaviest
cutting. The stand unit is also supported on the cross
slide with pre-tensioned roller elements for supporting and slide way (Fig. 4).

For supporting the vertical slide (Y axis) combined
flat and vertical guide ways are fixed by screws at the
front side of the stand unit.

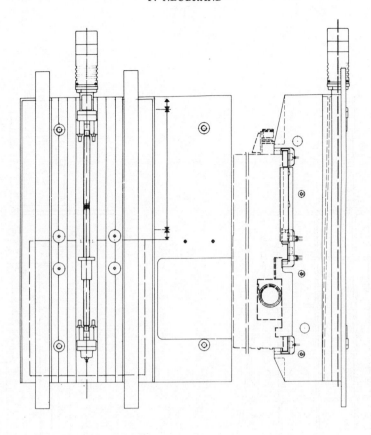

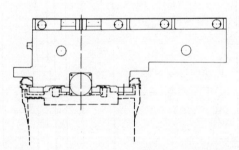

Figure 3. Cross slide.

The vertical slide is the intermediate element between main spindle gear with drive motor and the main spindle unit and at the same time the support of these units. It is also supported on roller elements. The whole vertical slide unit is driven via a pre-stressed ball rolling spindle.

The required balance of weight is obtained by a hydraulic cylinder due to the lack of space and for reasons of better dynamic properties.

Main drive and main spindle

In order to make an optimum use of the metal-cutting tools, the main spindle gear is equipped with three switching stages and a 2-step, switchable reduction gear. It is driven by a 30 kW d.c. motor. The driving motor is operated within a speed range of 1400 to 1960 rev/min with armature operation and 1960 to 3000 rev/min at field operation. The armature as well as the field-operation range are subdivided into 4 fixed speeds. In connection with the main spindle gear, 37 adjustable main spindle speeds with steps of 1·12 are thus obtained. The power reduction at lowest motor speed is max. 25 per cent.

The high power rating of the main spindle gear, however, can only become effective if the main spindle is adjusted accordingly.

As is known, the scope of operations comprises twist drilling, screwing down, rough milling, fine milling and fine cutting.

The resulting multitude of requirements on the main spindle cannot, thereby, be met optimally by a universal support (Fig. 5).

Therefore, the basic equipment of the machine includes a spindle head changing device which allows the appropriate spindle and supporting systems for the various operations to be used as alternatives.

The changing device thus allows as well the use of multispindle units for simultaneous machining at several points in one working rhythm similar to the machining on special and transfer machines.

The main part of this spindle-head supporting device is the spindle-head gripping unit mounted at the front side of the vertical slide (Fig. 6).

In this gripping unit, the spindle head is fixed in its position by four fixing bolts and is firmly clamped by a hydraulic gripping system. Thereby, the main

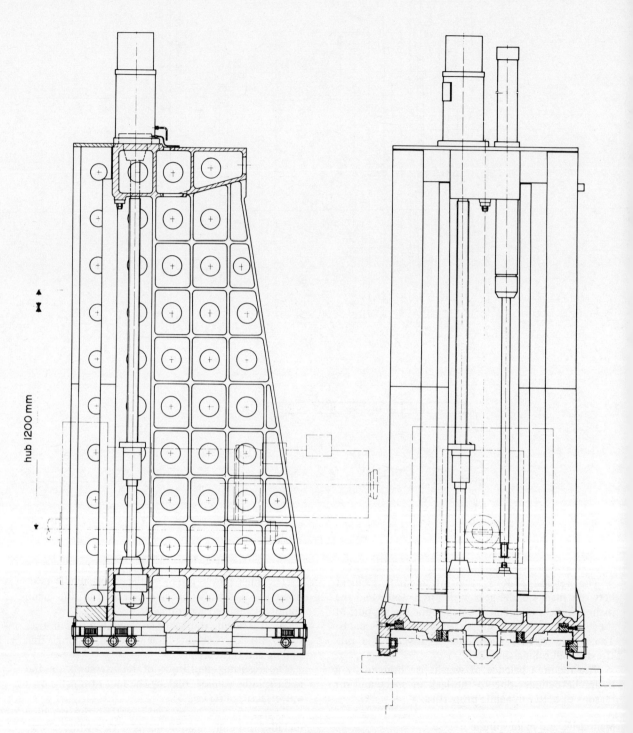

Figure 4. Moving column.

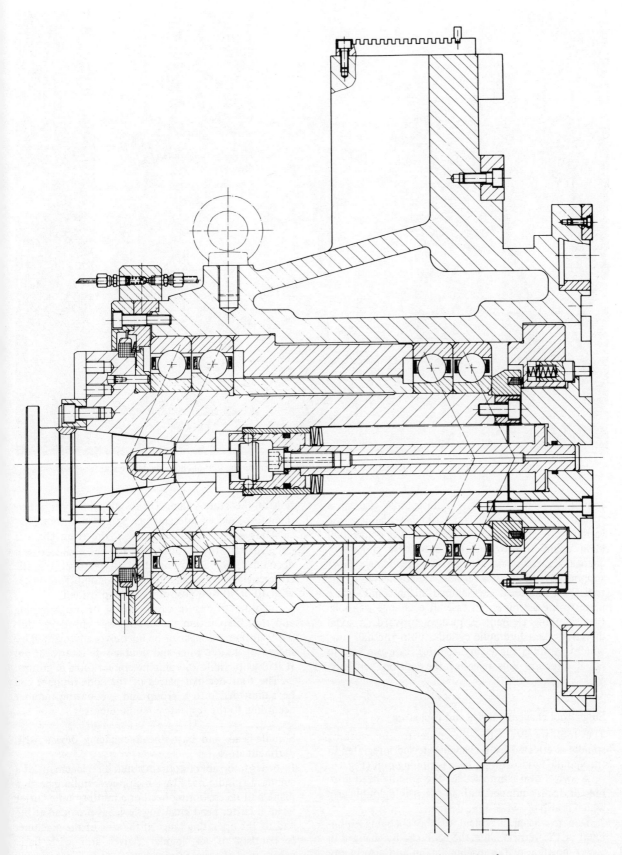

Figure 5. Interchangeable universal machining unit special roller bearing supported.

Figure 6. Arrangement of interchangeable main spindle unit with automatic single tool changing device and tool magazine.

spindle is connected to the main spindle gear via a special coupling device. In a similar manner, the various refrigerant pipes are automatically connected. Main spindles, provided for automatic single-tool change are equipped with an ISO steep-angle taper support. By means of a tension rod connected to the tool shaft, the tools are clamped in the spindle via a plate spring assembly. In case of a change of tools, the clamping element is pushed forward in axial direction via a hydraulic cylinder until the ball ring, which transmits the clamping force, can give way in radial direction thus releasing the tension bolt (Fig. 5).

Single-tool changing device and tool store

The tool transfer from the store to the working spindle is effected by a changing device integrated in the spindle-head (drilling head) gripping unit (Fig. 6).

A swivel arm supported in the spindle-head gripping device is equipped with an extensible double grip which is also of swivel-type design.

Due to the arrangement of the changing equipment at the vertical slide, the tools can be changed in any Y position of the machine stand and also in any Z position outside the collision range.

The tool store is designed as a chain-type store. It is fixed at the side of the machine stand. For selection of a programmed tool the chain is driven by a hydro-motor. The required tool is selected with the aid of code rings provided on the tool shafts which pass through an interrogation point during the revolu-

tion of the store chain. This coding of tools allows them to be placed into any free part of the store.

After a tool has been chosen, the store chain is coupled with the vertical slide and the chain drive is decoupled. By a movement of the vertical axis the store chain is compulsorily dragged along. Tool stores with 40 or 50 tool positions can be provided.

In order to ensure clear control of programming and tool preparation the coding was chosen so that 9999 different tools can be unequivocally marked by means of 9 code rings and double-side interrogation. It is thus possible to combine similar tools in groups.

The four decimal places of the code number can be subdivided into a group and a counting number according to the respective requirements.

Spindle-head and cutter-head changing device with pertinent store

In order to be able to obtain a quick replacement of a standard spindle head by a high-power milling head, a plane and recess-boring head or a multi-spindle cutter head, a cutter-head changing device is provided at the end of the operating range at the side of the machine. It consists of an angular swivel arm with basic capacity, at the same time to store up to 3 changeable spindle units.

In addition, a cutter-head inserting device is arranged beside the swivelling device. For carrying out a cutter-head change the angular swivelling arm, the machine stand (in direction Z) and the vertical slide (in direction Y) are moved to the fixed changing position.

Figure 7. Spindle head changing device (before exchange of multispindle boring unit).

The cutter head in the machine, after having been unlocked, is drawn into the angular swivel arm by means of the inserting device. In order to deliver the new cutter head, the vertical slide is to be adjusted to the changing position in accordance with the distance in the angular swivel arm. The cutter head carried in the angular swivel arm can then be pushed into the cutter-head gripping device (Figs 7 and 8).

If more than three cutter heads are required for machining a determined part spectrum, a cutter-head store is to be installed (Fig. 1).

The angular swivel arm serves then at the same time and in tipped position as a transfer equipment for the cutter-head store. The different cutter heads are taken out of the storing positions and transported to the angular swivel arm. The number of store positions can be optionally increased.

Similar to the individual tools, the cutter heads are provided with codes and can thus be called-off according to the program.

The whole changing operation is a fixed cycle which runs down automatically.

Reversible gripping device

The economy of the machine when processing large-size parts can be increased by supporting suitable workpieces for multi-side machining in a reversible gripping device (Fig. 9).

The rotation axis of this reversible gripping device is horizontal and arranged in parallel to the long-bed axis. The workpiece is clamped between the reversing device and the end support which can be moved in longitudinal direction. The reversing device is equipped with a positioning drive. In accordance with the equipment of the machine and the numerical control employed, the reverser can be operated as the fourth numeric axis (A axis).

If the reversing device is now used for turning the workpiece into a new machining position, the rotary

Figure 8. Spindle head changing device (after exchange of multispindle tapping unit).

table is disconnected after the position has been reached.

However, it is also possible to use the reversing device as a further feed axis. Rotation-symmetrical parts can thus be machined in circular direction. With the aid of a 3-axis guide control, curves and profiles can thus be cut into roller-type workpieces in connection with the X and the Y axis.

The end support can be moved and decoupled on a slide with an operating range of 2500 mm. The separation between end support and reversing device is determined by the shortest workpiece to be supported in the reversible gripping device.

SPECIAL FEATURES OF THE CONSTRUCTION

Feed drive independent from the bed length

Driving unit
Due to the possible working strokes of the X axis of 4000 to 10 000 mm it is not possible to drive the axis

via a ball rolling spindle. The spindle mass required for good dynamics for the axle drive would be too great and would simultaneously reduce the rigidity.

The arrangement of the driving unit on the cross slide makes the drive independent from the working stroke of the axis. Since it is difficult to realize the absence of play or even pre-stressing on rack-operated drives a hydrostatically supported worm travelling on the rack segments screwed to the long bed can be chosen as drive (Fig. 10). The flanks of the worm are on both sides provided with pockets. The required throttles are designed as capillary tubes which are radially arranged on the circumference according to the number of pockets. Only the hydrostatic pockets being meshed with the rack are triggered via a control reflector.

The hydrostatic worm is provided with threefold support in the drive housing. The end supports are designed as adjustable straight roller bearings. The central and the axial support of the worm are hydrostatic supports. The worm itself is driven by a tensioned gear.

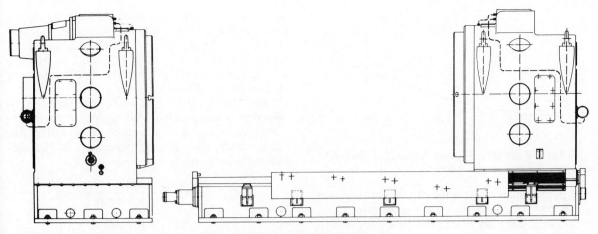

Figure 9. Reversible gripping device.

The drive shaft of the gear is driven by the axle driving motor, normally a thyristor-controlled d.c. motor, via a cross joint.

Calculation of the hydrostatic worm
When designing the hydrostatic system it was presupposed that the working pressure of the provided supply pump should not exceed 40 bars.

In order to ensure stable operating conditions, a temperature control was to be provided to ensure a temperature as constant as possible and thus also constant viscosity. For determining the bearing geometry it was therefore presupposed that the pump pressure P_P and the viscosity N were constant. With the already determined dimensions of the worm and a gap H_0 that could be realized in manufacture, and with a number z of pocket pairs as well as an escape length of 10 mm the following resulted (as shown in Fig. 11): ratio $N_R/N_P = 2.66$, rigidity for a disc according to a worm pitch (volute) of $c = 138.9$ kP/μm and the required oil quantity $Q = 1.7$ l/min.

With these values determined, the operating data for the hydrostatic worm were calculated.

Figure 12 shows the friction power N_R, the total power N_{ges} and the ratio friction power to pump power over the whole speed range of the worm. The ratio N_R/N_P decreases with reduced speed.

Figure 13 indicates the rigidity c, the load F, the required oil quantity, the pump, friction and total power including the ratio N_R/N_P for a disc with full circumference in function of the gap variation h.

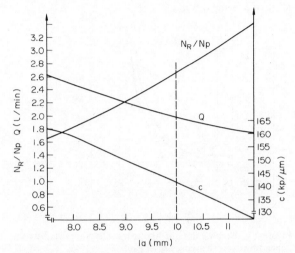

Figure 11. Determination of best geometry for hydrostatically supported worm with least amount of oil to be used at optimum N_R/N_P ($l = N_R/N_P = 3$).

The worm is not meshed with the rack on its whole circumference with the 18 pocket pairs.

Considering the start and stop for which the pockets must first be filled or can already be emptied, there remain 5 pocket pairs which are fully active. With a total of 8 threads of the worm, 40 meshed pocket pairs would result. A rigidity of $c = 138.9$ kP/μm was calculated for a full circumference with 18 pockets. The axial rigidity resulting with 40 pocket pairs being meshed is $c = 308$ kP/μm.

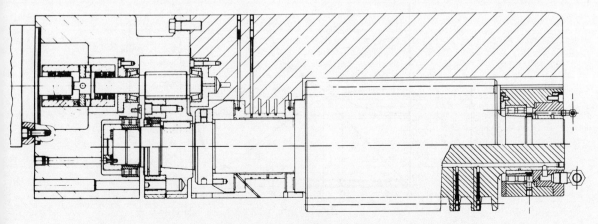

Figure 10. Drive unit with hydrostatically supported worm.

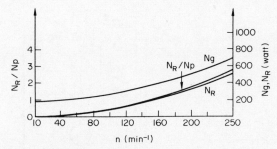

Figure 12. Amount of friction and efficiency of pump used on hydrostatically supported worm.

With maximum acceleration or retardation power of F is 2000 kP, which may occur at the driving unit in travelling direction of the machine axle, results, reckoned in terms of the calculation with 18 pockets corresponding to F being 900 kP in Fig. 13, an axial shift of h being 6·5 μm.

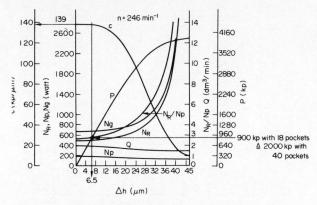

Figure 13. Working data of hydrostatically supported worm depending upon changing size of gap.

Calculation of the hydrostatic worm support
The calculation for determining the bearing geometry and the operating data for the radial and the axial bearing for supporting the worm shaft in the drive housing was made conforming to the calculation of the hydrostatic wrom. With a bearing diameter of D = 100 mm and bearing length of l = 90 mm, an escape length of l_u = 10 mm and l_a = 5 mm, a number of pockets z = 4 and an operating gap h_0 = 0·02 resulted in a radial rigidity c = 126 kP/μm.

For the axial bearing with the geometrical data D_a = 170 mm, D_i = 105 mm, l_a = 5 mm, a ring pocket

and a gap h_0 = 0·02 mm resulted in a bearing rigidity c = 342 kP/μm. Figure 14 shows the operating data in function of the gap variation h. With the maximum accelerative force F = 2000 kP results in this case a displacement of h = 5·9 μm. The rigidity is reduced to the value c = 326 kP/μm.

The total rigidity of the whole drive unit in travel direction is composed by the rigidity of the hydrostatic worm and the hydrostatic axial bearing. The series combination (series connection) of the individual rigidities results with the relation

$$c_{ges} = \frac{c_{sch} \times c_{ax}}{c_{sch} + c_{ax}} = \frac{308 \times 342}{308 + 342} \text{kP}/\mu\text{m}$$

$$= 162 \text{ (kP}/\mu\text{m)}.$$

This rigidity value could be confirmed at the machine with very small variations caused by the inaccuracies of manufacture.

By means of this hydrostatically supported transfer element, a total rigidity was achieved which is a prerequisite for the good dynamic behaviour of the axle drive. At the same time, the absence of play at minimum friction and high attenuation could be obtained in a simple manner.

Reversible gripping device as additional NC-axis A
If it is necessary to position a workpiece for multi-side machining in an optional angle to the initial position, a positioning system with continuous pitch is to be used.

The circular Inductosyn is suitable as an angle-measuring element for this application. In order to be easily accessible, it is arranged at the side in front of the rotary plate (Fig. 15).

A possible wobbling due to variations by large-size workpieces is prevented by the double supporting. Moreover, the measuring element can be aligned optimally to the rotary-plate centre by an adjusting device.

The support of the rotary plate is arranged as far as possible towards the outside in order to obtain a favourable load transmission on the reverser housing. The support is designed as a combined support for the absorption of radial and axial forces. The rotary plate is driven by a two-step, tensioned gear via a toothed wheel rim. The axle driving motor is coupled to the gear.

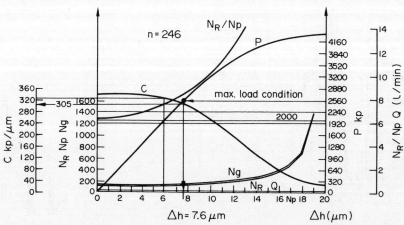

Figure 14. Working data of hydrostatic axial bearing depending upon changing size of gap.

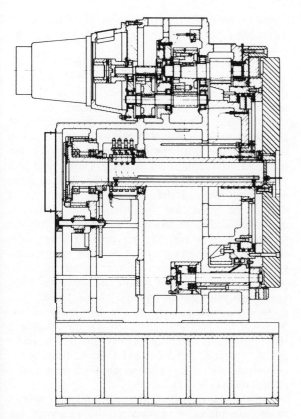

Figure 15. Drive and support of reversible gripping device.

If the reversing device is only used for positioning, the rotary plate is tensioned at a clamping disc with several clamping elements provided at the circumference. The tension is applied via spring assemblies. For release, the tensioning elements are to be triggered hydraulically.

Considering the concept of supporting and driving as well as the angle-measuring element, the reversing device can also be operated as a feed axle. In case of feeding operation, the hydraulic release remains continuously active.

With stationary tool axle it is possible to machine rotation symmetrical parts (Fig. 16) in circular direction. Curved parts can also be machined if a 3-axes way control is available at the machine.

ARRANGEMENT AND STRUCTURE OF A MACHINING PLANNING

A technical treatise of the discussed machine design enables, already in their basic conception and equipment and especially with their various possibilities of extension, the utilization of such machines to tool a very wide range of large workpieces and components of similar groups.

This possibility of utilization is exemplified on the production of large diesel engines.

The elaboration of the planning of this group of workpieces applies to the limit the full range of the technical possibilities of the machine design; the results are laid down in comparative computations of costs.

The planning begins with the examination of the drawings of the relevant workpieces to be machined and their grouping into families of components (Fig. 17).

Such an arrangement enables to discern similar machining processes and the recurrency of whole groups of machining operations. This enables subsequently to lay down criteria for the utilization of the tools.

The planning for a particular workpiece begins with the mapping out of machining faces and/or machining surfaces (Fig. 18).

At the same time the workpiece chucking fixture is elaborated and the means to realize this clamping and chucking are set down (Fig. 19).

To warrant a trouble free sequence of operation, the edges where the tools or the machine attachments could collide with the workpiece configuration as well as the chucking fixture and the clamping means should be checked in every working phase. If there

Fig. 16. Machining of big engine parts on machine type TC 22 SO.

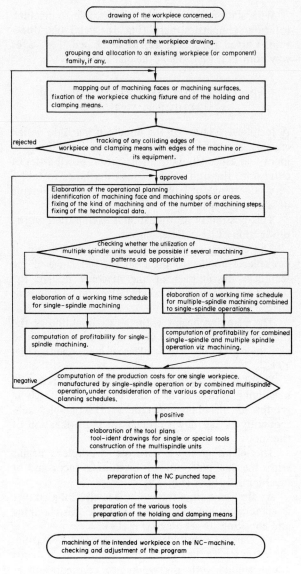

Figure 17. Diagram of operation sequence planned.

are some interfering edges, the arrangement of the workpiece or of the clamping means should be reconsidered and modified.

The next step involves the elaboration of the operating schedule. Every single spot to be machined should be identified continuously and the kind of machining operation and the number of steps required, laid down. This involves the selection of particular tools, as e.g. step tools or multiple-cutter tools. Together with the selection of the tools, the technical data concerned: cutting speed, feed and number of revolutions are fixed (Fig. 20); together with the tools and the sequence of operations, it should be examined whether some operation sequences and processes are repeated. As may be understood from Fig. 18, the cylinder bores areas show some machining operation with very similar geometrical configurations. If components or workpieces of the same family show the same machining operations, multispindle units may be provided.

As a preliminary step to the calculation of the economy in operation (profitability), the machining time required is computed a first time for a single-spindle operation (Fig. 21) and a second time for the combined machining with single tools and multi-spindle units (Fig. 22).

In such a way the costs of manufacturing one single workpiece are ascertained by means of the computation of profitability for every planning alternative considered. Such comparison of manufacturing costs enables to lay down the final decision criteria for the utilization of multispindle units for combined machining operations.

As the case may be, it will be necessary to proceed once again, or even several times, to reconsider the whole operation planning and to recalculate the time schedule and the profitability.

Once the optimum solution is decided upon, the second step will be to make up the tooling plans. In a parallel operation, the tool indent drawings will be

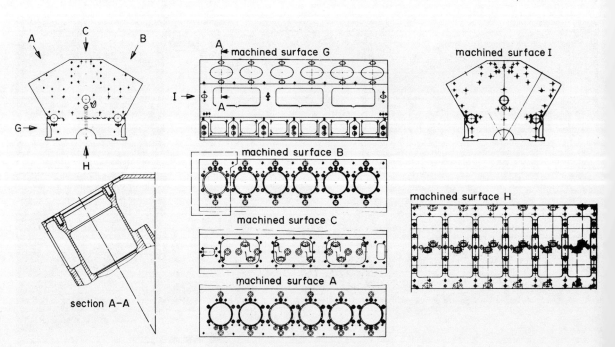

Figure 18. Component drawing.

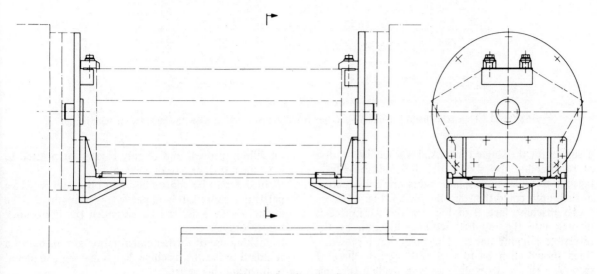

Figure 19. Clamping fixture.

bohrkopf–NR. werkzeug–NR. schneiden				bezeichnung	anzahl	bearbeitung	bearbeitungsdurchm. mm	bearb.–laenge mm	schnittgeschw. m/min	drehzahl u/min	vorschub mm/u	vorschub mm/min	op.zeit min	einzelzeit min	ges.zeit min
I	4	HSS				gewindebohren	M 16	25.00	10.0	200	1.99	398.0	0.15	2.18	4.28
I	37	HSS		90-92+97-1	8	spiralbohren	18.0 fase	50.00	31.6	560	0.36	200.0	0.25	2.00	2.00
I	38	HM		B 90-92	3	anstirnen	innensti. 26 + fase	10.00	65.3	800	0.10	79.4	0.16	0.47	0.47
I	39	HSS		B 97-101	5	gewindebohren	M 20 x 2	40.00	10.0	160	1.97	316.0	0.27	1.37	1.37
I	40	HSS		B 102	1	spiralbohren	24.50	40.00	30.7	400	0.35	141.0	0.28	0.28	
I	41	HSS				gewindebohren	R 3/4 zoll	35.00	10.2	125	1.79	224.0	0.33	0.33	
I	42	HM				anstirnen	innen+aussen st.	20.00	71.2	630	0.14	89.1	0.25	0.25	0.87
I	43	HSS		B 103-104	2	spiralbohren	8.5 fase	26.00	29.9	1120	0.20	224.0	0.12	0.23	
I	44	HSS				gewindebohren	M 10	18.00	9.8	315	1.42	447.0	0.10	0.20	0.43
				abgasseite											
I	1	HM	C	141-222-46	90	spiralbohren	10.2 fase	27.00	71.7	2240	0.16	355.0	0.08	6.85	
I	2	HSS				gewindebohren	M 12	20.00	10.5	280	1.79	501.0	0.10	8.99	15.83
I	46	HSS		B 223-230	8	tieflochbohren	10.00	90.00	25.1	800	0.16	126.0	0.79	6.33	6.33
I	47	HSS		B 231-238	8	tieflochbohren	28.00	80.00	24.6	280	0.36	100.0	0.83	6.63	
I	48	HM		KS 1-8	8	feinfraesen	kontrollschnitt	20.00	70.6	450	0.79	355.0	0.06	0.45	7.08
				steuerwellenseite											
I	40	HSS	D	B 247-248	2	spiralbohren	24.50	40.00	30.7	400	0.35	141.0	0.28	0.57	
I	41	HSS				gewindebohren	R 3/4 zoll	35.00	10.2	125	1.79	224.0	0.33	0.66	1.23
I	1	HM		B 249-428	180	spiralbohren	10.2 fase	27.00	71.7	2240	0.16	355.0	0.08	13.69	
I	2	HSS				gewindebohren	M 12	20.00	10.5	280	1.79	501.0	0.10	17.97	31.66
I	49	HSS		B 429-436	8	spiralbohren	2 stege 61 vorb.	150.00	30.6	160	0.79	126.0	1.19	9.52	
I	50	HM				aufsenken	1.steg 70 D.	38.00	69.2	315	0.80	251.0	0.15	1.21	
I	51	HM				vorbohren	2.steg 64 +fase	85.00	100.5	500	0.40	200.0	0.42	3.40	
I	52	HM				feinbohren	65 H8	70.00	114.3	560	0.20	112.0	0.62	5.00	19.13
I	46	HSS		B 437-444	8	tieflochbohren	10.00	90.00	25.1	800	0.16	126.0	0.79	6.33	6.33
I	47	HSS		B 445-452	8	spiralbohren	28.00	50.00	31.2	355	0.45	159.0	0.31	2.52	2.52
I	3	HM		453-488	36	spiralbohren	14.0 fase	35.00	70.3	1600	0.16	251.0	0.14	5.02	
I	4	HSS				gewindebohren	M 16	25.00	10.0	200	1.99	398.0	0.15	5.24	10.26
I	53	HM		B 489-506	18	aufsenken	25.00	50.00	70.6	900	0.25	224.0	0.22	4.02	
I	54	HSS				spiralbohren	22.00	50.00	31.1	450	0.35	159.0	0.31	5.66	
I	55	HSS				gewindebohren	M 24 x 2	45.00	10.5	140	2.01	282.0	0.34	6.10	15.78
				pumpenseite											

Figure 20. Part of an operation plan.

zeitzusammenstellung

bearbeitungszeit fuer	1702 operationen	438.69 min
werkzeugwechselzeit fuer einzelwerkzeuge in wechselstellung	64 x 0.15 min	9.60 min
zeit fuer vor-und ruecklauf in Z gei werkzeugwechsel	64 x 0.15 min	9.60 min
zeit fuer verfahrwege in XY (eilgang-geschwindigkeit 10 m/min)	315.80 m	31.58 min
positionierzeit in X-Y	1672 x 0.02 min	33.44 min
zeit fuer eilwege in Z (eilgang-geschwindigkeit 10 m/min)	53.55 m	5.35 min
positionierzeit in Z	3404 x 0.02 min	68.08 min
lesezeit		5.05 min
855 X anbohren m.halbem vorsch. je 3 sec.		42.75 min
gesamtzeit		644.146 min

stueckzeit bei 85 prozent maschinenauslastung 1 teil in 12.64 stunden

Figure 21. Time schedule for planned operations on 8-cylinder crankcase with single tool machining.

designed for the single tools. And if the computation of profitability has shown that it would be advantageous to utilize multispindle units, the construction of such units could then also be started (Fig. 23).

To machine face B of Fig. 18, the workpiece is pivoted into the required working location by the reversible gripping device. The operation representations should then be rotated 180° against these of face A. The multispindle units provided for this machining work should also be rotated 180° and placed in this position in the machine.

With the operation planning and the tool planning in hand the NC programming department will be able to produce the control punched tape for the NC machine—either by hand or by the appropriate device.

The tool plans enable to make up the lists of tooling preparation and setting

As far as composed of standard items, the holding and clamping means will be provided, together with the chucking fixture required.

This would be the end of the most important preparations for the machining of a particular workpiece.

SUMMARY

For a large range of workpieces which, because of their dimensions, their weight or their handling should remain stationary during the machining operations, a machine concept has been elaborated, in which the workpiece remains motionless during the

machining process and is only rotated or pivoted to machine a further face or surface.

In order to be adjustable in an optimum way to a particular range of workpieces or components, the main axis (x axis) can be extended (building-block construction).

Making use of further construction groups out of a standard series, the machine design may be completed to form another series.

In order to warrant excellent dynamic properties of the drive of the long bed axis, even with long travelling ways, a drive group equipped with a hydrostatic worm gear and completely independent from the length of the machine bed has been designed.

The drive group works without noticeable friction and is wear-resisting, the play required for proper dynamic properties is warranted by a very plain but effective solution.

The machine design offers the possibility to utilize multispindle units as well as the standard working spindle, and even to combine both modes; this enables to gain a very high grade of rationalization and scientific productivity, especially for families of workpieces or components.

No standard equipment can be quoted or indicated for the utilization of this machine concept. The whole machine must be determined as a machining plant for the intended range of workpieces; this requires in every case involved detailed investigations —technical and economical—as well as the elaboration of technical design-planning and the computation of profitability.

bohrkopf-NR.	werkzeug-NR. schneiden	bearb.-richtung bezeichnung anzahl	bearbeitung	bearbeitungsourchm. mm	bearb.-laenge mm	schnittgeschw. m/min	drehzahl u/min	vorschub mm/u	vorschub mm/min	op. zeit einzelzeit min	ges. zeit min
1	28 HSS	gewindebohren		M 12	20.00	10.5	280	1.79	501.0	0.10 0.90	1.58

zeitzusammenstellung

bearbeitungszeit fuer	1238 operationen	234.85 min
werkzeugwechselzeit fuer einzelwerkzeuge in wechselstellung	60 x 0.15 min	9.00 min
bohrkopfwechselzeit in wechselstellung	8 x 0.70 min	5.60 min
zeit fuer vor-und ruecklauf in Z bei werkzeugwechsel	56 x 0.15 min	8.40 min
zeit fuer vor-und ruecklauf in Z bei bohrkopfwechsel	8 x 0.20 min	1.60 min
zeit fuer verfahrwege in XY (eilgang-geschwindigkeit 10 m/min)	269.80 m	26.98 min
positionierzeit in X-Y	1212 x 0.02 min	24.24 min
zeit fuer eilwege in Z (eilgang-geschwindigkeit 10 m/min)	31.99 m	3.20 min
positionierzeit in Z	2476 x 0.02 min	49.52 min
lesezeit		3.70 min
663 x anbohren m.halbem vorsch. je 3 sec.		33.15 min
gesamtzeit		400.243 min

stueckzeit bei 85 prozent maschinenauslastung 1 teil in 7.85 stunden

Figure 22. Time schedule for planned operations on 8-cylinder crankcase with single tool machining and multispindle machining.

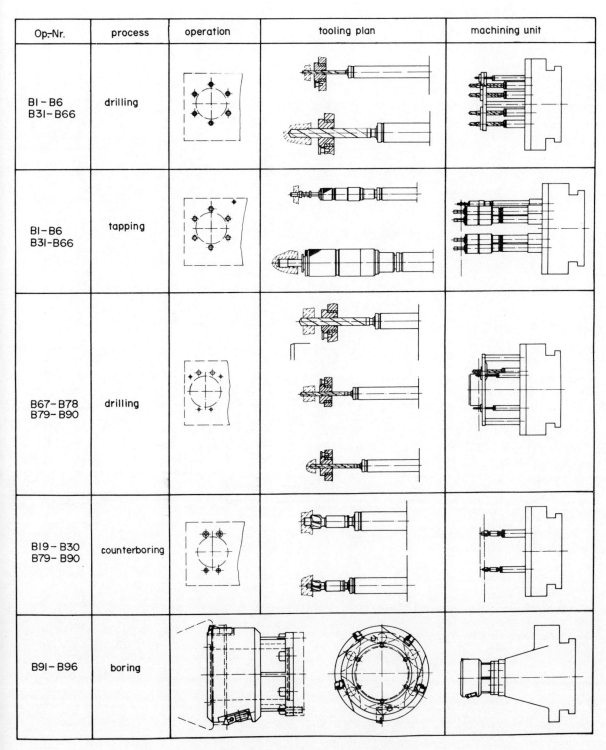

Figure 23. Tool layout for multispindle machining.

THE INFLUENCE OF THE MACHINE AND THE CONTROLLER ON THE SURFACE FINISH PRODUCED ON NUMERICALLY CONTROLLED MACHINE TOOLS

by

P. C. GALE*

SUMMARY

This paper discusses the effect of cyclic positioning errors on NC contouring machine tools on the resulting surface finish produced at the workpiece. Digital simulation is used to demonstrate that poor surface finish can result from cyclic positioning errors in one or more of the machine axes. Following this general introduction the role of the measuring system, the drive, and the slideways on the surface finish produced is examined in detail. Finally, techniques which can be used to solve such problems are discussed.

NOTATION

B angle of cut surface to the principal axis (radians)
D phase difference (radians)
F milling feed (mm per minute)
H period of X axis cyclic error (mm)
K period of Y axis cyclic error (mm)
N number of teeth in cutter
ΔN error normal to the cut surface (mm)
R spindle speed (revolutions per minute)
t time (minutes)
ΔT error tangential to the cut surface (mm)
U amplitude of X axis cyclic error (mm)
V amplitude of Y axis cyclic error (mm)
V_y velocity in the Y direction (mm per minute)
V_x velocity in the X direction (mm per minute)
V_T velocity in tangential direction (mm per minute)
ΔX error in the X direction (mm)
ΔY error in the Y direction (mm)

INTRODUCTION

The majority of NC machines in use in industry today are used on point-to-point applications such as boring, drilling and tapping. However, the number of milling machines used on contouring applications has increased sharply over the past five years.

Customers are now becoming more critical of the accuracy of parts produced on contouring NC machines. This paper will discuss a particular facet of workpiece accuracy, namely surface finish, and outline the measure taken by Cincinnati Milacron to improve the surface finish of parts produced on their range of NC products.

CAUSES OF POOR SURFACE FINISH

Surface undulations on a workpiece are a particular case of positioning errors produced on NC machine tools. Positioning errors fall into two basic types, random errors, which do not repeat over the travel of the machine, and cyclic errors which repeat with a regular pitch. It is the latter type of error which is the most troublesome from the surface finish point of view simply because the error pattern is so regular.

Consider a surface being milled, where P is the idealized cutter position assuming no positioning errors. In general, positioning errors will result and the actual cutter position will be at Q (see Fig. 1). If the surface is parallel to one of the machine principal axes then only a tangential error, ΔT, will result (Fig. 1a) and this cannot produce undulations on the workpiece. However, if the surface is inclined to the principal axes (Fig. 1b) normal errors, ΔN, are also produced which, if they are cyclic in nature, will cause cyclic undulations on the workpiece. This leads

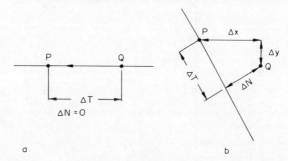

Figure 1. The effect of positioning errors on surface undulations.

* Cincinnati Milacron Ltd., Birmingham, England

us to the important general rule that *surface undulations can only result when machining surfaces inclined to the principal axes.*

In the case of two-dimensional contouring it is possible to calculate the effect of cyclic errors in each axis on the cutter position in a plane at any angle B to one of the principal axes[1]. Thus, if we have cyclic errors of amplitudes U and V and periods H and K with a phase difference D (Fig. 2), the normal error is given by:

$$\Delta N = \Delta y \cos B - \Delta x \sin B \qquad (1)$$

$$\Delta y = V \sin (2\pi V_y t/K + D)$$

and

$$\Delta x = U \sin (2\pi V_x t/H)$$

$$\Delta N = V \cos B \sin (2\pi V_y t/K + D)$$

$$- U \sin B \sin (2\pi V_x t/H) \qquad (2)$$

but

$$V_x = V_T \cos B \quad \text{and} \quad V_y = V_T \sin B$$

so that

$$\Delta N = V \cos B \sin (2\pi V_T t \sin B/K + D)$$

$$- U \sin B \sin (2\pi V_T t \cos B/H) \qquad (3)$$

Note that $V_T t$ is the distance of travel in the tangential direction, i.e. the distance along the cutter path, and this is a convenient variable for the calculation of the values of ΔN.

Equation (3) was programmed on a digital computer to determine the effect of the different parameters. Figure 3 shows a computer print-out for the case where the X and Y axes of an NC machine have equal cyclic errors of 2 mm pitch and 0·005 mm

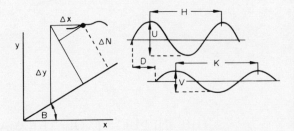

Figure 2. The effect of cyclic errors on a surface inclined to the principal axes.

amplitude, the imaginary surface cut being at 60° to the X axis. Two points are worthy of mention. Firstly the resulting error curve is non-sinusoidal as might be appreciated from examining the terms of equation (3). More important than this however, the peak-to-valley height of the surface generated is more than twice the amplitude of the cyclic errors in the individual axes. This is significant to the manufacturer of the machine tool because he must aim for cyclic errors of an order of magnitude better than the required peak-to-valley height allowable when contouring.

PHASE ANALOG SYSTEM

Cyclic positioning errors can arise from a variety of causes and these can best be understood by considering a modern phase analog measuring system (Fig. 4).

Pulses from the reference counter are converted to sine waves. These are split and one of them is shifted 90° before applying them to the synchro resolver stator. The sine wave output from the rotor of the resolver is converted back to a square wave and compared with the command counter at the phase

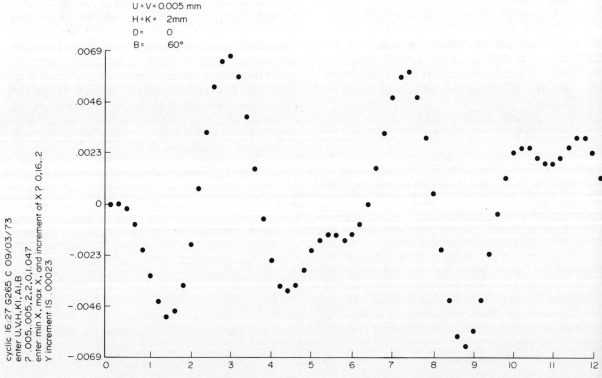

Figure 3. Computed surface undulations on a plane at 60° to one of the principal axes.

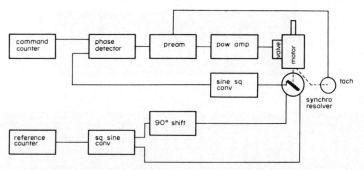

Figure 4. Simplified block diagram of a phase analog control system.

detector. Any difference in phase between these two square waves results in an output from the phase detector which is amplified and used to rotate the drive motor thereby altering the position of the resolver rotor until there is zero phase difference.

If the synchro resolver is directly attached to the leadscrew, errors can arise due to the drive and structure, since these are outside the position loop, and due to the measuring system. For slide feedback systems the errors are due to the measuring system. Both cases will be considered in this paper.

ERRORS IN THE MEASURING SYSTEM

Errors in both the transducer itself and the sinusoidal voltages which feed it can be grouped under the heading of measuring system errors.

The transducer inputs for both synchro resolvers and linear scales must consist of two sinusoidal waves with a 90° phase difference. For phase analog systems it is essential that these signals should have identical amplitudes and their phase difference should be precisely 90°. If this is not so, a cyclic positioning error which repeats once or twice every resolver revolution will result.

Figure 5 shows the result of a positioning check performed on a machine with this particular problem.

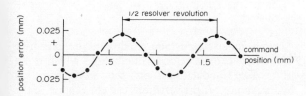

Figure 5. The effect of sine/cosine excitation voltage error.

Since the resolver inputs are common to all axes, similar cyclic errors will exist on all axes with this fault. Figure 6 shows a Talysurf trace of two surfaces milled at different feed rates at 60° to the X axis of the machine.

An important point to notice is that the pitch of the surface undulations is independent of feedrate. It will be seen that the surface undulations are of uneven pitch with irregular shapes. This confirms the findings mentioned earlier. Comparison of Figs 5 and 6 shows that if we are to limit the surface roughness, the cyclic error in each individual axis must be better than half the allowable amplitude of the undulations.

The mechanical and electrical properties of the synchro resolvers used must be carefully specified if freedom from cyclic errors is to be obtained.

The two stator windings which are fed with the sine and cosine voltages should be physically perpendicular to each other within ±10 minutes of arc. In addition, the transformation ratios between sine and rotor and cosine and rotor should be matched better

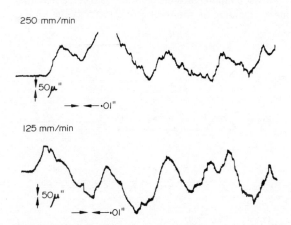

Figure 6. Talysurf trace of surface milled at 60° to one of the principal axes.

than 4 per cent. Reduced specification on either of these requirements results in a twice per resolver revolution cyclic error of the type already discussed.

The frequency of excitation of the resolver is an important parameter. Resolvers are available with or without brushes and, in general, the brush type allows a higher excitation frequency. The use of brushless resolvers well above their frequency limit results in a once per revolution cyclic error because of distortion on the rotor output.

Apart from maintaining a rigorous resolver specification, a representative cross section of each batch of resolvers is tested for cyclic error before being fitted to a machine.

Linear measuring devices are also subject to errors (Fig. 7). Mechanical errors in the device itself have been minimized by careful attention during manufacture. However errors can arise due to other causes, primarily:

(1) Imbalance between the currents into the two halves of the slider.
(2) Cross coupling between the stator and rotor windings or other interference effects as a consequence of the very low output voltages involved.

Figure 7. Example of Inductosyn® scale section and slider.

Cyclic errors due to imbalance of the currents can be minimized by the appropriate line balancing technique. Ideally, the input voltages should be compensated for both amplitude and phase. Lack of attention to this detail results in a second harmonic cyclic error pattern similar to that shown in Fig. 5.

Cross coupling effects can be corrected by adequate shielding of the wiring to and from the scales.

reduced by careful attention to the leadscrew alignment at the assembly stage. Errors can also arise due to leadscrew bearing faults resulting in a once per rev. or once per two revs. cyclic error.

Progressive leadscrew pitch errors rarely result in a poor surface finish, but cyclic pitch errors, by virtue of their regular nature, result in surfaces which are visually unacceptable.

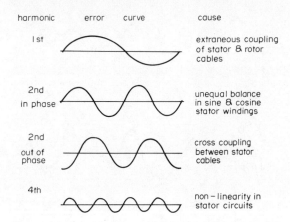

Figure 8. Types of error possible with Inductosyn® feedback.

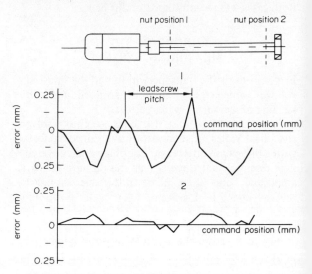

Figure 9. The effect of leadscrew misalignment on slide positioning.

A characteristic form of error curve is obtained with linear position transducers depending on the cause of the error. The different types of error curve are shown on Fig. 8[2].

FAULTS ON DRIVES AND SLIDEWAYS

Where the position feedback device is mounted on the end of the leadscrew, errors due to leadscrew misalignment or pitch errors can result in cyclic positioning errors since the leadscrew is outside the position loop.

Similarly any transmission errors in gearing between the leadscrew and the position feedback device will result in cyclic positioning errors. Repetitive errors in the gears such as base circle eccentricity will result in a once per gear revolution cyclic error at the machine slide.

Misalignment of the leadscrew results in cyclic errors of leadscrew pitch which increase in amplitude as the nut becomes closer to the leadscrew mounting bracket (see Fig. 9). Such errors can be

Even under closed loop conditions the axis drive motor can exhibit velocity fluctuations which result in surface undulations on the workpiece. On hydraulic motors the variations can correspond to the number of pistons or vanes in the motor itself but for d.c. motors velocity fluctuations arise due to firing of the silicon controlled rectifiers which control the motor. Velocity fluctuations such as those described can be reduced by an increase in the velocity loop gain around the motor[3]. This is acceptable for d.c. drives where the velocity loop gain is usually high but for hydraulic drives the velocity loop is required for linearization of the servo valve characteristics only and is thus usually of fairly low gain.

Position errors can arise due to stick slip phenomena on the slideways. The complete drive can be modelled in the form of a mass resting on a slideway attached to a spring of finite stiffness (caused by compliance of the drive) as shown in Fig. 10(a). Application of a force to the spring results in no movement of the mass until sufficient energy is generated to overcome the slide friction level. After movement does occur the energy has again to be increased to repeat the cycle. The velocity of the slide has the form shown in Fig. 10(*b*).

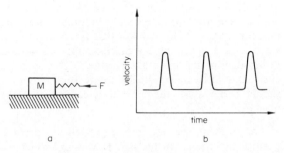

a   b

Figure 10. Stick slip phenomena.

The full mechanism of stick slip has only recently been understood[4,5] but it is now apparent that it can be reduced by increase in the lubricant viscosity, increase in the drive stiffness, or the addition of polar additives to the lubricating oil.

EFFECT OF DEFLECTIONS IN THE STRUCTURE

Vibrations in the machine structure can result in surface undulations on the workpiece. A feature of this type of fault is that the pitch of the undulations is dependent on the machine feedrate when cutting the workpiece.

Vibrations can arise from the transmission and spindle drive and are transmitted to the workpiece via the weakness in the machine structure.

EFFECT OF THE CUTTING PROCESS

It is beyond the scope of this paper to discuss the effect of chatter on the surface finish produced. However during normal cutting conditions surface irregularities occur because of the feed per revolution of the cutter. Thus for a milling cutter with N teeth, the pitch of the undulations due to the feed F mm/min and spindle speed R rev/min is

$$\frac{F}{NR}$$

Normally the pitch of feed-per-revolution undulations will be smaller than the pitch of undulation due to any of the forms of cyclic error; the pitch will, of course, be feedrate dependent.

TECHNIQUES USED TO SOLVE SURFACE FINISH PROBLEMS

All NC machines produced at Cincinnati Milacron undergo rigorous tests including both geometric, positioning accuracy, and machining tests.

As part of the positioning accuracy test the cyclic positioning error is checked at a number of positions along each axis. The test points are determined by dividing the leadscrew/scale/resolver pitch into ten

equal increments and measuring the error of each of these points over two pitches. Measurements are made using a Laser Interferometer.

A cutting test is also made to determine that the contouring accuracy and surface finish produced on the machine are satisfactory. The requirements of a test piece to check this are:

(1) It should be of such a shape as to enable simple measurements to be made.
(2) It should be small, light in weight, and relatively cheap.
(3) The cutting time should be short, i.e. the maximum information in the shortest time.

A test block which satisfies the requirements is shown in Fig. 11. It is aluminium and weighs only 1 lb 4 oz (0·56 kg).

Figure 11. Test block used for two axis contouring.

The circle is used to check roundness and smoothness of changeover points. The straight faces are used to check surface finish, the inclined ones for surface undulations due to the measuring system and drive, and the faces parallel to the principal axes are used to check for vibrations in the machine structure or spindle drive. Each face is milled at two feedrates to determine whether the undulations are feedrate dependent.

Some comments on the measurement of surface finish are necessary at this stage. Most surface measuring equipment has a fairly small 'cut off' length (in the case of the Talysurf this is 0·030 in (0·75 mm)) and this is the length over which the CLA value is determined by the instrument. However the undulations due to the causes discussed in this paper have a much greater pitch than this. Thus the CLA value indicates the short term surface finish rather than the longer wavelength surface undulations, hence the use of a CLA value to define the surface finish produced on an NC machine is inappropriate. This problem can be surmounted by obtaining a paper trace of the surface from the measuring equipment and measuring from this the peak-to-valley height of the undulations.

Experience allows the setting of an acceptable peak-to-valley height for each particular class of machine. We usually go further than this, however, and reject any machines which do not exhibit a random surface undulation pattern. In other words cyclic patterns of particular pitches (leadscrew/pitch/scale pitch) which can be defined for each type of machine are not acceptable.

REFERENCES

1. JAROMIR ZELENY (1967). Accuracy of automatically controlled machine tools, *8th M.T.D.R. Conference*, September.
2. ANON (1965). *Inductosyn Accuracy*, Farrand Controls Inc., January.*
3. S. K. CESSFORD and R. BELL (1971). The specification and performance of electro hydraulic motor drives for numerically controlled machine tools, *12th M.T.D.R. Conference*, September.
4. R. BELL and M. BURDEKIN (1967). The frictional damping of plain sideways for small fluctuations of the velocity of sliding, *8th M.T.D.R. Conference*.
5. R. BELL and M. BURDEKIN (1966-7). The dynamic behaviour of plain sideways, *Proc.* I. *Mech. E.*

---

* 'Inductosyn' is a trade mark of Inductosyn Corporation, Valhalla, New York, 10595, U.S.A.

DISCUSSION

Q. C. H. Thompson, Union Carbide, U.S.A. Have you used a laser interferometer to dynamically check slide positioning error? Do you feel that this is important, since it is a check of the machine slide in motion rather than in multiple static positions? This dynamic check should be more valuable in my opinion since it should better assess the contouring capability of the machine.

A. At present we are not equipped to perform positioning checks 'on the fly'. However, I will agree that this type of check would be a better assessment of the contouring capability of the machine.

Q. S. Taylor, University of Birmingham. Mr Gale referred to the principal axes of a machine and the fact that undulations cannot occur when cutting parallel to the principal axes. Would he please explain 'principal axes', and does he refer to the axes of vibration?

A. The principal axes are the x, y and z axes of the machine. These are not related to the axes of vibration.

SERVO SYSTEMS FOR THIRD GENERATION NC MACHINE TOOLS

by

JAROMIR ZELENÝ*

SUMMARY

Recently projected NC machines and machining centres for highly integrated and computerized production systems impose new requirements on performance and parameters of mechanical and control sub-groups. The paper presents some results and experience gained in VÚOSO, Prague during two years of systematic research and development work in position servo systems for above-mentioned applications. Servo systems have been solved as pulse-train controlled units processing step-changes of input pulse frequencies up to 40 kHz. One-increment sensitivity, infinite static stiffness and perfectly damped transient response are other important features of these units. Servo systems involve electronic circuits for fluent acceleration and deceleration of controlled movements and feedback units performing accurate measuring of output position for all basic applications in NC machine tools.

NOTATION

A. Time functions and their Laplace transforms

$x_H(t), X_H(s)$	'hard' command position signal
$x_S(t), X_S(s)$	'soft' command position signal
$\Delta x(t), \Delta X(s)$	position error signal
$x_C(t), X_C(s)$	'corrected' command position signal
$\Delta x_C(t), \Delta X_C(s)$	'corrected' position error signal
$x_T(t), X_T(s)$	position of the controlled mass for a system with translatory feedback
$x_R(t), X_R(s)$	position of the controlled mass for a system with rotary feedback
$x'_R(t), X'_R(s)$	angular position of the servo-motor output shaft
$v_I(t), V_I(s)$	input velocity signal
$v_{TF}(t), V_{TF}(s)$	tachometric feedback velocity signal
$\Delta v(t), \Delta V(s)$	velocity error signal

B. Constants and transfer functions

ϵ_{UL}	maximum angular acceleration of the unloaded servo-motor
ϵ_L	maximum angular acceleration of the loaded servo-motor
ω_R, ξ_R	natural frequency and damping coefficient of the servo-motor
ω_T	natural frequency of the translatory moving mass

K_{RT}	rotary to translatory movement transmission constant
K_T	torsional stiffness of the servo-motor
K_A	mechanical stiffness of the translatory mass
K_V	gain of the open position loop
K'_R	gain of the open velocity loop
T_{AD}	time constant of accelerating and decelerating circuits
T_{TF1}, T_{TF2}	time constants of tachometric feedback circuits
T_{CV}	time constant of command velocity correction circuits
T_{PE}	time constant of the position error detector
T_{PA}	time constant of the velocity error preamplifier
T_{CC}	time constant of the control current feedback
$K_{PA}, K_{CC}, K_{SV},$ $K_{HM}, K_{TF}, K_{CV},$ K_{TA}, K_{EM}	gain coefficients of individual servo sub-groups
$KG_{PA}(s), KG_{CC}(s),$ $KG_{SV}(s), KG_{HM}(s),$ $KG_{TF}(s), KG_{CV}(s),$ $KG_{TA}(s), KG_{EM}(s)$	transfer functions of individual servo sub-groups

INTRODUCTION

Two years ago, a systematic research and development work was started in VÚOSO, Prague in the field

* VÚOSO, Prague

of position servo systems for NC machine tools. The purpose of this effort was to prepare suitable feed drives for new generation of NC machines and machining centres designed for highly integrated and computerized manufacturing systems. Requirements imposed on individual components and sub-groups of position servo systems have been stated with respect to latest achievements in the field of integrated NC systems and developed servo systems make it possible to link new machines to most progressive control units. Main requirements were as follows:

(a) Minimum feed rate 1 mm/min

(b) Maximum working feed 5 m/min

(c) Rapid traverse 12 m/min

(d) Position increment 5 μm

(e) Fixed transmission ratio 1:1 between the servo-motor and the motion screw

(f) Full static output torque of the servo-motor at the position deviation of 5 μm

(g) Aperiodic transient response in closed loop condition

(h) Gain $K_V > 25$-50 sec for unloaded position servo system

(i) Start stop frequency of input pulses 40 kHz

(j) Steady state position error proportional to the programmed velocity

(k) High dynamic stiffness against the periodic external force

(l) Drift in closed position loop: $< 1 \mu$m

(m) Maximum static output torque: 50 N m

(n) Natural frequency of unloaded position servo system 35–50 Hz

(o) Maximum load moment of inertia on the motor shaft $J = 1$ kp cm sec^2

To fulfil these requirements, two different servo systems have been investigated and developed, these being the position servo system with hydromotor controlled by electrohydraulic servo-valve and the position servo system with thyristor controlled d.c. servo-motor. In the present stage of development, both servo systems are approaching the above stated parameters and have mutually comparable quality. They work with cyclic absolute measuring devices of 'resolver/inductosyn' type and have identical position-loop control circuits. Maximum interchange-ability of individual components enable their alternative use in individual machine tool applications.

SERVO SYSTEMS AND THEIR CONTROL CIRCUITS

General block diagram

Figure 1 shows a block diagram of input and control circuits and three servos for machine axes X, Y and Z. Servos are controlled only by command position pulses for + and − movement, without any additional auxiliary signals. Frequency and total number of input pulses express the command velocity and the command position, respectively. When no pulses are fed to the servo input, servo keeps its position automatically, by its active action. At zero command velocity, the output position coincides with the command position. At steady state command velocity, the output position lags behind the command posi-

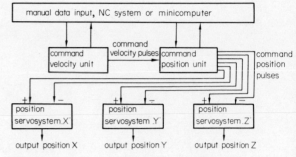

Figure 1. General block diagram of servo systems and their control circuits.

tion with an error proportional to the command velocity. In transient process, the output position follows aperiodically behind the command position. Servos are able to process sudden changes in input frequency of command pulses up to 40 kHz, which corresponds to the rapid traverse of 12 m/min at the position increment of 5 μm.

Command position pulses are generated by the command position unit. This unit emits pulses for individual servos either sequentially (for one servo at the same time) or simultaneously (for two or more servos at the same time). The first case means straight cut control, the second case may mean either a simultaneous automatic coordinate setting in two axes, when no geometric relation is kept between both movements or a continuous-path control, when there is a geometric relation, and the movement is controlled along a programmed path.

Velocity of movement is in all above-mentioned cases determined by the frequency of command velocity pulses on the output of command velocity unit. When the rapid traverse is commanded, the output frequency of command position pulses is constant and equal to 40 kHz. Working feed rates can be programmed either in mm per minute or in mm per spindle revolution. In the last case, the velocity unit contains a pulse generator mechanically connected to the machine spindle.

Both units accept programmed values of command velocity and command position from NC control system or from manual data input. Some auxiliary signals are sent back to the NC control system as for instance signals asking for new data.

Command velocity unit

Figure 2 shows a detailed block diagram of the command velocity unit. For rapid traverse, command velocity pulses, generated in clock and timing circuits are fed through synchronized gates directly to the output. Their frequency is adjustable up to the frequency of 400 kHz and is ten times higher than

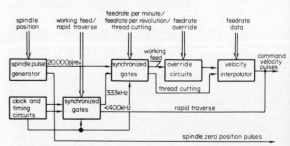

Figure 2. Block diagram of command velocity unit.

the nominal maximum frequency of command position pulses at the rapid traverse. The reason is evident from the structure diagram of the command position unit and will be explained in the following section.

For working feed per minute, a frequency of 333 kHz is generated and fed through synchronized gates to the input of the feed rate override circuit. This circuit serves as an adjustable pulse divider and can change the input frequency in 10 per cent steps.

Velocity interpolator accepts feed rate data either in mm per minute or in mm per revolution. For instance, when a working feed of 1000 mm per minute is programmed, the velocity interpolator obtains feed rate data '1000' and a frequency of 33 kHz appears on its output, commanding the demanded working feed. Similarly as in the case of rapid traverse, this frequency is ten times higher than the frequency of command position pulses for the position increment of 5 μm.

For working feed per spindle revolution, the spindle pulse generator with its output circuits emits 20 000 pulses per one spindle revolution. Pulses are fed through synchronized gates to the input of the override circuit and further to the velocity interpolator. When for instance the programmed value of '0500' is fed to the velocity interpolator, a number of 10 000 pulses appears on the output of the command velocity unit and a feed rate of 5 mm per spindle revolution is commanded.

For thread cutting, the function of the command velocity unit is similar as in the previous case with the difference, that the feed-rate override circuit is by-passed. The spindle pulse generator emits 'zero' pulse in the zero angular position of the spindle for synchronization of thread cutting operations.

Command position unit

Two variants of the command position unit are shown in Figs 3 and 4, these being the simple unit for straight-cut control and the complex unit for continuous-path control, respectively. Command velocity pulses with maximum frequency of 400 kHz at the rapid traverse (see previous section) are fed through synchronized gates to the pulse divider, which divides them by ten and puts each tenth pulse only through (Fig. 3.). In this way, any time-unevenness of the input pulse train, caused by the spindle pulse generator or velocity interpolator, is effectively compensated and reduced to one-tenth of its initial value. Pulses are fed to the input of the distance down-counter and simultaneously through a cascade of gates to the output of the command position unit. Before the movement is started by the start/stop synchronized gate, the distance counter is preselected into a state nominally corresponding to the programmed distance. During the movement the counter counts down to zero and the movement is stopped by the zero coincidence circuit as soon as the programmed position is reached. The pulse divider can be by-passed by the 'increased geometric scale' gate which opens the possibility of performing movements exactly ten times longer than programmed distance values. This feature can be used for instance for thread cutting up to slopes of 99·9 mm per spindle revolution. The output position pulses can be fed to

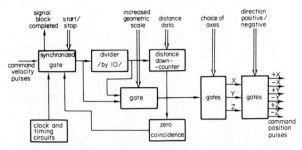

Figure 3. Block diagram of simple command position unit.

any one of the controlled axes X, Y and Z or even simultaneously to more of them. The last feature can be utilized for movements in 45°, 135°, 225° and 315° directions in any of XY, XZ and YZ planes.

Figure 4 shows schematically the complex command position unit for both the point-to-point and the command position control. For point-to-point and straight-cut operations, the command velocity pulses are fed through corresponding gates directly to the input of pulse dividers A and B.

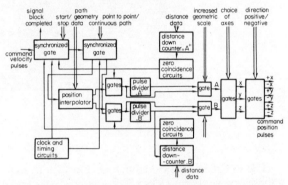

Figure 4. Block diagram of complex command position unit.

In point-to-point operations, the unit in Fig. 4 allows performing of simultaneous movements with identical velocity. This feature can be used for automatic coordinate setting along two different programmed distances in two controlled machine axes. For continuous-path operation, the command velocity pulses are fed to the input of the position interpolator. This interpolator, whether linear or circular, is an optional part of the command position unit. It accepts the path geometry data from the NC system and generates two trains of pulses with exact mutual geometric relation on its outputs A and B. Pulses are fed to the input of pulse dividers A and B as in the case of point-to-point operation. Function of both pulse dividers is similar as in Fig. 3. In case of a continuous-path operation, they compensate additionally the unevenness caused by the position interpolator. Also here, pulse dividers can be by-passed with the aid of 'increased geometric scale' circuits.

Output pulses of pulse dividers are fed to the distance counters A and B. Zero coincidence circuits A and B interrupt successively by means of corresponding gates the output channels A, B of the position interpolator as soon as individual down-counters A and B reach their zero stages. When zero coincidence condition is fulfilled in both counters, the start/stop circuit stops the interpolation and emits the 'block completed' signal back to the NC system.

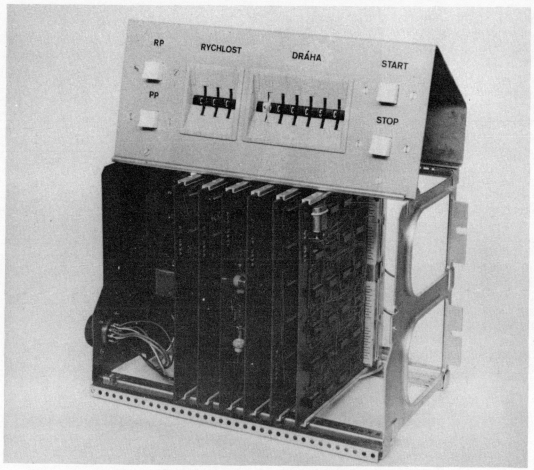

Figure 5. Experimental servo control unit.

Figure 5 shows an experimental control unit used in research work in VÚOSO. The unit generates series of pulses from the frequency of 3·33 Hz (1 mm/min) to maximum frequency of 33 kHz (10 m/min) and performs main functions for point-to-point operations shown in Figs 2 and 3.

POSITION SERVO SYSTEM WITH RESOLVER/INDUCTOSYN FEEDBACK

Block diagram
Figure 6 shows the block diagram of the position servo system for one of the controlled axes. As follows from Fig. 1, the servo system is of 'living servo' type and does not accept any auxiliary signals or additional data. 'Hard' command position pulses

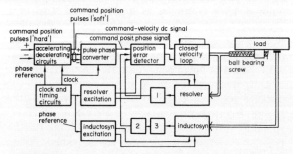

1 filter and shaper
2 pulse divider
3 filter and shaper

Figure 6. Block diagram of position servo system with resolver or inductosyn feedback.

coming from the command position unit are fed to the input of accelerating and decelerating circuits. These circuits suppress any sudden changes in frequency of command position pulses and produce 'soft' command position pulses with continuous and fluent changes of frequency. The circuit acts with respect to transferred frequency of pulses as an RC low pass filter with one time constant and generates exponential changes of output frequency in response to the step changes of the input frequency. The exact function of the circuit is described in the following section.

'Soft' command position pulses are fed to the input of the pulse-phase convertor. This convertor produces the command position phase signal with the carrier frequency of 2 kHz. The signal is rectangular and changes its phase with respect to the phase reference in increments of 0·18° in response to each command position pulse on its input. Evidently, 2000 input pulses are necessary to evoke a 360° change of output phase. Position error detector compares the command position phase signal with the actual position phase signal and produces the position error d.c. signal. On the input of the velocity error preamplifier, this signal is combined with the command velocity d.c. signal which is produced in accelerating and decelerating circuits and compensates for steady state error at constant velocity. The closed velocity loop unit involves the servo-motor, the amplifiers, the d.c. tachogenerator and the compensating circuits. The servo-motor output shaft is coupled directly to the ball-bearing screw with the pitch of 10 mm. Two variants of position feedback are considered, these being the direct coupled two-pole resolver and the linear inductosyn 2 mm/360°.

One revolution of the resolver corresponds to the distance of 10 mm and evokes a phase shift of 360° on its output. The resolver is excited by a frequency of 2000 Hz. At the feed rate of 600 mm/min (10 mm/sec) the output frequency of the resolver changes between values of 2001/Hz and 1999 Hz, depending on the direction of the movement.

The inductosyn evokes a 360° phase shift along a distance of 2 mm and is excited by a frequency of 10 000 Hz. Its output frequency changes in the above shown case between values of 10 005 Hz and 9995 Hz. Its output pulses are divided by the pulse divider (by 5) and, at the feed rate of 10 mm/s, we obtain values of 2001 Hz or 1999 Hz being identical with the case of rotary resolver.

This solution enables an alternative use of rotary or translatory measuring device. Both measuring devices can be also used in one servo system at the same time. This provides a possibility to synchronize the pulse divider by the resolver output pulses and to absolutize the inductosyn measuring system in the range of 10 mm.

Accelerating and decelerating circuits

Figure 7 in its upper part shows a detailed block diagram of the accelerating and decelerating circuits. 4 mHz clock pulses are fed to the synchronized one-pulse modulator together with ± 'hard' and 'soft' command position pulses with maximum frequency of 4 MHz. Each command position pulse is able to add or to take away one pulse from the clock train of pulses. In this way, the clock train of pulses is 'modulated' by the difference in the total number of 'hard' and 'soft' pulses.

The pulse divider divides the modulated train of pulses by 2000 and in this way, it converts the 'one pulse modulation' into a phase modulation with respect to the reference phase produced in clock and timing circuits. The carrier frequency of this signal is 800 Hz. The phase error detector produces pure digital signals on the carrier frequency of 2 x 800 Hz. The command velocity filter derives the d.c. component of the output digital signal of the phase error detector. This component is proportional to the commanded velocity and is used for compensation of steady state errors at constant velocity. Digital output signal of the phase error detector is fed to the input of gate circuits together with the clock frequency of 50 kHz. The clock frequency goes through the mentioned gates in time intervals, the total sum of which is proportional to the phase difference evaluated by

the phase error detector. Thus, the average frequency of output pulses is proportional to the value of phase error detected by the phase error detector. The last relation has in the Laplace transform the following form:

$$X_H(s) - X_S(s) = T_{AD} \cdot s \cdot X_S(s) \qquad (1)$$

Transfer function of the accelerating and decelerating circuits is:

$$G_{AD}(s) = \frac{X_S(s)}{X_H(s)} = \frac{1}{T_{AD} \cdot s + 1} \qquad (2)$$

At the maximum phase error of 12·5 mm the full frequency 50 kHz passes through the gate circuits. Evidently, for average 'soft' frequency of 40 kHz, an error of 10 mm is necessary. From this condition the time constant can be determined as follows (see equation (1)):

$$T_{AD} = \frac{10 \text{ mm}}{200 \text{ mm/s}} = 50 \text{ ms} \qquad (3)$$

This time constant, and simultaneously the steady state error at constant velocity can be reduced by raising the clock frequency beyond the shown value of 50 kHz.

Pulse-phase convertor, resolver and inductosyn excitation circuits

The pulse-phase convertor consists of two parts these being the synchronized one-pulse modulator and pulse divider (see lower part of Fig. 7). 'Soft' command position pulses are fed to the input of the synchronized one-pulse modulator together with clock frequency of 4 MHz. The 'modulated' train of pulses is fed to the input of the pulse divider which divides them by 2000 and produces the command position phase signal. It is evident that each 'soft' command position pulse causes a phase shift of 0·18° with respect to the reference phase generated in clock and timing circuits. At zero command velocity, the frequency of the command position phase signal is identical with the frequency of the phase reference signal and is equal to the value of 2 kHz. 2000 command position pulses will cause 360° phase shift and command the movement along a distance of 10 mm in increments of 5 μm.

The phase reference signal of 2 kHz is fed to the resolver excitation circuits where two rectangular waves with the mutual shift of 90° are derived.

Both rectangular waves serve directly as excitation of resolver windings. As they are rectangular and not sinusoidal, they contain not only the 1st harmonic but all higher odd harmonics. By means of a low pass filter on the resolver output, all higher harmonics are suppressed. Pure sinusoidal signal is shaped back to the rectangular form. Frequency of this resolver output signal depends on the actual angular velocity of the resolver rotor. One mechanical revolution of the rotor corresponds to the movement along the distance of 10 mm. At the rapid traverse of 12 m/min the resolver rotor makes 20 rev/s and the output signal frequency changes in limits of 2 kHz ± 20 Hz.

Inductosyn excitation circuits work with a frequency of 10 kHz. Also in this case, rectangular-form excitation signals can be used.

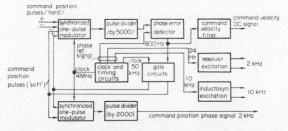

Figure 7. Block diagram of accelerating and decelerating circuits. Resolver and inductosyn excitation.

Electrohydraulic position servo system

Figure 8 shows the block diagram of the electro-hydraulic servo system. The output signal of the position error detector is filtered and fed to the input of the servo preamplifier together with the command velocity d.c. signal, the position error compensation signal and the velocity feedback signal. High-gain servo preamplifier controls through the servo amplifier the electrohydraulic servo-valve. The servo system involves five different feedbacks with the following functions.

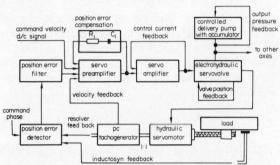

Figure 8. Block diagram of electrohydraulic servo system.

The position error feedback uses an RC compensation circuit across the servo preamplifier. Together with this derivative circuit, the servo preamplifier acts as an integrator and integrates the sum of all three input signals up to the frequency of f_{PA} given by the relation:

$$f_{PA} = \frac{1}{2\pi \, T_{PA}} = \frac{1}{2\pi \, R_1 C_1} \tag{4}$$

The value of f_{PA} in this case is 20 Hz. When no command velocity correction signal is used, the position error signal is equal to the d.c. tachogenerator feedback signal. In closed velocity and closed position loop condition, the servo system has constant error at steady state velocity and works with the gain K_V, given by the ratio of the output velocity to the corresponding position error. The value of K_V in our case is:

$$K_V = 50 \text{ s}^{-1} \tag{5}$$

which means that the servo has an error of 1 mm at the velocity of 50 mm/s (3 m/min). As the position error detector can work in a range of ±5 mm, the servo can theoretically process the feed rates up to 15 m/min. To increase the safety of its operation, the command velocity correction signal is used which can partly or fully compensate for the position error at constant velocity. When full compensation is used, the velocity feedback signal is completely balanced with the command velocity d.c. signal and the servo works with zero steady state error at constant velocity. At zero velocity, both the velocity feedback and command position signals are equal to zero and the servo preamplifier integrates the remaining position error until it fully disappears. Due to the integrating character of its preamplifier, the servo compensates fully for all drifts and external forces which may influence its output position. The servo has an infinite static stiffness and extremely high sensitivity. It reacts reliably on one-pulse (5μ) commands in both

directions and its repeatability is better than 1 μm. Its static output position can be influenced only by summing errors of resolver or inductosyn feedback devices.

The velocity feedback is derived from the d.c. tachogenerator giving output signal of 2 V at 1000 rev/min. This feedback stabilizes the position loop, damps all transient processes and keeps the steady state error exactly proportional to the command velocity. In open velocity loop condition, the constant input voltage of the servo preamplifier causes a constant angular acceleration of the servo motor. As the preamplifier input voltage, a sum of the position error and command velocity signal is considered here. For this case, the open velocity loop gain K_R' can be defined by the equation:

$$K_R' = \frac{\mathrm{d}v_{TF}/\mathrm{d}t}{\Delta v} \cdots \left[\frac{1}{\text{sec}}\right] \tag{6}$$

where v_{TF} and Δv are the tachogenerator output and the preamplifier input signals, respectively. In this case, the value of K_R' is about 800/s, which means for example that the tachogenerator output voltage changes with a rate of 8 V/s when a voltage of 10 mV is fed to the input of the servo preamplifier.

The control current feedback is applied across the servo amplifier and helps to reduce the time constant of the servo-valve control winding. With this compensation, the value of 1 ms of the time constant T_{CC} has been reached.

The valve position feedback used in the electro-hydraulic servo-valve compensates for friction forces and steady state flow forces acting on the spool. The feedback is a mechanical one and is performed by an elastic steel needle. By using three needles with different values of compliance, servo-valves for controlled flows of 25, 40 and 63 l/min can be completed.

The output pressure feedback is used with the controlled delivery pump. It keeps the output pressure constant by controlling the pump in full range of oil delivery. The unit has a built-in oil accumulator which helps to reduce the dynamic deviations of oil pressure during the transient processes. Typically a pump with maximum delivery of 40 l/min is used. After a sudden change in oil consumption from zero to maximum in 20 ms, the dynamic pressure deviation does not exceed ± 15 per cent of the nominal output pressure. The deviation reaches its maximum in 250 ms and the final value of the output pressure is re-established in other 350 ms. The output pressure is kept constant without any considerable deviations in the whole regulating range.

Figure 9 shows the complete mechanical part of the electrohydraulic unit developed in VÚOSO. The unit has static torque of 8 kP m at the inlet pressure of 140 bars. Its open-loop torsional stiffness is 260 kP m/rad. The feedback unit comprises the resolver and the tachogenerator.

Figure 10 shows the electrohydraulic two-stage servo-valve. It can control the flow of oil up to the value of 63 l/min at the pressure drop of 70 bars. It works with guaranteed performance in wide range of inlet pressures (from 20 to 160 bars) and temperatures (from 20°C to 70°C). The most important

Figure 9. Mechanical part of electrohydraulic drive unit.

Figure 10. Electrohydraulic servo-valve.

feature for servo applications is a very low hysteresis which does not exceed a value of 2 per cent in the whole working range. It has a perfectly damped transient response and its open-loop sensitivity is better than 0·5 per cent of the nominal deviation. For servo system synthesis, the servovalve can be considered as a unit with one time constant T_{SV}, the value of which is about 5 ms.

Thyristor controlled d.c. position servo system

The block diagram of the d.c. position servo system is shown in Fig. 11. The diagram is similar to the one shown in Fig. 8 for the electrohydraulic servo. Both servos use identical control circuits and their mechanical parts are easily exchangeable on the machine. Nevertheless, there are certain differences between internal parameters of both servos. The electric servo has much lower open-loop torsional stiffness and needs higher velocity loop gain K'_R. In

our case, a value of $K'_R = 2500$ (s^{-1}) has been used. On the other hand, the position-loop gain is lower and has the value of 25 (s^{-1}). In the electric servo, a velocity feedback compensation circuit is used additionally as shown in Fig. 11. The circuit works as a derivating link between the frequencies f_{TF1} and f_{TF2} given by the relations:

$$f_{TF1} = \frac{1}{2\pi T_{TF1}} = \frac{1}{2\pi(R_2 + R_3)C_2};$$

$$f_{TF2} = \frac{1}{2\pi T_{TF2}} = \frac{1}{2\pi R_2 C_2}$$ (7)

$$f_{TF1} < f_{TF2}$$

Frequencies lower than f_{TF1} and higher than f_{TF2} are transferred with constant amplitude through the circuit. In our case, the frequencies f_{TF1} and f_{TF2} are 50 Hz and 80 Hz, respectively.

A thyristor controlled servo has been completed with the type A 200 Lucas servo-motor and corresponding thyristor amplifier. Figures 12 and 13 show both units during their testing in VÚOSO. The servo-motor in Fig. 12 is equipped with the position feedback unit.

OPEN AND CLOSED LOOP TRANSFER FUNCTIONS

Figure 14(a) shows the general block diagram valid for both electrohydraulic and d.c. electric servo systems. The block diagram in Fig. 14(b) is mathematically identical with the diagram in Fig. 14(a). Its form is more suitable for the analysis of system transfer functions.

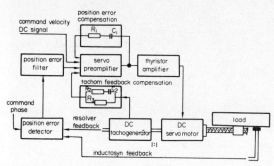

Figure 11. Block diagram of thyristor controlled servo system.

Figure 12. D.C. servo-motor Type Lucas A 200.

Figure 13. Thyristor amplifier.

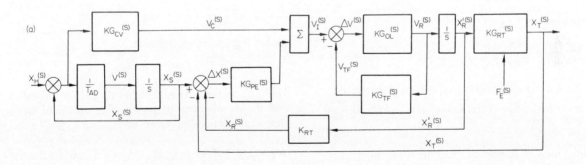

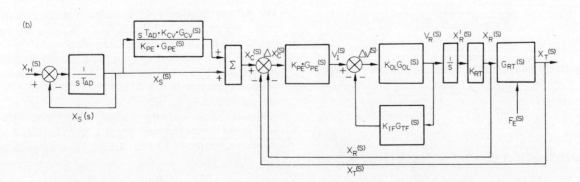

Figure 14. Block diagrams and transfer functions of position servo systems.

The following open and closed loop transfer functions can be defined:

Open velocity loop

$$KG_{OVL}(s) = \frac{V_{TF}(s)}{\Delta V(s)} = K_{OL} \cdot K_{TF} \cdot G_{OL}(s) \cdot G_{TF}(s)$$
$$(8)$$

Closed velocity loop

$$KG_{CVL}(s) = \frac{V_R(s)}{V_I(s)} = \frac{K_{OL} \cdot G_{OL}(s)}{1 + K_{OL} \cdot K_{TF} \cdot G_{OL}(s) \cdot G_{TF}(s)}$$
$$(9)$$

Open position loop with rotary feedback

$$KG_{OPR}(s) = \frac{V_R(s)}{\Delta V_C(s)} = K_{RT} \cdot KG_{PE}(s) \cdot \frac{1}{s} KG_{CVL}(s)$$
$$(10)$$

Open position loop with translatory feedback

$$KG_{OPT}(s) = \frac{X_T(s)}{\Delta X_C(s)}$$

$$= K_{RT} \cdot KG_{PE}(s) \cdot \frac{1}{s} KG_{CVL}(s) \cdot G_{RT}(s)$$
$$(11)$$

Closed position loop with rotary feedback

$$KG_{CPR}(s) = \frac{X_T(s)}{X_C(s)} = \frac{KG_{OPR}(s)}{1 + KG_{OPR}(s)} \cdot G_{RT}(s)$$
$$(12)$$

Closed position loop with translatory feedback

$$KG_{CPT}(s)$$

$$= \frac{X_T(s)}{X_C(s)} = \frac{KG_{OPT}(s)}{1 + KG_{OPT}(s)}$$

$$= \frac{K_{RT} \cdot K_{PE} \cdot (1/s) KG_{CVL}(s) \cdot G_{PE}(s) \cdot G_{RT}(s)}{1 + K_{RT} \cdot K_{PE} \cdot (1/s) KG_{CVL}(s) \cdot G_{PE}(s) \cdot G_{RT}(s)}$$
$$(13)$$

Closed loop of acceleration and deceleration circuits

$$G_{AD}(s) = \frac{X_S(s)}{X_H(s)}$$

$$= \frac{1}{s \cdot T_{AD}(1 + (1/s \cdot T_{AD}))} = \frac{1}{T_{AD} \cdot s + 1}$$
$$(14)$$

Transfer function of the command position correction

$$G_{CP}(s) = \frac{X_C(s)}{X_S(s)} = 1 + \frac{s \cdot T_{AD} \cdot K_{CV} \cdot G_{CV}(s)}{K_{PE} \cdot G_{PE}(s)}$$
$$(15)$$

Total closed loop transfer function for rotary feedback

$$\frac{X_T(s)}{X_H(s)} = G_{AD}(s) \cdot G_{CP}(s) KG_{CPR}(s) \qquad (16)$$

Total closed loop transfer function for translatory feedback

$$\frac{X_T(s)}{X_H(s)} = G_{AD}(s) \cdot G_{CP}(s) \cdot KG_{CPT}(s) \qquad (17)$$

Open-loop transfer functions (8), (10) and (11) determine the system stability and its internal damping. Closed-loop transfer functions express the system behaviour in steady state and transient conditions.

MAIN TIME CONSTANTS AND GAIN COEFFICIENTS

Block diagrams in Figs 8 and 11 show some differences between the electrohydraulic and the electric servo system. The open velocity loop of the electrohydraulic system has the following form:

$$KG_{OVL}(s) = K_{OL} \cdot G_{OL}(s) \cdot K_{TF}$$

$$= KG_{PA}(s) \cdot KG_{CC}(s) \cdot KG_{SV}(s) \cdot KG_{HM}(s) \cdot K_{TF}$$

$$= \frac{K_{PA}(T_{PA} s + 1)}{s \cdot T_{PA}} \cdot \frac{K_{CC}}{T_{CC} s + 1} \cdot \frac{K_{SV}}{T_{SV} s + 1}$$

$$\cdot \frac{K_{HM} \cdot K_{TF}}{\omega_R{}^2 s^2 + (2\xi_R s/\omega_R) + 1}$$
$$(18)$$

In this relation $KG_{PA}(s)$, $KG_{CC}(s)$, $KG_{SV}(s)$ and $KG_{HM}(s)$ are transfer functions of the preamplifier, the control current circuit, the servo-valve and the hydromotor, respectively. Typical values of corresponding time constants are:

$$T_{PA} = 8 \text{ ms}, \qquad T_{CC} = 1 \text{ ms}, \qquad T_{SV} = 5 \text{ ms}$$
$$(19)$$

The natural frequency ω_R of the hydraulic servomotor is given by the relation:

$$\omega_R = \sqrt{K_T/J_0} \qquad (20)$$

Here, K_T is the torsional stiffness of the servo-motor and J_0 its moment of inertia. For unloaded servomotor, corresponding values are:

$$K_T = 2 \cdot 8 \cdot 10^4 \text{ kP cm/rad}; \quad J_0 = 8 \cdot 7 \cdot 10^{-3} \text{ kP cm s}^2$$

and

$$\omega_R = 1800 \text{ s}^{-1} \quad (f_R = 300 \text{ Hz}) \qquad (21)$$

The natural frequency of unloaded hydraulic servomotor is extremely high but it falls rapidly by adding rotary masses to the servo-motor output shaft. For instance, for a ball-bearing screw with the diameter of 63 mm and length of 2800 mm, the added moment of load inertia is $J_0 = 3 \cdot 3 \cdot 10^{-1}$ kP cm s^2 and the natural frequency f_R falls slightly under 50 Hz.

Let us also evaluate the maximum angular acceleration ϵ for unloaded and loaded servo-motor. For maximum torque of 8 kP m, the corresponding values ϵ_{UL} and ϵ_L are:

$$\epsilon_{UL} = \frac{800}{8 \cdot 7 \cdot 10^{-3}} = 92 \cdot 10^3 \text{ s}^{-2} \qquad (22)$$

$$\epsilon_{\rm L} = \frac{800}{(330 + 8 \cdot 7) \cdot 10^{-3}} = 2 \cdot 3 \cdot 10^3 \text{ s}^{-2} \quad (23)$$

The damping coefficient $\xi_{\rm R}$ in equation (18) can be adjusted in a range of $0 \cdot 7$ to 3 by means of an auxiliary damping orifice, connecting both working spaces of the hydromotor. The minimum value of $0 \cdot 7$ corresponds to the totally closed orifice. Suitable adjusting of the damping orifice makes the damping coefficient $\xi_{\rm R}$ higher and independent of the output velocity. This allows application of much higher open velocity loop gain $K'_{\rm R}$ and leads to higher stiffness and lower drift of position servo systems. From equation (18) the velocity loop gain $K'_{\rm R}$ of the hydraulic servo is given by the relation:

$$K'_{\rm R} = \frac{K_{\rm PA} \cdot K_{\rm CC} \cdot K_{\rm SV} \cdot K_{\rm HM} \cdot K_{\rm TF}}{T_{\rm PA}} \text{ [s}^{-1}] \quad (24)$$

Typically, values of 800–1000 s^{-1} for $K'_{\rm R}$ of hydraulic servos with suitably adjusted damping orifice can be used in comparison with the value of 200 s^{-1} which represents a limit value for the case without damping orifice.

The transfer function of the open velocity loop for the above defined d.c. electric servo system has the form:

$$KG_{\rm OVL}(s) = KG_{\rm TF}(s) \cdot K_{\rm OL} \cdot G_{\rm OL}(s)$$

$$= KG_{\rm TF}(s) \cdot KG_{\rm PA}(s) \cdot K_{\rm TA} \cdot KG_{\rm EM}(s)$$

$$= \frac{K_{\rm TF}(T_{\rm TF1} \cdot s + 1)}{T_{\rm TF2} \cdot s + 1} \cdot \frac{K_{\rm PA}(T_{\rm PA} s + 1)}{s \cdot T_{\rm PA}}$$

$$\cdot K_{\rm TA} \cdot \frac{K_{\rm EM}}{(1/\omega_{\rm R}^2)s^2 + (2\xi_{\rm R}/\omega_{\rm R}) s + 1}$$

$$(25)$$

In this relation, $KG_{\rm TF}(s)$, $KG_{\rm PA}(s)$ and $KG_{\rm EM}(s)$ are transfer functions of the tachometric feedback compensation circuit, the preamplifier and the electric servo-motor, respectively. $K_{\rm TA}$ is the gain of the thyristor amplifier. Corresponding time constants have the following values:

$$T_{\rm TF1} = 3 \text{ ms}, \quad T_{\rm TF2} = 2 \text{ ms}, \quad T_{\rm PA} = 8 \text{ ms}$$

$$(26)$$

Open velocity loop gain $K'_{\rm R}$ of the d.c. servo is given by the relation:

$$K'_{\rm R} = \frac{K_{\rm PA} \cdot K_{\rm TA} \cdot K_{\rm EM} \cdot K_{\rm TF}}{T_{\rm PA}} \text{ [s}^{-1}] \quad (27)$$

In described type of d.c. electric servo systems, the value of 2500 s^{-1} for the open velocity loop $K'_{\rm R}$ has been applied. The natural frequency $\omega_{\rm R}$ and the damping coefficient $\xi_{\rm R}$ of the unloaded electric servo-motor are given by the relation:

$$\omega_{\rm R} = \frac{1}{\sqrt{T_{\rm M} \cdot T_{\rm E}}}; \quad \xi_{\rm R} = \frac{1}{2} \sqrt{\frac{T_{\rm M}}{T_{\rm E}}} \quad (28)$$

Here, $T_{\rm E}$ and $T_{\rm M}$ are the electric and mechanical time constants of the servo-motor. With unloaded A200 servo-motor, the corresponding values are:

$$T_{\rm M} = 8 \cdot 6 \text{ ms} \quad T_{\rm E} = 2 \cdot 2 \text{ ms} \quad (29)$$

Then, $\omega_{\rm R}$ and $\xi_{\rm R}$ have the following values:

$$\omega_{\rm R} = 228 \text{ s}^{-1}(f_{\rm R} = 36 \text{ Hz}); \quad \xi_{\rm R} \doteq 1 \quad (30)$$

The mechanical time constant $T_{\rm M}$ is proportional to the moment of inertia J_0 of the servo-motor, which in our case has the value of $0 \cdot 318$ kP cm s^2. This value is about 35 times higher in comparison with that of the hydraulic servo-motor. When the electric servo-motor is loaded with the load moment of inertia $J_{\rm L} = 3 \cdot 3.10^{-1}$ kP cm s^2, the natural frequency and the damping coefficient $\xi_{\rm R}$ change to the following values:

$$\omega_{\rm R} = 163 \text{ s}^{-1}(f_{\rm R} \doteq 23 \text{ Hz}); \quad \xi_{\rm R} = 1 \cdot 4 \quad (31)$$

Actually the damping coefficients in equations (30) and (31) are valid for servo-motor preloaded by higher torque. For lower output torque, the damping coefficients are substantially higher and may reach the value of 10 for fully unloaded servo-motor. This phenomenon is caused by changing of thyristor amplifier internal resistance in dependence of its output current. The natural frequency of the servo-motor and the system stability are not influenced by the mentioned feature of the thyristor amplifier.

Let us now evaluate the maximum angular acceleration for unloaded and loaded d.c. servo-motor. Nominal torque of the A200 servo-motor is 5 kP m, maximum torque in transient process reaches 30 kP m. Corresponding values of angular accelerations are:

$$\epsilon_{\rm UL} = \frac{3000}{318 \cdot 10^{-3}} = 9 \cdot 5 \cdot 10^3 \text{ s}^{-2} \quad (32)$$

$$\epsilon_{\rm L} = \frac{3000}{648 \cdot 10^{-3}} = 4 \cdot 6 \cdot 10^3 \text{ s}^{-2} \quad (33)$$

It is evident from equations (22), (23), (32) and (33) that unloaded d.c. servos of described type cannot reach the dynamic quality of unloaded hydraulic servos. On the other hand, at higher dynamic loads, both types of servos are equivalent or even d.c. servos can be better thanks to their higher maximum torque in transient processes. Their lower sensitivity to load variations makes their synthesis in individual applications more uniform and easier.

The position loop gain $K_{\rm V}$ is given for both types of servos by the relation:

$$K_{\rm V} = \frac{K_{\rm PE} \cdot K_{\rm RT}}{K_{\rm TF}} \text{ [s}^{-1}] \quad (34)$$

Values of $K_{\rm V}$ for unloaded hydraulic and electric servos are 50 s^{-1} and 25 s^{-1} respectively.

For higher dynamic loads, or for translatory position feedback, these values have to be reduced. In our case, steady state errors at constant velocity are partly or fully compensated by the command velocity correction circuit and reduction of the $K_{\rm V}$ does not influence the steady state accuracy of the system.

From the block diagram of Fig. 14(a), the condition for the full compensation of steady state errors can be written as follows:

$$K_{CV} = \frac{K_{TF}}{T_{AD} \cdot K_{RT}} \qquad (35)$$

When this condition is fulfilled, the position error detector process only minimum steady state errors at constant velocity and the system is controlled nearly exclusively by the command velocity correction signal. Because the input 'soft' pulses do not represent any sudden changes of command velocity, the system works all the time in more or less steady state condition, and its position error is concentrated in its accelerating and decelerating circuits as the difference between 'hard' and 'soft' pulses. From the external point of view, the system has a gain K_V' given by the relation:

$$K_V' = \frac{1}{T_{AD}} \ [s^{-1}] \qquad (36)$$

The velocity error preamplifier integrates the difference between the command velocity correction signal and the actual velocity signal which actually means that its output signal is proportional to the actual position error, both in steady and transient conditions. The position loop with its position error detector eliminates mainly the accumulated position error which may occur for instance due to imperfect fulfilment of equation (35). This can be performed effectively even with quite low values of K_V. As the value of K_V is critical for system stability, the shown way simplifies the situation in all cases, where the application of higher K_V would be difficult namely in cases with lower natural frequencies in the position loop.

Transfer function of the position error detector has the following simple form:

$$KG_{PE}(s) = \frac{K_{PE}}{1 + T_{PE} \cdot s} \qquad (37)$$

The time constant T_{PE} is about 2 ms but also higher values can be used in above-mentioned cases with lower natural frequencies in the position loop.

Transfer function of the command velocity correction circuit has similar form as equation (37). The corresponding time constant T_{CV} is also 2 ms.

Evidently, the ratio $G_{CV} : G_{PE}$ is approaching the value of 1 which simplifies the transfer function of the command velocity correction circuit in Fig. 14(b).

Transfer function $K_{RT} \cdot G_{RT}(s)$ of the controlled member of the machine has the form:

$$K_{RT}G_{RT}(s) = \frac{X_T(s)}{X_R'(s)} = \frac{K_{RT}}{(1/\omega_T^2)s^2 + (2\xi_T/\omega_T)s + 1}$$

$$(38)$$

The natural frequency ω_T of the translatory moving mass is given by the following relation:

$$\omega_T = \sqrt{\left(\frac{K_A}{M_T}\right)} \qquad (39)$$

Here, M_T is the translatory moving mass and K_A the axial stiffness of the drive. The K_{RT} coefficient of the rotary to translatory transmission, for the ball-bearing screw with the pitch of 10 mm has the value of:

$$K_{RT} = \frac{1}{2\pi} \ [cm] \qquad (40)$$

The damping coefficient ξ_T depends on the type of slideways and varies in a wide range. In most cases, the transfer function (38) involves nonlinearities as backlash, dead zone, hysteresis, etc.

In working conditions, the output member of the machine is exposed to the action of external cutting force $F_E(s)$.

The translatory output position $X_T(s)$ is then given by the relation:

$$X_T(s) = \frac{K_{RT}}{(1/\omega_T^2)s^2 + (2\xi_T/\omega_T)s + 1} \cdot X_R'(s)$$

$$+ \frac{1/K_A}{(1/\omega_T^2)s^2 + (2\xi_T/\omega_T)s + 1} \cdot F_E(s) \qquad (41)$$

Static stiffness of the servo-system with the rotary measuring device is equal to K_A. Static stiffness of the servo system with the translatory measuring device is infinite due to the integrating action of the velocity error preamplifier.

EXPERIMENTAL CHARACTERISTICS

During the development and testing of electro-hydraulic and d.c. servo system in VÚOSO, the following experimental characteristics have been investigated:

Open-loop characteristics
1. static characteristic (output velocity versus input voltage)
2. loading characteristic (output velocity versus load torque)
3. regularity of output velocity with and without dynamic load on the servo-motor
4. drift of the drive
5. transient response to the step of input voltage
6. open-loop frequency response

Closed velocity loop characteristics
1. static characteristics
2. loading characteristics
3. regularity of output velocity with and without dynamic load
4. drift
5. transient response to the step of input voltage
6. transient response to the step of load torque
7. frequency response to the input voltage
8. frequency response to the load torque

Closed position loop characteristics
1. static characteristic (output velocity versus position error)
2. static stiffness
3. sensitivity
4. regularity of output velocity at slow movements
5. drift
6. transient response to the step of input velocity
7. transient response to the step of load torque
8. frequency response to the load torque

9. available gain K_V with respect to varying dynamic load, damping of transient processes and regularity of slow movements

Some examples of above-mentioned characteristics will be shown in this chapter. Figures 15 and 16 show the open-loop static characteristics of electrohydraulic and d.c. electric drives, respectively. The hysteresis curve shown in Fig. 15 does not show any

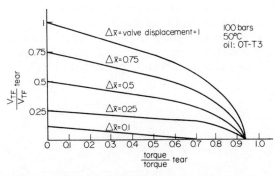

Figure 17. Open loop loading characteristic of undamped electrohydraulic servo-drive.

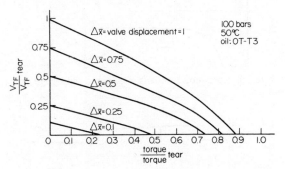

Figure 18. Open loop loading characteristic of damped electrohydraulic servo-drive.

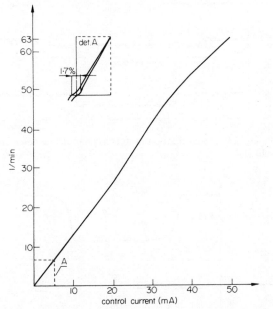

Figure 15. Open loop static characteristic of electrohydraulic servo-drive.

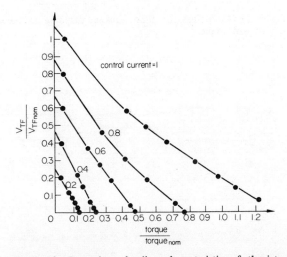

Figure 19. Open loop loading characteristic of thyristor controlled servo-drive.

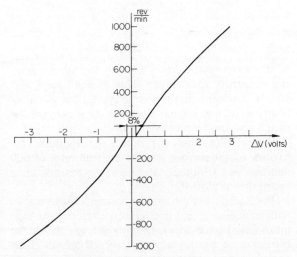

Figure 16. Open loop static characteristic of thyristor controlled servo-drive.

considerable dead zone and the hysteresis is smaller than 2 per cent. The curve in Fig. 16 has a dead zone, caused by passive friction forces of the d.c. servo-motor. No hysteresis has been observed in this case. Figures 17, 18 and 19 show the loading characteristics of both drives. Figures 17 and 18 show the differences in loading characteristics of the electrohydraulic servo-drive caused by adding of auxiliary damping orifice interconnecting both working spaces of the hydromotor (see equation (24)).

Figure 19 shows the corresponding loading characteristic of the d.c. servo-drive.

Nonlinear shape of characteristics shown in Figs 15–19 is for both servos fully eliminated by the action of velocity feedback with high values of velocity loop gain. The closed velocity loop and closed position loop static characteristics are linear and identical for both types of servos.

Figure 20(a) and (b) shows the open-loop transient response to the step of input voltage for unloaded (Fig. 20a) and loaded (Fig. 20b) d.c. servo-motor. The dynamic load was 3·3 kP cm s² (see equation (31)), the step change of velocity was 1000 rev/min.

From the closed-velocity loop characteristics, the transient response to the step of load torque and the frequency response to the input voltage are shown in Figs 21 and 22. Applied load torque in Fig. 21 is 32 N m, which is about 70 per cent of the nominal torque. The measurement has been performed at 1 rev/min and the maximum dynamic deviation of the velocity was about 25 rev/min.

The frequency characteristic in Fig. 22 relates to the unloaded d.c. servo system and shows its natural frequency about 33 Hz at the phase shift of 90°.

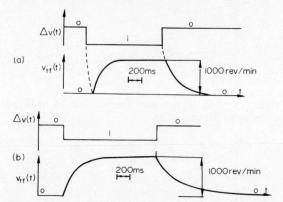

Figure 20. Open loop transient response to the step of input voltage.

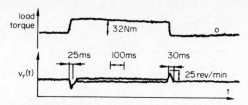

Figure 21. Closed velocity loop transient response to the step of load torque.

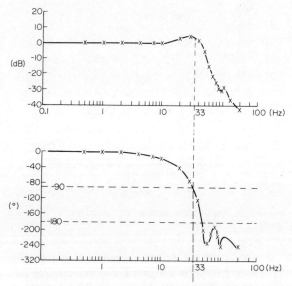

Figure 22. Closed velocity loop frequency response.

Figure 23(a), (b), (c), (d) and (e) show the behaviour of the servo system with closed position loop at small command velocities. The command position is changed in steps of 5 μm and the position error is plotted for feed rates of 1 mm/min (Fig. 23a), 2 mm/min (Fig. 23b), 5 mm/min (Fig. 23c), 10 mm/min (Fig. 23d) and 20 mm/min (Fig. 23e). Obviously, for feed rates lower than 10 mm/min, the servo acts as a stepping motor and follows individual steps of input position. For feed rates higher than 10 mm/min, the movement is fluent.

Graphical representation of servo movements at low velocities in Fig. 23 shows that the analogue part

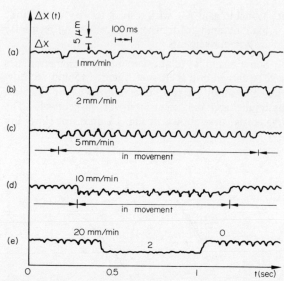

Figure 23. Closed position loop behaviour at low command velocity.

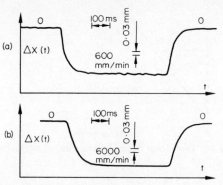

Figure 24. Closed position loop behaviour at high command velocity.

of the servo system has a much better repeatability and sensitivity than chosen value of input steps of 5 μm. It is evident that the servo system has potential ability to follow command position steps even smaller than 5 μm, say 2 μm or even 1 μm.

Figure 24(a) and (b) shows the aperiodic behaviour of the servo system at the feed rates of 600 mm/min and 6 m/min. Also here, the position error versus time is plotted.

For higher feed rates than 6 m/min, characteristics with accelerating and decelerating circuits have been measured. Figure 25(a), (b), (c) and (d) shows the behaviour at the rapid traverse of 10 m/min. Time constant of accelerating circuits was adjusted to 20 m/s (Fig. 25a). Position error (Fig. 25c) is proportional to the output velocity (Fig. 25b) during both accelerating and decelerating transient processes.

Figure 25(d) shows in detail the end of the decelerating transient process. It is evident that the position error has only one polarity and the decelerating process is completed without any overshoot. The total time for deceleration from 10 m/min rapid traverse to zero velocity is about 200 ms.

Figure 26(a), (b), (c) and (d) shows the transient response to the step of load torque in closed position loop condition. 80 per cent (40 N m) of the nominal load torque has been applied to the output shaft of the d.c. servo-motor (Fig. 26a,b) which corresponds

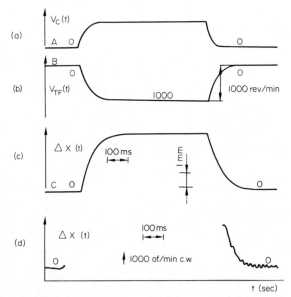

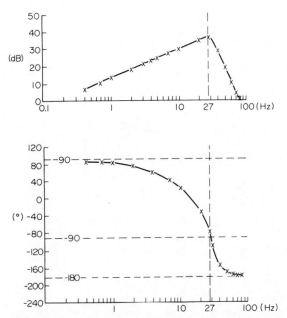

Figure 27. Closed position loop frequency response to the load torque.

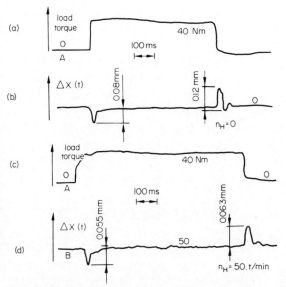

Figure 25. Closed position loop behaviour at rapid traverse of 10 m/min.

Figure 26. Closed position loop transient response to the step of load torque.

to the load force of 2500 kP. Curve 26(b) shows the dynamic deviation at zero output velocity. Maximum deviation corresponds to the dynamic stiffness of about 25 kP/μm. Curve 26(d) shows the dynamic deviation at the output velocity of 50 rev/min. The dynamic stiffness here is higher and has a value of about 40 kP/μm. Static stiffness is infinity in both cases. Time necessary for elimination of dynamic deviations is about 30 ms.

Figure 27 shows the frequency response to the load torque in closed position loop condition. Sinusoidal changes of the load torque were applied to the output shaft of the servo-motor in the frequency range of 0·1 Hz to 80 Hz. Amplitude of the load torque was about 20 per cent (10 N m) of the nominal torque. Also here, the servo system shows infinite stiffness at low input frequencies. Critical frequency is about 27 Hz. At this frequency, the servo has dynamic stiffness of about 10 kP/μm and the phase shift between the torque and deviation is

−90°. At higher frequencies, the dynamic stiffness rises rapidly with a slope of 80 dB/dek.

All above shown experimental characteristics relate to position servo systems with gain K_V equal to 50/s (electrohydraulic servo) or 25/s (d.c. servos). For higher dynamic loads, these values of gain K_V have to be reduced to values of about 10/s at load moments of inertia of 1 kP cm s^2. This reduction of gain does not influence the system accuracy and its static stiffness.

Drift of the position servo system has been checked at one month intervals and no zero position deviations higher than 1 μm have been observed.

CONCLUSION

Servo systems described in this paper can be characterized as pulse train controlled position servos with cyclic absolute position feedback, internal d.c. velocity feedback, digital built-in accelerating and decelerating circuits and integrating velocity error amplifiers. Special approach to some problems connected with their design and development has led to their satisfactory static and dynamic parameters including high accuracy, sensitivity, stiffness, maximum velocity and damping of transient processes. Modular design of developed servo systems enables alternative use of hydro-motors or d.c. electric servo-motors in combination with rotary or translatory measuring devices. Experimental tests have proved suitability of developed feed drive units for applications in NC machines and machining centres with highest parameters of accuracy and productivity.

REFERENCES

1. SKALLA (1973). Testing of electric D/C servo-drive HYPER-LOOP A 200, research report, VÚOSO, February.
2. VURM (1973). New line of electrohydraulic servovalves for nominal pressure of 160 bars, research report, VÚOSO, January.

EVALUATION OF THE CONTOURING PERFORMANCE OF ELECTROHYDRAULIC CYLINDER FEED DRIVES WITH THE AID OF DIGITAL SIMULATION

by

H. P. KHONG,* A. de PENNINGTON† and R. BELL*

SUMMARY

The effectiveness of digital simulation techniques to aid the design of feed drive control systems is demonstrated both in the sizing of the actuator and the performance of the complete system. The threshold and dynamic behaviour of a cylinder drive are modelled and a good correlation between experimental results and the simulation study is obtained. The extension of the simulation study to the prediction of the performance of a two axis contour machining system illustrates the possible use of simulation studies in the prediction of machine tool performance.

NOTATION

A	the effective area of the piston
B	the effective bulk modulus of the hydraulic oil
C	the linearized viscous friction coefficient
F_F	the non-linear friction term
F_{FC}	the coulomb friction
F_{FS}	the static friction
F_L	the load force
I	the servo-valve torque-motor current
i	the linearized servovalve torque-motor current
K_V	the servovalve flow gain
k_v	the linearized servovalve flow gain
k_p	the linearized pressure droop
k_l	the linearized crossport leakage coefficient
M	the load mass
P_L	the load pressure
P_S	the system supply pressure
p_l	the linearized load pressure
Q_L	the valve load flow
Q_{L1}, Q_{L2}	the load flow in each valve port
q_l	the linearized valve load flow
s	the Laplace operator
T_M	the servovalve torque-motor time constant
V_{DEL}	the threshold velocity of the non-friction characteristic
V_T	the effective trapped volume of the cylinder
X_0	the output position of the load mass
x	the linearized output position
ω_l	the angular natural frequency of the hydro-mechanical load
ω_v	the angular natural frequency of the servo-valve
ζ_l	the hydromechanical damping factor
ζ_v	the servovalve damping factor

INTRODUCTION

A major component of the effort required in the design and development of a new range of numerically controlled machine tools is concerned with the automatically controlled feed motions. The principal considerations in the design are the steady-state sizing of the drive elements, the provision of an adequate dynamic performance in the drives and the guarantee of compatibility of the numerical control system equipment and the feed drive elements.

A number of companies employ control system design aids based on empirical estimates of the axis dynamics required to meet particular servomechanism design procedures. The computer programs calculate the natural frequencies of the actuator-transmission subsystem. The design is considered to be effective when these natural frequencies fit into the ranges that have been considered to be essential by the control system manufacturer. This type of design procedure suffers from a limitation fundamental to all classical control system procedures, i.e., to relate the design end product based on, for example, the frequency response method to the specification of the control system application.

The establishment of a comprehensive specification for the controlled feed motions is not without

* Machine Tool Engineering Division, Department of Mechanical Engineering, U.M.I.S.T.
† Department of Mechanical Engineering, University of Leeds

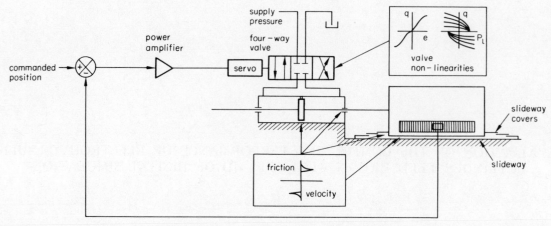

Figure 1. Schematic diagram of a cylinder feed drive.

problems. However, the most positive method to employ is for the designer to aim to meet the level of performance required to satisfy test standards for particular classes of machine tool. The most comprehensive range of specifications for NC machine tools is that published by the Aerospace Industries Association of America[1]. The study of the test workpieces described in these specifications emphasizes the gap that exists between the standard control system design procedures and the guarantee that they have met the required application specification. A more simple approach has been described by Evans *et al.*[2] of the General Electric Company (U.S.A.) for two axis continuous path machinery. The geometry of this test piece incorporates an adequate range of tests of the dynamic and threshold characteristics of the controlled axes. Other test procedures for the positioning accuracy of axes have been published by the NMTBA[3] and the VDI[4] and there is evidence of the growing use of these standards by machine tool manufacturers. If machine tool users and machine tool builders have confidence in such standard test procedures, then the design of NC systems can be given a definitive and systematic basis.

The aim of this paper is to indicate the value of digital simulation studies as a design aid to bridge the gap in the design of controlled axes that exists between design aids and procedures that are used by designers and the guarantee that the machine tool specification can be met. The optimum design process should produce the most effective and economic design that obviates the need for expensive effort by development engineers.

The use of a general purpose digital computer and one of the simulation languages now available makes it possible to handle both linear and non-linear functions equally well[5]. Several good survey articles[6,7] have presented a history of the development of simulation languages, and the solution of some control problems using the more widely used language is well illustrated in two books[8,9].

Among the advantages of digital simulation are greater accuracy, elimination of the difficulties associated with amplitude and time scaling, ease of generating mathematical functions. In contrast, by using an analogue computer there is the advantage of being able to see the influence of manually adjusting certain system parameters. However, the increasing

use of interactive graphics remote terminals offers the design engineer new flexibility and improved communication with the digital computer. In addition, as a result of the rapid advance in computer technology, the cost per simulation is comparable with presently available hybrid and analogue systems[10].

One of the earliest papers to demonstrate the potential of digital simulation in the area of electrohydraulic servomechanisms was presented by Gasich[11]. Further details on the practical significance of this work were given by McIntosh[12]. The system described in the paper was a high performance electrohydraulic positioning system for a computer rotary data storage file. More recently, Alpay and Mitchell[13] have described the problems and advantages in the application of the digital simulation package CSMP/360 to a problem in hydraulic control.

Over recent years, considerable understanding has been gained in the linear design of electrohydraulic cylinder control systems[14,15]. The practical relevance of this work has also been demonstrated[16]. However, the experimental work highlighted certain limitations of the linear approach and the need to study non-linear effects. Often the dominant non-linearities in this class of system arise from the servovalve and seal and load friction (Fig. 1). The more important characteristics of the servovalve are the variable flow gain about null, pressure droop due to orifice flow and flow saturation. Digital simulation offers a means of investigating the effect of these non-linearities on system design[17].

MODELLING OF THE ELECTROHYDRAULIC CYLINDER DRIVE

The model of the electrohydraulic cylinder drive is based on the parameters of a typical cylinder drive on a numerically controlled milling machine. A two stage electrohydraulic servovalve controls the flow of oil from a constant pressure supply at 1000 lbf/in² to a cylinder of equal area with the piston being rigidly coupled to the load. The load weighs 1800 lbf and is supported on recirculating roller bearings. Measured values of slide friction are typically 50 lbf static and 45 lbf sliding. Load displacement is detected via Inductosyn scales. The major drive parameters are specified in the Notation section above.

The various elements which constitute the drive

are shown schematically in Fig. 1. The sources of the non-linearities are indicated in the figure. The derivation of the linearized transfer function of the cylinder and load is readily available in standard reference works[18]. No formal development of the transfer function will be presented here, but the transfer function and describing equations are quoted in the Appendix. It can be seen from this analysis that the transfer function is an undamped second order term. Data on the dynamics of the servovalve are available from the manufacturers' publications. The model chosen for the servovalve is dependent on the load natural frequency and the capabilities of the simulation software. The overall transfer function of the actuator is formed by combining the servovalve and load dynamics.

The servovalve non-linear flow characteristic
A typical servovalve flow gain characteristic, Fig. 1, shows that within the region of ±5 per cent of the rated valve current, the valve flow can vary from 200 to 50 per cent of the nominal flow. Around null the internal leakage dominates the performance of the valve. It consists of two terms. One of them is due to leakage in the first-stage of the servovalve and is independent of spool displacement, representing a direct power loss but has no other significance. The other is second-stage valve leakage which affects drive damping and has a maximum value at null. The simulation of the servovalve flow gain is accomplished by using a function generator to specify a series of break-points on the curve.

The flow through the valve ports is assumed to obey the law of turbulent orifice flow while the leakage flow through the radial clearances of the valve spool are assumed to be laminar. The equations describing a four way spool valve with symmetrical port and non-linear flow gain, assuming zero return pressure, can be expressed as

$$Q_L = Q_{L1} - Q_{L2} \qquad (1)$$

where

$$Q_{L1} = K_V \cdot \left(1 - \frac{P_L(s)}{P_S}\right) \qquad (2)$$

$$Q_{L2} = K_V \cdot \left(1 + \frac{P_L(s)}{P_S}\right)^{1/2} \quad \text{for } I(s) < 0 \qquad (3)$$

$$Q_{L1} = Q_{L2} \qquad \text{for } I(s) = 0 \qquad (4)$$

and

$$Q_{L1} = K_V \cdot \left(1 - \frac{P_L(s)}{P_S}\right)^{1/2} \qquad (5)$$

$$Q_{L2} = K_V \cdot \left(1 + \frac{P_L(s)}{P_S}\right) \qquad \text{for } I(s) > 0 \qquad (6)$$

$$K_V = f\{|I(s)|, \text{sgn } I(s)\} \qquad (7)$$

These equations also apply for an overhauling load. The value of the load pressure must not exceed the supply pressure.

The non-linear friction characteristic
With a few exceptions, analogue computer circuits that have been developed to simulate the non-linear friction characteristic produce the effect of friction as a function of the sign of velocity only. These circuits remain correct so long as the system does not start from rest or becomes stationary. This is because it is necessary to consider both static and coulomb friction. Although coulomb friction is a function of the direction of velocity only, the point to note is that static friction depends on both the velocity and the applied force.

If we consider a free body initially at rest being subjected to an externally applied force, the static friction reacts in an equal and opposite direction to the applied force to prevent motion until the breakaway friction force is exceeded. Therefore, until the breakaway condition is reached, the static friction force only exists when there is an applied force and even then the magnitude will only be the same as that of the applied force. When motion has commenced the friction level does not change suddenly over to the coulomb level but remains at the breakaway level for a very small velocity. Only when the velocity of the body exceeds the threshold velocity do we get true coulomb friction.

The equations that describe the static and coulomb friction characteristic are

$$F_F(s) = F_L(s) \quad \text{for } |\dot{X}_0(s)| < V_{\text{DEL}}, |F_L(s)| < F_{\text{FS}}(s) \qquad (8)$$

$$F_F(s) = F_{\text{FS}}(s) \frac{\dot{X}_0(s)}{|\dot{X}_0(s)|}$$

$$\text{for } |\dot{X}_0(s)| < V_{\text{DEL}}, |F_L(s)| \geqslant F_{\text{FS}} \qquad (9)$$

and

$$F_F(s) = F_{\text{FC}} \frac{\dot{X}_0(s)}{|\dot{X}_0(s)|} \quad \text{for } |\dot{X}_0(s)| \geqslant V_{\text{DEL}} \qquad (10)$$

It should be noted that the friction model discussed above is an idealized one. In reality, the friction mechanism is extremely complicated owing to the interaction of numerous factors such as bearing design, type of lubricant, and quality of slideway alignment. Up to this point in time there is no definitive model that will effectively describe this phenomenon, although work in this direction is known to be in hand. Therefore, a more complicated model will not be attempted at this stage.

Verification of the model from experimental results
A selection of experimental responses obtained from the cylinder drive are shown in Figs. 2 and 3 for comparison with the simulated results. They were chosen to illustrate two points. First, it was necessary to demonstrate that for practical purposes the non-linear model effectively predicts the dynamic behaviour of the drive. Second, it is hoped that the result will enable a comparison to be made between the uncompensated and minor loop compensated responses. Figure 2 shows the slide velocity response of the open position loop system to a step input. The responses shown in Fig. 3 are those of the closed

P

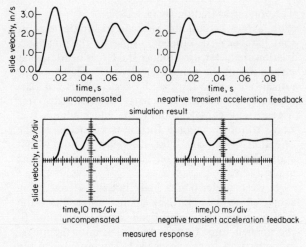

Figure 2.

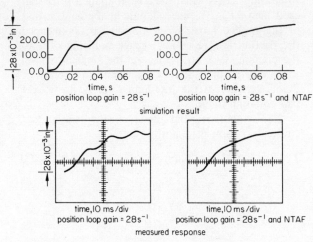

Figure 3.

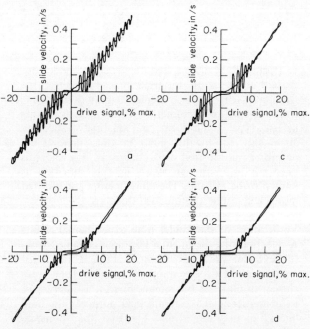

Figure 4. The prediction of actuator steady state characteristics.

position loop system. In both cases the damping effect of the minor loop compensation is evident.

In addition to the transient responses, the steady-

state characteristic of the cylinder drive can be simulated. Some of the simulated results are shown in Fig. 4. The ability to simulate the steady-state characteristic, especially about null, is an important feature of the model. This enables the drive deadband, which determines the maximum attainable accuracy of the closed loop system, to be obtained.

CLOSED LOOP SYSTEM DESIGN

There are several stages in the design procedure for a closed loop electrohydraulic cylinder drive system. The first stage normally involves working out the size of the cylinder. For a given application where the stroke and the inertia or mass of the controlled member are known, the effective area of the piston is determined by drive natural frequency and load force considerations. This, together with the velocity requirements, will in turn decide the type and capacity of the servovalve to be used.

The next step in the design procedure is to estimate an open position loop gain that will meet the specified resolution under steady state conditions. In order to do this the drive signal deadband must be known and this is obtained via a simulation of the open loop system. Various combinations of friction and valve leakage can be tried to ascertain the worst conditions of the deadband, Fig. 4. At this juncture it is also necessary to check that with this value of position loop gain the following error is not excessive, i.e. outside the limits of NC controller, when the mass is moving at full speed.

This so far has only sized the system for steady-state specifications. It is further essential to check

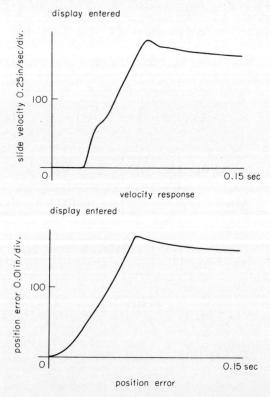

Figure 5. Simulation results for the transition from 0 to 24 ipm with an acceleration of 6 in/s².

dynamic response and stability with a given gain setting. If the dynamic response is not satisfactory, then compensation techniques such as acceleration feedback or phase lag compensation or both must be applied. However, a cautionary note must be made as to use of phase lag compensation in this type of system because an inappropriate choice of compensation parameters[17] can result in limit cycle oscillations. Again simulation can be relied on to provide effective answers. The error incurred by the system when it is accelerating up to speed can be large due to dynamic effects, such as that shown in Fig. 5. This also needs checking out as the system may fault if the error exceeds the limit permitted by the controller.

Up to this point, the design has been based on one axis and responses resulting from step or ramp inputs. The usual practice is to set up the controller of an axis by referring to its response to a step input. The criterion normally adopted is to aim for as fast a response as possible with minimum of overshoot. However, there are reasons to believe that a 'best' setting of one axis may not necessarily lead to a best contouring performance in terms of minimum path error. An example of this is that the results of one axis do not demonstrate the dynamic errors caused by the command signal such as that generated while contouring a circle. A two-dimensional system with matched axes would not show these errors when only required to traverse a straight line. It is, therefore, suggested that a two-axis simulation can give further insight into the behaviour of a contouring servo system.

of an individual servomechanism. A natural extension to this is to simulate a tool at the centre point of two independent axes, and to command the tool to move at a constant speed along a test path. The contouring performance of a machine can then be evaluated from a plot of the path error against time. The path error is defined by e_p in Fig. 7 and a sample plot is given in Fig. 8.

It is advisable that any test path should include as many practical features as possible. A plan view of the N.A.S. workpiece[1] for a 2-axis machine is shown in Fig. 6a. The dimension of the larger square is 12 in or 6 in depending on the size of the machine. It is not possible to simulate the full workpiece because the computing time needed to do this would be excessive. However, the most significant features of the N.A.S. workpiece can be combined in a two-dimensional[2] contour as shown in Fig. 6b. The test path includes a circular arc at one point of which the y-axis drive has to reverse direction slowly. With load friction present this slowly reversing command can produce a noticeable flat spot on the workpiece[19]. The test path also includes sharp corners where servo transients tend to produce larger path errors[20].

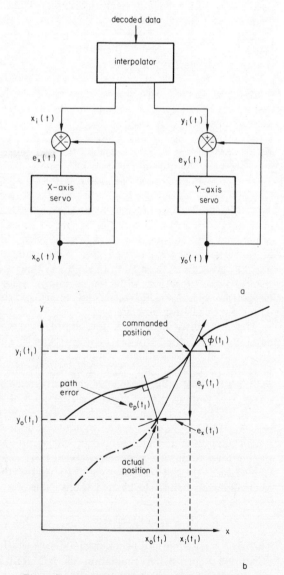

a

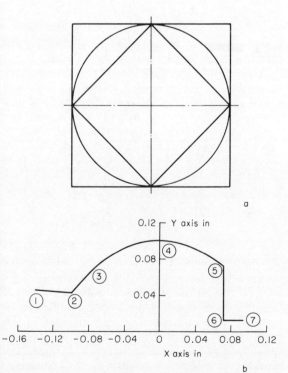

Figure 6. Representative two-axis machining test pieces.

TEST CONTOURS

The servomechanisms of a contouring machine are normally set up by tuning each axis in turn. Thus, analyses usually portray the response characteristics

Figure 7. Contour errors for two-axis machining.

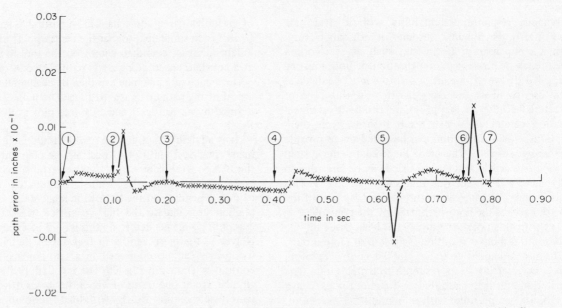

Figure 8. The predicted path error for the production of a testpiece produced to the geometry shown in Fig. 6b.

The work reported here was carried out using IBM's Continuous System Modelling Program (CSMP/360)[21]. The contouring control system consists of two independent electrohydraulic cylinder feed drives which move the tool parallel to the principal axes of a machine. Separate position command signals are supplied to the servos as continuous functions of time from a command generator, which coordinates the signals so as to command the tool to follow the desired two-dimensional path at the desired speed (25 ipm). For simplicity the two axes were assumed to be identical.

At an early stage it was realized that if the path error approach was to be acceptable, the handling of punched cards by the user had to be minimized. In order to solve this problem the CSMP operating system had to be changed. As part of the normal operating system it is possible to obtain a punched card output of variables listed using the PREPARE statement. The sub-program which controlled this output was, therefore, modified to write the variables out on online disc. As an additional check, time was also recorded together with the x and y values respectively. This was the format necessary for the path error calculation program.

The most difficult part of this approach is the development of the error calculation program. The test contour was defined using a programming system[22] written in the AED language[23]. After modification this program enabled the perpendicular distance of any point (x, y) from the true contour to be calculated. It was also necessary to test on which side (x, y) lay, so that the path error could be defined unambiguously as the cutter position moves around the contour. Additional statements coded in Fortran were added to the AED program to enable the calculated path error to be plotted against time.

CONCLUSIONS

The work reported in this paper has demonstrated the effectiveness of digital simulation as NC control system design aid. In particular, its value in the calculation of threshold performance must be emphasized.

It is not intended that the reader of this paper should feel obliged to have an accurate measure of the non-linearities present in a design but rather that the results of a simulation study should give guidance on the permissible ceiling values for these terms. For instance, in the class of drives discussed here, the machine tool builder should issue specifications for acceptable values of breakaway and running friction to the manufacture of hydraulic cylinders. These values being previously obtained from the simulation. Similarly, the slideways should also be so designed that the slideway friction characteristics satisfy the constraints indicated by the computer studies.

An important problem is seen to exist when the economics of the computer studies are evaluated within an industrial environment. The use of real time simulation when applied to particular lengthy and complex test workpiece geometries will result in expensive computer costs. There is a need for cooperative effort to establish optimum procedures for both the design, testing and evaluation of the accuracy of the controlled axes of NC machine tools.

ACKNOWLEDGEMENTS

The authors wish to thank Professor F. Koenigsberger for his permission to carry out this work in the Machine Tool Laboratories. The collaboration of the staff of the Control Systems Centre at U.M.I.S.T. and the Philips Research Laboratories, Eindhoven, is gratefully acknowledged. The work reported in this paper is part of a programme of feed drive studies sponsored by the Science Research Council. The provision of computing facilities to the Institute by the Science Research Council is also of immense help to this work.

REFERENCES

1. ANON. National Aerospace Standards 912, 913, 948, 960, 977, 978 and 979. Aerospace

industries Association of America, Washington D.C., U.S.A.

2. G. G. EVANS, H. E. VIGOUR and F. J. ELLERT (1966). An application of parameter optimisation of hydraulic servo design. 7th Joint Automation Control Conf., 506.

3. ANON (1968). Definition and evaluation of accuracy and repeatability for numerically controlled machine tools. National Machine Tool Builders Association, Washington, D.C.

4. ANON (1967). VDI 3227, 3228, 3229, 3230, VDI-Fachgruppe Betriebstechnik (ADB), Beuth Vetrieb, Berlin, Koln, May.

5. M. H. DOST (1971). Simulation languages: ideal analysis tools for the control engineer. IFAC Symp. on Digital Simulation of Continuous Processes, Hungary, September, Paper D6.

6. R. D. BRENNAN and R. N. LINEBARGER (1964). A survey of digital simulation: digital analog simulator programs. *Simulation,* **3**, No. 6, 22.

7. J. J. CLANCY and M. S. FINEBERG (1965). Digital simulation languages: a critique and guide. Fall Joint Computer Conf., Vol. 27, 23.

8. Y. CHU (1969). *Digital Simulation of Continuous Systems,* McGraw-Hill.

9. W. JENTSCH (1969). *Digitale Simulation Kontinuierlicher Systeme,* R. Oldenbourg Verlag, München.

10. B. A. CHUBB (1970). Application of a continuous system modelling program to control system design, 11th JACC, 350.

11. D. B. GASICH (1966). *Computer-Aided Design of Two Servosystems.* Wescon, Los Angeles, California.

12. R. P. McINTOSH (1967). A hydraulic servo positioner for a high-performance rotary data storage file, *Proc. Nat. Conf. Fluid Power,* 183.

13. S. A. ALPAY and T. MITCHELL (1971). Digital simulation of a hydraulic control system, *Hydraulics and Pneumatics,* **24**, No. 3, 81.

14. R. BELL and A. de PENNINGTON (1969–70). Active compensation of lightly damped electrohydraulic cylinder drives using derivative signals, *Proc. Instn. Mech. Engrs.,* **184**, Pt. 1, 83.

15. R. BELL and A. de PENNINGTON (1968). The design of active damping for electrohydraulic cylinder feed drives, *Proc. 9th M.T.D.R. Conf.,* 1309.

16. A. de PENNINGTON, D. W. MARSLAND and R. BELL (1971). The improvement of the accuracy of electrophydraulic cylinder drives for NC machine tools by the use of active feedback compensation, *Proc. 12th M.T.D.R. Conf.*

17. H. P. KHONG (1972). The digital simulation of electrohydraulic cylinder drives, PhD thesis, UMIST.

18. H. E. MERRITT (1967). *Hydraulic Control Systems,* John Wiley, New York.

19. G. AUGSTEN and D. SCHMID (1969). Einfluss von Spiel und Leibung auf die Konturfehler Hahngesteuerter Werkzeugmaschinen, *Steuerungstechnik,* No. 3, 103.

20. J. L. DUTCHER (1971). Servo drives for numerically controlled machines, *Numerical Control Society Proc.,* 371.

21. ANON, System/360 Continuous System Modeling Program (360A-CX-16X) User's Manual (H20-0367) IBM Corporation, Data Processing Division, White Plains, New York.

22. R. M. BURKLEY and W. B. BROADWELL (1972). Dynamic model for contouring NC devices. *Numerical Control Society Proc.,* 156.

23. D. T. ROSS (1967). The AED approach to generalised computer-aided design, *Proc. A.C.M.,* 367.

APPENDIX:

LINEAR SYSTEM TRANSFER FUNCTIONS FOR AN ELECTROHYDRAULIC CYLINDER DRIVE

(1) Electrohydraulic control valve:

$$\frac{q_1(s)}{i(s)} = \frac{k_v}{(1 + sT_M)\left(\dfrac{s^2}{\omega_v{}^2} + \dfrac{s2\zeta_v}{\omega_v} + 1\right)}$$

(2) Flow continuity equation:

$$q_l(s) = x(s) \cdot A + k_p p_l(s) + s p_l(s) + 4BV_T$$

(3) Thrust equation:

$$p_l(s) \cdot A = s^2 m x(s) + s c x(s) + \text{external load}$$

(4) The open loop transfer function:

The combination of the above equations gives rise to the overall transfer function

$$\frac{x(s)}{i(s)} = \frac{k_{v/A}}{s(1 + sT_M)\left(\dfrac{s^2}{\omega_v{}^2} + \dfrac{s2\zeta_v}{\omega_v} + 1\right)\left(\dfrac{s^2}{\omega_l{}^2} + \dfrac{2\zeta_l s}{\omega_l} + 1\right)}$$

if k_p and k_l and C are considered small.

DISCUSSION

Q. P. C. Gale. Could Dr Bell please comment on the methods he intends to use to check the test workpiece shown in Fig. 6?

A. The figure referred to by Mr Gale is included in the paper to show the typical output that is obtained from a digital simulation study. The error plot refers to the G.E. test contour and it follows that if a designer was to be asked to use this test contour to prove the acceptability of a machine then he could consider the computer data to indicate an acceptable or unacceptable drive design.

The problem posed by Mr Gale is important when international standards are established as the cost, complexity and measurement problems associated with acceptance test procedures have yet to be fully investigated.

The design aid discussed in this paper can be applied to any test workpiece geometry and do not influence, directly, the establishment of preferred standard test procedures.

WHY HIGH INERTIA DC SERVOS ARE SUPPLANTING HYDRAULIC DRIVES ON MACHINE TOOL SLIDES

by

ROGER GETTYS HILL*

INTRODUCTION

The latest thing in machine tool control is, of all things, a *high* inertia d.c. motor. Once considered an anathema to servo engineers, high inertia is proving to be very desirable when associated with the drives applied to machine tool slides. High inertia motors provide the necessary torque without gears. They assure a servo stability that makes high velocity loop gains easy to obtain, and they have the inherent ruggedness necessary for reliable performance. Remarkably it took twenty years to find this out. A short review of the history of the development of machine tool slide drives will make this quite clear.

HISTORY OF EVOLUTION OF MACHINE TOOL SLIDE DRIVES

Mechanical
Before World War 2 machine tool slide drives were largely mechanical. A single, constant speed a.c. motor was driving through an adjustable gear box most of the machines axes. In the more complex machines various mechanical linkages could provide programming movements through an automatic work cycle. Positioning was achieved by means of limit switches.

Servo developed
During the 1940s great strides were made in the development of servo mechanisms. Servos differed from earlier power-driven devices in that they would provide a much greater degree of control in terms of accuracy and capability. Smoothly adjustable speeds, for instance, replaced the stepped selection of changeable gears. Positioning by transducer rather than limit switch made possible programmed decelerations for more accurate control.

As servos evolved, two distinct approaches to the drive problem developed separately. One utilized the Ward-Leonard motor-generator concept with a d.c. motor as the machine driving means; while the other became the hydraulic systems developed during the war and used widely in the control of aircraft.

The hydraulic approach had a distinct advantage over the electrical for two reasons. First, the response time of a servo valve controlled hydraulic motor was substantially shorter than that of a conventional d.c. motor. Second, the size and weight of a hydraulic motor is substantially less than that of a d.c. motor with equal power rating.

Hydraulic servos favoured
The early work in numerical control seemed to dictate the highest possible response, so hydraulic drives were used almost exclusively by the American firms offering numerical control (NC) between 1960 and 1965.

Probably because that NC came to the European machine tool scene somewhat later than it did in the United States, European machine tool builders did not adopt hydraulics quite so completely. With many builders still using d.c. drives some development continued on this approach.

Conventional d.c. motors seemed to suffer from a high inherent inertia. They were limited to peak torques in the range of two-and-a-half to three times the rated torque. To obtain reasonable continuous torque ratings wound fields seemed essential, making the motors bulky, unwieldy, and inefficient.

In the other cases, hydraulics were found to suffer from inherent drawbacks in inefficiency and maintenance that made the development of an electrical equivalent seem very desirable.

Low inertia d.c. motor perfected
For this reason, Yaskawa in Japan perfected a new d.c. motor of radical design using a very small diameter armature without slots. This motor, called the Minertia motor (for minimum inertia), had mechanical and electrical time constants that rivalled the hydraulic servo valve motor combination. Many companies, including my own, immediately began to design controls around this motor. In 1963 silicon-controlled rectifiers (SCR's) were dropping dramatically in price, making the low inertia motor driven by inexpensive solid state components appear to be the

* President Gettys Manufacturing Co., Inc. Racine, Wisconsin

much sought after 'equivalent' to the high performance hydraulic units.

Many copies of the Minertia motor technique occurred. Siemens, General Electric, Reliance, and many machine tool builders themselves spent hundreds of thousands of dollars developing motors with low inertia and very high response. The disadvantage of a low inertia drive escaped the attention of all of us for some years because these disadvantages were shared by the hydraulic motors and we had, therefore, learned to live with them.

Early enthusiasm for the 'Minertia' approach is illustrated by Fig. 1 which is reproduced from a paper presented by Anderson and Fair of the Sundstrand

		Hydraulic system	D.C., SCR servo system
	Response	Very high but machine is limited to 0·2 g force	0·56 g available but limited to 0·2 g to protect machine. It is actually superior in the normal contouring range
	Drift	Subject to amplifier and servo valve drift	Superior. Subject only to amplifier drift
	Stability	Good	Superior. Our tests indicate a two to one advantage over our hydraulic system
	Gain	Good	Equal
	Power requirement	25 hp	8 hp
	Efficiency	Low	High. Better than 2 to 1 advantage with consequent savings in operating power cost
	Noise level	High	Low
	Warm-up period	Required	Not required
	Tachometer mounting	Less than ideal because in order to make it accessible for maintenance it must be geared to fluid motor. Angle type motor	Ideal. Directly coupled to d.c. drive motor
	Integration time in our plant	High. Hydraulic system must be flushed to minimize contamination of the oil which can affect servo valve operation. Approximately 32 h required to flush and adjust the hyd. system	Low
	Start-up time in customers plant	High (flushing of system required)	Low
	Floor space	Large (29·1 sq. ft)	Small (14·8 sq. ft)
	Machine heating problems	Hydraulic lines to servo valves must be thermally insulated from machine frame to prevent localized heating which could cause unequal expansion with consequent degradation of size control	Not required
	Heat exchanger	Required to maintain temperature of hydraulic system constant within 10°F	Not required
MAINTENANCE	Moving parts	Two a.c. motors and two hydraulic pumps plus one fluid motor and one servo valve with torque motors for each axis	One d.c. motor per axis plus one a.c. motor and hydraulic pump for vertical axis counter balance and tool changer operation
	Service specialists	Hydraulic, electrical and mechanical	Electrical and mechanical
	Servo problem diagnosis	Complex	Relatively simple
	Oil contamination problems	Very susceptible	Not susceptible
	Hydraulic leaks	Serious problem	Problem considerably reduced
	Filters	Filtering required. Filter cartridges must be changed periodically with consequent down time (8 h) for flushing of hydraulic system	Not required

Figure 1. Comparison of electric versus hydraulic servo systems on the OM2.

machine tool Company in 1966, at a machine tool conference. This chart is very interesting for while it demonstrates many advantages for the d.c. approach, it illustrates the ignorance all of us active in this area shared at that time.

POWER REQUIRED FOR SERVO DRIVES

The attention is drawn, particularly, to the line labelled 'power requirement'. The detailed specification of this machine is not known, however, from wide experience in applying drives on 'like' machines, it would be safer to guess that there was an *actual* peak average power requirement of some 10 000 pounds of thrust at maybe 10 inches per minute. This is roughly 0·25 h.p. being an astonishingly small part of 8 h.p., and much less the 25 h.p. needed for the hydraulic power supply.

There is a large discrepancy between actual power requirements and the power supplied, and so little engineering thought has been given to this anomaly. In the beginning machine tool slide drives were mechanical. A constant speed motor had to develop some maximum torque for maximum machine thrust ratings. The product of this torque and the constant speed was the motor's power rating. Since the acme lead screws that were used, plus the gear boxes through which they were driven, had a combined efficiency somewhere between 15 and 20 per cent, the actual a.c. motors chosen for prime mover would typically develop eight to ten times more power than was used in moving the slide against the cutting tool.

Now a servo motor driving directly into an efficient ballscrew faces an entirely different environment. This motor will typically provide high torques for high thrust, only at relatively low speeds since the highest cutting loads are always associated with low feedrates. At other times, for rapid traversing, the motor will operate at high speed; but will be devel-

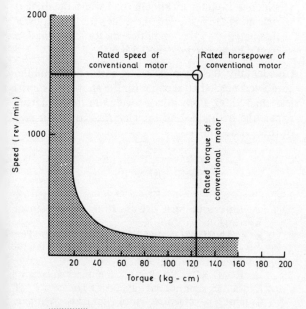

Typical continuous duty operating area of Servo Drive.

Figure 2. Comparison of conventional power rating, and actual power required of position servo drive.

oping, typically, less than 15 per cent of the rated torque.

Inspection of Fig. 2 will clarify the difference. The typical operating area of the servo used in positioning a machine tool slide lies in the relatively low power portion of the speed torque plane. High power is not what is needed, but rather high torque on the one hand and high speed on the other, but seldom, if ever, simultaneously.

NEED FOR HIGH ACCELERATIONS

To explain why the Sunderland hydraulic drive required 25 h.p., we must consider a further peculiarity of the servo drive. To deserve the term servo drive, the source of motion must be capable of very rapid starts and stops. Acceleration to full speed in less than 200 ms is essential. A really good servo could do this in less than 100 ms. Acceleration is, in turn, determined by the ratio of peak torque to the combined motor and load inertia as follows:

$$\text{Acceleration (radians/s}^2) = \frac{\text{Peak torque (lb in)}}{\text{Total inertia (lb in s}^2)}$$

Whilst the inertia of a hydraulic motor is so small as to be negligible, it is coupled directly to a lead screw which reflects a very large inertia. For this reason an acceptable torque to inertia ratio would dictate a high source pressure to provide high peak torques to overcome the high inertia. In addition, the hydraulic power supply must have a volumetric capacity to handle the rapid transverse requirements. Thus the hydraulic pump might have to provide pressures up to 1500 lb/in² at flow rates up to 20 gal/min. To be able to provide these *instantaneously,* this amount of power must be continuously generated whether used or not. For these reasons servo engineers became accustomed to the idea that a machine such as the Sundstrand OM2 needed a 25 h.p. slide drive. This was only about twice what the mechanical drives had used; and, after all, this was a servo so that 100 per cent increase in power required was not so much. All the while the maximum *power* we would ever require of this system was only one-quarter of a horsepower, or one-hundredth of that supplied by the power source.

It is clear now, as we look back, that the large source power required *both* from the mechanical and then the hydraulic servo drives kept us from questioning the inefficiencies inherent in the low inertia d.c. drive. It appeared to be three times more efficient than the hydraulic.

SEARCH FOR EQUIVALENT OF HYDRAULIC SERVO

Those who worked in the mid-1960s to find a d.c. drive replacement for hydraulics were looking for the d.c. 'equivalent' of a hydraulic servo rather than simply seeking the best overall solution. This, then, led many workers to low inertia d.c. motors as the 'equivalent' of low inertia hydraulic motors.

It was two years before GETTYS, began to realize that a better approach might be found. The initial impetus to looking elsewhere came from the cost and trouble experienced fitting motors with gear heads. Low inertia motors were very high speed (up to 5000 rev/min for rapid traverse) and, therefore, required gears. Since these were servo devices these gears had to be precision made. Each different size motor required its own special gears. The gear heads were costing more than the motors, and were subject to rapid deterioration. It seemed clear that what was needed was a motor with high torques and low speed so gears could be eliminated. The high torque, however, meant high inertia, and that meant poor response unless *very* high peak torques could be provided.

PEAK TORQUE LIMITED
BY MAGNETIC MATERIALS

Peak torque is limited in permanent magnet motors by the tendency to permanently demagnetize the field when high peak currents associated with the high peak torques are passed through the armature. (Motors with wound fields were too inefficient, and generally too large or bulky.)

The search led to a permanent magnet motor using a recently developed ceramic magnetic material which could retain its magnetism under very high flux densities. This motor employed a large armature, giving the high continuous torque rating sought, but it also tolerated peak torques ten to fifteen times rated without demagnetization. Thus the torque to inertia ratio far exceeded what was required to produce the necessary response.

Only after employing these motors to various machine tools was it realized that the high inertia so desperately avoided was actually a great asset.

WHY HIGH INERTIA
IS BENEFICIAL

To explain why this is so, thought must be given to what is involved in 'closing the loop' around the servo drive. It was known for years that high tachometer loop gains are very hard to obtain in a hydraulic servo because of the highly undampened nature of the natural resonance of the hydraulic system. Again, it was looked upon as 'natural' and tended to even overlook the added complication of reflected resonances from the machine slides, making each axis of each machine a new problem in stability. 'Tuning' and 'retuning' of each drive was an accepted procedure.

Placing our high inertia motor on the end of a drive screw was like adding a great flywheel to the system—a flywheel which can, of course, control beautifully because of the high torque to inertia ratio. What happens at the other end of the drive screw is virtually ignored by the high inertia motor. Not only does the d.c. motor not have a hydraulic resonance to trouble with, but it also so overwhelms the machine characteristics that each axis of each machine can be treated in exactly the same manner. In many cases our drives are sold without even a tachometer gain adjustment as this wants to be the same in each and every case. A servo loop with only one adjustment, position loop gain was now practical.

The high inertia armature brought with it yet another advantage, high thermal capacity. The Minertia motor, and all of its imitators, suffered from the fact that the high currents needed for their high response and which they were able to sustain without demagnetization were tolerable for only a matter of seconds before the armature would destroy itself by exceeding the rated temperatures. So fast is this destruction that it is virtually impossible to protect the motor with conventional overload relays, or even fuses.

The high inertia motor, on the other hand, has such thermal capacity that, for instance, operation at three times rated torque (current) is permissible for *thirty minutes* before armature temperatures reach dangerous levels. For this reason the large inertia motor provides a degree of ruggedness that matches the machine tool environment, allowing for operator error or misjudgment without destroying the drive.

Thus it is seen that the new high response, high inertia motors, provided something better than merely the 'equivalent' of the hydraulic servo. For it is concluded that:

(1) The simplicity of hydraulics in mounting, *no gear boxes.*
(2) The equal of hydraulics in *loaded* response.
(3) Greater stability, giving much higher velocity loop gains, and thus *greater* servo stiffness than hydraulics.
(4) Ruggedness equal to that of hydraulics.
(5) All the commonly understood advantages of electric over hydraulics; simplicity of installation, ease of maintenance, reliability, etc.
(6) The clincher—the *ultimate in efficiency*. For, using the Sundstrand OM2 machine once more as an example, the high inertia motor needed to meet these hypothetical specifications would have a conventional power rating of about 2·5 h.p. *Standby* power would be *negligible*. Power would be used *only* when it is supplied to the machine slide, and, under maximum load while the drive is delivering 0·25 h.p. to the slide, the total power dissipated by the amplifier–motor combination would be approximately 0·5 h.p.

(Incidentally, the complete drive, motor and amplifier combined needed to accomplish the above, now costs less than $800. *Both* drives covered in the Anderson paper, the hydraulic and the electric, cost in the area of three times this.)

REFERENCES

1.   GEORGE YOUNKIN (1969). Comparing machine tool feed drives, June. 33rd Annual Westinghouse forum.
2.   B. T. ANDERSON and D. G. FAIR (1966). The Sundstrand SCR servo drive, IEEE machine tool conference, October.
3.   H. E. MERRITT (1969). Cincinnati milacron; hydraulic component analysis, University of Cincinnati, machine tool controls seminar, September.
4.   G. S. ALBERS (1968). Lodge & Shipley Company; Stability analysis of N/C servos, IEEE machine tool conference, October.

DISCUSSION

Q. P. Brooks, Giddings & Lewis-Fraser. A number of servo driven machine tools rely on the static stiffness of the 'servo-lock' to hold a slide whilst one or more other axes are working. The emphasis of the paper is on the dynamic characteristic of high inertia d.c. drives. Does this design approach enable a better 'servo-lock' to be obtained over that available from the small inertia approach and associated drive, such that the designer need not be concerned with a back wind-up of the drive screw through its ball nut?

In asking this question it is appreciated that the frequency response of the static axis is a related but different consideration.

A. Recent tests by George Younkin of your American affiliate indicated that the static stiffness of a Gettys drive exceeded that of the hydraulic drives they have been using up to now.

Q. Sartorio, DEA, Torino. I would like to know the maximum speed (nominal) of this new motor and the maximum ratio between maximum and minimum speed under control.

A. The 2·5 HP motor referred to in the paper has a practical top speed of 2000 RPM.

With a special tacho a low speed of 0·1 RPM ±10 per cent has been accomplished.

OPTIMIZATION OF POSITION CONTROL LOOPS FOR NC MACHINE TOOLS

by

G. STUTE and D. SCHMID*

SUMMARY

A new criterion for optimizing position control loops is presented and discussed for some position control loop structures. By this optimization technique we obtain better results than by the performance criteria which are used in general. The functions of the performance index are also compared with the resulting path deviations at contouring.

NOTATION

A_e	undershooting error
A_{max}	maximum of A_e and $A_{\ddot{u}}$ (half tolerance bandwidth)
A_{ew}	undershooting error, measured along the bi-sector
$A_{\ddot{u}}$	overshooting error
D_0	damping ratio
f_0	natural frequency
I	index of performance
ISE	integral of squared error
IAE	integral of absolute error
$ISEV$	modified integral of squared error
$IAEV$	modified integral of absolute error
j	imaginary unit $\sqrt{-1}$
k_v	velocity gain $\dfrac{mm/s}{mm}$
t_0	group delay time
t_1	time displacement
v_B	feedrate
v_{xs}	velocity command signal (X-axis)
v_{xi}	velocity feedback signal (X-axis)
w	reference magnitude
x_a	position error signal
x_s, y_s, z_s	position command signals
x_i, y_i, z_i	position feedback signals
ω	frequency
ω_0	natural frequency
$\lrcorner$	step function
$\diagup$	slope function

INTRODUCTION

With the increasing demand for accuracy on one side and shorter machining time on the other, we frequently have to consider the dynamic behaviour of position control loops and their servodrives on numerically controlled machine tools.

In general the position control loop of an NC machine tool has a structure as shown in Fig. 1. The position command signals are generated in the interpolator and compared with the position feedback signals in the position controller. The difference between both signals, i.e. the error signal, is amplified and supplies in the form of the velocity command signal, the velocity control loop, comprising the feeddrive[1,2].

In the block diagram (Fig. 2) of a position control loop the dynamic behaviour is described by three blocks. The first one stands for the relation between the velocity command signal and the error signal of the position control loop. This relation, called velocity gain k_v, is one of the main parameters of a position control loop. The following block contains the dynamic behaviour of the feed-drive and the last one contains the integration between velocity and distance, i.e. position.

The problem is to find out the optimal velocity gain in relation to the dynamic parameters of the feed-drive and, respectively, the velocity control loop.

For parameter optimization there exist many techniques in control theory[3]. Unfortunately the

* Institut für Steuerungstechnik der Werkzeugmaschinen und Fertigungseinrichtungen der Universität Stuttgart

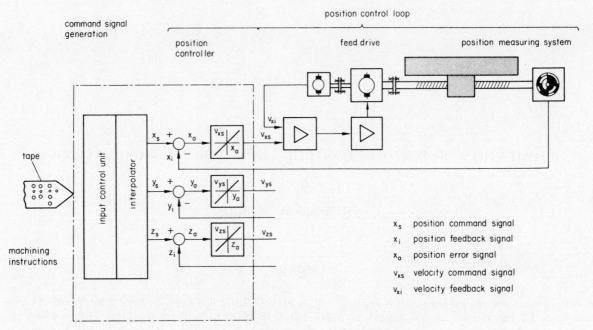

Figure 1. Block diagram of an NC machine tool.

x_s	position command signal
x_i	position feedback signal
x_a	position error signal
v_{xs}	velocity command signal
v_{xi}	velocity feedback signal

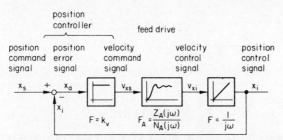

Figure 2. Block diagram of a position control loop.

conventional performance criteria do not provide results which are in good accordance with practical experience.

An index of performance, which seems more adequate to the special problem of optimizing position control loops, is now presented and discussed.

THE PERFORMANCE INDEX

For optimizing control loops we have many good optimization criteria, such as the integral of squared error criterion (*ISE*) or the integral of absolute error criterion (*IAE*) and similar integral criteria (Fig. 3). The base of these performance criteria is to minimize the area (control area) between the command signal (x_s) as the input signal and the controlled signal (x_i) as the output signal.

In the ideal case this control area becomes zero and we get an ideal transient response (Fig. 4a); it is to establish that this ideal transient response is directed at the performance criteria, mentioned above.

If we consider the specific problem at position controls, we see that in account of accuracy in machining a distortionless transient response, as shown in Fig. 4(b), is satisfactory. A time displacement between the output signal, i.e. the position control signal, and the input signal, i.e. the position command signal, causes no contour deviations at the workpiece. The time displacements must be, of course, identical in all machining axes.

According to the moving masses of a feed-drive,

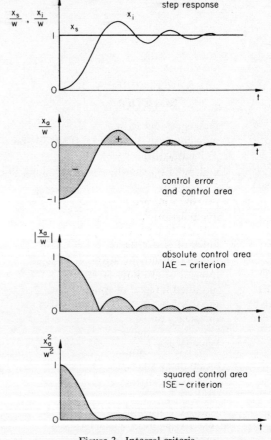

Figure 3. Integral criteria.

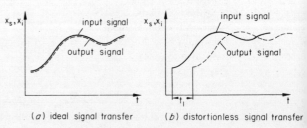

(a) ideal signal transfer   (b) distortionless signal transfer

Figure 4. Signal transmission.

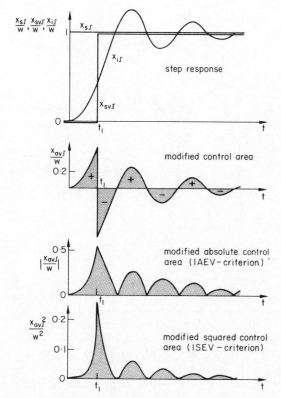

Figure 5. Modified integral criteria for a position control loop optimization.

the position control loop may never have a distortionless transient response. But an adequate performance criteria is found if the position control signal x_i is compared with a time-displaced 'virtual' position command signal x_{sv} (Fig. 5)[4]. So a modification of the usual performance criteria is obtained.

By squaring and integrating the difference between both signals we get the *ISEV* criterion. The integral of the absolute difference of both signals delivers the *IAEV* criterion.

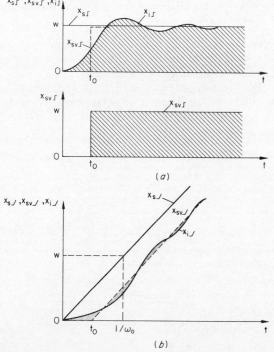

Figure 6. The meaning of the group delay time t_0 for step response and slope response.

In the ideal case, if the control area between the control signal x_i and the time-displaced command signal x_{sv} becomes zero, the transfer system would be really distortionless.

For the time displacement between the position control signal and the virtual position command signal the group delay time t_0 at the frequency $\omega \to 0$ is chosen.

The group delay time t_0 for position control loops is identical to the reciprocal value of the velocity gain k_v[4].

By this technique, it is also possible to apply the *ISEV* and *IAEV* criteria for step functions as well as for slope functions (Fig. 6).

THE OPTIMIZATION OF SOME STANDARD POSITION CONTROL LOOPS

A very simple position control loop is shown in Fig. 7. The dynamic behaviour of the servodrive is assumed now as a 'first order lag element' with the lag

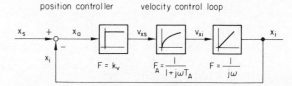

Figure 7. Block diagram of a position control loop. Velocity control loop (feed-drive) as a first order lag element.

time constant T_A or the corresponding natural frequency ω_0 being $1/T_A$ and the frequency response

$$F_A = \frac{1}{1 + j\omega/\omega_0}$$

The index of performance as a function of the velocity gain k_v, referred to the natural frequency ω_0, shows a minimum for $k_v \approx 0.8\ \omega_0$ (Fig. 8). The first formula for the relation between the dynamic

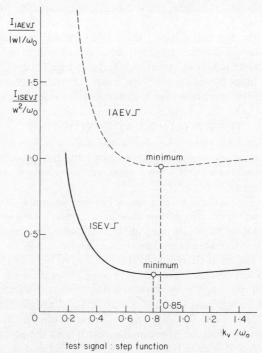

Figure 8. Index of performance as a function of the velocity gain (feed-drive as a first order lag element).

behaviour of the feed-drive and the gain of the position controller is now derived.

If the next step is calculated approximately, the dynamic behaviour of the servodrives has a second order lag element with the frequency response

$$F_A = \frac{1}{1 + 2D_0\,(j\omega/\omega_0) + (j\omega/\omega_0)^2}$$

we obtain for the *ISEV* index of performance a relief, as shown in Fig. 9. The lines deliver a constant index as a function of the damping ratio D_0 and the velocity gain k_v, referred to ω_0.

The minimum index of performance has values $D_0 \approx 0.5$ and $k_v \approx 0.4\,\omega_0$. So only half the velocity gain is achieved if the feed-drive is approximated by

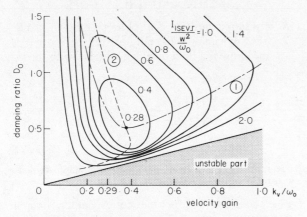

Figure 9. Lines of constant index of performance (feed-drive as a second order lag element).

a second order lag element instead of a first order lag element.

In addition if the relief has an increasing damping ratio the velocity gain has to be decreased.

By computation with a slope function as a position command signal similar diagrams are obtained.

On the other hand, the corresponding diagrams obtained by the generally used optimization techniques are very different to Fig. 8 and Fig. 9[4].

Many investigations with other transfer functions and less omissions have been successfully made[4,5].

THE DEVIATIONS AT CONTOURING

Contouring errors are also caused by the position control loop if the path vector changes its direction[1,2,6,7].

The influence on path deviations becomes very clear when contouring around a corner. An overshooting error $A_{\ddot{u}}$ and an undershooting error A_e (Fig. 10) have to be taken into account.

The path deviations often cause deviations in the workpiece contour; at a milling operation, for example on the workpiece inside contour, whereas the deviations at the outside contour may be neglected.

The undershooting error is measured as the shortest distance between the corner and the actual tool centre path. (On the contrary, the distance A_{ew}, measured along the bi-sector of the corner angle is often taken as the undershooting error. This distance

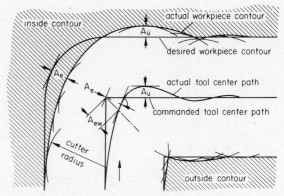

Figure 10. Undershooting error A_e and overshooting error $A_{\ddot{u}}$ at contouring around a corner.

has no relation to the workpiece tolerance bandwidth.)

By changing the velocity gain in relation to the dynamic parameters of the feed-drives the undershooting and overshooting errors change.

For position control loops with feed-drives corresponding to first order lag elements we obtain an overshooting and undershooting error, as shown in Fig. 11. The thick line A_{max} characterizes the maximum of both errors, respectively half of the

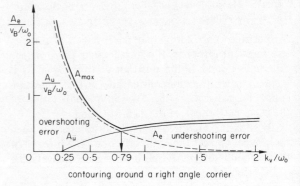

Figure 11. Path deviations as function of the velocity gain (feed-drive as a first order lag element).

tolerance bandwidth. A_{max}, as the function of the velocity gain, is similar in shape to the mentioned index of performance shown in Fig. 8.

For feed-drives with a dynamic behaviour of second order lag elements we obtain lines of constant A_{max}, depending on the damping ratio D_0 and velocity gain k_v/ω_0 (Fig. 12).

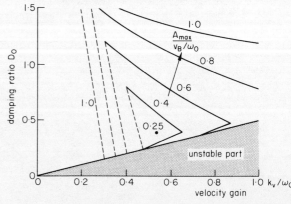

Figure 12. Lines of constant path deviations A_{max} (feed-drive as a second order lag element).

Comparing this relief with those obtained by the optimization (Fig. 9), an analogy can be stated.

CONCLUSIONS

We may recognize that the presented index of performance (*ISEV*) is suitable to optimize position control loops. Primarily, the resulting optimal parameters are mostly in accordance with practical experience. Secondly, the resulting index of performance, as functions of these parameters, is very similar to the corresponding functions of workpiece tolerance bandwidth, as these few figures have already shown.

REFERENCES

1. H. HEROLD, W. MASSBERG and G. STUTE. *Die numerische Steuerung in der Fertigungstechnik,* VDI-Verlag, Düsseldorf.
2. J. L. DUTCHER. *Servos and Machine Design for Numerical Control,* General Electric.
3. V. W. EVELEIGH (1967). *Adaptive Control and Optimization Techniques,* McGraw-Hill, New York.
4. D. SCHMID (1972). *Numerische Bahnsteuerung,* Springer-Verlag, Berlin, Heidelberg, New York.
5. D. SCHMID (1972). Dimensionierung von Lageregelkreisen bei Werkzeugmaschinen mit numerischer Bahnsteuerung, *wt-Z. ind. Fertig.* **62,** 554–9.
6. H. E. VIGOUR (1963). Effect of servomechanism characteristics on accuracy of contouring around a corner. *Trans. Amer. Inst. Electr. Electron. Eng.,* **82** (66), 120–4.
7. G. STUTE and D. SCHMID. Typische Konturfehler bei der Werkstückbearbeitung mit numerischer Bahnsteuerung, *Annals of the C.I.R.P.,* **18,** 531–40.

DISCUSSION

Q. H. P. Khong. In your paper you have only considered a linear feed drive and also only to optimize the velocity gain. My questions are, firstly, have you considered including non-linearities, such as static friction and backlash, in your study and the effect of these non-linearities on the results you have obtained? Secondly, have you considered optimizing the velocity loop compensation parameters, since the damping ratio of the velocity loop response affects the damping of the position loop?

A. In this short paper I could only present the main idea of the optimization technique which I have used and I have tried to demonstrate this with two simplified linear examples.

In answer to your first question: We have also considered more complex position control structures (see references 4 and 5) and have actually investigated non-linearities, like static friction and backlash.

Concerning your second question: With the performance index mentioned above you could take into account also the compensation parameters of the velocity loop in detail. But you cannot expect to obtain a brief and therefore practical formula for fixing the velocity gain, for example, if you try to take into account all the parameters which affect the position control loop.

MACHINE TOOL DESIGN

DESIGN CONSIDERATIONS FOR LIGHT-WEIGHT CONSTRUCTION: FATIGUE AND STIFFNESS OF FORMING AND CUTTING MACHINE TOOL STRUCTURES

by

R. UMBACH*

INTRODUCTION TO THE PROBLEMS OF LIGHT-WEIGHT CONSTRUCTION

General remarks on light-weight construction

For many years light-weight construction seemed to be reserved for aerospace projects, aeroplanes, and vehicles only. But nowadays the field of mechanical engineering also makes increasing use of light-weight construction, for instance in turbines, engines, pressure vessels, slide valves, etc.

Machine tools also exhibit certain features of light-weight design.

To obtain light-weight structures it is necessary to make use of classical design methods, based on static deformation and stress calculations including modern computer-aided design methods.

Success and failure of light-weight design depends on a balance between weight and costs on the one hand, and service life and safety on the other. Light-weight construction does not so much refer to the absolute weight of a structure as to the utilization of strength inherent in the material. This requires a knowledge of the load spectrum to be expected, the type of loading, and the internal force distribution.

To illustrate the point, a heavy duty metal forming press of 16 MN having a weight of about 180 tonnes can be designed in accordance with the principles of light-weight construction, very much as a railway truck having a weight of only 14 tonnes is capable of carrying additional goods weighing 25 tonnes. The specific differences between these two designs are not due to basically different design, but only to greatly differing working loads.

There are different types of light-weight structures, for instance cellular and box type structures, shell type structures, and tubular frames. Additional design features concern specifically the geometry of the cross-section, the wall thickness, and the number and shape of reinforcing ribs. Initially, light-weight design is not a problem of the material (be it for instance cast iron, globular cast iron, or welded steel design). Nevertheless there must be some knowledge available about the material to be used and its properties. Shape and strength are interrelated. When

this fact is disregarded, damage often occurs. Such is the case if insufficient attention is paid to shape and notch effects (slots, grooves, holes, etc.).

The designer has to take into account 'flow lines' of stress which are distributed so as to follow the shortest possible path. Thus sudden changes of cross-section should be avoided.

Types of loading and design features

The design of a structural member is determined by its use. There are three principal cases:

(1) The structural member is to be designed with respect to stiffness for which shape is the criterion. From the viewpoint of strength the member may then be overdimensioned. Several members in cutting machine tools, e.g. the bed and the column, belong to this group.

(2) The structural member is to be designed with respect to strength. Deformations must remain within allowable limits. These demands must be met with, particularly in the case of high performance vehicles.

(3) The structural member is to be designed with respect to both stiffness and strength. Forming machine tools, such as presses, belong to this group.

For those parts which are designed on the basis of stiffness, the dynamic behaviour is of special importance. This is the case with chatter in cutting machine tools—an important area in machine tool research.

Because of the high accuracy required on machined workpieces, and the low cutting forces when compared to forming, the design of cutting machine tools is based on stiffness, taking also thermal effects into consideration.

For those structural components which are to be designed on the basis of strength, loading, both static and fatigue, is the essential criterion[5,6].

Figure 1 shows the basic types of load spectra in relation to S–N curves. In the first case, the load is generally sinusoidal. The maximum stress amplitude is below the fatigue limit and the cumulative number of cycles is greater than 10^6.

---

* Rheinstahl AG, Forschungszentrum, Kassel

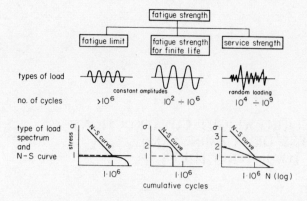

Figure 1. Strength based on stress spectrum[9].

In the second case, the stress amplitudes are greater than the fatigue limit; the number of load cycles lies in a relatively narrow finite life region, say between 10^2 and 10^6.

The third case represents random loading. Stress levels which occur very frequently are below the fatigue limit, whilst the highest stresses exceed the fatigue limit by a factor of about two and are stochastically distributed.

Fatigue resistance is mainly a question of the following factors: properties of material, design features of the construction, manufacture (quality of machining or welding, etc.) and surface condition, environmental influences, and types of load spectra. It should be mentioned that all these factors are not mutually independent. Quasi-static or dynamic loads may cause fatigue, resulting in cracks and fracture. In many cases such cracks reduce the strength and thus the fatigue life of the member concerned. Numerous failures of widely different structures are known, e.g. vessels, bridge structures, pressure vessels, etc. This is not different in the case of frames of presses and forming tools. Cutting tools for turning, drilling, and milling operations are equally prone to danger.

In Fig. 2 an example is shown of fatigue cracks between draw rod and tup of a 25 Mpm forging hammer. The rate of crack propagation was kept under observation.

In Fig. 3 the crack growth is plotted against the number of load cycles and shows good agreement between calculation and measurement. This is a case for 'how to live with cracks by taking calculable risk'. Additional information concerning fatigue strength was obtained in this case using fracture mechanics.

As mentioned before, every component is subjected to a definite load spectrum. The distribution of stress amplitudes and thus the type of spectrum determines the number of cycles of load. An example for this is given in Fig. 4 of a simple welded joint[10]. For several different spectra the S–N curve is recorded, showing a probability of failure of 90 per cent. With a decreasing proportion of the large stress amplitudes in the spectrum the number of cycles to failure increases, i.e. the S–N curve is shifted to the right.

All these results were obtained using conventional fatigue testing machines. Curve 'a' refers to a constant amplitude test identical to the S–N test. For testing

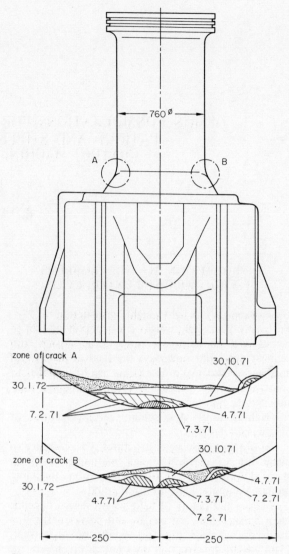

Figure 2. Fatigue cracks between draw rod and tup in a 25 Mpm forging hammer.

according to the remaining load spectra a block programme loading is employed.

Today, servohydraulic closed loop fatigue testing actuators enable new test methods to be applied, e.g. duplication of the service loading through magnetic tape control, randomized programme and random process tests.

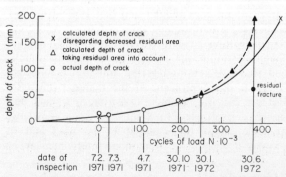

Figure 3. Rate of crack propagation referring to forging hammer in Fig. 2.

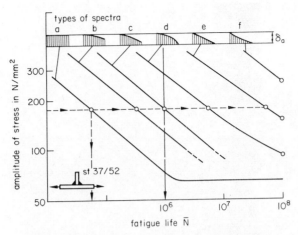

Figure 4. Fatigue life as a function of loading spectrum.

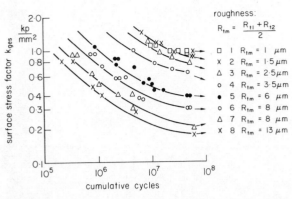

Figure 5. S–N curves for gears with different flank roughness[17].

For each individual case due consideration should be given to the suitability of the test; particularly so in case of more expensive and complicated design members. Often only a single specimen is available for testing. In such cases no statistical information can be obtained.

If there is just one single part available, as much information as possible must be extracted from it. Here the following aids can be used: theoretical analysis, static and dynamic tests, fatigue test, scale model tests, and the analogue computer.

Information obtained by using these aids may reduce the risk of failure and give greater assurance to the fatigue test result.

INTERDEPENDENCE OF STRENGTH AND GEOMETRY IN MACHINE TOOL DRIVES AND HYDRAULIC SYSTEMS

Several drive mechanisms in forming and cutting machine tools are highly stressed. This applies particularly to the following components: gears, shafts, crank shafts, spline shafts and keyways, clutches, brakes, lead screws, ball screws and anti-friction bearings[14,15].

Main drives and feed drives of cutting machine tools must exhibit appropriate bending and torsional stiffness. The stability of a machine tool will mainly depend on the behaviour of the drives[16].

In some cases this must be synthesized if it cannot be derived from a similar working unit.

The main factors influencing fatigue life of gears are applied loads, working accuracy, surface roughness, notch effects, heat treatment, and errors due to faulty assembly.

Figure 5 gives an example of the effect of surface roughness of the flanks of teeth, on fatigue strength. The surface stress factor is plotted against the number of cycles.

Hydraulic elements in machine tools do not always prove to have satisfactory fatigue strength. In most cases the reason may lie in an underestimation of the load spectrum applicable to these elements. Sometimes small corrections or changes in shape will lead to satisfactory fatigue strength.

The revolving cylinder in Fig. 6 turns a turret head through 180 degrees in less than 0·5 seconds. The rated pressure in the hydraulic system is about 70 bar. By a dynamic effect the pressure rises to 100 bar or more.

Too high a stress concentration factor α_k and notch factor β_k of the cylinder bushing (which is fixed in the outer cylinder and itself slotted to enable the revolving movement of the piston), was the reason for fracture. Larger radii and a smaller axial clearance of the bushing within the covers of the outer cylinder lead to a stress below the fatigue limit.

The power amplifier in Fig. 7 works as a clamping cylinder with a work cycle pressure of zero, 70 and 350 bar. The work cycle is repeated several times per minute. Cracks occurred for case 'B' after a relatively short duty time. The borehole for the oil supply was then altered as shown at 'A' involving a larger radius.

As a last example in this context a hydraulic control block is shown in Fig. 8. The visible crack at the front has its origin in a keyway-like groove, which runs under an angle into the left hole in order to provide a greater flow rate. The designer has paid no attention to the notch effect so created.

LIGHT-WEIGHT CONSTRUCTION IN CUTTING MACHINE TOOLS

As mentioned before, members of cutting machine tools are designed mainly on the basis of stiffness and stability. Thus, deflection and deformation of all components along the line of action of forces should be a minimum.

At the end of roughing, the workpiece should be as accurately-sized as possible.

The surface quality will depend on the static and dynamic behaviour of the machine tool. Cutting forces are far from constant, their distribution being a random function. Peklenik[21] gave explanations for this important fact. According to the type of load spectrum, the surface roughness is also stochastically distributed. The use of correlation techniques made the investigation of these interrelations possible: these techniques are already well known and indispensable in fatigue problems in aircraft and vehicles, for reducing weight on the one hand and to guarantee an appropriate safety factor on the other.

The total compliance of a machine tool is given by the load and the compliance of the single components

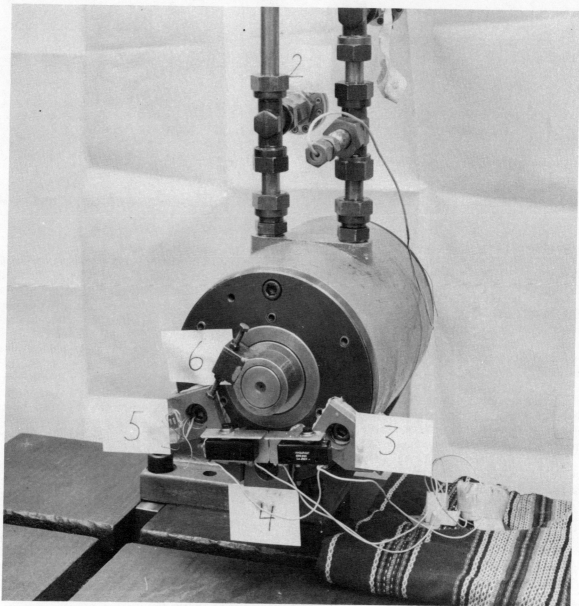

Figure 6. Rotary cylinder of a turret head.

(e.g. headstock, column, bed, etc.) along the line of action of the force. In many cases the foundation must be taken into consideration as well. In general, for every single component the loading can be reduced to bending moments or torques. Stiffness is, of course, affected by the position of tool and workpiece and the elastic properties of the material used: thus the maximum bending and torsional moments are fixed mainly by the working area of a machine tool, whereas the modulus of elasticity and the shear modulus are given by the chosen material. The designer has control over the rigidity of a component by fixing the second moment of area. This he obtains by selecting as large a cross-section as possible. In principle, this requirement is independent of the material, although certain properties may be of importance. This means that primarily light-weight construction is not tied to cast iron or steel. Nevertheless, the following factors should be considered:

(1) The value of the modulus of elasticity and its constancy, e.g. the change of modulus with wall thickness in the case of castings.

(2) The wall thickness and its constancy. For cast iron there seems to be an interrelation between minimum wall thickness and size of the component[23].

(3) The damping capacity of the material. In this respect the use of cast iron is advantageous, but it must be emphasized that the total damping ratio of a machine tool is about one to two orders higher than the damping capacity of the material.

With regard to casting the use of nodular iron with its higher modulus of elasticity pays dividends from the stiffness point of view. Only in welded structures can almost any wall thickness be realized. However there is a limit, because the wider the unrestrained area of thin steel plates, the more ribs must be added for stiffening to avoid forced vibrations occurring in the structure. As a matter of fact weight reduction is partly neutralized by the added weight of the ribs. Consequently, the rigidity of such a member must be related to its total weight. Several investigations were made to determine appropriate arrangements of ribs

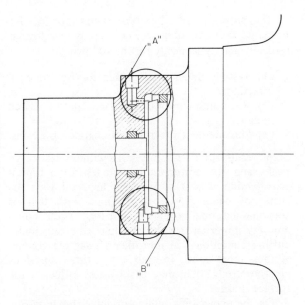

Figure 7. Power amplifying clamping cylinder ('A' improved design, 'B' original design which failed).

Figure 8. Fatigue crack in hydraulic valve block.

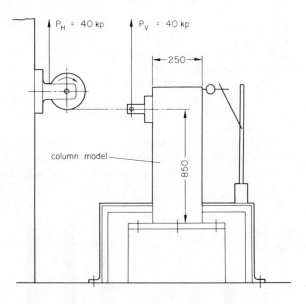

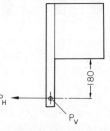

Figure 9. Test rig for static analysis of a box type column for a horizontal boring and milling machine[24].

TABLE 1 Comparison of rigidity ratios for a column with different cross-sections

model	sectional shape	weight ratio W (%)	rigidity ratio (%)		rigidity ratio per weight (%)	
			S_H	S_V	S_H/W	S_V/W
nr. 1		100	100	100	100	100
2		125	104	109	83	87
3		170	481	211	283	124
4		129	220	140	158	101
5		216	383	190	177	88
6		207	488	185	236	89
7		204	494	205	242	100
8		208	494	217	238	104

in relation to shape and size of a member and the loading case.

An example is given in Fig. 9 of a box type column in an open-sided boring and milling machine, showing the test rig for loading the column in two directions relevant to this type of machine tool. P_H acts in the horizontal direction on a cantilever arm, which stands on the milling head, P_V in the vertical direction on the same arm.

The more important results from these investigations are combined in Table 1 for eight different sectional shapes giving the ratio of rigidity per unit weight and the ratio of rigidity for both loading cases. The sectional shapes 4 to 7 refer to an internal counterweight type of column.

The longitudinal and horizontal ribs are arranged in such a way as to prevent sectional deformation, i.e. to maintain linearity of the sides and orthogonality of the corners.

An additional contribution to the recurrent controversy concerning the material is illustrated in Fig. 10 including approximate values for the main factors

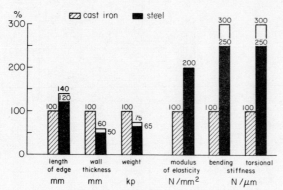

Figure 10. Stiffness comparisons concerning material and cross section of a box-type column[22].

of influence on the stiffness of box type sections. According to de Jong[22] the practical advantages of welded structures rest on certain premises. These include the co-operation of machine tool designers, steel construction engineers, and welding experts, and the use of semi-finished steel (bars, plates, etc.) and straight box-type sections (without slant planes, etc.).

The previously reported examples refer mainly to the static stiffness of machine tool members. However, the dynamic behaviour must also be taken into consideration[27]. This depends not only on the static stiffness, but also on:

(1) Ratio of masses.
(2) Ratio of frequencies.
(3) Ratio of damping.

The damping effect of a machine tool is determined mainly by the influence of relative vibration amplitudes in ways and joints and, of course, is not always proportional solely to the vibration velocity. Moreover in joints, under certain conditions, the system may change from a linear to a non-linear behaviour. This process is explained in Fig. 11[25].

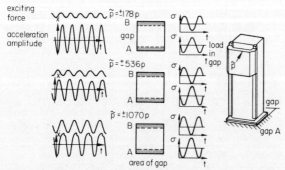

Figure 11. Linear to non-linear transition in a joint plane of a machine tool column.

The prototype of a box-type column for transfer machines corresponding to Fig. 12 was investigated statically and dynamically for the following items:

(1) Testing under static loads in the principal directions.
(2) Recording curves of resonance and of related modes.
(3) Measurement of cross-sectional deformations.

Some results are shown in Fig. 13 with the resonance peaks and the corresponding modes for a typical cross-section. The first mode is identical with the shape of bending, the third mode with that of torsional deformation under static load. These modes are very important for the dynamic behaviour under cutting conditions. In the other modes the column vibrates more or less in itself, but if these vibrations are too strong, they may nevertheless influence the surface finish.

On the planer type milling machine shown in Fig. 14 a static and dynamic analysis on the path of action of forces was carried out, the results of which should be taken into account for new designs. Figure 15 depicts the input and output instrumentation for dynamic tests. The milling machine was excited by a servohydraulic seismic force exciter using random signals from a random noise generator.

The output signals were stored on tape and analysed in the laboratory by a correlator and computer. Curves of amplitude and phase angles are the result, and thus compliance response loci.

The vibration modes were determined in the usual way. The input signal was a single sinusoidal frequency with constant amplitude. The corresponding output signals were picked up as amplitudes and phase angles at about 200 measuring locations. One of the modes so attained is shown in Fig. 16; the natural frequency was 55 c/s, the dominating vibrating components were the milling head, cross slide, and cross rail. The diagram in Fig. 17 demonstrates the dynamic behaviour during a rough milling operation on a cast iron block on the left of the figure, and on steel on the right. In both cases the position of the cross rail was fixed at a height of 1150 mm above the table of the milling machine. The dominant vibration frequency is plotted against the cutting frequency (number of cutter blades times rev/min of cutter).

During rough cutting on cast iron the dominant vibration frequency was usually close to the cutting frequency, whereas on steel it was almost independent of the cutting frequency. One or two of the dominating modes of the milling machine were excited in the latter case.

LIGHT-WEIGHT CONSTRUCTION IN PRESSES

Finally some examples are given which should indicate the significance of light-weight construction in presses. The importance of forming machine tools rises constantly. This has a special bearing, for instance, on hot and cold extrusion. In the latter case unheated workpieces of steel or other metals will be formed with high accuracy into finished or nearly

Figure 12. Welded box-type column for transfer lines.

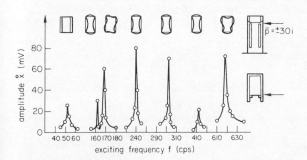

Figure 13. Resonance peaks and vibration modes in a fabricated box-type column.

completed components of various and often complicated shapes. Such operations not only require a high load capacity, but also a press frame that is sufficiently stiff and has high fatigue strength.

Because the load capacity of heavy duty presses ranges from about 1 MN to some 10 MN and because of the geometrical sizes, successful light-weight construction is highly desirable.

Attention must be paid in the design stage to the following factors:

(1) The load capacity.
(2) The working area, which has to carry full load

or a defined partial capacity, generally a circle or an ellipse.

(3) The mechanical presses have often to withstand the full load capacity 30 or 45 degrees ahead of the bottom dead centre.

(4) The dynamic peak load can exceed the static loading by 20 per cent or more.

(5) The load spectrum to be expected.

Figure 18 shows a hydraulically driven 15 MN press for hot extrusion. After a few years of service several cracks were found, one of which is seen in Fig. 19. Operating the press over a considerable period of time with three dies for three successive passes, these were mounted with large eccentricities on the bolster plate. Generally one or two of them received full load. Owing to these cracks, the load capacity of the press had to be limited to a maximum of about 5 MN. Because of the importance of this press for the above operations, several design alterations were proposed and model tests carried out. As a result the press was modified as follows:

(1) Additional ribs were inserted in the critical upper part; their shapes are shown in Fig. 20 together with tension stresses due to stress concentration in the corner;

(2) changed shape between column and table;

(3) four tie rods installed.

Figure 14. Planer type milling machine.

Figure 15. Instrumentation for dynamic analysis of a planer type milling machine.

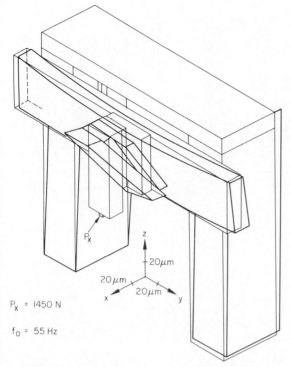

Figure 16. Vibration mode of cross slide system in a planer type milling machine.

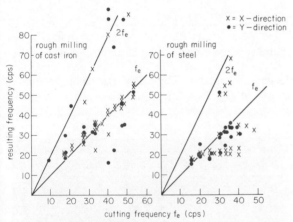

Figure 17. Dynamic behaviour during rough cutting operation of a planer type milling machine.

Figure 18. 4-column hydraulic press for 15 MN.

Figure 19. Fatigue cracks between table and column of a 15 MN hydraulic press.

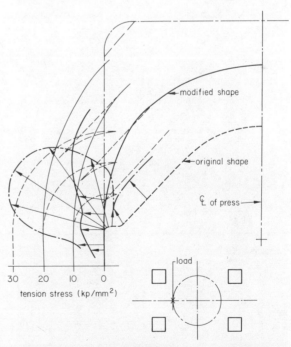

Figure 20. Effect of additional ribs on stress level.

With these modifications carried out, the press was again usable to full capacity. The stresses in the critical parts were measured by strain gauges and found to be within allowable limits.

Model testing in the early design stage is an essential aid for the design of presses. With the help of similarity transformation the actual behaviour of the press can be predicted[34].

Model tests for machine tool components are generally based on the law of elasticity (Hooke) and on Newton's law of motion. In the similarity equation for designing models, a linear scale for length is generally assumed (usually a scaled down version). The two other scale sizes, namely load and time, are directly connected by equations in combination with the above laws. All the influencing factors cannot always be taken into account to afford a complete degree of similarity. In some cases this is not even necessary.

Model tests enable quantitative predictions to be made in the design stage of the static, dynamic, and thermal behaviour. It is much cheaper to alter some sections of a model, if these are not satisfactory, than to have cracks or a breakdown later in service. This not only saves money but avoids loss of prestige too.

Further ease and economy in the use of models result from the fact that they can be made in dissimilar materials, provided that the model laws are adhered to. In this respect plastic materials are very useful.

Figure 21 shows the model of a straight-sided double crank press for 4 MN. These model tests were carried out during the design phase. The model was scaled down in the ratio 4 to 1. Critical locations for strain distribution were ascertained using 'stress coat' lacquer. Crack line density and orientation deter-

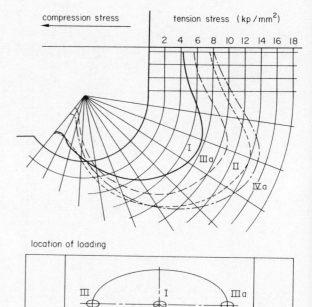

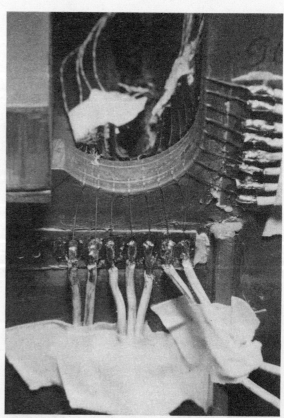

Figure 22. Stresses at the location 'column/table' of the model shown in Fig. 21.

mined the positioning of strain gauges. The results of these model tests for loads applied at points of an ellipse can be seen in Fig. 22 referring to a critical section of the press. The highest stresses occur when loaded at location 'IVa'. This particular section is shown in Fig. 23, with a strain gauge chain applied to it.

Figure 21. Model of a 4 MN straight-side double crank press.

Figure 23. Chain of strain gauges applied to location shown in Fig. 22.

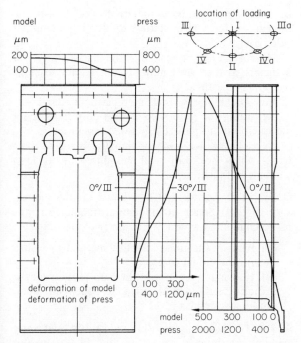

Figure 24. Deformation under static load of the press model shown in Fig. 21.

Figure 24 shows the distribution of deformation measured by means of dial gauges.

Valuable experience was gained with model tests of presses. In a few cases in connection with acceptance tests it was possible to check the results from model tests by measurement on the actual press. The results diverged by not more than 10 to 20 per cent, depending on the location of the load, which is certainly quite acceptable.

REFERENCES

1. A. LEYER (1972). Leichtbau als letzte Phase des Maschinenbaues, *Technica*, Nr. 24, Basel.
2. A. LEYER (1964). Kraftflußgerechtes Konstruieren, *Konstruktion*, **16** (10).
3. J. BROICHHAUSEN (1972). Strukturelles Versagen von Konstruktionselementen, *VDI-Z.*, **114** (4).
4. R. UMBACH (1972). Die Bedeutung von Gestalt- und Betriebsfestigkeitsuntersuchungen für die Konstruktion, *Rheinstahl Technik*, 1/72, Essen.
5. W. SCHÜTZ (1971). Werkstoffoptimierung für schwingbeanspruchte Bauteile, *Z. f. Werkstofftechnik*, **2** (4).
6. T. HAAS (1962). Loading statistics as a basis of structural and mechanical design, *Engineers Digest*, March, April, and May.
7. R. JARAUSCH (1972). Die Bedeutung bruchmechanischer Untersuchung für die Bauteilsicherheit, *Rheinstahl Technik*, 1/72, Essen.
8. H. HAHN. Spannungsverteilung an Rissen in festen Körpern, *VDI-Forsch.-Heft*, **542**.
9. W. SCHÜTZ and H. ZENNER (1973). Schadensakkumulationshypothesen zur Lebensdauervorhersage bei schwingender Beanspruchung, *Z. f. Werkstofftechnik*, **4** (1).
10. E. HAIBACH (1971). Probleme der Betriebs-

festigkeit von metallischen Konstruktionsteilen, *VDI-Z.*, **113**.
11. G. SCHULZE (1972). Einfluß des Schweißens auf das Zeit- und Dauerfestigkeitsverhalten von zwei schweißgeeigneten, niedriglegierten Vergütungsstählen mit Streckgrenzen über 600 N/mm², *Konstruktion*, **24** (12).
12. O. BUXBAUM (1966). Statist. Zählverfahren als Bindeglied zwischen Beanspruchungsmessung und Betriebsfestigkeitsversuch, Labor. für Betriebsfestigkeit, Darmstadt, Bericht Nr. TB65.
13. E. HAIBACH (1970). The allowable stresses under variable amplitude loading of welded joints, *Proceed. of Confer. on Fatigue of Welded Structures*, Brighton. The Welding Instit., Abbington, Cambridge.
14. H. PITTROFF (1965). Die Belastbarkeit von Wälzlagern, *Technica*, Nr. 9 and 10.
15. R. MUNDT and H. PITTROFF (1963). Riffelbildung bei Wälzlagern als Folge von Stillstandserschütterungen, *VDI-Z.*, **105** (26).
16. R. JARAUSCH and H. MARDER (1962). Berechnung erzwungener gedämpfter Drehschwingungen von Getrieben mit Hilfe elektronischer Rechenmaschinen, *Industrie-Anzeiger*, Nr. 63.
17. H. OPITZ and W. KALKERT (1965). Der Einfluß der Fertigungsgenauigkeit und der Schmierfilmausbildung auf die Flankentragfähigkeit, *Forschungsberichte des Landes Nordrhein-Westfalen*, Nr. 1476. Westdeutscher Verlag, Köln.
18. H. GROSS (1973). Ermittlung von Belastungskollektiven an installierten Getrieben, *Industrie-Anzeiger*, **95** (16).
19. A. A. BARTEL (1973). Stand und neue Entwicklungserfolge der Getriebeschmierung, *Maschinenmarkt, Forschung und Konstruktion*, **79**, 1.
20. W. FUNK and H. KREISKORTE (1973). Fatigue fractures as reason for failure of pressure vessels operating under cycling loads, *3rd Internat. Congress on Fracture*, Munich, VDEH-Verlag Düsseldorf.
21. J. PEKLENIK and T. MOSEDALE (1968). A statistical analysis of the cutting system based on an energy principle, *Proc. of the 8th Internat. MTDR Conference*, Manchester, 1967, Pergamon Press, Oxford.
22. H. DE JONG and P. FRECKMANN (1972). Neue Entwicklungen im Schwerwerkzeugmaschinenbau durch Anwendung der Stahlbauweise, *Industrie-Anzeiger*, **94** (77).
23. V. V. KAMINSKAYA et al. (1964). Bodies and Body Components of Metalcutting Machine Tools, National Lending Library for Science and Technology, Boston Spa, Yorkshire.
24. K. INOUE, T. MATSUMOTO and T. TSUJIMOTO (1972). A consideration on the rigidity of columns of machine tools. *Technical Review*, Mitsubishi Heavy Industries, October.
25. H. KRUG (1958). Leichtbau von Werkzeugmaschinen. *VDI-Berichte*, **28**.
26. A. HEISS (1950). Schwingungsverhalten von Werkzeugmaschinen-Gestellen, *VDI-forschungsbericht*, **429**.
27. R. UMBACH (1963). Dynamic behaviour of machine tool structures, *Proc. of the 3rd Internat. MTDR Conference*, Birmingham, 1962, Pergamon Press.

28. F. W. DREYER (1966). Über die Steifigkeit von Werkzeugmaschinen-ständern und · vergleichende Untersuchungen an Modellen. Dr.-Ing. Dissert., TH Aachen.
29. W. H. GROTH (1972). Die Dämpfung in verspannten Fugen und Arbeitsführungen von Werkzeugmaschinen, Dr.-Ing. Dissert., TH Aachen.
30. R. H. THORNLEY, R. CONOLLY, M. M. BARASH and F. KOENIGSBERGER (1964). The effect of surface topography on the static stiffness of machine tool joints, 5th Internat. MTDR Confer. Birmingham, *Int. Journal of Mach. Des. Res.,* 5.
31. K. LOEWENFELD (1955). Die Dämpfung bei Werkzeugmaschinen, 2. Fokoma München, Vogel-Verlag.
32. R. UMBACH (1966). Problems of stiffness and accuracy of large size machine tools, *Proc. of the 6th Internat. MTDR Conference,* Manchester, 1965, Pergamon Press.
33. H. OPITZ and M. WECK (1970). Determination of the transfer function by means of spectral density measurements and its application to the dynamic investigation of machine tools under machining conditions, *Proc. of the 10th Internat. MTDR Conference,* Manchester, 1969, Pergamon Press.
34. R. UMBACH (1962). Modellversuche als Hilfsmittel für die Werkzeugmaschinen-Konstruktion, *Industrie-Anzeiger,* Nr. 46.

DISCUSSION

Q. J. Peters, Leuven. May we ask you why you preferred to use the techniques of material modelling instead of computer aided design?

A. Until now we have preferred model tests, although we have now started to develop computer aided design.

Moreover, the model is more satisfactory for designers, as they get a better feeling for the machine part and its results.

For these model tests we used plastic material.

Regarding the results we are not only looking for deformations but particularly for stresses too.

Nevertheless, we hope to find the point of time, from which we only will compute, or at least combine both methods.

Q. H. Sato. Author had mentioned on the importance of the increase of the natural frequency of the torsional mode. I would like to ask two questions:

(1) What is the most effective method to realize this for the model analysed?
(2) How much increase of the natural frequency is sufficient, although it might be dependent upon the total system?

A. (1) To add suitable ribs and a top plate and avoiding apertures as far as possible. At least reduce these areas.

(2) Generally it is not so important to increase the natural frequency but to reduce the amplitudes. But the natural frequency may be of great importance on the dynamic behaviour of the total system, which is determined by:

(*a*) Ratio of masses of the combined parts (by bolted joints or slideways).
(*b*) Ratio of natural frequencies of these parts.
(*c*) Ratio of damping.

THE INFLUENCE OF THE CHARACTERISTICS OF MACHINE TOOL GUIDEWAYS CONCERNING THE DYNAMIC BEHAVIOUR OF MACHINE TOOL SLIDES

by

G. HAJDU*

SUMMARY

The vibrations of machine tool slides are dependent to a large extent on the peculiarities of the guideway system carrying the slide. This problem was analysed by applying a practically controlled mechanical–mathematical model and through its calculation by an Elliot-computer. Results are published showing design principles for convenient stiffness and damping relations of the guideway system and for suitable arrangement and measured ratios of this system. The analysis takes in consideration the different critical vibration directions at the diverse sorts of machine tools. A brief summary will be given concerning the influencing factors of guideway stiffness and damping.

INTRODUCTION

The slides of the metal-cutting machine tools moving in sliding or rolling guideways can effect various vibrating motions under the influence of a certain vibrating force. These motions can be classified in the following two fundamental groups:

(1) vibration of the slide, considered as an elastic body,
(2) vibration of the slide, considered as a rigid body.

The bending and torsioning vibrations, depending on the construction of the slide, can be placed in the first group. The extent of the vibrations can be influenced by selecting the slide and its supports, as the parameter of an elastic system.

The construction of the slide, has no influence—in case of the appropriate rigidity—on the vibrations of the second group, their formation and their extent will be influenced exclusively by the mass of the slide (inertia) the location of the supports, as well as by their stiffness and damping.

From the point of view of the machine tool working precision, the vibrations of the second group, effecting relative vibrations between the workpiece and the tool, perpendicular to the worksurface, are of great importance in the first instance. These vibrations influence the form-precision and the surface roughness of the products, made by machine tools. In Fig. 1 it is shown which coordinate direction they are critical, considering different types of machine tools. Our tests, carried out on an experimentally devised

mechanical–mathematical model, are aimed at the analysis of the factors influencing these vibrations[5].

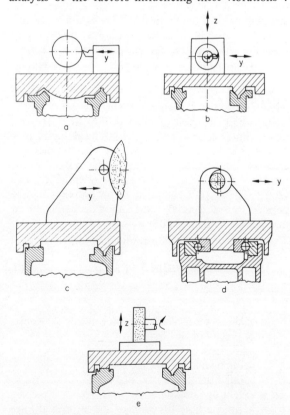

Figure 1. The critical directions of relative motions appearing between the tool and the work in case of: (*a*) turning, (*b*) fineboring, (*c*) cylindrical grinding, (*d*) internal grinding, (*e*) surface grinding.

* 'SZIMFI' Development Institute of the Machine Tool Works, Hungary, Pest megye.—H-2314 Halásztelek

Q

MECHANICAL-MATHEMATICAL MODELLING

In our findings the slide was replaced by a model having three degrees of freedom, as shown in Fig. 2. The slide is shown as a rigid body on the model, its symbolical mass is supported by three springs, loaded similarly by three dampers. The guideways can be simulated by a model acting along their lines of influence. In the case of flat and prismatic guideways, they can be modelled by a vertical spring and a

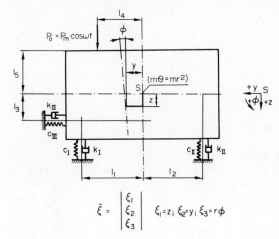

$$\bar{\xi} = \begin{vmatrix} \xi_1 \\ \xi_2 \\ \xi_3 \end{vmatrix} \quad \xi_1 = z_i \; \xi_2 = y_i \; \xi_3 = r\phi$$

Figure 2. Mechanical-mathematical model to examine the vibrations of slides.

vertically acting damper, while the prismatic guideways can be replaced by a spring respectively damping pairs of the vertical and horizontal influence lines. In the case of roller guideways, the two-way springing and damping can be modelled as a spring, respectively damping components of the guideway system.

When arranging the mass factor, the inertia relationships and the characteristic constants of the springs and dampers replacing the supports (assuming linear spring characteristics and damping proportional to velocity) into properly selected matrices and describing the force acting on the system, in the form of a harmonic function by a variable force vector. The motion equations of the system could be defined by a second order inhomogeneous linear matrix differential equation.

After adequate simplification we obtain:

$$\hat{M}\bar{\xi}'' + 2D\hat{K}\bar{\xi}' + \hat{C}\bar{\xi} = \bar{p} \tag{1}$$

where

$\bar{\xi}$ = general vector of motion interpreting the centre of gravity of the system,

$\hat{M}$ = mass matrix,

$\hat{K}$ = damping matrix,

$\hat{C}$ = spring matrix,

$2D$ = extent of damping,

$\bar{p}$ = harmonic time function of the impressed force vector,

(the prime means differentiating in accordance to the time scale without dimensions).

From the solution of the matrix differential equation system, we can determine the eigen fre-

quencies of the system, as well as the amplitudes of the forced vibrations at any frequency, referred either to the centre of gravity or to any triple coordinate system defining the motion of the body[1,2,3,4].

The latter affords the possibility of evaluating the vibrations of a given machine tool slide, from the point of view of the motion of its critical points, e.g. to test and determine the vibration amplitudes of a tool mounted on the slide, or the amplitudes of the toolpoints in different directions.

To work out the mechanical-mathematical model, a program was adapted for an Elliot-803 computer; in case of feeding a preselected combination of parameters, produces the data of the resonance characteristics related to the components of the motion vector $\bar{\xi}$ as well as the data of the respective phase shift curves for the whole resonance range. The points of the curves (which can also be made graphically by plotter), will be produced in adequately frequent steps (Fig. 1 and 3)[4].

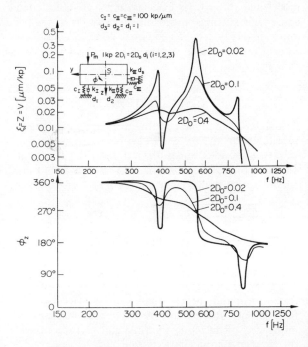

Figure 3. Calculated resonance and phase curve.

Experimental tests were carried out—not reported in detail in this paper—to control the model and the elaborate computer program by using an experimental slide mounted with adequate statistical and dynamical measuring transducers constructed specially for this purpose. The frequencies and amplitudes of the vibrations measured during the tests, carried out both with rolling and sliding guideway pads, corresponded with a good approximation to the values calculated by a mathematical model[5].

ANALYSIS OF PARAMETER EFFECTS

Using a lot of initial data, calculations were made by computer to examine the effects of traverse forced vibrations of the construction parameters, which are difficult to modify experimentally. The modified parameters were as follows:

(a) The extent and ratios of the stiffness of the supports or guideways.
(b) The extent and ratios of the damping of the supports or guideways.
(c) The distance of the guideways measured from each other.
(d) The height of the centre gravity of the slide above the planes of the guideways.
(e) The position of the point of application of the exciting force.
(f) The mass of the moving slide.

To evaluate the results of the calculations, the discussion of the so-called vibration form of the respective 3–3 natural vibrations—appearing according to the 3 degrees of freedom—by the systematizing figure

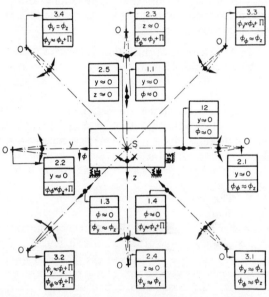

Figure 4. Systematizing of slide vibrations according to their centre of rotation in static condition.

shown in Fig. 4, was carried out. The vibratory motion of the slide is simulated here as a rotary vibration around a single fixed point[2]. The small tables of the figure show the conditions of the four possible translatorial motions (1.1 . . . 1.4) as well as of the five presumable special rotary motions (2.1 . . . 2.5) and finally of the four possible general rotary motions (3.1 . . . 3.4) inscribed.

The vibrations calculated by using the different combinations of parameters, were evaluated from the point of view of the following three machine groups:

(a) machine tools where the vibrations of forms 1.2, 2.3, and 2.4, containing large horizontal components, could become critical in respect of the surface quality of the workpiece, e.g. lathes for fine turning, cylindrical and internal grinding machines.
(b) machine tools, where the vibrations of form 1.1, 2.1, and 2.2 having large vertical components, can be dangerous, e.g. surface grinders.
(c) machine tools, in which the vertical and horizontal vibration components harmfully influence the quality of the machined surface, e.g. fineborers.

Vibrations of form 1.3, 1.4, 2.5, 3.1, 3.2, 3.3, and 3.4 are harmful and dangerous for all types of machine tool.

The results of the evaluation can be summarized as the following principal items.

(I) Generally only the two natural vibrations of lower frequencies are worthy of attention. The amplitudes of the natural vibrations of the highest frequency are mostly of negligible small values.
(II) In the case of sliding guideways and highly prestressed rolling guideways $p > 150$ kp/cm a specific roller loader having damping values of $2D > 0.25$. The vibrations transverse to the guideway direction, proved to have negligibly small values according to the tests. But with slightly prestressed roller guideways significant forced vibrations were found.
(III) The results of the tested combinations of guideway stiffness are shown in Table 1, where the occurrence of 'mark +' indicates the appearance of significant forced vibrations. It is apparent from the table that the stiffness condition of the guideway system ($C_I \approx C_{II} \approx C_{III}$) consisting of two crossroller chains and having the same stiffness both in vertical and horizontal directions, are

TABLE 1

Machine group		A			B			C		
Eigen frequency		ϵ_1	ϵ_2	ϵ_3	ϵ_1	ϵ_2	ϵ_3	ϵ_1	ϵ_2	ϵ_3
Stiffness ratios	$C_I < \dfrac{C}{2}$	+	+		+			+	+	
	$C_I \approx \dfrac{C}{2}$	+				+		+	+	
	$C_I > \dfrac{C}{2}$									
	$C_{III} < \dfrac{C}{2}$		+			$C_I < \dfrac{C}{2}$ +			+	+
	$C_{III} \approx \dfrac{C}{2}$	+				+		+	+	
	$C_{III} > \dfrac{C}{2}$	+				+ $C_I > \dfrac{C}{2}$			+ $C_I > \dfrac{C}{2}$	

+: the forced vibrations may be of significant values.

not the optimum with respect to the transverse forced vibrations. The optimum stiffness relationship exists, when the guideway is at the side of the exciting force, and is about 1·4 times stiffer in the vertical direction than in the horizontal, the horizontal stiffness value being equal to that of the smaller vertical stiffness. This relation can be realized by a guideway system consisting of a flat, and a prismatic guideway. A system of needle-rollers is advantageous in reducing the vibration sensitivity.
(IV) The eigen frequencies of the tested transverse vibrations can be altered significantly by changing:

(1) the guideway stiffness and the stiffness relationships,
(2) the mass of the slide,
(3) the damping conditions considerably.

A significant frequency change cannot be achieved by varying the distance of the guideways from each other.

(V) The natural vibration of the lowest eigen frequency consists mainly of the components of horizontal and oscillating vibrations. So they are damaging primarily to the machines belonging to groups A and C. Their amplitudes can be decreased very efficiently by the following procedure:

(1) by increasing the extent of the horizontal component of the vertical damping applied to the side effected by the vibrating force,
(2) by increasing the vertical stiffness of the guideway situated at the side affected by the vibrating force,
(3) by decreasing the stiffness of the horizontal direction,
(4) by arranging a wide guideway distance and making the centre of gravity low,
(5) finally by arranging that the exciting force acts, if possible, between the guideways.

The second natural vibration consists mainly of a component in the vertical direction and so it does not significantly affect machines of groups B and C, but it may contain dangerous horizontal vibrations even for the machines of group A. To decrease its amplitudes it is advisable:

(1) to increase the stiffness and damping of the guideway in the vertical direction at the side affected by the exciting force,
(2) to increase the vertical stiffness in case of the machines of group A, as well as in the machines of group B, if the stiffer guideway cannot be placed at the side of the vibrating force.

The above-mentioned general conclusions, deduced from the completed calculations can be applied as construction directives. The computer program at one's disposal makes the calculations of the forced vibrations to be expected feasible, for many possible design variations and so to help the designer in selecting the optimal solution.

INFLUENCING GUIDEWAY PROPERTIES

Further we intended by our experiments and by the theoretical analyses supporting our results to elucidate the dependence of the two fundamental properties stiffness and damping, of sliding and rolling guideways on certain design and working parameters.

Stiffness of sliding guideways

The experimentally measured variation of the stiffness of a flat guideway is shown in Fig. 5, plotted against the specific surface loading. The sliding velocity appears here as a parameter. The same guideway stiffness is shown on the right of the figure plotted against the velocity. In this case the specific loading was selected as the parameter. Our experience of the effects of lubricating oil is shown in Fig. 6, for the case of larger and smaller specific surface loading and varying sliding velocity.

It is very interesting to observe that the 'stick slip' inhibiting oil (Shell to 33) polarized chemically by activating additives, damages the stiffness conditions seriously in cases of larger specific loading and velocities and therefore it is advisable—when using such oils—to be very careful.

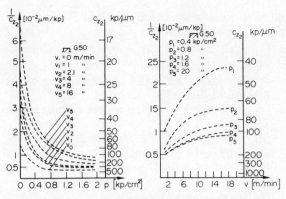

Figure 5. The measured variation of the sliding guideway stiffness plotted against the specific surface loading and the sliding velocity.

As a summary on the basis of the accomplished theoretical and experimental examinations, also considering the results[6] of Levit and Lurje, the following conclusions can be made[5]:

(1) The stiffness of the sliding guideways decreases progressively as oil viscosity increases. The guideway stiffness will not be influenced by oils of smaller viscosity than 20 cSt at 50°C temperature. The polarized oils cause the stiffness to increase in the case of small velocities ($v < 4$ m/min) and small surface loadings ($\bar{p} < 1\cdot2$ kp/cm^2) in comparison with the common oils of the same viscosity, but above the given limits they significantly affect the results in a contrary mode.

(2) The stiffness will be decreased by increasing the sliding velocity. This effect is changed progressively by increasing velocity. The phenomenon relates to the growing of the oil film thickness.

(3) The stiffness increases by the increased specific loading. This effect is also progressive.

(4) On the basis of the conclusions 1–3, it is therefore advisable in respect of the design of the machine tool sliding guideways:

(a) to use flat guideways in case of big loadings on prismatic guideways,
(b) not to choose a specific surface loading with a small value. It is advisable to keep its value above $0\cdot8$ kp/cm^2 at $v < 4$ m/min, and to keep it above $1\cdot2$ kp/cm^2 at greater velocities,
(c) the viscosity of the lubricating oil should be increased only as it is required by other aspects, e.g. wear and stick-slip, etc.,

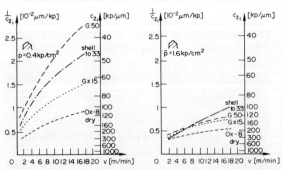

Figure 6. The measured variations of sliding guideway stiffness in case of different lubricants.

(d) polarized lubrication materials should be used only for small velocities ($v < 4$ m/min.)

Stiffness of rolling guideways

As was previously shown experimentally by Levina and Resetov,[7,8] the characteristic curve of the roller guideway deformation differs from the theoretic linearity at smaller surface loading; in this case, there is no steady contact between the guideway and the rolling element. The reasons of the deviation are to be found in the variance of sizes, and the form deviation of the ground surfaces. Furthermore deviation occurs from parallelism and in the form deviation of the rolling elements. The larger the specific loading, the easier it is to compensate for defects as deformations

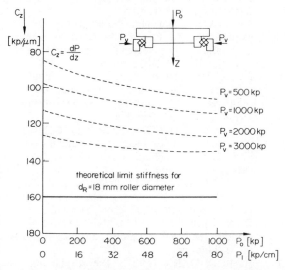

Figure 7. The measured variations of the rolling guideway stiffness at various prestresses.

are increased and a linear limit is approached. The system conforms to the linearity seen by the stiffness coefficient of the Bochmann formula[9]. From our experimental results, the stiffness of a guideway consisting of a crossroller chain, was formed and is shown in Fig. 7.

The continuously traced line shows the theoretical maximum stiffness limit. It is perceptible that this value will be approached by increasing the prestress load P_v. It can therefore be stated that the stiffness of the roller guideways depends on the loading of the rolling element, and it can therefore be increased by prestressing. Significant influences caused by rolling velocity as well as lubricants cannot be demonstrated.

The comparison of the static stiffness of rolling guideways and sliding guideways which can be arranged in the same volume, leads to the following conclusions:

(a) The roller guideways are only used in the case of small specific loading and at great slide velocities stiffer than the sliding guideways. When exceeding the limits $\bar{p} > 1$ kp/cm² and $v < 4$ m/min the sliding guideways are essentially stiffer.
(b) The use of rolling guideways is reasonable in spite of this fact—if we intend to eliminate the dependence on the operating conditions concern-

ing the stiffness, as the stiffness of the rolling guideways does not depend either on lubrication or on the slide velocity and is affected only slightly by loading.

Damping of guideways

The comparison of the damping conditions of the sliding and rolling guideways is shown in Fig. 8. In case of the same slide system the slightly prestressed

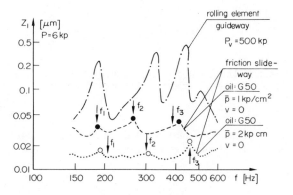

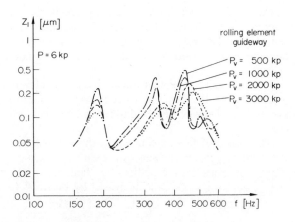

Figure 8. The measured resonance curves of the slide and rolling guideways.

rolling guideways have relatively high resonance amplitudes, but by increasing the prestress, these values could be decreased to the favourably high damping level of the sliding guideways.

Our initial theoretical examination[5] based on the work of Kimball and Lovell concerning material damping, showed that damping of rolling guideways increases proportionally to the square root of prestressing. A curve, made by such an assumption, shows the damping coefficient plotted Fig. 9.

According to the experience of the examinations the damping of rolling guideways can be increased to the same value as that of the slideway damping, by adequate, still permissible prestress. As a final result, the use of rolling guideways is not disadvantageous in comparison with sliding guideways considering resonance sensibility.

The results of further examinations proved that neither the rolling velocity nor the applied lubricants (oils and greases of different viscosity) influence the damping relations considerably. So the supposition of our theoretical examinations was verified, showing

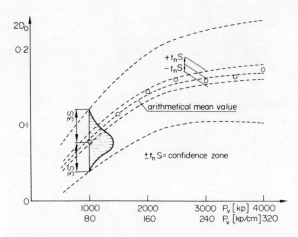

Figure 9. The variation of the experimentally determined damping factors of antifriction guideways plotted against the prestress.

that the material damping was the dominant factor in the case of rolling guideways.

CONCLUSIONS

In the foregoing sections we have pointed out, that in respect to transverse vibrations of machine tool slides, the flat, prismatic guideway pairs are very favourable, made either as sliding, or rolling guideways.

It is very advantageous to place the flat guideway at the side of the vibrating force as well as to make the distance between the guideways as large as possible.

The principal factors concerning the stiffness and damping of guideways were examined and the most important conclusion was, that the damping of the rolling guideways could be increased by prestress to the same level as in the case of sliding guideways.

ACKNOWLEDGMENTS

The author wishes to thank Prof. Dr. Ing H. Berthold; and T. U. Dresden for the scientific guidance of the research as well as for the 'SZIMFI' Developing Institute of the 'SZIM' Machine Tool Works, for the experimental work.

REFERENCES

1.   R. ZURMÜHL (1963). *Matrizen*, Springer Verlag, Berlin.
2.   S. FALK (1966). Numerical systems in technical mechanics. Manuscript, Society Bolyai Budapest.
3.   G. POPPER (1965). Solving algebraic equations of higher degree by electronic computers, *NIM IGÜSZI Communications of Computing Technics*, Budapest, March.
4.   Gy. HAJDU and Gy. POPPER (1970). The examination of forced vibrations of machine tool slides by electronic computer, *SZIM Journals*, Halásztelek, January.
5.   Gy. HAJDU (1970). Statische und dynamische Untersuchungen an Werkzeugmaschinenschlitten mit Gleit—und Rollenführungen. Dissertation, T.U. Dresden.
6.   G. A. LEVIT and B. G. LURJE (1962). Guideway calculation of feeding mechanisms on the basis of friction characteristic curves, *Sztanki i Insztrument*, January.
7.   Z. M. LEVINA (1961). Selection and calculation of the construction parameters of rolling guideways, ENIMSZ, Moscow, Study-aid.
8.   D. N. RESETOV (1951). Machine tool dimensioning for contact rigidity, *Sztanki i Insztrument*, January.
9.   H. BOCHMANN (1927). Die Abplattung von Stahlkugeln und Zylindern durch den Messdruck. Dissertation, T.H. Dresden.
10. A. L. KIMBALL and D. E. LOVELL (1929). Internal friction in solid element, *Physical Review*, December.

AN EXPERIMENTAL INVESTIGATION OF GRINDING MACHINE COMPLIANCES AND IMPROVEMENTS IN PRODUCTIVITY

by

W. B. ROWE*

SUMMARY

It was necessary to examine the nature of deflections in a centreless grinding machine in order to decide priorities in machine design. Static deflections were reported for controlled loading conditions. The results were analysed to reveal the importance of the various machine elements. Areas for design improvements arising from the investigation were suggested. Hydrostatic bearing spindles were substituted for the previous plain bearing spindles and grinding results demonstrated the improvements in sizing, roundness and reduced spark-out time. Particular attention is drawn to the importance of the dwell period for production rate and accuracy in plunge grinding operations. This work forms part of a Science Research Council supported programme.

INTRODUCTION

The results described in this paper were investigated as part of a more extensive programme having the following objectives:

(1) to determine critical areas in machine design,
(2) to achieve a better understanding of the important parameters in grinding.

REVIEW OF PREVIOUS WORK

Yonetsu[1] described the basic geometrical relationships and investigated the rounding process by harmonic analysis.

A digital simulation technique was developed[2] which allowed a more general investigation of the process and hence it was possible to obtain close agreement between experimental and theoretical workpiece shapes for sub-resonant effects. It was also shown at this time that even-order waviness may occur with increased workpiece support height due to geometric instability. A further conclusion reached was that a rigid machine will regenerate waviness more strongly than a flexible machine under conditions of geometric instability.

A chatter analysis[3] revealed the unusual nature of the limiting stability locus and confirmed the above conclusions. Chatter analysis allows measured machine dynamic characteristics to be taken into account when computing stability as shown by Sweeney and Tobias.[4] Unfortunately chatter stability charts for centreless grinding give little assistance as

they provide no indication of whether a particular combination of geometrical configuration and operating conditions is more favourable than another.

The geometric configuration has a strong influence on roundness accuracy and hence geometric stability charts were presented[5] which allow the geometric rounding tendency at any frequency to be examined. Thus by suitable selection of a workpiece speed[6] it is possible to achieve a condition where the centreless grinding geometric configuration has a stronger stabilizing effect for a particular frequency than the more basic grinding arrangement.

Schreitmüller[7] investigated the nature of the contact point between the workpiece and the grinding wheel. Experimentally this was difficult to achieve and it was not possible to achieve a high order of accuracy. However it was shown that compliance at the grinding point contributes significantly to the total compliance if the grinding machine is of a very rigid construction.

Attention was first drawn to the cyclic size variations in grinding by Loxham.[8] In a note to the Science Research Council Grinding Committee this feature was described in more detail.[9] By monitoring a number of parameters simultaneously it was shown that in a particular case a sudden reduction in size was accompanied by a reduction of the grinding forces, an increase in vibration level, and a deterioration of roundness accuracy. The sudden reduction in size was of opposite sense to the effect of wheel wear and was related to the reduction in grinding force which allowed the elastic deflections of the machine to be partly released.

---

* Lanchester Polytechnic, Coventry

Recently a simulation by hybrid computer has been described[10] which combines a digital simulation of the process with an analogue simulation of the machine. The hybrid computer gives the possibility of simulating complete grinding processes whilst monitoring roundness, size, vibration level, and deflections. These are parameters which are only measured with difficulty under experimental conditions. A further possibility is to vary the hybrid model to simulate an improvement in a particular part of the machine and hence determine the predicted change in grinding results.

One conclusion common to refs 2, 5, 9 and 10 is that, for quantitative prediction of grinding results, it is absolutely essential to include the static compliance and real time periods in the analysis. Forced vibrations and resonances are also important but not necessarily more so than static compliances.

The productivity of a centreless grinding machine is related to the ability to maintain shape and size. A machine which has a lower accuracy capability than another will require to be operated at lower removal rates and with more frequent attention in order to achieve given tolerances. The purpose of the present paper is to clarify some relationships between compliance, accuracy and effects on production rate by a practical demonstration of how machine modifications led to improvements in grinding.

THE COMPLIANT ELEMENTS OF THE MACHINE AND MODIFICATIONS

Deflections are conveniently described by compliances, i.e. deflections per unit force. In analysing the machine described in this paper the elements were classified as illustrated in Fig. 1(a). The force is shown reacted by radial forces on the two wheels which are linked by spindles, bearings, wheelheads and the tray. In practice there are both tangential and radial forces acting at the contact points with the two wheels and the work support plate. However the critical deflections are those which tend to affect the

depth of cut most directly. The investigation was therefore restricted to the force loop indicated.

Figure 1(b) summarizes the results of an analysis of compliances for the machine model specified in Appendix 1. The results indicate the effect at the grinding point due to the various elements. It will be apparent that a considerable instrumentation and careful analysis was required to determine these compliances. The methods have been employed over several years and have been improved to a stage where fair confidence has been achieved. Previously results were described for the control spindle unit and some of the instrumentation involved was indicated.[11]

The standard machine has plain phosphor bronze bearings each having an adjustable pad. The control wheel bearings which operate at low speed gave evidence of operation in the transition region between rubbing and full hydrodynamic lubrication, and hence the load/deflection characteristics are non-linear and subject to both speed and load changes.[11] The grinding wheel bearings appear to be fully hydrodynamic in operation and are stiffer in operation. Oil temperature measurements indicate operation at over 45 degrees centigrade.

The control wheel bearings on the standard machine were the most compliant element of the system under their normal operating condition for the grinding tests described. The maximum compliance measured was 13×10^{-6} in/lbf although measurements on an identical model produced much lower figures. This result led to the conclusion that it would be worth while to replace these bearings with a very much stiffer design. This result was conveniently achieved by substituting a hydrostatic bearing unit for the control wheel bearings on an identical model. Very high stiffness was ensured by diaphragm valve compensation.[12]

An interesting feature of the modifications was that they were applied to a worn machine that did not allow very good roundness accuracy. The control wheel unit replacement greatly improved the stiffness of the bearings, but made no noticeable difference to the roundness accuracy of the workpieces after grinding.

Subsequently hydrostatic bearings were substituted for the hydrodynamic grinding wheel bearings and improvements in roundness accuracy were then obtained. Another improvement which resulted was that the oil temperature could then be maintained at 24°C.

The rubber bonded control wheel was the second most compliant member of the standard machine and at some loads it was the most compliant element. This is another interesting feature of centreless grinding practice since it would not be difficult to employ much stiffer control wheels. It appears that some users skilled in the operation of centreless grinding machines claim that a flexible control wheel is necessary for good roundness accuracy and do, in fact, introduce softer wheels in certain circumstances. A possible explanation why control wheels are not harder may be due to the effect of forced vibrations. A compliant control wheel surface would not be expected to impress these vibrations on the workpiece surface as strongly as a rigid surface. However if

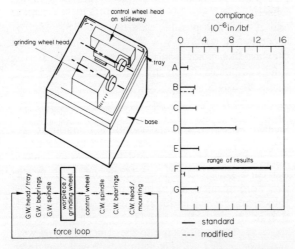

Figure 1. (a) The critical force loop in a centreless grinding machine. (b) Compliance at the grinding point due to various elements of a machine having standard bearings and a similar machine having modified bearings. A grinding wheelhead and tray, B grinding wheel bearings, C grinding wheel spindle, D control wheel, E control wheel spindle, F control wheel bearings, G control wheelhead and mounting.

the machine were very rigid the amplitude of forced vibrations would be lower, and it would then be expected that a stiffer control wheel would be acceptable. A very rigid research machine is in course of construction with a view to conducting controlled experiments on these mentioned effects. The control wheel does not act in a purely elastic manner and hence it was necessary to determine the load deflection curve by applying a force through a workpiece

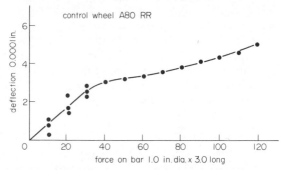

Figure 2. Deflections of a control wheel impressed with a bar of the same diameter as the workpieces.

for a short finite period. The result is illustrated in Fig. 2 which shows a low stiffness at light loads.[13]

GRINDING TESTS

In order to demonstrate the effect of compliance on the grinding performance the following measurements were made:

(1) In-process measurement of spindle deflections, workpiece rotation, infeed motion and time with various infeed rates up to 0·045 in/min. For these

tests a constant dwell period of 10 seconds was employed. The purpose of these tests was to investigate the magnitude of spindle deflections and the release of spindle deflections during the dwell period.

(2) The specimens resulting from the in-process measurements were subsequently coded and checked for size, roundness and surface texture. The order of measurements was randomized and the results were finally presented in the order in which the samples were ground.

(3) In-process and post-process measurements were made for varying dwell periods to determine whether productivity should be increased on the stiff machine by allowing a reduced dwell period. For these experiments a feed-rate of 0·032 in/min was employed for all the tests.

Grinding deflections and interpretation of the dwell period

The arrangement of the transducers for in-process measurements is illustrated in Fig. 3. A reference disc was attached to the end of each grinding spindle and 'trued' in position. A contactless inductive transducer was mounted from the diamond holder on each dressing attachment. The workpiece rotation was monitored by a similar transducer adjacent to a light disc attached to the workpiece. A further transducer monitored the infeed motion of the control slide. The transducers do not directly monitor spindle and bearing deflection at the grinding point since wheel-head distortions are also involved in these measurements. However the measurements are comparative for the two machines and illustrate some interesting points.

Figure 3. Arrangement of transducers for tests.

Two typical traces are illustrated in Fig. 4(a), which shows the deflection characteristics of the control wheel spindle for the two machines. The grinding cycle involves constant feed-rate followed by a constant dwell period as illustrated in Fig. 4(b). On the standard machine the maximum deflection measured is still increasing slightly after 8 seconds of infeeding. This means that the true depth of cut did not achieve a value equal to the infeed per half revolution of the workpiece as a consequence of machine deflections. The modified machine however

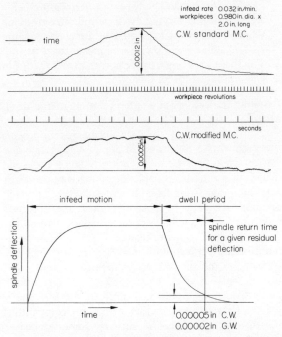

Figure 4. (a) Control spindle deflections during the grinding cycle. Grinding wheel A60 L5 V5. Control wheel speed 31 rev/min. (b) The idealized grinding cycle and a basis for comparing spindle return times.

achieved that condition in less than 4 seconds. The maximum deflection of the control wheel spindle on the standard machine was approximately two and a half times as large as the deflection on the modified machine and shows the improvement which had been achieved by the substitute bearings.

The deflection characteristic during the dwell period closely follows a geometric progression. Analysis of the common ratio 'r' of the geometric progression leads to two interesting parameters relating the resultant static compliance and the separating force due to grinding. The theory given in Appendix 2 shows that:

$$r = \frac{K_s}{\lambda_s} \qquad (1)$$

where

K_s = static normal grinding force coefficient
λ_s = static stiffness of the machine

and

$$K = 1 - r \qquad (2)$$

where

$$K = \frac{\text{True depth of cut}}{\text{Apparent depth of cut}}$$

Evaluating the common ratio for the examples illustrated in Fig. 4 yielded the following values

standard machine $K = 0.23$

modified machine $K = 0.44$

In previous measurements during fine grinding operations with a glazed wheel the lowest value recorded was $K = 0.01$.

Figure 5(a) shows the maximum deflections of the control wheel spindles during grinding at various infeed rates. For feed-rates varying between 0.009 in/min to 0.045 in/min the modified machine shows a linear relationship between deflection and feed-rate. The trend is also linear over most of the range for the standard machine but the slope is more than twice as

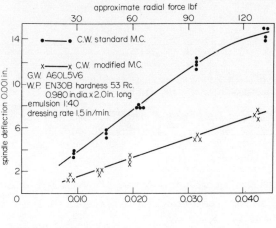

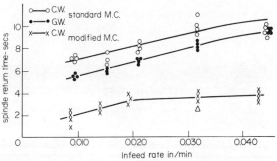

Figure 5. (a) Maximum control wheel spindle deflections for various infeed rates. (b) Spindle return times for various infeed rates.

great as the stiffer machine. At high infeed rates some stiffening up is apparent due to the characteristics of the plain bearing spindle. The maximum removal rate is 0.14 in³/in min which corresponds to 1.45 mm³/mm s. These rates compare with commercial practice for precision operations and were achieved with a 60 grit wheel. The corresponding spindle return times are illustrated in Fig. 5(b). It can be seen that the return times for the spindles on the standard machine vary between 5 and 10 seconds to achieve the accuracies specified in Fig. 4(b). However the stiffer machine achieved the same results within a time period 1 to 4 seconds. The results can be interpreted in two ways. Either the standard machine must be allowed a very much longer dwell period in production to achieve a close tolerance on size or alternatively if a reasonably short dwell period of 2 to 4 seconds is adopted on both machines it should be possible for the stiff machine to maintain a closer tolerance on size. The two succeeding sets of results support this view.

Workpiece accuracy with a ten-second dwell period
Figure 6 illustrates that varying infeed rates do not cause largely varying results if a sufficiently long dwell period is employed. It may be therefore that in some industrial applications only one grinding operation is necessary where two are customarily employed. However it is apparent that process variability is more evident in size variations with the standard machine than with the modified machine. After an initial phase of adjustment, apparent in the first few specimens, size was maintained within $\pm 7 \times 10^{-5}$ inches on the standard machine and within $\pm 4 \times 10^{-5}$ inches on the modified machine. The improved temperature stability of the modified machine may have assisted this result.

After a dwell period of 10 seconds the roundness errors illustrated in Fig. 6(b) shows that the modified machine produces slightly better accuracy than the standard machine. However this result has not been proved to be related to machine stiffness, since this accuracy was not achieved when the control wheel unit was replaced on the modified machine.

The surface texture values (shown in Fig. 6(c)) from the standard machine were slightly higher than the results from the modified machine. The difference is not great and insufficient evidence is available to give an explanation. Surface texture is strongly

dependent on average chip thickness which is directly related to the true depth of cut. A large value of average chip thickness corresponds to high surface texture values. There is also a secondary effect due to the depth of cut. This effect is that the wheel adjusts its surface topography according to the true depth of cut. If fine depth of cut values are sustained the grinding wheel grits have a greater tendency to glazing, whereas a redressing effect is experienced when a grinding wheel is plunged strongly into the workpiece.

Clearly the secondary effect requires a longer time period to become evident than the primary effect. It is possible that the secondary effect may explain the measured difference. The stiff machine impresses the grinding wheel more rigidly into contact with the workpiece while feeding and could conceivably lead to an improved redressing effect.

It is also possible that vibration levels have an influence. However it has been shown previously[14] that large reductions in the overall level of vibrations may tend to accentuate a tendency towards glazing in fine grinding operations. The lower level of vibrations on the modified machine do not appear to offer an explanation of the results in this case since the surface texture values were higher.

Workpiece accuracy with varying dwell periods
The size variations with increasing dwell periods, shown in Fig. 7(a), follow a geometric progression of

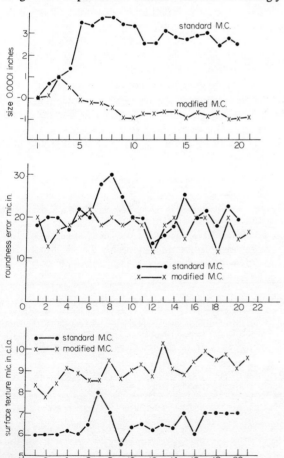

Figure 6. Specimens in grind order with various infeed rates and a 10-second dwell period. (a) Size variations; (b) roundness errors; (c) surface texture.

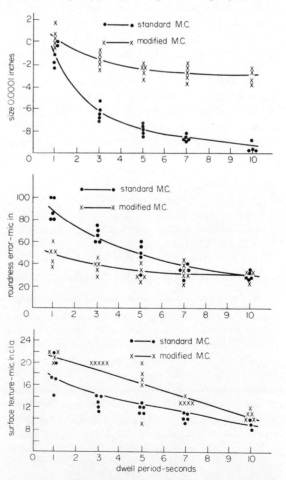

Figure 7. Workpiece accuracy for an infeed rate of 0·032 in/min and varying dwell periods. (a) Size variations; (b) roundness errors; (c) surface texture.

a similar shape to the size characteristic already discussed in detail. After 3 seconds dwell the modified machine is within 1.3×10^{-5} inches on size at 10 seconds compared with 2.4×10^{-5} inches on the standard machine.

The roundness results, Fig. 7(b), illustrate that there is an obvious improvement in roundness accuracy during the dwell period with both machines. On these tests the final roundness after ten seconds is almost the same for both machines. However roundness from the modified machine was found to be considerably better at the beginning of the dwell period. Thus while at the commencement of the dwell period the workpiece has achieved the same number of revolutions, on both machines, more material has been removed on the stiff machine. Both factors, number of workpiece revolutions and stock removed are important in the rounding process. Apparently the modifications were important for both roundness accuracy and productivity since the modified machine has almost obtained its ultimate degree of rounding-up after a 3-second dwell. The standard machine required 8 seconds to achieve a similar result.

The surface texture results obtained with various dwell periods, shown in Fig. 7(c), illustrate once more the higher values from the modified machine. The results are surprising since it has already been extensively demonstrated that after a short dwell period, the true depth of cut is lower with the modified machine. However a smaller average chip thickness should yield a lower surface texture result, which is the opposite of what was experienced. This leads one again to consider the secondary effect due to the adjustment of the wheel surface previously discussed. Perhaps the most outstanding conclusion from the surface texture measurements is however that surface texture is not a very reliable criterion of grinding machine capability. Surface texture for a given grinding wheel is strongly dependent on chip thickness and hence on dwell period. This is clearly evident in Fig. 7(c) for both machines. The values are of a reasonably similar magnitude at the commencement of the dwell period and reduce with time eventually approaching new values of similar magnitude. Surface texture is also strongly dependent on the grinding wheel surface and hence on the dressing procedure.[14] It is not always the grinding wheel condition which is most suitable for stock removal that gives the lowest surface texture values. This point is clearly illustrated in Fig. 8 which shows the relationships between grinding power, dressing traverse rate and surface texture for a finishing operation. The lowest surface texture values are obtained with a slow dressing traverse rate and correspond to a condition of low grinding efficiency and high power.

High removal rates and low surface texture values are not obtained at the same instant in grinding unless the number of active cutting grits which are brought into contact with the workpiece in unit time can be increased. Two techniques which have been successfully employed in this context are high-speed grinding and the increased grinding area approach[15].

However it appears that for some industrial applications a machine which combines high stiffness

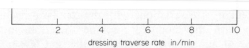

$$\begin{array}{ccccc} | & | & | & | & | \\ 2 & 4 & 6 & 8 & 10 \end{array}$$
dressing traverse rate in/min

Figure 8. Relationships between grinding power, surface texture and dressing traverse rate for a 0·0005 in depth of cut employing (a) an impregnated diamond dressing tool, (b) a standard diamond tool. Grinding wheel, A46/54 KS V50, 11·6 in dia, 1750 rev/min; control wheel, A80 RR, 6·7 in dia, 25 rev/min; workpieces, EN8, 1·0 in dia, 2·0 long; infeed rate, 0·004 in/min; emulsion, 1:40.

and rotation accuracy would allow a precise finishing operation to be eliminated. A rapid stock removal phase followed by an adequate dwell period should provide the accuracy required for many precision components.

CONCLUSIONS

(1) The critical areas which lead to roundness errors and deflections in grinding have been analysed. The conditions have been tested by modifying the most critical elements to effect an improvement in machine performance.

(2) A stiff machine allows a close size tolerance to be maintained with a short dwell period and hence allows increased productivity.

(3) Size, accuracy, roundness and surface texture all improve during the dwell period and hence the nature of the grinding cycle and number of operations to achieve required tolerances are worthy of further study in order to optimize productivity and reduce costs.

ACKNOWLEDGMENTS

To the Science Research Council who gave financial assistance towards the project.

To Mr P. J. Gardner for assistance with the grinding measurements and to Mr J. I. Willmore for assistance with compliance measurements.

REFERENCES

1. S. YONETSU (1959). Consideration of centreless grinding characteristics through harmonic analysis of out-of-roundness curves, *Proc. Fujihara Memorial Faculty of Engineering*, Keio University, **12**, (47), 8–26.

2. W. B. ROWE, M. M. BARASH and F. KOENIGSBERGER (1965). Some roundness characteristics of centreless grinding, *Int. J. Mach. Tool Des. Res.*, **5**, 203–15.

3. W. B. ROWE and F. KOENIGSBERGER (1965). The work-regenerative effect in centreless grinding, *Int. J. Mach. Tool Des. Res.*, **4**, 175–87.

4. G. SWEENEY and S. A. TOBIAS (1963). An algebraic method for the determination of the dynamic stability of machine tools, *Proc. Int.*

Production Engineering Research Conference, Pittsburgh (Sept.), 475–85.

5. W. B. ROWE and D. L. RICHARDS (1971). Geometric stability charts for the centreless grinding process, *J. Mech. Eng. Science,* **14**(2), 155–8.

6. D. L. RICHARDS, W. B. ROWE and F. KOENIGSBERGER (1972). Geometrical configurations for stability in the centreless grinding process, *Proc. 12th Int. Mach. Tool Des. Res. Conference,* Macmillan.

7. H. SCHREITMÜLLER (1971). Kinematische Grundlagen für die Praktische Anwendung des Spitzenloses Hochleistungsschleifens, Dissertation, T. H. Aachen.

8. J. LOXHAM (1960). The potentialities of accurate measurement and automatic control in production engineering, symposium on Machine Tool Control Systems, College of Aeronautics, Cranfield (August).

9. J. WILLMORE, W. B. ROWE and J. LOXHAM (1970). A note on the saw-tooth pattern in grinding, report to the Grinding Committee of the Science Research Council (December).

10. W. B. ROWE, J. I. WILLMORE and L. HULTON. A technique for simulation of cylindrical grinding processes by hybrid computation (to be published).

11. W. B. ROWE (1968). Experience with four types of grinding machine spindles, *Proc. 8th Int. Mach. Tool Des. Res. Conference,* Pergamon, 453–76.

12. W. B. ROWE and J. P. O'DONOGHUE (1970). Diaphragm valves for controlling opposed pad hydrostatic bearings, Paper 1, *Proc. I. Mech. E. Tribology Convention,* Brighton (May).

13. J. I. WILLMORE (1972). An investigation of the dynamic properties of a centreless grinding machine, M.Phil. thesis, Lanchester Polytechnic (January).

14. W. B. ROWE (1964). Studies of the centreless grinding process with particular reference to the roundness accuracy, Ph.D. thesis, University of Manchester.

15. W. B. ROWE (1971). A review of grinding process parameters, *Engineer's Digest,* **32** (10) (October), 41–8.

APPENDIX 1: SPECIFICATION OF THE CENTRELESS GRINDING MACHINE AND OPERATING CONDITIONS

The following specifications applied both to the standard machine and the modified machine which only differed because of the changed spindle bearings. The plain bearings on the modified machine were replaced by hydrostatic bearings and the control wheel drive arrangement had to be modified.

work-range (plunge-feed) grinding, 0·06 to 1 in
grinding wheel size (dia x width x bore), 12 x 3 x 4 in
control wheel sixe, 7 x 3 x 3 in
grinding wheel speed, 1780 rev/min
control wheel speed, 31 rev/min
control wheel dressing speeds:
     standard plain bearing machine, 491 rev/min
     modified hydrostatic bearing machine, 371 rev/min
grinding wheel Norton A60 L5V6 (unless otherwise stated)
control wheel Norton A80 RR51

workplate, Tungsten carbide 30° blade angle
workpieces, EN 30 B Hardness 53 Rockwell 'C'
     0·980 in dia x 2·0 in long
coolant, water-soluble oil 1:40
wheel dressing procedure:
     traverse rate 1·5 in/min
     2 passes with 0·001 in depth of cut
     2 passes with 0·0005 in depth of cut
     final pass without cut
stock removed, 0·004 in on diameter
dwell period, 10 s (unless otherwise stated).

APPENDIX 2: ANALYSIS OF THE SPARKING-OUT CHARACTERISTIC

Due to deflections in the machine the stock removed on diameter does not approach the value of the infeed motion per half-workpiece-revolution until a considerable dwell period has elapsed. If the difference between the applied infeed motion and the stock removed is defined as the apparent depth of cut A, and the true depth of cut at the same time is S, then a first-order relationship between these values may be assumed for the purposes of this analysis.

$$S_1 = K \cdot A$$

In a dwell period each subsequent half-revolution diminishes the apparent depth of cut so that:

$$S_2 = K(A - S_1) = (1 - K)S_1$$

and eventually

$$\frac{S_m}{S_n} = (1 - K)^n$$

which expresses a geometric progression of common ratio $r = (1 - K)$.

From the above description it follows that the apparent depth of cut A is the sum of the machine deflection x and the true depth of cut S

$$A = x + S$$

Both x and S to first-order approximation are proportional to the separating force P between the grinding wheel and the workpiece.

The relationships are

$$P = K_s \cdot S$$

where K_s is the static grinding force coefficient

$$P = -\lambda_s \cdot x$$

where λ_s is the resultant static stiffness of the machine, i.e. the reciprocal of the compliance.

It follows from the foregoing that

$$\frac{K_s}{\lambda_s} = -\frac{x}{S}$$

and

$$\frac{A}{S} = \frac{x}{S} + 1 = K$$

Hence

$$K = 1 - \frac{K_s}{\lambda_s} = 1 - r$$

and

$$\frac{K_s}{\lambda_s} = r$$

The significance of these equations is that the parameters K_2/λ_2 and K are dependent only on the shape of the magnetic properties during a particular grinding operation on a given machine and do not depend on which element is considered. The accuracy is not affected if one measures exactly at the grinding point or preferably at some other

possible to examine samples of the curve and hence the variation of λ during the dwell period. However this was not necessary in the present analysis.

HYDRAULIC SERVO SYSTEM AS APPLIED TO A PHOTOGRAPHIC PATTERN GENERATOR

by

J. J. 't MANNETJE*

SUMMARY

It is to be expected that within a few years the combination of a drawing table and a fix focus camera will not have the desired accuracy for artwork generation in the microelectronics field. The alternative solution: a photographic pattern generator, has been built at numerous places. The Philips Research Laboratories have realized one with the following specifications:

writing area	200×200 mm^2
smallest displacement command unit	0.5 μm
absolute accuracy	± 1 μm
repetitive accuracy	± 0.5 μm
maximum writing speed	10 mm/s
maximum acceleration	0.25 m/s^2
range of line widths	2–1900 μm (200 steps)
control computer	Philips P9201, 8K16 bits
input	papertape EIA or ISO or 9-track magnetic tape, ISO.

The most common patterns consisting of straight or curved parts can be drawn. The emphasis of this paper lies upon the mechanical parts of the electrohydraulic position servomechanism. For completeness short reference is made to the optical and electronic parts. The paper is complementary to another[2] which deals with the photographic pattern generator. The methods used to achieve both an accuracy of 0.5 μm and to meet the requirements of the specified velocity and acceleration are included, as well as the influence of the machine design upon the servo performance. The details of the slides and motors are first discussed and then the control loops.

NOTATION

G_s, G_r, G_f, G_v, H_c	transfer function
k	loop again
s	d/dt
t_1, t_2	bearing clearance
x	desired velocity
$\ddot{x}$	desired acceleration
x_1	displacement slide
$\beta_v, \beta_f, \beta_r$	damping ratio
$\omega_v, \omega_f, \omega_r$	resonance frequency
τ	time lag
ϵ^*, ϵ	error signal

DESCRIPTION OF THE SYSTEM

Basically there are two different optical methods to create a pattern on a photographic plate; one is by projecting a moving beam of light through a lens covering the entire field. The other is a coordinatograph with a moving projection system. We have chosen for the last method, as we believe that to be the only method to produce directly object plates with lines having an edge definition of 0.4 μm in a 40×40 mm^2 field, a size required for IC purposes.

Having made this decision our choice of the writing principle was determined by the other requirement; freedom of pattern geometry. In that respect we prefer continuous writing to the alternative solution, e.g. flash-projecting a series of characters and patterns chosen from a fixed set.

In a pattern generator of our type there is no serious benefits between speed or accuracy and writing area. Therefore the latter has been made larger than the object plate to make the instrument available for other purposes. Figure 1 shows a photograph of the machine.

* Philips Research Laboratories, Eindhoven, The Netherlands

Figure 1. The photographic pattern generator.

A small computer controls the position of the slides, the orientation and the dimensions of the rectangular writing aperture and the overall working of the instrument. A few geometric calculations describing higher level languages including the post-processors, such as APT are available to describe the masks and to generate the paper tape input.

In Fig. 2 a diagram is given of the servosystem. The two systems differ only in the applied load, i.e. the projection column (weight 30 kg) is mounted on one slide (weight 30 kg).

In the next sections the details of the mechanical components used are critically discussed. Then the mechanical components of the control system are summarized. After that a detailed discussion of the adjustment of the control loops is given.

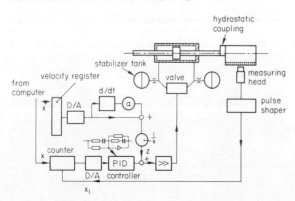

Figure 2. Schematic diagram of the servo.

THE MECHANICAL COMPONENTS

With regard to the high positional accuracy required, much attention must be paid to the mechanical construction as a whole.

To avoid additional movements in the drive system, the working line of the drive must pass through the centre of gravity of the load mass.

The machine resonances, frame and motor joints, must be chosen either to be very low, in order to compensate them by the servosystem itself, or to be very high (say three times the servo cut-off frequency).

The stiffness (weakness) of the machine frame is selected so that the resulting frequency will be much lower than the servosystem cut-off frequency.

The stiffness of the joints of the motor onto the frame must be very high because this will influence the dynamic characteristics of the servosystems. The resultant dynamics of the joints must be taken into account when the servosystem is stabilized.

In the following section the components are discussed with respect to the above considerations.

Two identical hydraulic slide ways are mounted perpendicular to each other on a slab of granite, one for the projection head, the other for the photographic plate. The slides with their hydrostatic bearings were developed earlier at the Philips Research Laboratories and are already applied in many high precision machines.

In all these applications they have proved their

reliability and usefulness. Full details of the construction and alternative forms are described elsewhere[1].

Of the applied slide design a cross-sectional view is given in Fig. 3. The slide contains three bearing blocks mounted onto the slide frame. Each block has three bearing sections with laminar restrictors.

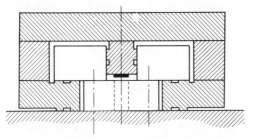

Figure 3. Cross-sectional view of a slide way that consists of three bars. This offers the opportunity to correct small deviations in the bars by bending and then fixing.

The resultant accuracy of this system was very high. The lateral deviations of the slide were hardly measurable and the rotations about a vertical axis during its travel over 200 mm did not exceed 0·5 seconds of arc.

The servomotors are symmetrical hydraulic rams. Their movement is controlled by a commercial hydraulic two stage four-way valve. The applied Moog valve delivers 3·8 litres/min at $70 . 10^5$ N/m^2 supply pressure at no load. The area of the piston was 25 cm^2.

The hydraulic rams (see Fig. 4) are equipped with bearings of tapered form. This bearing type has the highest stiffness value if $t_1 = 3t_2$ at a bearing length of 15 mm.

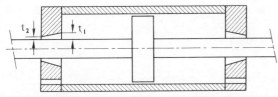

Figure 4. Schematic of hydraulic ram equipped with tapered bearings. $t_1 = 3t_2$.

The piston bar of the ram is connected to the slide via a hydrostatic coupling. This is done in order to eliminate the misalignments between the slide and the ram movements, so difficult alignments become unnecessary.

The coupling is shown in Fig. 5. Its stiffness is infinite in the direction of motion and zero in other directions.

At the end of the bar a plate is connected with a small clearance ($h = 10$ μm) on both sides to the housing. Because of the symmetrical form the bearing will allow movements in all directions except in the direction of motion. The stiffness of the bearing against rotations w_1 is zero, the stiffness against w_2 rotations is almost zero. A vertical shift of the components does not require any force.

The stiffness in the direction of the motion can be increased to an almost infinite value by using a

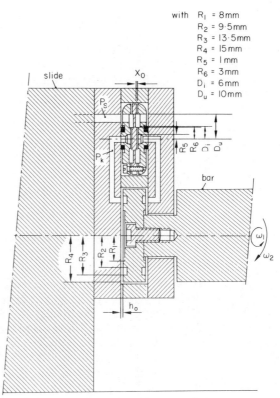

with R_1 = 8mm
R_2 = 9·5mm
R_3 = 13·5mm
R_4 = 15mm
R_5 = 1mm
R_6 = 3mm
D_i = 6mm
D_u = 10mm

Figure 5. Hydrostatic coupling.

variable flow resistance, called membrane double restrictors. A full description of the restrictor is given elsewhere[1]. The dimensions are selected so as to achieve the highest possible static stiffness. The frequency response characteristic, because of the small deviations ($<0·1$ μm) hardly measurable, shows only at high frequencies (≈ 500 Hz) very little amplitude rise. So this coupling is well suited for the purpose of this paper.

Two accumulator type dampers (called stabilizer tanks) are fitted via a laminar restrictor to the controlled volumes (Fig. 2). These dampers were added to adjust the relative damping of the motor to the desired level ($\beta_r \approx 0·5$).

The pressure source for the rams delivers a pressure of $70 . 10^5$ N/m^2 (1000 p.s.i.) and a maximum flow of 5 litres/min (5 c.i.s.). The pressure source to feed the slide bearings delivers a pressure of $28 . 10^5$ N/m^2 and a maximum flow of 5 litres/min.

The two supply units are completely separated and have their own temperature control units. This is done because the heat dissipation inside the hydraulic valve gives rise to a temperature rise of 3–4°C. Mixing of the two flows gives problems in maintaining the machine on one fixed temperature without gradients in the slides (and this is very important for IC purposes).

As a result it was observed that the temperature of the machine could be maintained at a level of 22°C ± 0·01°C.

The weakness of the frame, in the direction of the slides, resulted in a frequency of 9 Hz resp. 23 Hz and in a rotational resonance frequency of 40 Hz. The corresponding damping ratio was very large (0·3 → 0·9). The deviations occurring from these resonances

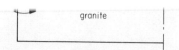

Figure 6. Schematic diagram showing the machine resonance.

The diameter of the bar of the ram is selected so that the corresponding frequency becomes approximately 3 times the motor resonance frequency. In the same way the thickness of the cylinders and the joints to the machine frame were selected. The resulting resonance frequency was approximately 400 Hz with a damping ratio of approximately 0·1, which is sufficiently high compared with the motor resonance.

THE ELECTRICAL AND MEASURING COMPONENTS

With regard to the high positional accuracy required much attention must be paid to the position measuring system. The best measurement is made when all the components of the drive, load and measuring system are situated in line. This is, measuring at the working line where the actual process takes place. It is clear that a perfect application of this principle cannot be performed. The used incremental measuring system is developed at the Philips Research Laboratories. The lines of the grating have a distance and a width of 8 μm. The measuring head operates according to the principle of interference of the grating with its image. A vibrating mirror, inside the head, is used for further interpolation until increments to 0·5 μm are obtained. The grating is placed as close to the working line of the process as possible; therefore it is mounted underneath the midbars (see Fig. 3) of the slides underneath the oil. This last point indicates that the grating is well stabilized against ambient temperature changes. The measuring head is mounted opposite the grating in the granite slab.

The measuring accuracy[3] of this combination equals 0·5 μm. An additional analogue interpolation technique improves the resolution to

displacement = $(k \cdot 0.5 \pm 0.1)$ μm with k = integer.

The number of pulses from this measuring system are compared (see Fig. 2) in a difference counter with the pulses coming from the control computer output interface. The output pulses are converted into an analogue signal by a D/A converter. The counter has a length of 2^5 pulses, which correspond to a displacement at ±16 μm.

THE SERVOSYSTEM

The open loop transfer function (see Fig. 7) is denoted shortly by $x_1/\epsilon = kG_s/s$ and consist of the hydraulic valve, motor and load. The frequency dependent part of this function G_s represents a cascade

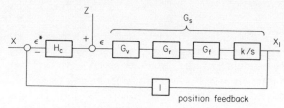

Figure 7. Block diagram of the servoloop.

connection of the dynamics of the valve, denoted by G_v, the dynamics of the ram G_r and the frame (weakness of joints) dynamics by $G_f \cdot s_o$.

$$G_s = G_v \cdot G_r \cdot G_f \tag{1}$$

(a) G_f

In a preceding paragraph we discussed the joint stiffnesses. It was found that the dominant resonance frequency equals 400 Hz with a damping coefficient of approximately 0·1. So the transfer function G_f appears to be

$$G_f = \frac{1}{1 + (2\beta_f/\omega_f)s + s^2/(\omega_f)^2} \tag{2}$$

$\omega_f = 400 \cdot 2\pi$
$\beta_f = 0.1$

(b) G_r

The type of the transfer function of a hydraulic ram is well known:

$$G_r = \frac{1}{1 + (2\beta_r/\omega_r)s + s^2/(\omega_r)^2} \tag{3}$$

The resonance frequency of the ram at midstroke equals $\omega_r = 210 \cdot 2\pi$ r/s.

The two accumulator type dampers connected to the controlled motor volumes offers the possibility to set the damping coefficient β_r to any desired value. For our purpose we set B_r between 0·4 and 0·5. The ω_r value will change slightly by this adjustment to 1·2 ω_r.

(c) G_v

The electrohydraulic valve appears to be the most critical and dominant component in the servo loop.

This valve contains several fundamental non-linearities, which are analyzed and discussed in detail in ref. 4. From this work it is recommended to tackle the valve dynamics by a third order linear model. So

$$G_v = \frac{1}{(1 + \tau s)(1 + (2\beta_v/\omega_v)s + s^2/(\omega_v)^2)} \quad (4)$$

The model constants τ, ω_v and β_v vary considerably, as given in Table 1.

TABLE 1   Variation of model constants with rated flow

Rated flow amplitude (%)	$\frac{1}{2\pi\tau}$ (Hz)	$\frac{\omega_v}{2\pi}$ (Hz)	β_v
≈ 9	1464	141	0·92 (selected values)
≈ 25	110·5	358	0·65
≈ 41	108	307	0·41
≈ 57	130	273	0·65
≈ 73	381	155	0·89
≈ 90	69	290	0·93

Because of the high required repetitive accuracy it is advisable to select the constants τ, ω_v and β_v at small rated flows for further analysis. The movement of the slide must be in accordance with the velocity profile generated by the control computer. A typical profile is shown in Fig. 9(a). From Fig. 2 we see that the velocity ($\dot{x}$) is transformed to a d.c. signal via a digital to analogue converter. Differentiation of this signal gives the acceleration ($\ddot{x}$) (see Fig. 9b) for a typical profile. The two signals are combined (z) and fed into the servo loop as shown in Figs. 2 and 7. From the latter we find

$$\epsilon^* = x - x_1$$
$$x_1 = (z + \epsilon^* . H_c)G_s . k/s \quad (5)$$

For optimum control the difference ϵ^* between the desired and actual positions must be equal to zero, so substituting $\epsilon^* = 0$ in the foregoing expression gives;

$$z = x \frac{s}{kG_s}$$

or with equations (2), (3), (4);

$$z = \left(x' + x''\left(\tau + \frac{2\beta_r}{\omega_r} + \frac{2\beta_f}{\omega_f} + \frac{2\beta_r}{\omega_r}\right) + \dots\right)\frac{1}{k} \quad (6)$$

Because of the software structure of the control computer it was only possible to derive the velocity and acceleration input commands with higher order corrections are neglected (see equation 6). The required feed forward signal can be written shortly as

$$z = \frac{1}{k}(x' + ax'') \quad (7)$$

with

$$a = \tau + \frac{2\beta_r}{\omega_r} + \frac{2\beta_f}{\omega_f} + \frac{2\beta_r}{\omega_r}$$

Now the position feedback loop will serve merely as a means to correct the deviations as expressed in the above approximation, disturbances and valve non-linearities.

As was indicated before, the hydraulic ram is a non-linear device since the resonance frequency varies with the position of the piston. The resonance frequency has its lowest value with the piston at mid-stroke. So the stability considerations must be performed according to the worst conditions and the resultant controller must have enough phase and amplitude margin to overcome the non-linearities of the valve.

At this stage a graphical technique might be applied to determine the controller actions. Whereas the graphical methods have the advantage of simplicity they do not give enough knowledge about the final gain, the dominant closed loop poles and the dependance of the final gain to changes in the values of the parameters. Therefore the method of root-loci is used.

From the foregoing expressions (2) (3) (4) the open loop poles are

$$\left.\begin{array}{l} -1245 \\ -504 \pm j238 \end{array}\right\} \text{valve}$$
$$\left.\begin{array}{l} -660 \pm j660 \\ -0·0 \end{array}\right\} \text{motor} \quad (8)$$
$$-312 \pm j2870 \;\} \text{joint}$$

The controller actions must result in a high static gain and a fast response with little overshoot caused by disturbances.

First of all an integrating action is necessary to achieve a high static gain. To improve the dynamic behaviour differentiating actions must be added.

To compare the different controller settings, use has been made of the root-loci technique.

As a first choice of the controller settings, the integrator time constant must be approximately 5 to 8 times the second system pole (first pole $s = 0$) and a differentiating action must have a time constant approximately equal to this second system pole. So an integrator action was selected with

pole: $- 3·33$ r/s
zero: $- 36·5$ r/s

Three different runs are made with three types of derivative control actions for comparison;

(a) No derivative action: the root-locus plot is shown in Fig. 8(a). The maximum allowable gain equals to approximately 3300 s⁻¹.
(b) Derivative action set to

pole: $- 17\,500$ r/s
zero: $- 555$ r/s

The root-locus plot is shown in Fig. 8(b).
The maximum allowable gain is approximately 4700 s⁻¹.

(c) Derivative action set to

poles: $-1500 \pm j700$
zeros: $- 500 \pm j230$

The root-locus plot is shown in Fig. 8(c).
The maximum gain equals approximately 5100 s⁻¹.

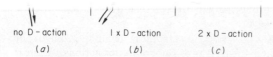

no D – action   I x D – action   2 x D – action
(a)   (b)   (c)

Figure 8. Root loci for different D-actions.

As can be seen from the plots it is advisable to add a derivative function to the controller both to achieve high static gain and acceptable dominant closed loop pole location. It was found that the actions of case (c) resulted in a worse behaviour around standstill, which may be caused by the valve non-linearities.

Finally for a controller with one derivative action, case (b) was selected.

PRACTICAL RESULTS

Some tests are performed at maximum speed (10 mm/s) and acceleration (0·25 m/s²).

15 μm and all lines have a width of 10 μm. The used acceleration (0·25 m/s²) and velocity 10 mm/s (if it is reached) are chosen.

CONCLUSION

A description of the hydraulic servosystems of the Philips pattern generator has been given. It is discussed that both a repetitive accuracy of the slides of less than 0·1 μm and a very fast dynamic response is possible.

The techniques used to achieve such a specification involve a careful mechanical design, anti-friction slideways, careful addition of control signals are discussed and found to be necessary in order to obtain the high performances, as listed in the introduction.

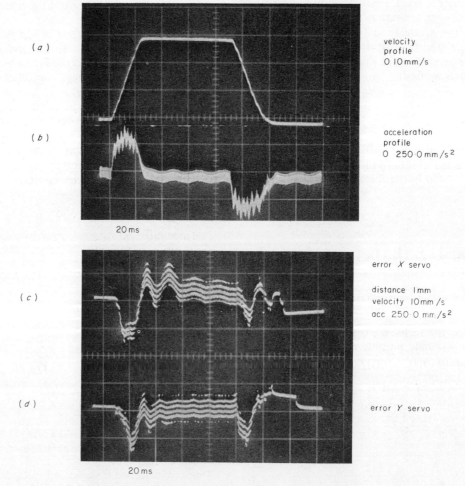

(a)   velocity profile 0 10mm/s

(b)   acceleration profile 0 250·0 mm/s²

20 ms

(c)   error X servo

    distance 1mm
    velocity 10mm/s
    acc 250·0 mm/s²

(d)   error Y servo

20 ms

Figure 9. Actual velocity, acceleration and error signals.

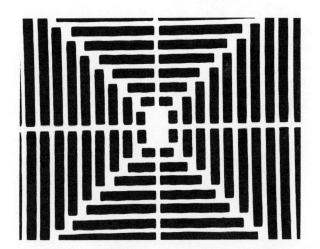

width 10 μm

shortest
line length 15 μm

Figure 10. Test pattern for the photographic pattern generator.

The work on the hydraulic servos is not finished yet, but further research is done to obtain faster and even more accurate servosystems, where the weakest point in the described system, viz. the electro-hydraulic valve, receives most of the attention.

REFERENCES

1. H. J. J. KRAAKMAN and J. G. C. DE GAST (1969). *Philips Tech. Rev.,* **30**, 121.
2. F. T. KLOSTERMAN and H. F. VAN HEEK (1971). Photographic pattern generator, paper presented at the topical meeting of the OSA. Optics in microelectronics, Jan. Copies on request from Philips Research Laboratories.
3. H. DE LANG, E. T. FERGUSON and G. C. M. SCHOENAKER (1969). *Philips Tech. Rev.,* **30**, 153.
4. A. DE PENNINGTON, J. J. 't MANNETJE and R. BELL. The modelling of electrohydraulic control valves and its influence on the design of electrohydraulic drives. Submitted for publication in the *Journal of Mechanical Engrs.*
5. F. T. KLOSTERMAN (1969), *Philips Tech. Rev.,* **30**, 57.

DISCUSSION

Q. M. Barash, Purdue University. What specific advantage does the 'miniature drafting machine' offer over the conventional method of making integrated circuit masks with the aid of standard NC drafting machines?

A. The advantage of the drafting machine over the conventional (cut and peel, etc.) method of making I.C. circuit masks concerns two factors. In the first place a higher accuracy is obtained, viz. a better outer size/accuracy ratio is achieved, combined with a much better line edge definition (≈ 0.1 μm). So finer details can be drawn. In the second place there is no intermediate step necessary to reduce the size to the standard form suitable for the photo repeaters. This is a saving in time, equipment and investment and no loss of accuracy or introduction of extra errors.

VIBRATION REDUCTION WITH CONTINUOUS DAMPING INSERTS

by

W. J. HAMMILL* and C. ANDREW†

SUMMARY

The case of uniform structural elements containing continuous damping inserts distributed between them is considered both analytically and experimentally with models representing machine spindles and cantilever bars. Continuous inserts are shown to be advantageous relative to the more conventional discrete ones.

INTRODUCTION

Damping inserts and vibration absorbers have been proposed for reducing the vibration amplitudes of engineering structures[1,2]. In general, however, the inserts considered have been of discrete lumped-parameter form, largely because the analyses of the structures to which they have been applied have used lumped-parameter models. In many structures there is an element, such as a spindle, which is essentially a continuous uniform element, and whose characteristics predominate in determining the important receptances of the structure. It would seem logical to damp such an element with a continuous, if not uniform, damping insert or vibration absorber. The purpose of this paper is to examine the effectiveness of such an insert.

PREVIOUS WORK

Taylor[3] and Cowley[4] have analysed lumped-parameter systems, and included damping as an approximation. Taylor suggested a modal damping assumption based on experience, while Cowley predicted modal damping using estimated real mode shapes and known damping source properties. The authors[1] discussed this work in detail, and proposed a combined eigenvalue and receptance solution, utilizing the basic method of Taylor for the bulk of the structure and combining this with discrete damping sources by receptance techniques. This represented a significant improvement in accuracy for a modest increase in effort.

None of these methods is applicable to the analysis of damping sources distributed over the length of a uniform element, without incurring penalties in terms of complication and loss of accuracy. However, the dynamic performance of damped 'sandwich beams' has received considerable attention in recent years. These have, in general, consisted of two purely elastic face plates with a linear viscoelastic core of specified complex shear modulus. Initial investigations were carried out by Kerwin[5], who determined overall loss factors for thin sandwich beams. Mead[6] has derived the sixth order differential equation of motion for transverse motion of a similar structure, and demonstrated the existence of damped normal modes, which can be used to determine the response to excitation. In these analyses the relative stiffnesses of the core and face-plates are assumed to be such that there is no relative transverse displacement between the face-plates at a particular cross-section. Damping can, therefore, only be attributed to shear strain in the core.

Such a model is not representative of the situation existing in heavy engineering structures like those of machine tools. In these the boundaries of the damping gasket or joint are likely to be of the same order of stiffness in the transverse direction as the gasket itself. Since the transverse stiffness of a gasket or oil film is considerably greater than the shear stiffness, this implies that the shear stiffnesses, and hence shear forces, are negligible in comparison with the other structural forces. Thus, the situation differs from that previously examined in that transverse

---

* CEGB, Berkeley Nuclear Laboratories
† University of Bristol, Department of Mechanical Engineering

effective vibration characteristics are to be modified, and an auxiliary beam, as shown in Fig. 1(a). This is developed to describe the characteristics of a pair of free-free beams of finite length as shown in Fig. 1(b).

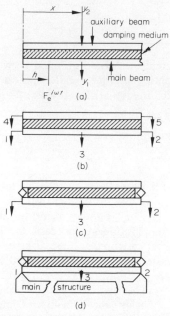

Figure 1. Schematic of beams with continuous inserts: (a) general case; (b) free–free beams of finite length–definition of receptance coordinates; (c) main and auxiliary beams pinned together–definition of receptance coordinates; (d) beam system coupled to main structure.

This system can then have various restrictions placed on it, or be connected into another system, by the method of receptances. Two special cases are considered:

(1) The two beams are pinned together and then pinned to a known structure at the ends, as indicated sequentially in Fig. 1(c) and (d)—a machine spindle configuration: the receptance at the mid-point is to be considered.

(2) The main beam is in the form of a cantilever and the auxiliary beam is connected to it only by the insert—a boring bar or like cantilever structure: the receptance at the tip is to be considered.

While the analysis described in the appendix is general, the cases now considered refer to the specific geometric configurations described in the experimental section; namely, 'spindle' 34 in long, 2 in square; and 'boring bar' 15 in long, 1 in square. The damping insert is in the form of a large number of butyl rubber pads, to facilitate control over the effective stiffness/unit length of the insert. These

rigid supports at the two ends, and it is assumed that the depth of the auxiliary beam and the effective stiffness/unit length of the insert are the variables. In the vicinity of the fundamental resonance of the main beam there are now two resonances. The insert stiffness (number of pads) can be adjusted such that these resonances are equal and this will be regarded as the optimum condition for the insert variable.

The optimum resonance amplitude is shown in Fig. 2(a), as a function of the depth ratio of the two beams, R. When R is small, the system is governed

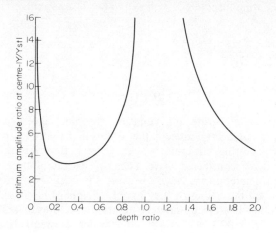

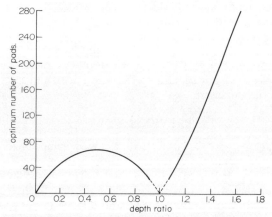

Figure 2. Theoretical results for pinned–pinned beams on rigid abutment: (a) variation of optimum amplitude ratio $|Y/Y_{st}|$ at beam centre with ratio of beam depths; (b) variation of optimum number of insert pads with ratio of beam depths.

largely by the undamped main beam, and its response approaches one infinite resonance as $R \to 0$. When $R = 1$ the two beams have the same natural frequencies, at which there is no relative motion and therefore no damping, and the resonant amplitude therefore again tends to infinity. Clearly there will be some intermediate values of R for which the two resonant amplitudes have a minimum value; from Fig. 2(*a*) this can be seen to be approximately $R = 0.3$ for the particular insert material and beam dimensions considered.

For values of $R > 1$, the optimum resonant amplitude decreases asymptotically to zero as R increases. It is interesting to note that R must be greater than approximately 2 if the optimum resonant amplitude is to be less than that corresponding to $R = 0.3$. The use of such a stiff auxiliary beam is not a practical solution since, for a given total depth of the two beams, the use of a stiffer main beam with $R = 0.3$ will give a similar resonant amplitude with a higher static stiffness.

The variation of the optimum stiffness of the insert is shown in Fig. 2(*b*), described in terms of the total number of pads forming the insert.

Cantilever with unpinned auxiliary beam (continuous absorber)

Calculations for this case were carried out to assess the influence of the flexibility of the auxiliary beam on its performance as a vibration reducer or 'absorber'. In practice, flexibility could be affected by the geometry or material of the beam. In the present case, the auxiliary beam dimensions were selected such that its first two non-zero modes of vibration straddled the fundamental frequency of the cantilever. The effect of flexibility was then assessed by including 2, 3 or 4 modal terms of the auxiliary beam in the series approximation for the cross receptances of this beam as described in the appendix. The resultant frequency responses for optimum damping insert conditions—defined as for the pinned–pinned beam case—showed negligible variation between the number of modal terms included, and hence the flexibility of the beam. This result was at first sight surprising. It can be explained in part by the near orthogonality of the flexural modes of the auxiliary beam and the first mode of the cantilever. Since the two beams are coupled via the damping insert, this implies that it is difficult for the cantilever to excite flexural modes of the auxiliary beam. Consequently these modes are not capable of producing large relative deformation of the insert, and hence energy dissipation. The opposite is true for the rigid body modes which are thus mainly responsible for the absorber's effectiveness.

The continuous absorber can be compared with a conventional rigid mass version. The rigid mass is assumed to have the same mass/unit length as the continuous absorber mass and to be connected to the cantilever by a discrete spring/damper at its centre of gravity. The optimum length of the rigid mass is approximately one-half of the cantilever length, noting that it must be contained within the overall length of the cantilever. In comparison with this system with an optimized spring/damper element, the continuous absorber shows an improvement of 15 per cent for the first cantilever mode. The advantage is smaller than was at first anticipated. This is primarily due to the small effect of the auxiliary beam near the abutment end of the cantilever and the diminishing returns of increasing mass ratio for high initial values

of, say, 0.25 or greater. It should be noted, however, that the continuous absorber is effective, if not optimally, for *all* modes of the cantilever.

EXPERIMENTAL

Pinned–pinned beams

The model used in this investigation consists of a main beam 2 in square and 34 in long mounted on a machine bed at each end. The mountings are assumed to provide no rotational constraint upon the main beam. Attached to the main beam is an auxiliary beam, also 2 in wide, and butyl pads are interposed between the two to simulate the continuous damping element. The pad characteristics were measured prior to incorporation into the model, and the characteristics are given in ref. 2. The main and auxiliary beams are attached to one another at each end by brackets with low bending stiffness, simulating a pinned fixing.

The direct and cross receptances of the support stiffnesses were determined separately over a frequency range from 30 to 200 Hz. This was accomplished by applying a harmonic force to each support in turn (without the main beam attached), transducing the force with a quartz dynamometer and the resultant motion at both support points with an accelerometer. In-phase and quadrature components of acceleration were measured at each frequency step with a Resolved-Components Indicator. The acceleration readings were then converted to displacement to define the receptances.

The receptance of the complete system was determined in a similar manner, by forcing and measuring acceleration at the mid-point of the main beam. During these tests the whole rig was enclosed in a polythene tent, and the temperature maintained at 78 ± 1 °F, the temperature at which the pad characteristics were obtained.

Cantilever with a continuous absorber

The model used in this investigation consists of a steel cantilever 1 in square and of 15 in overhang. This is clamped firmly to a very stiff I-section beam, in turn mounted on a massive concrete block. To the upper face of the cantilever overhang is attached an auxiliary beam of the same width, extending over its complete length and, interposed between the two, a number of cylindrical butyl pads. This rig was also temperature-controlled, as in the previous case.

The mass of the auxiliary beam was chosen so that it could be housed within the cantilever in the manner of a boring bar, without appreciable loss of main system stiffness. The stiffness of the 0.25 in deep auxiliary beam used initially was such that its first flexural frequency was considerably greater than that of the cantilever; hence only its rigid-body modes were likely to be effective as an absorber over the frequency range of interest.

Subsequent tests to examine the effect of auxiliary beam flexibility utilised a beam 0.125 in deep, to which were fixed seven steel blocks, intended to maintain the beam's mass constant but reduce its stiffness significantly. This ensured that the first flexural mode frequency of the absorber was less than that of the cantilever.

Pinned-pinned beams

The direct receptance at the centre of the pinned beam with its approximately optimal auxiliary beam and damping insert is shown in Fig. 3. Correlation between theory and experiment is good, and the dynamic magnification factor (Q) of the system has been reduced to about 3·25. The auxiliary beam

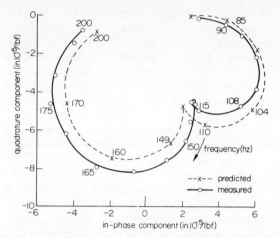

Figure 3. Predicted and measured receptance at centre of pinned–pinned beams coupled to main structure.

depth was chosen to be near the optimum value ($R = 0·30$) suggested in Fig. 2(a) for a beam with rigid end supports. The exact optimum for beam depth could be obtained by recalculating the curve of Fig. 2(a) including the influence of the finite end support stiffnesses. It should be noted, however, that the curves in both Fig. 2(a) and (b) are flat around the optimum values, which are therefore uncritical. A range of insert characteristics should also be included in these calculations to enable the optimum insert material to be selected, noting that other environmental limitations such as temperature, oil, etc., are also involved in this selection.

The use of the pinned end connections between the main and auxiliary beams would enable a precise location to be achieved between them, such as would be necessary in a balanced spindle assembly. The auxiliary beam could take the form of a sleeve inside or outside the spindle proper.

Cantilever with continuous absorber

The cantilever tip receptance is shown in Fig. 4. Again good correlation between theory and experiment has been achieved, and although not shown in these results, this also applies to correlation of predicted and measured phase angles. The two sets of experimental points for auxiliary beams of depths

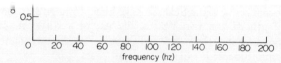

Figure 4. Predicted and measured receptance at cantilever tip for different auxiliary beam flexibilities.

0·25 and 0·125 in (but the same mass) confirm that auxiliary beam flexibility has negligible influence. The dynamic magnification factor has been reduced to about 3·5. Although in this case the improvement relative to an efficient discrete absorber is not marked, it should be noted that the continuous absorber has the advantage of versatility in affecting *all* modes of the system, and dissipates its energy over a wide area.

CONCLUSIONS

An analysis has been presented of the dynamic characteristics of uniform structural elements containing damping inserts distributed uniformly along them. Such inserts are shown to be advantageous for the two particular cases considered, representing a machine tool spindle and a cantilever boring bar.

ACKNOWLEDGMENTS

The authors wish to thank Professor J. L. M. Morrison, CBE, formerly Head of the Department of Mechanical Engineering, University of Bristol, and Mr A. E. DeBarr, Director of the Machine Tool Industry Research Association, for their advice and encouragement of this collaborative project; and the Machine Tool Industry Research Association for its sponsorship of one of the authors.

REFERENCES

1. W. J. HAMMILL and C. ANDREW (1971). Receptances of lumped-parameter systems containing discrete damping sources, *J. Mech. Eng. Sci.*, **13** (4), 296.
2. W. J. HAMMILL (1971). The dynamic analysis of structures containing damping sources, PhD thesis, University of Bristol.
3. S. TAYLOR (1966). The design of a machine tool structure using a digital computer, *Proc. 7th Int. Machine Tool Design and Research Conf.*, Pergamon Press, London, and The prediction of the dynamic characteristics of machine tool structures, PhD thesis, University of Birmingham.
4. A. COWLEY (1968). The prediction of the dynamic characteristics of machine tool structures, PhD thesis, University of Manchester.

5.   E. M. KERWIN, Jr. (1959). Damping of flexural waves by a constrained visco-elastic layer, *J. Acous. Soc. Am.*, **31**, 7.

6.   D. J. MEAD and S. MARKUS (1969). The forced vibration of a three-layer, damped sandwich beam with arbitrary boundary conditions, *J. Sound Vib.*, **10**(2), 163.

7.   R. E. D. BISHOP and D. C. JOHNSON (1960). *The mechanics of vibration*, Cambridge University Press.

APPENDIX: THEORY

Notation

a	Coefficient determined by end conditions of beam
M	Mass of beam
$y(x, t)$	Displacement of beam at a distance x from one end at time t.

For the above symbols the suffix 1 denotes the main beam, and suffix 2 the auxiliary beam.

C_i, D_i	Value of ith definite integral (defined in equation 5)
F	Vector of harmonic force $Fe^{j\omega t}$
I_{ij}	Value of definite integral (defined in equation 6)
k_1, k_2	Complex support stiffnesses
$k(x)$	Complex stiffness/unit length of damping medium
k^*	Complex stiffness/unit length of damping insert, assumed constant over insert length
l	Length of beam
P	Vector of harmonic force $Pe^{j\omega t}$
P_i	Vector of harmonic force at coordinate i
q_i	Vector of harmonic displacement of coordinate i
r	Number of modes
$u(x, t)$	Relative displacement of main and auxiliary beams at a distance x from one end, at time t
$U(x)$	Vector of harmonic relative displacement of main and auxiliary beams at a distance x from one end.
Y	Vector of maximum harmonic displacement
Y_{st}	Static displacement
α_{xh}	Cross-receptance of beam, measuring at a distance x from one end, forcing at a distance h from one end
β_{ij}, γ_{ij}	Cross-receptance of sub-system measuring at coordinate i, forcing at coordinate j
$\Theta(x)$	Phase lag between the relative displacement of main and auxiliary beams at a distance x from one end, and the applied force $Fe^{j\omega t}$
$\Phi_i(x)$	ith characteristic function of main beam
$\Psi_i(x)$	ith characteristic function of auxiliary beam
ω	Forcing frequency
ω_i	ith natural frequency of main beam
Ω_i	ith natural frequency of auxiliary beam

Analysis

The following analysis considers the dynamic response of two uniform elastic elements with a continuous spring/damping element interposed between them, as shown in Fig. 1(a).

The cross-receptance α_{xh} of a uniform elastic beam may be expressed in series form[7] as

$$\alpha_{xh} = \sum_{i=1}^{\infty} \frac{\phi_i(x)\phi_i(h)}{a_1(\omega_i^2 - \omega^2)}$$

For the main beam of Fig. 1(a) excited by an external force $Fe^{j\omega t}$ distant h from one end

$$y_1(x, t) = \sum_{i=1}^{\infty} \frac{\phi_i(x)\phi_i(h)F\, e^{j\omega t}}{a_1(\omega_i^2 - \omega^2)}$$

$$- \int_0^l \sum_{i=1}^{\infty} \frac{\phi_i(x)\phi_i(h)k(h)u(h, t)\, \mathrm{d}h}{a_1(\omega_i^2 - \omega^2)} \qquad (1)$$

Similarly for the auxiliary beam

$$y_2(x, t) = \int_0^l \sum_{i=1}^{\infty} \frac{\psi_i(x)\psi_i(h)k(h)u(h, t)\, \mathrm{d}h}{a_2(\Omega_i^2 - \omega^2)} \qquad (2)$$

Subtracting (1) and (2)

$$u(x, t) = \sum_{i=1}^{\infty} \frac{\phi_i(x)\phi_i(h)F\, e^{j\omega t}}{a_1(\omega_i^2 - \omega^2)} -$$

$$- \int_0^l \sum_{i=1}^{\infty} \frac{\phi_i(x)\phi_i(h)k(h)u(h, t)\, \mathrm{d}h}{a_1(\omega_i^2 - \omega^2)}$$

$$- \int_0^l \sum_{i=1}^{\infty} \frac{\psi_i(x)\psi_i(h)k(h)u(h, t)\, \mathrm{d}h}{a_2(\Omega_i^2 - \omega^2)} \qquad (3)$$

It will be assumed that r modes are sufficient to describe the characteristics of the two beams. It will also be assumed that $u(x, t) = U(x)\, e^{j(\omega t - \Theta(x))}$ and that the insert stiffness/unit length—$k(x)$—is constant. The latter assumption is made in order to simplify the analysis, although any specified distributions of the damping insert stiffness can be examined. Equation 3 may be written.

$$U(x)\, e^{-j\theta(x)} = \sum_{i=1}^{r} \frac{\phi_i(x)\phi_i(h)F}{a_1(\omega_i^2 - \omega^2)}$$

$$- k^* \int_0^l \sum_{i=1}^{r} \frac{\phi_i(x)\phi_i(h)U(h)\, e^{-j\theta(h)}\, \mathrm{d}h}{a_1(\omega_i^2 - \omega^2)}$$

$$- k^* \int_0^l \sum_{i=1}^{r} \frac{\psi_i(x)\psi_i(h)U(h)\, e^{-j\theta(h)}\, \mathrm{d}h}{a_2(\Omega_i^2 - \omega^2)} \qquad (4)$$

Now let

$$\int_0^l \phi_i(h)U(h)\, e^{-j\theta(h)}\, \mathrm{d}h = C_i$$

and

$$\int_0^l \psi_i(h)U(h)\, e^{-j\theta(h)}\, \mathrm{d}h = D_i$$

Both sides of equation 5 are now multiplied by $\phi_j(x)$ and each term integrated with respect to x between the limits $x = 0$ and $x = l$. Use is made of the orthogonality relationship

$$\int_0^l \phi_i(x)\phi_j(x)\,dx = \begin{cases} 0 \text{ if } i \neq j \\ l \text{ if } i = j \end{cases}$$

This gives

$$C_j = \frac{Fl\phi_j(h)}{a_1(\omega_j{}^2 - \omega^2)} - \frac{k^*lC_j}{a_1(\omega_j{}^2 - \omega^2)}$$

$$- k^* \sum_{i=1}^r \frac{D_i I_{ij}}{a_2(\Omega_i{}^2 - \omega^2)} \quad (j = 1, 2 \ldots r) \quad (6)$$

where

$$I_{ij} = \int_0^l \psi_i(x)\phi_j(x)\,dx$$

Similarly, multiplying both sides of equation 5 by $\psi_j(x)$ and integrating between the limits $x = 0$ and $x = l$ gives

$$D_j = F \sum_{i=1}^r \frac{\phi_i(h)I_{ji}}{a_1(\omega_i{}^2 - \omega^2)} - k^* \sum_{i=1}^r \frac{C_i I_{ji}}{a_1(\omega_i{}^2 - \omega^2)}$$

$$- \frac{k^*lD_j}{a_2(\Omega_j{}^2 - \omega^2)} \quad (j = 1, 2 \ldots r) \quad (7)$$

The $2r$ equations 6 and 7 can now be used to determine C_j/F, D_j/F ($j = 1, 2, \ldots r$).

Equation 1 may now be written:

$$y_1(x, t) = F e^{j\omega t} \sum_{i=1}^r \frac{\phi_i(x)}{a_1(\omega_i{}^2 - \omega^2)} (\phi_i(h) - k^* C_i/F) \quad (8)$$

The use of values of C_i/F, obtained from equation 6 and equation 7, in equation 8 permits calculation of the cross-receptance α_{xh} of the system of Fig. 1(b) at any frequency ω.

Two specific examples are now considered; firstly, a beam on finite support stiffnesses with an auxiliary beam pinned to it at either end and a continuous damping element interposed between the two, and secondly, a cantilever with a continuous free-free vibration absorber. These have practical significance with respect to spindle/bearing assemblies and long slender boring bars respectively.

Pinned-pinned beam
The required receptances can be derived in three stages. First, receptances of the free-free beams

describe the beam characteristics, equation 6 becomes

$$C_j = \frac{Fl\phi_j(h)}{M_1(\omega_j{}^2 - \omega^2)} - \frac{k^*lC_j}{M_1(\omega_j{}^2 - \omega^2)}$$

$$- \frac{k^*lC_j}{M_2(\Omega_j{}^2 - \omega^2)} \quad (j = 1, 2 \ldots r) \quad (9)$$

since both beams have the same characteristic function $\phi_j(x)$. Substitution into equation 1 gives

$$\frac{y_1(x, t)}{F e^{j\omega t}} = \sum_{i=1}^r$$

$$\left[\frac{\phi_i(x)\phi_i(h)(k^*l + M_2(\Omega_i{}^2 - \omega^2))}{M_1 M_2(\omega_i{}^2 - \omega^2)(\Omega_i{}^2 - \omega^2)} + k^*l(M_1(\omega_i{}^2 - \omega^2) + M_2(\Omega_i{}^2 - \omega^2)) \right] (10)$$

Similarly, equation 2 becomes

$$\frac{y_2(x, t)}{F e^{j\omega t}} = \sum_{i=1}^r$$

$$\left[\frac{\phi_i(x)\phi_i(h)k^*l}{M_1 M_2(\omega_i{}^2 - \omega^2)(\Omega_i{}^2 - \omega^2)} + k^*l(M_1(\omega_i{}^2 - \omega^2) + M_2(\Omega_i{}^2 - \omega^2)) \right] (11)$$

Equations 10 and 11 can now be used to determine any receptance of the system of Fig. 1(c).

If the beam system be denoted by B, and the main system by C, the derivation of the complete system receptances is as follows.

$$\begin{vmatrix} q_1 \\ q_2 \\ q_3 \end{vmatrix} = \begin{vmatrix} \beta_{11} & \beta_{12} & \beta_{13} \\ \beta_{21} & \beta_{22} & \beta_{23} \\ \beta_{31} & \beta_{32} & \beta_{33} \end{vmatrix} \begin{vmatrix} P_1 \\ P_2 \\ P_3 \end{vmatrix} \text{ beams}$$

and

$$\begin{vmatrix} q_1 \\ q_2 \end{vmatrix} = - \begin{vmatrix} \gamma_{11} & \gamma_{12} \\ \gamma_{21} & \gamma_{22} \end{vmatrix} \begin{vmatrix} P_1 \\ P_2 \end{vmatrix} \text{ main system}$$

These equations can be solved for q_3/P_3, the required receptance of the complete system.

The γ receptances will have been derived by the most convenient means available, such as experimental measurement, lumped-parameter model analysis, etc. It was pointed out in ref. 2 that this hybrid model analysis—for example, part uniform element, part lumped-parameter—can be economically advantageous irrespective of the inclusion of damping in the long uniform elements of the structure.

Cantilever with a continuous vibration absorber
It will be assumed that only the first mode of the cantilever is significant, while r modes are necessary to describe the characteristics of the auxiliary beam. It is required to calculate the tip-receptance of the composite system, and forcing and measuring both take place at the coordinate $x = l$. Equation 6 becomes

$$C_1 = \frac{Fl\phi_1(l)}{M_1(\omega_1{}^2 - \omega^2)} - \frac{k^* l C_1}{M_1(\omega_1{}^2 - \omega^2)}$$

$$- k^* \sum_{i=1}^{r} \frac{D_i I_{i1}}{M_2(\Omega_i{}^2 - \omega^2)} \qquad (12)$$

Equation 7 becomes

$$D_j = \frac{F\phi_1(l)I_{j1}}{M_1(\omega_1{}^2 - \omega^2)} - \frac{k^* C_1 I_{j1}}{M_1(\omega_1{}^2 - \omega^2)}$$

$$- \frac{k^* l D_j}{M_2(\Omega_j{}^2 - \omega^2)} \qquad (j = 1, 2 \ldots r) \quad (13)$$

Equation 12 and equation 13 can be combined to give

$$\frac{C_1}{F}\left[k^* l + M_1(\omega_1{}^2 - \omega^2) \right.$$

$$\left. - k^{*2} \sum_{i=1}^{r} \frac{I_{i1}{}^2}{k^* l + M_2(\Omega_i{}^2 - \omega^2)} \right]$$

$$= \phi_1(l)\left[l - k^* \sum_{i=1}^{r} \frac{I_{i1}{}^2}{k^* l + M_2(\Omega_i{}^2 - \omega^2)} \right] (14)$$

Equation 8 for this system becomes:

$$y_1(l, t) = F e^{j\omega t} \frac{\phi_1(l)}{M_1(\omega_1{}^2 - \omega^2)}\left[\phi_1(l) - \frac{k^* C_1}{F} \right] (15)$$

Equation 14 and equation 15 are now used to determine the required receptance over a suitable frequency range. The definite integrals $I_{i1}(i = 1, 2 \ldots r)$ are first evaluated once for all frequencies, the first two integrals (I_{11}, I_{21}) involving the rigid-body modes of the auxiliary beam. Also calculated are the appropriate natural frequencies $\omega_1, \Omega_1, \Omega_2 \ldots \Omega_r$.

Using these results, the value of the complex variable C_1/F can then be found at each frequency from equation 14, and substituted into equation 15 to determine the real and imaginary components of the tip-receptance of the composite system.

DISCUSSION

Q. A. Cowley. The paper indicates that the distributed type of damper does not require such critical tuning as the more conventional discrete (single auxiliary mass) damper, and this is an important advantage. Professor Andrew pointed out in his presentation that the distributed damper will have a desirable affect on all modes, which is not necessarily the case for the conventional damper. I would suggest that the single mass damper will have an effect on all modes except those which exhibit a node at the location of the auxiliary damper. The probability of the latter circumstance, however, is rather small.

The question I should like to ask is, therefore, what effect will the distributed damper have upon auxiliary modes compared with the single mass damper? I realize that such a question is too general to expect a complete answer but perhaps the authors could give some indication with reference to a specific example.

Q. G. J. McNulty. Third paragraph, Page 1, is not clear: sentence beginning 'Mead has derived'—Mead in his paper has analysed a continuous structure (i.e. with generalized mass, stiffness and damping). So he must have an infinite number of modes and it is not surprising that the sixth mode is excited (as stated by the authors).

The results produced by the authors for the inserts show good results at the lower modes, and I feel that because of the close proximity of the modes they will be suitable also for the higher modes.

Q. R. Umbach. (1) In Fig. 4 there are shown two peaks, which means two modes for the beam with the auxiliary system. This occurs if there are two systems connected with one another because of the ratio of masses, ratio of frequencies and ratio of damping. The question is whether there is only one peak in the particular range of frequencies for the cantilever arm without the damper or what explanation can be given?

(2) What happens with other polymers because of another damping?

A. The authors thank the contributors for their interesting and helpful comments.

In an attempt to answer Dr. Cowley's question, we have analysed the tip receptance of the cantilever at frequencies in the range of its second flexural mode, for the cases of the continuous and the discrete absorbers considered in the paper. With the continuous absorber, the worst dynamic magnification factor in this range is approximately 14, while the corresponding figure for the discrete absorber is approximately 360. It would be wrong to claim that these specific results prove the general case or have a general quantitative significance. The material characteristics were specific, and were assumed constant with frequency (it is interesting to note that the common practical rise in stiffness with frequency of rubber-like materials would improve the performance of both absorbers for the higher modes). The discrete absorber was optimized for the fundamental mode of the cantilever, and with the physical restrictions of containment within the cantilever the point of attachment has come near to the node of the second cantilever mode. A compromise between first and second mode performance could be attained, but this would, of course, cause the discrete absorber to compare even less favourably with the continuous one for the fundamental mode. The results do, however, bear out the point that the discrete absorber

order differential equation, and demonstrates the existence of damped normal modes. As with the fourth order equation for transverse vibration of simple uniform beans, this is an eigenvalue problem which has an infinite number of eigenvalues and associated eigenvectors (modes) as its solution. The mathematics, however, is complex, and in preference to such an exact classical approach we are proposing for our case a simpler series solution, which can be taken as far as is necessary for any desired accuracy. We did indeed derive the equations of motion in classical form[2], but only pursued the analysis for the simpler series solution.

Dr Umbach notes the two resonant peaks in the frequency range of the fundamental cantilever mode. We agree with him that this is approximately equivalent to the two peaks anticipated from a discrete cantilever fundamental as is shown in Fig. 4. This phenomenon is explained mathematically in ref. 2.

The use of different polymers for the damping element will modify the findings of the paper quantitatively. The optimum damping characteristic will vary from system to system, and would have to be determined by carrying out the analysis described, including optimization of the quantity of material used, for a number of different characteristics, and selecting the best. It is likely that, in practice, additional criteria such as oil or heat resistance would also have to be considered. It should be noted that no such material optimization has been carried out for the continuous absorber in the present case. The material used was selected on the basis of convenience of availability and known characteristics, and the results are therefore likely to be pessimistic relative to some other materials.

A CONTRIBUTION TO THE EFFECTIVE RANGE OF THE PRELOAD ON A BOLTED JOINT

by

YOSHIMI ITO*

SUMMARY

In order to clarify the static and dynamic behaviour of a bolted joint on a machine tool the effects of joint surfaces on the effective range of the preload are investigated in this paper.

Two basic experiments in which the interface pressure distribution driven from the preload has been measured by means of ultrasonic waves were carried out. From the results it is known that the effective range of the preload is quite different from the results presented already elsewhere, namely that the preload acts over a larger area of the bolted joint compared with the range calculated by Rötscher's pressure cone. For example, the half angle of pressure cone α given by our experiments is about 65 deg for the lapped joint surfaces, and furthermore this angle decreases with improving joint surface roughness.

We can conclude that the joint surfaces have large effects on the effective range of the preload and on the interface pressure distribution.

INTRODUCTION

With the advance of studies on the static and dynamic stiffness of machine tool structures the significance of the bolted joint has been pointed out, and many studies on the static and dynamic behaviour of the bolted joint (for example, on joint stiffness, on damping capacity and on the frequency response of the bolted joint) have now been done. However, the static and dynamic behaviour of the bolted joint have not yet been fully clarified, and further basic studies are required at this stage.

One of the reasons why we cannot get exact information on the bolted joint is that the interface pressure distribution, considering the effect of joint surfaces, has not been clarified. This interface pressure distribution is very important not only to clarify the static and dynamic behaviour of the bolted joint on a machine tool but also to calculate the spring constant of flange in a bolt-flange assembly found on other machines.

Concerning the fatigue strength of bolts, the relaxation of connecting bolts subjected to exciting forces, the leakage between the flanges, etc., many studies on the spring constant of flanges have so far been carried out because, as mentioned above, the bolted joint is one of the most common types of mechanical connections.

In most previous studies[1], however, the bolt-flange assembly has been dealt with as a three-dimensional problem in stress analysis, of an axi-symmetrical elastic body having finite length or infinite length, that is, the existence of the joint surfaces has not been considered in a stress analysis or in a stress distribution. If the stress distribution on the midplane of the finite hollow cylinder is related to the interface pressure distribution on the bolted joint, from the results of these previous studies it may be said that the interface pressure distribution is spread over within the range given by Rötscher's pressure cone.

According to our experience, from studies on the bolted joints on a machine tool, we cannot agree with the results already presented, because much interesting behaviour of bolted joints for which we could not provide a reason was frequently pointed out[2]. It was also known that the effects of joint surfaces on the normal stiffness of the hollow cylinder subjected to the compressed forces were very large; for example[3], the ratio between the normal stiffness of the jointed hollow cylinder and that of the equivalent solid is about $\frac{1}{3} \sim \frac{1}{5}$, though this ratio changes a little be the machining methods of joint surfaces and by the value of compressed force.

In this paper, therefore, consideration of the effects of joint surfaces, the interface pressure distribution on the bolted joint and on the joint surfaces between two contact elastic bodies have been measured by means of ultrasonic waves in order to

* Associate Professor: Department of Mechanical Engineering for Production, Tokyo Institute of Technology, 2-12-1, Ohokayama, Meguro-ku, Tokyo, Japan

In Fig. 1 the experimental set-up is shown. As shown in the figure the upper and lower test specimens having ground joint surfaces were connected to each other by using an M8 connecting bolt, and the

On the stem of the connecting bolt the strain gauges were fixed and by using these strain gauges the preload of the connecting bolt was controlled and kept at a constant value.

In all experiments, after the positioning of the holder at a certain point, the preloads of the connecting bolt were changed and the changes of the echo height on the C.R.T. were measured. The positions of the holder were changed at regular intervals along the circle and each position of the holder could be seen as the γ-θ coordinate by reading the graduation on the dividing plate. Thus the distribution of the echo height along the radius direction of the joint surfaces between the bolt-flange assembly can be measured and from this result it is easy to know the qualitative interface pressure distribution of the preload.

In Figs 2 and 3 some examples of measured results are shown. In these figures on the vertical axis the

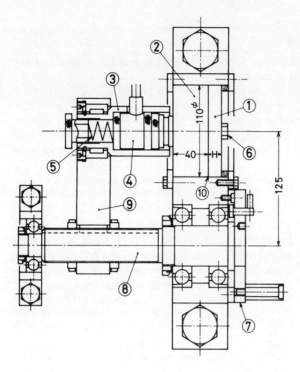

Figure 1. Schema of the experimental set-up for model test 1. (1) Upper test specimen, (2) lower test specimen, (3) holder, (4) crystal oscillator, (5) spring, (6) M8 connecting bolt, (7) dividing plate (8) threaded shaft, (9) arm, (10) joint surfaces to be investigated.

ultrasonic waves generated by the crystal oscillator were thrown into the joint surfaces through the bottom of a lower test specimen having a threaded hole. The interface pressure to be measured was detected by using the same crystal oscillator and was shown on the C.R.T. of the ultrasonic flaw detector as the echo height of the ultrasonic waves reflected at the joint surfaces.

According to previous studies published elsewhere[4] the echo height indicates the sound pressure of the reflected waves at the joint surfaces assuming that the material damping for the sound can be disregarded, and the contact pressure between the joint surfaces is approximately in linear relation to the echo height on the C.R.T. Therefore the interface pressure can be measured as the change of echo height.

In this measuring method the accuracy and reliability of measured values depend largely upon the

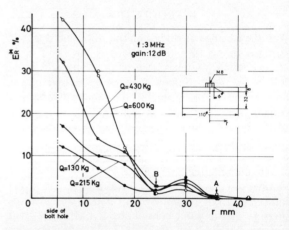

Figure 2. Changes of the qualitative interface pressure distributions with increasing preloads. Joint surfaces: ground, $H_{max} = 2 \cdot 0 \, \mu m$; jointed material: structural steel (JIS S45C); f: frequency of the crystal oscillator; Q: preload of the connecting bolt.

characteristic E_R^* is taken to show the measured results. Here E_R^* is equal to $(1 - E_R)$ and E_R given by h_e/h_i is the ratio of echo height[4], where h_e is the height of the reflected echo and h_i is the height of the initial pulse on the C.R.T.

It can be seen from the results shown in Fig. 2 that the half angle of the pressure cone α is 74 deg. Even if the interface pressure to be considered is only spread over within the point B, the value of α is about 63 deg. This latter result for the effective range of the preload is in close agreement with another experimental result presented already by R. Plock[5]. In that study the effective range of the preload was measured by putting a pressure sensitive paper between the

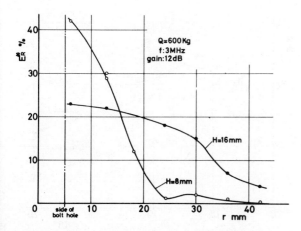

Figure 3. Changes of the qualitative interface pressure distributions by the thickness of upper flange. H: thickness of the upper flange.

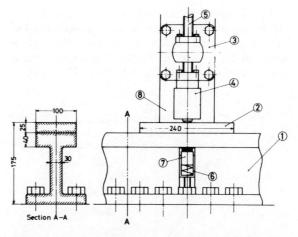

Figure 4. Schematic of the experimental set-up for model test 2. (1) Lower test specimen (guideway), (2) upper test specimen (saddle), (3) loading attachment, (4) load transducer, (5) loading screw, (6) holder, (7) crystal oscillator, (8) angle plate.

joint surfaces themselves, and from the results presented the value of α was calculated as about 65 deg. From the experimental results it can also be seen that the interface pressure distribution changes in relation to the preload and the thickness of the upper test specimen, namely the flange thickness, but the effective range of the preload is not influenced by the value of preload. The form of the interface pressure distribution itself becomes uniform with the increasing thickness of the flange.

THE INTERFACE PRESSURE DISTRIBUTION OF THE ELASTIC BEAM ON THE ELASTIC FOUNDATION (MODEL TEST 2)

On the flanges found on the bolted joints of machine tool constructions the bendings and warps can frequently take place easily when the connecting bolts are preloaded or the external loads caused by the cutting forces are applied, because the bolt-flange assemblies on machine tool constructions have the same shapes and dimensions as a thin plate subjected to a concentric load. It is, therefore, reasonable to consider that the most simple model of the bolted joint is an elastic beam on an elastic foundation subjected to a concentric load. Thus in the testing of this model the interface pressure distribution of the joint surfaces between the elastic beam and the elastic foundation has been measured by means of ultrasonic waves.

The experimental set-up which was used in the previous study on the pressure distribution of machine tool slideways[6] is shown again in Fig. 4. The experiments were carried out with a ground steel beam having the following dimensions, $b = 100$ mm, $L = 240$ mm and $h = 25$ mm, and with the ground steel foundation.

The experimental procedures are as follows. After the crystal oscillator was settled into the holder with the spring and pressed against the smoothed surface of the underside of the elastic foundation by using this spring, the concentric loads of measured values were applied on the elastic beam by using the load cell. Under this condition the interface pressures for the various concentric loads were measured as the

changes of the echo height on the C.R.T. By reason of the construction of the experimental set-up, it is difficult to measure simultaneously the interface pressure taking place at every point; therefore the measurements were taken at every loaded point and these loaded points were given by displacing the ground steel beam to a certain distance along the longitudinal axis only after the measurement of the interface pressure at the previous point.

From these measured results the echo height distribution, namely the qualitative interface pressure distribution, can be seen and one of the experimental results is shown in Fig. 5. It can be seen that the beam on the elastic foundation clearly indicates a local deflection with the increase of the concentric load. Therefore, the range affected by the concentric load, namely the effective range of the concentric load on the joint surfaces, may be regarded as the range within the distance a_0 shown together in Fig. 5.

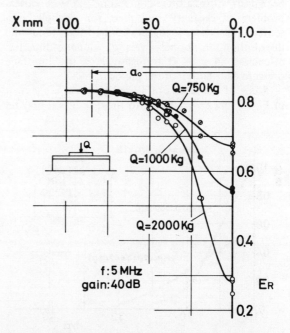

Figure 5. Measured result of the echo height distribution between the joint surfaces. Joint surfaces: ground, $H_{max} = 2\cdot0$ μm; jointed materials: mild steel (JIS SS41B).

R

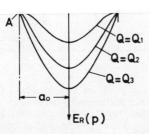

Figure 6.

the concentric load is between 500 and 1500 kg. This value of α is in close agreement with the results obtained by model test 1; therefore one can also confirm that the joint surfaces between the bolt-flange assembly or between the two contact bodies have a large effect on the interface pressure distribution and also on the effective range of the preload.

DISCUSSION

In the previous studies on the interface pressure distribution of the bolt-flange assembly there were very significant discrepancies concerned with the actual bolt-flange assembly, namely that the joint surfaces have not been taken into consideration in the theoretical analysis of the interface pressure distribution. In those studies the bolt-flange assembly has been dealt with as a three-dimensional axi-symmetrical problem of elasticity, and then, for example, presented in results by Shibahara[1] or by I. Fernlund[7] the distribution of normal stress σ_z along γ-direction calculated at $z = 0$ is regarded as the interface pressure distribution of the preload.

Some of the results presented by them are shown in Figs 7 and 8. According to these results it may be

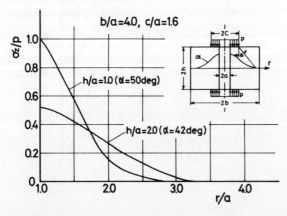

Figure 7. Theoretical interface pressure distribution presented by M. Shibahara and J. Oda. Poisson's ratio $\nu = 0.5$.

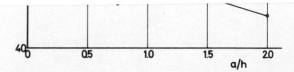

Figure 8. Theoretical values of α presented by I. Fernlund. Poisson's ratio $\nu = 0.25$.

said that the half angle of pressure cone α is nearly equal to 45 deg; however, as shown already in the previous chapter, the values of α obtained by our model tests and also R. Plock's experiments vary from 60 deg to 75 deg. From this result it is clear that the joint surfaces between the two flanges to be bolted have large effects on the interface pressure distribution and also on the effective range of the preload, and that the effective range of the preload spreads widely over the joint surfaces compared with the theoretical effective range of preload calculated in the previous studies in which the joint surfaces have not been considered.

As a result it can be concluded that the previous theoretical studies calculating the spring constant of the flange and the interface pressure distribution cannot be applied to solve the actual problem of the bolt-flange assembly, and thus the new proposals of how to calculate theoretically the interface pressure distribution of the bolt-flange assembly by considering the joint surfaces require the further development of the kind of studies described here.

However, as shown in Figs 7 and 8 interesting results can be pointed out from the theoretical results calculated already for the interface pressure distribution by using the theory of elasticity. Although the quantitative correlations between the actual and the theoretical results relating to the range of preload and to the form of interface pressure distribution are different from each other, the experimental results show that the effective range of the preload is always constant, even if the preload changes considerably, and that the interface pressure distribution shows a gentle curve with the increasing thickness of the flange. These results are qualitatively in good relation to the theoretical results presented already elswhere.

To clarify further the effects of the joint surfaces on the effective range of the preload, model test 2 was carried out with the various joint surface conditions. In Table 1 the values of angle α calculated from the measured results of this test are shown. For all that the values of α vary with the condition of each joint surface, they are generally scattered within 60 deg to 70 deg, and the half angle of the pressure cone α indicates small values when the joint surfaces are good, having a small roughness, or

TABLE 1 Measured results of α for the various joint surfaces conditions. Nomenclature in brackets are JIS Nos. for metallic materials

Conditions of Joint Surfaces						Half Angle of the Pressure Cone α deg
Lower Test Specimen			Upper Test Specimen			
Material	Machining Method	Surface Condition	Material	Machining Method	Surface Condition	
Cast iron (FC25)	Lapped	Flame Hardening (Depth:1mm) H$_B$:450~470 H$_{max}$:20μm	Cast iron (FC35)	Scraped	30/1in²	67
					60/1in²	63
			Cast iron (FC25)	Ground	H$_{max}$:15μm	69
					H$_{max}$:40μm	72
					H$_{max}$:2.8μm H$_{RC}$:40 (F.H)	58
				Lapped	H$_{max}$:1.7μm	63
			Case hardening Steel	Ground	H$_{max}$:2.5μm H$_{RC}$:59	61
Cast iron (FC35)	Scraped	30/1in²	Cast iron (FC35)	Scraped	30/1in²	67
					60/1in²	63
			Cast iron (FC25)	Ground	H$_{max}$:20μm	70
					H$_{max}$:32μm H$_{RC}$:30 (F.H)	63
				Lapped	H$_{max}$:1.7μm	63
Mild steel (SS41B)	Ground	H$_{max}$:20μm	Mild steel (SS41B)	Ground	H$_{max}$:20μm	73

the joint surfaces hardened, namely when the normal contact stiffness of the joint surfaces is marked.

In Fig. 9, in one of the results which was investigated, the effect of the joint surfaces roughness on the values of α is also shown. In this experiment for reasons of easy control of the surface roughness the joint surfaces lapped with each other have been investigated. From this experimental result it can also be seen that the joint surfaces roughness has a large effect on the values of α, and that the values of α decrease and approach 45 deg with the improving in the surface roughness.

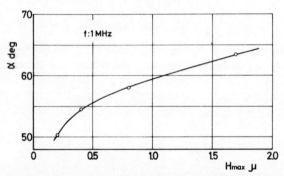

Figure 9. Effects of the surface roughness on the half angle of pressure cone. Joint surfaces: lapped; jointed materials: flame hardened cast iron–cast iron (JIS FC25).

These facts mean that the effective range of the preload may be under the control of the distribution of the real contact points between the joint surfaces; therefore if two perfectly flat surfaces having no surface roughness are in contact with each other, the effective range of the preload may be equal to the theoretical range of the preload.

From the results mentioned above the significance of the joint surfaces between the bolt-flange assembly can also be confirmed.

CONCLUSION

In order to clarify the influence of the joint surfaces on the effective range or on the interface pressure distribution of the preload, in this study two model tests, in which the interface pressure or the contact pressure was measured by means of ultrasonic waves, were carried out with the simple bolt-flange assembly and with the elastic beam on an elastic foundation.

Contrary to the theoretical analysis presented elsewhere, in which the bolt-flange assembly has been dealt with as three-dimensional axi-symmetrical problems of elasticity and the joint surfaces have not been taken into consideration in its calculation, the experimental results indicate clearly that the joint surfaces have large effects on the effective range and on the interface pressure distribution of the preload.

From the experimental results it is known that the effective range and the interface pressure distribution are closely related to the joint surfaces condition, and that the half angle of pressure cone decreases by improving the joint surfaces roughness. The half angle of pressure cone seen in all these experiments varies from 60 deg to 70 deg notwithstanding the various joint surfaces conditions. The half angle of pressure cone calculated theoretically is about 45 deg, and therefore we can conclude that the preload of connecting bolt at the actual bolt-flange assembly spread over a wide range of joint surfaces to compare with the range of preload at the ideal bolt-flange assembly.

It may also be said that there are some discrepancies in the previous studies, calculating theoretically the interface pressure distribution by using theory of elasticity; therefore the significance of further experimental studies, investigated such as here, should be pointed out, and theoretical studies, especially on how to determine the boundary condition at the joint surfaces, are furthermore of importance.

ACKNOWLEDGMENTS

The author wishes to thank Prof. Dr M. Masuko for his advice and help during this study. The financial support of the Scientific Research Fund from the Ministry of Education is also gratefully acknowledged.

REFERENCES

1. M. SHIBAHARA and J. ODA (1969). *Journal of the Japan Society of Mechanical Engineers*, 72, 1611 (in Japanese).
2. For example, Y. ITO and M. MASUKO, *Proc. 12th Inter. M.T.D.R. Conf.*, p. 97, Macmillan.
3. R. H. THORNLEY et al. (1965). *Int. Journal of Mach. Tool Des. Res.*, 5 (1/2), 57.
4. M. MASUKO and Y. ITO (1969). *Annals of the C.I.R.P.*, 17 (3), 289.
5. R. PLOCK (1971). *Industrie-Anzeiger*, 93 (27), 571.
6. M. MASUKO and Y. ITO (1970). *Proceedings of the 10th Inter. M.T.D.R. Conference*, p. 641. Pergamon Press, Oxford.
7. I. FERNLUND (1970). *Konstruktion*, 22 (6), 218.

A CRITICAL EVALUATION OF THE PROBLEMS OF SLIDEWAY COVERS ON HIGH TRAVERSE SPEED MACHINE TOOLS

by

H. NEUREUTHER*

SUMMARY

This report presents a descriptive and pictorial evaluation of the problem of protecting high traverse speed machine tools. The possibilities of providing adequate protection for the slideways in lathe-bed grinders, milling machines with grinding capability and planing machines, are given particular consideration.

INTRODUCTION

Due to the increasing demand for precision in machine tools, it has become imperative that thought be given to increasing their useful life, and to the possibility of reducing their wear. This is of particular importance in this age of mechanization. Through environmental effects, such as contamination of the ambient air, attacks by chips, grinding dust, etc, the susceptibility to wear has increased. Generally, the threat posed by these unwelcome occurrences is met through the installation of so-called protective covers, so as to protect the slideways. Basically, one can differentiate between three types of coverings.

(1) Roller-blind covers (see Fig. 1);
(2) folded bellows (see Fig. 2);
(3) telescoping steel way covers (see Fig. 3).

Naturally, all types of covers have advantages and disadvantages. In particular, in the case of modern, high-capacity machines which are used at operating speeds of up to 120 m/min, due to the high accelerative forces, vibration and acceleration thrusts are produced, which can be conveyed to the surface being ground, (e.g. in the case of grinding machines), and therefore make some types of covers appear as unsuitable. This problem will be examined and an endeavour will be made to explain the problems posed by the individual covers, and to present suggestions for their solution.

Roller-blind type covers (Fig. 1)
This type of cover provides the simplest possibility of protecting the slideways against contamination. However, to a certain extent it is a compromise, and should be used only when, for reasons of space, it is not possible to use another type of cover.

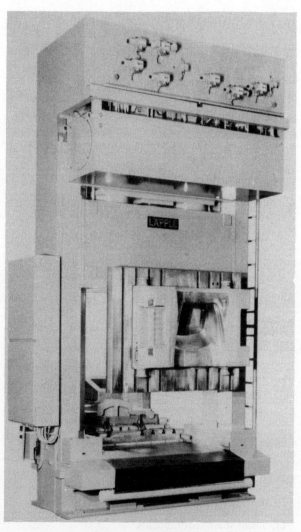

Figure 1. Roller blind cover.

---

* Gebruder Hennig, Ismaning, Bavaria

Figure 2. Planing machine with grinding capability. Travel: 16 m; max speed; 80 mmin⁻¹. Roof-type folded bellows with extension control: overall width 2.4 m.

Figure 3. Telescoping steel way cover on a boring and milling machine. Overall width: 2·6 m; travel: 32 m; max. speed: 6 m min⁻¹.

Its main advantages are

(*a*) lesser space requirements
(*b*) ease of movement
(*c*) lower accelerative forces due to lesser dead weight
(*d*) economical in price.

As disadvantages, we can mention

(*a*) insufficient protection of the slideway (see Fig. 4) (Protects only the upper part of the slideway)
(*b*) low stability
(*c*) low useful life due to contamination by chips
(*d*) undulation of the roller-blind due to high operating speeds (see Fig. 5).

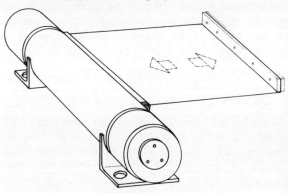

Figure 4.

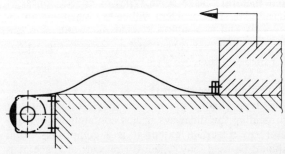

Figure 5.

Since this type of covering is generally well known, it is not necessary to discuss this aspect further. Insofar as a recommendation is concerned, it can only be stated that this type of cover represents only a compromise, and should be used only as a possible solution for problems of a minor nature.

Folded bellows (Fig. 2)

This is probably the best known and most widely disseminated type of cover for the protection of slideways in machine tools. Here, the important innovations are only considered, and particularly in detail, the problems of high traverse speed lathe-bed ways grinding machines with travel limits of more than three metres in length. The main problem in this type of machine, is that caused by the enormous acceleration forces which occur in the process of reversal. These forces must be taken up by the internal bellows armature in relation to the bellows' self-acceleration, taking into account the weight of the bellows. In the path of motion depicted in Figs 6(a) and 6(b), it can be seen that at high operating

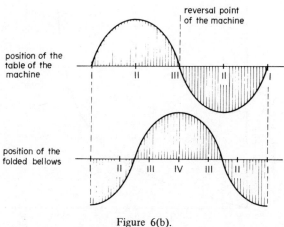

Figure 6(b).

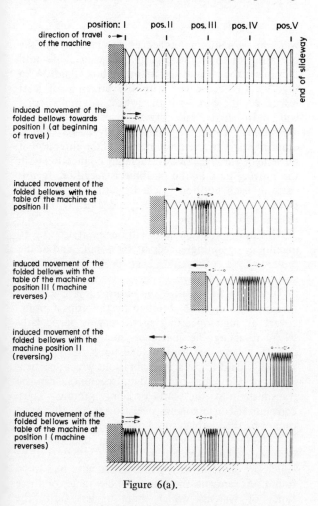

Figure 6(a).

speeds, definite 'natural oscillations' will appear in the bellows in the form of horizontal waves. These waves, generated by the machine, travel continuously through the entire length of the bellows. Thus, it can happen that the machine is already accelerating in one direction, whilst the centre section of the bellows is still moving in the opposite direction. As a result, constructive precautions must be taken to prevent a tearing of the bellows, particularly at the moment of reversal. Through the use of bellows, the need for what is called 'extension controls', which are used

mainly in the centre section, can be avoided. From the diagram (Fig. 7), the operating characteristics of an 'extension control' of this type is noted.

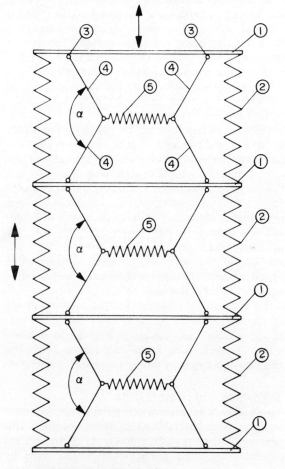

1. stiffeners (metal or plastic)
2. folded bellows section
3. pivot bolts welded to the stiffeners
4. linkage rods
5. spiral springs

Figure 7.

Operating characteristics of the 'extension control'

It is possible to completely subdue the occurring acceleration and deceleration forces in the middle of the elastic joints depicted. From the diagram shown in Fig. 8, it is obvious that, through an increase in the opening angle α, the increase in force in the direction

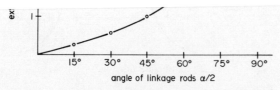

angle of linkage rods $\alpha/2$

Figure 8.

of movement can be expressed as a direct tangential function: $P = P_F \tan(\alpha/2)$. This proves that by means of a suitable combination of linkage rods, pivot bolts and springs, a progressive extension control, which prevents, with absolute dependability, any damage to the covers, particularly at high operating speeds, can be set up. As a positive secondary feature, one can mention the adequate parallel construction of the stiffeners. Additionally, through a suitable design of the extension control, tilting of the stiffeners in the vertical axis can be prevented.

Similar covers will be put into operation primarily in grinding and planing machines, so as to protect the horizontal slideways. However, and specially in the case of milling machines with grinding capability, and that of planing machines, further difficulties can arise due to contamination by chips. In particular, it is possible that chips may fall into the centre of a fold, and that through the joint movement of the cover (at the compressed position of the bellows), the folds will be cut through. Thus, after a relatively short period of time, the cover will become unserviceable, and must be renovated with considerable expense.

To counter this disadvantage, the machine manufacturer will specify in such cases that the cover should be used only for grinding work. In all other types of work, such as milling, planing, etc., the machine must operate without slideway protection.

Telescoping steel covers (Fig. 3)

These allow a solution which provides total protection for the slideways in all modes of operation. It is generally known that the slideways in chip-producing machines are being protected more and more by means of telescoping steel covers. This type of cover has the advantage of being exceptionally sturdy and resistant to damage and wear. A further advantage is that with this type of cover the operator can step or stand on it, regardless of the position of the machine, without fear of damaging it. Regretfully, the telescoping steel covers which have been manufactured up to now can be used only with low operating speeds. Normally, their applicability limit lies in the range of operating speeds from 6 to 10 m/min. In special cases, the possibility exists, due to a suitable 'Pantograph-type' extension control system inside the cover, to distribute evenly the induced motion of the

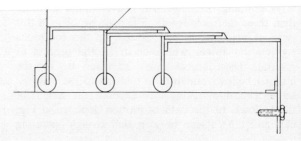

Figure 9.

will, in a random sequence, be racked by the operation of the moving parts of the machine until they meet the stop of the following box (Point 'A' in Fig. 9). In accordance with the principle of least resistance, the box with the lowest internal friction will move first. The larger boxes constitute an exception, because they will normally be fastened to a moving part of the machine. Here, the direction of movement through the force-locking connection with the moving part of the machine is specified. In this case, in each position reached by the machine, the smallest box is in a state of rest, and is firmly fixed to the end of the machine.

This system can be used without hesitation up to maximum operating speeds of 10 m/min, and at the present time offers the best protection for the machine tool slideways.

It is easy to visualize the enormous acceleration forces which must be absorbed by the stops in heavy covers at high operating speed ranges, particularly when reversing the machine. Firstly, these forces impose a large mechanical load upon the stop and secondly, hard impacts and vibrations which, particularly in the case of grinding machines, can be transmitted to the surface being ground. Thus, conventional telescoping steel covers of this type cannot be used in high speed machines.

Therefore, new ways of locking the individual boxes together through the use of other systems which are more effective and secure have had to be found. At present there is a whole range of these types of systems, which more or less fulfill the specifications claimed for them. A description of the most important versions are:

(1) 'Pantograph' systems
(2) Elastic shock absorbers
(3) Hydraulic shock absorbers

(1) 'Pantograph' systems

These consist of a system of linkage rods, which have the effect of controlling the motion of the individual telescoping boxes when the machine is in operation.

A representation of a 'pantograph' system of this type is shown in Fig. 10. In this example, each

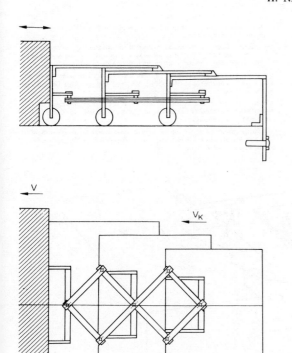

V_K Relative speed of one box to the next

n   Number of boxes in the cover

k   Successive numbers of the boxes (the box connected to the moving part of the machine is the number I box)

v   Operating speed of the machine

Figure 10.

individual box is bound by means of joint bolts to the pantograph system. This has the effect of distributing the relative speeds to all of the boxes in the cover, attaining therefore an adequate uniformity of movement. The distribution of speeds in the individual boxes results from the following formula:

$$V_k = \frac{1}{n}(n-k) \cdot V$$

where V_k = relative speed of one box to the next

   n = number of boxes in the cover

   k = successive numbers of the boxes (the box connected to the moving part of the machine is the Number 1 box)

   V = operating speed of the machine

This formula confirms the fact that in a system of this type, the relative speed of the individual boxes to each other decreases with the increase in the distance of the boxes from the moving parts of the machine. Still, relatively high initial speeds occur only in the first one-third of the cover. This further confirms the fact that extension controls of this type can only be employed up to maximum operating speeds of 30 m/min. In the case of very large covers (large dead weight), this limit must again be considerably reduced. The reason for this lies primarily in that the pivoting parts (the pantographs), are not able to absorb the increasing forces which occur with greater masses and higher accelerative forces. In such extreme types of stress, one can only guarantee a relatively short useful life for the cover.

(2) Elastic shock absorbers

This system is similar to the one previously described. However, in this system there exists an uncontrollable locking-in of the individual boxes. The principle of an 'elastic' extension control of this type is depicted in Fig. 11.

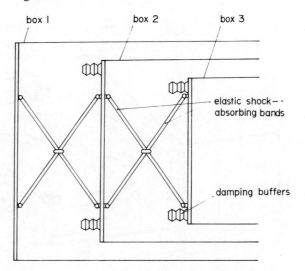

Figure 11.

This system can be employed in most cases. However, it has the disadvantage that there is no damping at all when the individual boxes are thrust together. In the case of the boxes moving away from each other, the damping is likewise uncontrolled, and it can happen that the individual boxes may suffer hard acceleration shocks, which may be transmitted to the surface being ground. In the case where, after long use one of the elastic bands should rupture, there would also be the danger that individual pieces of this extension control could fall on the slideway, and cause considerable damage to the machine. Consequently, it is obvious that this must also be considered only as a temporary solution, and that efforts must be made to perfect this principle.

(3) Hydraulic shock absorbers as damping elements and extension controls for steel covers

In order to provide also adequate protection in the form of telescoping steel covers for the slideways of the larger machines with high operating speeds, it became necessary to find entirely new ways of doing this. First of all, oil-hydraulic shock absorbers of normal commercial pattern were built in. After a relatively short testing period, it appears however, that these damping elements do not provide the desired results. Particularly when squeezed together, the individual boxes cause a clattering noise. This occurs due to the impact of the individual boxes against each other. In this position, the shock absorbers produced insufficient damping, since due to the particular setting of the shock absorbers, there was an unfavourable distribution of forces in the nearly compressed position of the cover (Fig. 12).

It can be seen in the diagram, that in the case of a relatively high transversal force 'Q', there is only a very small horizontal force 'H' which depends upon the opening angle of the shock absorbers. However, it

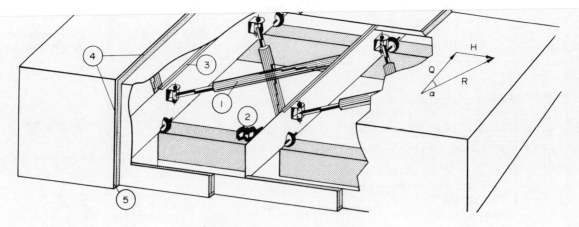

Q = transversed force
H = horizontal force (extension preventive force)
R = damping force of the shock absorbers

Figure 12(a).

would be desirable to reach a potentially large horizontal force. In particular in the maximum and minimum positions, the horizontal force 'H' must be strong enough to impede impact of the individual boxes against each other.

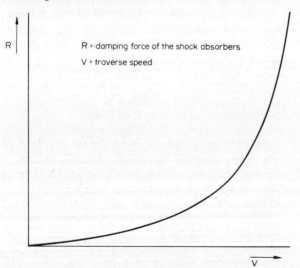

R = damping force of the shock absorbers
V = traverse speed

Figure 12(b).

It appears that in these extreme positions, a relatively small amount of additional damping is sufficient to impede the impact of the individual boxes. This problem could be resolved by means of two additional changes in the system. Firstly, special rubber shock absorbers with progressive spring characteristics could be built in. Additionally, the damping characteristics of the shock absorbers would be changed to such an extent, that in the area of the minimum position, damping could reach its greatest effect. All further improvements could, through changes in the damping tubes inside the shock absorbers, attain a progressive damping which would be dependent upon the speed (see Fig. 12b).

The tests show that at a low operating speed (10 m/min), almost no damping occurred, and that the damping force would rise rapidly along with a corresponding rise in operating speeds. This result coincides exactly with the requirements presented, and the endurance tests showed that this system can be utilized without hesitation with operating at speeds up to 80 m/min.

CONCLUSIONS

It can be stated that this last type of slideway cover for high-speed, metal-cutting machine tools, provides at the present time the best protection for the slideways. All alternative solutions constitute only compromises, and in most cases will be used only for reasons of economy. The type of cover to be used in a specific machine depends above all on the stress criteria and ambient conditions prevailing. Regretfully, no general rule can be given here. However, it would be advisable, particularly in view of new developments in machinery, to contact the slideway cover manufacturers. Thus, a specific recommendation to fit the machine can be obtained. In this manner, a designer is in a position to spare himself considerable development costs, and possibly, modification costs as well.

COMPUTER AIDED DESIGN

THE ROLE OF COMPUTERS IN MACHINE TOOL DESIGN

by

F. M. STANSFIELD*

SUMMARY

The applicability of computers to machine tool design is determined in practice not only by technical advantages but also by administrative and economic factors. Techniques by which computer programs can be made to meet the needs of designers are being developed and will be described in this paper. Reference will be made to features of computer programs developed by the Machine Tool Industry Research Association as examples of these techniques.

INTRODUCTION

It cannot be disputed that there are tasks encountered in machine tool design which can be undertaken more thoroughly and more quickly with the aid of computers than by any other methods. It is obvious in only a few cases, however, that the applications of computers to machine tool design are economic. In many situations the advantages of using computers are of a nature than cannot yet, if ever, be stated directly in terms of financial benefit. Nevertheless there are some requirements which, clearly, must be satisfied in order that the potential economic usefulness of computers in machine tool design can be fully realized. Computer programs must be highly user-oriented with respect to data preparation, presentation of results and the ease with which they might be integrated into the normal processes of design. The paper describes the requirements for computer programs which are to be used in design and explains some techniques which are being developed to improve the applicability of computers in machine tool design.

SOME OBJECTIVES OF THE DESIGNER

Some of the general objectives of the designer of a machine tool are

(1) a high rate of metal removal;
(2) accuracy of both size and geometry of work-piece;
(3) good surface finish;
(4) consistent performance;
(5) low selling cost.

These general objectives may be expressed in terms of the more specific requirements:

(1) high static stiffness of the fixed structure, sliding joints, journal and thrust bearings, mechanisms and drives;
(2) avoidance of unacceptable natural frequencies of the whole machine;
(3) acceptable patterns of vibration of the whole machine;
(4) adequate damping capacity of the whole machine to resist vibration;
(5) high speeds of operation;
(6) low rates of wear in sliding joints, journal and thrust bearings, mechanisms and drives;
(7) low thermal distortion in the whole machine;
(8) insensitivity of performance to variable factors such as the quality of fit of adjacent parts, surface finish, site conditions, relative thermal expansion;
(9) low design and development cost;
(10) low manufacturing cost.

(It is necessary to consider also the strength of a structure in the design of a few types of machine tools, but in most cases the satisfaction of the requirements with respect to stiffness ensures that a structure will have ample strength.)

THE DESIGNER'S TASK

It is fair to say that the concept that the activity of design involves a process of optimization is a relatively recent one, and that its emergence as a maxim for design is due largely to the increasing use of computers. In the past there were only bad, good,

* Machine Tool Industry Research Association

(*a*) experience of the relationships which exist between performance and design is being embodied in many computer programs to which the designer has access; and

(*b*) the numerical data which are being acquired concerning the properties of an ever-increasing variety of materials can be processed by these programs rapidly and effectively.

There is also much theoretical knowledge, established for many years, which is being employed in design for the first time because it required the speed of computers to make its application feasible.

Many of the requirements which must be fulfilled are mutually conflicting and, therefore, in order to determine the most suitable design of machine for a specified purpose it is necessary to investigate the performance of a wide range of possible designs. For example, a suitable compromise must be found between a design of bearing which is very stiff but in consequence generates much heat, and a design of bearing which, at the other extreme, operates at temperatures only a little above ambient temperature but may be large dimensionally, or have a low load-bearing capacity or low stiffness. The designer's task of searching for the optimum design may be lightened by the use of computers in proportion to the extent to which the problem can be defined in terms of hard facts and proven relationships between design and performance—because this determines how much of the problem can be expressed in the form of computer programs.

The designer's work involves calculation and decision making, both of which may be based partly on established information and partly on tentative assumptions. Several types of decision making may be required.

(1) The application of determinate logic—logic which is applied to just-sufficient data and leads to unambiguous conclusions.

(2) The application of indeterminate logic—logic which is applied to insufficient or excessive, and therefore ambiguous, or even conflicting data, and leads to a reduction in the number of alternatives.

(3) Subjective decisions, based on the experience of the company and on personal experience and intuition.

Computers, which are essentially devices capable of performing logical operations at great speed, may therefore be used to assist the designer in performing calculations and those parts of decision making referred to in items 1 and 2 which involve the application of formal logic.

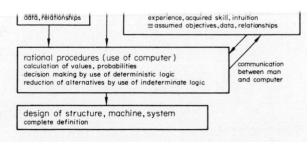

Figure 1. A graphical representation of the design process.

SPECIFIED OBJECTIVES represents the initial specification which, alone, contains insufficient information for design purposes. The block labelled FACTS represents additional information concerning

(1) established theoretical relationships between, for instance, dimensions and the performance of the machine;

(2) quantitative data—for instance, material properties.

Another source of information is the experience of the individual designer and of the company: this is represented in Fig. 1 by the block labelled EXPERIENCE. The term 'experience', in this context, is intended to refer only to that kind of information which has not been verified sufficiently well to merit being called factual information.

The part of the design procedure involving calculation and the exercise of logic is represented in Fig. 1 by the block labelled RATIONAL PROCEDURES. This is the part which might possibly be transferred to the computer. The designer would usually wish to examine the effect on the design of alternative hypotheses and of alternative values of data wherein some doubt might exist about where precisely the truth lies. In order to determine the most suitable design it is necessary for the designer to carry out the following sequence of steps repetitively.

(*a*) Supply tentative information (EXPERIENCE).
(*b*) Supply established information (FACTS).
(*c*) Apply calculation and logic (RATIONAL PROCEDURES).
(*d*) Obtain and consider the results.

Hence, provision must be made for a two-way flow of information between the designer who supplies the experience and the computer which applies the rational procedures; subjective decisions are required from the designer at intermediate stages in the execution of calculations and logical operations. Facilities which make it possible to use computers in this way are referred to as interactive facilities. The term is sometimes applied with particular reference to the use of time-sharing computers from teletype terminals and sometimes to the use of computers having graphical display and light-pen facilities. The former system aids the interactive use of computers by reducing the cost of time spent in making decisions whilst using the computer. The latter system aids the interactive use of computers by facilitating communication with the computer in terms of visual images—these are more readily understood than strings of numerical values.

THE DEVELOPMENT OF USER-ORIENTED TECHNIQUES

Means of communication with computers have improved greatly since the first electronic digital computer, ENIAC, was completed in 1946. Advances have been brought about by a variety of developments:

— high level languages such as ALGOL and FORTRAN which facilitate the writing and reading of computer programs;
— conversational programming languages such as JEAN, TELCOMP and BASIC, which facilitate the development of computer programs;
— time-shared computing services which enable the costs of a central processor to be shared and which make the use of conversational languages and programs economic;
— digital graph plotters for producing diagrammatic and graphical output;
— graphical display computer terminals with light pens, to facilitate direct interaction between the user and the program.

The recent development of graphical-display computers with light-pen facilities is, however, one of the most significant contributions to easier communication with computers and one which will greatly facilitate their use in the design process.

Essentially, graphical-display computers are computers to which have been added facilities by which diagrammatic displays can be generated on the screen of a cathode-ray tube and by which the operator can cause the display to be modified, as for instance by a light pen. The computer might provide all the services required for computation and for control of the display equipment, in which case it is referred to as a stand-alone system. Alternatively, it might be linked to a much larger time-sharing computer and referred to as a satellite graphics terminal.

The display screen presents data which originate from computations carried out by the program or which are supplied directly by the user of the program by means, for example, of a keyboard or a paper-tape reader; in both cases the data are processed by the program for presentation in graphical or diagrammatic form.

A light pen is a particularly convenient means of communicating with a graphical-display computer. It is a hand-held device shaped like a pen and detects light emitted locally at the point on the display at which it is directed. It thus enables the computer to identify any particular element of the display at which the user of the program points it. Subsequent

actions within the computer depend on the instructions embodied in the computer program and on the instructions given by the user of the program by his selection of the positions of a number of keys, known as sensing keys. The program responds by performing further calculations or by modifying the display or both, or by printing, punching or plotting the results. It should be noted that such a system is very versatile with respect to the ways in which it can be adapted by the programmer to provide the user with facilities for controlling the running of a program.

THE GRAPHICAL DISPLAY COMPUTER AT M.T.I.R.A.

The combination of a graphical-display computer with a large time-sharing computer is a particularly convenient one. Access to a large time-sharing computer fulfils the occasional need for powerful computing facilities without incurring unnecessarily high overheads for the less extensive computations which probably constitute the greater part of design. The system at M.T.I.R.A. is of this type and consists basically of an Elliott 905 computer and an Elliott 928 graphical-display unit with a light pen, linked by a high-speed line to the ATLAS 2 time-sharing system at the C.A.D. centre in Cambridge.

Generally, the procedure for use of the Elliott 905/928 graphical-display terminal is to use the terminal as a simple device for the input and output of data and to write programs which are executed in the ATLAS computer. Some routines are executed locally within the Elliott 905/928 graphical-display terminal, but these are mainly standardized procedures for interpreting data flowing in both directions between the actual input and output devices and the ATLAS computer and for maintaining the picture on the screen.

The facilities offered by the use of a screen and a light pen enable programs to be written so that troublesome procedures of data preparation are avoided. For instance, it is not necessary to enter data on forms in a prescribed order or preceded by prescribed symbols, to punch tapes (or cards) from the data on the forms, to edit the tapes to correct punching errors, to operate the paper-tape reader and the computer console in order to enter the data tapes into the computer and to initiate computations, or to repeat all these procedures each time it is required to correct an error in the original data or to modify the data in the light of the results computed by the program. The program may be written so that the foregoing procedures are replaced by the following procedures: identify with the light pen, in turn and in any order, each variable depicted on the display and enter its value at a keyboard; create, modify, move or delete lines or portions of a display by simple and direct actions with the light pen and sensing keys; observe, on the display, errors incurred in data whilst entering it at the keyboard and correct them immediately by similar procedures; modify data for further runs of the program by similar procedures; initiate computations by directing the light pen at a word on the display such as 'calculate'. Whereas procedures involving the use of paper tape and cards

broader process of design, including also decision-making. The facilities of the C.A.D. centre's time-sharing system enable several programs and sets of data to be maintained on file simultaneously; a file may be arranged to hold data required for entry into several programs and to hold results from all these programs. When the results from one program are to be used as the data for another, the file may be used as a means of conveying the data from one to the other. Hence, groups of components whose design requirements are interrelated may readily be optimized, each component being designed by one or more individual programs. The inconvenience of transferring data by the medium of printed results and paper tape is avoided, and hence there is a great reduction in the time required to investigate the influence of the designs of related components upon each other. The elimination of much of the manual handling of data enables individual design procedures to be incorporated into a smooth, continuous process of design.

The success of such a system depends upon the ease with which a user can learn how to use the particular programs which are relevant to his needs. This in turn depends on the selection by the programmer of the most suitable techniques by which the user may communicate with the program—the selection and devising of appropriate techniques involves principles of systems design and of ergonomics and is a main objective of M.T.I.R.A.'s work on computer-aided design.

A SUITE OF M.T.I.R.A. INTERACTIVE GRAPHICS PROGRAMS FOR THE DESIGN OF MACHINE TOOL SPINDLES AND BEARINGS

A suite of six graphics programs has been developed, with the cooperation of the C.A.D. centre at Cambridge, for the design of machine tool spindles and bearings, and they are known by the titles

INITIAL SPINDLE DESIGN
ROLLING ELEMENT BEARINGS
HYDROSTATIC BEARINGS—STAGE 1
HYDROSTATIC BEARINGS—STAGE 2
DEFLECTION ANALYSIS OF SPINDLES
DATA CONTROL FACILITY

The last of these programs greatly simplifies the use, in association, of the other five applications programs.

These graphics programs are much easier to use than the earlier batch-processing versions from which programs have to be run repeatedly, as is usually the case for purposes of design as distinct from purposes of analysis of the performance of a given design.

The graphics programs are easier to use individually and are also more convenient for use in association with each other because of additional facilities which have been provided. Any program may be called into use simply by identifying it, with the aid of the light pen, amongst a displayed list of programs. Provision is made for the filing of complete sets of data and results from each program, for reference on subsequent occasions. Any such 'Data Sets' which are regarded as final may be assembled together as 'Design Records'. The contents of the Data Sets and Design Records may be examined at any stage in design. Selected data from any one program may be transferred readily, as initial data, to any other program.

All operations for using the programs are carried out by one or two basic types of operation using the light pen and keyboard—these operations have been standardized throughout the programs as far as the system permits.

The program DATA CONTROL FACILITY

The suite of graphics programs for the design of spindles is entered by this program. Its first response is to request the user to state whether or not a name has previously been entered for a file in which the user may accumulate the data and results. If so, then the file name must be identified; otherwise a new name must be assigned to a file.

The second response of the program is to display a list of names of programs from which the user is requested to identify the particular program which he wishes to use.

Facilities are provided to display for inspection data which may have been accumulated during previous use of the programs. There are two categories of such data, termed Data Sets and Design Records respectively. Data Sets are complete arrays of self-consistent data and results, associated with particular applications programs. When the design of a component has been developed to a stage beyond which it cannot be developed further using only the current applications program, then its associated Data Set may be embodied in a larger array of more permanent data, termed a Design Record. The latter may comprise an accumulation of data for related components. Thus Data Sets and Design Records reflect different degrees of finality of data, the Design Records containing the more completely developed data.

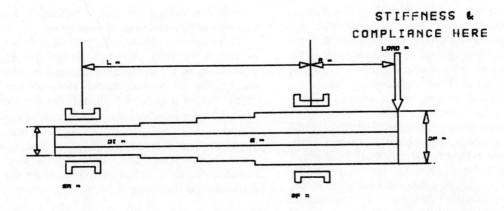

Figure 2. INITIAL SPINDLE DESIGN: program entered.

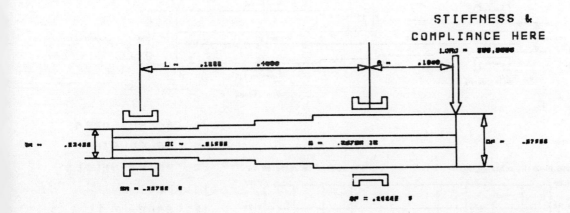

Figure 3. INITIAL SPINDLE DESIGN: data entered.

stiffness determined for a bearing by the use of the program ROLLING ELEMENT BEARINGS must be equal to, or nearly equal to, the stiffness assumed in designing the spindle by use of the program INITIAL SPINDLE DESIGN. Also, the diameter of the bearings must be consistent with the diameter of the spindle. The fulfilment of these requirements is the duty of the user of the programs.

The user can request Data Sets and Design Records to be displayed so that he might compare the values of variables in related sets of data. If it should be necessary to improve the consistency between values relating to adjacent components, then the user must consider which applications program to enter with a sequently to drive a plotter to reproduce the displayed data as hard copy.

The program INITIAL SPINDLE DESIGN (Figs 2, 3, 4 and 5)

This program is intended for use in optimizing the size and main proportions of a spindle for high stiffness at the cutting zone. For the purposes of this program, a spindle is defined in terms of a greatly simplified mathematical model. Hence a spindle is defined in terms of the values of only eight variables. The program computes only results which are essential for considering its stiffness characteristics. In view of the ease with which the small quantity of data can

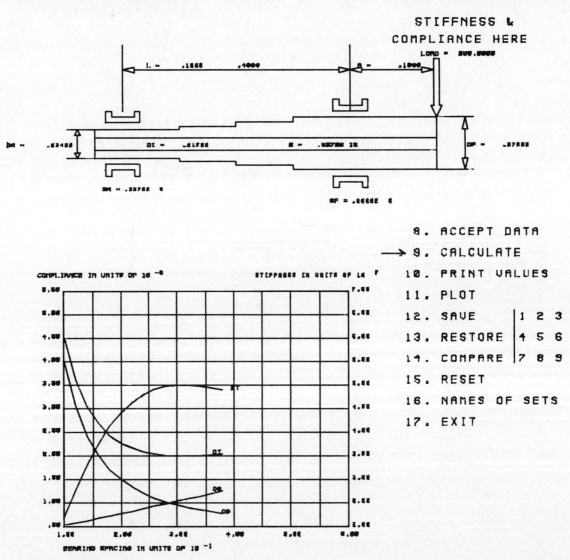

Figure 4. INITIAL SPINDLE DESIGN: command 9. CALCULATE initiated—a range of values assigned to length *L*.

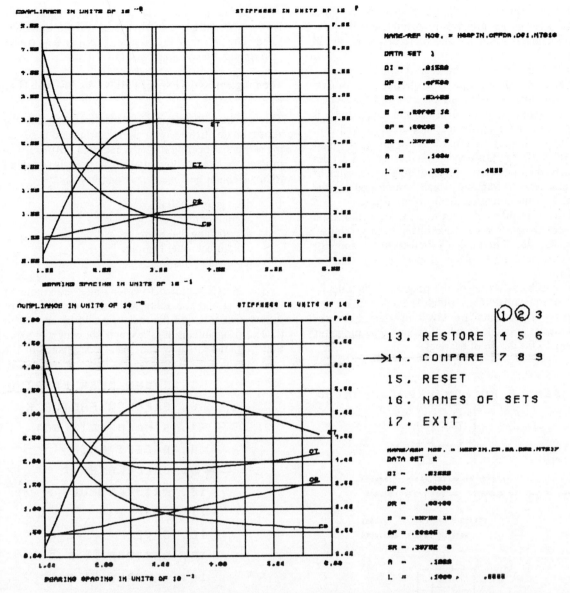

Figure 5. INITIAL SPINDLE DESIGN: command 14. COMPARE initiated to compare Data Sets 2 and 3.

be entered and of the small amount of calculation performed by the program, it may readily be used in a repetitive manner to optimize the size and main proportions of a spindle.

Although the program is nominally concerned with the design of spindles supported in two bearings, it is recommended for determining the position of two of the bearings of a three-bearing spindle: when two bearings are optimally spaced, the addition of a third bearing is unlikely to increase the stiffness at the cutting zone by more than 5 per cent. The third bearing may therefore be sited solely according to the needs for support at points at which driving loads are exerted on the spindle.

Data are entered for each variable in turn by

(1) indicating, with the light pen, one of the variables displayed on a diagrammatic representation of the spindle,
(2) entering its value at the keyboard. The values which have been entered are displayed for inspection and may be changed if a mistake is made.

To facilitate optimization, graphical displays may readily be obtained of the variation of stiffness and compliance with changes in any one of the eight variables defining a spindle. The four curves which are plotted relate to the following performance characteristics:
— the contribution to radial compliance at the cutting zone of deflections only within the spindle;
— the contribution to radial compliance at the cutting zone of deflections only within the bearings;
— the total radial compliance at the cutting zone;
— the net radial stiffness at the cutting zone.

As a further aid to optimization, pairs of such graphs (of four curves) may be displayed one above the other. The (vertical) scales for compliance and stiffness are then identical in the two graphs. The graphs may be plotted against the same variable, in which case the horizontal scales are also made identical, or they may be plotted against different variables. These facilities make it possible to compare, readily, the effects of changing the values of any two variables.

the independent variable which is indicated by the position of the cursor on the horizontal scale.

Various facilities are provided for inspecting Data Sets and Design Records which have been previously compiled, for obtaining printed results, and for obtaining hard copy of any display at any stage of design.

A subsidiary facility of the program is the calculation of the radial loads which are exerted on the bearings. These data are saved in Data Sets, for subsequent reference, when using the programs relating to the design of bearings.

rolling element bearings, singly and in pairs:

>   deep groove ball bearings
>   angular contact ball bearings
>   self aligning ball bearings
>   cylindrical roller bearings
>   taper roller bearings.

The user is required to enter some dimensions of the bearing and the rolling elements, the number of rolling elements and the radial load to be applied to it. The program displays a graph showing the variation of radial stiffness and radial compliance of the

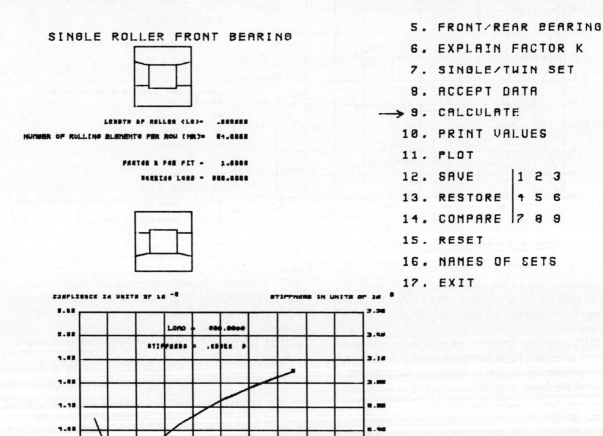

Figure 6. ROLLING ELEMENT BEARINGS: command 9. CALCULATE initiated.

bearing with load. A vertical line is displayed on the graph at the value of the load which is applied to the bearing. The stiffness at this load is also displayed numerically.

The facilities for entering data, saving sets of data, comparing graphs for different sets of data, obtaining tabulated data, and obtaining hard copy of graphs are similar to those provided in the program INITIAL SPINDLE DESIGN. As in the case of the latter program, exit from the program causes entry to be made, automatically, into the program DATA CONTROL FACILITY.

The programs HYDROSTATIC BEARINGS–STAGE 1 and HYDROSTATIC BEARINGS–STAGE 2 (Fig. 7)

These programs are intended for the design of oil hydrostatic bearings in the form of cylindrical, conical or thrust bearings. The same criteria apply for the optimization of the design of oil hydrostatic bearings as for the optimization of the design of spindles generally: the values of dimensions and other design parameters for a particular bearing must be determined carefully so as to achieve a delicate balance between the desired properties of high speed, high stiffness and low temperature while satisfying constraints imposed by practical requirements.

The design can be carried out in either one or two stages. In the first stage, with is optional, a range of possible geometrical forms of bearing are computed which all have the same stiffness, or alternatively, the same load capacity. The program for the first stage is thus intended to be used as a starting point for the design of a hydrostatic bearing.

The program for the second stage computes a number of performance characteristics for a range of values of any one of a number of design variables. Input data for this program may readily be transferred, via the DATA CONTROL FACILITY, from the output of the first stage. In this case, the performance characteristics are first displayed as curves which are plotted against variation of the landwidth and length of the bearing. The data may then be changed readily to cause the program to display the curves plotted against any other variable.

There are too many performance characteristics of interest to be displayed with sufficient clarity on one graph: hence two graphs may be displayed. On one of these, curves show the variation of friction power, pumping power, total power consumption, and rate of oil flow. On the other, curves show the variation of stiffness, resting load capacity and ultimate load capacity.

The facilities for entering data, saving sets of data,

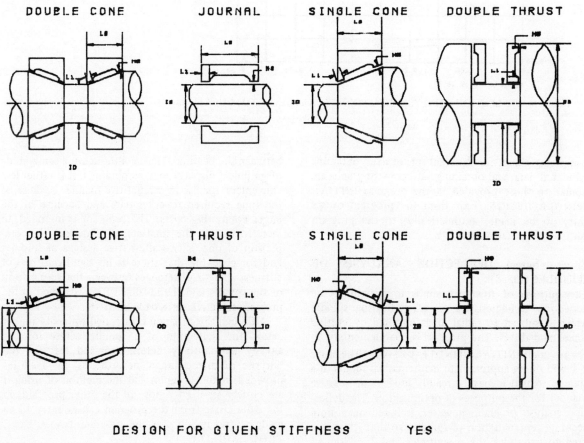

Figure 7. HYDROSTATIC BEARINGS–STAGE 1: program entered.

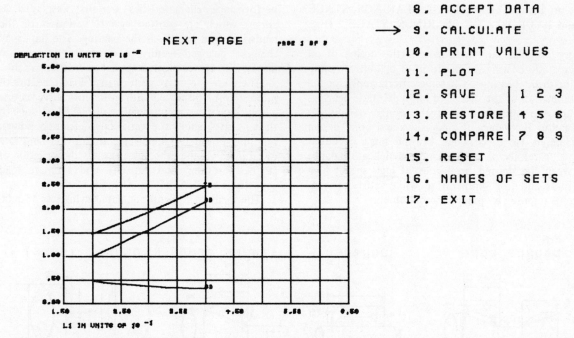

8. ACCEPT DATA
→ 9. CALCULATE
10. PRINT VALUES
11. PLOT
12. SAVE | 1 2 3
13. RESTORE | 4 5 6
14. COMPARE | 7 8 9
15. RESET
16. NAMES OF SETS
17. EXIT

Figure 8. DEFLECTION ANALYSIS OF SPINDLES: command 9. CALCULATE initiated—a range of values assigned to length L1.

comparing graphs for different sets of data, obtaining tabulated data, and obtaining hard copy of graphs are similar to those provided in the program INITIAL SPINDLE DESIGN. Exit from the program causes entry to be made, automatically, to the program DATA CONTROL FACILITY.

The program DEFLECTION ANALYSIS OF SPINDLES (Fig. 8)

The purpose of this program is to determine the deflection characteristics of a two-bearing spindle which is subject to radial loads imposed by cutting forces and drives. The program is complementary to the program INITIAL SPINDLE DESIGN. The latter is based on an approximate mathematical model of a spindle so that it may be readily used in an iterative manner for the purposes of optimization. The deflection analysis program, however, is based on a more accurate representation of spindles so that it may be used to determine the magnitudes and directions of deflections and of deflection-gradients at the bearings and at the cutting zone.

Considerably more data must be entered into the program to define the spindle. For example, in this case, the spindle may have any number of steps

between the bearings. Hence, although the same kinds of graphical displays may be obtained, it is rather less convenient for use in a repetitive manner because of the time required to enter data and because of the larger computing costs. The program is intended, in fact, for use in the final stages of design, for comparison of one or two alternative designs of spindles and for checking that there is no significant loss of stiffness due to differences between the actual form of the spindle and the idealized form assumed in the program INITIAL SPINDLE DESIGN.

Although this program is intended for analysing, in detail, the deflection of a spindle whose form is already substantially determined—and less for the purpose of direct design—nevertheless, the facilities provided in the program and the method of using it are closely similar to those of the other programs in the suite. Exit from the program causes entry to be made, automatically, to the program DATA CONTROL FACILITY.

CONCLUSIONS

It has been shown how the interactive facilities of graphical display computer terminals, linked to a large time-sharing computer, can be used to provide highly user-oriented procedures for the design of machine tool components and of assemblies of components.

ACKNOWLEDGMENTS

This paper is published by courtesy of the director of the Machine Tool Industry Research Association.

DISCUSSION

Q. J. Cyklis. The optimizing of the high stiffness is not always in agreement with demand of the high quality of the other parameters, such as geometrical accuracy. Is it possible to introduce into the program another factor? I think that would be difficult and computation should be performed for the global criterion including the other factors.

A. The relative importance of different criteria of good performance of a component, or machine, depends upon the particular requirements of the machine and must often be decided subjectively in the absence of quantitative rules. For instance, for heavy cutting conditions it might be particularly important that a spindle should have high stiffness, whereas for light cutting conditions it might be particularly important that it should have the smallest possible run-out.

Designers should generally, therefore, be provided with several programs, each having a clearly defined purpose and each having the capability of being used readily in association with the others. In sufficiently simple circumstances, several performance criteria might reasonably be computed in one program.

The suite of graphics computer programs for the design of machine tool spindles, as described in the paper, is designed to enable new programs to be added, if there is sufficient justification for them, and is designed to enable all the programs in the suite to be used in an integrated manner.

ANALYSIS OF MACHINE TOOL JOINTS BY THE FINITE ELEMENT METHOD

by

N. BACK†, M. BURDEKIN* and A. COWLEY*

SUMMARY

This paper presents in the first place some remarks on the normal and shear stiffness of machine surfaces. It is shown that the shear stiffness can be calculated from the parameters that define the normal stiffness of the machine surfaces.

In the second place, the deformations and pressure distributions are calculated for several examples of joints, taking into account the interface surface and structural compliance by using the finite element method. The joints considered are typical examples of sliding and bolted joints.

Finally, examples are presented to show the influence of the flatness deviation and shear stiffness of the surfaces upon the joint deformation and pressure distribution at the interfaces.

Close correlation between computed and measured results demonstrates the validity of the computational models and procedures.

NOTATION

c	Constant
E	Modulus of elasticity of the material
f	Coefficient of friction
G	Shear modulus of the material
h	Asperity height peak to valley
K_n	Normal compliance of the machine surfaces
K_s	Shear compliance of the machined surfaces
m	Constant
p_a	Apparent interface pressure
p_n	Normal pressure
p_s	Shear pressure
p_{sf}	Limit friction pressure
R	Constant
S	Constant
Z	Number of spots of contact per square inch
λ_n	Normal deformation of the surface
λ_s	Shear deformation of the surface
μ	Poisson's ratio
Δ	Flatness deviation

INTRODUCTION

The performance of a machine tool can be significantly influenced by the static and dynamic stiffness of the sliding and bolted joints which are incorporated in the structural design. For a precise analysis of these factors it is necessary to know the deformations of the components forming the joint as well as the pressure distribution at the contacting surfaces. To calculate these values it is necessary to take into consideration not only the surface compliance but also the structural components surrounding the surfaces in contact.

The surface compliance has been analysed by many researchers such as Dolbey[1], Levina[2,4] and Ostrovskii[3]. More details of the analysis and factors affecting the normal compliance of the machined surfaces can be seen in a survey presented by the authors[5]. The relationship between the normal pressure and the approach of the surfaces is of the form represented in Fig. 1(a), which is based upon experimental data from Ostrovskii[3]. Mathematically the surface characteristics can be represented as

$$\lambda_n = c p_n{}^m \qquad (1)$$

where λ_n is the approach of the surfaces or the deformations of the asperities in μm, p_n the interface pressure in kgf/cm^2, and c and m are parameters depending upon the pair of materials and surface finish.

From the experiments of Levina[2,4] Dolbey[1] and Ostrovskii[3] the values of c and m have been obtained for cast iron, as shown in Table 1.

---

* Division of Machine Tool Engineering, Department of Mechanical Engineering, U.M.I.S.T.
† Department of Mechanical Engineering, U.F.S.C., Brazil

$h = 15\text{–}20\ \mu m$, $Z = 5\text{–}12$ spots/in²	2·0–2·6	0·5	1·5–2·0	0·5
Hand-scraped, $h = 6\text{–}8\ \mu m$, $Z =$				
$Z = 15\text{–}18$ spots/in²/Ground,	1·0–1·3	0·5	0·8–1·0	0·5
$h = 1\text{·}0\ \mu m$ CLA				
Peripheral ground/Peripheral ground				
$h = 1\text{·}0\ \mu m$ CLA	0·8–0·9	0·5	0·6–0·7	0·5
Finish Planing/Finish Planing	0·78	0·5	0·6	0·5

The shear compliance was analysed by Kirsanova[7] and for repeated loadings the relationship between shear pressure and shear deformation is given by

$$\lambda_s = K_s p_s \qquad (2)$$

where λ_s is in μm and p_s in kgf/cm².

Kirsanova[7] found that the shear compliance of cast iron surfaces is dependent upon the surface finish and that it decreases with the increase of the normal pressure. From the results obtained by Kirsanova[7] and by the authors[6,8] it can be seen that the relationship between the shear compliance and the normal interface pressure can be written as

$$K_s = \frac{R}{(p_n)^S} \qquad (3)$$

where R and S are again parameters dependent upon the pair of materials and surface finish.

The authors[6,8] measured the shear deformations of cast iron machined surfaces and obtained results, as shown in Fig. 1(b). It can be seen that for the first

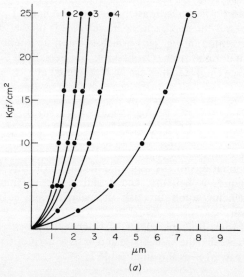

Figure 1 (a) Relation between the interface pressure and the deflection for cast iron. 1–Peripheral grinding. 2–Fine scraping, $Z = 24$ to 36 spots/in². 3–Finish planing. 4–Conventional scraping. 5–Coarse scraping, $Z = 5$ to 10 spots/in².

loading there are permanent deformations but for repeated loading and unloading the relationship between the shear pressure and deformation is linear as is represented by equation (2). From these measurements and from the measured normal compliance of the surfaces, the normal and shear compliance, obtained as a function of the normal interface pressure, are given in Fig. 1(c). The influence of surface finish on the measured shear compliance of cast iron surfaces is shown in Fig. 1(d). From the experimental results of Kirsanova[7] and from the authors[6,8] as well as from theoretical considerations[8] the ratio of the shear and normal stiffness can be written as follows:

$$\frac{dp_n}{d\lambda_n}\Big/\frac{dp_s}{d\lambda_s} = \frac{R}{cm}\,p_n^{(1-m-s)} \qquad (4)$$

The value of m can be considered equal to $0\text{·}5$, and in the same form $S = 0\text{·}5$; therefore, the ratio (4) becomes independent from the normal pressure. By analysing the experimental results[6-8] and from theoretical considerations[8] the relationship (4) can be written as follows:

$$\frac{R}{cm} = \frac{E}{G} = 2(1 + \mu)$$

where E and G are the modulus of elasticity of the material and μ the Poisson's ratio. Consequently the parameter R can be calculated by the following equation:

$$R = 2cm(1 + \mu) \qquad (5)$$

The values of R calculated by equation (5) for cast iron surfaces of different surface finishes are given in Table 1.

For the precise calculation of the deformations and pressure distribution in the joints it is necessary to consider the compliance of the structural components surrounding the contacting surfaces. The authors[6,9,10] have developed three methods to calculate the deformations and pressure distribution in the joints, namely the hydrostatic, plate and spring methods. In the plate and spring method the structural components of the joints are divided into finite

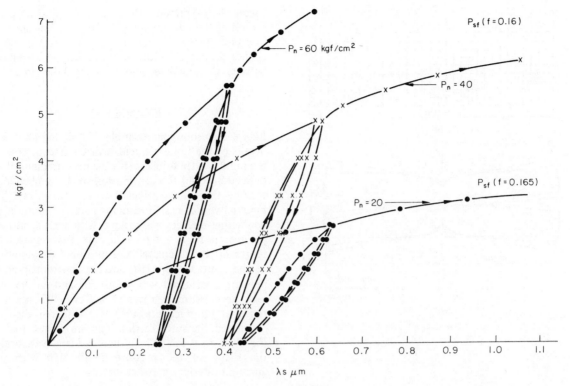

Figure 1 (b) Relationship between shear pressure and deflection of cast iron surfaces (Ground/Ground).

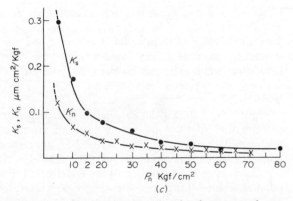

Figure 1 (c) Relationship between the shear, normal compliance and normal pressure for cast iron surfaces (Ground/Ground).

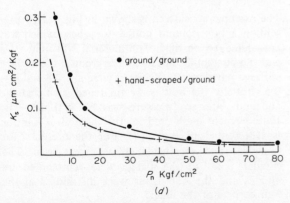

Figure 1 (d) Comparison of the shear compliance of cast iron surfaces. Ground/Ground + Hand-scraped/Ground.

elements (either rectangular or triangular plates). At the contacting surfaces the nodes of the finite elements must be coincident, forming pairs of nodes. A

detailed description of these methods is given in the references[6,9,10]. A brief outline of the spring method is given below. Each pair of nodes, one node on each contacting surface, is connected by a spring. These springs are defined initially such that only the normal stiffness is considered, but the shear stiffness of the surfaces can also be included. The stiffness of these springs is defined as a function of the surface stiffness and the normal pressure.

When the components of the joint are connected by the springs then the deformations of the joint can be calculated as a general finite element method solution[11-13]. The solution involves an iterative procedure; for the first iteration the normal interface pressure distribution must be assumed, and from the first finite element solution a new pressure distribution is calculated and the spring connections are redefined. This procedure continues until a final convergent solution is obtained. This solution gives the deformations of the joint and the pressure distribution is calculated as a function of the compression of the springs.

For joints similar to those presented in this paper, the three analytical methods were compared and have shown to give very similar results in the majority of cases. However, the spring method is the most powerful since this permits the inclusion of the correct shear stiffness of the surfaces.

EXAMPLE 1

Figure 2(a) shows two identical cast iron beams which were loaded by a central load. The surface finish at the contact was hand-scraped and the corresponding surface stiffness parameters determined experimentally were $c = 0.25$ and $m = 0.7$. These

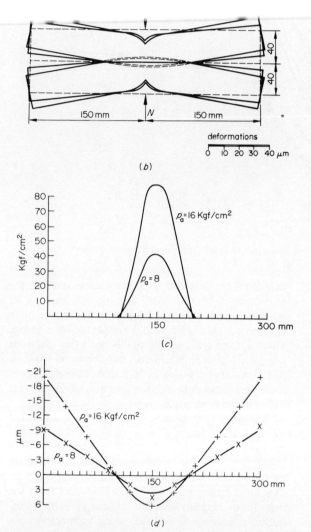

(b)

deformations
0 10 20 30 40 μm

(b)

(c)

(d)

Figure 2 (a) Finite element division of the beams and longitudinal position of the probes for the measurement of the contact deformations. (b) Deformations of the beams for p_a = 8 kgf/cm² and p_a = 16 kgf/cm². (c) Pressure distribution for p_a = 8 kgf/cm² and p_a = 16 kgf/cm². (d) Contact deformations for p_a = 8 kgf/cm² and p_a = 16 kgf/cm². Calculated + Measured.

values differ from the results in Table 1 because the range of interface pressure was larger. The modulus of elasticity (E) of the cast iron was 9500 kgf/mm². The thickness of the beams was 40 mm but for the calculations a plane state of stress was assumed and, therefore, for the theoretical solution it was taken as 10 mm. The applied load on the joint was measured by a load washer and the contact deformations by inductive transducers fixed along the beam[6]. The deformations of the beams are presented in Fig. 2(b) for apparent interface pressures of 8 and 16 kgf/cm².

the inductive transducers (for the measurement of the deflections) and the distribution of the applied loads for the theoretical calculation are also shown. The material was of cast iron and the surface finish of the contacting surfaces was ground with surface stiffness parameters of c = 0·69 and m = 0·5. Figure 3(b) represents the calculated and measured deformations of the joint. On the left side the apparent interface pressure considered was 4 kgf/cm² and on the right side the pressure used was 16 kgf/cm². When a large range of interface pressures is considered then the calculated pressure distributions are represented by Fig. 3(c). As it can be seen there is a large variation of the shape of the pressure distribution when the apparent interface pressure increases.

EXAMPLE 3

Figure 4(a) shows again the finite element division and the position of the inductive transducers of a dovetail slideway. The material is cast iron and the contacting surfaces were hand-scraped. For the theoretical solution it was assumed that the surface stiffness parameters c and m were 0·25 and 0·7 respectively.

Figure 4(b) shows the calculated and measured deformations of the joint for an apparent interface pressure of 8 kgf/cm². The value of 2·51 μm represents the surface compression at the point A and −1·95 μm gives the separation of the interfaces at the point B. Figure 4(c) gives the corresponding pressure distribution for several loadings.

EXAMPLE 4

The example considered is shown in Fig. 5(a), representing a box column bolted to a base which was considered to be rigid. The material was mild steel and the contacting surfaces were ground; the surface stiffness parameters considered were c = 0·3 and m = 0·5. Initially the bolts were tightened such that the apparent interface pressure was 67 kgf/cm². The finite element mesh used for the theoretical solution is represented in Fig. 5(b). For the present case, triangular and rectangular plates were used. The circular holes in the flange were represented by squares and the bolt loads were distributed at the four corners of the holes.

Figure 5(c) shows the deformed column for an applied load of 200 kgf. This figure shows the results of two different assumptions; one when the flange is assumed to be rigid and the other including the effect of the flange. The difference between these two solutions when considered at point A is large and the

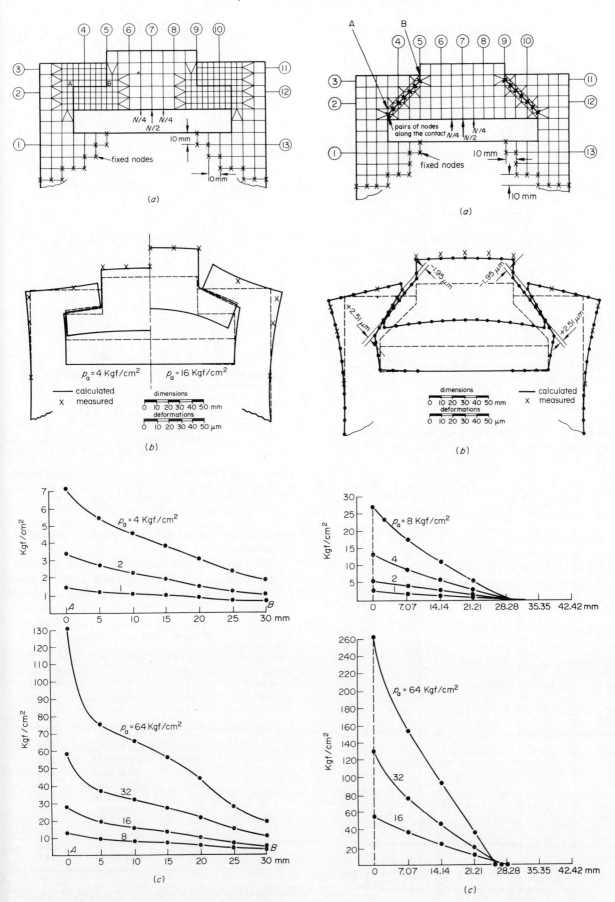

Figure 3. (*a*) Finite element division of the joint and position of the probes for the measurement of the deflections. (*b*) Deformations of the joint. Left side for $p_a = 4$ kgf/cm². Right side for $p_a = 16$ kgf/cm². (*c*) Pressure distribution at the contacting surface of the joint for several interface pressures.

Figure 4. (*a*) Finite element division of the joint and position of the probes for the measurement of the deflections. (*b*) Deformations of the joint for $p_a = 8$ kgf/cm². (*c*) Pressure distribution at the surface in contact for several interface pressures.

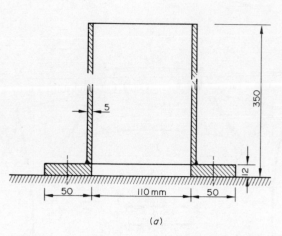

Figure 5. (*a*) Dimensions of the column model used for the calculations and tests. The base is assumed to be rigid.

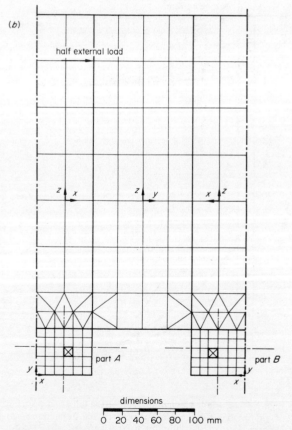

Figure 5 (*b*) Finite element division of the box column represented in a plane.

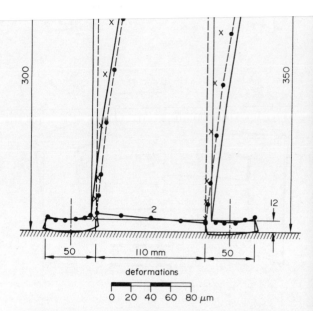

Figure 5 (*c*) Deformations of the column with a bolted flange. x Calculated for the column built in at the base; • when the flexibility of the flange is included; + measured deflection.

effect of the joint on the total deflection is equal to 43 per cent.

The calculated and measured deflections at the point *A* are also represented in Table 2. As it can be seen there is good correlation between the calculated and measured values. A further consideration with

TABLE 2   Calculated and measured deflections of the box column

Int. pressure kgf/cm^2	Applied load kgf	Method used	Deflection at the point A μm
50	200	measured	51·7
	400	measured	102·0
67	200	Built in column	22·7
	200	Joint flexible	40·5
	200	measured	46·0
	400	Built in column	45·4
	400	Joint flexible	94·5
	400	measured	95·0
90	200	measured	46·3
	400	measured	94·5
110	200	measured	45·8
	400	measured	94·0

this column was the analysis of the effect of the bolt pre-load, and the results for this are shown in Table 2. When the pre-load was increased from 50 to 67 kgf/cm^2 the deflection decreased by a considerable amount, but for larger pre-loads the improvement was small.

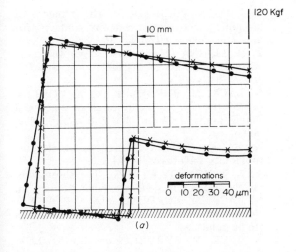

(a)

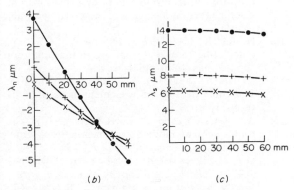

(b)   (c)

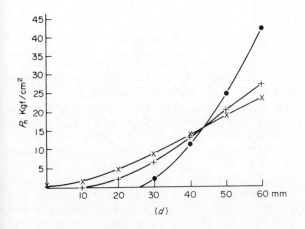

(d)

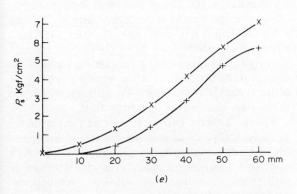

(e)

Figure 6. (a) Finite element division and calculated deformations of the half of the joint. (b) Normal contact deformations, · f = 0, + f = 0·2 and x f = 0·3. (c) Shear deformations, · f = 0, + f = 0·2 and x f = 0·3. (d) Normal pressure, · f = 0, + f = 0·2 and x f = 0·3. (e) Shear pressure, + f = 0·2 and x f = 0·3.

For this example the pressure distribution at the interfaces can be calculated and the results for the pre-load of 67 kgf/cm² can be seen in ref. 6.

EXAMPLE 5

The example shown in Fig. 6(a) was specially chosen to analyse the effect of the shear stiffness of the surfaces. As it can be seen the shear deformation at the contact is large, and the inclusion of the shear forces at the contact therefore have a considerable effect upon the deformations. For the theoretical analysis the material was assumed to be of cast iron with stiffness parameters $c = 0·8$ and $m = 0·5$. Figure 6(b) shows the contact deformations for the values of the coefficient of friction $f = 0$, $f = 0·2$ and $f = 0·3$. The corresponding shear displacements are given in Fig. 6(c). According to the authors[6,8] it is possible to verify whether the shear displacements are in the elastic range or if the surfaces are slipping. In the present example the shear displacements are in all cases larger than the elastic limit; the shear pressure represented in Fig. 6(e) can therefore be calculated from the normal pressure represented in Fig. 6(d) multiplied by the corresponding friction coefficient.

EXAMPLE 6

For the analysis of the effect of the flatness deviation upon the pressure distribution and deformations in joints the simple model shown in Fig. 7(a) was chosen. The beam was assumed to be of cast iron with surface stiffness parameters $c = 1·0$ and $m = 0·5$.

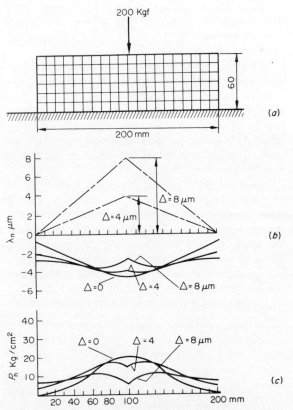

Figure 7. (a) Finite element division of the beam mounted on a rigid base. (b) Contact deformations. (c) Pressure distribution.

the present example the flatness deviation considered was very simple, but any shape possible in practice can be considered and the theoretical procedure described by the authors[6] is the same for all cases.

CONCLUSIONS

The examples presented in the paper cover all types of joints used in machine tools. The correlation between the theoretical and experimental results is very good, and the work has demonstrated that the stiffness and pressure distribution in machine tool joints can be calculated precisely. The small differences between the calculated and measured deflections, are dependent upon several factors.

The finite element meshes used for examples 2, 3 and 4 will possibly not represent exactly the actual flexibility of the joints. In particular, for the box column the assumption of a rigid base can introduce some errors.

The machining errors that always are present are sources for differences in the results. The machining errors were verified for all test models with the blue calibrator and maintained as small as possible.

The material used for the test models was controlled such that major casting failures of the cast iron were absent. In the tests no error was observed that could be due to casting failures.

A further source of error could be in the evaluation of the parameters c and m. When the deformations of the joints are analysed it can be seen that some differences in the values of c and m would give small errors in the total deflections. The authors[6,9,10] verified that it is not necessary to know precisely the values of c and m for a sufficiently accurate solution.

When the deflections of the joints are calculated by assuming rigid components, the results are several times smaller than the results obtained in the examples given. Even for very low interface pressure the errors obtained when assuming rigid components are large and increase with the increase of the interface pressure. For the calculation of the stiffness and pressure distribution in the joints it is not therefore possible to assume that the structural components are rigid.

REFERENCES

1. M. P. DOLBEY and R. BELL (1970). The contact stiffness of joints at low apparent interface pressure, *Annals of C.I.R.P.*

2. Z. M. LEVINA (1965). Calculation of contact deformations in slideways, *Machines and Tooling*, **36**.

7. V. N. KIRSANOVA (1967). The shear compliance of flat joints, *Machines and Tooling*, **38**.

8. N. BACK, M. BURDEKIN and A. COWLEY. Normal and shear stiffness of machine surfaces (to be published).

9. N. BACK, M. BURDEKIN and A. COWLEY. Pressure distribution and deformations of machined components in contact (to be published).

10. N. BACK, M. BURDEKIN and A. COWLEY. Calculating local deformations in machine tool connections (to be published).

11. S. HINDUJA (1971). Analysis of machine tool structures by the finite element method, PhD thesis, U.M.I.S.T.

12. A. COWLEY and S. HINDUJA (1970). The finite element method for machine tool structural analysis, *Annals of C.I.R.P.*, **18**.

13. O. C. ZIENKIEWICZ and Y. R. CHEUNG (1970). *The Finite Element Method in Structural and Continuum Mechanics*, McGraw-Hill.

DISCUSSION

Q. J. R. Barber. In carrying out experiments to deduce the normal compliance parameters c, m as in the authors' figure 1(a), there is one point which cannot be established experimentally—i.e. the origin. It is impossible to apply a zero load between the solids. However, in using a logarithmic plot to find the exponent m, the position of the origin will be crucial to the result. Would it not be preferable to use a representation based on a finite preload reference point? Indeed, there is a school of thought which maintains that the load/compliance curve should be asymptotic to the compliance axis with a merely arbitrary cut off point at the highest asperity. If such a representation was used, it would have the added advantage of making the curve continuous through the origin, thus simplifying the numerical analysis. All points on the interface could be regarded as being 'in contact' except that at some the stiffness and hence the contact stress would be infinitesimal.

Finally, have the authors considered the possibility of applying their method to the analysis of 'limits and fits' problems—e.g. the stress distribution between a shaft and a shrunk-on collar? The characteristic of such problems is that there are now two opposed deformable surfaces.

A. We accept the point raised by Mr Barber on the use of logarithmic plots when determining the empirical constants. In practice however, one carries out a procedure which is effectively the same as he proposes i.e. emphasis is placed upon the higher pressure range of the deflection curve where the uncertainty of the results is negligible.

At this stage we cannot see that a different representation of the surface characteristics would have any significant advantage.

We hope that we have demonstrated that the finite element technique is a powerful engineering tool which can be applied to components having contacting surface. We see no reason why one should not apply this technique to the analysis of limits and fits but as yet we have not encountered any real problems in this field and have therefore not considered a detailed analysis.

S

COMPUTER AIDED DESIGN OF TOOLS

by

H. K. TOENSHOFF*

SUMMARY

Cost movements and lack of personnel compel the rationalization of the design office. The paper reports on the investigations made at a machine tool manufacturer in order to find out the possibilities of applying CAD in the design office. A cost-to-savings analysis thereby resulted in favour of the area of tool design for tools required in multistage machines. The structure of the programming system CADOT is explained and the possible directions of its expansion are discussed.

INTRODUCTION

In the last few years it has become evident that rationalization efforts of a company must not be restricted to the production process and the organization. In fact the possibility to rationalize must be investigated for all spheres of action[1]. Quite a few research projects and numerous publications of recent date pay special attention to the application of electronic data processing methods in production planning and design office, with a view to utilizing the computer for those activities which are repetitive and which have solutions in the form of algorithms. These activities have different levels of suitability to automatic processing and, of course, depict a wide range of cost of implementation.

The computer as a thinking machine may be considered as a challenge to the intellectual abilities of mankind. It is worth while and interesting to compare the effectiveness of natural and artificial intelligence in a two-dimensional field (Fig. 1) with the degree of complexity in the x-axis and the working speed in the y-axis. The computer is able to solve simple problems at high speed. The human being is superior when complex problems are involved. The design problems are of complex nature requiring a high speed of computation and it is obvious that a dialogue system with graphics display gives itself to effective use[2].

This paper reports on the application of computer aided design (CAD) in the machine tool industry to design boring and turning tools. Work is being done at present on this project at IFW of TU Hanover. The general method of introducing CAD will be explained taking the problem of tool design as

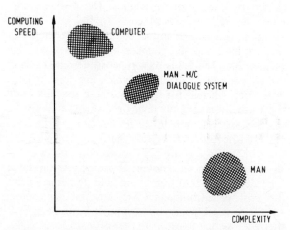

Figure 1. Computing speed vs complexity for different systems.

an example. The following three subjects are discussed:

(1) selection of suitable design problems out of a wide range of design activities.
(2) structure of the design program for lay-out planning, design of boring tools, determination of machining conditions and the cost calculation for tools.
(3) the possible extensions of the system.

SELECTION OF SUITABLE DESIGN PROBLEM

The design activities of a machine tool manufacturer have been investigated in this project. The production programme of this manufacturer consists of turret

* Institut für Fertigungstechnik und Spanende Werkzeugmaschinen Technische Universität, Hannover

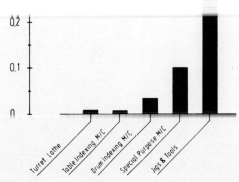

Figure 2. Ratio of design costs to turnover for different products.

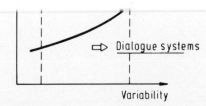

Figure 3. Software costs by design through variations.

in the amount of work load of the design office caused by the various machines. The ordinate contains the ratio of design costs to the cost of the product. This ratio is influenced by the design life of the product, e.g. the number of machines sold for each design. This leads to very small design costs per product for those machines which are sold in large quantities, whereas for special purpose machines and tools usually made only once or in small numbers the design costs can sum up to 33 per cent of the total cost of the product. It can therefore be expected, that the use of a computer for design purposes in the latter case brings in substantial savings. At the same time during the selection of suitable design problems the cost of introducing CAD must be considered. Apart from the necessary computer equipment it is thereby essential to estimate the software costs.

To appraise the amount of software to be developed it is useful to categorize the design activities in various phases and to determine the possibility of EDP for each phase. The following categories must be considered.

(a) Determination of the functional and working principle;
(b) determination of the design arrangement and calculation of design elements;
(c) drafting.

The design process proceeds in the above sequence from the abstract to the concrete concept. The work relief of the designer through computer can also be expected to ascend in the same order. The use of a computer has, therefore, not been considered at present for the first design phase of the project at hand. The functional and working principle are viewed as given.

The second phase promises to be the more suitable for the design of tools with CAD. The solution of problems in this phase can be tackled with the method of 'design by variation'. To estimate the

A dialogue system proves to be inevitable for problems having high variability. Should the programming system for such a problem be made up of only algorithms then the processing tends to be complex and uneconomical. The degree of variability of the two design problems discussed here is shown in Fig. 3. Even if the functional and working principles of some of the assemblies of special purpose machines are known, there exist numerous possibilities of design arrangements. These considerations lead to the choice of the design of tools for the CAD application, even though for the design of special purpose machines at least an equal amount of rationalizing effect could be expected.

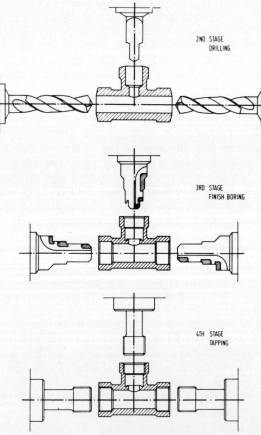

Figure 4. Layout plan for a typical application.

An idea about the problems encountered during the design of tools and about the extent of design work needed a typical example for a workpiece and the related tools to manufacture it are illustrated in Fig. 4. The design work involves essentially lay-out planning to determine the different stages shown in the figure, the tool design and the design of jigs and fixtures.

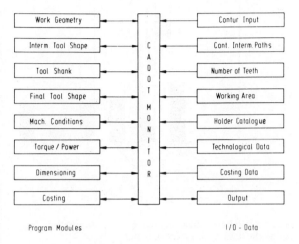

Figure 5. Program structure of computer aided design of tools.

In Fig. 5 the structure of the programming system CADOT has been explained. The required sub routines are shown on the left hand side of the figure, whereas an account of the input/output data and other storage mediums is given to the right of the monitor. The system is essentially developed in form of program modules, which can be changed or replaced at ease. The contour information of the finished workpiece, the intermediate passes and the blank are put into a suitable format. This can be carried out with the help of 'form elements'. The module 'work geometry' builds up the necessary tables to register the end points and the points of tangency in case of circular shapes of all the above-mentioned contours. For each of the 'contour of the intermediate paths' a separate tool must be designed. At present it is intended to put the data on intermediate paths. Provisions are made however to calculate this data at later stages of the project.

The subroutine 'Torque/Power' has access to the technological data like specific cutting force stored on external mediums and computes the torque developed during the boring process depending upon the material to be removed and the number of teeth of the cutter. The suitable diameter of the tool shank is then chosen from the holder catalogue specified by the design office and also stored in the system. Depending upon the available working space the length of the tool shank is determined. These computations are carried out by the module 'Tool Shank'.

At this stage of the process the finished shape of the tool has been defined and can be displayed through 'Output' either on a plotter or on a graphic display. The module 'dimensioning' undertakes apart from outputting the lengths and diameters also the problem of selection of tolerances of the tool dimen-

sions. Several facts must be considered to achieve correct tolerances. For example, because of unavoidable radial runout the boring tools produce slightly oversized bores. On the other hand thin-walled workpieces tend to warm up easily resulting in undersized bores. To guarantee the fulfilment of the specified tolerances it is necessary to profit from the practical experiences of the design office. Such know-how is proposed to be included in the module 'dimensioning'. It is a great advantage of CAD to be able to converse with the designer and at the same time store the know-how in readily accessible form for the future. Such a system would assure a stable design quality of the company in spite of fluctuations of personnel in the design office.

An additional work carried out by the module 'dimensioning' is the calculation of the core diameter of holes with threads. This is done by storing the information on the usually used thread standards or by analytical calculations.

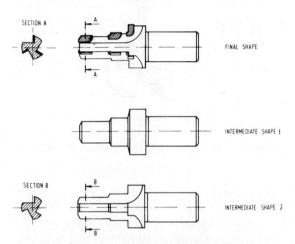

Figure 6. Finished and intermediate tool shapes.

The drawing of the boring tool can now be plotted. The upper part of Fig. 6 shows the drawing produced by CADOT, where the final shape of the tool has been drawn. This drawing can be used for the finish grinding of the tool.

The manufacture of tools proceeds, however, in stages. At first the tool material is turned to intermediate shape 1. The grooves to accommodate carbide tips and to provide for chips must then be milled (intermediate shape 2). The subroutine 'intermediate tool shape' delivers drawings as shown in Fig. 6 to manufacture the tool shape 1 and 2. The production of paper tapes to control NC lathes (for tool shape 1) and NC milling machines (for tool shape 2) are reserved for the later phase of development of the CADOT project. It is then possible to realize an integrated information processing system starting from the design office and ending with the production.

The CADOT system delivers in addition to the drawings mentioned above also lists containing technological information to the assembly department. This information consists of values of speeds and feeds to be set up and the time required to manufacture the given workpiece. Torque and power

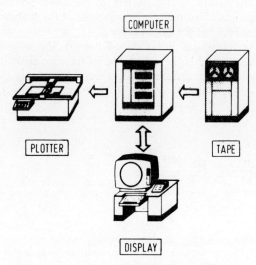

Figure 7. Computer configuration for design office.

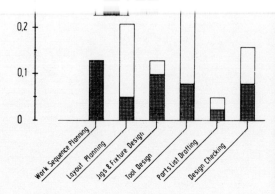

Figure 8. Relative time requirements for different activities during the design of jigs, fixtures and tools.

The CADOT system outlined so far operates with the following computer configuration (see Fig. 7). The computer has 64 k 16 bit words of memory, out of which 8 k words are present in core and the rest in form of virtual memory on a fixed head disc. The program library and other data can be stored in additional 64 k storage on the same disc. Further a cassette unit serves as a scratch tape. The source program containing job specifications is read through a card reader. CADOT is initiated from the control teletype. The graphic display forms the important part of the I/O equipment, with which the designer pilots the execution of CADOT. Because of the small degree of variability of the design process and because of the fact that most of the modules are made up of algorithms a storage oscilloscope type of display is adequate. The much costlier alternative—a display working in pick-and-trace mode which needs a minimum core storage of 4 k as a support for display repetitions—is not necessary. The storage type display is however slow in operation due to the need to generate the picture every time a change is desired. The drawings are made on a plotter. This configuration could be rented for a sum of about DM 4500, per month. Additional costs for programming and operating personnel are involved, so that in total nearly DM 7000 per month must be brought forth to pay for the upkeep of the equipment.

It is useful to compare these costs with the expected savings. As CADOT has not yet been sufficiently tested in practice only approximate values based on the experiences collected so far can be given

in this case. CADOT may be used in later stages of development to draw up quotations as explained elsewhere and especially to choose from a number of possible sequences. A considerable relief of the designer can be expected in the phase of 'Lay-out Planning'. It is usual practice of the design bureau to develop a lay-out plan at the beginning of tool design, so that the designer can decide about the working space needed and check up the danger of collision of tools. With CAD, however, the lay-out plan is made at the end of the design process. The necessary drawings of the various tools can be plotted from the information stored in the computer. The shape of the workpiece can also be retrieved from the system. It thus results in 75 per cent savings.

The design of work holders is little influenced by CADOT. The tool design process itself is essentially reduced in cost and time. Savings are to be expected also in the fields of design checking, preparation of part lists etc. All in all a reduction of about 50 per cent in design time can be achieved with CADOT. The reduction of design time is especially desirable for tool design activities, because the permission to proceed with design work is often given by the purchaser only at stages when the manufacture of the machine tool is nearly complete. A longer design time would mean that the capital intensive machine tool must wait with the manufacturer till the tools are ready. A flexible and quick design office contributes a lot towards the ease of selling the machine tools.

Apart from time savings CADOT proves also to be economical. According to the first applications of CADOT 20 per cent savings in design cost are expected. This may seem to be modest, but other advantages like more reliable design, storage of know-how and reduced design time when considered are convincing arguments which speak for the introduction of CADOT.

POSSIBLE EXTENSIONS OF THE PROGRAMMING SYSTEM

More savings can be expected when CADOT is extended to include other fields of activities of the firm. As already reported CADOT can be extended with little effort to include calculations leading to quotations. The preparation of NC tapes to produce the intermediate tool shapes would also result in cost reduction.

So far the design of boring tools has been dealt with. It is planned to extend CADOT to the design of turning tools and holders. The degree of variability of the turning problem is however greater and as shown in Fig. 3 the designer should work essentially in a dialogue mode. The method of menu technique promises to be most suitable in this connection[3]. The dimensions and the shape of standard tools are stored in the system. The designer can make these tools appear on the display. Depending upon the application different sets of menu can be made available. The designer can then decide upon the suitable tool. In case no such tool is present in the menu, he can alter the standard tool to suit the requirements.

A high degree of sophistication can be reached, when the described programming system is extended to include the design of whole multistage machine tools. Substantial work has been carried out at IFW of TU Hanover[4] to analyse a given workpiece spectrum and to develop a multistage machine operating with minimum costs.

CONCLUSIONS

The use of electronic data processing equipment can be expected to result in a rationalization of the work carried out in the design office also. However the hardware and software costs tend to increase rapidly for complex design problems. A CAD project, which is now being tested in practice, has therefore been developed to solve design problems having a small degree of variability with a view to demonstrate the advantages of this concentption on the example of the design of boring tools. This system named CADOT consists of various modules which are easily changed or altered and which take into account the know-how of the design office. CADOT produces drawings of the finished tools and their intermediate shapes. The preliminary estimates of costs and time savings are promising. By extending CADOT to include other activities of the company the efficiency of the programming system can be increased.

REFERENCES

1. Computer Aided Design, NEL-Report 242, August 1966, Ministry of Technology, U.K.
2. H. K. TÖNSHOFF and D. SANKARAN (1971). Konstruieren von Werkzeugmaschinen mit Unterstützung von Rechnern, *Konstruktion*, **23** (9), 333-8.
3. H. OPITZ, H.-P. WIENDAHL and U. BAATZ (1970). Bildschirmunterstütztes Konstruieren—Prinziperarbeitung, Gestaltung und Detaillierung mit der Menutechnik, *Ind. Anz.*, **92** (98), 2371-4.
4. D. SANKARAN (1973). Zur Lösung des Zuordnungsproblems in der Arbeitsvorbereitung mit Hilfe pseudoboolescher Methoden, Diss., TU Hannover.

DISCUSSION

Q. R. Umbach. Does the program take into consideration standardized tool tip measures?

A. Yes. First of all the shape of the chip groove is determined by considering the amount of material to be removed and the space required for the transport of the chip. In case carbide tipped tools are used the standard measurements are referred to. With the further extension of CADOT it is planned to choose the most suitable tip among several alternatives stored in the system.

DEVELOPMENT OF THE FINITE ELEMENT METHOD FOR VIBRATION ANALYSIS OF MACHINE TOOL STRUCTURE AND ITS APPLICATION

by

H. SATO, Y. KURODA and M. SAGARA*

SUMMARY

A standard type of lathe is adopted for an example of the analysis of the vibration characteristics of machine tool structures. As the computation method the finite element method, which will be referred to below as FEM, is used. The effectiveness of the method is studied using simplified thin plate perspex model structures. The fundamental structure of the lathe is composed of elements such as the bed and its supporting columns. It is shown that the method is powerful for analysing the structural elements and for studying the effects of ribs on the vibration characteristics. A method for obtaining the vibration characteristics of the integrated fundamental structure, by describing each structural element as an equivalent beam having the same fundamental natural frequency as the FEM model, is developed. This makes it possible to perform the computer aided analysis by combining the function of the large-scale computer with that of the medium one for the principal mode of vibration of complicated machine tool structure. FEM is applied to the extensive analysis of the actual lathe structure. Again the effectiveness of the method is proved. It is also noticed that there are some difficulties in evaluating the boundary condition of supporting points. The computation method itself has wide versatility for the structure of machine tools in general. The adequacy of the method is examined by comparing the results with those obtained experimentally for both the small size model and the actual structure.

INTRODUCTION

The natural frequencies and the modes of vibration of a machine tool have been studied in order to improve its machining performance and to resist adequately the onset of self-excited vibration[1-3]. Various methods to compute these vibration characteristics have been developed, and experiments to identify the dynamic properties of the actual machine tools have been established[4-8]. However, it does not seem that the development of a computation method capable of evaluating the vibration characteristics at the design stage has been completely successful. The models often used for the computation are generally too simple and cannot take into account the effects of plates and ribs.

Recent advances of the finite element method in structural engineering make it possible to obtain not only the static behaviour but also the dynamic characteristics of complicated structures, the limitation being connected with the memory size of the computer[9,10]. However, its application to machine structures has not yet been fully exploited, especially for structural vibration.

The method used here adopts a finite element procedure using FEM to identify the natural frequencies and the vibration modes. Even with a large-scale computer with a memory of 65k words, the number of elements cannot be made large enough to perform the accurate computation of the simplest type of general lathe structure. Thus a new method to estimate vibration characteristics is developed. The machine tool is usually broken down into several fundamental structural elements, such as the bed and its support in the case of a lathe. It is proposed that the analysis of the vibration characteristics of each structural element is made firstly using FEM which simulates the original structure as exactly as possible. Then each structural element is represented by an equivalent beam structure, the fundamental natural frequency of which is equal to that obtained from the FEM analysis. The total structure is composed of these equivalent beam systems. The analysis for such beam structures has often been made for such machine structures by various methods. The method is studied first using rather simple small size perspex models[11], and its effectiveness is verified.

Then it is applied to the analysis of a lathe structure. The bed is reduced to a simplified plate structure to which beams corresponding to slideways and the like are attached. The bed structure is supported on rubber pads for experimental measure-

* Institute of Industrial Science, University of Tokyo

The rectangular thin plate element is used for the analysis. Comparison with the model experiment makes it obvious that the estimate obtained using FEM for the natural frequency of the torsional mode is higher than that of the actual system. This is improved by introducing an angular deformation in plane rotation in the plate finite element description.

ANALYSIS ON PERSPEX MODEL STRUCTURE

Analysis of bed structure

FEM is first applied to the analysis of the vibration characteristics of a model structure made of methacrylic resin perspex plate. It is composed of substructures such as the bed and its supporting columns. The perspex model is easily made and the vibration experiments can be readily performed, the results of which should be compared with those by the analysis.

Figure 1(a) shows a general view and dimensions of the model structure which is used for the analysis. Although it is a simplified structure which does not have a gear box, spindle system, tail stock and other auxiliary items, the structural configuration is based on a standard type of lathe shown in Fig. 1(b). The configuration is idealized and simplified, the slides are

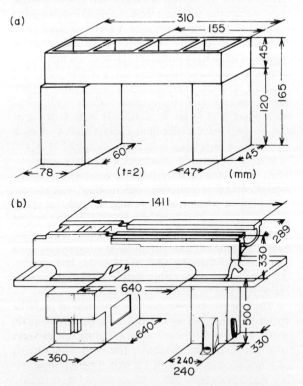

Figure 1. General view and dimensions of the fundamental structure of the lathe: (a) small-size perspex model; (b) actual structure.

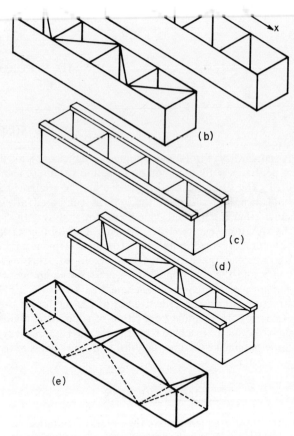

Figure 2. General view of various types of bed structures.

only the bending but also the torsional stiffness are added in Fig. 2(b). The models shown in Fig. 2(c) and (d) have beams corresponding to slideways.

Table 1 gives the natural frequencies for these bed-like structures. The analysis is made with the beds cantilevered. For the type of structure shown in Fig. 2(a) the experiment is conducted using a model made from steel plate of 0·6 mm thickness. This aims at eliminating the effect of the frequency dependent property of Young's modulus of perspex in verifying the results by the analysis.

Taking into account the fact that the estimate of the natural frequency by FEM gives an upper bound solution for the correct value, both results agree well with respect to the bending mode of vibration about the y axis. Looking at the torsional mode about the x axis, the result of the computation is about two times higher than that of the experiment. This could be caused by the assumption in the FEM procedure that the in-plane angular displacement about the z axis (θ) is neglected in constructing the stiffness matrix. This results in a constraint about the x axis and the higher estimate of the torsional vibration frequency is obtained by the analysis. It was difficult to carry out

TABLE 1   The natural frequency of the various types of be structures in Fig. 2 (Hz)

Model in Fig. 2	(a)*		(a)		(b)		(c)	(d)
Dominant mode	exp.	comp.	exp.	comp.	exp.	comp.	comp.	comp.
Bending around y	21·0	25·3	29·8	28·2	155	150·9	39·3	160·7
Bending around y	68·0	77·6	116	85·9	520	586·0	129·0	649·9
Bending around y	117	132·0	190	154·3	–	714·9	235·0	–
Torsion around x	210	446·4	90·0	176·4	–	653·0	175·7	580·0
Bending around z	–	473·0	–	159·6	–	171·9	167·2	179·5

* Steel plate.

TABLE 2   The natural frequency for the equivalent beam integrating the fundamental structure by the perspex model (Hz)

Mode		Rubber	Bearing at 8 points	Bearing at inside 4 points	Bearing at outside 4 points	Computation
Bending around y	1st	284	275	269	277	144
	2nd	491	467	484	500	402
Bending around z	1st	362	335	340	361	426
Torsion around x	1st	212	195	202	210	220
	2nd	643	–	–	635	504

the measurement of the natural frequency of the bending mode about the z axis.

The difference in the results seen between the experiments and the computation is in some cases large and is not systematic for the perspex model; however, good agreement can generally still be found. Comparing column (a) with (b) in Table 1, the effect of the diagonal rib plate shown in Fig. 2(b) is obvious. The natural frequencies of the bending mode about z and the torsional mode in column (b) becomes remarkably higher than those in (a).

The effect of attaching beams which simulate slideways is to give a slight rise in the natural frequencies of the corresponding modes. From the details of the modal shapes for Fig. 2(c) and (d), the bending and the torsional modes of vibration are no longer independent but coupled to each other, although the amount of the coupling is small. The natural frequencies for the model of Fig. 2(e) are not listed in Table 1 because of the difficulty of the mode comparison. These are obtained as 150·2, 155·4, 176·9 Hz and so on.

The computations by the FEM are made by using an element for one rib plate and a side panel between the rib plates, so that the number of nodes for the structure in Fig. 2(a) is twenty, four of which are used as fixed points.

As for the supporting columns, both analytical and experimental investigations indicate that the results agree well. The configuration of the structure is simpler than that of a bed. They are columns with rectangular and square cross section. The boundary condition at the bottom is fixed and the top of the column is open and free. In the analysis each side panel is divided into four rectangular elements.

Analysis of the integrated structure

The analyses made so far indicate that the FEM for plate structures would also be effective for the integrated total structural system as far as the memory capacity of the computer is concerned. However, it will be difficult to carry out the computation for complicated and large size machine tool structures without some simplification. In this study a simplified method is developed that allows substructure to be represented by an equivalent beam which has the same natural frequencies (first order bending modes around two axes and the torsional mode around the longitudinal axis). The equivalent beam is characterized by second cross-sectional moments and shear moduli which give respective fundamental natural frequencies equal to the results obtained by the FEM for the thin plate structure. Then the total integrated structure is composed of beam systems, and we can therefore investigate the vibration characteristics using a computer with medium-size memory capacity.

Table 2 shows the natural frequencies of the equivalent beam of each structural element which are used for integrating the total structural system. The experimentally and the analytically computed results are compared in Fig. 3. The computation is performed by the beam structure system as mentioned above; the modes are shown on the perspective of the model.

As for the first natural frequency, both the frequency and the modal shape agree well. Some discrepancy is found in the amplitude at the left end of the bed, that is, it is almost zero for the experiment and a small amplitude is observed for the computation. This would be inevitable because the computation is carried out under a different boundary con-

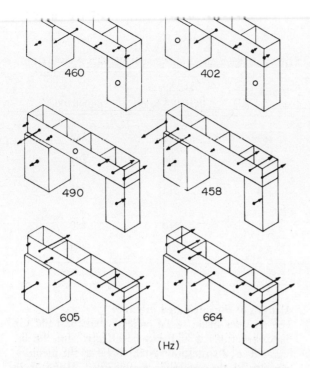

Figure 3. The natural frequency and the vibration modes by computation and experiment for the integrated fundamental structure of the perspex model.

dition around the joints. Taking this into account, the mode for the second frequency also shows good agreement although the difference in the frequency is greater.

The third mode which is regarded as the mode accompanying torsional deformation of the bed also shows good agreement. In the actual system this type of mode is often observed as the second one. This would be a result of the difference of the nature of the material and that of the boundary condition at the bottom of the column. As for the latter, the complete fixed end is not usually realized for the actual system.

As for the fourth mode, the similarity of the mode shape is disturbed at the left end, which indicates a weakness in the equivalent beam model. The disagreement is greater for the modes of higher natural frequency than this. Although the natural frequencies in the higher frequency range are successfully obtained from the computation, it is difficult to compare these properly with the experimental results

The investigation shows that the FEM is a powerful tool for identifying the vibration characteristics of the basic structure. It is also obvious that the analysis by the equivalent beam structure is also effective for the integrated system when the equivalent beam is identified adequately.

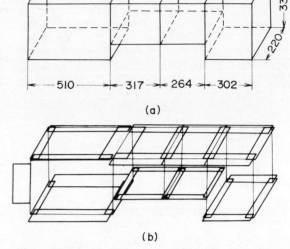

(a)

(b)

Figure 4. The composition of the bed for computation by the FEM: (a) part described by the plate element; (b) part described by the beam element.

of supporting the bed are considered. Rubber pads and newly developed bearings are used for the support. The latter is composed of a steel ball and a base which has a spherical seating, the radius of which is a little larger than that of the ball. Since the support is realized by point contact, the constraint of the support is very flexible and seems closely to resemble a free-free boundary condition. The constraint is flexible for all directions of the coordinate axes, which are shown in Fig. 5.

Figure 5 compares the experimental results of the natural frequencies and the normal modes with those obtained by the analysis. From the experimental results, the lowest natural frequency occurs at 212 Hz and is a torsional mode around the x axis. The lowest natural frequency for the bending mode around the x axis is observed at 284 Hz. The second order natural frequency for the same sort of mode which has three nodal points appears at 491 Hz. The natural frequency for the bending mode around the z axis, which has scarcely been reported in work done so far and was not expected to be obtained, is clearly evident at 362 Hz. The results obtained by the computation are also shown in Fig. 5. The skeleton of the FEM model is shown on the original drawing of the structure. The parts which have curved and complex shape and slanted panels are all straightened and simplified. The natural frequencies and the modes are obtained under the free-free boundary condition.

The results indicate modes which look similar to the experimental ones. However, the modal points are

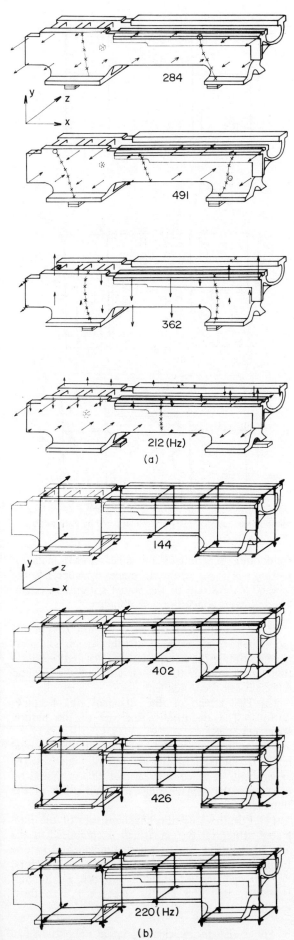

Figure 5. The natural frequency and the vibration mode of the bed: (*a*) by experiment, (*b*) by the computation.

insufficient for an accurate comparison. The general shapes express the mode characteristics comparable with the experimental results. As for the natural frequency at 144 Hz for the bending mode around the *y* axis, this is about half as much as that of the experiment. The natural frequencies for the first mode about the *x* axis, the first bending mode around the *z* axis and the second bending mode around the *y* axis show quite a good approximation, considering the discrepancy of the configuration between the actual system and the model used for the analysis and of the boundary conditions.

As for the torsional mode around the *x* axis, a rather large difference in the natural frequencies was found in the study using a perspex model. A method to improve this by compensating the effect of in-plane[1 2] angular displacement is used. Further investigation is necessary to acquire better agreement. In addition to these natural frequencies some others are computed in the analysis which were difficult to find in the experiment. The panel mode with the front and the back panels vibrating in the *x* direction is an example. The deformation is caused by the shear force applied to each panel.

Table 3 shows the natural frequencies for the experiment with various supporting conditions and for the analysis. The variance of the natural frequencies according to the change of the supporting condition is not great. It can be said that the difference is within 10 per cent. The change in the modal shapes can hardly be discerned.

The fundamental structure consists of the bed and the supporting columns. In the actual system they have large holes to accommodate the power motor and its access, and they are rather complex. The results of the excitation experiment are shown in Fig. 6. A rocking mode is observed for the lowest natural frequency. In the second mode both ends of the bed move in reverse phase. The frequencies of these two modes are 55 and 110 Hz. The natural frequencies corresponding to these modes for the complete structure including the tail stock, gear trains, spindle system and all other auxiliary parts were 48 and 80 Hz[3]. This proves that the auxiliary structural system causes the natural frequencies of the fundamental structure to decrease.

The FEM analysis for the fundamental structure is made by the aforementioned analytical bed model and the box type column supporting the bed, which is expected to be stiffer than the actual one. The bottom of the supporting column is fixed. The mode shapes which are comparable with those for the actual system obtained by experiment were found; however, the frequencies are much higher than those

TABLE 3   The natural frequency of the actual bed structure under various supporting conditions, and of the computation for the structure under a free-free condition (Hz)

Dominant mode element str.	Bending	Bending	Torsion
Bed	26·7	154	158
Thick supporting column	418	556	715
Slender supporting column	546	546	842

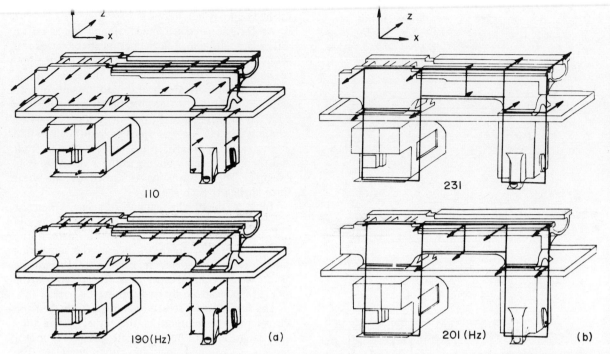

Figure 6. The natural frequency and the vibration mode of the fundamental structure: (*a*) by experiment, (*b*) by computation.

of the experiment. The turn of the second and the third mode is reversed. This means that if the support is pure box and the boundary condition at the foot is fixed, the structure is estimated to be much stiffer than the actual system.

At the moment it is difficult to formulate the supporting column in a more realistic way than this. Thus the results are considered to be the stiffest natural frequency which can be realized. The boundary condition at the foot in the actual system is never fixed, so that an appropriate flexible support should be incorporated into the analysis. According to the analysis natural frequencies were found in addition to those shown in Fig. 6. These again were not observed in the experiment.

The basic study to obtain the natural frequency and the mode shape by applying the FEM to the fundamental structure of a lathe is made. Further refinement for adjusting the boundary conditions making the appropriate model seems necessary, especially for the estimation of the integrated structural system.

CONCLUSIONS AND ACKNOWLEDGMENT

The FEM is applied to identify the vibration characteristics of machine tool structures. In particular, a lathe structure is taken for the main analysis. The effectiveness of the method is verified at first by conducting experiments on small models made of perspex. A basic study is made by applying the method extensively to the actual structural system. The following conclusions are obtained.

(1) Application of the FEM to various types of model bed structures made of thin perspex and steel plate to obtain the natural frequencies and the vibration modes verifies the effectiveness of the method especially for the bending modes.

(2) The effect of the diagonal ribs becomes obvious. The quantitative increase of the natural frequency is given.

(3) A higher natural frequency for the torsional mode is obtained by the computation than by the experiment. It is considered that this is caused by neglecting the inplane angular deformation of the finite elements.

(4) For the vibration characteristics of the integrated structural system, which is composed of the equivalent beam models whose fundamental natural frequencies are the same as those of each structural element such as the bed and its supporting column, it is shown that the results agree well with those obtained experimentally.

These are the conclusions based upon the small size models. In these studies the configurations of the model structure and of the FEM analysis are similar and the boundary condition which is adopted in the experiment is realized in the FEM analysis. The conclusions described below are derived by the extensive application to the actual system.

(5) Even large-scale digital computers with 64K words of memory are not adequate to model the bed structure accurately when using the FEM. Quite good agreement is obtained especially for the modal shapes in spite of the discrepancy in the boundary conditions and the configuration.

(6) The natural frequencies which cannot be easily excited in the experiment are found in the analysis. It seems that they might be caused by the assumption that the actual system behaves as a thin plate structure. Further investigation is needed.

(7) The compensation for the in-plane angular deformation works well. Good agreement is found as for the natural frequency of the torsional mode between the experiment and the analysis.

(8) The analysis for the integrated fundamental structural system by modelling the supporting column as a box-type column provides a higher natural frequency than that by the experiment. Modal shapes which resemble closely those obtained for the actual system are found.

Further refinement in estimating the boundary conditions and the modelling of the actual system would provide a better estimation of the structural vibration characteristics when using the FEM.

The authors express their sincere gratitude to Professors N. Takenaka, A. Watari and S. Fujii at the University of Tokyo for their fruitful discussions and suggestions. They also owe much to Messrs K. Suzuki, M. Komazaki and M. Ohori for their assistance in this work. This study is partly supported by the special research fund provided by Ministry of Education and the JSME project. HITAC 5020, FACOM 270-30 at University of Tokyo and UNIVAC 1108 in UNICON are used for the computation.

REFERENCES

1. S. A. TOBIAS (1961). *Machine-Tool Vibration*, Blackie.
2. H. E. MERRIT (1965). Theory of self-excited machine-tool chatter, *Trans. ASME*, Ser. B, 11.
3. H. SATO and T. AKUTSU (1971). A study of identification of dynamic characteristics of machine tools by means of micro tremor, *Proc. 12th MTDR*, Macmillan.
4. S. TAYLOR and S. A. TOBIAS (1964). Lumped-constants method for the prediction of the vibration characteristics of machine tool structures, *Proc. 5th MTDR*, Pergamon.
5. J. C. MALTBAEK (1964). Classical beam method for the prediction of vibration characteristics of machine tool structures, *Proc. 5th MTDR*, Pergamon.
6. T. SATA and N. TAKASHIMA (1971). Analysis on dynamic stiffness of machine tools by the finite element method, *Proc. Autumn Meeting JSPE* (in Japanese).
7. N TAKAHASHI and S. OHNO (1971) On free vibration in lathe bed, *J. IIS*, Univ. of Tokyo (in Japanese).
8. M. YOSHIMURA and T. HOSHI (1972). Computer approach to dynamically optimum design of machine tool structures, *Proc. 12th MTDR*, Macmillan.
9. O. C. ZIENKIEWICZ (1971). *The Finite Element Method in Engineering Science*, McGraw-Hill.
10. H. C. MARTIN (1966). *Introduction to Matrix Methods of Structural Analysis*, McGraw-Hill.
11. F. M. STANSFIELD (1965). Some Notes on the use of perspex models for the investigation of machine tool structures, *Proc. 6th MTDR*, Pergamon.
12. O. C. ZIENKIEWICZ et al. (1968). Arch dams analysed by a linear finite element shell solution program, *Symp. on Arch Dams*, Inst. Civil Engineers.

APPENDIX

The formulation of the FEM and the methods of obtaining the eigenvalues and eigenvectors which correspond to the natural frequencies and the normal modes are summarized below. Details are given in the references.

The displacement function for the basic rectangular element is given as

$$u = \alpha_1 + \alpha_2 x + \alpha_3 y + \alpha_4 xy \tag{1}$$

$$v = \alpha_5 + \alpha_6 x + \alpha_7 y + \alpha_8 xy \tag{2}$$

$$w = \beta_1 + \beta_2 x + \beta_3 y + \beta_4 x^2 + \beta_5 xy + \beta_6 y^2 + \beta_7 x^3 + \beta_8 x^2 y + \beta_9 xy^2 + \beta_{10} y^3 + \beta_{11} x^3 y + \beta_{12} xy^3 \tag{3}$$

where u, v and w is the displacement of the element shown in Fig. 7. This is a very common description of the function.

Based upon the relation of the strains, the stresses and the modal forces with the displacement function, the stiffness and the mass matrices of the element $[K]^e$ and $[M]^e$ are respectively given as

$$[K]^e = \int [B]^t [D] [B] \, \mathrm{d(vol)} \tag{4}$$

$$[M]^e = \int [N]^t \rho [N] \, \mathrm{d(vol)} \tag{5}$$

where $[B]$, $[N]$ and $[D]$ are obtained from the characteristic of the displacement function for the element and the relation between stress and strain, and ρ is weight per unit area. These can be made for in-plane and out-of-plane deformation. The stiffness and the mass matrices for the total structure $[K]$ and $[M]$ can be obtained by summing those of the individual elements according to the configuration of the structural system.

The equation of motion for free vibration is given through the general equilibrium equations as

$$K \{\delta\} = -M \{\ddot{\delta}\} \tag{6}$$

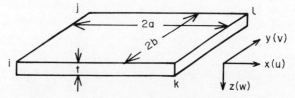

Figure 7. Rectangular element and its coordinate system.

model structure made of perspex while Householder's method is used for the analysis of the actual structure.

The time consumed for the computation of the bed shown in Fig. 5, which has 20 nodes, is about 750 s when using Jacobi's method. It takes about 180 s for the first 11 natural frequencies and modes by Householder's method. A UNIVAC 1108 with 64k words is used for both.

DISCUSSION

Q. Dr S. Taylor, University of Birmingham. The frequency errors showing in the paper could arise from three sources.

(a) Mathematical errors in the computation, particularly as I understand large elements were used.
(b) Errors in the experiments due to construction

analysis is not the only source of error. I have not paid much attention to analyse the source of error so far. The investigation is mainly made from the viewpoint of how effective the finite element method works to identify vibration characteristics of the machine tool structure as the first stage of the application of the method. Then the roughest element division is taken for the limit of the memory size of the computer and for the evaluation of the method under the worst condition.

(a) Mathematical errors should be generally investigated by further study using fine mesh.
(b) Errors in the experiments can be easily introduced by the causes Dr Taylor pointed out. However, as for the experiment reported here the ambiguity of the material constant of the perspex seems the most responsible cause.
(c) As for the analysis of the actual structure system it would be one of the most important sources.

ANALYSIS OF A MILLING MACHINE: COMPUTED RESULTS VERSUS EXPERIMENTAL DATA

by

J. A. W. HIJINK and A. C. H. VAN DER WOLF*

SUMMARY

The paper describes the transformation of a structure of a horizontal milling machine into a beam model. From the comparison of the static results of the model and the experimental values of the milling machine some corrections in the model are carried out. The dynamic results are compared by means of the direct and cross receptances between table and spindle of the machine tool. To this end an arbitrary value for the overall relative damping is introduced into the model.

INTRODUCTION

In order to calculate the static and dynamic properties of a machine tool structure, it is necessary to define a suitable topological model of that structure. The details of such a model depend on the structure itself, on the available program facilities and on the person who defines the model.

The machine tool structure under discussion is a universal milling machine, equipped for horizontal milling. A computer program based on the 'finite-element' method was available. A description of this program is given in ref. 1.

When computing the static properties of the model, any kind of load and combination of loads may be placed in any station point of the model. The displacements and rotations of all the station points can be calculated. For the dynamic properties the natural frequencies with corresponding modal shapes are computed. It is also possible to calculate the modal flexibilities and—according to Cowley[2]—the frequency response between two points of the structure. Viscous damping is assumed to occur.

THE MODELLING

In the computer program the relevant stiffness quantities of the beam elements are—in general—calculated from the length of the element, the cross-sectional area, the second moments of area with respect to the principal axes, the second polar moment of area and the material properties. The program also offers the possibility to characterize some elements by direct input of the stiffness quantities. Thus, it is possible to put in other elements than 'beams'.

Figure 1 shows how the Jaspar milling machine is reduced to a mathematical model of elastic and stiff elements. The length axis of the elastic elements is taken through the centre of gravity of the machine-parts cross-section. For the connection between the

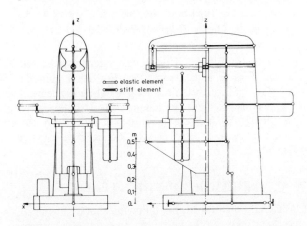

Figure 1. The Jaspar milling machine and its mathematical model.

elastic elements, stiff elements can be used. A stiff element is defined as an element for which the rotations and displacements of one of the two station points of the element depend upon those of the other station point. For the computer program the numbering of the station points is of importance. To decrease the bandwidth of the stiffness matrix, and with it the computing time, the difference between two independent station points of an element must be as small as possible. It has to be noticed that the dependent point of a stiff element is not important

---

* University of Technology, Eindhoven, The Netherlands

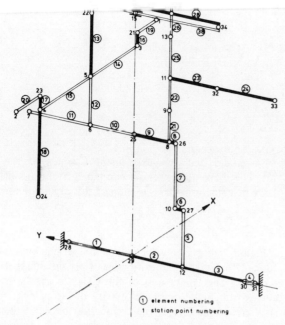

Figure 2. The numbering of the elements and the station points in the mathematical model.

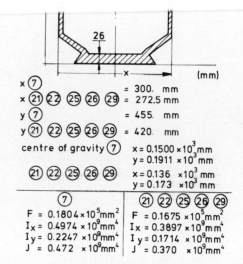

	$x \textcircled{7}$					= 300. mm
	$x \textcircled{21} \textcircled{22} \textcircled{25} \textcircled{26} \textcircled{29}$					= 272.5 mm
	$y \textcircled{7}$					= 455. mm
	$y \textcircled{21} \textcircled{22} \textcircled{25} \textcircled{26} \textcircled{29}$					= 420. mm

centre of gravity $\textcircled{7}$ $x = 0.1500 \times 10^3$ mm
  $y = 0.1911 \times 10^3$ mm

$\textcircled{21}\ \textcircled{22}\ \textcircled{25}\ \textcircled{26}\ \textcircled{29}$ $x = 0.136 \times 10^3$ mm
  $y = 0.173 \times 10^3$ mm

$\textcircled{7}$	$\textcircled{21}\ \textcircled{22}\ \textcircled{25}\ \textcircled{26}\ \textcircled{29}$
$F = 0.1804 \times 10^5 \text{mm}^2$	$F = 0.1675 \times 10^5 \text{mm}^2$
$I_x = 0.4974 \times 10^9 \text{mm}^4$	$I_x = 0.3897 \times 10^9 \text{mm}^4$
$I_y = 0.2247 \times 10^9 \text{mm}^4$	$I_y = 0.1714 \times 10^9 \text{mm}^4$
$J = 0.472 \times 10^9 \text{mm}^4$	$J = 0.370 \times 10^9 \text{mm}^4$

Figure 3(a). Cross-section of elements 7, 21, 22, 25, 26 and 29.

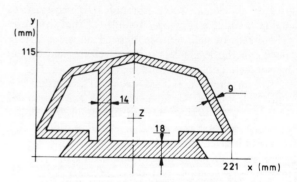

centre of gravity $x = 0.106 \times 10^3$ mm
  $y = 0.430 \times 10^2$ mm

$F = 0.7600 \times 10^4 \text{ mm}^2$
$I_x = 0.1066 \times 10^8 \text{mm}^4$
$I_y = 0.2500 \times 10^8 \text{mm}^4$
$J = 0.2011 \times 10^8 \text{mm}^4$

Figure 3(b). Cross-section of element 33.

of the station points. This global system is indicated in the Figs 1 and 2 as XYZ. For each element separately a local system of axes is defined. The local X-axis coincides with the length axis of the element. The local Y- and Z-axis must coincide with the principal axes of the cross-sectional area of the element.

PROPERTIES OF THE ELEMENTS

As already mentioned, the following characteristics of every element are to be known in order to compose the stiffness matrix of that element:

- the length and the cross-sectional area of the element,
- the second moments of area with respect to the local Y- and Z-axis,
- the effective second polar moment of area with respect to the local X-axis,
- the modulus of elasticity and the shear modulus of the material.

To calculate the second moments of area we can use auxiliary programs. Some examples of cross-sections whose characteristics are calculated with one of these programs are shown in Figs 3(a), (b), (c) and (d). In general, the shape of the cross-sections is obtained from the drawings of the machine tool. The auxiliary program used in this case, calculates the second

moments of area with respect to the principal axes of the cross-section. For this the cross-section is fed into the program by means of coordinates. For closed thin-walled cross-sections the polar moment of area is calculated according to Bredt's relation and for thin-walled open cross-sections De St Venant's relation is used. The output of the auxiliary program also gives the coordinates of the centre of gravity and the magnitude of the cross-sectional area.

Finally, for calculating the dynamic behaviour of the structure, the uniformly distributed mass per element and the lumped masses in the station points are to be known.

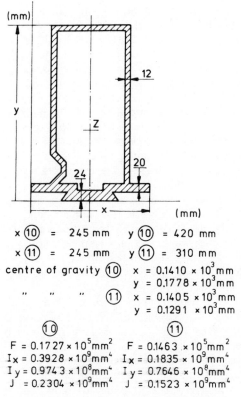

$x \, \textcircled{10} \;\; = \;\; 245 \text{ mm} \qquad y \, \textcircled{10} \;\; = \;\; 420 \text{ mm}$

$x \, \textcircled{11} \;\; = \;\; 245 \text{ mm} \qquad y \, \textcircled{11} \;\; = \;\; 310 \text{ mm}$

centre of gravity $\textcircled{10}$ $x = 0.1410 \times 10^3 \text{mm}$
                         $y = 0.1778 \times 10^3 \text{mm}$

"   "   " $\textcircled{11}$ $x = 0.1405 \times 10^3 \text{mm}$
                         $y = 0.1291 \times 10^3 \text{mm}$

$\textcircled{10}$ $\textcircled{11}$

$F = 0.1727 \times 10^5 \text{mm}^2 \qquad F = 0.1463 \times 10^5 \text{mm}^2$

$I_x = 0.3928 \times 10^9 \text{mm}^4 \quad I_x = 0.1835 \times 10^9 \text{mm}^4$

$I_y = 0.9743 \times 10^8 \text{mm}^4 \quad I_y = 0.7646 \times 10^8 \text{mm}^4$

$J = 0.2304 \times 10^9 \text{mm}^4 \quad J = 0.1523 \times 10^9 \text{mm}^4$

Figure 3(c) Cross-section of elements 10 and 11.

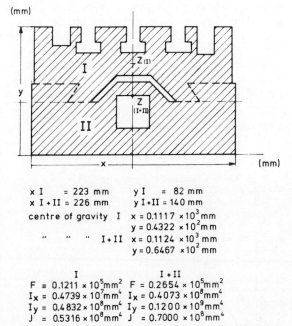

$x \, \text{I} \;\;\; = 223 \text{ mm} \qquad y \, \text{I} \;\;\; = 82 \text{ mm}$

$x \, \text{I+II} = 226 \text{ mm} \qquad y \, \text{I+II} = 140 \text{ mm}$

centre of gravity   I $x = 0.1117 \times 10^3 \text{mm}$
                        $y = 0.4322 \times 10^2 \text{mm}$

"   "   "   I+II $x = 0.1124 \times 10^3 \text{mm}$
                        $y = 0.6467 \times 10^2 \text{mm}$

     I   I + II

$F = 0.1211 \times 10^5 \text{mm}^2 \quad F = 0.2654 \times 10^5 \text{mm}^2$

$I_x = 0.4739 \times 10^7 \text{mm}^4 \quad I_x = 0.4073 \times 10^8 \text{mm}^4$

$I_y = 0.4832 \times 10^8 \text{mm}^4 \quad I_y = 0.1200 \times 10^9 \text{mm}^4$

$J = 0.5316 \times 10^8 \text{mm}^4 \quad J = 0.7000 \times 10^8 \text{mm}^4$

Figure 3(d). Cross-section of elements 14 (I + II), 15 (I + II), 19 (I) and 20 (I).

THE STATIC RESULTS

First of all, the measurements of the milling machine in the laboratory are used as a feedback for the original model by comparing the results of the measurements with those of the calculations. Actually, this feedback was necessary in

two places. The first correction was carried out in the part where the column is connected with the basis of the machine-tool frame. In the first model this connection was considered to be stiff. However, the measurements showed relative large rotations of the column with respect to the basis. This is corrected by adapting the stiffness of the elements 1 and 4 in the basis of the frame, in a way that the calculated rotations are in agreement with the measurements. The second point of correction is element 12. In the beginning this element was also considered to be stiff because of the fact that during the measurements

TABLE 1   Examples of static loading

Loading case	Station point	Loading	Direction
1	36	1000 N	$+X$
2	36	1000 N	$+Y$
3	17	1000 N	$+X$
	22	1000 N	$-X$
	22	50 Nm	$-Y$
4	20	1000 N	$+X$
	7	1000 N	$-X$
	7	600 Nm	$-Y$
	7	55 Nm	$-Z$
5	20	1000 N	$-Y$
	20	60 Nm	$+Z$
	22	1000 N	$+Y$
	22	50 Nm	$-X$
	22	60 Nm	$-Z$
6	17	1000 N	$+Z$
	22	1000 N	$-Z$
7	20	1000 N	$+Z$
	7	1000 N	$-Z$
	7	55 Nm	$+X$

the supports are fixed. However, it appeared that the table exhibited relative large displacements and rotations. The reason for this flexibility is the possibility to rotate the longitudinal carriage with respect to the cross carriage. These carriages are fixed with two bolts and this joint happened to be very flexible in horizontal direction. This effect is only partially taken into account in the latest model. Table 1 shows the static loading cases. The deflections of the station points caused by these loadings are listed in Table 2. In addition to this, Table 2 gives the deviation of the calculated deflection with respect to the measured value if the latter is larger than 5 μm.

A detailed analysis of the results[3] shows that the causes of the deviations originate mainly from:

- the stiffnesses of the connection between column and basis of the structure,
- the stiffnesses of the connection of the carriages,
- the stiffnesses of the spindle-bearing system,
- the stiffnesses of the connection between over-arm and column.

Within the scope of this paper we accept the model and conclude from Table 2 that the average deviation is about 24 per cent.

	34	+X	(3)	(0·7)	–
	35	+X	(4·5)	(6·6)	–
	36	+X	5·5	5·4	1·8
	17	+X	35	31	11·4
	20	+X	19	23	21·1
	39	+X	16	20	25
4	7	–X	20	3·1	84·5
	13	+X	(4)	(3·1)	–
	14	+X	(3·5)	(4·1)	–
	34	–X	(2)	(1·5)	–
	35	+X	8	8·3	3·8
	36	+X	7·5	5·4	28
	17	+X	21	23	9·5
	20	+X	53	48	9·4
	39	+X	41	38	7·3
5	35	–Y	(4·5)	(1·8)	–
	36	–Y	(4·0)	(2·4)	–
	39	–Y	5·0	6·2	24
	39	–Z	20	14	30
6	17	+Z	40	23	42·5
	20	+Z	19	12	36·8
	20	+Y	(1)	(0·3)	–
	39	–Y	(1·5)	(1·6)	–
7	7	–Z	8	1	87·5
	17	+Z	20	12	40
	20	+Z	60	36	40
	20	+Y	15	14	6·7
	39	–Y	(2·5)	(0·0)	–

In order to obtain a correct impression of the transfer function in a certain frequency range, it is necessary to introduce a value for the damping ratio ζ. Especially at an initial stage of the analysis, it is proposed to introduce for every mode the same overall damping ratio. In our case, we chose for this arbitrary value $\zeta = 0.03$. After this, the transfer function can be calculated for a number of frequencies.

If the experimental transfer curves are available, it is possible to obtain from these the natural frequencies and the damping ratios of the several modes. With the aid of the modal flexibilities, the calculated natural frequencies can be matched with the experimental ones. In this way we can attach to a number of modes an experimental damping value, while the remaining modes can be suppressed by means of a large ζ-value. Again the transfer function can be calculated.

The results of the analysis of the Jaspar milling machine are shown in Figs 4(a), (b) and (c). These figures show the result of the calculation of the receptances between point 17 (the tool) and 22 (the workpiece). The experimental values are also plotted in these figures. From the results we might conclude that the natural frequencies can be found with this analysis. Furthermore, the

THE DYNAMIC RESULTS

The natural frequencies with corresponding mode shapes are calculated with the aid of the latest model. To this end, the mass distribution per

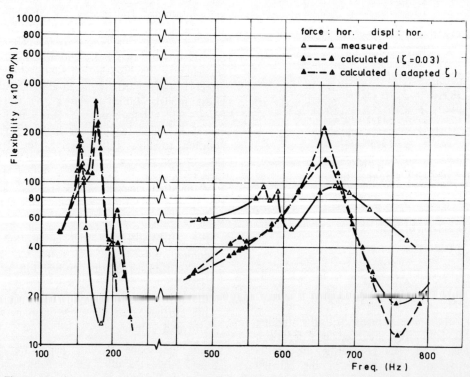

Figure 4(a). Direct receptance between points 17 (the tool) and 22 (the workpiece) for X-direction.

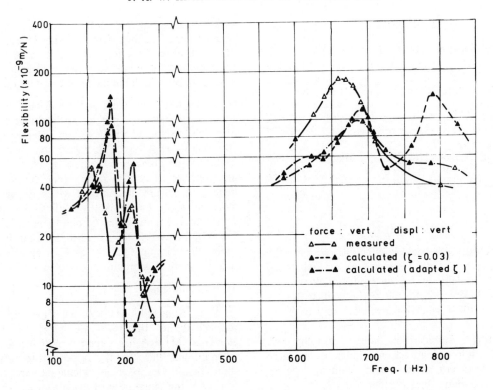

Figure 4(b). Direct receptance between points 17 (the tool) and 22 (the workpiece) for Z-direction.

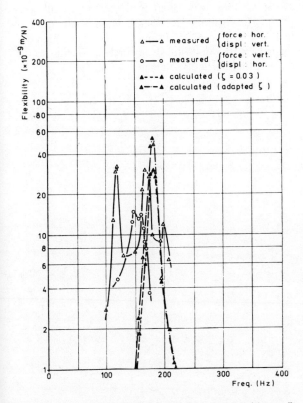

Figure 4(c). Cross receptance between points 17 (the tool) and 22 (the workpiece) for X- and Z-direction.

calculated values give a fair indication of the order of magnitude of all important modal flexibilities. Finally, it has to be remarked that carefully adapting the ζ-value per mode has not so many advantages for this machine tool; the introduction of an arbitrary value for the overall damping in the model seems to be equally satisfactory.

CONCLUSIONS

In spite of the complicated nature of a machine tool structure such as a horizontal milling machine, it is possible to transform that structure into a relatively simple beam model which can be considered to be representative to a certain extent of that structure.

At this moment feedback to the model via measurements cannot be avoided and is essential in order to obtain more insight into the technique of modelling.

In order to give this computer aided design method more reliability in the design stage of machine tools, it is of vital importance to have more fundamental information concerning the stiffness and the damping of machine elements such as bolted joints, spindle-bearing systems etc. This information should be made available in such a numerical form, that those specific machine elements can be introduced into the model as elements with well defined stiffness and damping matrices.

ACKNOWLEDGMENT

The authors wish to express their gratitude to Mr W. J. T. Bouwman and to Mr P. R. M. van Dijk for their help in carrying out the calculations and the experiments.

DISCUSSION

Q. G. J. McNulty, Sheffield Polytechnic. Why did the authors use a uni-model viscous damping, when the structure (especially at higher modes) will be predominantly hysteretic? Certainly the analysis is simplified using the former by accuracy may be sacrificed.

The authors stated that modal clamping was obtained by suppressing adjacent modes using a large damping coefficient. This would seem to neglect modal cross coupling: please comment.

A. Because the damping ratios in machine tools are relatively low ($S = 0.03$ satisfies often), it does not matter whether you use viscous or hysteretic damping in order to calculate the frequency response. Furthermore, using Cowley's method for calculating the dynamic response (see ref. 2), only viscous damping can be applied.

not possible to deduce from experimental measurements the nature of damping which is present in a machine tool. Consequently either assumption is satisfactory for purposes of computer aided design. This comment is valid when, as is usual in machine tools, the damping in any work is appreciably less than the critical damping.

A. No comment.

Q. F. M. Stansfield, MTIRA. What method was used for calculating the shear flexibility of the beam elements? MTIRA has experience of using a similar program and finds that it is important to estimate the shear flexibilities of beam elements carefully—they can be of the same order of magnitude as the bending flexibilities.

A. No shear flexibility, other than caused by torsional loading, was taken into account in this program. The authors feel that it is only necessary to introduce shear when using a fairly short element, loaded by two opposite shear forces, working at both ends of that particular element.

INITIAL APPLICATIONS OF DYNAMIC STRUCTURAL ANALYSIS TO COMPUTER-AIDED DESIGN OF MACHINE TOOLS

by

TETSUTARO HOSHI and MASATAKA YOSHIMURA*

SUMMARY

Procedures for computer-aided design of machine tool structures are described. They represent applications presently under way in co-operation with several machine tool industries. The design process discussed consists of a computer analysis based on the receptance method and experimental identification of the dynamics of the prototype, followed by the design modification in pursuit of the dynamically optimum design from the viewpoint of the mass and stiffness distribution. An example is presented to illustrate the use and effectiveness of the procedure.

INTRODUCTION

Three categories of dynamic structural analysis techniques are known at present. They are the lumped-mass beam method, the distributed-mass beam method, and the finite element method, each of these is capable of estimating the dynamic performance of an elastic structure when its form is known. When the design is to be developed for a given purpose, such analysis techniques do not give an optimum solution *a priori*, but repeated use is necessary resulting in an empirical design process as illustrated in Fig. 1.

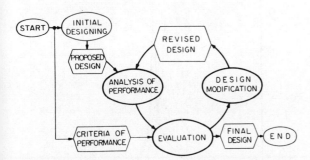

Figure 1. Cut and try approach of designing with repeated use of analysis technique.

The basic requirement of the analysis technique is to analyse the performance of the proposed design. Additionally the following two functions are also essential requirements in order to satisfy the machine designer:

Evaluation aid function
The computed result should indicate the error function deviation from the optimum design.

Design modification aid function
This function should point out the part of the structure to be modified and also indicate the change to be made.

The receptance method which the authors have been developing for seven years[1] comes under the category of a distributed-mass beam method, and it is an intermediate solution between the other two categories with respect to the complexity of the mathematical model, computing requirements such as time and cost. The research has been pursued in the three following sections.

(1) Development of computing technique
A system of computer programs have been developed based on the receptance principle.

(2) Initial applications
The technique is applied for the design improvement of the prototype machine tools or machines under current use, and its effectiveness has been and continues to be assessed for different applications.

(3) Extensive applications
The technique is applied for the development of new models.

Several cases of the initial applications are currently under way. The process of computer-aided design by use of the receptance method is introduced and discussed with reference potentials and limitations.

OUTLINE OF COMPUTER-AIDED DESIGN WITH THE RECEPTANCE METHOD

Principle of performance analysis
The undamped frequency response of the total

* Kyoto University, Japan

grinding wheel is dressed by a diamond tool mounted at the tip of the arm A. The arm, hinged by a pair of rolling bearings, assumes a horizontal position for dressing, supported by the arm rest which is a cantilever beam extended fom the axle bracket. The structure is approximated by the model as illustrated in Fig. 3. The complete system is divided into sixteen sub-systems, which are joined by a series of nine computations (syntheses) programmed consecutively.

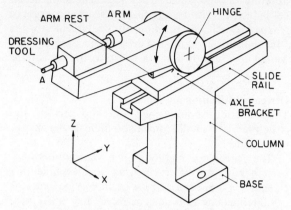

Figure 2. Illustration of the dressing unit.

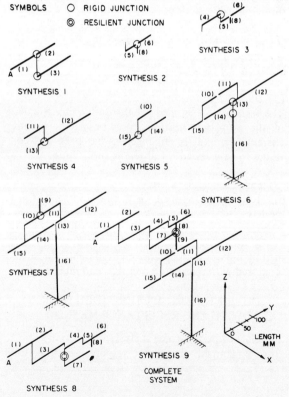

Figure 3. Computation model of the dressing unit.

is indicated. Resonant frequencies up to the Nth order are identified in this manner.

The modal shapes of the structure are then computed at each of the resonant frequencies thus chosen. The synthesis computations used at the mode shape computing stage is so designed that the receptances at all ends of every sub-system are obtained. In addition the dynamic forces exerted on the sub-system through these ends are stored. After all the synthesis computations are finished, the dynamic force data is used to define the boundary conditions to compute displacements at an arbitrarily given number of equidistant points for every sub-system. The modal shape is thus analysed in detail, and also at this stage, the maximum values of the kinetic and potential energies retained by each sub-system are computed. The modal flexibility and energy distribution are computed based upon the energy data as discussed in the following sections.

System evaluation by means of modal flexibility

For the prototype design of the dressing unit shown in previous figures, two of the computed frequency responses are depicted in Fig. 4, in comparison with the experimental data measured by the exciting test of the actual model. It is seen that each of the direct

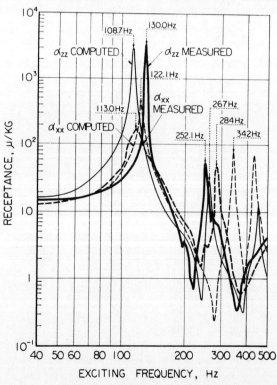

Figure 4. Frequency responses at the dressing tool point A measured and computed on the prototype model.

receptances in the X and Z directions show a remarkable resonance at its lowest order mode. The mode shapes at these two modes by both analysis and experiment are shown in Fig. 5. From the dominantly higher response peaks at the lowest two modes it was concluded that the structure can possibly show improvement with respect to these modes. This point, however, is more clearly apparent when the modal flexibilities are compared. The modal flexibilities are the equivalent static flexibilities attributable to individual modes.

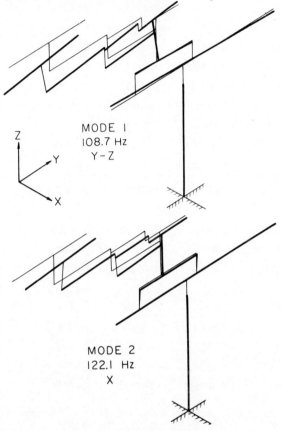

MODE 1
108.7 Hz
Y-Z

MODE 2
122.1 Hz
X

Figure 5. Computed mode shapes at lowest two resonance frequencies.

The theory of modal flexibility[2] discusses a lightly-damped structure vibrating at a resonance by a negligibly small exciting force. It is shown theoretically, that the receptance $(\alpha_{i,j})_m$ between two co-ordinates i and j, and at a mth order resonance mode is expressed by the following equation.

$$(\alpha_{i,j})_m \equiv \left(\frac{X_i}{F_j}\right)_m = \sin\beta \, \frac{\pi(f_{i,j})_m}{\delta_{Am}} \qquad (1)$$

In the equation, β designates the phase lag of the response X_i to the dynamic force F_j, and δ_{Am}, the logarithmic decrement of the total system at the mth order mode. The modal flexibility $(f_{i,j})_m$ is defined by the formula*:

* According to A. Cowley[3], modal flexibility is defined as $(f_{i,j})_m$ in the following formula which expresses the dynamic flexibility $\alpha_{i,j}$ for a frequency ω:

$$\alpha_{i,j} = \frac{X_i}{F_j} = \sum_{m=1}^{\infty} \frac{\omega_m^2}{\omega_m^2 - \omega^2}(f_{i,j})_m$$

Since this equation reduces to the same as equation (3) when $\omega = 0$ is substituted, it is understood that the two definitions of the modal flexibility are essentially the same.

$$(f_{i,j})_m = \frac{X_i X_j}{2E_m} \qquad (2)$$

meaning that it is the product of the displacement amplitudes in the two coordinate directions divided by twice the total maximum vibration energy E_m of the complete system. E_m is the total sum over all sub-systems of the maximum values of the kinetic energies (or of the potential energies), both sums being equal when the system is at a resonance. Since E_m is proportional to the square of the displacement amplitude (X) irrespective of the damping, the modal flexibility is independent of δ_{Am} which indicates the damping ability of the system. Modal flexibility is a parameter which reflects allocation of masses and stiffnesses throughout the system. When modal flexibilities are computed for all modes of vibration and added together, the sum equals the static flexibility $(f_{i,j})_{st}$, that is:

$$(f_{i,j})_{st} = \sum_{m=1}^{\infty} (f_{i,j})_m \qquad (3)$$

According to the above theory, a dynamically optimum design can be defined as such that has a minimum value of static flexibility, which is shared by the modal flexibilities of many modes in an even manner or in a selective manner suitable for the anticipated purpose of its use, and sufficient damping is in effect at each mode. When the computed modal flexibility shows a dominant concentration at a particular mode, it is known that the system is remote from the optimum design and is possible to be improved with respect to that mode by some modified allocation of the mass and stiffness.

For example: e.g. the dressing unit, the direct modal flexibilities at the dressing tool point A are computed up to the fourth order mode and the results are shown in the upper three rows of Table 1. It is immediately seen that at the first order mode, the modal flexibilities in the Y and Z directions are large and they share more than 98 per cent of the corresponding static flexibilities. In the second order mode, the modal flexibility in the X direction shares 94 per cent of the static flexibility. Thus, it is understood that the structure can be greatly improved by improving the rigidity with respect to these first two modes.

Design modification suggested by energy distribution
In order to realize the improvement expected from the computed modal flexibility, the parts of the structure must be specified where design modifications are most effective. This fact is established, by analyzing the distribution of the vibration energy in the system. Both the kinetic and potential energies consist of those associated with the bending, longitudinal, and torsional deformations of the subsystems, and these values are computed with the modal shapes. In the middle and lower parts of Table 1, the results for the dressing unit are presented. For the first two resonance modes derived, those parts of the system where the concentrated energy distributions exist are of interest. Design of these parts and their vicinities are to be modified in such a way that the energy concentrations are diminished. Con-

Ratios of kinetic energy taken by							
Dressing tool and micro adjuster				0·353	0·326	0·0302	0·0421
Arm	sub-system 3			0·472	0·414	0·0465	0·0643
	sub-systems 4, 5, 6, 8			0·130	0·188	0·0630	0·289
	axle bracket	sub-system 9		0·000387	0·00769	0·0141	0·0238
		sub-systems 10, 11		0·00118	0·00739	0·0413	0·0271
arm rest				0·0100	0·0137	0·00822	0·00777
slide rail				0·00986	0·0644	0·819	0·541
column and base				0·000139	0·000341	0·00208	0·00754
Total				1·0	1·0	1·0	1·0

Ratios of potential energy restored in								
dressing tool and micro adjuster		bending	0·000197	0·00109	0·000235	0·000361
		torsion	–	–	–	–
arm	sub-system 3	bending	0·00733	0·00292	0·00292	0·00293
		torsion	–	0·00115	–	0·000236
	sub-systems 4, 5, 6, 8	bending	0·00241	0·0110	0·00203	0·0207
		torsion	–	0·06°/	–	0·0331
axle bracket	sub-system 9	bending	0·0456	0·0796	0·00283	0·0111
		torsion	–	0·169	–	0·0796
	sub-systems 10, 11	bending	0·103	0·0392	0·0471	0·00346
		torsion	–	0·173	–	0·0395
arm rest		bending	0·600	0·000190	0·0937	0·000216
		torsion	–	–	–	–
slide rest		bending	0·178	0·0679	0·721	0·296
		torsion	–	0·212	–	0·0859
column and base		bending	0·0572	0·111	0·127	0·317
		torsion	–	0·0629	–	0·121
Total			1·0	1·0	1·0	1·0

sidering the results listed for the first mode, it is seen that 60 per cent of the potential energy is stored in the arm rest. This means that the strain is dominantly concentrated at the arm rest; therefore, an increase in its stiffness is an effective modification. Also it is seen that the dressing tool-micro adjuster set and the end part of the arm (sub-system 3) retain 35·3 and 47·2 per cent of the kinetic energy respectively; therefore, it is also effective to reduce the cross-section areas and consequently the weights of these parts. Reductions in the stiffness of these parts also occur but they have no significant effect because they have practically no share in the potential energy. Considering the results for the second mode, the kinetic energies retained by the dressing tool micro-adjuster set and the end part (sub-system 3) of the arm are high at 32·6 and 41·4 per cent respectively. Weight reduction of these parts are again effective to reduce this. The potential energy at the second mode does not show particular concentrations, rather it is distributed mainly in the following modes 21·2 per cent due to the torsion of the slide rail, 17·3 per cent due to the torsion of the axle bracket (sub-systems 10 and 11) and 16·9 per cent due to the torsion of the other part of the axle bracket (sub-system 9).

As a consequence of the analysis of Table 1, two modifications are suggested on the dressing unit. The first is to increase the stiffness of the arm rest: effective for improvement with respect to the first mode. The second is to reduce the weight of the arm: effective in the first and second modes. The first modification has been carried out; namely, the design of the axle bracket has been changed such that the flexural rigidity of its connecting arm rest is about ten times as much as in the original design.

Excitation tests have been carried out on this new model: the results are shown in Fig. 6. Compared to

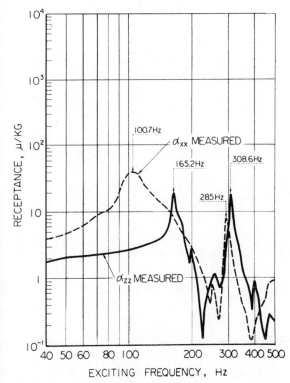

Figure 6. Frequency responses at the dressing tool point A measured on the modified model.

the response curves for the original model shown in Fig. 4, drastic improvements are apparent. Considering the receptance in the Z direction, static rigidity is increased by about eight times, the resonance frequency of the first mode is increased from 130 to 165 Hz, and the dynamic rigidity at the first resonance is increased by about sixteen times. For the receptance in the X direction, increase factors of three times in the static rigidity and four times in the dynamic rigidity at the second mode resonance are obtained. Further improvements are anticipated if the other modification is made to reduce the weight of the arm.

When the modal flexibility is reduced by modified structural designs as illustrated in the foregoing example, it is also important to have a large logarithmic decrement δ_{Am} according to equation (1). Theoretically[2], this parameter is known to depend on the distribution of the potential energy according to the following formula:

$$\delta_{Am} = \pi \sum_{\substack{\text{summation} \\ \text{over all} \\ \text{sub-systems}}} \left(\begin{array}{l} \text{loss factor which rep-} \\ \text{resents the damping} \\ \text{ability of the sub-} \\ \text{system material} \end{array} \right) \frac{\left(\begin{array}{l} \text{maximum} \\ \text{potential} \\ \text{energy of} \\ \text{the sub-} \\ \text{system} \end{array} \right)}{E_m} \quad (4)$$

Therefore, optimum result in damping is attained by studying the energy distribution, and by trying to allocate greater latent damping at the parts of the structure where greater potential energy is stored.

PROCESS OF DYNAMIC STRUCTURAL ANALYSIS BY SYNTHESIS OF RECEPTANCE PROGRAM SYSTEM

Structure of the synthesis of receptance program system

The program system consists of a manual, a group of sub-programs and an auxiliary program for computation of cross-section values. The sub-programs are kept in the disc-pack file of the FACOM 230-60 dual processing system at the Data Processing Center, Kyoto University. They are accessed by executing a problem-oriented main-program through the batch-job counter at the Center, or through a time-sharing-system terminal. Table 2 illustrates the names of sub-programs to be used for the computation of the frequency response and the mode shape. Areas of the core memory as listed are secured at the CPU, in addition to the core area for the main-program.

Data preparation

The following data are prepared: a set of design drawings; weight data of main sub-assemblies of the machine; modulus of elasticity, shear modulus, and mass density of the materials; radial and axial stiffness of the main spindle bearings; type of guide ways; when rolling or hydrostatic ways are used, stiffness of the feed-drive system, and normal and lateral stiffness of the guiding system; and the design of foundation and mounting stiffness data.

Modelling

The structure is approximated by a model as illustrated in Fig. 2 and the sequence of the syntheses are designed.

Preparation of sub-system data

Sub-system data are determined from the drawings by use of the cross-section value program. The data consist of the axial moments of inertia around the two principal axes of the cross-section, the torsional stiffness of the cross-section, the cross-section area, the length of the sub-system, and the length of the imaginary rigid mass-less beam assumed at an end of the sub-system, in order to cover the offset between the neutral axes of adjacent sub-systems.

Main-programming

Main programs for computations of the frequency response and the mode shape are written according to the sequence of the syntheses. The two programs are very similar having but minor differences.

Preparation of the input data

The input data are arranged in accordance to the main-programs. The data are only partially different for computations of the frequency response and the mode shape.

Error checking

Before the computation, checks should be made to find and eliminate errors in the main-programs and the input data. The programs are run at a few different frequencies and the computed results are checked for conformity to the following rules. The

response	SYNRF	8940		
Synthesize receptance of two or more sub-systems for mode shape			SYNRM	5756
Compute energy distribution called by SYNRM			ENER1	12778
Inversion of a matrix	INV	816	INV	816
Assign free condition (zero force) to specified ends after synthesis	ZER	510		
Scan for resonance frequencies	SCAN	1524		
Table output of computed receptance	OUT1	674		
Calculate sub-system receptance for 3-dimensional uniform beam (When two ends are assumed on a sub-system incl. fixed spring)	SUR45F	3062	SUR45M	4406
(When three ends are assumed on a sub-system)	SUR55F	6267	SUR55M	10526
Size of Common areas (High speed core memory)		2738		5902
(Large size core memory)		28000		51248
Total when all are used		66380		106950

computed results of the mode shape should not contain a discontinuity at the joints between sub-systems; reciprocals, the receptances should hold $\alpha_{i,j} = \alpha_{j,i}$ when $i \neq j$; orthogonality, some elements of the receptance matrix should be zero by principle; and equality, some outputs of the computations of the frequency response and modal shape should be equal when computed for a common frequency.

Computation of frequency response
Several computing jobs are processed for the frequency response. Starting with coarsely selected frequencies, the vicinity of the resonance frequencies is scanned more closely. The output data is tabulated by the line printer at the Data Processing Center. Response curves at the tool point A and the work point B are plotted manually, and the static flexibility and the resonance frequencies are identified. A method providing graphical output by use of a digital plotter is under preparation.

Computation of mode shape
Mode shapes are computed at the resonance frequencies thus identified, and the results are tabulated by the line printer, and then plotted manually. The use of a digital plotter is under preparation.

Evaluation
Modal flexibilities and energy distributions are computed manually and tabulated as illustrated in Table 1.

APPLICATIONS

General approach of applications
The dynamic structural analysis is under trial uses at the present moment and initial applications are being attempted using the machine tools under current use or prototype machines with the intention of improving their performance through design modifications. In this type of application, the design is analysed by computation, and simultaneously by dynamic tests of the real machine. The computed and measured results are compared altogether and related to the cutting performance experienced in the field test, evaluations are made and design modifications proposed. In the case of the aforementioned dressing unit, the modification was immediately made on the real machine because it was a small structure. However, in many cases it is more economical to make intermediate tests, and study the effect of the modification experimentally on scale models. When the effect of the modification is proved both by computation and tests on the scale models, it will be incorporated in the actual machine. Initial applications are being made following the above concept. For example some of the projects are: the development of an extension quill for an internal grinder wheel head; designing of production models for a machining centre having a portal column and several knee-type milling machines; and prototype design of an NC drill press.

Computing time and cost
Total CPU time needed to analyse a machine struc-

ture ranges from 1 800 to 11 000 s depending on the complexity of the structure and the completeness of the analysis. The computer cost is estimated at 560 to 3400 dollars. By using faster and larger computers, however, the figures quoted are anticipated to reduce substantially.

CONCLUSIONS

One of the techniques for the dynamic structural analysis, e.g. the receptance method, has been discussed with respect to its application to the computer-aided design of machine tools. Although the analysis is confined to the undamped system at the present time, the method is efficient for pursuit of the optimum design with respect to the allocation of mass and stiffness in the structure. Evaluation and the design modification of the structure is greatly benefited by the computed results of the modal flexibilities and the energy distribution. Analysis and design modification of the dressing unit have illustrated the use and effectiveness of the computing technique.

Initial applications are being pursued on several machines with the intention to prove the validity of the analysis technique. When it is proven, use of the technique at the principal design stage of the machine structure will bring about a profound impact for the design optimization.

One of the present limitations is the accuracy of estimating the torsional rigidity of a sub-system beam, when it has end plates, partitions, and/or cutouts. Use of a static structural analysis by the finite element method is now being considered as a solution to this difficulty.

Computing time and cost are often beyond reasonable costing at the present moment. Use of a faster and larger computer, however, will resolve this problem.

REFERENCES

1. M. YOSHIMURA (1970). Analysis of structural dynamics by digital computer-study for analysis of structural dynamics of machine tools (1st Report), *J. of the Japan Society of Precision Engineering*, **36**, No. 3 (March), 212–18 (in Japanese).
2. M. YOSHIMURA and T. HOSHI (1972). Computer approach to dynamically optimum design of machine tool structures, *Proc. 12th Int. M.T.D.R. Conf.*, Macmillan, 439–46.
3. A. COWLEY (1972). Co-operative work in computer aided design in the C.I.R.P., report submitted to T.C.Ma. of C.I.R.P. (January).

DISCUSSION

Q. G. J. McNulty. In Fig. 4. The two frequencies 108·7 and 130·0 Hz are for the same modal configuration. As the theoretical peak is finite, a δ_{AM} value must have been used in equation (1). These two peaks show a widely different damping value— assessing this on their half power points.

A. Computations for present study have been made assuming no damping, namely $\delta_{AM} = 0$. Computed

response curves in Fig. 4 should always soar up to infinity at every peak.

Q. A. Cowley. Figure 1 shows a systematic procedure for the application of C.A.D. In this diagram the authors indicate that an important feature of this procedure is the establishment of 'criteria for performances' in items of computed parameters. I should like to ask the authors what formal efforts have made in their laboratory or in other organizations in Japan to establish such criteria.

A. In general principles, the criteria should consist of the admittable static flexibility, the way in which the static flexibility is shared by modal flexibilities, avoidance of resonant frequencies from the vicinities of possible disturbing frequencies, and possibility of effective latent damping existing at members sharing potential energy. The specific values of these criteria, however, should be left entirely to the designer's judgement as he would make a decision considering the purpose and environment of the particular structure he is designing.

Q. Professor C. Andrew: It is interesting to note that each of the foregoing three papers describes a different analytical technique, each one optimum in terms of accuracy/cost-efficiency for particular types of machine tool element. Thus: finite elements for intricate shapes, joints and systems subject to torsion; connected beam analysis for slender, quasi-uniform elements; and lumped-parameters for the larger components away from the critical parts of the strain loop. Instead of trying to use solely one method or another, would it not be better to use all three, each for its own optimum application, and to combine the separate results by a receptance matrix? This can show cost-saving for a one-off analysis; the savings can be considerably greater in a synthesis for which the characteristics of one component are to be varied while the rest of the system is kept fixed. Such a technique has been described in (1), with particular reference to systems with heavy damping. It is also used in (2) for including uniform (damped) beam elements in a structure for which the other dynamic characteristics have been measured experimentally.

1. W. J. HAMMILL and C. ANDREW (1971). Receptances of lumped-parameter systems containing discrete damping sources, *J. Mech. Eng. Sci.*, **13**(4), 296.
2. W. J. HAMMILL and C. ANDREW (1973). Vibration reduction with continuous damping inserts, 14th Int. M.T.D.R. Conf.

A. Yes, it would be better to use all three, each for its own optimum application. As for the analysis of the perspex small model, the FEM analysis was applied for each element structure. If the integrated structure is analysed by FEM, it would not be cost saving, but time and memory of computer consuming. In this study the characteristics of the integrated system are obtained as a structure composed by equivalent beams the characteristics of which is based on the FEM analysis for each element structure. The results shows that this simplified and saving approach is justifiable.

grinding method. Those three different methods should be used for modelling various parts of a machine. The computation on each component and assembling of the results for simulation of the total machine composite is possible within reasonable cost and CPU time by the use of a recent large-scale computing system.

Dr S. Taylor, University of Birmingham: I agree with Dr Cowley that flexible modes, rather than low frequency modes, can be computed best but that stiff modes are difficult or at least very expensive to compute.

With reference to Professor Andrew's comment, I agree that a confirmation of techniques is the most effective computing system. The relatively expensive finite element method or experimental or other data can be used for difficult and complex elements to establish the static conditions. The dynamic analysis can then be made for the complete structure using a cheaper method such as the lumped mass flexibility technique.

Dr. Ing. R. Noppen, Aachen/Germany. In the final discussion on the foregoing three papers the following statement was made: 'the beam method should preferably be used for the analysis of machine tool structures rather than the finite element technique, because both methods yield wrong results at higher

either method is based is whether the actual problem *can* be matched by either procedure or whether it is beyond their respective limitations[1].

It is not wise to base the choice of one or other method on computing costs because the expense of wrong conclusions arrived at due to inadequate analyses can be much higher than those for additional calculation time.

(1)   H. OPITZ and R. NOPPEN. The Evaluation of Computer Aided Methods for the Design of Machine Tool Structures. To be published in C.I.R.P. Annals, 1973.

Dr Lowel Wilcox, Gleason Works, USA: Two developments should be considered that may make finite elements more appropriate in the dynamic analysis of machine tool structures in the near future. The first of these, is, that the problem of considerable data storage, necessary to accurately model a machine tool structure, can be reduced dramatically through the use of the 'substructuring' approach now available. The second point is, that as experience is accumulated using the finite element method, there will be more efficient ways of using a given number of elements to obtain satisfactory answers, consequently, tending to reduce the number of elements required.

ELECTRICAL FORMING AND MACHINING

A NEW LOOK AT ELECTRO-CHEMICAL FORMING IN PRACTICE

by

H. J. BILLING* and P. LAWRENCE*

SUMMARY

Is electro-chemical forming (machining) a viable process for general engineering?
Do the specialists give industry what it needs?
Healy experience indicates that little has been done to cater for the largest potential users. (Some examples will be given which illustrate a new approach towards future developments in this area.)

INTRODUCTION

Our company has been associated with electro-chemical machining or, as we prefer to call it, electro-chemical forming, since 1966. We were attracted to the process by the ability to machine hard and tough metals as easily as soft metals—comparatively simple production of shapes and forms which would be difficult by conventional machining, a burr free job, and minimal distortion—qualities which many other production engineers must have found interesting.

With such attributes could we, as tool makers, plagued by the curse of distortion during hardening and the difficulty of correcting moulds and tools in the hardened state, ignore this panacea which was being offered? Why then had other manufacturers who must have had problems similar to ours not accepted the hand of the fair damsel who was born in 1927 and must surely by now be reaching maturity? There must have been many interested suitors but we were not able to find a proportionate number of marriages. Our early investigations told us that the process had been invented by a Russian gentleman named Gusseff. Were politics then the reason for non-acceptance? We could not believe this and being sure that there must be something more fundamental, we commenced our investigations and research to find the answer.

In presenting this paper we aim to illustrate the reasoning which led to the conclusion that manufacturers of equipment have, in general, concentrated their efforts in a direction which has led to over-elaboration. We believe that if the process is to receive wide approval greater attention to simplification is necessary.

We have concluded that ECF offers the greatest advantages when applied to mass production and where the amount of metal to be removed is small. At the same time, we are very much aware that there are applications where bulk metal removal by ECF is the most economical—and sometimes the only practical solution—but we submit that these cases are in the minority and by themselves do not represent a large percentage of the total potential.

PART I: BASIS OF ARGUMENT

Consideration of the way in which ECF developed helps to explain some of the reasons for the present state of the art.

Two of the properties of the process which had immediate appeal were the ability to machine metals irrespective of their hardness and to produce configurations which were difficult to make by orthodox machining methods. The introduction of the process coincided with a requirement in the aerospace industry where production engineers were called upon to machine the new exotic metals to shapes which did not lend themselves to conventional production methods.

Effort was, therefore, directed by the machine manufacturers towards an industry where the primary consideration was to achieve an objective set down by

* Healy of Leicester Limited

T

necessary to consider the process in some detail.

On the positive side we have, in addition to the properties referred to earlier, two other major advantages in comparison with normal machining.

(1) Absence of tool wear.
(2) Virtually no temperature effect on the job.

Against this, on the negative side, we have:

(1) Low metal removal per unit of power consumed.
(2) Need to supply a consistent film of electrolyte over the complete working surface of the tool and the job.
(3) The production (using neutral salt electrolyte) of very hygroscopic minute particles of metal hydroxides which contaminate the electrolyte.

In general the cost of removing metal in chip machining does not increase in proportion to the amount of metal removed.

With ECF the cost is approximately in direct proportion to the metal removed. The 'chip' size is always the same irrespective of the accuracy required or volume to be removed.

ECF is most likely to be competitive when the volume to be removed is small rather than large.

Electrolyte has to pass between the interface of the tool and the job and has three essential functions:

(a) To allow electrolytic dissolution of metal;
(b) to remove products of reaction, i.e. hydroxide of the metal and hydrogen;
(c) to remove the heat generated by the electrical current.

In practice the electrolyte has to pass at speeds up to 100 ft/s to satisfy these requirements. Any failure to obtain a consistent homogeneous film will result in variations in the electrolytic process.

It will be readily obvious that the further electrolyte has to travel from entry to exit of the job, the greater is the chance of the film breaking up. To obtain the flow rate, the pressure of the electrolyte has to increase proportionately with the increase in distance from entry to exit. The longer the path the greater is the attention required in design and manufacture to avoid areas of electrolyte starvation.

The cost of tool development and plant will, therefore, be least where the distance the electrolyte has to travel is minimal, i.e. where the jobs are small rather than large.

The amount of hydroxide produced which has to be removed from the electrolyte, is therefore directly proportional to the volume of metal machined; on

With large volumes of metal removal, however, this is seldom the case.

From the foregoing it can be seen that ECF is most suited to applications where small parts and large production are encountered. This is contrary to the areas where most effort has been concentrated up to the present time. We have very many examples of general engineering applications where the best features of the process are demonstrated without the major disadvantage arising.

At the commencement of this paper I referred to the interest which prompted our company to investigate the potential of ECF. We were looking for a means of eliminating distortion of moulds and tool parts during hardening. Regretfully, I must say that the more we delve into ECF the less likely we considered it as a suitable process for this application.

In most cases the number of cavities or parts which are required is insufficient to justify the tool development costs. There are of course cases where the process is viable, one example being with forging dies which have to be cut regularly and others where the manufacture by normal methods is extremely difficult, but these are the exception rather than the rule. An area of application which meets all the requirements of ECF is deburring. There has been a limited usage in this field, but the scope and potential is very large. Excellent results and economies can be obtained, but most applications in use still employ the unnecessarily sophisticated equipment developed to meet the requirements of the aerospace industry and as a result, in our opinion, are too complicated and costly. We believe that if more effort had been expended on tool design this would have been a far more fruitful area. Just two examples will illustrate this point. Most design parameters are still determined on a rule of thumb basis and there is no satisfactory insulating material which can be applied as a thin coat. Comparatively simple equipment with good tool design will open up many markets which are at present just potential. It may not be out of place at this point to recall a comment by Mr Wedgwood Benn, a former Minister for Trade and Industry, who, when asked to give his opinion as to whether Mr Nixon's aide Dr Kissinger had been successful in a deal with the Russians because of his 'Whizz Kidd' ability, replied that whilst he did not in any way deny the justification for the 'Whizz Kidd' reputation he felt that the main reason for success resulted from a 'Real Common Interest' as without this no deal could succeed.

It has been our aim to probe the area of 'Real Common Interest' between ECF and industry because we believe that just trying to show how clever we are at making novel products will be of little real interest to an industry that wants 'Nuts, Washers, and Screws'.

The following part of this paper gives some examples of the practical application of our philosophy and indicates the development we anticipate by following this course in the future,

PART II: APPLICATION

Feed system

The first major obstacle which confronted us as potential users of ECF was the high capital outlay for equipment available at the time. If we wished to explore the potential of the process for our particular requirement the only alternative was to construct a simplified version ourselves. In doing this the first requirement was to evolve a suitable tool feed system. The most common form of precision machine tool movement is the traditional leadscrew driven dovetail slide. Perhaps inevitably this system was carried over to ECF machines, if only because the first machines were converted conventional machine tools. As ECF requires a tool feed system which has no backlash, low feed rate, and capability of operating in a harsh environment, the traditional solution was less than ideal. Improvements, such as recirculating ball screws and roller slides have been used successfully, but on the face of it the best system would appear to be a hydraulic ram. The advantages are that it is very stiff, and when filled with oil under pressure, is totally resistant to ingress of electrolyte and can easily be made backlash free. The chief difficulty experienced with such rams when used on ECF machines has been their failure to provide the constant tool feed rate required, either because of stick-slip, or metering problems in continuous flow systems. In a simple system the speed of the ram depends on the pressure and viscosity of the oil and the metering orifice. Both changes in oil temperature and load on the ram can alter its speed. The amount of oil being metered may only be of the same order as the internal leakage in other valves in the circuit, which also makes control more difficult. These difficulties are overcome if an 'integrated' or step feed system is employed since it is relatively simple to inject a constant volume of oil per unit time if a positive displacement pump is connected directly to the ram cylinder. The stability of the feed rate depends on maintaining the frequency of the pump strokes constant, which is easily done electronically. Naturally, the steps by which the tool advances must be sufficiently small so as not to interfere with the operation of the process.

Experience has shown that steps equal to 30 per cent of the end gap are in no way detrimental to the cutting action or to the resultant finish on the workpiece. Control of the feed rate can be achieved either by altering the frequency of the pump strokes or the volume swept by the pump.

Machine construction

Most machine tool manufacturers started from the premise that ECF machines need to be large rigid structures with good damping characteristics. This is the experience of many years of evolution of conventional machine tools, but is it relevant to ECF equipment? Certainly there can be considerable separating forces between tool and work—normally in the region of 200 lb/in^2—but there is none of the 'hammering' effect very often prevalent in conventional machining.

ECF tends to be carried out inside a box because of the necessity to confine electrolyte spray. A box can be made as a very rigid structure and may be used as the basic machine tool element which sustains all the machining forces in the process. The rest of the machine tool then only provides a means of supporting the box at the correct height and housing all the other components of the plant. Bearing this in mind, an excellent material for the construction of the casework and general support structure is laminated glass fibre since it is inexpensive, completely corrosion free, tough, and may be moulded easily thus eliminating many joints. Metal or even wooden stiffeners can easily be moulded in where necessary, giving a rigid yet much lighter structure than alternative methods of construction.

The incorporation of all the ancillary equipment into one case, results in a much more economical plant than the alternative collection of individual units.

Power unit and protection devices

Short circuit damage means that the machine must be stopped, and the tool and workpiece examined, the tool dressed in position if the damage is minimal and possibly a cusp removed from the workpiece. In more severe cases it will be necessary to remove the tool and have it remachined and insulated, etc., and the workpiece may be scrapped out.

In an attempt to prevent short circuit damage, spark detection devices have been fitted to large plants, the object being to sense sparks before short circuit actually occurs and shut the equipment down before any serious damage is caused to the tool. Whilst the notion is a good one, most users report that although damage may be reduced by an indeterminate amount, certainly all damage is not prevented. Additionally the device must by its nature be extremely sensitive and cannot differentiate between sparks and other transients. Thus it is prone to spurious shutting down of the plant which can be frustrating. The whole problem of tool damage, as we have previously argued, is amplified by working with larger components. Both the likelihood and cost of down time and rectification are increased. The spark detection device, as we see it, is yet another expensive complexity in the ECF process which has come about as a result of treating symptoms instead of causes.

Handling electrolyte

ECF seems to demand that considerable quantities of electrolyte are available, but is this necessary? Large volumes of electrolyte are expensive to prepare, both in material value and time. Valuable space is taken up and time and money can again be wasted in raising the working temperature of the electrolyte. The effect of these factors is obviously nullified if the volume of electrolyte in use can be kept to a minimum. The salt in the solution is not normally

than chemical. The sludge tends to coagulate on filter elements causing high pressure differentials, and eventually may coagulate in the work gap leading to short circuiting. There are four solutions to the electrolyte dirt problem at present:

(1) Dump when dirty;
(2) Settle and draw off sludge;
(3) Separate by centrifuge—either full continuous flow or by-pass;
(4) Separate by filter press in line—not usually practical.

For small plants (1) and (2) are the most economic solutions. In the interests of keeping the volume of electrolyte to a minimum and implementing cost savings on expensive electrolytes (3) and (4) come into vogue.

We feel that there should be an even better method of sludge separation: is electronic filtration a possibility? Is there an additive which will promote rapid settling and compaction of the hydroxide without detriment to the cutting process? Our experience indicates that there is such a material, the use of which enables electrolytes to be loaded with sludge concentrations up to an order of magnitude greater than is presently thought possible. Reference is often made to the necessity for maintaining electrolyte temperatures within close limits whilst machining. Our experience is that for dimensional tolerances of ±0·004 in a variation of 10°C causes no embarrassment.

Moving tools
The chief cause of short circuits is starvation of electrolyte at some point between the job and the tool.

Experiments have shown clearly that the incidence of this fault is very much reduced if there is a relative motion between tool and workpiece in addition to the normal tool feed. A speed of 10 ft/min or even less is all that is required.

Stagnation points in the electrolyte flow no longer occur, or at least do not remain over the same point of the workpiece. Minor tool damage does not halt production since the defect which would be caused by the damaged area is averaged out over the whole of the machined surface.

It is not always possible to use this technique, but it does apply to any job having rotational symmetry.

Simple machining can be done without fixtures by utilizing a rotating disc. Components may be offered up to the disc as on grinding machines, the rotation of the disc being the means of introducing a uniform moving film of electrolyte.

required for forming may often be used since it is not always necessary to feed the tool during machining, and control of the parameters is not as vital.

The chief fault observed with existing ECF deburring equipment is excessive short circuit damage to tooling. Ideally, when tooling is designed, the gap between the tool and work is set so as to give clearance with the largest expected burr. It is not possible to take the apparently expedient course of using a very large gap to build in a safety factor, since selectivity of metal removal would be lost and cycle time increased. It follows that if burrs larger than those catered for in the initial design occur then a short circuit will result. This situation often occurs. This also applies when any loose materials, swarf, etc., adhere to the work. The ways in which the problem can be overcome are:

(a) start with a much larger cutting gap and feed the tool during machining as in the normal forming process;
(b) average out the burrs in some way so that the height of the burrs is not a matter of chance.

Either course results in some of the attractive simplicity of the EC deburring process being lost, although we feel that to guarantee its reliability this is essential. Burr averaging could be achieved by an operation prior to the actual deburring operation, but it would be far neater to incorporate it in one cycle. It would be necessary to design the deburring tool so that it would be capable of crushing the outstanding burrs, probably making use of a rotary action where possible, e.g. when deburring the ends of gear teeth.

We have outlined in the foregoing what we consider to be the major reasons for limited acceptability of the process to date, and some of the solutions which we think will assist in changing this situation. There still remain many aspects, perhaps less important, which nevertheless seem to us to call for attention. When some of these are solved, the correctness or otherwise of our philosophy will have been determined.

DISCUSSION

E. R. Brealey. The two processes, ECF and spark erosion can be seen as complementary processes. Spark erosion with its high rate of electrode wear, used for roughing at the rate of one electrode for two components. ECF with its high quality of surface finish and lack of electrode wear.

High dimensional accuracy can be achieved by removing metal at a high rate by ECF followed by spark erosion. The resulting heat affected zone can be removed by a short acid etch.

The importance of a name. The use of the word 'machining' rather than 'forming'. Machining is associated in industry with higher wage rates than are normally paid to operators in the process departments.

Q. N. Hodgson. Mr Lawrence makes a good case for a simple small machine but obviously this does not eliminate the need for large machines for large components and sophisticated controls for accurate components.

From a commercial point of view one 20 000 amp machine may be equal to ten 2000 amp machines. If, by optimum size, Mr Lawrence means one generating the maximum turn over then I suggest that the answer is a variety of sizes with complex controls or with very simple controls.

A. Mr Hodgson's observation is obviously correct since in no sphere of metal working can one size or type of machine satisfactorily cope with every type and size of component to be machined.

Our view point is that ECF, by its nature, is most suited to medium and large batch production. It follows that the cost benefits to be obtained by using the process are most likely to be achieved in this area. There is a loose relationship between the quantity of parts produced and their size. The machine we are advocating will cope with a large proportion of the sizes of components which are made in large batch quantities.

We believe that a machine of this size, which is sophisticated only as far as its function demands and is thus modestly priced, is more likely to propagate the use of Electro-Chemical Forming.

A CONTROL SYSTEM FOR ELECTROCHEMICAL MACHINING USING A MINI-COMPUTER

by

K. SEIMIYA* and S. ITO*

SUMMARY

A control system has been developed with a view to improving both the accuracy and efficiency and the presentation of new possibilities in electrochemical machining. A mini-computer was employed as a central performance unit to memorize a control program, to process data, to judge on optimal conditions and to give instructions in feed rate and power source voltage. Specific resistance of electrolyte and gap voltage are measured on-line and given to the central performance unit to calculate an adequate feed rate for the required gap. Some experiments on program control of ECM were performed by using this system, in which the taper of machined holes was controlled.

INTRODUCTION

With its merits electrochemical machining seemed to be promised a good future when it came on the market more than ten years ago. It has been developed in this decade to such an extent that units are frequently available for sale in manufacturing shops. But now its destination is clouded by some defects mainly involving lack of accuracy. The inaccuracy results from difficulty in controlling electrode gap thickness affected by many parameters. Many problems that have never been studied must be solved to achieve perfect improvement of accuracy. Among these are the problem of systemization of ECM, including control of the process. Objectives of the systemization necessarily include accuracy improvement as well as productivity gain. An ECM control system has been developed to initiate studies on systemization and some experiments have been performed on basic program control using this system.

BASIC CONCEPT OF ECM CONTROL

Control of machined depth

The accuracy of electrochemical machining has been discussed on the basis of equilibrium gap given as,

$$h = k\eta(E - E_0)/10 . vR \qquad (1)$$

where h is front gap thickness in mm, k specific removal rate in mm^3/A min, η current efficiency, E gap voltage in volts, E_0 decomposition voltage in volts, v feed rate in mm/min and R specific resistance of electrolyte in Ω cm. The equation was originally derived for the case with an electrolyte whose current efficiency is constant, involving NaCl solution. It is, however, applicable to another case like $NaNO_3$ solution on condition that correlations between current efficiency and other parameters are given. Control of ECM at the first stage aims at maintenance of the above gap h through a whole machining process in spite of changes in other machining parameters. The gap control is realized principally with the correction of either gap voltage or feed rate in accordance with changes in other variables. Adoption of gap voltage as a control parameter, however, is not advantageous for the following reasons. Firstly, it cannot be controlled directly but depends on power source voltage E_s. Secondly, quick and precise control of E_s is very difficult. Feed rate should therefore be taken as the control variable. Gap voltage E and specific resistance R are measured on line and passed to a central performance unit through an A-D converter.

Different ways are selected to control feed rate, depending on the kind of motors used for machine feed. Feed-back-control is inevitably needed by a standard type electrochemical machine to control the induction motor, while a stepping motor of a digital type electrochemical machine can automatically be driven in proportion to the number of control pulses from a central performance unit.

Length of feed T must be related to the calculated machined depth H as,

$$H = T + h - h_0 \qquad (2)$$

where h_0 is initial gap and h is given by equation (1). A digital type electrochemical machine is advantageous since it can use the soft-ware of counting command pulses in place of direct measurement of table or tool displacement, compared with a standard type electrochemical machine which cannot avoid the employment of a detecting system of feed length.

*Mechanical Engineering Laboratory, Japan

depth should be controlled such as parallel-flow machining with plane electrode, optimization can be limited to the final stage. On the other hand, earlier machining stages can be concerned with the shape of machined product in two- or three dimensional ECM. Two-dimensional machining practically refers to the form that has almost the same cross section at any depth.

The switchover of machining conditions from one stage to the next is proceeded with in respect of feed length or machined depth regarded as a substitute for time variable. Each stage of a control program comprises a set of feed length T, front gap h and power source voltage E_s as follows.

$$T_1, T_2, \ldots T_n, h_1, h_2, \ldots h_n, E_{s1}, E_{s2}, \ldots E_{sn}$$

These represent an ECM process that starts at electrode gap h_1 and power source voltage E_{s1}, changes the conditions to h_2 and E_{s2} respectively at feed length T_1, repeats similar change $(n-1)$ times and ends at T_n.

Adaptive control

Adaptive control is the expansion of optimal control and includes the whole concept of the latter. The two controls differ in whether optimal conditions are on-line determined or not. Figure 1 shows the

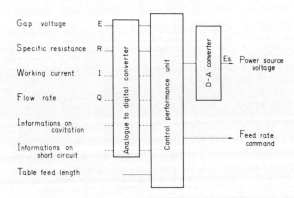

Figure 1. Schematic diagram for adaptive control of ECM.

schematic diagram representing a comparatively advanced control for ECM. Working current, flow rate, information on cavitation and short circuit are required for a judgement of optimal conditions. The central performance unit refers to a mini-computer or a similar device when the system works independently. It may be substituted for a terminal device of a large-scale computer if it works as a subsystem belonging to an integrated manufacturing system.

CONTROL SYSTEM FOR ECM

Central performance unit

Figure 2 shows the block diagram of the prototype control system for electrochemical machining developed by the authors. A mini-computer of 8 kilo-word

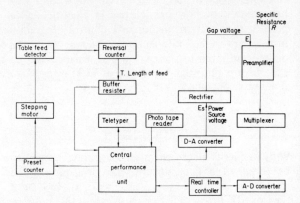

Figure 2. Block diagram of the prototype control system for ECM.

core-memory was employed as the central performance unit. The quantity of memory can be increased by the additional unit of 4 kilo-words within the total limit of 32 kilo-words. A photo tape reader and a teletyper for input and output of data are used in association with the mini-computer. The central performance unit functions to memorize a control program and input data, to process data, to calculate, to judge on optimal conditions and to give instructions on feed rate and power source voltage.

Devices concerning input data

Analogue input data, involving gap voltage and specific resistance at the present stage, are given to the central performance unit through preamplifiers, multiplexer (multi-parameters meter) and analogue to digital converter. Only two of eight channels of the multiplexer are used now and others are preserved for future expansion. A reed-relay is used in channel selection of the multiplexer. Gap voltage is measured at terminals of the electrochemical machine. The difference between gap voltage and power source voltage refers to the drop in resistance known as Ohm's drop through leading-wire. Specific resistance of electrolyte is detected with a detector consisting of a kind of cell and its protective metal tube, on the basis of electro-magnetic induction. Since the detector is sunk into an electrolyte-reservoir, measured values of specific resistance may be somewhat different from that at the inlet of the electrode gap. The lower limit for measuring specific resistance with the detector is placed around 3 Ω cm. The real time controller, mainly comprising a timer and squarewave oscillator, controls the sequence of time for the A-D converter and the preset counter. The timer works with the aid of the clock pulses of 1 kHz from the oscillator. Though the timer can provide the A-D converter with four kinds of sampling period of 1 to 0·01 s, 1 s has exclusively been selected and preferred.

A magnetic-scale was attached to the electrochemical machine in order to detect table feed length, which is given to the central performance unit through reversal counter and buffer resister. Values of table feed length from the detector were typed out in comparison with what were calculated on the basis of counting the total number of pulses given to the stepping motor. Differences between the two values did not exceed ±0·01 mm.

Devices concerning output command
The number of pulses per second for the realization of an aimed gap is temporarily set in the preset counter and then given to the stepping motor in accordance with the command from the central performance unit. The clock pulses of 10 kHz from the oscillator are used for controlling the preset counter. For the authors' electrochemical machine, the frequency of some 760 pulses per second refers to the feed rate of 1 mm/min.

Since a stepping motor cannot follow a rapid change in frequency of driving pulses, some countermeasures should be prepared in the control program when the change is inevitably required. Consequently, a sub-program has always been prepared for the gradual acceleration at the starting time. Attention should necessarily be paid to possible disorders, due to noises, that occur from both the a.c. power-source line and the connecting cable between the central performance unit and the stepping motor. A stepping motor can easily stop owing to noises during machining that apparently induce a rapid change in frequency.

A silicon-rectifier of 1000 amperes, used as the machining power source, is instructed in its output voltage E_s by the central performance unit. Digital instruction values are converted by the D-A converter and given to the input-device of the rectifier. The input-device regulates the rectifier's output voltage according to the instructed voltage.

Reliability of the system
Much attention should be focused upon the reliability of the system when it is introduced into commercial manufacture. But it does not seem sensible to discuss the reliability of the present control system with so much room left for improvement. Problems often arose with analogue data sampling due to malfunction of A-D converter and mini-computer in respect of the interface. Weakness of the stepping motor to noises was also notable. Gradual improvement has been made, however, and our system now works well in the laboratory. Further development depends on the need of users and manufacturers for an electrochemical machine.

APPLICATION EXAMPLES
Some experiments were performed to represent application examples of the control system on program-control of ECM for the purpose of controlling machined hole's taper.

Experimental method
A couple of the same mild-steel-bar of 25 x 40 x 50 mm were used to form an anode-workpiece of 40 x 50 x 50 mm. Electrochemical machining was performed along the centre-line of the interface so that the machined shape can easily be inspected. A cylindrical, brass tool-electrode was used for the machining. The outer and inner-diameter of the electrode was 25 mm and 10 mm, respectively. The working area of the electrode was limited to the front plane and the edge of radius of 2 mm by means of a proper insulation. The electrolyte was 10 wt per cent NaCl solution. Electrolyte flow rate was manually adjusted to a proper value for each case by watching the flow-meter. The power source voltage was fixed at 10 volts for simplification.

Results
The calculated value of feed rate plotted in Fig. 3 to Fig. 7 inclusively were shown together with the related values of feed length, specific resistance and

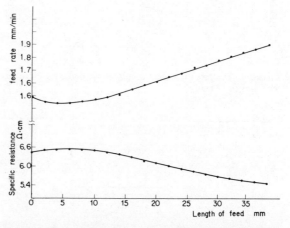

Figure 3. Dependence of feed rate and specific resistance on feed length as the result of control to maintain a constant gap.

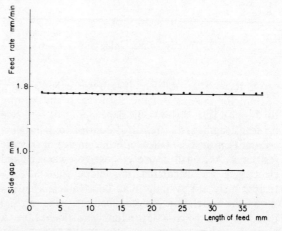

Figure 4. Correlation between feed rate, side gap and feed length in the case of machining a straight hole.

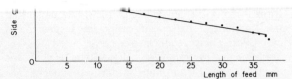

Figure 5. The case of conical cavity whose diameter decreases with length of feed.

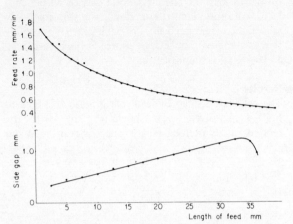

Figure 6. The case of conical cavity whose diameter increases with length of feed.

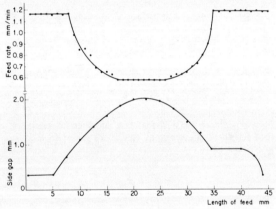

Figure 7. The case of machining convex cavity.

gap voltage. The total feed length was established at 38 mm for the cases of Fig. 3 to Fig. 6 and 45 mm for Fig. 7. Figure 3 shows the dependence of feed rate on feed length, as the result of control to maintain a constant gap in accordance with the change of specific resistance. As a small electrolyte-reservoir was used and electrolyte was circulated, the temperature of electrolyte increased by more than ten degrees. Figure 4 shows the correlation between feed rate, side gap and feed length in the case of machining a straight hole. A large electrolyte-reservoir was used in the cases of Fig. 4 to Fig. 7 inclusively, to minimize temperature rise

Figure 7 refers to the case of machining of a convex cavity, or a hole of expanded body. Though the aimed gaps of the first and last stage were the same, the resulting gaps were different. This suggests the necessity for modification in consideration of the effect of flow rate etc.

CONCLUSIONS

A control system has been developed with a view to improving both the accuracy and efficiency and the presentation of new possibilities in ECM. A mini-computer was employed as a central performance unit to memorize a control program, to process data, to judge on optimal conditions and to give instructions in feed rate and power source voltage. Specific resistance and gap voltage are on-line detected and given to the central performance unit to calculate an adequate feed rate for the keep of aimed gap. Some experiments performed by using this system under control programs have shown its usefulness, though much room has been left for improvement to minimize troubles in respect of both synthetic adjustment and unit devices' improvement.

ACKNOWLEDGMENT

The authors wish to thank K. Kikuchi, S. Shida and H. Matsumoto for considerable assistance in experiments.

DISCUSSION

Q. P. Lawrence, Healy of Leicester. The claim that this adaptive control system, which modifies the tool feed to compensate for change in electrolyte conductivity, is novel is refuted. A similar system, known as the 'Barmax' system was developed and used by Rolls-Royce approximately 10 years ago.

(This paper was presented by C. F. Noble and the undermentioned is his reply to the above contribution.)

A. I accept the point about the Barmax machine but what makes the Japanese research novel is the fact that a computer is used to judge on optimal conditions and to give instructions on initial feed rate and power source voltage. This same computer then monitors the machining process by giving instructions to vary the feed rate as conditions demand. A further important point is that the research reported is part of a program to monitor further variables by computer control. Whether or not this long term aim of the research can be considered industrially viable is for the delegates to judge.

THE CONIC WHEEL FOR ELECTROCHEMICAL SURFACE GRINDING

by

M. M. SFANTSIKOPOULOS and C. F. NOBLE*

SUMMARY

The introduction of the conic wheel to vertical spindle electrochemical surface grinding is discussed and experimentally assessed. It is shown that the scope of the process could be widened by permitting large depths of cut at reasonably high table feed rates. Conclusions are drawn about wheel geometry, supply of electrolyte solution, process efficiency, accuracy and limiting conditions.

NOTATION

To ensure continuity the symbols used in previous papers[1,2] have been retained as follows.

b = radial width of wheel
β = dimensionless factor
C = electrochemical constant = $k(V - \Delta V)\epsilon/\sigma F$
D = wheel diameter
D_1 = constant
D_2 = constant
e = grit protrusion height
f_1 = set depth of cut relative to grit surface
f_{1a} = actual depth of cut
f_{20} = critical table feed rate
f_{20B} = critical table feed rate, technique B
F = Faraday
i = current density
$I_{d.c.}$ = d.c. current
k = conductivity
K_3 = depth of cut ratio
R = material removal rate
R_A = material removal rate, technique A
R_B = material removal rate, technique B
V = applied voltage
V_S = volume of material removed per stroke
V_{SF} = volume of Faradaic removal per stroke
w = component width (transverse to table feed direction)
ϵ = chemical equivalent
θ = conic wheel angle
σ = density of workpiece.

INTRODUCTION

Two methods[1,2] can be adopted for electrochemical surface grinding when using a vertical spindle machine with work-table traverse. In technique 'A' the periphery of a straight cup wheel is used but, because the 'critical table feed rate' is determined by the limitation imposed by the small size of the 'equilibrium gap', only low feed rates are possible. To increase removal rate the working area must be enlarged by employing a very wide workpiece (and consequently large diameter wheels) or by enlarging the depth of cut. Use of the latter does not constitute a surfacing operation and, in any case, creates difficulties for maintaining continuity of electrolyte solution within the machining zone. A further disadvantage is that any susceptibility to time-dependent side effects (overcut, grooving, edge erosion, staining) is increased by the low feed rates.

In technique 'B' the face of the cup wheel is used and the set depth of cut is limited to a fraction of the height of abrasive grit above the wheel bond. However, relatively high table feed rates are possible because, instead of a limitation imposed by an equilibrium gap, the critical feed condition is determined by the need to prevent overcut—i.e. no part of the wheel must behave in an entirely electrochemical manner. Unfortunately, the absence of an equilibrium gap condition makes process optimization a difficult and complicated procedure, especially for shop-floor use. The technique is extremely sensitive to distribution characteristics of the electrolyte solution, grinding wheel conditioning, machine rigidity and geometric accuracy and forced vibration level. These complications result in a much reduced capacity for grinding large workpieces, but the proneness to electrochemical side effects is much reduced.

The present authors believe that use of a conically shaped wheel offers the opportunity to combine the advantages of both techniques and to provide a means to combat some of their individual disadvantages. Although conic wheels have been referred to in leaflets from some ECG equipment manufacturers

---

* Machine Tool Division, Mechanical Engineering Dept., U.M.I.S.T.

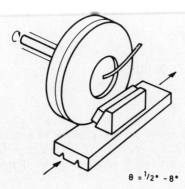

$$\theta = \frac{1}{2}^\circ - 8^\circ$$

Figure 1. The Anocut grinding principle with conic wheel and horizontal spindle.

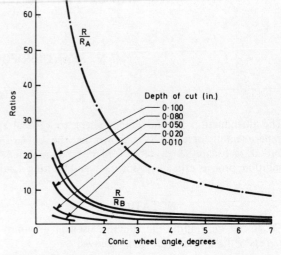

Figure 2. The theoretical model: (a) and (b) geometry, (c) removal rate characteristics.

grinding is impractical. This paper is devoted to use of the conic wheel for surfacing operations with a vertical spindle where the working area is relatively large. Therefore, provided the d.c. power unit has sufficient capacity, it offers faster removal rate.

THEORETICAL CONSIDERATIONS

The theoretical model is given in Fig. 2(a) and (b). The critical table feed-rate for maintaining a certain electrochemical gap will be

$$f_{20} = \frac{C}{\beta e \sin \theta} \tag{1}$$

The necessary mechanical action of the abrasive grit for maintaining process continuity, accuracy and quality is represented by the mechanical action factor β which, in principle, should be <1.

It can be seen from equation (1) that employment of a conic wheel results in a considerably faster table feed and, therefore, also with material removal rate, compared with deep-cut technique A where $\theta = 90^\circ$. When compared with the analysis of shallow depth technique B[1,2] the following table feed ratio can be formulated

$$\frac{f_{20}}{f_{20B}} = \frac{K_3(2 - K_3)}{2\beta} \cdot \frac{e}{b \sin \theta} \simeq \frac{0.000618}{\sin \theta} \tag{2}$$

where typical parameter values have been taken (i.e. $K_3 \simeq 0.2$, $\beta \simeq 0.7$, $e \simeq 0.0015$ in, $b = 5/8$ in).

With the above values, any angle greater than $3'$ necessitates slower table feed rates. However, when removal rate is considered, the following ratio is revealed

$$\frac{R}{R_B} = \frac{2 - K_3}{2\beta} \cdot \frac{f_1}{b \sin \theta} \simeq 2.06 \frac{f_1}{\sin \theta} \tag{3}$$

Equation (3) is plotted in Fig. 2(c) for different depths of cut, together with the equivalent ratio for deep-cut technique A, i.e. R/R_A. It can be seen that for greatest benefit from the conic wheel it is necessary to apply it to deep cuts and employ small angles. However, many practical applications could involve depths of the order of 0·01 to 0·02 in and the results shown in Fig. 2(c) suggest that, for a range of angles up to 3° to 4°, removal rate is comparable or higher than with the normal cup wheel technique B (i.e. $R/R_B \geqslant 1$).

A further consideration is component size in relation to wheel size. With wide workpieces the wheel contact arc may be sufficiently greater than the width to cause undue variation in the effective cone angle. To limit this effect the authors suggest that the mean contact arc should not be greater than the width by more than 5 per cent. To satisfy this geometric condition it can be shown that

$$w_{max} \leqslant \frac{D - b}{2} \tag{4}$$

Any further restriction on width will depend on the current capacity of the supply unit and on the mean current density specified for a particular component material or wheel.

Hence

$$w_{max} \leqslant \frac{I_{d.c.,max}}{i_{max} b} \leqslant \frac{D - b}{2} \tag{5}$$

Therefore, a large diameter wheel and/or a small rim width may be necessary.

EXPERIMENTAL INVESTIGATIONS

I.S.O.K20 $3/4 \times 3/4 \times 3/16$ in^3 tungsten carbide workpieces were ground using a 10 per cent concentration of a $NaNO_2 + NaNO_3$ electrolyte, with an Abwood SG4HE vertical spindle grinding machine. The hydraulics for table motion were then modified in order to accomplish the necessary creep feed rates with the conic wheel. Instrumentation made it possible to record continuously the performance during grinding. The electrolyte system was redesigned to supply the fluid both through external nozzles and through the hollow machine spindle. Electrolyte from the latter was guided to the machining zone by means of a flow shaper and a distributor (see Fig. 3). Total electrolyte flow rate was 6 l/min. The 8 in diameter, 5/8 in rim width grinding wheel was manufactured to the authors' specification by Messrs J. K. Smit & Sons using 100 diamond grit size, 50 concentration. Maximum peripheral surface speed was 6600 ft min^{-1}. Four wheel angles were tried and Table 1 shows the established limiting operating conditions for each.

For the particular grinding operation, lower voltages were necessary with the larger angles because it was found difficult to obtain satisfactory electrolyte flow along the full path of the cut. The table feed rates were also significantly lower than those for the smaller angles. Given that the main electrolyte supply into machining gap is from the machine spindle (the side nozzle supply plays a rather secondary supporting role since most of the electrolyte is dispersed by the centrifugal forces developed

TABLE 1.   Limiting operating conditions

θ	$f_{1,max}$ (in) (geometrical limit)	$f_{1,test}$ (in) (close to $f_{1,max}$)	V (volts)	f_{20} (in/min)
1°	0·0080	0·0075	7	2·100
1½°	0·0123	0·0100	8	1·750
2°	0·0164	0·0140	6	1·225
2½°	0·0204	0·0175	4	0·500

by the wheel rotation), the wheel geometry (angle, rim width) is of high importance for distribution characteristics. Wheel geometry, therefore, indirectly influences the grinding conditions in addition to its primary role described by equation (1). The larger the angle, the more strained are the electrolyte conditions (see Fig. 4a) and it is important to realize that the rim width determines the electrolyte path length.

Appropriate wheel design consequently necessitates consideration of these aspects in conjunction with the attainable range of depths of cut to be applied, workpiece size and electrolyte supply system.

There may exist strong reasons for objecting to electrolyte flooding in the vicinity of the working area because it encourages stray-machining and ECG side-effects. It would seem, nevertheless, that for heavy grinding, where the main target is high rate of material removal, a directed electrolyte jet under low pressure is reasonable (see Fig. 4b).

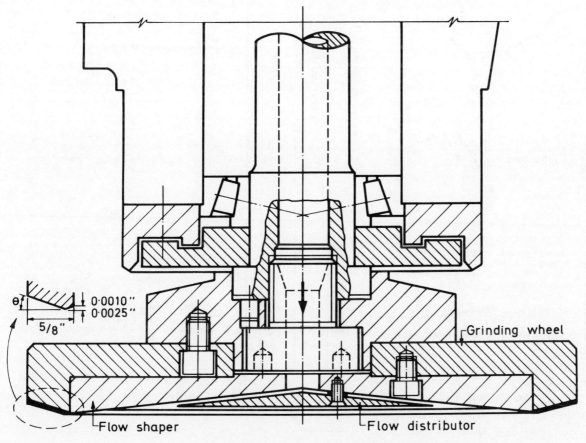

Figure 3. Conic wheel with flow shaper and distributor fitted to spindle.

Electrolyte Supply System

b

Figure 4. Path of electrolyte from (a) central supply and (b) external nozzles.

ficult, an increased proportion of evolved gas and possibly local boiling followed, and led to a reduction in equivalent conductivity. This in turn led to a reduction in the gap and some electrolytic action was replaced by increased mechanical activity as illustrated by the spindle load current. Passivation of the workpiece surface combined with adherence of electrochemical precipitates just before actual

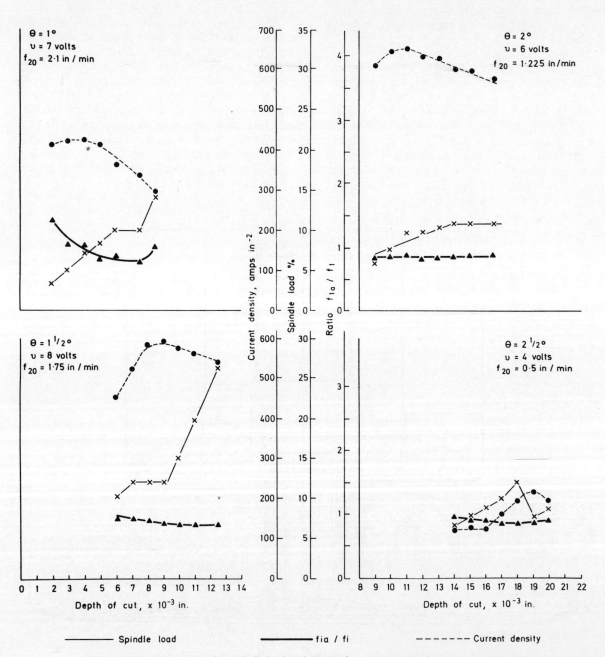

Figure 5. Mean current density, spindle load and depth of cut ratio results for varying depths of cut.

grinding starts (see Fig. 6) is held as responsible for the lower current densities with the smaller depths of cut. It is thus confirmed that the method should preferably be applied to deep cuts matched to the specific wheel angle and rim geometry. However, developed current densities exceed the 400–500 A in^{-2} of technique B which were obtained at the expense of a substantial contribution from sparking[2]. In the present case 1°, 1½° and 2° cone angle grinding was effected entirely free of any sparking apart from one exception at maximum depth of cut with $\theta = 1°$. (The combination of maximum depth of cut, maximum voltage, and maximum table feed rate.) Increased sparking and unstable machining occurred,

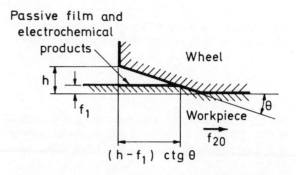

Figure 6. Operating dimensions showing possible passive region.

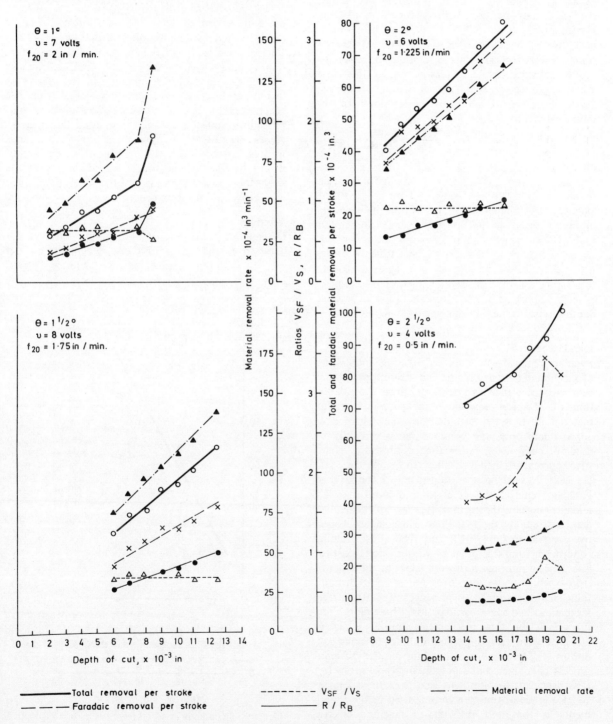

Figure 7. Material removal characteristics vs. set depth of cut.

lationship with the set depth of cut, whereas the unstable conditions associated with electrical sparking in the case of $\theta = 2\frac{1}{2}°$ have resulted in an exponential variation. A similar deviation from linearity is met also—as already mentioned above—for the case of the maximum depth with $\theta = 1°$. It is to be noticed, nevertheless, that even for zero depth of cut there is some electrochemical removal, tending, however, to vanish as wheel angle increases due to the large average gap so created. Actual depth of cut is, therefore, a linear function of the set depth of cut with a general form

$$f_{1a} = D_1 f_1 + D_2 \qquad (6)$$

D_1 takes into account elastic deflections caused by cutting, evolved gas pressure and hydrodynamic forces. It should be independent of the set depth of cut over a reasonably wide range. From the diagrams of Fig. 5—even for the case $\theta = 2\frac{1}{2}°$—it can be seen that within a good approximation and for the experimental set-up used, $D_1 \simeq 0.85$. It is, therefore, possible to assess the number of (single) deep cuts (close to the maximum, $D_2 \ll f_1$) required to remove a certain stock, and, also, to decide upon the final depth of cut for the finishing pass(es).

Faradaic removal

Faradaic removal was evaluated by time integration of machining current records and assuming a 100 per cent current efficiency. However, since the calculation of chemical equivalent for such a complex material as tungsten carbide must be regarded as approximate only, the obtained values can only be used for comparison rather than as absolute values. Also, current efficiency is likely to be less than 100 per cent. Nevertheless, as seen in Fig. 7 the ratio of Faradaic removal to actual total removal V_{SF}/V_S remains constant throughout the range of the applied depths of cut for the first three wheel angles, respectively equal to 0.64, 0.70 and 0.88. Best grinding conditions, from this point of view at least, exist for $\theta = 2°$. The ratio has its lowest value—of the order of 0.55—with $\theta = 2\frac{1}{2}°$ but does not remain constant, thus demonstrating the presence of both strong mechanical and spark participation. These plots allow the conclusion that mechanical removal is always included when operating under limiting operating conditions, the proportion depending on the different process parameters but especially on wheel geometry. It is necessary to point out that the almost straight WC-Co grade of tungsten carbide used in the experiments is the most difficult to machine electrochemically and consequently the mechanical contri-

the angles $1\frac{1}{2}$ and $2°$. This fact comes in support of the conclusions already made that the conic wheel technique is best suited to deep cuts. These should be close to the maximum possible with the particular wheel geometry, provided that situations such as those developed with $\theta = 2\frac{1}{2}°$ are avoided. The latter condition requires, above all, a suitable design for the electrolyte supply. On the other hand, it should be remembered that spark-erosion and substantial mechanical removal are inherent in the results of technique B. The conic grinding wheel offers the advantages of a one cut, instead of multiple cut, procedure and the capability to grind larger workpieces. Because an electrochemical equilibrium gap is formed, the conic wheel especially offers the opportunity for determining operating conditions on the shop-floor, instead of by an otherwise complicated laboratory optimization procedure.

Accuracy

In general, roughness of the ground surfaces lay between 6 and 11 μ in CLA, depending on the specific grinding conditions. No serious flatness

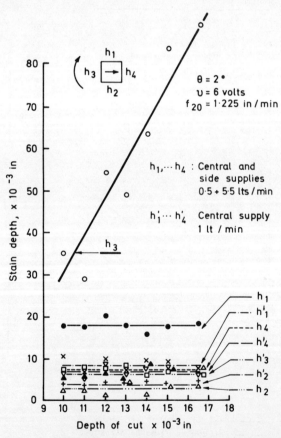

Figure 8. Stain depth and workpiece orientation vs. depth of cut with full electrolyte supply and with spindle supply only.

deviations were observed with the exception only of the $\theta = 1\frac{1}{2}°$ case were 'steps' of the order of 400–750 μin were produced. These are attributed to problems with electrolyte flow associated with the high voltage (8 V) applied.

The steps emphasize, once again, the importance of the electrolyte flow conditions, as especially influenced by the wheel geometry. They further emphasize the role that voltage and feed rate play with respect to ECG side-effects, which fact equally well applies to the observed edge erosion and staining. Both appeared in a form which has been described previously[1] but, due to the higher voltages and considerably lower table feed rates employed in the present investigations, stain depth was much larger although, as before, strongly dependent on the orientation of the particular side of the workpiece with respect to direction of wheel rotation and table traverse. Nevertheless, as Fig. 8 clearly demonstrates, it was proved possible to reduce its size substantially by operating with electrolyte supply through the grinding spindle only.

CONCLUSIONS

Introduction of the conic wheel to vertical spindle grinding provides a more satisfactory approach to solving some of the particular problems associated with the process. Nevertheless, it requires that parameters directly associated with the machine tool, power unit, grinding wheel component size and material are correctly interrelated. Removal with a substantial mechanical contribution under near limiting conditions can be conducted under controlled electro-chemical action with the 2° cone angle. This was made apparent by its smooth and spark-free operation. It is thus possible to predict actual depth of cut within a good approximation.

The surface finish produced is satisfactory but electrolyte flow and distribution can so easily develop undesirable effects on surface flatness and on edge geometry and appearance.

It is advisable to effect a finishing-to-size operation by employing the shallow depth of cut, high table feed-rate technique, even under non-optimized conditions. The electrolyte supply system is a crucial factor in any of the techniques so far used, and this fact must be borne in mind if further attempts are made to improve use of the conic wheel. Limiting the supply and its direction has been shown to considerably reduce undesirable edge features on tungsten carbide components but it has not yet been found possible to completely eliminate the problem. A concentrated effort on this aspect could prove a valuable topic for further research.

ACKNOWLEDGMENTS

The authors are pleased to record thanks to Abwood Machine Tools Ltd for the loan of one of their electrochemical surface grinding machines. Thanks are also due to Hoy Carbides Ltd for the provision of tungsten carbide workpieces and to Keelavite Hydraulics Ltd for the valves which permitted use of low table feed rates during the investigations.

UMIST is acknowledged for providing facilities for conducting the work and the National Technical University of Athens for the S. Niarchos Scholarship awarded to the first named author.

REFERENCES

1. M. M. SFANTSIKOPOULOS and C. F. NOBLE (1972). Vertical spindle electrochemical surface grinding, *Proc. 12th Int. M.T.D.R. Conference,* Macmillan.
2. M. M. SFANTSIKOPOULOS and C. F. NOBLE (1973). Dynamic and geometric aspects of vertical spindle E.C.G. *Proc. 13th Int. M.T.D.R. Conference,* Macmillan.
3. Electrochemical grinding: Anocut Engineering Co. Paper No. WAL-3, May 1968. Functions and applications of the wheel.
4. J. C. WILSON. *Electrochemical Surface Grinding.* Thompson Grinder Co., Springfield, Ohio, U.S.A.
5. *Electrolytic grinding.* The Cincinnati Milling Machine Co., Cincinnati, Ohio, U.S.A.
6. M. M. SFANTSIKOPOULOS (1973). Studies in electrochemical surface grinding, PhD Thesis, UMIST, Jan.

DISCUSSION

Q. E. Kaldos, Hungary. Mr Noble's interesting paper is welcomed as one of the studies aimed to extend the electrochemical grinding process to other fields than resharpening of cutting tools.

Conclusions drawn from the experiments are valuable, and, in my opinion, satisfy industrial demands. When listening to Mr Noble's lecture, some questions arise.

(1) Has Mr Noble carried out experiments with wheels having smaller rim width? I propose a smaller width than applied during the experiments presented, and not greater than 10 mm, which promises better electrolyte supply along the gap. In my opinion, the side nozzle supply should be negligible in this case.

(2) Looking at Fig. 7, on the top left, a breaking point can be found on the diagram representing the total removal rate. At this comparatively high feed rate, i.e. 2 in/min., the material removal is forced by the table feed, and the mechanical contribution to total removal is high; therefore this curve should be linear. Could Mr Noble give the answer, why this deviation exists between theoretical consideration and results obtained from experiments?

(3) Has comparison between attainable grinding ratios of the different kinds of surface grinding methods been made? If yes, does the conic wheel technique promise a better result than the others, or not?

(4) Has the proposed equation 4 concerning the maximum component size been proved experimentally or even theoretically?

In my opinion—and let me propose this to Mr Noble—by means of the method of dimension analysis, if reasonable basic units are chosen, more correct equations could be obtained.

with this value.

(3) We have only made comparisons with other ways of using a vertical spindle, table traverse technique. These all offer the opportunity to grind large surface areas, and in this respect compare very favourably with the horizontal spindle, workpiece-under-pressure method which has become accepted practice for EC tool grinding purposes. Perhaps I am biased towards use of the conic wheel, but if delegates will study two previous papers in this field, refs 1 and 2, I think they will also become enthusiastic.

(4) Equation 4 is a simple expression derived from the suggestion that the mean horizontal arc of contact should not exceed the component width by more than 5 per cent. With the wheel dimensions used a nominal $2°$ cone angle would effectively reduce to approximately $1\frac{3}{4}°$ at the position determined by equation 4, and in view of our results on cone angle I think this is a reasonable limitation. However, it is a suggestion only.

As far as dimensional analysis is concerned, although I accept it to be a powerful tool for analysis purposes its use has to be carefully evaluated and I do not believe it would be of any value in respect of equation 4.

Q. D. A. Lindenbeck, De Beers Diamond Research Laboratory. It is very important to convey the fact that diamond grinding wheels require in all types of grinding processes a specific profile. This applies to diamond tool producers and especially to diamond

'natural' value and only then are the data obtained valid for practice, keeping in mind that in a correctly chosen grinding process generally the later retrieving or redressing should be necessary during the life of the diamond wheel.

A. In ECG, diamond wheel wear is much less important than when grinding by mechanical means only. Consequently rather a lot of workpiece material has to be consumed if wheel wear is to be investigated, and this will be both expensive and time-consuming. I accept that wear will occur especially, presumably, at rapid changes in profile, but I would not expect angles on an ECG wheel of the type proposed in this paper to alter noticeably unless the requirements for maintaining electrolytic action are ignored. In Refs 1 and 2 we emphasized the importance of wheel surface topography, maintenance of grit protrusion and removal of debris and products of electrolysis from the wheel. Periodic or perhaps continuous, conditioning of the wheel is extremely important for efficient use of ECG, and the procedures adopted for conventional grinding with diamond wheels should be altered.

Q. N. Hodgson. How did Mr Noble measure the grit protrusion of the grinding wheel?

A. We used a capacitive transducer having a 0·1 in diameter active area. I would refer Mr Hodgson to our previous paper, ref. 2, for more information on this aspect.

PRINCIPLES AND APPLICATIONS OF ELECTROFORMING

by

W. R. WEARMOUTH*

SUMMARY

The principles involved in electroforming are discussed, followed by a review of the advantages of electroforming as a production process. The main types of electroforms are classified under three headings: tools; actual components for engineering service or consumer goods; foil and mesh products; and the many diverse applications under each heading are examined in detail. Finally the future role of electroforming is considered in the light of the rapid diversification in applications over the last few years.

INTRODUCTION

Electroforming has been used for more than a hundred years to produce shapes by the deposition of metals in an aqueous bath with direct current. It has been reported that a Professor Jacoby of the Academy of Science in St. Petersburg, Russia, reproduced parts by electroforming as far back as 1838, and in 1907, Thomas Edison patented a process for making articles by electroplating[1]. Electroforming has been used in industry for forty years, yet today the process is being improved and developed more rapidly than ever before and, as a result, is offering new and cost-effective answers to many design and production problems. It is estimated that electroforming uses about 4 per cent of the nickel consumed in all electrodeposition operations.

Electroforming is the production of a component by an electrodeposition technique, the operation of which is basically simple. The required metal is deposited on a mandrel immersed in a suitable electrolyte, until it is strong enough to be self-supporting. The removal of the mandrel then virtually completes the production of the electroform. The mandrel will have the required surface shape and finish, and although there may be very little adhesion of the metal deposited, the surface contour and finish of the mandrel will be faithfully reproduced on the contact face of the electroform.

In commercial electroforming, expertise is needed, particularly in some applications, and this paper has three objectives: Firstly the advantages of electroforming as a production process will be discussed. Secondly the process itself will be explained in more detail particularly with respect to electrolytes, deposit properties and mandrel requirements. Thirdly

the present uses of electroforming will be considered in detail, and the future role of electroforming in industry discussed, in the light of rapid diversification in the applications of electroforming over the last few years.

ADVANTAGES

In some instances, e.g. gramophone record stampers, electroforming is the only practicable method of fabricating a product. In others, electroforming is used as a method of production because its technical advantages enable that production to be carried out at lower cost.

Simplification

An object of intricate form can be produced in a single stage by electroforming. This is perhaps exemplified by the mass production of bellows which could not be readily produced in a single stage in any alternative way (see Fig. 1). Another field where this

Figure 1. Miniature metal bellows electroformed in nickel. (Cuniform Engineering Ltd.)

---

* International Nickel Limited, Birmingham

signal is recorded on a track which is about 25 μm across and approximately 1 km long on a 300 mm

Figure 2. Press fitted with electroformed record stampers. (Decca Record Co.)

diameter L.P. record, within a tolerance of ±2 μm (see Fig. 2). Other examples are the reproduction by electroforming of diffraction gratings and the production of electroformed wave guides where dimensional accuracy is essential if the electronic requirements are to be satisfied.

Surface variety
A variety of surface textures can be reproduced simultaneously on the initially-deposited electroformed surface. For example, a leather-grained finish can be reproduced (see Fig. 3), or a mirror finish may be provided, or both surfaces may be formed together on one part.

Figure 3. Mould for plastic case with leather grained finish. (Fa Simon, Western Germany.)

Figure 4. Parabolic electroformed mirror, rhodium-plated finish. (Cuniform Engineering Ltd.)

multiple tools for fabricating materials such as plastics as in the use of record stampers or multi-cavity moulds for plastic gears. These moulds are frequently produced by assembling together several electroformed cavities, all of which are identical and capable of use in producing an exactly formed gear from plastic.

ELECTROFORMING SOLUTIONS AND DEPOSIT PROPERTIES

Nickel is the metal most used for electroforming and is followed in importance by copper, and, to a much lesser extent, by iron and other metals. Nickel is strong and tough and highly resistant to corrosive attack, erosion, and abrasion. Also the mechanical properties of electrodeposited nickel, such as stress, hardness and ductility may be varied at will between wide limits.

Nickel
Nickel can be deposited over a wider, controlled range of mechanical properties than other metals commonly used in electroforming. Most electroforming of nickel is carried out from the Watts solution, the conventional sulphamate solution (containing about 300 g/l nickel sulphamate), and the concentrated sulphamate solution used in the Ni-Speed process[3] which contains 600 g/l nickel sulphamate. These solutions are used alone or with addition agents which may be organic or inorganic.

Deposits from the Watts solution are matt, and have a hardness of 150–200 HV. Their tensile strength is approximately 410 MN/m², elongation about 30 per cent, and the deposit stress 150 MN/m² tensile at 5 A/dm². Deposits from conventional sulphamate solutions are also matt but not so dull as the deposits from Watts type solutions. They have a hardness of 180 HV, tensile strength 410 MN/m², elongation of 18 per cent and a deposit stress of 14 MN/m² tensile. The main difference is the markedly lower internal stress in the deposits from the conventional sulphate solution.

Additional agents are commonly made to these solutions, especially to the Watts solution, in order to achieve lower deposit stress, higher hardness and strength, to increase deposit lustre and to prevent pitting. Additives such as p-toluene sulphonamide, saccharin, and naphthalene 1,3,6-trisulphonic acid cause stress to become more compressive, an increase in deposit lustre, and a decrease in grain size, giving an increase in hardness up to values of 600 HV.

However, the concentrations of these substances in solution, and hence the exact effects that they have on deposit properties, are difficult to control. Moreover, these addition agents give rise to incorporation of sulphur in the deposits which causes embrittlement on heating to temperatures in excess of 200°C[4]. The exact embrittling temperature is determined by factors such as sulphur content, the time at a given temperature, the rate of cooling, the stress distribution, the distribution and forms of sulphur etc. As well as causing incorporation of sulphur, in many instances, organic addition agents form breakdown products in use, and, as these accumulate, the control of deposit properties becomes more difficult and necessitates regular re-purification of the solution.

The development of the 'Ni-Speed' process[3] and the use of INCO 'S' Nickel* permits under appropriate conditions, the use of high deposition rates and the control of internal stress from a compressive state to a tensile state, without the use of sulphur-containing addition agents. 'Ni-Speed' was developed as a high-speed process and, under appropriate conditions where current distribution is uniform, current densities up to 80 A/dm² can be used, giving 1 mm of nickel per hour. At the other extreme, as in the production of complex moulds and dies where current density is non-uniform, the superior throwing power of the Ni-Speed process allows the use, in many instances, of current densities of 2-3 A/dm² instead of the 1 A/dm² commonly the maximum with other plating processes. This permits a large reduction in production time. Deposit hardness is 220-250 HV, tensile strength approximately 800 MN/m², elongation 10 per cent, and deposit stress may be varied between 105 MN/m² compressive and 140 MN/m² tensile by selecting the deposition conditions. Zero stress can be attained at current densities up to 30 A/dm² giving deposition rates up to 375 μm/h[4].

A development of the Ni-Speed process is the use of cobalt as a hardening agent[6,7]. Compared with organic additives, cobalt has the two advantages in that there are no breakdown products and the deposits remain free from sulphur and therefore do not become brittle when heated. However the maximum hardness obtainable with cobalt is not as high as that achieved using organic addition agents, the highest value is 525 HV using a solution with 6 g/l cobalt, so as to give a deposit of 65 per cent

nickel-35 per cent cobalt. With 1 g/l or a greater quantity of cobalt in the solution, the deposit stress is always tensile at deposition rates between 50 and 350 μm/h, although the level of stress is lower than that in deposits from a Watts bath. However, by appropriate choice of deposition rate and cobalt concentration in the solution, hard alloys of nickel-cobalt can be achieved with zero internal stress. Such low values of stress ensure freedom from distortion during electroforming of precision parts. The maximum hardness obtainable is approximately 350 HV, corresponding to a rate of deposition of 62 μm/h.

In addition to the high as-deposited hardness of the nickel-cobalt alloys, substantial differences both in room temperature hardness and in hot hardness, between nickel and nickel-cobalt alloys are observed after heat treatment over the temperature range 200-600°C. The high initial hardness and hardness-retention properties of the alloys encourages their exploitation for operation at elevated temperatures, such as electroformed moulds and dies for zinc alloy die-casting[8].

Nickel-manganese deposits have been developed by Stephenson[9] as high strength electroforming alloys. Deposits having elongation exceeding 6 per cent and hardness up to 400 HV were obtained from a nickel sulphamate bath containing approximately 30 g/l manganous ions. An advantage of this patented process is that the rate of manganese depletion from the solution is slow, as the deposits only contain approximately ½ per cent manganese. As with nickel-cobalt alloys, the nickel-manganese alloys exhibit improved room temperature hardness and hot-hardness than nickel.

Nickel-iron alloys have been electroformed in order to take advantage of the particular properties of certain compositions. Firstly the electroforming of 'Permalloy'* which has the composition 80 per cent nickel-20 per cent iron finds application as a magnetic screening material for cathode ray tubes, transformers, relays and other sensitive devices. Secondly the electroforming of 36 per cent nickel-64 per cent iron, which has a low coefficient of linear expansion for use in high tolerance wave guides.

Copper

Less is known about the control of deposit properties of copper than of nickel, and the degree to which those properties can be varied is less for copper. Copper is used for the inside surfaces of electroformed wave guides and similar applications where high electrical conductivity is required, for spark erosion electrodes, and as a backing material for nickel electroforms. Copper deposition is mostly carried out from acid sulphate copper, cyanide copper, or pyrophosphate copper electrolytes.

The hardness of copper from the acid solution is normally about 80 HV, tensile strength 115-410 MN/m², elongation up to 50 per cent and deposit stress usually less than 14 MN/m² tensile. Various additions are made to the solutions including glue, molasses, Rochelle salt, or phenol sulphonic acid in

---

* INCO electrolytic un-wrought 'S' Nickel Rounds. INCO is a Trade Mark.

* Trade Mark.

elongation up to 50 per cent and deposit stress about 70 MN/m².

Pyrophosphate solutions are used particularly for spark erosion electrodes, since the sparking properties of the copper deposited appeared to be better than the properties of copper deposited from either the cyanide copper or the acid copper solutions. Reported values of hardness are 90–120 HV, elongation 10–12 per cent, and deposit stress about 30 MN/m². Addition agents are often used to modify the basic properties of the pyrophosphate copper deposits.

Aluminium
Work has been carried out by Brenner in the USA[10] and Heritage[11] in the UK on electroforming in aluminium. However the solutions used are non-aqueous, toxic and flammable and therefore involve complex plants. Thus electroforming of aluminium parts from such solutions is normally beyond the scope of the industrial electroformer and is done for special applications only.

Iron
It is possible to electroform iron which is strong, dense and relatively ductile. However, one disadvantage of iron for electroforms is its tendency towards surface oxidation or rusting. Iron plating solutions are comparatively stable, but since the iron is in the ferrous state difficulties sometimes occur when oxidation produces insoluble ferric compounds which precipitate and cause roughness. Moreover pitting has also been a common problem.

Precious metals
Electroforming in silver and gold is increasing as a result of the great shortage of skilled craftsmen in the Goldsmiths' and Silversmiths' industry. A specialized and highly important use of a composite of silver and gold and either copper or nickel is in the electroforming of high-frequency radar wave guides. Electroformed gold parts have also found use in high-reliability electronics where despite its cost in comparison with other metals, the inertness to corrosive environments and the excellent electrical properties make the gold parts the preferred material. The baths mainly used for gold deposition are the cyanide and acid electrolytes with certain addition agents. Electroforms of silver are desirable where the highest electrical conductivity is required in the electrical and electronics field. The electrolytes used for electroforming silver are mainly variants of the silver cyanide–potassium cyanide type with addition agents to promote dense deposits.

precision in manufacture. Mandrels may be permanent or temporary, depending on the shape of the article to be electroformed. If no re-entrant angles are present then rigid mandrels can be used which can be withdrawn intact from the finished electroform and are generally re-usable. If re-entrants are present, however, it is necessary to use a mandrel which can be collapsed in some way from within the electroform. Such mandrels are generally used only once although frequently the materials from which they are produced can be recovered. Shapes of electroforms produced from permanent mandrels could be cylinders or hemispheres whereas temporary mandrels would be used for articles such as bellows and heat exchangers.

Permanent mandrels
Both metallic and non-metallic materials can be used for the production of this type of mandrel.

(a) Metallic: Mandrels having a natural passive film that facilitates separation of the electroform and are normally made from stainless steel, nickel, or chromium-plated steel. If the mandrel is used repeatedly, the passive film needs reinforcement, and this may be done by preliminary cleaning followed by treatment in a dichromate solution. An example of a complex holloware article produced by electroforming onto a polished stainless steel mandrel is shown in Fig. 5. The component parts, that is, the body, spout etc. were electroformed into separate mandrels and subsequently assembled as shown.

Mandrels can also readily be machined from copper and brass, and have the great advantage that they are cheaper than stainless steel. They are also easily worked and therefore tend to be used where

Figure 5. Electroformed coffee maker.

intricately engraved or textured surfaces are required. These metals may be passivated in a solution of sodium sulphide, or alternatively given a thin electroplated coating of nickel and/or chromium.

(b) *Non-metallic:* Permanent mandrels are also made from acrylic resins, e.g. 'Perspex'*, epoxy resins, and vinyl plastics. In all cases, the surface of the plastic must be made conducting, and this is usually done by coating with a reduced silver film applied preferably from a twin jet spray gun in order to achieve a uniform film with economy of operation. Plastic mandrels tend to be less expensive than metal mandrels but do have inferior mechanical properties and cannot be produced to such close dimensional tolerances nor with quite such a high quality surface finish.

Temporary mandrels

There are basically three techniques by which a mandrel can be removed from a complex shaped electroform with re-entrant angles: these are by chemical dissolution; by fusion; or by collapsing.

(a) *Soluble mandrels:* Electroforms having exact dimensions are frequently made with mandrels of aluminium or zinc. The mandrel is removed after the electroforming operation by dissolution in caustic soda or acid as required. Such solid mandrels are relatively expensive however and their removal by

Figure 6. Left: Mandrel in 'ZAM' prepared by blow moulding. Right: Nickel electroform made on 'ZAM' mandrel by the Ni-Speed process. (Mandrel supplied by the Royal Military College of Science.)

dissolution can be a lengthy operation. The recent development of super-plastic zinc-based alloys may advance this field. One such material is a zinc-aluminium-magnesium alloy (ZAM) developed by the Royal Military College of Science[12]. This may be blow moulded at 250°C into a shaped die and when cooled has a yield strength exceeding that of mild steel. The moulding possesses a high finish and gives good reproduction of the surface details of the die with no expensive machining operations. The thin, light, strong conducting mandrel can easily be dissolved away from the electroform with no apparent attack on the nickel electroform[13] (see Fig. 6). Another super-plastic zinc-based alloy that can be

---

* Trade Mark.

blow moulded is 'SPZ' developed by Imperial Smelting Corporation Ltd.

(b) *Fusible mandrels:* Materials used to produce fusible mandrels must have relatively low melting points so that they can be removed from the electroform without causing damage or distortion. The two most commonly used classes of materials are low melting point alloys and waxes. Fusible alloy mandrels have potentially great advantages for electroforming in that they are conducting but unfortunately most formulations contain small amounts of bismuth which, when the alloy is in the

Figure 7. Electroformed heat exchanger element. (Efficient Heat Exchangers Ltd.)

molten state during removal, causes embrittlement of the nickel or copper electroform by penetration at grain boundaries. A bismuth free, eutectic tin-zinc alloy has recently been developed[15,16] specifically as a mandrel material, which can be die-cast to the required shape, with a high surface finish. The main use for this mandrel material so far has been for the production of spark erosion tools. It has been used for the production of compact heat exchangers[13,14] (see Fig. 7), and could obviously be extended to the production of holloware.

Waxes have been used for many years to produce mandrels for electroforming. However a number of

Figure 8. Left: Mandrel in conducting wax. Right: Copper electroformed spark erosion tool as used in shaping blades for the Rolls-Royce RB 211 engine. (Electro Formers Ltd.)

becoming coated with a thin layer of wax and being rendered non-wettable in the plating solution. However a new procedure for producing conducting waxes has been developed by Electro Formers Limited using processing techniques involving pressure so that no shrinkage occurs and reproduction of the surface of the mould is good[13]. Figure 8 illustrates a mandrel in conducting wax for electroforming a copper spark erosion tool.

APPLICATIONS

These applications may be divided into three broad groups: tools for the subsequent shaping of other materials; actual components for engineering service or consumer goods; and foil and mesh products for a variety of uses.

Tools

Gramophone Record stampers provide the traditional application of an electroformed plastic mould and are the second biggest use for electroforming. The initial, cut, disc is coated with electroformed nickel to give a nickel master, which is passivated and used to form several electroformed 'mothers'. From each of these, several identical stampers are electroformed for use on the record presses (see Fig. 2). The versatility of moulds and dies that can be produced as nickel electroforms is a boon to the plastics industry, where output ranges from toys to sophisticated engineering components which exploit the merits of electroforming to produce both complex shapes and accurate dimensions. The good thermal conductivity of such dies allows rapid cooling, and the parting lines, undulating if required are very precise which minimizes flashing. Tools for applying leather graining or other textures to soft metals, unpolymerized

Figure 9. All-nickel moulds for ice lollies electroformed using the Ni-Speed process.

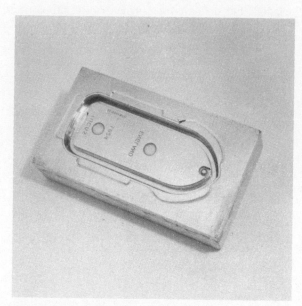

Figure 10. Electroformed nickel–cobalt alloy die insert for zinc die-casting.

Electroformed dies are used for pressing sheet materials, such as steel, brass or plastics and usually take the form of electroformed shells backed up with filled epoxy resin, a cast iron bolster or special concrete. By electroforming, the dies can be made more quickly and, if extensive contour machining and hand finishing are required in preparing the conventional die or mould, the cost is less.

Masks for spraying paint where an exact fit over a contoured surface is necessary are readily made by electroforming, as well as printing plates for high quality printing of paper. Bank notes are printed in many countries in this way, using printing cylinders made by replication from a hand-engraved master plate. For this application the Ni-Speed process is used extensively in conjunction with INCO 'S' Nickel in order to achieve high deposition rates and reduce electroforming time. Spark erosion tools (see Fig. 8) electroformed in copper are used to accurately form hard materials that are difficult to machine by conventional procedures. Also diamond cutting tools can be made using a working surface of electroformed hard nickel containing incorporated diamonds.

A recent development is the use of foundry patterns electroformed in nickel-cobalt alloy (see Fig. 11). In a four-year trial such a set of patterns have been shown to have a life comparable to and probably greater than, that of conventional machined cast-iron patterns. The electroforming method of

Figure 11. An experimental foundry pattern electroformed in nickel–cobalt alloy. (Rotoplas Ltd.)

manufacture offers the possibility of making production patterns from wooden master patterns without any of the costly machining and hand-finishing needed when making conventional tooling. On the basis of the limited experience gained from this trial it is thought that cost savings could amount to 25–30 per cent in some cases[17].

Components

The aerospace industry has realized the merits of nickel electroforms for the production of nose cones and other parts for rockets and satellites, nozzles for wind tunnels, mirrors for satellites and other equipment, bulk heads for fuel tanks in rockets; and, on aircraft, erosion shields for engine cowls, helicopter rotors and other parts. Even when the dimensions are only a few centimetres, electroforming can be more economical and Green[18] showed how electronic screening cans could be electroformed on a stainless steel mandrel costing £120, whereas conventional press tools would have cost £1000. Where components are subject in service to high operating temperatures electroforming offers the opportunity of incorporation of cooling channels of complex form.

The production of electroformed bellows has already been briefly mentioned (see Fig. 1), and these can be produced in a large variety of sizes and forms for various applications such as pressure sensitive devices and flexible couplings. Such bellows tend to be expensive compared with the value of the electroformed metal itself, largely because the mandrels used are individually machined, usually from aluminium. However for a large number of identical designs there is no reason why cheaper mandrels requiring only a minimum of finishing should not be produced from a suitable mould or die.

Patents published during the last few years have shown how electroforming can be used to manufacture highly efficient heat exchangers[19,20]. Such heat exchangers are built up from electroformed elements (see Fig. 7), and for a given cost, give an equivalent heat exchanger of only 10 per cent of the volume, compared with a conventional shell-and-tube-type exchanger made of mild steel.

Wave guides are mostly made by electroforming (see Fig. 12), using a permanent or expendable mandrel depending on the geometry of the section. A finish of 3–5 μ in is readily achieved. Whereas wave guides were at one time electroformed entirely in copper, interest has arisen in thin copper backed with nickel, in order to combine the high conductivity of the initial copper layer with the strength of the nickel, to give a lighter and equally strong structure.

Figure 12. Twin track wave guide bend. Wall thickness 0·25 mm copper, 1·27 mm nickel. (Plessey Electronics Group.)

In the production of surface roughness standards, the benefits to be obtained by electroforming, such as high fidelity, simultaneous reproduction of a variety of surface textures, and the ability to mass produce products, are well illustrated.

The manufacture of metal foams by electroforming onto highly porous urethane foams was described by Cohen, Power and Fabel[21] in 1968. Possible commercial applications include electrochemical anodes and electrodes, filters, heating panels for walls and floors, flame arresters, home vaporizers, and as a material for absorbing sound and impact[22].

The electroforming of holloware[2] has immense potential in terms of economy of manufacture and a previously unobtainable freedom of design. The coffee maker electroformed by General Electric in the USA (see Fig. 5) was made on permanent stainless steel mandrels and sold many thousands before the manufacture was discontinued. Other items produced commercially include fruit bowls, jugs, switch plates, and condiment sets. However full realization of the potential will depend on the development of a suitable fusible or flexible, i.e. a temporary mandrel system.

Foil and mesh products

The section of the electroforming industry devoted to the production of mesh products is large and diverse, and indeed the largest single application of nickel electroforms is in the production of the seamless screen cylinders used in textile printing machines (see Fig. 13). These permit faster printing and give better colour register than is obtainable in flat-bed printing.

Figure 13. Textile printing machine fitted with electroformed screen cylinders. (Gebr. Stork N.V., Holland.)

The original application was for textile printing but has now been extended to printing of tufted carpets and of wallpaper. The screens are of two types: those electroformed with a pattern built into the screen, and, those electroformed as plain mesh on which the pattern is applied by means of photo-sensitive stopping-off lacquers. Plain flat screens are also made in large quantities for a great variety of applications, and the biggest producer in Europe, VECO Zeefplatenfabriek NV (Holland), can supply standard 1 metre squares in over five hundred combinations of thickness, hole size, and hole shape (circular or slotted, conical or venturi section).

Electroformed nickel mesh is used in one large group of alkaline batteries for supporting a sheet of porous sintered nickel powder which is impregnated with nickel salts or with cadmium salts which are converted to the hydroxides to form respectively the

Figure 14. Support mesh for alkaline batteries. (Electro Formers Ltd.)

wide foils. Cigarette machine tapes are electroformed in nickel as continuous bands with no welded join on cylindrical mandrels, and mesh inserts for electric razor heads are electroformed in nickel in quantities greater than a million. For the latter a mandrel technique is used that causes a slight protrusion around the holes to give a cleaner shaving action.

Electroformed nickel filter segments are curved and assembled in centrifuges used continuously in sugar refineries for separating sugar from the mother liquor. The holes in the sieve are conical in section as a result of the growth of the deposit, and the sieves are assembled with the cones pointing inwards, thus ensuring that the sieve does not become blocked in use. Large numbers of filters are also made for handling fruit juices and recent advances in mandrel techniques have enabled seamless conical filters to be electroformed for centrifugal separators.

THE FUTURE

The merits of electroforming have already been established in the versatile applications described and illustrated. During the last few years, in particular, there has been a rapid diversification in the applications, and the future of the industry will undoubtedly be one of continued expansion. Some expansion will come through the use of newer and more convenient mandrel materials such as super-plastic alloys, conducting waxes, bismuth-free fusible alloys, and improved plastics. Further expansion will stem from the exploitation of novel, rapid, backing techniques for those electroforms such as mould inserts which have to be supported in service. Expansion will also arise through improvements in the properties of the electroforms themselves, by the development of alloys of nickel with metals such as cobalt and manganese, and by the use of dispersion-hardened materials made up from nickel containing ceramic, cermet, or nickel powder particles[23,24].

The most immediate expansion, however, will be realized through broadening of the market for applications of those techniques in which the industry is already skilled. The market increase will be stimulated by publicity efforts from the electroforming companies and others who have an interest in the industry.

ACKNOWLEDGMENT

The author wishes to thank International Nickel Limited for permission to publish this paper, and also the companies who provided the electroformed articles which are illustrated in the text.

REFERENCES

1. T. A. EDISON (1906). U.S. Patent 821,621. (1907). U.S. Patent 865,688.
2. A. C. HART (1973). Electroform '73, Connaught Rooms, London, 13–14 February.
3. R. J. KENDRICK (1964). *Trans. Inst. Metal Finishing,* **42,** 235–45.
4. W. H. SAFRANEK (1966). *Plating,* **55,** 1211.
5. G. L. J. BAILEY, S. A. WATSON and L. WINKLER (1969). *Electroplating and Metal Finishing,* **22,** 21–34.
6. K. C. BELT, J. A. CROSSLEY and R. J. KENDRICK (1968). *Proc. 7th Int. Met. Fin. Conf.,* May, 222–9.
7. K. C. BELT, J. A. CROSSLEY and S. A. WATSON (1970). *Trans. Inst. Metal Finishing,* **48,** 132–8.
8. W. R. WEARMOUTH and K. C. BELT (1973). Electroform '73, Connaught Rooms, London, 13–14 February.
9. W. B. STEPHENSON (1966). *Plating,* **53,** 183–92.
10. D. E. COUGH and A. BRENNER (1952). *Jnl. Electrochem. Soc.,* **99,** 234.
11. R. J. HERITAGE (1955). *Trans. Inst. Metal Finishing,* **32,** 61.
12. (1971). *The Times,* 29 January.
13. S. A. WATSON (1972). *Jahrbuch der Oberflachentechnik,* **28,** 187–97.
14. H. G. E. WILSON (1973). *Electroplating and Metal Finishing,* **26,** 29.
15. (1970). German Patent Application 1,944,092, April.
16. J. J. MARKLEW (1972). *Machinery and Production Engineering,* 10 May, 653.
17. (1972). *Inco Nickel,* No. 36, 7.
18. A. F. GREEN (1968). *Plating,* **55,** 594–9.
19. (1965). U.K. Patent 1,009,178, Published Nov.
20. (1968). U.K. Patent 1,115,988, Published June.
21. L. A. COHEN, W. H. POWER and G. A. FABEL (1968). *Materials Engineering,* April, 44–6.
22. L. C. LUNDSTROM (1965). *S.A.E. Journal,* November, 22–6.
23. S. J. HARRIS and P. J. BODEN (1973). Electroform '73, Connaught Rooms, London, 13–14 February.
24. E. C. KEDWARD (1972). Electroplating and Metal Finishing, **25,** 20.

DISCUSSION

Q. P. Lawrence. The paper does not give much indication of the time involved in electroforming various components. I have the impression that it is a slow process, likely to be restricted to one off or small quantity production. Does the author see any hope for large volume production by this technique?

A. The speed of electroforming is largely governed by the geometry of the part to be made, and under appropriate conditions where current distribution is uniform, nickel can be deposited at a rate of 1 mm per hour. However for the electroforming of moulds and dies where the current distribution is non-uniform slower rates of deposition have to be employed. Thus it may be necessary to electroform for a period of days to deposit the required thickness, but this requires non-skilled labour and does not depend on the availability of skilled tool makers and a limited number of cutting machines. The only restriction is the amount of spare tank capacity. Of course it is possible to electroform more than one object at a time in a particular tank, and where these objects are of a similar geometry, e.g. spark erosion tools, they can indeed be made on a large volume production basis, the restriction on the actual number being the size of electroforming tank and the availability of power.

Q. C. F. Noble, UMIST. To my knowledge this is the first time that the subject of electroforming has appeared in the MTDR Conference and I believe this indicates a general lack of appreciation of its potential in industry at large. More than a decade ago Metachemical Processes Ltd., Crawley were producing very high strength nickel electroforms for anti-rain-erosion applications in aircraft and as can be seen from the present International Nickel paper possible applications now include very intricate components and tools. Designers and production engineers should give electroforming the consideration it deserves during their process selection studies.

A. I was pleased to accept the invitation to present a paper on electroforming at the 14th MTDR Conference and agree fully with Mr Noble that there is a lack of appreciation of its potential in industry at large, particularly in the tool making industry. However during the last few years the number of new applications has increased and there is no doubt that the future of electroforming will be one of continued expansion. The reluctance shown by engineers and designers to consider electroforming probably results from their lack of knowledge of the process, but it is to be hoped that publicity efforts from the electroforming companies and others who have an interest in the industry will bring to these people the realization that electroforming is indeed a competitive process, to be given important consideration to during their process selection studies.

THE PERFORMANCE OF LASERS AS MACHINE TOOL SYSTEMS

by

A. F. D. S. TAYLOR*

SUMMARY

Although lasers are now becoming accepted in the manufacturing industries, the machine tool industry does not yet seem to be fully aware of their potential. A brief introduction to lasers and the laser machining process is given, together with details of laser types (including output power and mode of operation) that are likely to be of interest to the machine tool industry. The machining capabilities of various types is given and typical cutting speeds for commercially available industrial CO_2 lasers is included, together with the reasons for considering the laser system as a useful extension of the range of advanced machine tools. Laser systems can be numerically controlled or part of a computerized complex. A brief discussion on some adverse aspects of lasers today, and the likely future remedies is included, together with a short discussion on laser safety requirements.

INTRODUCTION

The laser is an optical amplifier of light energy and produces coherent radiation at a fixed wavelength. This radiation can be produced in short duration high intensity energy pulses (with relatively low average power), or continuously, and can be focused with a mirror or lens to a very small diameter spot having a very high energy density. In principle the spot size cannot be smaller than the particular wavelength of the laser and in practice is usually two or three times this size. The power of a laser is often qualified as being multi mode or single mode (Sometimes referred to as uniphase or fundamental mode) which governs the minimum size of spot to which the laser beam can be focused. The single mode laser, having a beam conforming to a Gaussian energy distribution, permits the smallest possible spot size and hence is likely to be the most useful laser for most machining purposes where the need is to remove material as efficiently as possible.

Although from time to time lasers giving very high output powers are reported (>60 kW CW (continuous wave) and mainly from the USA, it must be stressed that these lasers are not generally available as industrial lasers. Although in the very near future lasers producing output powers of 10 kW CW will probably become commercially available, 2 kW CW is the most powerful laser that is on offer in the UK at the present time. Only lasers available in the UK are discussed.

THE LASER AS A TOOL

The machining process is one of rapid heating where the material is melted, vaporized, or ablates. In the case of non-metals the material absorbs almost all the laser energy and machining can be achieved with modest laser power. However most metals reflect some of the laser beam and in addition thermal conduction removes heat from the vicinity of the cut; there is a threshold level of a few hundred watts of laser power below which the machining of metals is impracticable. In some metal cutting applications a reactive gas is introduced to the cutting region to create an exothermic reaction whereby heat is generated in addition to that supplied by the laser.

Some of the potential advantages of a laser as a machining tool are

(1) the possibility of high tool velocities with rapid stopping and starting (because a laser beam has zero inertia)
(2) there is no wear on the cutting tool
(3) the workpiece need not be rigidly held
(4) difficult materials can be cut
(5) small wastage because of the narrow kerf width
(6) intricate shapes can be cut at high speed automatically
(7) accurate and repetitive cuts are possible
(8) narrow heat-affected zone adjacent to the cut
(9) deep, very small holes are possible

---

* Laser Applications Group, UKAEA Culham Laboratory, Abingdon, Berks.

The two types of lasers most likely to be of practical use in the Machine Tool industry are the solid state lasers and the gas lasers. Solid state lasers based on ruby, neodymium doped glass (Nd glass), and yttrium aluminium garnet (YAG) are physically small and operate in pulsed mode at low mean powers and are generally used for working components of thin section or small size. A ruby laser emits radiation at the wavelength of 0.6943 μm and the other two at 1.06 μm. Broadly speaking glass lasers produce high energies per pulse ($\sim$60 J) at pulse repetition frequencies (PRF) of up to five per second and are useful for making spot welds and drilling small holes. Because of its better characteristic the Nd glass is tending to displace the ruby laser. The YAG laser operates at a lower peak power but a higher PRF and in some cases continuously.

Gas lasers fall into two categories. The He–Ne laser which is very low power (0.5–5 mW) continuous wave and emits radiation at 0.6328 μm, is used extensively in the field of measurement and alignment. The CO_2 laser on the other hand operates at continuous powers up to 2 kW at a wavelength of 10.6 μm, is used for cutting a vast range of materials and for welding a somewhat smaller range. CO_2 lasers can also operate in the pulsed mode where they are now being used in the field hitherto dominated by the truly pulsed lasers.

Historically, pulsed lasers were the first to be applied commercially and will be found in highly automated industries. Typical applications are: drilling of diamonds to be used as wire drawing dies, piercing of synthetic ruby bearings to be used in watches, dynamic balancing where imbalance in high speed rotating parts is corrected by removing material with a laser pulse automatically fired by vibration detecting equipment (possibly an He–Ne laser) while the machinery is running, trimming both thin film and thick film resistors used in the micro-electronic circuits industry where the resistance film may be only a few Angstrom units thick (1 Å $\equiv 10^{-8}$ cm), scribing ceramic substrates (e.g. alumina) on which certain types of micro-circuits are laid so that individual circuits can be separated from a larger block of units.

The He–Ne laser (the first of the gas lasers to be used commercially) is now used extensively for alignment purposes in the engineering field and in surveying, where, for example, it may be used for measuring the roundness and diameter of hot steel rods on line in steel mills, for detecting surface blemishes in white paper (which could have an adverse psychological effect for example) in the food industry, as fire

Material	Thickness (mm)	Cutting speed (cm/min)
quartz	2	100
asbestos board (dense)	3	180
resin bonded fibreglass	3	300
rubber sheet (dense)	3	500
rubber sheet (sorbo)	3	1000
plywood	18	30
paper	Newsprint	>60 000
art board	1	6000
formica	1.5	550
acrylic sheet	1.5	1500
ABS plastic	2.5	850
melinex film (mylar)	0.025	>30 000
P.T.F.E.	6	100
titanium	1	750
stainless steel	1	450
hardened tool steel	3	170
mild steel	1.5	450
textiles	15/16 oz suiting	5000
leather	5	250

Depending on the laser, optical components, gas jet, etc., cutting speeds can vary by a factor of two, and should not be taken as the criteria when comparing a laser with a conventional machine tool.

range of materials that can be cut with a CO_2 laser together with cutting speeds that can be achieved. It is unwise to try and assess one laser with respect to another on published cutting speeds alone because usually not all the relevant facts are published at the same time. For example, it is necessary to know the beam characteristics (i.e mode order, diameter, power, divergence), the type of optics used to focus the beam, cutting environment (i.e. if gas assistance was used and if so which gas and at what pressure) quality of cut achieved, width of cut (kerf) and heat affected zone.

Probably only the He–Ne lasers are available 'off the shelf', with the machining lasers being made to order. This does not imply that every laser is specially designed, but because of the specialized nature of a particular application it may be desirable to change some features of the laser, without altering the basic design. Table 2 indicates the range of lasers currently available in the UK together with the name of the manufacturer, some of whom have parent companies outside the UK. The international market for lasers, obviously, is much larger than the home market.

LASER SYSTEMS

Conventional machine tools today can be very sophisticated and though they can be very precisely controlled they rely for their effectiveness on well

TABLE 2 'Machine tool' lasers that are available in the UK

Type		Power	Mode	Manufacturer	Parent Co.
CO_2		450 W CW	Single	Ferranti Ltd, Dundee	UK
CO_2		500 W CW	Multi	Marconi-Elliott Avionic Systems Ltd Boreham Wood, Herts.	UK
CO_2	(a)	50 W CW	Single	Coherent Radiation Labs	USA
	(b)	100 W pk pulse	1–100 PPS	Royston, Herts.	
CO_2	(a)	250 W CW	Single	Coherent Radiation Labs Royston, Herts.	
	(b)	500 W pk pulse (1–10 msec)	1–100 PPS	Coherent Radiation Labs Royston, Herts.	
	(c)	25 KW Q-switch (50 msec)	50–400 PPS	Coherent Radiation Labs Royston, Herts.	
CO_2	(a)	500 W CW	Single	Coherent Radiation Labs Royston, Herts.	
	(b)	100 W pk pulse (≤1 msec)	1–1000 PPS		
Ruby Nd glass YAG		Various powers and PRF		J. K. Lasers Ltd, Rugby, Staffs	UK
Ruby N_2		Various powers and PRF		I.R.D. System Computers Ltd Newcastle upon Tyne	UK
Ruby		Various powers and PRF		Barr and Stroud, Glasgow	
Ruby Nd glass		Various powers and PRF		KORAD Roditi International Corp Ltd, London	USA
YAG		Pulsed 12 W CW	Multi		
CO_2		300 W CW also long pulse up to 500 W	Multi		

This is not a comprehensive list and laser assessment labs are aware of other less publicized sources both in the UK and on the continent of Europe. He–Ne lasers are not included because the list would be too long, but most of the sources above also supply this type of laser.

established machining techniques albeit with machining elements manufactured from superior materials. Even so there are areas of industry where materials are required to be machined that cause rapid wear on the tools or cause excessive down time because tools become clogged. The more sophisticated this machinery becomes the more desirable it is to reduce the down time to a minimum.

In the context of laser systems the term system implies the sum total of a laser, optical interface and work handling equipment. The high power laser is usually supplied by the manufacturer with all its control equipment and to a specification, but the interface is the responsibility of the user. Some laser manufacturers will supply the interface but not many are willing to consider a complete system. The optical interface will normally consist of certain basic components such as beam dump, power monitor, focusing elements and gas jets where needed. With fittings and interlocks this could increase the basic laser cost by 20 per cent. From here on the system can be as simple or as sophisticated as the manufacturing process demands. In the simplest form one might imagine the laser and optics projecting a beam on to a constantly moving strip of material to be trimmed, or a more advanced work handling device might be a numerically controlled coordinate table programmed to be automatically producing pre-determined profiles. At the extreme, one could envisage a total system, computer controlled, where material loading, unloading and collating is accomplished, where the material is scanned for defects or perhaps pattern recognition before being transferred to the cutting

area where, after being cut it is either accepted or rejected. Whether the system is simple or complex, total integration is essential for success and maximum economic advantage. The successful application of commercially available lasers now depends not so much on the reliability of the laser but on the design of the complete system. Because a laser beam does not exert force on the material to be worked, the work-handling equipment can be significantly simplified in design and construction. For example, it is necessary only to position and support sheets of metal, plastics, wood, textiles, rubber, quartz, etc; or their composites, for trimming or for profile cutting. Intricate shapes can be cut with a minimum of waste and certain foods and confectionery can be cut without risk of contamination from knives or saws and certain mouldings, castings and pressing can have flashings or rough edges removed cleanly. Tubes can be cut, the edges of which may be profiled in a manner that eliminates the need for further operations.

Normally the systems engineer will have three options open to him when considering the layout of a production line based on a laser: (1) the laser can be stationary and project a fixed beam on to a moving workpiece, (2) the workpiece can remain stationary while the laser head traverses the workpiece, (3) both workpiece and laser can remain stationary while the laser beam is moved over the workpiece via a low-inertia scanning mirror. All three options will allow either programming or numerical control. However it would be prudent for the designer to consider the laser as an extension to the range of machine tools

microscope and closed circuit television monitor for close viewing of the workpiece that can be manipulated along three axes by micro-adjustors, either manually or automatically.

Systems for CO_2 lasers, because of the larger size of the material to be processed, and because the hazard to the eyes can be designed out usually allow better direct viewing, bearing in mind other safety aspects. Figure 1 shows, schematically, one of the possible processing systems in which it would be possible to use a cutting head shown diagrammatically in Fig. 2. A more sophisticated system which will allow much faster linear speeds employing a system of mirrors to exploit the inertia-less pro-

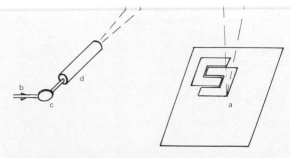

Figure 3. Schematic diagram of a laser beam steering system. a, workpiece; b, laser beam; c, metal mirror; d, zoom lens; e, scanning mirror.

perty of the laser beam (shown schematically in Fig. 3), can be linked to a completely integrated and computerized manufacturing process. Such a system has the disadvantage that gas jet cutting is not easily achieved and in addition the optical components are required to be made to a very high standard in order to achieve the aberration-free small focused spot size that gives the laser its unique machining capability.

Laser systems engineers are still a rarity in this country and in the main are to be found within the laser manufacturers' laboratories and the laser assessment laboratories. This is not necessarily a good thing since a system will then tend to be built around the laser rather than the laser being built into an integrated system. It does seem a little surprising that the machine tool industry itself with its expertise in the manufacturing industry should be slower than individual manufacturers to accept this new tool. No doubt it may be considered in some quarters that the laser is in some respects still a laboratory toy, but practice has shown that the laser can be a very rugged device and that the inadequacy has been the system itself.

Possibly the first systems to be used industrially in which a CO_2 continuous wave laser was incorporated were those in both the UK and the USA paper converting industries. In the UK, a Ferranti laser was used by BOC Ltd in their 'Falcon' system for cutting slots in 'Form boards'[1] into which steel cutting and creasing blades were set, to create dies. The laser and cutting head was fixed and the material to be cut was manipulated by a coordinate table controlled by an electro-optic scanning device, locked on to a line drawing facsimile of the profile to be cut. Although the system is well engineered, it does rely on very accurate facsimiles. In contrast to the 'Falcon' system is one designed by the Atlas Steel Rule Die Company of America. The system was built around a Coherent Radiation Labs laser, and was designed to produce a similar end product to the British system. A more futuristic attitude was, however, adopted and the

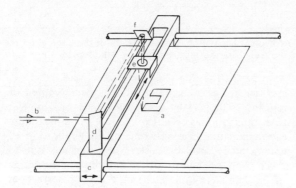

Figure 1. Schematic diagram of a coordinated laser cutting head. a, workpiece; b, laser beam; c, saddle for X movement, d, mirror (moving in X direction); e, saddle for Y movement (carried on X saddle and holding focusing lens); f, mirror (moving in Y direction).

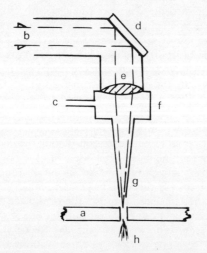

Figure 2. Diagram of a laser cutting head. a workpiece; b, laser beam; c, inlet for cutting gas to pressure chamber f and nozzle g; d, metal mirror; e, focusing lens; h, escaping gas and debris.

stationary material was traversed by a coordinated mobile cutting head and optically aligned with a stationary laser. The entire system was numerically controlled by punched tape and offered a flexible system together with fast setting up times.

This then, typifies the differing philosophies. On the one hand is a machine tool and on the other hand is a system, and such a system can be extended or adapted to take account of changing demands.

It should be clearly understood that lasers are in a somewhat different category to conventional machine tools and as such should be considered as one part (sometimes the essential part) of a complete system and for this reason it would be prudent to purchase in the UK when considering laser equipment unless the objective has been clearly defined and the laser and system very tightly specified. Indeed for the designer with no laser expertise, 'do it yourself' can be costly. For those designers who might consider the laser as a possible solution to a machine tool problem it is worth noting that in the UK there is, in addition to certain interested Research Associations, at least one independent laser laboratory[2] where advice can be sought and problems discussed in confidence on the topic of laser applications. Feasibility studies can be undertaken which, when concluded, would result in unbiased advice on all the aspects of a total system.

Obviously, the laser is a specialized piece of equipment and is not fully understood by everyone outside the laser industry but this should not be a reason for failing to exploit its potential. After all, one does not need to understand the intricacies of modern electronic circuits in order to use a desk calculator.

ADVERSE ASPECTS AND REMEDIES

In general a laser cannot be regarded as a multi-purpose machine tool to be installed in a machine shop for processing a varying range of materials by all grades of operatives because, safety considerations apart, considerable experience is necessary to obtain the optimum conditions for machining different materials. Undoubtedly this constraint will diminish in time but for the foreseeable future a laser will be part of a system. The expertise that is required to optimize a laser system is at present only to be found in the laboratories of a few laser manufacturers and in the larger assessment laboratories, but these skills are not inborn and will eventually extend to the machine tool industry as lasers are used more in industry. Manufacturing industries in general would seem to be more aware of this than the machine tool industry itself, and as an example 'Plascut'[3] have installed a Ferranti 400 W CO_2 laser to cut sheet material on a jobbing basis and to gain the necessary expertise.

It is frequently claimed that lasers are not powerful enough to undertake the machining tasks that are problematical for conventional machine tools; this is mainly true only in the metals field where sheets greater than, say, 6 mm thick are required to be cut or where welding is required to the standards attained by electron beam welding. However, we are likely to see lasers in the near future that will satisfy these criteria; a CO_2 laser is at an advanced stage of development at Culham Laboratory that will produce

10 kW CW of output power and studies are in hand to extend this breed of laser to produce very much higher powers. These lasers cannot be portable however but they are extremely robust and compact.

Commercially available CO_2 lasers have tended to be somewhat large and at the same time fragile by machine shop standards but this situation is rapidly changing and it is now possible to buy a 400-500 W laser weighing about 100 lb and measuring about $150 \times 50 \times 50$ cm that is robust enough to be mounted directly on to a motorized coordinate table. Because of their long length lasers have been subject to vibration which causes stability drift due to the critical alignment of the optical cavity mirrors and here again matters have improved because of the reduction in size and the unique and novel construction techniques now employed in the manufacture of high power lasers. Indeed, it is possible to lift a laser from its stand when it is producing power, without affecting the output.

Because a laser is an optical device some components need to be handled with great care. There are encouraging signs that, with the advances in laser technology, new materials and components are appearing that will withstand the full rigours of industrial life.

One further aspect of lasers that is constantly aired is laser lifetime. This is difficult to predict with accuracy because of the limited number of CO_2 lasers in the field. However, enough is known about individual components to consider amortization over five years, but during this period it is likely that some inexpensive components will need to be replaced. This need not be an inconvenience since the replacements can be made during routine service arrangements.

SAFETY ASPECTS

Laser equipment can present three hazards to personnel, all of which can be guarded against. Firstly a laser is a high voltage device and as such must be installed in such a way that personnel cannot tamper with it. The dangerous parts of the equipment will be adequately protected by the laser manufacturer and mandatory requirements under the Factories Act will have been complied with.

Secondly, lasers used for machining emit coherent radiation at wavelengths that cause burning and their beams cannot always be seen, therefore adequate protection must be arranged to prevent exposure of parts of the body to the beam. Simple fixed screens are normally adequate and, depending on the type of laser, these can be transparent. In the immediate vicinity of the cutting or processing area however (where the laser beam is focused) more stringent guarding must be arranged that is preferably interlocked with the laser beam switch. A focused continuous wave laser can sever a finger just as effectively as a rotating saw, although the wound will be self-cauterized and so will not bleed!

The third hazard is directed against the eyes. Laser radiation should not, and need not be allowed to enter the eye. There are certain threshold levels of intensity above which permanent damage can occur.

straight along the beam.

It is inadvisable to look directly into any laser beam with the unprotected eye even though the power output may be less than 10^{-3} watt and all laser systems can be designed in a way that will strictly minimize the hazards mentioned above.

Lasers are not dangerous to use if a sensible code of practice is followed and indeed can be significantly safer than some machine tools. British Specification BS 4803/1972 sets out a code of practice and recommendations which could form a basis for safety procedures that could ensure that a laser system was the safest machine tool available to industry.

CONCLUSIONS

Experience suggests that lasers are now becoming reliable in operation in an industrial environment, but the number of systems in use is still comparatively handling equipment that can result from laser machining, but a much greater liaison between laser systems laboratories and machine tool designers is necessary if the manufacturing industries are to be supported effectively. A great opportunity now exists for the machine tool designer to exercise his intellect and past experience of specialized work-handling techniques in this modern expanding area of machine tool technology.

REFERENCES

1. Die Making and Die Cutting: First Annual Symposium, March 1971, York (Sponsor: W. Notting Ltd. Bowling Green Lane, London EC1).
2. UKAEA Culham Laboratory, Abingdon, Berks.
3. Optics of Laser Technology, February 1972.

WIRE SPARK EROSION TECHNIQUES FOR THE TOOL ROOM AND MODEL SHOP

by

D. G. HUGHES and J. L. SHELDON*

SUMMARY

The 'Agiecut' NC Wire Spark Erosion machine and the 'Agiepac' programming system are described in their application to the manufacture of fine blanking and conventional press tools and the production of prototype and one-off components. Organizational and operational changes to take maximum advantage of the new system are considered.

INTRODUCTION

The competitive pressure for progress means that, in engineering, we must always strive to make things quicker, cheaper and better than before. This is an account of the results of one such attempt. A new process and new methods developed to go with it have brought results which will affect press-toolmaking and, eventually, many other engineering activities.

The development work took place in a factory which was concerned, primarily, with the manufacture, in large, small or very small batches of flat or bent metal components made from strip. Sizes ranged from 7 to 300 mm long, and materials from aluminium and beryllium copper to stainless steel, in thicknesses from 0·1 to 3 mm. Quantities needed could be anything from two to two million, with a significant part of the output falling within the 500 to 10 000 a year range. Tight limits, intricate profiles and many pierced holes abounded.

A factory engaged in such work is, in effect, a jobbing shop *par excellence*, and problems of production and effective control are particularly intractable.

In the factory, fine blanking presses (25-250T capacity) were in use. This process minimizes further operations, gives clean fully sheared edges and maintains close control over hole and profile tolerances. But these triple action presses need expensive, super-accurate tools. For instance, punch/die clearances are, typically, ·008 mm (·0003 in). Variations from this of as little as ·005 mm (·0002 in) either way cause trouble. Few toolmakers can be found who can handle such work with predictable and unfailing success, and toolmaking errors and delays can be very costly, even in a shop such as this one, with many years of fine-blanking experience.

THE ENGINEERING BACKGROUND

For a number of years the writers of this paper have been considering, in a general Production Engineering context, a problem which lies at the heart of such high cost and manufacturing delay. It is a problem in applied craftsmanship. How to make one-offs quickly and with an accurately known shape and tolerance at all points.

However much we may tool-up subsequently, there usually has to be a prototype. And prototypes are one-offs. One or many sophisticated and accurate machine tools may be employed to make the prototype, but it generally involves also a large element of human craftsmanship, judgement and skill. Humans are fallible, the human hand and eye were not really designed to work in units of ·002 mm. If a shape is really complicated, it is difficult for an inspector to say what a man *has* made, let alone for the man to be sure at the time. The real cost of making one of something in a model shop or toolroom is an eye-opener to the uninitiated.

But somehow prototypes are made, modified, checked, tested and approved in a working context. Now we have to make a tool which will output a quantity product exactly like the prototype. A tool is a one-off, only this time the craftsman is working with thicker, harder, more intractable material, which, when the toolmaker has done his best, will probably have to be heat-treated, causing further problems of shrinkage and distortion. The probability is very low that the product from a tool will be exactly like a model shop prototype, or that either of them will be close, in all details, to the draughtsman's specification.

Satisfactory tools for a fine-blanking press can take several months to produce the tool components,

* International Computers Limited

cutting two-dimensional shapes out of metal. Conventional' spark erosion is a technique now well known and much used for toolmaking and other purposes. In wire spark erosion, the usual electrode is replaced by a thin wire, which, as it moves forward cuts through any metal in its path, just as a cheese wire or a band saw does. Wire spark erosion machines have been available for some time, the Russians being the pioneers. The novelty of the machine made by AGIE Ltd. is that the movement of the cutting wire in relation to the work is regulated by a simple but flexible numerical control system.

Any metal, hard or soft, can be cut, to any two-dimensional shape which can be programmed, and the programmable steps in each axis are ·002 mm (·00008 in). The copper wire electrode, being continuously renewed does not wear; unlike a normal electrode, a grinding wheel or milling cutter. One of these machines was purchased, and, in order to evaluate its performance was assigned to the most difficult task available, the cutting-out of dies, punches and other components for fine-blanking tools.

A problem with all NC machines is that they have to be provided with a control tape to tell the cutter where to go. This, in the past has usually involved programmers and other people with special skills, and long delays in making and proving the tapes. And still we find errors built into the tapes. This is one of the major causes of the slow general adoption of NC machines. The Agiecut machine is simpler to program than most because it works only in two dimensions and because it uses only one dimensionally stable tool—the copper wire. Very simple shapes can be manually programmed by the operator, but awkward shapes will take him days and difficult shapes defeat him utterly. To achieve its full potential the machine tool needs a simple, cheap and foolproof method of programming. This is now available.

Programming Innovation

The machine-tool—the 'Agiecut'—can be located anywhere on the tool room floor, although as with jig-borers and other high-precision equipment it is advisable to screen it from draughts and temperature changes. Beside, or near to, the 'Agiecut' machine is located a teletypewriter terminal (which looks like a typewriter on a stand). Any intelligent and versatile toolmaker can be taught to operate an Agiecut machine in a few days. And the same man can also be taught in a few days to use the teletypewriter and the Agiepac programming system. He is then able to produce his own fully edited control tapes, simple ones in minutes and really complicated ones in an hour or two. All he works from is the component drawing.

system was begun as soon as the Agiepac machine was ordered. Control tapes produced by Agiepac have driven the original machine for several thousand operating hours. The system is now available commercially, it is producing tapes successfully at an increasing number of installations throughout the country, and worldwide.

Important features of the system are:

(1) No special programmers are used. Agiecut control tapes, complete edited and ready to run, can be produced—after minimal instruction—by anyone who can read the relevant engineering drawing. Typically the programs are made by the ultimate user, the Agiecut machine operator.
(2) Built-in checks and safeguards make program errors extremely unlikely.
(3) The time to make, and have available, a program tape is minutes only—typically twenty to one hundred—depending on the complexity of the job.

Operating Characteristics of the Machine Tool

The instructions which reach the Agiecut machine from the control cabinet can cause the table (and workpiece) to do one of three things in relation to the electrode:

(1) To move in a straight line in any direction for a specified distance.
(2) To describe an arc with a specified radius and centre for a given number of whole degrees.
(3) To describe an ellipse or a portion of an ellipse.

The third facility is not essential, since any two dimensional shape can be described to any desired accuracy by a suitable combination of straight lines and arcs. The unitary programming and movement steps of the Agiecut being ·002 mm (·00008 in) this is accurate enough for the most precise work.

The movement instructions issued from the control cabinet are, in one important respect, not the same as the statements on the control tapes. The tape statements describe the nominal size and shape of the article to be cut. But the wire cuts a path ·3 mm (·012 in) wide approximately, and therefore any male shape cut would, if no action were taken, be ·15 mm small all round. A female shape would be correspondingly enlarged. By means of a dial-in-control on the control cabinet, the cutter path can be offset by any desired amount (up to 1 mm). In principle, using this facility, male and female shapes can be made from the same tape such that they match exactly or have any desired clearance. In practice, for technical reasons, there has to be a short 'lead-in', and this must of course be from opposite sides of the path according to whether a male or female form is required. Operating experience shows that the precise clearances needed for fine-blanking tooling are fully within the capabilities of the machine.

Using the Agiepac Programming System

The teletypewriter is, in fact, a send and receive keyboard terminal connected to a remote computer. It can be located virtually anywhere. Placed in the toolroom next to the Agiecut machine it becomes just another machine tool or facility available to the toolmakers. (The computer does not have to be in the factory or anywhere near it. Baric, on their Manchester computer, provide Agiepac connection facilities for the whole of Britain. You hire computer time as and when you need it). The operator does not need to know anything about computers, only what facilities are available to him and how to make use of them. He also has to be able to type with at least one finger, but most of us can do that.

Few engineers, as yet, know how to use an interactive computer program. A brief description will help to put it in its proper place and show how its use leads logically to further innovations.

The first operation stop is to get connected to the computer and call up the Agiepac program—technically called 'logging-in'. After that the program itself takes charge. The program—and now that the program is active, the computer—contains within itself all the information needed to produce a valid Agiecut program-tape for any job which is properly described to it (and which is within the capacity of the Agiecut machine). 'Agiepac' stands for Agiecut *Interactive* Programming and Calculation routine, and the *interaction* between the computer and the operator now begins. The computer causes a question to be typed out by the teletype, and the operator is required to answer this question. The answer may be 'Yes', 'No', or a simple command such as 'Start'; or it may be a dimension or an angle. The computer considers whether the answer is appropriate, and permissible. If accepted, the data is utilized to further the programming activity and the next question is put.

Before the shape of the job is discussed, there are certain preliminary questions to be answered, such as whether the input data is in inches and what minimum radius is allowable. The program works in inches or metric, as commanded, and the operator is warned if any input statements or data derived therefrom will result in radii which are too small. When the preliminaries are complete, the program asks for details of the profile geometry. The operator must select a suitable datum point. Usually, on any drawing, there is one point of reference from which a number of the major dimensions are derived. If there is, then this would be the datum point. A start point must also be selected on each profile, this point being the place where cutting the profile will begin on the Agiecut machine. The operator selects the order in which the profiles will be dealt with (leaving, naturally, the external profile until last) and also selects the start points.

It is unnecessary to go into further detail except to make the following points clear. Component geometry is by reference to X and Y axes passing through the datum point, with angles described in the conventional fashion with zero degrees along the X axis. Any valid geometric description of a line or arc can be used and will be understood by the computer; for instance, a line can be described as joining two points, passing through a given point at a stated angle, etc. Start and ends of lines or arcs do not have to be fixed by the operator, the computer will see that each element fits appropriately between the one before and the one after it. Suitable blend radii are put in automatically by the computer, which also makes appropriate adjustments for the fact that the Agiecut machine can only operate rotationally on whole degrees. When the computer has accepted what is intended to be the last statement for a profile, the operator types 'Finish' and the computer then indicates whether the end of the profile will link up with the start point. (If it does this is a pretty sure indication that there is no error). The total of the deviation between the intended input shape and the shape which the Agiecut program will specify is then stated (the error, due mainly to the automatic interpolations which take care of fractional degree angles, is rarely more than ·003 mm) and the profile length. When all profiles have been dealt with the computer indicates the total profile length and also—very useful for fine-blanking press-tooling—the centre of shear for the sum of the profiles.

There are, of course, a number of facilities not covered in this brief description. Special sections of the program can be called up to deal with the problems of gear teeth and of cam faces; and any trigonometrical or other calculations can be done by the computer by calling up the appropriate calculating sub-program at any time. Both sides of the computer/operator conversation are recorded on the teletype printer and also all data to be punched into the Agiecut tape. The computer having been told whether a punch or a die-form (i.e. a male or female) is required can now work out the appropriate lead-ins that will be required. Therefore, on demand, it now produces—from the tape punch on the side of the teletypewriter—a fully edited tape complete with lead-ins, control stops and moves from one profile to another. Coordinates for the centres of start holes are also outputted and these are related not to the original datum but to the centre of shear (which of course will be the centre of a fine-blanking die and is therefore now the most generally useful datum point).

The time from calling up the computer to output of the tape runs from about 20 minutes for a simple shape to 100 minutes or so for a really complicated one with, say, a number of gear teeth and several internal forms.

CAPABILITY OF THE AGIECUT/AGIEPAC SYSTEM

Operating experience with the Agiecut and programming experience with Agiepac soon showed that

out in this manner one or several true prototypes and hand these to the customer for him to measure or test under working conditions. Sometimes experience showed that modifications were desirable—not because the component was different from specification but because people do not always specify what they really want. New prototypes of amended shape were easy to program and cut.

Furthermore, if a prototype was considered satisfactory, it was easy to cut ten, twenty or a hundred as a preproduction batch. Not only could we be sure that these would be identical with the first prototype, but—because the same tapes would be used—we were able to do something else that had never been possible before, guarantee that when a tool was made the product from it would be identical to the prototype (within ·01 mm everywhere if due care was taken). We have, in effect, provided samples from the tool before the tool is made.

Note also that, for all practical purposes, drawing tolerances no longer matter. The programming system programs on nominal dimensions and the machine tool cuts on nominal. The actual tolerances (variability) are the same every time—in effect they are the accuracy of the cutting process. 'Spot-on' and quickly available prototypes are a great boon to the Development Engineer, and it is good to be able to answer a prospective customer's enquiry with 'this is what it will cost you, this is our delivery time, and here is a sample of what you will get'.

MAKING BEST USE OF THE SYSTEM

Facilities and Procedures

When we had got rid of the small problems and mysteries—the 'bugs'—which are inevitable in anything new, and when we had fully assured ourselves that the machine tool, carefully used, really would work to the tolerances stated, and the program system really would deliver the right tapes, then we were faced with a further problem—how best to make use of our new capability. Accuracy being taken for granted, speed is the great advantage conferred—quick reaction time and short lead time. So the next step was to organize to reduce other delays in the toolmaking system.

It was soon realized that most of the work on a tool could be done before the shape of the component was known. Die, punch guide and ejector blanks were hardened, ground, fitted into appropriate die sets and placed in stock. Also various sizes and shapes of tool steel for punches and piercing punches.

conventional spark erosion machine, located near to the Agiecut machine, and operated by the Agiecut operator.

For dies, punch guides and ejectors this system worked perfectly, but there was a potential problem with punches. A punch has to be very firmly anchored otherwise it will break or pull out as it is withdrawn from the strip. For this reason it is normally made integral with a thicker shank into which its form is carefully blended, and it is impossible to cut such a punch on the Agiecut machine. There are two ways out of this dilemma; experience and the geometry of the job determined which sould be used in any given case.

One way was to make straight sided punches on the Agiecut and find some means of holding them. Having, over the years, had plenty of experience of broken or damaged fine blanking punches, we tackled this expedient very gingerly. We were agreeably surprised at the lack of trouble. Two factors seem to help; punch/die clearances were now nearer to optimum, and more uniform all round the periphery than we could achieve in the past; and the Agiepac program accurately placed the true centre of shear at the geometric centre of the punch. Together these two factors virtually eliminated bending stresses and considerably reduced tension stresses during the stripping cycle. A combination of pinning and araldite techniques proved remarkably successful.

Inevitably, however, there were occasions when it was not considered possible to use a straight sided punch, and in these cases female electrodes (of graphite or copper) were cut on the Agiecut and a normal punch was made on the conventional spark eroder. This practice is not quite so accurate as the direct method, since two sets of tolerances are involved, those of the Agiecut and those of the conventional eroder (where electrode wear is one of the factors). So it became established practice to make the punch, measure it across some easily established dimension, and then cut the die and other components to suit the punch. This meant that accurate clearances were maintained and the true geometrical form preserved but the product from the finished tool might be a little larger or smaller than the optimum (or than the prototype). The error thus introduced could, with care, be kept down to less than ·01 mm (·0004 in) radially; and it is important to note that this error is detectable, and was measured, as soon as the punch was made. If the measured error was liable to affect some particularly tight tolerance, new electrodes and a new punch were made before starting work on the die and other tool components.

Changes in Organization and Attitudes

Inevitably, the introduction of new ways of doing things will have repercussions throughout an organization. We cannot enter fully into this here, but some points from our experience in handling the Agiecut/Agiepac system are relevant. In an established toolroom we were able to bring the system into use alongside the existing system with minimal disturbance. This was a great advantage, not only because it permitted careful and realistic assessment of the cost and time saving, and the improvement in quality of the product; but also because the workpeople both on and off the project itself were able to see that, though their jobs might change the changes would mean greater job satisfaction, greater personal responsibility and release from much tool-fitting, tool modification and tool repair work which, though demanding and troublesome is often mere manual drudgery.

We took four men of wide toolmaking experience, three for shifts and one as relief, and gave them a brief training and the fullest possible explanations of what we were trying to do. They quickly became an experienced and enthusiastic team. Shift working presents problems with conventional fine blanking toolmaking unless each shift works on different tools; but it is the natural way to use an Agiecut machine, which cuts metal relatively slowly. Since, however, all the essential variable data is on tape there is no difficulty in handing a half-finished job on from one operator to another.

It was soon obvious that tool design, planning and progress in the old sense of the term were no longer necessary; and even estimating was much simpler (and more accurate!) A potential new job, if accepted as practicable, was simply given to the man on day shift, and he sat down at the teletypewriter and made a tape. If there were difficulties which had not so far been noticed, he soon found them. Drawing errors, ambiguities, inconsistencies or onerous and expensive requirements such as too tight radii were all discovered by the man himself or revealed by the computer as he went methodically round the profiles. If there was an unresolved problem he could not proceed. If he got his tape, all is well. And because the rate of working of the Agiecut machine—and the standard spark eroder—were pretty accurately known for all conditions and thicknesses of material, it only took a few minutes to say how many hours of work would be needed to make the tool components or the prototype parts. A comparison with, and if necessary rearrangement of, the loading schedule for the Agiecut gave a realistic completion date for the work. Thus estimates of cost, time and completion dates were available quicker and more accurately than ever before, merely as a by-product of an hour-or-so's programming activity.

The job, if accepted, was now firmly on the Agiecut team, and stayed there, right up to the stage of tool assembly. They had made the tapes, and they made the job. There were no arguments about whose fault it was if there was an error. And it was easy to check the exact progress of any one job. The men too, gained pride and responsibility in knowing that they made a whole, complete entity instead of just performing one operation on it.

OUTPUT

What can be done

A compound blanking tool such as is used in fine blanking has always been costly and taken a long time to make, including, all too frequently, much time and expense on modification of faulty parts, and careful adjustment and fitting. For presses in the 25 to 100 tons range typical tool prices run from £750 to £2000 plus. What is often much more troublesome is that it usually takes from three to six months to make a tool even when there is no delay in starting the work. It is difficult to compare one toolroom with another, one toolmaker with another or one tool with another. It will not, therefore, be profitable to enter into a detailed and possibly misleading comparison of our own experience with the old and the new methods of making compound press tools of the fine blanking type. Typical output data may be more informative. Consider a component of complicated form and tight external tolerances, with three internal holes—a slot, a keyhole slot and an irregular shaped hole. The total periphery is about 200 mm (8 in) and the material 1·5 mm (·060 in) M.S. It is not of any significance how complicated the shape is, in any of the subsequent operations, except that if it is *very* difficult to describe geometrically, the programming time might be increased by up to an hour or so. Production will be on a 25/40 ton press and a tool for a component of this size would normally be worth £750 to £1200, according to the shape of the product and the tolerances on it.

For such a component it will take about an hour to get a tape; eight hours machine time to make start holes and cut ten prototypes on the Agiecut machine; and thirty hours total machine time to cut the tool components. Note that at no time is more than one man involved, and that while he is programming or making start holes there will be other work going forward on the Agiecut machine. Assuming that high priority is given to the work (i.e. there is no queuing problem) there is no difficulty in having the tool assembled and running on production in two weeks from first sight of the component drawing.

Or consider a job of similar complexity with 500 mm (20 in) of periphery and made from 3 mm (·120 in) mild steel—a job suitable for a 100/160 ton press. The tape will still take an hour to make, but this time six prototypes would need 20 hours of machine time and the tool components about 100 hours. This job could be fully tooled and operational in three weeks.

Further Progress

The Agiecut machine was acquired, and the Agiepac programming system developed, for the general purpose of evolving a better—because more accurate and predictable—method of producing two dimensional one-offs. A fine blanking toolroom was chosen for the initial experiments because an old established one happened to be available, and because the compound

This paper is primarily a report of work that has been done, and of the progress made. But some further consequences should now be noted. The first is that fine blanking tools can, for the first time, be made with truly interchangeable parts. Articles coming from different tools can be literally identical, and what is more important new component parts for a broken tool—perhaps many miles away—can be supplied with confidence that they will fit. And, because tools are cheaper and quicker to make, it becomes economic to make such tools for a smaller total requirement of components. So the work content of the press shop itself will be altered. Naturally, also the manufacture of conventional blanking tools, both simple and progression, is facilitated.

The Agiecut machine and Agiepac are both new, and there is a capability for further development to give each, and the joint system, greater capacity and flexibility. To take a simple illustration, Agiepac already allows for the rotation of a component so that the data can be fed in from the as-drawn position and the tape will be appropriate to the predetermined strip-layout position. Soon Agiepac itself will be able to calculate the optimum strip layout angle. From this it is but a step to the incorporation of machine tool technology, press technology and materials data. The computer handles with ease and accuracy all the data which men so easily forget or miscalculate. Most of what is important about the component and the way it is to be made is already being told to the computer to get the tape. Ultimately, we hope, a few extra questions, posed by the computer and answered by the operator, will produce optimum settings for the machine tool and the presses; and details of material requirements and scrap percentage for future press operations.

CONCLUSIONS

This paper is in the nature of a progress review, since we are dealing here with a new type of machine tool and a new development in programming techniques. It is obvious that, as time goes on, there will be expansion in the capabilities and improvements in the performance of both. Already, however, from our now extensive workshop experience of the operation of the combined system we can draw some pretty solid conclusions:

(1) Numerically controlled wire spark erosion is shown to be an outstandingly accurate and flexible method of cutting metal—equally adaptable to prototype manufacture and to the production of already hardened press tool components.

Q. C. E. Noble, UMIST. I understand that the control system for the Agiecut machine only permits straight and radial lines to be cut and that it only has a limited capacity to withdraw the wire from the machining zone during operation. Have these limitations caused your Company any serious problems?

A. The Agie wire control system also copes with elliptical curves, but we have never used the facility. Any 2D shape can be described to any appropriate degree of accuracy using radial and straight lines. Our Agiepac programming system, with automatic incorporation of suitable blend radii and facilities for calculating routines as well as toothed-gear and cam-form routines, ensures that all departures from nominal form are within the minimum Agiecut movement increment of ·002 millimetres, so there is, in effect, no limitation.

The reversible memory capacity of 511 Bit is, we find, fully adequate for all operating conditions.

Q. E. S. Moore, National Research Council Canada. Does decay of profile accuracy occur with increase in workpiece thickness?

A. However carefully the flow of dielectric is adjusted there is some tendency for the width of the slot cut by the wire to be greater at the middle than at the top and bottom of the material section. The effect is negligible on a section less than 15 or 20 mm, but is of the order of ·010 mm on 30 mm material and progressively more on thicker material. Thus, though the profile shape is accurately maintained, the profile size varies across the section, the variation being radially half the above figures. Using wire spark erosion methods all fine blanking dies are in one piece and all punches easily replaceable. Therefore 20 mm, or at most 30 mm material is now adequate for punches, dies and other tool components. For other types of tooling these very small variations of size across the cut section would, of course, be of no practical significance.

Q. Dr J. R. Crookall. The surface produced by a wire-electrode E.D.M. machine has the usual appearance of an electro-discharge machined surface made at a low discharge energy level. Whilst the roughness amplitude may be fairly low, the surface has a non-specular, rough appearance, and has slopes higher than many other types of surface. Are there any limitations in the use of such a surface in fine blanking particularly, or is this 'disadvantage' more apparent than real?

A. Despite the apparently 'poor' surface finish produced by the wire spark erosion machine the fine blanking tools work very well, probably because clearances are always near to optimum and the surface irregularities are aligned in the direction of punch/die movement, which assists lubrication. The fine vertical lines are observable on the component edge, but the all-round component surface finish and appearance is extremely good by any realistic stan-

dards. If any particular face of a component needs to have a really top-grade line-free appearance, stoning or polishing of the appropriate die face will ensure this. The flexibility of the offset control operation on the Agie machine makes it simple to leave any appropriate (say ·008 mm) extra amount of metal on that face only.

INDUSTRIAL APPLICATIONS OF GLOW DISCHARGE ELECTRON BEAM TECHNOLOGY

by

J. CONDLIFFE* and R. A. DUGDALE*

SUMMARY

The new technique of generating electron beams from the Glow Discharge is briefly described and the present stage of development is indicated. A description of electron guns for welding, strip processing, vapour deposition, and other applications is then given. The main industrial interest at present lies in welding applications and the paper will discuss in some detail single shot tube welding and linear welding. In addition developments in the more conventional 'point focus' welding are described. The experience so far gained in relation to strip processing, vapour deposition, and other processes depending on the precise application of controlled heating, is discussed.

INTRODUCTION

Glow discharge was employed in the early days of ion and electron physics to generate experimental beams of particles.[1] As high vacuum technique developed more sophisticated sources tended to replace it. This was particularly evident in the field of electron beam technology (e.g. cathode ray tubes, electron microscopes, etc.). There is today however a growing need in industry for electron beam sources which do not require high vacuum for their operation and which are moreover compatible with frequent admission of air at atmospheric pressure. This is particularly the case in the field of material processing, e.g. the application of electron beams to welding[2], machining, vapour deposition, and other purposes[3]. The most commonly used electron source in this field at present depends on thermionic emission from a tungsten filament in high vacuum. However, a return to the early techniques of using glow discharge to generate beams can be undertaken with advantage[4].

Glow discharge beam sources have therefore become a new technology and the authors have aimed for some time now to develop it for engineering purposes.

BASIC PRINCIPLES AND PRESENT STATE OF DEVELOPMENT

Figure 1 shows the main features of a simple glow discharge electron gun. The discharge takes place between two electrodes in an atmosphere of gas at a specific reduced pressure (usually in the pressure range 20 mtorr to 200 mtorr). A plasma formed on the anodic side supplies energetic ions to bombard the cathode where they liberate electrons by a secondary emission process. These electrons are accelerated in the field away from the cathode to form the electron beam. The shape and power of the electron beam is a function of the electrode geometry

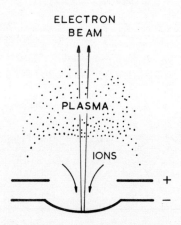

Figure 1. Main features of a simple glow discharge electron gun.

and the working gas pressure. Thus the diagram of Fig. 1 could represent a gun generating a thin pencil beam with the electrode structure having a cylindrical symmetry about the beam axis, or it could represent a section through a thin, sheet-like beam extending a considerable distance perpendicular to the section. Again, rotation of the section about any axis lying in its plane would form a structure giving either a converging or a diverging, sheet-like beam. Other

---

* A.E.R.E., Harwell, Berks.

unsophisticated processes requiring either a concentrated beam input (e.g. up to about 10^3 W/cm^2) or a dispersed beam. Automatic regulation of gas pressure can then be employed to control beam power. This can be done simply and cheaply by means of a closed loop system sensing gun current and governing the admission of gas by means of an electrically operated leak valve. The main restriction is that the rate of gas evolution during processing should not exceed the

emerging beam can be refocused or otherwise magnetically manipulated as required by the process.

Figures 2, 3 and 4 show examples of practical guns based on these principles which are currently under development for specific applications. These guns work well in various gases, including air, at voltages in the range 10 kV to 40 kV and at powers of several kW. We have operated experimental guns at power levels up to about 50 kW and expect development to

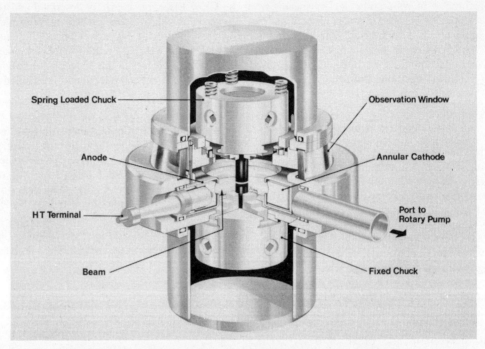

Figure 2. Ring focus gun for 'single shot' tube welding (schematic).

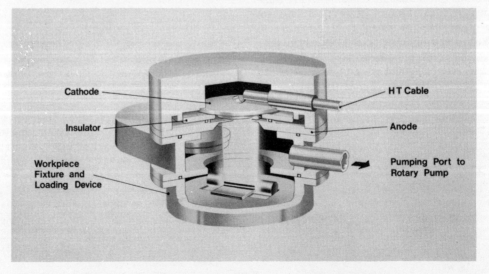

Figure 3. Line focus gun for 'single shot' welding (schematic).

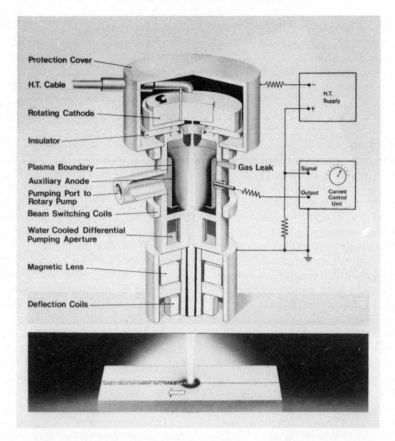

Figure 4. Point focus gun for general purpose welding, etc. (schematic).

higher powers and energies to be straightforward. Glow discharge guns are characterized by their ruggedness, durability, and reproducibility. They can be quite efficient (70-90 per cent conversion to beam power), and will operate for long periods with little attention in crude vacuum conditions ($10^{-2}-1$ torr). With the aid of rotating cathodes and magnetic lenses point focus beams can be produced with high power densities in the range 10^6-10^7 W/cm^2.

These guns are being developed and assessed for a range of industrial applications; some of these will now be outlined in detail.

WELDING

Considerable interest is currently being shown in a new welding technique known as 'one-shot' welding, in which an electron beam is shaped so that the entire weld line is heated simultaneously (i.e. in a single shot) without any movement of the beam or the workpiece. Machines of this type are now developed up to 10 kW power at 15 kV with automatic control. They can weld steel and other metal tubes up to about 3 cm O.D. and wall thicknesses up to about 3 mm (see Fig. 5), and techniques for welding thicker tubes are under development. Dissimilar metals can be welded (e.g., tubes of molybdenum and stainless steel) and silica tubes have been joined. A special achievement was the butt welding of stainless steel tubes 1 in (25 mm) outside diameter by 0·006 in (0·15 mm) wall thickness.

The two tubular components to be welded are loaded into upper and lower chucks (Fig. 2). Vacuum

Figure 5. Examples of 'single shot' tube welds.

sealing covers are then brought into position and the chamber pumped down to about 50 mtorr in 10-15 s. The pressure is automatically adjusted for the welding operation, the gun fired and a butt weld is made in about 1 s, without rotation of either the beam or workpiece. With simple modifications to

during the welding operation. One of the chucks is spring loaded so that when the weld occurs the chuck is free to move slightly and in doing so switches off the beam.

An additional refinement recently developed is the use of a 'joule meter' to terminate the 'shot'. This, as its name implies, integrates the product of voltage and current and improves weld reproducibility by effectively compensating for variations in operating conditions.

Flexible seals have been developed which allow the technique to be used for welding fittings to long tubes, and the increased powers now available will mean that larger components can be handled.

A novel use for the tube welder shown in Fig. 2 is its use for sealing tubes. If the 'shot' is given an extended duration the ring of molten metal contracts to close the tube (Fig. 6). Tubes of up to 1 in (25 mm) O.D. have been sealed in this way.

The straight line focus glow discharge guns (Fig. 3) can also be used for butt welding components, employing the one-shot welding technique just described. A 'double line' gun, i.e. one to weld both sides of the butt simultaneously, is under development.

Point focus welders based on the concept illustrated in Fig. 4 have advantages over the now established high vacuum thermionic point focus welders can be just as readily used for the precise welding of thin section components with fine focus and low beam power. In recent applications components were fabricated in thin section molybdenum and stainless steel (0·005 in to 0·010 in (0·13 to 0·25 mm) thickness). It has been necessary in some of these applications to ensure that the melted region did not penetrate the thin section stainless steel. Owing to the precision available in beam power control this was accomplished without difficulty.

A point focus welder can also be applied to insulating materials such as ceramics. An alumina encapsulated tungsten–rhenium thermocouple for high temperature applications in corrosive environments has been satisfactorily developed which depends on this technique for its fabrication.

Low power point focus guns can be designed with high brightness and very fine focus. A microfocus gun of this type was developed and achieved 30 micron molten tracks in stainless steel at a working distance of 2 in (50 mm) from the single lens employed. A beam current of about 0·3 mA at 30 kV was focused onto the steel. This type of gun is suitable for electron probe applications, micro-welding and micromachining.

STRIP PROCESSING

A variety of laboratory trials have been carried out on electron beam processing of moving strip using an unfocused, 15 cm width, line-gun operating at power levels up to about 8 kW (10 kV to 25 kV). Most interest lies in the field of electron beam curing of thin lacquer coatings. Conventional lacquers are cured in an electron beam exposure of the order of 5×10^{-3} coulomb/cm^2. For many applications this is too high a dose, and current activity in the field is aimed at developing lacquers which can be cured by a much reduced exposure.

There are two basic types of E.B. curing processes, viz. 'in vacuum' and 'out of vacuum' processes. The 'out of vacuum' process requires a high voltage (e.g. 300 kV) owing to losses in the 'window' (heating of the window also requires a limitation on beam current and therefore process rate). The 'in vacuum' process is free from these disadvantages and a high process rate at a voltage to suit the coating thickness is possible. Our experience is presently limited to 10 micron coatings at a voltage of 25 kV, but development should lead to increases in these parameters.

Experimental air to air equipment for the continuous processing of 10 cm wide strip is now under construction. The strip will be run from a supply coil at atmospheric pressure into the soft vacuum process-

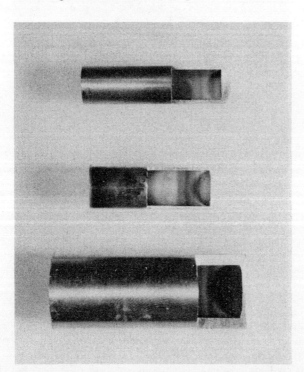

Figure 6. Stainless steel tubes sealed in single shot ring focus gun.

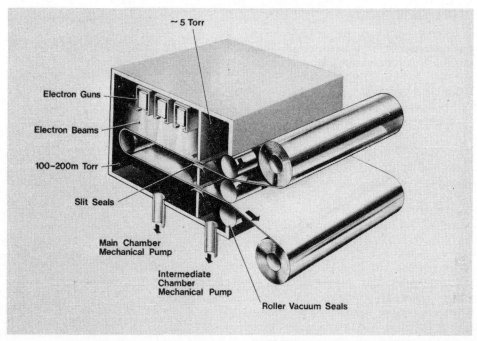

Figure 7. Plant for continuous processing of thin sheet (schematic).

ing chamber and then out again, as indicated in Fig. 7. The equipment will be fitted with line beam guns employing long-life rotating cathodes. This plant is designed as a test facility for the development of electron beam processes on a wide variety of materials. In addition to lacquer curing it will be suitable for heat treatment of moving strip and for continuous coating processes by vapour deposition (even bright aluminium can be put down under soft vacuum if the process rate is high enough).

VAPOUR DEPOSITION

The differentially pumped, point focus electron gun provides an excellent heating source for high-rate evaporation and deposition of ceramics and metals.

A system of this type is shown in Fig. 8. A rotating cathode point focus gun generates a beam of about 1 kW power at voltages in the range 15 kV to 25 kV. This is brought through a 2 mm aperture into the work chamber. The air pressure to work the gun is about 40 mtorr, but, owing to the differential pumping provided by the two roughing pumps (each of 8 l/s capacity), the chamber may be operated at an air pressure of about 1 torr. A very flexible system of electronic beam programming has been developed to govern the operations performed by the beam in the work chamber. This is based on a principle of beam-time sharing and contains six time slots in each of which the beam current, position, and motion are independently controlled. The sequence of operations contained in the six time slots may be cycled at up to 200 c/s. This form of beam programming permits a wide variety of operations to be carried out by the single gun. In effect the gun becomes equivalent to up to six independently controlled guns simultaneously focusing sharply to a point, line, loop, or other figure, or scanning over a rectangular area. We are especially interested in using it to study ceramic coatings put down under controlled conditions of deposition rate,

Figure 8. Laboratory apparatus for investigating E.B. processing in soft vacuum. A, electron gun; B, work chamber; C, gun control unit; D, beam programmer, E, waveform display.

substrate temperature, and composition. The beam then carries out all the necessary operations including evaporation by point or line focus on one or more sources, as well as substrate heat-treatment by scanning. The equipment is rugged, durable, and automatically controlled. It has operated for periods of at least 100 hours 'beam on' time without maintenance or change of cathode.

Most of the research so far has been devoted to the deposition of oxides. It has been shown possible to deposit, for example, dense amorphous alumina at

FUTURE DEVELOPMENTS

Our laboratory work has shown a wide variety of processes to be feasible. These include various types of localized, distributed, and fast heat-treatments; sintering (e.g. of metal powder to make sheet); various types of welding including the novel single shot line and ring welds; fast and clean brazing; ceramic welding, including ceramic to metal joints and welds between sapphires; evaporation techniques for powder making and coating (including both ceramics and metals); evaporation under a point focus beam as a method of machining and perforating; chemical processing; etc.

There are obviously many possibilities for the future. Any single application development will depend on market opportunities, economic factors, and particularly the availability and cost of capital. For the majority of industrial applications glow discharge beam equipment is simpler and cheaper to engineer than high vacuum thermionic equipment. It is ideally suited to long term operation by automatic control in gassy applications.

The most developed applications at the present time are in welding and we expect to see the capability range in this field extended in the near future, particularly in terms of increased power for the ring and line focus welders. We hope also to raise the working voltage for point focus welders to 60 kV and similarly to increase the voltage of lacquer curing guns so that thicker coatings of lacquer and other organic coatings can be treated. The ceramic vapour deposition technique is nearing the stage of practical application and simplified custom built equipment may soon be engineered.

There is a growing interest in E.B. processing of materials[7] and the glow discharge source has considerable merit for application in a number of appropriate fields.

2. A. H. MELEKA (Ed.) (1971). *Electron Beam Welding,* McGraw-Hill.
3. See for example, R. BAKISH (Ed.) (1970). Electron and ion beam technology, *4th Int. Conf. Electrochemical Society.*
4. R. A. DUGDALE (1971). *Glow Discharge Material Processing,* Mills and Boon.
5. J. T. MASKREY and R. A. DUGDALE (1972). *J. Phys. E.,* **5**, 881.
6. J. D. F. RAMSAY and R. G. AVERY (1972). Effects of compaction on the surface areas and pore structures of oxide ceramic powders, *Proceedings of 1st Int. Conf. on the Compaction and Consolidation of Particulate Matter,* Brighton, September. R. G. AVERY and J. D. F. RAMSAY (1972). Compaction and calcination studies on ultrafine ceramic powders prepared by vapour condensation, Presented at Brit. Ceram. Soc. Basic Science Section Meeting on Electrical, Magnetic and Optical Ceramics, London, December.
7. R. M. SILVA (Ed.) (1972). *Proceedings of the 2nd Electron Beam Processing Seminar,* Frankfurt, Universal Technology Corporation.

DISCUSSION

Q. C. F. Noble, UMIST. We heard from a previous speaker that the machine tool industry was reluctant to take up the potential of the laser. Has Dr Dugdale experienced similar problems in attracting *both* machine tool user and machine tool designer in the development of his experimental work for commercial use.

A. There are some remarks at the end of the paper on this point. We are now interesting both machine tool user and designer in the commercial development of welding equipment and we hope in the near future to similarly attract commercial investment in lacquer curing and in vapour deposition.

GRINDING

SELECTION OF GRINDING WHEELS FOR USE AT HIGH SPEED

by

K. B. SOUTHWELL*

SUMMARY

The already established use of resinoid bonded wheels at speeds up to 16 000 s.f.p.m. is briefly discussed, outlining the methods of reinforcement which enable these wheels to operate at the higher speeds. The paper goes into greater detail concerning vitrified wheels for use in high speed precision grinding at speeds up to 12 000 s.f.p.m. describing the selection of wheels in relation to their performance characteristics, with mention being made of the refinements necessary in the process of grinding at high speed and the potential benefits to the overall production operation. Emphasis is placed on the safety aspects which generally restrict precision grinding operation to a maximum operating speed of 12 000 s.f.p.m. Reference is made to changes that would be necessary before speeds of 16 000 s.f.p.m. and higher could become a production reality.

INTRODUCTION

The need for new terminology in the grinding industry, such as Abrasive Machining, High Efficiency Grinding and High Speed Grinding, only emphasize the change in image which is required. For too long the view has been held that grinding is purely a finishing operation which has to be tolerated as an additional operation if close tolerances on size, form, geometry and surface finish are required.

Although the precision side of the industry in general has been slow to grasp the opportunities offered by the increased efficiency of higher peripheral speeds, it is true to say that in certain areas of high batch production, bearing industry, etc., the improvement achieved at 12 000 s.f.p.m. has been sufficiently dramatic to encourage the call for even higher speeds and the demand for sheels to operate at 20 000 s.f.p.m. is fast becoming a reality.

The steel and foundry industries were the first to utilize greater power and increased speeds, these moves taking place in the application of resinoid wheels on fettling and cutting-off operations.

GRINDING WHEEL PARAMETERS

Resinoid wheels
Fettling wheels are at present being used up to 16 0C0 s.f.p.m. Such speeds are made possible and safe by not only using wheels of adequate strength—with bursting speeds over 30 000 s.f.p.m.—but also protecting the operator adequately in case of wheel breakage through any cause.

The high bursting speed is achieved by wheel design and method of manufacture. Hot pressed wheels in resin bonds achieve very low porosity and the resultant increase in bond strength, together with a high strength annulus of fine abrasive mixing, incorporating a high tensile steel ring, around the bore, achieves the required structural strength to give a high bursting speed and the required margin of safety.

The part of the wheel adjacent to the bore is the most highly stressed part of a rotating wheel, and therefore requires most attention in terms of reinforcement.

More porous wheels made by cold pressing processes require a different method of reinforcement. Both cut-off wheels, which are supplied for operation up to 20 000 s.f.p.m., and cold pressed fettling wheels, are reinforced by incorporating one or more layers of glass cloth within the wheel, and special weaves and thickness of glass cloth have been developed for this purpose. Not only do glass cloths add strength to a wheel but, particularly in the case of cutting off wheels they add a measure of impact strength and an ability to flex laterally before actual breakage.

Requirements for vitrified grinding wheels
Under freely rotating conditions bursting occurs in a disc or cylinder as a result of centrifugal stresses.

In the case of a grinding wheel additional stresses are super-imposed on the structure during the grinding operation. To accommodate this and also to

---

* Universal Grinding Wheel Co Ltd.

and bond used, with wheel structure, and dimensions. In particular white aluminous abrasive wheels have slightly higher bursting speeds than brown aluminous abrasive wheels, whilst large wheels have lower bursting speeds than small ones.

Safety factors

The present safety factors which are used for vitrified wheels to be operated at 12 000 s.f.p.m. are defined in terms of the ratio of wheel strength to stress, or bursting speed of the wheel to its maximum operating speed, which means that we are applying a safety factor of 4 : 1 (stress being proportional to the square of the velocity).

The factory testing of these wheels is carried out below the design strength, and is set at an intermediate speed which ensures an adequate safety factor whilst, of course, not having too high a percentage of breakage on test.

A review of the standards required by different countries and bodies does not show complete agreement. The German D.S.A. regulations require a factory speed test after certification of a wheel specification, of a minimum of 1·4 times M.O.S.

The generally accepted level of speed testing in this country for wheels to be operated at 6000 s.f.p.m. is believed to be 1·5 times M.O.S., which relates to a stress loading of 2·25 times that encountered at M.O.S. For vitrified wheels supplied for use at 12 000 s.f.p.m. we at Universal use a test speed of 1·73 times M.O.S. which equates to a stress loading of 3 times that imposed at M.O.S. (free running).

The U.S.A. regulations states that speed testing of 1·5 times M.O.S. is an adequate standard for normal stresses imposed during use, and do not include a design factor. They, therefore, consider testing at 2·25 times the stress level encountered at M.O.S., a sufficient safety level.

Wheels for use at 12 000 s.f.p.m. which do not meet the required strength level, can be reinforced but their manufacture does cause problems in large wide wheels and in very coarse and soft grades.

An alternative way of looking at this safety factor problem is by considering strength margins. The strength to stress ratios which are now used increase the strength margin required as the wheel speed increases (see Fig. 5). Strength margin can be defined as the product strength above that required to withstand the maximum rotational stresses in the wheel when it is running freely at the recommended operating speed.

This approach would enable a wider range of wheels, both reinforced and unreinforced to become

ring after the wheel has been speed tested to the required level. The onus is on the machine manufacturer to guard the machine adequately to protect against such contingencies, and of course the user must take all possible care to ensure the wheel is sound, correctly mounted and properly used.

GRINDING PARAMETERS

Metal removal rate

The primary advantage to be gained in operating a wheel at high speed is its ability to reduce machining times. This is possible because of the high metal removal rates obtainable. Also, these increased removal rates are achieved without deterioration in the main grinding parameters, i.e. finish, form holding and grinding ratio.

Results from trials carried out in our Abrasive Engineering Department have confirmed that removal rates of 0·5 in^3/min at 6000 s.f.p.m., 3·0 in^3/min at 12 000 s.f.p.m. and 5·0 in^3/min at 16 000 s.f.p.m. can be obtained without affecting the grinding parameters—Figs 1, 2 and 3.

Increases in removal rates can be utilized, either by increased output or by eliminating prior machining operations. Reducing the grinding time per component can only be a viable proposition if the actual

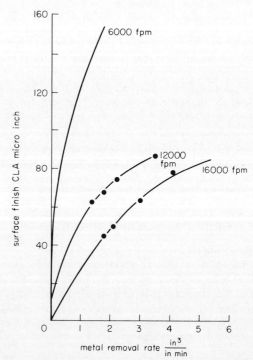

Figure 1. Influence of metal removal rate and wheel speed on surface finish.

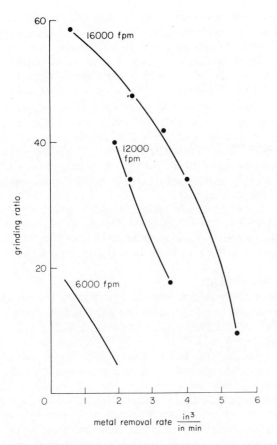

Figure 2. Influence of metal removal rate and wheel speed on grinding ratio.

grinding time is a sufficiently high percentage of the floor to floor time to make it economical. The elimination of prior machining operations has been proved technically and viably by the major Machine Tool Manufacturers, but due to restriction on capital investment only a limited number of machines have at the time of writing been installed, these being mainly in the cam and crankshaft area.

Forces

At constant metal removal rates, as the wheel speed is increased, the grinding forces fall[1]. As previously stated the benefit of high speed grinding, is the increased metal removal rate potential, therefore to take full advantage of uplifted wheel speeds, it follows that wheel-head power should be increased proportionally to the increase in speed in order that grinding force availability remains equal

$$\text{Tangential Force} - \frac{\text{HP x 33 000 ft.lb/min}}{\text{Wheel Vel. Ft/min}}$$

If grinding forces are allowed to fall, either from low infeed rates or insufficient power, loss of grinding efficiency is inevitable. At high speeds and low force loadings, abrasive wear is mainly attritious, which leads to an all-round drop in the volume of metal removal before surface chatter occurs.

A case history that illustrates this problem was encountered at a customer who was grinding case hardened shafts at a wheel speed of 6000 s.f.p.m. removing 0·20 in. from diameter in a grinding time of 23 seconds, and obtaining an average of 30 shafts per

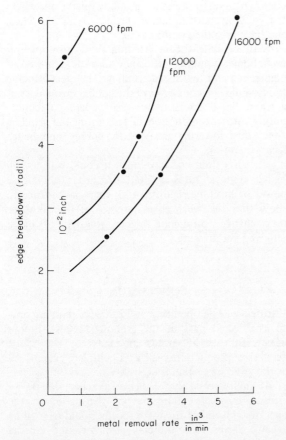

Figure 3. Influence of metal removal rate and wheel speed on form holding.

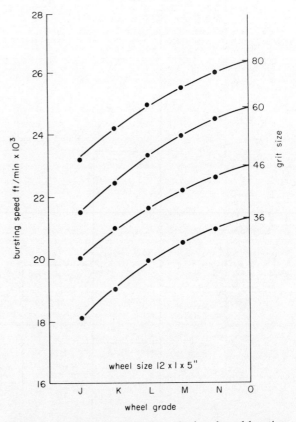

Figure 4. Relationship of grit size, wheel grade and bursting speed.

Wheel selection

The main criteria when making a wheel selection for use at 12 000 s.f.p.m., are—wheel strength, metal removal rate, finish and form tolerance, number of components per dressing.

The strength of a grinding wheel decreases as the grit size is increased (i.e. coarser) and/or the bond size (i.e. grade) is reduced. The metal removal rate potential at 6000 s.f.p.m. is usually linked with the surface finish and form holding requirements, and the selection is made by choosing a wheel of the coarsest grit and softest grade that will meet these two tolerances. At 12 000 s.f.p.m. it has been established that for maximum efficiency one grit size finer is required, but due to the changes required in wheelhead spindle design for use at high speed, a change to one grade softer is usually necessary.

Our experience has shown that initial requests for trial wheels have usually been for a coarser grit size

0·45 in^3, taking the first 0·010 in of the wheel depth as the working zone.

It can be clearly seen that such a wheel has the ability to achieve high removal rates and is capable of accommodating the material in its structure, all this being possible only under correct grinding conditions.

Grinding fluids

Over the past few years a considerable amount of work has been carried out on grinding fluid application. A comprehensive programme of work has been completed at M.T.I.R.A. by Dr G. Sweeney under the sponsorship of D.T.I. This highlights the advantages to be gained by correct grinding fluid application, and compares the increase in all-round grinding efficiency obtained by different low pressure delivery systems illustrating that such systems using inexpensive attachments give equivalent grinding performances to those attainable with high-pressure fluid-delivery systems.

Despite all the publicity on this subject, it seems to have had little effect in industry, and it is still a common sight to see some grinding operations where the grinding spark-stream is carried around the wheel and back into the grinding zone.

With possible future grinding fluid development, even higher grinding efficiency may be achieved at high speed and low speed grinding. The ideal grinding fluid would, of course, have the cooling properties of a water based coolant, but also capable of contributing into the reduction of friction in the grinding zone, thus lowering the specific energy and hence heat generation.

If such a fluid was developed it could lead to still higher removal rates without the need to increase power or machine rigidity. Also, there is evidence to show that the wear characteristics of the abrasives may change, producing higher volumes of metal removed per dressing.

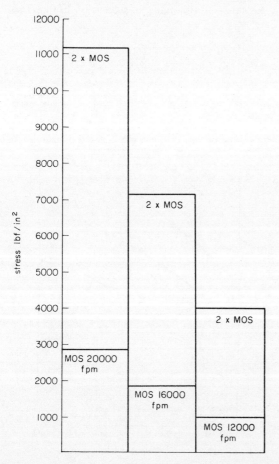

Figure 5. Diagram of strength margins of various wheel speeds, using a safety factor of 4 : 1.

CONCLUSION

Vitrified wheels operating at 12 000 s.f.p.m. are now an established part of the grinding process. Before wheels for use at speeds in excess of this speed become a production reality, a complete reappraisal of the grinding system is necessary.

At such speeds the chance of a wheel failure by misuse, must be taken out of the hands of the operator. Machines for these speeds would have to be capable of absorbing the energy on breaking and the containment of all fragments must be guaranteed.

With machines designed to this level of safety, a

review of wheel safety levels may be possible. Without these changes, wheel development must be capable of producing a wheel with vastly improved strength characteristics, and, of course, must achieve this without affecting the wheel's grinding performance.

REFERENCES

1.   H. OPITZ and K. GUHRING. *High Speed Grinding*. Annals of C.I.R.P., Vol. XVI.

THE FACTORS INFLUENCING THE SURFACE FINISH PRODUCED BY GRINDING

by

C. RUBENSTEIN*

SUMMARY

The hypothesis that the surface roughness in grinding is a function both of the grit depth of cut and of the average plateau area per grit is proposed and evidence is cited which is consistent with this hypothesis. Deductions from this hypothesis, both qualitative and quantitative, are shown to be in accordance with published data. In this way, a large body of apparently unrelated factors are shown to be consistent with a single, coherent picture of the surface roughness generated in grinding.

INTRODUCTION

An exact understanding of the mechanism by which the surface finish is generated in the grinding process is yet to be acquired. The nearest approach that has been made, which holds any prospect of enabling a prediction of the factors influencing the surface finish and their effect on surface finish, is the indication by Yang and Shaw[1] and by Opitz, Ernst and Meyer[2] of the existence of a correlation between the grit depth of cut and the surface finish in grinding. There are, however, two limitations to progress from this point, viz. the surface finish is not solely dependent on the grit depth of cut and while expressions for the mean grit depth of cut in reciprocating or cylindrical plunge grinding exist, very little exists for the more generally used traverse grinding process. In the present paper, an attempt will be made to remove these limitations and to compare the results of an analysis of surface finish in grinding with published data. It is emphasized at the outset, however, that this work does nothing towards offering an explanation of the physical mechanisms by which the surface finish in grinding arises. Again, it is assumed throughout that grinding is occurring in the absence of chatter, surface burn or wheel loading.

ANALYSIS

Hypothesis

As mentioned above, the surface finish is not solely dependent on the grit depth of cut. The hypothesis is that the surface finish in grinding is a function both of the grit depth of cut, t_1 and of the plateau area per grit, which arises as a result of attritious wear of the grits during grinding, i.e.

$$CLA = f(t_1)\,\phi\,(a)$$

Only when the plateau area per grit is a constant will there be a time independent correlation between surface roughness and the grit depth of cut.

To examine the validity of this hypothesis, consider the observation[3] that the average plateau area per grit increases, in general, with grinding time; and that the force components in grinding are a function, *inter alia*, of the average plateau area per grit[18] so that, when the average plateau area per grit increases with grinding time, the force components will be increasing functions of grinding time. It follows, therefore, if the hypothesis be true, that when the average plateau area per grit is an increasing function of time *both* the surface finish and the grinding force components will be increasing functions of time.

Data confirming this suggestion have been obtained by McKee, Moore and Boston[4] who have examined the changes in both horse-power and surface roughness with grinding time when hardened SAE 52100 steel was cylindrically ground with alumina wheels all having grade J hardness. They showed that over 60 passes using wheels of 46, 60, 80 and 150 grit size, both the horse-power and the surface roughness were increasing functions of time. Further, their data show that the rate of increase of horse-power was greatest with the 46 grit wheel and decreased in the order 46, 80, 150. Similarly, the rate of increase of surface roughness with grinding time was greatest with the 46 grit wheel and decreased in

---

* Department of Mechanical Engineering, University of the Negev, Beersheba, Israel

cesses can be derived as particular cases, cylindrical traverse grinding will be considered subject to the simplifying assumption that the wheel speed V is very much greater than the work speed v.

The length of contact between wheel and work is

$$l \simeq \sqrt{\frac{Dd}{1 \pm (D/D_w)}}$$

where D is the wheel diameter, D_w is the work diameter, and d is the infeed/revolution of the workpiece (this is equivalent to the wheel depth of cut in surface grinding).

(In external grinding the positive sign is applicable, in internal grinding the negative sign.)

Consider one revolution of the wheel. If w is the average width of cut per grit, and t_1 is the average depth of cut per grit, then the removal by one grit in traversing the length of contact between wheel and workpiece is

$$wt_1 l$$

In the time required for the wheel to complete one revolution, the amount of workpiece removal is

$$wt_1 l n_c W \pi D$$

where n_c is the number of active grits per unit area of wheel surface and W is the width of the wheel.

The time for one revolution of the wheel is $\pi D/V$, hence the removal rate is

$$wt_1 l n_c W V$$

But the removal rate can also be expressed as

$$vXd$$

where X is the traverse rate (equivalent to the cross-feed in surface grinding).

Hence, the average uncut chip cross-section, wt_1, is given by

$$wt_1 = \frac{v}{V} \cdot \frac{X}{W} \cdot \frac{1}{n_c} \cdot \frac{d}{l}$$

$$\propto \frac{v}{V} \cdot \frac{X}{W} \cdot \frac{1}{n_c} \sqrt{\left(\frac{d}{D_e}\right)} \qquad (1)$$

where D_e is the equivalent diameter given by

$$\frac{1}{D_e} = \frac{1}{D} \pm \frac{1}{D_w}$$

Particular cases

(i) Plunge cylindrical grinding, $X = W$

$$wt_1 = \frac{v}{V} \cdot \frac{1}{n_c} \sqrt{\left(\frac{d}{D_e}\right)} \qquad (1a)$$

Equation (1) gives

$$t_1 = \left\{ \frac{v}{V} \cdot \frac{X}{W} \cdot \frac{1}{rn_c} \sqrt{\frac{d}{D_e}} \right\}^{1/2} \qquad (2)$$

Equation (1a) gives

$$t_1 = \left\{ \frac{v}{V} \cdot \frac{1}{rn_c} \sqrt{\frac{d}{D_e}} \right\}^{1/2} \qquad (2a)$$

Equation (1b) gives

$$t_1 = \left\{ \frac{v}{V} \cdot \frac{X}{W} \cdot \frac{1}{rn_c} \sqrt{\left(\frac{d}{D}\right)} \right\}^{1/2} \qquad (2b)$$

and equation (1c) gives

$$t_1 = \left\{ \frac{v}{V} \cdot \frac{1}{rn_c} \sqrt{\left(\frac{d}{D}\right)} \right\}^{1/2} \qquad (2c)$$

It should be noted that equation (2c) is identical with that derived by Backer, Marshall and Shaw[6], who obtained their equation using a different argument. In contrast, equation (1b) is at variance with the equation derived by Lindenbeck[7] who, in the present nomenclature, obtained the expression

$$wt_1 = \frac{v}{V} \cdot \frac{1}{n_c} \sqrt{\left(\frac{2Xd}{WD}\right)}$$

Special case; traverse grinding; a is constant

In this case when a is independent of grinding time, the surface roughness will be a function only of the grit depth of cut, t_1. Suppose that the peak to valley roughness height, h, is proportional to t_1. Rubert[8] has shown that

$$h \propto (CLA)^{1/n}$$

and that

$$1/n = 0.85, \text{ i.e. } n = 1.2$$

Hence

$$CLA \propto h^n$$

and if our supposition is correct,

$$CLA \propto (t_1)^n$$

But from equation (2)

$$t_1 = \left\{ \frac{v}{V} \cdot \frac{X}{W} \cdot \frac{1}{rn_c} \sqrt{\frac{d}{D_e}} \right\}^{1/2}$$

whence CLA is proportional to

$$\left\{ \frac{v}{V} \cdot \frac{X}{W} \cdot \frac{1}{rn_c} \sqrt{\left(\frac{d}{D_e}\right)} \right\}^{n/2} \qquad (3)$$

If grinding is performed at constant removal rate $vXd = Q$ then expression (3) can be rewritten

CLA is proportional to

$$\left(\frac{1}{WVn_cr}\right)^{n/2}\left(\frac{Q}{D_e}\right)^{n/4}(vX)^{n/4} = \left(\frac{Q}{WVn_cr\sqrt{D_e}}\right)^{n/2}d^{-n/4}$$

For comparison, if Lindenbeck's expression for t_1 is used (equation 1d) CLA is found to be proportional to

$$\left(\frac{Q}{Vrn_c}\right)^{n/2}(WD_eXd)^{-n/4} = \left(\frac{1}{Vrn_c}\right)^{n/2}\left(\frac{Q}{WD_e}\right)^{n/4}v^{n/4}$$

Thus, the contrast between the present expressions and Lindenbeck's expression for t_1 is that at constant removal rate in, say, surface traverse grinding, the present expressions predict a linear relation between

$$\log CLA \text{ and } \log d$$

while Lindenbeck predicts a linear relation between

$$\log CLA \text{ and } \log v$$

As will be seen below, experimental data exists which enables these predictions to be tested.

DEDUCTIONS

Qualitative
In terms of the model proposed above, in conjunction with previously published considerations of the factors controlling machinability[9] and their application to grinding[10] the following deductions may be made:

Work hardness
In ref. 9 it was shown that as the hardness of a workpiece increases it will tend to become less ductile. This will tend to reduce the adhesion between workpiece and cutting tool or abrasive grit. As an immediately obvious consequence, the tendency to wheel loading in grinding will reduce if, with the workpiece in a softer state, there was any tendency for this phenomenon to occur. Such a change will effect an improvement in surface finish (cf. cutting in the presence of built up edge). A further consequence, however, is attendant on a reduction in grit/workpiece adhesion, namely, the lower the adhesion between workpiece and abrasive grit (or cutting tool), the more negative may be the rake angle at which chip formation ceases to occur (this has been termed the 'critical rake angle'[10]). This means that, considering the random array of rake angles which grits in a grinding wheel present to the workpiece, as the critical rake angle becomes more negative, the proportion of contacting grits which participate in metal removal will increase, i.e. n_c will increase. (For further discussion of this point, see ref. 10). An increase in n_c will produce a reduction in surface roughness (see expression 3), i.e. as the workpiece hardness increases, the surface finish should improve.

Krabacher[11] presents data which show a slight improvement in surface finish obtained when cylindrical grinding steel as the workpiece hardness increases. Lindenbeck[12] shows that a considerable improvement in surface finish occurs as the workpiece hardness increases from about 12 Rockwell C to 55 Rockwell C when EN6 steel is ground either by a metal-bond or by a resin-bond diamond wheel. It will be shown below, however, that an increase in workpiece hardness could conceivably produce a worsening of surface finish.

Grinding fluid
A fluid may influence the grinding process either by acting as a coolant or by acting as a lubricant and, in general, both mechanisms will occur simultaneously. If the fluid is capable of lubricating the grit/workpiece contact region, this will produce a reduction in grit/workpiece adhesion, an increase in n_c and an improvement in surface finish.

As a coolant, a fluid can influence the grinding process in two ways.

(a) the workpiece will be cooled causing it to be less ductile which, as has been seen, produces an improvement in surface finish;
(b) the grits can suffer a thermal shock causing grit fracture which will tend to limit the size of the average plateau area per grit, again producing an improvement in surface finish.

Thus, irrespective of the mode of action, the surface finish should improve as either the lubricating ability or the cooling ability of a grinding fluid improves. Krabacher[11] detected only a slight improvement in surface finish as the concentration of a water-soluble grinding fluid was increased. McKee, Moore and Boston[4] also detected an improvement in surface finish as the concentration of grinding fluid in water was increased from 2 to 16·7 per cent.

Average plateau area per grit, a
Although, above, the average plateau area per grit has been introduced as a factor influencing the surface finish, no attempt has been made to suggest a mechanism by which an increase in a could cause a deterioration in surface finish. It is quite feasible that the plateau area per grit influences surface finish in more than one way. However, one way in which a can influence surface finish can be visualized, immediately, from the realization that the plateau area per grit is exactly analogous to the flank wear land which appears as cutting tools wear. In ref. 13 it has been shown that the temperature in the flank contact region of a tool increases as the flank wear land increases. In ref. 14 evidence is quoted to show that in grinding, under conditions such that the grit depth of cut is constant, provided no hindrance to continued growth of plateau area per grit exists, the wear rate increases rapidly and the workpiece exhibits 'burn' when the plateau area reaches such a value that a critical temperature is reached. Thus, as a increases, the temperature in the grit/workpiece contact zone will increase resulting in an increase in workpiece ductility, a decrease in n_c and a deterioration in surface finish. Furthermore, with increased ductility,

hardness grade of a wheel increases, its modulus of elasticity increases. Hence, under given grinding conditions and with a given workpiece (i.e. at a given grinding force) the deflection of a grit within the bond material due to the tangential grinding force component will increase as the hardness grade decreases. Thus, considering a grit which in the absence of any force applied to the grit presents a rake angle less negative than the critical rake angle, the greater deflection of the grit (in the direction of cutting), the greater the chance of the grit presenting a rake angle more negative than the critical rake angle. Whence, it can be seen that as the hardness grade decreases the proportion of cutting grits i.e. n_c will reduce and the surface finish will deteriorate. (Similarly, as the strength properties of the workpiece material increase, more grit deflection will occur at a given wheel hardness and surface finish will deteriorate.) Thus, with *increase* in wheel hardness, the surface finish will *improve.*

(2) *Bulk wheel deflection*: For the same reasons, as the modulus of elasticity of the wheel decreases, the bulk deformation of the wheel will increase (for given grinding conditions and a given workpiece) and the effective diameter in the wheel/workpiece contact region will increase causing an improvement in surface finish (see expression 3). Thus, from this mechanism, with an *increase* in wheel hardness the surface finish will *deteriorate*.

(3) *Grit retention*: Peklenik and Opitz[17] have shown that as the wheel hardness grade increases, the force needed to detach a grit from the bond material i.e. the grit retention force increases. Since the force acting on a grit depends on the grit depth of cut and on the plateau area per grit[14,18,19], it follows that, provided the grinding conditions are not so severe as to cause catastrophic wheel wear throughout the grinding period (because t_1 is too great), a grit will remain held by the bond until the plateau area grows to such a value that the force acting on the grit becomes equal to the grit retention force—at which point, the grit will be pulled out of the bond. Accordingly, the higher the wheel hardness, the greater the average plateau area (after a sufficient grinding time has elapsed after dressing) and the worse the surface finish to be expected. Thus, from this mechanism, an *increase* in wheel hardness will result in a *deterioration* in surface finish *after a sufficient grinding time has elapsed.*

From these considerations, it is clear that a monotonic variation of surface roughness with wheel hardness is not to be expected. There is not a great deal of experimental evidence available on the influence of wheel hardness on surface finish. The results in ref. 4

...grits or pyramids of fixed geometry (grit size having no effect on grit shape) then, obviously because of geometric similarity considerations, r will be constant, independent of grit size. Suppose, however, the grits resemble spheres of diameter δ (Fig. 1);

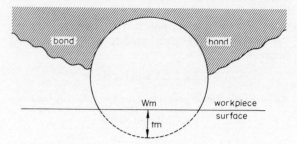

Figure 1. Diagrammatic representation of a spherical grit taking a maximum depth of cut t_m.

it is shown in the appendix that at a fixed removal rate by wheels of fixed concentration of grit/unit volume of wheel (i.e. wheels having a given structure), the width-to-depth ratio r is constant in this case, too.

Indirectly, however, with wheels of fixed grit concentration, a change of grit size influences the grit depth of cut via the concomitant change in n_c, the number of cutting grits per unit area of wheel surface ($n_c \propto C/\delta^3$). Thus as grit size increases (i.e. δ increases) n_c reduces and *CLA* increases (expression 3).

Via its influence on the average plateau area per grit, grit size can influence surface finish because of the better grit retention of large grits. For a given bond material and a given porosity of wheel bond, the number of bond posts per grit will be a function of the area of grit embedded in the bond, so that with a given dressing (and hence, a given grit protrusion) larger grits will have more bond posts per grit than smaller grits, i.e. the grit retention force will increase with the size of the grit.

This is seen to be true from the measurement of the average force required to dislodge grits from various grinding wheels when the grit retention force is found to increase as grit size increases[17]. The effect of an increase in bond retention force on surface finish has already been considered when it was shown that an increase in bond retention force produces a deterioration in surface finish because of an increase in the average plateau area per grit, a, hence, the surface finish will *deteriorate* as the grit size *increases*. Concomitantly, the grinding force will increase as the grit size increases at a fixed wheel structure and a fixed removal rate. As mentioned above experimental data consistent with these suggestions exist[4]. Makino[20] also finds that the surface finish improves as the grain size decreases when hardened steel is ground by alumina wheels. Conradi[21] quotes data obtained by grinding tungsten carbide with diamond wheels of 100 concentration, the surface finish improving monotonically from 44 μm with 40/60 grit to 24 μm with 200/230 grit.

It is, perhaps, interesting to note that the tendency for surface finish to improve as abrasive particle size decreases is not restricted to abrasive removal by fixed abrasive particles. Matsunaga[22] found that the surface finish deteriorated as the particle size increased during the lapping of steel; similarly in both the lapping[23] and polishing[24] of glass, surface finish improves as particle size decreases.

Wheel width

From expression (3) it follows that with a given wheel, a given workpiece and given dressing treatment (i.e. r, n_c, D constant), if the wheel speed V, the wheel depth of cut d, the work speed v and the cross-feed, X, are all held constant, then as the wheel width W increases, the surface finish will improve. This conforms with the experimental results obtained by Sagarda[25] when grinding steel with cup-type diamond wheels. (In addition, Sagarda found that at fixed cross-feed, wheel width and wheel speed, with a given wheel/workpiece combination the surface finish deteriorated as the normal force component, which, at constant cross-feed, is a function of the product vd (see ref. 18) increased. This, again, is consistent with expression (3).)

Quantitative

Nagi[26] has measured the roughness of steel specimens, surface ground at constant removal rate when

   (i) the cross-feed X was held constant
   (ii) the workpiece speed v was held constant and
   (iii) the wheel depth of cut d was held constant

His data are presented in Fig. 2(a), (b) and can be represented as

$$CLA = 1.3\, d^{-0.3}\text{ when } X \text{ is held constant and}$$

$$CLA = 1.3\, d^{-0.35}\text{ when } v \text{ is held constant,}$$

where *CLA* is expressed in μin and d in inches.

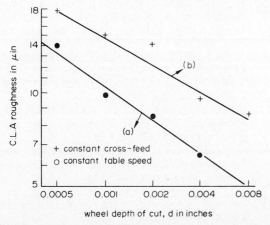

Figure 2. The centre-line-average roughness presented against wheel depth of cut at constant removal rate in surface grinding (*a*) with table speed held constant, (*b*) with cross-feed held constant. Data from ref. 26.

The corresponding values of n are 1.2 and 1.4 which are to be compared with Rubert's value of $n = 1.2$.

If, as in (iii) above, the depth of cut is held constant at $d = 0.002$ in. then the equations representing the data in Fig. 2(*a*), (*b*) suggest that the *CLA* would lie between $1.3\,(0.002)^{-0.3}$ and $1.3\,(0.002)^{-0.35}$, i.e. between 8.5 and 12.8 but, in any case, would be constant, independent of the individual values of X and v.

In the following table, Nagi's data can be seen to be constant at 11.7 ± 0.7 as predicted above.

X in	v ft/min	CLA μin
0·00625	34	11·2
0·0125	17	12·5
0·025	8·5	11·5
0·05	4·25	11·0
0·10	2·125	12·5

In contrast, Lindenbeck's expression, equation 1(d) would suggest that at constant removal rate with the table velocity held constant, the *CLA* would be constant: a conclusion which differs from the data shown in Fig. 2(*a*) where the *CLA* varies from 14 μin to 5·3 μin while the work speed is held constant at 8·5 ft/min.

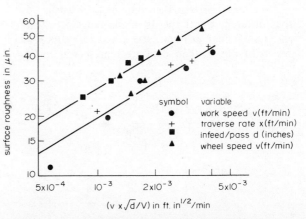

Figure 3. The centre-line-average roughness presented against $(vX\sqrt{d}/V)$. External cylindrical traverse grinding. Data from ref. 11.

In Fig. 3, Krabacher's data[11] obtained in cylindrical traverse grinding are presented as *CLA* against $vX\sqrt{d}/V$ in accordance with expression (3). The data fall, surprisingly, about two distinct lines and from the slopes of these lines, which equal 0·6, the value $n = 1.2$ is obtained, again in agreement with Rubert's value. No explanation can be offered for the fact that the data lie about two distinct lines with the separation shown in Fig. 3. The attribution of such a separation to a change in wheel diameter would necessitate the wheel diameter changing from 24 in to 8 in.

Opitz, Ernst and Meyer[2] have investigated plunge cylindrical grinding at high wheel speed under conditions such that the wheel speed to work speed ratio was constant.

$$f(t_1) = \left\{ \frac{q}{rn_c} \sqrt{\frac{d}{D_e}} \right\}^{n/2}$$

At constant removal rate per unit wheel width $Q' = vd = qVd$

$$f(t_1) = \left(\frac{1}{rn_c} \right)^{n/2} \left(\frac{qQ'}{VD_e} \right)^{n/4}$$

$\therefore CLA$ is proportional to

$$\left(\frac{1}{rn_c} \right)^{n/2} \left(\frac{qQ'}{VD_e} \right)^{n/4} \phi(a)$$

Thus for a given wheel/workpiece combination and a given dressing technique (i.e. r, n_c, constant), a given speed ratio (q constant) and given wheel and workpiece diameter (D_e constant)

$$CLA \text{ is proportional to } (Q'/V)^{n/4} \phi(a)$$

Opitz *et al.* present their data as CLA versus Q' at four different wheel speeds in the range 30–60 m/s and obtain four distinct curves, according to wheel speed. In Fig. 4, their data are represented as CLA

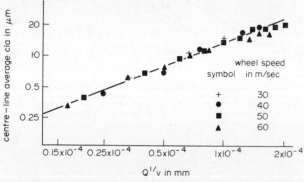

Figure 4. The centre-line-average roughness presented against removal rate per unit wheel width per unit wheel speed, Q'/V. External cylindrical plunge grinding. Data from ref. 2.

versus Q'/V and all their data now fall about a single line (log-log) as expected. The slope of this line, however, is 0·78. This implies that if $\phi(a)$ is constant, independent of (Q'/V) then $n = 3·1$, i.e. a very different value from the previously obtained values of 1·2. Alternatively, of course, $\phi(a)$ may be dependent on (Q'/V) in this case—there is no way of determining from their results whether or not this is true.

CONCLUSIONS

Deductions from the simple hypothesis advanced above have been shown to be in accord with published information concerning the influence of a

data lend support to the hypothesis, the subsequent analysis and Rubert's observation that the peak-to-valley roughness, h, is proportional to $(CLA)^{0·85}$ in grinding.

REFERENCES

1. C. T. YANG and M. C. SHAW (1955). *Trans. Am. Soc. mech. Engrs.*, **77**, 645.
2. H. OPITZ, W. ERNST and K. F. MEYER (1965). *Proceedings of the 6th Int. M.T.D.R. Conference*, Pergamon Press, 581.
3. H. TSUWA (1964). *Trans. Am. Soc. mech. Engrs.* Series B, **86**, 371.
4. R. E. McKEE, R. S. MOORE and O. W. BOSTON (1947). *Trans. Am. Soc. mech. Engrs.*, **69**, 891.
5. R. S. HAHN and R. P. LINDSAY (1969). *Proceedings of the 10th Int. M.T.D.R. Conference*, Pergamon Press. 95.
6. W. R. BACKER, E. R. MARSHALL and M. C. SHAW (1952). *Trans. Am. Soc. mech. Engrs.*, **74**, 61.
7. D. A. LINDENBECK (1970). Schleifen von Eisenwerkstoffen mit Diamantschleifscheiben, Dr-Ing. Dissertation, Technical University, Hannover.
8. M. P. RUBERT (1959). *Engineering*, **188**, 393.
9. C. RUBENSTEIN (1967). *Machinability* (Iron and Steel Institute Conference on Machinability, London 1965), Iron and Steel Inst. Special Report Series No. 94, 49.
10. C. RUBENSTEIN, F. K. GROSZMANN and F. KOENIGSBERGER (1967). *Science and Technology of Industrial Diamonds* (Proceedings of the International Industrial Diamond Conference, Oxford), vol. 1, 161. Industrial Diamond Information Bureau, London.
11. E. J. KRABACHER (1959). *Trans. Am. Soc. mech. Engrs.*, **81** (Series B), 187.
12. D. A. LINDENBECK (1971). *Ind. Diamond Rev.*, **36**, 400.
13. J. N. GREENHOW and C. RUBENSTEIN (1969). *Int. J. Mach. Tool Des. Res.*, **9**, 1.
14. V. PACITTI and C. RUBENSTEIN (1972). *Int. J. Mach. Tool Des. Res.*, **12**, 267.
15. P. D. SINGHAL and H. KALIZER (1965). *Proceedings of the 6th. Int. M.T.D.R. Conference*, Pergamon Press. 629.
16. G. PAHLITZSCH and E. O. CUNTZE (1965). *Proceedings of the 6th Int. M.T.D.R. Conference*, Pergamon Press, 507.
17. J. PEKLENIK and H. OPITZ (1962). *Proceedings of the 3rd Int. M.T.D.R. Conference*, Pergamon Press, 163.
18. C. RUBENSTEIN (1972). *Int. J. Mach. Tool Des. Res.*, **12**, 127.
19. C. E. DAVIS and C. RUBENSTEIN (1972). *Int. J. Mach. Tool Des. Res.*, **12**, 165.

20. H. MAKINO (1966). *Bull. Japan Soc. Prec. Eng.*, **1**, 281.
21. V. CONRADI. Diamond abrasive wheel technology, Diamond Information L. 13, De Beers Industrial Diamond Division.
22. M. MATSUNAGA (1966). Report of the Institute of Industrial Science, University of Tokyo, vol. 16, no. 2, Serial No. 105.
23. O. IMANAKA (1966). *Annals CIRP*, **13**, 227.
24. Y. ISHIDA (1964). *J. Mech. Lab. Japan*, **10**, 76.
25. A. A. SAGARDA (1969). *Machines and Tooling*, **40**, no. 7, 47.
26. H. M. NAGI (1972). MSc dissertation, University of Manchester Institute of Science and Technology.

APPENDIX: THE WIDTH TO DEPTH RATIO FOR SPHERICAL GRITS

Figure 1 shows that the chord of width ω_m subtends an angle θ at the centre of the diametral section of the sphere. Then the area of the segment bounded by the chord and the smaller arc is $(\delta^2/8)(\theta - \sin\theta)$.

$$\sin\theta = \theta - \theta^3/3! + \ldots$$

The area is $\dfrac{1}{48}\delta^2\theta^3$

Now if t_m is small $\theta \simeq 2\omega_m/\delta$, i.e.

$$\text{area of sector} \propto \frac{\omega_m^{\,3}}{\delta}$$

If the concentration of abrasive grit/unit volume of the wheel C, is fixed, then when the grit size is such that the average grit diameter is δ the number of grits/volume of wheel $\propto C/\delta^3$ and the number of grits/area wheel surface $\propto 1/\delta^2$, hence, the number of cutting grits/area wheel surface $n_c \propto 1/\delta^2$.

Suppose the removal rate $= Q$, then removal rate per grit $\propto Q\delta^2/WD$, and the cross-sectional area of workpiece removed, per grit, is proportional to $Q\delta^2/WDV$.

But this has been seen to be expressed as $\omega_m^{\,3}/\delta$, i.e.

$$(Q\delta^2/WDV) \propto (\omega_m^{\,3}/\delta)$$

hence, at a fixed removal rate, at fixed wheel speed with given wheel dimensions and a fixed concentration of grit per unit volume of wheel

$$\omega_m \propto \delta$$

Now $r \sim \dfrac{\omega_m}{t_m}$

and for small t_m, $t_m \propto \dfrac{\omega_m^{\,2}}{\delta}$

$r \sim \dfrac{\delta}{\omega_m}$ i.e. constant.

DISCUSSION

Q. J. Peters. May I make a general comment: Looking at the paper by Dr R. Hahn and Dr Lindsay, the one of Dr Cegrell and the one of Professor Rubenstein, I like to insist that notwithstanding a different terminology and symbolism they finally emphasize a very similar idea.

In Figs. 5 and 6 Dr Hahn and Dr Lindsay show that the different curves of Fig. 3 and Fig. 4 can be replotted in a single curve with respect to a fundamental parameter they call T_{ave}—similarly Dr Rubenstein shows in Fig. 4 that surface finish can be plotted in a single curve with respect to a parameter Q'/V. Also Dr Cegrell showed a slide not reproduced in his paper with a similar conclusion.

I would further like to quote the cooperative work done in C.I.R.P. where the formulae proposed by prominent authors all include a similar factor, precisely the same as above. This factor can be called the *equivalent chip thickness* or the *equivalent cutting depth,* using the standard symbol h_{eq}.

The physical meaning of h_{eq} is obvious and must be explained:

$$h_{eq} = \frac{av_w}{v_s}$$

where a is the infeed/min, v_w is the workspeed, v_s is the wheel speed.

Consequently $av_w = Z'_w$ = stock removal rate.

When divided by the wheel speed it just means the thickness of the cut material if this is measured over the displaced path of the grinding wheel circumference:

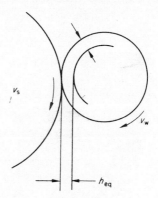

Using this parameter, roughness, forces, stock removed, etc. can be expressed in a graph quite conveniently.

I would suggest to refer to *C.I.R.P. Annals* where this method has been described, and suggest the authors insist more on their common points than on the second order differences that separate their theories.

A. While I would not disagree with the very laudable aim expressed by Professor Peters that, wherever possible, those points on which we are all agreed should be clearly indicated, I cannot agree that h_{eq} is, in all circumstances, the relevant parameter against which roughness variation and force variation, etc., are to be compared.

One obvious deficiency of h_{eq} is the absence of the cross-feed—a factor known to influence the surface finish in grinding. Thus, the parameter h_{eq} could only be applicable to a plunge grinding operation. In fact, as shown in the paper, the analysis offered here suggests that for a plunge grinding operation the surface finish is related (by a power relation) to $(v_w/v_s)a$ (to use the same notation as Professor Peters).

Where we can agree—and this is, I think, the essential

Rubenstein in Fig. 4 plots some results from Opitz   1. H. Opitz and K. Guhring.

THE EFFECT OF HIGH DEPTH OF CUT ON GRINDING PERFORMANCE OF ALLOY STEELS

by

T. MATSUO*

SUMMARY

This paper reports the results of a systematic test program designed to determine the effect of high depth of cut on grinding performance on conventional surface and cylindrical grinding machines. The grinding tests have been carried out on several soft and hardened alloy steels under the operating conditions of extremely high depths of cut and high table speeds. The optimum operation condition is discussed for application to actual abrasive machining, from the standpoints of the metal removal rate, wheel wear, grinding ratio, power, surface finish, and dimensional error.

INTRODUCTION

Recently, abrasive machining has attracted special interest. Grinding is normally associated with small rates of metal removal, but there has been a recent trend to extend this rate to higher values and into the range of other machining processes using single point tools by increasing the depth of cut or feed rate.

It is certainly true that in certain applications this abrasive machining offers economical advantages over other methods using single point tools. However, the selection of proper grinding conditions is of substantial practical significance in the abrasive machining. There is a considerable number of papers on grinding[1-3], but hitherto very little systematic research[4-6] has been done on abrasive machining. According to the work of Krabacher[7], who studied the influence of operating variables on the grinding performance, using a cylindrical grinder, it was shown that an increase in depth of cut increases wheel wear, decreases grinding ratio, deteriorates the surface finish, and increases the dimensional error. Many more negative effects are expected in abrasive machining, but the increase in the metal removal rate may be effectively obtained.

However, no detailed information is available on the consideration that we should apply to the execution of abrasive machining. Thus, it is intended that this paper will give an appreciation of the extent to which desirable abrasive machining may be successfully applied in surface or external cylindrical grinding.

EXPERIMENTAL PROCEDURE

The conventional surface grinder (PSG 6E, 1·5 kW)

TABLE 1 Work materials and grinding conditions

	Surface	Cylindrical
Work materials	SKD 11 steel (H_v 200) SKH 4 steel (R_c 67) Maraging steel (R_c 52)	SCM 3 steel (H_S 290) SNCM 26 steel (R_c 35)
Grinding wheel	WA46KmV, WA60JmV PA60J8V	WA60KmV, DA60KmV 19A46KmV
(Wheel size)	(205 × 19 × 50·8)	(405 × 75 × 152·4) (355 × 50 × 127·0)
Depth of cut	0·01 to 0·20 mm	0·01 to 0·20 mm
Total depth of cut	1 to 4 mm	0·4 to 1·0 mm in dia.
Table speed	3 to 15 m/min	0·3 to 2·0 m/min
Workpiece rotating speed	–	10 to 65 m/min
Wheel speed	2000 to 2400 m/min	2000 to 2600 m/min
Dressing	(15 to 25 μ) × 2	(15 to 25 μ) × 2
Coolant	Dry, Emulsion HSG 55	1·3 per cent solution (TL 131) Soluble

and external cylindrical grinder (45MGU28, 5·5 kW) were used in the research. The cylindrical grinding was of traverse type, and no feed was used for the surface grinding. The work materials and grinding conditions used in the study are given in Table 1. The SKD 11 steel is a 1·4 per cent C, 13 per cent Cr tool steel, while the SKH 4 steel is a typical high speed steel. The SCM 3 and SNCM 26 steels are a Cr-Mo and a Ni-Cr-Mo low alloy steels, respectively. The 6 mm width, 100 mm length plate was used for the surface grinding experiments, and the 60 to 120 mm

* Department of Production Engineering, Kumamoto University, Japan

×

DISCUSSION OF RESULTS

1. Surface grinding results

A strict linear relationship is seen between the wheel wear and grinding time under almost all of the operating conditions. The metal removal rate is also shown to be proportional, to some extent, to the depth of cut. The results obtained in the surface grinding on SKD 11 steel are presented in Fig. 1,

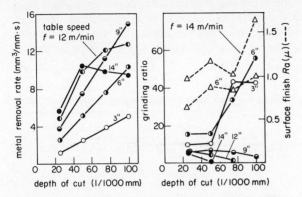

Figure 1. The effect of depth of cut and table speed on grinding performance for SKD 11 steel.

which shows the variation in actual metal removal rate, grinding ratio, and surface finish with increasing depth of cut and table speed. It is evident from the illustration that the metal removal rate is increased linearly with the increase in depth of cut, at table speeds lower than 9 m/min. It has been generally understood[8] that the grinding ratio decreases with the increase in metal removal rate. However, this the present experiment. Also, no special vibration appeared. However, a large distortion was found when the SKH 4 steel was ground at a depth of cut of 0·05 mm. The amount of coolant seems to be largely effective for controlling the distortion of work.

Cylindrical grinding results

Table 2 lists part of the results obtained in the external cylindrical grinding experiments on the SCM 3 steel, where the grinding performance is shown for two table speeds. It is clearly seen from the table that the wheel wear increases and the grinding ratio decreases with increasing depth of cut. However, it should be noted here that a relatively good tolerance is held even after being ground at high depths of cut. Grinding force is also not so high. Accordingly, it is expected to execute grinding at higher depths of cut than are usually used.

Table 3 shows the results obtained from grinding the SNCM 26 steel under the same operation conditions. Two types of wheel have been used for the study. The result is generally the same as that for the SCM 3 steel. The effect of table speed in the performance is seen in Fig. 2 where the SCM 3 steel was tested at a wheel speed of 2400 m/min, using the WA60K8V wheel. Table speed effect was also studied in the grinding of SNCM 26 steel. As a result, the optimum operating condition is considered a combination of about 0·05 mm of depth of cut and a table speed of about 0·7 m/min. The change in horsepower with grinding time is presented in Fig. 3. It is understood for the illustration that a proper wear of the wheel results in the presence of a steady state of power consumption.

TABLE 2   The effect of depth of cut in cylindrical grinding of alloy steel

Work material: SCM 3 (60 mm in diameter), wheel: WA60KmV (405 × 75 × 152·4), wheel speed: 1850 m/min, work rotating speed: 21 m/min, coolant: soluble (×80)

Depth of cut (1/1000 mm)	Table speed (m/min)	Grinding time (min)	Metal removal rate (mm³/mm . s)	Wheel wear (mm³/sec)	Grinding ratio	Power (kW)	Surface finish (Ra, μ)	Diametral error (μ)
10	0·5	40	0·124	0·166	45·2	0·55	0·30	2
25	0·5	16	0·317	0·724	26·3	1·29	0·35	2
50	0·5	8	0·686	2·39	17·2	1·89	0·75	3
100	0·5	4	1·17	9·15	7·40	2·32	0·92	2
150	0·5	3·2	1·41	90·8	0·92	2·66	1·20	6
10	1·0	20	0·226	0·365	37·1	0·79	0·45	1
25	1·0	8	0·467	5·67	10·5	1·54	0·85	1
50	1·0	4	1·02	17·1	2·8	2·69	1·50	3
100	1·0	2	1·95	93·9	1·3	2·74	5·30	5

TABLE 3   The effect of depth of cut in cylindrical grinding of alloy steel

Work material: SNCM 26 (120 mm in diameter), wheel speed: 2600 m/min, work rotating speed: 25 m/min, coolant: 1·3 per cent solution TL131

Depth of cut (1/1000 mm)	Table speed (m/min)	Grinding time (min)	Metal removal rate (mm³/mm.s)	Wheel wear (mm³/sec)	Grinding ratio	Power (kW)	Surface finish (R_{max}, μ)	Diametral error (μ)	Wheel specification
10	1·0	37·1	0·66	0·465	53·0	1·42	3	>5	
25	1·0	14·8	1·78	1·42	45·3	2·98	6	10	WA60KmV
50	1·0	7·41	2·90	9·52	10·9	3·55	9	10	(355 × 50 × 127)
75	1·0	5·06	3·72	27·5	4·6	3·76	10	20	
100	1·0	3·81	4·02	45·5	3·2	4·33	10	25	
150	1·0	3·04	4·07	93·1	1·7	6·15	11	30	
200	1·0	2·24	4·76	145	1·2	7·18	12	–	
25	1·0	15·7	1·32	1·42	36·9	2·21	4	–	
40	1·0	14·6	1·83	6·68	10·8	2·32	10	–	DA60KmV
50	1·0	11·5	2·28	10·5	8·6	2·97	–	–	(355 × 50 × 127)
75	1·0	9·15	3·04	22·4	5·3	3·46	–	–	
100	1·0	7·58	3·29	39·5	2·1	3·56	20	–	
150	1·0	6·14	3·53	81·3	1·7	5·88	21	–	

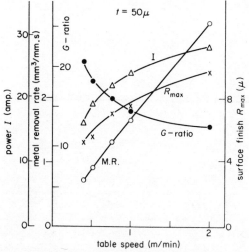

Figure 2. The effect of table speed in cylindrical grinding performance for SCM 3 steel.

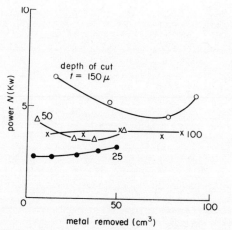

Figure 3. The variation in horsepower with the metal removed.

CONCLUSION

From the above-mentioned experimental work it may be concluded that the high depth of cut grinding has

promising possibilities for an abrasive machining method, depending on the operating condition. However, additional work is necessary for the application of experimental results to actual abrasive machining operation. Especially, more research should be done to investigate fully the limit of the application in this grinding from the standpoints of the surface quality and dimensional error. Studies are continuing to establish a definitive operating condition.

ACKNOWLEDGMENT

This work was done as a part of the co-operative research in the Sectional Committee on Heavy Grinding of Steels, Jap. Soc. Prec. Eng.

The author is grateful to the members of the committee for their help.

REFERENCES

1. R. E. McKEE, R. S. MOOR and W. BOSTON (1948). An evaluation of cylindrical grinding performance *Trans. Amer. Soc. Mech. Engrs.*, **70**, 893.

2. R. S. HAHN and R. P. LINDEAY (1969). The influence of process variables on material removals, surface integrity, surface finish and vibration in grinding, *Advance in Machine Tool Design and Research*, p. 95.

3. H. OPITZ (1967). High efficiency grinding, *Proc. of Seminar on European Development in Horizons in Manufacturing Technology*, April, 84.

4. L. P. TARASOV (1967). Factors affecting grindability of highly alloyed steels, *Proc. in International Conference on Manufacturing Technology, Proc. A.S.T.M.E.*, Sept., 689.

5. C. POLLOCK (1967). The effect of some operating variables in vertical spindle abrasive machining of mild steel, *Trans. Amer. Soc. Mech. Eng., Series B, J. Eng. for Ind.*, May, 323.

SURFACE FINISH IN DIAMOND GRINDING OF STEEL

by

D. A. LINDENBECK*

SUMMARY

It has already been shown in many publications that diamond grinding of steel, in terms of stock removal, is viable. This paper shows that by using different techniques and machine parameters, high quality surface finishes can be achieved without sacrificing stock removal rates and adversely affecting the economics of a process. It also indicates that the most effective way of obtaining a better surface finish is by dressing the diamond wheel with a diamond impregnated dresser, improving the surface finish from R_a 2·2 μm to R_a 0·2 μm.

NOTATION

a	downfeed
a_M	downfeed applied to the machine
$a_{1.1}(t)$	effective downfeed for section 1 during the first grinding pass
$a_{2.1}(t)$	effective downfeed for section 2 during the first grinding pass
B	rim width
b	number of relative extrema
D	wheel diameter
d_d	total downfeed necessary for wheel dressing
F_q	average undeformed chip cross-section
H	wheel profile
l_a	length of workpiece on which one circumference of the grinding wheel is delineated
N_D	number of diamonds per unit area of wheel surface
N_{SP}	number of cutting processes per unit area of workpiece surface
n_s	wheel speed
R_a	workpiece roughness (average)
R_{a0}	minimum possible workpiece roughness
R_p	wheel roughness
$\Delta R_{1.1}$	radial wear of section 1 during the first grinding pass
S	compression factor for delineated length of wheel circumference
s	crossfeed
u	table speed
v_s	peripheral speed
W_W	wheel wear per unit workpiece material removed
α,β,γ	constants

INTRODUCTION

During the last five years much has been published about the grinding of ductile materials, mainly steel, with diamond wheels. Most of these papers were concerned with the economics of the grinding process, and few mentioned the surface finish of ground ductile materials. Very often, roughness is the major factor in deciding the economics of a specific grinding operation and whether or not it is acceptable.

One of the major differences between conventional and diamond grinding wheels is the number of cutting points on the wheels. In a normal 100 concentration diamond grinding wheel, diamond particles make up 25 per cent by volume of the impregnated layer, whereas in conventional wheels the proportion of abrasive particles is approximately 80 per cent. A conventional wheel will therefore have a significantly higher number of grinding points on the surface than a diamond wheel of the same dimensions and containing the same grit size, resulting in a lower surface roughness on the workpiece.

Because a conventional wheel (Al_2O_3 or SiC) is easily dressed, the grinding process is usually divided into two parts.

(a) rough grinding, for stock removal
(b) finish grinding, for surface quality.

The low primary cost, and the rapidity with which the conventional wheel can be dressed, permit a dressing operation to be carried out before the start of every rough grinding process, leaving a high degree of roughness on the grinding wheel for relatively rapid stock removal. On completion of the rough

---

* De Beers Diamond Research Laboratory

methods of improving the workpiece surface quality using diamond wheels.

THE WEAR PROCESS IN SURFACE GRINDING WITH PERIPHERAL WHEELS

In order to understand the evolution of the workpiece surface, it is necessary to analyse the effect of the wear process on the wheel geometry.

Figure 1(a) shows the initial situation when the workpiece is completely even and the grinding wheel can be represented by a cylinder.

Figure 1(b) shows the geometric conditions when a crossfeed of s is applied. During the first table traverse the effective downfeed changes because of the wear which occurs in section 1 of the grinding wheel. The geometric conditions are shown in Fig. 1(c).

Depending on contact time, $a_{1.1}(t)$ is the effective downfeed in section 1 of the grinding wheel width during the first grinding pass. a_M is the downfeed applied to the wheel. $\Delta R_{1.1}(t)$ is the decrease of the wheel radius in section 1 during the first grinding pass.

The effective downfeed for section 2 of the wheel is:

$$a_{2.1}(t) = \Delta R_{1.1}(t) - \Delta R_{2.1}(t) \tag{2}$$

For the third section of the wheel it is:

$$a_{3.1}(t) = \Delta R_{2.1}(t) - \Delta R_{3.1}(t) \tag{3}$$

If this process is continued many times, a step profile shown in Fig. 1(e), will be obtained.

For finite surfaces and normal wear values, the influence on the last section is relatively low, so that, for the crossfeed in the opposite direction, a cylindrical wheel can be assumed up to about the middle of the wheel rim. After the workpiece has been ground with crossfeed in both directions, the profile shown in Fig. 1(f) will be obtained.

During successive grinding passes, the effective downfeed is increased by the amount of material which was not removed by the centre part of the wheel during the previous pass. Therefore, the effective downfeed on each section of the wheel changes, not only because of the wear of the adjacent section of the wheel during the current pass, but also because of the wear of the centre part of the wheel during the previous pass.

Taking these *effects* into consideration and assuming an *exponential function* (4) as a model for the interrelation between *radial wear* and effective *downfeed*, the effective downfeed for each section of the wheel can be calculated in relation to the grinding time. The formula was derived from practical grinding results carried out at the DRL.

$$W_W = A \exp(\alpha a + \beta s + \gamma u) \tag{4}$$

Figure 2 shows an example of the effective downfeed for three sections on the wheel width in relation

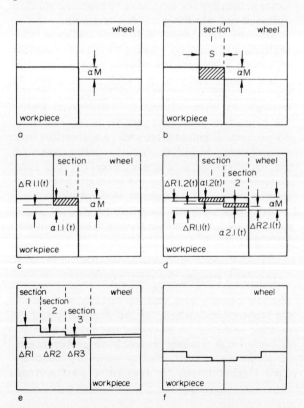

Figure 1. Diagrammatic representation of the development of the grinding wheel profile.

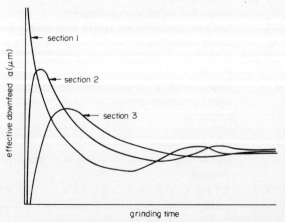

Figure 2. The interrelation between the effective downfeed and the grinding time for three sections of a grinding wheel.

to the number of passes for crossfeed in one direction only. Initially, section 1 feeds the total downfeed a_M. The effective downfeed decreases quickly and the load is placed on the second section, where, at that time, it is higher than that of the first section.

After several passes, however, these differences equalize and the effective downfeed becomes the same for all sections of the wheel, i.e. the profile of the wheel develops into steps of equal height.

In practice, the profile does not develop in definite steps because the application of the crossfeed is not very accurate. The corners of the steps developed and the outside edges of the wheel break away, while development from both sides overlaps in the centre of the wheel face profile, tending to flatten the centre. Thus, a profile which can be described as roof-shaped is formed, with the peak of the roof and the edges rounded. Furthermore, this profile is superimposed with a roughness which depends on the grinding conditions and on the grinding wheel parameters.

DESCRIPTION OF WHEEL SURFACE

From this wear process, it follows that the wheel surface developed can be measured and described by the base profile and the superimposed roughness. The measurement is effected by a depth gauge with a knife edge of 2 mm x 0·1 mm, in steps of 0·1 mm over the width of the wheel. The points are plotted and connected by straight lines. A typical profile measured in this way is shown in Fig. 3. According to the principle of the profile development, the base profile is calculated as follows.

The profile of a wheel consists of a series of measurements; of which one point is the absolute maximum and one point the absolute minimum. The absolute maximum is generally near the centre of the wheel face, the absolute minimum near one of the two sides. The measurements also consist of a number of relative maxima b and relative $b + 1$ or $b - 1$ minima. These extremes are analysed and the average of the coordinates of each of the adjacent extremes calculated. These averages represent the so-called first average line of the wheel profile. The same procedure is then carried out with the extremes of the first average line to get the second average line until only three or four points are left. The two lines drawn through these points represent the base profile of the wheel. If the last average line is drawn through only two points, the second last average line is drawn by two straight lines which are calculated by the system of the least square. The meeting point of the two straight lines (tip of the profile) is the reference point for further calculations. To obtain the slope of the two base profile lines, the heights between the tip of the profile and the two meeting points of the profile lines with the outside edges of the wheel are calculated. Reference is made to H, as the average of the two values. Faults in diamond distribution can be detected when the two values differ significantly.

The superimposed wheel roughness on the base profile is referred to as R_p (levelling depth value), and is calculated from the sum of the distance of all profile measurements on either side of the centre from the parallel to the base profile through the highest point, divided by the number of measurements (Fig. 3).

EVOLUTION OF THE WORKPIECE SURFACE

The wheel surface already described comes into contact with the workpiece surface. However, only about half of the wheel width will cut at any one time when it has a profile similar to that shown in Fig. 3. For each rotation of the grinding wheel the circumference of the wheel is delineated by length l_A on the workpiece. This length depends on the table speed and wheel speed and is given by

$$l_A = \frac{u}{n_s} \qquad (5)$$

where u = table speed, n_s = wheel speed.

The compression factor S_u is defined as the ratio of the circumference of the wheel to the length l_A:

$$S_u = \frac{n_s \cdot \pi \cdot D}{u} \qquad (6)$$

D = diameter of the wheel.

The workpiece surface is also affected by the number of cutting processes N_{SP} which occur per square unit on the workpiece surface.

$$N_{SP} = S_u \cdot N_D = \frac{N_D \cdot n_s \cdot \pi \cdot D}{u} \qquad (7)$$

N_D is the number of diamond particles per square unit on the surface of the grinding wheel. If N_{SP} is high, the super-imposition of the individual cutting processes will also be high, so that a low surface roughness may be expected.

The roughness on the workpiece depends on two components.

(1) A constant as a minimum value, called R_{a0}, for the super-imposition of all wheel profiles on one line.

(2) A part which is reciprocal to the number of cutting processes per square unit.

$$R_{a0} + h\left(\frac{u}{N_D \cdot n_s \cdot \pi \cdot D} \cdot R_{a0}\right) \qquad (8)$$

The minimum value R_{a0} depends consequently on the wheel roughness R_p and can be expressed as

$$R_{a0} = f(R_p)$$

The foregoing facts relate to the roughness of the spherical surface created on the workpiece by the

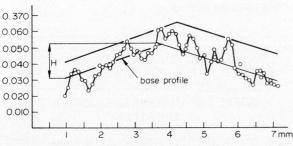

Figure 3. The measured wheel profile.

$$f(R_p) + h\left(\frac{u}{N_D \cdot n_s \cdot D \cdot \pi} \cdot R_p\right) \qquad (9)$$

Therefore, the workpiece surface will automatically consist of parallel lanes, each of width s, which are compressed mappings of the centre section of the wheel surface and, because of base profile, are not flat. Thus, waves are formed on the workpiece surface in addition to the roughness within each parallel lane.

The average roughness on a workpiece surface is usually expressed by the value R_a, which depends on the roughness measured within the single lanes only, depending on R_p, and the wave pattern which is a function of the wheel profile expressed by H. (Waviness can furthermore be caused on the workpiece because of machine vibration, spindle deflection etc. These effects are, however, not considered in this paper.) When wider wheels are used, the influence of the wave value of H diminishes. Therefore, the influence of the wheel profile in all formulae is expressed by H/B. (B = rim width.)

The workpiece roughness R_a can then be formulated in the following way:

$$R_a = f(R_p) + g\left(\frac{H}{B}\right) + h\left(\frac{u}{N_D \cdot n_s \cdot \pi \cdot D} R_p\right) \quad (10)$$

This formula represents the surface roughness obtained when the operating process is carried out under constant conditions for wheel roughness R_p and wheel profile H, which means that there was neither a change in grinding parameters at the end of the grinding process nor any spark-out.

With two further subtractive elements in the equation, formula (11) extends to include the change of machine parameters:

$$R_a = f(R_p) + g\left(\frac{H}{B}\right) + h\left(\frac{u}{N_D \cdot n_s \cdot \pi \cdot D} R_p\right)$$
$$- i\left(\frac{H}{B} \cdot \Delta s\right) - k\left(\frac{H}{B} \cdot \Delta a\right) \qquad (11)$$

Δa and Δs are the change of grinding parameters between the initial rough grinding process (a_1 and s_1) and the finish grinding process (a_2 and s_2) with $a_1 > a_2$ and $s_1 > s_2$.

The functions i and k only exist while there is no significant change of the values of R_p and H. This can be assumed for light brief finish grinding conditions using diamond wheels, since the wheel wear is low. R_p and H are determined during the time of rough grinding, when considerably more stock is removed.

The functions i and k require further explanation. By changing the machine parameters for a short time for finish grinding, functions h, i and k show that it is possible to improve surface finish on the workpiece. By decreasing table speed and increasing wheel speed,

For the function k, two main points have to be taken into account, Firstly, the geometrical cutting conditions for the diamond particles in the centre of the wheel do not change when a smaller downfeed is applied, because, provided H is almost constant, only the outer edges of the wheel are not in contact with the workpiece. Because R_p is constant, the roughness within the parallel lanes is again unaffected. However, because the normal force is considerably lower at smaller downfeeds, the overall geometrical accuracy of the workpiece is improved.

The second aspect is mainly related to spark-out processes where the effective downfeed changes after each grinding pass. If, under these conditions, the crossfeed position can be arranged in such a way that the wheel does not travel along the same lanes as during the previous grinding pass, but is offset by half the crossfeed width, then the peaks of the waves created during the previous pass are removed by the centre of the wheel. When the effective downfeed—and with that the normal force—is sufficiently decreased, the wheel profile will only remove the peaks of the waves with less than the width of the crossfeed, thus decreasing the wave depth by leaving at least double the number of parallel lanes. As the decrease of the downfeed under certain conditions mainly changes the wave pattern which is caused by the wheel profile H, the function k is expressed by $(H/B) \cdot \Delta\alpha$. The required off-set of the crossfeed position can be assumed because of the very erratic functioning of the crossfeed system on most grinding machines.

From the detailed discussion above, the following suggestions for improving workpiece finish can be made. On a grinding operation, with constant grinding conditions, the roughness depends on the wheel profile H, which gives a wave pattern on the workpiece, and the wheel roughness R_p, which gives the roughness within the wave, H and R_p being dependent on the grinding conditions. Equation (11) shows that a brief second operation with an increased wheel speed and a decreased downfeed, crossfeed and table speed will improve the workpiece finish, each of them in a different manner.

INTERRELATIONSHIP BETWEEN WHEEL PROFILE, WHEEL ROUGHNESS, WORKPIECE ROUGHNESS, GRINDING CONDITIONS AND WHEEL PARAMETERS IN GRINDING D2 DIE STEEL

To ensure the development of the characteristic profile and roughness on the diamond wheel, the measurements were taken after a grinding time which ensured radial wear on the wheel of more than the depth of the downfeed applied.

During the tests, no attention was paid to obtaining a good surface finish, i.e. the workpiece was ground under specific conditions without spark-out so that the measured workpiece, the wheel roughness and wheel profile are characteristic for those conditions.

Influence of downfeed, crossfeed and table speed

Figures 4, 5 and 6 show the influence of the machine parameters on the wheel base profile H, the wheel roughness R_p and the workpiece R_a. The parameters are chosen to cover a reasonable range of theoretical stock removal rates within the capabilities of the grinding machine. It is interesting to note that the number of diamonds per surface area is approximately 400/cm² without showing any definite trend. The grinding conditions were as follows.

	Figure 4	Figure 5	Figure 6
downfeed	varied	0·100 mm	0·025 mm
crossfeed	2 mm	2 mm	varied
table speed	16 m/min	varied	4 m/min
wheel speed	30 m/s	30 m/s	30 m/s
wheel type	D1A1 Resin Bond 80/100 mesh		175 x 10 mm 100 conc.

workpiece material D2 die steel (AISI)

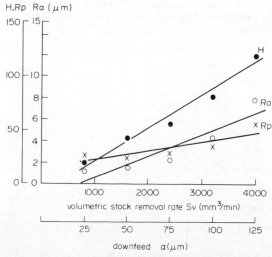

Figure 4. Interrelation between wheel roughness, base profile, workpiece roughness and downfeed. ● Ra, x Rp, ○ H. H = wheel profile; Rp = wheel roughness; Ra = workpiece roughness.

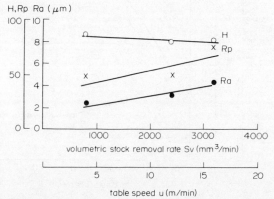

Figure 5. Interrelation between wheel roughness, base profile, workpiece roughness and table speed. ● Ra, x Rp, ○ H. H = wheel profile; Rp = wheel roughness; Ra = workpiece roughness.

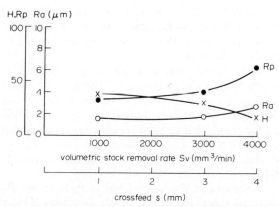

Figure 6. Interrelation between wheel roughness, base profile, workpiece roughness and crossfeed. ● Ra, x Rp, ○ H. H = wheel profile; Rp = wheel roughness; Ra = workpiece roughness.

The various machine parameters each have differing influences on the value H. A change of the downfeed has a very pronounced effect on H (Fig. 4) which is always slightly less than the downfeed applied, confirming theoretical considerations discussed previously. When the table speed is changed, but downfeed is maintained constant at 0·1 mm, a constant value of H might be expected. However, the increased wear of the wheel at higher stock removal rates also has an effect on the base profile (Fig. 5). This is understandable if one considers a grinding process on an infinitely wide workpiece. Eventually the downfeed, which is applied at the start, is worn off the grinding wheel, returning the wheel to a straight profile parallel to the axis. At intermediary stages, H has values between downfeed a and zero as the wear progresses. For ease of explanation, a higher wear at higher table speeds can be considered as grinding of a wider workpiece, thus explaining the slight drop of H with increased table speed.

An even more pronounced drop in H can be observed when crossfeed is increased (Fig. 6), a situation which decreases the number of steps across the wheel width. Thus the centre of the wheel has to do more work from the beginning. When the crossfeed equals half the wheel width, apart from edge chipping, an almost straight profile, parallel to the wheel axis, should be obtained. With a crossfeed of 4 mm in the test, this situation almost pertains (Fig. 6).

Figures 4 to 6 show the wheel roughness R_p, which increases at higher stock removal rates. This increase is most pronounced with an increase in the table speed.

According to equation (12), an increase of the stock removal rate obtained by an increase of table speed (u), crossfeed (s) or downfeed (a) increases the average undeformed chip cross-section cut by one diamond F_q.

$$F_q = \frac{u \cdot \sqrt{2a \cdot s}}{v_s \cdot N_D \cdot \sqrt{D \cdot B}} \qquad (12)$$

A bigger chip cross-section means that larger steel chips have to be moved along between workpiece and wheel surface, consequently increasing the bond erosion and leaving deeper grooves, which are measured in a higher R_p value on the wheel surface.

H, has the most pronounced influence on the R_a value as shown in Fig. 4.

Although, in the case of higher table speed, R_p increases more rapidly, the slightly decreasing H value is the more important factor in limiting the increase of the workpiece roughness R_a (Fig. 5).

This effect can be seen again when crossfeed is increased. Although the wheel roughness increases, as it did when changing the downfeed, the base profile H decreases, so that, in total, a higher crossfeed results in the least increase of the workpiece roughness mainly because of a reduced influence of wave pattern.

It can therefore be concluded that the most economical way to get a reasonable finish at high stock removal rates is the application of low downfeed but high table speeds and maximum possible crossfeeds.

Influence of wheel parameters

Figures 7 and 8 show the relationship between diamond concentration, diamond particle size, base profile, wheel roughness, number of diamonds and workpiece roughness.

The following were the grinding conditions:

downfeed	0·05 mm
crossfeed	1·5 mm
table speed	16 m/min
wheel speed	30 m/s
wheel type	D1A1 resin bond 125 x 6·4 mm
workpiece material	D2 die steel (AISI)

An increase of the concentration from 60 to 120, at constant grit size, doubles the number of diamonds per unit area of the surface and improves, according to equation (11), the workpiece surface finish. However, there are two other factors which have the opposite effect, i.e. the wheel roughness R_p and the base profile H. These increase with the concentration, because the considerable increase in wear at lower concentrations affects the wheel profile H in the same way as increasing the crossfeed or table speed, H decreasing at higher wear rates.

The increased number of diamonds N_D at higher concentrations decreases the average chip cross-section per diamond, which means that the diamond can protrude more from the bond before breaking out, thus increasing the wheel roughness R_p.

The result of the counteracting effects of N_D, H and R_p is that the workpiece roughness R_a remains practically unchanged at different concentrations (Fig. 7). However, the better overall accuracy and the more economic grinding at lower wear rates suggest the use of high concentrations.

A change of diamond grit size theoretically raises the number of diamonds N_D by the power of two.

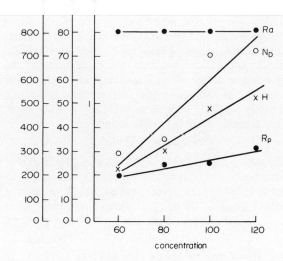

Figure 7. Interrelation between wheel roughness, base profile, workpiece roughness, number of diamonds and concentration.

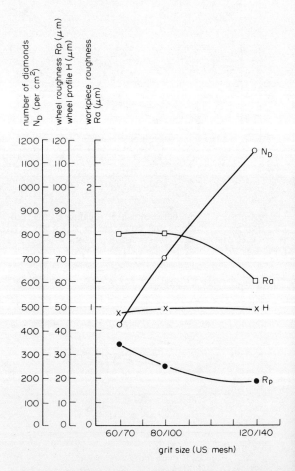

Figure 8. Interrelation between wheel roughness, base profile, workpiece roughness, number of diamonds and grit size.

However, because smaller diamonds are not held as securely in the bond, the increase of N_D is below the theoretical value and a higher wear can be expected with smaller grits. But the change of the wear rate in the investigated range was not very pronounced so that the wheel profile H remained practically constant, and did not influence the workpiece surface.

Because the smaller diamond particles protruded less, the wheel roughness R_p decreased. This, together with the increased number of diamonds per unit surface area, improves surface finish (lower value of R_a) (Fig. 8).

From the foregoing discussion, it can now be stated that a higher concentration and a small grit size give the required geometrical accuracy and the lowest surface roughness possible under the specific conditions.

PROPOSALS FOR SURFACE QUALITY IMPROVEMENT

The previous sections discussed the influence of machine and wheel parameters on the workpiece finish under constant grinding conditions, showing that improved surface finish can only be obtained when smaller grit sizes are used with lighter grinding conditions, both of which increase grinding costs. In addition, brief changes in grinding conditions after rough grinding, equation (11), involve higher costs because of lower stock removal rates over and above the fact that the best improvements obtainable are limited by R_p and H. Therefore, the following suggestions are put forward to enable R_p and H to be reduced even at high stock removal rates.

Reduction of wheel roughness
The explanation of the development of the base profile and consequently the workpiece surface, suggest that special attention should be paid to the centre section of the wheel, because this is the last part of the wheel in contact with the workpiece and therefore determines the workpiece roughness within the parallel lanes. If the wheel roughness within the centre portion of the wheel can be reduced, while retaining a high degree of roughness on the outer edges of the wheel, then a good stock removal rate can be assured with the edges of the wheel, and a good surface finish by the centre portion. A special type of grinding wheel was therefore designed. It contained coarse grit (60/70 mesh) in the two edges of the wheel rim, giving high wheel roughness, and a fine grit size (170/200 mesh) in the centre portion for low wheel roughness, the three sections having equal width.

A 127 x 6·4 mm D1A1 resin bond wheel designed in this way and containing DXDA-MC diamond grit was used under the following grinding conditions:

cutting speed	26 m/s
table speed	16 m/min
downfeed	0·050 mm
crossfeed	1·25 mm
workpiece material	D2 die steel (AISI)

A wheel containing only 60/70 mesh grit was tested under the same conditions and gave a workpiece roughness of $R_a = 1·5\ \mu m$. With the three-layer wheel the wear was only slightly higher but the workpiece roughness was reduced to $R_a = 1·2\ \mu m$.

Reduction of the base profile
As shown above, the base profile has a decisive influence on the R_a value. Without reducing the downfeed, the effect of H can only be reduced when a wider wheel is used. This involves higher initial wheel costs and the use of wider wheels is not always physically possible. However, the same effect can be derived when a wheel of reduced width is used with a crossfeed applied in one direction only. Then the base profile, as shown in Fig. 1(e), is spread over the whole wheel width, so that all calculations are as for a wheel of double the rim width. The base profile becomes 'flatter' as the quotient H/B becomes smaller so that a lower waviness occurs on the workpiece surface, reducing the R_a value. Furthermore, according to equation (12), the average undeformed chip cross-section decreases by the factor $\sqrt{2}$. On the other hand, for the same stock removal, every particle comes into contact twice as often. Therefore, the actual wear behaviour depends on the application.

The grinding process with crossfeed in one direction only should be carried out on grinding machines which can be programmed in a suitable way so that no significant time loss is encountered, only approximately 3 per cent.

The logical consequence of the considerations in the two previous sections is the use of a two-layer wheel with crossfeed in one direction only, the two layers containing a coarse and a fine grit respectively.

Dressing of diamond wheels
When the above procedures do not give the required surface finish, the wheel roughness and base profile must again be reduced.

On a wheel with a profile as shown in Fig. 3, the roughness of the centre portion of the rim must be reduced and have a base profile parallel to the wheel axis, after the rough grinding process is carried out in the usual way. This can be achieved by dressing the diamond wheel with a diamond dressing tool. In this particular experiment a bronze bond diamond dressing block, containing 80/100 mesh De Beers MDA-S at 150 concentration, was used. A 177 x 10 mm resin bond diamond wheel containing DXDA-MC, 80/100 U.S. mesh at 100 concentration, was dressed with four downfeeds each of 0·010 mm.

Thereafter, four finishing passes, each with 0·010 mm downfeeds were carried out on the rough ground workpiece, giving a surface finish of $R_a = 0·2\ \mu m$.

The question immediately arises concerning the cost of diamonds in such a dressing procedure. In an industrial application an impregnated round disc should be used as a dressing tool. This disc is inserted in the dressing apparatus of the grinding machine and applied in the conventional way. The total downfeed, d_d, of the dressing tool after touching the wheel can be calculated in the following way.

$$d_d = \frac{s \cdot a \cdot 2}{B}$$

$$
\left.
\begin{array}{l}
s = \text{crossfeed} \\
a = \text{downfeed} \\
B = \text{rim width}
\end{array}
\right\} \text{during rough grinding}
$$

THE PRODUCTION OF FINE SURFACE FINISHES
WHILE MAINTAINING GOOD SURFACE INTEGRITY
AT HIGH PRODUCTION RATES BY GRINDING

by

ROBERT S. HAHN and RICHARD P. LINDSAY*

NOTATION

$\bar{v}_f$ cross slide advancement rate
$\bar{v}_w$ rate of recession of the work radius
F_n induced force normal to the wheel-work
F_t tangential force to the interface
V_s wheel peripheral speed
V_w wheel-work speed
K_w spring representing work support system
K_s spring representing wheel support system
K stiffness of wheel-work contact
Z'_s wheel wear rate per unit width
Z'_w work wear rate per unit width
D_w diameter of work
D_s diameter of wheel

INTRODUCTION

The production of fine surfaces at low cost underlies our industrial economy and relates to the efficiency of manufacture of many products. In recent years the grinding process has been developed to the point where stock removal rates, surface finish, surface integrity, and the correction of initial geometrical errors can be predicted and better controlled through the recognition of basic grinding parameters. These parameters relate the stock removal, wheel wear, surface finish and surface integrity to the grinding machine operating variables such as wheel speed, work speed, feed rate etc. Using these relationships the optimum grinding conditions can be found for producing workpieces at minimum cost while maintaining good surface finish and surface integrity.

BASIC RELATIONSHIPS IN
PRECISION GRINDING

In grinding metals three distinct processes[1] take place at the interface of the abrasive grain and the workpiece: (a) Rubbing—e.g. where the grain rubs on the work causing elastic and/or plastic deformation in the work material with essentially no material removal. (b) Ploughing—e.g. where the grain causes plastic flow of the work material in the direction of

sliding, extruded material being thrown up and broken off along the sides of the groove, resulting in low rates of stock removal. (c) Cutting—e.g. where a fracture takes place in the plastically stressed zone just ahead of the rubbing grain, causing the formation of a chip and resulting in fairly rapid stock removal rates.

To illustrate this behaviour and the basic relationships between grinding parameters, consider Fig. 1. The cross slide is advancing at a rate $\bar{v}_f$. Simultaneously the work radius is receding at a rate $\bar{v}_w$,

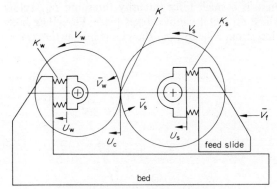

Figure 1. Plunge grinding system.

while the wheel radius is wearing at a rate $\bar{v}_s$. Induced forces exist during the steady-state: F_n, normal to the wheel-work contact surface and F_t, tangential to the interface. The wheel peripheral speed is V_s and the work speed is V_w. The work support mechanism is represented by a spring K_w, while the wheel support system is shown schematically as K_s. The wheel-work contact also has some stiffness, shown as K.

In explaining the grinding process it has been customary to report results in terms of downfeed (on surface grinders) or infeed rate (on cylindrical grinders). Under these conditions the grinding performance is dependent on the rigidity of the machine tool. For example, a downfeed of 0·002 inch on a very stiff machine will induce a much larger wheel-work contact force than the same downfeed on a more compliant machine. Accordingly, grinding performance (surface finish, wheel wear, stock

* Cincinnati Milacron—Heald Division, Worcester, Massachusetts

removal rates:

$$\bar{v}_f = \bar{v}_w + \bar{v}_s \qquad (1)$$

The total material removal rates in cubic inches per minute per inch of width are

Work: $Z'_w = \pi D_w \bar{v}_w \times 60 \ (\text{in}^3/\text{min per inch})$ (2a)

Wheel: $Z'_s = \pi D_s \bar{v}_s \times 60$ (2b)

where

$$D_w = \text{diameter of work}$$
$$D_s = \text{diameter of wheel}$$

Figure 2 is a plot of Z'_w and Z'_s, the material removal rates for different total applied force intensities; also shown are the surface finish and horsepower. From this graph we can see that for this work material, there is a small force intensity threshold F'_{no} below which no metal removal takes place. Ploughing takes place from F'_{no} to F'_{pc}, the ploughing cutting transi-

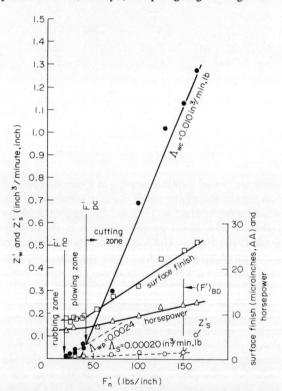

Figure 2. Wheel–work characteristic chart showing metal removal rate surface finish, wheel-wear rate and horsepower versus interface normal force intensity (force per unit width). Test conditions: wheel: A80K4V; dress lead: 0·003 in/rev; dress depth: 0·0002 in; wheel speed: 12 500 ft/min; work speed: 250 ft/min; coolant: Cimcool 5 star; material: AISI 52 100, $R_c = 60$; equivalent diameter: $D_e = 2\cdot0$ in; external plunge climb grinding.

linear equations:

Work: $Z'_w = \Lambda_{wc}(F'_n - F'_{pc})$ (3a)

Wheel: $Z'_s = \Lambda_s F'_n$ (3b)

Where the constant of proportionality in each case is the Greek letter lambda (Λ) and has the units: 'cubic inches of material being removed per minute, per pound of interface normal force intensity'. Λ_{wc} is the 'Metal Removal Parameter' and Λ_s is the 'Wheel Removal Parameter'.

Using equation (2), equation (3a) for the work removal can be written two ways:

Work: $\bar{v}_w = \dfrac{\Lambda_{wc}(F'_n - F'_{pc})}{\pi D_w 60}$ (4a)

$(F'_n - F'_{pc}) = \dfrac{\pi D_w 60 \bar{v}_w}{\Lambda_{wc}}$ (4b)

Wheel: $\bar{v}_s = \dfrac{\Lambda_s F'_n}{\pi D_s 60}$ (4c)

Equations (4a) and (4c) apply to controlled-force grinding where the force intensity is prescribed while equation (4b) applies to feed-rate grinding where the feed rate is prescribed. In practice, the wheel wear velocity, $\bar{v}_s$ is generally small compared to $\bar{v}_w$ so that $\bar{v}_f$ can be substituted for $\bar{v}_w$ in equation (4b) giving:

$$(F'_n - F'_{pc}) = \dfrac{\pi D_w 60 \bar{v}_f}{\Lambda_{wc}} \qquad (5a)$$

Equation (5a) gives the induced force caused by applying a feed rate $\bar{v}_f$ on to the grinder. It can be seen that the induced force is directly proportional to the workpiece diameter and width and inversely proportional to Λ_{wc}. Thus doubling Λ_{wc} will cause the induced force to be halved if the feed rate $\bar{v}_f$ is held constant.

From the standpoint of gross metal removal the cutting region and Λ_{wc} are significant. From the standpoint of precise sizing and producing smooth surface finishes the region around F'_{no} and F'_{pc} is very important.

The data shown in Fig. 2 presents in a graphical way the complete grinding action and is called a 'Wheel–Work Characteristic Chart'. This chart is helpful in developing fast efficient grinding cycles.

SURFACE FINISH

For the production of workpieces with good surface finish at low cost it is desirable to obtain relationships between surface finish and the other pertinent grind-

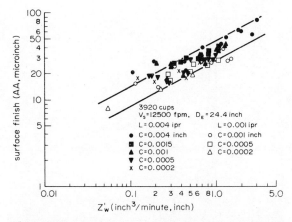

Figure 3. Surface finish versus metal removal rate. Test conditions: wheel: 12A801M6VFMD2; wheel speed: 12 500 ft/min; work speed: 800 ft/min; wheel diameter: 3·25 in; work diameter: 3·75 in (mean); dress lead: 0·001 and 0·004 in/rev; dress depth: 0·0002, 0·0005, 0·001, 0·0015, 0·004 in; coolant: heavy-duty soluble oil w/water; work material: AISI 4610, R_c = 62 (roller bearing cup #3920); internal, climb, plunge grinding.

ing variables. The most obvious grinding variable is the stock removal rate Z'_w. Figure 3 shows surface finish plotted against stock removal rate, Z'_w, in grinding No. 3920 roller bearing outer cups for various dressing conditions (L = dressing lead of single point diamond, c = diamond depth of cut, D_e = $D_w D_s/(D_w-D_s)$ the equivalent diameter). It will be seen that high stock removal rates generally produce rougher surface finishes. Also considerable scatter will be noted. Figure 4 shows surface finish for different values of Z'_w for two values of equivalent diameter D_e and at two wheel speeds (3500 ft/min and 12 000 ft/min). The surface finish data clearly cluster into two groups. At a given metal removal rate the finish will be improved by increasing D_e and/or increasing the wheel speed. In view of Figs 3 and 4 it would be desirable to find a unique relationship which would reduce the scatter and predict the surface finish for various dress conditions, equivalent diameter, wheel speed, and metal removal rate. Reichenbach[2] and other investigators have suggested the underformed chip thickness is an important grinding variable.

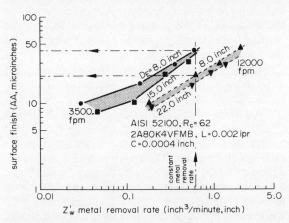

Figure 4. Surface finish versus metal removal rate. Test conditions: wheel: 2A80K4VFMB; wheel speed: 3500 ft/min and 12 000 ft/min; work speed: 250 ft/min; wheel diameter: 2·5 and 3·25 in; equivalent diameter: 8·0, 15·0 and 22·0 in; dress lead; 0·002 in/rev, dress depth: 0·0004 in on dia.; coolant: chemical emulsion w/water Mfg. A; work material: AISI 52 100 at R_c = 62; internal, climb, plunge grinding.

Based on this work, a partially empirical relation giving the average chip thickness, T_{ave}, in terms of the dressing and grinding conditions has been developed[3]:

$$T_{ave} = K \left(\frac{V_s}{V_w}\right)^{3/27} \frac{(dL)^{16/27}\left(1 + \dfrac{c}{L}\right)}{D_e^{8/27}} \left(\frac{Z'_w}{V_s}\right)^{19/27} \quad (8)$$

where: V_s is the peripheral wheel speed (in/s)

V_w is the peripheral work speed (in/s)

K = 85, a dimensional constant (in$^{-16/27}$)

d is the average diameter of abrasive grain in the wheel (in)

L is the dressing lead (inch of axial diamond motion per wheel revolution)

c is the diamond depth of penetration into the wheel measured on diameter (in)

D_e is the equivalent diameter (in)

Figure 5 shows the data of Fig. 3 replotted against T_{ave}. It will be seen that the scatter has been considerably reduced in spite of the wide variation of

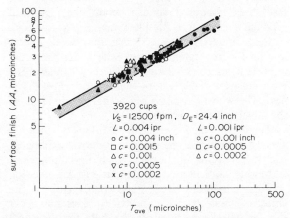

Figure 5. Surface finish versus theoretical chip thickness. Test conditions: same as Fig. 3.

dressing conditions. Figure 6 shows the data of Fig. 4 replotted against T_{ave}. The two scattered parts of the data have been united into one fairly narrow band. Also included in Fig. 6 are data for the 0·003 in/rev dress lead. It is clear that by using T_{ave} as the independent variable the widely scattered data are unified. It will be observed that below 10 μin the slope is less than that above 10 μin. This region

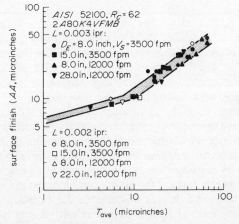

Figure 6. Surface finish versus theoretical chip thickness. Test conditions: same as Fig. 4.

ficorder instruments, involves roughness in the circumferential direction. It has been shown that local wheel hardness variations around the circumference of the grinding wheel[5] cause periodic roughness imperfections in the ground workpiece.

SURFACE INTEGRITY

In addition to producing workpieces with good surface finish and geometry, it is often very important to produce satisfactory surface integrity. The problem is to define the operating variables so that fast stock removal rates can be obtained and still produce good surface finish and integrity at low cost.

It is important to recognize that work speed, interface force intensity and wheel sharpness are important variables relating to thermal damage. The most elusive variable among these is the wheel sharpness. The wheel sharpness can be defined in terms of the metal removal parameter Λ_w.

The slope of the Z'_w versus force intensity curve has already been seen to be the metal removal parameter, Λ_w. It has also been shown[6] that if a grinding wheel is dressed and then used continuously under precision grinding conditions, wear flats will develop on the abrasive grains. If gross wheel wear does not occur (i.e. the wheel wear is attritious) these flat areas will continue to grow to create some stable characteristic real area of contact. If in addition, these tests be performed under conditions where the interface normal grinding force between the wheel and work is held constant, it will be found that the developed grinding rate $\bar{v}_w$ (rate of change of work radius) will diminish with time. Thus the wheel is losing its ability to remove material through the dulling action. Now since the growth of the real area of contact causes $\bar{v}_w$, at some force intensity, to diminish, then the metal removal parameter must be decreasing with time.

Figure 7 shows the variation of Λ_w with time under a controlled-force precision internal grinding conditions. Λ_w then is an indicator of the sharpness of the wheel as measured by the ability of it to remove stock: when Λ_w is large the wheel is acting sharply, when dulling occurs Λ_w diminishes.

Since Λ_w is then a measure of wheel-work grinding capability, we have a quantitative means of describing wheel sharpness. It has been shown that variations in the dressing ratio C/L can affect Λ_w. Thus by dressing a wheel differently, its sharpness can be changed, but as long as Λ_w can be measured, one can determine changes in sharpness due to changes in dressing methods. Wheel sharpness will be seen to be an important parameter in determining the surface integrity of ground surfaces.

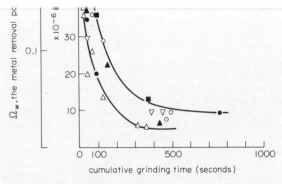

cumulative grinding time (seconds)

Figure 7. Metal removal parameter versus cumulative grinding time. 60 lb/in (10 500 N/m) wheel A60L5V; 30 lb/in (5250 N/m) wheel A60L5V; 120 lb/in (21 000 N/m) wheel A60L5V; 60 lb/in (10 500 N/m) wheel 2A90P6; 20 lb/in (3500 N/m) wheel 2A90P6; 100 lb/in (17 500 N/m) wheel 2A90P6; wheel speed: 7600 ft/min (38 m/s); work speed: 1100 ft/min (515 m/s); wheel diameter: 1·87 in (47·5 mm); work diameter: 2·37 in (60·2 mm); dress lead: 0·003 in/rev (0·076 mm/rev). Climb work rotation: coolant: Flow Rex 100; work material: AISI 52100 $R_c = 60$; internal grinding.

In order to establish threshold limits for force intensities beyond which thermal damage will occur, grinding tests were made on AISI-E52100 steel rings, at Rockwell C 60.

The test procedure was to increase the interface force at each work speed and to examine the workpiece for cracks after hydrochloric acid immersion. A representative run is shown in Fig. 8 which presents photographs of the ground and etched parts taken at 15x magnification through a microscope. As the force intensity is increased, cracking begins at the edges of the workpiece and spreads across the piece, until, at high force intensities, cross cracking (i.e. parallel to the direction of the grinding grit path) occurs. The limiting force intensity is defined as that which causes the first appearance of cracking.

Figure 9 shows this critical force intensity for different work speeds for two sharpness values: Λ_w equal to 88×10^{-6} in³/s lb ($0\cdot323$ mm³ N⁻¹ s⁻¹)—a sharp wheel, and equal to 25×10^{-6} in³/s lb ($0\cdot0918$ mm³ N⁻¹ s⁻¹)—a dull wheel. As work speed is increased, the tolerable force intensity rises, for each wheel sharpness. The importance of knowing the degree of wheel sharpness, or Λ_w, may be illustrated by a hypothetical grinding operation, using skip dressing (avoidance of dressing for some time). With work speed at 500 ft/min, ($2\cdot5$ m/s) a force intensity of 180 lb/in (31 N/mm) is permissible (see Fig. 9) and as long as the wheel remains sharp, thermal tensile stressing of the work will not occur. However if the wheel dulls (as it will with continual usage) so that Λ_w is reduced to say 25×10^{-6} in³/s lb ($0\cdot0918$ mm³ N⁻¹ s⁻¹) the force intensity of 180 lb/in (31 N/mm)

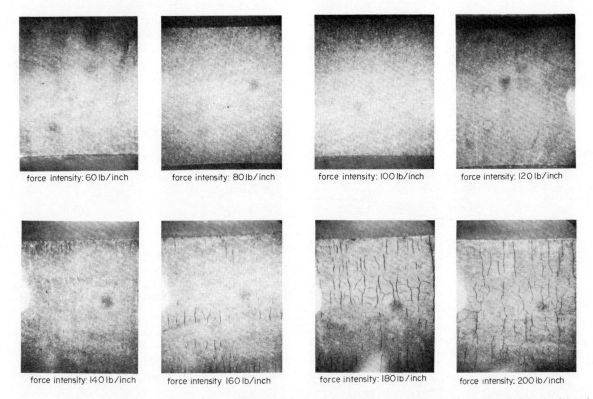

force intensity: 60 lb/inch   force intensity: 80 lb/inch   force intensity: 100 lb/inch   force intensity: 120 lb/inch

force intensity: 140 lb/inch   force intensity 160 lb/inch   force intensity: 180 lb/inch   force intensity: 200 lb/inch

Figure 8. Workpieces ground under various force intensities for constant wheel sharpness. Wheel: 97 A1001L6VFM; wheel speed: 7700 ft/min (38·5 m/s); work speed: 180 ft/min (0·9 m/s); wheel diameter: 2·0 in (51 min); work diameter: 2·31 in (59 mm); dress lead: 0·004 in/rev (0·10 mm/rev). climb work rotation: coolant: Cimperial 20:1, 110 lb/in^2 (7·7 kg/cm^2); work material: AISI 52100 R_C = 60; internal grinding; hydrochloric acid etch after each grind−10 min at 150°F (65°C).

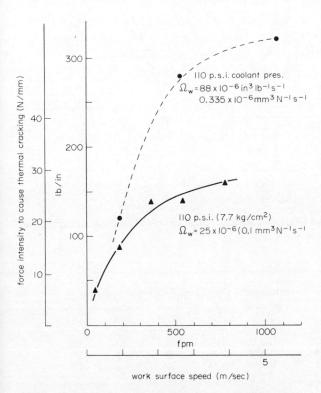

Figure 9. Threshold force intensity to produce thermal cracking. Wheel: 97A1001L6VFM; wheel speed: 7700 ft/min (38·5 m/s); work speed: 180 ft/min (0·9 m/s); wheel diameter: 2·0 in (51 mm); work diameter: 2·31 in (59 mm); dress lead: 0·004 in/rev (0·10 mm/rev) sharp, 0·0001 in/rev (0·025 mm/rev) dull. Climb work rotation: coolant: Cimperial 20:1, 110 lb/in^2 (7·7 kg/cm^2); work material: AISI 52 100 R_C = 60; internal grinding, hydrochloric acid etch after each grind−10 min at 150°F (65°C).

is above the permissible level for that Λ_w sharpness value, and the work will have thermally induced tensile stresses. To be conservative and set a force intensity below that permissible for a dulled wheel, here say 120 lb/in (20·6 N/mm), will cause grinding time to be lost since the penetration rate, $\bar{v}_w$, will be slow, even with a sharp wheel.

High speed surface grinding of René 80 specimens was performed under various conditions and the residual stress pattern in the workpieces was determined by Metcut Research Associates. The operating conditions are given in Table 1.

TABLE 1

Test no.	Wheel condition	V_s (ft/min)	V_w (ft/min)	Avg. Z_w (in^3/min)	Procedure
2	Dull	11 000	60	0·075	Remove 0·010 in at 0·075 in^3/min
3	Dull	11 000	500	0·204	Remove 0·010 in at 0·204 in^3/min
7	Sharp	11 000	500	0·347	Remove 0·010 in at 0·347 in^3/min
6	Sharp	11 000	500	0·620	Remove 0·008 at 1·48 in^3/min; remove 0·002 at 0·347 in^3/min; dwell for 10 s
Low stress ref.	—	2 000	60	0·015	Remove 0·008 at 0·018 in^3/min; remove 0·0008 at 0·0144 in^3/min; remove 0·0012 at 0·0072 in^3/min

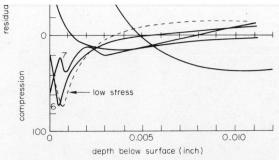

Figure 10. Residual stress pattern in surface grinding René 80 with Sulfochlorinated Oil [courtesy of Metcut Research Associates].

Test no.	Wheel speed (ft/min)	Work speed (ft/min)	Avg Z_w (in³/min)
2	11 000	60	0·075
3	11 000	500	0·204
7	11 000	500	0·347
6	11 000	500	0·620
Low stress	2 000	60	0·015

The residual stress patterns for the four high-speed and low-stress reference grinds are shown in Fig. 10. Grinds 2 and 3 using a dull wheel produced maximum tensile residual stresses of 200 000 and 60 000 lb/in² (1375 and 410 N/mm²) respectively. Tests 6 and 7 using a sharp wheel, yielded maximum compressive stresses of 70 000 and 60 000 lb/in² (482 and 410 N/mm²) respectively which compared favourably with the 60 000 lb/in² compressive stress of the low-stress reference grind. It is important to realize that one can produce either tensile or compressive residual stresses at 11 000 ft/min using oil by judiciously altering the grinding procedure.

In addition to the residual stress information, Metcut Associates also performed fatigue studies on

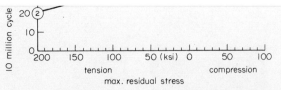

Figure 11. Endurance limit–10⁷ cycles–versus maximum residual stress for René 80. Surface ground with Sulfochlorinated Oil [courtesy of Metcut Research Associates]. See conditions in Fig. 10.

Since the goal is to increase the endurance limit, this graph shows that a residual compressive stress will do this best. Figure 12 implies that whenever a surface finish of 19 μin AA (0·48 μm) or less is produced,

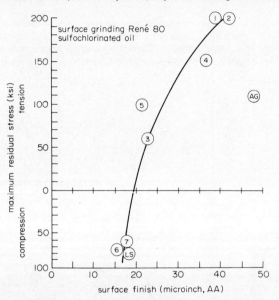

Figure 12. Residual stress versus surface finish. See conditions in Fig. 10.

TABLE 2

Test no.	Wheel condi-tion	Λ_w (in³/min lb)	V_s (ft/min)	V_w (ft/min)	Z_w (in³/min)	Stock removed (in³)	Average Z_w (in³/min)	Residual stress (kSi)	10⁷ cycle stress (kSi)	Finish to grind (AA)
2	Dull*	0·00288	11 000	50	0·075	0·338	0·075	+200	20	42
3	Dull	0·00288	11 000	500	0·204	0·338	0·204	+60	35	23
7	Sharp*	0·0058	11 000	500	0·347	0·338	0·347	−60	Not done	18
6	Sharp	0·0058	11 000	500	1·48	0·270				
					0·347	0·068				
					0·100	0·016	0·620	−70	40	16
Low stress			2 000	60	0·018	0·270				
					0·014	0·027				
					0·007	0·041	0·0148	−74	43	19

* Note: Dull wheel produced by dressing with 0·001 in/rev lead and 0·0005 in depth of dress on diameter. Sharp wheel produced by dressing with 0·004 in/rev lead and 0·001 in depth of dress on diameter. All dressing performed with single point diamond.

under these conditions, the maximum residual stress will be compressive: thus the low metal-removal-rate conditions necessary to produce a smooth finish evidently cold-work the metal sufficiently to create compressive stresses. Other data shown in Fig. 12, grinds 1, 4 and 5 were performed under identical or intermediate conditions and are included to supplement the residual stress-surface finish conclusion. Condition 'AG' (Abusive Grind) was a reference grind performed at a metal removal rate of 0·072 in³/min (1·18 mm³/min) with no coolant.

Figure 13 shows the practical productivity benefits available using these results. It is most desirable to

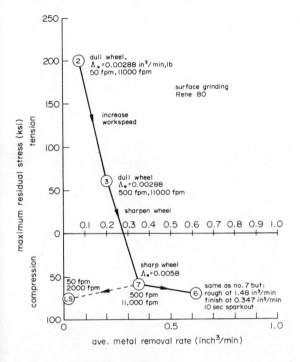

Figure 13. Maximum residual stress versus average metal removal rate. See conditions in Fig. 10.

produce compressive residual stresses for good endurance performance and high average metal removal rates to lower machine costs: therefore in Fig. 13, moving 'down' (tensile to compressive stresses) and to the 'right' (higher stock removal rates yields lower machining costs) is most desirable.

Condition 2 is a poor result: 200 000 lb/in² (1375 N/mm²) tensile residual stress, and Z_w being 0·075 in³/min (1·23 mm³/min). It was caused by a low work speed of 50 ft/min (0·25 m/s) and a dull grinding wheel giving both poor quality and high machining cost. If the work speed is increased from 50 to 500 ft/min (2·5 m/s) conditions improve to point 3: 60 000 lb/in² (410 N/mm²) tensile residual stress, and Z_w being 0·204 in³/min (3·342 mm³/min). This is still a poor stress condition, but the metal removal rate has been raised 172 per cent. Keeping the high work speed and sharpening the wheel improves the situation to point 7: Z_w of 0·347 in³/min (5·685 mm³/min) and a 60 000 lb/in² (410 N/mm²) compressive stress. Now, keeping a sharp wheel and high work speed and grinding most of the stock out at a rate (0·008 in of the 0·010 in stock at a Z_w of 1·48 in³/min (24·25 mm³/min)) and then

finish grinding with a low Z_w rate and a 10 s sparkout, conditions further improve to point 6: a 70 000 lb/in² (482 N/mm²) compressive residual stress has been produced at an average Z_w of 0·620 in³/min (10·15 mm³/min). Thus, from condition 2 which had a poor quality surface and was very expensive to produce, we have proceeded to condition 6, a quality surface efficiently produced. By increasing the work speed from 50 to 500 ft/min, grinding with a sharp wheel and removing 80 per cent of the stock at a high metal removal rate, the surface stress was changed from 200 000 lb/in² tensile to 70 000 lb/in² compressive and the productivity increased 725 per cent ((0·620-0·075)/0·075).

The Low Stress condition (LS), produced at 60 ft/min (0·3 m/s) work speed and 2000 ft/min (10·1 m/s) wheel speed at an average Z_w of 0·0148 in³/min (4·0 mm³/s), yielded at 74 000 lb/in² (510 N/mm²) compressive residual stress. Condition 6 had the same compressive stress (70 000 lb/in²) but was achieved at a production rate of factor of 40·9 higher, i.e. ((0·620-0·0148)/0·0148).

CONCLUSION

The grinding process, when properly controlled provides a method of removing unwanted stock, generating good size and geometrical accuracy, producing good surface finish and favourable surface residual stress and fatigue characteristics.

The development of efficient grinding cycles can be accomplished with the aid of 'Wheel-Work Characteristic Charts' to determine optimum rounding up, roughing and finishing conditions.

The effect of dress lead and diamond depth of cut, wheel speed, work speed and stock removal rate on surface finish can be described by using T_{ave}.

Sub-surface compressive stresses can be produced by judicious control of work speed, stock removal rate and wheel sharpness.

The endurance limit is related to the maximum sub-surface residual stress.

A relationship exists between the produced surface finish and the maximum residual stress level. Finish grinding with sharp wheels at high work speed to produce good surface finish will also produce compressive sub-surface stress.

High stock removal rates can be combined with good surface integrity by finish grinding following the rough grind.

REFERENCES

1. R. S. HAHN (1964). Controlled force grinding—a new technique for precision internal grinding, *Trans. ASME, J. of Eng. for Ind.*, Series B, **68**, 287-93.
2. G. S. REICHENBACH *et al.* (1956). The role of chip thickness in grinding, *Transactions of ASME*, **78**, 847, ASME, New York.
3. R. P. LINDSAY (1972). On the surface finish-metal removal relationship in precision

VARIABLE WHEEL SPEED–A WAY TO INCREASE THE METAL REMOVAL RATE

by

GUNNAR CEGRELL*

SUMMARY

This investigation was made by the Metal Cutting Research Department at Volvo Flygmotor, Trollhättan, Sweden. In this paper the results of an experimental study of grinding at variable wheel speed is described. The grinding wheel speed was changed in several steps during grinding. The purpose was to find out if this method has industrial applications. Parameters such as surface finish, vibration amplitude, and feed efficiency were studied. The ordinary G-ratio was modified. A direct comparison between laboratory results and production results can be made.

INTRODUCTION

Regenerative chatter is generally accepted as one of the main sources of self-excited vibrations. It is caused by the chip thickness variation and depends upon the time delay between two successive cuts. The chatter can be represented by a closed-loop model. Many theories and explanations have been made. Unfortunately it is difficult to put these theories into practice so that they can be used in an economic way.

The grinding processes that nowadays are used in production plants are in general not controlled in any way, i.e. when the grinding process is started it is given predetermined data. During the grinding time all these data are constant.

The purpose of the experimental study was to investigate the grinding process when the grinding wheel speed is changed. Parameters such as surface finish, vibration amplitude, and feed efficiency are of the greatest interest. Each parameter is studied as a function of the metal removal rate. The changing of the wheel speed was controlled by a program box with predetermined step variations. Thus the control system used was an open-loop system.

Previous studies[1] have shown that the chatter amplitude can be decreased by using a variable wheel speed.

Grinding investigations have been carried out in an ever-increasing volume, and a standardized grinding test is needed. This study was carried out in accordance with the standardized grinding test developed at Volvo Flygmotor. The test has been so designed that it can be directly applied to industrial testing, and parameters such as metal removal rate, grinding wheel, workpiece material, and coolants can be tested in a systematic way.

TEST CONDITIONS

The method
The external cylindrical grinding machine is equipped with a traversing table. Two workpieces of equal diameter are ground parallel and the sequence of operations for the test is given below:

(1) measuring surface finish
(2) grinding; total radial working allowance 0·15 mm
(3) measuring surface finish and workpiece diameters
(4) grinding; total radial working allowance 0·15 mm
(5) Measuring surface finish and workpiece diameters.

This sequence is repeated with feeds (s_a) of 2·5, 5·0, 7·5 and 10·0 μm for each pass. See Fig. 1.

Moreover the above-mentioned tests were carried out at three different workpiece speeds (v_d), 55, 85 and 145 rev/min. The traversing speed (s_l) is changed in proportion to the workpiece speed so that the over-

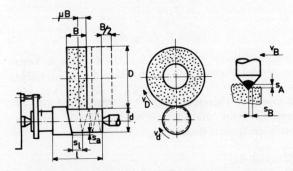

Figure 1. Schematic view of the machine parameters.

* Metal Cutting Research Department at Volvo Flygmotor, Trollhättan, Sweden

with the traversing table.

The machine

For the experiment a Kellenberger 50 MU grinding machine equipped for external cylindrical grinding was used. This machine is equipped with a traversing table, and the wheel motor assembled on the machine is an a.c. motor. In order to change the wheel speed a static frequency transformer was connected between the wheel motor and the main circuit connection. It was possible to change the power frequency continuously between zero and 87 Hz.

The headstock stiffness was 43 N/μm.

The stiffness between centres was 25 N/μm.

The workpiece

The set-up between centres is conventional. The shape of the workpiece was chosen so that its stiffness far exceeded the stiffness of the most common tailstocks (Fig. 2). The purpose was to study the

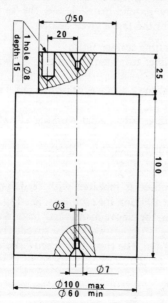

Figure 2. The workpiece.

process without any weakness in the workpiece. Tests in several materials would be very time-consuming and preparatory tests showed that this was not necessary. Therefore only two different types of material were studied, one with austenitic and one with martensitic structure.

Material data:

Grade: Martensitic steel AMS 5616 (Greek Ascoloy).
Analysis: 0·15% C; 0·50% Si; 0·50% Mn; 14% Cr; 2·2% Ni; 3·5% W; 0·50% Mo.
Hardness: 36 HR$_c$.

materials, but it is used as a reference wheel in Volvo's standardized test.

To dress the wheel a single point diamond was used. The diamond was turned so that the maximum contact width between wheel and workpiece did not exceed 0·2 mm. The feed of the diamond was $s_A = 0·02$ mm at each pass and the dressing feed rate $v_B = 0·25$ m/min.

The coolant delivery was kept at a constant level. As the wheel was working as an air pump, the coolant jet was sprayed on the wheel in a radial direction. This arrangement made it possible to omit the air scraper on the wheel periphery.

Coolant data:

Castrol, Syntilo 9, chemical solution; concentration: 1:100.

Test programs

A preliminary investigation was made to find suitable programs to control the grinding wheel speed. Four programs were chosen, as shown in Fig. 3. Program 0 represents constant wheel speed.

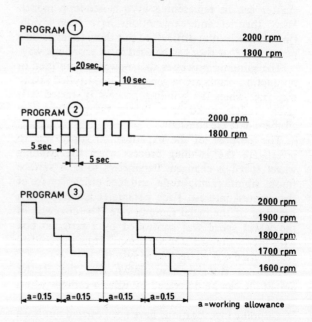

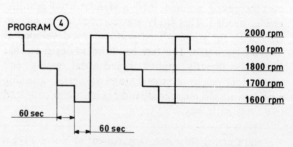

Figure 3. Test programs.

PARAMETERS

Surface finish

A Perth-O-Meter W3B was used to measure the surface finish. The surface finish was measured at six different places evenly spaced over half the periphery of the workpiece. The mean value was calculated.

The surface finish can be described as follows:

$$O_y = K_1 \cdot V^{K_2} \rightarrow \log O_y = K_2 \log V + \log K_1$$

where O_y = surface finish,
          V = stock removal,
          K_1 and K_2 are arbitrary constants,

i.e. the surface finish is a straight line curve on a log–log diagram. From this diagram the surface finish can be plotted against the metal removal rate.

Vibration amplitude

Accelerometers were used to measure the vibration amplitude, and were glued directly onto the back centre. The normal and tangential amplitudes were measured. The accelerometers were connected to a piezo-amplifier and an oscilloscope; the oscilloscope was calibrated in acceleration due to gravity.

The vibration amplitude can be described as follows:

$$\log A(t) = K_1 \cdot V + K_2$$

where $A(t)$ = amplitude of vibration at time t,
          V = stock removal,
          K_1 and K_2 are arbitrary constants.

i.e. the amplitude is a straight line in a lin–log diagram. From this diagram the amplitude can be plotted against the metal removal rate.

Feed efficiency

Differences appear in the cutting zone when the wheel speed is changed. For instance the chip thickness changes; variable wheel speed affects the stock removal rate and wheel wear. The G-ratio is the volume of material removed divided by the volume of grinding wheel consumed in removing it. It is an indication of grinding wheel cost per part ground.

Another cost component is, of course, time. Generally in any cost analysis[2] the wheel cost is the most insignificant, so setting machine parameters to achieve a high G-ratio in order to 'save wheel' is a fallacy.

The ordinary G-ratio is not practically usable for larger test series, and generally it is difficult to analyse the difference between theoretical and practical stock removal, as regards workpiece deflection and wheel wear.

Therefore the modified G-ratio is defined by

$$G_{\text{mod}}\text{-ratio} = \sum \frac{\Delta d_p}{\Delta d_t} = \frac{\text{practical change of diameter}}{\text{theoretical change of diameter}}$$

where $0 < G_{\text{mod}}\text{-ratio} < 1$,

and can be described as follows

$$\log G_{\text{mod}} = K_1 \cdot V + K_2$$

where V = stock removal,
          K_1 and K_2 are arbitrary constants.

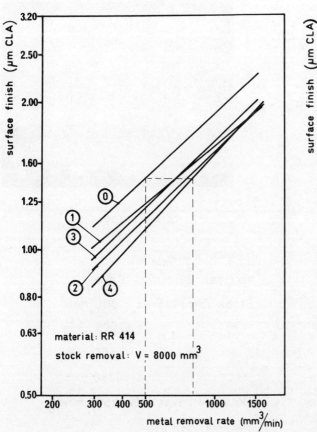
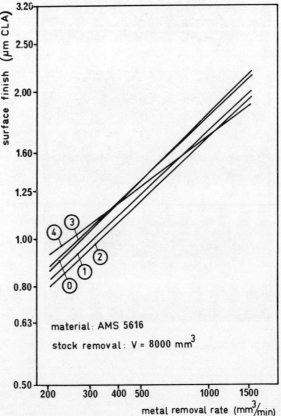

Figure 4. The surface finish as a function of the metal removal rate.

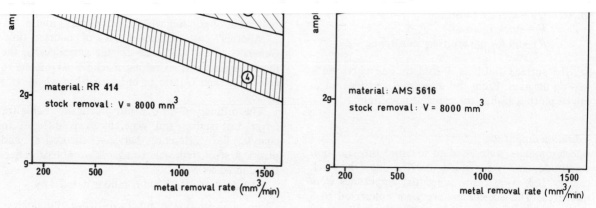

material: RR 414
stock removal: V = 8000 mm³

material: AMS 5616
stock removal: V = 8000 mm³

metal removal rate (mm³/min)

metal removal rate (mm³/min)

Figure 5. The amplitude of vibrations in the normal force direction.

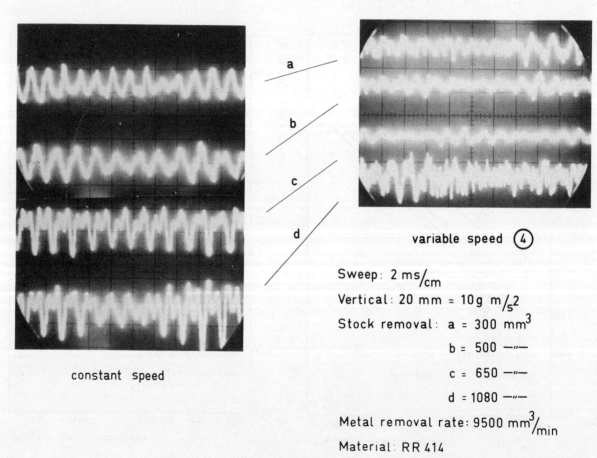

a

b

c

d

constant speed

variable speed ④

Sweep: 2 ms/cm
Vertical: 20 mm = 10g m/s²
Stock removal: a = 300 mm³
b = 500 —"—
c = 650 —"—
d = 1080 —"—
Metal removal rate: 9500 mm³/min
Material: RR 414

Figure 6. Oscilloscope pictures.

There are many resemblances between G_{mod}-ratio and the ordinary G-ratio. Of course G_{mod}-ratio is a function of the machine stiffness, but when it is used for relative measurements the advantages are obvious. For diamond and borazon wheels the wheel cost is significant and the ordinary G-ratio is used.

EXPERIMENTAL RESULTS

Figure 4 shows the surface finish as a function of the metal removal rate for both materials. For an austenitic material (RR 414) it is suitable to use a step variation like program 4. It is possible to increase the metal removal rate from 500 to 800 and still maintain the same surface finish (This is shown by the dashed line). For a martensitic material (AMS 5616) the advantages increase at a higher rate of metal removal when program 4 is used. Measurements on the workpiece have also confirmed a reduction in waviness.

Figure 5 shows the amplitude of vibrations in the normal force direction. The amplitude corresponding to each program is contained in the dashed zone. For the austenitic material (RR 414) the amplitude increases when the wheel speed is constant (program 0), but it decreases greatly when program 4 is used. For the martensitic material (AMS 5616) the amplitude decreases to a lower level when variable speed is used. Figure 6 shows pictures from two tests which have been carried out under the same conditions. The left-hand picture shows the vibration amplitude for different stock removals at a constant speed. The right-hand picture shows the same test with the wheel controlled by program 4.

Figure 7 shows the G_{mod}-ratio for both materials. G_{mod}-ratio is the practical change of diameter divided by the theoretical change of diameter. If G_{mod}-ratio is expressed as a function of the metal removal rate, it can also be expressed in terms of the feed efficiency by

$$G_{mod}\text{-ratio} = \frac{W_p}{W_t}$$

where W_p = the practical metal removal rate,
W_t = the theoretical metal removal rate.

The metal removal rate on the X-axis is the theoretical one.

CONCLUSIONS

In order to evaluate the results it is necessary to compare all increases and all decreases with a reference value of the metal removal rate. A representative value for this parameter is 500 to 1500 mm^3/min (for a wheel width of 18 mm).

By applying a suitable step variation to the grinding wheel speed it is possible to attain the following results:

The metal removal rate can be increased by 25–50 per cent while still maintaining the same surface finish. The metal removal rate can be increased by

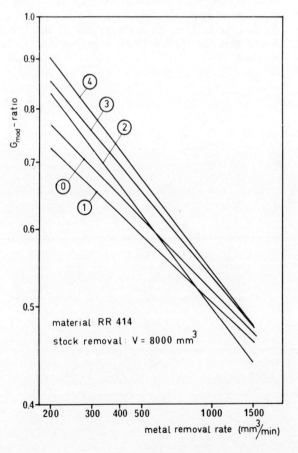
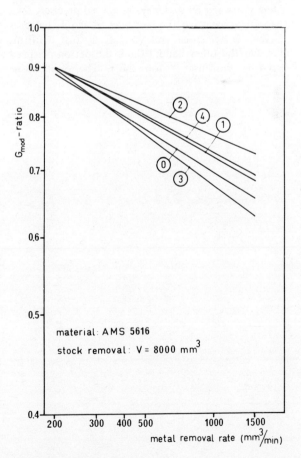

Figure 7. The G_{mod}-ratio as a function of the metal removal rate

not exceed the critical force of the wheel.

(2) There must be a time delay, as it takes about 1 to 10 seconds before the process is adapted to the new speed.

REFERENCES

1. B. BARTALUCCI, G. G. LISINI, P. C. PINOTTI (1970). Grinding at variable speed, *Proc. 11th M.T.D.R. Conf.*, Birmingham, Pergamon.
2. R. P. LINDSAY (1971). On the metal removal and wheel removal parameters, surface finish, geometry, and thermal damage in precision grinding, Worcester Polytechnic Institute.

DISCUSSION

Q. Professor J. Peters. I am astonished when you claim that it is difficult to put theories into practice—a few years ago Dr A. Lweys predicted precisely on a theoretical base that the best way to avoid chatter is to vary wheel speed and your study just confirms this. On the other hand I think the period of speed variation you use is too short. The oscilloscope that they can be used in a *production plant* in an *economic way*.

I agree that chatter still exists, but the chatter amplitude is reduced a lot. For instance the waviness of the workpiece can be reduced to half the value by applying program 4 to the wheel. You claim that the period of speed variation is too short. The preliminary investigation showed less results with a longer period of variation. On the other hand the period must be a function of the metal removal rate, so my suggestion to adapt a vibration sensitive instrument to the process would give an optimal variation of the period time, and thereby optimal grinding conditions.

In Volvo's standardized grinding test the surface finish is measured at six different places evenly spaced over half the periphery of the workpiece. For every single *observed* value of the surface finish, the dispersion will be: 8 per cent lower, or 12 per cent higher. The mean value of six measurements are then calculated. The test can easily be reproduced and then the dispersion will be very low. An example will describe the reproducibility. A year ago I got a deviation of approximately 8 per cent in our surface finish curves. After an investigation of the test I found that the surface finish apparatus had to be repaired.

HEAT AFFECTED ZONE IN GRINDING OPERATIONS

by

M. MARIS and R. SNOEYS*

SUMMARY

The temperature rise, due to the passage of a grinding wheel, may have detrimental effects on the surface integrity of the outermost layers of the ground workpiece. In this contribution, a comparison has been made of various models, worked out for evaluating the heat affected zone. The predicted temperature fields and theoretically determined heat affected zones have been compared with experimental data.

From this evaluation, some factors have been put forward, which are relevant for practical grinding work. Of special interest in this respect is the work speed. In some cases, especially cut-off operations, also the infeed may become quite important. It might be possible indeed to increase significantly the efficiency of the operation, with respect to thermal damage by evacuating a large quantity of heat, simply by removing fast enough the heated-up material so as to minimize the heat conduction into the workpiece bulk material.

This kind of analysis may indicate further trends in cut-off operations, in high-speed grinding and in abrasive machining operations.

NOTATION

Symbol	Dimension	Definition
A	°C	maximum surface temperature
a	m	depth setting
B	1/m	exponential decay factor
b	m	width of cut-off wheel
C	°C	critical temperature limit
C_v	J/m³ °C	specific heat per unit volume
c	J/kg °C	specific heat per unit mass
D_{eq}	m	equivalent diameter $= \dfrac{d_s d_w}{d_w \pm d_s}$
d_s	m	grinding wheel diameter
d_w	m	workpiece diameter
E_{sp}	J/m³	specific grinding energy
F_t'	N/m	tangential force per unit width
H_v	N/m²	Vickers hardness
h	J/m² s °C	convective cooling coefficient
k	J/m s °C	thermal conductivity
L	—	$= l v_w / 2\alpha$
l	m	half width of heat source
l_c	m	theoretical contact length
M_s	°C	martensitic transformation temperature
P	W/m	power per unit width
Q	J/m² s	total grinding heat flux
q	J/m² s	heat flux to workpiece metric contact area
Q'	J/m² s	heat flux to wheel per unit geo-
q'	J/m² s	heat flux to grain
T	°C	temperature
T_m	°C	maximum temperature
t	s	time
v_f	m/s	infeed rate
v_s	m/s	wheel speed
v_w	m/s	work speed
Z	—	$= z v_w / 2\alpha$
z	m	depth
z_H	m	depth at which hardness variations vanish
z_T	m	theoretical determined depth at which the M_s temperature is reached
z_s	m	depth at which residual stress vanish
α	m²/s	thermal diffusivity, $= k\rho c$
Φ	—	decimal fraction of grinding energy entering the workpiece
ρ	kg/m³	specific mass

INTRODUCTION

In the first part a critical survey of the literature is presented concerning the heat in grinding process and its effect upon the workpiece surface integrity. It is widely accepted that the induced damage in the uppermost layer of a ground workpiece is mainly of thermal origin. Metallurgical transformations, residual tensile stresses and hardness changes show that the heat input adversely affects the machined surface. In

---

* Instituut voor Werkplaatstechniek en Industrieel Beleid, Universiteit Leuven, Belgium.

where every cutting edge may be considered separately.

Various experiments have been carried out to measure workpiece temperatures when grinding in various conditions[2,5,6,8,25]. Due to the use of small sized thermocouples and other refinements of the measuring techniques the experimental results have become more reliable. However, whether making a theoretical analysis or an experimental investigation, it is essential to make a clear distinction between the various temperatures in the contact area[3,22].

(a) The temperature rise at the chip-grain interface affects the grinding wheel and strongly influences the chemical reactions between chip and grit;
(b) a temperature increase in the shear area is an important factor in terms of chip formation physics[7,10];
(c) finally, the temperature rise in the sub-ground surface, which is usually only a fraction of the increase previously mentioned, may be important in terms of surface integrity.

THERMAL MODELS FOR GRINDING OPERATIONS

Thermal balance

Almost all the described models start from Jaeger's analysis[1] of a moving slider on an adiabatic surface, either for the continuous band heat source model or for a single grain heat source.

N. Des Ruisseaux[4] has extended this analysis for cases where cooling is present. He also included in his approach a cylindrical geometry taking care of cylindrical plunge grinding conditions. He showed that the continuous band heat source model describes adequately the global action of the grains on the workpiece. He concludes that cooling is effective only when the grinding fluid would be able to enter the interference zone of the heat source area. He further emphasized the difference between the temperature in the chip shear plane and workpiece surface temperature.

Sauer[5] reaches similar conclusions. His model moreover takes into account the deformation energy induced in the workpiece. Although the amount of heat going into the workpiece material obviously is quite an important feature, only a few of the models described in the literature attempted to determine this quantity.

An energy balance of the grinding operation indicating the energy flux going to the chip, to the wheel, to the coolant and finally to the workpiece is

grinding 30 per cent is immediately taken away again by cooling mainly in the interference zone.

TABLE 1 Heat balance based on experimental work or analytical derivations by various authors

Author	Ref.	Per cent to workpiece	Per cent to chip and wheel	Comments
Sauer	5	30–70	70–30	experimental
Lee	9	80	20	experimental
Malkin	15	60–80	40–20	experimental
Sato	16	84	16	experimental
Outwater and Shaw	2	35	65	analytical (shear energy)
Eshghy	19	10	90	analytical (abrasive cutoff)

Outwater and Shaw[2] only treat the shear energy in their theory, disregarding the rubbing of the grains. In abrasive cut-off about all the energy which is conducted into the workpiece will be taken away again in the chips. This is a good approximation of reality when the side losses into the remaining material are small. Eshghy estimates these losses at 10 per cent for an infeed rate of 8 mm/s.

It is generally recognized that the ratio of normal force to tangential force is roughly 3/1 to 5/1 in grinding. Even when a small effective coefficient of friction is assumed $(0·2, \ldots, 0·3)$ almost the entire tangential force amplitude may be attributed to a friction action. The amount of heat to melt a unit volume of chip requires the heating up to, let us say, the melting temperature of carbon steel, $1500°C$ $(7·5 . 10^9$ J/m$^3)$, as well as the melting heat $(2·10^9$ J/m$^3)$. Comparing this extreme value of $9·5 . 10^9$ J/m^3 to the usual specific grinding energy of carbon steels $(30–60 . 10^9$ J/m$^3)$, it turns out that a maximum of 15 to 30 per cent would be immediately carried away by the chip. From this point of view the mentioned values of Lee, Malkin and Sato seem to be reasonable.

Another part of the grinding energy can be directly conducted to the grinding wheel. R. S. Hahn[26] found a relation between the rate of wheel wear and the amount of heat injected into the grain. R. Snoeys[27] measured mean surface temperatures on very small areas (1 mm^2) of a bakelite bonded diamond wheel at several distances (15 to 35 mm) from the contact area. Using an infrared radiation meter, temperatures of 130–180°C were found and cooling rates of $3 . 10^4$ to $6 . 10^4$ °C/s, indicating the importance of the cyclic heating-up of the grains.

The amount of heat conducted into the grain cannot be calculated anymore by a continuous model as is done for the workpiece[4], because only a small part of the contact area is occupied by grains (Fig. 1). Hahn and Lindsay[27] measured the real area of contact. From their measurements one can conclude that 1 to 3 per cent of the geometric contact area is in contact with the grains. For an upper bound analysis, let us assume that the surface temperature of the grain is about the melting temperature of the metal to be ground, e.g. $T_0 = 1500°C$.

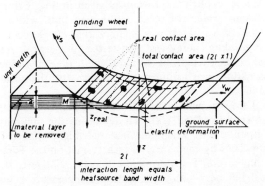

Figure 1. Geometrical or total contact and real contact area in grinding. The elastic flattening of the wheel causes an increase of interaction length.

A one-dimensional model yields the upper bound temperatures

$$T = T_0 \, \mathrm{erfc}\,(z/2\sqrt{\alpha t}) \qquad (1)$$

The heat flux q into the aluminium-oxide grain is proportional to the mean temperature gradient for a given contact time ($t = 2l/v_s$).

$$q = 2T_0 \sqrt{k\rho c/\pi t} \qquad (2)$$

As an example, the contact time t is about $35 . 10^{-6}$ s and using the thermal properties of corumdum at $1500°C$ one obtains $q = 10^7/\sqrt{t} = 1.7 . 10^9$ J/m^2 s. To compare this value with the total energy flux in the contact area, one should reduce the mentioned value of q proportionally to the real contact area of the grains, say maximum $0.03\, q = 4.5 . 10^7$ J/m^2 s. Usual total energy fluxes ($F_t' v_s$/contact length) range from $1.5 . 10^8$ to $3 . 10^8$ J/m^2 s which is about 3.5 to 7 times the maximum heat input into the grains. The value of $4.5 . 10^7$ J/m^2 s may be considered to be some maximum value corresponding to a dull wheel. For a sharp wheel, where the real contact area is roughly 1 per cent of the total contact area, this value is three times smaller and 5 to 10 per cent of the total energy would enter the grain.

Real contact length

Another parameter required for a thermal analysis and which might yield some difficulties is the real contact or interaction length of the wheel and workpiece. Sometimes it is assumed that this length differs little from the theoretical value $l_c = \sqrt{aD_{eq}}$ (e.g. ref. 7), but other authors[11-14,28], show by combined theoretical and experimental investigations that there might be a considerable difference between theoretical interaction length and practical values. The combined interaction depth in plunge grinding

composed of the actual depth of cut and the elastic deformation of the grinding wheel surface (Fig. 1), measured by Okamura and Nakajima[12] amounts to 5 to 10 times the depth setting for depth settings smaller or equal to roughly 5 μm. Taking this combined depth, instead of the depth setting, to calculate the effective contact length, a relative increase of the heat source band width of roughly 100 per cent has be taken into consideration. This point of view is also emphasized by some direct experimental evidence.

Makino et al.[14] and Brown[13] have tried to measure the effective contact lengths directly. The experimental value turned out to be about two to three times the geometrical contact length. These values have actually been predicted as a result of measuring the contact stiffness of the grinding wheel[11]. Some Hertzian elastic flattening of the grinding wheel surface occurs due to the normal force, creating a considerable increase of the contact area relative to the theoretical one especially at low depth settings. This effect is still amplified in internal grinding.

Hahn and Lindsay[28] measured contact lengths in internal grinding by pressing a non-relating grinding wheel against a non-rotating polished workpiece. The scratches produced by moving the workpiece in axial direction showed that the contact length increased with force intensity (F_t'(N/mm)) and amounted up to ten times the theoretical contact length.

As a conclusion it may be stated that a great deal of interesting work has been carried out with regard to thermal aspects in grinding. However, the energy balance yielding the various quantities of heat entering the material surface, the chips and the wheel may be critically discussed. In view of the mentioned evidence it seems reasonable to accept that about 80 per cent of the grinding energy enters the workpiece.

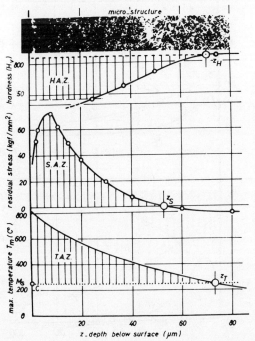

Figure 2. Surface damage of a ground surface in terms of microstructure change, hardness variation and residual stresses. Also the theoretical maximum temperature distribution has been plotted. (Experimental results due to Decneut.)

terms of residual stresses and microhardness changes, experimental results by Decneut[23] in plunge grinding have been used. Those results yield the definition of the Stress Affected Zone (SAZ) and the Hardness Affected Zone (HAZ) (Fig. 2). These two experimental appreciation functions may be related to what can be called the Thermal Affected Zone (TAZ), which is a theoretical defined value corresponding to the area between the maximum temperature distribution function and a transition temperature C (Figs. 2, 3). The three mentioned functions, describing the effect of residual stresses, hardness variations and temperature both in absolute value and in space, may thus be defined as

$$SAZ = \int \sigma \, dz$$

$$HAZ = \int |\Delta H_v| \, dz \qquad (3)$$

$$TAZ = \int_{C}^{T_{max}} z \, dT$$

A more explicit expression is given in Fig. 6 for a constant value of heat input per unit width P ($P = F_t'(v_s + v_w)$). It was assumed that 80 per cent of the grinding energy was transferred to the work. The real contact length based upon the earlier mentioned discussion, was assumed to be twice the theoretical value.

The workpiece material has been ball-bearing 100 Cr6 steel (SAE 52 100). To determine the minimum temperature at which the material might change its properties by metallurgical transformation the Time–Temperature–Transformation diagram relative to this steel has been used. The critical temperature C was assumed to be $250°C = M_s$. Although the contact time for a point at the surface is about 1 to 10 ms, the time during which a point remains above 250°C is about five times larger for high work speeds (1·5 m/s) and small depth settings (1 μm) and twenty times larger for low work speeds (0·25 m/s) and high depth settings (25 μm), so that each point at the ground surface remains for 5 to 200 ms above the 250°C limit. The fairly good correlation (Fig. 4) between thermal influence and both surface integrity evalu-

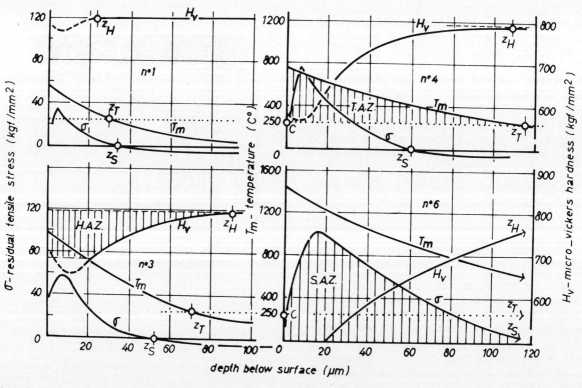

Figure 3. Four particular grinding cases for which the surface damage has been evaluated. The reference numbers refer to Table 2. Definitions are given of the hardness affected zone (HAZ), the stress affected zone (SAZ), the thermal affected zone (TAZ). (Experimental results due to Decneut.)

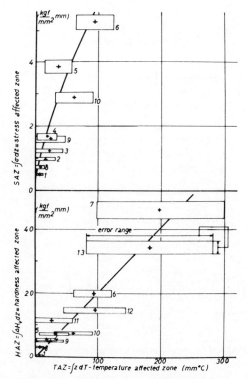

Figure 4. Correlation between the thermal affected zone and both the stress affected and hardness affected zone. Numbers refer to Table 2.

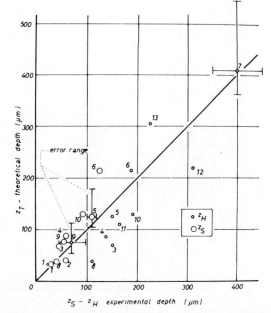

Figure 5. Comparison between the theoretical predicted depth of thermal influence and the experimental values in terms of hardness observations and stress measurements. The work conditions are represented in Appendix 2, numbers refer to Table 2.

ating functions (HAZ and SAZ) shows that temperature effects are the dominant factors influencing residual stresses and metallurgical changes. The predicted values of the depth at which the material regains its initial hardness and the depth at which the residual stress vanishes are of the same order of magnitude of the real values. Table 2 compares the practical values (z_H and Z_S) to the theoretical Z_T value. Figure 5 shows the same result, together with an estimated error range.

It must be noticed that for higher depth settings (a) the predicted values are larger than the real ones, whereas for low values of a the former tend to underestimate. This may be due either to a systematic error in the analysis of the model or to the choice of the model itself. Takazawa's formula is accurate only for small dimensionless values for L and Z (say $L, Z < 10$), as also stated by Lee[17]. Calculating Jaeger's

accurate solution for the same experiments of Table 2 shows that the 250°C limit is situated at 1·2 times the depth, predicted by Takazawa's formula. This would still increase the discrepancies with experimental observations. Using Jaeger's model for large depth settings yield thus predicted heat affected depths which are too large. Werner[7] therefore introduces the effect of thermal resistance of the material layer (M in Fig. 1) just in front of the wheel and which will be ground away. The amount of heat, conducted into that layer, will partly be taken away in the chips and prevented from penetrating into the ground surface. This explains to a certain extent why the temperatures of the ground surface decrease with increasing stock removal rate. Increasing the work speed lowers also the specific energy, but not to such an extent as to explain the measured temperature decrease[8].

Werner's model can be approximated by using Jaeger's analysis, but instead of subtracting a fixed amount, say 20 per cent, from the total grinding

TABLE 2   Analysis of Decneut's experimental work

No. test	V_s m/s	F'_t (N/mm)	v_w (m/s)	a (µm)	A^* (°C)	B^* (m⁻¹)	SAZ (kg/mm)	HAZ (kg/mm)	TAZ (m°C)	Z_S (µm)	z_H (µm)	Z_T (µm)
1	30	4·9	1·5	0·83	575	24 605	0·49	–	0·0064	28	25	34
2	30	8·96	1·5	2·08	845	20 763	0·97	–	0·016	42	–	59
3	30	11·16	1·5	4·16	895	18 265	1·22	–	0·0216	50	150	69
4	30	4·06	0·5	2·5	595	10 043	1·67	5·65	0·0176	60	125	87
5	30	6·09	0·5	5	760	8842	3·85	7·25	0·032	110	150	126
6	30	12·5	0·5	12·5	1260	7467	5·25	20·1	0·0893	125	190	217
7	30	12·2	0·25	25	1430	4242	–	46	0·198	–	400	412
8	60	2	1	1·25	505	17 661	0·73	3·04	0·0064	38	110	39
9	60	3·9	1	2·5	824	15 536	1·6	5·02	0·0216	53	70	77
10	60	8·12	1	6·25	1390	13 131	2·9	7·64	0·061	82	190	130
11	60	2·71	0·5	5	670	8842	–	11·5	0·0256	–	160	112
12	60	6·43	0·5	12·5	1282	7460	–	14·85	0·095	–	310	220
13	60	10·81	0·5	25	1832	6562	–	34·4	0·182	–	225	304

* Equation (4): $T = A \exp(-Bz)$

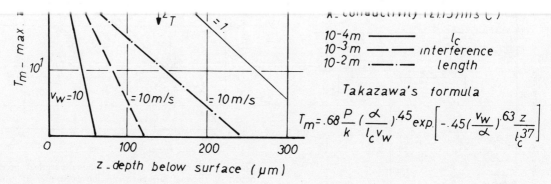

Figure 6. Maximum temperature below ground surface according to Takazawa's approximation for constant power input and various values of work speeds and contact lengths.

energy to take account for the heat removal in the chip, one should subtract an amount of energy which is a function of the depth setting. To describe the thermal resistance of the layer M in Fig. 1, one should count the real depth z_{real} from the real surface during grinding and not from the ground surface (z).

The more complex models discussed above will give a more accurate analysis, but compared to the relatively very simple formula given by Takazawa some rules of thumb are rather difficult to derive.

As a conclusion, using Takazawa's formula with a given heat input per unit wheel width ($0\cdot8$ $F_t'(v_s + v_w) = 10^6$ w/m), (see Fig. 6) higher contact lengths (or depth settings) yield lower temperature at the surface but a slower decay. An increase of the work speed gives rise to both lower surface temperature and quicker decay. Knowing that the temperature is proportional to the heat input, one can use the graphs of Fig. 6 to estimate grinding temperatures or to determine the maximal allowable heat input per unit wheel width to avoid thermal damage. For instance, for a work speed of 1 m/s and a contact length $2l_c$ of about 2 mm the maximum power input which can be allowed without fearing thermal damage is about 100 w/mm. Obviously, for a given stock removal rate, a low heat input is obtained by lowering the specific energy. This, of course, can be realized by properly selecting the grinding wheel, using adequate dressing conditions and possibly by reducing the amount of friction in the cutting process by using doped lubrifiant coolants and fillers.

HEAT BALANCE IN CUT-OFF OPERATION

The ideal kinematic condition for removal of heated up material is realized in the abrasive cut-off operation. Indeed, the direction of infeed v_f coincides with the main direction of heat flow, Fig. 7(a), and in steady state all the energy entering the workpiece is

carried away by the chips. Obviously, in that case the total grinding energy will partly go to the chip and the other part to the wheel. Because the total grinding energy can easily be measured, the main point is to determine the percentage going to the wheel in order to assess its thermal load and the other part going to the chip.

Eshghy[18,19] uses a one-dimensional thermal model to determine the temperature distribution in the wheel and in the layer of width b of the material which will be removed. He assumes a constant contact temperature at the contact surface of the workpiece and the abrasive wheel. Further the wheel is supposed to be in contact with the workpiece in the entire geometric contact area and he uses thermal constants presumably for a mixture of grain and bond material, including also some percentage of pores.

The heat balance, based on this model, predicts that 70 per cent of the total grinding energy would be immediately transferred to the wheel. By a limit analysis Eshghy also assesses the side losses into the workpiece which have been neglected in his model. For an infeed rate of 8 mm/s and a wheel width of 4·7 mm he finds that these losses would account for 10 per cent of the total energy flow. Moreover, an increase of the infeed rate and of the wheel thickness would lower this value. Therefore, in a first approximation, for most cut-off operations those side losses may be neglected. A new analysis which is presented here takes into account the variation of the specific energy with infeed rate, which has been neglected by Eshghy. Also the influence of the magnitude of the real contact area is emphasized.

The effect of infeed rate

To analyse the maximum temperature distribution in the material layer which will be ground away, the thermal model of Fig. 7(b) has been used. Instead of a constant contact temperature a constant heat flux was used, because the total heat flux is an easily

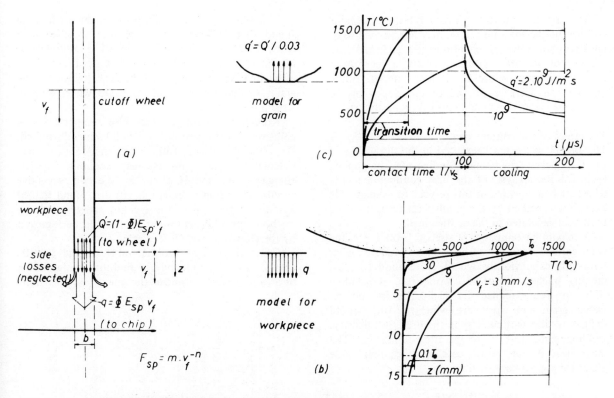

Figure 7. (a) Abrasive cut-off operation and thermal model. The total heat input per unit area is given by $Q = E_{sp}v_f$, which partly goes to the wheel (Q') and partly to the chip q ($n = 0.35$). (b) Steady state temperature distributions in the workpiece show the dependence of infeed rate. (c) Heat pulse on grain cutting tip with transition period, contact time and cooling.

measurable quantity. Moreover, at the material-grain interface temperatures larger than the mean workpiece material surface temperature occur[4,5]. This heat flux represents the amount of heat conducted into the material which is subsequently removed as a chip. The exact analytical solution is given in Appendix 1. The steady state temperature distribution

$$T_m = \frac{q}{h + v_f\rho c} \exp\left(-v_f z/\alpha\right) \qquad (6)$$

is reached after a time t:

$$t = 36\, \alpha/v_f^2 \qquad (7)$$

Calculations for various values of the cooling show that the latter has only a minute influence on this transition time. For plain carbon steel this transition

TABLE 3   Transition periods in seconds in cut-off operation

Down feed rate v_f (mm/s)	$\dfrac{\text{Transition temperature}}{\text{Steady state temperature}} \times 100 = \dfrac{T}{T_m} \times 100\%$		
	100%	90%	50%
3	70	7	0.7
30	0.7	0.07	0.007

period t amounts to $0.6/v_f^2$ ms, which is inversely proportional to the square of the downfeed rate v_f. For the usual ranges of v_f the transition times are obtained (Table 3). Ninety per cent of the steady state is reached after 20 mm depth for slow down feed and after 2 mm for high down feed rate.

Equation (6) shows that a large infeed rate limits the temperature rise in the workpiece material. The decay of the temperature ahead of the grinding wheel is such that 10 per cent of the maximum temperature at the surface will exist at a depth of 12 mm at $v_f = 3$ mm/s and at a depth of 1.2 mm for $v_f = 30$ mm/s (Fig. 7b).

In cut-off operations usually no coolant is used. The maximum temperature at the surface is then

$$T_0 = \frac{q}{v_f C_v} \qquad (7)$$

where q represents the heat flux going into the chip,

$$q = \Phi \cdot E_{sp} \cdot v_f \qquad (8)$$

Φ represents the decimal fraction of the total power per unit area ($E_{sp} \cdot v_f$) going into the chip. $C_v = \rho c$ J/m³ °C and is the specific heat per unit volume. This yields that the surface temperature T_0 is proportional to the specific energy and inversely proportional to the specific heat per unit volume:

$$T_0 = \Phi \cdot E_{sp}/C_v \qquad (9)$$

T. Storm[21,30] who experimented with ultra high infeed rates (15 to 75 mm/s) and a cutting speed $v_s = 100$ m/s showed that the specific energy E_{sp} for plain carbon steel may be written as an exponential function of the infeed rate:

$$E_{sp} = m \cdot v_f^{-n} \qquad (10)$$

m is a constant and $n = 0.35$. This value of the exponent n agrees well with Shaw's[29] ($n = 0.25$–0.5).

Y

again for the analysis of the heat conducted to the wheel. It is interesting in this respect to evaluate the heat flow absorbed by the grains with respect to the total heat generated in the contact area.

A maximum temperature rise of 1500°C for the grain contact surface may be assumed because higher temperatures obviously would correspond to local melting phenomena of the workpiece. A suchlike situation would reduce drastically the friction between grain and workpiece which in turn would reduce at that particular spot the heat flux density, yielding some sort of temperature limitation. The mean heat flux q' in the case of aluminium oxide grains for a given contact time t is

$$q' = 10^7/\sqrt{t} \qquad (11)$$

Using experimental results reported by Shaw[29] and Storm[30] specific energies of 15 to 5 J/mm³ for a 24 grit size wheel are found. The heat flux q' may be determined for practical cases where a cutting speed v_s of the order of 80 m/s, downfeeds from 3 to 75 mm/s and contact lengths between wheel and workpiece ranging from 10 to 100 mm are used. Furthermore, it is assumed that for slow downfeed rates 3 per cent of the geometrical contact area is in contact with the wheel and only 1 per cent in the case of high downfeeds, where the self sharpening effect is more pronounced.

In order to be able to compare the amount of energy per unit time flowing to the wheel Q' and the total grinding power Q, both these values are calculated for 1 m² geometric contact area:

for the wheel: $Q' = 0.03$ to $0.01 q'$
for the workpiece: $Q = E_{sp} \cdot v_f$, using equation (8) with $\Phi = 1$.

These values are represented in Table 4. At low infeed rates v_f a considerable part of the grinding energy flows to the wheel. At larger values for v_f the amount of heat going into the wheel decreases and becomes more and more negligible.

TABLE 4 Maximum power flux per unit geometric contact area to wheel (Q') compared to total grinding power flux per unit geometric contact area (Q) for two values of contact length l_c

v_f (mm/s)	$Q'(10^7$ J/m² s) $l_c = 10$ mm	$l_c = 100$ mm	$Q(10^7$ J/m² s)
3	2·5	1	4·5
6	1·5	0·6	9
75	0·8	0·25	37·5

Calculating the temperatures assuming specific energies of 15 and 10 J/mm³ for the two respective speeds (according to equation 10) it is found that at 3 mm/s 40 per cent of the grinding energy would go to the chips whereas 60 per cent would be removed by the chip at 9 mm/s infeed rate.

Considering the absolute value of the specific energy E_{sp} (Fig. 8) it may be said that at high values of E_{sp} the fraction above the energy of 10 J/mm³

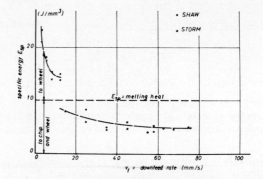

Figure 8. Specific grinding energy versus infeed rate according to Shaw's and Storm's results[29,31]. Shaw: $V_s = 60$ m/s; 25 × 25 mm rod hot-rolled, AISI 1020. Storm: $V_s = 100$ m/s; $d_w = 20$ mm, CK45 and 9S20.

(required to melt a unit volume of chips) necessarily has to be absorbed by the wheel, whereas 10 J/mm³ goes partly to the wheel and partly to the chip. Even if almost the entire grinding energy would be transmitted to the wheel it is obviously favourable to use a relative high infeed rate in order to lower the thermal load on the grain by decreasing the specific energy.

However, the importance of the self sharpening effect becomes more dominant when increasing the infeed. Therefore the decrease of real contact area will yield a decrease of this thermal load even for a constant specific energy. When increasing the downfeed v_f it may be favourable to increase the cutting speed v_s too in order to maintain the same mechanical load on the grain. This again agrees with Shaw[29] and also with Rüggeberg's[31] recommendations. Rüggeberg claims higher temperatures occur when increasing the wheel speed only. Early failure of the wheel may be obtained indeed because a larger real contact area would result in suchlike conditions, allowing more energy to flow into the wheel.

Thermal shock load of the wheel
In assessing the maximal heat flux to the wheel it was assumed that the surface temperature of the grain immediately raised to 1500°C at the instant of contact. The time, required to reach that high temperature is, however, sometimes an appreciable part

of the contact time, even if all the energy would be transferred to the wheel. As a result, the assumption of a maximum surface temperature at the surface of the wheel of 1500°C during the whole contact time yields maximal values for the heat conducted to the grain.

Using the model of an infinite half space with a constant flux q' acting at the border (Fig. 7c), the temperature distribution is found to be:

$$T = \frac{q'}{k}\left[2\sqrt{\frac{\alpha t}{\pi}}\exp\left(-\frac{z^2}{4\alpha t}\right) - z\,\mathrm{erfc}\left(\frac{z}{2\sqrt{\alpha t}}\right)\right] \tag{12}$$

The maximum temperature existing at the boundary is equal to:

$$T_0 = \frac{2q'}{k}\sqrt{\frac{\alpha t}{\pi}} \tag{13}$$

Using average values for the thermal characteristics of aluminium $(k = 20\ \mathrm{J/m\,s\,°C},\ C_v = 5 \cdot 10^6\ \mathrm{J/m^3\,°C},\ \alpha = k/C_v)$, the time necessary to reach 1500°C at the surface is given by

$$T = \frac{\pi}{4}kC_v\left(\frac{1500}{q'}\right)^2 \tag{14}$$

The mean total heat flux $Q(Q = E_{sp}v_f)$ ranges from 40 to 400 J/mm^2 s. If all the heat is conducted to the wheel, the flux flowing through the grains should be equal to Q divided by 0·03 if 3 per cent of the geometric contact area is in contact with the grains. Grain heat fluxes from 10^9 to 10^{10} J/m^2 s are obtained which are comparable with power densities observed in spark erosion processes. The transition periods are found to be of the order of 1 to 100 μs.

In reality the heat source is acting on a rather limited area and not on an infinite half space. The transition times are therefore minimum values. Compared to transition times obtained by Van Dijck[32] for a finite heat source area, the calculated values amount to ±75% of the transition times obtained with a circular heat source of 100 μm diameter.

The transition times, being inversely proportional to the square of the heat flux to the grain (equation 14) are in reality larger than the values obtained here, because not all the grinding energy is entering the wheel.

The magnitude of the real contact area is also quite important. Here 3 per cent of the geometric contact area has been postulated. The smaller this percentage, the more these transition times are reduced.

The rate of temperature decrease of the grain after leaving the contact area can be described by the same theoretical model. In addition a negative heat source is introduced, acting on the boundary when the grain is leaving the contact area. When no convective cooling or radiation losses are assumed (adiabatic surface) the surface temperature can be described by:

$$T = \frac{2q'}{k}\sqrt{\frac{\alpha}{\pi}}(\sqrt{t} - \sqrt{t - t_0}) \tag{15}$$

where t_0 stands for contact time. The rate of cooling is:

$$\frac{\partial T}{\partial t} = \frac{q'}{\sqrt{\pi k C_v}}\left(\frac{1}{\sqrt{t}} - \frac{1}{\sqrt{t - t_0}}\right) \tag{16}$$

For a particular value of grain heat flux $q' = 10^9$ J/m^2 s the results (Table 5) after a contact time $t_0 = 20\ \mu$s are obtained. These examples of ultra

TABLE 5   Cooling rates of grain cutting edges

$t - t_0$	$-\dfrac{\partial T}{\partial t}$	T
(μs)	(°C/s)	(°C)
0	∞	440
10	$6\ 10^7$	230
100	$4\ 10^5$	90
1000	0·2	4

high cooling rates and the frequency heating up (10–100 times per sec) which the grain is subjected to, might indicate that the wear at high temperatures is caused by thermal fatigue.

CONCLUSION

Using a relative simple analysis, based on Takazawa's[22] approximation for Jaeger's[1] solution of a moving heat source on a half space, a good correlation was found between the theoretical predicted depth of surface damage and the experimental data concerning hardness affected zone, residual stress affected zone and microstructure variation. Using the graphical representation of this analysis in Fig. 6 a good estimation of surface quality can quickly be obtained.

The total heat input per unit width acting in the contact area between wheel and workpiece is given by the product of tangential force per unit width and wheel speed $(F_t' \cdot v_s)$. In common grinding operations 70 to 80 per cent of this heat is transferred immediately to the workpiece. About 15 to 20 per cent is going to the chip and the remaining part conducted to the wheel. Cooling is effective only when the coolant can penetrate into the contact zone in order to reduce the high local temperatures in the interface.

An important factor in grinding is the specific energy, which should be kept as low as possible. Appropriate choice of the grinding wheel, of the dressing conditions and the use of lubrifiants and fillers may reduce it considerably.

In the abrasive cut-off operation, compared to usual plunge grinding conditions, very high infeed rates are possible without thermal damage of the workpiece because all the heat flowing in the direction of the workpiece, is removed by the chip.

Higher infeed rates yield specific energies and smaller thermal influenced zones in the workpiece. Thermal pulses, comparable to those occurring in spark erosion processes are active on the grain cutting edges yielding thermal loads which may be a major reason of grit wear.

and *Proc. of the Royal Soc. of N. South Wales*, **76** (3).

2. J. OUTWATER and M. C. SHAW (1952). Surface temperatures in grinding, *Trans. ASME*.

3. R. S. HAHN (1956). The relation between grinding conditions and thermal damage in the workpiece, *Trans. ASME*, May.

4. N. DES RUISSEAUX (1968). Thermal aspects of the grinding process, Ph.D. thesis, Univ. of Cincinnati.

5. SAUER (1971). Thermal aspects of grinding, Ph.D. thesis, Carnegie-Mellon Univ. Pittsburgh.

6. K. TAKAZAWA (1966). Effects of grinding variables on the surface of hardened steel, *Bull. Jap. Soc. of Prec. Eng.*, **2** (1).

7. G. WERNER (1971). Kinematik und Mechanik des Schleifprozesses, Ph.D. thesis, T. H. Achen.

8. G. WERNER and M. DEDERICHS (1972). Spanbildungsprozess und Temperatur beeinflüssung des Werkstücks beim Schleifen, *Industrieanzeiger*, **94** (98).

9. D. G. LEE, R. D. ZERKLE and N. R. Des RUISSEAUX. An experimental study of thermal aspects of cylindrical plunge grinding. *Trans. ASME*, paper 71-WA/Prod. 4.

10. G. W. BOKUCHAVA (1963). Cutting temperatures in grinding, *Russian Engineering Journal*, **43** (11).

11. SNOEYS and I.-CHIH-WANG (1969). Analysis of the static and dynamics stiffness on the grinding wheel surface, *M.T.D.R. Conference*.

12. OKAMURA and NAKAJIMA (1972). The surface generation mechanics in the transitional cutting process, *Proc. of the Int. Grinding Conference Pittsburgh*, April (ed. M. C. Shaw).

13. BROWN, SAITO and SHAW (1971). Local elastic deflections in grinding. *CIRP Annals*, **19** (1).

14. MAKINO, SUTO and FOKUSHIMA (1966). An experimental investigation of the grinding process, *Journal of Mech. Laboratory of Japan*, **12** (1), 17.

15. MALKIN (1968). The attritious and fracture wear of grinding wheels, Sc.D thesis, MIT.

16. SATO (1961). Grinding temperatures, *Bull. of Jap. Soc. of Grinding Eng.*, nr. 1.

17. D. G. LEE (1971). An experimental study of thermal aspects of grinding, Ph.D. thesis, Univ. of Cincinnati.

18. S. ESHGHY. Thermal aspects of the abrasive cut-off operation, Part 1, Theoretical analysis, *Trans. ASME*, 66-WA/Prod. 21.

19. S. ESHGHY (1967). Thermal aspects of the abrasive cut-off operation? Part 2, Partition functions and optimum cutoff, *Trans. ASME*.

20. V. V. KOROLEV. Temperature distribution calculations during abrasive machining, *Machines and Tooling*, **42** (4).

21. T. STORM (1970). Adaptive control in cutoff grinding, *M.T.D.R. Conference*, vol. A.

27. R. SNOEYS (1965). Het onderzoek van de slijpbewerking, *Het Ingenieursblad*, **34**, (11–12).

28. R. S. HAHN and LINDSAY (1967). On the effects of real area of contact and normal stresses in grinding, *CIRP Annals*, **15**.

29. M. C. SHAW (1967). Mechanics of the abrasive cutoff operation, *Trans. ASME*, paper 11-WA/Prod. 13.

30. T. STORM. Internal report of the University of Technology, Delft, The Netherlands.

31. T. RUGGEBERG (1972). High-speed and high-efficiency abrasive cutting, a challenge to circular cold sawing, *Proc. of the Int. Grinding Conference. Pittsburgh*, April (ed. M. C. Shaw).

32. F. VAN DIJCK (1973) Physico-Mathematical analysis of the EDM Process, Ph.D. thesis Univ. of Leuven.

APPENDIX 1

Analytical solution of the thermal model (Fig. 7)
The formulation of the problem:

$$\text{energy balance: } \alpha \frac{\partial^2 T}{\partial z^2} = \frac{\partial T}{\partial t} - v_f \frac{\partial T}{\partial z}$$

boundary conditions:

$$-k \frac{\partial T}{\partial z}\bigg|_{z=0} = q - hT(0, t)$$

$$T(z, 0) = 0$$

$$T(\infty, t) = \text{finite}$$

$T(z, t)$ represents the temperature rise above ambient. Introducing the quantities

$$d = \frac{v_f}{2\sqrt{\alpha}}$$

$$a = \frac{h\sqrt{\alpha}}{k} + d$$

The solution of the problem can be obtained by Laplace transformation. In the case of cooling:

$$T(z, t) = \frac{q\sqrt{\alpha}}{k}\left[\frac{1}{2(a-d)} \, \text{erfc}\left(\frac{z'}{2\sqrt{\alpha t}} + d\sqrt{t}\right)\right.$$

$$+ \frac{1}{2(a+d)} \cdot \exp\left(-\frac{zv_f}{\alpha}\right) \cdot \text{erfc}\left(\frac{z}{2\sqrt{\alpha t}}\right.$$

$$\left. - d\sqrt{t}\right) - \frac{a}{a^2 - d^2} \cdot \exp\left[\left(\frac{zh}{k}\right)(a^2 - d^2)t\right]$$

$$\left. \cdot \text{erfc}\left(\frac{z}{2\sqrt{\alpha t}} + a\sqrt{t}\right)\right]$$

In the case of no cooling:

$$T(z, t) = \frac{q\alpha}{2kv_f}\left[-\left(1 + \frac{v_f z}{\alpha} + \frac{v_f^2 t}{\alpha}\right) \cdot \mathrm{erfc}\left(\frac{z}{2\sqrt{\alpha t}}\right.\right.$$

$$+ d\sqrt{t}\right) + 2v_f\sqrt{\frac{t}{\pi\alpha}}\exp\left(-\left(\frac{z}{2\sqrt{\alpha t}} + d\sqrt{t}\right)^2\right)$$

$$\left. + \exp\left(-v_f\frac{z}{\alpha}\right) \cdot \mathrm{erfc}\left(\frac{z}{2\sqrt{\alpha t}} - d\sqrt{t}\right)\right]$$

In both cases

$$\lim_{t\to\infty} T(z, t) = \frac{q}{h + v_f\rho c}\exp\left(-v_f z/\alpha\right) \quad h \geqslant 0$$

APPENDIX 2

CIRP, grinding conditions
External cylindrical plunge grinding without spark-out
Workpiece
  ball bearing steel 100 Cr 6 (52100 AISI), $d_w = 80$ mm
  heat treatment quenching from 850°C in oil at 130°C and
  subsequent tempering at 150°C for 3 h
  hardness: HRC 62–63
  ultimate strength: 220–230 kg/mm²
  width of contact workpiece–grinding wheel: 12 mm
Grinding wheel—EK60L7VX
  $d_s = 700$ mm
  E modulus: 43·7 kN/mm²
  density: 2·09 g/cm³
Coolant
  emulsion (soluble oil BP energol SB-C) 3 per cent
  rate: 40 l/min
Dressing
  last pass of dressing determined by
  dressing lead $S_d = 0·2$ mm/wheel revolution
  dressing depth of cut $a_d = 0·05$ mm
  tool: 'diamant fliese-Winter' type FA/G 180 ND
Grinding machine
  type: CC
  power: 20 kW
  horizontal static stiffness: 49 N/µm
  horizontal dynamic stiffness: 10·0 N/µm
  free workpiece resonance: 690 Hz

DISCUSSION

Q. Barber, Univ. Newcastle upon Tyne. The authors remark that the actual contact length between the grinding wheel and the work will be greater than the geometrical length because of the Hertzian elastic deformation of the wheel. Although a Hertzian stress distribution may be produced at the start of the process, the metal removal rate will be higher in the regions of high stress and this will tend to redistribute the normal force. In the steady state, the stress distribution will accommodate itself to that necessary to produce the imposed metal removal rate at all points in the contact region and this is highest near the edge of the uncut surface, tapering to zero at the cut surface. Hence, I would expect the normal stress to be highest at the left hand end of the contact region in the author's figure. One effect of such a redistribution of pressure would be to cause more of the heat generated to be removed in the chip material. Also, the equilibrium interaction length will probably differ from that predicted by the Hertzian theory. Both effects would be expected to influence the depth of the heat affected zone considered by the authors.

A. The contact between grinding wheel and workpiece is fairly complicated and it should be stressed that the interaction length cannot merely be deduced from a Hertzian theory. One of the main reasons for this is the very limited contact area between grains and workpiece. The flattening of the grinding wheel is mainly caused by the displacement of the separate grits in contact, because they are elastically mounted by the bond. For theoretical considerations and practical evidence in this respect we refer to refs. 11 and 13. One might expect, however, a difference between the contact lengths in the beginning of the process and in steady state. In this paper we are only concerned with the latter.

The ratio of tangential to normal force in grinding shows that most of the force is due to a friction action and not to metal removal. This is why the authors believe that the stress relief, as expected by Mr Barber, will exist, but to a limited amount. However, it is true that for a more detailed analysis of the process one should consider the increase of forces as is described by Mr Barber. This would result in a heat source (in the thermal model), composed of a predominant uniform distributed source and a triangular distributed one, superposed on the former, so that the highest power density may be expected near the edge of the uncut surface.

GRINDING WITH CUBIC BORON NITRIDE

by

J. TRIEMEL*

SUMMARY

Beginning with a comparison of various abrasives and their mechanisms of wear, the specific problems when grinding high-speed steel with cubic boron nitride cup wheels are discussed. The necessary truing and dressing of the grinding section before the first usage is given special attention, because these preparations have an important effect on ground finish and wheel wear. With the help of theoretical consideration and practical measurements the process leading to the formation of the shape of the grinding section is investigated. This phenomenon is observed during the initial application of a trued and dressed wheel and has a great importance for the deep grinding process. By measuring the grinding force components, the average temperature and the surface roughness, the mechanism of wear in the contact zone between workpiece and tool can be explained.

INTRODUCTION

With the development of wear-resistant, i.e. heavy-to-machine materials, like high-speed steel, tungsten carbide and ceramics and because of the increased requirements of quality and tolerance of the workpiece the abrasive engineering practice has increased its importance within the production technology. The necessity of machining these stable materials led to an intensive development in the last decade. Due to concentrated research the usual abrasives were improved and synthetic industrial diamonds produced for the first time in the mid-fifties.

Since 1969 a new synthetic abrasive has been available, the cubic crystalline modification of boron nitride, produced under high temperature and pressure conditions. For machining hard, ductile materials, the usual grinding wheel with aluminium oxide and silicon carbide can be used, however there is the alternative of using diamond and boron nitride tools which are relatively more expensive. On the basis of their different physical properties there are

Table 1. Properties of various abrasives.

	aluminium oxide	diamond	cubic boron nitride
production	1. natural mineral 2. electrically molten from bauxite	1. natural mineral 2. crystallization from carbon under high pressure and temperature conditions	crystallization from hexagonal boron nitride under high pressure and temperature conditions
thermal stability up to K	1430	900......1900	1650
Knoop microhardness, 100g	2100	6000...8000	4700
wear phenomenon	essentially mechanical destruction when grinding hard materials	graphitization when grinding ferrous materials with low carbon content	decomposition by hydrolysis when using coolants containing water
economical range of application	all materials with low to medium hardness	hard, brittle materials e.g. tungsten carbide, glass, ceramic materials, stones, steels rich in carbon	hard, ductile ferrous materials e.g. high speed steels, tool steels low in carbon,

* IFW, Tu Hannover

kp/mm² and boron nitride with a Knoop hardness of 4500 kp/mm² [1] have definitely superior hardness and wear resistance when compared with other abrasives, but on the other hand their manufacture is very costly. The selection of the correct abrasive for a specific grinding application depends mainly on economic considerations and the necessary technological requirements. Although boron nitride has a lower hardness than diamond it is at the same time slightly more expensive. However it is economically viable to use boron nitride under specific grinding conditions, where its wear resistance property is superior to that of diamond.

The machining of hard, brittle materials with diamond wheels presents no special problems, whereas the grinding of ferrous materials leads to an increased wheel wear. This is probably due to a chemical/thermal reaction in the contact zone. The high temperature existing at the work-point favours in the microscopically small layers of the diamond a change in grating to the hexagonal carbon. The carbon then diffuses in the workpiece materials and chemically reacts with the alloying elements like tungsten, vanadium and chromium to form their respective carbides. The affinity of the carbon in diamond to diffuse is especially high when grinding ferrous materials low in carbon content[2,3].

The wear phenomenon described above has not been found in similar form in boron nitride wheels. But when grinding with water-based coolants at high temperatures a decomposition by hydrolysis has been experimentally shown[3]. Dry grinding or usage of suitable coolants can help to avoid high tool wear.

ally HSS, tool and alloy steels (Fig. 1). The smaller contact zone temperatures developed during grinding with boron nitride reduce the tendency to form soft skins in the surface layers of the material structure[5]. This aspect, and the reduced wear lead to better accuracy of shape and size of ground workpieces.

STRUCTURE OF THE GRINDING SECTION

To achieve optimal machining results it is necessary to prepare the grinding wheel carefully before initial usage. The axial and/or radial runouts of the wheel should be reduced by truing the grinding section. During this process grit and bond materials are removed to prepare the shape of the wheel.

Truing is necessary before initial usage and this costly and time consuming operation is required in only exceptional cases during a production run. After truing, the grinding wheel has to be sharpened in a dressing operation. During the dressing process bond material is removed from the plane surface of the grinding section so that there is enough space for chips produced during machining. These 'chip channels' are formed between the grits in the peripheral direction of the wheel. The other bond material not removed as 'chip channels' fulfil the function of supporting grit in the direction of the cutting forces.

Truing

The truing of diamond and boron nitride grinding wheels presents a difficult problem because of their high wear resistance and because of the necessity to remove as small an amount of grit as possible (due to it being a costly material) and at the same time to achieve sufficient grit sharpness. In IFW different truing devices have been investigated experimentally and the experience has shown that; (a) the Norton truing device and (b) the tungsten bonded diamond truing device are suitable. In the following paragraphs the working principles of these devices are explained and their relative merits discussed.

The Norton truing device is suitable for boron nitride wheels. This device works on the principle of a centrifugal brake with adjustable setting of the relative velocities of the driving boron nitride grinding wheel and the driven silicon carbide truing wheel. The bond is removed in such a manner that the grits fall out of the grinding section without mechanical destruction.

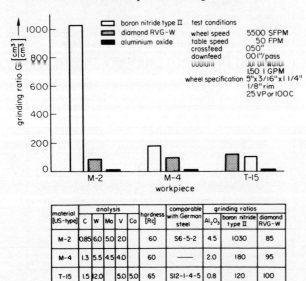

Figure 1. Grinding ratios when grinding various high-speed steels[4].

material [US-type]	analysis					hardness [Rc]	comparable with German steel	grinding ratios		
	C	W	Mo	V	Co			Al₂O₃	boron nitride type II	diamond RVG-W
M-2	0.85	6.0	5.0	2.0		60	S6-5-2	4.5	1030	85
M-4	1.3	5.5	4.5	4.0		60	——	2.0	180	95
T-15	1.5	12.0		5.0	5.0	65	S12-1-4-5	0.8	120	100

* The grinding ratio is defined as the ratio of the volume of workpiece material removed to the volume of grit bearing bond worn from the grinding wheel.

The second method of truing boron nitride wheels is the application of the tungsten bonded diamond truing device. The truing operation with this device results in mechanical damage of the grit.

The surfaces of the grinding sections resulting from these devices are entirely different. The grinding section after truing with the Norton device is shown in Fig. 2. It has a sharp cutting surface with a high roughness. These conditions allow a short subsequent

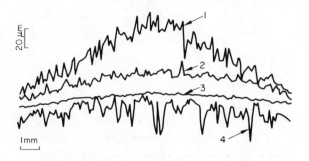

1. stabilized profile
2. profile after truing with a SiC-truing wheel (Norton truing device)
3. profile after truing with a diamond truing device
4. profile after dressing with an Al_2O_3 – dressing stick

Figure 2. Grinding section profile after truing and dressing.

dressing. The profile of the cutting section is roof-shaped, which shall be explained later. This roof-shaped profile accelerates the formation of the final profile required for the production runs. This method is therefore suitable for machine grinding although 5 to 10 times greater truing time as that required for diamond truing is necessary.

As can be seen from Fig. 2 the grinding section after truing with the diamond truing device is plain and smooth. This demands a longer subsequent dressing, usually 3 to 5 times than that for the Norton device. Diamond truing is effective when high accuracy requirements of the shape of the grinding section exist. Therefore this method of truing can be recommended for wheels used for off-hand grinding of tools, for plunge-grinding operations and for form truing applications.

Dressing

The cutting quality of the grinding wheel is influenced by the roughness of the grinding section which has been superimposed on the shape of wheel by the dressing operation. The surface roughness can be recorded with the help of a truing device. The shape of the grinding section in the radial direction is defined as the reference profile. The depth of smoothness R_p is used as a measure for the roughness factor of the surface. R_p is determined by taking the average of the axial distances between the actual (recorded) profile and the reference profile. For good grinding conditions it has been found that the value of R_p must be a minimum of 10 μm for the grinding wheel described below.

Tests have been conducted on resin bonded grinding wheels with the FEPA standard wheel shape 6A2 and the specification 200-15-6-50-100/120 and 18 vol % boron nitride. It has been found that the following properties are essential for the dressing materials:

high porosity to absorb coolants, high grit hardness and soft bonding material. It is usual practice to use aluminium oxide sticks for dressing. The tests with other materials indicate that lime sandstone is equally suitable and at the same time much cheaper.

WEAR MECHANISM

By the investigation of the wear mechanism of boron nitride grinding wheels it is useful to differentiate between macrogeometric and microgeometric wear.

During the initial stages of grinding the formation of the grinding section shape takes place[6,7]. At the end of this process the wheel wear is constant without affecting the shape of the grinding section. This mechanism is denoted as the macrogeometric wear. Apart from this the grit and the bond material are subjected to damage. The mechanism of the wear of the single grit particles and bond is called the microgeometric wear.

Macrogeometric wear

The theoretical calculations to determine the macrogeometric wear assume a plane grinding section at the beginning of the grinding process (Fig. 3). During the first grinding pass only the outermost edge of the

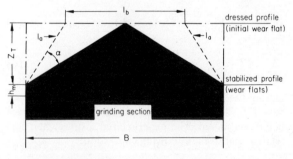

B   width of grinding section
h_m average wear depth
l_a length of wear flats
l_b length of centre plane wear flat
Z_T infeed
a angle of wear edges

Figure 3. Formation of the grinding section profile.

grinding section comes into action whereas the rest of the wheel touches the surface of the workpiece which has already been ground. The length of the cutting edge equals the setup depth of grinding.

The deep grinding process described is characterized by high depths of cut and the corresponding macrogeometrical formation of the grinding section shape is critical for the results achieved in the process.

The boron nitride particles at the edge of the grinding section are subjected to a heavy load during the initial stages of the first cut. This is due to the fact that only a small number of grits are available to remove the entire volume of workpiece material with the setup conditions predetermined. The wear at this stage is very high. The grits which lie in the top edge of the grinding section have an unstable seating due to the one sided support of the bond. Because of the low bonding forces the wear begins at the edge and the worn surface makes up an angle a to the face of the grinding section. The backpasses cause a similar removal of grits from the smaller diameter of the

related to an equal and minimum amount of material to be removed. These elements have a number of grits to do the task. The grinding section reduces gradually at a constant rate. The reduction per number of passes or time interval is

$$A_S = h_m \cdot \sqrt{z_T^2 + \left(\frac{B}{2}\right)^2} \qquad (2)$$

Now the shape of grinding section after each pass and the number of passes z_e required to reach the ultimate profile can be calculated based on the model in Fig. 3. The wear volume V_S^* is determined by the average circumference $D_m \cdot \pi$ of the grinding section multiplied by the wear area A_S^*.

$$A_S^* = \frac{B - l_b}{2} \cdot z_T \qquad (1 \leqslant z \leqslant z_e) \qquad (3)$$

$$V_S^* = D_m \cdot \pi \cdot \frac{B - l_b}{2} \cdot z_T \qquad (1 \leqslant z \leqslant z_e) \qquad (4)$$

During the formation of the wear profile the wear per pass V_{SH}^* depends upon the actual number of passes being considered. The experimental results when correlated lead to the following expression:

$$V_{SH}^* = f(z) = a \cdot z^b \qquad (5)$$

The volume of wear V_S^* after a definite number of passes can be calculated through the integration of V_{SH}^* from 1 to z_e

$$V_S^* = F(z) = \int_1^{z_e} a \cdot z^b \, dz = D_m \cdot \pi \frac{B - l_b}{2} \cdot z_T \qquad (6)$$

By equating the expression 6 and 4 l_b can be determined. Knowing l_b the values of α and z_e are derived:

$$l_b = B - \frac{2a(z^{b+1} - 1)}{(b+1)D_m \cdot z_T} \qquad (1 \leqslant z \leqslant z_e) \qquad (7)$$

$$\alpha = \arctan \frac{(b+1)D_m \cdot \pi \cdot z_T^2}{a(z^{b+1} - 1)} \qquad (1 \leqslant z \leqslant z_e) \qquad (8)$$

$$z_e = \left[\frac{1}{2a}(b+1)D_m \cdot \pi \cdot B \cdot z_T + 1\right]^{1/(b+1)} \qquad (9)$$

Test conditions to determine the wear characteristics are shown in Fig. 4. Measurements were initially carried out after each pass and subsequently after a small number of passes. The coefficients of the

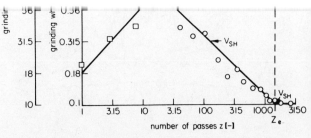

Figure 4. Wheel wear and grinding ratios during formation of the grinding section profile.

expression (5) were computed with the help of a program of least squares fit.

$$V_{SH}^* = 3\cdot638\, z^{-0\cdot506} \qquad (10)$$

Figure 4 shows the relation between V_{SH}^*, G_H^* and number of passes in log plot. The number of initial passes to reach the ultimate profile was found to be 1458. The corresponding workpiece material removed sums up to 105 cm³.

Microgeometric wear

The wear mechanism during the formation of the profile as described above can be observed in the microscopic region, too. The grits on the active wear flats, which take part in the actual machining work, split off and fall out of the bond. The increased mechanical and thermal load of the wheel edges during the first passes lead to a heavy wear of the bond, perceptible in its pungent smell.

The boron nitride particles have very sharp edges and smooth crystal surfaces and therefore receive a metal coating when used with resin bonds to achieve better holding properties. The adhesion between the grit and the coating is not quite satisfactory. It has been observed that the grits fall out of the coating before they are fully used up.

The middle section of the grinding profile, being a remainder of the original dressed profile plays only a subordinate role in the grinding process. This part is distinctly visible to the naked eye as a dark circumferential element. This dark shade is caused by small chip particles with diameters of about 1 μm which are either accumulated in the chip channels or adhere to the bond. This phenomenon is due to the sliding of the middle zone grits on the ground work surface. These grits smooth the small summits left by the other parts of the grinding section. As these grits are exposed to small cutting forces only, they get blunt rather quickly. This results in larger areas of the flanks leading to higher heat generation in the wheel and workpiece zones. Microscopic observations show that chips and grits get welded in the middle zone.

THERMAL AND MECHANICAL LOAD

During the formation of the grinding section the cutting force components and the prevailing average temperatures have been determined. The forces were measured with the help of a piezo-electric system. For temperature measurement the method of Büttner[6] was used. A constantan wire is embedded between two work piece plates which welds with the workpiece surface during the first passes so that a thermo-couple is formed. Figure 5 shows that the mechanical and thermal loads are extremely high during the initial stage of the profile formation and they remain constant after reaching the ultimate profile of the grinding wheel.

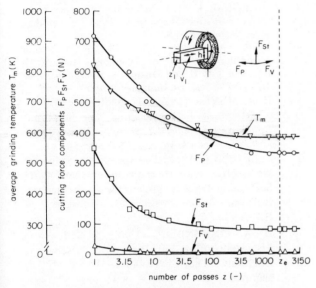

Figure 5. Cutting force components and average temperature during formation of the grinding section profile.

SURFACE FINISH

The shape of the profile influences the surface finish of the ground workpiece. During the initial stages of the formation of the grinding section the amount of the grinding section not taking part in active machining is quite high. This portion of the profile has a smoothing effect just as in the fine grinding process.

Figure 6 shows the depth of roughness R_t and the average value of the roughness R_a of the ground workpiece surface as a relation to the number of passes of the grinding wheel after truing and dressing. These curves show that the roughness of the ultimate profile increases with the number of passes until the ultimate profile of the grinding wheel is reached. At the initial stages a roughness of about 1 μm can be achieved. This superior surface quality is quite desirable for many applications. However, an accompanying structural change in the outer layers of the workpiece surface is not always avoidable.

In order to check the reproducibility of this effect, the middle portion of the grinding section was trued to defined heights with the diamond truing device. The surface finish after these grinding experiments are shown in Fig. 7. Both R_a and R_t have similar values as for the profile formation process for corresponding lengths l_b. To achieve good surface

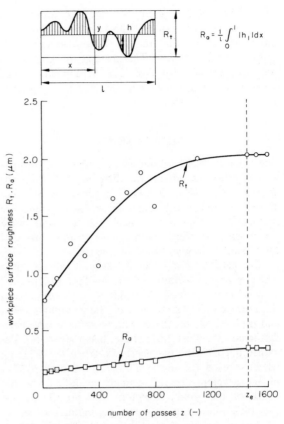

Figure 6. Workpiece surface roughness during formation of the grinding section profile.

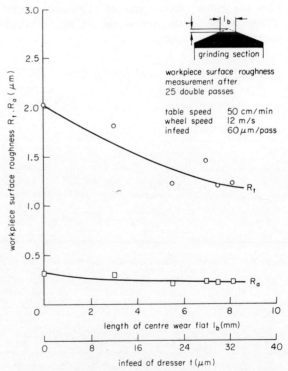

Figure 7. Workpiece surface roughness when using definite dresses grinding section profiles.

finish it is suggested that after attaining the ultimate profile a truing process is undertaken to produce a middle flat and work with reduced depth of infeed.

The experimental investigations of the profile

The machining of hard ductile materials present quite a few problems which are different for different abrasives. Due to their special mechanisms of wear, aluminium oxide and diamond grinding wheels are restricted in their economical use to grind these materials. It is often advantageous to machine hard ductile materials with boron nitride tools.

The careful truing and dressing of the grinding wheel is very important in achieving optimum results. Two different methods of truing were presented and the corresponding structures of the grinding section and their effects on the ground workpiece have been discussed. Practical hints have been given for the choice between the two methods. When the grinding wheel is used for the first time, a grinding section profile develops depending upon the infeed and the width of the grinding section. This process can be very time consuming for the deep grinding process.

During the initial application of the wheel high mechanical and thermal loading are developed leading to rapid wear. The middle zone of the grinding zone has a smoothing effect on the surface produced. After the grinding section reaches its ultimate profile corresponding to the machining set up, constant working conditions prevail.

1. (1970). Borazon-Scheiben, *Schliff und Scheibe*, Norton GmbH 1, S. 3/6.

2. T. N. LOLADZE and G. V. BOKUCHAVA (1967). The wear of diamonds and diamond wheels (English translation of Russian original). *Izdatelstvo 'Mashinostroeniye'* Moscow.

3. D. BORSE (1972). Schleifen mit Borazon—Fortschritt und lohnende Aufgabe TZ f. prakt., *Metallbearb.* **66** 1, S. 1/8.

4. N. P. NAVARRO (1970). The technical and economic aspects of grinding steel with Borazon Typ II and Diamond, *Technical Paper MR 70-198 Soc. of Manufacturing Engineers.*

5. H.-R. MEYER (1971). Technologie und Wirtschaftlichkeit beim Schleifen von Stahl mit Borazon, *Jahrbuch der Schleif-, Hon.-, Lapp- und Poliertechnik* **44**. Ausg. S. 130/154.

6. A. BÜTTNER (1968). Das Schleifen sprödharter Werkstoffe mit Diamanttopfscheiben unter besonderer Berücksichtigung des Tiefschleifens, Doctorate Thesis, Technical University of Hanover.

7. D. A. LINDENBECK (1970). Schleifen von Eisenwerkstoffen mit Diamantschleifscheiben, Doctorate Thesis, Technical University of Hanover.

GENERATION OF SURFACE TOPOGRAPHY ON A GROUND SURFACE

by

H. KALISZER* and G. TRMAL*

SUMMARY

This paper analyses the types and the formation of surface topography generated on the workpiece periphery due to a conventional plunge grinding operation. The parts of the work covered in this paper are concerned with analysing the sources responsible for the formation of surface roughness and waviness. The paper has been written as a general guide and contains practical examples.

INTRODUCTION

The topography of a surface generated during grinding depends to a large extent upon the superimposed action of several factors such as the cutting action of abrasive grains, the relative vibratory motion between the wheel and the workpiece etc.

To study the effect of machining upon the generation of surface irregularities a plunge grinding operation has been selected with a workpiece located in dead centres. The practical conclusions from this type of grinding will be generally the same as for any other type of grinding processes.

The topography of a ground surface can be basically measured in two main directions i.e.

(i) In the direction normal to the main cutting motion (motion of the cutting grits). Such direction corresponds to a topography 'across the lay'.
(ii) In the direction of the main cutting action or 'along the lay'.

To evaluate the effect of machining upon the generation of surface topography a stylus method was employed since such method gives the best representation of the surface in the selected section. The investigated topographical parameters were measured by three different types of stylus instruments, i.e. Talysurf with a sharp stylus traversing across the lay on the surface tested. Rotary Talysurf with a sharp stylus traversing along the lay and detecting the measured irregularities located on the periphery of the machined component and finally with a Talyrond measuring also peripherally, but with a blunt chisel type stylus.

FACTORS INFLUENCING THE GENERATION OF SURFACE TOPOGRAPHY

Cutting action of the abrasive grains

In any grinding operation the number of cutting grains taking part in the generation of surface topography is very large. In spite of the random shape of the grains and their random distribution it is possible to predict a geometrical relationship between the work of a single grain and the shape of its penetration into the workpiece material. To establish such a relationship it is assumed that a single grain which is built into a matrix of a wheel generates a circular trajectory in its motion relative to the workpiece[1]. (The work speed because of its minimal effect upon the grain trajectory can be neglected.) On the basis of the above assumption it is possible to determine approximately the length of a single scratch produced by a particulrar grain as shown on Fig. 1, i.e.

$$L = 2 \sqrt{\left(\text{PVH} \times \frac{D_s \times D_w}{D_s + D_w} \right)}$$

where L is the length of the scratch, D_s is the wheel diameter, D_w is the workpiece diameter, and PVH is the peak-to-valley height.

If we consider that an average surface roughness obtained in a grinding operation $R_a = 0.6\ \mu m$ (PVH = 0.0035 mm) and we select a very common wheel diameter $D = 300$ mm the max. length of a single scratch $L \approx 1$ mm.

If we assume that the cutting part of the grains is

* Department of Mechanical Engineering, University of Birmingham

Figure 1. Length of a single scratch produced by an abrasive grain.

etc. As a result of the dressing operation the irregularities 'across the lay' caused by the abrasive grains are superimposed upon the shape generated by the dresser (see Table 1).

ball shaped with a very small radius ($r = 5.5 \times 10^{-3}$ in)[4] or represents a conical shape with a width to height ratio of 15[5] then the width of a single scratch will be approximately 0.06 mm. In reality grains may contain more than one cutting edge and may be distributed at different wheel depth. For this reason the ground surface can be approximate to a surface consisting of many randomly distributed scratches

The relative motion of the wheel-workpiece

The effect of the wheel-workpiece relative motion can be divided into two main groups, i.e. the motion of the workpiece in relation to its angular position and the relative vibratory motion between the wheel and the workpiece.

The relative motion of the workpiece in relation to

TABLE 1. Types of surface profile generated during grinding.

direction	type of stylus	recorded profile	technological sources
across the lay	sharp	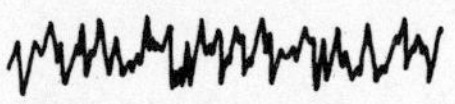	cutting action of the grains. note: expressed most often by surface roughness parameters.
		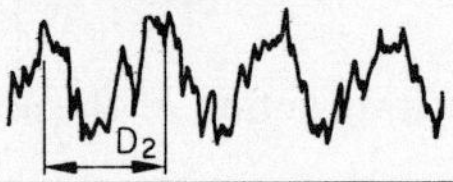 D_2	basic shape caused by the dressing feed and sharp single point dresser. grain cutting action. superimposed. note: included into surface roughness. wave length too short for a normal cut-off.
		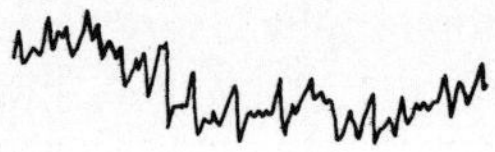	deviation from straightness in dresser movement. vibration of the dresser. non uniform wear alongside the wheel width. note: cross sectional waviness. included or cut off from surface roughness parameters, depending upon the wave length.
along the lay	sharp		cutting action of grains. note: expressed sometimes as surface roughness surface grinding. not widely used.
	chisel shape	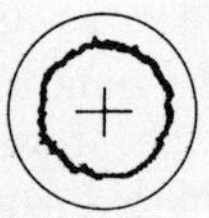 filter N I+450 waves on the periphery	note: the envelope of grain traces due to stylus shape. sometimes considered as waviness.
		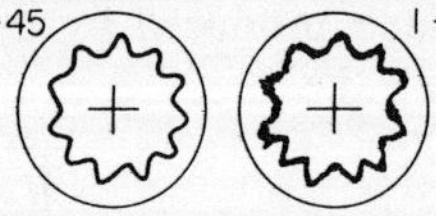 I-45 I+450	relative vibrations of low frequency (wheel speed). note: waviness.
	sharp	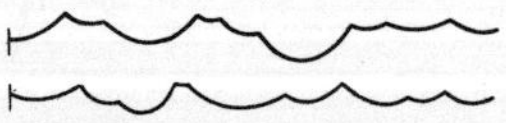	"chatter," usually self exciting vibrations of high frequency. note: can be easily detected by unaided eye. the waves are of similar (or shorter) wavelength as scratches caused by individual grains and of about 10 times smaller amplitude.
	calculated	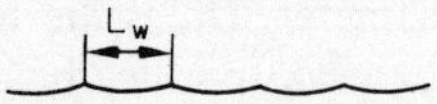 L_w	
	chisel shape	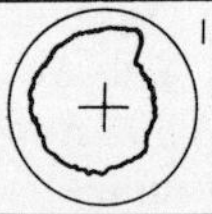 I-450	fast withdrawal of the wheel. "step" equal to the depth of cut. note: error of shape.
		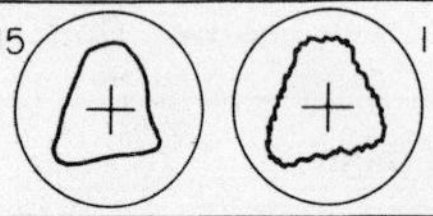 I-15 I-450	workpiece motion in relation to its angular position. inaccurate centres or work spindle. *

its angular position is mainly due to the inaccuracy of the workpiece centres or work spindle motion. As a result of this angular positioning the grinding process produces an error of form (for example, ovality). A relative vibratory motion between the wheel and the workpiece can develop due to the presence of external forces (for example centrifugal forces excited by the wheel unbalances) or can be generated by the internal forces produced by the grinding process itself (chatter vibration). In the first case (low frequency vibration) the superposition of the waves which are generated during subsequent revolutions of the workpiece must be taken into consideration. Such superposition creates generally a wave shift explained in detail elsewhere[2]. In the most unfavourable case when the number of waves generated on the workpiece periphery forms an integral number the amplitude of the waviness will be equal to the amplitude of vibration. In all other cases when the wave shift $\phi \neq 0$ the amplitude of the resultant waviness will be reduced. In cases when the frequency of vibration increases, the wheel diameter must also be taken into consideration (Fig. 2).

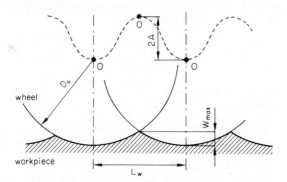

Figure 2. Effect of wheel diameter upon the amplitude of waviness.

Regardless of the amplitude of vibration, the maximum waviness (PVH) generated on the workpiece periphery cannot exceed a certain value W_{max}, which can be approximately calculated as follows:

$$W_{max} = L_w^2/4D_s = V_w^2/4f^2D_s$$

where L_w is the wavelength on the workpiece periphery, f is the frequency of vibration, V_w is the peripheral speed of the workpiece, and D_s is the diameter of the grinding wheel.

A more exact determination of maximum waviness as a function of frequency has been investigated by Peters[3].

In case of high frequency self-excited vibration, the chatter marks (chatter waviness) although easily detected with the naked eye cannot be traced by means of conventional stylus type instruments. This can be demonstrated by considering a very common grinding operation ($V_w = 20$ m/min., $D_s = 300$ mm and frequency of self-excited vibrations $f = 600$ Hz). In such a case the calculated amplitude of waviness cannot exceed 0.2 μm and the wavelength falls within the range of 0.5 mm. Such waviness is undetectable in view of the irregularities generated by the grain

cutting action which are 10 to 15 times higher than those caused by the chatter vibration itself.

A brief review of various types of surface irregularities generated by the grinding process and their respective technological reasons are given in Table 1.

Non-uniformity in surface topography

On certain occasions strips of higher surface roughness may appear on the workpiece periphery. These strips correspond to one revolution of the wheel. Analysis of this phenomena shows that part of the wheel periphery generates a surface roughness higher than the remaining part of the wheel. To find the reason for such a behaviour all obvious factors effecting the surface topography have to be analysed, i.e. the rate of material removal, the time of grinding, the wheel hardness, and the dressing technique. A non-uniformity of any of these factors may lead to non-uniform surface topography.

The rate of material removal has a pronounced effect upon the resultant surface roughness. Due to the fact that the rate of material removal (Z^1) is a function of the product of depth of cut and work speed, the relationship between the depth of cut versus surface roughness can be analysed by considering a constant work speed. A typical relationship between surface roughness and depth of cut for a constant work speed is shown on Fig. 3.

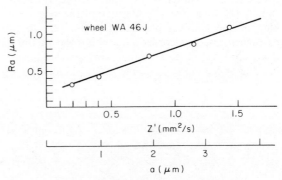

Figure 3. Surface roughness as a function of depth of cut.

The time of grinding or alternatively the volume of material removed (V^1) after completing the wheel dressing operation has also a strong influence upon surface roughness. Such a relationship for three different wheel hardnesses is shown on Fig. 4.

The effect of wheel hardness is also shown on the same Fig. 4. As can be seen from the figure the

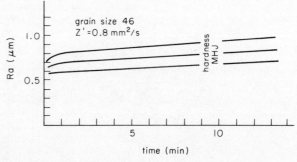

Figure 4. Surface roughness as a function of the volume of material removed.

and the wave shift $\phi = \pi$. The geometrical configuration for such conditions is shown on Fig. 5. As can be seen from the figure. The rate of material

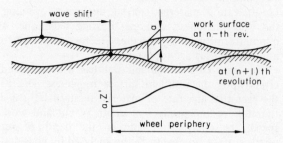

Figure 5. Change in the rate of metal removed due to the wave shift.

removal during one wheel revolution does not remain constant and consequently the surface roughness should follow a similar pattern from the very beginning of the grinding process. The waviness in this particular case as recorded on a Talyrond, should be relatively small because of the existing wave shift $\phi = \pi$. The experimental results after 2 minutes of

uniformly and independently upon the developed run-out. Subsequently, the generation of waviness on the workpiece periphery will develop as well. As a result, softer parts of the wheel will produce a rougher surface on the peaks of the workpiece waves. This phenomenon develops gradually with time during a grinding operation.

In the case when the shift $\phi = \pi$, the harder parts with a tendency to form a run-out are loaded more and, therefore, they also wear more intensively, producing a rougher surface. These two opposed processes have an overall minimizing effect.

Figure 7 shows a practical example of the resultant waviness and surface roughness generated in such a situation by a non-uniform wheel.

CONCLUSIONS

On the basis of the results obtained, the following conclusions can be drawn:

(i) Due to the action of cutting grains the workpiece periphery represents a surface consisting of

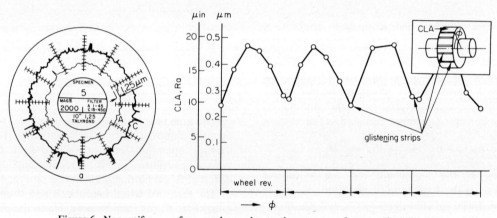

Figure 6. Non-uniform surface roughness due to the pressure of non-uniform load.

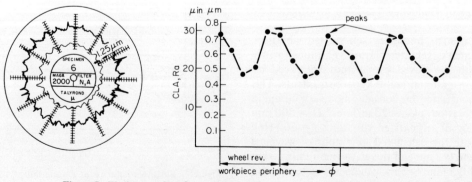

Figure 7. Waviness and surface roughness generated by a non-uniform wheel.

many randomly distributed scratches with a length to width ratio of approximately 20.

(ii) The low frequency workpiece waviness generated on the workpiece periphery depends to a large extent upon the wave shift. In case of high frequency vibration, the amplitude of waviness cannot exceed a certain maximum value which is a function of several factors such as the frequency of vibration, size of the wheel, etc.

(iii) In case of high frequency vibration the chatter waviness cannot be easily traced by a conventional stylus instrument.

(iv) To explain the generation of strips with higher surface roughness generated on the workpiece periphery, the uniformity of hardness and uniformity of loading must be considered.

ACKNOWLEDGMENT

The authors wish to thank Professor S. A. Tobias for providing facilities and the Science Research Council for their financial support.

REFERENCES

1. G. K. LAL and M. C. SHAW (1972). Wear of single abrasive grains in single grinding. *Proc. of the International Grinding Conference,* Pittsburgh.

2. H. KALISZER and G. TRMAL (1972). Apparent run-out of wheel periphery and its effect on surface topography, M.T.D.R., Birmingham.

3. J. PETERS and D. VAUHERCK. Unbalance of electromotors and their influence on the surface geometry in surface grinding.

4. K. NAKAYAMA and M. C. SHAW (1967–68). Study of the finish produced in surface grinding. *Proc. Inst. Mech. Engrs.,* p. 171.

5. M. C. SHAW (1972). Fundamentals of grinding *Proc. of the International Grinding Conference,* Pittsburgh.

METAL CUTTING

THE PRESENT STAGE AND FUTURE OF THE MACHINABILITY DATA SERVICE IN JAPAN

by

T. SATA*, N. FUJITA*, Y. HIRAMATSU†, K. KOKUBO† and H. TAKEYAMA‡

SUMMARY

In the paper the requirements for the machinability data service in the field of automated manufacturing, the desirable types of the service and several approaches to attain the data service system are discussed. The organization and the present research activities to develop the national machinability data centre in Japan are described. Technological aspects such as evaluation and processing of the machining data are also presented. Lastly some ideas on the future development of the technological problems such as the simulation of the cutting process and the diagnosis program for the machining troubles are proposed to increase the usefulness of the data service and to ease the data processing.

INTRODUCTION

As in any industrial country, the present tasks of Japanese industries are attended to the rationalization of small lot production and the displacement of manual work from the workshops as much as possible. Those aims can be achieved by introduction of numerical controlled machine tools and building up a highly automated and integrated manufacturing system. For the rational design and the effective operation of such a manufacturing system, the industry needs not only highly reliable hardware such as machine tools, computers and other automated equipment but also preparation of the computer programs for production design, scheduling and production of numerical control tapes which work well with the aid of good machining data. Thus the needs for the central information centre to collect, to process and to supply machining data have been growing up among Japanese industries. Three years before, the Japan Society for the Promotion of Machine Industry started the project work for establishing the Machining Data Center with cooperation of the industries and the academic society.

This paper first describes the organization and the activities of the project work. In the latter parts of the paper emphasis is put on the description of the technological aspects of the machining data service such as preparation of service programs and fundamental studies of the data generation.

TASKS OF THE MACHINING DATA CENTER

In order to meet the demands from the industries for the data and information service, it would be desired that the Machining Data Center performs the following tasks:

(1) to render the information service on the technology of the conventional machining processes and development of new manufacturing process,

(2) to recommend the optimum or standard machining condition in various kinds of the machining processes depending on the circumstances of the workshop and to supply other related data such as cost and machining time,

(3) to prepare the data file required for specific automatic program system of numerical control,

(4) to render the consulting service to solve the troubles taking place in any machining process,

(5) to establish the standard testing method for evaluation of the performance of work, tool and machine tool and to deliver the standard test pieces of work or tool material if required.

The jobs of the Machining Data Center to execute these tasks are classified into the following sections:

(1) collection of data and information,
(2) processing of data and information,
(3) execution of various services above mentioned by using the data and information stored,

* Faculty of Engineering, University of Tokyo
† Technical Research Institute, Japan Society for the Promotion of Machine Industry
‡ Mechanical Engineering Laboratory

by the classification codes or the key words, and delivered to the user.

The machining data to be collected at the Data Center can be classified into the following three categories: (a) the data obtained at work shops, (b) at laboratories and (c) extracted from the technical documents, which are also given classification codes and stored. This workshop data would be retrieved with the technical codes and delivered without any modification to the user for his reference. The laboratory data would be usually processed by a proper statistical method such as regression analysis or factorial analysis into the empirical formula whose parameters are stored in the form of specified data file. At user's special request such as the optimum machining condition in a specified case or specified data file, this data or the data file are proposed with the proper service programs.

When the user's request is not satisfied with workshop data or laboratory data, the literature data would be searched.

If the machining troubles at the workshops and the remedies for them are studied in many cases and if the data related to these cases are properly processed, the diagnosis of machining troubles and recommendation of the remedies would be possible at the Machining Data Center.

In some cases the Machining Data Center may be requested to evaluate the performance of newly developed tool materials, work materials, or machine tools. For these requests the Machining Data Center would be necessary to standardize the cutting test and to supply the standard work materials or tool materials to the user.

RESEARCH ACTIVITIES FOR ESTABLISHMENT OF THE MACHINING DATA CENTER (MDC)

In view of the necessity for establishment of the machining data service mentioned above, the research project for establishment of the National Machining Data Center started at the Japan Society for the Promotion of Machine Industry (JSPMI) in 1970. First the survey study on the foreign machining data centres and the domestic private data services was carried out[1,2]. To investigate the needs for the machining data services in Japanese industries the extensive surveys were conducted several times by distributing the questionnaires among the industries and by visiting their factories[3]. Of the 370 answers to the questionnaires, the kinds of service requested are shown in Fig. 1. It is found that recommendation of the optimum machining condition and information of workshop data, technical consulting and literature service on the machining technology are mainly desired.

In 1971, the executive committee and the PR committee were established under the leadership of Dr Tomonaga, Director of Technical Research Institute, JSPMI. The executive committee is in charge of the system design, data collection, and data evaluation, on the other hand the PR committee is in charge of communication and PR activities.

Based on the results of the survey research, it was decided that the Machining Data Center would start the following five services:

(1) recommendation of cutting conditions,
(2) the laboratory data service,
(3) the workshop data service,
(4) literature service,
(5) consulting.

It was also decided that the data service would start first on turning operation followed by milling operation, drilling operation and other operations.

For collection of machining data the executive committee prepared the laboratory data format and the workshop data format, and decided the specification of four kinds of the standard work materials (carbon steel, Cr–Mo steel, stainless steel and cast iron), and two kinds of carbide tool tips and ordered these materials.

Now four senior industrial research institutes are conducting the turning tests to collect the laboratory data.

Concerning the workshop data service, two systems, for the private and the public data service, are now under consideration. The private data service means that each company has its own account to deposit and draw the machining data occasionally. On the other hand all workshop data collected by the Machining Data Center are opened to the public for the data service.

The computer programs for the machining data processing and information retrieval are under development.

The executive committee has started to list manufacturing engineers who are willing to cooperate with the Machining Data Center in digesting the technical documents and in consulting services.

These data services are scheduled to start in 1974. The organization of the project team for establishing the Machining Data Center is illustrated in Fig. 2 and the time-schedule of the project is shown in Fig. 3.

To promote the project work on the machining data services, various fundamental studies on machinability, analysis of machining processes and characteristics of machining data are necessary to be done. The work is undertaken by the Special Committee on Machinability in the Japan Society of Precision Engineering with the financial support from JSPMI. The special committee consists of three

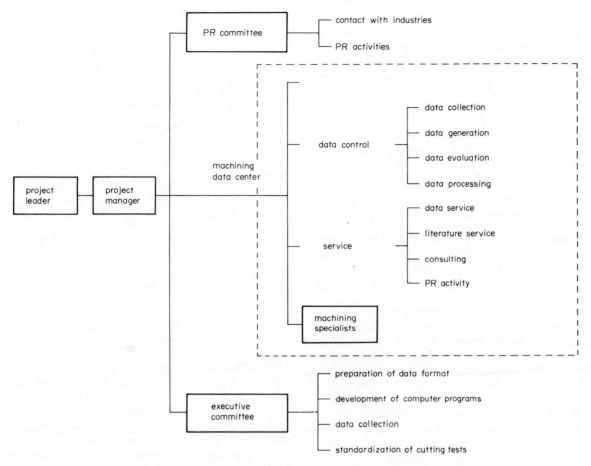

Figure 2. The organization of the project team for establishment of MDC.

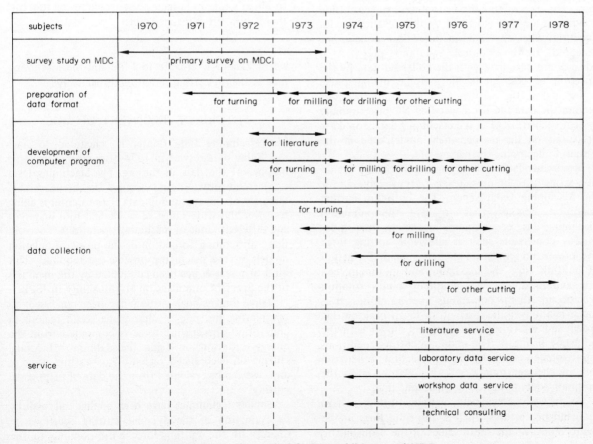

Figure 3. The time schedule for establishment of MDC.

materials on machinability.

TECHNOLOGICAL ASPECTS OF MDC

To start the data service, many technological problems related to evaluation and processing of the machining data should be studied. The following methods of the data evaluation are now being examined.

First, it is necessary to evaluate the machining data collected to eliminate any mistake or error in data acquisition. Manual inspection of the data sheets must be done to find any mistakes in filling in the sheets. Second, by utilizing redundancy of the data described in the sheets, some parts of data description can be checked by the computer program. Third, when the variance analysis on the machining data is possible, the extremely deviated data can be automatically picked up by the computer program and then examined manually to find any cause of deviation.

When machining data belonging to a certain category are collected, the mathematical model may be formulated to describe the characteristics of the data. The data of tool-life or tool-wear versus cutting conditions are well described by the so-called Taylor equation. When any fixed mathematical model is not accepted, many empirical equations prepared in advance are examined with the collected data by the criterion of best model fit and the fitting model is selected as the mathematical model. If the users require the data file for a specified NC programming system such as EXAPT, the machining data stored are processed on the mathematical model used in the system to determine the parameters.

Recommendation of the optimum cutting condition is one of the important service types required for the Machining Data Center. The optimum cutting condition denotes that to give the minimum machining cost or minimum machining time under various constraints such as allowable cutting force and power, no generation of chattering, no formation of built-up edge, the maximum and minimum feed rate and maximum and minimum spindle rotation will occur. All the constraints may be expressed as linear relations in the space of the logarithmic cutting speed and logarithmic feed rate as shown with the solid line in Fig. 4. The optimum cutting conditions approximately given by the intersection A between one of the constraints and the relation given the optimum tool life shown by thick solid line.

As is well known, any mathematical data have probabilistic features, that is, the solid lines in the figure represent the mean values of the probabilistic

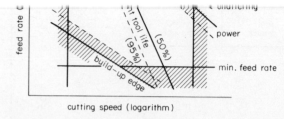

Figure 4. Probabilistic determination of the optimum cutting condition.

distributions, or the values attained with the probability of 50 per cent. In practice, however, the recommendation of data are required to have higher reliability such as 95 per cent. Therefore the constraints and tool-life relation should be shifted as shown with the broken lines in Fig. 4, and the optimum cutting condition for recommendation is displaced from the intersection A to B[4].

Because the cutting conditions usually change in some range in 'even cutting' process with a single tool in practical machining operation, it becomes very important to estimate the tool-life under varying conditions. It has been found that a similar relation to Miner's rule in fatigue is applicable to the tool life, that is, tool life is reached when

$$\Sigma\, \tau_i/T_i = 1$$

where T_i and τ_i are the tool life and the machining time at the ith cutting condition respectively.

FUTURE DEVELOPMENT OF MDC

The Machining Data Center is scheduled to start machining data service in 1974 after storing the large number of the data in turning. The Machining Data Center, however, should acquire the immeasurable quantity of the machining data if the Center is going to cover the wide range of work and tool materials and different kinds of machining operations. To avoid these difficulties, formation of the proper simulation models of the machining process are required. This kind of research has been conducted by the members of the Special Committee on Machinability in JSPE.

Three-dimensional cutting has been analysed by considering the pile of thin plane strain regions in chip flow. The cutting force is computed from the energy criterion by using the data in orthogonal cutting on each sliced region. Thus cutting force in any case can be evaluated from the data of orthogonal cutting[5].

Similar techniques have been applied successfully to evaluation of the dynamic cutting coefficient[6]. Change of the cutting force corresponding to the inner and outer modulation is computed on the model of the sliced plane strain region. The dynamic cutting coefficient thus obtained is used in the stability analysis of the three-dimensional cutting system, to give the stability chart as shown in Fig. 5.

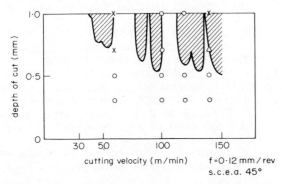

Figure 5. The stability limit against chattering ○: stable cutting; ×: chattering.

The process of formation of surface roughness in cutting has been well studied and the block diagram representing the formation of surface roughness is pictured as shown in Fig. 6[7]. Since the surface

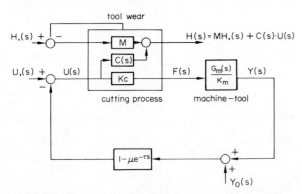

Figure 6. The block diagram of formation of surface roughness.

roughness depends on the geometrical component determined by the tool nose and feed rate, swelling up of work material in cutting, eccentric rotation of the spindle, relative displacement or vibration between tool and work and the change of the tool shape due to the progress of tool wear, it is necessary to derive the mathematical expressions of these components and to store the data concerning each process in order to evaluate the roughness of the machined surface.

In spite of the great effort made, the processes of wear and failure of cutting tool have not been well elucidated. The processes, however, may be pictured as being constituted by diffusive wear, abrasive wear, mechanical fatigue and thermal fatigue as shown in Fig. 7. Examination of wear debris of tool material on the chip surface by an X-ray micro-analyser has shown that the diffusive wear is predominant in the range of high cutting speed while the abrasive wear becomes larger with the decrease of the cutting speed[8]. By the local temperature measurement on the tool face it has been disclosed that the rate of the diffusive wear depends not only on temperature but also on the distance from the cutting edge probably due to the formation of thin film between tool and chip[9]. Furthermore extensive studies will be necessary to understand more thoroughly individual processes involved in tool wear phenomena and to find the influence of various conditions of tools, work and cutting on the wear processes.

It is scheduled that the consulting work at the Machining Data Center starts by many machining specialists with good technical experience. Such an approach must eventually become more difficult and must be changed due to the increasing difficulty to get new specialists and possible unwillingness of people to do such tedious work as consulting. New approaches such as computerized diagnosis of the machining troubles may solve the difficulty, while the task is left in future.

CONCLUSION

The project work to establish the Machining Data Center is now proceeding in Japan to aim at the start of the data service in 1974. Cooperation of the

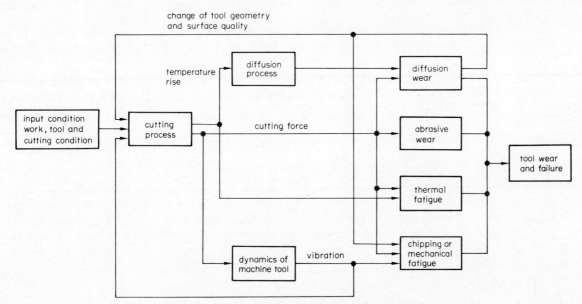

Figure 7. The process of wear and failure of cutting tool.

techniques.

REFERENCES

1. T. SATA and K. KOKUBO (1971). Investigation on machining data bank (Report No. 1), *Report of Technical Research Institute of JSPMI.*

2. Y. TOMONAGA, Y. HIRAMATSU, K. KOKUBO and S. SHINYA (1972). Investigation of machining data bank (Report No. 2), *Report of Technical Research Institute of JSPMI.*

(1973). Analysis of three-dimensional cutting dynamics, *Ann. CIRP,* **22**, to be presented.

7. T. SATA and M. KUBO (1966). Surface roughness formation in metal finishing, *Jour. Jap. Soc. Prec. Engg.,* **32**, 118.

8. K. UEHARA (1972). On the mechanism of crater wear of carbide cutting tool, *Ann. CIRP,* **21**, 31.

9. M. HIRAO and T. SATA (1973). Measurement of local temperature of the tool face in cutting, *Jour. Jap. Soc. Prec. Engg.,* **39**, to be presented.

SATISFYING INDUSTRY'S NEEDS FOR IMPROVED MACHINING DATA

by

A. I. W. MOORE*

SUMMARY

This paper contains a review of the background and developments leading to the setting up of a machining data bank facility in the UK. Also described is the more recent development whereby companies participate in a 'Club' specially formed to provide subscribers with continuous supplies of 'proven' machining data. Details are given of the way the 'Club' operates.

INTRODUCTION

For companies engaged in machining practice, overall machining performance has a significant influence on profitability, and in particular, the general levels of feeds and speeds used are an important factor. Although this simple fact is appreciated by many companies, a large proportion of them feel unable to take positive overall steps to raise their machining rates, because they do not have the necessary machining data in a form for which they can make comparisons. As a consequence, the actual feeds and speeds used are left to the discretion of the machine operator. Usually, the only constraint is a 'blanket' machining time allowance agreed between management and operator which also includes the time taken to load, and unload workpieces into the machine, in-process measurements, etc., as well as the actual cutting time.

Although some managements can expect to run into resistance from machine operators whenever they attempt to implement improved machining data, it is essential that they should have access to a source of reliable data so that they can make informed judgements about the steps they should take to remedy undesirable situations.

In many other companies where such obstacles do not exist, machining speeds and feeds are often set with respect to a variety of data sources, most of which are simplified or generalized to a degree which makes them suitable only as a starting point for establishing the speed and feed which should be used in an individual machining situation.

Partly in recognition of managements' needs for reliable machining data, and partly to stimulate companies to seek to improve their general levels of speeds and feeds, a series of steps have been taken over the past few years which have culminated in the setting up of a machining data system in the UK which it is believed will set the pattern for similar developments in other countries throughout the world.

A feasibility study undertaken on behalf of the Ministry of Technology[1] provided a sound basis for the development of the UK system, by establishing (i) companies' needs in respect of machining data, and (ii) the potential financial benefits which could be expected to result to companies and the nation as a whole from industrial implementation of improved machining data.

Subsequently, PERA set up a computerized machining data bank, and this forms the backbone of the PERA Machining Data Club activity which has now been firmly established in the UK.

In this paper, an outline is given of the more important aspects of these developments, together with an indication of the current activities and developments of the above-mentioned 'Club'.

THE NEEDS OF COMPANIES

Two significant features of the feasibility study for the Ministry of Technology were, firstly, an industrial survey of a typical cross-section of British industry comprising some fifty companies, and secondly, a careful analysis of the levels of industrial machining performance.

Industrial survey

A questionnaire was compiled which comprised some sixty-eight questions. Each company was visited, and the interviewer guided the discussion by referring to the questionnaire, and completed it himself. In this way, it was possible to ensure that a consistent

* R & D division, PERA

sources of machining data used.

For about one-quarter of the companies, machining conditions were specified by shop floor personnel, and in the remainder by planning methods or ratefixing staff. Apparently, most machine operators complied with the machining conditions specified, although in five cases it was stated that they never complied.

The principal sources of machining data were 'experience', experimental results or trial runs, cutting tool manufacturers, company internal standards, and general published technical literature. Some companies obtained data from associate companies or from other companies—often through contacts between individuals on a personal level.

Two questions in particular produced answers which apparently contradicted each other. Whereas thirty-seven companies said they obtained the data they required from their regular sources of information, forty said the data they used could be improved. Nevertheless it was encouraging to find that over forty companies believed that improved data would be implemented on the shop floor. Only five companies thought that union attitudes would create an insurmountable obstacle to implementing improved data.

Only six of the fifty-one companies co-operating in the survey were not positively interested in using a machining data centre as a source of information, almost all of the remainder envisaging no difficulty in getting their employees to use it.

When companies were asked about the kind of information they would expect to obtain from a machining data centre, as expected, almost everyone required cutting speeds, feeds, etc. Likewise, they specified a need for cutting tool details, and two-thirds of them required tool life data and coolant details.

On being asked which factors they would specify when requesting information from a centre, about half of the companies indicated that a particular cutting tool would be stipulated. Virtually everyone, as expected, would specify the workpiece material and type of machining operation.

About two in every three companies expected to receive merely starting-point data, but one in three expected precise data.

Although it was clear that a machining data centre would be of interest to almost all companies, the same high proportion of companies were also interested in one or other of various kinds of handbooks, with the emphasis on those that could operate on the basis of regular updating or revision of data.

data available from other sources.

By comparing industrial data with data from various other sources, it became feasible to formulate a basis for assessing the overall potential for improving machining practice through the implementation of data which would be provided by a machining data service. It was also recognized that it would be necessary to identify and assess the main sources of data for a data bank. Therefore, the feasibility study included a thorough search to bring to light all published and unpublished sources of data and information on metal cutting.

Some 6000 published articles and 250 books were scrutinized for data about specific machining situations, and about 150 articles and twenty-seven books were considered to be worthy sources of data. Of the latter, eight were particularly suitable.

PERA's research reports—some 450 in all—were estimated to contain in the order of 10 000 data points in the 'proven data' category, and 30 000 in the 'estimated data' category. Manufacturers of cutting tools were recognized as another important source of data, but although virtually all such companies were well able to provide useful data in terms of feeds and speeds they would recommend for different machining situations, actual records comprising 'proven data' were comparatively scarce.

The more important items revealed by the survey were treated as a 'primitive' data bank, and comparisons were made between some 550 items of industrial data on the one hand, and corresponding appropriate data from the bank. In effect, each item was treated as a synthetic enquiry for data.

Analysis of the comparative data showed conclusively that the recorded industrial data relating to actual machining situations was consistently inferior in terms of machining rates. Specifically, it was concluded that there was scope for increasing machining rates on average by about 150 per cent ranging from 50 per cent in turning operations to 600 per cent in end milling operations.

Potential benefits to industry

By treating the aforementioned data comparisons as representative of industry as a whole, an analysis was undertaken to determine the economic benefits of implementing improved machining data in the UK industry. The analysis was based on the potential application of improved data in six common machining operations (turning, drilling, etc.). Because only the time actually spent cutting metal would be affected, an average machine utilization of 20 per cent was assumed. Expression of benefits in financial terms was based on an average machining cost of £1 per hour.

Extrapolation of the survey data to represent the overall national situation required estimates to be made of the total time spent throughout the country in carrying out the six selected machining operations on the various types of machine tools used. By making use of published data[2,3] estimates were made of the total effective numbers of machines performing the selected machining operations, and hence, the total effective time spent on these operations.

By combining the results of the analysis with average potential increases in machining rates of *half* those indicated by the data comparisons referred to previously, it was concluded that the indicated overall saving to industry would be in the order of £70 million per annum. The corresponding respective savings for each of the six selected machining operations are as follows.

Drilling:	£21·4 million	Tapping:	£6·0 million
Turning:	£20·2 million	End milling:	£4·2 million
Reaming:	£8·4 million	Face milling:	£10·8 million

Expressed in terms of a typical individual company using twenty-five production machine tools, the equivalent potential saving is £4000 in one year.

ESTABLISHMENT OF MACHINING DATA SERVICE

Computerized data bank

Having estimated the scale of the potential savings to industry through implementation of improved machining data, it was quickly recognized that it would be essential to establish a source of such data in a form which would facilitate immediate implementation. As a first step, therefore, PERA designed a computerized system for storing machining data and proceeded to establish a data bank facility by encoding thousands of items of machining data taken from actual records of machining situations.

This basic concept ensured that certain essential features were built into the data bank system. Firstly, because the cost of sifting and sorting individual items of data is prohibitive, and secondly, because any criterion which might be used to determine which would be the better of alternative items of data might not apply in different companies, the decision was made to include virtually any item of data provided that it represented a record of an actual machining situation. Thus, the computer was treated merely as a vast storage facility which could sort data rapidly. Data is stored on magnetic tapes—each one can accommodate some 50 000 data points—and the storage capacity is of course unlimited. The system embodies a well-developed 'search' program so that data for up to twenty-four individual enquiries can be sought in a single run of a data bank tape. Other programs enable tapes to be up-dated at intervals as required.

It is important to understand that the computer is not used in any way to carry out calculations—for example, to determine minimum costs or maximum production rates—nor is it capable of being linked to a data transmission system of any kind. Most of the data on tape is in encoded form and would be unintelligible to anyone outside PERA.

Provision has been made in the design of the system for storing on tape virtually every detail of a machining situation, including tool life, tool life criterion, etc. and the following are examples of the 'data fields' included.

(a) type of machining operation.
(b) workpiece material—specification, form, condition, etc.
(c) cutting tool—design, geometry, material, size, etc.
(d) machine tool—type, horsepower, condition, etc.
(e) cutting conditions—speed, feed, etc.
(f) cutting fluid—type, dilution, etc.
(g) performance criteria—tool life, mode of tool failure, accuracy, surface finish, etc.

The computerized data bank provides the means for dealing with any company's machining data requirements. If necessary, a search can be made quickly against a specific requirement and a number of companies make use of this facility. However, because companies do not readily accept that good machining data is a commodity which is worth paying for directly, means have had to be devised of using the data bank facility to supply those companies which recognize the value of good data with data in a form which is acceptable to them and at a price they are prepared to pay. The answer has proved to be a unique Club facility which is believed to be the first one set up anywhere in the world to provide companies with machining data.

PERA machining data club

The Club has been founded on the principle that companies individual needs could be met by indirect financing of the facilities required. Nevertheless, to establish and maintain the resources required, the cost to each company had to be low and also the support of a sufficiently large number of companies was essential.

The Club provides two main services to its subscribers but, in addition, there are a number of other unique features which are designed to provide companies with what they increasingly require. The two main services cover:

(a) regular monthly supplies to subscribers of twenty specially prepared machining data bulletins which are inserted in loose-leaf binders to build up into successive volumes of the PERA machining data manual,
(b) limited free access to PERA's machining data service whereby companies can request data concerning specified machining situations.

Each data bulletin contains one or more items of proven machining data extracted from the computerized data bank, that is, data which constitutes a record of an actual machining situation. The form of presentation is such that, when filed in a binder, all bulletins relevant to a particular machining situation

bulletins is that data can be provided about any individual design of cutting tool which users may be called upon to use. For example, to take the field of reaming, it is quite feasible to include data about the many different designs of reamers available commercially, whether they be of the floating blade, expanding blade, or conventional types; whether they have 'quick-spiral', 'slow-spiral', or straight flutes; whether they have a low bevel lead angle, or a special tooth geometry such as the PERA reamer; or whether they be of the long-flute, chucking, or piloted type. It has been recognized that users require data concerning any or all of the many different types of cutting tools available, not merely about so-called 'conventional' tools. The PERA loose-leaf manual system has established the means whereby users can be provided with data about all the wide variety of cutting tools used.

If a Club subscriber cannot locate the data he requires from the manual, as will often be the case in the early stages, he can make use of the data service by contacting PERA. If—again (as sometimes will happen)—PERA does not have proven data to meet the enquirer's needs, he can be provided with starting point data while attempts are made to locate what is required from other subscribers. If the latter fails to reveal proven data, PERA can take steps to acquire it by undertaking machining tests or by assisting the original enquirer in generating and recording what is required at the company's factory.

Another feature of the manual is that up-dating is no problem. As new or improved data becomes available, it becomes a simple matter to replace obsolete bulletins. Such new or improved data can arise from machining tests specifically designed for the purpose, or as a result of feedback from companies who have applied data from the manual, or again, from cutting tool manufacturers and others willing to have data disseminated to users. Notwithstanding the source of data, it must however be of the 'proven' variety. The insistence on proven data is essential—acceptance or inclusion of any other type of data in the computerized data bank for publication in the manual would detract from the long term value of the manual when several volumes have been established, so that it would acknowledge a continuing need for users to experiment on the shop floor to develop 'proven' data.

One more feature of the manual is that the loose-leaf system enables a wide range of data to be included concerning each type of tooling, and thus leaving the user to select that system which best suits his own performance criteria, whether this be related to small, medium or large quantity production overall

number of companies joining has ensured a solid foundation of support, and there is every reason to suppose that the Club will continue to expand in the future. It is also encouraging to note that the Club qualifies for grant support by the Department of Trade and Industry.

Already, enquiries have been received from various organizations in several overseas countries, and it is expected that there will be no obstacles to overseas companies being able to join the PERA Club.

As may be expected, initially, capacity has been concentrated on issuing data bulletins to UK Club subscribers, but as support has grown increasing attention is being given to fuller development along the lines already indicated.

CONCLUSIONS

Although the PERA machining data Club has been in operation for only a comparatively short time, it is already clear that the degree of industrial support shown so far is such that the Club is providing the kind of service which companies need in respect of machining data. With increased industrial support, there is no reason why the range of facilities available to Club subscribers should not be broadened and the transmission of data through the Bulletin system accelerated.

In the longer term, it should be possible to develop methods of rationalizing and streamlining the presentation of proven data, but the present form of bulletin data should be adequate to meet the needs of most companies for the immediate future.

ACKNOWLEDGMENTS

The author wishes to thank the Director and Council of PERA for permission to publish this paper.

REFERENCES

1. Feasibility study on the need for a machining data service. Ministry of Technology, November 1969.
2. PERA Report 142: Industrial machining practice—a survey of machining operations.
3. Census of machine tools in Britain, *Metalworking Production*, July 1966.

DISCUSSION

Q. J. Fisher, Wickman Wimet Ltd. Despite savings accruing from the use of improved machining Data, have any of the speakers experienced any reluctance to pay the economic cost of producing such data remembering that the cost per enquiry for computerized data could be as much as £20?

A. A. I. W. Moore. PERA has found by experience that, in general, although companies are usually willing to pay for machining data to be generated to provide solutions to machining problems, there is some resistance to charges made for searching or retrieving data which already exist. This is probably due to a number of reasons, one of which may be an unwillingness to recognize machining data as a kind of 'commodity'. The development of the PERA Machining Data Club described in the author's paper appears to be an effective way of solving this problem, inasmuch that subscribers can make requests for data within the overall subscription charge.

APPLICATION OF MACHINABILITY DATA BANKS IN INDUSTRY

by

JOHN F. KAHLES* and MICHAEL FIELD*

SUMMARY

Suggestions are presented for developing a national numerical machinability data bank based upon the *Machining Data Handbook*.[1] Modifications and additions which are indicated by our growing technology must be limited by salability considerations of data bank products. Sources of input data are identified and several computerized machinability data bank systems are referenced. It is believed that an international numerical machinability data bank is not required at this time.

INTRODUCTION

At the international meeting of the American Society for Information Science in October 1972, the INTELSAT IV communications satellite provided a link from the Shoreham Hotel in Washington, D.C. to Europe. Participants at this meeting were able to engage in on-line searching of data bank files located in Darmstadt, Germany. This was a dramatic demonstration of the progress which has been made in the software and hardware of communication systems and of the interest which has been shown not only in national but in international information systems. It should be pointed out, however, that the more than casual participant in 'international file searching' at this meeting would have recognized that the data banks being searched were *document* rather than *numerical* files and that there were negative indications concerning the economic feasibility of international satellite searching of these document files. Other comments were made regarding inadequacies of documentation systems in relation to specific industry needs, but those concerning *economics* and the need for *numerical data banks* stand out as being of prime significance in relation to machinability data banks.

In the United States, a number of information analysis centres sponsored by the U.S. Government have been developed in support of industry and government, and for a number of years the services of these centres were made available to qualified users, essentially at no cost. Beginning in January 1972, these information centres were required to recover costs for all services and data products. This action placed all centres in an entirely different economic framework as a result of 'cost recovery' contract provisions. In effect, all information analysis centres, including the Machinability Data Center (MDC) which is operated by Metcut Research Associates Inc., were required to give high priority to the salability of services. MDC's present contract with the U.S. Government calls for selling sufficient publications and services to collect dollars equivalent to 50 per cent of the initial contractual funding. None of the income is returned to the U.S. Government, but rather all income is added to the original contract, thereby increasing the scope of MDC's effort. The experience gained in managing MDC in two different types of economic environment—namely, total Government funding and cost recovery—has necessitated considering new and improved concepts of machining information technology in a very practical way.

The above background information is provided because it forms the basis for MDC's present outlook concerning numerical machinability data banks, both national and international. In this paper, therefore, principal consideration will be given to concepts which can lead to salable products and services from machinability data banks in our present economic environment. Consideration will be given to MDC's present capabilities, to the types of machinability data which are currently available and salable, and to improvements which appear essential to keep pace with our growing technology without seriously affecting marketability.

MDC's PRESENT STATUS

At the present time, the Machinability Data Center maintains a computerized document file composed of approximately 25 000 items. These documents are retrievable using *material identification* and *machining operation*. In addition, uniterms, such as tool geometry, cost and production, cutting force, safety, etc., are used to identify pertinent document

---

* Vice-President and President, Metcut Research Associates Inc.

z

processed by the Machinability Data Center since it became operative in late 1964. These are retrievable from the Specific Inquiry File. Specific inquiries contain carefully evaluated information from MDC's Document File and supplemental information from industry. They also contain information which reflects the experience of the machining data analyst handling the inquiry. As in the case of the Document File, the Specific Inquiry File is computerized and in particular is a valuable source of information for handling duplicate type requests for machining information. It is important to call attention to the fact that MDC has always retained the right to reuse information contained in its Specific Inquiry File, but in order to protect the proprietary rights of inquirers, individuals or companies are not identified without permission.

At the present time, MDC has no *numerical* machinability data bank. However, a number of data formats and computer programs have been developed to accommodate a bank of machinability data. One such format is shown in Fig. 1. This format provides a basis for output of final technical evaluated data. Other programs which were developed in order to assist output requirements of MDC include a program for cost and production, calculation of metal removal rates, and computation of residual stress. Of the above programs, the one for cost and production is currently being sold as a package to individual companies. A sample printout shown in Fig. 2 enables one to make a quick analysis of his machining situation and to select machining conditions for achieving highest production and least cost.

In the paragraph above, it was stated that the Machinability Data Center has no numerical machinability data bank. This statement was made with some reservations in light of the fact that MDC has been producing and publishing the well-known *Machining Data Handbook,* now in its second edition. This handbook truly represents a type of numerical machinability data bank, and the fact that to date it has not been computerized and implemented as a national machinability data bank in a time-sharing service is a matter of choice based primarily upon economic considerations. In addition to the *Machining Data Handbook,* MDC has collected a large quantity of what A. W. J. Chisholm[2] has characterized as 'absolute test data'. The above-mentioned data are derived from tool life testing, and to date two publications have been made available from MDC:

(1) Machining Data for Numerical Control, AFMDC 66-1, December 1966.

choice of parameters than provided in the *Machining Data Handbook.* (See section on Negative or Less than Optimum Data.)

In summary, the experience gained in the operation of the Machinability Data Center since 1964 provides a good basis for evaluation of national and international needs for machinability data banks. In particular, a year's operation in a 'cost recovery' climate has had an important effect upon data bank design considerations.

CONCEPTS FOR A NATIONAL NUMERICAL MACHINABILITY DATA BANK

The concepts which are currently being pursued at MDC in relation to a national numerical machinability data bank are based upon what might be considered as practical and economical modifications of the *Machining Data Handbook.* These modifications are now under study, and therefore the ideas which are presented and proposed below are subject to change. Hopefully, the ideas expressed in this paper will bring forth helpful information required for the design of an operating system, one which will also meet the required cost recovery goals. At this time, it is important to note that a number of national machinability data bank designs have been developed previously by MDC, but all previous design concepts are considered economically unsound in the present day 'cost recovery environment.' This does not imply that these previous designs might not be totally adequate to provide a new order of magnitude of machining technology in industry, both technically and economically, in a different economic climate. At present, all sorts of social, psychological, and other factors mitigate against salability of data to industry from highly sophisticated national data banks. Furthermore, the U.S. Government has chosen to evaluate performance and judge need for funding of data centres to a large extent on output sales. This choice is made largely because of the inability of anyone to assess total value of national information services in actual dollars.

Figure 4 shows a typical page from the *Machining Data Handbook* which can be and has been computerized for study purposes. In reviewing this format, it would appear reasonable to consider making modifications and additions as suggested below in order to keep pace with industry's requirements and to broaden the scope of MDC's services. Underlying all future design considerations are the limits to which any national system is effective in intervening or relating to individual plants. Under

```
 *    MACHINING        *    MATERIAL          *   HEAT TREAT   *  MATERIAL *   A F M D C   DATA   INDEX   *
 *                     *                      *                *           *                             *
 *    OPERATION        *   GROUP  DESCRIPTION *   CONDITION    *  HARDNESS *   SOURCE   * STATUS * CLASS  *

      TURN, SINGLE PNT     301  INCO718          SOLUTIONED       ROCK C29                               INQUIRY

      TURN, SINGLE PNT     301  INCO718          SOLUTIONED       ROCK C29    950002 660019      EB       AFMDC


 **************        TOOL      * BACK * SIDE * END *SIDE* ECEA * SCEA *  NOSE   *     CHIP   BREAKER     *
 *TOOL GEOMETRY *      STYLE     * RAKE * RAKE * RELF*RELF*      *      *  RADIUS  * TYPE * WIDTH * DEPTH  *
 **************(SEE CHART)*       * DEG  * DEG  * DEG *DEG * DEG  * DEG  *  INCHES *      * INCH  * INCH   *

                      11 161       00     0 05   05   05    15     15     0.032    *****   ******  ******  ******  AFMDC


 **************        TOOL  *            CUTTING FLUID                          CUTTING FLUID
 *TOOL MATL-FLUID*          *    TRADE  NAME                                     CONCENTRATION
 **************        MATERIAL*                      DESCRIPTION

                      C2 K68   NOT  REPORTED      WATER SOLUBLE OIL -LIGHT DUTY              1 TO 020   AFMDC


 **************        CUT   * FEED   * DEPTH * TOOL * WEAR * SURF * UNIT *
 *NUMERICAL DATA*      SPEED *        *  CUT  * LIFE * LAND * FIN  *  H P *
 **************        FT/MIN* IN/REV * INCH  * MIN  * INCH * RMS  * HP/CUIN*

                      125.     0.009    0.060   10.    0.015  *****  2.000  *******  ******  AFMDC
                      110.     0.009    0.060   15.    0.015  *****  2.000  *******  ******  AFMDC
                       98.     0.009    0.060   30.    0.015  *****  2.000  *******  ******  AFMDC
                       90.     0.009    0.060   45.    0.015  *****  2.000  *******  ******  AFMDC
```

Figure 1. Output of final technical evaluated data.

COST AND PRODUCTION RATE FOR TURNING

BRAZED CARBIDE TOOLS

DATA* SET* NO *	WORK MATERIAL	*HARD* *NESS* *	TOOL* MATL* *	*CUT* *SPD* *F/M*	FEED IN/REV	*TOOL* *LIFE* MIN*	*FEED* *COST* * $	*RAPD* *TRAV* * $	*LOAD* *UNLD* * $	SET* UP * $	*TOOL* CHNG* * $	*TOOL* DEPR* * $	*TOOL* SHPN* * $	RE * BRAZ* * $	TIP COST* * $	*GRIND* *WHEEL* * $
1	AISI 4340	300	C-7	470	0.0100	15	0.55	0.04	0.34	0.15	0.18	0.05	0.55	0.06	0.01	0.01
2	AISI 4340	300	C-7	400	0.0100	30	0.65	0.04	0.34	0.15	0.10	0.03	0.32	0.03	0.00	0.01
3	AISI 4340	300	C-7	360	0.0100	45	0.72	0.04	0.34	0.15	0.08	0.02	0.24	0.02	0.00	0.00
4	AISI 4340	300	C-7	325	0.0100	60	0.80	0.04	0.34	0.15	0.06	0.02	0.20	0.02	0.00	0.00

COST AND PRODUCTION RATE FOR TURNING

THROWAWAY CARBIDE TOOLS

DATA* SET* NO *	WORK MATERIAL	*HARD* *NESS* *	TOOL* MATL* *	*CUT* *SPD* *F/M*	FEED IN/REV	*TOOL* *LIFE* MIN*	*FEED* *COST* * $	*RAPD* *TRAV* * $	*LOAD* *UNLD* * $	SET* UP * $	*INDX* *INST* * $	*HLDR* *DEPR* * $	*INSERT* COST * $
1	AISI 4340	300	C-7	470	0.0100	15	0.55	0.04	0.34	0.15	0.01	0.00	0.04
2	AISI 4340	300	C-7	400	0.0100	30	0.65	0.04	0.34	0.15	0.00	0.00	0.02
3	AISI 4340	300	C-7	360	0.0100	45	0.72	0.04	0.34	0.15	0.00	0.00	0.01
4	AISI 4340	300	C-7	325	0.0100	60	0.80	0.04	0.34	0.15	0.00	0.00	0.01

COST AND PRODUCTION RATE FOR TURNING

SOLID HIGH SPEED TOOLS

DATA* SET* NO *	WORK MATERIAL	*HARD* *NESS* *	TOOL* MATL* *	*CUT* *SPD* *F/M*	FEED IN/REV	*TOOL* *LIFE* MIN*	*FEED* *COST* * $	*RAPD* *TRAV* * $	*LOAD* *UNLD* * $	SET* UP * $	*TOOL* CHNG* * $	*TOOL* DEPR* * $	*TOOL* SHPN* * $	*GRIND* *WHEEL* * $
5	AISI 4340	300	T-1	77	0.0100	15	3.39	0.04	0.34	0.15	1.13	0.08	2.26	0.03
6	AISI 4340	300	T-1	63	0.0100	30	4.14	0.04	0.34	0.15	0.69	0.04	1.38	0.01
7	AISI 4340	300	T-1	54	0.0100	45	4.83	0.04	0.34	0.15	0.53	0.03	1.07	0.01
8	AISI 4340	300	T-1	45	0.0100	60	5.80	0.04	0.34	0.15	0.48	0.03	0.96	0.01

Figure 2. Cost and production printout.

MATERIAL	CONDITION & MICROSTRUCTURE	BHN	TOOL MATERIAL: TRADE NAME	TOOL MATERIAL: INDUSTRY GRADE	TOOL GEOMETRY: BR°	SR°	SCEA°	ECEA°	RELIEF°	NOSE RADIUS in.	CUTTING FLUID Code Appendix Page A-2	DEPTH OF CUT in.	FEED ipr	TOOL LIFE END POINT in.	TOOL LIFE - minutes vs SPEED-feet/minute R=Recommended Speed Appendix Page A-2				
HIGH TEMPERATURE ALLOYS - NICKEL BASE WROUGHT - *(cont.)* WASPALOY	SOLUTION TREATED & AGED	388	883	C-2	5	0	30	30	7	.032	11 1:20	.060	.009	.015	16				
	AUSTENITIC														120				
"	"	"	"	"	"	"	60	60	"	"	"	"	"	"	30				
	"														120				
"	"	"	"	"	"	"	75	75	"	"	"	"	"	"	60				
	"														120				
WASPALOY	SOLUTION TREATED & AGED	388	883	C-2	5	0	75	75	7	.032	11 1:20	.060	.009	.015	5	15	30	60	
	AUSTENITIC														190	147	130	120	
RENE' 41	SOLUTION TREATED	321	-	T15 HSS	0	15	0	5	5	.032	11 1:20	.060	.009	.060	15	30	45	75	
	AUSTENITIC														19	17	15	12 R	
RENE' 41	SOLUTION TREATED	321	K6	C-2	-5	-5	15	15	5	.032	00	.060	.009	.016	5	15	24		
	AUSTENITIC														67	55	50		
RENE' 41	SOLUTION TREATED	321	K6	C-2	0	5	15	15	5	.032	00	.060	.009	.016	5	15	20	25	
	AUSTENITIC														90	74	67	60	
"	"	"	"	"	"	"	"	"	"	"	11 1:20	"	"	"	16	20	30	38	
	"														85	80	73	70	
RENE' 41	SOLUTION TREATED & AGED	365	-	T15 HSS	0	15	0	5	5	.032	52	.060	.009	.060	15	30	45	81	
	AUSTENITIC														14	13	12	12 R	
RENE' 41	SOLUTION TREATED & AGED	365	K6	C-2	-5	-5	15	15	5	.032	00	.060	.009	.016	5	15	25		
	AUSTENITIC														80	58	40		

Figure 3. Turning data for numerical control.

MATERIAL	HARD-NESS BHN	CONDITION	DEPTH OF CUT inches	SPEED fpm	FEED ipr	TOOL MATERIAL	SPEED fpm	FEED ipr	SPEED - fpm BRAZED	SPEED - fpm THROW-AWAY	FEED ipr	TOOL MATERIAL
4. ALLOY STEELS, WROUGHT (cont.)	175 to 225	Hot Rolled, Annealed or Cold Drawn	.150	90	.015	M2, M3	110	.015	300	400*	.020	C-6
			.025	120	.007	M2, M3	140	.007	400	500	.007	C-7
	225 to 275	Annealed, Normalized, Cold Drawn or Quenched and Tempered	.150	75	.015	T15, M33, M41 Thru M47	95	.015	280	350*	.020	C-6
			.025	100	.007	T15, M33, M41 Thru M47	125	.007	360	460	.007	C-7
	275 to 325	Normalized or Quenched and Tempered	.150	60	.015	T15, M33, M41 Thru M47	75	.015	260	330	.015	C-6
			.025	80	.007	T15, M33, M41 Thru M47	95	.007	325	400	.007	C-7
	325 to 375	Normalized or Quenched and Tempered	.150	50	.010	T15, M33, M41 Thru M47	65	.015	215	280	.015	C-7
			.025	65	.005	T15, M33, M41 Thru M47	80	.007	280	350	.007	C-7
Medium Carbon 1330 50B40 6150 1335 50B44 81B45 1340 5046 8630 1345 50B46 8637 4032 50B50 8640 4037 5060 8642 4042 50B60 8645 4047 5130 86B45 4130 5132 8650 4135 5135 8655 4137 5140 8660 4140 5145 8740 4142 5147 8742 4145 5150 9254 4147 5155 9255 4150 5160 9260 4161 51B60 94B30 4340	375 to 425	Quenched and Tempered	.150	35	.010	T15, M33, M41 Thru M47	55	.010	175	225	.015	C-7
			.025	45	.005	T15, M33, M41 Thru M47	60	.005	225	275	.007	C-7
	45R$_c$ to 48R$_c$	Quenched and Tempered	.150	30	.010	T15, M33, M41 Thru M47	40	.010	160	210	.010	C-7
			.025	40	.005	T15, M33, M41 Thru M47	45	.005	190	240	.005	C-8
	48R$_c$ to 50R$_c$	Quenched and Tempered	.150	25	.010	T15, M33, M41 Thru M47	35	.010	130	160	.010	C-8
			.025	35	.005	T15, M33, M41 Thru M47	40	.005	160	180	.005	C-8
	50R$_c$ to 52R$_c$	Quenched and Tempered	.150	20	.007	T15, M33, M41 Thru M47	30	.007	105	140	.010	C-8
			.025	30	.003	T15, M33, M41 Thru M47	35	.003	125	160	.005	C-8
	52R$_c$ to 54R$_c$	Quenched and Tempered	.150	-	-	-	-	-	80	90	.010	C-8
			.025	-	-	-	-	-	90	100	.005	C-8
	54R$_c$ to 56R$_c$	Quenched and Tempered	.150	-	-	-	-	-	70	80	.005	C-8
			.025	-	-	-	-	-	80	85	.005	C-8
High Carbon 50100 51100 52100 M-50	175 to 225	Hot Rolled, Annealed or Cold Drawn	.150	95	.015	M2, M3	110	.015	340	425*	.020	C-6
			.025	125	.007	M2, M3	140	.007	440	525	.007	C-7

See Sections 2.1 and 2.2 for Tool Geometry.
See Section 3 for Cutting Fluid.
* Reduce Cutting Speed 25% for .300 inch Depth of Cut.

Figure 4. Typical page from *Machining Data Handbook*.

most circumstances, even the finest complement of machining information which can be furnished may only serve as 'starting data'.

Materials

An examination of Fig. 4 reveals the need for an obvious modification in the machining data for materials. It is evident that the 52 steels contained in a specific grouping such as 'Alloy Steels Wrought–Medium Carbon' do not have equivalent machining characteristics. Lack of absolute data, the fact that the *Machining Data Handbook* contains starting recommendations only, and economic reasons largely account for having to resort to excessively large groupings of materials throughout the entire book. In the future, industry will have increased need for smaller groupings and perhaps need for absolute data for individual alloys. This will be especially true when greater quality control is achieved from one alloy heat to another. MDC's experience in generation of absolute data and collection of data has shown that accomplishing the objective of regrouping of materials presents a difficult problem. In all likelihood, mathematical correlation and extrapolation procedures would be needed to effect this data improvement.

Material designations

Various sectors of industry in the United States have their own preferences for designating materials. In handling steels alone, extensive use is made of AISI, SAE, AMS and ASTM specifications. It is important to note that an engineer using AMS numbers may be totally unfamiliar with an AISI equivalent. Currently, the Unified Numbering System (UNS) is under consideration by ASTM and SAE. Until standards become available, a machinability data bank must provide for identification of equivalent materials. In addition to the standards referred to above, many companies have their own internal standards which are used corporate wide. For example, one U.S. firm finds it advantageous to use numbers such as 10325PA-2, 13219BP, and 11720AA to represent AISI 4140, X45, and AISI 309, respectively. In this firm, the more commonly used AISI designations, etc., are not used and are even unfamiliar in conversations concerning material problems among shop labour and supervision. With regard to such unique in-house methods of designating materials, responsibility for making conversions rests with the plant itself. In some instances, however, MDC could conceivably be supportive.

Machining parameters

Significant and extensive modifications of the data contained in the *Machining Data Handbook* are required for a national numerical machinability data bank in relation to feeds, depths and widths of cut, hole depths in operations such as drilling, boring, etc. In examining Fig. 4, only two depths of cut are provided for an operation such as turning. An additional note at the bottom of some of the appropriate turning pages states 'Reduce Cutting Speed 25% for ·300 inch Depth of Cut.'

It appears unlikely that absolute data can ever be generated or collected economically for all of the useful variations in machining requirements. It appears necessary, therefore, to select existing and/or develop new mathematical procedures for calculation of machining conditions. Calculation procedures have been used in the past with varying degrees of success and, to the best of our knowledge, not in application to a national data bank. In order to qualify for the national numerical machinability data bank being contemplated, the validity of calculation procedures must be tested. The Machinability Data Center expects to devote its major effort in improving its present 'bank'—namely, the *Machining Data Handbook*—to selecting, testing and using interpolation and extrapolation procedures for expanding data bank capabilities.

Modification of data by industry grouping

The *Machining Data Handbook* is a collection of data of varying quality in that it consists of data derived from absolute tests, from industry surveys including time standards data, from the periodical literature and trade journals, from engineering evaluations, etc. Experience has shown that data derived from industry vary considerably depending upon the type of industry, production quantities, and many other factors. It has become apparent that the machining data required for various industries should be tailored. It is common to find wide variations in machining parameters used for turning in industries such as automotive, machine tool, screw machine, etc. Accordingly, it appears important to study the possible improvement of MDC's data by making a detailed study of machining data characteristics through industry grouping of data requirements. Below are listed a few typical industries with differing needs for machining data:

Industry
Automotive and Farm Equipment
Screw Machine Products
Aerospace
Foundry
Instrument and Small Precision Machine Shop
Machine Tool
Shipbuilding
Steel Mill—Billet and Roll Turning Shops

Hopefully, the specific needs of industries such as noted above can be characterized and identified through modifications and addition of commonly used machining parameters. For example, recommendations might be tailored as indicated in the typical examples below:

(1) Data for large production lots where tool changes can be made quickly can be differentiated from tool room or small lot production by selecting machining parameters which provide for variations in economical tool life.
(2) Data requirements for roll turning in steel mills and other heavy equipment shops can be accommodated by adding recommended parameters for very heavy chip loads.

The above examples are just a few of the approaches which may be found applicable. How far these approaches can be pursued will be a function of their cost.

Negative or less than optimum data

The *Machining Data Handbook* makes single recommendations for specific materials. For example, the handbook recommends using premium type high speed steels and heavy duty oils for drilling of high strength wrought steels in the hardness range of $48R_C$ to $50R_C$. For many machining situations, the above data is less than adequte in that the data provides for a specific type of HSS tool material, namely, T15, M33 or one of the super-hard high speed steels, and a heavy duty oil. Many companies are not concerned with optimization of individual machine tools but rather with the optimization of costs and production for an entire job or production area. Therefore, a manufacturing engineer needs machining data for a steel such as mentioned above using other tool materials and other fluids. In relationship to the example noted above, he may want to know what the effect will be of using an M1, M7, or M10 drill and possibly a soluble oil. Such information will enable him to determine whether he will find it essential to inventory the premium type tool materials or cutters and also whether he must necessarily take time and spend money to change the cutting fluid in a large system supplying a given machine tool.

It is quite possible that extrapolation procedures may be helpful in supplying a data bank with less optimum data. In any event, it would appear logical to expand data bank capability by preserving all negative data which ordinarily accumulates during absolute testing programs.

Tool geometry and cutting fluid

In the present handbook, tool geometry and cutting fluid recommendations are contained in separate sections. Computer printouts from a machinability data bank should contain those recommendations coupled closely to feed, speed, depth of cut, and other parameters. An examination of the tool geometry section will show that material groupings are also too broad and the validity of these broad groupings needs further questioning. Cutting fluids pose a very complex data problem, especially for machining situations related to difficult-to-machine materials, stress corrosion cracking, surface integrity, and anti-pollution requirements. At the present time, there is no product standardization among cutting fluid manufacturers. Pricing of raw materials and new developments result in cutting fluid formula modifications,

3 Oils—heavy duty
4 Emulsifiable oils—light duty (general purpose)
5 Emulsifiable oils—heavy duty
6 Chemicals and synthetics—light duty (general purpose)
7 Chemicals and synthetics—heavy duty
8 Specials—light duty
9 Specials—heavy duty

A typical major group, Oils—medium duty, is divided into the following subgroups:

Sulfurized mineral oil
Compounded sulfurized mineral oil
Chlorinated mineral oil
Sulfo-chlorinated mineral oil
Sulfurized mineral oil + fatty oil
Chlorinated mineral oil + fatty oil
Mineral oil with fatty compounding + sulfur and chlorine additives
Sulfo-chlorinated mineral oil + fatty oil
Compounded sulfo-chlorinated mineral oil + fatty oil

At present, the *Machining Data Handbook* supplies recommendations using only major cutting fluid groups. No preferences are indicated among subgroups such as noted above, and yet such distinctions should be recognized, at least for certain machining situations, in a viable machining data bank.

Power and forces

The second edition of the *Machining Data Handbook* provides considerable data on forces and horsepower for machining of most of the important classes of alloys and for some of the widely used machining operations. A national machinability data bank should provide for computer output of both forces and horsepower for specific machining situations in a format which is far more convenient than presently displayed in the *Machining Data Handbook*.

Surface finish and accuracy

At the present time, supplementing a machinability data file with surface finish and accuracy data, along with the other data amplifications mentioned previously, provides a considerable challenge in the economical operation of a machinability data bank. Surface finish data could be stored easily in a data bank; however, it must be recognized that for a specific operation and material a very wide range of surface finishes can be achieved. It is doubtful at this time whether industry will look to a data bank for finish data except in very general terms.

Deflection and cutter breakage data

Programming of adaptive control units requires setting of deflection limits for cutters as well as parameter limits which when exceeded would endanger breaking or seriously chipping the cutting edge of a tool. Availability of this latter data is limited. At present, it appears that while data banks are essential for adaptive control, further study is needed to determine if and how they should relate to a national data system.

Types and sources of data

As noted previously, the *Machining Data Handbook* consists of data of varying quality from many different sources and is sometimes downgraded so as to provide starting recommendations for users of varying needs and capabilities. Serious consideration must be given to the question of imposing such limitations in the design of a national machinability data bank. Perhaps some help in this connection will be obtained if data is modified as a result of different industry groupings. (See section on Modification of Data by Industry Grouping.) Common types of data available for collection purposes are well known, and most types have been used in the production of the *Machining Data Handbook*. They include:

(1) Data from testing programs previously referred to as data from absolute tests.
(2) Data collected from industrial plants engaged in machining and grinding operations. This data is of varying quality depending upon management's cognizance of the importance of machining technology, size of plant, type of product, extent of competition, purchasing policies, and labour relations. In any case, experience has shown that no valid assumptions can be made regarding quality of data as a function of the overall reputation of companies.
(3) Calculated data. Data derived from mathematic models has great potential for development of a national machinability data bank. Extrapolation procedures have been used in the past, but they have often been misinterpreted and overextended. Simple proportional calculations, for example, have been used to determine the machinability index of a material and to select machining parameters, particularly cutting speed. The use of machinability indices is not recommended for improvement or expansion of a machinability data base.
(4) Machinability correlations—Hardness, Chemical composition, Microstructure and Short-time tests. It is anticipated that various types of correlation procedure will find usefulness in evaluating input for a machinability data bank. Such procedures have been used previously, but it is anticipated that additional care needs to be exercised in order to upgrade the quality of data presently contained in the *Machining Data Handbook*. Microstructure, for example, has been extremely useful in categorizing machinability data for grey, ductile, and malleable irons for wide variety of machining operations. Hardness has been extremely useful for the quenched and tempered martensitic steels and some short-time tests have been found correlative for free machining steels.

(5) Standardized machining tests. Various organizations, including the International Standards Organization (ISO), have investigated the feasibility of using standardized tests in order to create a valid body of machinability data. This question has been addressed by Chisholm, Mills and Redford in their paper on 'The Assessment of Machinability.'[2] It is apparent that while many people refer loosely to the machinability of a particular material, no agreement has been reached concerning such a material property. Based on efforts expended to date in defining machinability, it is very probable that no useful definitive group of basic material properties will be found to define 'machinability', except very generally. Fortunately, it does not appear practicable or in fact even necessary to define or assess 'the machinability of a material'. Rather, industry is primarily interested in obtaining those machining parameters which will enable one to assess the 'machinability characteristics of a system' as defined by productivity, cost, and in some instances the integrity of a surface of a machined part through an evaluation of mechanical properties and service performance (surface integrity). Presently, it is our opinion that standardized machining tests are limited in their usefulness unless new and improved mathematical models can be developed which will enable one to extrapolate with confidence to machining operations other than the one used in the standardized test—most frequently turning.

From the comments above, it is quite apparent that many important improvements are possible and required for converting the *Machining Data Handbook* to a numerical machinability data bank. It will be necessary to make changes selectively and with utmost care in planning so as to reflect the effect of all changes on product costs for customers. In addition to cost, there are other circumstances affecting design. One of these concerns the inability of any national system to be sufficiently relevant to individual industry needs. Specifically, individual plants already maintain machinability data banks of various types. Most of these are tailored to meet particular needs in a given plant. Every plant, for example, has some type of tooling data bank and a machine tool data bank to support its manufacturing operations. At present, most of these are not computerized. In the future, individual systems in operation at companies will become more sophisticated. However, it can be assumed with confidence that companies will always rely in part on outside sources for their data requirements.

In addition to relevance factors, many other conditions mitigate against extensive intervention of national systems. These include many human relations problems such as the NIH factor (not invented here), competitive and proprietary considerations, and management and purchasing policies which allow for in-house expenditures far in excess of the cost of system. It has been said that when a publication in the United States reaches a cost of $25, then 75 per cent of sales potential to individuals disappears. This trend has definitely been noted in comparing sales of the 1st and 2nd editions of the *Machining Data Handbook*. Presently, the *Machining Data Handbook* in single copy quantities is selling for $35 and as of July 1973, 7731 copies of the 2nd edition have been sold. This response has been gratifying; however, considering the potential market in the United States alone and considering the value of the information contained in this handbook, one might have anticipated many more sales. It has been concluded that pricing has had a strong influence, especially because considerable exposure has been given to this book through selective direct mail advertising, with mailings totalling over 700 000. From this experience, MDC further concludes that marketing of data products would become very difficult if pricing fell in the range of $50 to $500 per data product.

Up to this point, no specific reference has been made as to the form of output from a national machinability data bank. MDC at present favours some type of publication or publications similar to the *Machining Data Handbook*. The present costs of direct computer intervention in answering specific questions are high, and at this point in time it is very doubtful whether a system can be operated economically in this fashion. Dumping of large portions of files for distribution to various industries appears reasonable, and of course emergency conditions will always provide situations where some users will resort to individual file intervention.

A national machinability data system under present conditions will not be totally supported by the U.S. Government, and therefore the question of salability of products looms as a key consideration in system design.

A LIMITED NATIONAL NUMERICAL MACHINABILITY DATA BANK

In view of the plans outlined above for an improved national numerical machinability data bank, it is important to take note of an existing capability in the USA. Currently, the Carboloy® Systems Department of General Electric operates a computerized machinability program as a part of the General Electric time-sharing service network. Essentially, the program used for input and output is based upon a mathematical model, including all significant machining parameters. This system is described in detail in *N/C Machinability Data Systems*[3] and also in Carboloy® Systems Computerized Machinability Program

tion furnished costs a subscriber about $1.00 above the prorated minimum costs. When a potential customer chooses to interrogate a bank such as the Carboloy® CM System, he is facing annual minimum costs of the order of $1200 to $2400 per year. Realistically, these costs are being compared in the marketplace with handbook information such as available in the *Machining Data Handbook* at $35 per copy. Cost confrontations of this type must be resolved before a sophisticated machinability data bank can become a reality at MDC which must meet contractual cost recovery goals.

IN-HOUSE COMPUTERIZED MACHINABILITY DATA SYSTEMS

MDC has the task of establishing the characteristics of a national machinability data bank in contrast with existing and developing 'in-house' data systems for handling machining information. Practically all companies engaged in machining technology have some type of machinability data system. For lack of a better description, these have been classified as 'in-house' systems in contrast with the national system such as discussed herein. In-house systems have inherent advantages in that they serve specific and limited needs, are more accessible, and furnish far more detailed requirements than could be supplied from a national system. In contrast, a national system has the capability of supplying reliable and timely machining data, especially for materials and machining operations which are new or with which a specific plant has had limited or no experience. For the Government, national system output has the great advantage of eliminating duplication of effort and upgrading national competence in machining technology on a broad scale. All indications to date are that intervention by means of a national data bank in most plants would be limited at least initially to the type of information currently carried in the *Machining Data Handbook* but improved as suggested in this paper. Of course, other special services would be available, such as special computer programs, cost analyses, etc.

It is appropriate in this presentation to recognize some of the types of improved systems, mainly in-house, which are in development in various companies in the U.S. and abroad. For example, in addition to the GE system, there are computerized systems such as FAST, EXAPT, and ABEX. These are described in detail in ref. 3.

Briefly, FAST, a proprietary system of International Business Machines (IBM), is written in FORTRAN and covers those machining operations being performed by N/C machining centres. FAST is designed to select proper cutting tools and appropriate feeds and speeds, but variations of FAST can also provide improved productivity by variations of tool geometry, and calculate times for N/C machining operations.

EXAPT, a proprietary system of Exapt–Verein, Aachen, Germany, interfaces with APT and has been designed to select appropriate cutting tools and cutting conditions for given materials, calculate tool paths, and determine an appropriate machining operation sequence.

ABEX, a proprietary system of Abex Corporation, Mahwah, New Jersey, USA, is a computerized approach for making machining analyses of cost and production rates. The system is designed to use plant machine tool data, plant cost data, and machinability data from laboratory tests as well as from the shop. Output formats provide for a quick assessment of tooling and cutting parameters for obtaining maximum production and minimum cost. The ABEX system also provides for using other strategies, including parts per tool, surface finish, accuracy, etc. Also, data analyses can be made to assess the relationship of costs of time in cut compared with setup, tool tip, load and unload, and rapid traverse costs. This information enables planners to give primary attention to those operations which contribute the highest costs.

With regard to machining data at any level of sophistication, it is clear that national and in-house data banks have specific functions which complement one another very well. It is possible that certain large companies will eventually be able to totally justify on-line computer services.

AN INTERNATIONAL MACHINABILITY DATA BANK

Many of the design criteria for a national machinability data bank, as outlined in this paper, are also applicable for the development of an international counterpart. Based upon the Machinability Data Center's experience with the *Machining Data Handbook*, there is no reason for one to believe that a similar data publication might not be salable in all countries of the world which are actively involved in machining technology. Acquisition of data could be made in a fashion similar to that used for development of the *Machining Data Handbook*, thereby reflecting any differences in user needs for particular countries. Assuming that various countries of the world develop their own machining data books, the first international data bank would simply be a collection of the various books of each of the countries, with additional capability for translation of language, units, and materials.

Based upon MDC's experience, it is recommended that design of an international numerical machinability data bank be postponed in favour of developing and publishing handbooks financed by government and industry in individual countries. Common Market participating countries may find added incentives for a cooperative approach. We believe that handbook type machining data projects will naturally lead into more sophisticated national machining data systems. As in the case of a national numerical machining data bank for the U.S., our recommendations relative to an international system are premised upon an assumption of need to recover the cost of output products in order to demonstrate usefulness. This assumption has many ramifications, not all of which are subscribed to by those of us involved in the operation of the Machinability Data Center. Nonetheless, 'cost recovery' is a way of life in the United States among information analysis centres—the longevity of which can presently be prolonged best through sales of 'best seller' type machining data publications. Perhaps other countries with different approaches to the justification of funding will provide us all with better insights into the science and technology of machining information.

REFERENCES

1. MACHINABILITY DATA CENTER (1972). *Machining Data Handbook*, Metcut Research Associates Inc., Cincinnati, Ohio.
2. A. W. J. CHISHOLM, B. MILLS and A. H. REDFORD (1972). 'The assessment of machinability', *SME* Paper No. MR72-163, Society of Manufacturing Engineers, Dearborn, Michigan.
3. *N/C Machinability Data Systems*, 1971. Society of Manufacturing Engineers, Dearborn, Michigan.

DISCUSSION

Q. J. Fisher, Wickman Wimet Ltd. Despite savings accruing from the use of improved machining data, have any of the speakers experienced any reluctance to pay the economic cost of producing such data, remembering that the cost per enquiry for computerized data could be as much as £20?

A. The Machinability Data Center of the USA has definitely experienced a reluctance on the part of users to pay for technical inquiries. The data presented in the table below substantiates this fact.

Machinability Data Center (MDC)
specific inquiry services

1965–1972–before cost recovery (no charge for enquiry services)

Total inquiries	6842
Inquiries per year (average)	977

1 January 1972–31 July 1973–after cost recovery

Total inquiries	192
Inquiries per year	121
No. of paid inquiries	94
Paid inquiries per year	64
No. of free telephone inquiries	98
Total income from 94 paid inquiries	$7524
Average charge per inquiry	$80

SYSTEMATIC DETERMINATION OF CUTTING DATA SUPPORTED BY AN INFORMATION CENTRE

by

W. KÖNIG*, W. EVERSHEIM* and D. PFAU*

SUMMARY·

Starting from the difficulties in the determination of cutting data, the support of this planning task by an information centre for cutting data is presented. The way of acquisition of cutting data is shown as well as the information service given by the centre to the industrial users. After presenting the stage of development carried out by the WZL, an example proves the successful use of the information centre.

INTRODUCTION

In the past, organizational measures and advances in automation were mainly responsible for the increase in the productivity of chip removal machining. With the increasing demand for industrial goods, coupled with only a small increase in the number employed, shorter working hours and sharply rising costs, the price level and thus the competitiveness and growth of industrial production can be maintained only through increased productivity.

In West Germany there are 1·1 million machine tools employed in the production process. Of these 430 000 are used either for turning, drilling or milling. Several studies on the utilization of machine tools have determined that only 25 per cent of the total available time is actually used directly for machining work. 75 per cent of the time is lost through lack of utilization intensity, numerous non-productive jobs, and inappropriate application.

Ways and means of improving the amount of the usable time of the production facilities relative to machine tools, and for shortening the turn-round time with respect to the workpiece are shown in Fig. 1. The following methods can be used to achieve improvements in production:

(1) Improvement of machining technology and its application.
(2) Automation of the machining flow.
(3) Automation and integration of machining, transportation and storage.
(4) Generally a far reaching and precise planning of production.

Designers and management are responsible for the latter improvements. Efforts in this direction show up—as indicated on the right side of Fig. 1—with a computer controlled adaptive manufacturing system, which integrates the machining process and the flow of information and material. Technologists are responsible for the first measure—the improvement of technology.

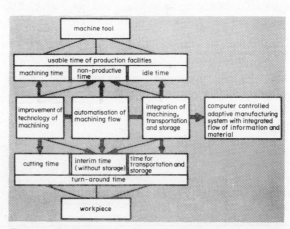

Figure 1. Possibilities for the improvement of the utilization of production facilities and shortening of the turn-round time in manufacturing.

The significance of the improvement of the machining technology is shown by the following consideration. A reduction of just 1 per cent of the utilization losses of the machine tools used in West Germany would mean an annual saving of 162 million DM in 1972 and 297 million DM in 1975. If this study is limited to lathes, drilling- and milling-machines, it would still mean a saving of 53·8 million DM in 1972 and 98 million DM in 1975.

* Laboratorium für Werkzeugmaschinen und Betriebslehre, TH Aachen

compared to the optimum ones, whereby the economy and the productivity are influenced unfavourably.

On the other hand reliable and qualified information on the working process is not available in sufficient quantity. A general survey carried out among eighty-six firms from ten branches of industry for the setting up of a general materials-data bank showed that 83 per cent would like to obtain data on machinability, especially on ferrous metals.

Further sources of uncertainty are the variations in the machinability of the materials, variations in the wear resistance of the cutting tools and of the behaviour of the machine tool-tool-workpiece system. Today variations of up to 100 per cent are to be expected in machining of workpieces of the same type of steel with similar heat treatment but from different charges. Such uncertainties are the cause of the selection of lower than optimum cutting values. This naturally results in a poor utilization of machine tools.

The trend of production cost, especially the increase of machine hour cost rate, demands a reconsideration of the cutting values in use today. In connection with lowered tool cost this trend involves decreasing tool lives for optimum cost-machining, as is shown in Fig. 2 for an example of machining a cylindrical shaft. These values fast approach those valid for the minimum machining time. Corresponding to the reduction of tool life the cutting speed for optimum cost must be considerably forced up.

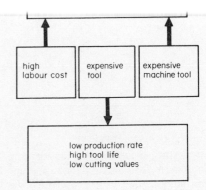

Figure 3. Influence of cost factors on cutting data recommendation.

The influence of the cost factors on the selection of cutting data, e.g. for working on a cost optimum basis, is shown in Fig. 3. A rise in the cost of labour and machine tools will require a corresponding increase in production rates, thus leading to a reduction in tool life and a simultaneous increase in the cutting values until such time as the reduction of machine tool and labour cost per workpiece are compensated by the increase in tool costs. Rising costs of tools have the opposite effect on the cutting values.

CUTTING VALUE DETERMINATION SUPPORTED BY THE INFORMATION CENTRE

The determination of cutting values, relative to the technological and economic factors is an important task of the work planning department, which is thus responsible for the production resources available. The work plans, in conventional single piece and small batch production, contain more or less detailed information regarding the cutting values to be used. The machine tool operator thus has the opportunity to alter the recommended cutting values to his advantage.

The employment of highly automated production equipment, such as NC-machine tools and conventional automatics presumes a detailed planning of each step of work. The checking and alteration of information carriers, such as punched-tapes and cams, should be avoided, as far as possible, because of the time and expense involved. Thus, it is necessary to have reliable information on cutting values for the production of such data carriers.

Suitable information, regarding the optimum ranges of the machining parameters, can contribute towards the economic utilization of the machine tool even if the machining of the workpiece is carried out on a machine tool with adaptive control.

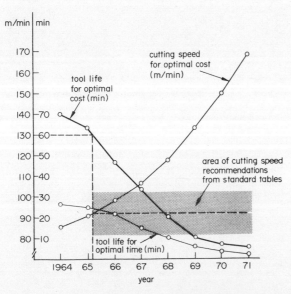

Figure 2. Adaptation of cutting data for turning to the development of cost factors in conventional manufacture.

The determination of recommended cutting values by the work planning department can be supported by an information centre for cutting data (INFOS). The information centre collects cutting values from industry and from research institutions. Data from these sources of information are checked, evaluated and stored in a concentrated form (Fig. 4). Through

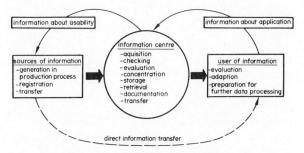

Figure 4. Information flow and distribution of tasks between the information centre and its 'information markets'.

electronic data processing, it is possible to have permanent access to large stocks of recorded data in the information centre. At the same time, it is possible to evaluate these data in accordance with criteria based on the latest developments in research and industry. This information will be made available to the user of the centre in a form suited to his needs.

It is anticipated that the user gives information about the application of the cutting data to the centre. This feedback allows to control the centre's effectiveness. On the other hand the centre gives information about the usability of the data coming from the sources of information.

The demand for information from the centre's customers will vary in the different stages of technological development. These stages can be characterized, as shown in Fig. 5, by research, development, pilot application, distribution, application, and substitution. The demand is largest when a new technology is distributed and generally applied in industry.

As the supply of information to a company is largest when the distribution and general application have come to an end, a temporary and quantitative information gap will be left. It is the task of the information centre to close this gap for industry. Hence, industry itself must carry out the most important task: it must cooperate with the information centre and supply it with cutting values.

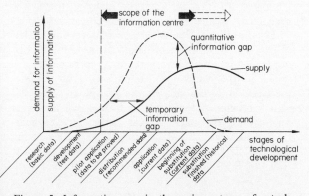

Figure 5. Information gap in the various stages of a technology and the scope of the information centre.

ACQUISITION OF CUTTING VALUES

For data acquisition it must be ascertained that the data characterizing the corresponding cutting process are described with sufficient accuracy. For example, in turning it is not sufficient to register the cutting depth, the feed and the cutting speed. Without the tool life and the criteria for tool wear, it is not possible to judge the utilization intensity of the production equipment. Thus an evaluation of these data is impossible, as well as its employment to similar cases.

In addition, the conditions under which the values are determined must also be registered. The designation of a material alone is not enough. A more precise description of the heat-treatment and the tensile strength etc. is also required.

The importance of this requirement was shown by a survey of cutting data in six companies for the turning of the material C35 with cutting material P20. The feeds and cutting speeds, recommended by the different companies, were recorded for a tool life of 60 min and a flank wear of 0·8 mm. In order to allow a comparison, reductions of the cutting values caused by the equipment used, have not been taken into consideration.

In spite of the fact that the same machining operation was being compared, differences in cutting speed of almost 100 per cent were noted with respect to the lower limit. The differences in the feed varied from 200 to 250 per cent. It was not possible to trace the reasons for these variations since no additional information on the material or further influencing factors was given. In addition, the registered data represented values which were influenced by the experience of the work planner. Such values, however, are not always objective.

The following requirement for the information centre can thus be derived: only those values should be registered which have actually been used, together

INFOS WZL,TH-Aachen D 5I Aachen Wüllnerstr. 5	data registration		reg.-no.	
			department	
	cutting operation:		date	
material	name(1)		DIN–no. (2)	
	manufacturer			
	heat treatment (3)			
	tensile strength	kp/mm²	hardness HRC(4)	
	remarks			
workpiece	name			
	drawing–no.		lot size	
	dimension		surface quality/tolerance	
	surface (5)		stability (6)	
machine tool	type		power	kW
	manufacturer			
	year		stability(6)	
tool	name			
	manufacturer		setting angle	degree
	dimension (8)		clearance angle (16)	degree
	cutting material (9)		rake angle	degree
	manufacturer		inclination angle	degree
	tip-type (10)		included angle (7)	degree
	tip-fixture (11)		back relief angle	degree
	cutting edge (12)		side relief angle	degree
	chip former (13)		helical angle (19)	degree
	no. of edges			
	corner (14)		stability (6)	
	kind of sharpening (15)		oil channel (20)	
cutting conditions	interrupt cut			
	coolant (21)			
	stability of clamping (6)		general stability (6)	
cutting values	spindle speed	min⁻¹	depth of cut (22)	mm
	speed	m/min	swath (23)	mm
	feed s	mm/rev	distance at entry (24)	mm
	feed u	mm/min	limitations (25)	
cutting results	tool life	min	total tool path	mm
	tool life criterion (26)		chip form (27)	
	remarks(28)			
sketch of the workpiece with dimension,clamping,tool, and cutting area				

Figure 6. INFOS data-sheet for acquisition of cutting values from production.

production. The number of influencing factors here were held to a minimum. This is necessary in order to limit the work in the workshop and to guarantee the flow of the large amounts of data necessary for the successful functioning of the information centre.

In order to use the collected data for other machining problems, even those parameters have to be described which cannot be quantified or whose influence cannot be quantified. In addition, standardized names, definitions, testing and measuring methods, dimensions, etc. must be used. For this purpose comments on data registration are given on the back of the acquisition forms.

SERVICES OF THE INFORMATION CENTRE FOR CUTTING DATA

The machining data, provided and registered by industry and research, must be transformed into a suitable data service by the information centre. The most important fields of the centre's activities are shown in Fig. 7.

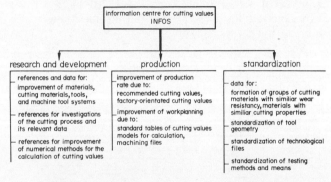

Figure 7. Prospective fields of application of the information centre for cutting data.

Research and development departments can recognize trends by using information on materials and their machinability. Furthermore the centre can give hints for detailed investigations of the cutting process and as yet unknown parameters, or on parameters whose influence is so far uncertain, and for the development of numerical methods for the determination of cutting values.

A data bank offers national and international institutions the facility of a faster, well founded producing of standards, e.g. in the grouping of materials and cutting materials and standardization of tool geometries, technological files and testing methods.

The main field of application of the information centre, however, is no doubt in the field of industrial

		depth of cut (mm)					
tensile strength___60–70							
hardness_____180–200HV		2	4	6	8	10	
surface_____roughed	0.25	260 115 5 65 130	260 223 10 65 260	260 335 15 65 390	260 450 19 65 520	260 560 24 65 645	v f n o r
carbide_____GC 125; STI 25 gold;							
STI 15 gold; TT 25 extra; TT 15 extra	0.315	240 135 6 75 150	240 265 11 75 300	240 395 16 75 445	240 525 21 75 600	240 655 26 75 745	v f n o r
ISO–quality_____							
particularities____coated	0.4	220 155 6 85 175	220 310 11 85 345	220 465 17 85 515	220 615 23 85 685	220 770 28 85 855	v f n o r
tool life exponent__–0,04							
tool geometry							
α _____5°	0.5	200 180 6 95 195	200 360 11 95 395	200 540 18 95 590	200 720 24 95 790	200 900 30 95 990	v f n o r
γ _____6°							
λ _____0°	0.63	185 215 7 110 225	185 425 13 110 455	185 635 20 110 680	185 845 27 110 910	185 1060 33 110 1035	v f n o r
ϰ _____70°							
speed			5 kW		15 kW	25 kW	
vmax____260							
vmin____180							

v_____	speed	(m/min)
f_____	main cutting force	(kp)
n_____	power required	(kW)
o_____	surface rate	(mm²/min)
r_____	volume rate	(mm³/min)

Figure 8. Table of standard values for continuous longitudinal turning.

Figure 8 shows a table collated by the information centre, with the values recommended for the continuous longitudinal turning of MR CK 45. The details of the material are given on the left hand side of the figure together with trade names of the coated cutting materials, for which this table is valid. On the right different depths of cut and feeds with the corresponding cutting speeds are given calculated for a tool life of 20 min and a flank wear of 0·4 mm. Further machining data given here are the main cutting force, the cutting power required and cutting rates to be achieved.

Such general tables of recommended cutting values do not contain factors due to machine tool, tool or workpiece. Although giving the work-planner only a rough idea, they are a substantial improvement when compared to the information material previously used.

Besides the tables with generally recommended values, it is possible with the aid of the computer, to set up tables in which many more parameters are taken into consideration. Figure 9 shows a table, prepared with the aid of the computer, with the values recommended for the turning of the steel 16MnCr5. The material has been forged, heat-treated-BF, and is to be machined with cutting material P10 on a 23 kW lathe. The optimum cost tool life is 10 min. The cutting depths possible with the tool are given in the first column. The corresponding feeds and cutting speeds are given in the next columns. The following columns contain the

material:	16 MnCr 5, forged, heat treated – BF
cutting material:	P 10
machine:	power 23 kW, torque 350 mkp, cost 40 DM/h
tool:	nose radius 1mm, length of cutting edge 16mm
tool life:	10 min
wearland:	0,2mm

bearbeitungskennwerte bei kappa = 90 grad, schnittloenge = 1000mm, drehdurchm = 300mm							
schnitt-tiefe	vor-schub	schnitt-geschw.	tatsaechl. standzeit	genutzie leistung	schnitt-kraft	zeit / schnitt	kosten/schnitt
1	,667	202	10	3,9	118	2.6	3.9
2	,800	185	10	8,2	271	2.6	3.6
3	,800	181	10	12,0	406	2.7	3.7
4	,800	178	10	15,8	541	2.7	3.7
5	,800	177	10	19,5	677	2.7	3.8
6	,800	173	10	23,0	812	2.8	3.8
7	,800	149	19	23,0	947	3.1	4.1
8	,768	134	20	23,0	1050	3.5	4.7
9	,657	134	20	23,0	1050	4.0	5.4
10	,571	134	20	23,0	1050	4.6	6.1
11	,503	134	20	23,0	1050	5.2	6.9
12	,447	134	20	23,0	1050	5.7	7.6

| depth of cut | feed | speed | actual tool life | power used | main cutting force | time per cut | cost per cut |

Figure 9. Computer edited table of standard values for turning.

expected tool lives, the power used and the main cutting forces. In this example, the power limit of the lathe is reached with a cutting depth of 6 mm. A further increase in the depth of cut first causes the cutting speed to be reduced to the lowest permissible value and then leads to a reduction of the feed. This orientation of the cutting values to machine tool power available, leads to longer tool lives than the cost-optimum one. This is shown in the column for the actual tool life.

Technologically orientated programming systems are being increasingly applied for programming NC-machines. An example is EXAPT, in which the technological data of the machining process, such as depth of cut, feed and speed, are determined automatically. For this the characteristics of the cutting behaviour must be presented to the NC-processor for the material–cutting material combinations to be used, in the form of a so-called 'material file' which is produced once only. The production of such machining files is a further achievement of the information centre, which can call on its ample store of cutting data for this purpose.

The industrial benefits of the information centre are not, however, limited to the supplying of such information for machining as standard tables and machining files. Whilst accounting for the increasing application of data-processing equipment, the information centre can also provide the necessary software for computerized cutting data calculation. In production planning, cutting data calculations can be carried out on a factory's own computer, saving time and expense.

Figure 10 shows schematically the sequence of the cutting data calculation produced within the factory. For each operation, only those segments to be cut must be defined. The proposed machines and tools must be given by means of an identification number, and the intention of the cutting data calculation (e.g. optimum cost or optimum time values) must be noted. The characteristics of the machines and tools are put into the program in the form of a machine-tool file and tool file, which must be produced once only. The machining file with the characteristics of the cutting behaviour of the material–cutting material–combination used can also be obtained from the information centre. Apart from the actual

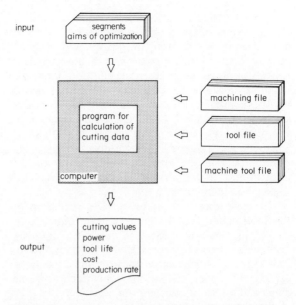

Figure 10. Computer aided calculation of cutting data in workshop planning.

machining values information which would otherwise be obtained manually such as, for example, machining times and costs can also be determined automatically.

ORGANIZATION AND METHODS OF THE MACHINING INFORMATION CENTRE

The range of operations of the information centre can be divided into three groups:

(1) acquisition and storing of data
(2) processing and retrieval
(3) documentation and transfer of information.

The procedure for carrying out these tasks can be subdivided into internal and external information processing (Fig. 11). The internal part includes all

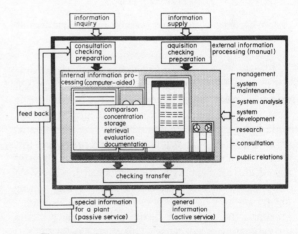

Figure 11. Organization of the information centre.

the tasks which are carried out by the computer. This comprises, in particular, the comparison of newly obtained data with existing data, the concentration and storage of obtained data, the retrieval of data from the data bank and the evaluation of data with corresponding documentation of the results.

effectiveness of the information system and making appropriate improvements.

The following is a synopsis of the operations of the information centre. Before new information is stored in the data bank, it must be formally examined, checked for plausibility and evaluated. This extremely time-consuming work must partly be carried out by experts, as the computer cannot be employed for valuation, for example.

Routine tasks such as the formal check, checking for completeness or converting the numerical data into a clear and helpful form can, however be extensively allocated to the computer.

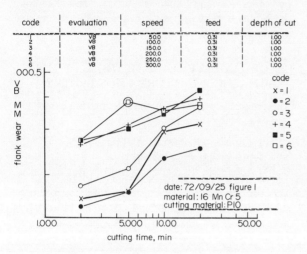

Figure 12. Wear–cutting time diagram (printer–plotter output).

Figure 12 shows, for example, a combined line printer–plotter output for the registration of measured wear values in conjunction with the cutting time. As can be seen from the diagram and the underlying details for cutting speed, feed and depths of cut, greater wear figures were determined in this case with the low cutting speed of 50 m/min (heavily traced curve), than with the higher cutting speed of 100 m/min. As long as no errors in measuring, writing or transposing are responsible for these characteristics, the cause can lie in built-up edges in the lower cutting-speed range. This must be taken into account in evaluating these data. The point marked by a circle obviously is out of line, as the flank wear becomes smaller again with increasing cutting time.

A further check of data to be stored is made in comparison with those already stored. Figure 13 represents a machining data comparison program output, in which for this example two sets of machining-data are compared for the turning of the

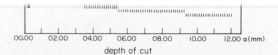

Figure 13. Comparison of cutting values 16MnCr5–P10.

steel 16MnCr5 with a cutting material P10. The feeds to be used for various depths of cut are represented by the 'A'-curve for both sets. The corresponding cutting speeds are given by curves '1' and '2' respectively and in both cases relate to the same tool life of 20 min and flank wear of 0·3 mm. As can be seen, cutting speeds differ by about 70 per cent. The differences in this case are primarily derived from the different heat-treatment of the material and its resulting differences in hardness and strength. With the aid of such comparisons of cutting values, factors having an influence on the cutting process can be analysed and quantified. Similarly, groups of comparable operating conditions can be determined, which are necessary for the drawing up of standard tables.

The cutting data obtained are ready in the data bank, after being checked, evaluated and stored. They can be used for determining machining values at any time, for example when requested by the industry. For defining inquiries for cutting data specific inquiry-sheets were developed. The results of the cutting data calculations can be advised to the enquiring party by mail, telephone or even telex.

INFOS—WORKING GROUP

In the drafting phase of the information centre for cutting data a working group was formed in the machine-tool laboratory to which manufacturers of machine-tools, tools, cutting-materials and materials belong, as well as various users of these products (Fig. 14). In addition, associations are represented in this group which work in similar project areas; that is, in the forming of technological information systems or in related fields as the EXAPT association. One purpose of the working group is to supervise the tasks and thus ensure that the information system is built up on actual practice; another purpose is to act as a representation to future industrial users of the machine-tool laboratory.

The preparation of cutting data for building up a satisfactory data store is at present the task of the companies represented in the working group. Today about 300 extensive data records from systematic wear–time experiments have been collected for turning of about fifty different materials. Additional single point data acquisition from workshop has been undertaken. On the other hand the working group

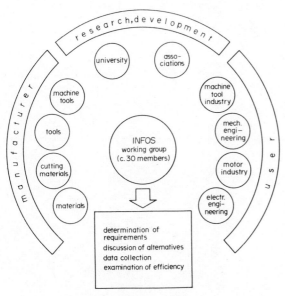

Figure 14. INFOS working group.

workpiece: spindle
material: 16 MnCr 5, forged, heat treated—BF
cutting material: P 10
operation: roughing

	before	now		
	64	16	min	tool life
	80	140	m/min	speed
	0.5	0.76	mm/rev	feed
	12.8	7.6	min	machining time
	9.73	7.37	DM	machining cost per workpiece

Figure 15. Workshop rationalization by adapted tool lives.

members have in return the right of access to the existing benefits of the information centre.

At first the data acquisition and the information service was restricted to turning operations. Since the beginning of 1973 milling and drilling operations have been included. Cooperation in this working group is basically open to anyone who is ready to provide machining details.

CONCLUSION

In conclusion, an example will demonstrate the successful employment of the information centre for cutting data. The relevant operation was roughing on a NC-lathe of twenty forged main-shafts in material 16MnCr5. The forging crust of the workpieces had already been removed and the operation was carried out with a tool fitted with P10 carbide.

The result of the cutting values obtained from the information centre is shown in Fig. 15. The machining values previously used in the production are shown by the upper rectangles, whilst the new machining values applied by the information centre are represented by the lower rectangles. Because of the considerably shorter tool life of 16 min as opposed to 64 min previously, and correspondingly higher cutting speed and feed values, the output was increased with simultaneous reduction of production costs. A total saving of about 25 per cent of production costs could be achieved.

The presentations in this paper have shown that the prerequisites for the setting up of a comprehensive information centre of cutting data exist in industry or were created by the machine-tool laboratory. Successful operation of the information centre is now largely dependent on the willingness of industry to cooperate with the centre and to provide up-to-date machining data in sufficient quantity and quality.

REFERENCES

VDI (1965). Richtwerte für das Längsdrehen im nichtunterbrochenen trockenen Schnitt, *VDI-Richtlinien,* **3206**, Düsseldorf.

W. KÖNIG (1971). Leistungssteigerungen bei spanenden und abtragenden Bearbeitungsverfahren, *Verlag Girardet,* Essen.

AUTORENKOLLEKTIV (1971). Informationszentrum fur Schnittwerte Berichtsheft zum 14. Aachener Werkzeugmaschinen-Kolloquium.

D. PFAU, G. WERNER and H.-H. WINKLER (1971). Informationszentrum für Schnittwerte—Anleitung zur Verschleiß-Schnittzeit-Untersuchung beim Drehen, *Industrie-Anzeiger,* 93. Jg., Nr. 105.

H.-H. WINKLER, (1972). INFOS—ein Informationszentrum für Schnittwerte, *Refa-Nachrichten,* Nr. 1.

W. KÖNIG, D. PFAU and H.-H. WINKLER (1973). INFOS—Informationszentrum für Schnittwerte, RKW-REFA Betriebstechnische Reihe.

DISCUSSION

Q. J. Fisher, Wickman Wimet Ltd. Despite savings accruing from the use of improved machining data, have any of the speakers experienced any reluctance to pay the economic cost of producing such data remembering that the cost per enquiry for computerized data could be as much as £20?

A. W. König. We have no experience in this matter, as in the present stage of development of our project, the transfer of improved machining data is limited to the members of the INFOS working group and is free of charge within this group. I think, however, there will be no great difficulties, if the cost per inquiry is adequate to its complexity. For simple inquiries the cost then might be less than £20.

THE COLLECTION AND DISSEMINATION OF MACHINING DATA

by

J. FISHER and J. HARGREAVES*

SUMMARY

This paper discusses the views of a cutting tool manufacturer with regard to the collection, storing, and dissemination of machining data.

The many difficulties in obtaining adequate data and transferring it into a form suitable for calculating machining recommendations are discussed giving particular emphasis to the problems of predicting tool life.

The means of storing and disseminating the data which was obtained are also discussed and reasons for the ultimate choice of methods given.

INTRODUCTION

Each year a great deal of money is spent on cutting metal. In 1969 Dr Merchant estimated that 5 per cent of the gross national product of most industrial countries was spent in removing metal[1]. Based on this estimate the cost in the United Kingdom of cutting metal at that time could therefore be put at £2000 million per annum and today would be nearer £2500 million per annum.

It is generally accepted that the production performance of machine shops could be improved. Developments in casting and forming processes have reduced the levels of stock to be removed from components and in some cases have entirely eliminated machining. However, the vast majority of components still require some form of machining operation.

Developments in machine tool technology have increased the stability and power available which in turn could exploit recent developments in metal cutting tools and materials, but the general level of cutting conditions have not yet been fully exploited with the older standard machines.

POTENTIAL APPLICATIONS FOR MACHINING DATA

For a machining data system to be successful account has to be taken of many more factors than workpiece material specification and type of operation being performed. This is because cutting tools are used on many types of machines operating under different economical and practical circumstances. For instance the speeds and feeds used for machining carbon steel on a roll turning lathe would be very different from those used on a 7 kW turret lathe machining the same material. The heaviest cut on the turret lathe would be regarded as a light cut on the roll turning lathe.

When preparing a machining data system for carbide tools the main groups of machines to be considered varies widely, but nevertheless can be broadly grouped according to their own economical and practical circumstances.

Relatively simple machines such as centre lathes, turret lathes, and vertical and horizontal milling machines have used tungsten carbide tools for many years. Over the last few years they have been increasingly applied to complex operations such as copy turning. Cutting conditions for all of these machines are often determined by machine operators or planning engineers who have limited assistance from published data, and consequently tend to select conservative machining conditions. The capacity and power of these machines vary widely, and some of these factors have to be taken into account to give a good machining recommendation.

The automatic lathe is another type of machine which uses a large number of tungsten carbide cutting tools. These machines have special significance because they are often left unattended whilst in operation and consequently it is essential to avoid unscheduled interruptions. Unexpectedly rapid wear or catastrophic failure of tools in any section could lead to a catastrophic failure of tools in the following section. Therefore machining conditions have to be selected so that any interruption for tool re-setting occurs at scheduled intervals. A further complication on the automatic lathe system can be the lack of choice of

---

* Wickman Wimet Limited

machining conditions have to be adjusted accordingly. When several tools are cutting simultaneously the disposal of swarf is yet another problem which can be met on the automatic lathe, consequently it may be necessary to adjust the feed rate to give a chip form which can be easily removed from the cutting zone.

In many ways the problems which apply to automatic lathes also apply to numerically controlled lathes. They too are sometimes left unattended for periods whilst machining, and so premature tool failure and swarf disposal problems have similar significance. Therefore, a predictable and reliable tool performance is again of the utmost importance. However one advantage of some numerically controlled systems is the facility to manually override the machining conditions embodied in the NC programme, therefore the recommended cutting conditions can be refined after production starts.

Another situation where the selection of machining conditions is particularly important arises with the transfer line. A breakdown on any one station could lead to the shut down of the complete line, and so a predictable and reliable tool performance is very important. In these circumstances the machining conditions are usually selected to give longer tool life compared with simple inexpensive machines.

FACTORS TO BE TAKEN INTO ACCOUNT WHEN PREPARING MACHINING DATA

The applications above indicate the widely varying circumstances under which cutting tools are used and underline the problems in providing accurate machining data. Taking account of the workpiece material and its conditions will only give approximate machining conditions, and if better and more reliable machining data is to be recommended, the power available to cut metal, the rigidity of the set-up and the workpiece itself, and the type of machine on which the work is being carried out must all be considered.

The workpiece material alone raises large problems when selecting machining data, especially if tool life is to be predicted. In fact, after careful consideration it was found that it was not possible to evaluate the tool life both because of the lack of a full understanding of the mechanics of metal cutting, and the difficulty of obtaining sufficient reliable information from industry about the applications under consideration. Variations of the workpiece material can arise from composition, microstructure, hardness, work hardening properties and surface characteristics.

which varies with tool material, C is a constant depending on work and tool material.

It is now generally accepted that there are many exceptions to this premise. For example, it is well known that the machining characteristics of cold drawn low carbon steel bars are superior to similar bars in the annealed condition, whilst recently developed materials for the aerospace industry have proved difficult to machine although they are not specially hard.

Inclusions in the workpiece can have widely and dramatically different effects upon machinability. For example, inclusions such as sulphides or lead are deliberately made to steel to assist machinability, but other inclusions such as alumina resulting from de-oxidation can drastically reduce tool life.

Composition, additives and heat treatment all affect the microstructure of the material. Studies on steels and cast irons have shown which structure will enhance tool life or surface finish, and which will cause these to deteriorate[2,3]. Tolerances on composition of materials of nominally the same specification can cause differences in microstructure which can lead to wide variations in tool life[4].

Another difficult feature to assess is the stability of the tool-machine-workpiece system. Users of tungsten carbide cutting tools are familiar with the detrimental effect that 'chatter' can cause on the performance. The present state of knowledge is such that the effects of stability cannot be quantified and any attempt to define degrees of stability would therefore depend largely on the subjective judgement of individuals.

TOOL LIFE

All the factors already outlined will produce a large amount of scatter when plotting tool life curves for given machining conditions and workpiece specifications.

Some workers have pointed out the difficulties in predicting tool life from the Taylor Tool Life equation:

$$VT^n = C$$

where V = cutting speed ft/min, T = tool life, C = constant
or in its modified form

$$VT^n a^x s^y = C$$

where s = feed/rev, a = depth of cut, n, x and Y are exponents.

Ideally the above formulae would determine tool life for any given set of cutting conditions. They also form the basis of procedures from which cutting conditions to give minimum cost or maximum production are derived.

Values of n are assumed to be well established and constant for a given range of machining conditions. For example, for tungsten carbide tooling they are considered to lie between 0·2 and 0·3. However a recent review by Pilafidis[5] has shown that n can vary as much as 3 to 1 in tests conducted in different laboratories using identical tool materials, workpiece materials and tool geometries. In all cases the value obtained for 'n' was higher than the values commonly assigned. Work by Barrow[6] and others has also shown that the log V/log T plot is non-linear and that a constant value of n only applies under certain conditions. Barrow has suggested that non-linear relationships are likely to be found under the following conditions:

(a) When machining high strength heat resistant materials.
(b) Machining under conditions to give a long tool life.
(c) At high metal removal rates ($a > 0·150$ in, $s > 0·020$ in).
(d) Under the finish machining conditions indicated by Kronenberg[7].

Despite the straightening techniques which are available for tool life curves, the predictions of tool life could still be considerably incorrect. Consequently it was concluded that the Wimet machinability charts linked with experience in the field would form the best basis for our machining data system.

PREPARATION OF MACHINING DATA SYSTEM

Having carefully reviewed all the problems mentioned above it was considered that the most satisfactory system would generate the data listed below:

(1) Cutting speed, feed rate and depth of cut which would give a good performance.
(2) The grade of carbide tool to be used together with cutting rake and size of corner radius.
(3) The power the recommended cut would require.
(4) Whether or not a coolant should be used.

To enable these recommendations to be built up the following information was required:

(1) Work material and its hardness.
(2) Stock to be removed.
(3) Surface condition of workpiece.
(4) An indication of the rigidity of the system.
(5) If the cut was continuous or interrupted.
(6) The horsepower available on the machine.

The problem was to decide how to combine the effects of all the factors listed above in as simple a form as possible. The machinability charts devised by Dr Trent[8,9] formed an ideal basis to work from. These charts showed the range of cutting speeds and feed rates at which the cutting tool material-workpiece combination functioned satisfactorily and the types of tool failure which occur outside the

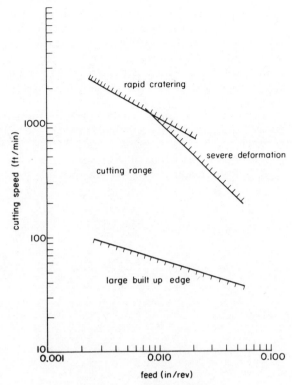

Figure 1. Wimet machinability chart.

cutting range. These charts were compiled from short time tests carried out in the laboratory (see Fig. 1).

Data points extracted from field reports which represented good machining practice in industry were then plotted on these charts. It could then be seen that while there was a wide spread of conditions on the cutting speed scale, the spread on the feed scale was less. These also showed that within the usable

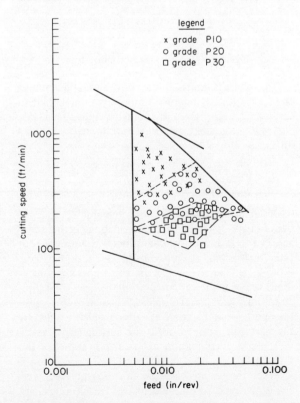

Figure 2. Effect of cutting speed and feed rate on grade of hardmetal used.

chart, a typical example being shown in Fig. 3.

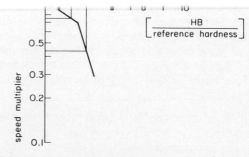

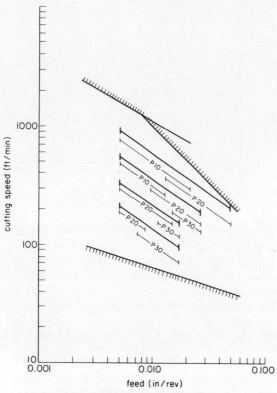

Figure 3. Composite machinability chart.

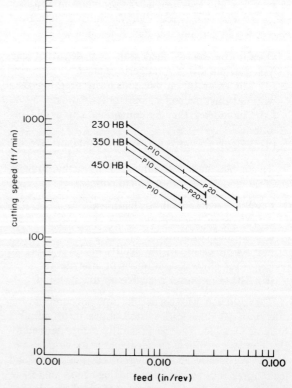

Figure 4. Relationship between work material and cutting hardness, speed and power required.

under group headings (e.g. plain carbon steels) which made the practical development of the system much easier than if a separate chart had been necessary for each workpiece material.

Within a material group the hardness could also vary substantially. In order to allow for this variation, information concerning the relation of workpiece hardness to speed and power was augmented by industrial experience and an extensive programme of laboratory testing. This enabled a relationship to be established to allow a graph (shown in Fig. 4) to be produced from which a correction factor could be obtained.

This chart is similar to the conventional machining chart except that it caters for more than one grade of carbide. It also gives the cutting speed-feed rate curves for four conditions of set-up, these being based on:

  (i) Surface condition of workpiece.
  (ii) Rigidity of set-up.
  (iii) Whether the cut was smooth or interrupted.

If all three conditions were favourable, then the upper curve would be used for selecting cutting speed and feed. If only one condition was unfavourable then the second curve would be used. If two conditions were unfavourable the third curve would be used, whilst the lower curve would be used if all three conditions were unfavourable. It also proved possible to superimpose the grade of carbide which would give the best performance on to the cutting speed versus feed curve. Examination of practice in industry also showed that the feed rates were limited according to whether the set-up was rigid or not, and these were also superimposed on the composite machining chart.

Experience with this approach showed that it was possible to combine several workpiece materials

Figure 5. Relationship between material hardness and speed, and grade of hardmetal.

Sometimes the effect of increasing workpiece hardness was to restrict the feed rate and require a change in the grade of carbide used. For example, the effect of increasing the hardness of alloy steels from 230 HB to 350 HB and 450 HB resulted in changes in the top curve of the composite machining chart for this particular material group. This is illustrated in Fig. 5.

The most commonly machined materials were arranged into nineteen separate groups for each of which a composite machining chart was prepared. The nineteen groups are shown in Fig. 6.

The generalized formula used to calculate the cutting conditions from the power available is of the form:

$$S = \left(\frac{P \cdot e}{C \cdot a}\right)^{1/Z}$$

where S = feed/rev. (in),
P = power available at motor (hp),
e = machine tool drive efficiency,
a = depth of cut,
C = constant $\big\}$ depending on tool geometry,
Z = exponent $\big\}$ workpiece material and hardness.

The derivation is discussed in Appendix 1.

It is now possible to prepare a machining recommendation if the following information is provided:

(1) Work material type.
(2) Work material hardness.
(3) Surface condition of workpiece (i.e. clean, drawn, rolled, forged or cast).
(4) An indication of the set-up rigidity (i.e. rigid or flexible).
(5) If cut was continuous or interrupted.
(6) Depth of stock to be removed.
(7) Horsepower available on the machine.

Having obtained this information the following procedure is necessary:

(1) Determination of the degree of 'favourability' of the system. This means the selection of which line on the composite machining chart applied to the system under consideration and noting the constraints on feed rate which the chosen line imposes.
(2) Calculate the correction factor for cutting speed and power to allow for the effect of workpiece hardness.

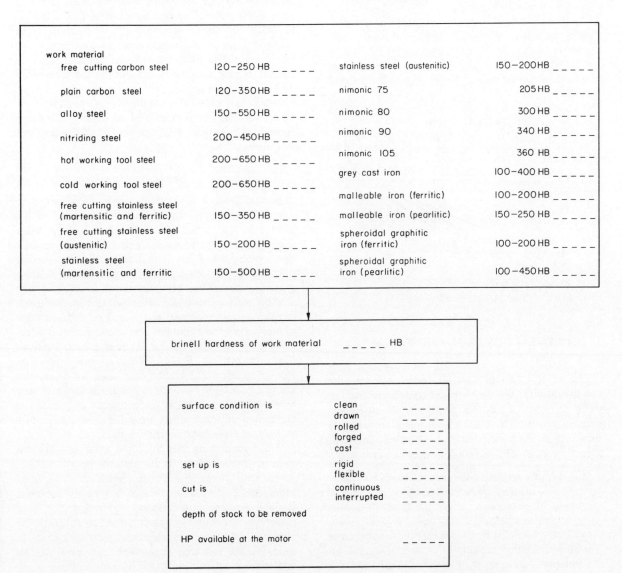

Figure 6. Questionnaire for Wimet machining data system (stock removal cuts).

feed rate exceeds the upper limit the feed rate chosen will automatically assume the value of this upper limit. Slight adjustments to speed may be made to use more power but generally under these circumstances not all of the power will be utilized.

(4) The determination if the depth of cut divided by feed rate falls within the ratios which are normally considered to constitute good machining practice. Kronenberg[7] has indicated typical depth of cut-feed ratios which at the time of developing this system constituted current machining practice.

(5) From the feed rate, calculate the cutting speed to be used, using the appropriate curve on the composite machining chart.

(6) Reduce cutting speed to compensate for workpiece surface conditions, stability of set-up and severity of cut.

(7) Determine the grade of tungsten carbide, from the material, hardness, feed rate, favourability, and cutting speed.

(8) Determination of the minimum recommended corner radius from the feed rate.

From these considerations all the following data are obtained:

(a) Feed rate.
(b) Number of cuts.
(c) Depth of cut.
(d) Cutting speed.
(e) Grade of tungsten carbide.
(f) Minimum corner radius.
(g) Rake angle.
(h) Power required.
(i) Cutting fluid (where applicable).

DISSEMINATION OF MACHINING DATA

When considering a machining data system it is important to consider carefully who will use the system and what will be the most convenient form in which the data should be presented. Extensive experience provided by cutting tool manufacturers has shown that machining data is required for four different purposes by different categories of users.

(1) Machine setters setting up machines for a new operation. Quite often they have a reasonable amount of information about the types of cuts to be taken, condition of the machine and power available for cutting. However they sometimes have difficulties because of the inadequate information concerning the workpiece material which is available to them.

(3) Sub-contract shops, who often have minimal information about the workpiece material, type and condition of machine to be used, who do need guidance to enable floor-to-floor times to be estimated on which they could base cost data.

(4) Application engineers involved in improving machining practices and productivity.

The form of the reply to a query depends very much on its source. On one hand there is a need to give a quick answer when an unexpected problem arises when setting up a machine. On the other hand if a company is considering the purchase of a new machine a more detailed recommendation, taking longer to prepare, would be more suitable. The data system must therefore be organized in such a way as to cater for these varying types of enquiry. One solution is to refer the enquiries to a central point where the data is held and the recommendation transmitted by post, telex or telephone. Alternatively data can be condensed into a form which can be included in handbooks, catalogues or on slide rules. These can be freely circulated to industry, and are consequently available at the source of the problem or enquiry.

DISCUSSION

The performance of cutting tools is affected to varying degrees by many factors, most of which cannot be quantified. Naturally this leads to wide scatter of tool performance under everyday production conditions. Thus it is not possible to predict optimum conditions but only to give a good starting point for machining. If optimization is required then this has to be based on cutting data collected from the job under consideration. A well established technique to find the minimum cost or maximum production is by the use of the Machining Cost Analysis Sheet shown in Fig. 7. This requires more information about the system than has to be given for a starting point. Often much of this information is only available to the financial departments of companies concerned and underlines again the problems in providing comprehensive machining data.

Good machining data collected from industry was important to enable the determination of realistic speed/feed curves, but it was found that the comprehensive information required was rarely available. In some cases, particularly on the lesser used materials, no suitable information at all could be obtained. In such cases laboratory tests were made to generate the required data. Whenever possible this was checked against industrial practice and any necessary adjustments made.

Wickman Wimet Limited
machining cost analysis sheet

customer			purpose of analysis	cost reduction	
component	spindle	*	machine hour rate/min	5d	
material	EN 31	*	operators' rate/min	2d	compiled
machine	Dean Smith and Grace lathe	*	setters' rate/min		date
operation	turn O.D.	*	grinders rate/min		checked

tool data			
tool no.	econotool S112 – insert SNU 433		
grade	XL2B		
speed (ft/min)	450	550	600
feed (in/rev)	0.015	0.015	0.015
depth of cut (in)			
type of failure	normal wear	normal wear	normal wear
tool life (mins)	11.5	7.9	4.14

(a) cutting cost			
*(1) cutting time (mins)	0.32	0.29	0.24
(2) cutting cost (A1) x O/Hd total rate	2.24d	2.03d	1.68d

(b) tool cost/piece	82/6	82/6	82/6
*(1) initial cost of tool	990d	990d	990d
*(2) no. of regrinds or tip index's per tool or holder	300	300	300
(3) depreciation cost per regrind or index (BI ÷ B2)	3.3d	3.3d	3.3d
(4)*(a) reservicing time (min)			
*(b) reservicing total O/Hd			
(c) reservicing cost (B4a x B4b)			
(5) tip cost	77d	77d	77d
(6) tip edge cost (B5 ÷ no. of edges)	9.6d	9.6d	9.6d
(7) total tool edge cost (B3 + B4 or B6)	12.9d	12.9d	12.9d
*(8) pieces prod. edge	36	27	17
(9) tool edge cost/piece (B7 ÷ B8)	0.36d	0.48d	0.76d

(c) tool changing cost			
*(1) tool changing time (min)	1 min	1 min	1 min
(2) tool changing cost (C1 x rate)	7d	7d	7d
(3) tool changes/piece (1 ÷ B8)	1/36	1/27	1/17
(4) tool changing cost/piece (C2 x C3)	0.19d	0.26d	0.41d

(d) time per piece			
(1) tool changing time/piece (min)(C1 x C3)	0.03	0.04	0.06
(2) total tool changing and machining time/piece (A1 + D1) mins	0.35	0.33	0.30

total tool and machining cost/piece (A2) + (B9) + (C4)	2.79d	2.77d	2.85d

note *enter for cost analysis purposes example 1 1

Figure 7. Machining cost analysis chart.

Several possible methods of holding the machining data were considered and these included tables, graphs and a library of data points. The graphical technique was the most promising because it allowed automatic interpolation between data points; and the data could be expressed in empirical terms which lent themselves to operation by computer. This technique also lent itself to being operated very quickly (either manually or by computer) and with minimum training of the operator.

In June 1968 a trial system was launched at the International Machine Tool Exhibition in London when a terminal was connected to an ICL 1905 computer several miles away. Visitors were invited to feed their machining problems into the computer and they received a printed recommendation in the form shown in Fig. 8 in a matter of seconds. The response

Wickman Wimet Ltd.– computerised machining data

go
4 350 1 4 1 2 .1 25

ack

feed rate	0.020 in/rev
number of cuts	1
depth of cut	0.100 in
cutting speed	110 ft/min
use Wimet grade XL3	
minimum corner radius	0.045 in
use negative rake tools	
power required	3.14 HP

Figure 8. Typical recommendation from Wimet computer machining data system.

was disappointing and this influenced future thinking concerning the commercial viability of such a service. Instead attention was turned to producing the required data in a simpler and cheaper way, such as in tables, in a handbook dealing with tool application, and on a slide rule. Industry has responded to this approach much more enthusiastically.

ACKNOWLEDGEMENTS

The authors wish to express their thanks to the Directors of Wickman Wimet Limited for permission to publish this paper.

REFERENCES

1. Editorial, *Metalworking Production,* April 1970.
2. H. J. SIEKMANN and R. G. BRIERLEY (1964). *Machining Principles and Cost Control,* McGraw-Hill.
3. E. J. A. ARMAREGO and R. H. BROWN (1969). *The Machining of Metals,* Prentice-Hall.
4. PERA Report 179: Preliminary Study of Variation in the Machinability of Carbon Steels.
5. E. J. PILAFIDIS (1971). Observations on Taylor '*n*' values used in metal cutting, *Annals of CIRP,* 24.
6. G. BARROW (1971). Tool life equations and machining economics, *Proc. 12th MTDR Conference.*
7. M. KRONENBERG (1966). *Machining Science and Application,* Pergamon Press.
8. E. M. TRENT (1963). Cutting iron and steel with cemented carbide tools—Part 1, An analysis of tool wear, *Jnl. ISI,* 201, Oct.
9. E. M. TRENT (1959). Tool wear and machinability, *Jnl. I. Prod. E.,* March.

where V = cutting speed (ft/min),

$\quad\quad s$ = feed/rev (in),

$\quad\quad C_1$ = constant,

$\quad\quad n_1$ = an exponential constant.

This work material is of a given reference hardness H_r.

Power to remove 1 cu in of metal/min is related to the feed rate by an expression:

$$u = C_2 s^{-n_2} \quad\quad (2)$$

where u is the power to remove 1 cu in of metal/min,

$\quad\quad C_2$ a constant,

and $\quad n_2$ an exponential constant.

The depth of cut a is a given parameter normally the depth of stock or in the case of 2 cuts—$\frac{1}{2}$ depth of stock etc.

The basic equation relating power at the motor and the machining variables above is:

$$P \times e = 12 \times V \times a \times s \times u \quad\quad (3)$$

where P = horsepower available,

$\quad\quad e$ = efficiency of driving mechanisms including motor.

where H_B is the Brinell hardness number of the workpiece material,

$\quad\quad H_r$ is the Brinell hardness number of the reference material,

$\quad\quad g$ and f are constants depending on the tool, work material, tool geometry and hardness value.

$\quad\quad p$ and q are exponents depending on the tool, work material, tool geometry and hardness value.

Note: The values g, f, p and q are defined and constant over a limited range of hardness only.

Substituting equations (1) and (2) and the hardness correction factors in (3) give an equation of the form:

$$Pe = 12 C_1 s^{-Z_1} . a . s . C_2 s^{-Z_2} k . m.$$

$$= C . a . s^Z . k . m.$$

or

$$S = \left(\frac{P . e}{C . a} \right)^{1/z} \quad \text{where} \quad \begin{array}{l} C = 12 C_1 . C_2 . k . m. \\ Z = (1 - z_1 - z_2). \end{array}$$

MACHINABILITY TESTING AND QUALITY CONTROL CONSIDERATIONS IN THE DEVELOPMENT OF HARDMETAL CUTTING TOOLS

by

M. J. HENTON* and G. A. WOOD*

SUMMARY

Over the years many machinability and machining tests have been developed but these have been primarily for the evaluation of the machinability of work piece materials rather than cutting tool materials. Organizations confronted with the problem of developing or evaluating cutting tool materials have frequently adopted very similar tests. However we have found that such tests are not ideal even for comparing two existing materials, and they are completely unsatisfactory when the object is to develop new materials. It is essential that the mechanisms that result in eventual tool failure, and the effect of the carbide variables, should be recognized from the testing procedure. The way these observations are utilized in machining tests, especially in relation to the hardmetal variables such as grain size and composition, is described in this paper.

In the system which we have adopted, hardmetal grade development is based upon the results of a series of comparative laboratory machining tests of short duration, and considerable backing to these results is obtained from carefully conducted field testing. This type of development procedure is discussed with reference to some specific cases, including both homogeneous and coated tips.

Development done in this way automatically relates the performance of the grade to the hardmetal parameters and immediately indicates the important factors for quality control. The way in which satisfactory quality control of the hardmetal cutting tool grades is effectively and economically obtained is described.

INTRODUCTION

Most machinability tests have been designed to define the machinability of the work material and the literature is full of proposed machining tests for its measurement[1-7]. Many of these tests are based upon the Taylor equation, $VT^N = C$, and generally involve long and costly procedures. The results may only relate to the given machining conditions of that test, making their transposition to a range of applications on the shop floor fraught with serious difficulties.

These machining tests generally emphasize the rate of flank wear and it is frequently assumed that this is the only form of wear that should be considered. When cutting conditions are good, flank wear can be the dominant form of wear, but there are other forms of wear that can be more important in determining the useful life of a tool[8]. All the following factors can have a dominant effect under the appropriate cutting conditions:

(1) Mechanical chipping of the cutting edges.
(2) The presence or absence of a built up edge.
(3) Cratering on the rake face.
(4) Deformation of the tool due to high cutting forces and temperatures.
(5) Rake face grooving.
(6) Thermal fatigue.

Tests are also required to enable cutting tool materials to be evaluated. All too frequently it has been assumed that the same machinability tests would be satisfactory. This procedure has frequently been used with some limited success by those who have to choose between cutting tools for different jobs; but when the problem becomes one of formulating new cutting tool materials the limitations of this approach become evident.

For such work it is essential that the wear mechanisms that result in tool failure in different applications are recognized and then related to the chemical, physical and mechanical properties of the tool material. The way in which this can be done is described in this paper.

HARDMETAL GRADE DEVELOPMENT

A new hardmetal cutting tool grade could be introduced to offer an improvement in performance over an existing grade, or to extend the range of applications already provided. New hardmetal grades may

---

* Wickman Wimet Limited

evidence concerning the wear mechanism. In order to carry out such an examination it is necessary to remove any adhering work material from the insert by immersion in hydrochloric acid so that the detail on the contact area can be observed. The inserts are then examined by various methods to determine the types of wear that are predominating and causing short tool life.

Premature failure of a hardmetal cutting tool does not generally require the development of a new hardmetal grade. Examination of the used inserts from a particular application may often show that the problem could be solved by using another standard grade from the range available, or by the use of a different tool geometry.

Examination of the used inserts in this way enables the inadequacies of the existing grade to be identified, and a new grade can be formulated by altering the composition or structure in a manner which will reduce these inadequacies. For instance if a grade is failing predominantly by attrition wear, the new grade would probably be of finer grain size. Initially several variations may be considered that would provide improvements. Inserts are then prepared from experimental batches of powder for laboratory machining tests.

TESTING METHODS

Any proposed new cutting tool grade would be evaluated by two types of machining test. Initially the new grades would be subjected to short term comparative laboratory tests to establish that an adequate improvement in performance over the existing grade had been obtained. These would be supplemented by extensive field tests to verify the laboratory findings and establish the useful application range of the new grade.

The laboratory machining tests are carried out using accurate inserts made from the proposed new grades and the best existing grade. The inserts are flat polished on all faces using 6–12 μm diamond powder, so that small critical variations in the behaviour of the cutting edge can be accurately observed and measured.

The following tool geometry is usually adopted:

Top rake	$+8°$
Side clearance	$6°$
Approach angle	$15°$
Nose radius	0.8 mm

Test inserts, of both the experimental and standard grades, are then subjected to a series of short machining tests which are designed to generate the test is terminated when a measurable difference is observed in the amount of the particular wear under investigation. Differences in the rates of cratering, rake face grooving, thermal fatigue, mechanical chipping or deformation can be established fairly quickly. However, to establish significant differences in flank wear rates, longer tests are generally required in order to produce a 0.010 in or 0.015 in flank wear land.

When the optimum new grade has been established in the laboratory machining tests, a large batch of the powder is produced and suitable styles of insert manufactured for field tests. The tests are again conducted on a comparative basis using the existing grade and the new grade.

It is essential that the normal shop criteria of performance are reinforced by laboratory examination of the development of wear and failure during the life of the insert. At the end of the practical life of a cutting tool insert, the wear mechanisms involved are often completely obscured and misleading results are then obtained. To avoid this, the test inserts are removed at various stages during their life, so that the progression of cutting edge deterioration can be observed. The test is extended to give complete economic tool life data only if such observations are satisfactory.

DEVELOPMENT OF A GRADE TO COMPETE WITH HIGH SPEED STEEL

Hardmetal requires higher cutting speeds than high speed steel if it is to be used successfully and economically. On many machines that use high speed steel tools it is not possible to raise the cutting speeds sufficiently for hardmetal to be used. A detailed examination of hardmetal inserts used at low cutting speeds has shown that they generally fail by attrition wear. That is, discrete particles of carbide are plucked out of the insert by the action of the built-up edge or chip. The grade best suited to resist this type of wear is a fine grain size tungsten carbide–cobalt alloy, such as Wimet H, which is a 6 per cent Co grade with an average carbide grain size of 0.8 μm. When used under high speed steel cutting conditions on the shop floor, this grade has excellent attrition wear resistance but fails because of edge chipping or massive fracture, Therefore, to develop a grade to withstand these conditions it was necessary to maintain the fine tungsten carbide grain size for resistance to attrition wear, but increase the toughness to withstand the chipping and fracture.

It is known that toughness and strength are increased by an increase in cobalt content. In terms of transverse rupture strength an increase from 6 to 10 per cent cobalt gives a 50 per cent increase in strength, and this was considered sufficient to significantly improve the resistance to edge fracture. To evaluate the proposed fine grain 10 per cent Co grade (10F) in laboratory machining tests, it was necessary to demonstrate improvements in its resistance to chipping and fracture over H grade, while maintaining the resistance to attrition wear. This therefore involved two basic tests to investigate both mechanisms of failure.

The resistance to attrition wear of the two grades was compared in low speed turning tests. Typical machining conditions for these tests were:

Test 1

Work material	Meehanite GA
Cutting speed	90 ft/min
Feed	0·0104 in/rev
Depth of cut	0·080 in
Cutting time	12 min
Cutting conditions	Dry
Grades	H and 10F

Figure 1. H grade insert after laboratory test No. 2.

Under these cutting conditions a built up edge covered the contact zone of both inserts. After the tests this was dissolved off with hydrochloric acid. Examination showed that very little attrition wear had occurred on either grade.

The relative resistance of the two grades to edge chipping was investigated with a turning test using a bar containing a 0·5 in wide longitudinal slot. This gave an interruption in cut once every revolution. Typical machining conditions were:

Test 2

Work material	En8, containing 1 slot 0·5 in wide
Cutting speed	145 ft/min
Feed	0·0044 in/rev
Depth of cut	0·080 in
Cutting conditions	Dry
Cutting time	3 min
Grades	H and 10F

After the test, the contact zone on both inserts was completely covered by a built up edge. This was removed so that the cutting edge of the inserts could be examined. The H grade insert was badly chipped around the nose and some chipping had also occurred on the rake face (Fig. 1). The 10F insert was still in very good condition (Fig. 2). These results were confined over a range of cutting conditions.

If the two grades had been compared in the BSI

Figure 2. 10F grade insert after laboratory test No. 2.

machinability test[3] the results would have indicated that H grade was superior to 10F because its higher hardness would have given lower flank wear rates. However in this instance flank wear was not the critical failure mechanism to be studied.

Indexable inserts and brazed blades were then produced for field tests. These were carried out on applications where hardmetal had previously been unsuitable because of failure by chipping or fracture. The purpose of the tests was to demonstrate that 10F could be used below the previously accepted cutting speed range for hardmetal, and to establish its useful range of cutting conditions on a wide range of applications and work materials.

Field tests confirmed that 10F is a tougher grade than the 6 per cent Co fine grain grade because it suffered significantly less chipping and massive fracture, and also resisted attrition wear. They showed that it performed well on both cast irons and steels up to cutting speeds of 200/250 ft/min and under these conditions generally outperformed high speed steel tools by a substantial margin.

DEVELOPMENT OF TITANIUM CARBIDE COATED GRADES

The introduction of titanium carbide coated grades provided inserts with considerably lower flank wear and crater wear rates than conventional hardmetal grades when machining cast irons and plain carbon to medium alloy steels. This was readily demonstrated in laboratory turning tests (Fig. 3). These improvements are associated with the inherent properties of high hardness and resistance to diffusion wear of the very fine grain size titanium carbide layer, even though it was only 0·0002 in thick.

The subsequent field testing of these grades showed improvements in the life of the inserts of 200–400 per cent over the uncoated grades (Fig. 4). Many field tests were conducted so that the useful range of applications could be established and the test inserts were returned to the laboratory for examination. It was evident that the titanium carbide coated grades performed considerably better than the uncoated grades on many applications, and offered increased tool life or higher cutting speeds for similar tool life.

Industrial usage soon showed that there could be a substantial difference in performance between various

the titanium carbide layer was found to occur mainly with variations in depth of cut and cutting speed, such as in copy turning.

From this observation a short term simulated copy turning test was developed for laboratory use which enabled differences in the adhesion of the titanium carbide layer to be demonstrated. The test is a turning operation carried out on a prepared bar of Meehanite GA cast iron. A series of grooves 0·040 in deep and with a width equivalent to 10 s turning at the feed and speed selected are cut into the bar. The test is then carried out with a total depth of cut of 0·080 in resulting in a change in the depth of cut between 0·040 in and 0·080 in every 10 s. This provides fluctuating thermal and mechanical stresses in the tip on the outer half of the depth of cut, and gives a sensitive test of layer adhesion.

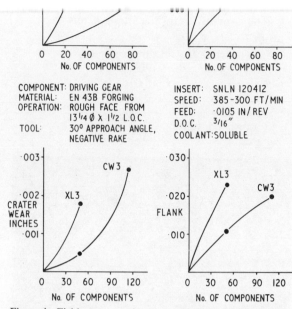

COMPONENT:	DRIVING GEAR	INSERT:	SNLN 120412
MATERIAL:	EN 43B FORGING	SPEED:	385–300 FT/MIN
OPERATION:	ROUGH FACE FROM	FEED:	·0105 IN/REV
	13¼ Ø X 1½ L.O.C.	D.O.C.	3/16″
TOOL:	30° APPROACH ANGLE,	COOLANT:	SOLUBLE
	NEGATIVE RAKE		

Figure 4. Field test results on titanium carbide coated grades.

Typical machining conditions for this test are:

Work material	Meehanite GA
Cutting speed	230 ft/min
Feed	0·0104 in/rev
Depth of cut	0·040 in for 10 s
	0·080 in for 10 s
Lubricant	Dry
Cutting time	2 min 20 s

The difference in titanium carbide layer adhesion properties between two different coated inserts is illustrated in Figs 5 and 6.

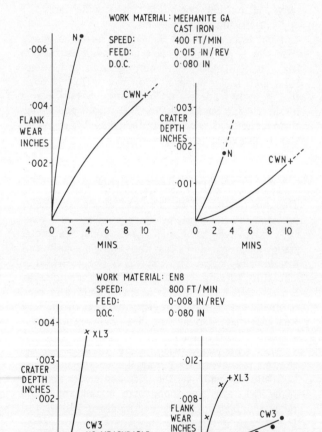

Figure 3. Laboratory machining tests with titanium carbide coated grades.

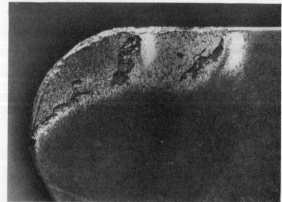

Figure 5. Insert with poor titanium carbide layer adhesion. After laboratory simulated copy turning test.

The machining test makes it possible to study the effect of coating parameters upon the layer adhesion properties and develop new coated grades or coating

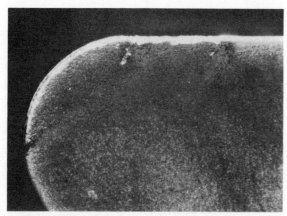

Figure 6. Insert with good titanium carbide layer adhesion. After laboratory simulated copy turning test.

Figure 8. CW620 after machining 100 cast iron components.

processes using short time laboratory tests, rather than lengthy field tests.

Using the test it was possible to define those variables in the coating process which influence layer adhesion and a modified process was developed which gave a substantial improvement in this respect. The difference between the behaviour of the two products was clearly seen from the following field test. The coated grade, initially produced for cast iron applications CWN, and the improved coated grade CW620, were used to machine cast iron pressure plates under the following conditions:

Material	Cast iron BS 1452 Grade 17
Cutting speed	280/180 ft/min
Feed	0·032 in/rev
Depth of cut	0·0625 in
Conditions	Dry
Grades	CWN and CW620
Components	CW620—initially 100—
produced	ultimately 200
	CWN—initially 44—
	ultimately 100

The two inserts were not taken to their ultimate life in the first instance but were removed for examination at a fairly early stage. The wear that occurred is illustrated in Figs 7 and 8.

Figure 7. CWN after machining 44 cast iron components.

If this had not been done, the vital evidence showing that the failure was due to the layer being plucked off, would have disappeared. In fact the ultimate life of CWN on this application was 100 components, while CW620 after 100 components, as shown in Fig. 8, went on to machine a further 100 components before indexing was necessary.

To date, field testing has shown that the coated grades will operate over a wider application range than the uncoated grades. For example the cast iron grade CW620 is suitable for the ISO K10–K30 range and the steel cutting grade CW540 for the ISO P10–P40 range. Coated grades will generally provide a longer life than the uncoated grade under the same cutting conditions, or similar tool life at increased cutting speeds.

PRODUCTION AND QUALITY CONTROL CONSIDERATIONS

The performance of hardmetal cutting tools depends on a complex combination of physical and chemical properties that are functions of the composition and structure. Many of these properties are not immediately apparent. For example, the important physical properties are not simply hardness and density. Strength, both at room temperature and elevated temperatures, and in compression and tension, and the coefficient of thermal expansion and thermal conductivity are also important. In the case of the chemical properties of the material, resistance to oxidation, and diffusion wear are important. These properties can be considered to be either structure sensitive or composition sensitive. For instance, resistance to diffusion wear is related to titanium carbide content and is therefore composition sensitive, whilst resistance to attrition wear is related to the grain size of the tungsten carbide and hence is structure sensitive. Some properties however, such as hardness, are sensitive to both structure and composition.

In addition, these properties can be influenced by defects of various types which may affect the strength, wear resistance or structure of the sintered tip. A low level of porosity is particularly important if strength and wear resistance are not to be impaired.

The aim in hardmetal manufacture is to keep porosity at the lowest possible level. The industry has now advanced to the stage where porosity in cutting tool grades rarely has an inhibiting effect on tool life.

The other major defects which have to be avoided are etaphase and uncombined carbon. These are related to the carbon content of the hardmetal, and can only be avoided by precise carbon control at all manufacturing stages. A deficiency of carbon causes etaphase to be produced on sintering. This has a serious embrittling effect on a material which at best has little toughness. In addition, this defect is usually associated with a finer structure than normal and this

During the development stage of a new grade of hardmetal it is normal practice to carry out extensive metallurgical testing to provide background information for the laboratory and field tests. In this way, the limits of both composition and structure, and the properties which are dependent on these, can be established. The problems of introducing a new grade into production are minimized by using standard materials and manufacturing processes whenever possible. However when the new grade is finally launched and becomes commercially available, the intensive metallurgical testing carried out initially is not economically possible.

QUALITY CONTROL PARAMETERS

The earlier discussion concerning the difficulties of machining tests, particularly the need for several types of test, shows quite clearly that there is no uncomplicated, cheap machining test that is directly related to the performance of a cutting tool tip which could be used for the purpose of routine quality control. In consequence, quality control has to use indirect methods based on easily measured properties of hardmetal. The evidence acquired at the development stage is used in establishing the limits of composition and grain size and the easily measured properties which are dependent on these—hardness, density and coercive force. The quality control system is therefore designed to ensure that:

(1) The composition and grain size of the tips manufactured conform to the requirements established during the development stage.
(2) The tips are not porous and are free from either etaphase or uncombined carbon.
(3) The tips are dimensionally correct and have a satisfactory surface finish.

QUALITY CONTROL PRACTICE

It is impossible to divorce quality control from economic considerations and with hardmetal these play an unusually important role. The raw materials and the processing to the finished product are most expensive, yet if the finished product has to be scrapped it is almost valueless. There is no melting pot into which the scrap hardmetal can be thrown and then be re-melted and re-cast. In these circumstances the hardmetal manufacturer has a strong incentive to institute a comprehensive system of quality control, not simply to ensure that he has satisfied customers, but also to minimize his own scrap costs. There is a greater incentive in the hardmetal industry to be 'right first time' than in many other industries.

degree of involvement in quality on the part of all personnel supervising and operating the process.

An important feature of quality control at Wickman Wimet is in the degree of involvement of the Quality Control Department in discussions concerning design and manufacturing processes. It is the responsibility of the Quality Control Department to issue detailed instructions on every process that is carried out. These instructions are issued to all production supervisors. In addition, the Quality Control Department issues a Quality Control Manual detailing the tests, analyses and specification of all materials purchased, or manufactured within the Company.

In the case of Wimet hardmetal, manufacture starts from Wolfram Ore. The process can be divided conveniently into three stages.

(1) The extraction of tungsten from Wolfram Ore.
(2) The manufacture of 'ready to press' powders from the extracted tungsten.
(3) The manufacture of sintered tips from the powders.

1. The extraction of tungsten from Wolfram Ore

In this stage, tungsten is extracted from the ore and converted to an extremely pure tungsten compound, ammonium paratungstate. As very little purification takes place in subsequent manufacturing stages, it is important that the ammonium paratungstate does not contain significant amounts of impurities likely to cause porosity in the final hardmetal. Impurities which may affect the grain size of materials made in later manufacturing stages must also be avoided. To ensure that these requirements are met the extraction process has to be carefully monitored. This involves extensive chemical analysis of samples from intermediate stages of the process and of the ammonium paratungstate[9].

2. Manufacture of 'ready to press' powders

The ammonium paratungstate is firstly reduced to tungsten metal and then mixed with carbonblack and heated to produce tungsten carbide. In order to satisfy the grain size requirements of the sintered hardmetal, several different grain sizes of tungsten metal and tungsten carbide are produced. Control of the grain size of the cutting tool tip starts at the tungsten metal stage, where a wide variation in grain size can be readily obtained by alteration of the reduction conditions.

The importance of carbon control in hardmetal manufacture cannot be over-emphasized. This depends to a very large extent on the control exercised over the carbon content of tungsten carbide and other carbides used in the manufacturing process. Complete assurance of freedom from etaphase and uncombined carbon in the sintered material, is only achieved by carefully monitoring the carbon content of each batch and ensuring that it is within the close limits specified.

The grain size of the tungsten carbide also has a significant effect on the grain size of the final hardmetal. Thus it is particularly important that the tungsten carbide is manufactured within closely controlled limits of grain size. To ensure this, each batch of tungsten carbide is checked on a routine basis.

The same criteria of close carbon control and grain size control applies to the other carbides used in the manufacture of the complex alloys used for machining steel and the method for testing tungsten carbide is also used for these materials.

Alloy powders are made by wet ball milling together those carbides and cobalt required to give the final composition. After ball milling the powders are dried and sieved and from each batch a sample is taken and analysed for carbon, cobalt, titanium and tantalum. A sintered test piece is also made from each batch and the following tests are carried out.

Physical properties:

(1) Density.
(2) Hardness.
(3) Coercive Force.

Metallographic examination:

(1) Appearance of fracture.
(2) Assessment of porosity on polished section.
(3) Structure.

The powder is only passed for use if the general quality of the test piece is satisfactory and the physical properties, structure and composition are within the limits specified for the particular grade. Powder in this condition is not normally suitable for consolidation and it is necessary to add a pressing lubricant, usually paraffin wax. The powder is then either sieved or granulated depending upon how it is to be used.

A point not generally appreciated is the difficulty in obtaining accurate hardness measurements on hardmetal. This occasionally leads to customers requesting impossibly tight hardness limits for particular applications. The commonly used hardmetal alloys have hardnesses of 1000 to 1800 HV and even with the most accurate commercial machines available for measuring hardness, the results are accurate to only 1-2 per cent. Errors in measurement can account for variations of the order of 10-35 HV and this is probably exceeded under most practical conditions. The same error in measurement corresponds to insignificant variations in hardness with the softer metals generally used in the engineering industry.

Coercive force is an indirect method of measuring average hardness and can be carried out with much greater accuracy than a hardness test. Coercive forces also give some indication of abnormal deviations in carbon content.

3. Production of sintered tips
Two major methods of producing sintered tips are used. For small quantities simple shaped blocks are first pressed using waxed powder and finally shaped by machining after a dewaxing-presintering operation. Quantity production of simple shapes is normally carried out on automatic tabletting presses. For this process waxed and granulated powders are used and the tips are pressed as closely as possible to finished shape. Regular tests are taken during long runs to ensure there has been no alteration in press setting and any machining necessary is carried out after a dewaxing-presintering operation.

The large amount of contraction which takes place on final sintering with hardmetal, approximately 20 per cent on linear dimensions, has to be taken into account during soft machining and in the design of press tools.

The tips are inspected before final sintering to ensure that they are dimensionally correct and undamaged. Unsatisfactory tips are scrapped at this stage when their reclamation is more economical than after sintering. The tips are then sintered, usually in batch vacuum furnaces. After sintering, tests are carried out on each furnace charge before unloading, to ensure that there has been no adverse temperature gradient during sintering.

If the tests are satisfactory each order is sent to the Inspection Department for final inspection. The standard tests carried out are:

(1) Measurement of hardness, density and coercive force on at least one tip from each order.
(2) 100 per cent inspection for chipping, cracks and surface defects which might indicate possible metallurgical trouble.
(3) Dimensional inspection of a sample from each order.

Only tips which have the correct physical properties and are dimensionally correct and undamaged are finally passed for despatch to customers.

4. Dimensional control
Grinding hardmetal is costly and the use of unnecessarily large sintered tips exerts a high economic penalty. Thus the general aim in hardmetal manufacture is to obtain the best possible dimensional accuracy directly from sintering. There has been considerable discussion recently on the possibilities and problems involved in manufacturing hardmetal components to precise tolerances and a good deal of information has been published on the subject[10].

In general the alloys used for machining applications and the shapes required are not those which cause serious problems in dimensional control. This may be the reason for cutting tool tips becoming the first group of components to be covered by external specifications. At the moment there are two, BS 4193 part I 1967 relating to throwaway inserts, and ISO Recommendation R242 relating to turning tool tips.

ACKNOWLEDGMENTS

The authors wish to acknowledge the valuable assistance contributed by their colleagues from many departments of Wickman Wimet Limited that has enabled this paper to be produced. They also wish to express their thanks to the directors of Wickman Wimet for permission to publish the paper and to Mr

3. BRITISH STANDARDS INSTITUTION (May 1972). *Fifth Draft Proposal—Tool Life Testing with Single Point Turning Tools*, B.S.I. 72/33047.

4. (1956). Life tests for single point tools of sintered carbide, *A.S.M.E. ASA*, **B5**, 34.

5. N. HALLBERG (1968). Standard machining test, *Cutting Tool Engineering*, Jan./Feb., pp. 25–28, Mar./Apr., pp. 5–7.

6. C. MOORE (1968). The steelmakers approach to machinability, MTDR Conference.

7. W. B. HEGINBOTHAM and P. C. PENDEY (1968). A variable rate machining test for tool life evaluation, MTDR Conference.

8. E. M. TRENT (1969). Tool wear and machinability, *Journal of the Institution of Production Engineers*, Mar.

9. G. A. WOOD (1970). Quality control in the hard-metal industry, *Powder Metallurgy Symposium Institute of Metals*.

10. E. LARDNER, G. E. SPRIGGS and G. A. WOOD (1972). Dimensional control in hard-metal manufacture, *Powder Metallurgy Symposium, Institute of Metals*.

DISCUSSION

Q. J. Tlusty. Which are the special features of your test in which you detect the ability of the TiC layer to adhere to the substrate.

A. The laboratory simulated copy turning test, used to compare the adhesion properties of the titanium carbide layer onto the substrate, creates fluctuating thermal and mechanical stresses within the layer and immediate substrate and coupled with the fact that the machining is carried out under built up edge conditions, shows up any weaknesses by plucking the layer off the insert. The test was developed after the observation that variations in speed, feed and depth of cut, as can occur in copy turning of steel components, and machining cast iron components

related to the high hardness of the titanium carbide and its inherent resistance to diffusion reactions with the chip: another feature is that the temperature at the work piece/tool interface is lower for a given set of machining conditions with the coated tip, than for an uncoated tip. This is important because the diffusion wear rate is related exponentially to the temperature, so that small temperature rises give large increases in diffusion wear rates. Therefore the titanium carbide layer initially acts as a barrier between the chip and carbide tool, and even when the 0·0002 in thick layer is penetrated, and a crater and flank wear land forms, the titanium carbide around the edge helps to provide lower wear rates than with upcoated tips.

Q. E. R. Brealey, Trent Polytechnic. (1) Many of the machining tests quoted in the papers have been performed on the high strength materials developed for the Aerospace industry. Have the authors examined the surface condition of the test pieces after machining?

2. Some materials have very shallow cracks in the surface after machining. This cracked surface must be removed to restore the fatigue life of the component. Extra costs of manufacture are therefore created. I suggest that information about cutting the more difficult materials at high rates should be given together with a warning that possible surface damage may be created.

A. In our paper we set out to describe our techniques of grade development, which involves understanding of principles of alloy development, and requirements of good quality control backup. With a cutting tool grade the initial machining tests are carried out in the laboratory but eventually the need for testing, under industrial conditions, arises. When these field tests are carried out, on cast irons, mild steel or the high strength alloys, our initial concern is the performance of the cutting tool. However, the condition of the finished work piece has to be acceptable to the customer for the trial to have been successful.

THE STRESS-TEMPERATURE METHOD OF ASSESSING TOOL LIFE

by

I. YELLOWLEY* and G. BARROW*

SUMMARY

A method of assessing tool life by considering the normal stress acting on the flank face of a turning tool and the cutting temperature is described. The results presented indicate that when using carbide (Wc–Co only) tools and high strength thermal resistant workpiece materials, the technique gives satisfactory results with considerable savings in both cost and time. The technique can be used to predict the life of new workpiece materials, or for monitoring variations in tool life from batch to batch of one workpiece material.

It is proposed that relating tool life to normal stress and cutting temperature is a more fundamental method of assessing both tool and workpiece materials in that a so-called 'basic wear resistance' of tool materials can be considered. It is evident, however, that a great deal of fundamental study is required before a complete physical model of the wear process in cutting may be evolved. It is hoped that the results presented will aid the search for this model.

NOTATION

V	Cutting velocity
h	Orthogonal depth of cut
h_c	Chip thickness
h_e	Equivalent chip thickness in the turning operation
γ_o	Orthogonal rake angle
ϕ_i	Shear angle
λ_c	Chip thickness ratio (h_c/h)
γ_c	Shear strain in primary deformation zone
μ_γ	Apparent coefficient of friction on tool rake face
K	Shear yield strength of work material at metal cutting conditions
σ_0	Maximum normal stress on rake face
σ_n	Normal stress on tool flank at the tool edge
a_e	Ratio of work done in the primary deformation zone to work done in the secondary deformation zone
V_B	Flank wear land
t	Time
T	Tool life
θ_c	Cutting temperature
θ_s	Stress modified temperature
α_1	Secondary temperature multiplier
β_1	Primary heat division parameter

INTRODUCTION

The increasing use of sophisticated machine tools has stimulated the interest in obtaining reliable tool life data. While the application of reliable tool life data can lead to reduced machining costs, any gains achieved must be balanced by the cost of obtaining such data.

One method of obtaining reliable tool life data is the systematic collection and dissemination of data from industry, resulting in the so-called machinability data banks. Whilst this development is of significant importance, there are problems of financing such a system and also of determining the reliability of data obtained from sources not under direct control of the data banks. The latter problem is extremely important since, whilst data from industry may be 'reliable data' in the sense that they have been used successfully in practice, there is no proof that they are optimal data for universal application. In view of this, methods of assessing tool life under laboratory or controlled industrial conditions are required.

The conventional method of obtaining tool life data under laboratory conditions is to undertake actual tool life tests over a range of cutting conditions in order that tool life may be expressed in an empirical form using an equation such as the well-

---

* Machine Tool Engineering Division, Department of Mechanical Engineering, U.M.I.S.T.

many of which (e.g. radioactive, taper turning and facing methods) are well known and will not be discussed here. In some cases the techniques have not resulted in considerable cost saving due to lengthy preparation time etc., whilst in other cases the results have not proved reliable enough for general application. In view of this, none of the techniques have been used extensively (except by the originators). Most of the methods used to date have been completely empirical in nature and it is felt that a technique with a more fundamental basis will be accepted with greater confidence.

REQUIREMENTS OF A RAPID TOOL LIFE TESTING TECHNIQUE

The main uses of such a technique can be summarized as follows:

(a) The testing of new workpiece materials.
(b) The testing of new tool materials.
(c) The monitoring of variations in tool life which can occur in different batches of a workpiece material of the same nominal composition.

These objectives must be achieved both quickly and cheaply with reasonable degree of accuracy. It is, of course, difficult to determine what accuracy should be aimed for, since in some cases extreme accuracy is not required and the attainment of extreme accuracy can only be obtained with a sacrifice in cost reduction. It is the authors' opinion that even on tool and workpiece materials which have been subject to close quality control in manufacture, the assessment of tool life to ±50 per cent of the mean value is acceptable for a rapid test method.

THE INFLUENCE OF TEMPERATURE ON TOOL LIFE

As long ago as 1907 Taylor[1] noted the influence of cutting temperature on tool life. However, it was left to Schallbroch et al.[2] to demonstrate that tool life could be related to temperature by the equation

$$T\theta^n = \text{Constant} \qquad (2)$$

where the cutting temperature θ was measured at various values of cutting speed and feed using the worktool thermocouple method of temperature measurement. The work of Schallbroch et al. was undertaken on high speed steel (HSS) cutting tools, but several other workers have demonstrated the validity of equation (2) using carbide tools with both flank and crater wear as tool life criteria. Experience has shown that a unique relationship between temperature and tool life only exists for one tool-

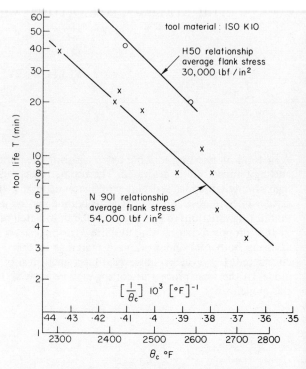

Figure 1 Tool life against the inverse of cutting temperature.

reasonable relationship holds for one tool workpiece combination, cutting temperature cannot be used to predict tool life when more than one workpiece material is being machined with one carbide. The form of the relationship depicted in Fig. 1 is compatable with a rate process, this type of relationship has been observed by Takeyama.[3]

THE INFLUENCE OF STRESS ON TOOL LIFE

In attempting to evaluate the influence of stress on tool life it is necessary to consider briefly the wear mechanisms which are thought to contribute to the wear of cutting tools.

Adhesive wear
It is now accepted that pressure welding can exist between the tool on one hand and the chip (and possible workpiece) on the other. This pressure welding may be cyclic in nature, thus providing the opportunity for the 'plucking out' of particles of tool material.

Abrasive wear
It is thought that this type of wear can be of two types, two body abrasion and three body abrasion. In the former case the wear is caused by the work material abrading to the tool material, whilst in the

latter case separated particles of tool material enter the contacting region between work and tool.

Diffusion wear

At high cutting temperatures diffusion reactions between the carbide tool and the work material are possible. A thorough investigation of the reactions which occur under static conditions between carbide tools and a steel workpiece has been carried out by Opitz and Konig[4]. Whilst the actual conditions of stress and temperature in the metal cutting situation are extremely complex, estimates of mass diffusion have been made by Bhattacharyya and Ghosh;[5] from their calculations they infer that at high cutting speeds the main influence of diffusion processes on the wear rate is through the local weakening of the tool interface, the proportion of wear which may be attributed to mass diffusion being very small.

It is evident that the extent of diffusion is limited by the particular work material, tool material combination.

Chemical wear

This occurs through the interaction between tool and workpiece material in an active environment, e.g. cutting fluid or air. At high temperatures oxidation of the tool material can occur leading to the formation of weak complex carbides.

Flank wear

Experience has shown that in general, tools fail in practice due to the flank wear land reaching a limiting value. In all the tests reported in this paper on flank wear criterion for tool life of $V_B = 0.015$ in or $V_{B \max} = 0.030$ in was used.

It is assumed that the total volume (V) of tool material removed, is due to both 'primary' wear particles being removed and secondary mechanical wear taking place as the primary particles traverse the wear land. Both these types of wear are related as indicated below.

It is assumed that in unit time a volume V_1 of primary particles is produced. It is likely that the temperature will exert a strong influence on the primary wear either through its influence on the properties of the tool material, the thermal stressing of the tool material or by the facilitate of diffusion processes. Hence

$$V_1 = f(\theta_c, \sigma_n) \qquad (3)$$

The secondary wear value V_2 can be represented by the expression

$$V_2 = A.V_B.V_1 \qquad (4)$$

where $A = f(\sigma_n)$.

As the time required for a primary wear particle to traverse the wear land is extremely small, then in unit time the total wear volume is approximately given by

$$V = V_1 (1 + A.V_B) \qquad (5)$$

i.e.

$$dV_B/dt = \text{Const. } V_1 (1/V_B + A)$$

If one assumes that at reasonable values of wear land the secondary wear forms by far the largest proportion of the total wear then the slope of the secondary region of the flank wear land time characteristic will be approximately given by

$$dV_B/dt_{\text{secondary}} = \text{Const. } f_1 (\theta_c, \sigma_n) f_2 (\sigma_n) \qquad (6)$$

As the length of the wear 'land' increases to relatively high values the temperature gradient down the wear 'land' will decrease and more and more primary particles will be formed at a greater distance from the tool edge, additionally the same phenomenon will facilitate the formation of secondary wear, i.e. an avalanching effect will bring about the tertiary region of the tool wear land–time relationship.

It is evident, therefore, that flank wear of cutting tools is influenced by both temperature and the normal stress (σ_n) which act on the flank face of the cutting tool. (This point has been demonstrated by Chao and Trigger[6] for the case of the rake face. They showed that a reasonable correlation exists between wear rate, normal stress and interface temperature. Their analysis was based on the combination of simple adhesion wear and a rate process.)

It is now necessary to consider methods of assessing the values of stress and temperature which exist in the wearing regions.

DETERMINATION OF NORMAL FLANK STRESS AND CUTTING TEMPERATURE

Normal flank stress

While there have been several cases in the past of research workers measuring stresses on a cutting tool by means of photoelasticity, the technique is limited in its application, and theoretical methods of assessing the normal stress in the flank are required when several tests are envisaged, or when using material other than light alloys or pure low strength material, e.g. lead.

In view of this, an analysis of metal cutting was undertaken[7], which was aimed at obtaining reliable expressions for the contact length between chip and tool and the normal stress on the flank face at the tool edge.

The model of cutting makes assumptions regarding the distribution of the rake face stress, then by considering the well-known relationships which may be derived from force equilibrium and energy considerations in the primary and secondary deformation zones, show that it is possible to express the apparent coefficient of friction in the mode suggested by Zorev (8), i.e.

$$\mu_\gamma = \frac{\lambda_c \cos \gamma_o}{1 + a_e - \lambda_c \sin \gamma_o} \qquad (7)$$

The natural contact length between chip and tool can then be expressed as

$$KB_0 = \frac{h \, 2 \gamma_c \, (1 + a_e - \lambda_c \sin \gamma_o)}{a_e \cos \gamma_o \, (1 + \sin 2 \, (\phi_i - \gamma_o))} - \frac{1}{2 \cos \gamma_o}$$

$$(8)$$

The maximum normal stress on the rake face σ_0 is given by

$$\sigma_0 = K \, (1 + \sin 2 \, (\phi_i - \gamma_o)) \qquad (9)$$

It is obviously difficult to arrive at a temperature characterizing the conditions at the flank, however, as one may infer that the greatest influence of temperature on the process is in the facilitation of primary particle removal the authors came to the conclusion that the average rake face temperature should provide a reasonable approximation to the temperature acting at the tool edge.

While methods for measuring cutting temperatures were considered, it was felt that no available technique was acceptable for the use in mind since several results were required with a reasonable degree of accuracy. In view of this several methods of calculating cutting temperature were considered. The method suggested by Boothroyd[9] was chosen since it was recognized that in order to obtain reasonable results, the frictional heat source must be assumed to have finite thicknesses. Boothroyd found that, in general the thickness of the secondary deformation zone was approximately 0·2 of the chip thickness. This would also seem to be the case for nickel alloys (from photomicrographs obtained by Mutze[10]. This value has been used by the authors for all materials tested.

The average temperature at the chip tool interface can, therefore, be calculated as

$$\theta_c = \frac{K\gamma_c}{\rho C_s} \; (1 - \beta_1) + \frac{\alpha_1}{a_e} \qquad (11)$$

DETERMINATION OF TOOL LIFE-STRESS MODIFIED TEMPERATURE PLOT

In order to obtain a reasonable relationship between temperature, stress and tool life for one material it is advantageous to undertake tests on materials which give a reasonably wide range of stresses and temperatures. In order to obtain this, tests were conducted with one tool material and the following workpiece materials:

N901 (nickel alloy)
H50 (5 per cent Cr hot work die steel)
IM1 550 (titanium alloy)
Nimonic 115 (nickel alloy)
En40C (Ausformed Cr-Mo-Va nitriding steel)

It was found that under the conditions tested, temperatures between 2300°F and 2800°F and flank stresses between 5000 lbf/in² and 70 000 lbf/in² were obtained. Figure 2 shows the relationship between temperature–flank stress and tool life. Using Fig. 4 it is possible to obtain a reasonable relationship between flank stress and temperature for a constant tool life by drawing a series of straight lines. Whilst it is not certain at this stage whether the relationship between flank stress and temperature is linear, it is a reasonable approximation since the influence of flank

cutting temperature, θ_c °F

Figure 2. Cutting temperature–flank stress–tool life plot.

stress is much less than temperature as indicated in Fig. 1. (For example in Fig. 1 a change in flank stress from 30 000 to 54 000 lbf/in², i.e. 80 per cent is equivalent to a change in temperature from 2430°F to 2570°F, i.e. 5·8 per cent for a constant tool life of 20 minutes.)

Using Fig. 2 it is then possible to plot tool life against the stress modified temperature as shown in Figs 5 and 6. The tool life-stress modified temperature plot could be considered to indicate the 'basic wear resistance' of a tool material, in that the stress modified temperature for a given tool life could be compared for different tool materials. However, for assessing tool life on different workpiece materials or under different cutting conditions on one carbide it is better to construct a cutting temperature–flank stress–tool life plot as shown in Figs 3 and 4.

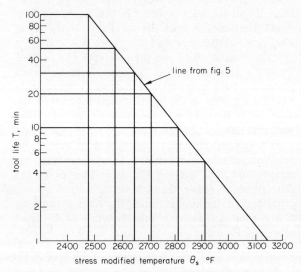

Figure 3. Construction of stress–temperature–tool life plot.

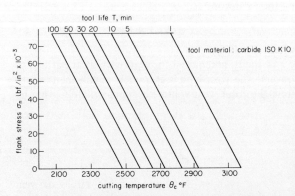

Figure 4. Cutting temperature–flank stress–tool life plot.

Having obtained a diagram of the form shown in Fig. 4 for one tool material the tool life can be assessed on any workpiece material and any cutting conditions from a simple force test. It should be realized however, that the technique will not be valid under conditions where built-up edge occurs and where the mode of tool wear changes drastically, e.g. plastic deformation and brittle failure.

It is reasonable to expect that different tool materials will exhibit different cutting temperature-flank stress–tool life characteristics, i.e. have different wear resistance. While this is undoubtedly the case, it would appear that cutting tools of similar composition exhibit similar wear characteristics when both stress and temperature is taken into account. This point is considered in Fig. 7 where tests on ISO K10 carbide from a different manufacturer are plotted together with the stress modified temperature line taken from Fig. 5. The results shown in Fig. 7

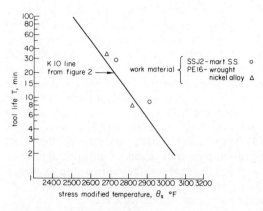

Figure 7. Comparison between line for carbide ISO K10 from manufacturer A and carbide ISO K10 from manufacturer B.

indicates that one tool life stress modified temperature line can be used for several 'similar' grades of tool material. However, if the grade of carbide is significantly different, e.g. an ISO P grade carbide, a completely different line must be used as indicated in Fig. 8. As shown in Fig. 8 the line for ISO PO5 carbide

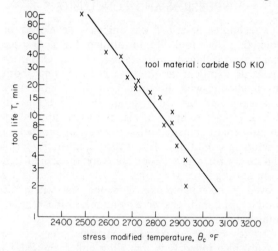

Figure 5. Tool life against stress modified temperature.

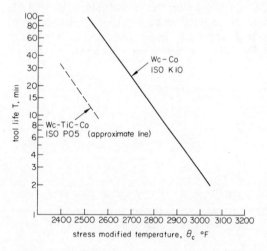

Figure 8. Comparison between the modified temperature-tool life relationship of Wc-Co and Wc-TiC-Co Carbide Grades.

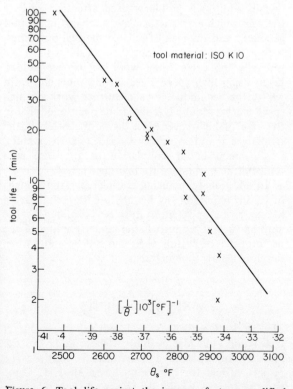

Figure 6. Tool life against the inverse of stress modified temperature.

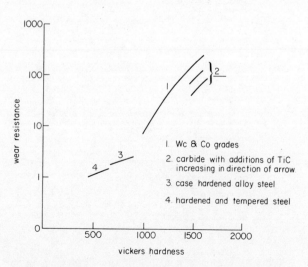

Figure 9. Relative wear resistance as a function of vickers hardness.

of expressing the wear resistance of cutting tools than comparing results under standardized machining conditions, e.g. constant speed, feed, etc.

TEST PROCEDURE

In order to establish the cutting temperature–flank stress–tool life plot (Fig. 4) it is, of course, necessary to undertake several conventional tool life tests using one tool material and preferably several workpiece materials. It is normal procedure of the authors to use the concept of equivalent chip thickness and to plot the results in the form of a V-T-h_e plot[12]. The use of the equivalent chip thickness enables the tool life tests to be undertaken at any speed, feed, depth of cut, tool approach angle and tool nose radius, provided the corresponding cutting force test is

```
work material INC 718
tool material ISO K 10

?00.00
*E
*G
: 100
: 0.0053
: 5
: 426
: 220
: 2.26
: 0.150
: 0.032
: 7.0
shear angle is = 24.7034 coeff friction is = 0.6321
equiv shear strain is = 2.5383 energy ratio is=2.7581
shear yield stress is = 146632.00
contact length is = 0.0198
sticking length is = 0.0102
: 146632
: 2.7581
max norm stress is = 239479.00
max shear stress is = 113492.00
max norm stress (flank) is = 35538.700
thermal no is = 22.1440
beta par is = 10.1514
beta is = 0.9056
alpha par is = 14.1875
alpha is = 3.0141
int temp is = 2485.9400
*E
*G
: 150
: 0.0053
: 5
: 399
: 180
: 2.10
: 0.150
: 0.032
: 7.0
shear angle is = 26.4097 coeff friction is = 0.5604
equiv shear strain is = 2.4115 energy ratio is = 2.9157
shear yield stress is = 154969.00
contact length is = 0.0169
sticking length is = 0.0080
: 154969
: 2.92
max norm stress is = 260040.00
max shear stress is = 113910.00
max norm stress (flank) is = 31763.900
thermal no is = 31.4366
beta par is = 15.5568
beta is = 0.9352
alpha par is = 20.6801
alpha is = 3.3721
int temp is = 2610.5400
*
```

Figure 10. Computer print out of stress and temperature calculations.

nents of the cutting force F_v and F_f and the chip compression ratio λ_c, from which the cutting temperature and normal stress can be calculated using the appropriate equations. In view of the number of calculations required a computer program has been developed (a typical print out is shown in Fig. 10).

DISCUSSION AND CONCLUSIONS

The technique described was originally developed for predicting the tool life in turning for aero-space materials machined with carbide tools. On these materials excellent results have been obtained both for predicting what life to expect under laboratory conditions and also under production conditions. In the latter case samples of material were supplied by an industrial company with a request for production machining conditions. The cutting conditions (speed and feed) for a given tool life were then predicted from cutting force tests only[13]. The predicted values proved to be reliable except in one case where the production conditions were by no means ideal (heavy interrupted cutting).

While other workpiece materials than those already mentioned (e.g. low alloy and carbon steel) have been used, it would appear at this stage that certain additional problems need attention such as the role which crater wear plays in changing the tool geometry and the lack of reliable thermal properties, before the technique can be considered to be generally applicable. Some work has also been undertaken using high speed steel cutting tools and again whilst the technique is valid further work on high speed steel is required. It should be mentioned that in the case of high speed steel the normal stress in the flank face is not as influential as it is with carbides, i.e. a much better correlation of tool life with temperature is obtained for high speed steel.

In addition to reducing considerably the amount of testing required in evaluating tool life, relating tool life to stress and temperature is considered by the authors to give a much better understanding of the performance of cutting tools and can be useful when developing new tool materials.

ACKNOWLEDGMENTS

The work described formed part of the contract 'Prediction of Machinability' sponsored by the Ministry of Defence Procurement Executive. The project officer for the contract was Mr J. L. Caine and the authors would like to record their thanks to him for his continued interest and encouragement.

Considerable help was given by Rolls Royce (1971) Ltd. and in this connection particular thanks go to Mr B. Pegg, Manufacturing Methods Development Department, Derby, for his support and many valuable discussions.

REFERENCES

1. F. W. TAYLOR (1970). On the art of cutting metal, *Trans. A.S.M.E.*

2. H. SCHALLBROCH, H. SCHAUMANN and R. WALLICHS (1938). Testing for machinability by measuring cutting temperature and tool wear, *Vort. Dtsch. Ges. Metalk.*, 4.

3. H. TAKEYAMA (1967). Wear of cutting tools, *Bull. J.S.P.E.*, **2** (3).

4. H. OPITZ and W. KONIG (1967). Basic research on the wear of cutting tools, *Machinability I.S.I. Special Report*, No. 94.

5. A. BHATTACHARYYA and A. GHOSH (1964). Diffusion wear of cutting tools, *Proc. 5th M.T.D.R. Conference.*

6. K. J. TRIGGER and B. T. CHAO (1956). The mechanism of crater wear of cemented carbide tools, *Trans. A.S.M.E.* **78**, July.

7. I. YELLOWLEY and G. BARROW The evaluation of contact length in orthogonal metal cutting. To be published.

8. N. ZOREV (1966). *Metal Cutting Mechanics*, Pergamon Press.

9. G. BOOTHROYD (1963). Temperature in orthogonal metal cutting, *Proc. I. Mech. E.*, No. 177.

10. H. MUTZE (1967). Dissertation T. H. Aachen. (English translation B.I.S.I. 6132.)

11. Sandvik Ltd. *Cemented Carbide as Tool and Design Material.*

12. I. YELLOWLEY and G. BARROW. The assessment and application of machinability data, *Fertigung* 1/71.

13. I. YELLOWLEY. Stress temperature exercise carried out at U.M.I.S.T. for Rolls Royce (1971) Ltd., Unpublished Report, May 1971.

DISCUSSION

Q. J. Tlusty. How do you determine the stess on the flank? I believe that you determine it for the tip of the tool and derive it from the stress on the rake side of this tip. Does this apply to the whole flank wear area?

A. The normal stress on the flank is estimated from the Mohr circle at the tool tip, which is derived by a consideration of rake face stresses.

Whilst the authors would concede that this is an approximation (i.e. they do not imagine that the normal stress acting over the flank will be constant and equal to the calculated value). It is evident that if the theoretical value of stress gives a guide to relative stress levels on various materials then the empirical procedure employed to determine the stress modified temperature—tool life plot will ensure that valid results will be obtained when using this plot for the prediction of tool life.

Q. A. H. Redford, University of Salford. (1) Has the stress-temperature method been applied to more common workpiece materials and if so was reasonable correlation obtained?

(2) Why do the results obtained using 'sharp' tools give a reasonable estimate of tool life when for most of a tool's useful life the tool is not sharp?

(3) How can the method be used to estimate tool life when some form of chip breaking device is used?

(4) For the exceptional circumstances where under practical conditions the radial force is very high, does the method give a reasonable prediction of tool life.

A. (1) As mentioned in the paper, work has been carried out on low alloy and carbon steels. The results on these materials have been encouraging although the scatter of life values has been larger than those presented in the paper. (In some cases discrepancies of up to 4 to 1 have been found.) The authors would attribute most of this error to the lack of reliable thermal properties in these materials.

(2) In the authors' opinion the results imply the following:

a. The initial wear rate of the tool is a unique function of the initial temperature and stress.

b. The tool life is a function of the initial wear rate, the initial temperature and the initial stress (i.e. the shape of the tool wear curve is dependent on the initial conditions of stress and temperature).

(3) The method has not been applied to tools with chip breakers. In the opinion of the authors, however, it would be possible to predict the life of tools using 'sintered in' chip control grooves, however, as one would normally expect the life of such tools to be slightly higher than that for a tool with a plane rake face then the effort required may not be justified.

(4) The reasoning behind this question is not at all clear, it would appear that the discusser is associating increasing radial force with departure from orthogonal cutting conditions, this is obviously fallacious. It should be possible, using the method to predict tool life for all practical conditions because the method assumes that the equivalent chip thickness in the practical situation results in the same values of temperature and stress as has been found for an equal uniform feedrate in the force tests. The validity of the method thus depends on the validity of the concept of equivalent chip thickness when complex tool shapes are considered.

Q. G. A. Wood, Wickman Wimet. In the paper test inserts were used from different manufacturers and grouped together according to their I.S.O. classification.

It is important to note that the I.S.O. classification is based on applications and does not define hardmetal grades. Inevitably different manufacturers will use different compositions and grain sizes to meet the needs of the the range of applications within each group, based upon their own experience. In these circumstances, a PO5 insert for example, from one manufacturer may differ sufficiently in grain size and composition, from a PO5 insert supplied by another manufacturer to have a measurable effect on any test results obtained.

EXPERIMENTAL ANALYSIS OF THE CORRELATION BETWEEN CUTTING FORCES VARIATION WITH TIME AND CUTTING DATA

by

R. IPPOLITO*, G. F. MICHELETTI* and R. VILENCHICH†

INTRODUCTION

A systematic study to correlate tool wear with cutting forces and to utilize force measurement as an indirect measure of cutting edge wear, has been continued for many years at the Institute of Mechanical Technology of Turin Polytechnic. The following remarks were derived from the results obtained[2,3,5,8,9]:

Cutting forces change very markedly with time (up to 100 per cent); these changes are attributable to cutting edge wear.

In modern carbide tips, both back and crater wear of the tool generally increase cutting forces.

The effects of different wear on cutting forces cannot in general be separated.

When the tool is close to collapse (very advanced wear) the force versus time ratio is flattened; an inverse ratio is sometimes observed.

Forces must be measured with great accuracy to prevent accidental variations of other parameters (such as machining stock and hardness) from being interpreted as variations due to cutting edge wear. Continuous recording allows accurate control of random phenomena.

When cutting parameters and tool wear change, the law by which force changes with time obviously changes.

Because this law is represented by a straight line, to a good degree of approximation[2,3,5,8], when cutting conditions change, the slope of the line changes. Its constant will also change of course, but this is not known and has no relevance to the problem under review. This slope is defined as the 'Rate of Variation' of cutting force. It is therefore a function of overall tool wear.

If a concept of *equivalent wear* is introduced as follows: 'two tools shall have equivalent wear when cutting force changes by the same amount', variations of cutting forces can be correlated with Taylor's Law.

A value ΔF of increased cutting force can in fact be taken as a criterion of tool life. Recent tests[6] in cooperation with Aachen support this assumption by showing good correlation between increase ΔF of cutting force, in a defined time interval and under given cutting conditions, and machinability of a steel.

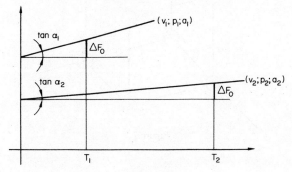

Figure 1.

From this hypothesis it follows (Fig. 1) that:

$$T_1 \tan \alpha_1 = T_2 \tan \alpha_2$$

whence

$$\frac{T_1}{T_2} = \frac{\tan \alpha_2}{\tan \alpha_1} = \left(\frac{v_2}{v_1}\right)^{1/n}\left(\frac{p_2}{p_1}\right)^{y/n}\left(\frac{a_2}{a_1}\right)^{x/n} \quad (1)$$

having introduced the well-known generalized law:

$$T^n = \frac{C}{vp^y a^x}$$

From (1) above it can immediately be deduced that:

$$\tan \alpha = \frac{dF}{dt} = C_0 v^q p^r a^s \quad (2)$$

where $q = 1/n; r = y/n; s = x/n$

Equation (2) can be made linear by taking the logarithms of both members as follows:

$$\ln \tan \alpha = \ln C_0 + q \ln v + r \ln p + s \ln a \quad (3)$$

Equation (3) makes it possible to obtain the response surface by using the multiple linear regression method.

---

* Politecnico di Torino, Istituto di Tecnologia Meccanica, Torino, Italy
† Windsor University, Windsor, Ontario, Canada

constant time intervals. These values, processed by the linear regression method have made it possible to evaluate the equation of straight line:

$$F = C_1 + C_2 t$$

with C_1 being a constant. C_2 being $\tan \alpha$, the slope.

compares experimental data with values obtained by using the mathematical model (data corresponding to dF/dt). Figures 4 and 5 show the calculated trend of dF/dt as a function of variables v, a and p (the latter being indicated as a parameter for each surface).

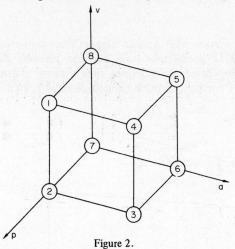

Figure 2.

TABLE 1

No.	a (mm/rev)	p (mm)	v_t (m/min)	Other conditions
				Workpiece-material
1	0.142	2	300	UNI 34 Cr Mo 4
2	0.142	2	200	Tool-material
3	0.285	2	200	(1) Carbide P 35
4	0.285	2	300	coated with
5	0.285	1	300	titanium
6	0.285	1	200	(2) Carbide P 10
7	0.142	1	200	Tool rake-angle
8	0.142	1	300	$\gamma = +6°$
				Type of cut:
				orthogonal

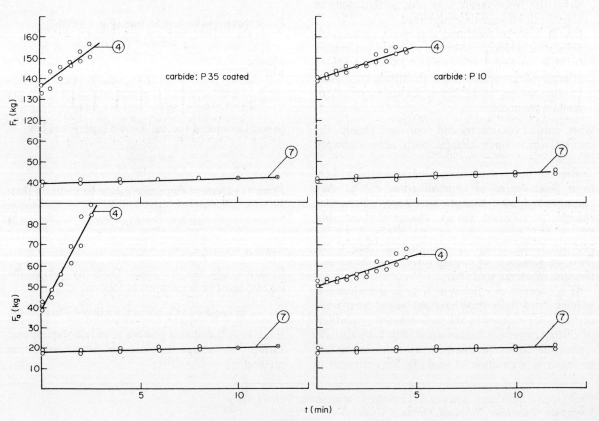

Figure 3.

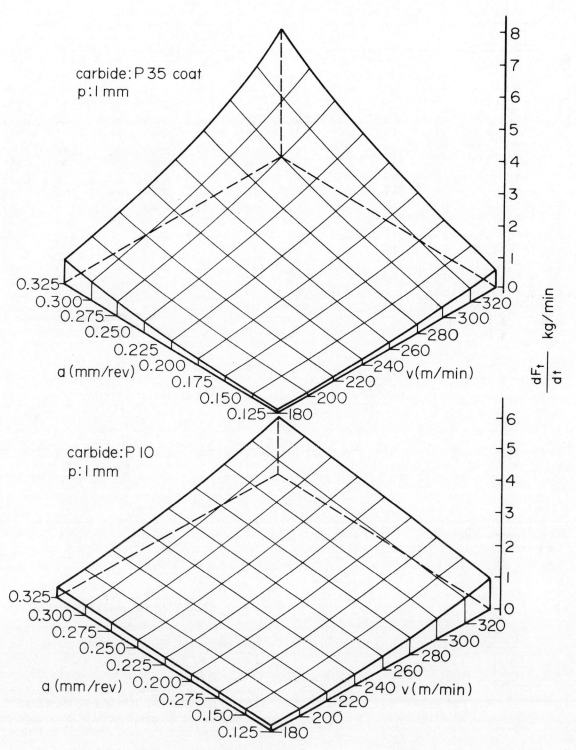

Figure 4.

3	127·5	0·75	0·94	44·9	0·54	0·99	134	1·07	0·973	50·5	0·54	0·985
	123·8	1·44	0·95	42·2	0·8	0·983	132	0·65	0·980	51·5	0·38	0·981
4	120·9	10·63	0·993	39·9	15·9	0·983	130	3·43	0·970	50	2·53	0·912
	129·4	4·62	0·912	33·2	25	0·980	129	3·22	0·989	49	3·95	0·968
5	66·2	1·72	0·983	19·8	3·5	0·996	68	1·17	0·953	21	1·25	0·941
	63·8	1·64	0·988	18·5	2·5	0·996	67·5	2·25	0·970	19·5	2·75	0·97
6	66·5	0·69	0·99	22·5	0·51	0·99	70	0·45	0·949	25	0·21	0·983
	67	0·67	0·975	21·5	0·73	0·99	69·5	0·35	0·940	23·5	0·21	0·980
7	38·5	0·31	0·97	18·2	0·19	0·90	41	0·30	0·981	17	0·17	0·969
	40·5	0·14	0·88	17·2	0·21	0·98	42·5	0·21	0·979	19·5	0·12	0·938
8	36·2	0·65	0·979	16·5	0·37	0·976	43	0·39	0·938	20·5	0·19	0·98
	38·9	0·36	0·924	17·4	0·25	0·978	41	0·54	0·985	20	0·16	0·93

TABLE 3

	Carbide P 35 coated						Carbide P 10					
	C_0		q	r	s	M.C.	F	C_0		q	r	s	M.C.	F
$\dfrac{\mathrm{d}F_t}{\mathrm{d}t} = C_0 V^q a^r p^s$	4·07	10^{-6}	2·77	2·10	0·973	0·94	28·82	4·45	10^{-7}	2·78	0·70	1·60	0·94	29·17
$\dfrac{\mathrm{d}F_a}{\mathrm{d}t} = C_0 V^q a^r p^s$	2·39	10^{-9}	4·37	2·94	0·92	0·92	20·52	2·77	10^{-9}	3·76	1·33	1·58	0·91	17·94

M.C. = multiple correlation
F = F value

TABLE 4

	Carbide P 35 coated		Carbide P 10	
No.	C_2 computed	C_2 exper.	C_2 computed	C_2 exper.
1	0·96	0·76	2·17	2·77
		0·71		2·47
2	0·31	0·34	0·70	0·88
		0·30		0·83
3	1·36	0·75	1·15	1·07
		1·44		0·65
4	4·17	4·62	3·53	3·43
		10·63		3·22
5	2·12	1·72	1·17	1·17
		1·64		2·25
6	0·69	0·69	0·37	0·45
		0·67		0·35
7	0·16	0·14	0·23	0·30
		0·31		0·21
8	0·49	0·65	0·72	0·39
		0·36		0·59

ANALYSIS OF EXPERIMENTAL RESULTS

The analysis of results obtained enables the conclusion that good correlation exists between the theoretical law and experimental results. The mathematical model postulated can in other words explain the real phenomenon. It might be interesting to try and explain the difference found between coefficients of equations for the two materials and those for measured forces F_t and P_a.

Exponents for machining depth of cut p and feed a are tied to the exponent of cutting speed. This remark will be useful later on in the paper. The analysis of the two series of data can show first of all that in the case in point, Taylor's index corresponds with the inverse exponent of cutting velocity; this holds true for both materials tested, namely

0·365 . . . 0·230 for P 135

0·365 . . . 0·265 for P 10

The above values were obtained from the formula for $\mathrm{d}F_t/\mathrm{d}t$ and for $\mathrm{d}F_a/\mathrm{d}t$ respectively. These values are

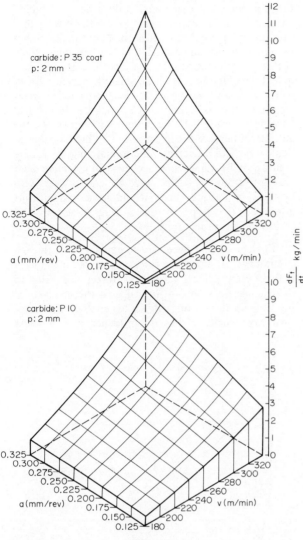

Figure 5.

Carbide P 135 under test conditions shows substantially different wear from that found traditionally; greatly reduced 'wear land', practically a non-existent crater, bevel formation on the main cutting edge which decreases the rake angle. The latter phenomenon reaches maximum values at higher feed. This means that: at low feed at which wear land formation predominates, increase is less than for Carbide P 10; at high feed, at which the edge is formed, variation is greater. This explains Carbide P 135 has greater sensitivity to feed.

The above findings make some conclusions possible, as follows: the experimental investigation on a defined combination of tool and machined material can allow mathematical relations which cannot be generalized and are only valid for combinations of materials under review; the only possible generalization is on the mathematical formula of the response surface.

Monte Carlo method was applied to the determination of independent variable with 10, 30, 50 and 70 per cent variation, and compared with the values obtained by RSM.

RSM provides relatively poor confidence limits for the independent variable, which is comparable only to 70 per cent variation obtained by Monte Carlo. Clearly, the combined method of RSM with Monte Carlo methodology provides much more meaningful confidence limits for practical purposes, because it takes into account the probabilistic nature of the machining variables. The combined method of RSM with Monte Carlo methodology has been successfully applied to other machining operations such as determination of tool-life in turning and milling and on G-ratio determination in grinding[11,12,13] from which optimum conditions can be assessed[14,15].

Computer programs connected to this work are not included for the sake of brevity but can be supplied on request.

CONCLUSIONS

The original hypothesis and mathematical model chosen to simulate the phenomenon's real behaviour are confirmed. There follows a significant step forward in utilizing measurement of cutting forces and their variations for measuring the degree of wear of a cutting edge.

It must however be pointed out that this method can be used advantageously in the absence of marked machining stock variations. These can be interpreted as alterations to cutting edge geometry. Such a measuring technique could be used in long finishing operations (such as turning and milling) which are in any case the only ones in which the use of cutting edge wear condition adaptive control is meaningful.

The opportunities as outlined in this article could enhance traditional adaptive techniques in which force measurement is used to change tool feed so as to compensate for variations in machining stock on the machined part.

In other terms, during finishing operations, the derived signal from the cutting force gauge will be an input for the adjusting block, acting on cutting speed or machining feed to keep the 'Rate of Variation' of the cutting force constant.

in good agreement with conventional data. Greater sensitivity to speed is however noted for feed forces. This is explained by remarking that as speed increases, tool crater wear phenomena also increase, which increases friction angle. Consequently, with reference to the classical shear plane theory, there is an increase of the angle $(\tau - \gamma)$ between resulting force and the main cutting component. This causes a more rapid increase of normal force.

In so far as individual materials are concerned, exponents x and y remain substantially constant for dF_t/dt and dF_a/dt. Variations of r and s are due to the variations of q mentioned earlier. Remarkable differences can however be noted between the two materials. Carbide P 10 was found to be more sensitive to depth of cut and less sensitive to feed than P 135.

The above is justified by the following two phenomena:

Carbide P 10 shows marked decrease of cutting edge; since this is to be considered as a decrease of depth of cut, it has a greater influence at low values of p so that variation of force is influenced with time. This phenomenon is practically absent when machining with a titanium plated tip. This justifies carbide P 10's greater sensitivity to depth of cut.

tool wear relationship for P10, P20, P30 Carbides, *CIRP General Assembly*, Nottingham, September.

4. A. DE FILIPPI and R. IPPOLITO (1968). Studio delle forze agenti nel piano dorsale dell'utensile nella tornitura ortogonale, *Macchine*, Milano, Giugno.

5. A. DE FILIPPI and R. IPPOLITO (1969). Valutazione dell'usura dell'utensile nella fresatura frontale per mezzo della forza di taglio, *Macchine*, Milano, Dic.

6. G. F. MICHELETTI (1970). Work on machinability in the co-operative group C of CIRP and outside this group, *Annals of the CIRP*, **18** (1), 13-30.

7. G. F. MICHELETTI, DE FILIPPI and R. IPPOLITO (1967). Détermination de l'indice d'usinabilité par une corrélation entre les forces de coupe et l'usure de l'outil, *Rapport C.E.C.A.* Gennaio.

8. A. DE FILIPPI and R. IPPOLITO (1972). Analysis of the correlation among: cutting force variation (vs. time), chip formation parameters, machinability, *CIRP General Assembly*, Stockholm.

9. R. IPPOLITO (1973). Analisi della correlazione tra la variazione delle forze di taglio nel tempo ed i parametri di formazione del truciolo, *Macchine*, Milano, Gennaio.

10. A. DE FILIPPI (1972), Analisi della correlazione tra incremento della forza di taglio e lavorabilità, *Macchine*, no. 9.

11. R. VILENCHICH, K. STROBELE and R. VENTER (1972). Tool-life testing by response surface methodology coupled with a random strategy approach, Paper 10, *Proceedings of the MTDR Conference*, Birmingham, September.

12. G. F. MICHELETTI C. BOER and R. VILEN-CHICH (1973). A statistical model of Taylor's equation for tool-life and computer optimization of unit costs and production rates in front milling, *CIRP Annals*.

13. A. DE FILIPPI, R. VILENCHICH and G. F. MICHELETTI (1973). A statistical approach to grinding tests, *CIRP Annals*.

14. K. IWATA, Y. NUROTSU, T. IVATSUBO and S. FUJII (1972). A probabilistic approach to the determination of the optimum cutting conditions, *Journal of Engineering for Industry*, 1099-1107, November.

15. R. VENTER, R. VILENCHICH and M. DE MALHERBE (1973). Statistical techniques and their application to some manufacturing processes, *CIRP Annals*.

DISCUSSION

Q. A. H. Redford, University of Salford. It has been would not be valid since a change in effective wear criterion from say flank wear to crater wear with increase in cutting speed would have a significant effect on the slope of the cutting force-turn and thrust force-turn relationships.

A. First of all it is necessary to explain the influence of crater wear on cutting forces. Crater wear can have two different effects on cutting forces according to its distance from the cutting edge. If the distance 'x' Fig. 1 from the cutting edge is big enough, the crater wear rake angle 'φ' has no influence on the shear

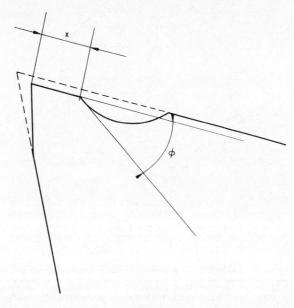

Fig. 1

zone so the only influence that the crater wear has on the chip formation is an increase in the contact length chip-tool and this phenomenon, according to many authors, produces an increase of the friction angle and consequently as cutting forces increase (this variation can be demonstrated by the variation of the F_t/F_a ratio with the time when the crater wear is present).

The explained condition is the more usual with the modern carbide tips and with normal values of the feed, working alloy steel. If the 'x' value becomes so small that the cutting zone 'feels' the crater wear rake angle, generally the cutting edge feals.

Working stainless-steel at very high cutting speed and with low feed value it is possible that the 'x' value is so small that the effective rake angle becomes φ so that it grows with time more quickly than friction angle and it is possible to obtain a cutting forces decrease with time.

We obtained similar results in the past working alloy steel with H.S.S. tools at high cutting speed.

Considering flank wear effect it can be said that in the condition previously described, its effects are smaller compared with those of crater wear.

We can conclude that, working in the standard condition on alloy steel, cutting force always increases with time till the end of the tool life.

METAL FORMING

COMPARISON OF GEAR PRODUCTION BY ROLLING WITH TRADITIONAL TECHNIQUES

by

A. N. NOWICKI*

SUMMARY

This paper deals with methods of forming of gears by rolling from a blank or finish rolling. Only processes which are on the verge of becoming a production proposition or are already in production are considered.

INTRODUCTION

Examination of gear cutting methods in the high volume industry shows that in most instances the following procedure is adopted:

(1) Gear blanks e.g., in the form of forgings, are prepared for gear cutting.

(2) Machined blanks are then loaded into Gear Hobbing machines or, in the case of shouldered gears, into Gear Shaping machines and are machined to within approx. 0·3 mm of finished size, when measured over pins.

(3) The gears are then transferred into Gear Shaving machines where they are finish machined.

(4) The gears are then heat treated.

(5) They may then be loaded into Gear Honing machines where nicks and burrs are removed. This process is optional and in a number of cases is not used.

Of the above mentioned only (2) and (3) are associated with the actual gear cutting process and it is these that gear rolling techniques are attempting to replace.

Gear rolling processes have received a great deal of attention, especially in the last decade. A large number of gear rolling methods were examined; however, of the total only a few processes were developed to the stage where it could be said that they are a practical production possibility and it is with these, that this paper is mainly concerned.

Forming of gears by rolling is a process whereby the tooth form is produced by displacement rather than by cutting away the unwanted metal and on average the reaction force and, therefore, distortion to which the rolling dies and the rolled component are subjected, is higher than is the case with conventional processes in current use. This distortion usually has to be compensated for by applying corrections to forming dies, hence of necessity the cost of dies tends to be greater than that of, say, shaving cutters. Since rolling dies tend to have a relatively long life span, in order to recover tooling costs, large quantities of components need to be produced. It is, therefore, suggested that gear rolling processes are most applicable to high gear production industries, e.g., the automobile industry.

HOT ROLLING OF GEARS FROM AN UNGASHED BLANK

To the writer's knowledge, the only hot rolling to full depth process that is in full industrial use has been developed in the USSR[1]. The machine consists basically of two dies which, in most cases, are located between side plates and surface induction heating equipment, as shown in Fig. 1. The principle of operation is as follows:

(1) A blank is positioned between the dies which are fully retracted.

(2) It is then induction heated to a temperature of 850–1200°C.

(3) With the rotational speed of the blank synchronized to that of the dies, either one or both dies are radially fed to a dead stop, thus producing the gear.

To allow for metal flow, the blank outside diameter is below that of the finished gear outside diameter, and the size is suitably chosen so that at the completion of rolling, the displaced metal fully fills the space enclosed by the die and the side plates, thus forming a gear which does not require subsequent machining.

Basically, the same process is used for hot forming of bevel gears. In this instance, however, only one die in the form of a bevel gear is used.

* Charles Churchill Ltd

induction preheater

Figure 1. Principle of hot rolling of gears from an ungashed blank.

It is claimed that gears produced by the hot rolling process have a 25 per cent increase in wear resistance and 15–20 per cent increase in tooth strength over gears produced by conventional methods.

Typically, 4000–5000 components can be produced per regrind and the dies can be reground up to four times.

Accuracies which can be achieved, together with typical required accuracies, are given in Table 1.

TABLE 1. Typical accuracies for hot rolled gears in the truck transmission range

Parameter	Typical req't (mm)	Hot rolled (mm)
Tooth alignment per flank	±0·034/100	0·035–0·05
Profile deviation from true form	0·022	0·055
Concentricity T.I.R.	0·063	0·1–0·15
Tooth thickness variation	0·045	0·06–0·09
Stock left for final machining per flank	0·1	0·1–0·15 of module

As can be seen, gears produced by the hot rolling process are far less accurate than are the present day requirements of, say, the truck industry. Due to the relatively large residual errors, it is necessary to leave excess stock of the order of 0·4 mm per flank on the gear. This stock is far above that which can be tolerated for gear shaving or gear rolling, thus prior to these an intermediate operation is required, e.g., finish hobbing.

It can, therefore, be seen that in cases where gears of low accuracy are required, e.g., for some mining or earth moving equipment applications, the process can prove to be economical; however, for higher quality gear production where in order to achieve the required accuracy additional machining operations are required, the process will prove to be uneconomical.

COLD ROLLING OF GEARS FROM AN UNGASHED BLANK

Although there are a number of processes which potentially are capable of producing finished gears in the range 1–1·5 mm module, to the writer's know-

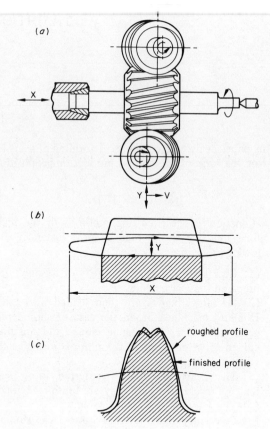

Figure 2. Principle of cold rolling of gears from an ungashed blank.

synchronized, the work is axially fed between the worms. At the end of each pass, the worms are slightly retracted while the work is returned to the start position. The worms are then fed in ready for the next pass, the cycle is indicated in Fig. 2(b). The ratio of axial work strokes per one revolution of the worms can be adjusted, hence the profile of the formed teeth can be controlled. During the roughing operation this ratio is 2, and the profile thus produced is inscribed by three lines as indicated in Fig. 2(c). For the finishing operation the ratio is automatically altered to a non integer number between 1·5 and 2·5 which has the effect of increasing the number of enveloping planes, which causes the generated profile to approach a true involute.

Although acceptable quality gears can be produced, it has been found that in order to achieve this, the blanks have to be machined to complex forms, thus increasing blank preparation costs. Field trials are currently being carried out to establish the economics of the process, and there are already indications that the cost of production is higher in relation to conventional methods.

FINISH ROLLING OF GEARS

Finish rolling of gears has now been talked of for a number of years and as far back as 1968 Ford Motor Company of USA have successfully finish rolled sun gears of approximately 1·4 mm module for their automatic transmission gearboxes.

It is only recently that reasonably successful results have been achieved when rolling gears for the manual change transmission gearboxes, the gears being in the 2–2·5 mm module range.

Single, twin and three die rolling techniques have been investigated, but lately most effort has gone into the development of single die and twin die rolling.

Although there are some basic differences between the single die and the twin die rolling processes, experiments have shown that the pattern of behaviour in each case is somewhat similar, thus many of the problems associated with controlling the process are common to both methods.

Single die finish gear rolling

This technique was developed in Europe[3] and is now well established in the Ford Europe plants.

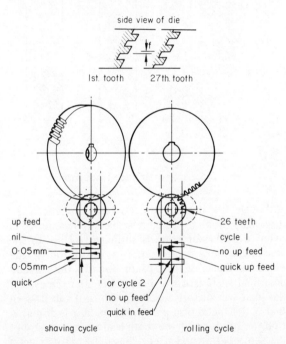

Figure 3. Principle of shaving and single die finish rolling of gears.

The rolling die often looks similar to a shaving cutter, the major difference being that in shaving the axis of the work and the shaving cutter are 'crossed', whereas in rolling the axis of the work is always parallel to the axis of the rolling die, as can be seen in Fig. 3. Both the shaving cutter and the rolling die teeth are serrated, thus when shaving, a combination of crossed axes and serrations results in the latter acting as cutting edges, whereas in rolling no cutting takes place and the whole purpose of the serrations is to reduce the contact area between the die and the work and, therefore, reduce the separating force between the two. This force is much greater when finish rolling than it is when shaving and this is due to the fact that in rolling line contact occurs between

the teeth on the work and the teeth on the die, whereas in shaving, due to the 'crossed axis' between the work and the shaving cutter, theoretically only point contact exists.

In order to achieve a satisfactory surface finish on the flanks of the rolled gear, great care has to be taken during the design and manufacture of the die. A correctly designed rolling die, as is also the case with shaving cutters, should have a number of teeth which do not have common factors with that of the workpiece, in which case every tooth on the die should work every tooth on the gear once only before the cycle is repeated. The serrations on the rolling die teeth have also to be arranged so that each successive time a tooth on the gear is rolled, the serration pattern moves axially a certain distance, usually referred to as the feed of serration per tooth, shown in Fig. 3 as f.

Serration feeds as low as 0·03 mm were tested and it was found that surface finish deteriorated with increase in serration feed; however, the lower the serration feed, the longer the cycle time required for completion of rolling. A satisfactory compromise between surface finish quality and cycle time has been achieved with the serration feed of 0·46 mm. Typically, a surface finish of the order of 0·3 μm is obtained.

Dies with varied ratios of land width to serration width were tested and it was found that this variant had little effect on the surface finish; however, as one would expect, increase in the serration width had the effect of reducing the separating load between the die and the work, shown in Fig. 4.

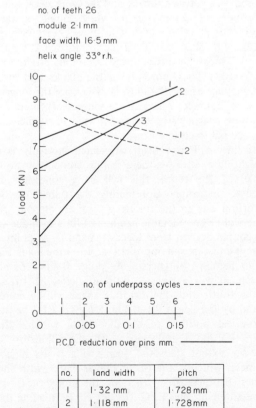

Figure 4. Effect of serration proportions and rolling passes on the separating force.

no.	land width	pitch
1	1·32 mm	1·728 mm
2	1·118 mm	1·728 mm
3	0·864 mm	1·728 mm

was investigated and the results are summarized in Fig. 4. It is of interest to note that the reduction in the separating force is minimal even after five feed passes. It can, therefore, be concluded that nearly all metal displacement occurs during the first feed pass; a very similar pattern was observed during shaving.[4]

As can be seen from Fig. 4, for a typical gear of approx. 2·1 mm module, the separating force during single die rolling process can be of the order of 9 kN, whereas, for the same type of gear during shaving, the corresponding force is only approx. 0·9 kN at 0·025 mm depth of cut[4].

The relatively large rolling force has a marked effect on the accuracy of the finished product and in practice it is necessary to grind complex corrections into the die in order to achieve leads and profiles of acceptable values.

In Fig. 5(a) is shown a helical gear tooth which is in mesh with the rolling die, rotating in the direction shown. Considering contact between one pair of teeth, L and T indicate the leading and the trailing points of contact respectively and if there is zero crossed axis between the die and the workpiece, these are the only conditions in which point contact occurs; in all other conditions line contact occurs. Checking a rolled gear it will be found that at points L and T most metal is removed. As rolling progresses, the contact lines move down the flanks and at a certain stage are A–B and C–D, which are the longest lines of contact. Along these lines least metal is removed, particularly points A and C, which in practice fail to clean up.

If line APB is considered, which normally is a straight line, because there is line contact on both flanks of the tooth, this will be a condition of heaviest distortion. Considering the purely elastic distortion due to the rolling first of all, at point A, there will be considerable bending of the tooth due to the couple derived from loads at A and D. At point P there is an equal and opposite load from the load at Y so no bending will occur. At point B little or no movement will take place, this point being the stiffest of the tooth, thus the distorted shape will be represented by the line A'PB in Fig. 5(b). Let EFG be the theoretical position of the die tooth, x being the theoretical metal displacement. The die will distort to a lesser extent than the gear owing to its higher stiffness, which is due to the longer flank width and higher hardness than that of the rolled gear, as represented by line EFG'. Under elastic loading the gear will take the form of the die, i.e., EFG' which will become G'FA" as soon as the load is removed. As can be seen, distance A'E is small compared to x, which explains why the corner A fails to clean up.

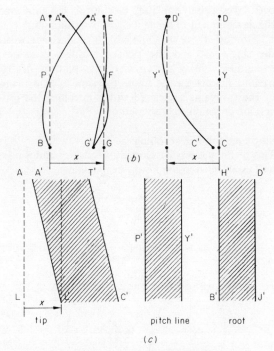

Figure 5. Analysis of single tooth error.

Although the line of contact does not strictly go diagonally from tip to root as has been assumed, in practice this assumption is sufficiently close to the true behaviour. Consideration of distortion of the opposite flank of the tooth follows the already described analysis where flank CYD after removal of the load will distort to C'Y'D'. Thus, it can be seen that if the theoretical amount of metal x is displaced from points L and T the tooth lead trace at the tip of the tooth before and after rolling will approximate to Fig. 5(c). When contact is at a point such as P and Y, which are on the pitch line, there will always be an equal and opposite reaction from the contact on the other flank of the tooth, preventing any bending moment distorting the final shape and an amount x is removed at all points. Similarly, because the root of a tooth is stiff compared with other points, there is little deviation from the theoretical displacement. In practice points towards L and T have progressively more metal removed probably due to increased pressure as the area of contact is reduced. In addition there is a tendency to 'fall off' at the edges of the tooth due to metal extrusion on the unsupported parts of tooth A–H and C–J shown in Fig. 5(a). From the above analysis it is apparent that in order to achieve satisfactory tooth lead graphs it is necessary to correct for 'single tooth errors' or as it is sometimes called 'tooth twist'. This correction can be achieved either by grinding into the rolling die a certain amount of 'tooth twist' or, alternatively, the gears to be rolled can be deliberately hobbed off lead. Whichever method is used, it is essential to strictly control the amount of lead error during hobbing in order to achieve a repeatable final result.

During investigations carried out by Ford Motor Company[5] on twin die semi-rough rolling of gears, it was found that the flow of metal on the flanks of a rolled tooth assumes a complex pattern. However, when finish rolled gears with approx. 0·18 mm stock removal on diameter over pins were examined, no evidence of either work hardening or complex metal flow could be detected, i.e., neither etching in nitral nor macrohardness survey proved positive. This suggests that if there is any work hardening effect, the hardened layer must be less than, say, 2 μm. Although the metallurgical tests were unsuccessful, profile graph checks of a finish rolled gear, when rolled with an uncorrected die, with single direction of rotation, clearly indicated a deposit of metal near the meshing line on the driven flank and a hollow on the coasting flank, which confirm the Ford findings.[5] It can thus be seen that in order to achieve straight and similar profiles on both flanks of a tooth, complex profile corrections may have to be ground into the rolling die, in addition corrections to the profile of the die may be different on each flank. The unbalance between the two profiles exists irrespective of whether single direction of the die rotation is used or reversal of rotation half way through the rolling cycle occurs. This behaviour can be explained by the fact that most of the metal flow takes place in the first pass and when the die rotation is reversed for the return pass, only partial balancing in profiles occurs.

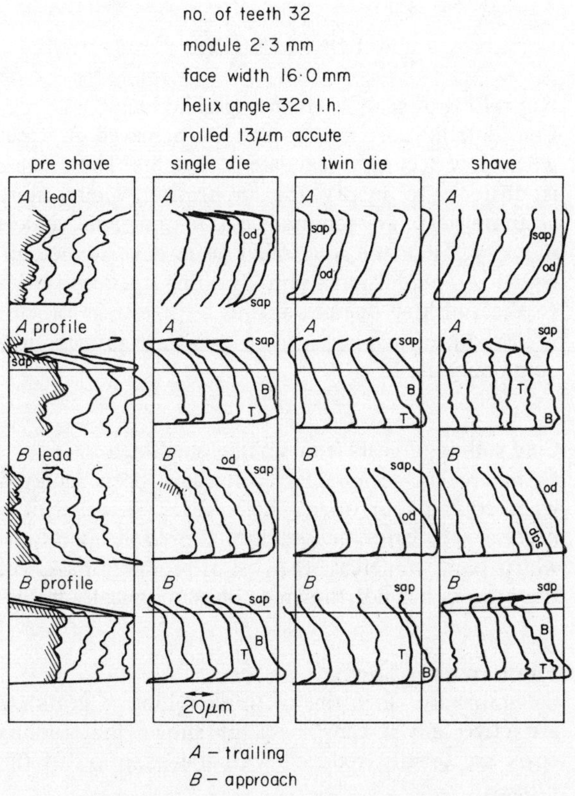

Tests with reversal of die rotation also produced much larger scatter in the profile results than did the tests with single direction of rotation of the die.

In practice, die tooth twist can be readily ground, e.g., on a generating type of gear grinder this can be achieved by sideways offset of the pitch block. This has the effect of constantly changing the base diameter thus progressively changing the profile of the ground tooth. Complicated profile corrections present a greater problem and investigations are still being carried out to resolve this problem.

In Fig. 6 are shown typical lead and profile graphs for a 2·3 mm module passenger car gear after shaving, single die finish rolling and twin die finish rolling.

As already mentioned, gear rolling techniques are most applicable to large through flow industries and, therefore, for a new process to be generally accepted, basically two conditions have to be met: firstly, the quality of the finished product has to be at least as good and preferably better than that presently achieved; secondly, the cost of manufacture has to be equal to or lesser than the existing cost. In the table below is shown a typical tooling cost comparison of shaving versus single die finish rolling and as the results show, by changing to finish rolling considerable savings can be achieved.

TABLE 2. Tool cost comparison for shaving and single die finish gear rolling

	Shaving	Rolling
Cost per tool	£75	£100
Pieces per grind	2000	25 000
Number of grinds	10	3
Hours per grind	4·8	22
Minutes per piece grind	0·1440	0·0528
Grind cost per piece (£2 per hour)	0·48 p	0·18p
Tool life (pieces per tool)	20 000	75 000
Tool cost per piece	0·37p	0·13p
Total tool cost per piece (tool and grind cost)	0·85p	0·31p

Additional savings can be obtained due to the rolling process being faster, for example

A maindrive gear, shaving time 35 s, is finish rolled in 9 s.

An idler gear, shaving time 40 s, is finish rolled in 8 s.

An intermediate gear, shaving time 45 s, is finish rolled in 10 s.

On the basis of the above times, it is estimated that if initially fifty-six shaving machines were required to achieve the required output, the same output can be achieved with only twenty-two single die finish rolling machines, which in turn results in approximately 300 m² saving in shop floor area.

Twin die finish gear rolling

Although single die finish rolling of gears has many advantages, it also has a number of disadvantages, e.g., the rolling die is serrated, hence can easily be damaged in use due to either excessive loads or careless handling; the rolling load has to be directly

no. of teeth 32
module 2·3 mm
face width 16·0 mm
helix angle 32° l.h.
rolled 13μm accute

A — trailing
B — approach

Figure 6. Typical lead and profile graphs.

as closely as is practically possible, in order to eliminate side loads on the work mounting spindle.

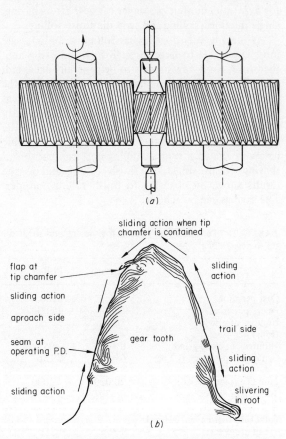

(a)

sliding action when tip chamfer is contained

flap at tip chamfer

sliding action

aproach side

seam at operating P.D.

sliding action

gear tooth

sliding action

sliding action

trail side

sliding action

slivering in root

(b)

Figure 7. Principle of twin die finish rolling of gears.

Normally unserrated dies are used, although in practice there is no reason why serrated dies, or a combination of the two, cannot be utilized. However, it should be remembered that serrated dies may be more expensive to produce; in addition the serrations may create stress raisers, which in turn will reduce the life of the dies.

As previously mentioned, with reference to single die rolling, the twin die rolling process also causes uneven flow of metal on the opposing flanks of a tooth and a typical pattern[5] is shown in Fig. 7(b). The flow pattern is most apparent with increased metal displacement and is hardly visible with stock removal over pins of, say, 0·18 mm.

Because unserrated dies are commonly used, the separation force between the work and the dies is greater than is the case with single die finish gear rolling and a typical separating force for an approx. 2·3 mm module of 16 mm face width gear can be of the order of 27 kN for approx. 0·18 mm over pins stock removal.

is not possible to correct the dies accordingly.

An additional problem with large stock removal, which is indicated in Fig. 7(b), is that of metal overlap. On the driven side, as the metal is rolled towards the meshing contact point the two 'streams' of flow can overlap and form a stress raising point. Similarly, an overlap can occur in the root of the trailing flank as metal is forced into the fillet. The latter of the two can create more serious problems, since the stress raising fault occurs at the maximum stress point on the tooth.

Die life is very long and when rolling 20 teeth, 20° helical pinion gears of 1·37 mm module and 22 mm face width outputs as high as 1 250 000 components per regrind have been achieved[6]. Thus for the full life of a set of dies 2 500 000 components can be finish rolled, as against approximately 80 000 components per corresponding shaving cutter life. Tool cost per piece is claimed to be much lower for finish rolling, e.g., 0·15 cents against 2·5 cents for shaving.

As yet, there is no cost information available for twin die finish rolling of gears in the 2·5 mm module range, since it is only recently that reasonably acceptable lead and profile accuracies have been achieved; typical results are shown in Fig. 6.

CONCLUSIONS

Hot rolling of gears from an ungashed blank

Considerable cost savings can be achieved in cases where low accuracy gears are acceptable. The process is difficult to justify for application where more accurate gears are required, e.g., commercial vehicle industry. Here the hot rolled gears have to be subjected to an additional semi-finishing operation prior to the finishing operation, thus a three-stage process is introduced in place of the two-stage process now commonly used.

Cold rolling of gears from an ungashed blank

Commercially acceptable results have been achieved when rolling spur or low helix gears. However, it is necessary to carry out additional pre-roll operations which tend to increase the cost of production. As yet the process has not shown to be economically viable.

Finish rolling of gears

Economically, all forms of finish rolling of gears are attractive and it can be readily shown that tooling costs are greatly reduced with the adoption of the process.

There are, however, areas where certain doubts

may still exist, e.g., it is possible to produce better quality tooth leads and profiles when shaving than when finish rolling.

ACKNOWLEDGMENTS

The writer wishes to thank the Ford Motor Company and in particular the Transmission Plant at Halewood for their cooperation by providing some of the information presented in this paper.

REFERENCES

1. U.K. Patent Specification 1134450 (7 October 1966), 'VNIIMETMASH'—U.S.S.R.
2. U.K. Patent Specification 1293412 (18 October 1972), MAAG Gear Wheel & Machine Co. Ltd.
3. U.K. Patent Specification 1207776 (7 October 1970), Carl Hurth Maschinen und Zahnradfabrik; and U.K. Patent Specification 1224140 (3 March 1971), Carl Hurth Maschinen und Zahnradfabrik.
4. D. B. WELBOURN and J. D. SMITH (1968). Summary of work on shaping hobbing and shaving, presentation to B.G.M.A. on 26 September.
5. U.K. Patent Specification 1164512 (17 September 1969), Ford Motor Co. Ltd.
6. Finish cold roll forming of gears at Ford Motor Company, (1971). Ford—U.S.S.R. technical exchange forum.

DISCUSSION

Q (1) W. T. Chesters, David Brown Gear Industries Ltd. I would query the claim that hot rolling increases tooth strength by as much as 15–20%. Work on the hot forging of coarse pitch teeth in very large gears has shown that although ideal metal flow was obtained, the increase in the bending fatigue strength of teeth, after casehardening and grinding, was only marginally improved, the gain being zero to 50 per cent.

Q (2) W. T. Chesters, David Brown Gear Industries Ltd. With the single die rolling process, there is a tendency for deformed metal to produce a lap in the centre of the tooth crest. In your illustration of the twin die formed tooth (Fig. 7) this is not shown. Is a lap formed but sealed and hidden by further plastic flow of metal?

A (1) The 15–20 per cent claimed increase in tooth strength refers to a comparison of a hot rolled gear with an equivalent finish hobbed gear.

It is true to say that if a hot rolled, or forged, gear is subsequently processed, e.g., hobbed or ground, the increase in the bending strength will be smaller than the quoted value.

A (2) Both single die and twin die rolling processes tend to produce a lap in the tooth crest.

During finish rolling, where the volume of the displaced metal is low, this lap is usually 'lost' in the tip chamfer, as would be in the illustrated case.

During rough rolling, with root relief in the dies, the lap is clearly visible at the tooth tip. When full form dies are used it is possible to seal this lap.

Q. Professor J. M. Alexander, Department of Mechanical Engineering, Imperial College. Would it be possible to have hot rolling followed by cold rolling of gears, perhaps using the same machine, thereby to achieve an economical production method?

A. Provided that acceptable accuracy hot rolled gears are produced, there is no reason why this approach cannot be adopted; however, present-day hot rolled gears appear to be outside the required pre-finish roll gear accuracy.

Q. N. R. Chitkara, Department of Mechanical Engineering, U.M.I.S.T., Manchester. Has the author sectioned any of the rolled gears to investigate whether there was any evidence of axial cracks or cavities being developed?

A. Two types of rolled gears were sectioned and tested for cracks etc:

(1) Finish cold rolled automobile gears of approximately 2·1 mm module with approximately 0·15 mm of displaced stock on diameter, when measured over pins, i.e., approximately 0·025 mm per flank.

(2) Rough cold rolled automobile gears of approximately 1·4 mm module, rolled from a pre-gashed blank where the gash area was approximately 50% of the finished gash.

In both cases no axial cracks were detected.

THE COLD ROLL FORMING OF SHAPES OF REVOLUTION WITH PARTICULAR REFERENCE TO BALL-BEARING INNER AND OUTER RACES

by

PAUL F. EGAN*

SUMMARY

This paper deals particularly with the cold roll forming of I.S.O. Ball Journal Bearings in EN31 steel. It explains the basic principles of both internal and external cold roll forming and illustrates how the process is being practically introduced into the bearing industry using available production equipment. The paper enumerates on various important production and economic considerations and concludes with references to other potential applications for the basic process of cold roll forming.

INTRODUCTION

The development of cold metal roll forming as later described in this paper very much involves new techniques, providing the Production Engineer with a new and exciting manufacturing technology. Detailed study of specific applications, particularly bearing race rolling, indicate substantial cost savings in material, production times and operator costs, when compared with conventional processes. Additionally, quality levels and physical property improvements are extra bonuses. More specifically, some of the main advantages of cold roll forming are listed as follows:

Some process advantages

(1) Higher production rates can be achieved when compared with conventional methods.

(2) Higher and more consistent quality can be expected from cold rolling when compared with conventional methods.

(3) An extremely high tool life can be expected from cold rolling.

(4) A higher utilization of material is inevitable by using rolling techniques.

(5) Rolling techniques give improved strength and improved metallurgical characteristics.

(6) There is a higher machine utilization, and a considerable reduction in floor space requirement.

(7) The capital investment is substantially reduced.

Anticipated further process advantages not taken into consideration in the above statement are as follows

(a) There should be less distortion after heat treatment.

(b) Less grinding stock allowance should be required.

The object of this paper is to outline the development of two particular cold rolling applications and show how it is possible to apply the basic techniques to a whole number of components suitable for cold rolling. Some typical components similar to those under consideration are illustrated in Fig. 1.

It took only a brief examination of the traditional methods of bearing race manufacture to appreciate that a roll forming operation could drastically reduce production times, improve the quality of the end product and significantly improve metal utilization. In fact detailed analysis suggested that in some cases up to 40 per cent of metal could be saved when compared to conventional processes. Subsequent practical experience has proved this point beyond doubt.

During the development of the process, particular attention was paid to the material used for the rolling dies. The problem was to strike the optimum balance between wear resistance and brittleness to give long die life. Much pioneering work has been completed in particular with research into heat treatment effects on the rolling dies. Also a great deal of work has been done in connection with the development of a suitable cold rolling lubricant which can withstand the relatively high rolling pressures and still maintain a barrier of lubricant between the bearing race and the rolling die which is forming it.

The paper is not meant to be a deep, technical treatise, but more an historical report of what has been practically achieved in the development of cold roll forming applied to bearing race production.

---

* Managing director Formflo Limited

Figure 1. Typical inner and outer bearing races.

BASIC PROCESS DESCRIPTION
OF THE COLD ROLL FORMING
OF BALL-BEARING INNER RACES

The general principles of inner race cold roll forming are illustrated in Fig. 2. An annular ring, which comprises the starting blank, is located in the core by a split mandrel. It is initially located axially between two sets of collapsible springs (omitted for clarity).

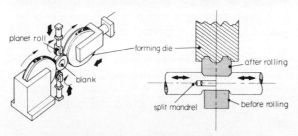

Figure 2. Basic principles inner race rolling.

The springs accurately locate the end faces and hold the starting blank in the correct position relative to the bore radius form on the split mandrel and the external forming dies (rolls). The split mandrel is capable of free rotation but is in no way actually power driven.

There are two identical main forming dies of approximately 10 inches in diameter. The dies are mounted on precision spindles and each spindle is driven rotationally by means of universal spindles from a gear box and electric motor drive. One of the spindle assemblies is sensibly fixed and is not capable of axial movement. The other spindle assembly is attached to a hydraulic cylinder, which provides the axial forming load and movement, relative to the fixed spindle. A double planet roll system is provided, which constrains undesirable excessive deformation of the blank during rolling. The planet rolls are capable of free rotation and are in no way power driven. Each of the two separate planet roll assemblies is attached to an independent hydraulic cylinder, such that a planet roll force can be applied simultaneously with the forming operation. The planet rolls do not substantially contribute to the actual forming operation.

The method of cold roll forming under discussion comprises loading a starting blank on to the split mandrel as previously described and initiating a rolling cycle. The right hand die, which is rotationally driven, is fed forward by means of the forming cylinder. The blank is eventually engaged and carried forward, together with the split mandrel assembly, until the other axially static, rotating forming die is reached. The blank and split mandrel now start to spin under the frictional forces applied. During this operation the planet roll cylinders are actuated under low pressure and engage the blank. The planet rolls are now normally rotating by means of the frictional forces applied. At this stage the blank is effectively constrained externally at four points. The planet roll pressure is switched to a high value at the appropriate point in time after the main forming dies start to deform the blank. The forming dies have an annular form equivalent to the external form being rolled and the split mandrel has a form equivalent to the bore radius configuration. During the actual rolling operation, the blank grows in both diameter and width. It is because of the latter effect that it is possible to simultaneously roll the internal form with the external form. After rolling, the bore of the finished bearing race is a very loose fit on the split mandrel and can easily be removed. At the end of the actual infeed of the forming cycle, the rolls are allowed to dwell for a short period before retraction, the forming rolls and planet rolls retract simultaneously, and the cycle repeats. During the process of cold forming a copious supply of lubricant is sprayed directly onto the part being rolled as well as the rolling dies.

STARTING BLANK CONSIDERATIONS FOR INNER BEARING RACE ROLLING

A typical inner race blank is shown with the ideal required tolerances in Fig. 3. The inner race, which is cold roll formed from this blank, is also shown. The

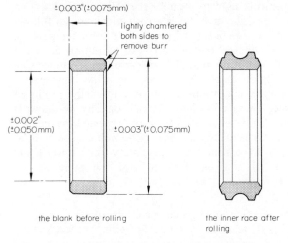

Figure 3. Typical inner race and starting blank.

form of the starting blank is an annular ring with the corners lightly chamfered to remove excessive burr. It is recommended that either hot rolled or cold reduced tube produced to normal standard commercial specification is used as a starting material. It is recommended that if used, the hot rolled tube is peeled on the outside diameter which simply leaves a bore sizing and part off operation to produce the blank. It is necessary to bore the tube, to remove eccentricity, which is usually inherent in either hot rolled or cold reduced tube. In the case of the former, it also assists in removing a decarburized layer, detrimental to subsequent processes. There are numerous ways of manufacturing the blank including multi-spindle automatics, which provides a very economical manufacturing process. Work has also been carried out using a single spindle machine in conjunction with a slitting saw, with very encouraging results. Interesting experimental programmes currently being evaluated include tube shearing, tube fatigue shearing and disc slicing.

Alternative material forms
It is possible to use other forms of starting material such as cold or hot formed rings or even bar stock. ERW tube is a possibility for certain applications. Most bearing manufacturers require that rings should be made in BSEN31 (SAE 52100) and most application work has been carried out on this material. However it is equally possible to roll most case hardening steels. Ideal materials for rolling exhibit low sulphur and phosphorous contents, and are usually fine grained with a ductile crystal structure exhibiting high per cent elongation and low work hardening properties.

Most of the application work so far completed on hot rolled or cold reduced tube has utilized a starting blank with an initial hardness of between 220 and 300 Vickers. There is not a significant change in hardness after rolling, when the correct process procedures are maintained. Ultimately it is thought that sinter forged rings may prove to be the ideal starting blank. However there is much work to be done on this particular aspect, both technically and commercially.

Process accuracy related to blank tolerance
It was found during the early experiments, that variations in the starting blank dimensions and hence variations in the blank volume, had a marked effect on the accuracy of the rolling process with common rolling machine settings, i.e. mechanical slide stops, rolling feed rate, dwell, etc. The initial solution to this problem involved using an electronics gauging system to measure the inside and outside diameter and thickness of the starting blank. The volume could then be automatically calculated and any variance from mean related to a mechanical stop position on the rolling machine. The stop was automatically set by means of a stepping motor, which received an output signal from the gauging/computer unit. The actual computation is not as straightforward as simple volume variations converted to a stop setting. Individual variations of each of the three basic blank dimensions had a different effect on the rolling diameter accuracy. A size variation on the outside diameter for instance had a different effect on equal size variation in the bore diameter. Similarly, an equal size variation in the width had another effect again. It was however, possible to determine the mathematical relationship between the variables and use it for the purpose previously described. Volumetric variations in the blank are to a large extent controlled into the rolled width with an equal spread either side of the centre line of the ring. This can be easily sized after heat treatment with a double disc grinding operation. The equipment to carry out the gauging and computation is necessarily a significant cost element of the total system and whilst fully developed and completely practical obviously adds to the general complication of what is a fairly sophisticated system.

It was decided to look for a more simple solution. It was found that by fitting a simple caliper type in process gauging device the problem was solved. When the ring grows to the correct size during the rolling process the forming infeed cylinder is stopped such that it never reaches the mechanical stop, it dwells and then retracts. This is a very simple, practical and inexpensive solution to a complex problem and is accurate to the limits specified under dimensional quality, if the blank tolerances previously stated are maintained.

QUALITY OF COLD ROLLED INNER BEARING RACES

Dimensional quality
It will be appreciated from the foregoing remarks regarding starting blank dimensional accuracy, that this aspect has an important effect on the dimensional accuracy of the finish rolled bearing inner race.

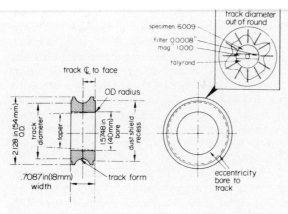

Figure 4. Nomenclature regarding inner race dimensional quality.

Typical 40 mm bore inner race

Bore	±0·0025 in (±0·062 mm)
Track	±0·0025 in (±0·062 mm)
O.D.	±0·0025 in (±0·062 mm)
Dust shield recess	±0·0025 in (±0·062 mm)
Width	The as-rolled thickness is a function of the starting blank volume. Most excess metal is displaced sideways equally about the track centre line. The excess metal, however, should be comfortably removed with a double disc grinding operation
Track to face difference	0·0025 in (0·062 mm)
Taper	0·0015 in max (0·037 mm)
Out of round	0·0025 in max (0·062 mm)
Eccentricity bore to track	0·002 in (0·050 mm)
Track form including dust shield recess	This is a function of roll design and should not alter once established correctly.

It is undesirable to generalize regarding quality levels and each application should be studied on its own merits. The specific example is an I.S.O. 40 mm bore inner race with a normal dust shield and bore radius/chamfer configuration. This application is quite representative of what can be expected, and comparable results could be achieved for geometrically similar inner races from 15 mm bore to, say, 60 mm bore, light, medium and heavy series (size of race cross section). The dimensional stability of the light series races needs extra care, as does the rolling of particular soft material—i.e. 220 Vickers, the rolling

axis of the tube. This means that during a machining operation used to produce the ball track and related features, the grain takes a path which almost follows the surface contour of the external form being rolled. From limited fatigue tests carried out to date, the indications are that this will result in the inner race having a longer life under the same comparative working conditions as a conventionally produced race. The assumption that can logically be drawn is that (a) the increased life can be expected as a quality improvement, (b) the bearing section can be reduced to give the same life as previously, (c) a lower grade material can be used to give the same life as previously.

Surface finish and grinding considerations
It is a feature of the cold rolling process under discussion that under normal conditions the as-rolled surface finish of the part being processed is usually almost identical to the surface finish of the rolling dies. It is usual in similar rolling configurations, to aim for a surface finish in the order of 5 RMS on the finish rolled component. There are numerous applications in the bearing industry for 'commercial bearings' (conveyor, office furniture, castors, etc.), where it is not necessary after finish rolling to grind the race. In most cases the elimination of grinding would not be possible with a turned race. By improving the surface finish on this type of bearing an extra bonus is gained. Features such as operating noise level, surface pitting and hence longer life are improved and additionally production costs are not so high. A typical, practical production example to illustrate this point is given later in this paper.

Distortion after heat treatment
Rolled races distort less after heat treatment than turned races. This is in part related to the previous subheading as far as elimination of grinding is concerned. This is a subject currently receiving considerable attention with the aim of eliminating most grinding requirements in due course. There is, however, much work to be carried out in this direction and at present it is necessary to grind normal I.S.O. bearings.

It is obvious that if rolled races distort less after heat treatment when compared to turned races, the possibility exists to reduce the required grinding allowances that are necessary to allow for cleaning up after heat treatment.

This has the effect of promoting a reduction in the grinding cycle time, the exact value of which is dependant of course upon the exact magnitude of the grinding allowance reduction. Equally important however is the effect that this smaller allowance has on grinding wheel wear. It should be possible to grind many more races within the required grinding tolerance prior to wheel redress. This will improve grinding quality and hence assist in one of the next stages of production, namely assembly. Selective assembly is normally used, which requires the stocking of numerous dimensional grades. The number of grades can be reduced if closer tolerances can be held, thus resulting in a substantial cost saving.

Consistency of quality

Another quality aspect worth considering is the question of consistency. As there is negligible rolling tool wear in the cold rolling process, very consistent dimensional characteristics are apparent as well as consistent surface finish. This is not so in the case of turning, where the tools dull and wear and have to be reground and replaced regularly.

Rolling tool considerations for cold rolling bearing inner races

It is undesirable to generalize regarding tool life, and the tool lives specified in this paper are always related to a particular example. Many factors influence tool life, including material being rolled and its condition, tool material of the rolling die and its condition. The type of cold rolling lubricant used can have a dramatic effect on tool life, as can the rate at which the actual rolling is performed. Under reasonably controlled conditions, however, an estimated rolling die life in the order of 750 000 parts should be comfortably achieved when considering an I.S.O. shieldless 40 mm bore inner race in BSEN31 steel. Depending upon the exact configuration of dust shields, an estimated rolling die life in the order of 500 000 parts should be comfortably achieved when rolling the equivalent bearing race with dust shields. In both cases the rolls could be reground at least six times. An important consideration of this long tool life is the effect on the overall process efficiency. It is not necessary to adjust the machine regularly for tool wear, and hence a high utilization is possible. As a bonus, higher, more consistent quality levels can be expected. Another important tooling factor is the simplicity and speed at which the rolling machine can be changed from one size of race to another. On average it takes less than one hour to change the tooling set up completely and be rolling another part.

Cycle times for cold rolling bearing inner races.

Once again it is considered undesirable to generalize regarding cycle times, and hence the cycle times specified in this paper are related to a particular example. There is, however, one interesting general comparison between rolling and turning. Rolling times are normally substantially shorter than equivalent turning times. In addition, when considering the previously mentioned range of inner race sizes 15–60 mm bore, there is only a small numerical change in the cycle time when rolling, the numerical change, however, when considering turning is quite substantial. The floor to floor times for the above range when rolling would be in the range 6–8 s. In the case of the 40 mm bore I.S.O. inner race the cycle time from floor to floor would be about 7 s. The actual rolling cycle is only about 2 s, the other time is taken up mainly with material handling. The turning time for the equivalent bearing race would be in the order of 30 s.

Material utilization for cold rolling bearing inner races

Once again generalization is considered to be possibly misleading when material utilization is being discussed. In the case of the 40 mm bore inner race at least 30 per cent material will be saved when compared to turning the equivalent race conventionally. The percentage could be more, depending on what material is used in the comparison (i.e. hot rolled or cold reduced tube). This represents a real saving, not only in direct material costs, but in less obvious factors. Expensive swarf handling is not necessary when considering cold rolling. Additionally substantially lower stock levels can be held of basic raw material, resulting in reduced inventory and storage space requirements.

PRODUCTION MACHINERY FOR COLD ROLLING BEARING INNER RACES

Description of inner race cold rolling machine model IR/1000

The machine is clearly illustrated in Fig. 5, and a specification can be found at the end of the description.

The basic inner race rolling machine comprises a rigid fabricated base carrying one fixed head and one moving head. The moving head is of the quill feed type with a maximum stroke of 6 in to allow easy access for die changeover. The working stroke of the piston, however, is controlled by micro switches mounted on the side of the head casting. The piston is of the 'through rod' type and the head is additionally supported by means of upper and lower support blocks.

Identical spindle housings are mounted on both the fixed and the moving head and these carry the die support spindles which are mounted in four taper roller bearings, pre-loaded to give minimal run-out. The spindle surfaces are hardened as necessary to maintain life and accuracy. The bearings are lubricated by means of an oil bath and are sealed for life. Each spindle is driven via a universal joint assembly and a co-axial gear box. The two gear boxes are linked and driven via a tooth belt and pulleys from a single 30 hp motor. The whole of the drive assembly is mounted on a fabricated base situated towards the rear of the roll head assembly. Nominal ring size control is achieved manually by means of an easily set stop nut mounted on the rear piston rod. Individual ring size control is afforded by means of a unique system in which the stop is automatically set according to the volume of the particular blank presented. Included in the basic machine is a patented 'ring growth control' system necessary for successful rolling of thin section rings.

An hydraulic power pack is provided for driving main head cylinder, 'growth control roll system' and the Auto Load unit. The pack comprises all necessary control valves and interconnecting piping between pack and machine for the above stated functions. The

Figure 5. Production machine model IR/1000.

pack would be mounted on an extension of the machine base, in order to avoid dismantling of the hydraulic lines for machine shipment. The basic electrical control system consists of electrical control cabinet, operators push button station and all necessary relays and switches, motor contractors, push buttons, indicator lights, timers and interconnecting wiring for control of the basic machine functions.

A rolling lubricant system is provided comprising 50 (227 litres) imp. gallon tank mounted in the machine base, together with a pump rated at 6 gal/min (27 l/m). The pump suction line would be fitted with a suction filter followed by a micro-suction filter complete with gauge indication of the filter condition. The pump and micro filter are accessibly mounted on the side of the lubricant tank.

An automatic loading fixture complete with entry and exit chutes is provided. These chutes can be connected to complete automation systems if required.

Specification for IR/1000

Maximum rolling force	50 000 lbf	22 700 Kg f
Maximum die diameter	10 in	254 mm
Minimum die diameter	Dependent on ring size	
Maximum die width	3 in	76 mm
Spindle speed range	46–88 rev/min	
Maximum ring O.D.	3·75 in	95 mm
Minimum ring O.D.	1 in	35·5 mm
Machine weight	12 000 lb	5440 kg
Floor area		
(incl. ancillaries)	10·5 x 8·5 ft	3·2 x 2·59 mm

N.B. Larger rings can be rolled on this machine depending upon exact configuration.

BASIC PROCESS DESCRIPTION OF THE COLD ROLL FORMING OF BALL-BEARING OUTER RACES

The general principle of outer race cold roll forming will be readily appreciated by reference to Fig. 6. An annular ring which comprises the starting blank is very loosely located on the outside diameter in a housing which can be either simple or a two-piece split assembly as illustrated. It is initially located

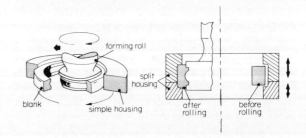

Figure 6. Basic principles outer race rolling.

axially between two sets of collapsible springs (omitted for clarity). The springs locate against the end faces and hold the starting blank in the correct positional relationship to the forming die, and related tooling components. The housing can be hydraulically prerotated to impart a similar peripheral speed to the blank, as that of the rolling die, such that indentation on initial rolling contact is avoided.

There is one cantilever type forming die, the actual rolling diameter of which is as large as possible, conducive with clearance considerations. The die is mounted into a precision spindle. The whole spindle assembly is mounted vertically and is capable of a vertical linear motion to bring the rolling die clear of the housing. The housing is mounted in precision bearings and comprises part of a slide assembly whose line of action is at 90° to the spindle. The slide assembly is connected to a hydraulic cylinder arrangement, which provides two basic movements: (a) a transfer motion, (b) a forming motion. Although not mandatory, the production machine described later, has two identical housings at 10 in centres such that rolling can take place at one housing whilst loading and unloading is taking place at the other.

The method of cold roll forming under discussion comprises loading a starting blank into the housing and initiating a rolling cycle. The housing slide is fed forward with the housing and blank, rotating by means of the forming cylinder. The blank is eventually engaged by the forming die and the rotational, frictional forces imparted, take over from the prerotator hydraulic motor. The blank will start to grow radially until it forms an interference fit into the housing. At this point further metal deformation is sideways, usually equally about the centre line of the bearing outer race. The forming die has an annular form equivalent to the internal form being rolled. The housing as previously stated, can be either split or simple. In the case of the former it is possible to simultaneously form external corner radii/chamfer configurations. In the case of the latter only plain, straight external surfaces can be produced and any additional requirement must be incorporated either as a second operation or previously at the blank preparation stage.

After rolling the O.D. of the race is an interference fit in the housing and has to be ejected by means of a hydraulic cylinder. At the end of the actual infeed of the forming cycle, the roll is allowed to dwell for a short period of time before retraction. During the process of cold forming a copious supply of lubricant is sprayed directly onto the part being rolled as well as the rolling die.

GENERAL NOTE

Many of the remarks previously made with regard to inner bearing race rolling apply to outer race cold rolling. There are however, certain fundamental differences and these are listed under the various appropriate sub-headings as follows:

Starting blank considerations for outer bearing race rolling

A typical outer race blank is shown with the ideal required tolerances in Fig. 7. The outer race which is cold formed from this blank is also shown. All other principles are the same as for inner race rolling although the application detail may be a little different.

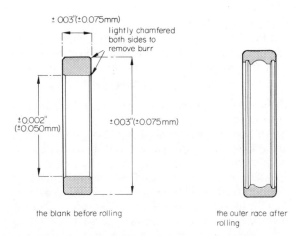

Figure 7. Typical outer race and starting blank.

Quality of cold rolled outer races

Dimensional quality
Considerations here are very similar to inner race rolling although the nomenclature is somewhat different. Reference should be made to Fig. 8 which clearly illustrates the various important dimensional elements.

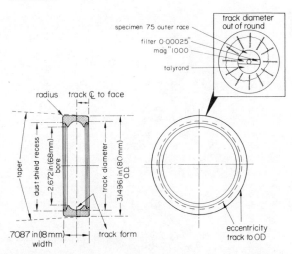

Figure 8. Nomenclature regarding outer race dimensional quality.

Typical outer race

O.D.	±0·0025 in (±0·062 mm)
Track	±0·0025 in (±0·062 mm)
Bore	±0·0025 in (±0·062 mm)
Dust shield recess	±0·0025 in (±0·062 mm)
Width	The as-rolled width is a function of the starting blank volume. Most excess metal is displaced sideways equally about the track centre line. The excess metal however, should be comfortably removed with a double disc grinding operation
Track to face difference	0·0025 in (0·062 mm)
Taper	0·0015 in (0·037 mm)
Out of round	0·0015 in (0·037 mm)

sponding outer race for an I.S.O. shieldless 40 mm bore inner race. Depending on the exact configuration of dust shield an estimated die life in the order of 150 000 parts should be comfortably achieved when rolling the equivalent bearing race with dust shields. The material of the die for outer race rolling presented many problems during development. A completely different set of conditions, due to the cantilever configuration, exist than those found when rolling inner races. The main problem was to strike a happy balance between wear resistance and brittleness under bending stresses to give long die life. Much pioneering work has been completed and particular attention has

order of 9 s. The turning time for the equivalent bearing race would be in the order of 40 s.

PRODUCTION MACHINERY FOR COLD ROLLING BEARING OUTER RACES

Description of outer race cold rolling machine model OR/1000

The machine is clearly illustrated in Fig. 9 and a specification can be found at the end of the description.

The basic outer race machine comprises a rigid,

Figure 9. Production machine model O/R 1000.

fabricated base carrying four basic assemblies, namely die head assembly, moving cylinder assembly, roll head assembly and transfer slide assembly.

The die head assembly carries two die housings, each vertically mounted in heavy duty taper roller bearings, each housing being pre-rotated by means of an hydraulic motor. The whole assembly is slidably mounted on two hardened and ground horizontal columns rigidly attached to the machine base.

The moving cylinder assembly is a double cylinder arrangement and is attached to the die head assembly and the machine base. One cylinder causes the index movement of the die housing and the other cylinder is used during the forming operation.

The roll head assembly is rigidly bolted to the machine base during the normal operation but it is also pivotably attached so that by removing the fixing screws, the whole assembly can be hydraulically tilted approximately 40° to afford easy access to the forming roll shaft, integral with a special purpose heavy duty casting carrying a hydraulically driven forming rool shaft, integral with a special purpose hydraulic cylinder for lowering and raising the forming roll to and from the rolling position.

The roll head casting also carries two vertically mounted eject cylinders for ejecting the rolled ring into the transfer slide assembly.

The transfer slide assembly is attached to the underside of the die head assembly and carries two slide mechanisms each mounted on the centre line of the die housings. A load cylinder is also mounted below each die housing and this lifts the blank ring into the rolling die and also acts as a face location during the rolling operation. Also included in the basic machine are two automatic loader units, and an exit chute is situated at the rear of the machine. The machine is fitted with short lengths of inlet and outlet chuting for connection to automatic handling devices if required.

An hydraulic power pack unit is provided with all the necessary valves, controls and interconnecting piping between the pack and the machine.

An electrical control system is provided, consisting of electrical control cabinet, operator's push button station and all the necessary relays, switches, motor contactor push buttons, indicator lights, timers and interconnecting wiring between the panel and machine.

A rolling lubricant system is provided, comprising 40 imp. gallon (182 litres) reservoir integral with the machine base, gear pump rated at 3 gal/min (13·6 l/m) complete with suction filter and micro suction filter including gauge indication of the filter condition. Both the pump and micro filter are accessibly mounted on the side of the machine base.

Specification for OR/1000

Maximum rolling force	51 000 lbf	23 100 kgf
Maximum stroke of cylinder	0·625 in	15·9 mm
Spindle speed range	50–500 rev/min	
Maximum ring O.D.	3·375 in	85 mm
Minimum ring O.D.	1·5 in	38·1 mm
Maximum ring width	1 in	25·4 mm
Machine weight	12 000 lb	5450 kg
Floor area	10·5 x 8·5 ft	3·2 x 2·59 m

SOME INTERESTING COMBINATION ROLLING APPLICATIONS

Conveyor bearing

Figure 10 shows an overhead conveyor bearing outer race, which involves both separate internal and an

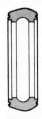

Figure 10. Conveyor bearing outer race.

external forming operations. In this particular case the following main advantages when compared to conventional manufacturing methods were gained:

(1) 43 per cent material saving.

(2) 8 s external operation + 12 s internal operation floor to floor times, compared with 70 s multispindle automatic floor to floor time.

(3) Ball track grinding eliminated.

Self lubricating outer race

Figure 11 shows a self lubricating type bearing outer race, which can be formed in one operation. Because

Figure 11. Self lubricating outer race.

of the relatively thin cross section, a combination of actual plastic deformation and bending can be used to form both the internal and external profiles simultaneously. A basic model OR/1000 machine is used with a split housing technique.

CONCLUSION

This paper records only the first chapters in the story of cold metal roll forming. The basic principles have already been applied to the cold rolling of gears as well as a number of other internal and external shapes of revolution. Fan and water pump spindles, ball rod ends and fractional house power electric motor shafts are typical examples of the scope of the process. It is confidently expected that cold roll forming will be applied to an ever increasing list of applications in the future. In fact new components are being and will be designed with cold roll forming particularly in mind as a production method. To reflect that this paper looks to the future, is perhaps to be more than ordinarily prognastic.

might one not just as well machine the ball track at the same time and thus avoid an additional handling operation and additional process costs? In other words, is one likely to be incurring increased conversion costs in order to save material costs, and is the total cost of the product really going to be cheaper by changing to the rolling process?

A further point that occurs to me is concerned with the important operation of blank preparation. Is any special purpose machinery envisaged to deal specifically with this operation?

Lastly, reference is made in the paper to the possible use in the future of blanks produced from metal powder other than tube. Could Mr Egan please comment on this?

A. 1. Blank preparation is probably one of the most important aspects of the cold roll forming of ball bearing inner and outer races. The economics of the process are very much tied up in blank preparation.

There are, of course many other methods of blank preparation in addition to the one mentioned, i.e. using tube and multi-spindle automatics. I do not consider that the multi-spindle automatic route is necessarily the most economic method of blank preparation, however an important consideration is the fact that the bearing companies will have redundant multi-spindle automatics available if they in fact decide to adopt cold roll forming. It would seem sensible to use existing capital plant if possible until such times as this plant becomes worn out and needs replacing.

With regard to the specific points raised, 'might one just not as well machine the ball track at the same time as the other features and thus avoid an additional handling operation?' This is a question which I am continually asked—the answer is categorically 'no'. Whilst it is dangerous to generalize, the indication of some of the arguments proposed in favour of the economics of cold roll forming are described in my paper. The two primary points are that a substantial reduction in material used can be made and secondly substantially faster cycle times can be achieved. A specific example is quoted in the paper. There are a number of very important other factors listed in the paper but I would like to stress that care should be taken when comparing economics and consideration given to compare identical configurations. Our appreciation to date when considering the current method of cold forming and blank preparation on multi-spindle automatics is that there is a break-even point on approximately 20 mm bore diameter inner race. Any increase in size after this shows progressively better economic justification.

the rolling machine one for one.

This method of blank preparation would presuppose the use of either hot rolled or cold reduced tube as a starting material.

It should be borne in mind, however, that there are other roots of blank preparation such as hot or warm forging, sinter forging or maybe even the use of bar stock material. Each of these alternative methods would have its own problems and would need different consideration, both technically and economically.

3. This is probably one of the most interesting methods of blank preparation but I would like to emphasize in the long term, whilst we have carried out experiments of rolling powder metal blanks in EN31 material, there are still many technical problems to be overcome. My personal opinion is that one day the technology will be developed to a sufficient degree to enable sinter forged rings to be successfully rolled.

Our experiments to date have indicated that a very high density blank is required, something in the order of 99·6. One of the main problems is in obtaining a blank with the correct metallurgical characteristics. EN31 has, as I am sure you are aware, a relatively high percentage of chromium and this is not one of the most ideal constituents from the point of view of sintering.

I think, in conclusion, that tube and bar stock material are with us for a long time yet, but sinter forged blanks certainly should not be disregarded in the long term, which will depend on the ultimate prices of the powder. In the meantime, a lot will depend on the successful implementation of warm forging of bar stock material which is receiving a great deal of emphasis in several quarters.

Q. Professor J. Peters Leuwen. Do you immediately get the final form, or do you require further grinding?

A. 1. When considering I.S.O. bearings it is necessary to grind after heat treatment. Typical achievable tolerances in relation to a specific application are quoted in the paper. These tolerances would not be good enough for final use, however we have found that due to the fact that there is less distortion after heat treatment the consistency of dimensions is quite constant. We are currently carrying out a substantial work programme on improving the tolerances we are presently obtaining on I.S.O. bearings. This to a large extent is tied up with the accuracy of the starting blank. At the moment, however, we envisage grinding after heat treatment.

2. On more commercial type bearings, such as

conveyor bearing, which I mentioned in my paper, the grinding of the ball track and profiled outside diameter have been eliminated with considerable economic advantage. The surface finish, even after heat treatment, of these two features is in the order of 5 micro inches RMS, and hence assembly can be effected without any further work at all.

It should be borne in mind, however, that this is a slow speed application where great precision is not required.

INJECTION MOULDING MACHINES

by

D. G. DAY*

In past years the injection moulding machine has been almost entirely disregarded by the machine tool manufacturers and, at present, is still not classified as a machine tool due to the rigid definition of this class of machine.

For many years the market has been extremely competitive resulting, in many cases, in lowered quality standards in order to be competitive, and in some cases, poor after-sales services have aggravated the situation.

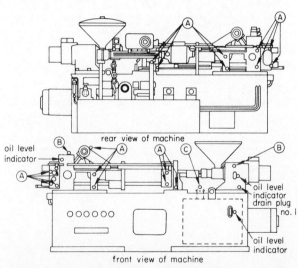

Figure 1. Anker Werk single-screw injection moulding machine.

Generally U.K. built machines lack engineering finesse but manufacturers will usually comply with safety requirements and electrical standards requested by the purchaser whilst most European built machines (mainly Germany and Switzerland) are well engineered but fall far short on safety and, until such time as the International Standards are generally accepted, electrically as well.

It has, therefore, largely been left to individual users to set out standards of their own, derived largely as a result of experience. This to some extent is detrimental to the reliability of the basic machine because quite obviously no manufacturer will immediately incorporate a user's requirements into the basic design without a guarantee of further orders. Consequently, the additional features called for by the user invariably have the appearance of being afterthoughts of a temporary nature.

This situation is changing slowly. As manufacturers redesign their ranges, they are tending to improve the engineering quality and the safety features. These, coupled with the advances made with solid state control to the point, with one manufacturer, where a change of sequence is made simply by changing a key card, have raised these machines to an extremely sophisticated level requiring advanced technical expertise to maintain them.

This being the case, it is logical therefore to expect them to be as safe and reliable as it is humanly possible to make them. For this reason they become expensive to purchase and can experience long delivery times.

Despite all these advances there are still areas in which the makers are unwilling to assist the purchaser. For example it is still unusual to see a machine proved to be capable of the performance quoted by the manufacturer. This applies in particular to the die locking unit where it is extremely rare to see tests carried out to ensure that the specified maximum locking force is being achieved or even simple checks to ensure equal distribution of that force between the tie bars. From a production point of view it is unusual to see a die offered up to demonstrate the machine's capability to fill the die at the rated capacity.

We have evolved over the past years our own techniques for ensuring that, as far as it is practical and economic, these machines perform to the manufacturer's specification, comply with the relevant company, national and international standards on engineering and meet our demanding requirements on safety.

SAFETY

This is of paramount importance on these machines for the following reasons:

(a) The relatively short distances travelled by the dies usually between 6 in and 9 in but dependent upon the size and shape of the component.

* Joseph Lucas (Electrical) Ltd

Figure 2. Horizontal injection moulding machine—danger points.

(b) The speed at which the dies close to the cushion point approaches that of a small power press.

(c) The requirement of some components for inserts to be introduced usually at a point between the extremes of die movement which reduces the margin of safety available to an operator should an unauthorized movement of the dies occur.

(d) The high locking forces incurred in stopping the dies being forced apart during injection. Even so called die sensing pressures which prevent full lock being applied if a foreign body is present, are sufficient to cause considerable damage.

(e) Injection pressures of 1000 lb/in^2 are quite common.

(f) The relatively high material temperatures, 300°C regularly occurring.

(g) The use of band heaters on the barrel to achieve the material temperatures exposing the operator to the danger of burns.

Generally, all moving and/or heated parts of the machine are enclosed either by fixed or moving guards made from heavy gauge sheet metal or steel mesh.

On an injection moulding machine having the locking unit mounted horizontally with the injection unit in line with its axis, it is usual to fit sliding guards to the front and rear of the locking unit with a hinged guard over the top, to facilitate tool changes. The two sliding guards must be interlocked electrically with the control system in such a manner that the whole of the machine control system is inactive when either gate is opened.

This interlocking is usually achieved by the use of two limit switches on each gate activated by opposite movements (i.e. one is depressed and one released) and they must be operated directly by the gate with no intermediate linkages. Furthermore, these switches must be connected in series so that the failure of any one will deactivate the circuit.

The top guard may or may not be electrically interlocked since it is for use by the setter only and is fastened by captive screws. However, on small models, it would be possible to work in the tooling area and to avoid confusion, it is preferable to have a case electrically interlocked by one limit switch in series with the sliding guard switches.

The front sliding guard must also be interlocked by another medium since this has most usage during

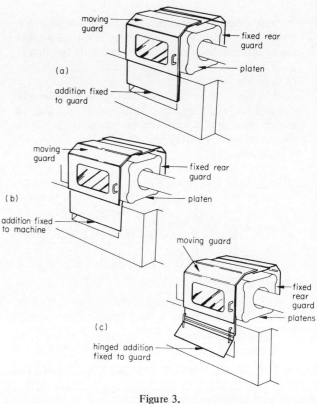

Figure 3.

normal working and this is usually achieved either hydraulically or mechanically.

When hydraulic interlocking is utilized, this again must be a fail-safe system incorporating two valves connected in series, and activated in the same manner as the limit switches (i.e. one depressed and one released). It can be connected into the circuit in many ways but its ultimate purpose must be to prevent closure of the dies with the gate open. An automatic die opening sequence may be incorporated but it is not regarded as mandatory at present.

Mechanical interlocking should preferably act directly upon the main control valve to prevent its use with the gate open and also stop the gate being opened whilst the machine is cycling. Alternatively, the action of the gate can be used to insert or withdraw a scotch to physically halt the die, should an unauthorized movement occur. But this feature

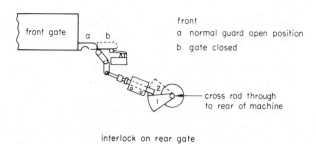

front
a normal guard open position
b gate closed

interlock on rear gate

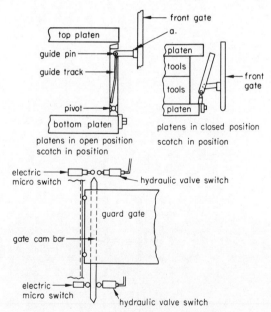

Figure 4. Mechanical interlock.

may not be used on all machines. It must obviously be extremely robust and should preferably be progressive in nature (i.e. it must be operational at any die position).

Figure 5. Vertical model moulding machine.

Machines, having a vertically arranged locking unit and usually utilizing split line injection are interlocked in a similar manner with a preference for mechanical interlocking utilizing a scotch. This is because hydraulic valves are naturally subject to wear leading to creep on a downstroking press.

Where dies are required to stop part way through the closing sequence for the introduction of inserts, it is not sufficient to rely on the normal interlocks alone. This has led to the use of a large port dumping valve capable of passing full pump output back to the reservoir with absolute minimal back pressure. This again is activated by the front gate but since back

pressures can be introduced by lengthy pipework a linkage is permissible.

Since high injection pressures are fairly common, it is essential to interlock the injection system with the locking system and the system which holds the injection nozzle against the sprue bush. On machines employing pure hydraulic locking and hold on mediums, this is usually achieved by pressure switches and where mechanical locking mediums are used (i.e. toggle arrangements) the signal is usually initiated by a micro-switch which is depressed by one of the toggles on achieving full lock.

Any viewing windows in the safety gates must be able to satisfy BS 3757, Rigid PVC Sheet. Part 1. To achieve this standard, the window needs to be $\frac{1}{4}$ in thick using such materials as Davic Clear 024 or Oroglass.

To prevent interference with interlocks, it is essential that these should be enclosed by sheet metal guards. This is to protect them from physical damage as well as to stop interference because operators will attempt to marginally reduce cycle times in order to increase their virtual work output.

RELIABILITY

General

Once in production, these machines are worked hard for extended periods with little or no time allowed for planned maintenance. In order to be economically viable they are usually worked 24 hours a day for 5/7 days a week until they eventually break down. This is basically because a large percentage (approximately 60 per cent) of the cycle time, the machine does no actual work, namely during the cooling period, although during this period the pump is working continuously and for part of it the injection cylinder is being returned to its start position and the material is being plasticized ready for the next shot.

Since the return on the capital cost of these machines is low when producing high quality components, stand-by plant cannot readily be justified, thus making reliability highly desirable although not always achievable.

Locking units

The writer's preference is for direct hydraulic locking machines because they have fewer mechanical moving parts. This is a prime factor on maintenance because the high pressures produced between link pins and bushes during locking makes satisfactory lubrication virtually impossible to achieve at an economic price. Inevitably, this leads to accelerated wear made worse if offset tooling is used, resulting in the necessity to replace toggle sets at an average of $1\frac{1}{2}$ to 2-year intervals. When one considers that they cost between £300 and £500 for the small and medium range machines, the preference for direct hydraulic locking, which will probably last for 10 years before overhaul, is understandable. Also, with hydro-mechanical systems, there is the need for some form of adjustment to compensate for variations in closed die height thus introducing another mechanical feature. The direct hydraulic lock unit is of course self compensating.

the need to cushion both ends of the stroke to prevent damage to both the machine and dies. The design of toggles generally provides an inbuilt acceleration and deceleration thus reducing the cushioning period.

Hydraulic systems

Probably the most common fault on any hydraulic system is that of leakage generally due to the use of rigid pipework and fittings which are not genuinely re-usable. A large range of coupling sleeves are in use which cut into the pipe to provide a seal. This coupled with vibration from inadequately secured pipework will lead to leakage as will the removal and reconnection of pipes for replacement of valves, etc.

Vibration must, therefore, be reduced to a minimum. As well as reducing transient pressure pulses physical layout, eliminating sharp elbows and similar features must be planned. Whilst it does not always produce a neat layout, the use of flexible pipework does reduce vibration transmission and also relieves the necessity to produce intricate and accurate forms in rigid pipework.

Pump units are a section of the system which is particularly vulnerable if misused. Cavitation, which can damage a pump in a very short space of time, should obviously be eliminated when developing a new model but can be inadvertantly reintroduced after a period of time if for example a very fine mesh suction filter is used resulting in a premature partial clogging of that filter. Also, a low oil level, inadequate sealing of the suction line, high oil temperature and the use of the wrong type of oil, can produce the same effect.

The rating of the pump also requires careful consideration. If, for example, a pump is rated at 1000 lb/in^2 continuously, there is little point in expecting to achieve 1000 lb/in^2 on injection (i.e. assuming the pump pressure control valves are set correctly). After the settling down period, one could expect a drop of approximately 50 lb/in^2 at this point. Therefore, in order to achieve a reasonable pump life, it is preferred that the machine manufacturer should use a higher rated pump governed down to the desired line pressure. Whilst slightly more expensive, the increased pump life is sufficient compensation.

Efficient oil cooling also plays an important role in reliability and the first factor to be considered is the size of the reservoir. On a medium size machine of approximately 200 tons locking force, the reservoir would usually contain 120 to 150 gallons. The surface area should be kept as large as possible for natural cooling with additional cooling provided by

previously, of producing pump cavitation due to reduced viscosity and, also, damaged seals due to temperature. If the oil is maintained at too low a temperature, inefficiency results due to increased electrical power consumption to drive the pump. From this, it will be readily appreciated that an oil temperature gauge is of prime importance yet surprisingly it is very rarely fitted.

As a general requirement, BS 4575 should be complied with on hydraulic systems.

Lubrication

Adequate and correct lubrication at regular intervals must be one of the most important facets of reliability. Fortunately, on direct hydraulic lock machines, very little lubrication is required with only two areas requiring attention, these being the tie bar bushes and the injection unit slides.

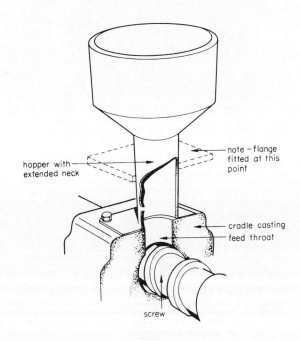

Figure 6.

Hydromechanical lock machines are extremely difficult to lubricate satisfactorily, due to the high pressures produced between the toggle pins and bushes. Unfortunately, toggle arrangements preclude the use of recirculatory systems and, since a positive displacement system is necessary to circulate the lubricant properly, losses can be excessive. Grease is not really acceptable because of the possibility of contamination by the moulding powders which would produce a strongly abrasive mixture. Excessive losses are undesirable due to:

(a) economic grounds,
(b) possible contamination of hydraulic oil,
(c) possible spillage on to the floor causing dangerous conditions,
(d) possible contamination of the components.

Pneumatic systems

Compressed air is not generally used for prime moving functions because of the requirement for high pressures. It is, however, used in some cases for auxiliary functions such as component ejection and, in one case, as an intermediate control medium between the electrical control and the hydraulic system.

Since the systems encountered are usually very simple, there is little chance of failure. Possibly, the main points to ensure are that an air service unit (i.e. filter/lubricator unit) is fitted and that pressure requirements are compatible with those available on site.

Electrical system

The basic requirement of the electrical system should be BS 2771 used in conjunction with the purchaser's own company standards. Since BS 2771 is almost identical in content to IEC 204 produced by the International Standards Organization, machines purchased from continental countries should stand a reasonable chance in future of compliance with our own basic standards. This should help to keep additional costs for any company requirements down to a minimum. Until such standardization is achieved, any machine which is new to an organization should be examined carefully and additional features or alterations determined and quoted for before ordering. To attempt modifications after it is installed can be costly and in some cases impractical.

The type or make of equipment is usually a matter of individual preference but it is obviously advantageous if the engineer concerned can standardize on one or two makes to reduce his spares holding.

On this type of machine it is essential that the temperature control instruments, timers and general control circuit components should be housed in a cabinet entirely separate from the machine. There are a number of reasons for this, a few are quoted:

(a) To prevent the ingress of moulding powders, oil contamination etc.
(b) To remove delicate equipment from the effects of vibration.
(c) To provide better accessibility for maintenance purposes.

MECHANICAL

It is difficult in this field to say which features are more important, and the following points are not in any order of priority.

(a) Toggles (the problems encountered with these have already been discussed).
(b) Die height adjustment which is normally fitted in conjunction with toggle units. This is usually a screw jack arrangement which is subjected to heavy loading. Correct lubrication is important.
(c) The cams which operate limit switches, hydraulic valves and mechanical interlocks should be case hardened and the ones which operate interlocks must be welded to the safety gates. This gives a longer working life and prevents their removal.
(d) Gate runners and brackets should be dowelled in position for ease of repositioning after removal for major work.
(e) The whole of the injection unit should be pivoted to simplify screw removal.
(f) All units or sub-assemblies bolted to the main chassis should be fitted with secondary locking mediums (e.g. lock washers or lock nuts).
(g) Nozzle tip radius should conform to that on which the purchasers company has standardized. At present, there is no standard within the trade.
(h) All guards and covers should be retained by captive screws which can be unfastened only with a tool. The consumption of nuts and bolts which are not captive can reach fantastic proportions.
(i) The design of the hopper feed throat should be such that it is impossible for an operator to insert his hand into the screw area. Allied to this is the method of shutting off the flow of powder to the screw. It is quite common to see operators hammering these to get them shut.

As you will already appreciate, these are mostly minor detail points, but if they are not correct, they will certainly cause annoying maintenance problems like a safety gate movement stiffening up because somebody accidentally hit the runner or the final slow close limit switch works loose and allows the die faces to meet at full speed.

GEOMETRIC ALIGNMENTS

At present it is extremely rare for the purchaser to be supplied with an alignment acceptance chart and it would be fair to say that most manufacturers rely solely on manufacturing tolerances to provide some semblance of accuracy on assembly. Also, many manufacturers make no final inspection of the assembled unit in this respect.

It is to be hoped that the eventual acceptance of injection moulding machines as part of the machine tool group in the British Standards will resolve this problem to the satisfaction of all concerned.

PERFORMANCE TESTING

This, again, is an area which manufacturers are reluctant to demonstrate to the purchaser.

Two functions in particular should be proved by the manufacturer to the satisfaction of the purchaser. They are:

(1) The ability to achieve the specified locking force and the equal distribution of that force in all tie bars. On horizontal machines, this can be measured fairly simply with a dial indicator to measure the extension of each tie bar. Vertically arranged locking units cannot usually be checked that simply because invariably, there is nothing

HYDROSTATIC EXTRUSION: THE STATE OF THE ART

by

J. M. ALEXANDER*

INTRODUCTION

The subject of hydrostatic extrusion has been written and talked about over the past ten years or so almost to the exclusion of any other metal working topic. Notable reviews have been given by Dr H. Ll. D. Pugh of the National Engineering Laboratory, his most recent paper being published in the *Annals of the C.I.R.P.*[1], but perhaps his most important contributions are summarized in the Bulleid Memorial Lectures for 1965[2]. Recently he has published also a chapter in a book on this subject[3], and I and my colleague Dr Lengyel have written a book on *Hydrostatic Extrusion*[4] as well as a number of review articles[5,6,7,].

Therefore in this paper I do not intend to go back again over all the ground which has been covered by these reviews. In the above references I have selected only those articles which are of a review nature—an enormous amount of work has been carried out on this process both in the USA and in the USSR as well as in other countries all over the world such as Japan, France, Sweden and elsewhere. I believe that much of the work which has been done remains unpublished because this is such a versatile and important process that I believe many of its advantages are already being exploited commercially and, for obvious reasons, no great public airing is being given to it.

I intend briefly to try and point out where I believe we have got to in studying this very important process, and I will concentrate mainly on the apparent difficulty of getting the process into industrial production. I have no doubt that this process will displace many of the major continuous production processes such as rolling and drawing which exist today, for *continuous* hydrostatic extrusion is already with us and has been demonstrated to work in practice. The main advantage of a *continuous* hydrostatic extrusion process over other *continuous* processes, such as rolling and drawing, is the enormous reductions which can be obtained in one die without excessive wear of that die. In the extrusion of commercially pure aluminium, for example, it is possible to reduce the area in one die by an extrusion ratio of 10 000:1, that is, to reduce the diameter of a round bar by a factor of 100 times so that a bar, entering the die at a diameter of 8 cm, will emerge from the die with a diameter of 0·8 mm. It is interesting to imagine the sequence of rolling and drawing operations which would be necessary to achieve that result. There are a number of alternative ways of doing it but, however you do it, it involves a lot of costly machinery. The hydrostatic extrusion process performs the operation in one die! Pressure required to achieve this enormous reduction is really quite moderate—less than 15 kb, which is well within the capacity of extrusion press containers.

I believe one reason that this exciting process has not yet come into industrial use is that there is much money invested in equipment which manufacturers know works extremely well and efficiently, and has already been amortized against its original capital cost. It would be replaced by equipment which is new, possibly with operating difficulties because of the nature of the high pressure involved; and generally there is a reluctance to embark on something with which the manufacturers are not familiar. However, I believe that we will shortly *have* to recognize this process for the major contributions it can make, particularly with regard to its ability to achieve reductions on materials which are otherwise either extremely difficult or impossible to deform.

I shall therefore discuss in some logical order the various problems which are preventing us from adopting this process in production. First I will consider the difficulty of making this process continuous.

CONTINUOUS HYDROSTATIC EXTRUSION

The first effort to make the process continuous seems to have been my own attempt in 1965, in collaboration with Dr Bela Lengyel; we devised and patented a system for the semi-continuous hydrostatic extrusion of a continuous length of feed stock (Fig. 1). The idea rests on the use of a clamp which both clamps

*Department of Mechanical Engineering, Imperial College, London

container to be hydrostatically extruded by an augmenting plunger containing the die. This advances over the clamped feed stock which is surrounded by fluid at a pressure slightly below that required to cause hydrostatic extrusion. The product is accordingly forced up the die through the centre of the augmenting plunger to be wound on to a spool or drawn out of the front of the press, either by the control of the augmenting ram, or by a controlled pull from the coiler or other means of pulling out the product.

Although this is ostensibly an intermittent process, there is no reason why the product which goes in and leaves the press should not be in a continuous length; the fact that it is intermittent is quite immaterial. The press could be fed from a continuous casting machine in which the continuous casting mould can move either forward or backward to accommodate the stopping and starting of the clamp feed stock. Alternatively, the intermittent feed would be absorbed by elastic bending of the feed stock if it were in the form of a long arc. The issuing material can be either sawn or cut into lengths which are required for production, by a flying saw or by a static saw when it is stationary; or it can be coiled on to an accumulator block of the kind customarily employed in the wire drawing industry while the machine is extruding wire, and paying off the accumulator block while the press is stationary.

A number of difficulties can be envisaged in the operation of such a system. First, the operation of the clamp presents difficulties. Clearly, it is difficult to envisage feed stock of varying dimensions being satisfactorily clamped and sealed against pressures of the order necessary for the process. It is therefore necessary to 'size' the feed stock entering the press. This can be done either by providing a drawing or shaving die to act on the feed stock before it goes into the press, or by employing a sizing tube which has been described previously by my colleagues and myself. This then enables accurately sized feed stock to enter the clamp so that the clamp itself can have a continuous cylindrical surface and not be of the 'collet' type, which might mark the feed stock leading to marks on the surface of the product.

Then again, the continual raising and lowering of the pressure inside the container poses fatigue problems for the container and seals and other equipment subjected to it. Actually the cycle time of such a press would be of the order of 0·5 to 1·5 min, and this could involve a large number of cycles over the operating period of the press. These fatigue problems are being currently investigated in my department, and we feel perfectly confident that we can design

This is not aesthetically very satisfying, but what we are concerned with at the moment is how to achieve a viable continuous hydrostatic extrusion process, and not the aesthetics of getting it to operate in the most satisfying manner. Also it is possible, by altering the design of the die itself—in particular the position of the sealing rings on the outside of the die—to arrange for it to open up slightly upon release of the fluid pressure or possibly not to change its dimensions at all, although this is quite difficult. We are using finite element methods of analysis of the die to see whether we can devise a die shape which retains the dimensions as accurately as possible, of the material passing through it.

The semi-continuous method of operation has been investigated at the UKAEA Laboratories in Risley by Mr Derek Green and his colleagues, on the Fielding 'Hydrostat' press. They had some difficulties with marking from the clamp and with marking of the product, but I am sure that these difficulties can be overcome and that the process could be made a viable commercial reality. What is important is that it has been demonstrated both at the UKAEA and at Imperial College that this process can be made to work.

Doubtless it would be better if a fully continuous process could be developed—this would avoid the continual stopping and starting of the feed stock and the emerging product and this would certainly ease the situation at both ends of the press. The first attempt at a continuous process seems to be that embodied in the UKAEA Patent No. 1177223, which describes a double clamp system in which the clamps are arranged in tandem, one in front of the other and alternately grip and feed the feed stock forward. This is an extremely difficult system to achieve in practice—we are trying to clamp, seal and move forward the feed stock against a pressure of about 11 kb, and this involves extremely large forces, together with the problems of dynamic sealing against those forces and the difficulty of handling a feedstock which is often extremely soft by comparison with the enormous pressure inside the container. Nevertheless such a system can be made to work, although there are no details published about whether the UKAEA ever succeeded with their system. At Imperial College we are certainly trying to achieve this form of continuous process.

Recently we have seen an elegant idea from Dr Fuchs of Western Electric Ltd, USA[8], describing a continuous extrusion process in which the feed stock is urged into the container by the shearing drag of a very viscous fluid which is axially circulated around many channels along the length of the feed stock in

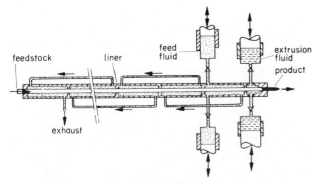

Figure 2.

such a way that it creates a rising pressure towards the entry to the container (Fig. 2). This is an extremely difficult system to achieve, and although Dr Fuchs has demonstrated it to work on a laboratory scale, I believe that the production version of this equipment which has been installed in Western Electric's factory in Atlanta, Georgia, is not yet operational, although there is every hope that it will be, of course. There is a much better illustration of his process in the review by Dr Pugh[1].

At the NEL/AIRAPT Conference held in June 1973, a second ingenious invention by Dr. Fuchs[13] was described, for continuous hydrostatic extrusion. This features a segmented container made up into four endless tracks, as shown in Fig. 3, which are driven by gears of the rack and pinion type in an axial direction over the die. After passing over the die the container tracks divide and loop back to the entry point into which the feed stock is introduced. The feed stock is round bar, accurately sized and encased in a polymer which is picked up before entry into the container. This polymer serves to 'glue' the feed stock to the walls of the moving container within which pressure is built up until the die is reached, through which hydrostatic extrusion takes place, the polymer then serving as the extrusion fluid. The four tracks are pressed together by externally pressurized support pads and the prototype machine has been designed for a pressure of 150 000 psi (10·33 kb).

It was stated at the Conference that the endless chamber extruder had operated on aluminium wire rod of $\frac{3}{8}$ in diameter sized to 0·365 in diameter with a reduction ratio of 27 : 1 at rod speeds of 12 feet per minute. A second endless chamber extruder for

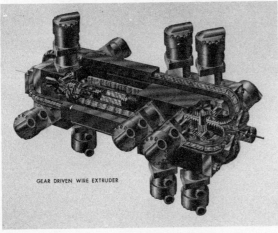

Figure 3.

copper wire which will operate at double the present pressure is being designed.

Also important in this context is a process devised by Derek Green of the UKAEA, called 'Hydrospin'[9] (Fig. 4). This process has been described extensively both in my book with Dr Lengyel and also in the review by Dr Pugh; it is basically a process for the hydrostatic extrusion of a billet into tube, with the issuing material from the tube then virtually machined by a rotating tool in such a way that the

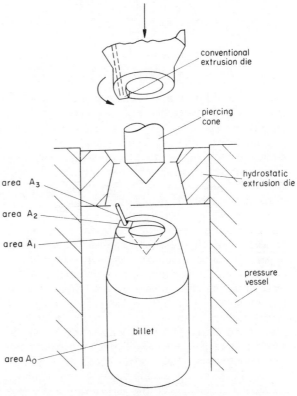

Figure 4.

chip is forced to pass through a small die in the centre of the face of the 'cutting tool' by the process of conventional extrusion to form the wire product. This is a very ingenious system but it is not really a continuous process. Billets of finite size only can be accommodated in the container, although it is maintained that billets can be virtually pressure-welded together by introducing a following billet before the first one has been completely extruded out. However, there are other problems with this process, I believe—for one thing it must be very difficult to achieve reasonable lubrication of the machining face of the rotating tool, and I am sure this will give problems in applying this process to harder materials, such as steel. Nevertheless, it may have applications in certain facets of production.

Of more interest, possibly, is another recent ingenious idea from Derek Green for the continuous extrusion forming of wire sections[10] (Fig. 5). Although this process is not hydrostatic extrusion, I mention it because it is relevant in the context of our trying to find a continuous process of extrusion for metals. The billet in this case is square or rectangular in cross-section and is continuously fed over the surface of a large diameter roll containing a groove. (Soft stock of circular cross-section can be fed into

contact areas between feedstock and groove shown dotted

Figure 5.

the groove.) The die is carried on an arcuate surface which fits into the groove, and the general mechanics of this process is such that the material is forced forward by the frictional contact of the billet in its groove through the die by a conventional extrusion process. It is very ingenious, but again I think the actual equipment involved is going to be quite difficult to engineer in practice.

It is interesting that we have had for many years the continuous process of the screw extrusion of plastics in which the virgin material is put into a hopper in powder form and eventually continuously extruded out through the die by either single, double or triple screws of Archimedean form, thus achieving a continuous product. It is rather surprising that mechanical metallurgists have not really tackled the problem of making the extrusion process for metals continuous in the same way. After all, we have the continuous casting process which is now developed to a great degree of sophistication, and the metal which comes out of that is moving very slowly and could nicely be introduced into the back end of a continuous extrusion machine. I am sure this is the way that industry must develop in the future and that many articles such as girders and railway lines, even in hard steels, will be made this way by hydrostatic extrusion.

VIABLE COMMERCIAL PRODUCTS

In considering the application of hydrostatic extrusion to give a viable industrial process one immediately turns to the problem of finding a suitable product. In my view the most important products at the moment which could be made better by a continuous hydrostatic extrusion process than by any other are those involving more than one material because of the high interfacial pressures which can be developed. One of the most interesting products of this type is copper-covered aluminium for electrical conductors such as are used in electrical machinery in industry or in the wiring of buildings and so on. When we consider the product itself, it is attractive because of the relative cheapness of aluminium as compared with copper. Aluminium has approximately 61 per cent of the electrical conductance of copper, so that a pure aluminium conductor has to be very little bigger than its equivalent copper conductor of the same

surfaces where electrical arcing may take place causing the hazard of fire. Further, aluminium does not have the resistance to creep which copper possesses; therefore in clamped contacts the pressure between the aluminium and the clamping member may gradually relax and contact again be lost, which is rather unfortunate. Then again there is some opposition to using aluminium for house wiring because of the fire hazard. The fact is that aluminium melts at a temperature of 666°C compared with copper at 1083°C, so that if there is a fire in a house the electrical wiring is much more likely to melt and give rise to even more dangerous situations than with copper. Personally I think this last problem is exaggerated, and that if you have a fire in a house you have got problems anyway without worrying about the likelihood of the electrical conductors melting.

In spite of all these problems with aluminium, there is no doubt in my mind that, from the point of view of the conservation of natural resources, we must move towards introducing copper-covered aluminium electrical conductors. The covering of copper (or alternatively a thin coating of nickel could suffice) is there to enhance the corrosion resistance of the aluminium and to enable lengths of it to be soldered together. The creep strength of the aluminium itself can doubtless be improved by introducing some alloying element such as iron, but there seems to be little or no work being carried out on this at present, and the subject is one which metallurgists might care to pursue. The actual static strength of the conductor seems to be less important for household wiring—in fact the more ductile the wire the better, as far as I can make out. Therefore the softness and ductility of commercially pure aluminium is a desirable characteristic for house wiring. It is only the creep strength which needs to be improved, but we shall obviously have to lose some of the softness and ductility of the material if we are going to alloy it with creep strengthening elements.

For overhead conductor or cables and underwater submarine cables it seems to me that strength is important. It is possible to have cheap strength, cheap electrical conductivity and cheap corrosion resistance by having a composite cable comprising a steel core surrounded by aluminium covered in copper or nickel. There is no fundamental problem in producing such a composite rod or wire by hydrostatic extrusion. The interesting feature of the process is that when you combine soft and hard materials together and extrude them together they both suffer the same area reduction during the extrusion process, even though one of the metals may be much harder than the other.

For example, at Imperial College we have been extruding stainless steel and copper together as well as copper and aluminium, and stainless steel and mild steel. In general these materials will happily extrude together through the same extrusion ratio at pressures very much less than those required for the harder of the materials concerned. It is a process which is analogous to the ancient sandwich rolling technique with which most of us are familiar. The softer material tries to extrude out faster than the harder one, and therefore by the shearing action between the two materials it introduces compressive stresses in the longitudinal direction in itself and tensile stresses in the harder material over which it is trying to extrude. Thus the plasticity of the two materials (or rather the propensity for plastic flow) is brought nearer together.

What I am trying to say is that by introducing an all-round compression in the softer material its plastic flow is inhibited, while by introducing a tensile stress in one direction together with the compressive stress in the other in the harder material it is encouraged to flow plastically. Because of this the two materials end up by having roughly the same flow stress and therefore they extrude out through the same extrusion ratio in spite of the fact that the pressure in the fluid is very much less than that required for the harder material.

A paper on this subject was presented by myself and Dr C. S. Hartley at the NEL/AIRAPT conference held in Stirling University last June. In that paper you will find a very simple theory which describes the mechanics of this process quite adequately on the basis of the equation

$$p = K + w \ln R \qquad (1)$$

which we have often used to describe extrusion processes of this type. For hydrostatic extrusion in which both friction and redundant work are almost absent, it is sufficent to use an expression of the form

$$p = Y_m \ln R \qquad (2)$$

where Y_m is the mean yield stress or flow stress of the material being deformed. By using this equation and combining it in proportion to the areas being extruded for a composite billet, very realistic estimates of the pressure required to extrude these composite billets can be obtained.

Although at first sight it appears that very hard metals could be extruded through large ratios by encasing them in softer metals in this way, this is not always the case. In a recent investigation of the hydrostatic extrusion of copper sheathed in aluminium (there being 16 per cent by volume of copper), the harder core of copper was found to break up into sausage-like segments for extrusion ratios exceeding about 3·6. The investigators, Osakada, Limb and Mellor[14] have developed an upper bound analysis capable of predicting possible deformation modes such as uniform deformation (as being discussed here), cladding (i.e. no deformation of core material) and fracture of either core or sheath material.

Thus a note of caution must be sounded—it appears that when the harder material forms the

sheath there is less likelihood of fracture at large extrusion ratios, so that combinations such as copper on aluminium and stainless steel on mild steel stand more chance of success than if the harder metal forms the core.

Regarding other interesting products that can be made by hydrostatic extrusion, I refer back to the various review articles describing the interesting variants of hydrostatic extrusion, such as fluid-to-fluid extrusion which makes possible the extrusion of very hard materials such as cast iron and beryllium without the problem of having to encase them in a soft material incidentally, and also the production of shaped products of various kinds. Readers will no doubt be familiar with this work. It would be of more interest to describe some of the theoretical work which we are doing at Imperial College.

THEORETICAL BACKGROUND

Apart from the rather simple approach I have just described, which enables the prediction of extrusion pressure in extruding composite billets, what we are interested in is the interfacial stress strain conditions that apply in the interface between the copper and the aluminium, for example—and also, of course, in the general shape of the deforming zone. We are developing finite element methods for analysing the flow conditions in large deformation problems such as this. We have managed to adapt a very simple idea, suggested originally by Cornfield and Johnson of the Electricity Council[11], which is simply to take the elastic finite element solutions or programs and adapt them for plastic flow conditions by replacing strain with strain rate and displacement with velocity. The Young modulus and Poisson ratio are both replaced by analogous plasticity parameters. Unfortunately if one replaces Poisson's ratio with the value one half, then the matrices involved in the finite element formulation are singular and cannot be inverted. The trick to overcome this problem, devised by Cornfield and Johnson, is to replace Poisson's ratio not with a half but with a number nearly equal to one half such as 0·495. This gives a very realistic solution—more realistic than the value one half because it does give some allowance for the fact that the material is not incompressible. The work I am referring to has already been published in the NEL/AIRAPT Conference[12]. Figure 6 shows the comparison

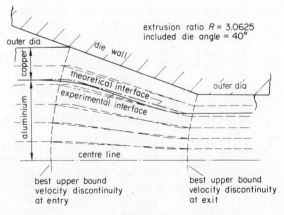

Figure 6.

overcome by the ingenuity of the metallurgical engineer so that hydrostatic extrusion can at last become a viable commercial reality. At present I am trying very hard to introduce the process in industry. For the record, Imperial College has received more than £100 000 from the Science Research Council specifically for the industrial development of hydrostatic extrusion. My own research efforts have been directed towards that end for the last seven or eight years, and I feel that we have fulfilled our commitments. Further, the former Ministry of Technology awarded £300 000 to the UKAEA who, jointly with Fielding and Platt, have done much of the initial spade-work in trying to get hydrostatic extrusion accepted by industry. It has been somewhat disappointing that more of their machines have not been sold; they are not all that expensive. Then again a considerable amount of government money has been poured into the National Engineering Laboratory researches under Dr Pugh to advance the process of hydrostatic extrusion. He has devoted his whole life to this topic, and he has made great contributions which will undoubtedly enable the process eventually to be put into production. There is no limit to the possibilities which one can foresee for this versatile process, particularly in special applications such as the development of more economic materials in the sense both of finance and of the proper, efficient utilization of natural resources.

REFERENCES

1. H. Ll. D. PUGH (1972). Hydrostatic extrusion—A review, *Annals. of the C.I.R.P.*, **21**/2, 167–86.
2. H. Ll. D. PUGH (1965). Recent developments in cold forming, Bulleid Memorial Lectures, University of Nottingham.
3. H. Ll. D. PUGH (1971). The application of hydrostatic pressure to the forming of metals, chapter 9 of *Mechanical Behaviour of Materials under Pressure*, Elsevier Publishing Co., Amsterdam, London, New York.
4. J. M. ALEXANDER and B. LENGYEL (1971). *Hydrostatic Extrusion*, Mills and Boon.
5. J. M. ALEXANDER (1969). Hydrostatic extrusion, Report MW/FRB/2/69, BISRA Intergroup Laboratories of the British Steel Corporation.
6. J. M. ALEXANDER (1972). Critical comments on continuous hydrostatic extrusion, *Materials Science and Engineering*, **10**, 70–4 (American Society for Metals and Elsevier).
7. B. LENGYEL (1968). A semi-continuous hydrostatic extrusion production process, *Metals and Materials* (January).

in hot rolling including the effect of various temperature distributions, Electricity Council Research Centre, Report No. ECRC/M427.
12. J. M. ALEXANDER and C. S. HARTLEY (1973). On the hydrostatic extrusion of copper-covered aluminium rods, Int. Conf. on Hydrostatic Extrusion at Stirling University, NEL/AIRAPT (June).
13. F. J. FUCHS and G. L. SCHMEHL, Continuous hydrostatic extrusion, Proceedings of NEL/AIRAPT International Conference on Hydrostatic Extrusion, June 1973, Vol. 2, pp. 334–341.
14. K. OSAKADA, M. LIMB and P. B. MELLOR, Hydrostatic extrusion of composite rods with hard cores, *Int. J. of Mechanical Sciences*, **15**(4), 291, 1973.

DISCUSSION

Q. P. A. Woodrow, NRDC. Could continuous hydrostatic extrusion cope with the output speeds from continuous casting plants (e.g. Southwire process)?

A. I am not sure of the actual output speeds which are developed from the continuous casting plant of the Southwire process. The average input speed which would be accepted by a 'clamp' system of semi-continuous or continuous hydrostatic extrusion might be rather slow (I would estimate 1 inch per second as being a possible figure). However, it can be envisaged that the input speeds which could be accepted by either Dr Fuch's continuous extruders, or Mr Green's 'grooved roll' could be much higher than that. It seems to me that the 'clamp' system would be more suitable for larger diameter feed stock being continuously cast fairly slowly, whilst the systems of Dr Fuchs and Mr Green would be more suited to smaller diameter feed stock being cast at higher speed.

Q. J. W. Bassett, Delta Metal B.W. Ltd., Birmingham. A continuous extrusion facility fed from a continuous casting plant appears to offer great possibilities in reducing the capital cost of the equipment employed when compared with the high capital cost of conventional casting, re-heating and extrusion methods.

It would appear that to produce say 1 in diameter bar one might only require to continuously cast say 1 $\frac{1}{4}$ in diameter bar, and continuously extrude it down to 1 in diameter to give dimensional accuracy and required properties due to working.

My company would certainly be very interested in hearing of developments along these lines.

A. These comments are very interesting and mirror my own thoughts about possible developments of hydrostatic extrusion. In discussions I have had with metal manufacturers, however, there always seems to be reluctance to putting processes such as continuous casting and continuous extrusion in series because of the difficulties which would be introduced by the breakdown of any one of the many 'links' in the chain of operations. Whilst appreciating these problems I feel sure that the economic advantages to be gained would make the necessary expense of such a development worthwhile.

Q. S. E. Rogers (Chairman), DFRA, Sheffield. In your paper Professor Alexander, you have said you 'believe one reason that this exciting process has not yet come into industrial use is that there is much money invested in equipment which manufacturers know works extremely well and efficiently and has already been amortized against its original capital cost. It would be replaced by equipment which is new, possibly with operating difficulties because of the nature of the high pressure involved; and generally there is a reluctance to embark on something with which the manufacturers are not familiar'. Can you see this problem being solved in the not-too-distant future?

A. I think it will be solved in the near future, mainly because a number of firms are beginning to introduce hydrostatic extrusion processes, often with small presses carrying out operations for which hydrostatic extrusion has special advantages. As operating experience is gained I feel sure this will stimulate development of the more sophisticated operations and equipment such as that required for continuous extrusion which I have been describing. Increased knowledge and data will be available, hopefully to reveal economic advantages of hydrostatic extrusion which will give the necessary impetus for investment in this equipment. Also a number of large hydrostatic extrusion presses have already been sold to industry and these will clearly provide much of this initial operating experience, at least within those firms which have purchased them.

CROPPING OF ROUND BAR UTILIZING PLASTIC FATIGUE

by

HIDEAKI KUDO* and KIMIO TAMURA†

SUMMARY

As an alternative to existing bar and tube cropping techniques, a cropping method is described which utilizes plastic fatigue fracture and consists of roller grooving and subsequent rotary bending. Solid billets thus cropped at an appropriate number of rotations to fracture have flat and smooth cut ends and show no deterioration in ductility in a subsequent side pressing test. The present technique is most suitable for relatively brittle materials, such as high-speed steel, since only several tens of rotations are required to obtain a good cut end. Preliminary experiments in applying this method to tube cropping give promising results. Directions to shorten the cropping time for less brittle materials are suggested.

INTRODUCTION

Billets and slugs to be forged or extruded in subsequent operations are at present mostly cut from a bar stock by sawing, parting-off or shearing. Press shearing is widespread in mass production because of material saving and high productivity.

In order to avoid cut end defects and distortion which appear more or less in the conventional shearing operation, novel bar shearing processes which utilize a high hydrostatic pressure have been developed in Hungary and Japan[1,2]. Although they achieve excellent cut end surfaces, their inherent drawback is a relatively short tool life, especially in cropping steels and other high-strength metals and alloys.

A very different process of chipless cropping has been reported in the USSR[3]. In this a knife-edged or screw-shaped roller die is pressed to a rotating bar or tube to form and enlarge a circumferential groove until material separation occurs.

We have tried to confirm the validity of this process by indenting simultaneously three knife-edged rollers to a circular rotating bar. But it was unsuccessful, since material separation took place only when the groove depth approached the bar radius and the lips produced on both sides of the groove became considerably thick and high. Another disadvantage of this technique is the considerable axial elongation that occurs during the process.

Here we report some experimental results obtained in developing a new cropping technique that consists of roller indentation for forming a circumferential

groove on the cylindrical surface of bar stock and subsequent rotary bending to cause plastic fatigue fracture. The necessary number of rotations to fracture and cut end aspects are investigated in relation to some process parameters with bars of three kinds of steel.

Application of the present technique to tube cropping is also attempted. Finally, directions to obtain satisfactory cut ends at a moderate cycle time of operation are discussed and suggested.

EXPERIMENTAL EQUIPMENT

A used turing lathe was modified into an experimental fatigue cropping machine (Fig. 1). A 20 mm diameter bar stock (1) to be cropped was clamped by the scroll chuck (2) attached to the main spindle. At the first stage of cropping operation, a knife-edged roller (3) was pressed into the rotating bar by the action of a pressure-oil cylinder to form a V-shaped groove circumferentially. The bending arm (4) which

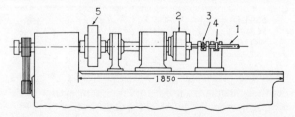

Figure 1. Schematic of experimental machine. (1) Bar stock to be cropped; (2) scroll chuck; (3) knife-edged roller; (4) bending arm; (5) electrical torque transducer.

* Yokohama National University
† Mechanical Engineering Laboratory, Japan

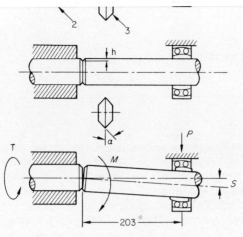

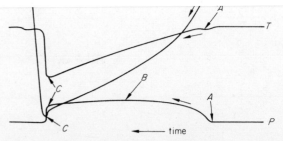

Figure 3. Typical oscillogram of bending load P, bar end deflection S and rotating torque T against time t. (A) Start of rotary bending; (B) approximate point of crack initiation; (C) start of final unstable fracture.

Figure 2. Sequence of new cropping process. (1) Ball bearing contained in bending arm; (2) jaws of scroll chuck; (3) knife-edged roller. (a) Roller indentation to form circumferential V-shaped groove; (b) rotary bending to cause plastic fatigue; (c) fatigue fracture and separation.

The roller indenting and bar bending loads were detected through load cells of a wire strain-gauge type, while the travel of the bending arm was detected by a dial gauge which included a potentiometer. The bar rotating torque was measured with an electrical torque pick-up (5 in Fig. 1). The maximum attainable speed of the main spindle was 400 rev/min.

TEST MATERIALS AND PROCEDURE

The bar materials tested were cold-drawn plain carbon steel (SS41-D), Cr–Mo alloy steel (SCM21) and high-speed steel of Co type (SKH 9). Their mechanical properties are summarized in Table 1.

TABLE 1   Mechanical properties of test materials

Material notation	Tensile strength kg/mm²	Elongation %	VPN
SS41-D	55·2	15·5	192
SCM21	59·7	34·1	208
SKH 9	72·9	17·1	263

In the present experiment, the effects of groove depth h ranging from 0·5 to 2 mm, semi-wedge angle α ranging from 15 to 45° of groove profile* (see Fig. 2c), and bending moment M were investigated in relation to the number n_f of rotations to complete fracture and cut end aspects. The number of rotations n was determined from the rev/min and the running time of the machine. The cut end aspects were

---

* A 0·1 mm radius is provided at the groove bottom.

nucleation was markedly retarded, apparently owing to a compressive stress developed continuously at the groove bottom by the roller edge. A predetermined constant bending load was therefore applied to the specimen after withdrawal of the grooving roller at a distance of 203 mm from the groove. Figure 3 illustrates a typical diagram of bending load P, bar end deflection S and rotating torque T against time t, recorded on an oscillogram. The standard rev/min of bar chosen in the experiment was 250.

RESULTS AND DISCUSSION

Nominal bending stress versus number of rotations to fracture

The nominal bending stress σ_n is calculated at the groove bottom disregarding the notch effect, normalized by dividing by the tensile strength σ_B and plotted in Fig. 4 against logarithm of the number n_f of rotations to fracture.

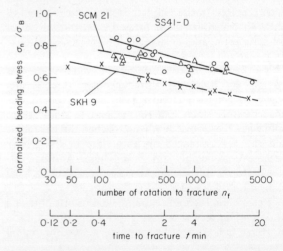

Figure 4. Relationship between normalized bending stress, i.e., ratio of nominal bending stress σ_n at groove bottom to tensile strength σ_B, and a number of rotations n_f of bar stock to fracture ($h = 0.5$ mm, $\alpha = 45°$).

An approximately linear relation is observable. The n_f value decreases, for a given σ_n/σ_B, as the tensile strength σ_B increases, that is, as the bar material is harder. In another series of experiments, n_f was found to decrease with increasing groove depth h and less markedly with decreasing groove angle α. An increase of the machine rev/min from 20 to 400 did not result in any appreciable change of n_f.

Because of the low rigidity of the present experimental machine, no attempt was made to determine the minimum length of cut billet that was attainable in the present cropping process.

Cut end aspects

Figure 5 illustrates a typical cut end appearance, which consists of areas of initial cracking A, steady fatigue fracture B and final unstable fracture C. The initial cracking started at about the middle of the cropping duration (at B in Fig. 3), while the unstable fracture was accompanied by a rapid increase in the bar end deflection.

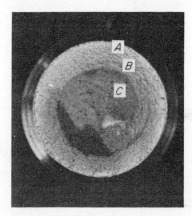

Figure 5. Typical end appearance cut at relatively small number of rotations to fracture. (A) Area of initial cracking; (B) area of steady fatigue fracture; (C) area of final unstable fracture.

The initial crack zigzagged from the groove bottom to a depth of about 0·1 to 0·2 mm. It was found that this area was closely related to the work-hardened zone developed by the roller indentation. In a subsidiary experiment, in which the specimen groove was machined, the initial crack showed a straight path. But the narrow area of the initial cracking would cause no serious problems for billets subsequently being processed.

The area of steady fatigue fracture was fairly smooth and flat, while the area of unstable fracture for SS41-D and SCM21 exhibited a rough and, in many cases, severely curved surface. Scanning electron micrographs of them are reproduced in Fig. 6. Figure 6a, taken from area B, is a typical fatigue fracture surface exhibiting the striation pattern. Figure 6b is a typical surface of ductile fracture characterized by the dimple pattern. The area of unstable fracture of SKH9, however, was smooth and flat, and revealed traces of brittle fracture. For all test materials, the area of unstable fracture narrowed as the bending moment M, and accordingly the number n_f of rotations to fracture, increased (Fig. 7a, b). The n_f necessary to obtain a smooth and flat surface was lower for the more brittle material.

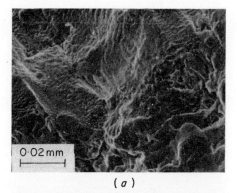

0·02 mm

(a)

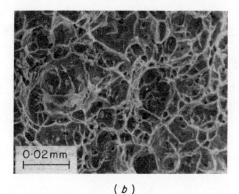

0·02 mm

(b)

Figure 6. Scanning electro micrographs of cut end. (a) Area B of Fig. 5; (b) area C.

Through the side pressing test down to 50 per cent reduction in height of cropped billets, the ductility of the smooth end surface was not inferior to that of the machined surfaces. This confirmed that the billets properly cropped through plastic fatigue could be successfully processed in subsequent forming operations without intermediate annealing. Since the necessary groove depth to attain satisfactory cut end was so small that no appreciable axial elongation took place in the bar stock, we expected to obtain billets of a precise length.

Preliminary experiment in tube cropping

Using bearing steel tubes, a preliminary experiment was conducted to apply the present fatigue fracture method to tube cropping. Figure 7c illustrates some of the cut end appearances. In tube cropping, the provision of circumferential grooves on both the outer and inner surfaces of tube was found favourable for obtaining a good cut end.

Further experimental work in tube cropping is now being done to improve the present technique.

Productivity problems

It was found that the main problem which could arise in the application of the present method to industrial production was how to shorten the cycle time of the operation without losing surface quality. A high rotating speed would be one solution to this.

Since the final unstable fracturing was thought to be accelerated because of overload in the experiments described above, some preliminary experiments were conducted in which the speed of bar deflection was controlled with a stopper at an intermediate stage of cropping process.

(a)

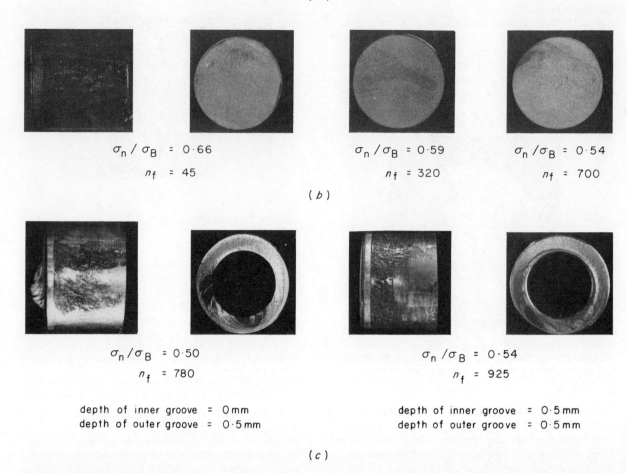

σ_n / σ_B = 0·66 σ_n / σ_B = 0·59 σ_n / σ_B = 0·54

n_f = 45 n_f = 320 n_f = 700

(b)

σ_n / σ_B = 0·50 σ_n / σ_B = 0·54

n_f = 780 n_f = 925

depth of inner groove = 0 mm   depth of inner groove = 0·5 mm
depth of outer groove = 0·5 mm   depth of outer groove = 0·5 mm

(c)

Figure 7. Cut end appearance of bars and tubes (h = 0·5 mm, α = 45°). (a) Cold-drawn carbon steel bar SS41-D; (b) high-speed steel bar SKH-9; (c) bearing steel tube.

The diagrams adopted of deflection S against number n of rotations are shown by the solid lines in Fig. 8 together with the point of complete fracture (x) and photographs of cut end appearance.

In comparing Figs 7a and 8, it will be seen that a cut end appearance of SS41-D similar to that obtained with an n_f of 2100 and a constant bending load was attainable with an n_f of 800 when S was controlled to decrease the bending load towards the termination of cropping. A tentative diagram of S against n to minimize n_f without losing surface quality is suggested by the broken line in Fig. 8.

An optimum control of bar end deflection or bending load will be attained if a suitably designed machine is constructed equipped with automatic or adaptive control devices.

CONCLUSION

The possibility and usefulness of a new bar and tube cropping technique based on plastic fatigue fracture in billet production was investigated for the purpose of overcoming drawbacks inherent to the existing cropping techniques. The use of this technique was found promising in obtaining smooth and non-brittle cut ends of billet at a moderate production rate, especially for relatively high strength materials such as high-carbon, tool, bearing and high-speed steel bars. Successful application to tube cropping was also anticipated.

There still, however, remains much to be solved before the present technique is successfully put to use in industrial production.

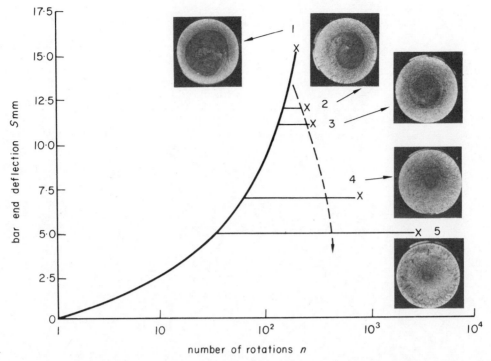

Figure 8. Bar end deflection versus number of rotation diagram and cut end appearance in deflection controlled fatigue cropping.

ACKNOWLEDGMENTS

We thank Messrs H. Fujii and S. Takamura, Fujikoshi Co., and Mr A. Kuroda, Kuroda Instruments Co., for offering experimental equipment and materials. We are indebted to Mr M. Terasaki, Mechanical Engineering Laboratory, for carrying out the experimental work, and to Miss Y. Harada, Yokohama National University, for preparing and typing the manuscript.

REFERENCES

1. (Anon.) (1965). Cold-flow shearing, *Metal-working Production* (27 Oct.), 59.
2. T. NAKAGAWA and K. MIYAMOTO (1972). Tool life test in cramp shearing of steel bar or wire. *Proc. 12th Int. MTDR Conf.,* Macmillan, 567.
3. (Anon.) (1967). Component production by transverse rolling. *The Engineer* (10 Nov.), 611.

THE EFFECT OF TEMPERATURE AND STRAIN-RATE ON THE COLD AND WARM WORKING CHARACTERISTICS OF ALLOY STEELS

by

P. F. THOMASON, B. FOGG and A. W. J. CHISHOLM*

SUMMARY

The dynamic yield stress–strain relations for a selected range of alloy steels were obtained with the aid of a plane–strain cam plastometer at temperatures in the range 20 to 600°C and at constant plane-strain rates up to 52/s. The tests were carried out to 90 per cent compression (a natural strain of 2·3), which makes the data suitable for estimating extrusion and forging loads in cold or warm working processes. The results are presented in the form of mean equivalent yield stress curves to facilitate this type of application.

In addition to providing data on the dynamic yield stress–strain relations, the results also provide information on the effects of temperature on the basic ductility of the alloy steels.

INTRODUCTION

There is evidence to suggest[1] that tool pressures in the cold forging of some carbon and alloy steels can be reduced by as much as 50 per cent if the billets are pre-heated to temperatures ranging from 200 to 600°C. These pressure reductions should enable materials normally used in cold forging to be worked to even greater deformations without violating prescribed safe limits of tool stress. Alternatively they should enable some of the higher strength alloy steels, which have previously been excluded from the process because of tool strength limitations, to be formed on a commercial basis. This presupposes that lubrication and tool wear problems can be overcome.

There is the additional limiting factor of fracturing which occurs in such processes as upsetting, flanging, coining and, in certain cases, in the backward extrusion of cans. It is obvious that in the general context of cold forging, the influence of temperature and strain rate on the 'ductility' of these materials might be of equal importance to their effects on the yield stress–strain relations.

It has long been established[2] that in an extrusion process the punch or die pressure for a particular tool geometry, area reduction and friction condition is directly proportional to the mean yield stress of the material over the relevant range of equivalent strain. Preliminary work has shown that this reasoning is applicable to a process such as combined upsetting and radial forging[3], provided that a good estimate is made of the mean equivalent strain.

It follows that reasonably good estimates of the reduction in forging pressure should be possible by measuring the yield stress–strain characteristics of the steels at the appropriate values of pre-heat temperature, strain rate and total equivalent strain. It is realized, however, that the estimated reductions in pressures will only be valid if satisfactory methods are available for effectively lubricating the billets for deformation at the elevated temperatures.

In the present work, the dynamic yield stress–strain curves were obtained by means of a cam plastometer fitted with a plane-strain compression sub-press and in situ furnace, as previously described[4]. The compression tests were carried out to 90 per cent reduction (a total equivalent strain of 2·66) at various temperatures from ambient up to 600°C and at constant plane-strain rates from 3 to 52/s. The plane-strain yield stress curves were then converted to mean equivalent yield stress curves for direct use in calculating cold and warm forging loads and tool pressures.

MATERIALS AND SPECIMENS

The materials for the plane-strain compression tests were supplied to BS 970 (1955) and the chemical analyses are shown in Table 1. Before machining the materials into plane-strain compression specimens, tests were carried out to ensure that they were free from internal defects and to measure the initial anisotropy of the yield stress. The tests, which are described elsewhere[5], showed the materials to be sound, with a fine grain size and a negligibly small initial anisotropy.

The materials were machined into plane-strain compression specimens of 0·2 in thickness, 1·25 in

---

* Department of Mechanical Engineering, University of Salford, Salford M5 4WT

En 34	0·17	0·21	0·47	0·016	0·019	1·92	0·20	0·25	
En 56D	0·32	0·39	0·34	0·013	0·022	0·43	13·12		
En 58B	0·050	0·63	0·85	0·018	0·029	9·05	18·08		Titanium 0·63
Comm. grade mild steel	0·15	0·24	0·65	0·035	0·010	0·20			Aluminium 0·009 Nitrogen 0·010

TABLE 2   Annealing treatment and resulting hardness, for each steel

Steel (B.S. 970 1955)	Heat treatment	Vacuum furnace	Hardness HV 30 after annealing
En 2A	1 h at 890°C	Furnace cooled	121
En 8	1 h at 800°C	Furnace cooled	172
En 11	1 h at 870°C	Furnace cooled	245
En 19C	1 h at 870°C	Furnace cooled	230
En 30B	1 h at 680°C	Furnace cooled	270
En 34	1 h at 920°C	Furnace cooled	162
En 56D	1 h at 950°C $\frac{1}{2}$ h at 700°C $\frac{1}{2}$ h at 600°C	Furnace cooled	197
En 58B	1 h at 1050°C	Water quenched	150
Comm. grade mild steel	1 h at 920°C	Furnace cooled	115

width and 8 in length, and were ground to 10 μin CLA index. After machining, the specimens were annealed in a vacuum furnace (for details see Table 2) and then treated with a colloidal graphite–cadmium oxide lubricant[5,6], to give controlled friction conditions between die and specimen during the dynamic plane-strain compression tests.

EXPERIMENTAL METHODS AND ANALYSIS OF RESULTS

A detailed description of the methods of operating the cam plastometer and performing the tests has been given elsewhere[4,5].

The cam plastometer test results were obtained in the form of load–strain oscilloscope photographs, which were then converted to plane-strain yield stress curves and corrected for the effects of die-face friction[5,6]; the results being repeatable to within ±2 per cent. The plane-strain yield stress curves were then transformed into equivalent yield stress curves and the first step consisted of converting the yield stress in plane-strain P to an equivalent uniaxial yield stress $\bar{Y}$, with the aid of the von Mises yield criterion in the form

$$\bar{Y} = P/1·155 \tag{1}$$

The plane-strain ϵ_p was then converted to an equivalent uniaxial strain ϵ by the equation

$$\bar{\epsilon} = 1·155\,\epsilon_p \tag{2}$$

which is based on the hypothesis that the state of work-hardening is a function only of the total plastic work[7].

For direct application in metalworking calculations the equivalent yield stress curves must be transformed into mean-equivalent yield stress curves, and this is done by systematically calculating the total work done per unit volume of plastic material W (i.e., the area under the equivalent yield stress curve) at various total equivalent strains $\bar{\epsilon}$. The mean-equivalent yield stress Y_m at the strain $\bar{\epsilon}$ is then given by the relation

$$Y_m = W/\bar{\epsilon} \tag{3}$$

The advantage of having the results in the form of mean-equivalent yield stress curves is that the frictionless extrusion pressure P_0 in a metalworking process is equal in magnitude to the mean work done per unit volume of plastic material. Hence, if an expression for the mean-equivalent strain $\bar{\epsilon}_m$ in an extrusion process is available, the mean work done per unit volume will be given by $Y_m\bar{\epsilon}_m$, where Y_m is the mean-equivalent yield stress at an equivalent strain $\bar{\epsilon} = \bar{\epsilon}_m$. The frictionless extrusion pressure P_0 is thus given by the equation[8]

$$P_0 = Y_m\bar{\epsilon}_m \tag{4}$$

The extrusion pressure P_0 can be corrected for the effects of friction between the extrusion-billet and tooling[8]; the tool loads to carry out the extrusion process can then be determined directly. The primary requirement of this method for estimating extrusion pressures is an expression for the mean equivalent strain $\bar{\epsilon}_m$ in the extrusion process, which includes redundant plastic straining; such expressions are available for a wide range of extrusion geometries.

EXPERIMENTAL RESULTS

Dynamic plane-strain compression tests were performed on the range of steels listed in Table 2, and the complete set of plane-strain yield stress curves are given elsewhere[5]; however, a typical set of results for En 8 medium carbon steel is shown in Fig. 1. In all cases the plastic flow stress, at large strains, was found to decrease with increased strain rate, due to a reduction in conduction heat-losses from the deforming metal[4,5]. Hence, for reasons presented in a previous paper[5], the results obtained at the highest compression strain rate were taken as those most appropriate for application to metal-working calculations; the high strain rate data is therefore, presented in the form of mean-equivalent yield stress curves (Fig. 2). Also plotted, for comparison with these results, are a number of quasi-static isothermal yield stress curves. As heat conduction losses were relatively small at the higher strain rates, the dynamic curves are described as 'adiabatic' or 'polytropic'[4]; it needs to be emphasized, however, that true adiabatic conditions were not achieved in the present experiments. This was particularly the case towards the end of the compression stroke where the yield stress curves displayed an upward sweep (see Fig. 1); the results in Fig. 2 were corrected for this effect[5]. An indication of the general effect of an increase in test temperature on the dynamic yield stress characteristics of the complete range of steels is given by the composite graph in Fig. 3, where the 'adiabatic' mean-equivalent yield stress Y_m at an equivalent strain of 2·66 is plotted against the specimen pre-heat temperature. This graph indicates that all the steels exhibit an appreciable reduction in yield stress with increasing temperature, although some of the steels

show a slight retardation in the thermal softening effect in the temperature range 400 to 500°C, notably the En 2A mild steel, the En 8 and En 30B steels and the En 56D stainless steel.

An important feature of the plane-strain compression tests was the intervention of edge-cracking and distortion effects (plastic instability) for certain steels in the intermediate range of pre-heat temperatures. These effects, which are summarized in Table 3, became so severe in certain cases that the test results had to be discarded.

DISCUSSION

Effect of temperature on yield stress

All materials tested in this programme show an appreciable overall reduction in yield stress with increasing pre-heat temperature, although the rate at which the yield stress decreases is different for each material (Fig. 3). The En 58B austenitic stainless steel and the En 2A mild steel exhibit the greatest amount of thermal softening as the pre-heat temperature increases from 20 to 300°C; for temperatures in the range 300 to 400°C, the En 2A mild steel and the En 8 medium-carbon steel show little further reduction in yield stress, and this low thermal softening sensitivity is thought to be the result of accelerated strain ageing effects[5]. A similar reduction in thermal softening sensitivity has been observed[1] in this pre-heat temperature range when performing warm extrusion operations on En 2A mild steel billets.

The rest of the alloy steels show a fairly gradual decrease in yield stress as the temperature is raised to about 500°C, followed by a large reduction between 500 and 600°C (Fig. 3). The behaviour in this region is thought to be due to accelerated spheroidization effects[5].

Effect of strain rate on yield stress

The plane-strain yield stress curves for all the steels tested in the present work were found to exhibit a low strain-rate sensitivity for pre-heat temperatures up to 600°C. In fact, the yield stress curves at plane-strains in excess of 0·5 were lowered by an increase in strain rate, and the results for En 8 in Fig. 1 are typical in this respect. The most likely explanation for this inverse strain-rate effect, at large plastic strains, is that when high strength materials are plastically deformed at high strain-rates a large increase in temperature occurs within the metal, in direct proportion to the total plastic work. Hence, the single compression, or polytropic yield stress curves, for metals of high thermal-softening sensitivity and low strain-rate sensitivity, are expected to display a reduction in yield stress with increasing strain-rate, once the internal temperature rise is sufficient to outweigh the effect of initial strain-rate sensitivity. It follows from this, that metals with a high thermal softening sensitivity and a low strain-rate sensitivity should have yield stress curves which are reduced in magnitude as the strain-rates are increased towards a limiting value corresponding to adiabatic conditions. This effect is clearly demonstrated for the En 8 steel in Fig. 1, where it can be

TABLE 3 Relationship between specimen pre-heat temperature and fracture effect for the plane-strain compression tests.

steel (B.S.970 1955)	initial temperature of specimen				
	condition of specimen				
	no fractures	edge fractures	separation	distortion	remarks
En 2A	23°C 100°C 200°C 300°C 600°C	400°C 500°C			cracks very small at 400°C and 500°C
En 8	600°C	21°C 100°C 200°C 300°C 400°C 500°C			cracks large at 300°C and 400°C small at 500°C
En 11		300°C 400°C 500°C 600°C		400°C at 52/s 500°C at 26/s and 52/s	cracks very small at 600°C
En 19C		300°C 400°C 500°C 600°C		400°C at 52/s 500°C at 26/s and 52/s	cracks very small at 600°C
En 30B	600°C	21°C 100°C 200°C 300°C 400°C 500°C			cracks small at 500°C
En 34		23°C 100°C 600°C	200°C 300°C 400°C 500°C load trace distorted		cracks very small at 600°C
En 56D		23°C 100°C 200°C 300°C 400°C 500°C 600°C			cracks small at 500°C very small at 600°C
En 58B	23°C 100°C 200°C 300°C 400°C 500°C 600°C				
comm. grade mild steel		23°C 100°C 600°C	200°C 300°C 400°C 500°C load trace distorted		

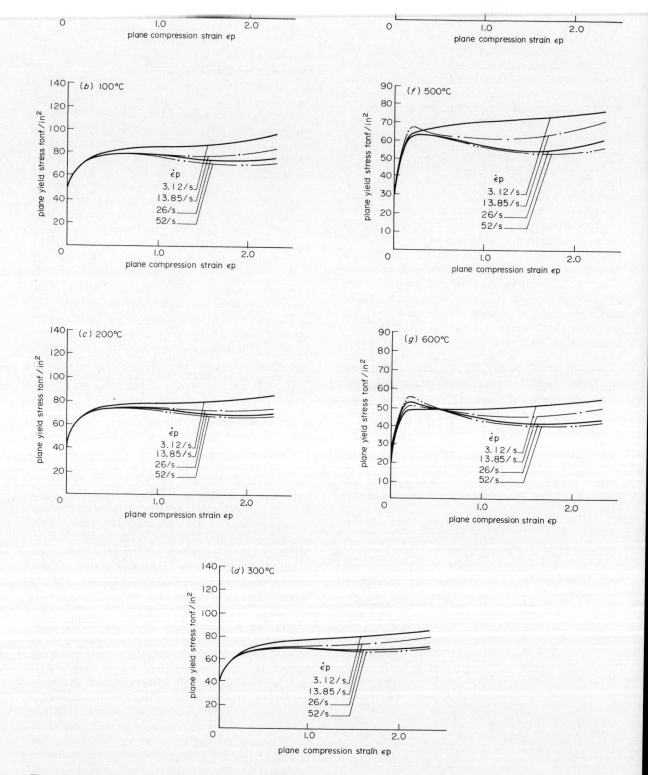

Figure 1. Plane strain yield stress curves for En 8 medium carbon steel at pre-heat temperatures from 21°C to 600°C and at various constant plane strain rates.

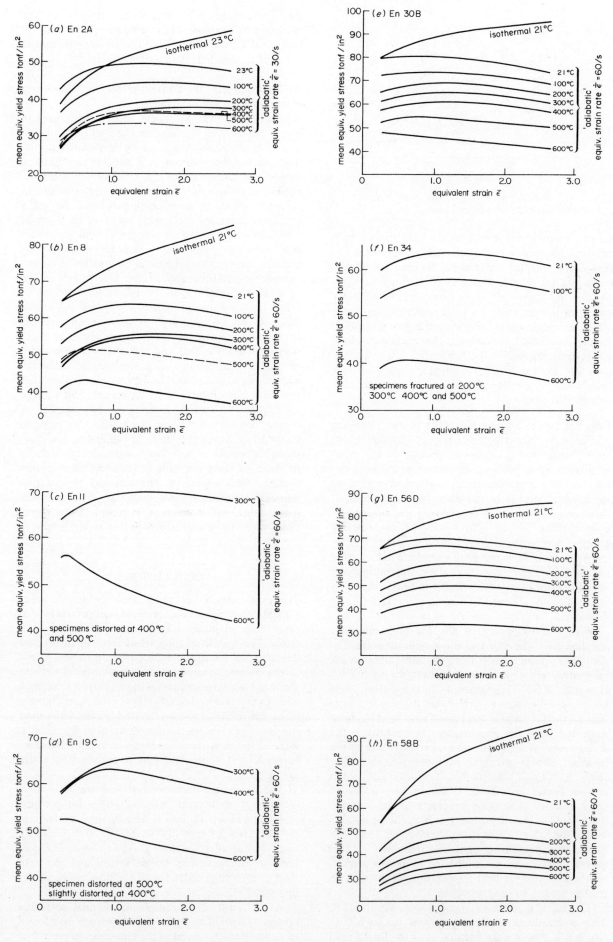

Figure 2. Mean equivalent yield stress Y_m curves for each steel at various pre-heat temperatures.

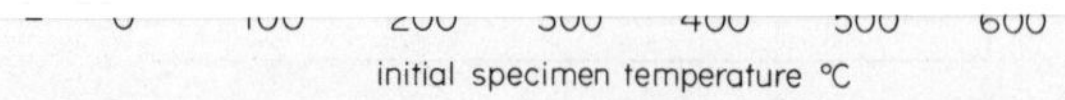

Figure 3. 'Adiabatic' mean equivalent yield stress at an equivalent strain of 2·66 for various steels.

seen that very little lowering of the yield stress curves can be expected beyond a plane-strain-rate of 52/s. Hence, for practical purposes results obtained at this strain-rate can be considered as approximately 'adiabatic' plane-strain yield stress curves, and such results are directly applicable to those metalworking processes which are carried out at speeds giving approximately 'adiabatic' conditions. The mean-equivalent yield stress curves for the 'adiabatic' strain-rate are therefore given in Fig. 2, for direct application to the calculation of metalworking loads by the methods described above.

Edge fracture as a guide to ductility in warm working
The edge fracture effect, which was observed in most steels, with varying degrees of severity in the intermediate range of pre-heat temperatures (Table 3), can be used as a semi-quantitative method of estimating the forging ductility of the alloy steels[5]. The basis of this method consists of comparing the severity of edge-cracking in the alloy steel, at a particular pre-heat temperature, with that observed in the En 2A mild steel. Since the En 2A steel is a cold forging steel of high ductility, any alloy steel displaying edge-fractures of a severity similar to or less than that material can be assumed to have an acceptable forging ductility.

With the exception of the En 58B austenitic stainless steel, the edge-fracture results show that the alloy steels have a much lower ductility than En 2A mild steel at pre-heat temperatures up to 500°C. However, at 600°C the ductility of these alloy steels has increased to such an extent that it virtually equals that of the En 2A mild steel (see Table 3); it has been suggested[5] that the improvement in ductility is the direct result of an accelerated spheroidization of the lamellar carbide particles in these pearlitic alloy steels. This result suggests that, if pearlitic alloy steels are pre-heated to 600°C, it should be possible to carry out warm working processes to deformation limits only achieved normally at lower temperatures when the alloy steels have received a long and expensive spheroidizing annealing treatment[9]. These general conclusions are supported by recent work[10] on the effect of pre-heat temperature on ductility in cold and warm heading processes.

The En 58B austenitic stainless steel was found to have a ductility, at all temperatures up to 600°C, which is probably greater than that of the En 2A mild steel. With the exception of the En 58B steel, all the

methods described earlier, it is possible to obtain an immediate indication of the pre-heat temperature needed to reduce the tool loads for alloy steels to an acceptable level, by an inspection of Fig. 3. Since the extrusion pressures for a wide range of En 2A mild steel components give an acceptable tool life under cold working conditions[1,2] (i.e. billets initially at ambient temperature) it should be possible to warm-extrude successfully any of the alloy steels if the pre-heat temperature of the billets is sufficient to bring their mean-equivalent yield stress below the value for En 2A mild steel at ambient temperature; i.e. below the line $A-A$ in Fig. 3. The results in Fig. 3 suggest, therefore, that any one of the present range of medium-carbon and alloy steels could be warm-worked with acceptable tool pressures, if the billet pre-heat temperature is in excess of 550°C. The success of the warm working process would, of course, depend on the use of a satisfactory elevated temperature lubricant and, in addition, the tool materials would require hot-hardness characteristics in excess of these normally present in cold extrusion tooling. With regard to the elevated temperature lubrication problem, the colloidal graphite–cadmium oxide lubricant[6] used in the present experiments might prove to be satisfactory in this application.

CONCLUSIONS

The cam plastometer yield stress curves show, in general, that moderate pre-heat temperatures can cause large reductions in the yield stress of alloy steels. The 'pearlitic' alloy steels show relatively low ductility for pre-heat temperatures up to 400°C but as the pre-heat temperature approaches 600°C these steels show a large reduction in yield stress combined with an increase in ductility; these effects are thought to be the result of the accelerated spheroidization of the lamellar carbides. In the region of 400 to 500°C some of the alloy steels are found to exhibit increased hardness and brittleness due to accelerated strain ageing effects.

The austenitic stainless steel and the mild steel show large reductions in yield stress for pre-heat temperatures up to 300°C and further increases in pre-heat temperature up to 600°C give relatively small additional reductions in yield stress. The ductility of the austenitic stainless steel and the mild steel is of a high order for all pre-heat temperatures up to 600°C.

As a result of predominant thermal softening effects, the yield stress of the alloy steels was found to depend on the speed of the compression tests. The faster the rate of deformation, the lower the amount of heat transfer from the plastically deforming metal, and the lower the yield stress; this effect should only occur, however, when the rate of deformation is less than that required for adiabatic deformation.

ACKNOWLEDGMENTS

The authors gratefully acknowledge the very helpful advice and encouragement given by Mr G. H. Townend, Research Manager, GKN Group Research Laboratories, during this programme of work.

REFERENCES

1. F. HOWARD and H. A. J. DENNISON (1966). *Sheet Metal Industries*, **43** 183.
2. H. Ll. D. PUGH, M. T. WATKINS and J. McKENZIE (1961). *Sheet Metal Industries*, **38**, 253.
3. P. E. WATSON (1968). *Forging of slabs in closed dies, undergraduate project report*, University of Salford, March.
4. P. F. THOMASON, B. FOGG and A. W. J. CHISHOLM (1968). *Advances in Machine Tool Design and Research*, Pergamon Press, London, 287.
5. P. F. THOMASON, B. FOGG and A. W. J. CHISHOLM (1969). University of Salford, Mechanical Engineering Research Report no. 69/33.
6. P. F. THOMASON and B. FOGG (1970). *Tribology in iron and steel works,* Iron and Steel Inst. publication no. 125, 142.
7. R. HILL (1950). *The Mathematical Theory of Plasticity*, Oxford University Press, 172.
8. W. JOHNSON and H. KUDO (1962). *The Mechanics of Metal Extrusion*, Manchester University Press, 108.
9. H. D. FELDMANN (1961). *The Cold Forging of Steel*, Hutchinson.
10. P. F. THOMASON (1969–70). *Proc. Instn. Mech. Engrs.*, **184**, 885.

THE FORMING OF AXISYMMETRIC AND ASYMMETRIC COMPONENTS FROM TUBE

by

M. E. LIMB*, J. CHAKRABARTY*, S. GARBER† and P. B. MELLOR‡

SUMMARY

A hydraulic tube-forming machine has been designed and manufactured to investigate the cold forming of axisymmetric and asymmetric components from circular tube. Radial expansions of 50 per cent have been achieved with axisymmetric components, and tee-pieces have been successfully formed in a wide range of materials.

INTRODUCTION

In the last twenty years there has been increasing interest in the cold forming of components from cylindrical tube. It has been common practice for some years to form copper tee-pieces for domestic water systems and more recently similar methods have been applied to steel tee-pieces, having the main branch of the tee up to 12 inches in diameter. The cold forming from steel tube appears to have originated in Japan and has been reported by Ogura, Ueda and Takagi[1].

The forming of a tee-piece is a relatively simple matter when compared with axisymmetric tube expansion. In both cases however the aim is to change a process from one of stretch-forming to one of drawing so that new material is continuously fed into the working region. Both processes depend on the combined loading of an internal fluid pressure and an axial compressive load, but the relationship between the fluid pressure and the axial load is much more critical in the case of axisymmetrical expansion.

Fuchs[2] has achieved uniform radial expansions of 100 per cent without any thinning in copper tubes. The operation was carried out in a high fluid pressure environment which meant that axial frictional forces were very small. For the general run of axisymmetric components a smaller radial expansion is generally required and there is a need for simpler tooling. Wallick[3] has described a simple process for the radial expansion of tubes of up to 2 in diameter in which more modest but useful expansions were obtained.

The present work is concerned with the development of a machine to investigate the optimum conditions and the final strains obtained when forming components from tube. Various materials with widely different properties have been tested.

THE TUBE BULGING MACHINE

The layout of the machine can be appreciated from Fig. 1. An axial compressive end load is applied to the tubes by two horizontally opposed rams each of 30 tonf capacity. The bottom half of the die block is bolted onto the bed of the machine and the top half is lowered and held in position by a vertical hydraulic ram of 30 tonf capacity. All the rams are double acting.

The horizontal rams are driven by a small Madan air-hydro pump and the internal fluid pressure is generated by a Madan double acting air-hydro pump which will give a fluid pressure up to 25 000 lbf/in². This latter pump is located beneath the bed of the machine. The machine is manually operated from the control console which also contains gauges indicating the pressure at the rams.

A requirement of the present process is that the horizontal rams move in by equal amounts. This cannot be achieved by applying equal pressures to the horizontal rams since the frictional resistance at the two ends of the tubes can be very different, resulting in unequal and uncontrollable movements of the rams. The problem was solved by designing an equal volume chamber (Fig. 2). Oil from the pump is

* Department of Mechanical Engineering, University of Birmingham
† Department of Industrial Metallurgy, University of Birmingham
‡ School of Mechanical Engineering, University of Bradford

Figure 1. The tube bulging machine.

delivered to the underside of piston A which in turn drives two identical but isolated pistons B and C. The connections from the equal volume chamber to the horizontal rams are made of equal lengths of high pressure steel piping. In practice the movements of the horizontal rams were controlled to within 0·010 in of each other. During continuous operations it was necessary at intervals to bring both rams and the equal volume chamber to a common datum. This operation was made simple by the addition of a four-way compensating valve through a non-return valve (Fig. 3).

During the forming operation the tubes have to be sealed at both ends to contain the fluid and arrangements have to be made to pre-pressurize the tubes before the start of the main forming operation. Oil is pumped into the tubes through holes drilled in the stems of the horizontal rams and the tubes are sealed by the plugs (Fig. 4), which are screwed into the ends of the rams. The outside diameters of the plugs are a loose fit inside the bore of the tube and sealing at low pressures is achieved by neoprene standard 'O' ring seals. At pressures exceeding $1500 \, \text{lbf/in}^2$ a seal is achieved by the pressure of the shoulders of the sealing plugs against the ends of the tube. The plug insert length in the tube is kept to a minimum so that the plugs do not interfere with the flow of the metal during the forming operation. The inlet plug (Fig. 4a) has a central hole through which oil is fed into the tube, and the plug at the opposite end (Fig. 4b) has in addition a drilled hole connecting the central hole to the top surface of the plug so that air can escape from the tube via a valve prior to forming. This valve also serves as a relief valve during the forming operation. It can be preset to any required pressure.

During the forming operation the pressure at the

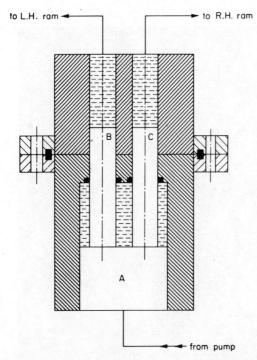

Figure 2. Schematic arrangement of the equal volume chamber.

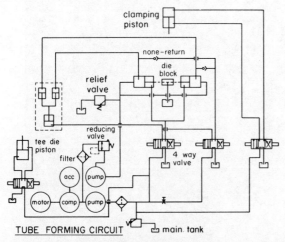

Figure 3. The hydraulic circuit.

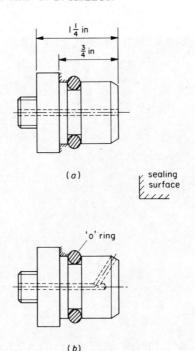

Figure 4. Sealing plugs.

input side of the ram is measured with a Kistler pressure transducer and a linear displacement transducer measures the movement of the horizontal rams. The signals from three instruments are fed via an amplifier to a U.V. recorder.

Tube materials
The initial dimensions and the mechanical properties of the tube materials are given in Table 1. All tubes had initial outside diameters of $1\frac{1}{2}$ in. Axisymmetric components were made from steel, aluminium and aluminium alloy tubing, and for the tee-pieces the additional materials copper and brass were also used.

The commercially pure aluminium tube had alternate longitudinal zones of coarse and fine grains (Fig. 5). This is a processing defect originating from the pattern of metal flow around the bridge dies during the hot extrusion. In this mechanical working process there are widely different amounts of work being imposed on the flowing metal depending on its

position relative to the fixed die supports. The subsequent recrystallization in the extruded tube depends on the particular local processing combinations of strain rate, strain and temperature and results in these longitudinally oriented zones of coarse and fine grain size. The behaviour of this material in subsequent cold forming operations depends on the stress system and this will be discussed later with reference to the two types of component.

TABLE 1

Material	Thickness t (in)	Tensile strength (tons/in^2)	Hardness (V.P.N.)
Steel	0·080	26·5	137
Annealed copper	0·048	15·2	39
	0·064	15·2	38
Com. pure aluminium EIC	0·064	5·3	31·1
	0·080	5·8	31·4
Aluminium alloy HV90	0·064	6·3	33·0
	0·084	6·5	33·0
Brass 70/30 annealed	0·048	21·0	85
	0·064	24·0	109

Forming of axisymmetric components
Fig. 6 illustrates the type of failure that is likely to occur in the forming of an axisymmetric component. Fig. 6(a) shows the original tube and Fig. 6(d) the successful component with 50 per cent radial expansion, which is now much shorter in length than the original tube. Fig. 6(b) illustrates the small amount of expansion that is possible if an increasing internal pressure is applied without any axial movement of the rams, and Fig. 6(c) the case when there is insufficient movement of the rams. The most common type of failure is the fracture shown in Fig. 6(b) and (c) but gross buckling of the tube can occur if the internal pressure is too small. The component shown in Fig. 6(d) was formed before the equal

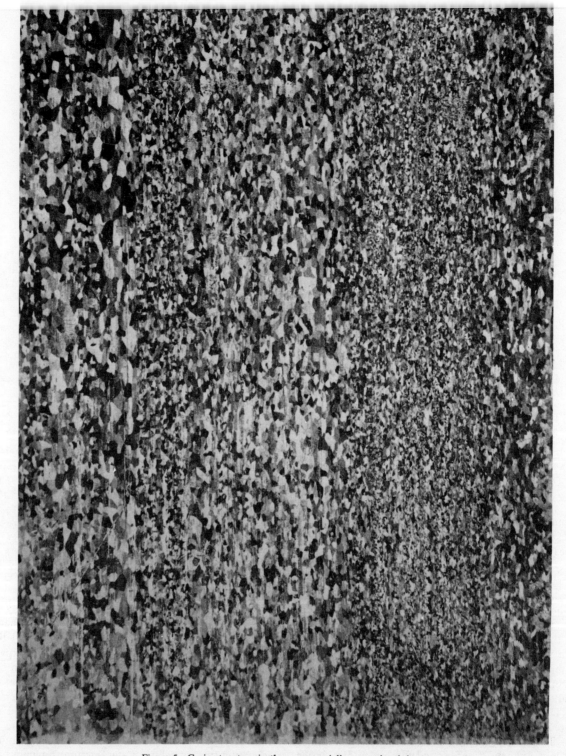

Figure 5. Grain structure in the commercially pure aluminium.

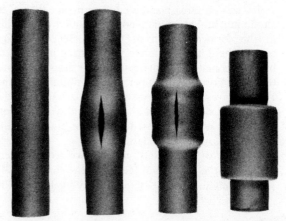

Figure 6. The forming of an axisymmetric component in steel.

volume chamber was fitted. Although the component was successful as regards radial expansion, it is off centre in the axial direction. The maximum thinning strain in this particular component is 0·24 and the axial shortening of the tube is 18 per cent.

The bursting pressures of the original tubes were determined for each material by increasing the internal pressure while keeping the axial load just large enough to form an effective seal. It was found that the well-known mean diameter formula predicted the bursting pressure p_u with reasonable accuracy. This has the form

$$p_u = 2 \,(\text{T.S.}) \frac{(K-1)}{(K+1)}$$

where T.S. is the tensile strength of the material and K is the ratio of the outside diameter to the inside diameter of the original tube.

The sequence of the axisymmetric tube forming was as follows. A tube was placed in the die and the axial rams brought into position to seal the tube. The vertical ram was then lowered and the die closed around the tube. Oil was then pumped into the tube until a certain pre-set pressure was reached. Finally the axial rams were advanced to compress the tube and the combination of axial force and increasing internal pressure formed the tube. It should be noted that in the forming of the axisymmetric components the internal volume increased and that therefore it was necessary to keep supplying oil to the inside of the tube throughout the process.

Under the tensile hoop stresses generated in the simple axisymmetric forming, plastic deformation in the commercially pure aluminium commenced in a non-uniform manner at the distinct coarse-grained zones because the yield stresses there were significantly lower than in the adjacent fine-grained structure. The circumferential variation in wall thickness observed in the component after straining originated from this inhomogeneous plastic behaviour; also, instability followed by fracture invariably occurred at these locations.

Experiment showed that the most satisfactory method of forming thin material was to increase the internal pressure as a step function during the ram input stroke (Fig. 7). By adopting this procedure premature instability was avoided. This method was

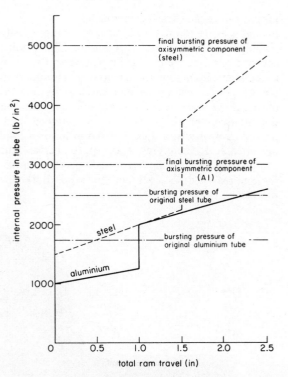

Figure 7. Variation of internal pressure with ram travel when forming axisymmetric component.

also used for steel but here the technique of forming was not as critical. Fig. 8 shows the wall thickness variation along a steel component formed to a radial expansion of 50 per cent. This component was formed after the equal volume chamber had been fitted and the expansion is situated centrally along the axis of the formed component. Notice that in this case the maximum thinning strain is only 0·06.

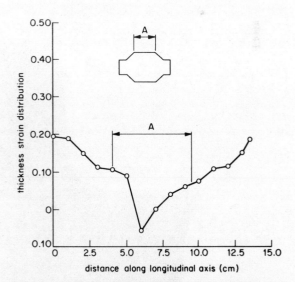

Figure 8. Variation of thickness strain for an axisymmetric steel component, at a radial expansion of 50 per cent.

There appears to be no published theoretical analysis of this type of process. Chakrabarty[4] has considered the radial expansion of a tube, with axial shortening, for the particular case when there is no change in wall thickness.

the oil seal and to continue the axial shortening process.

existing in the forming of tee-pieces the commercially pure aluminium exhibited appreciably higher ductility

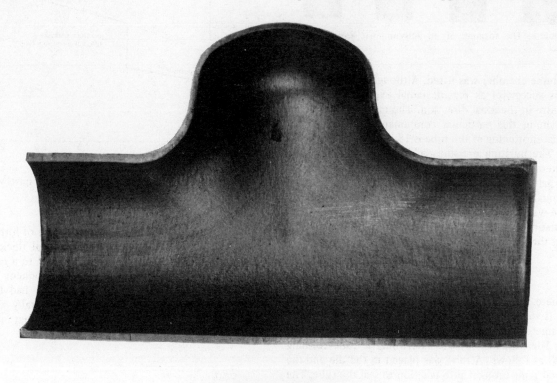

Figure 9. Sectioned tee-pieces in (top) copper and (bottom) aluminium alloy.

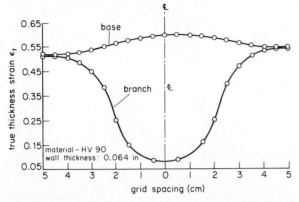

Figure 10. Thickness strain variation in tee-piece formed from aluminium alloy.

and performed satisfactorily during the large shape changes. The ductility limitations imposed by the combination of hoop tensile stresses and the coarse grained zones when forming axisymmetric components was here circumvented.

The amount of axial shortening that occurs in the forming of tee-pieces is illustrated in Fig. 11 for the aluminium alloy. The original length of 6·87 in has

Figure 11. Aluminium alloy tee-piece.

been reduced to 3·75 in and this reduction is taken up by an increase in wall thickness. It was again found convenient to increase the internal pressure at chosen positions of ram travel (Fig. 12).

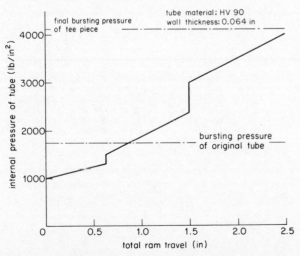

Figure 12. Variation of internal pressure with ram travel when forming tee-piece.

In all cases the die radius leading from the main branch to the tee was based on $\frac{3}{8}$ in, in a plane through the dome and parallel to the centre axis of the original tube. Great care was taken to blend this radius when manufacturing the die.

A variety of surface lubricants were used in forming tee-pieces: PTFE film, colloidal graphite, Rocol R.T.D. spray and Rocol A.S. paste. Lubrication assists the flow of the metal and changes the geometry of the dome. Without lubrication the dome produced is much more pronounced because of the large amount of stretch forming. Using a surface lubricant between the tube and die, the dome of the tee becomes much flatter, the branch piece increases 20 per cent in length, and the wall thickness of the branch becomes thicker.

The most effective lubricant is the PTFE film, but it would be too expensive to use on a production basis. Colloidal graphite and Rocol A.S. paste give the next best results and the latter apparently adheres to the forming die much longer than the colloidal graphite. These observations will be reported in detail at a later stage.

CONCLUSIONS

The machine has been found to operate well, within its load capacity of 30 tonf, and successful components have been formed. Much useful information has been obtained concerning the pressure and axial load requirements and the relationships between them. The forming of axisymmetric components presents a more difficult problem than the forming of tee-pieces. It would seem desirable when forming axisymmetric components to have a control system that will follow a preselected relationship between internal fluid pressure and ram travel.

Experience with the commercially pure aluminium emphasizes the fact that material properties are less important when forming tee-pieces than when attempting a large radial expansion.

ACKNOWLEDGMENTS

This investigation was carried out at the University of Birmingham and supported by a grant from the Science Research Council. The authors would like to thank Mr A. H. Cole for useful discussions on the equal volume chamber and Mr R. K. Jackson for carrying out mechanical and metallurgical tests. The authors would also like to thank Mr D. E. Burton, Technical Manager of Rocol Ltd, for useful discussions and for supplying the various lubricants.

REFERENCES

1. T. OGURA, T. UEDA and R. TAKAGI (1966). The hydraulic bulging process, *Industrie-Anzeiger*, 88, 770, 1001.

2. F. J. FUCHS (1965). Production metal forming with hydrostatic pressures, A.S.M.E. Paper, 65—Prod. 17.

3. C. R. WALLICK (1968). Tube bulging with axial pressure, *Metalworking Production*, 52.

4. J. CHAKRABARTY (unpublished). Axisymmetric tube expansion.

PRODUCTION CONSIDERATIONS FOR THE HIGH SPEED FORGING OF SPUR GEAR FORMS

by

A. R. O. ABDEL-RAHMAN and T. A. DEAN*

SUMMARY

The contents of this paper contain some of the results from a wide investigation into the commercial production of spur gear forms on high speed hammers. The design of tooling for this process is discussed and the performances of die sets made to empirically derived principles are determined. Several component shapes which have been successfully formed are illustrated and an examination of their quality is made.

INTRODUCTION

Gears are extensively used in power transmission for conveying torque, changing rotational speeds and the line of action and direction of those speeds. Gears of increased strength and reduced weight are needed in modern high speed mechanisms to cope with the high speeds employed and the greater loads that they will be subjected to. This situation is exemplified by the aero space and aircraft industries. A Rolls-Royce report[1] states that aero engine gears are now required to transmit greater loads, and Lavoie[2] mentions that gears are now forged for use in truck differentials and in helicopters. The design of gears is therefore critical in striking a balance between size to minimize weight and strength to avoid failure. Wear resistance and beam strength are the usual factors influencing the design of gears[3]. In most cases the design is based on the beam strength. To increase the beam strength requires either the use of stronger materials or the use of larger teeth. While stronger materials add to the cost of components, the use of larger teeth increases tooth sliding velocities and hence the rate of wear.

Gears are normally made by cutting methods which form the teeth either from bar stock, for low quality gears, or from forged blanks, for high quality gears. Forging a complete gear form can offer two main advantages:

(a) A saving in material can be made if gears can be forged with integral teeth.
(b) The forging offers an increase in the beam strength due to the contoured grain flow. Miller[4] reported that forged gears ran twice as long as machined ones in fatigue tests; and Parkinson[5] states that forged gears had seven times as much fatigue strength as machined ones as indicated by the average number of cycles to failure.

Gears forged with integral teeth are of two types, namely:

(1) Gears forged to finished dimensions without the need for a post forging machining process on the tooth profile. The majority of these are bevel gears.

A production process in Great Britain[6] enables the tooth profile of bevel gears to be completely forged to BSS 545 (1949) Class C, and Class B gears are made using a subsequent cold coining process. Tests[1] made on gears formed in Germany by this process showed that compared with a master form:

(a) the tooth profiles were identical
(b) the tooth sizes varied by a maximum of 0·0005 in
(c) the pitch line of the forged gears varies by 0·0005 in the axes.

(2) According to Kobyskovsky[7] a machining allowance of 0·020 in to 0·030 in is made on the profiles of spur gears forged in the USSR.;

A comparison of gears forged by the Western Gear Corporation, with American Standard Full Depth, shows the outer diameters to be oversize by 0·055 in and the root diameters to be oversize by between 0·017 in and 0·069 in. More than one grinding operation would be required to remove these amounts of excess material.

Spur gear forms have a profile section which is basically rectangular, compared with the conical sections of bevel gears. This makes the shapes far more difficult to form for the following reasons:

(a) The complete filling of vertical sided cavities requires considerable forging effort.
(b) The lack of natural draft (as exists for bevel gears) renders very difficult the removal of the components from the cavity.

---

* Department of Mechanical Engineering, University of Birmingham

of gear forging dies have been variously reported to be 300 components for forging[4] and in the case of finishing dies for bevel gears 4000 components[6]. Erosion occurs in the case of gear dies on the tooth profile. This is because during forging the teeth are surrounded by large volumes of hot billet stock and tend to soften. To reduce the overheating of the die insert, it is important to minimize the time for which the billet is in contact with the dies. An obvious way of doing this is to use a high speed forging machine. The Petro-Forge machine used in the investigations reported in this paper is attractive in this respect as it has a combined deformation and dwell time of the order of 5-10 ms and ejection of the component can be concurrent with the withdrawal of the ram thus minimizing the total contact time of hot stock and die.

DIE DESIGN

Preliminary die

A sound die design is of great importance to the production of components and must be such that a large quantity of components within specified limits of size may be produced. To obtain a preliminary assessment of the performance of a die constructed to a traditional pattern for HERF machines, using several bolted inserts, the dies shown in Fig. 1 were

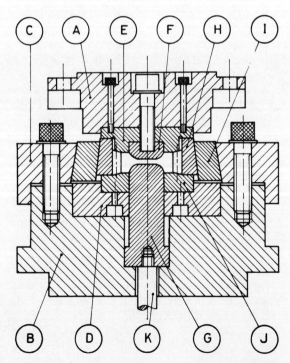

Figure 1. Preliminary die set (traditional HERF design).

(S1).
(c) The punch (E) is bolted to the top die holder (A) by screws (S2). The punch has an insert (F) shaped to produce the geometry required on the component's topface. The insert (F) is held to (E) by means of screw (S3).
(d) Ejection is effected by the cylindrical ejector (G) which also acts to form the central part of the lower face of the gear. The ejector is seated in a counter bore on the bolster (B), which also guides it during operation. The ejector is connected to the ejection cylinder (bolted to the machine below the bed) by the ejector rod (K) incorporating screw (S4).

Forging tests using this die set revealed the following features of the process:

(a) Scale and lubricant deposits accumulated in the lower part of the die cavity. These deposits prevented the complete filling of subsequent forgings. No easy method for clearing this detritus was available.
(b) A combination of the ejection force, acting in the centre of the component, and the frictional restraints between the sides of the die cavity and the gear caused the gear to bend about its axis.
(c) The cooling of the flash, which was sometimes formed between the punch (E) and the die insert (H), caused the component to stick to the punch and great difficulty was experienced in removing the forging from the tools.
(d) The screws S1, S2 and S3 were plastically deformed and frequently needed replacement.

It was also found that a rigid connection between ejector (G) and the ejector cylinder could not be adequately designed, for if the length of the ejector rod K was such that the ejector was seated before the hydraulic cylinder had reached the end of its stroke then the screwed portion was subject to a tensile stress of nominally about 50 T/in², (770 MN/m²) which together with the shock load from forging was sufficient to cause fracture. However, if the ejector rod was of a length to allow the ram of the ejector cylinder to reach the bottom of its stroke then the situation arose where the ejector was not fully seated. The full forging load was taken on the ejector rod which buckled.

(e) After a small number of forgings were made, the die insert cracked and later completely failed by breaking into four sectors. Ring (1) was designed to produce initial compressive stresses to counter the forging pressure. However, due to the relaxation of the clamping ring (C) when the screws slackened; the prestressing ring was rendered ineffective. It appeared that the screws slackened due to the alternating stresses produced by the clashing of the dies.

(f) The top of the tooth profile in the die insert became distorted due to the high loads engendered by the formation of a thin ring of hot forging stock which was tapped between punch and insert. The damage could be reduced by increasing the billet volume and hence increasing the thickness of this ring. However, increasing this flash reduces the utilization of the stock material.

Survey of design requirements

A survey of the observations obtained from the preliminary die showed that for a successful, commercially practicable gear forging process to be established, the following conditions have to be met.

(a) A provision for easily removing the deposit of scale and lubricants.

(b) A means of removing components from the dies without distorting them.

(c) A design of a cavity which avoids the sticking of the gear to the punch.

(d) A use of as few bolts as possible and a minimum number of die inserts.

(e) An elimination of the clashing of the punch and the die insert.

(f) A provision of an adequate and sustained compressive stress on the die. Insert to counter the forging pressure inside the die cavity.

(g) An avoidance of excessive waste of material.

Improved die design

The above information was made use of to design the dies illustrated in Fig. 2. The main features of this design are:

(a) The die insert is a duplex cylinder. This provides sustained compressive stresses on the inner

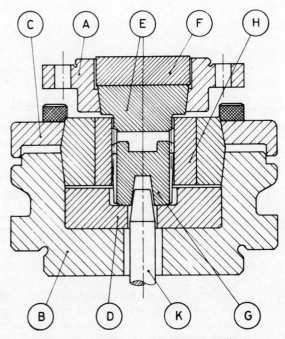

Figure 2. Improved design of spur gear die set.

ring and therefore offers advantages over a tapered clamping ring which depends for its effect on the tightness of the bolts.

(b) The die insert is tapered at its outer surfaces which mate with the die pot (B) and the clamping ring (C). These tapers are designed to provide a further support from the bolster and the clamping ring. Furthermore, with the use of the proper tapers, adequate frictional forces can be attained to prevent the rotation of the die insert about its axis.

(c) The die insert is not supported at its bottom surface (note the gap between H and D). This allows the insert to plunge deeper into the bolster during the first few blows after which a further tightening of the bolts is necessary. No further tightening was required.

(d) The ejector is a gear form that mates with the die insert. This has the following advantages:

(i) It gives complete support to the forged component and this eliminates distortion during ejection.

(ii) During its upward stroke, it clears away the scale and lubricant deposits. These deposits can be blown away after removing the forging. A further clearing action is obtained as the ejector returns.

(iii) It transmits the forging load to the bolster (B) over a wide area through the ejector seat (D).

(e) The connection between the ejector and ejector rod is an 8° taper which replaces the previous screw S4. The taper allows for the alignment of the ejector during its travel. It also provides a means of replacing the previous rigid construction of the ejector–ejector rod and ejection cylinder. The ejector rod is shown in Fig. 2 with the ejection cylinder at its bottom stroke. When the ejector cylinder is actuated, the rod locates in the taper bore of the ejector. When the rod moves downwards, it pulls the ejector along with it until the ejector is firmly seated on its support. The contact is then broken and the rod is free to travel downwards until the ejector cylinder reaches the end of its stroke.

(f) The punch is a single unit held within the punch holder by a 7° taper (14° inclusive) that prevents rotation of the punch about its axis. This eliminates the use of bolts. Three specific types of punches were designed, as shown in Fig. 3. They were:

(i) Flat punch.

(ii) Stepped punch where the step diameter is smaller than the root diameter of the gear die and the step height hp is long enough to clear the top part of the die teeth when the punch reaches its maximum displacement.

(iii) Gear shaped punch.

The section of this punch is a gear profile which mates with the die insert.

Die dimensions

The sizes of the punch and ejector were limited by the gear to be forged. The maximum outer diameter

Figure 3. Forging punch profiles.

of the duplex cylinder was restricted to 7·25 in by the size of the bolster. The ring dimensions and the interference were calculated using the method suggested by Parsons and Cole[8], which is based on the Mohr criterion of failure.

Die materials

Except for the bolster and the top punch holder which were made from En 24, all other pieces were made from a 5 per cent chromium hot-working steel. The die insert, punch and ejector were heat treated to Rc 52–56. The ejector seat was heat treated to a hardness of Rc 36–40.

Die manufacture

Points of interest in the manufacture of the dies are as follows:

(a) The two rings of the die insert were shrunk together subsequent to the spark machining of the tooth profile. This was because the available spark machine was of insufficient size to accommodate the assembled rings.

(b) After the punches and ejector had been machined, desired clearances between these and the die insert were obtained by placing them in aqua regia. A clearance between punch and die of 0·005 in was obtained. The clearance between ejector and die, in operation, was also 0·005 in. The dimension of this feature proved to be of great importance to the satisfactory working of the die.

PERFORMANCE OF DIES

Forging tests

In order to test the production viability and performance of this improved die design, a series of forging tests were performed. Various dies together with different sizes of cylindrical billets were used. Billets were coated with graphite and heated to 1200°C in an electric muffle furnace either in air or in a protective atmosphere of cracked ammonia.

The forging machine used was a Mk II C Petro-Forge[9] with a rated energy of about 10 000 ft lbf at a closing speed of 30 ft/s. Tests showed the following:

(a) The problem of accumulation of scale and lubricant deposits was eliminated by using a gear shaped ejector.

(b) Distortion of the component during ejection was eliminated by providing a complete surface support.

... ... the bolster using the ejector.

(f) The previous failure of the insert was not experienced.

(g) Distortion at the top of the teeth depended on the type of punch used, the required width of the gear and the positional level of the ejector top face. This is further explained below.

Effect of punch shape

(i) Flat punch

Using sufficient stock material to theoretically just fill the gear portion of the die severe damage of the top part of the tooth profile occurred together with underfilling of the lower corners of the cavity. The use of excess material improved the filling and slightly reduced the above-mentioned damage to the die, but material waste was increased.

(ii) Stepped punch

This punch was primarily designed in order to reduce the amount of wasted material associated with using the flat punch. The flash here is in the form of a ring compared to the disc obtained previously. The ring has a radial width just larger than the total tooth depth of that particular die.

With a thickness of flash ring of at least 3/8 in die filling could be achieved without distortion of the die tooth profile. If the step height on the punch (and consequently the thickness of flash that can be formed) is less than this value of 3/8 in damage to the dies similar to that obtained using the flat punch occurred.

(iii) Gear shaped punch

The need to avoid deformation of the top of the die teeth and to avoid wastage of material suggested the use of a punch that clears the teeth of the die and eliminates gross flash formation. It was possible to completely fill the die cavity using this type of punch. This punch shape therefore offers the following advantages:

(1) No excess material is required in order to achieve die filling and no material need be wasted in order to get the required gear width.

(2) Because no flash is formed over the die teeth, the axial tooth loading which occurred with previous punches was eliminated. With this design the whole of the forging load is applied to the component.

(3) All that is required to form gears of different width is that different billet volumes be used. This is a feature which is not possible using the other two punch shapes.

However, this punch requires the following:

(1) Extreme accuracy of alignment which apparently requires more care than the other two punches.

(2) The level of energy required has to be the exact amount to fill the die cavity. Excess energy would cause the radial overloading of the die insert.

QUALITY OF THE FORGED COMPONENTS

Shapes of forged gears

A number of the gears forged are shown in Fig. 4. The tooth sizes are 16 D.P., 10 D.P. and 5 D.P. A selection of these gears were examined to determine their metallurgical properties.

Figure 4. Typical forged spur gears.

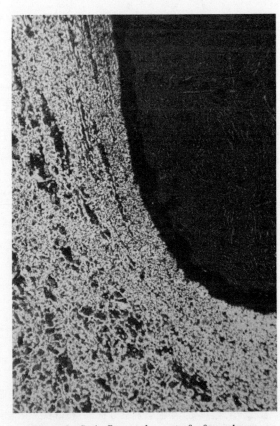

Figure 5. Grain flow at the root of a formed gear.

Grain structure

Figure 5 illustrates the form of metal flow around the root and tooth profile. It can be seen that an uninterrupted grain line has been attained which would be unimpaired by surface grinding. The general pattern of metal flow into the teeth of all the forged gears is as shown in Fig. 6.

Visual appearance

On some gears longitudinal marks were to be seen on the teeth profiles particularly at the junction of the involute curve and the fillet radius. These were caused by contact with the die on ejection. Some gears had blemished surfaces due to scale falling off after forging.

Surface oxidation

Beside the surface roughness of the dies, the surface finish of the tooth profile was to a great extent dependent on the scaly oxidation of the billet both before and after forging.

The oxidized surface of the billets heated in air was removed prior to forging either by banging the

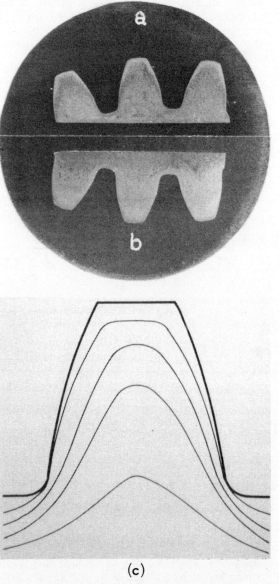

Figure 6. Pattern of material flow in a tooth.

was found to detach itself from the billet during furnace heating. Atmosphere controlled heating required a fast transfer time from the furnace to the dies in order to prevent oxidation before forging. After forging scale obviation methods were the same as those used with atmosphere heating. Very slight oxidation was noticed at the tips of the teeth on gears forged in this manner but the profile was free from oxidation.

Decarburization

Decarburization of billets heated in atmosphere was evident around the complete gear profile. It had a maximum depth of about 0·007 in at the tips of teeth and the layer decreased in thickness to nearly zero at the base of the tooth flank. A similar depth of

tively, are shown in Fig. 7. It can be seen that although the general level of the quenched component has increased, the pattern of hardness variation present in the air cooled component still persists.

Carburization

Billets carburized to a depth of 0·020 in were forged after heating in both open and controlled atmospheres. Figure 8 shows some hardness values taken on these gears. It can be seen that for the gears heated in air, the pattern of hardness remains the same as before due to the loss of carbon from the surface. For gears heated in a controlled atmosphere the pattern is reversed to a desirable one of higher hardness values on the outer surface decreasing towards the centre of the tooth.

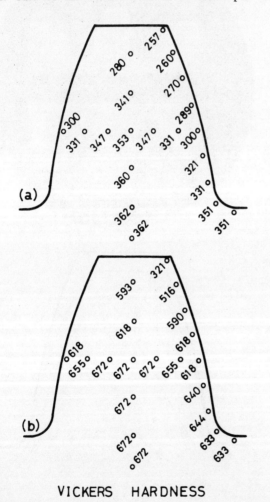

VICKERS HARDNESS

Figure 7. Pattern of hardness on a formed tooth heated in atmosphere. Material En 26, die 16 D.P. 40 teeth. (a) sand cooled, (b) oil quenched.

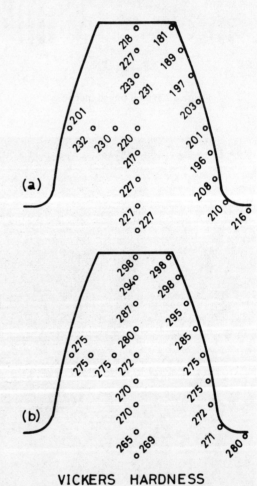

VICKERS HARDNESS

Figure 8. Hardness patterns of gears formed from carburized billets. Carburization depth 0·020 in, material En 8, die 10 D.P. 28 teeth. (a) billet heated in air, (b) billet heated in controlled atmosphere.

CONCLUSIONS

The die design principles advocated in this paper appear adequate for the production of gear forms in quantity.

Spur gears of various shapes can be produced to finished form although some surface blemishes are evident on teeth profiles. Metallurgical examinations of components indicate that desirable patterns of grain flow and hardness can be obtained.

REFERENCES

1. Rolls-Royce Limited, Bristol Engine Division, Ministry of Technology Contract Number KC/3N/20/CB. 12(a). February 1968.
2. F. J. LAVOIE (1968). High velocity forming 'new force in gear technology', *Machine Design*, 5 Dec., 146–51.
3. *Machinery's Handbook*, seventeenth edition, pp. 714.
4. R. C. MILLER (1967). Pneumatic–mechanical forging–'state-of-the art', *The First International Conference of the Center for High Energy Forming, Estes Park, June 1967*, p. 10.
5. F. L. PARKINSON, *Evaluation of High Energy Rate Forged Gears with Integral Teeth*, Western Gear Corporation, Lynwood, California, U.S. Army Contract, DA44-177-AMC-321(T), p. 50.
6. K. LACY (1967). Hot forged gears to fine tolerances, *Metal Working Production*, 4 January, p. 34.
7. N. F. KOBYSKOVSKY, G. D. SHOROCHOV and J. P. SOGRISHIN (1967). Forging of spur gears on a high speed forge, *Kuznechno-Shtampovochnoe Proizvodstvo*, 8
8. B. PARSONS and B. N. COLE (1967–68). A generalised approach to the optimum design of short composite cylinders, *Proceedings of the Institute of Mechanical Engineers*, 182 (3c).
9. L. T. CHAN and S. A. TOBIAS (1969–70). Performance characteristics of Petro-Forge Mk I and Mk II machines. *Proc. I. Mech. E. (London)*, 184 (1).

DISCUSSION

Q. E. R. Brealey, Trent Polytechnic. Were the tools and dies designed for a particular type of press?

Did the authors consider the use of the closed die technique employing an additional ram?

A. Yes; the tools were designed for high speed forging hammers and in particular Petro-Forge machines.

It is not practicable, on these devices, to employ movements of the dies other than those afforded by the main ram.

SCALE MODEL TESTING IN MAGNETIC FORMING OF TUBES

by

S. T. S. AL-HASSANI*

SUMMARY

A non-dimensional analysis of the magnetic forming process is developed. In situations where only the geometrical scale factor n_1 and the capacitance scale factors n_2 are employed, the analysis predicts that the discharge energy required to obtain similar deformations varies as $n_1^{3.5}/n_2^{0.5}$ and that the peak value of current varies as $n_1^{1.25}/n_2^{0.25}$. The present magnetic forming unit only allows experiments for the case when $n_2 = 1$ and experiments on tube expansion and reduction, using three different sizes, have shown that these predictions are reasonable.

INTRODUCTION

Magnetic forming is a technique for shaping metals by means of high strength transient magnetic fields. The fields are produced by a sudden discharge of a bank of capacitors through a robust coil adjacent to the part to be formed. Part of the field penetrates the workpiece inducing circulating currents in it which alter the original field producing large impulsive forces. These forces could be large enough to accelerate the material to velocities of the order of a few hundred feet per second in tens of microseconds duration. In forming tubular workpieces[1], solenoidal coils are used and the magnetic field is an axial one and this induces a circumferential current in the workpiece. The workpiece current can be considered as a solenoidal sheet current in the tube which is concentric with the coil and the direction of this current is opposed to the coil current during almost the whole of the discharge period. The resulting magnetic force is radial, tending to push the tube wall away from the coil. Tubular components may thus be formed or assembled in a similar manner to other high rate processes which employ explosive charges, gas detonation or high energy underwater sparks. Typically, these other processes employ water or gas as a medium to deliver the energy to the workpiece. In magnetic forming the energy is delivered by induction and, consequently, there are no contact forces.

The magnetic forming process has a number of applications in industry. Generally, the parts formed or assembled are relatively small and capacitor banks with storage energies of 10–30 kJ are used[2]. Larger magnetic forming units up to 240 kJ storage energy have been used in secondary forming operations on the large light alloy fuel vessels used in the moon rockets[3].

For forming relatively large components, the capital cost of capacitor banks and the development costs for coils and workpieces for full-scale tests would be quite high and therefore methods which would permit reasonable extrapolation of small scale studies would be most valuable.

The performance of geometrically similar workcoils to expand aluminium tubes was investigated theoretically and experimentally by Al-Hassani et al.[4]. They found that, in expanding geometrically similar tubes, the discharge energy required to produce geometrically similar final shapes increased with $n^{3.5}$ where n is the geometrical scale factor. They maintained the capacitance of the bank unaltered in all their experiments. In their analysis they used expressions for the equivalent lumped circuit parameters and obtained expressions for the pressure on the tube and the radial velocity. Scaling was then made by adjusting the appropriate geometrical terms in the parameters.

Notwithstanding the close agreement between the theoretical and experimental results of ref. 4, Lawrence[5] presented a non-dimensional analysis, which did not include circuit equations and suggested that the energy ratio should be n^3. His work was unsupported by experimental evidence and reference was often made to strain energy and to the speed of sound in the metal undergoing deformation. These

---

* Applied Mechanics Division, Mechanical Engineering Department, University of Manchester Institute of Science and Technology

thirdly to explain the reason for the disagreement between the analyses of Al-Hassani[4] and Lawrence[5].

In the present investigation, thin-walled tubes of a constant diameter/thickness ratio of the same aluminium alloy were formed by means of geometrically similar workcoils. The tubes had nominal diameters of 1, 2 and 4 in. The coils were solenoidal, having a length/diameter ratio of about one, and were placed inside or outside of the tubes. The tubes were then freely expanded, or reduced by discharges through the coils and measurements of velocity, circuit current waveform, and the final deformed shapes were obtained. These were then compared with the theoretical results.

VARIABLES OF THE PROCESS

The non-dimensional analysis can be used to scale variables of the magnetic forming process when different materials and different geometries are used at different discharge conditions.

The variables which influence the magnetic forming of a tube using an air-cored solenoid are divided into independent and dependent variables as listed in Tables 1 and 2. The dimensions of each are indicated in terms of $MLT\mu$ where M, L, T and μ denote mass, length, time and permeability respectively.

Other independent variables such as the thermal conductivity, temperature coefficient of resistivity and elastic constant are neglected.

TABLE 1   Independent variables

J_0	discharge energy, joules	ML^2T^{-2}
C	capacitance of the bank, farads	$\mu^{-1}L^{-1}T^2$
L_i	internal inductance of the bank, henries	μL
R_i	internal resistance of the bank, ohms	μLT^{-1}
N	number of turns per unit length of the coil	L^{-1}
l	length of coil or tube, metres	L
d_o	outside diameter of coil, metres	L
d_i	inside diameter of coil, metres	L
ρ_c	resistivity of coil material, ohm/metres	μL^2T^{-1}
Y	yield stress of the workpiece, Newtons/(metres)^2	$ML^{-1}T^{-2}$
D	diameter of workpiece, metres	L
h	wall thickness of workpiece, metres	L
γ	density of workpiece, kilograms/(metres)^3	ML^{-3}
ρ	resistivity of workpiece material, ohm-metre	μL^2T^{-1}
μ	permeability	μ

ω	angular frequency of discharge, radians/second	T^{-1}
β	damping factor, per second	T^{-1}
δ_c	skin depth in coil, metres	L
δ	skin depth in workpiece, metres	L

There are many dependent variables but those indicated in Table 2 are useful and frequently used in magnetic forming analysis.

DIMENSIONLESS GROUPS IN MAGNETIC FORMING

Since there are four dimensions, M, L, T and μ, and fifteen independent variables, there should be eleven dimensionless groups. Choosing variables γ, C, D and μ as those which cannot make a dimensionless group, we have the following dimensional equations.

$$[L] = [D], \qquad [M] = [\gamma D^3], \qquad [T] = [\mu CD]^{1/2}$$

and $\quad [\mu] = [\mu]$ \hfill (1)

where square brackets denote dimensional equality.

Using the (π)-theorem[6] and equation (1), the following dimensionless groups are obtained,

$$\pi_1 = \frac{\Delta D}{D}, \qquad \pi_2 = \frac{J_0}{ML^2T^{-2}} = J_0\frac{\mu C}{\gamma D^4}, \qquad \pi_3 = \frac{L_i}{\mu D},$$

$$\pi_4 = \frac{R_i}{D}\left(\frac{CD}{\mu}\right)^{1/2}, \qquad \pi_5 = ND, \qquad \pi_6 = \frac{l}{D},$$

$$\pi_7 = \frac{d_o}{D}, \qquad \pi_8 = \frac{d_i}{D}, \qquad \pi_9 = \frac{\rho_c}{D^2}\left(\frac{CD}{\mu}\right)^{1/2},$$

$$\pi_{10} = \frac{Y\mu C}{\gamma D}, \qquad \pi_{11} = \frac{h}{D} \quad \text{and} \quad \pi_{12} = \frac{\rho}{D^2}\left(\frac{CD}{\mu}\right)^{1/2}.$$

(2)

The final hoop strain $\Delta D/D$ is related to the other parameters by some function f_1 and we have,

$$\frac{\Delta D}{D} = f_1\left[\frac{J_0\mu C}{\gamma D^4}, \frac{L_i}{\mu D}, \frac{R_i}{D}\left(\frac{CD}{\mu}\right)^{1/2}, ND, \frac{l}{D}, \frac{d_o}{D}, \frac{d_i}{D},\right.$$

$$\left.\frac{\rho_c}{D^2}\left(\frac{CD}{\mu}\right)^{1/2}, Y\left(\frac{\mu C}{\gamma D}\right), \frac{h}{D}, \frac{\rho}{D^2}\left(\frac{CD}{\mu}\right)^{1/2}\right]. \quad (3)$$

Similar equations to (3) may be obtained when considering other dependent variables. However, before considering any of the laws of similitude it is worth while to investigate the nature and meaning of these dimensionless groups. Evidently, groups π_1, π_6, π_7, π_8 and π_{11} are ratios of dimensionally identical quantities. The other groups, however, are not quite as obvious. Group π_2 is simplified by writing

$$\pi_2 = \frac{J_0 \mu C}{\gamma D^4} = \frac{J_0}{\gamma D^3 (D/\mu C)}$$

The factor $(D/\mu C)$ has dimensions of $(L/T)^2$ so that π_2 represents the ratio of stored electrical energy to kinetic energy.

In situations where the internal inductance and resistance have an appreciable influence on the forming, groups π_3 and π_4 provide a method by which scaling of L_i and R_i could be achieved. They are also useful in combining with other dimensionless groups to produce new useful products. However, in an efficient magnetic forming machine, L_i and R_i are small and π_3 and π_4 can, therefore, be justly ignored.

π_5 defines the number of turns in the coil and is useful in determining the variation of coil pitch with the geometrical scale factor.

Products π_9 and π_{12} are not directly interpreted. However, noting that $[\omega] = [\mu CD]^{-1/2}$ and rearranging, we have

$$\pi_{12} = \frac{\rho}{\mu \omega D^2}$$

or in terms of the skin depth δ,

$$\pi_{12} = \left(\frac{\delta}{D}\right)^2 \quad \text{and similarly} \quad \pi_9 = \left(\frac{\delta_c}{D}\right)^2$$

It is evident that these groups indicate the distribution of current and magnetic field within the conductors and consequently play an important role in situations where similitude of these distributions are required, such as in the induction heating[7].

Lastly, π_{10} may be written in the form

$$\pi_{10} = \frac{Y \mu C}{\gamma D} = \frac{Y D^2}{(\gamma D^3 / \mu C)},$$

which represents the ratio of the yield force to the inertia force.

SIMILARITY CONSIDERATIONS

The principles of complete physical similitude indicate that a prototype must be scaled from a model such that all the dimensionless products remain unchanged. Since there are four dimensions, $MLT\mu$, the laws of similitude can be determined in terms of any four scale factors. Selecting scale factors to represent the geometry, the electrical system, the workpiece strength and the workpiece material, we have

$$n_1 = \frac{D_n}{D_1} \tag{4}$$

$$n_2 = \frac{C_n}{C_1} \tag{5}$$

$$n_3 = \frac{Y_n}{Y_1} \tag{6}$$

and

$$n_4 = \frac{\gamma_n}{\gamma_1} \tag{7}$$

where subscripts n and 1 refer to prototype and model respectively.

Assuming that the materials involved in the process are non-magnetic, the permeability will remain the same for both the prototype and model and is equal to the permeability of free space, μ_0.

Complete similitude of the final deformed shape requires each group in equation (3) to be the same for both model and prototype. Hence, in combining equations (4) to (7), with the equality of each group, we obtain the following scaling laws

$$\frac{\Delta D_n}{\Delta D_1} = n_1 \tag{8}$$

$$\frac{J_{0n}}{J_{01}} = \frac{n_3 n_1^4}{n_2} \tag{9}$$

$$\frac{N_n}{N_1} = \frac{1}{n_1} \tag{10}$$

$$\frac{l_n}{l_1} = \frac{d_{on}}{d_{o1}} = \frac{d_{in}}{d_{i1}} = \frac{h_n}{h_1} = n_1 \tag{11}$$

$$\frac{\rho_{cn}}{\rho_{c1}} = \frac{\rho_n}{\rho_1} = \frac{n_1^{3/2}}{n_2^{1/2}} \tag{12}$$

and from π_{10}

$$n_3 n_2 = n_4 n_1 \tag{13}$$

It is clear from equation (12) that if the materials of the model and prototype coils are the same, the materials of the model and prototype workpieces must also be the same. Only when the coil core is not air (i.e. $\mu_n \neq \mu_1$) can the materials be different, but this introduces complexities to the problem.

Similarity of forming

If the same materials are used, $n_3 = n_4 = 1$, equation (13) becomes

$$n_2 = n_1 \tag{14}$$

which reduces equation (9) to

$$\frac{J_{0n}}{J_{01}} = n_1^3 \tag{15}$$

Also from equation (12) we have

$$n_2 = n_1^3 \tag{16}$$

which reduces equation (9) to

$$\frac{J_{0n}}{J_{01}} = n_1 \tag{17}$$

Equations (15) and (17) present a dilemma and a complete physical similitude is not possible when using the same material.

Interest in forming is usually directed towards the amount of plastic deformation achieved in the workpiece. Similarity of forming is said to exist when the final formed shapes of the model and prototype are the same.

Equation (3), however, suggests that forming similitude may be obtained by selecting any other value for n_2 provided that the appropriate change in

In terms of dependent variables the final deformation is clearly a function of the total magnetic impulse delivered to the workpiece, the equivalent circuit frequency, damping factor, and the yield pressure of the workpiece. Equation (3) can now be simplified as follows.

Using the total magnetic impulse delivered to the workpiece, F, instead of the discharge energy in π_2, the group becomes

$$\pi_2' = \frac{F}{lD}\left(\frac{\mu C}{D}\right)^{1/2} \tag{18}$$

where

$$F = \int_0^\infty P \, dt \tag{19}$$

and P the instantaneous mean pressure on the workpiece.

The strength term $Y(\mu C/\gamma D)$ can be written in terms of the yield pressure, $P_y = 2Yh/D$, as

$$\pi_3' = P_y\left(\frac{\mu C}{\gamma D}\right) \tag{20}$$

Replacing the electric terms by the damping factor β and the frequency ω, we obtain new dimensionless groups

$$\pi_4' = \beta(\mu CD)^{1/2} \tag{21}$$

and

$$\pi_5' = \omega(\mu CD)^{1/2} \tag{22}$$

Since the geometrical parameters are included in the above groups they may be ignored and equation (3) becomes

$$\frac{\Delta D}{D} = f_2\left[\frac{E}{\gamma D}\left(\frac{\mu C}{D}\right)^{1/2}, \frac{P_y}{\gamma}\left(\frac{\mu C}{D}\right), \beta(\mu CD)^{1/2}, \right.$$
$$\left. \omega(\mu CD)^{1/2}\right] \tag{23}$$

The magnetic impulse F

In real situations the circuit parameters change during the deformation process and the integral of equation (19), which defines the total impulse delivered to the workpiece, cannot be analytically determined. If, however, we use a simplified theory which assumes constant circuit parameters with zero internal resistance and inductance, we may write the discharge current in the coil as[8]

$$i = -\sqrt{\frac{2J_0}{L}}\, e^{-\beta t}\sin \omega t \tag{24}$$

tube exerted by an internal coil is[8]

$$P = \frac{i^2}{\pi D}\frac{\partial L}{\partial D} \tag{27}$$

which combines with equations (24) and (25) and gives

$$P = \alpha j\, e^{-\beta t}\sin^2 \omega t \tag{28}$$

where

$$\alpha = \frac{D^2 d_i^2}{(D^2 - d_o^2)(D^2 + d_i^2 - d_o^2)} \tag{29}$$

and j, the energy discharged per unit volume contained by the tube to be formed and is given by

$$j = \frac{4J_0}{\pi D^2 l} \tag{30}$$

Equation (28) also holds for tube reduction using an external coil, but in this case we have

$$\alpha = \frac{1}{1 - (D/d_i)^2} \tag{31}$$

Integrating equation (28) and assuming constant α, β and ω, we have the total impulse

$$F = \int_0^\infty P\, dt = \alpha j \int_0^\infty e^{-2\beta t}\sin^2 \omega t \, dt$$
$$= \frac{\alpha j}{2}\cdot\frac{\omega^2}{\beta^2 + \omega^2} \tag{32}$$

In most magnetic forming applications we have $\omega^2 \gg \beta^2$ and equation (32) reduces to

$$F \approx \frac{\alpha j}{2\beta} \tag{33}$$

which shows that for high frequency operations the total magnetic impulse is directly proportional to the energy discharged and inversely proportional to the damping factor.

Thus group π_2' of equation (18) may now be written as

$$\phi = \frac{\alpha j}{\gamma D\beta}\cdot\left(\frac{\mu C}{D}\right)^{1/2}\cdot\frac{\omega^2}{\beta^2 + \omega^2} \tag{34}$$

or for high frequency

$$\phi = \frac{\alpha j}{\gamma D\beta}\left(\frac{\mu C}{D}\right)^{1/2} \tag{35}$$

which is an impulse number representing the ratio of the total magnetic impulse delivered to the momentum of the workpiece.

The significance of ϕ is that it combines all the circuit and geometrical parameters in a simple form which is directly related to the forming. It is clear that ϕ can be kept constant by various changes in the factors of the product. However, only when scaling is made such that $n_1 = n_2$ will the impulse number indicate a condition of dynamic similitude.

For operations where the strength of the workpiece is small compared with the applied peak pressure it is convenient to assume that the yield pressure of the tube is negligible and the total magnetic impulse, F, may be used as an approximate indication of the extent of diametral displacement.

Conditions for similar forming

Similarity of forming is obtained when the condition $(\Delta D/D)_n = (\Delta D/D)_1$ is satisfied. This is possible, for geometrically similar arrangement, if approximately

$$\frac{\Delta D_n}{\Delta D_1} = \frac{F_n}{F_1} = n_1 \tag{36}$$

From equations (33) and (36), and noting that α is dimensionless we have

$$\frac{j_n}{j_1} \cdot \frac{\beta_1}{\beta_n} = n_1 \tag{37}$$

which for

$$\frac{\beta_n}{\beta_1} = \left[\frac{C_1 D_1}{C_n D_n}\right]^{1/2} = (n_1 n_2)^{-1/2}$$

(viz. equation 21), this gives

$$\frac{j_n}{j_1} = \left(\frac{n_1}{n_2}\right)^{1/2} \tag{38}$$

or in terms of J_0

$$\frac{J_{0n}}{J_{01}} = \frac{n_1^{7/2}}{n_2^{1/2}} \tag{39}$$

If $n_2 = n_1$, $J_{0n}/J_{01} = n_1^3$ which satisfies the condition of dynamic similitude where $\phi_n = \phi_1$, but if $n_2 = n_1^3$, $J_{0n}/J_{01} = n_1^2$, which shows that, if scaling is made according to the electric similarity, similar forming is possible although dynamic similitude is not satisfied. If, in this case, plastic deformation was absent, i.e. $F_n = F_1$, electric similarity gives $J_{0n}/J_{01} = n_1$ as before.

Equation (39) also shows that for $n_1 = 1$, i.e. forming tubes of the same size, the energy required to produce the same amount of plastic deformation decreases with increasing capacitance. This agrees well with the published experimental results[1]. Equation (39) may be used to determine the effect of scaling on any other dependent variable.

The peak magnetic pressure, P_1, occurs at $t = \pi/2\omega$ and is given, from equation (28), by

$$P_1 = \alpha j \, e^{-\beta\pi/\omega} \tag{40}$$

Hence the scaling law for the peak pressure for $\beta_1/\omega_1 = \beta_n/\omega_n$ from equations (21) and (22), is

$$\frac{P_{1n}}{P_{11}} = \frac{j_n}{j_1}$$

which combines with equation (38) and gives,

$$\frac{P_{1n}}{P_{11}} = \frac{n_1^{1/2}}{n_2} \tag{41}$$

In the same way the peak current scales such that

$$\frac{I_{1n}}{I_{11}} = \frac{n_1^{5/4}}{n_2^{1/4}} \tag{42}$$

The velocity at the end of the first half cycle scales such that

$$\frac{V_n}{V_1} = \frac{(P_1 \pi/\gamma h\omega)_n}{(P_1 \pi/\gamma h\omega)_1} = \left(\frac{n_1}{n_2}\right)^{1/2} \frac{(n_1 n_2)^{1/2}}{n_1} = 1 \tag{43}$$

Equations (39), (41), (42) and (43), therefore, represent the most general scaling laws for similar forming in magnetic forming applications in which the capacitance and geometry are the only varying independent parameters.

TABLE 3 Scaling laws for similar forming

	Scaling laws			
n_2	1	n_1	n_1^2	n_1^3
J_{0n}/J_{01}	$n_1^{3 \cdot 5}$	n_1^3	$n_1^{2 \cdot 5}$	n_1^2
P_{1n}/P_{11}	$n_1^{1/2}$	1	$n_1^{-1/2}$	n_1^{-1}
I_{1n}/I_{11}	$n_1^{5/4}$	n_1	$n_1^{3/4}$	$n_1^{1/2}$
V_n/V_1	1	1	1	1
ω_n/ω_1	$n_1^{-1/2}$	n_1^{-1}	$n_1^{-3/2}$	n_1^{-2}

Table 3 gives the scaling laws for some variables of the process when geometrically similar coils are used and selection is made for $n_2 = 1$, $n_2 = n_1$, $n_2 = n_1^2$ and $n_2 = n_1^3$.

EXPERIMENTAL RESULTS AND COMMENTS

To test the above theoretical results, an extremely large multi-capacitor capacitor bank with almost negligible internal inductance and resistance is required. The bank at Manchester has a limited number of capacitors and it is only convenient to test scaling laws for situations when $n_2 = 1$.

The theoretical results for $n_2 = 1$, however, agree well with the theoretical and experimental results of previously published work[4], for expanding tubes, although the method of derivation of the scaling laws are different from the above. Tubes of 1, 2 and 4 in outside diameter were expanded and reduced in diameter by discharging the capacitor bank through internal and external coils of 1, 2 and 4 in nominal diameters. The external coils are encased in split steel casings as shown in Fig. 1 and typical aluminium tubes formed by these coils are shown in Fig. 2. The internal coils, however, are wound on wooden or tufnol formers, as shown in Fig. 3 and typical tubes expanded by them are shown in Fig. 4.

Letting $n_1 = 1$ to represent the medium size, $2 \cdot 0$ in diameter, the large size will be defined by $n_1 = 2$ and the small size by $n_1 = \frac{1}{2}$.

Figure 1. Geometrically similar external coils used for reducing the diameter of 4, 2 and 1 in outside diameter aluminium tubes.

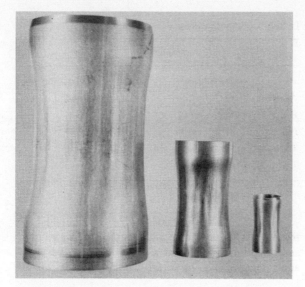

Figure 2. Geometrically similar tubes of 4, 2 and 1 in o.d. reduced in diameter by the coils of Fig. 1.

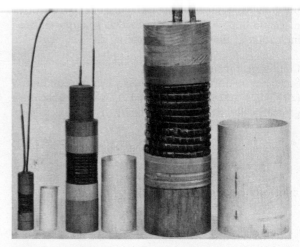

Figure 3. Geometrically similar internal coils used for expanding the diameter of 4, 2 and 1 in diameter aluminium tubes.

Figure 4. Geometrically similar tubes of 4, 2 and 1 in o.d. expanded in diameter by the coils of Fig. 3.

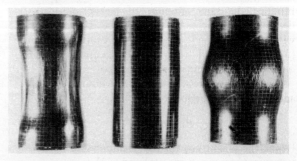

Figure 5. 2·0 in o.d. and 0·064 in wall thickness aluminium tubes with printed grids. Right, expanded by a 2·0 in nominal diameter internal solenoidal coil and Left, reduced by a 2·0 in nominal diameter external solenoidal coil.

Final profiles and strains of the tubes were measured by means of grids printed on the cylindrical surface, as shown in Fig. 5. Figure 6 shows the final profiles of tubes of three different sizes reduced to a maximum diametral reduction, $(\Delta D/D)_{max}$, of 7·8 per cent. In the same figure the results for tubes expanded to $(\Delta D/D)_{max}$ of 14 per cent are also shown. These curves show that, for constant capacitance $C = 80$ μF, the final deformed shapes are geometrically similar and that the maximum value of $\Delta D/D$ for a tube is a suitable parameter of overall deformation.

From a number of tests on tubes of each size, the energy required to produce 4 to 14 per cent reduction, or expansion, for each size is obtained and a logarithmic plot of J_{0n}/J_{01} versus the scale factor n_1 is shown in Fig. 7.

Similarly, the ratios of the first peaks of current I_{1n}/I_{11} for $(\Delta D/D)_{max}$ of 4 to 14 per cent are plotted in Fig. 8 and compared with equation (42) for $n_2 = 1$.

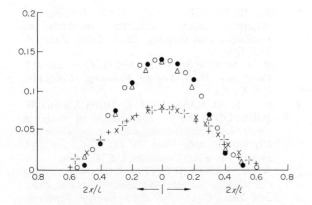

Figure 6. Final deformation profiles, $\Delta D/D$ along the tube plotted non-dimensionally for the same $(\Delta D/D\text{max}) = 14$ per cent expansion and $(\Delta D/D_{\text{max}}) = 7\cdot8$ per cent reduction.

+ reduction   △ expansion for 1·0 in dia.
× reduction   ● expansion for 2·0 in dia.
-⊦- reduction   ○ expansion for 4·0 in dia.

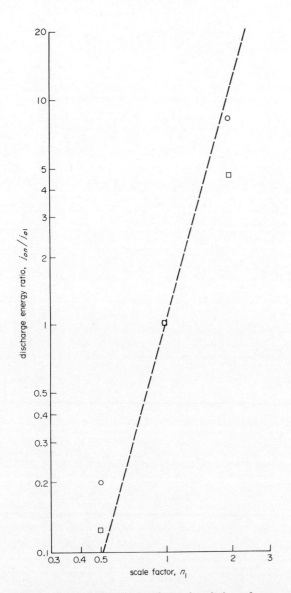

Figure 7. Theoretical and experimental variation of energy ratio with geometrical scale factor for constant capacitance, $C = 80$ μF. □ Tube reduction, ○ tube expansion, - - - theoretical.

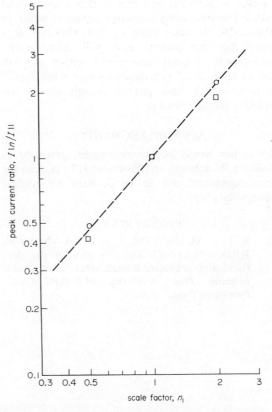

Figure 8. Theoretical and experimental variation of peak current ratio with geometrical scale factor for discharges using $C = 80$ μF and producing the same $(\Delta D/D_{\text{max}})$.
□ Tube reduction, ○ tube expansion, - - - theoretical.

The velocity for each size was measured using the pin-contact method. The results show that, like in tube expansion, tubes reduced to the same $(\Delta D/D)_{\text{max}}$ have nearly the same velocities at the end of the first half cycle. This is in agreement with equation (43).

CONCLUSIONS

A non-dimensional analysis which incorporates the circuit parameters of the magnetic forming process has been developed. It has been shown that complete physical similitude of forming between the model and prototype is not possible. However, the analysis identifies various possibilities of similitudes by simply selecting the appropriate values of the geometrical scale factor, n_1, and the capacitance scale factor, n_2. Dynamic similarity is achieved when selecting $n_2 = n_1$, thus, giving a prototype to model energy ratio equal to n_1^3, which agrees with Lawrence's prediction[5]. On the other hand, forming similarity, using constant capacitance in which case $n_2 = 1$, the energy ratio becomes $n_1^{7/2}$ which agrees with Al-Hassani's prediction[4]. Other similitudes, such as those required for induction heating where $n_2 = n_1^3$, are easily deduced.

Experimental results on aluminium tubes of various sizes were performed using constant capacitance and the general level of agreement between the analysis and the experimental results demonstrates the usefulness of the present non-dimensional

The author would like to express his gratitude to Professor W. Johnson and Professor J. L. Duncan for their suggestions and to Dr S. Reid for reading through the script.

REFERENCES

1. S. T. S. AL-HASSANI, J. L. DUNCAN and W. JOHNSON (1967). The influence of the electrical and geometrical parameters in magnetic forming, *Proc. 8th Int. M.T.D.R. Conf.*, Pergamon Press.

5. W. N. LAWRENCE (1967). Scale modelling calculations for electro-magnetic forming, *Proc. 2nd Int., Conf. for High Energy Forming*, Colorado, U.S.A., June.

6. E. BUCKINGHAM (1914). On physically similar systems, *Phys. Rev.*, Ser. 2, **4**, 345.

7. P. G. SIMPSON (1960). *Induction Heating*, McGraw-Hill.

8. S. T. S. AL-HASSANI (1969). The electro-magnetic forming process, Ph.D. Thesis, University of Manchester.

AN INVESTIGATION INTO THE EFFECT OF PUNCH SHEAR AND CLEARANCE IN THE PIERCING OF THICK STEEL PLATE

by

F. W. TRAVIS*

SUMMARY

Primarily, the paper examines the effect of providing different degrees of double-shear for the punch, and also different punch-die clearances, on the punch load-displacement characteristics and on the quality of the hole, in the piercing of 60 mm ϕ holes in mild steel plate, ranging in thickness from $\frac{1}{4}$-$1\frac{5}{8}$ in (6·35-41 mm). The alleviating effect of punch double-shear on the rapid unloading of the blanking press following separation of the plug from the stock is also considered.

A line diagram and a photograph of the set-up used are presented, along with typical punch load-displacement records, photographs of typical specimens, and graphs showing the effect of variation of the experimental parameters upon the maximum punch force and the associated penetration.

INTRODUCTION

The blanking, or piercing, operation is of long standing and, as is well recognized, is one of the major manufacturing processes. In recent years interest has centred on the use of very low punch-die clearances—of the order of 0·0005 in—to secure a close tolerance component of good edge finish and requiring no subsequent finishing processes. The additional advantages of a slight coining action around the periphery of the (eventual) hole, secured by use of a blankholder of slight conical form[1], or having an indenting ridge[2,3], are well appreciated, and the latter technique is used extensively in practice. A further technique with the same aim, is 'finish piercing' or 'finish blanking'[4], in which a chamfer is provided on either the edge of the punch or the edge of the die, respectively, depending upon whether the stock or the blank constitutes the component and thus requires the better edge finish and profile. With these techniques, the main 'edge finish' effect is to delay the fracture of the blank from the stock, so that the plastically sheared region—which has a desirable smooth burnished appearance—extends over almost the whole depth of the hole, instead of being limited to a comparatively small depth, followed by a matt irregular fractured surface.

A drawback to the use of fine blanking techniques is that very accurate alignment and guidance of the punch and die is required, which is restrictive, and furthermore, as the punch-die clearance is reduced, the maximum force that the punch is called upon to deliver, increases slightly[5,6,7]. Thus for situations in which the pierced hole is provided, say, primarily, to reduce the weight of a structural member, or to provide access, where the major requirement of the hole is simply that it should be free of cracks, lips, burrs and torn edges, corner rounding and other dimensional aspects being relatively unimportant, low clearances are neither necessary nor desirable.

One such case is in the piercing of the stiffening frames of a ship, to provide drainage holes; the thickness of frames, currently, can be as much as $1\frac{1}{2}$ in, and the order of punch load involved is 10^3 tonf. Thus it is highly desirable to explore means of reducing punch loads so that current frame thicknesses can be pierced more easily, and also so that any future increased frame thicknesses can be accommodated, while at the same time maintaining conditions which will ensure the required standard of finish of the pierced hole.

In the work presently reported, the percentage radial clearance—defined as the radial clearance expressed as a percentage of the plate thickness—was increased progressively up to a value of 30 per cent, as opposed to the order of 0·1-0·2 per cent of fine-blanking techniques, and in addition, two different degrees of punch double-shear were used.

The reader is referred to ref. 8 for a general survey of the conventional blanking operation and to ref. 9 for an examination of the mechanism of fine blanking.

* University of Strathclyde, Glasgow

finitely variable from 0 to 1 in/s and was maintained constant throughout the tests at 0·6 in/s. As shown in Figs 1a and b, the punch assembly is mounted on the (stationary) top platen of the press and the die assembly is mounted on the bottom platen, along with clamps for holding the stock.

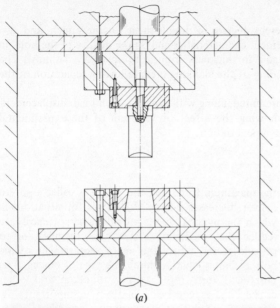

(a)

(b)

Figure 1. (a) a line diagram, and (b) a photograph, of the piercing assembly.

ground to an internal diameter of 63·8 mm—to provide a radial clearance of 1·9 mm or 0·075 in—and to test increasing thicknesses of plate, after which the procedure was repeated with the punch reground to 10° and then 20° shear. The die was then reground to an internal diameter of 67·6 mm—to provide a radial clearance of 3·8 mm or 0·150 in—with the punch shear starting at 20° and then being reground to 10° and finally flat. Thus, the most severe conditions were left to the end of the series of tests, and regrinding of the punch and dies was minimized.

Instrumentation

The displacement of the punch was measured by a linear potentiometer mounted between the top and bottom platens of the press as shown in Fig. 1b, the resistance changes of the potentiometer being translated by appropriate circuitry into the deflection of a galvanometer of a UV recorder. Slip gauges were used to calibrate the system.

The punch load was measured by fitting a pressure transducer into the main hydraulic line at the point where it enters the cylinder of the press, the resistance changes of the transducer being used to drive a second galvanometer of the UV recorder. Calibration of the load system was accomplished using a columnar load cell of 350 tonf capacity set directly between the platens of the press, the load cell itself having been calibrated previously in a universal testing machine. While calibration is carried out under essentially static conditions, with the hydraulic pressure supplied to the ram being limited by a manually operated and infinitely variable by-pass valve, it is felt that the pressure transducer is sufficiently close to the cylinder for flow pressure losses to be neglected during an actual piercing operation. Likewise it is believed that the inferring of the press load—rather than its direct measurement—from the cylinder supply pressure will not lead to serious error, as care was taken during calibration to ensure that pressure rises were progressive, so that friction forces between the press cylinder and bore would act in the same sense, as would occur in an actual piercing operation.

EXPERIMENTAL RESULTS

Throughout the figures, round and diamond shaped points refer to tests employing 0·075 in and 0·150 in radial clearance respectively, and hollow, half solid and solid points refer to 0°, 10° and 20° punch double-shear respectively.

Figure 2 presents a selection of the punch load records, showing how the latter vary with plate thickness, radial clearance and punch double-shear.

Figure 3a shows how the maximum punch load, and Fig. 3b the penetration of the punch at maximum load, vary with plate thickness.

In Fig. 4a the maximum punch load is divided by the nominal area to be sheared, i.e. the circumference of the punch multiplied by the plate thickness, and is thus presented as a nominal shear stress, while in Fig. 4b the penetration is normalized by dividing it by the plate thickness, both quantities being plotted against percentage radial clearance.

An attempt is made in Fig. 5 to correlate the results from tests using different degrees of punch double-shear, by plotting the nominal shear stress against the amount by which the leading edge of the punch double-shear is in advance of the trailing edge, i.e. $r \tan \phi$, the latter quantity being normalized by dividing it by the plate thickness t.

Photographs of typical results are presented in Fig. 6.

DISCUSSION

Maximum punch load
While the maximum punch load is seen from Fig. 3a to increase progressively with plate thickness, and the trend is noted of lower load with increasing shear, particularly at lower plate thickness, results are more easily interpreted when presented as in Fig. 4a.

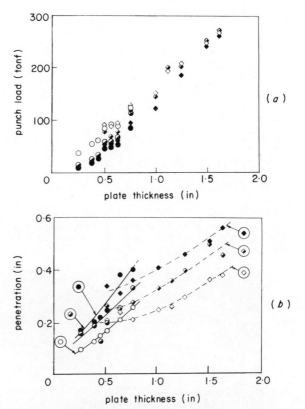

Figure 3. Graphs of (a) maximum punch load, and (b) penetration at maximum load, against plate thickness, upper and lower respectively.

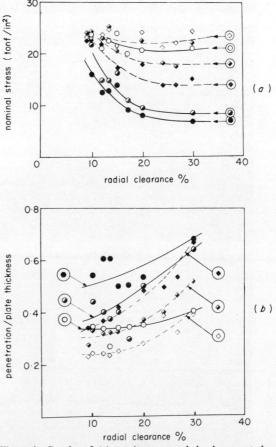

Figure 4. Graphs of (a) maximum punch load presented as nominal shear stress, and (b) penetration normalized by division by plate thickness, against percentage radial clearance, upper and lower respectively.

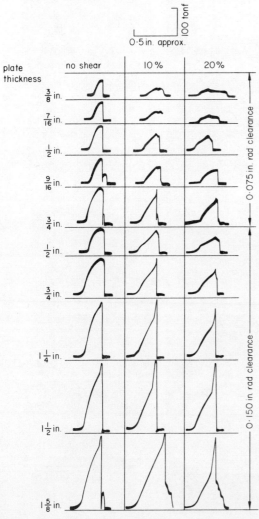

Figure 2. Punch load–approximate displacement characteristics.

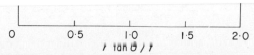

Figure 5. Nominal shear stress against amount leading edge of punch shear is in advance of rear edge, the latter quantity normalized by division by plate thickness.

ran from the edge of the punch to the edge of the die at all points around the circumference; instead, in one area the line of separation ran down more steeply from the edge of the punch, to leave a lip of stock material extending beyond the edge of the die. Thus, at this order of clearance, the sharp edge of the die is becoming ineffective in promoting separation, and

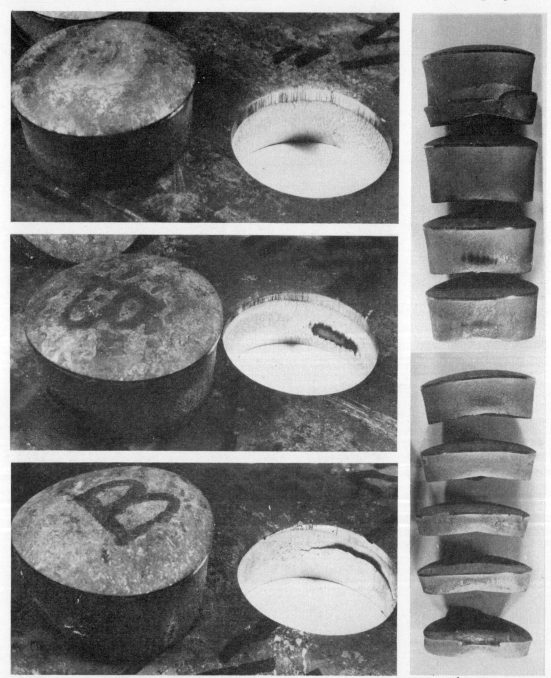

Figure 6. For a punch–die clearance of 0·150 in, showing on left, the stock and plug from tests using $1\frac{1}{4}$ in thickness plate where the punch shear was top: 0°; middle: 10°, and bottom: 20°; and on right, the whole range of plugs, in order, from tests using 10° double-shear.

this could account for the slightly greater punch load to secure separation.

It will be noted from Fig. 4a that the nominal stress for tests using the larger radial clearance is distinctly higher than that for lower radial clearance tests, although the percentage radial clearances cover the same range; this can be considered on the basis of the punch diameter to plate thickness ratio, D/t, which for the large radial clearance tests is always half of that of the smaller radial clearance tests, at any particular percentage radial clearance. This aspect has been examined by Chang and Swift[10] and Chang[11], who carried out tests into the slow-speed piercing of $\frac{1}{2}$ in thick mild steel plate at radial clearances of 0, 5, 10 and 20 per cent and D/t ratios of 2, 5, 8 and infinity, the latter value being provided by use of a straight-edged tool. It was found that at 0 per cent radial clearance the nominal shear stress for D/t ratios of 2, 5, 8 and infinity was approximately 22·8, 22·6, 21·7 and 20·9 tonf/in[2] respectively, while at 20 per cent radial clearance the figures were 21·3, 18·7, 18·9 and 20·1 tonf/in[2] respectively. Chang explained the results in terms of the nature of the surface of separation; as the curvature of the tool is increased there is a tendency for the formation of more cracks, with production of multiple tongues, which interfere, and thus result in increased strain, and hence strain hardening, before separation occurs. The single crack occurring in bar shearing using a straight-edge tool was claimed to be due to the absence of constraint, and in support of this Chang carried out tests constraining the ends of the bar, finding that the nominal shear stress then rose to the level of that found when using circular punches, and furthermore that multiple cracks were encountered.

Further work on this topic was carried out by Maeda and Tamura[12] employing radial clearances of up to 30 per cent and D/t ratios of from 2 to 300. For each of the three materials used—mild steel, copper and aluminium—the nominal shear stress was almost independent of the D/t ratio for ratios above 5–6, but below a figure of about 5, the nominal shear stress was highly dependent. In the present work the D/t ratios range from 1·45 to 4·72 for the larger radial clearance tests and 3·15 to 9·44 for the smaller radial clearance tests, and thus place the nominal shear stress in the highly dependent category. The author is of the opinion, however, that if curvature of the material in the shear zone is the major factor when considering varying punch diameters, then, for tests at constant percentage radial clearance, a more representative and meaningful quantity would be the ratio of the punch diameter to the radial clearance (this cannot be examined for the present work as only two values of the latter ratio were used). It is unlikely, however, that a single parameter can be found to completely define the punch load in an operation as complex as piercing; this is supported by inspection of all previous published work, when, although general trends may be noted, no such universal parameter is indicated.

When considering the double-shear curves of Fig. 4a, there is seen to be a substantial drop in the nominal shear stress, as the radial clearance is increased beyond about 10 per cent, and particularly for the lower radial clearance results, when for a radial clearance of 30 per cent, the shear stress falls to about one-third of that of the no double-shear tests. For the same percentage radial clearance, the plate thickness of the lower radial clearance—0·075 in—tests is a half of that of the greater radial clearance—0·150 in—tests, and thus the effect of the punch shear will be more pronounced. To attempt to allow for the latter, the nominal shear stress is plotted in Fig. 5 against the amount by which the leading edge of the punch double-shear is in advance of the training edge, normalized by division by the plate thickness.

While there is still some separate grouping of points, the general trend of reduction of nominal shear stress with increasing values of $r \tan \theta/t$ will be noted, from an average of about 23 tonf/in[2], to an asymptotic value of about 7 tonf/in[2]. The proportionally greater effectiveness of $10°$ double shear over that of $20°$ double-shear, is seen from the generally lower curve, particularly at $r \tan \theta/t$ values of less than unity. With the greater punch double-shear, there was found, as shown in Fig. 6 and discussed later, to be a pronounced tendency to the formation of inter-penetrating tongues along the surface of separation, with, as already considered in respect of punch curvature, greater strain, and hence strain hardening, before the plug is pushed clear of the stock.

It would appear that there is no advantage in providing more than $10°$ double-shear and, for radial clearances below about 10 per cent, no advantage in providing double-shear at all, on the basis of reduction in maximum punch load. The latter observation is consistent with the results of the earliest reported[13] systematization investigation of punch load-displacement characteristics, in which when piercing mild steel plate at radial clearances of 2·7, 4·6 and 10·3 per cent, virtually no difference in maximum nominal shear stress was observed in the use of flat-ended or (single) shear punches.

Punch penetration at maximum load

From Fig. 3b, there is seen to be an increase in the penetration at maximum punch load with increase in the degree of double-shear. This can be attributed largely to the purely geometric aspects of progressively commencing to pierce the stock from the apex of the punch, with separation of the plug and stock in this vicinity being inhibited by the restraint of the lesser deformed regions. Increased penetration in this case is not associated with increased plastic shearing prior to separation, such as occurs when using low clearances with flat-ended punches, and which desirable feature forms the objective of fine-blanking and similar techniques. The amount of plastic shearing at the order of percentage radial clearance involved in the present tests is very slight, as shown by the comparatively small plastically sheared rims to the holes of the specimens of Fig. 6. Maximum punch load occurred immediately prior to complete metallurgical separation (i.e. the creation of a complete surface of separation between the plug and stock) while complete physical separation in some instances, particularly when using smaller plate thicknesses and

normalized basis of Fig. 4b, there is seen to be little difference between the results for the two different radial clearances employed, at a radial clearance of 30 per cent, but with a reduction in radial clearance, the normalized penetration is higher for the smaller radial clearance tests. An explanation for this is that the latter, as discussed earlier, have D/t ratios of a half of those of corresponding larger radial clearance tests and the 'punch curvature' effect for these tests is enhanced; the material is thus effectively more constrained, with the result that separation, and hence the realization of maximum punch load, is delayed.

Surface of separation

From visual examination of the pierced holes, the following general pattern was observed:

(i) No shear. All surfaces were free from flaws over the central range of radial clearances involved, but at radial clearances below 10 per cent interpenetrating tongues (such as displayed in the $20°$ punch double-shear specimen of Fig. 6, but continuing completely around the surface) were formed, while at radial clearances of 30 per cent, the line of separation, as discussed earlier, did not run from the edge of the punch to the edge of the die, resulting in a small lip extending beyond the edge of the die in one place.

(ii) $10°$ double-shear. Interference of plug and stock, initially resulting in abrasion, but subsequently with the production of deep tongues, was encountered with decrease of radial clearance below 10 per cent, while above 15 per cent tearing of material to produce a small lip, as above, was noted, both of these features occurring at the apex of the punch. (See bottom specimen on right of Fig. 6.)

(iii) $20°$ double-shear. In this case the lower radial clearance defects of the $10°$ double-shear tests progressed into the higher radial clearance defects, at a radial clearance of about 15 per cent, with no completely satisfactory surfaces being produced.

The main result of punch double-shear on the surface of separation seems to be effectively, to reduce the radial clearance at the apex of the punch, and increase it at the trailing edge. Thus, by reference to Fig. 6, it can be noted that increase of punch double-shear has the same result, at the apex, as reduction in radial clearance, in the production of the 'low clearance' defects of the no shear tests, while at the trailing edge, increase of double-shear results in decrease of the depth of the plastically sheared rim, which has been confirmed in many publications as being associated with increase in radial clearance.

This argument is entirely reasonable, as when the

and punch load at separation, particularly with smaller plate thicknesses. However, when considering greater plate thicknesses, where the maximum loads for double-shear and no double-shear tests become increasingly large and close, the very desirable (from the point of view of press operation) feature of the punch loading falling to a low value before separation of the plug, is associated with overcoming the interference of interpenetrating lips at the surface of separation. Thus, from the point of view of securing a pierced hole of satisfactory surface finish, no advantage can be gained in the reduction of the punch load at separation by the use of double-shear. However, if further operations were proposed to finish the hole to a required standard, or if the surface finish of the hole were immaterial, then decided advantages can be gained.

CONCLUSION

In the production of holes of acceptable surface finish, the following conclusions can be drawn.

(1) Punch double-shear beyond $10°$ is unadvisable.

(2) Provision of double-shear restricts the range of radial clearance over which a satisfactory hole can be produced; in the present work from 10 to 30 per cent for no shear, to 10–15 per cent for $10°$ double-shear.

(3) The reduction in maximum punch load secured by use of double-shear, becomes appreciable with increasing radial clearance, and over the range in which $10°$ double-shear can produce satisfactory results (i.e. 10–15 per cent), is from about 10 to 50 per cent, depending upon the ratio of hole diameter to plate thickness ratio employed.

(4) There are limited advantages in double-shear in reducing the punch load at separation.

ACKNOWLEDGMENTS

The author wishes to thank Mr I. McInnes for carrying out the piercing tests on which the present paper is based.

REFERENCES

1. M. MEYER and E. KIENZLE (1962). *C.I.R.P., Ann.,* **11**, 2.
2. A. GUIDI (1962). *Metalworking Production,* **106**, 40.
3. Fr. BOSCH and A. STAGER (1963). *Werkstatt u. Betrieb,* **96**, 10.

4. Prod. Eng. Res. Assn. Reports nos. 27, 57 and 64.
5. W. JOHNSON and F. W. TRAVIS (1965-6). *Proc. Inst. Mech. Engrs.*, **180,** Part 3I.
6. W. JOHNSON and R. A. C. SLATER (1965-6). *Proc. Inst. Mech. Engrs.*, **180,** Part 3I.
7. W. JOHNSON and F. W. TRAVIS (1968). High-speed blanking of steel, *Engineering Plasticity*, Cambridge University Press.
8. W. JOHNSON and R. A. C. SLATER (1967). *Proc. Int. Conf. Man. Tech. C.I.R.P.-A.S.T.M.E.*, Ann Arbor, Michigan.
9. R. JOHNSTON, B. FOGG and A. W. J. CHISHOLM (1968). *Proc. 9th Int. Mach. Tool Des. Res. Conf.*, Pergamon.
10. T. M. CHANG and H. W. SWIFT (1950-51). *J. Inst. Metals*, 78, 119.
11. T. M. CHANG (1950-51). *J. Inst. Metals,* **78,** 393.
12. T. MAEDA and K. TAMURA (1960). *Bull. Jap. S.M.E.*, **3,** 312.
13. G. C. ANTHONY (1911). *Trans. A.S.M.E.*, **33,** 369.
14. R. BALENDRA and F. W. TRAVIS (1970). *Int. J. Mach. Tool Des. Res.*, **10,** 249.

IMPROVEMENT OF DEFORMABILITY OF SHEARED EDGE BY APPLICATION OF ADDITIONAL COMPRESSION

by

VLADIMIR CUPKA*, TAKEO NAKAGAWA† and TERUFUMI MACHIDA‡

SUMMARY

To achieve high deformability in sheet metal forming it is necessary that the sheared edges of sheet metal do not crack during the stretch forming operation. The present investigation details improvements that were achieved in the deformability of the sheared edge by additional compression. A marked improvement in deformability was also obtained by performing a small coining operation on the sheared surfaces before stretching. As a combined method, a reverse stretching is newly developed and possible applications to actual performance are discussed.

INTRODUCTION

Stretchflanging occupies an important position next only to deep drawing and punch stretching in sheet metal forming. In actual performance, most of the troubles encountered in stretchflanging are due to the fractures that occur along the sheared edges of the sheet material. Several investigations[1-3] have recently been carried out to study this type of fracture, and some progress has been made in its prevention. The role played by certain metallurgical properties of the material has also been carefully assessed[2-7], and new commercial quality hot rolled steel sheets, which allow high deformability of the sheared edge, have also been developed. The effect of shearing and forming conditions on stretchability is now well correlated and it is less difficult to select appropriate pressing conditions to meet particular requirements.

Among the various techniques employed for improvement of the deformability in sheet metal forming, it is well known that cut-off punching[3] sharply improves the deformability limit. The harmful sheared zone is removed by cut-off punching before stretchforming in a manner similar to shaving or machining. The present investigation aims primarily at further improvement of the quality of the sheared edge and its deformability in subsequent stretchforming operations. It is supposed that to prevent crack formation in stretch forming an application of addition compression into the sheared surface is very important. Several different ways of doing so are proposed. At first, the effect of additional compressive forces, applied to the sheared surface at the time of stretching, on deformability is treated experimentally. Then coining of the sheared surface before stretching is proposed as a further alteration, and results of experimental investigation are presented. Based on positive results of above-mentioned investigations a new stretchflanging method, hereafter called reverse stretching is suggested. This method makes use of the beneficial effects of both the additional compressive forces applied to the sheared surface as well as coining and represents a promising solution to the problem under study. Finally, possible applications of these methods are discussed.

EXPERIMENTAL

Experimental equipment and procedure

For detecting the deformability of the sheared edge, the hole expanding test (Fig. 1) was adopted, using a double action hydraulic sheet metal forming machine. The hole expanding limit λ was determined when the crack had developed fully through the thickness of the sheet and is expressed as the average stretch of the hole diameter, given by

$$\lambda = (d_B - d_i)/d_i$$

where d_i is the initial hole diameter and d_B is the fractured hole diameter. In experimental work, the initial hole diameter d_i was standardized at 10 mm and pierced by a simple die-set punching tool with an approximate clearance of 10 per cent on each side.

---

* Slovak Technical University, Bratislava, Czechoslovakia; Visiting Researcher, University of Tokyo, 1971–73
† Institute of Industrial Science, University of Tokyo, Japan
‡ Nippon Institute of Technology, Saitama, Japan; Research Fellow, University of Tokyo

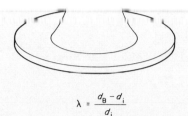

$$\lambda = \frac{d_B - d_i}{d_i}$$

Figure 1. Hole expanding test.

Material

Hot rolled mild steel sheet 150 mm in diameter and 3·2 mm thick, commonly employed for stretchflanging in car industries, was used for specimens. Its mechanical properties are as follows: tensile strength 34·5 kg/mm², elongation 48·4 per cent, $C = 54·1$ kg/mm² and $n = 0·22$ in, $\sigma = C\epsilon^n$, and r-value 0·84.

RESULTS AND DISCUSSION

Additional compression during stretching

To affect the quality of the sheared surface, an appropriate additional compression introduced to the sheared surface is directed in two ways in accordance with Fig. 2a and b. Compression applied along the thickness to the sheared surface, that is, perpendicularly to the plane of the sheet, represents the first case. More powerful is the case with radial compression acting perpendicularly to the sheared surface. Unfortunately, both of these methods will meet difficulties in actual performance. Despite this fact, the discussion is carried out to make clear the effect of hydrostatic compression on the deformability limit in stretchflanging. For experimental confirmation the above-mentioned testing machine, with an additional hydraulic cylinder attached, was used.

From the experimental results presented in Fig. 2c it can be seen that additional compression causes deformability limit to increase considerably if compared with that without compression. This tendency is more obvious in case of stretching with radial compression in which the hole expanding limit is three times that obtained in conventional stretching process. The difference between hole expanding limits λ in terms of the effect of additional compression parallel and perpendicular to the sheared surface comes from a different mechanism to be achieved in both cases. The former does not include a direct action of additional compression on the sheared surface which results only in slight improvement of the additional compression, and it is also accompanied with a remarkable thickness reduction at the

(a) Compressive stretching with round counter punch

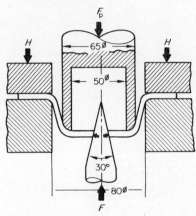

(b) Compressive stretching with conical counter punch

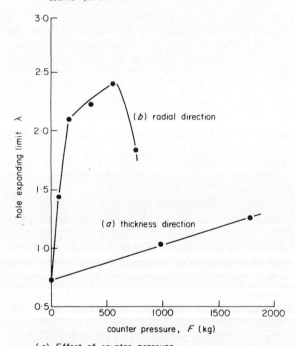

(c) Effect of counter pressure

Figure 2. Effects of additional compression during stretching.

inner hole periphery. On the other hand, the compression applied in radial or perpendicular direction to the sheared surface is responsible for its wholly new quality and, as will be revealed later, this together with radial pressure is the leading factor influencing deformability of the sheared edge in a new stretching process. With increasing magnitude of counter force in radial compression the generated bending moment exceeds the resistance of the inner hole periphery which is bent inside, and the sheared surface becomes free of radial compression. Consequently the hole expanding limit λ shows decreasing tendency in the range of excessive counterforce.

Coining operation before stretching

As a pre-treatment of the sheared edge a small coining operation was applied before stretching and the subsequent effect on the deformability of the sheet was investigated. To coin, depress and burnish the fracture zone contained in the sheared surface a polished conical punch of the type shown in Fig. 3a was used. After coining the hole was expanded, with a coined burr side of the sheared surface facing expanding die-cavity. Figure 3b shows the variation

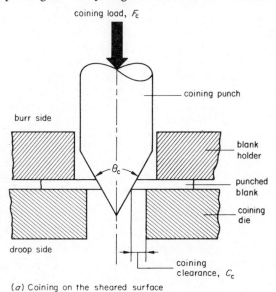

(a) Coining on the sheared surface

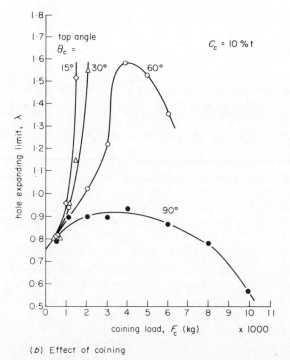

(b) Effect of coining

Figure 3. Application of coining and its effects on deformability of sheared edge.

of the hole expanding limit for a different top angle θ of a conical coining punch and a variable coining load. It can be pointed out that, by coining before stretching, the deformability of the sheared edge also recovers to a great extent. It is clear that in general, the smaller top angle θ of the conical coining punch results in a bigger hole expanding limit. This phenomenon occurs because for the smaller angle θ the component of coining load perpendicular to the sheared surface is bigger and, also, the contact area between conical punch and sheared surface increases. On the other hand the optimum coining load exists because its exceeding causes the additional workhardening. This can be seen in Fig. 3b for conical coining punches provided with 60° and 90° top angle. In the case of 60° top angle, the optimum coining load nearly corresponds to the load at which the fracture zone of sheared edge disappears by coining, as shown in Fig. 4.

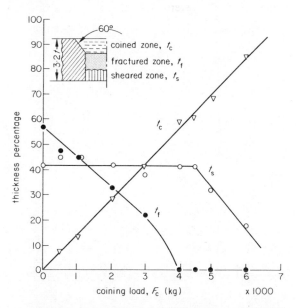

Figure 4. Constitutions of coined specimen section.

In the case of 15° and 30° top angles, however, a larger coining load cannot be applied to the sheared edge in so far as the conical coining punch may contact the coining die. Nevertheless, it can be expected that under such smaller angle conditions (small angle θ) the optimum load will also exist because of the increasing workhardening generated near the sheared surface. And detailed observations in the case of smaller angle reveal that with increasing coining load behind the limit required to remove fracture zone, deformability of the sheared edge still increases. Further, comparing the hole expanding limit after coining and that obtained after the fracture zone of the sheared edge was removed by machining providing the sheared surface with the same shape as it was after coining, the following can be noted. The deforming limit of the machined edge is a little less compared to that of the sheared hole with a coined edge. From the above results, it is concluded that the improvement of deformability by coining results mainly from the disappearance and burnishing of fractured zone, while the compressive pre-strain effect seems to be a secondary factor.

In this section, a process of reverse stretching is proposed as a new method including positively the two above-mentioned effects. The schematic explanation of this reverse stretching process is shown roughly in the left of Fig. 5a. At first a pre-stretching is performed to some extent at the state that the burr side of the sheared edge faces to the die cavity. At the following stage, the pre-stretched sheet is stretch-formed from the reverse direction. Consequently, at

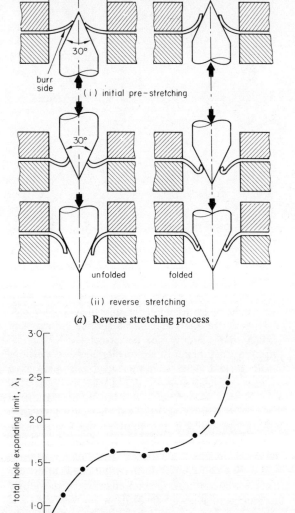

(a) Reverse stretching process

(b) Effect of reverse stretching

Figure 5. Effects of reverse stretching on the deformability of the sheared edge.

the total expanding limit drops down because of crack initiation along the sheared edge during pre-stretching.

As the different type of reverse stretching method, by making larger the difference between the diameter of a coining die and the hole diameter of sheet at the time of coining, means using the larger coining clearance (see Fig. 3a), the initial pre-stretching as well as coining become simultaneously possible. Figure 6 shows the relations among the total hole expanding limit λ_t, the initial hole expanding ratio λ_i at coining and the coining clearance C_c. In this trial the deforming limit of sheared edge is improved again better than any of those in previously mentioned methods, as shown in Fig. 6b.

It must be remarked, however, that the biggest deformability was obtained in the case of the folded-type reverse stretching. This folded-type reverse stretching is schematically shown on the right of Fig. 5a and the cross section of flanged product is also shown in Fig. 6b. If the coining clearance, the coining load and the top angle of stretching conical punch exceed their limits, then the inner hole periphery is bent sharply inside at the time of coining or pre-stretching and consequently the bent lip is folded in the stage of reverse stretching. In the folded-type reverse stretching the fragile sheared edge is covered by the sheet and stretched at the highly compressed state between the sheet and punch. As the results of this hydrostatic compression on the sheared edge, its deformability at the optimum condition reaches almost the same value as in the case of machined edge, that is five times the deformability of conventional stretching. It is a matter of course that the application of this special folded-type stretching will be limited in actual performance, but it is also useful for getting the round and smooth edge of the flanged products.

Returning to the ordinary reverse stretching without any folded lip shown in Fig. 6, an adequate value of coining clearance should be selected in order to get the bigger deformability. Judging from the experimental results, it exists at about 150–200 per cent of the sheet thickness. But much higher deformability is expected by making the shapes of coining punch and die more suitable, such as a smaller top angle of punch and a conical die or a die with a round edge.

POSSIBLE APPLICATIONS OF COMPRESSION STRETCHING IN ACTUAL PERFORMANCE

It is considered that the stretching after coining as well as the reverse stretching mentioned above are

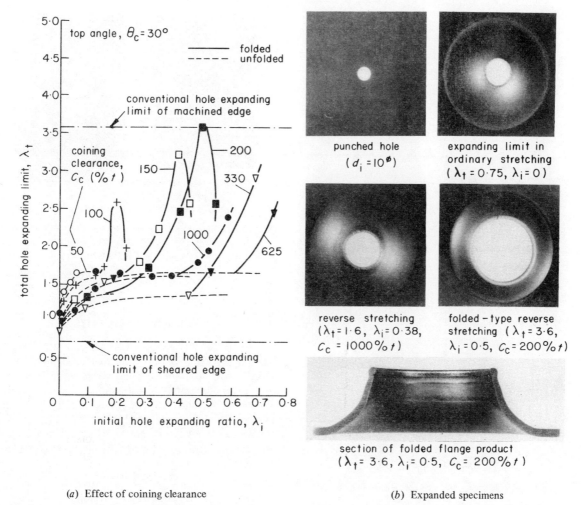

(a) Effect of coining clearance

(b) Expanded specimens

Figure 6. Effect of coining clearance in reverse stretching.

directly applicable to the actual performance. Several features about these improvements and the possible applications in actual performance are discussed here.

General features of compression stretching

As the features of this compressive stretching, the following can be pointed out.

(1) Fracture occurrence can be delayed and the hole expanding limit as well as the flange height of stretched parts can be improved to a great extent.

(2) It is effective to prevent the crack origination from the burr side. Until now there was no successful method to solve this problem except cut-off punching.

(3) More uniform wall thickness can be obtained by the application of radial compression.

(4) In the reverse stretching, instead of the first stretching, the punch stretching before the piercing is applicable in order to get higher wall height.

Tooling in compression stretching

As a disadvantage of this compression stretching the necessity of additional compressive operation is pointed out. This problem can be solved considerably by the simply modified tooling in the case of hole flanging, or burring as shown in Fig. 7.

A. Application of stretching with coining

(1) Piercing + coining + burring (separately)

(2) (Piercing + coining) + burring
Piercing and coining are done at a single stroke, but this is practically impossible.

(3) Piercing + (coining + burring)
Coining and burring are performed at the same time.
See Fig. 7a–(1) (2) (3).

B. Application of reverse stretching with pre-stretching

(1) Piercing + pre-stretching + reverse-stretching (separately). See Fig. 5a.

(2) (Piercing + pre-stretching) + reverse-stretching. See Fig. 7b–(1) (2).

(3) Piercing + (pre-stretching + reverse-stretching). See Fig. 7b–(3).

C. Application of reverse stretching with pre-bulging

In this application a higher stretched wall can be obtained.

(1) Pre-bulging + piercing + reverse-stretching (separately). See Fig. 7c–(1).

(2) (Pre-bulging + piercing) + reverse-stretching. See Fig. 7c–(2) (3).

(3) (pre-bulging + piercing) + reverse-stretching) (all together). See Fig. 7c–(4) (5).

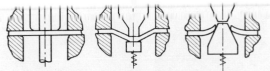

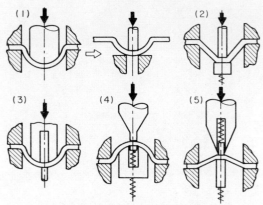

(b) Reverse stretching with pre-stretching

(c) Reverse stretching with pre-bulging

Figure 7. Possible applications of compression stretching.

(a) Without coining
(b) After coining ($\theta_c = 60°$, $F_c = 4000$ Kg)

Figure 8. Application of coining to bending. Material; 0·45 per cent C steel sheet, t = 3·0 mm. Bending tool; 90° V-bending, r_p = 10 mm. Burr is tension side.

is depressed by coining into compression side of bending.

ACKNOWLEDGMENTS

We thank Mr Toyoharu Takano and Mr Kiyoshi Suzuki, Institute of Industrial Science University of Tokyo, for their invaluable assistance and cooperation in experimental investigations. We would also like to thank the Kawasaki Steel Co. Ltd. for the supply of steel material used in the investigations. We extend our thanks to Dr N. R. Chitkara, University of Manchester Institute of Science and Technology, for his help in preparing the manuscript.

Application of coining to the crack prevention in bending

Till now burring operation is exclusively discussed as the typical example of stretchflanging, but as a natural consequence this improvement of deformability by additional compression can be applied to the other stretch forming of the sheared edge as in bending.

In bending the coining of fractured zone as it was confirmed experimentally is very effective for fracture prevention originated from the burr side. Figure 8 shows some of the experimental results. In this case a further beneficial effect can be expected as well as the previously mentioned effects—namely that the fragile burr and fractured zone of the sheared surface

REFERENCES

1. T. NAKAGAWA and K. YOSHIDA (1968). *Reports I.P.C.R.*, **44**, 150.
2. T. NAKAGAWA and K. YOSHIDA (1971). *Proceedings ICSTIS, Suppl. Trans. Iron and Steel Institute of Japan*, **11**, 813.
3. T. NAKAGAWA and K. YOSHIDA (1970). *J. Japan Soc. Tech. of Plasticity*, **11**, 665.
4. J. KUBOTERA et al. (1969). *Tetsu-to-Hagane*, **55**, S539.
5. T. FUJIOKA et al. (1970). *Tetsu-to-Hagane*, **56**, S120.
6. T. NAKAGAWA, H. KAWASE and K. YOSHIDA (1969). *Tetsu-to-Hagane*, **55**, 648.
7. T. NAKAGAWA, M. TAKITA and K. YOSHIDA (1970). *J. Japan Soc. Tech. of Plasticity*, **11**, 142.

AUTHOR INDEX

SUBJECT INDEX